T0140427

Advances in Soft Computing 40

Advances in Soft Computing

Editor-in-Chief

Prof. Janusz Kacprzyk
Systems Research Institute
Polish Academy of Sciences
ul. Newelska 6
01-447 Warsaw
Poland
E-mail: kacprzyk@ibspan.waw.pl

Further volumes of this series can be found on our homepage: springer.com

Marek Kurzynski, Edward Puchala, Michal Wozniak, Andrzej Zolnierek (Eds.)
Computer Recognition Systems, 2005
ISBN 978-3-540-25054-8

Abraham Ajith, Yasuhiko Dote, Takeshi Furuhashi, Mario Köppen, Azuma Ohuchi, Yukio Ohsawa (Eds.)
Soft Computing as Transdisciplinary Science and Technology, 2005
ISBN 978-3-540-25055-5

Barbara Dunin-Keplicz, Andrzej Jankowski, Andrzej Skowron, Marcin Szczuka (Eds.)
Monitoring, Security, and Rescue Techniques in Multiagent Systems, 2005
ISBN 978-3-540-23245-2

Frank Hoffmann, Mario Köppen, Frank Klawonn, Rajkumar Roy (Eds.)
Soft Computing Methodologies and Applications, 2005
ISBN 978-3-540-25726-4

Mieczyslaw A. Klopotek, Slawomir T. Wierzchon, Kryzysztof Trojanowski (Eds.)
Intelligent Information Processing and Web Mining, 2005
ISBN 978-3-540-25056-2

Abraham Ajith, Bernard de Bacts, Mario Köppen, Bertram Nickolay (Eds.)
Applied Soft Computing Technologies: The Challenge of Complexity, 2006
ISBN 978-3-540-31649-7

Mieczyslaw A. Klopotek, Slawomir T. Wierzchon, Kryzysztof Trojanowski (Eds.)
Intelligent Information Processing and Web Mining, 2006
ISBN 978-3-540-33520-7

Ashutosh Tiwari, Joshua Knowles, Eral Auineri, Keshav Dahal, Rajkumar Roy (Eds.)
Applications and Soft Computing, 2006
ISBN 978-3-540-29123-7

Bernd Reusch, (Ed.)
Computational Intelligence, Theory and Applications, 2006
ISBN 978-3-540-34780-4

Miguel López-Díaz, María ç. Gil, Przemysław Grzegorzewski, Olgierd Hryniewicz, Jonathan Lawry
Soft Methodology and Random Information Systems, 2006
ISBN 978-3-540-34776-7

Ashraf Saad, Erel Avineri, Keshav Dahal, Muhammad Sarfraz, Rajkumar Roy (Eds.)
Soft Computing in Industrial Applications, 2007
ISBN 978-3-540-70704-2

Bing-Yuan Cao (Ed.)
Fuzzy Information and Engineering, 2007
ISBN 978-3-540-71440-8

Bing-Yuan Cao (Ed.)

Fuzzy Information and Engineering

Proceedings of the Second International Conference of Fuzzy Information and Engineering (ICFIE)

Editor

Prof. Bing-Yuan Cao
School of Mathematics and
Information Science Institute
Guangzhou University
Guangzhou 510006
P.R. China
E-mail: bycao@gzhu.edu.cn
caobingy@163.com

Library of Congress Control Number: 2007923716

ISSN print edition: 1615-3871
ISSN electronic edition: 1860-0794
ISBN-10 3-540-71440-5 Springer Berlin Heidelberg New York
ISBN-13 978-3-540-71440-8 Springer Berlin Heidelberg New York

Springer is a part of Springer Science+Business Media
springer.com

Printed in Germany

Typesetting: by the authors and SPS using a Springer LaTeX macro package

Printed on acid-free paper SPIN: 11930303 89/SPS 5 4 3 2 1 0

Preface

This book is the proceedings of the Second International Conference on Fuzzy Information and Engineering (ICFIE2007) from May 13-16, 2007 in Guangzhou, China. The conference proceedings is published by Springer-Verlag (Advances in Soft Computing, ISSN: 1615-3871).

This year, we have received 257 submissions from more than 10 countries and regions. Each paper has undergone a rigorous review process. Only high-quality papers are included. The overall acceptance rate is approximately 42%, including the extended papers and post papers.

The ICFIE2007, built on the success of previous conferences, the ICIKE2002 (Dalian China), is a major symposium for scientists, engineers and practitioners in China as well as the world to present their latest results, ideas, developments and applications in all areas of fuzzy information and knowledge engineering. It aims to strengthen relations between industry research laboratories and universities, and to create a primary symposium for world scientists in fuzzy fields as follows:

1) Fuzzy Information;
2) Fuzzy Sets and Systems;
3) Soft Computing;
4) Fuzzy Engineering;
5)Fuzzy Operation Research and Management;
6) Artificial Intelligence;
7) Rough Sets and Its Application;
8) Application in Fuzzy Mathematics and Systems, etc.

In addition to the large number of submitted papers, we are blessed with the presence of nine renowned keynote speakers and several distinguished panelists.

At the same time, we shall organize three workshops including Fuzzy Geometric Programming and Its Application, Operation Research and Management, and Research for Fuzziness.

On behalf of the Organizing Committee, we thank Guangzhou University in China, Fuzzy Information and Engineering Branch of China Operation Research Society for sponsorship; Japan Society of Fuzzy Theory and Systems (SOFT), Singapore International Institute of Systems Science, Fuzzy Information and Engineering Branch of International Institute of General Systems Studies China Branch (IIGSS-GB) and International Fuzzy Mathematics Institute in USA for Co-Sponsorships. We are grateful to the supports coming from National Natural Science Foundation of China, Guangzhou Science and Technology Association, Education Bureau of Guangzhou City, Guangdong Provincial Association of Science and Technology, California Polytechnic State University in USA, and Industrial Engineering and Operations Research of North Carolina State University in USA and J. of Fuzzy Optimization and Decision Making (FODM). We are showing gratitude to the members of the Organizing Committee, the Steering Committee, and the Program Committee for their hard work. We wish to express our heart-felt appreciation to the keynote and panel speakers, workshop organizers, session chairs, reviewers, and students. In particular, I am thankful to my doctoral student, Yang J.H, and my masters students, Tan J, Zhang Y.F, who have contributed a lot to the development of this issue. Meanwhile, we thank the publisher, Springer, for publishing the ICFIE 2007 proceedings as J. Advances in Soft Computing. Finally, we are appreciated for all the authors and participants for their great contributions that made this conference possible and all the hard work worthwhile.

May 2007

Bing-yuan Cao
Guangzhou, P.R. China

Organization

ICFIE 2007 is organized by

Guangzhou University

Fuzzy Information and Engineering Branch in the Operations Research Society of China

technically co-sponsored by

Japan Society of Fuzzy Theory and Systems (SOFT)

Singapore International Institute of Systems Science

Fuzzy Information and Engineering Branch of IIGSS-GB

International Fuzzy Mathematics Institute in USA

and supported by

National Natural Science Foundation of China

Guangzhou Science and Technology Association

Education Bureau of Guangzhou City

Guangdong Provincial Association for Science and Technology

California Polytechnic State University in USA

Industrial Engineering and Operations Research of NCSU in USA

J. of FODM

Organizing Committee

Conference Chair :

Cao Bing-yuan (China)

Honorary Chairs :

Lotfi A. Zadeh (USA) Jian-she Yu (China)

Steering Committee:

J.C.Bezdek (USA)
Z.Bien (Korea)
D. Dubois (France)
Gui-rong Guo (China)
M.M. Gupta (Canada)
Xin-gui He (China)
Abraham Kandel (USA)
G.J.Klir (USA)
L.T. Koczy (Hungary)
Ying-ming Liu (China)
E.Mamdani (UK)
R.P.Nikhil (India)
M. Sugeno (Japan)
Hao Wang (China)
P.P. Wang (USA)
Pei-zhuang Wang (USA)
H.J. Zimmermann (Germany)

Program Committee:

Chair :
Hong-xing Li (China)
Co-Chair :
Shu-Cherng Fang (USA)

Members

K.Asai (Japan)
Guo-qing Chen (China)
Ovanes Chorayan (Russia)
Yinjun Feng (China)
Cheng-ming Hu (USA)
Li-min Jia (China)
Jim Keller (USA)
N.Kuroki (Japan)
T.Y. Lin (USA)
D.D.Majumder(Indian)
M.Mukaidono(Japan)
Jin-ping Ou (China)
D.A. Ralescu (USA)
V.V.Senkevich(Russia)
Enric Trillas (Spain)
Guo-jun Wang (China)
B.L.Wu(China Taiwan)
R.R.Yager (USA)
J.P. Barthelemy (France)
Mian-yun Chen (China)
H.P.Deng (Australia)
Si-cong Guo (China)
Chong-fu Huang (China)
Guy Jumarie (Canada)
E.E. Kerre (Belgium)
D.Lakov (Bulgaria)
Bao-ding Liu (China)
M.Mizumoto (Japan)
J.Mustonen (Finland)
Witold Pedrycz (Canada)
Da Ruan (Beglium)
Qiang Shen (UK)
L.P.Wang (Singapor)
Xi-zhaoWang (China)
Cong-xin Wu (China)
T. Yamakawa (Japan)
Tian-you Chai (China)
Shui-li Chen (China)
M. Fedrizzi (Italy)
Ming-hu Ha (China)
Hiroshi Inoue (Japan)
J.Kacprzyk (Poland)
K.H.Kim (USA)
Tsu-Tian Lee (China Taiwan)
Zhi-qiang Liu (Hongkong)
J.Motiwalla (Singapore)
Shohachiro Nakanishi (Japan)
H. Prade (France)
E.Sanchez (France)
Kai-quan Shi (China)
Guang-yuan Wang (China)
Zhen-yuan Wang (USA)
Yang Xu (China)
Bing-ru Yang (China)

Local Arrangements Chair : Guang-fu Cao (China)

Co-Chair : Zeng-liang Liu (China)

Secretary : Yu-bin Zhong (China)

Member : Yu-chi Zhong (China)

Publicity Chair : Kai-qi Zou (China)

Member : Da-bu Wang (China)

Publication Chair :
Cao Bing-yuan

Co-Chair :
Michael Ng (Hong Kong)

Accounting:
Yun-lu Cai (China)

Treasurer:
Ji-hui Yang (China)

Reviewers

Shi-zhong Bai
Klaas Bosteels
Bing-yuan Cao
Tian-you Chai
Shui-li Chen
Mian-yun Chen
Xiu-min Chen
Yong-yi Chen
Jie Cui
Guan-nan Deng
He-pu Deng
Ren-ming Deng
Yijun Fan
Shu-cherng Fang
Jia-li Feng
Robert Fullr
Si-cong Guo
Kai-zhong Guo
Ming-hu Ha
Bao-qing Hu
Chong-fu Huang
Zhe-xue Huang
J.Kacprzyk
E.E.Kerre
J.Keller
K.H.Kim
Hong-xing Li
Qing-guo Li
Lei Li
Tian-rui Li
Zu-hua Liao
Bao-ding Liu
Xiaodong Liu
Zeng-liang Liu
Marc Lavielle
Sheng-quan Ma
E.Mamdani
M.Mizumoto
Michael.Ng
R.Pal Nikhil
W.Pedrycz
Lech Polkowski
H.Prade
Germano Resconi
Da Ruan
M.Sanchez
Yu Sheng
Hui-qin Shi
Kai-quan Shi
Yun Shi
Ye-xin Song
Da-bu Wang
Guo-yin Wang
Li-po Wang
P.P. Wang
Xue-ping Wang
Xi-zhao Wang
Zhi-ping Wang
Ber-lin Wu
Ci-yuan Xiao
Kai-jun XU
Yang Xu
Ze-shui Xu
Bing-ru Yang
Ji-hui Yang
Xue-hai Yuan
Ai-quan Zhang
Min Zhang
Qiang Zhang
Xin-ai Zhang
Yu-bin Zhong
H.J. Zimmermann
Wu-neng Zhou
Kai-qi Zou

Contents

Part I: Fuzzy Information

Part III: Soft Computing

Part IV: Fuzzy Engineering

Part I

Fuzzy Information

Contrast Enhancement for Image by WNN and GA Combining PSNR with Information Entropy

Changjiang Zhang[1] and Han Wei[2]

[1] College of Mathematics, Physics and Information Engineering, Zhejiang Normal University, Jinhua, postcode:321004 China
zcj74922@zjnu.cn

[2] College of Mathematics, Physics and Information Engineering, Zhejiang Normal University, Jinhua, postcode:321004 China
weihan0627@126.com

Abstract. A new contrast enhancement algorithm for image is proposed combing genetic algorithm (GA) with wavelet neural network (WNN). In-complete Beta transform (IBT) is used to obtain non-linear gray transform curve so as to enhance global contrast for an image. GA determines optimal gray transform parameters. In order to avoid the expensive time for traditional contrast en-hancement algorithms, which search optimal gray transform parameters in the whole parameters space, based on gray distribution of an image, a classification criterion is proposed. Contrast type for original image is determined by the new criterion. Parameters space is respectively determined according to different contrast types, which greatly shrinks parameters space. Thus searching direction of GA is guided by the new parameter space. Considering the drawback of trad-tional histogram equalization that it reduces the information and enlarges noise and background blutter in the processed image, a synthetic objective function is used as fitnees function of GA combing peak signal-noise-ratio (PSNR) and in-formation entropy. In order to calculate IBT in the whole image, WNN is used to approximate the IBT. In order to enhance the local contrast for image, dis-crete stationary wavelet transform (DSWT) is used to enhance detail in an im-age. Having implemented DSWT to an image, detail is enhanced by a non-linear operator in three high frequency sub-bands. The coefficients in the low frequency sub-bands are set as zero. Final enhanced image is obtained by adding the global enhanced image with the local enhanced image. Experimental results show that the new algorithm is able to well enhance the global and local contrast for image while keeping the noise and background blutter from being greatly enlarged.

Keywords: contrast enhancement; wavelet neural network; genetic algorithm; discrete stationary wavelet transform; in-complete Beta transform.

1 Introduction

Traditional image enhancement algorithms are as following: point operators, space operators, transform operators and pseu-color enhancement [1]. H.D. Cheng gave a kind of algorithm for contrast enhancement based on fuzzy operator [2]. However, the algorithm cannot be sure to be convergent. Lots of improved

B.-Y. Cao (Ed.): Fuzzy Information and Engineering (ICFIE), ASC 40, pp. 3–15, 2007.
springerlink.com

histogram equalization algorithms were proposed to enhance contrast for kinds of images [3]. The visual quality cannot be improved greatly with above algorithms. Tubbs gave a simple gray transform algorithm to enhance contrast for images [4]. However, the computation burden of the algorithm was very large. Existing many contrast enhancement algorithms' intelligence and adaptability are worse and much artificial interference is required.

To solve aboveproblems, a new algorithm employing IBT, GA and WNN is proposed. To improve optimization speed and intelligence of algorithm, a new criterion is proposed based on gray level histogram. Contrast type for original image is determined employing the new criterion. Contrast for original images are classified into seven types: particular dark (PD), medium dark (MD), medium dark slightly (MDS), medium bright slightly (MBS), medium bright (MB), particular bright (PB) and good gray level distribution (GGLD). IBT operator transforms original image to a new space. A certain objective function is used to optimize non-linear transform parameters. GA is used to determine the optimal non-linear transform parameters. In order to reduce the computation burden for calculating IBT, a new kind of WNN is proposed to approximate the IBT in the whole image. Combing IBT, GA with WNN enhances global contrast of an image. Local contrast of an image is enhanced by DSWT and non-linear gain operator. Final enhanced image is obtained by adding the global enhanced image to the local enhanced image.

2 IBT

The incomplete Beta function can be written as following:

$$F(u) = B^{-1}(\alpha,\beta) \times \int_0^u t^{\alpha-1}(1-t)^{\beta-1}dt, 0 < \alpha, \beta < 10 \tag{1}$$

All the gray levels of original image have to be unitary before implementing IBT. All the gray levels of enhanced image have to be inverse-unitary after implementing IBT. Let x shows gray level of original image, g indicates unitary gray level. We have:

$$g = \frac{x - \min(x)}{\max(x) - \min(x)} \tag{2}$$

Where $\min(x)$ and $\max(x)$ shows the minimum gray level and the maximum one in original image respectively. g is mapped to g':

$$g' = IB\,(a, b, g) \tag{3}$$

Let x' shows gray level of enhanced image, we have:

$$x' = [\max(x) - \min(x)]\, g' + \min(x) \tag{4}$$

3 Contrast Classification for Image Based on Histogram

Based on gray level histogram, contrast classification criterion can be described in Fig.1:

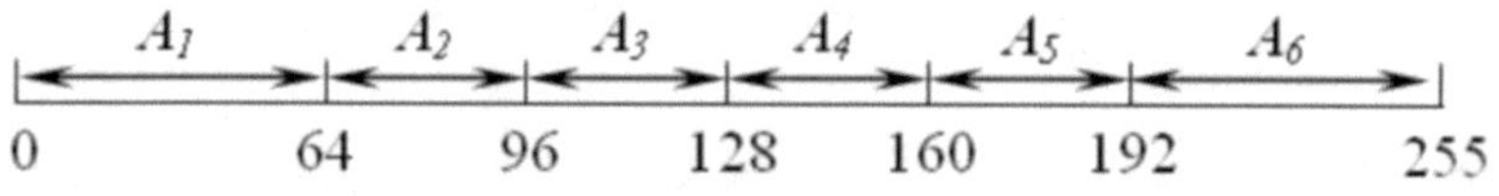

A_1 A_2 A_3 A_4 A_5 A_6

0 64 96 128 160 192 255

Fig. 1. Image classification sketch map based on gray level histogram

Given that original image has 256 gray level values (gray level value ranges from 0 to 255), the whole gray level value space is divided into six sub-spaces:$A_1, A_2, A_3, A_4, A_5, A_6$, where $A_i (i = 1, 2, \cdots 6)$ is the number of all pixels which is in the ith sub-space. Let,

$$M = \max_{i=1}^{6} A_i, B_1 = \sum_{k=2}^{6} A_i, B_2 = \sum_{k=2}^{5} A_i, B_3 = \sum_{k=1}^{5} A_i$$
$$B_4 = A_1 + A_6, B_5 = A_2 + A_3, B_6 = A_4 + A_5$$

The classification criterion in the following can be obtained:
$if\ (M = A_1)\ \&\ (A_1 > B_1)$
Image is PB;
$elseif\ (B_2 > B_4)\ \&\ (B_5 > B_6)\ \&\ (B_5 > A_1)\ \&\ (B_5 > A_6)\ \&\ (A_2 > A_3)$
Image is MD;
$elseif\ (B_2 > B_4)\ \&\ (B_5 > B_6)\ \&\ (B_5 > A_1)\ \&\ (B_5 > A_6)\ \&\ (A_2 < A_3)$
Image is MDS;
$elseif\ (B_2 > B_4)\ \&\ (B_5 < B_6)\ \&\ (A_1 < B_6)\ \&\ (A_6 < B_6)\ \&\ (A_4 > A_5)$
Image is MBS;
$elseif\ (B_2 > B_4)\ \&\ (B_5 < B_6)\ \&\ (A_1 < B_6)\ \&\ (A_6 < B_6)\ \&\ (A_4 < A_5)$
Image is MB;
$elseif\ (M = A_6)\ \&\ (A_6 > B_3)$
Image is PB;
else
Image is GGLD;
end
where symbol & represents logic "and" operation.

4 Transform Parameters Optimization by GA

GA, originally proposed by Holland, are a class of parallel adaptive search algorithms based on the mechanics of natural selection and natural genetic system, which behave well in searching, optimization, and machine learning. GA can find the near-global optimal solutions in a large solution space quickly. GA has been used extensively in many application areas, such as image processing, pattern recognition, feature selection, and machine learning. We will employ the GA to

Table 1. Range of transform parameters

Parameter	PD	MD	MDS	MBS	MB	PB
α	[0,2]	[0,2]	[0,2]	[1,3]	[1,4]	[7,9]
β	[7,9]	[1,4]	[1,3]	[0,2]	[0,2]	[0,2]

optimize transform parameters [5]. If the algorithm is used directly to enhance the global contrast for an image, it will result in large computation burden. The range of α and β can be determined by Tab.1 so as to solve above problems.

Let $\mathbf{x} = (\alpha, \beta)$, $F(\mathbf{x})$is the fitness function for GA, Where $a_i < \alpha, \beta < b_i (i = 1, 2)$, a_i and $b_i (i = 1, 2)$ can be determined by Tab.1. GA consists of the following steps (procedures):

A. Initialization. An initial population size P for a genetic algorithm should be generated, which may be user-specified or randomly selected.

B. Evaluation. Fitness of each chromosome within a population will be evaluated. The fitness function maps the binary bit strings into real numbers. The larger (or smaller) the fitness function value is, the better the solution (or chromosome) will be.

C. Selection. Based on the fitness values of bit strings in the current population, pairs of "parent" bit strings are selected which undergo the genetic operation "crossover" to produce pairs of "offspring" bit strings forming the next generation. The probability of selection of a particular chromosome is (directly or inversely) proportional to the fitness function value.

D. Crossover. Crossover exchanges information between two parent bit strings and generates two offspring bit strings for the next population. Crossover is realized by cutting individually the two parent bit strings into two or more bit string segments and then combining the two bit string segments undergoing crossing over to generate the two corresponding offspring bit strings. Crossover can produce off-springs which are radically different from their parents.

E. Mutation. Mutation is to perform random alternation on bit strings by some operations, such as bit shifting, inversion, rotation, etc. which will create new offspring bit strings radically different from those generated by the reproduction and crossover operations. Mutation can extend the scope of the solution space and reduce the possibility of falling into local extremes. In general, the probability of applying mutation is very low.

F. Stopping criterion. There exists no general stopping criterion. The following two stopping criteria are usually employed: (1) No further improvement in the fitness function value of the best bit string is observed for a certain number of iterations or (2) A predefined number of iterations have been reached. Finally, the best bit string obtained is determined as the global optimal solution. Before running GA, several issues must be considered as follows.

A. System parameters. In this study, the population size is set to 20 and the initial population will contain 20 chromosomes (binary bit strings), which are randomly selected. The maximum number of iterations (generations) of GA is set as 60 and100 respectively in our experiment.

B. Fitness function. In this paper, Shannon's entropy function is used to quantify the gray-level histogram complexity [7]. The entropy function traditionally used in thermodynamics was first introduced by Claude Shannon in 1948 for the measurement of uncertainly in the information theory. Given a probability distribution $P = (p_1, p_2, \ldots, p_n)$ with $p_i \geq 0$ for $i = 1, 2, \ldots, n$ and $\sum_{i=1}^{n} p_i = 1$, the entropy of is:

$$E_{ntr} = -\sum_{i=1}^{n} (p_i \cdot \log p_i) \tag{5}$$

Where $p_i \cdot \log p_i = 0$ by definition for $p_i = 0$. Since p is a probability distribution, the histogram should be normalized before applying the entropy function. Considering noise enlarging problem during enhancement, peak signal-to-noise (PSNR) is used to quantify the quality of an enhanced image:

$$P_{snr} = 10 \cdot \log \left(\frac{MN \cdot \max\left(F_{ij}^2\right)}{\sum_{i=1}^{M} \sum_{j=1}^{N} \left(F_{ij} - G_{ij}\right)^2} \right) \tag{6}$$

Where F_{ij} and G_{ij} are gray level value at (i, j) in original image and enhanced image respectively. M and N are width and height of the original image respectively.

The fitness (objective) function is used to evaluate the goodness of a chromosome (solution). In this study, the fitness function is formed by Equation (7):

$$F_{ctr} = -E_{ntr} \cdot P_{snr} \tag{7}$$

Less F_{ctr} is, better contrast of enhanced image is.

C. Genetic operations. For GA, the three genetic operations, namely, reproduction, crossover, and mutation, will be implemented. In this study, a multi-point crossover is employed. For the multi-point crossover, the crossover-point positions of the bit string segments of pair-wise bit strings are randomly selected. Mutation is carried out by performing the bit inversion operation on some randomly selected positions of the parent bit strings and the probability of applying mutation, P_m , is set to 0.001.

The GA will be iteratively performed on an input degraded image until a stopping criterion is satisfied. The stopping criterion is the number of iterations is larger than another threshold (here they are set as 60 and 100 respectively in order to enhance two images). Then the chromosome (the solution) with the smallest fitness function value, i.e., the optimal set of IBT for the input degraded image, is determined. Using the optimal set of IBT enhances the degraded image. Totally there are two parameters in the set of IBT (α and β). The two parameters will form a solution space for finding the optimal set of IBT for image enhancement. Applying GA, the total two parameters will form a chromosome (solution) represented as a binary bit string, in which each parameter is described by 20 bits. We will employ the GA to optimize continuous variables [5]. If the algorithm is used directly to enhance image contrast, it will result in large computation cost.

5 IBT Calculation with WNN

IBT is calculated pixel-to-pixel. Operation burden is very large when pixels in original image are large. Different IBT have to be calculated to different α and β. Different IBT need to be calculated one time in every iterative step during optimization. To improve operation speed during the whole optimization, a new kind of wavelet neural network is proposed.

Let $f(x) \in L^2(R^n)$, WNN can be described approximately as follows:

$$Wf(x) = \sum_{i=1}^{N} w_i \psi[(a_i x - \tau_i)] \tag{8}$$

where τ_i is translation factor, a_i is scale factor, $Wf(x)$ shows the output of WNN. The translation factor, scale factor and wavelet basis function, which are on the same line, is called wavelet unit.

Parameters to be estimated are w_i, a_i, τ_i, $i = 1, 2, \cdots, N$ (where N is the number of wavelet unit). "Forgetting factor" algorithm is used to train weight of WNN. Iterative prediction error algorithm is employed to train translation factors and scale factors. Weight, translation factors and scale factors are trained iteratively and mutually with above two algorithms [6].

Let,

$$\theta = [a_1, a_2, \cdots, a_N, \tau_1, \tau_2, \cdots, \tau_N] \tag{9}$$

$$z_i(x) = \psi(a_i x - \tau_i) \tag{10}$$

Equation (8) is rewritten as:

$$Wf(x) = \sum_{i=1}^{N} w_i z_i(x) = \phi_t^T \mathbf{W}_t \tag{11}$$

Definition:

$$\phi_t = [z_1(t), z_2(t), \cdots, z_N(t)]^T \tag{12}$$

$$\mathbf{W}_t = [w_1(t), w_2(t), \cdots, w_N(t)]^T \tag{13}$$

Where $z_i(t)$ and $w_i(t)$ show the output of ith wavelet unit at time t and corresponding to weight respectively. T shows transpose of matrix. "Forgetting factor" algorithm can be written as following:

$$\mathbf{W}_t = \mathbf{W}_{t-1} + \mathbf{K}_t [y_t - \phi_t^T \mathbf{W}_{t-1}] \tag{14}$$

$$\mathbf{K}_t = (\alpha + \phi_t^T \mathbf{P}_{t-1} \phi_t)^{-1} \mathbf{P}_{t-1} \phi_t \tag{15}$$

$$\mathbf{P}_t = [\mathbf{P}_{t-1} - \mathbf{K}_t \phi_t^T \mathbf{P}_{t-1}]/\alpha \tag{16}$$

Where α is forgetting factor, $0 < \alpha \leq 1$. Parameter matrix θ can be estimated by following iterative prediction error algorithm:

$$\theta_t = \theta_{t-1} + \mathbf{R}_t [y_t - \phi_t^T \mathbf{W}_{t-1}] \tag{17}$$

$$\mathbf{R}_t = \frac{\mathbf{S}_{t-1}\mathbf{T}_t}{1 + \mathbf{T}_t^T \mathbf{S}_{t-1}\mathbf{T}_t} \tag{18}$$

$$\mathbf{S}_t = [\mathbf{S}_{t-1} - \mathbf{R}_t \mathbf{T}_t^T \mathbf{S}_{t-1}] \tag{19}$$

Where $\mathbf{T}_t = \partial W f(x)/\partial \theta$ shows gradient vector of output for WNN. Weight, translation factors and scale factors are trained iteratively and mutually with above two algorithms.

IBT can be calculated by the above WNN. Parameters α, β, g are input to trained WNN and output g' for IBT is obtained directly. 100000 points are selected as sample sets. Parameter and , which are between 1 and 10, are divided into 10 parts at the same interval. Parameter , which is between 0 and 1, is divided into 1000 parts at the same interval. 25 wavelet units are selected. The dimension number of input layer and output layer are determined according to the dimension number of input samples and output samples. Mexican hat wavelet is selected as mother wavelet:

$$\psi(x) = (1 - x^2)e^{-x^2/2} \tag{20}$$

The "forgetting factor" $\alpha = 0.97$ in the WNN. Mean square error is selected as error index and set as 0.00001.

6 Local Contrast Enhancement by Non-linear Operator

Based on DSWT, a non-linear enhancement operator, which was proposed by A. Laine in 1994, is employed to enhance the local contrast for image [8]. For convenience, let us define following transform function to enhance the high frequency sub-band images in each decomposition level respectively:

$$g[i,j] = MAG\{f[i,j]\} \tag{21}$$

Where $g[i,j]$ is sub-band image enhanced, $f[i,j]$ is original sub-band image to be enhanced, MAG is non-linear enhancement operator, M, N is width and height of image respectively. Let $f_s^r[i,j]$ is the gray values of pixels in the $\mathbf{r}th$ sub-band in the $\mathbf{s}th$ decomposition level, where $s = 1, 2, \cdots, L$; $r = 1, 2, 3$. $\max f_s^r$ is the maximum of gray value of all pixels in $f_s^r[i,j]$. $f_s^r[i,j]$ can be mapped from $[-\max f_s^r, \max f_s^r]$ to $[-1, 1]$. Thus the dynamic range of a, b, c can be set respectively. The contrast enhancement approach can be described by:

$$g_s^r[i,j] = \begin{cases} f_s^r[i,j], & |f_s^r[i,j]| < T_s^r \\ a \cdot \max f_s^r\{sigm[c(y_s^r[i,j] - b)] - \\ sigm[-c(y_s^r[i,j] + b)]\}, & |f_s^r[i,j]| \geq T_s^r \end{cases} \tag{22}$$

$$y_s^r[i,j] = f_s^r[i,j]/\max f_s^r \tag{23}$$

In order to obtain detail image, inverse discrete stationary wavelet transform is done to reconstruct detail image. Final enhanced image can be obtained by following equation:

$$\mathbf{I_F}[i,j] = \mathbf{I_G}[i,j] + C * \mathbf{I_L}[i,j] \tag{24}$$

Where $\mathbf{I_F}$ represents final enhanced image, $\mathbf{I_G}$ is global enhanced image, and $\mathbf{I_L}$ indicates local enhanced image, $C \geq 1$ represents a constant to be used to adjust the magnitude of local enhancement.

7 Algorithm Steps

Steps of contrast enhancement algorithm for image can be described as follows:

Step1. Determine contrast type of input image by the criterion in section 3;

Step2. Global enhancement is implemented by combing IBT, GA and WNN;

Step3. DSWT is made to the original image;

Step4. The non-linear operator in the DSWT domain is used to obtain the detail image;

Step5. Add the global enhanced image with the detail image to obtain final enhanced image;

Step6. Evaluate the quality of enhanced images by Equation (7) in section 4.

8 Experimental Results

Fig.2 is relationship curve between number of evolution generation and Best, where $Best = F_{ctr} = -E_{ntr} \cdot P_{snr}$. Fig.3 (a)-(b) are an infrared tank image and its histogram respectively.

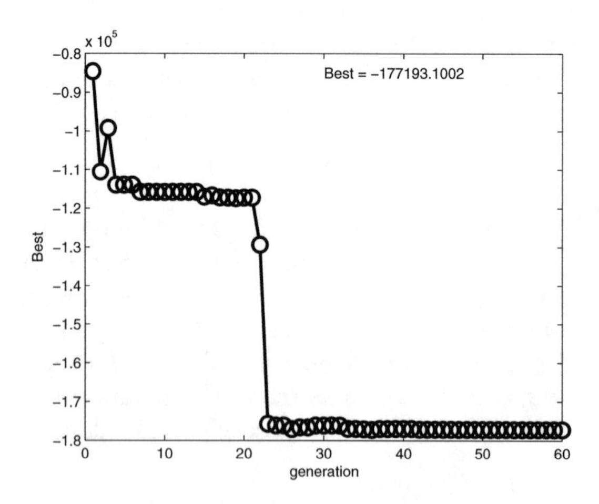

Fig. 2. Relationship curve between generation and best individual fitness

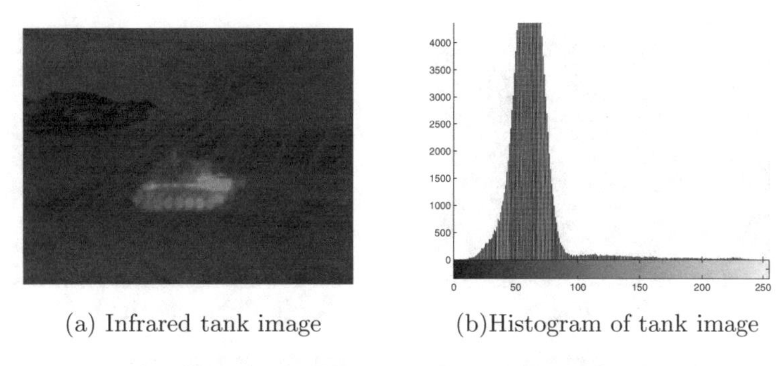

(a) Infrared tank image (b)Histogram of tank image

Fig. 3. Tank image and its histogram

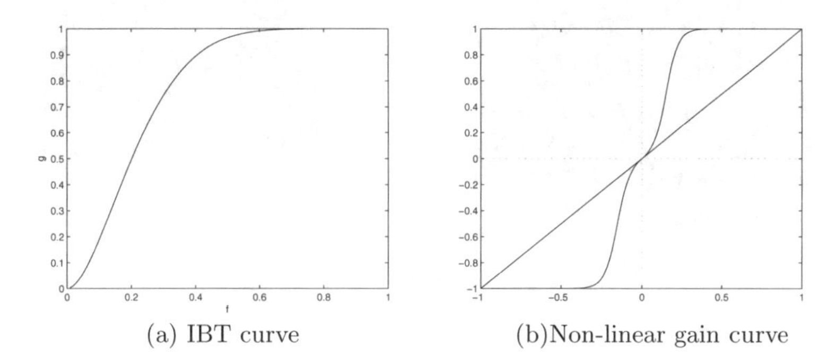

(a) IBT curve (b)Non-linear gain curve

Fig. 4. IBT curve and non-linear gain curve

Fig.4 (a) represents transform curve obtained by GA, where $alpha = 1.9971$, $beta = 7.0000$. Fig.4 (b) is a non-linear gain curve used to obtain the detail image, where $b = 0.15, c = 25$. Fig.4 (a)-(b) are used to enhanced the contrast in Fig.5 (a). Fig.5 (a) is a noisy image. Two traditional contrast enhancement algorithms are compared with the new algorithm. They are histogram equalization (HE) and unsharpened mask algorithm (USM) respectively.

Fig.5 (a) is a noisy image, which is added white noise (standard variance of noise is 4.6540). Fig.5 (b)-(d) are enhanced images by using HE, USM and the new algorithm respectively.

Although the global contrast is good when HE is used to enhance Fig.5 (a), background clutter and noise is also enlarged while some detail information also lost. Although local detail is well enhanced when USM is used to enhance Fig.5 (a), the global contrast is bad and noise is greatly enlarged in Fig.5 (c). The new algorithm can well enhance the global and local contrast in Fig.5 (d), and the background clutter and noise is also well suppressed. It is very obvious that the new algorithm is better in visual quality than HE and USM.

Fig.6 is relationship curve between number of evolution generation and Best. Fig.7 (a)-(b) are a fox image and its histogram respectively.

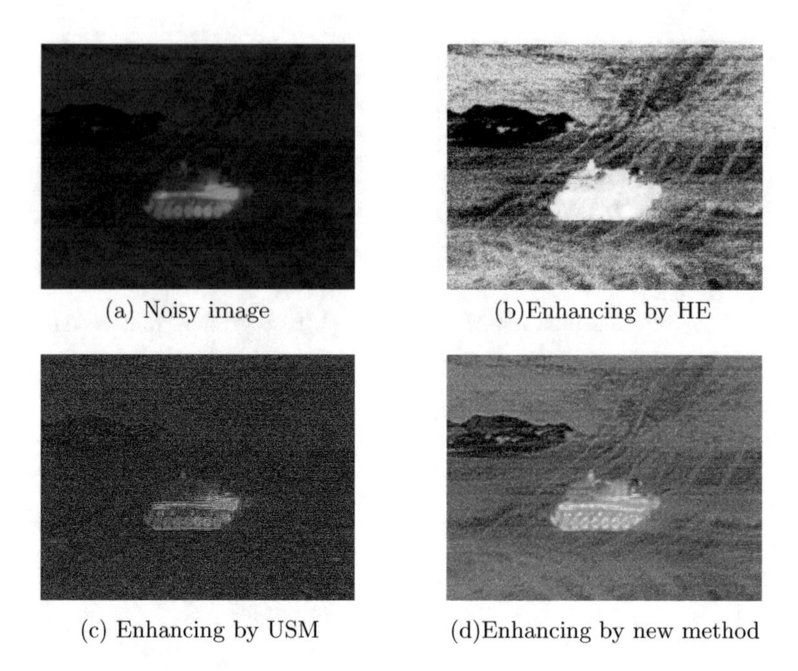

(a) Noisy image (b)Enhancing by HE

(c) Enhancing by USM (d)Enhancing by new method

Fig. 5. Enhancement by three methods

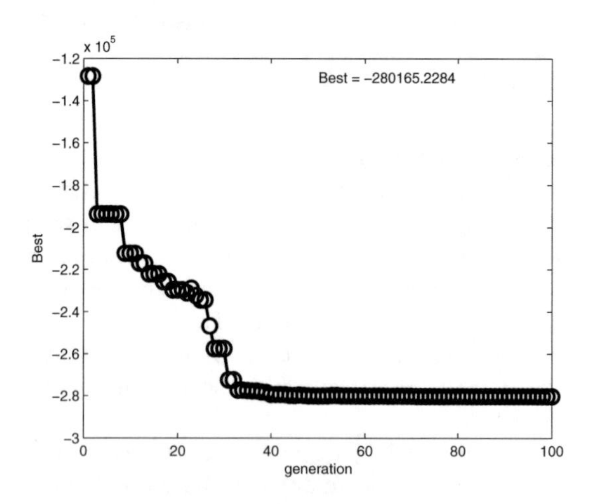

Fig. 6. Relationship curve between generation and best individual fitness

Fig.8 (a) represents transform curve obtained by GA, where $\alpha = 7.9991$, $\beta = 1.9980$. Fig.8 (b) is a non-linear gain curve used to obtain the detail image, where $b = 0.15$, $c = 35$.

Fig.8 (a)-(b) are used to enhance the global contrast and obtain the detail image in Fig.9 (a) respectively. Fig.9 (a) is a noisy image, which is added white noise (standard variance of noise is 6.3147). Fig.9 (b)-(d) are enhanced images

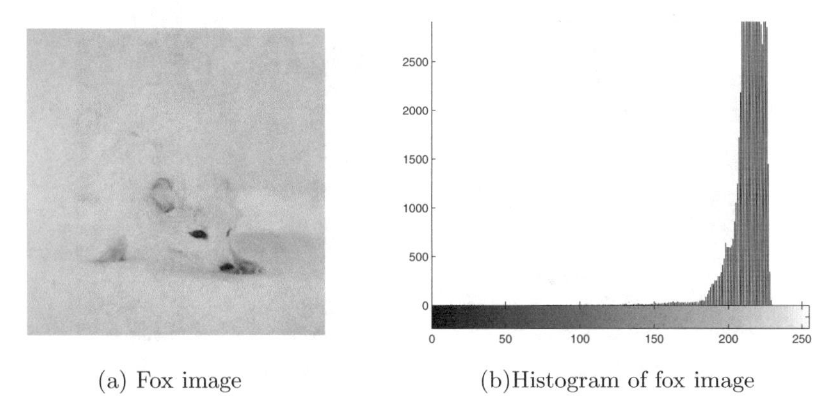

(a) Fox image (b)Histogram of fox image

Fig. 7. Fox image and its histogram

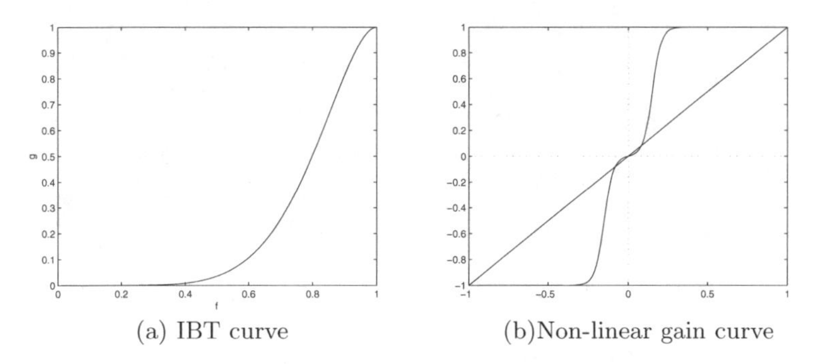

(a) IBT curve (b)Non-linear gain curve

Fig. 8. IBT curve and non-linear gain curve

using HE, USM and the new algorithm respectively. From Fig.9, the noise is greatly enlarged when HE is used to enhance contrast in Fig.9 (a). Some detail information in Fig.9 (b) is lost, for example, mouth and foot of fox cannot be distinguished. Although some detail information in Fig.9 (c) is well enhanced, however, noise in Fig.9 (c) is greatly enlarged. Some detail is also lost in Fig.9 (c), for example, the back of the fox has been submerged in the background and noise. Compared with HE and USM, global and local contrast is well enhanced while noise is less enlarged in Fig.9 (d).

In order to show the efficiency of the new algorithm, Equation (7) is used to evaluate the enahnced image quality. Evaluation results are 1.3653e+006, 9.9277e+005 and -1.7730e+005 respectively when USM, HE and the new algorithm are used to enhance the contrast in Fig.5. Similarily, the evaluation results are 1.1450e+007, 8.1576e+005 and -4.7548e+005 respectively in Fig.9. This can draw the same conclusion as in Fig.5 and Fig.9.

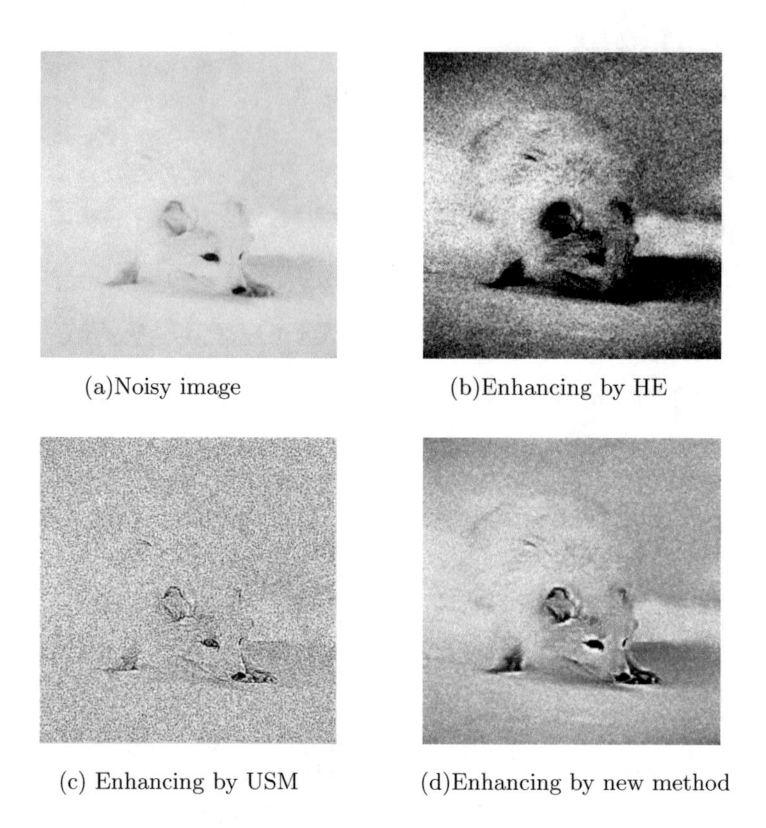

(a)Noisy image (b)Enhancing by HE

(c) Enhancing by USM (d)Enhancing by new method

Fig. 9. Enhancement by three methods

9 Conclusion

Experimental results show that the new algorithm can adaptively enhance the global and local contrast for an image while keeping the noise in an image from being greatly enlarged. The new algorithm is better than HE and USM in visual quality.

References

1. Azriel Rosenfield, Avinash C K (1982) Digital Picture Processing. Academic Press, New York
2. H.D. Cheng, Huijuan Xu (2002) Information Sciences 148:167–184
3. Stark, J.A (2000) IEEE Transactions on Image Processing 9:889–896
4. Tubbs J D (1997) Pattern Recognition 30:616–621
5. Ming-Suen Shyu, Jin-Jang Leou (1998) Pattern Recognition 31:871–880
6. Meiling Wang, Changjiang Zhang, Mengyin Fu (2002) Journal of Beijing Institute of Technology 22:274–278

7. J.C. Fu, H.C. Lien, S.T.C. Wong (2000) Computerized medical imaging and graphics 24:59–68
8. Laine, A., Schuler, S (1993) Part of SPIE's Thematic Applied Science and Engineering Series, Newport Beach, California 24:1179–1182

A Generalized Decision Logic Language for Information Tables*

Xiaosheng Wang

School of Computer Engineering, Jimei University, Xiamen 361021, China
davidwangxs@yahoo.com

Abstract. In this paper, we put forward a generalized decision logic language (*GDLL*) for information tables. The language is a generalization of the decision logic language (*DLL*) proposed by Z. Pawlak. Although it is an effective logic tool for analyzing the data in information tables, the *DLL* has a disadvantage in analyzing the data in inconsistent information tables. As an improvement of the *DLL*, the *GDLL* extends the range of the truth values of logic formulas from 0 and 1 to the interval [0, 1], which considers the degree of a logic formula's being true. The decision rules obtained from decision tables have the truth values that represent the credibility degrees of the decision rules by using the *GDLL*. Therefore, the *GDLL* is useful for dealing with the data in inconsistent information tables.

Keywords: Rough Sets, Decision Logic Language, Information Tables, Decision Rules, Credibility.

1 Introduction

Rough set theory, originally proposed by Z.Pawlak in Poland at the beginning of 1980s [1], has developed into a powerful mathematical method of data processing. Since inducing decision or classification rules is one of the major problems investigated by rough set theory, the theory is being widely used for data mining and machine learning.

In rough set theory, using the granular computing approach, the decision or classification rules are induced from information tables, a form of representing information and knowledge. Granular computing is a general principle of utilizing granules with appropriate size to resolve problems in terms of the practical necessity in problem solving. A granule is a set of elements that are drawn together by indistinguishability, similarity, proximity or functionality [2]. In rough set theory proposed by Z.Pawlak, the elementary granules are formed by equivalence relations and the other granules are studied based on the elementary granules. For inducing rules by using the granules, Z.Pawlak proposed a decision logic language (*DLL*) that combines classical logic with set theory to deal with the data in information tables [3]. The facts have proved that the *DLL* was a highly effective tool for analyzing the data in information tables [4-7]. However, the past explorations of the *DLL* mostly were

* Project supported by the Science Foundation of Jimei University, China.

B.-Y. Cao (Ed.): Fuzzy Information and Engineering (ICFIE), ASC 40, pp. 16–21, 2007.
springerlink.com

restricted within consistent information tables because using the *DLL* for inconsistent information tables is not fairly successful. But we had better not give up using the *DLL* when explore inconsistent information tables since the *DLL* is very useful for analyzing the data in information tables. For improving the application of the *DLL* to inconsistent information tables, we propose a generalized decision logic language (*GDLL*) and apply it to inconsistent information tables. The result shows that the *GDLL* is helpful for analyzing the data in inconsistent information tables. The structure of the article is the following. In Section 2, we introduce the decision logic language proposed by Z.Pawlak. In Section 3, we define a generalized decision logic language, and apply it to deal with the data in inconsistent information tables. Section 4 includes some conclusions and further research problems.

2 Decision Logic Language

In rough set theory, the information table is defined as the turple: S=(U, At, $\{V_a \mid a\in At\}$, $\{I_a \mid a\in At\}$), where U is a finite nonempty set of objects, At is a finite nonempty set of object attributes, V_a is a finite nonempty set of values for $a\in At$, I_a: $U\rightarrow V_a$ is an information function. Each information function I_a is a total function that maps an object of U to exactly one value in V_a [3, 5]. The following is an information table:

Table 1.

Objects	Attributes			
patients	headache	muscle ache	temperature	cold
e_1	yes	yes	high	yes
e_2	yes	yes	normal	no
e_3	yes	yes	very high	yes
e_4	no	yes	normal	no
e_5	no	no	high	no
e_6	no	yes	very high	yes
e_7	no	no	high	no
e_8	yes	no	very high	yes
e_9	no	yes	very high	yes

In Table 1, U=$\{e_1, e_2, e_3, e_4, e_5, e_6, e_7, e_8, e_9\}$, At={headache, muscle ache, temperature, cold}, $\{V_a \mid a\in At\}$={{yes, no}, {yes, no}, {normal, high, very high}, {yes, no}}, $I_{headache}(e_1)$=yes, $I_{muscle\ ache}(e_2)$=yes, $I_{temperature}(e_4)$=normal, $I_{cold}(e_5)$=no, etc.

Usually we denote the value of an object x over attribute subset $A\subseteq At$ by $I_A(x)$, which is a vector with each $I_a(x)$, $a\in A$, as one of its components [4, 6].

In [3], Pawlak defined a decision logic language (*DLL*) for information tables as follows: each (a, v) is an atomic formula, where $a\in At$ and $v\in V_a$; if φ and ψ are formulas, then so are $\neg\varphi$, $\varphi\wedge\psi$, $\varphi\vee\psi$, $\varphi\rightarrow\psi$, and $\varphi\leftrightarrow\psi$. The semantics of *DLL* are defined through the model of information tables. The satisfiability of a formula φ by an object x in information table S, denoted by $x \models_S \varphi$ or for short $x \models \varphi$ if S is understood, is defined by the following conditions [6-7]:

(1) $x \models (a, v)$ iff $I_a(x)=v$,
(2) $x \models \neg\varphi$ iff not $x \models \varphi$,
(3) $x \models \varphi \wedge \psi$ iff $x \models \varphi$ and $x \models \psi$,
(4) $x \models \varphi \vee \psi$ iff $x \models \varphi$ or $x \models \psi$,
(5) $x \models \varphi \rightarrow \psi$ iff $x \models \neg\varphi \vee \psi$,
(6) $x \models \varphi \leftrightarrow \psi$ iff $x \models \varphi \rightarrow \psi$ and $x \models \psi \rightarrow \varphi$.

If φ is a formula, then the set $m_S(\varphi)=\{x \in U \mid x \models_S \varphi\}$ is called the meaning of φ in S. We simply denote $m_S(\varphi)$ as $m(\varphi)$ if S is understood. Therefore, the meaning of a formula φ is the set of all objects having the property expressed by φ. At the same time, φ can be viewed as the description of the set of objects $m(\varphi)$. Obviously, the following properties hold [3, 7]:

(a) $m((a, v))=\{x \in U \mid I_a(x)=v\}$,
(b) $m(\neg\varphi)=\sim m(\varphi)$,
(c) $m(\varphi \wedge \psi)=m(\varphi) \cap m(\psi)$,
(d) $m(\varphi \vee \psi)=m(\varphi) \cup m(\psi)$,
(e) $m(\varphi \rightarrow \psi)=\sim m(\varphi) \cup m(\psi)$,
(f) $m(\varphi \leftrightarrow \psi)=(m(\varphi) \cap m(\psi)) \cup (\sim m(\varphi) \cap \sim m(\psi))$.

In the *DLL*, a formula φ is said to be true in S, written $\models_S \varphi$, if and only if $m(\varphi)=U$, i.e., φ is satisfied by all objects in the universe. We can infer easily that $\models_S \neg\varphi$ iff $m(\varphi)=\varnothing$, $\models_S \varphi \rightarrow \psi$ iff $m(\varphi) \subseteq m(\psi)$, and $\models_S \varphi \leftrightarrow \psi$ iff $m(\varphi)=m(\psi)$.

The axioms of the *DLL* include two parts: the classical propositional logic calculus system's axioms and some specific axioms on information tables. The specific axioms include:

(1) $(a, v) \wedge (a, u) \equiv 0$, where $a \in A$, $u, v \in V_a$, $u \neq v$ and "0" represents "false";

(2) $\bigvee_{v \in V_a} (a, v) \equiv 1$, where $a \in A$ and "1" represents "true".

The MP rule is the only inference rule of the *DLL*, i.e., if $\models_S \varphi \rightarrow \psi$ and $\models_S \varphi$, then $\models_S \psi$.

In the *DLL*, an implication $\varphi \rightarrow \psi$ is called a decision rule in information table S, where φ and ψ are two formulas in S; a decision algorithm is a finite nonempty set of decision rules. If $\models_S \varphi \rightarrow \psi$, then $\varphi \rightarrow \psi$ is called a consistent decision rule in S. And we state that a decision algorithm A is consistent if every decision rule contained in A is consistent.

Usually we divide the attributes into condition attributes and decision attributes in information table S, and call S a decision table, written $S=(U, At=C \cup D)$, where C is the set of condition attributes and D the set of decision attributes. For example, we can view Table 1 as a decision table, where C={headache, muscle ache, temperature}, D={cold}. In the rest of this paper, we restrict each decision rule mentioned to the decision rule in decision tables, such that the premise of the rule is a description on C, and the conclusion is a description on D. Decision table S is called a consistent decision table if the descriptions on D are identical in the event that the descriptions on C are identical. For consistent decision tables, we have the following determining theorem. First we present a definition.

Definition 1: [8] Let R_1 and R_2 be two equivalence relations on the set U. If for $\forall x, y \in U$, xR_1y implies xR_2y, we say that R_1 is finer than R_2, or R_2 is coarser than R_1, denoted by $R_1 \leq R_2$.

In decision table $S=(U, At=C \cup D)$, an equivalence relation induced by attribute subset $A \subseteq At$, written R(A), is given by: for $x, y \in U$, $xR(A)y \Leftrightarrow I_A(x)=I_A(y) \Leftrightarrow (\forall a \in A)Ia(x)=Ia(y)$ [4].

Obviously, if $A \subseteq B \subseteq At$, then $R(B) \leq R(A)$.

Theorem 1: [9] Given a decision table $S=(U, At=C \cup D)$, then S is consistent iff $R(C) \leq R(D)$.

According to Theorem 1, we know that Table 1 is consistent.

3 Generalized Decision Logic Language

It is clear that every decision rule in consistent decision tables is consistent. However, in inconsistent decision tables, there must exist more than one inconsistent decision rule.

For example, Table 2 is an inconsistent decision table since $R(C) \leq R(D)$ does not hold. In fact, in Table 2, rule γ: (headache, no)$\wedge$(muscle ache, no)$\wedge$(temperature, high)$\rightarrow$(cold, yes), is inconsistent. So is rule δ: (headache, no)$\wedge$(muscle ache, no)$\wedge$(temperature, high)$\rightarrow$(cold, no).

It is the common way of dealing with inconsistent decision tables to keep consistent decision rules and discard inconsistent decision rules. However, sometimes this way is not sensible because valuable information may be lost. For instance, the γ is a useful decision rule since it is correct under two-third cases. Therefore, for more effectively dealing with inconsistent decision tables, we propose a generalized decision logic language (*GDLL*).

Firstly, the formulas in the *DLL* are all the formulas in the *GDLL*. The meaning of formula φ in the *GDLL* is the same as that in the *DLL*. In the *GDLL*, an implication $\varphi \rightarrow \psi$ is said to be α $(0 \leq \alpha \leq 1)$ true in S, written $\models_{S(\alpha)} \varphi \rightarrow \psi$, if and only if $|m(\varphi) \cap m(\psi)|/|m(\varphi)|=\alpha$; a formula σ is said to be α $(0 \leq \alpha \leq 1)$ true in S, written $\models_{S(\alpha)} \sigma$, if and only if $|m(\sigma)|/|U|=\alpha$, i.e., σ is satisfied by $|U|*\alpha$ objects in the universe. In fact, since σ is equivalent to implication $1 \rightarrow \sigma$ and $m(1)=U$, the second definition can be inferred from the first one. Obviously, $\models_{S(\alpha)} \neg\sigma$ if and only if $\models_{S(1-\alpha)} \sigma$. In the *GDLL*, we define the inference rule: if $\models_{S(\alpha)} \varphi \rightarrow \psi$ and $\models_{S(\beta)} \varphi$, then $\models_{S(\alpha\beta)} \psi$, where $0 \leq \alpha, \beta \leq 1$. In addition, the axioms of the *GDLL* are the same as those of the *DLL*.

Table 2.

Objects	Attributes			
patients	headache	muscle ache	temperature	cold
e_1	no	no	high	yes
e_2	yes	yes	normal	no
e_3	yes	yes	very high	yes
e_4	no	yes	normal	no
e_5	no	no	high	yes
e_6	yes	no	very high	yes
e_7	no	no	high	no
e_8	yes	no	very high	yes
e_9	no	yes	normal	yes

Here, α and β denote the truth degree of formulas. The larger α and β are, the higher degrees of truth the corresponding formulas have. Obviously, when α=β=1, the *GDLL* is turned into the *DLL*. Hence, the *GDLL* is the generalization of the *DLL*, and can be applied to a wider scope, such as inconsistent decision tables. For simplicity, we denote the decision rule(or implication) $\varphi\rightarrow\psi$ with truth α as $\varphi\rightarrow_{\alpha}\psi$. Using the *GDLL* for the decision rule γ, we have:

(headache, no)$\wedge$(muscle ache, no)$\wedge$(temperature, high)$\rightarrow_{2/3}$(cold, yes).

Furthermore, we obtain the following decision rules by applying the *GDLL* to Table 2 :

γ_1: (headache, no)$\wedge$(muscle ache, no)$\wedge$(temperature, high)$\rightarrow_{2/3}$(cold, yes),

γ_2: (headache, no)$\wedge$(muscle ache, no)$\wedge$(temperature, high)$\rightarrow_{1/3}$(cold, no),

γ_3: (headache, yes)$\wedge$(muscle ache, yes)$\wedge$(temperature, normal)$\rightarrow_1$(cold, no),

γ_4: (headache, yes)$\wedge$(muscle ache, yes)$\wedge$(temperature, very high)$\rightarrow_1$(cold, yes),

γ_5: (headache, no)$\wedge$(muscle ache, yes)$\wedge$(temperature, normal)$\rightarrow_{1/2}$(cold, no),

γ_6: (headache, no)$\wedge$(muscle ache, yes)$\wedge$(temperature, normal)$\rightarrow_{1/2}$(cold, yes),

γ_7: (headache, yes)$\wedge$(muscle ache, no)$\wedge$(temperature, very high)$\rightarrow_1$(cold, yes).

So there are 3 consistent decision rules and 4 inconsistent decision rules in Table 2. Among the 4 inconsistent decision rules, γ_1 and γ_2 are more credible than γ_5 and γ_6 since they are less uncertain. In general, the closer to 1/2 the value of α is, the more uncertain the corresponding decision rule gets.

After inducing the decision rules from inconsistent decision tables by using *GDLL*, we always keep the decision rules that we believe credible and discard the others based on some criteria. One commonly-used way is to give a threshold, and then keep the decision rules whose truth degree is above the threshold. For example, in Table 2 if the threshold is set as 0.6, then decision rule γ_1, γ_3, γ_4 , γ_7 are kept, while γ_2, γ_5 , γ_6 are discarded. In other words, we believe that for Table 2, decision rule γ_1, γ_3, γ_4 and γ_7 are credible and informative. It is a practical way of dealing with inconsistent decision tables.

4 Conclusions

The decision logic language (*DLL*) proposed by Z. Pawlak was designed for analyzing the data in information tables. Although it is a strong logic tool of dealing with the data in information tables, the *DLL* needs to be improved for analyzing the data in inconsistent information tables. Therefore we propose a generalized decision logic language, written *GDLL* for short. In *GDLL*, the truth value of a logic formula is not limited to 1 (true) or 0 (false), but any value between 0 and 1. The truth value of a logic formula indicates the degree of the formula's being true. As the specific logic formula, each decision rule obtained from decision tables has the truth value that represents the credibility degree of the decision rule. Thus the *GDLL* is appropriate for analyzing the data in inconsistent information tables. In this article, we focus on analyzing the truth values of the decision rules induced from inconsistent information tables with the *GDLL*. However, some other problems, such as how to induce or reduce decision rules and decision algorithms from inconsistent information tables using the *GDLL*, need to be studied.

References

[1] Pawlak Z (1982) Rough sets. International Journal of Computer and Information Sciences 11: 341-356

[2] Zadeh LA (1997) Towards a theory of fuzzy information granulation and its centrality in human reasoning and fuzzy logic. Fuzzy Sets and Systems 90: 111-127

[3] Pawlak Z (1991) Rough sets-theoretical aspects of reasoning about data. Kluwer Academic Publishers, London

[4] Yao YY (2004) A partition model of granular computing. Lecture Notes in Computer Science 3100: 232-253

[5] Yao YY (2002) Granular computing using information tables. Data Mining, Rough Sets and Granular Computing. Physica-Verlag, Heidelberg: 102-124

[6] Yao YY, Yao JT (2004) Granular computing as a basis for consistent classification problems [Online]. Available: http://www2.cs.uregina.ca/~yyao/grc_paper/

[7] Yao YY, Zhong N (1999) Potential applications of granular computing in knowledge discovery and data mining. Proceedings of World Multiconference on Systemics, Cybernetics and Informatics, Computer Science and Engineering, Orlando, U.S.A: 573-580

[8] Zhang B, Zhang L (1992) Theory and applications of problem solving. North-Holland, Amsterdam

[9] Wang X (2006) Inducing decision rules: a granular computing approach. Proceedings of the 2006 IEEE International Conference on Granular Computing, Atlanta, U.S.A: 473-477

New Similarity Measures on Intuitionistic Fuzzy Sets

Jin Han Park[1,⋆], Jong Seo Park[2], Young Chel Kwun[3,⋆⋆], and Ki Moon Lim[1]

[1] Division of Mathematical Sciences, Pukyong National University, Pusan 608-737, South Korea
jihpark@pknu.ac.kr
[2] Department of Mathematics Education, Chinju National University of Education, Chinju 660-756, South Korea
parkjs@cue.ac.kr
[3] Department of Mathematics, Dong-A University, Pusan 604-714, South Korea
yckwun@dau.ac.kr

Abstract. Intuitionistic fuzzy sets, proposed by Atanassov, have gained attention from researchers for their applications in various fields. Then similarity measures between intuitionistic fuzzy sets were developed. In this paper, some examples are applied to show that some existing similarity measures are not effective in some cases. Then, based on the kind of geometrical background, we propose several new similarity measures of IFSs in which we consider three parameters describing a IFS. Finally, we apply these measures to pattern recognition.

Keywords: Intuitionistic fuzzy set, Similarity measure, Pattern recognition.

1 Introduction

The theory of fuzzy sets proposed by Zadeh [15] has showed successful applications in various fields. In fuzzy set theory, the membership of an element to fuzzy set is a single value between zero and one. But in reality, it may not always be certain that the degree of non-membership of an element in a fuzzy set is just equal to 1 minus the degree of membership. That is to say, there may be some hesitation degree. So, as a generalization of fuzzy sets, the concept of intuitionistic fuzzy sets (IFSs) was introduced by Atanassov in 1983. Bustince and Burillo [5] showed that this notion coincides with the notion of vague sets proposed by Gau and Buehere [10]. Measures of similarity between IFSs, as an important content in fuzzy mathematics, have attracted many researchers. Chen [6, 7], Chen and Tan [9] proposed several similarity measures for calculating the degree of similarity of vague sets; Li and Cheng [11] also proposed similarity measures of IFSs and applied these measures to pattern recognition. But Liang and Shi [12] and Mitchell [13] pointed out that Li and Cheng's measures are not always

⋆ This work was supported by the Korea Research Foundation Grant funded by the Korean Government(MOEHRD) (KRF-2006- 521-C00017).

⋆⋆ Corresponding author.

B.-Y. Cao (Ed.): Fuzzy Information and Engineering (ICFIE), ASC 40, pp. 22–30, 2007.
springerlink.com

effective in some cases and made some modifications, respectively. Szmit and Kacprzyk [14] gave a geometrical representation of a IFS and introduced three parameters to describe the distance between IFSs. In this paper, some examples are applied to show that some existing similarity measures are not effective in some cases. Then, based on the kind of geometrical background, we propose several new similarity measures of IFSs in which we consider three parameters describing a IFS. Finally, we apply these measures to pattern recognition.

2 Basic Notions and Definitions of Intuitionistic Fuzzy Sets

In the following, we firstly recall basic notions and definitions of IFSs which can be found in [1, 2, 3, 4].

Let X be the universe of discourse. An intuitionistic fuzzy set A in X is an object having the form

$$A = \{(x, \mu_A(x), \nu_A(x)) : x \in X\}$$

where $\mu_A, \nu_A : X \to [0,1]$ denote membership function and non-membership function, respectively, of A and satisfy $0 \leq \mu_A(x) + \nu_A(x) \leq 1$ for every $x \in X$.

Obviously, each fuzzy set A in X may be represented as the following IFS: $A = \{(x, \mu_A(x), 1 - \mu_A(x)) : x \in X\}$.

For each IFS A in X, we call

$$\pi_A(x) = 1 - \mu_A(x) - \nu_A(x)$$

the intuitionistic index of x in A. It is hesitation degree of x to A (see, [1, 2, 3, 4]). It is obvious that $0 \leq \pi_A(x) \leq 1$ for each $x \in X$. For example, let A be an IFS with membership function $\mu_A(x)$ and non-membership function $\nu_A(x)$, respectively. If $\mu_A(x) = 0.7$ and $\nu_A(x) = 0.1$, then we have $\pi_A(x) = 1-0.7-0.1 = 0.2$. It can be interpreted as the degree that the object x belongs to the IFS A is 0.7, the degree that the object x does not belong to the IFS A is 0.1 and the degree of hesitation is 0.2. Thus, IFS A in X can be expressed as

$$A = \{(\mu_A(x), \nu_A(x), \pi_A(x)) : x \in X\}.$$

If A is an ordinary fuzzy set, then $\pi_A(x) = 1 - \mu_A(x) - (1 - \mu_A(x)) = 0$ for each $x \in X$. It means that third parameter $\pi_A(x)$ can not be casually omitted if A is a general IFS, not an ordinary fuzzy set. Therefore, this representation of an IFS will be a point of departure for considering the our method in calculating the degree of similarity between IFSs.

For $A, B \in \text{IFS}(X)$, Atanassov [1, 2] defined the notion of containment as follows:

$$A \subseteq B \Leftrightarrow \mu_A(x) \leq \mu_B(x) \text{ and } \nu_A(x) \geq \nu_B(x),\ \forall x \in X.$$

As we mentioned above, we can not omit the hesitation margin in the representation of IFS and so we redefine the notion of containment as follows:

$$A \subseteq B \Leftrightarrow \mu_A(x) \leq \mu_B(x),\ \nu_A(x) \geq \nu_B(x) \text{ and } \pi_A(x) \geq \pi_B(x),\ \forall x \in X.$$

Similarity measure is a term that measures the degree of similarity between IFSs. As an important content in fuzzy mathematics, similarity measures between IFSs have gained much attentions for their wide applications in real world, such as pattern recognition, machine learning, decision making and market prediction.

Definition 1. (Li and Cheng [11]) Let S : IFS$(X) \times$ IFS$(X) \rightarrow [0,1]$ be a mapping. $S(A,B)$ is said to be the degree of similarity between $A \in$ IFS(X) and $B \in$ IFS(X) if it satisfied the properties (P1)-(P4):
$(P1)$ $0 \leq S(A,B) \leq 1$;
$(P2)$ $S(A,B) = 1$ if $A = B$;
$(P3)$ $S(A,B) = S(B,A)$;
$(P4)$ $S(A,C) \leq S(A,B)$ and $S(A,C) \leq S(B,C)$ if $A \subseteq B \subseteq C$, $A,B,C \in$ IFS(X).

But this definition has some limitations. So, Mitchell [13] gave a simple modification of it by replacing $(P2)$ with a strong version $(P2')$ as follows.

$(P2')$ $S(A,B) = 1$ if and only if $A = B$.

This definition was proved to be more reasonable that Li and Cheng's.

3 New Approaches to Calculating Similarity Measures

In this section, we firstly recall similarity measures proposed by Li and Cheng [11] and Liang and Shi [12]. Then we give an example to show that the Liang and Shi's method is not always effective in some cases. So we propose several new similarity measures on IFSs and compare proposed method with Li and Cheng's and Liang and Shi's methods.

Assume that A and B are two IFSs in $X = \{x_1, x_2, \ldots, x_n\}$, respectively.

Li and Cheng defined similarity measure between IFSs A and B in X as follows: Let $\varphi_A(i) = (\mu_A(x_i) + 1 - \nu_A(x_i))/2$, $\varphi_B(i) = (\mu_B(x_i) + 1 - \nu_B(x_i))/2$ and $x_i \in X$. Then

$$S_d^p(A,B) = 1 - \frac{1}{\sqrt[p]{n}} \sqrt[p]{\sum_{i=1}^{n} (\varphi_A(i) - \varphi_B(i))^p}\ , \tag{1}$$

where $1 \leq p < \infty$.

From $S_d^p(A,B)$, we note that $\varphi_A(i)$ is the median value of the interval $[\mu_A(x_i), 1 - \nu_A(x_i)]$. Then we consider the following fact: If median values of each subinterval between IFSs are equal respectively, similarity measures between two

IFSs are equal 1. Thus, $S_d^p(A,B)$ does not satisfy the condition $(P2')$ of Michell's definition. To solve this problem, Liang and Shi proposed another definition of similarity measure between IFSs as follows: Let $\varphi_{\mu_{AB}}(i) = |\mu_A(x_i) - \mu_B(x_i)|/2$, $\varphi_{\nu_{AB}}(i) = |(1-\nu_A(x_i)) - (1-\nu_B(x_i))|/2$ and $x_i \in X$. Then

$$S_e^p(A,B) = 1 - \frac{1}{\sqrt[p]{n}} \sqrt[p]{\sum_{i=1}^{n} (\varphi_{\mu_{AB}}(i) + \varphi_{\nu_{AB}}(i))^p}\ , \tag{2}$$

where $1 \leq p < \infty$.

Although in most cases the similarity measure $S_e^p(A,B)$ gives intuitively satisfying results, there are some cases in which this is not true. The following example shows one such case.

Example 1. Assume that there are two patterns denoted with IFSs in $X = \{x_1, x_2, x_3\}$. Two patterns A_1 and A_2 are denoted as follows:

$$A_1 = \{(x_1, 0.2, 0.6), (x_2, 0.2, 0.6), (x_3, 0.2, 0.5)\};$$
$$A_2 = \{(x_1, 0.4, 0.6), (x_2, 0.2, 0.6), (x_3, 0, 0.3)\}.$$

Assume that a sample $B = \{(x_1, 0.3, 0.7), (x_2, 0.3, 0.5), (x_3, 0.1, 0.4)\}$ is given. Then, it is obvious that $S_e^p(A_1,B) = S_e^p(A_2,B)$. So, the patterns cannot be differentiated using the above Liang and Shi's method. Hence, we cannot obtain correct recognition results.

In order to deal with this problem, based on the geometrical representation of an IFS, we propose another definition of similarity measure between IFSs by considering the hesitation degrees of IFSs in Liang and Shi's definition as follows:

Definition 2. Let $\varphi_{\mu_{AB}}(i) = |\mu_A(x_i) - \mu_B(x_i)|/2$, $\varphi_{\nu_{AB}}(i) = |\nu_A(x_i) - \nu_B(x_i)|/2$, $\varphi_{\pi_{AB}}(i) = |\pi_A(x_i) - \pi_B(x_i)|/2$ and $x_i \in X$. Then

$$S_g^p(A,B) = 1 - \frac{1}{\sqrt[p]{n}} \sqrt[p]{\sum_{i=1}^{n} (\varphi_{\mu_{AB}}(i) + \varphi_{\nu_{AB}}(i) + \varphi_{\pi_{AB}}(i))^p}, \tag{3}$$

where $1 \leq p < \infty$.

From Definition 2, we can obtain the following Theorems 1 and 2.

Theorem 1. *$S_g^p(A,B)$ is a degree of similarity in the Mitchell's sense between two* IFSs *A and B in $X = \{x_1, x_2, \ldots, x_n\}$.*

Proof. Obviously, $S_g^p(A,B)$ satisfies $(P1)$ and $(P3)$. As to $(P2')$ and $(P4)$, we give the following proof.

$(P2')$: It is obvious that $\mu_A(x_i) = \mu_B(x_i)$, $\nu_A(x_i) = \nu_B(x_i)$ and $\pi_A(x_i) = \pi_B(x_i)$ from Eq. (3). Therefore, $A = B$.

$(P4)$: Since $A \subseteq B \subseteq C$, we have $\mu_A(x_i) \leq \mu_B(x_i) \leq \mu_C(x_i)$, $\nu_A(x_i) \geq \nu_B(x_i) \geq \nu_C(x_i)$ and $\pi_A(x_i) \geq \pi_B(x_i) \geq \pi_C(x_i)$ for any $x_i \in X$. Then we have

$$\begin{aligned}
&\varphi_{\mu_{AB}}(i) + \varphi_{\nu_{AB}}(i) + \varphi_{\pi_{AB}}(i) \\
&= |\mu_A(x_i) - \mu_B(x_i)|/2 + |\nu_A(x_i) - \nu_B(x_i)|/2 + |\pi_A(x_i) - \pi_B(x_i)|/2 \\
&\leq |\mu_A(x_i) - \mu_C(x_i)|/2 + |\nu_A(x_i) - \nu_C(x_i)|/2 + |\pi_A(x_i) - \pi_C(x_i)|/2 \\
&= \varphi_{\mu_{AC}}(i) + \varphi_{\nu_{AC}}(i) + \varphi_{\pi_{AC}}(i).
\end{aligned}$$

So we have

$$\sqrt[p]{\sum_{i=1}^{n}(\varphi_{\mu_{AC}}(i) + \varphi_{\nu_{AC}}(i) + \varphi_{\pi_{AC}}(i))^p} \geq \sqrt[p]{\sum_{i=1}^{n}(\varphi_{\mu_{AB}}(i) + \varphi_{\nu_{AB}}(i) + \varphi_{\pi_{AB}}(i))^p}.$$

Therefore, $S_g^p(A, C) \leq S_g^p(A, B)$. In the similar way, it is easy to prove $S_g^p(A, C) \leq S_g^p(B, C)$. □

Theorem 2. *Assume that $S_d^p(A, B)$, $S_e^p(A, B)$ and $S_g^p(A, B)$ are given in Eq. (1), Eq. (2) and Eq. (3), respectively. Then we have $S_d^p(A, B) \geq S_e^p(A, B) \geq S_g^p(A, B)$.*

Proof. Let $x_i \in X$. Since $((\mu_A(x_i) - \mu_B(x_i))/2 - (\nu_A(x_i) - \nu_B(x_i))/2 \leq |\mu_A(x_i) - \mu_B(x_i)|/2 + |\nu_A(x_i) - \nu_B(x_i)|/2 \leq |\mu_A(x_i) - \mu_B(x_i)|/2 + |\nu_A(x_i) - \nu_B(x_i)|/2 + |\pi_A(x_i) - \pi_B(x_i)|/2$, we have $(\varphi_A(i) - \varphi_B(i)) \leq \varphi_{\mu_{AB}}(i) + \varphi_{\nu_{AB}}(i) \leq \varphi_{\mu_{AB}}(i) + \varphi_{\nu_{AB}}(i) + \varphi_{\pi_{AB}}(i)$. Therefore, $S_d^p(A, B) \geq S_e^p(A, B) \geq S_g^p(A, B)$. □

Example 2. We consider the two patterns A_1 and A_2 and the sample B discussed in Example 1. Let $p = 1$. Then, using Eq. (3), we find $S_g^1(A_1, B) = 0.8333$ and $S_g^1(A_2, B) = 0.8667$ and thus obtain the satisfying result $S_g^1(A_1, B) \neq S_g^1(A_2, B)$. Hence, we obtain correct recognition results.

When the universe of discourse is continuous, we can obtain the following similar result. For IFSs $A = \{(x, \mu_A(x), \nu_A(x)) : x \in [a, b]\}$ and $B = \{(x, \mu_B(x), \nu_B(x)) : x \in [a, b]\}$, set

$$S_k^p(A, B) = 1 - \frac{1}{\sqrt[p]{b-a}} \sqrt[p]{\int_a^b (\varphi_{\mu_{AB}}(x) + \varphi_{\nu_{AB}}(x) + \varphi_{\pi_{AB}}(x))^p dx}\ , \qquad (4)$$

where $1 \leq p < \infty$.

Similarly, $S_k^p(A, B)$ has the following result.

Theorem 3. *$S_k^p(A, B)$ is a degree of similarity in the Mitchell's sense between two* IFSs *A and B in $X = [a, b]$.*

Proof. The proof is similar to that of Theorem 1. □

However, the elements in the universe may have different importance in pattern recognition. So we should consider the weight of the elements so that we can obtain more reasonable results in pattern recognition.

Assume that the weight of x_i in X is w_i, where $w_i \in [0, 1]$, $i = 1, 2, \ldots, n$ and $\sum_{i=1}^{n} w_i = 1$. Similarity measure between IFSs A and B can be obtained by the following form.

$$S_{gw}^{p}(A, B) = 1 - \sqrt[p]{\sum_{i=1}^{n} w_i(\varphi_{\mu_{AB}}(i) + \varphi_{\nu_{AB}}(i) + \varphi_{\pi_{AB}}(i))^p}, \tag{5}$$

where $1 \leq p < \infty$.

Likewise, for $S_{gw}^{p}(A, B)$, the following theorem holds.

Theorem 4. *$S_g^p(A, B)$ is a degree of similarity in the Mitchell's sense between two IFSs A and B in $X = \{x_1, x_2, \ldots, x_n\}$.*

Proof. The proof is similar to that of Theorem 1. □

Remark 1. Obviously, if $w_i = 1/n$ $(i = 1, 2, \ldots, n)$, Eq. (5) becomes Eq. (3). So, Eq. (3) is a special case of Eq. (5).

Similarly, assume that the weight of $x \in X = [a, b]$ is $w(x)$, where $0 \leq w(x) \leq 1$ and $\int_a^b w(x)dx = 1$.

$$S_{kw}^{p}(A, B) = 1 - \sqrt[p]{\int_a^b w(x)(\varphi_{\mu_{AB}}(x) + \varphi_{\nu_{AB}}(x) + \varphi_{\pi_{AB}}(x))^p}, \tag{6}$$

where $1 \leq p < \infty$.

Theorem 5. *$S_{kw}^{p}(A, B)$ is a degree of similarity in the Mitchell's sense between two IFSs A and B in $X = [a, b]$.*

Proof. The proof is similar to that of Theorem 1. □

Remark 2. It is obvious that Eq. (6) becomes Eq. (4) if $w(x) = 1/(b-a)$ for any $x \in X = [a, b]$. So, Eq. (4) is a special case of Eq. (6).

4 Application to Pattern Recognition Problem

Assume that the question related to pattern recognition is given using IFSs. Li and Cheng proposed the principle of the maximum degree of similarity between IFSs to solve the problem of pattern recognition. We can also apply the principle of the maximum degree of similarity between IFSs to solve the problem as well as Liang and Shi.

Assume that there exist m patterns which are represented by IFSs $A_i = \{(x_i, \mu_A(x_i), \nu_A(x_i)) : x_i \in X\} \in \mathrm{IFS}(X)$ $(i = 1, 2, \ldots, m)$. Suppose that there be a sample to be recognized which is represented by $B = \{(x_i, \mu_B(x_i), \nu_B(x_i) : x_i \in X\} \in \mathrm{IFS}(X)$. Set

$$S^p_{gw}(A_{i_0}, B) = \max_{1 \le i \le n} \{S^p_{gw}(A_i, B)\}, \tag{7}$$

where $X = \{x_1, x_2, \ldots, x_n\}$. When $X = [a, b]$, set

$$S^p_{kw}(A_{i_0}, B) = \max_{1 \le i \le n} \{S^p_{kw}(A_i, B)\}. \tag{8}$$

According to the principle of the maximum degree of similarity between IFSs, we can decide that the sample B belongs to some pattern A_{i_0}.

In the following, two examples are given to show that the proposed similarity measures can overcome the drawbacks of some existing methods and can deal with problems more effectively and reasonably than those methods.

Example 3. Assume that there are three patterns denoted with IFSs in $X = \{x_1, x_2, x_3\}$. Three patterns A_1, A_2 and A_3 are denoted as follows:

$$A_1 = \{(x_1, 0.2, 0.6), (x_2, 0.1, 0.7), (x_3, 0.0, 0.6)\};$$
$$A_2 = \{(x_1, 0.2, 0.6), (x_2, 0.0, 0.6), (x_3, 0.2, 0.8)\};$$
$$A_3 = \{(x_1, 0.1, 0.5), (x_2, 0.2, 0.7), (x_3, 0.2, 0.8)\}.$$

Assume that a sample $B = \{(x_1, 0.3, 0.7), (x_2, 0.2, 0.8), (x_3, 0.1, 0.7)\}$ is given. For convenience, assume that the weight w_i of x_i in X is equal and $p = 1$.

By Eq. (1), we have

$$S^1_d(A_1, B) = 1,\ S^1_d(A_2, B) = 1,\ S^p_d(A_3, B) = 0.983.$$

By Eq. (2), we have

$$S^1_e(A_1, B) = 0.900,\ S^1_e(A_2, B) = 0.867,\ S^p_e(A_3, B) = 0.883.$$

By Eq. (3), we have

$$S^1_g(A_1, B) = 0.800,\ S^1_g(A_2, B) = 0.733,\ S^p_g(A_3, B) = 0.767.$$

From the above result, it is obvious that $S^p_d(A, B) \ge S^p_e(A, B) \ge S^p_g(A, B)$. We see that the degree of similarity between A_2 and B is the greatest and so we can say A_2 should approach B.

Example 4. Assume that there are two patterns denoted with IFSs in $X = [1, 5]$. Two patterns A_1 and A_2 are denoted as follows:

$$A_1 = \{(x, \mu_{A_1}(x), \nu_{A_1}(x)) : x \in [1, 5]\};$$
$$A_2 = \{(x, \mu_{A_2}(x), \nu_{A_2}(x)) : x \in [1, 5]\},$$

where

$$\mu_{A_1}(x) = \begin{cases} \frac{4}{5}(x-1), & 1 \le x \le 2, \\ -\frac{4}{15}(x-5), & 2 \le x \le 5, \end{cases} \quad \nu_{A_1}(x) = \begin{cases} -\frac{1}{10}(9x-19), & 1 \le x \le 2, \\ \frac{1}{10}(3x-5), & 2 \le x \le 5, \end{cases}$$

$$\mu_{A_2}(x) = \begin{cases} \frac{1}{5}(x-1), & 1 \le x \le 4, \\ -\frac{3}{5}(x-5), & 4 \le x \le 5, \end{cases} \quad \nu_{A_2}(x) = \begin{cases} -\frac{1}{10}(3x-13), & 1 \le x \le 4, \\ \frac{1}{10}(9x-35), & 4 \le x \le 5. \end{cases}$$

Consider a sample $B = \{(x, \mu_B(x), \nu_B(x)) : x \in [1,5]\}$ which will be recognized, where

$$\mu_B(x) = \begin{cases} \frac{3}{10}(x-1), & 1 \le x \le 3, \\ -\frac{3}{10}(x-5), & 3 \le x \le 5, \end{cases} \quad \nu_B(x) = \begin{cases} -\frac{1}{5}(2x-7), & 1 \le x \le 3, \\ \frac{1}{5}(2x-5), & 3 \le x \le 5. \end{cases}$$

For convenience, let $w(x) = 1/(b-a)$ for any $x \in X = [a,b]$ and $p = 1$. By applying Eq. (4) or (6), we have $S_k^1(A_1, B) = 0.3638$ and $S_k^1(A_2, B) = 0.6407$. Thus we can say A_2 should approach B. Obviously, this result is coincide with the result of Li and Cheng's.

5 Conclusions

The similarity measures between IFSs were developed by some researchers. Although some existing similarity measures can provide an effective way to deal with IFSs in some cases, it seems to obtain unreasonable results using those measures in some special cases. Therefore, new similarity measures are proposed to differentiate different IFSs. The proposed similarity measures can overcome the drawbacks of some existing methods and can deal with problems more effectively and reasonably than those methods.

References

1. Atanassov K (1984) Intuitionistic fuzzy sets. in: V. Sgurev (Ed), VII ITKR's Session. Sofia (June 1983 Central Sci and Techn Library Bulg Academy of Sciences).
2. Atanassov K (1986) Intuitionistic fuzzy sets. Fuzzy Sets and Systems 20: 87–96.
3. Atanassov K (1989) More on intuitionistic fuzzy sets. Fuzzy Sets and Systems 33: 37–46.
4. Atanassov K (1994) New operations defined over the intuitionistic fuzzy sets. Fuzzy Sets and Systems 61: 1377-142.
5. Bustince H, Burillo P (1996) Vague sets are intuitionistic fuzzy sets. Fuzzy Sets and Systems 79: 403-405.
6. Chen SM (1994) A weighted fuzzy reasoning algorithm for medical diagnosis. Decis Support Systems 11: 37-43.
7. Chen SM (1995) Measures of similarity between vague sets. Fuzzy Sets and Systems 74: 217-223.
8. Chen SM (1997) Similarity measure between vague sets and elements. IEEE Trans Systems Man Cybernt 27: 153-158.
9. Chen SM, Tan JM (1994) Handling multi-criteria fuzzy decision-making problems based on vague sets. Fuzzy Sets and Systems 67: 163-172.
10. Gau WL, Buehere DJ (1994) Vague sets. IEEE Trans Systems Man Cybernt 23: 610-614.
11. Li D, Cheng C (2002) New similarity measures of intuitionistic fuzzy fuzzy sets and applications to pattern recognitions. Pattern Recognition Lett 23: 221-225.

12. Liang Z, Shi P (2003) Similarity measures on intuitionistic fuzzy fuzzy sets. Pattern Recognition Lett 24: 2687-2693.
13. Mitchell HB (2003) On the Dengfeng-Chuitian similarity measure and its application to pattern recognition. Pattern Recognition Lett 24: 3101-3104.
14. Szmit E, Kacprzyk J (2000) Distances between intuitionistic fuzzy sets. Fuzzy Sets and Systems 114: 505-518.
15. Zadeh LA (1965) Fuzzy sets. Inform and Control 8: 338-353.

Information Source Entropy Based Fuzzy Integral Decision Model on Radiodiagnosis of Congenital Heart Disease

Yecai Guo

Anhui University of Science and Technology, Huainan 232001, China
guo-yecai@163.com

Abstract. Based on fuzzy mathematical principle, Information source Entropy based Fuzzy integral decision Model(IEFM) on radiodiagnosis of congenital heart disease is created. In this paper, attaching function, quantum standard, weight value of each symptom, which causes disease, are determined under condition that medical experts take part in. The detailed measures are taken as follows: First, each medical expert gives the scores of all symptom signs of each symptom based on their clinic experience and professional knowledge. Second, based on analyzing the feature of the scores given by medical experts, attaching functions are established using the curve fitting method. Third, weight values of symptoms are calculated by the information source entropy and these weight values modified by optimization processing. Fourth, information source entropy based fuzzy integral decision model(IEFM) is proposed. Finally, the relative information is obtained from the case histories of the cases with radiodiagnosis of congenital heart disease. Fuzzy integral value of each case is calculated. Accurate rate of the IEFM is greater than that of the Single Fuzzy Integral decision Model(SFIM) via diagnosing the anamneses.

1 Introduction

The mathematical model of microcomputer medical expert is used as analyzing the scientific ideals of medical experts and a mathematical description of dialectic diagnosis process.

From mathematical views, the diagnosis process of medical experts on a diseases may be viewed as a rule of correspondence between symptom sets (including physical sign and laboratory indexes) and disease type sets and decribed as follows:

Assume that symptom set is denoted as

$$\mathbb{Z} = \{\mathbb{Z}_1, \mathbb{Z}_2, \cdots, \mathbb{Z}_j, \cdots, \mathbb{Z}_n\} \tag{1}$$

and disease type set is written as

$$M = \{M_1, M_2, \cdots, M_i, \cdots, M_m\} \tag{2}$$

where $\mathbb{Z}_j$ denotes the j th symptom of the case, and M_i is the i th disease.

Accordingly, the diagnosis process of medical experts may be regarded as the rule of the correspondences between $\mathbb{Z}$ and M and this mapping is written as $\mathbb{Z} \to M$.

B.-Y. Cao (Ed.): Fuzzy Information and Engineering (ICFIE), ASC 40, pp. 31–40, 2007.
springerlink.com

From information theory, the diagnosis process of medical experts is a process of information processing. In this process, the patients are information sources and the medical expert is a receiver of the given information obtained from the given patients. A diagnosis process is just a communication between the medical experts and the patients. After a communication, the valuable information that redounds to the accurate diagnosis upon the disease can be obtained by the medical experts from the patients and this information capacity can be measured by the information source entropy. In informatics, it is well-known that the information source entropy measures the degree of great disorder or confusion of system. As for God's image, under normal physiological condition, a zoetic organism is an organized or ordered open system. As a rule, under pathological condition, the entropy of system will increase with increasing disorder or confusion of system. So, in order to improve the accurate diagnosis of clinist on a disease, it is very appropriate to employ the information source entropy for analyzing the value of clinic symptoms.

The scientific ideals and diagnosis process of medical experts, which include logic thinking and visual thinking, are most complicated. The logic thinking may be divided into formal logic thinking and dialectic logic thinking. The formal logic thinking and the dialectic logic thinking integrate and infiltrate each other, and the diagnosis process of medical experts on a disease incarnates hominine subjectivity. Apparently, symptoms of sufferers and diagnostic estimation of medical experts are forceful fuzzy. For processing fuzzy information, it is very necessary to combine statistical analysis, information theory, and fuzzy mathematics to insure the rationality of the medical expert diagnosis system model.

The organization of this paper is as follows. In section 2, we establish the model on radio-diagnosis of congenital heart disease. Section 3 gives an example to illustrate the validity of this model and optimization processing.

2 Medical Expert Diagnosis System Model

Based on medical theory and clinic experiences of the medical experts and concerned references[1][2], the symptom set of radiodiagnosis upon congenital heart disease is written as

$$\mathbb{Z}=\{\mathbb{Z}_1,\mathbb{Z}_2,\mathbb{Z}_3,\mathbb{Z}_4,\mathbb{Z}_5,\mathbb{Z}_6,\mathbb{Z}_7,\mathbb{Z}_8,\mathbb{Z}_9,\mathbb{Z}_{10},\mathbb{Z}_{11},\mathbb{Z}_{12}\} \tag{3}$$

where $\mathbb{Z}_1,\mathbb{Z}_2,\mathbb{Z}_3,\mathbb{Z}_4,\mathbb{Z}_5,\mathbb{Z}_6,\mathbb{Z}_7,\mathbb{Z}_8,\mathbb{Z}_9,\mathbb{Z}_{10},\mathbb{Z}_{11},\mathbb{Z}_{12}$ present cardiac type, size of heart, left shift of heart, right atrium, right ventricle, left ventricle, left shift of ascending aorta, aortic knob, pulmonary artery segment, blood vessel around pulmonary, pulmonary hypertension, pulsation of hilus of lung, respectively. The jth symptom $\mathbb{Z}_j$ is denoted by

$$\mathbb{Z}_j=\{\mathbb{Z}_{j1},\mathbb{Z}_{j2},\cdots,\mathbb{Z}_{jl},\cdots,\mathbb{Z}_{iK}\} \tag{4}$$

where $\mathbb{Z}_{jl}$ represents the lth symptom sign of the jth symptom $\mathbb{Z}_j$ (See formulas of attaching function), $j=1,2,\cdots,12$, $l=1,2,\cdots,K$, and K is the number of symptom signs.

Assume that the type set of congenital heart diseases is written as

$$M = \{M_1, M_2, M_3, M_4\} \tag{5}$$

where M_1, M_2, M_3, M_4 present in turn atrial septal sefect, ventricular septal sefect, patent arterial duct, pulmonary stenosis.

2.1 Conformation of Attaching Function of Each Symptom

Attaching function of each symptom sign is given by following equations

$$\mu_i(\mathbb{Z}_1) = \begin{cases} \frac{a_{11}+10}{62}(normal, \mathbb{Z}_{11}) \\ \frac{a_{12}}{112}(mitral\ valve, \mathbb{Z}_{12}) \\ \frac{a_{13}+10}{62}(aortic\quad valve, \mathbb{Z}_{13}) \\ \frac{a_{14}+10}{90}(mixed, \mathbb{Z}_{14}) \\ \frac{a_{15}+10}{62}(sphere, \mathbb{Z}_{15}) \end{cases}, \tag{6a}$$

$$\mu_i(\mathbb{Z}_2) = \begin{cases} \frac{4a_{21}-15}{15}(normal\quad size, \mathbb{Z}_{21}) \\ \frac{20a_{22}-15}{234}(larger, \mathbb{Z}_{22}) \\ \frac{a_{23}}{14}(much\quad larger, \mathbb{Z}_{23}) \end{cases}, \tag{6b}$$

$$\mu_i(\mathbb{Z}_3) = \begin{cases} \frac{2a_{31}+1}{21}(no\quad, \mathbb{Z}_{31}) \\ \frac{a_{32}+10}{50}(yes, \mathbb{Z}_{32}) \end{cases}, \tag{6c}$$

$$\mu_i(\mathbb{Z}_4) = \begin{cases} \frac{2a_{41}+1}{21}(normal, \mathbb{Z}_{41}) \\ \frac{a_{42}+10}{34}(low\text{-}grade\quad accretion, \mathbb{Z}_{42}) \end{cases}, \tag{6d}$$

$$\mu_i(\mathbb{Z}_5) = \begin{cases} \frac{10a_{51}+35}{15}(normal\quad size, \mathbb{Z}_{51}) \\ \frac{20a_{52}-3}{234}(low\text{-}grade\quad accretion, \mathbb{Z}_{52}) \\ \frac{a_{53}+11}{112}(mid\text{-}grade\quad accretion, \mathbb{Z}_{53}) \end{cases}, \tag{6e}$$

$$\mu_i(\mathbb{Z}_6)=\begin{cases}\dfrac{4a_{61}-3}{40}(normal\quad size,\mathbb{Z}_{61})\\ \dfrac{a_{62}+10}{234}(low\text{-}grade\quad accretion,\mathbb{Z}_{62})\\ \dfrac{a_{63}+10}{38}(mid\text{-}grade\quad accretion,\mathbb{Z}_{63})\end{cases}, \tag{6f}$$

$$\mu_i(\mathbb{Z}_7)=\begin{cases}\dfrac{a_{71}+1}{21}(no\quad,\mathbb{Z}_{71})\\ \dfrac{a_{72}+10}{20}(yes,\mathbb{Z}_{72})\end{cases}, \tag{6g}$$

$$\mu_i(\mathbb{Z}_8)=\begin{cases}\dfrac{a_{81}+10}{36}(lessening,\mathbb{Z}_{81})\\ \dfrac{a_{82}+10}{24}(normal,\mathbb{Z}_{82})\\ \dfrac{a_{83}+10}{80}(low\text{-}grade\quad accretion,\mathbb{Z}_{83})\\ \dfrac{a_{84}+10}{30}(mid\text{-}grade\quad accretion,\mathbb{Z}_{84})\end{cases}, \tag{6h}$$

$$\mu_i(\mathbb{Z}_9)=\begin{cases}\dfrac{a_{91}+5}{22}(normal\quad size,\mathbb{Z}_{91})\\ \dfrac{2a_{92}-3}{20}(low\text{-}grade\quad accretion,\mathbb{Z}_{92})\\ \dfrac{2a_{93}-3}{20}(mid\text{-}grade\quad accretion,\mathbb{Z}_{93})\end{cases}, \tag{6i}$$

$$\mu_i(\mathbb{Z}_{10})=\begin{cases}\dfrac{a_{10,1}+10}{23}(lessening,\mathbb{Z}_{10,1})\\ \dfrac{a_{10,2}+4}{17}(normal,\mathbb{Z}_{10,2})\\ \dfrac{a_{10,3}+10}{43}(low\text{-}grade\quad accretion,\mathbb{Z}_{10,3})\\ \dfrac{a_{10,4}+10}{35}(mid\text{-}grade\quad accretion,\mathbb{Z}_{10,4})\end{cases}, \tag{6j}$$

(6k)

$$\mu_i(\mathbb{Z}_{11})=\begin{cases}\dfrac{a_{11,1}-5}{100}(no,\mathbb{Z}_{11,1})\\ \dfrac{a_{11,2}+11}{70}(yes,\mathbb{Z}_{11,2})\end{cases}, \tag{6l}$$

$$\mu_i(\mathbb{Z}_{12}) = \begin{cases} \frac{a_{12,1}}{12}(no, \mathbb{Z}_{12,1}) \\ \frac{a_{12,2}+10}{24}(yes, \mathbb{Z}_{12,2}) \end{cases}.$$

where a_{jl} is a score of the lth symptom sign of the j th symptom $\mathbb{Z}_j$ and given by medical experts. After a_{jl} is given by medical experts, the attaching function value of the symptom $\mathbb{Z}_j$ can be calculated by Eq.(6).

If the symptom signs of the given sufferer are indicated by $\{\mathbb{Z}_{1u}, \mathbb{Z}_{2v}, \cdots, \mathbb{Z}_{nw}\}$ $(u, v = 1, 2, \cdots)$, it is necessary to substitute the score a_{jl} given by medical experts into Eq.(6) to obtain the attaching function values of symptom signs.

2.2 Information Analysis of Radiodiagnosis Value and Weight of Each Symptom

Based on the statistical analysis of case histories, the prior probability of the ith disease is denoted by $P(M_i)$ and $i = 1, 2, 3, 4$, the questionably diagnosis result given by medical expert before distinguishing or identifying the suffering disease of the sufferer may be described by the information source entropy [3], i.e.,

$$H(M) = -\sum_{i=1}^{4} P(M_i) * \log_{10} P(M_i) \, (i = 1, 2, 3, 4) \tag{7}$$

The medical experts can obtain the symptom sign set $\{\mathbb{Z}_{j1}, \mathbb{Z}_{j2}, \cdots, \mathbb{Z}_{jl}, \cdots, \mathbb{Z}_{iK}\}$ via examining the jth symptom $\mathbb{Z}_j$ of the sufferer with congenital heart disease. For example, if the medical expert examines the size of heart of the sufferer, the expert can get the radiodiagnosis symptom sign of the normal or low-grade accretion or mid-grade accretion of heart and $K = 3$. After the medical expert gets the radiodiagnosis symptom sign $\mathbb{Z}_{jl}$ of the sufferer, the diagnosis uncertainty that the medical expert distinguishes or identifies the disease type of the sufferer will reduce from $H(M)$ to $H(M \mid \mathbb{Z}_{jl})$

$$H(M \mid \mathbb{Z}_{jl}) = -\sum_{i=1}^{4} P(M_i \mid \mathbb{Z}_{jl}) * \log_{10} P(M_i \mid \mathbb{Z}_{jl}) \, (i = 1, 2, 3, 4). \tag{8}$$

where $P(M_i \mid \mathbb{Z}_{jl})$ is a probability of the disease M_i under the condition that the symptom sign $\mathbb{Z}_{jl}$ of the sufferer has been determined. Accordingly, the radio- diagnosis symptom sign $\mathbb{Z}_{jl}$ provides the information $T(M, \mathbb{Z}_{jl})$ for the medical expert, i.e.,

$$T(M, \mathbb{Z}_{jl}) = H(M) - H(M \mid \mathbb{Z}_{jl}) \, . \tag{9}$$

Based on the statistical analysis of case histories, the probability $P(\mathbb{Z}_{jl})$ of the radio-diagnosis symptom sign $\mathbb{Z}_{jl}$ of the sufferer can be obtained, $j = 1, 2, \cdots, 12$,and $l = 1, 2, \cdots, K$. Averagely, the entropy $H(M)$ of the sufferer will decrease to $H(M \mid \mathbb{Z}_j)$.

$$H(M \mid \mathbb{Z}_j) = \sum_{i=1}^{4} P(\mathbb{Z}_{jl}) * H(M \mid \mathbb{Z}_{jl}) \cdot \tag{10}$$

The average information $T(M, \mathbb{Z}_j)$, which provides for the medical expert by the radiodiagnosis symptom $\mathbb{Z}_j$, is written as

$$T(M, \mathbb{Z}_j) = H(M) - H(M \mid \mathbb{Z}_j) \cdot \tag{11}$$

Suddenly, if a symptom sign provides more information for medical expert, the symptom sign has larger diagnosis value. In like manner, if a symptom provides more information for medical expert, the symptom has larger diagnosis value.

The initialization weight value g_{j0} of the j th radiodiagnosis symptom $\mathbb{Z}_j$ can estimated using Eq.(13)

$$g_{j0} = T(M, \mathbb{Z}_j) / \sum_{j=1}^{n} T(M, \mathbb{Z}_j) \cdot \tag{12}$$

$$g_0 = \{g_{10}, g_{20}, \cdots, g_{j0}, \cdots, g_{n0}\} \cdot \tag{13}$$

where g_0 is a weight set of the symptoms and $j = 1, 2, \cdots, n$. In actual application, g_0 must be modified into g ,i.e.,

$$g = g_0 \pm \Delta g \cdot \tag{14}$$

or

$$g_j = g_{j0} \pm \Delta g_j \cdot \tag{15a}$$

Where

$$\Delta g = \{\Delta g_1, \Delta g_2, \cdots, \Delta g_j, \cdots, \Delta g_n\} \tag{15b}$$

Where Δg_j is a modified factor of weight value g_{j0} of symptom $\mathbb{Z}_j$.

2.3 Fuzzy Integral Decision-Making

For λ-Fuzzy measure $g(\mathbb{Z}_i)$ [4],[5], we have

$$g(\mathbb{Z}_1) = g_1, \tag{16a}$$

$$g(\mathbb{Z}_j) = g_1 + g(\mathbb{Z}_{j-1}) + \lambda \cdot g_j \cdot g(\mathbb{Z}_{j-1}) \cdot \tag{16b}$$

When $\lambda = 0$, according to the Eq.(16), we also have

$$g(\mathbb{Z}_1) = g_1, \tag{17a}$$

$$g(\mathbb{Z}_j) = g_1 + g(\mathbb{Z}_{j-1}) \cdot \tag{17b}$$

Using Eq.(17), fuzzy measure may be computed.

In limited ranges, fuzzy integral is denoted by the symbol "E" and written as

$$E_i = \int \mu_i(\mathbb{Z}_j) \circ dg(\mathbb{Z}_j) \,. \tag{18}$$

where $\mu_i(\mathbb{Z}_j)$ is an attaching function of the symptom $\mathbb{Z}_j$ and monotone order. Weight value g_i is arrayed according to the depressed order of $\mu_i(\mathbb{Z}_j)$. "$\circ$" represents fuzzy integral operator or *Zadeh* operator, i.e., $M(\vee,\wedge)$ operator[6]. However, it is improper to employ *Zadeh* operator for computing fuzzy measure due to losing some usable information using *Zadeh* operator. So a new operator that combines *Einstain* operator "$\overset{+}{\varepsilon}$" and outstretched *Zadeh* operator "$\bar{V}$" is defined in this paper. In this case, Eq.(18) may be written as

$$E_i = \mathop{\bar{\mathrm{V}}}_{j=1}[\mu_i(\mathbb{Z}_j)\overset{+}{\varepsilon}\, g_j] \,. \tag{19}$$

where

$$\mu_i(S_j)\overset{+}{\varepsilon}\, g_j = [\mu_i(S_j)+g_j]/[1+\mu_i(S_j)\times g_j],$$

$$\bar{V} = [\mu_i(S_j)\overset{+}{\varepsilon}\, g_j) + \mu_i(S_{j-1})\overset{+}{\varepsilon}\, g_{j-1})]/2 \,.$$

Let $E = \max_{i=1}^{4}\{E_i\}$, then the disease corresponding to E is just the diagnosed result.

Eq.(18) or (19) is called as fuzzy integral decision-making model. Moreover, g_j in Eq.(18) or (19) is determined by the information source entropy, so Eqs.(18) or (19) and (12) are called as Information Entropy based Fuzzy integral decision-making Model (IEFM). The IEFM of diagnosing congenital heart disease is not established until now.

3 Example Analysis and Optimization Processing

3.1 Example Analysis

The symptom set of the sufferer with the congenital heart disease was obtained by

$$\mathbb{Z} = \{\mathbb{Z}_{12},\mathbb{Z}_{22},\mathbb{Z}_{31},\mathbb{Z}_{42},\mathbb{Z}_{52},\mathbb{Z}_{61},\mathbb{Z}_{71},\mathbb{Z}_{81},\mathbb{Z}_{93},\mathbb{Z}_{10,1},\mathbb{Z}_{11,1},\mathbb{Z}_{12,1}\} \,. \tag{20}$$

The radiodiagnosis process of the Information Entropy based Fuzzy integral decision-making Model(IEFM) to the sufferer is as follows:

Step 1: The scores of each symptom sign is given by medical experts and initialization weight value g_{j0} is calculated using Eq.(12) and shown in Table 1. Substituting the scores shown in Table 1 into Eq.(6), the attaching function value of the symptom sign $\mathbb{Z}_{jl}$ of the symptom $\mathbb{Z}_j$ is calculated and shown in Table 2. Using Eqs.(10) and (12), the diagnosis values of the symptom $\mathbb{Z}_j$ and its symptom sign $\mathbb{Z}_{jl}$ can be quantized in Table 2.

Table 1. Symptoms and their signs, weight initial value

	$\mathbb{Z}_1$	$\mathbb{Z}_2$	$\mathbb{Z}_3$	$\mathbb{Z}_4$	$\mathbb{Z}_5$	$\mathbb{Z}_6$	$\mathbb{Z}_7$	$\mathbb{Z}_8$	$\mathbb{Z}_9$	$\mathbb{Z}_{10}$	$\mathbb{Z}_{11}$	$\mathbb{Z}_{12}$
sign	$\mathbb{Z}_{12}$	$\mathbb{Z}_{22}$	$\mathbb{Z}_{31}$	$\mathbb{Z}_{42}$	$\mathbb{Z}_{52}$	$\mathbb{Z}_{61}$	$\mathbb{Z}_{71}$	$\mathbb{Z}_{81}$	$\mathbb{Z}_{92}$	$\mathbb{Z}_{10,1}$	$\mathbb{Z}_{11,1}$	$\mathbb{Z}_{12,1}$
g_0	0.03	0.10	0.04	0.08	0.14	0.06	0.05	0.07	0.14	0.09	0.10	0.09
g	0.04	0.10	0.05	0.08	0.15	0.07	0.05	0.08	0.15	0.08	0.05	0.10

Table 2. Scores of symptom signs and Attaching function values of the given case

	M_1		M_2		M_3		M_4	
$\mathbb{Z}_{jl}$	a_{jl}	$\mu_i(\mathbb{Z}_j)$	a_{jl}	$\mu_i(\mathbb{Z}_j)$	a_{jl}	$\mu_i(\mathbb{Z}_j)$	a_{jl}	$\mu_i(\mathbb{Z}_j)$
$\mathbb{Z}_{12}$	10	0.9079	7	0.5495	9	0.7258	10	0.9005
$\mathbb{Z}_{22}$	10	0.7650	4	0.2786	3	0.0234	5	0.3368
$\mathbb{Z}_{31}$	8	0.7430	9	0.7858	10	1.0000	10	1.0000
$\mathbb{Z}_{42}$	6	0.6154	-10	0.0000	-10	0.0000	4	0.4298
$\mathbb{Z}_{52}$	10	0.9137	5	0.3515	7	0.4958	8	0.5834
$\mathbb{Z}_{61}$	10	1.0000	4	0.2785	3	0.2109	10	1.0000
$\mathbb{Z}_{71}$	7	0.5000	10	1.0000	10	1.0000	10	1.0000
$\mathbb{Z}_{81}$	9	0.7336	7	0.4878	-10	0.0000	8	0.5714
$\mathbb{Z}_{92}$	6	0.4125	3	0.1633	3	0.4091	5	0.3354
$\mathbb{Z}_{10,1}$	-10	0.0000	-10	0.0000	-10	0.0000	9	0.7006
$\mathbb{Z}_{11,1}$	10	0.8472	9	0.8095	9	0.8322	10	0.9852
$\mathbb{Z}_{12,1}$	1	0.0952	9	0.5806	9	0.5806	10	1.0000

Step 2: Based on clinic experience and relative literatures[4], the initialization weight value g_0 is modified into g ,and g is shown in Table 1.

Step 3: The attaching function $\mu_i(\mathbb{Z}_j)$ of the case is put into a monotone degressive order. Weight values corresponding to the attaching function $\mu_i(\mathbb{Z}_j)$ are also arranged with attaching function $\mu_i(\mathbb{Z}_j)$.

Step 4: Using Eqs. (19), we can get the fuzzy integral values as follows:

$$E_1 = 0.94, E_2 = 0.88, E_2 = 0.90, E_4 = 0.97. \quad (21)$$

So the sufferer is the case with pulmonary stenosis. This diagnosed result of the IEFM is in agreement with that of the medical experts.

2.2 Optimization Processing

The information entropy based fuzzy integral decision-making model(IEFM) on microcomputer medical expert system has entire theory, detailed characterization and computational simplification. It is feasible to apply the IEFM to clinic diagnosis.

In order to illustrate the performance of the IEFM compared with the Single Fuzzy Integral Model(SFIM), we have gone on the following researches.

The diagnosed 165 cases with congenital heart disease, were retrospectively tested by the IEFM and the SFIM, respectively. The diagnosed results were shown in Table 3.

Table 3. Diagnosed results results for previous cases

Method	165 Cases		Coincidental number	Accurate rate (%)
IEFM	Atrial septal sefect	(45 Cases)	42	93
	Ventricular septal sefect	(33 Cases)	28	88
	Patent arterial duct	(21 Cases)	18	86
	Pulmonary stenosis	(17 Cases)	15	88
SFIM	Atrial septal sefect	(45 Cases)	38	84
	Ventricular septal sefect	(33 Cases)	27	81
	Patent arterial duct	(21 Cases)	16	76
	Pulmonary stenosis	(17 Cases)	13	76

Table 3 has shown that the diagnostic accurate rate of the information entropy based fuzzy integral decision-making model(IEFM) is higher than that of the Single Fuzzy Integral Model(FIM). Based on the results, we think that it is valuable to employ the IEFM for diagnosing the congenital heart disease. However, from applied view, it is very necessary to carrying out the optimum diagnosis result of the IEFM. Thereby, the following problems must be processed commendably and further.

(1) Weight values in the IEFM must be determined with reason

The first, weight values with reason is a key of improving accurate rate of the IEFM. Under condition of lots of case histories, the initialization weight value g_{j0} can be computed by the information source entropy. The second, it is very necessary to modify the initial weight value g_{j0} whereas modified value Δg_j is given via experiment-ing again and again. Finally, weight values of high diagnostic accurate rate can be expressed as $g_{j_0} \pm \Delta g_j$

(2) Operator in the IEFM must be reasonably selected

Due to losing some available information for diagnosing the disease using *Zadeh* operator, the appropriate operator must be given to improve the computational precision in the IEFM. However, the work load of selecting operator or modifying weight values is very large. For decreasing the work load, the self-educated system, which can set up the modified factor of weight values and select the appropriate operator, has to be established using the artificial intelligence method. As for this self-educated system, its inputs are the case histories recoreded by medical experts. According to comparison the diagnosis results using the IEFM with medical expert, we must modify the weight values using the modified factor $\pm\Delta g_j$, and fuzzy integral value E_i are computed using the different operators. The self-educated process doesn't finish until that the enough high accurate rate of the IEFM are obtained for previous cases.

(3) Attaching function, quantum standard, and weight value of each symptom, must be determined under condition that medical experts take part in.

(4) The IEFM is not modified again and again according to a great many of history cases until that its diagnostic accurate rate is enough high and it can be directly applied to the clinic diagnosis.

Acknowledgements

The research has been Supported by Natural Science Fund of Anhui Province (050420304) and the Science Foundation of Educational Office, Anhui Province, People's Republic of China(2003KJ092, 2005KJ008ZD), and by Doctor Fund of Anhui University of Science and Technology(2004YB005).

References

1. Zhou Huaiwu(1990) Medical biology mathematics. Beijing, Renmin Sanitary Book Concern.
2. Zhou Huaiwu (1983) Mathematical medicine. Shanghai, Shanghai Science and Technology Press.
3. Guo Yecai(1992) Fuzzy integral method based coronary heart disease prediction. Journal of Mathematical Medicine. 5(4): 79-81.
4. Yecai Guo,Wei Rao,Yi Guo (2005) Research on predicting hydatidiform mole canceration tendency by fuzzy integral model. LNAI 3613, Fuzzy Systems and Knowledge Discovery, FSKD2005, 1 Part1:122~129.
5. Xi-zhao Wang, Xiao-jun Wang (2004) A new methodology for determining fuzzy densities in the fusion model based on fuzzy integral. 2004 International Conference on Machine Learning and Cybernetics. Shanghai, China, Aug. 26-29:1234-1238.
6. Zhang Deshun (1993) Synthetical application of outstretched *Zadeh* operator fuzzy operator. Journal of Mathematical Medicine. 6(supplement):180~181.

The Design and Research of Controller in Fuzzy PETRI NET

Yubin Zhong

School of Mathematics and Information Sciences, Guangzhou University,
Guangzhou Guangdong 510006, P.R. China
Zhong_yb@163.com

Abstract. This paper designs a class of Fuzzy PETRI NET controller based on the character of place invariant in PETRI NET. This controller is simple in operation and is easy to put into practice. It has satisfying effects and can provide methods to do research in designing and optimizing the performance of system in solving the dead lock phenomenon in concurrent system.

Keywords: Place invariant; Controller; Fuzzy PETRI NET; Dead lock Phenomenon; Optimize;

1 Introduction

PETRI NET is an important implement in building models and analyzing discrete events dynamic systems. It was widely used in controller design and in building models for many systems. Further more, it does not only help to research in the performance of systems, but also controls and optimizes it based on models. However, the models built were normally huge because of the complexity of the real system. So if general PETRI NET is applied in analyzing the performance of system directly, it can lead to dimension disaster on computer. In order to solve this problem, this article proposes a new controller design based on PETRI NET, which generalizes control designing to Fuzzy PETRI NET. A new design method for maximum feedback allowed controller is given. This controller is composed of variance place which is connected to Fuzzy PETRI NET. The dead lock phenomenon can be solved and can ensure system to avoid access in some forbidden state according to given constraints if this method is put to use.

2 The Definition of Fuzzy PETRI NET

2.1 The Definition of Fuzzy PETRI NET

Fuzzy PETRI NET（FPN for short）is a knowledge model, which is based on traditional Fuzzy generating regulation and PETRI NET and is composed of place, variance, reliability, and valve value.

FPN is defined as an eight-tuple, FPN=（P, T, I , O, C_f ,a , b, M_0）, in this eight-tuple p=(p_1 ,p_2 ,……,p_3) is limited place set ; T=（t_1 ,t_2 ,…..,t_n）is

B.-Y. Cao (Ed.): Fuzzy Information and Engineering (ICFIE), ASC 40, pp. 41–49, 2007.
springerlink.com

limited variance set, $P \cap T = \phi$; I is input function, place to directed arc set of variance $I(t_j)=p_i$; O is output function, the varied place directed arcs set $O(p_i)=t_j$; C_f : T→[0,1] is the reliability of variance ; a : P→[0,1] is reliability of place ; b : T→[0,1] is valve value which is activated in varying. About $p_i \in I(t_j)$, iff $a(p_i)>b(t_j)$, the variance t_j can start up, M_0 is the initial Fuzzy token.

2.2 The Rules of Activation

In Fuzzy PETRI NET, in order to activate the variant, the following two conditions must be satisfied:

a. There must exist a token, which represents a Fuzzy inputting variable in the place where you input ;

b. It must satisfy Fuzzy regulation condition that associates with variant, that reliability in inputting place must be greater than its valve value。

2.3 Place Invariant

Based on one of the structure characteristics of PETRI NET, which depends on its topological structure but not its initial label, place invariant is the place set that holds a fixed number of tokens and it can be expressed by an n-dimensional column vector X. Its nonzero elements are corresponding to special invariant places, while zero elements are on the contrary. A place invariant is defined as an integer vector X, and satisfies:

$$M^T X = M_0^T X \tag{1}$$

M_0 is the initial label of the network, M is the next label, the meaning of equation (1) is that in invariant place the summation of all weight, which signs its token, is constant and the summation is determined by the initial label of PETRI NET. Place invariant of network can be deduced by the integral solution of the following equation :

$$M^T D = 0 \tag{2}$$

D is connection matrix with n×m dimensions, D=I O, n and m are the place number and the variant number of network, respectively. We can observe that the arbitrariness linear combination of place invariant is place invariant in network. Place invariant is the key value for analyzing PETRI NET, because it allows analyzing the structure of network independently and dynamically.

3 Design of Controller

Suppose the controlled object system is a Fuzzy PETRI NET model which is composed of n places and m variants. Here b equal to 0. The aim of controlling is force process to obey the constraint of the following inequality :

$$l_1 M_i + l_2 M_j \le b_j \tag{3}$$

M_i and M_j are the labels of the places p_i and p_j in Fuzzy NET. l_1, l_2 and p_j are integer constants. After laxity variable M_s is brought, the constraint of inequality can be transformed into equation form :

$$l_1 M_i + l_2 M_j + M_s = b_j \tag{4}$$

M_s represents a new place p_s, it has additional Fuzzy token which is needed to balance two sides of the equation. It ensures that the summation of Fuzzy token weight in place p_i and p_j is less or equal to b_j all the time. The place p_s ought to belong to the controlling network. To calculate its structure we can bring in laxity variable M_s, then bring in a place invariant according to whole controlled system which is defined by equation (4). Obviously, the number of constraints in inequality (3) is the same as the number of controller places. So the size of controllers is in proportion to the number of constraints in inequality (3).

Because of bringing in a new place in network, the connection matrix D of controlled system changes into n×m dimensional matrix D_p, Plus one row corresponding to the place brought by the laxity variableM_s. This row, which is calledD_s, belongs to the connection matrix of the controller. To calculate the arc which connects controller place to system initial Fuzzy PRTRI NET, we use equation (2) of place invariant. When vector x_j is place invariant defined by equation (4), unknown values are the elements of the new row in matrix.

We can describe the above problem as follows : All constraints in inequality (3) can be written into matrix forms:

$$L M_p \leq B \tag{5}$$

Here M_p is label vector of Fuzzy PETRI NET, L is n_c×n dimensional integer matrix, n_c is the number of constraints in inequality (3), B is n_c×1 dimensional integer vector.

We can change (5) into equation by bringing in laxity variable:

$$L M_p + M_c = B \tag{6}$$

M_c is $n_c \times 1$ dimensional integer vector which indicates controlling place label. Because every place invariant defined by equation (6) must satisfy equation (2),we can deduce :

$$X^T D = 0$$

$$[L, I][D_p, D_c]^T = 0 \tag{7}$$

Here I is n_c×n_c dimensional identity matrix, because the coefficients of laxity variable are 1, it includes arcs which connect controlling place to Fuzzy PETRI NET in matrix D_c. So according to the given Fuzzy PETRI NET model D_p and constraints L and B which process must be satisfied, Fuzzy PETRI NET controllerD_c can be defined as follows:

$$D_c = -L D_p \tag{8}$$

We must notice that the initial label in controller should be included. The initial label of controller Mc_0 is the label that can satisfy place invariant equation (6) and depends on the initial label of Fuzzy PETRI NET. As to the Given the equation (1), initial label vector equation (6) can be defined as following form:

$$LMp_0+Mc_0=B \tag{9}$$

Which is

$$Mc_0=B-LMp_0 \tag{10}$$

Below is an example of controller design, as shown in fig.1. The connection matrix of network is :

$$D_p=\begin{pmatrix}-1 & 0 & 1\\ 1 & -1 & 0\\ 0 & 1 & -1\end{pmatrix}$$

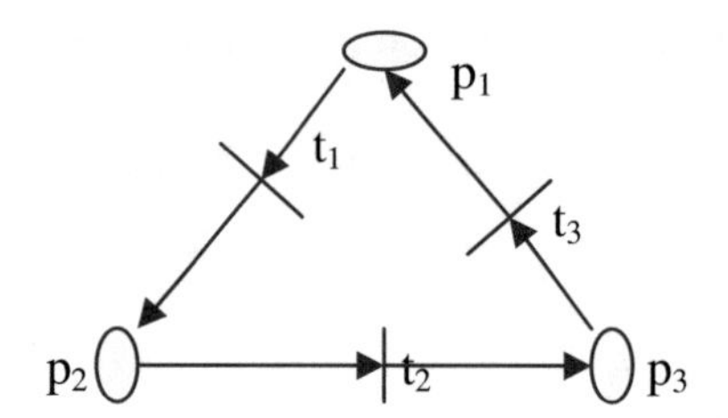

Fig. 1. Primary PETRI NET

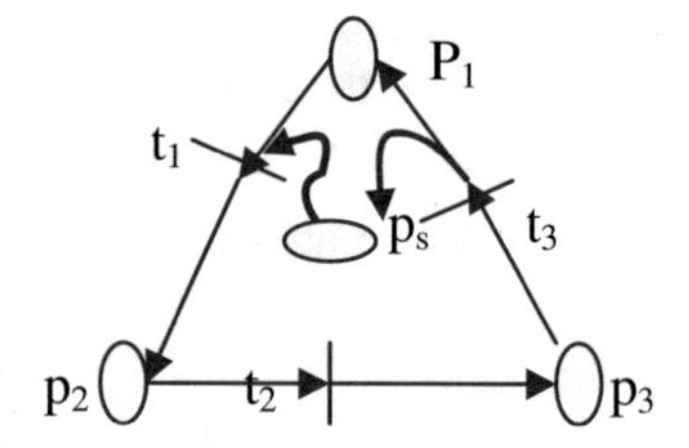

Fig. 2. Network contain controller

Its initial Fuzzy label is :

$$Mp_0=[M_1,M_2,M_3]^T=[1,0,0]^T$$

Assume the aim of controlling is that place p_2 and p_3 do not contain Fuzzy token at the same time, which is:

$$M_2+M_3\le 1 \tag{11}$$

$$L=[0,1,1],\ B=1$$

Because [1,0,0] is not a place invariant of network, the primary network does not satisfy given constraints. we change equation (11) into equation constraint by bringing in laxity variableM_s :

$$M_2+M_3+Ms=1 \tag{12}$$

Laxity variable M_s shows Fuzzy label of the place p_s in controller, the label X^T =[0,1,1] of equation (12) is the place invariant of the controlled network. We can calculate the connection matrix of the controlled net by equation (8):

$$D_c = -L D_p = [-1,0,1]$$

The initial condition of controller place is given by equation (10):

$$Mc_0 = B - L Mp_0 = 1$$

Fig.2 shows the place p_s in PETRI NET and its connection which belongs to feedback controller.

4 A Synthetical Algorithm Aiming at the Dead Lock Phenomenon

Now, we utilize the model of Fuzzy PETRI NET to provide a synthetical algorithm aiming at the dead lock phenomenon based on Fuzzy PETRI NET model controller. Detailed steps are :

Step 1: Produce its reachable state fig. RMG(Σ) from PETRI NET model (Σ) of primary system ;
Step 2: Produce reachable state fig. RMG(ΣΘC) of object system from reachable state fig. RMG(Σ) and control criterion (Spec)of primary system;
Step 3: Produce reachable state fig. RMG(C) of controller from reachable state fig. RMG(ΣΘC) of object system;
Step 4:Produce PETRI NET model (C)of controller from reachable sate fig. RMG(C)

The detailed processes of every step above are:

Algorithm 1

Input : RMG(ΣΘC)=(V, E, F), output : RMG(C)=(V_c, E_c, F_c)

Step1. Let $V_c=\{\Gamma P\rightarrow P_c(M) | (M\in V)\wedge(\Gamma P\rightarrow P_c(M)$ is projection sub-vector of M ιυ P_c },

Step2. Suppose M, M"∈V, (M ,M")∈E, and f((M, M"))=t, then we can let $E_c \leftarrow E_c \cup \{(\Gamma P\rightarrow P_c(M), \Gamma P\rightarrow P_c(M''))\}$, $Fc(\Gamma P\rightarrow P_c(M), \Gamma P\rightarrow P_c(M''))\leftarrow t$, because ΣΘC is compounded synchronously fromΣand C actually, for t∈TΣΘC, we have cases as follows:

Case 1. If t∈TΣ-TC, then M[t]>M", iff ΓP→PΣ(M)[t]>ΓP→PΣ(M"),and $\Gamma P\rightarrow P_c(M)=\Gamma P\rightarrow P_c(M'')$;

Case 2. If t∈TΣ-TC, then M [t]>M", iff $\Gamma P\rightarrow P_c(M)$[t]>ΓP→Pc(M"),and ΓP→PΣ(M)=ΓP→PΣ(M") ;

Case 3. If t∈TΣ∩TC, then M [t]>M", iff ΓP→PΣ(M)[t]>ΓP→PΣ(M"), and $\Gamma P\rightarrow P_c(M)$[t]>$\Gamma P\rightarrow P_c(M'')$;

Thus ,RMG(C)=(V_c,E_c,F_c) is obtained by RMG(ΣΘC)=(V,E,F) projecting in C actually.

Algorithm 2

Now suppose input:RMG(C), output :C=(P_c,T_c,F_c,Mc_0)

Step1. Let F_c← ,mark Mc_0,push Mc_0 in stack ;

Step2. If stack is not empty, do Step3. Otherwise finish ;

Step3. If there exists a node M_c, which is adjacent to top but has not been marked, do Step4. Otherwise go to Step5 ;

Step4. Let $\Delta M_c = M_c$-stack(top), mark f((stack(top), M_c))=t, do Step4.1 ;

Step4.1 For $p_c \in Pc$, if $\Delta Mc(p_c)>0$, then let $F_c \leftarrow F_c \cup \{(t, p_c)\}$; if $\Delta Mc(p_c)<0$, then let $F_c \leftarrow F_c \cup \{(p_c,t)\}$,
Step4.2. Mark M_c and push M_c in stack, go to Step3;

Step5. Heap stack, go to Step2. According to the state equation of the network, we can know:

$$M_c\text{-stack(top)}=A^T t$$

Here A^T is the transposed matrix of connection matrix of net C, t=f((stack(top),M_cc)). In this way, we can get the structure (P_c,T_c,F_c) of net C from every two nodes of RMG(C) and correlative edges. Then according to the state equation mentioned before, we can get initial label Mc_0 of C from label of initial node of RMG(C), thereby we get net C=(P_c,T_c,F_c, Mc_0).

5 Dealing with Dead Lock Phenomenon

Dead lock phenomenon is an abnormal phenomenon which takes place easily in concurrent system. If it is not eliminated, it will lead to the paralysis of the whole system. Now, we discuss the problem of dead lock control. Σ1 is a PETRI NET in fig.3(a), fig.3 (b) is reachable state fig. RMG(Σ1) of Σ1. It is easy to see that there are two dead lock states: (02000) and (00200). In order to eliminate the dead lock states we bring in controlling place s1 and s2, and set control criterion Spce={$M(p_2)<2$, $M(p_3)<2$, $M(p_2)$+M(s1)=1, $M(p_3)$+M(s2)=1|M∈R(Mo)}.

For $Mo(p_2)=Mo(p_3)=0$, so Mo(s1)= Mo(s2)=1. Change RMG(Σ1) into RMG(Σ1ΘC1)(see fig.3(c)). We can get RMG(C1)(see fig.3(d)) and C1(see fig. 3(e)) according to algorithm 1and 2, respectively. C1 that we get in his way is PETRI NET model of controller. We must notice that the structure of C1 may not be unique.

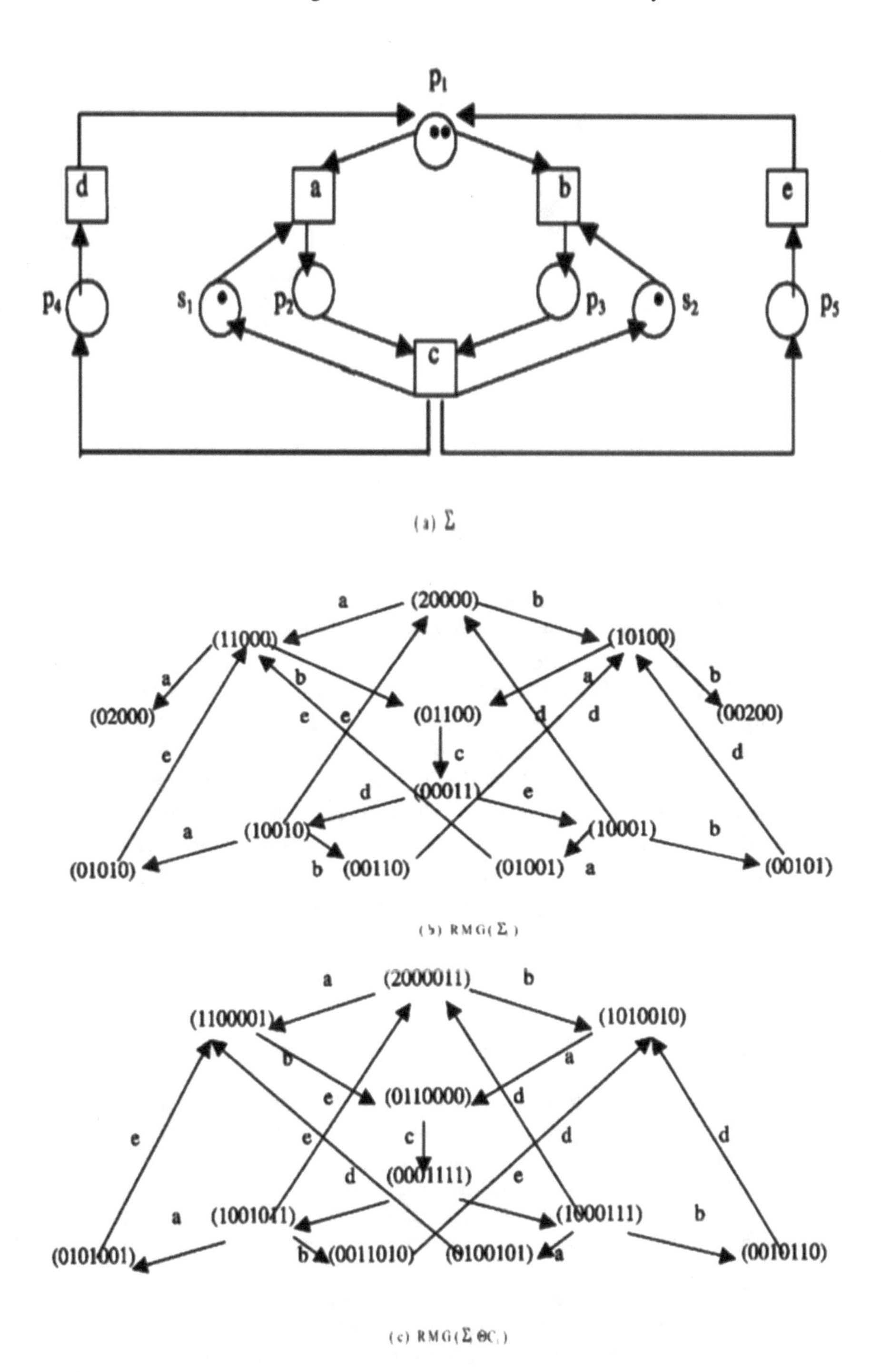

Fig. 3. Synthetical picture of deadlock controller

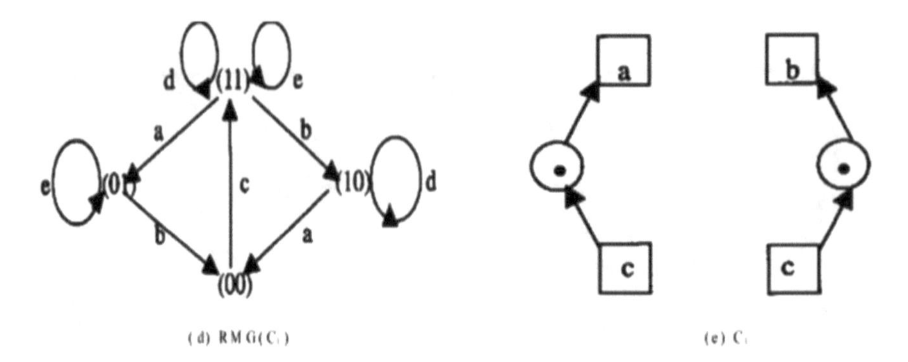

(d) RMG(C) (e) C

Fig. 3. (*Continued*)

6 Summary

This paper gives a method of controller design based on Fuzzy PETRI NET. It points out that this is an algorithm to maximum feedback allowed controller of Fuzzy PETRI NET. The controller is composed of places and arcs. Its size is in proportion to the number of constraints. This algorithm can be put into practice easily because it only involves the vector operation and the matrix operation. We can utilize this method to solve dead lock phenomenon in concurrent system. Then, our research can sum up the controller design of Fuzzy PETRI NET as a mathematical problem, which is a matrix equation or a matrix inequality problem. Does the solution to mathematical problem exist doubtless? What are the conditions assure it of existing? How to get the special solution if it exists? If there exist many solutions, we can raise the design problem of the most optimized controller under some certain condition.

References

1. Looney C. (1988) Fuzzy Petri Nets for Rule_based decisionmaking. IEEE Trans. on SMC,18(1):638-646
2. Holloway L E and Krogh B H(1990) Synthesis of Feedback logic for a class of Control Petri Nets. IEEE Trans. On Automatic Control, 35(5): 514-520.
3. Qinyou Wang and Zhu jun(1998) A synthetical method of feedback control for controlled PETRI NET. Information and Control , 27(5):326-330
4. M.C.Zhou, F. Dicesare and A.A. Desrochers(1992) A Hybrid Methodology for Symthesis of Petri Net Models for Manufacturing system. IEEE Trans. Robotics & Automation, 8(3): 356-361
5. Chen Zhanrong and Zhong Yubin(2003) Design of adaptive inserting rule fuzzy controllers. Chinese Control Theory & Applications. 20(1):129-132
6. Zhong Yubin(2005) The FHSE model of sofeware system for synthetic evaluating enterprising. Journal of Guangzhou University. 4(4):316-320

7. Zhong Yubin(2006) The researching and realizing in ERP system based on C/S structure mode mixed with B/S. Chinese Journal of Information of Microcomputer. 11:31-33
8. Zhong Yubin(1994) Imitating and forecast of economic systems of district industry of central district of city. Chinese Journal of Decision Making and Decision Support Systems. 4(4):91-97.

Fuzzy Tracing the Source of Net-Virus

Lansheng Han[1], Shuxia Han[2], Bing Peng[1], and Varney Washington[3]

[1] School of Information Security Computer Science Department , Huazhong University of Science and Technology,Wuhan 430074, China
hanlansheng@hotmail.com

[2] School of Mathematics, Huazhong University of Science and Technology,Wuhan 430074, China

[3] Chelae Vocational Training Institute, Tubman Boulevard P.O.816, Monrovia Liberia

Abstract. The paper proposes the original node of the spreading path of the virus must be the source of the virus. However most network virus have more than one spreading channels and the hints of the virus' spreading are also unclear. The paper establishes a fuzzy source tracing equations for the network virus. Working out the equations repeatedly, the paper gets the spreading path of the virus; thus the start node of the path should be the source of the virus in the network. Finally, the paper carries out the simulation test. The results of the test verify our tracing model and methods of working out the equations. Thus the paper opens a theoretic fuzzy way to tracing the source of net viruses.

Keywords: Net-virus, fuzzy trace, network security.

1 Introduction

Net-virus constitutes an important part of network-security [1]. Seeking for the propagation path and then working out the source of the virus has become an indispensable subject for blowing the virus-embed and preventing the propagation of the virus. However, currently most works of combating the virus only focus on the superficial technical aspect and have not yet established a mature theory [2].

In fact, source tracing is the inverse-process of virus' spreading and also be considered a kind of inverse problem, yet different from inverse problems in strict definition [3][4]. Firstly, it does not have definite condition as strict as inverse problems. Secondly, its propagation medium is not continual but a concrete, logical digital network. Thirdly, many properties of the medium of the inverse problem are not known even by the modern techniques, but the net viruses and its medium are both digital application whose propagation information can be recorded. Therefore, the inverse-process of the net virus' spreading is objective. In other words, the solution of this kind of inverse problems does exist if we get enough spreading information of the virus.

B.-Y. Cao (Ed.): Fuzzy Information and Engineering (ICFIE), ASC 40, pp. 50–58, 2007.
springerlink.com

Unfortunately, as seeking the source of the virus is retracing what already happened, some information might be unclear which is also part reason why there is no mature theory on it.

First when new virus appears, people are unclear its spreading features. Some modern advanced virus may have several spreading channels; such as CodeRed II and Sadmind II both have five spreading channels [5].

Second when people get know the spreading features, some propagate information may getting uncertain such as the connection log is too long, we can can't be sure which connection caused infection.

Third even if people know the spreading features immediately, its spreading styles are still unclear, for example most macro virus are hided in many kind of files, those file may transmitted on the Internet by many kind of connections such as FTP, MSN,QQ or Email attachment.

But this unclear information might be enough for us to trace the virus using fuzzy method. Table1 lists the spreading channels and their spreading match degree of Macro virus "OurCode" which is also a sample in our simulation test.

Table 1. Spreading style and their match degree of Macro virus "OurCode"

Connection style	Possibility to spread virus	Match degree
FTP	0.445	0.445
MSN	0.371	0.371
Email	0.637	0.637
QQ	0.215	0.215
Downloading	0.452	0.452

Net-virus, as an application program, usually propagates in a logic sub-net according to its characteristics [6]. Most sub-nets are relatively stable in a given time. Therefore, we adopt division methodology: tracing the source in a sub-net and judging whether it is the original source, if it is the original source, the job is done; or else, the virus must be spread from another sub-net, thus tracing the source in that sub-net. To be strict, the number of the sub-nets and the computers in the sub-nets are both finite, therefore the steps of source tracing is finite too.

Now we define the source of a certain virus in a sub-net as the commencement of the propagation path of the virus within the sub-net. According to this definition, the virus sources may be not unique, which is also congruent to the reality.

2 Source Tracing Modeling

The computer system in the sub-net is considered a node $v_i (i = 1, 2, \cdots, n)$. The connection of node v_i to v_j is considered a directed edge denoted by (v_i, v_j), which corresponds to the spreading direction of the virus. Thus a directed graph

$D(V, E(t))$ is formed where is the set of nodes in the sub-net, $E(t)$ is the binary relation on V . Because the connection is dynamic with the time, we let $E(t)$ be the function of the time. Therefore, the connection among the nodes in the sub-net can be described by a mutative Boolean matrix$C_t = (c'_{ij})_{nn}$ called connection matrix, where

$$c_{ij}^t = \begin{cases} 1 \ (v_i, v_j) & \text{in} \quad \text{E(t)}, \\ 0 \ (v_i, v_j) & \text{notin} \quad \text{E(t)}. \end{cases}$$

We know that if virus spread from v_i to v_j in time t, then$c_{ij}^t = 1$. However, if $c_{ij}^t = 1$, the virus does not necessarily spread from v_i to v_j in time t. It depends on the connection match degree for the certain virus. For example, an email virus has the match degree 0.35 to spread on a FTP connection.

We present the match degree function$M(C_t) = (a_{ij}^t)_{nn}$ on the basis of the above connection matrix, where $M(c_{ij}^t) = a_{ij}^t$ denotes the match degree of the c_{ij}^t to spread the virus. For example$M(c_{ij}^t) = 0.7$ denotes the connection at time t from v_i to v_j is thought to have the probability 0.7 to spread the virus. Vertices are prefer to preserve its original state until methods taken place to change its state, so we let $M(c_{ii}^t) = a_{ii}^t = 1$ for any time t. Of cause if $a_{ij}^t = 1$, the virus is certainly propagated from v_i to v_j.

As narrator above, in most case, the source tracing are retracing what already happened, so the states of some vertices are clear, but some are fuzzy which depend on people's fuzzy judgment or deduce. So fuzzy variable x_{it} is introduced to denote the state of node v_i in time t. For example $x_{it} = 0.8$ denotes v_i is thought to have the probability 0.8 to be infected in time t . Therefore, vector$(x_{1t}, x_{2t}, \cdots, x_{nt})$ denotes the state of all the nodes in the subnet in time t.

To look for the source of the virus, the time when the nodes are infected by the virus is very important hint, and the capture of these hints is usually by discrete time. Therefore, we divide the time into time intervals whose size usually depended on the propagation characters of the specific virus and the type of the specific net connection. For example, within a time interval at least one connection or one infection of virus can be finished or the node state can change once. Of course, the interval shouldn't be too large in case of necessary information be missed thereby source tracing will be more difficult.

According to the protocols and abilities of the net nowadays, it is known that connections among nodes are random. Hence in one time interval, node v_i is set to be able to receive connections from more than one node, and during the same time interval, v_i can send information to more than one node (such as broadcasting, multicasting). By the definition of the source of virus, we need to find out which nodes have the large probability in the sub net infects v_i . Consequently, the vector $(a_{i1}^t, a_{i2}^t, \cdots, a_{in}^t)$is used to denote the acceptance vector of v_i in time t . Also, the state vector of the subnet in time t is $(x_{1t}, x_{2t}, \cdots, x_{nt})$. Therefore, after the connection of this time interval, the impact of the subnet on v_i can be expressed by the formula below:

$$(a_{i1}^t, a_{i2}^t, \cdots, a_{in}^t) \cdot (x_{1t}, x_{2t}, \cdots, x_{nt}) = a_{i1}^t \cdot x_{1t} \vee a_{i2}^t \cdot x_{2t} \vee \cdots \vee a_{in}^t \cdot x_{nt} \quad (1)$$

Operators $\cdot$ in (1) are Boolean multiply. Taking v_i as an example, if in time t, $x_{it} = 0.2$ but in time $t+1$, $x_{i(t+1)} = 0.9$ then there must exist $j \in N$ such that $a^t_{ij} \cdot x_{jt} \geq 0.9$ in (1), and in fact those v_js are the suspicious virus -carriers. As those are fuzzy state, we give a threshold $\tau \geq 0$, and called the set of those nodes Virus-Carries Node (VCN) of $v_{i(t+1)}$ denoted by $S(v_{i(t+1)}) = \{v_{jt} | a^t_{ij} \cdot x_{jt} \geq \tau\}$. Among those VCN the node with the biggest $a^t_{ij} \cdot x_{jt}$ infecting v_i is called Pre-Infected Node (PIN) of $v_{i(t+1)}$ denoted by $P(v_{i(t+1)}) = v_{jt}$. In converse $v_{i(t+1)}$ is called Post-Infected Node (POIN) of v_{jt}. It is clear that $P(v_{i(t+1)}) \in S(v_{i(t+1)})$. In theory, if we get the state of all the infected nodes, we can find the virus source in the subnet just utilize PINs. However in actual, some useful information of connection or the state of the nodes may be lost, be sure of not miss any information we also use $S(v_{i(t+1)})$.of the infected nodes to trace. So the next focus on how to work out PIN or $S(v_{i(t+1)})$ of every infected node.

Besides virus infecting, virus disinfecting can also change the state of v_i such as running anti-virus software or reload the infected computers, and such disinfecting is not fuzzy but a clear process. Actually, virus scanning and cleaning are also important means to obtain the states of the nodes. We let δ_{it} denotes cleaning variable of node v_i in time t, for example, $\delta_{it} = 1$ shows that v_i is infected by virus and the virus is cleaned in time t, $\delta_{it} = 0$ shows that v_i is scanned in time t and find no virus, $\delta_{it} = \Theta$ shows v_i has not been scanned. Yet we know disinfecting the virus of v_i before or after its' connection in any time interval may have different influences on its' adjacent nodes. For instance, v_j is first disinfected then connected to v_i , then v_{jt} is not in $S(v_{i(t+1)})$; but v_j first connects to v_i then is disinfected, it is possible that $v_{jt} \in S(v_{i(t+1)})$. Thus we set δ'_{it} as the variable of the virus scanning or disinfecting on node v_i after the connection in the time interval t. So the change of v_i from time t to time $t+1$ is:

$$(a^t_{i1}, a^t_{i2}, \cdots, a^t_{in}) \cdot ((x_{1t}, x_{2t}, \cdots, x_{nt})^T + (\delta_{1t}, \delta_{2t}, \cdots, \delta_{nt})^T) + \delta'_{it} = x_{i(t+1)} \quad (2)$$

Where operator " + " is Boolean add. Considering the change of the states of all the nodes in the subnet from time t to time $t+1$, we get state changing equation or in another words source tracing equation of the sub net:

$$\begin{bmatrix} a^t_{11} & a^t_{12} & \cdots & a^t_{1n} \\ a^t_{21} & a^t_{22} & \cdots & a^t_{2n} \\ \cdots & \cdots & \cdots & \cdots \\ a^t_{n1} & a^t_{n2} & \cdots & a^t_{nn} \end{bmatrix} \cdot \left[\begin{bmatrix} x_{1t} \\ x_{2t} \\ \cdots \\ x_{nt} \end{bmatrix} + \begin{bmatrix} \delta_{1t} \\ \delta_{2t} \\ \cdots \\ \delta_{nt} \end{bmatrix} \right] + \begin{bmatrix} \delta'_{1t} \\ \delta'_{2t} \\ \cdots \\ \delta'_{nt} \end{bmatrix} = \begin{bmatrix} x_{1(t+1)} \\ x_{2(t+1)} \\ \cdots \\ x_{n(t+1)} \end{bmatrix}$$

This equation can be simplified as:

$$A_t \cdot (X_t + \delta_t) + \delta'_t = X_{t+1} \quad (3)$$

In which A_t, δ_t, δ'_t, X_{t+1} are known or estimated and part of X_t may also known. Hence the problem of source tracing is the process of working out the whole vector X_t. In other words we need get PIN or $S(v_{i(t+1)})$ of all the infected nodes in time $t+1$ and the state of all the nodes in time t. If the two problems are solved, then the state in time $t-1$ can also be worked out based on the state in time t. Repeated as above, the initial state of this sub-net and the infecting chain formed by the infected node next to its PIN will be obtained. Then the virus source in the sub-net is found out.

3 Solutions to the Equations

As analyzed above, the process of source tracing to the virus is the process of working out the source tracing equation (3) repeatedly, which is the inverse-process of the propagation of the virus: from $t+1$ to t , and then to $t-1$, until the initial time t_0 when this subnet is firstly infected by the virus. For this reason, two problems should be solved during working out equation (3): First to work out the x_t of all the nodes according to their x_{t+1} especially the infected nodes except for nodes whose state maybe known or estimated by the cleaning methods; Second to find out the PIN or $S(v_{i(t+1)})$ of each infected node. Because equation (3) is a formula constructed by fuzzy variable and Boolean operations, besides cleaning variables δ_{it} orδ'_{it} of some nodes may be unknown, it is not suitable to use elimination method. Considering the actual network and after several experiments, we achieve the following methods for the solutions:

Step 1, we know in the actual net, nodes involve in connection in usual time are only a small part of the complete graph, in other words, A_t is often a sparse Boolean matrix. Therefore, the nodes without any connection are to be picked out first. For those nodes, only disinfecting can change its' states. As $a^t_{ii}=1$, so those nodes change according to: $x_{i(t+1)}=x_{it}+\delta_{it}+\delta'_{it}$. The concrete algorithm is: if $x_{i(t+1)}>0$, then $x_{it}=x_{i(t+1)}$,$S(v_{i(t+1)})=\{P(x_{i(t+1)})=x_{it}\}$. That is the state has not changed, the PIN of the node is itself; if $x_{i(t+1)}=0$, and δ_{it} or $\delta'_{it}=1$, then $x_{it}=1$, or else $x_{it}=0$, for this case $S(v_{i(t+1)})=\{P(x_{i(t+1)})=Null\}=\phi$ this is to say they have no suspicious infectee nor PINs .

Step 2, picking up all the nodes with $x_{i(t+1)}=0$, as in this case $S(v_{i(t+1)})=\{P(x_{i(t+1)})=Null\}=\phi$. Their states change as: $\vee^n_{j=1}(a^t_{ij}(x_{jt}+\delta_{jt})+\delta'_{jt})\vee(x_{it}+\delta_{it}+\delta'_{it})=0$. Then according to $\delta'_{jt}=1$ or $\delta_{jt}=1$we get $x_{jt}=1$; if $\delta_{jt}=0$ or $\delta'_{jt}=0$ then $x_{jt}=0$. Therefore, the rest nodes fulfill δ'_{jt} or $\delta_{jt}=\Theta$, that is, the states of those nodes have not been changed: $\vee^n_{j\in\{1,\cdots,n\}}a^t_{ij}x_{jt}=0$. So $x_{jt}=0$.

Step 3, after the above two steps, the left nodes with $x_{i(t+1)}>0$ have the state changing as: $\vee^n_{j=1}(a^t_{ij}(x_{jt}+\delta_{jt})+\delta'_{jt})\vee(x_{it}+\delta_{it}+\delta'_{it})=x_{i(t+1)}>0$. If $\delta'_{jt},\delta_{jt}=1$ then $x_{jt}=1$, if $\delta'_{jt},\delta_{jt}=0$ then $x_{jt}=0$, for the two cases $x_{jt}$is not in $S(x_{i(t+1)})$ except for $\delta'_{jt}=1$. And now the left are those nodes with $\delta'_{jt},\delta_{jt}=\Theta$ who change as the equation: $\vee^n_{j=1}a^t_{ij}x_{jt}\vee x_{it}=x_{i(t+1)}>0$. Nodes in this formula satisfy $a^t_{ij}>0$, if the state of all such j-vertices were known, then $S(x_{i(t+1)})=\{x_{jt},\text{where}a^t_{\text{ij}}\text{x}_{\text{jt}}>\tau\}$ and $P(x_{i(t+1)})=x_{jt},(\text{where}a^t_{\text{ij}}\text{x}_{\text{jt}}=\max(\text{S}(\text{x}_{\text{i(t+1)}})))$. If the state of such j-vertices were uncertain, we can decrease the number of items by adjusting the size of time interval, Obviously, the more items the formula has, the more difficult to make sure of the state of each item. It is best that only one item left in $\vee^n_{j=1}a^t_{ij}x_{jt}$, then most of x_{jt} and $P(x_{i(t+1)})$

can be achieved. However we cannot exclusive the possibility that the formula has the following case:

$$\begin{cases} x_{i(t+1)} = x_{it} \vee a_{ij}^t x_{jt} > 0 \\ x_{j(t+1)} = x_{jt} \vee a_{ji}^t x_{it} > 0 \end{cases}$$

Which means v_{it} and v_{jt} sending information to each other in the same time interval and then the both get infected. We called this case SS (Symmetric State). In theory, we can solved SS of the following case: As $x_{it} \geq 1$ and $x_{jt} \geq 1$ so if $a_{ij}^t < x_{i(t+1)}$ then $x_{it} = x_{i(t+1)}$ and $S(x_{i(t+1)}) = \{P(x_{i(t+1)}) = x_i\}$; similarly if $a_{ji}^t < x_{j(t+1)}$ then $x_{jt} = x_{j(t+1)}$ and $S(x_{j(t+1)}) = \{P(x_{j(t+1)}) = x_{jt}\}$. However, if $a_{ij}^t \geq x_{i(t+1)}$or $a_{ji}^t < x_{j(t+1)}$, we can't solve such equation in theory.

Fortunately, in practice, this case rarely happens. For example, it has not been found in our simulation tests. However, if it happens in practical work, it means that since the time interval t, until the time when the virus begins to be traced, the states of these nodes have not changed by virus disinfecting. We can use virus-scanning tools on those SS nodes. Because most virus-scanning tools can record the infection information of the computer system, such as the infected files and the infecting time, therefore, the states of these nodes in time t can be achieved, and the problem is solved.

By three steps above, the states of all the nodes $x_{1t}, x_{2t}, \cdots, x_{nt}$ in the subnet in time t are achieved; and at the same time, the PINs $(P(V_{i(t+1)}))$ and $S(v_{i(t+1)})$ of all the infected nodes in time $t+1$ are also achieved. Thus the steps can be moved to work out the following state equation of the previous time interval:

$$A_{t-1}(X_{i(t-1)} + \delta_{i(t-1)}) + \delta'_{i(t-1)} = X_{it}$$

Repeat this procedure until PINs and $S(v_{i(t+1)})$ of all the infected nodes in the subnet do not exist in this subnet, that is $P(x_{it_0}) = Null$ or $S(v_{it_0}) = \phi$. Hence x_{it_0} must be a commencement of the propagation path of the virus in this subnet, and v_{it_0} is a source of the virus in the subnet. Usually, the virus source in a subnet is not necessarily unique; for example, there may be another $P(x_{it_1}) = Null$; thus v_{it_1}may also be a source of the virus.

4 Simulation Tests

As all the computers in our laboratory are connected in physical layer, in order to simulate our test in the application layer based on a logic sub net, we select 30 computers with each creates new email box, FTP and QQ connections that can be only used in our tests. Every user in our tests can select his partners from the rest 29 users. To avoid doing any harm to the computers, we select Macro virus "OurCode" .

To avoid any moderate interfere by the test and keep the spreading information fuzzy as in actual, we let every user scan or disinfect virus by his own habit, and only a special manager can collect the connecting information by accident. Considering the connection frequency is not very high, we select 24 hours as the time interval and each run last 8-10 days. For the length limited, we only list the source-tracing paths of the infected nodes. The following three tables are the results got by solving the tracing equation respectively by VCN with $\tau = 0.61$, VCN with $\tau = 0.71$ and PIN of the infected nodes.

For convenience, from Table 2 to Table 4, we use the following notation:

$$E = 8th \quad \text{day}, Se = 7th \quad \text{day}, Si = 6th \quad \text{day}, Fi = 5th \quad \text{day},$$

$$Fe = 4th \quad \text{day}, T = 3rd \quad \text{day}, S = 2nd \quad \text{day}, F = 1st \quad \text{day},$$

$$sv8 = S(v_{i8}), sv7 = S(v_{i7}), sv6 = S(v_{i6}), sv5 = S(v_{i5}), sv4 = S(v_{i4}),$$

$$sv3 = S(v_{i3}), sv2 = S(v_{i2}), Pv8 = P(v_{i8}), Pv7 = P(v_{i7}), Pv6 = P(v_{i6}),$$

$$Pv5 = P(v_{i5}), Pv4 = P(v_{i4}), Pv3 = P(v_{i3}, Pv2 = P(v_{i2}).$$

Table 2. Source-tracing table using VCN with $\tau = 0.61$, the integer in the table denotes the No. of the computer, $\delta_i = 1$ denotes the vertex get cleaning on that day

E	sv8	Se	sv7	Si	sv6	Fi	sv5	Fe	sv4	T	sv3	S	sv2	F
						$\delta_3 = 1$	3	3	3	3	5			
4	4	4	9											
				$\delta_5 = 1$	5	5	5	5	5	5	5	5	5	5
								$\delta_7 = 1$	7	7	7	7	5	
9	9	9	9	9	9	9	9	9	7,3					
12	12	12	12	12	26	12	9							
14	4,12													
			$\delta_{18} = 1$	18	18	18	18	3						
22	12	22	26											
26	26	26	26	26	26	26	26	26	7,3					

From Table 2 we can see that v_5 is the only source of the virus. The nodes v_3, v_5, v_7, v_{18} are clean respectively on 5th, 6th, 4th,7th day. If we raise the threshold, let $\tau = 0.71$ then we get the following tracing Table 3.

From Table 3 we can see that increasing the threshold may decrease the difficulty to solve the tracing equation but may lost some useful information, for instance in Table 3 we know v_4 get infected on 7th day, but we can't get its $S(v_4)$, so by our definition of the source it must be one of the source of virus, but by Table 2 it was infected by v_9 on 6th day.

From the Table 4, we can see that the fifth node is only source of the virus. It is interesting that v_5 get infected again on 7th day by v_6 that was infected by v_5 on the second day.

From the test we get some experience and conclusions:

(1) Using VCN to trace the source, the higher the threshold, the less the information to deal with, but the source is less reliable; the lower the threshold, the more the information to deal with, but the source is more reliable.

Table 3. Source-tracing table source using VCN with $\tau = 0.71$, the integer in the table denotes the No. of the computer, $\delta_i = 1$ denotes the vertex get cleaning on that day

E	sv8	Se	sv7	Si	sv6	Fi	sv5	Fe	sv4	T	sv3	S	sv2	F
						$\delta_3 = 1$	3	3	3	3	5			
4	4	4												
				$\delta_5 = 1$	5	5	5	5	5	5	5	5	5	5
								$\delta_7 = 1$	7	7	7	7	5	
9	9	9	9	9	9	9	9	9	7					
12	12	12	12	12	12	9								
14	4, 12													
22	22	22	26											
26	26	26	26	26	26	26	26	26	7					

Table 4. Source tracing table-using PIN of the infected nodes, the integer in the table denotes the No. of the computer

E	Pv8	Se	Pv7	Si	Pv6	Fi	Pv5	Fe	Pv4	T	Pv3	S	Pv2	F
5	5	5	6					$\delta_5 = 1$	5	5	5	5	5	5
		δ_6	6	6	6	6	6	6	6	6	6	6	5	
				δ_9	9	9	11							
11	11	11	11	11	11	11	11	11	11	11	5			
14	14	14	11											
17	17	17	11											
					δ_{21}	21	21	21	21	21	21	6		
		δ_{23}	23	23	9									

(2) Using PIN to trace the source, as the information to deal with is not very much, the solving is not very difficult, but the source is no more reliable than using VCN with lower threshold.

(3) The smaller of the time interval, the easier source-tracing equation be worked out. But more times of equations are worked out repeated.

(4) The more times of virus scanning in the sub net, the more information we will get and then the easier to work out the equation.

So, in order to find the source of the virus immediately, we'd better scan or clean virus often, which can supply us more information to get the solutions to the equations.

5 Conclusion and Acknowledgements

The paper defined the source of the virus as the commencement node of the propagation path. As the spreading information of the virus is not certain or clear, the paper constructed a fuzzy source-tracing equations and presented the main steps to work out the equations. Finally simulation tests were carried out and the test result verified the valid of source-tracing equation and the ways of working out the equation. Thus the paper opened one path to construct theoretic fuzzy ways to tracing the source of viruses.

References

1. C. Wang, J. C. Knight, and M. C. Elder (2000) On computer viral infection and the effect of immunization. In Proceedings of the 16th ACM Annual Computer Security Applications Conference, December.
2. Steve R. White (1998) Open Problems in Computer Virus Research Virus Bulletin Conference, Munich, Germany, October
3. Albert Tarantola (2004) Inverse Problem Theory and Model Parameter Estimation. Society of Industrial and Applied Mathematics(SIAM):5-67.
4. Richard C. Aster, Brain Borchers, Clifford H. Thurber.(2005) Parameter Estimation and Inverse Problems. Elsevier Academic Press:3-23.
5. Jose Roberto C. P., Betyna F. N., Liuz H. A. M.(2005) Epidemiological Models Applied to Viruses in Computer Networks. Jounal of Computer Science 1(1): 31-34.
6. Cliff C.zou, Don Towsley, Weibo Gong (2003) Email Virus Propagation Modeling and Analysis, Technical Report:TR-CSE-03-04.

Uncertain Temporal Knowledge Reasoning of Train Group Operation Based on Extended Fuzzy-Timing Petri Nets

Yangdong Ye[1], Hongxing Lu[1], Junxia Ma[1], and Limin Jia[2]

[1] School of Information Engineering, Zhengzhou University
Zhengzhou, 450052, China
yeyd@zzu.edu.cn, iehxlu@zzu.edu.cn, mjx232@yahoo.com.cn
[2] Beijing Jiaotong University, Beijing 100044, China
jialimin@jtys.bjtu.edu.cn

Abstract. In order to analyze quantitatively uncertain temporal knowledge in train operation, this paper proposes a temporal knowledge reasoning method, which introduces fuzzy time interval and computation of possibilities to Extended Fuzzy-Timing Petri net(EFTN) in existed train group operation Petri net models. The method can represent the temporal firing uncertainty of each enabled transition, and thus is convenient for modeling in conflict. It also provides a technique for quantitative analysis of possibilities of conflict events. The train group behaviors model shows that the method can deal with temporal uncertainty issues during the train's movement including multi-path selecting in conflict, train terminal time and the possibility of train operation plan implementation, thus provides the scientific basis for the reasonable train operation plan. Compared with existed methods, this method has some outstanding characteristics such as accurate analysis, simple computation and a wide range of application.

Keywords: Uncertainty, knowledge reasoning, EFTN, train group operation.

1 Introduction

Efficient analysis of time parameters is the key issue of train group behavior modeling in Railway Intelligent Transportation System(RITS) [1]. For the stochastic feature of train group operation system and the influence of objective uncertainty factors and unexpected incidents, the time information in train operation has significant uncertainty.

The processing of uncertain temporal knowledge is one of the important issues in Petri net modeling [2, 3, 4, 5, 6, 7, 8, 9, 10, 11, 12]. Murata [2] introduced fuzzy set theory to handle temporal uncertainties in real-time systems and proposed Fuzzy-Timing High-Level Petri Nets(FTHNs) which features the fuzzy set theoretic notion of time using four fuzzy time functions. Zhou and Murata presented a method to compute the possibility [6].

This paper proposes a method of temporal knowledge reasoning and validating by introducing fuzzy time interval [7] and computation of possibilities [6] to EFTN [5] in present train group operation Petri net models [8, 9, 10, 11] to

B.-Y. Cao (Ed.): Fuzzy Information and Engineering (ICFIE), ASC 40, pp. 59–64, 2007.
springerlink.com

analyze quantitative temporal uncertainty related to conflict in train operation. Combined with possibility computation, we analyzed quantitatively the uncertain temporal knowledge of train group operation.

In next section, we extend EFTN and then propose new formulas correspondingly. Formulas for possibility computation are also presented. We illustrate the analysis of temporal uncertainty in section 3. Comparison of different temporal knowledge reasoning methods for temporal uncertainty is given in Section 4. In Section 5, we provide summary and concluding remarks.

2 EFTN and Relative Computations

An EFTN model [5] is a FTN with the default value of fuzzy delay being [0,0,0,0], and each transition is associated with a firing interval with a possibility p in the form of p[a,b]. For more accurate representation and analysis of temporal uncertainty related to transition, we extend the interval to fuzzy time interval [7] in the form of p[a,b,c,d]. Then, an EFTN system is an ordinary net and is defined as a 7-tuple $\{P, T, A, D, CT, FT, M_0\}$, where $\{P, T, A, D, FT, M_0\}$ is a FTN and CT is a mapping function from transition set T to fuzzy firing intervals, $CT : T \rightarrow [a, b, c, d],\ 0 \leq a \leq b \leq c \leq d.$

A fuzzy firing interval p[a,b,c,d] indicates time restriction of a transition. If a transition is enabled, it must fire at time interval [b,c]. Firing at [a,b] or [c,d] is also possible, but it is impossible to fire beyond [a,d]. Possibility p is 1 if the transition is not in conflict with any other transition, and p can be less than 1 when we want to assign different chances to transitions in conflict.

2.1 Updating Fuzzy Timestamps

To handle uncertain temporal information, Murata proposed four fuzzy time functions: fuzzy timestamp, fuzzy enabling time, fuzzy occurrence time and fuzzy delay [2]. The detailed introduction of fuzzy time functions and corresponding computations of latest and earliest are given in [2,5,6]. We use the same methods to compute fuzzy enabling time and new fuzzy timestamps.

About fuzzy occurrence time, if there is no transition conflict for transition t, which has a fuzzy firing interval $CT(t) = p[f_1, f_2, f_3, f_4]$, then the fuzzy occurrence time of t is

$$o(\tau) = e(\tau) \oplus p[f_1, f_2, f_3, f_4]. \tag{1}$$

where $p = 1$ because of no conflict exists. $\oplus$ is extended addition [2]. Assume that there are m enabled transitions, t_i, $i = 1, 2, \cdots, m$, whose fuzzy enabling times being $e_i(\tau)$, $i = 1, 2, \cdots, m$ and whose fuzzy firing intervals being $CT(t_i) = p_i[f_{i1}, f_{i2}, f_{i3}, f_{i4}]$, $i = 1, 2, \cdots, m$. Then the fuzzy occurrence time of a transition t_j with its fuzzy enabling time $e(\tau)$ and fuzzy firing interval $CT(t_j)$, is given by

$$o_j(\tau) = Min\{e_j(\tau) \oplus p_j[f_{j1}, f_{j2}, f_{j3}, f_{j4}],\ earliest\{e_i(\tau) \oplus p_i[f_{i1}, f_{i2}, f_{i3}, f_{i4}],\ i = 1\ ,2, \cdots,\ j, \cdots,\ m\}\}. \quad (2)$$

where Min is the intersection of distributions [2].

2.2 Possibility Computation

Zhou and Murata proposed a method to compute the possibility that a FMTL formula is satisfied on a given transition firing sequence in FTN [6]. Suppose that e and f are fuzzy time with membership function π_e and π_f, respectively. Then $e \leq f$ means that e takes place before f. The possibility of the relation is defined below:

$$Possibility(e \leq f) = \frac{Area([E, F] \cap \pi_e)}{Area(\pi_e)} \quad (3)$$

when f is a precise date, $f = [f', f', f', f']$, we have

$$Possibility(e \leq f) = \frac{Area(The\ part\ of\ \pi_e\ where\ \tau \leq f)}{Area(\pi_e)} \quad (4)$$

3 Analysis of Temporal Uncertainty

Next, we compute some possibilities in train group operation to illustrate the specific process of temporal uncertainty. The relative railway net and train operation plan are given in [10]. Figure 1 shows the corresponding EFTN model and the descriptions of places and transitions are given in [10] too. For the convenience of computation, the fuzzy timestamps in initial state are all set to be [0,0,0,0]. Now we analyze following issues of imprecise knowledge reasoning in train Tr1 operation.

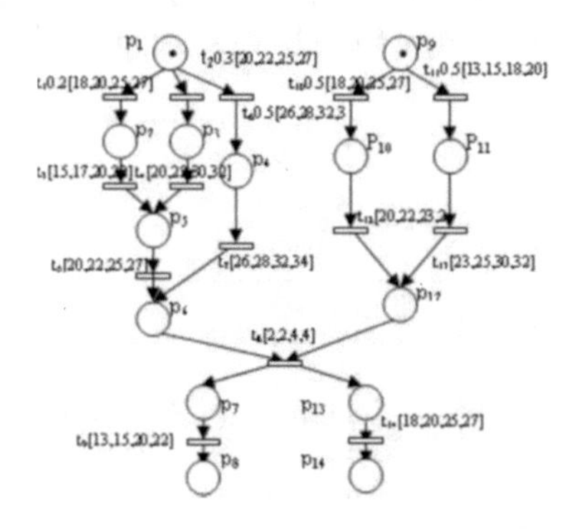

Fig. 1. EFTN model of train operation

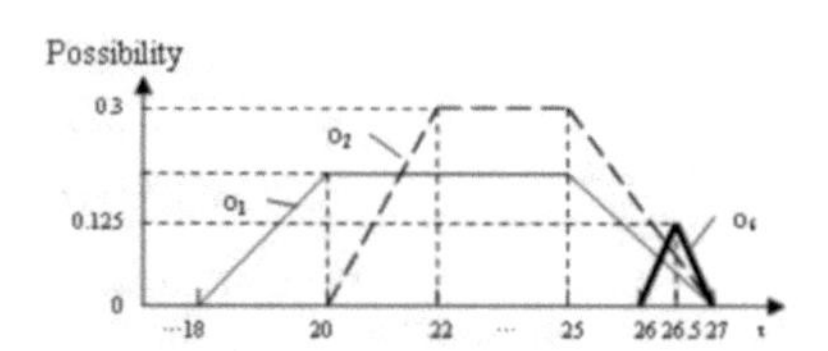

Fig. 2. Fuzzy occurrence times of t_1, t_2,t_6

1. Which path should train Tr1 choose among three alternatives? The first section of each path correspond transitions t_1, t_2, t_6 in conflict respectively. Suppose that their fuzzy occurrence times are o_1, o_2, o_6 respectively. These times can be calculated by formula (1) (2) and the results are shown in Figure 2.

   From Figure 2 transition t_6 has the lowest possibility to fire, though its fuzzy firing interval contains the highest possibility 0.5 among the three. Therefore the path which corresponding transition sequence doesn't containing t_6 should be main priority.
2. How much possibility does the train Tr1 by way of station S5? Tr1 stay in S5 is described by token in p_5. There are two corresponding transition sequences, t_1t_3, t_2t_4, exist when token moves to p_5. Our computation produces two fuzzy timestamps: 0.2[33,37,45,49] and 0.3[40,44,55,59]. Therefore the possibility of train Tr1 via the station S5 is $0.2 + 0.3 = 0.5$. On the other hand, from the fuzzy timestamp 0.125[26,26.5,26.6,27] in place P_4 the possibility of Tr1 via the station S4 is only 0.125, though there is a higher possibility 0.5 that Tr1 runs on Se14 according to plan.
3. Whether is train Tr1 sure to run on Se25 during the period from 7:30am to 7:40am? And what is the possibility? The train Tr1 running on Se25 is represented by transition t_3 which fuzzy occurrence time $o3 = 0.2[33, 37, 45, 49]$. We have $Possibility = 0.2 \times Possibility(30 \leq o_3 \leq 40) = 0.2 \times 0.4167 = 0.0833$, thus train Tr1 is possible to running on Se25 during the period from 7:30 to 7:40, but with a very low possibility.

4 Comparing Analysis

Existed temporal knowledge reasoning methods analyze time parameters either according to sprouting graph created with finite system states in EFTPN [10] or TPN [11,12] model or by computation of fuzzy time functions based on FTPN [8, 9] model. We analyze the example in section 3 using other temporal knowledge reasoning algorithms and the specific contrasts with the above analysis can be seen in Table 1.

Table 1. Temporal knowledge reasoning contrast

	TPN	EFTPN	FTPN	EFTN
Representation of temporal uncertainty	Time interval	Fuzzy time interval	Fuzzy time function	Fuzzy time function
Processing of temporal uncertainty in conflict	No quantitative analysis	Quantitative analysis	No quantitative analysis	Quantitative analysis
Data structure	Dynamic sprouting graph	Dynamic sprouting graph	Timestamp	Timestamp
Computation procedure	4	4	2	2
Analysis of possibility	No	Yes	Imprecise	Yes

The following conclusions can be deduced from Table 1:

1. The descriptions of temporal uncertainty of all methods are consistent.
2. Our model can precisely describe the different possibility which the conflict event occurs and analyze quantitative uncertain temporal knowledge in conflict.
3. Our computation procedure is simple. By simple repeated addition and comparison of real numbers, our method just needs two procedures, system states searching and time parameters calculating and analyzing when token moves, thus enhanced the efficiency.
4. Associating transition with fuzzy firing interval can represent firing uncertainty of an enabled transition and make it possible to distinguish transitions involved in the same conflict relation by their firing possibility, and thus is very important to precisely describe and analyze quantitative conflict relation.
5. Time analysis is easier in our method. Our token structure of fuzzy timestamp is much simpler than sprouting graph structure.

EFTN has been used for system modeling in different areas [3, 4, 5, 10]. The introduction of fuzzy time interval makes EFTN model more accurate.

5 Conclusion

In this paper, we proposed a method of uncertain temporal knowledge reasoning and validating to handle temporal uncertainty of train group operation. Our contribution can be summarized as: (1)We associated each transition with a fuzzy firing interval, which describes accurately the temporal knowledge of this transition.(2)We promote the efficiency of reasoning and making our method suitable for time-critical applications.(3)Oriented to temporal uncertainty of train group operation, we analyzed quantitatively several issues of imprecise knowledge reasoning.

Quantitative analysis of temporal uncertainty of conflict events gives a direction for the research of new method of train group operation adjustment, and also enriches the research of expert system of other categories in knowledge engineering. To apply the method to more complicated systems is our next work in future.

Acknowledgement

This research was partially supported by the National Science Foundation of China under grant number 60674001 and the Henan Science Foundation under grant number 0411012300.

References

1. Limin Jia, Qiuhua Jiang (2003) Study on essential characters of RITS [A]. Proceeding of 6th International Symposium on Autonomous Decentralized Systems, IEEE Computer Society: 216-221
2. Tadao Murata (1996) Temporal Uncertainty and Fuzzy-timing High-Level Petri Nets. Lecture Notes in Computer Science 1091: 11-28
3. T Murata, J Yim H Yin and O Wolfson (2006) Petri-net model and minimum cycle time for updating a moving objects database. International Journal of Computer Systems Science & Engineering 21(3): 211-217
4. Chaoyue Xiong, Tadao Murata, Jason Leigh (2004) An Approach for Verifying Routing Protocols in Mobile Ad Hoc Networks Using Petri Nets. IEEE 6th CAS Symp. on Emerging Technologies: Mobile and Wireless Comm 2: 537-540
5. Yi Zhou, Tadao Murata (2000) Modeling and Performance Using Extended Fuzzy-Timing Petri Nets for Networked Virtual Environment. IEEE Transactions on System, Man, and Cybernetics-part B: Cybernetics 30(5): 737-755
6. Yi Zhou, Tadao Murata (1999) Petri Net Model with Fuzzy-timing and Fuzzy-Metric Temporal Logic. International Journal of Intelligent Systems 14(8): 719-746
7. Didier Dubois, Henri Prade (1989) Processing Fuzzy Temporal Knowledge. IEEE Transactions on Systems, Man and Cybernetics 19(4): 729-744
8. Yang-dong Ye, Juan Wang, Li-min Jia (2005) Processing of Temporal Uncertainty of Train Operation Based on Fuzzy Time Petri nets.Journal of the china railway society 27(1): 6-13
9. Yang-dong Ye, Juan Wang, Li-min Jia (2005) Analysis of Temporal Uncertainty of Trains Converging Based on Fuzzy Time Petri Nets. Lecture Notes in Computer Science 3613: 89-99
10. Yan-hua Du, Chun-huang Liu, Yang-dong Ye (2005) Fuzzy Temporal Knowledge Reasoning and Validating Algorithm and Its Application in Modeling and Analyzing Operations of Train Group. Journal of the china railway society 27(3): 1-8
11. Ye Yang-dong, Du Yan-hua, Gao Jun-wei, et al (2002) A Temporal Knowledge Reasoning Algorithm using Time Petri Nets and its Applications in Railway Intelligent Transportation System. Journal of the china railway society 24(5): 5-10
12. Woei-Tzy Jong, Yuh-Shin Shiau, Yih-Jen Horng, et al (1999) Temporal Knowledge Representation and Reasoning Techniques Using Time Petri Nets. IEEE Transaction on System, Man and Cybernetics, Part-B: Cybernetics 29(4): 541-545

An Improved FMM Neural Network for Classification of Gene Expression Data

Liu Juan[1], Luo Fei[1], and Zhu Yongqiong[2]

[1] School of computer, Wuhan University, Hubei 430079
liujuan@whu.edu.cn, luofei_whu@126.com
[2] The key lab of multimedia and network communication engineering, Wuhan University, Hubei 430079
zyqzhuyongqiong@126.com

Abstract. Gene microarray experiment can monitor the expression of thousands of genes simultaneously. Using the promising technology, accurate classification of tumor subtypes becomes possible, allowing for specific treatment that maximizes efficacy and minimizes toxicity. Meanwhile, optimal genes selected from microarray data will contribute to diagnostic and prognostic of tumors in low cost. In this paper, we propose an improved FMM (fuzzy Min-Max) neural network classifier which provides higher performance than the original one. The improved one can automatically reduce redundant hyperboxes thus it can solve difficulty of setting the parameter θ value and is able to select discriminating genes. Finally we apply our improved classifier on the small, round blue-cell tumors dataset and get good results.

Keywords: Classification, Gene Selection, FMM Neural Network.

1 Introduction

cDNA microarray experiment is an attracting technology which can help us monitor the expression condition of thousands of genes simultaneously. By analyzing genes expression result, the active, hyperactive or silent states of genes in different samples or tissues can be reflected. Furthermore, it provides a way to discriminate normal and disease tissues or different subtypes of tumors. Many cases [1,2] have proven that large-scale monitoring of gene expression by microarrays is one of the most promising techniques to improve medical diagnostics and functional genomics study. In practice, in virtue of gene microarray analysis, accurate classification of tumor subtypes may become reality, allowing for specific treatment that maximizes efficacy and minimizes toxicity. Many machine learning methods have been applied to exploit interesting biological knowledge, such as clustering, classification, feature selection and so on. Clustering analysis includes clustering samples into groups to find new types according to their profiles on all genes, and clustering genes into groups according to their expression profiles on all samples[3,4]. Classification analysis aims to build a classifier from the expression data and the class labels of training samples. Especially, Classification analysis of gene expression data [5-7] is useful in categorizing sub-types of

B.-Y. Cao (Ed.): Fuzzy Information and Engineering (ICFIE), ASC 40, pp. 65–74, 2007.
springerlink.com

tumors, which may be difficult to be distinguished for their similarity of clinical and histological characteristics. Gene selection [8-10] is critical in filtering redundant genes so that the key genes can be found out. It is particularly important in practice, for availability of a small number of samples compared to a large number of genes.

Fuzzy sets theory introduced by Zadeh [11] has been widely applied to many fields, such as fuzzy decision system, fuzzy expert system, fuzzy control system and so on. Compared to traditional set, which uses probability theory to explain the occurrence of crisp events and determines whether the events as occurring or not without middle ground, the fuzzy sets theory can provide a mechanism for representing linguistic concepts such as "high", "middle" and "low" and measure the degree how membership of an event is. Attentively, the conception of "degree" to which how membership an event is in fuzzy sets is different from the conception of "probability" that an event occurs in probability theory. For example, when describing a sample belonging to a certain class, fuzzy sets theory simply states the belonging degree as fully belonging to, quite belonging to, a little belonging to and so on, while traditional sets theory states unequivocally that the sample belongs to the class or not. The combination of fuzzy sets and pattern classification has been widely studied by many people [12,13]. Simpson[14] provided the combination of fuzzy set theory and neural network to construct the Fuzzy Max-Min neural network classifiers. It has been compared with several other neural, fuzzy and traditional classifier on Fisher iris data and shown good results in his work. However, there are still some drawbacks of the Fuzzy Max-Min neural network, such as lacking a definite instruction to set proper value for key parameter and so on.

In this paper, we will focus on the problems of classification and genes selection based on FMM (Fuzzy Max-Min) neural network [14]. Due to there being huge amount of genes and small number of samples, our work consists of two steps: selecting informative genes and building classifiers. Basically, our gene selection method is a kind of ranking based filtering methods. But traditional filtering methods just evaluate the discriminant ability of genes according to some criteria and choose the top ones, which lead to that there may exist many redundant (similar) genes in the selected gene sets. Differently, our method first clusters genes according to their similarity and choose the non-redundant genes, then choose those genes with most discriminant abilities from them. When constructing the classifier, we analysis some drawbacks of original FMM neural network and present an improved method. The proposed method has been evaluated on the SRBCT data set and has shown good performance.

This paper is organized as following. Section 2 first introduces the FMM neural network, then analyzes some drawbacks of it, and presents our improved FMM neural network at last. In section 3, we introduce how to do genes selection work with the improved FMM neural network. Section 4 is the experiment based on the dataset of the small round blue-cell tumors. The paper ends with conclusion and in section 5.

2 Improving FMM Neural Network Classifier

2.1 Original FMM Neural Network

Fuzzy Max-Min neural network is a three layer neural network. The first layer is the input layer.$F_A = (x_1, x_2 \ldots x_n)$ has n processing elements, one for each of the n genes of one training sample (or pattern). The second layer is the hyperboxes layer. Each node b_i in $F_B = (b_1, b_2 \ldots b_m)$ represents a hyperbox. The third layer is the output layer $F_C = (c_1, c_2 \ldots c_p)$ and the node c_i represents a class label. The structure of FMM neural network shows in Fig.1.

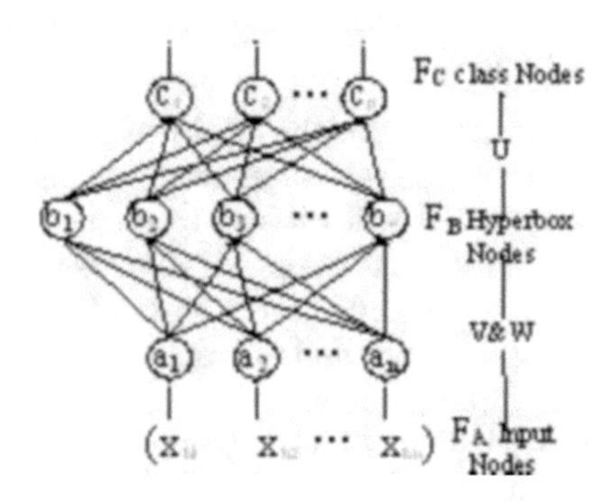

Fig. 1. Three layers structure of FMM neural network

Each node b_j in F_B is connected with all nodes in F_A, and there are two sets of connections between each input node and each of the m hyperbox fuzzy set nodes, the min vector ν_j and the max vector ω_j, which represent the minimal and the maximal point of the hyperbox respectively. A hyperbox determines a region of n-dimension hyper-space and a fuzzy set and it is defined by its minimal vector ν_j and maximal vector ω_j as formula (1):

$$B_j = \{X_k, \nu_j, \omega_j, f_j(X_k, \nu_j, \omega_j)\} \forall X_k \in R^n \tag{1}$$

X_k is the kth input sample in the training sample set R^n,usually the training sample set is normalized, $f_j(X_k, \nu_j, \omega_j)$ is the membership function and defined as

$$f_j(X_k, \nu_j, \omega_j) = \frac{1}{2n} \sum_{i=1}^{n} [max(0, 1 - max(0, \gamma min(1, X_{ki} - \omega_{ji}))) + max(0, 1 - max(0, \gamma min(1, \nu_{ji} - X_{ki})))] \tag{2}$$

γ is the sensitivity parameter that regulates how fast the membership values decrease as the distance between X_k and B_j increases.The aggregate fuzzy set that defines the kth sample class C_k is

$$C_k = \bigcup_{j \in K} B_j \tag{3}$$

where K is the index set of those hyperboxes associated with class k. Note that the union operation in fuzzy sets is typically the max of all of the associated fuzzy set membership functions. The maximal size of any hyperbox is bounded by $\theta \in [0, 1]$,a user-defined vale, and must meet the constraint:

$$n\theta \geq \sum_{i=1}^{n} [max(X_{ki}, \omega_{ji}) - min(\nu_{ji}, X_{ki})] \tag{4}$$

The learning process of FMM neural network includes three steps: expansion, overlap test and contraction.

1. When a new sample is input into the FMM neural network, the hyperbox which have highest membership with the sample and its class is the same with the sample's class label will expand to include it under the maximal size constraint shown in formula (4) met. If there is no expandable hyperbox found, a new hyperbox will be added into the FMM neural network. In this case, the new hyperbox is a point and its minimal vector and maximal vector are the same with the input sample.
2. When addition of the new pattern leads to expansion of some hyperbox, the expansion may cause the overlap with other hyperboxes. Therefore, it is necessary to test overlap between hyperboxes that represent different classes.To determine if this expansion created any overlap, a dimension by dimension comparison between hyperboxes is performed. If, for each dimension, at least one of four cases summarized in [14] is satisfied, then overlap exists between the two hyperboxes.
3. If the overlap between hyperboxes from different class exists, the overlap must eliminate by adjusting the size of the hyperboxes according to the contraction principle in [14].

The learning process lasts until all samples are put into the FMM neural network to train.

2.2 Drawbacks of FMM Neural Network

Studying the FMM Neural Network, we found that there are some drawbacks in it. One is that it is difficult to set a proper value to the parameter θ in the formula (4), which is used to restrict the maximal size of hyperboxes set. θ influences not only the resolution of the sample space, but also the performance of the classifier. On one hand, small value of θ will contribute to high classification accuracy, especially when there are large amount of overlapping parts among different classes in the dataset, however, too much small θ may lead to too many trivial hyperboxes produced during the learning process, which complicates the structure of the classifier and also decreases the performance of the classifier. On the other hand, too large θ will reduce hyperboxes number to simplify structure but result in large size of hyperboxes, which may contain samples from different classes. This case decreases the classification accuracy. Therefore, how to determine an optimal value of θ is very difficult for users, for the concept of "proper" θ varies greatly according to different datasets. There isn't standard quantitative description for it.

The second drawback is that it neglects some useful information. In original FMM neural network, a hyperbox is defined only by minimal vector ν and maximal vector ω . All hyperboxes have two kinds of information after construction of the classifier. The class it belongs to and the size of itself. Besides these information, it is natural to evaluate which hyperbox is better when there are several hyperboxes belonging to one sample class. The original FMM neural network has no way to evaluate those hyperboxes due to lacking of useful information. Obviously, other information should also be considered during designing the FMM neural network.

2.3 Improvement of FMM Neural Network

Based on the analysis of drawbacks of the original FMM neural network, we proposed an enhanced FMM neural network. One enhancement is that we let the hyperbox have more information. With regard to those hyperboxes belonging to the same class, a new parameter ρ_i is defined to reflect the sample density of hyperbox B_i

$$\rho_i = \frac{N_i}{V_i}, V_i = \prod_{j=1}^{n}(\omega_{ij} - \nu_{ij}) \tag{5}$$

In B_i, N_i is the number of samples whose class label is the same with the hyperbox B_i's. V_i is the volume of hyperbox B_i. Obviously, although a class may have several hyperboxes, the samples belonging to this class couldn't averagely lie in each one. Therefore, the parameter ρ_i is able to approximately reflect distribution of samples of one class. The more correct samples lie in the hyperbox B_i, the higher the value of parameter ρ_i is and the more confidently the hyperbox B_i is to represent its sample class. So this parameter can be used to evaluate those hyperboxes belonging to the same class.

In order to guarantee enough accuracy of classification, parameter θ always is set to small value. As described above, small θ value makes size of hyperboxes small, thus the number of hyperboxes is large. It makes the structure of FMM neural network complicated due to production of trivial hyperboxes since overlapping between hyperboxes belonging to the same class is neglected in the original FMM neural network. Therefore, ρ_i can also helps to define homogenous hyperboxes to eliminate trivial hyperboxes. Here homogenous hyperboxes are those ones that have similar ρ_i, belong to the same sample class and overlap.

As mentioned above, the proper value of θ is very important for the performance of FMM neural network. However, it isn't easy for most of users to estimate a perfect θ for the classifier. In order to solve this problem, another improvement of FMM neural network is made. Depending on definition of

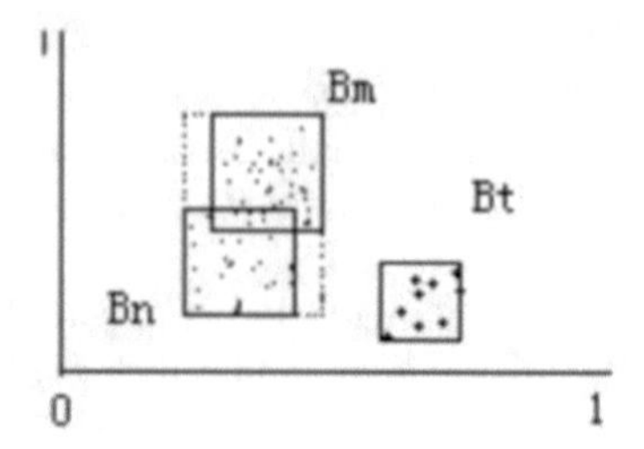

Fig. 2. two homogeneous hyperboxes can be aggregated into one

homogenous hyperboxes, we just fix θ to small value to guarantee accuracy, and then reduce some trivial hyperboxes by aggregating those homogeneous after the process of training. Assume that the hyperboxes B_m and B_n belong to the same class and B_t belongs to different class, just as shown in Fig.2. We define those hyperboxes as homogeneous ones only if they meet requirements below:

1. $B_m \cap B_n \neq \emptyset$
2. $\frac{max\{\rho_{B_m}, \rho_{B_n}, \rho_{(B_n \cup B_m)}\}}{min\{\rho_{B_m}, \rho_{B_n}, \rho_{(B_n \cup B_m)}\}} - 1 \leq \lambda$
3. $(B_m \cup B_n) \cap B_t) = \emptyset$

Requirement 1 makes sure two hyperboxes overlap. The intersection operation of two hyperboxes means to find out the overlapping part between them. The case indicating overlapping existence is $\nu_m \leq \nu_n \leq \omega_m \leq \omega_n$ or $\nu_n \leq \nu_m \leq \omega_n \leq \omega_m$. Requirement 2 makes sure there is a similar sample density among two hyperboxes and union of them. If the sample density of hyperboxes differs greatly, as though two hyperboxes belong to the same class and overlap, they couldn't be aggregated, for much unknown sample space is included and it will cause misclassification. The parameter λ controls the different degree of hyperboxes' sample density. Requirement 3 makes sure there is no overlapping between the new hyperbox produced by union operation and any other hyperbox representing different class. If above three requirement are met, we aggregate B_m and B_n and produce a new hyperbox B_q(shown as the dashed line hyperbox in Fig.2.). Meanwhile, delete B_m and B_n from FMM neural network and add B_q. Its minimal and maximal vectors decide as $\nu_q = min(\nu_n, \nu_m)$,$\omega_q = max(\omega_n, \omega_m)$. Although we introduce another parameter λ, its meaning is more clear than θ for any dataset, so it is more easy to set proper λ for most average users.

3 Gene Selection

Gene selection in classification aims to find optimal genes subsets which provide good discrimination of sample classes by identifying the genes that are differentially expressed among the classes when considered individually. It is an indispensable process for the number of genes is much greater than that of samples. Therefore, redundant genes should be eliminated. The merit of implementing gene selection lies in that through evaluation of contribution to classification of each gene, we can find out high-performance genes and eliminate low performance ones so that not only decrease the dimensions but also avoid influence of noisy genes. In this paper, we perform two steps to implement the gene selection. The first step is to cluster genes into the groups using KNN clustering method, and then select the representative ones from each group to form the new candidate genes. This step is to reduce redundant similar genes since their discriminant ability is regarded same as their representatives. The second step is to choose top-ranked genes from the candidates. If the value of a gene varies little within same sample class and great between different sample classes, the gene is expected to have good class discrimination ability and has high rank. FMM neural network is still used here to rank the genes left after step 1.

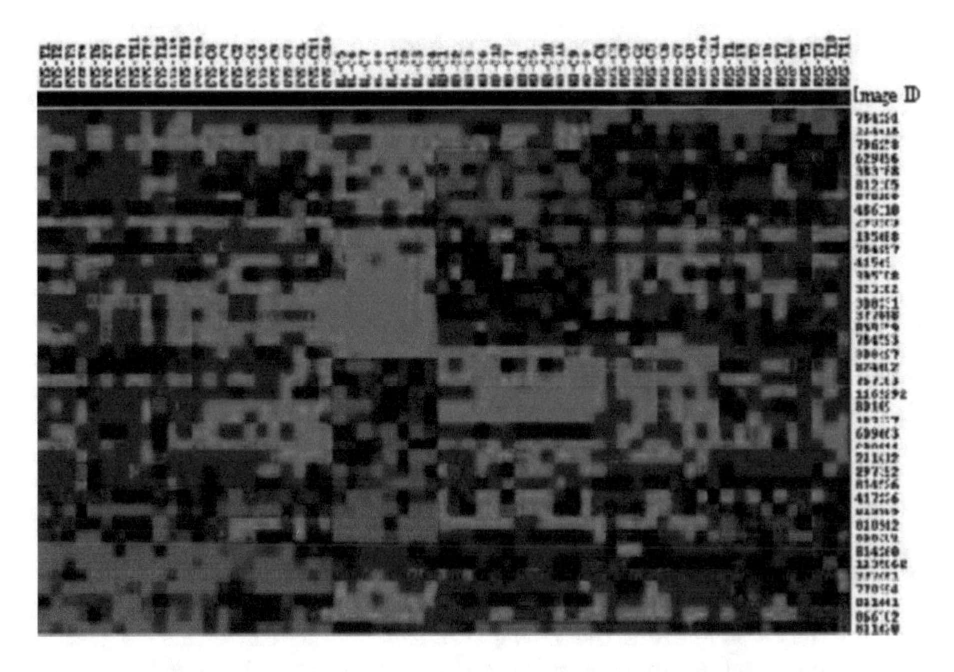

Fig. 3. the expression of top ratio 40 genes in 63 samples

We add a new connection η_{ij} between the jth input node in first layer and the ith hyperbox node in the improved FMM neural network. Initially, $\eta_{ij} = 0$. During the training process, η_{ij} will update as:

$$\eta_{ij}^{new} = \eta_{ij}^{old} + (X_{kj} - \overline{X}_{Cj})^2, \overline{X}_{Cj} = \frac{\sum_{k=1}^{n} X_{kj} * I(X_k, C)}{n} \tag{6}$$

$$I(X, C) = \begin{cases} 1 & \text{X belongs to class C} \\ 0 & \text{otherwise} \end{cases}$$

X_{kj} is the jth dimension value of the kth sample and belongs to the ith hyperbox by membership function. Here n is the total number of samples.

After training, total m hyperboxes are produced and we analyze how well each gene discriminates samples by calculating its ratio of between-groups to within-groups sum of squares.

$$Ratio(Gene_j) = \frac{\sum_{k=1}^{p}(\overline{X}_{kj} - \overline{X}_{.j})}{\sum_{i=1}^{m}\sum_{k=1}^{p} \eta_{ij} * Class(B_i, k)}, \overline{X}_{.j} = \frac{\sum_{k=1}^{n} X_{kj}}{n} \tag{7}$$

$$Class(B_i, C) = \begin{cases} 1 & \text{hyperbox } B_i \text{ belongs to class C} \\ 0 & \text{otherwise} \end{cases}$$

In formula (7), m is the total number of hyperboxes and p is the total number of classes. The numerator part is to compute between-groups sum of squares of the gene j and the denominator part is to compute its within-groups sum of squares. Obviously, the bigger the Ratio ($Gene_j$) is, the better the jth gene can discriminate different sample classes.

4 Experiment Evaluation

We applied our method on the dataset of SRBCTs (the small, round blue-cell tumors)published by Khan in [15]. The SRBCTs of childhood, which include neuroblastoma (NB), rhabdomyosarcoma (RMS), non-Hodgkin lymphoma (NHL) and the Ewing family of tumors (EWS), are so named because of their similar appearance on routine histology. In Khan's experiment, expression levels from 6567 genes are measured for each of the 88 samples. In order to reduce number of redundant genes, some genes were filtered out by measuring their significance of the classification and PCA in [15].

Based on the same SRBCTs dataset with Khan's, which includes 2308 genes whose red intensity more than 20 in 88 samples, we used 63 samples as training set from the 88 samples and the other 25 samples were used as testing set. During learning process, the parameter γ donated 4 in the formula (2), for it is moderate for learning rate. The parameter θ was set as 0.0375 in the formula (4)and λ was 0.1.

Table 1 is the detailed testing result on 25 testing samples with our improved FMM neural network combining with gene selection. Except five normal samples T1, T2, T3, T6 and T7, other 21 tumorous samples were 100 percents accurately classified by forty genes with highest value of ratio of between-groups to within-groups sum of squares. Fig.3. is the color boxes diagram of these forty genes in 63 samples training dataset. From Fig.3, we can see these forty genes are able to discriminate four types of SRBCTs well.

Table 1. Classification 25 testing data by improved FMM neural network

Test sample	Improved FMM	Histological Label	Source Label
T1	BL	Sk. Muscle	SkM1
T2	EWS	state Ca.-C	PC3
T3	EWS	Sarcoma	A204
T4	NB	NB-C	SMSSAN
T5	RMS	ERMS-T	ERDM1
T6	RMS	Sk. Muscle	SkM2
T7	EWS	Osteosarcoma-C	OsA-C
T8	NB	NB-C	IMR32
T9	EWS	EWS-C	CHOP1
T10	RMS	ERMS-T	ARMD1
T11	BL	BL-C	EB1
T12	EWS	EWS-C	SARC67
T13	RMS	ERMS-T	RMS4
T14	EWS	EWS-C	9608P053
T15	EWS	EWS-C	ES23
T16	EWS	EWS-C	9903P1339
T17	RMS	ERMS-T	ARMD2
T18	BL	BL-C	GA10
T19	RMS	ERMS-T	ERMD2
T20	NB	NB-C	NB1
T21	NB	NB-C	NB2
T22	NB	NB-C	NB3
T23	NB	NB-C	NB4
T24	BL	BL-C	EB2
T25	EWS	EWS-C	ET3

5 Conclusion

In this paper, we proposed an improved FMM neural network to classify the gene expression data. The main contributions include introducing more information along with the hyperboxes that helps to overcome dilemma in selecting a proper parameter θ; New gene selection method consisting of clustering and FMM neural network based filtering. The experiment results on SRBCT show that the proposed method have good performance.

References

1. Sorlie T, Perou CM, Tibshirani R, et al. (2001) Gene expression patterns of breast carcinomas distinguish tumor subclasses with clinical implications. Proc Natl Acad Sci USA; 98:10869-10874.
2. M.J.van de Vijver, Y.D.He, L.J.van't Veer, et al.(2002)Gene-Expression Signature as a Predictor of Survival in Breast Cancer. N. Engl. J. Med., December 19; 347(25): 1999 - 2009.

3. LUO F, KHAN L, BASTANI F, et al. (2004)A Dynamical Growing Self-Organizing Tree (DGSOT) for Hierarchical Clustering Gene Expression Profiles. Bioinformatics, 20(16):2605-17.
4. Dembele, D. and Kastner, P.(2003)Fuzzy c-means method for clustering microarray data. Bioinformatics, 19, 973-980
5. Golub T. R., Slonim D. K., Tamayo P., Huard C., et al.(1999)Molecular classification of cancer: class discovery and class prediction by gene expression monitoring. Science (Wash. DC), 286: 531-537
6. Alizadeh, A.A., et al.(2000) Distinct types of diffuse large B-cell lymphoma identified by gene expression profiling. Nature, 403, 503-511
7. Ramaswamy, S., et al.(2001)Multiclass cancer diagnosis using tumor gene expression signatures. Proc. Natl Acad. Sci. USA,98, 15149-15154
8. Model, F., Adorjan, P.(2001)Feature selection for DNA methylation based cancer classification. Bioinformatics 17 Suppl 1, S157-S164
9. Park, P.J., Pagano, M.(2001)A nonparametric scoring algorithm for identifying informative genes from microarray data. Pac. Symp. Biocomput. 52-63
10. Guoyon,I., Weston,J., Barnhill,S. and Vapnik,V.(2003)Gene selection for cancer classification using support vector machines. Machine Learning.46, 389-422
11. L. Zadeh.(1965)fuzzy set, Inform, and Control, vol. 8, pp. 338-353
12. Shinn-Ying Ho, Chih-Hung Hsieh, Kuan-Wei Chen.(2006)Scoring Method for Tumor Prediction from Microarray Data Using an Evolutionary Fuzzy Classifier. PAKDD2006,Proceedings. pp. 520 - 529.
13. Chakraborty, D., Pal, N.R.(2004).A neuro-fuzzy scheme for simultaneous feature selection and fuzzy rule-based classification, IEEE Transactions on Neural Networks, pp: 110- 123, Volume: 15, Issue: 1
14. Simpson, P.K. (1992).Fuzzy min-max neural networks: 1. Classification.IEEE Transactions on Neural Networks, 3, 776-786
15. Khan, J., Wei, J. S., Ringner, M., et al.(2001)Classification and diagnostic prediction of cancers using gene expression profiling and artificial neural networks. Nat. Med. 7, 673-679

Using Assignment Matrix on Incomplete Information Systems Reduction of Attributes

Songyun Guo[1] and Zhi Tao[2]

[1] Shenyang Artillery Academy, Shenyang, 110162, China
Gsy1968@tom.com

[2] College of Sciences, Civil Aviation University of China, Tianjin, 300300, China
T86543213@163.tom

Abstract. This paper presents an assignment matrix based heuristic attribute reduction method, states the significance of attributes in the incomplete decision table by introducing assignment matrices and their distances from one another, which is then used as heuristic information for selecting attributes. This algorithm has a polynomial time complexity. Seeking the minimal relative reduction in a decision table is typically a NP-hard problem, and its complexity can be reduced by using the algorithm provided in this paper. An example shows this algorithm can achieve the minimal relative reduction of incomplete decision table.

Keywords: assignment matrix, incomplete decision table, rough set, attribute reduction.

1 Introduction

Rough set theory was developed by a Polish Scholar Pawlak in 1982. As a new mathematic tool that deals with incomplete and uncertain problems, it has become a new, hot academic topic in the artificial intelligence area. Its main idea is deriving decision-making or classification rules of a problem by attribute (knowledge) reduction on the premise of keeping the classification capability of knowledgebase. At present, rough set theory has been successfully applied to machine learning, decision-making analysis, process control, pattern recognition and data mining, etc.

Attribute reduction is one of the core issues in rough set theory. For an information system, people always expect to find all reductions or the minimal reduction. But unfortunately, seeking all reductions and the minimal reduction are both NP complete problems. Recently some researchers have worked out many attribute reduction algorithms, most of which studied on complete information systems. However, the incompleteness of information (missing attribute values of objects) exists widely in the real world. For example, in the intelligent decision-making system for medical diagnostics, since some patients are unsuitable to one instrument examination, it is impossible to obtain some symptoms (attributes) of the patients (objects). Therefore, how to obtain useful knowledge from incomplete information systems is appearing more important.

B.-Y. Cao (Ed.): Fuzzy Information and Engineering (ICFIE), ASC 40, pp. 75–82, 2007.
springerlink.com

This paper defines the assignment matrix based on tolerance relation (reflexivity, symmetry), characterizes the relative significance of attributes by using the distance between the assignment matrices, uses this significance as heuristic information for selecting attributes, and thereby presents the relative reduction algorithm of decision table attributes, and points out it has a polynomial time complexity. Furthermore, an example in this paper shows that it is possible to find the relative reduction of an incomplete information system by using this algorithm.

2 Incomplete Information System and Tolerance Relation

Definition 1 The quadruplet $S=(U, A=C\bigcup D, V, f)$ is called an information system, where U is a nonempty finite set of objects, called discussion domain; $A=C\bigcup D$ is a nonempty finite set of attributes, C is a conditional attribute set, D is a decision attribute set, and $C\bigcap D=\phi$; $V=\bigcup_{a\in A} V_a$, V_a is the value domain of the attribute a; f is an information function of $U\times A\rightarrow V$, which assigns an information value to each attribute of each object, that is, $\forall a\in A. x\in U, f(x,a)\in V_a$. If $D=\phi$, the information system is called a data table; otherwise, it is called a decision table. If there exist $x\in U, a\in C$, such that $f(x,a)$ is unknown (noted as $f(x,a)=*$), the information system is incomplete; otherwise, it is complete.

For an incomplete information system, Kryszkiewicz M defined a tolerance relation T_P on U:

$$\forall\phi\subset P\subseteq A,$$

$$T_P(x,y)\Leftrightarrow\forall a\in P, f(x,a)=f(y,a)\vee f(x,a)=*\vee f(y,a)=* \qquad x, y\in U.$$

Obviously, T_P meets reflexivity and symmetry. The tolerance relation determined by the attribute $a\in A$ is briefly noted as T_a.

Let $T_P(x)$ denote the whole objects which are tolerant with x under P, i.e. $T_P(x)=\{y\in U \mid T_P(x,y)\}$. For P, $T_P(x)$ is the maximum set of objects that may be undistinguishable from x.

For $B\subseteq C$, $U/T_B(x,y)=\{T_B(x)\mid x\in U\}$ is used to represent a classification of the discussion domain U. Any element in $U/T_B(x,y)$ is called a tolerance class. Generally, the tolerance classes in $U/T_B(x,y)$ don't make up of a division of U, but a coverage of U.

For the decision table $S=(U, A=C\bigcup D, V, f)$, we can assume the decision attribute value is not empty because experts can make decisions even if the information is incomplete. Based upon decision attribute values, the discussion domain can be divided into nonintersecting equivalent classes, noted as $U/D=\{Y_1,Y_2,\cdots,Y_m\}$. For $B\subseteq C$, note $D_B(x)=\{f(Y_i,D)\,|\,T_B(x)\bigcap Y_i\neq\phi,\ 1\leq i\leq m\}$, where $f(Y_i,D)$ represents the decision values of the equivalent class Y_i, i.e., $D_B(x)$ represents all possible decision values of B tolerance class of x., called B assignment set of x.

Definition 2. For an incomplete decision table $S=(U,C\bigcup D,V,f)$, $b\in B\subseteq C$, if $\forall x\in U$, and $D_{B\setminus\{b\}}(x)=D_B(x)$, then b is unnecessary in B with respect to decision attributes; otherwise, b is relatively necessary in B. If each attribute in B is relatively necessary, then B is relatively independent; otherwise, B is relatively dependent.

According to Definition 2, if removing an attribute from the conditional attribute set doesn't result in any new inconsistency in the decision, then the attribute is unnecessary in the attribute set with respect to decision; otherwise, it is necessary to decision.

Definition 3. For an incomplete decision table $S=(U,C\bigcup D,V,f)$, all the relatively necessary attributes in $B\subseteq C$ are called the relative core of B, noted as $Core_D(B)$.

Definition 4. For an incomplete decision table $S=(U,C\bigcup D,V,f)$, $R\subseteq B$ is called a relative reduction of B, if

1) $D_R(x)=D_B(x)$;
2) R is relatively independent.

It can be proved that the relative core is the intersection of all relative reductions. That is:

Property 1. $Core_D(B)=\bigcap Red_D(B)$, where $Red_D(B)$ represents the relative reduction of B.

3 Assignment Matrix and Measurement of Conditional Attributes' Significance to Decision

First we provide the concept of assignment matrix:

Definition 5. For an incomplete decision table $S=(U,C\bigcup D,V,f)$, $U=\{x_1,x_2,\cdots,x_n\}$, $U/D=\{Y_1,Y_2,\cdots,Y_m\}$, the assignment matrix of the conditional attribute subset $B\subseteq C$ is defined as:

$$M^B = (m_{ij}^B)_{n\times m}$$

Where, $m_{ij}^B = \begin{cases} 1 & f(Y_j, D) \in D_B(x_i) \\ 0 & f(Y_j, D) \notin D_B(x_i) \end{cases}$

Using the assignment matrix, we can easily evaluate the assignment set of each object under any conditional attribute subset. At the same time, the assignment matrix can be also used to determine whether an attribute is necessary or not: if the resulting assignment matrix after removing an attribute becomes unchanged, then the attribute is unnecessary; otherwise, it is necessary. For an incomplete decision table, people always expect to find its minimum reduction by removing the unwanted attributes. For this reason, the distance between two assignment matrices is defined as following:

Definition 6. For an incomplete decision table $S = (U, C \cup D, V, f)$, M^E, M^F are the assignment matrices of the conditional attribute subsets E, F respectively, then the distance between M^E, M^F is defined as:

$$d(M^E, M^F) = \sum_{i=1}^{n}\sum_{j=1}^{m} | m_{ij}^E - m_{ij}^F |$$

Where, $| m_{ij}^E - m_{ij}^F | = \begin{cases} 0 & m_{ij}^E = m_{ij}^F \\ 1 & m_{ij}^E \neq m_{ij}^F \end{cases}$

The distance between assignment matrices can be used to describe the relative significance of an attribute to the attribute subset (relative decision attribute D).

Definition 7. The relative significance of an attribute $b \in B \subseteq C$ in B is defined as:

$$sig_{B\setminus\{b\}}(b) = d(M^B, M^{B\setminus\{b\}}) = \sum_{i=1}^{n}\sum_{j=1}^{m}(m_{ij}^{B\setminus\{b\}} - m_{ij}^B)$$

The definition above shows that the significance of attribute $b \in B$ in B with respect to decision D can be measured by the variation of assignment matrix which is caused by removing b from B. Consequently, we can easily get the following conclusion.

Property 2. An attribute $b \in B$ is relatively necessary in B if $sig_{B\setminus\{b\}}(b) > 0$.

Property 3. $core_D(B) = \{ b \in B \mid sig_{B\setminus\{b\}}(b) > 0 \}$.

Definition 8. The relative significance of attribute $b \in C \setminus B$ to B is defined as:

$$sig_B(b) = d(M^B, M^{B \cup \{b\}}) = \sum_{i=1}^{n} \sum_{j=1}^{m} (m_{ij}^B - m_{ij}^{B \cup \{b\}})$$

Definition 8 shows that the relative significance of one attribute to one attribute set is measured by the variation of assignment matrix which is caused by adding this attribute. So it can be used as heuristic knowledge to find the minimum reduction by continuously adding the attributes with the maximum relative significance.

4 Assignment Matrix Based Attribute Reduction Algorithm

From Property 3 we can easily evaluate the relative core $core_D(C)$ of an incomplete decision table. Use the relative core as the starting point of evaluating the minimum relative reduction, and according to the attribute significance defined in Definition 8, select the important attributes to add to the relative core one by one until the distance between its assignment matrix and that of the original conditional attribute is zero.

The following is the steps of the algorithm:

Input: incomplete decision table $S = (U, C \cup D, V, f)$.

Output: the minimum relative reduction of $S = (U, C \cup D, V, f)$.

Step 1: calculate the assignment matrix M^C of conditional attribute set C.

Step 2: for each $c \in C$, seek $sig_{C \setminus \{c\}}(c)$, take

$$core_D(C) = \{ c \in C \mid sig_{C \setminus \{c\}}(c) > 0 \}.$$

Step 3: let $B = core_D(C)$, repeat:

(1) Determine if $d(M^C, M^B) = 0$ is true. If true then go to Step 4; otherwise, go to (2).

(2) Select an attribute $c \in C \setminus B$ which is fit to $sig_B(c) = \max_{c' \in C \setminus B} sig_B(c')$, and combine it to $B = B \cup \{c\}$, go to (1).

Step 4: the output B is one relative reduction of $S = (U, C \cup D, V, f)$.

Below we'll analyze the time complexity of the above algorithm.

The time complexity of Step 1 for seeking the assignment matrix is $O(|C||U|^2)$; the time complexity of Step 2 is $O(|C|^2|U|^2)$; the time complexity of (1) in Step 3 is $O(|C||U|^2)$, the time complexity of (2) is $O(|C||U|)$, at worst this step needs to loop $|C|$ times. Thus the time complexity of the above algorithm is $O(|C|^2|U|^2)$.

5 Example and Analysis

In order to show the availability of the algorithm, we try to find the minimum relative reduction of the incomplete decision table on cars that was provided in Literature [5] (See Table 1).

Table 1. Incomplete decision table

Car	Price	Mile-age	Space	Max speed	Conclu-sion
1	high	high	big	low	good
2	low	*	big	low	good
3	*	*	small	high	bad
4	high	*	big	high	good
5	*	*	big	high	excellent
6	low	high	big	*	good

In Table 1 the conditional attribute set *C*={Price, Mileage, Space, Maximum speed}, and the decision-making attribute set *D*={Conclusion}. The following is the process of finding its relative reduction by using the assignment matrix based attribute reduction algorithm (where $Y_1 = \{3\}, Y_2 = \{1,2,4,6\}, Y_3 = \{5\}$):

Step 1: evaluate $$\boldsymbol{M}^C = \begin{bmatrix} \mathbf{0} & \mathbf{1} & \mathbf{0} \\ \mathbf{0} & \mathbf{1} & \mathbf{0} \\ \mathbf{1} & \mathbf{0} & \mathbf{0} \\ \mathbf{0} & \mathbf{1} & \mathbf{1} \\ \mathbf{0} & \mathbf{1} & \mathbf{1} \\ \mathbf{0} & \mathbf{1} & \mathbf{1} \end{bmatrix}$$

Step 2: because $M^{C\backslash\{Price\}} = M^{C\backslash\{Mileage\}} = \begin{bmatrix} 0 & 1 & 0 \\ 0 & 1 & 0 \\ 1 & 0 & 0 \\ 0 & 1 & 1 \\ 0 & 1 & 1 \\ 0 & 1 & 1 \end{bmatrix}$,

$M^{C\backslash\{Space\}} = \begin{bmatrix} 0 & 1 & 0 \\ 0 & 1 & 0 \\ 1 & 1 & 1 \\ 1 & 1 & 1 \\ 1 & 1 & 1 \\ 1 & 1 & 1 \end{bmatrix}$, $M^{C\backslash\{Max\ speed\}} = \begin{bmatrix} 0 & 1 & 0 \\ 0 & 1 & 1 \\ 1 & 0 & 1 \\ 0 & 1 & 1 \\ 0 & 1 & 1 \\ 0 & 1 & 1 \end{bmatrix}$,

we can separately calculate $sig_{C\backslash\{Price\}}(Price) = sig_{C\backslash\{Mileage\}} = 0$,

$sig_{C\backslash\{Space\}}(Space) = 5$,

$sig_{C\backslash\{Max\ speed\}}(Max\ speed) = 2$.

So the relatively core is $core_D(C) = \{\text{Space}, \text{Max speed}\}$.

Step 3: let $B = core_D(C)$, because $M^B = M^C$, $d(M^C, M^B) = 0$ is true. Go to Step 4.

Step 4: output the minimum relative reduction $B = \{\text{Space}, \text{Max speed}\}$ of the original incomplete decision table.

The above calculating result is exactly the correct result of this example. From the example analysis, it is concluded that we can find the minimum relative reduction of incomplete decision table by using the assignment matrix based attribute reduction algorithm provided here.

6 Summary

Rough set theory derives the classification or decision-making rules of problems by means of attribute reduction, but attribute reduction is mostly based on complete information system. Regarding the tolerance relation based incomplete decision tables, this paper introduces the concept of the significance of conditional attributes to decision by defining assignment matrices and the distances between them, which is then used as heuristic information to help set up the attribute reduction algorithm for incomplete decision table. The example shows this algorithm can find the minimum relative reduction of incomplete decision table.

Ackowledgements

This work described here is partially supported by the grants from National Science Foundation Committee of China and General Administration of Civil Aviation of China (60672178), and Doctor's Launch Foundation of Civil Aviation University of China (05qd02s).

References

[1] Pawlak Z (1991) Rough sets—theoretical aspects of reasoning about data[M]. Dordrecht: Kluwer Academic Publishers. 9-30.
[2] Han Z X, Zhang Q, Wen F S (1999) "A survey on rough set theory and its application[J]." Control Theory and Application, 16(2): 153-157.
[3] Pawlak Z (1998) "Rough set theory and its application to data analysis[J]." Cybernetics and Systems, 29(9): 661-668.
[4] Hu X H (1996) "Mining knowledge rules from databases-a rough set approach[C]." Proceedings of IEEE International Conference on Data Engineering, Los Alamitos: IEEE Computer Society Press, 96-105.
[5] Kryszkiewicz M (1998) "Rough set approach to incomplete information systems[J]." Information Sciences, 112: 39-49.

An Image Compression Algorithm with Controllable Compression Rate

J.L. Su[1,2], Chen Yimin[1], and Zhonghui Ouyang[2]

[1] School of Computer Engineering and Science,
Shanghai University, 200027, shanghai, P.R. China
[2] TSL School of Business and Information Technology,
Quanzhou Normal University, 362000, Quanzhou, P.R. China
su_jinlong@yahoo.com.cn

Abstract. A method of fuzzy optimization design based on neural networks is presented as a new method of image processing. The combination system adopts a new fuzzy neuron network (FNN) which can appropriately adjust input and output, and increase robustness, stability and working speed of the network. Besides FNN, wavelet transform also be applied to compression algorithm for a higher and controllable compression rate. As shown by experimental results, the communication system can compress and decompress image data dynamically which proves that it is a more practical and more effective method than traditional methods for communication in wireless networks, and it is especially better at video data transmitting.

Index Terms: FNN, wavelet transform, compression rate.

1 Introduction

The image compression algorithm is designed for an wireless communication system, which is aim to control communication image data timely and automatically for congestion control. When the detecting algorithm gets the result that the network is under high risk of congestion, the system will try to higher the compression rate to avoid the coming congestion, with the cost of image quality losing.

The compression algorithm is including a wavelet transform subprogram and FNN vector classification sub-algorithm. The wavelet transform is function to spilt the image data gradually, and control the amount of processing data which is relate to the compression rate, while keep the important detail of the original image as much as possible. In additionally, the compression task is handled by FNN algorithm, which is aim to compression the data and retain detail information efficiently and effectively.

The overall organization of the paper is as follows. After the introduction, in Section 2 we present fundamental issues in fuzzy core neural classification, the theory and network structure of FNN. Some technical details are explained in Section 3. In this section, practical issues of image data split based on wavelet transform are touched upon. And the whole process of the compression algorithm is clarified at the same time. Some of our experiments and results are presented on Section 4. Finally Section 4 concludes the paper.

B.-Y. Cao (Ed.): Fuzzy Information and Engineering (ICFIE), ASC 40, pp. 83–88, 2007.
springerlink.com

2 The Fuzzy Neural Network [5, 6]

Fuzzy core neural networks have emerged as an important tool for classification, which includes the Bayesian classification theory, the role of posterior probability in classification, posterior probability estimation via neural networks, and the relationships between neural networks and the conventional classifiers. As we know, the neural-network-based control of the dynamic systems compensates the effects of nonlinearities and system uncertainties, so the stability, convergence and robustness of the control system can be improved. The fuzzy neural network possesses the advantages of both fuzzy systems and neural networks. Recent research activities in fuzzy core neural classification have established that FNN are a promising alternative to various conventional classification methods.

This paper describes a fuzzy optimization model we have developed, which allows us to control the compression rate of sequences of images derived from Multi-sensors-fusion, and to relieve the congestion in GPRS communication. The model is a fuzzy core neural network (FNN).

The fuzzification procedure maps the crisp input values to the linguistic fuzzy terms with the membership values between 0 and 1.The Rule Base stores the rules governing the input and output relationship for fuzzy controller. The inference mechanism is responsible for decision making in the control system using approximate reasoning. The operations involved in this study are “AND”. For CFNN, the inference mechanism includes pessimistic-optimistic operation layer and compensatory operation layer. The pessimistic-optimistic operation layer includes two types of cells: pessimistic compensatory cells and optimistic compensatory cells. Pessimistic compensatory cells always come up with a solution for the worst situation, while optimistic compensatory cells always choose a solution for the best situation. The compensatory fuzzy neuron can map the pessimistic input as well as the optimistic input to the compensatory output, which may lead to a relatively compromised decision for the situation between the worse case and the best case. The defuzzification procedure maps the fuzzy output from the inference mechanism to a crisp signal.

3 The Technical Details [8-10]

The special compression algorithm introduced in this paper is a FNN based compression algorithm, whose compression rate can be easily and timely controlled, that means, we discard some of the less important information if necessary, in order to achieve very high compression rate when GPRS transmission ability is low. Thus, we can maintain a smooth connection and avoid the data traffic congestion at the cost of image quality. The image compression algorithm showed in this paper has two characteristics:

1. controllable compression rate
2. better image quality under the same compression rate than other compression algorithms

The first characteristic is achieved by wavelet transform for splitting, and the second one is done by FNN (fuzzy core neuron network) for vector-classification-compression algorithm.

Data compression is one of the most important applications of wavelet transform [10]. Wavelet transform can be generated from digital filter banks. Wavelet transform hierarchically decomposes an input image into a series of successively lower resolution images and their associated detail images.

Discrete-wavelet-translation of digital images is implemented by a set of filters, which are convolved with the image in rows and columns. An image is convolved with low-pass and high-pass filters and the odd samples of the filtered outputs are discarded resulting in down sampling the image by a factor of 2. The wavelet decomposition results in an approximation image and three detail images in horizontal, vertical, and diagonal directions. Decomposition into L levels of an original image results in a down sampled image of resolution 2^L with respect to the original image as well as detail images.

Images are analysed by wavelet packets for splitting both the lower and the higher bands into several sub-images at a time. A set of wavelet packets is gained. The following wavelet packet basis function $\{w_n\}(n = 0,1,\cdots\infty)$ is generated from a given function w_0.

$$w_{2n}(l) = \sqrt{2}\sum h(k)w_n(2l-k) \tag{1}$$

$$w_{2n+1}(l) = \sqrt{2}\sum_{k} g(k)w_n(2l-k) \tag{2}$$

Where the function $w_0(l)$ can be identified with the scaling function ϕ, and $w_1(l)$ with the mother wavelet ψ, $h(k)$ and $g(k)$ are the coefficients of the low-pass and high-pass filters, respectively. Two 1-D wavelet packet basis functions are used to obtain the 2-D wavelet basis function through the tensor product along the horizontal and vertical directions. (see Fig.2.)

In this paper, we use Mallat algorithm as the wavelet transform algorithm.

The reconstruction expression of Mallat can be seen as the following:

$$\begin{aligned} C_{K-1}(n,m) = \frac{1}{2}\Big[& \sum_{k,l\in Z} C_K(j,l) + \\ & \sum_{k,l\in Z} d_k^1(j,l)h_{n-2j}g_{m-2l} + \sum_{k,l\in Z} d_k^2(j,l)g_{n-2j}h_{m-2l} \\ & + \sum_{k,l\in Z} d_k^3(j,l)g_{n-2j}g_{m-2l}\Big] \end{aligned} \tag{3}$$

Fig.1 shows that original image splits into a series of sub-images. D_j^1, D_j^2, D_j^3 in Fig.1 are represented as sub-images with high frequency characteristics in horizontal, vertical and diagonal directions respectively. The more times we use wavelet

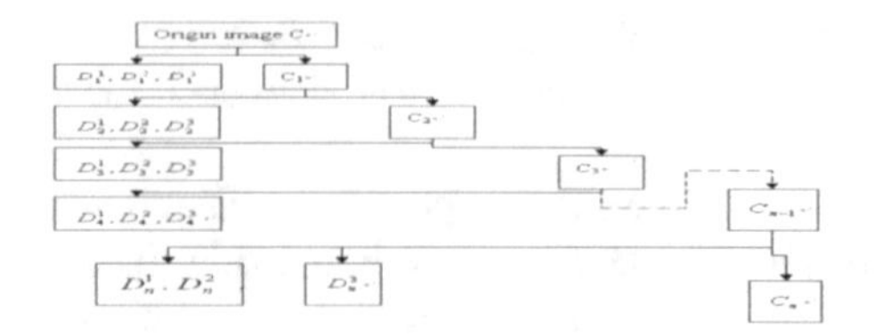

Fig. 1. Splitting High and Low Frequency Image by Wavelet Transform

transform, the more sub-images we get. The more sub-images we acquire, the less image information we lose. But if we want to get more information by decomposition we need to compress more useful information derived from the original image. For this reason, we hope the compression result can be controlled; therefore the congestion control algorithm may avoid transferring too much data while the wireless communications network being jammed. And if data traffic does not seem to happen, we may hope to send more information through the channel to transfer as high quality images as possible. The combination algorithm, as can be seen from Fig.2. , based on FNN and wavelet transform, is right for GPRS communications. It can send smallest image data continuously, with sound image quality; it can control the compression rate timely and appropriately, while taking efforts to avoid data congestion.

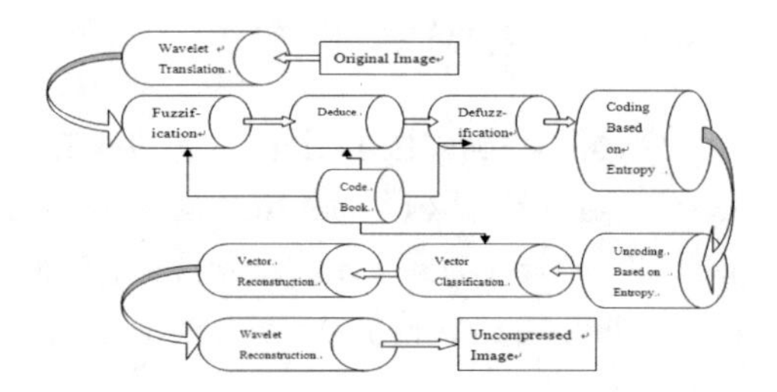

Fig. 2. Image compression and Decompression process

4 Performance

To compare our compression algorithm with other popular & traditional algorithms, another two typical compression algorithms, WENRAR transform and fractal image compression in DCT domain based on adaptive classified weighted ISM [12], are introduced and adopted to test the same pictures whose average performances are shown in table 1. As can be seen from Table1, our image compression algorithm, which is based on FNN and wavelet transform, has shown much better performance than others in our experiments.

Table 1. Comparison of Some Typical Compression Algorithm

	WENRAR Translation	Fractal Image Compression in DCT Domain Based on Adaptive Classified Weighted ISM [12]	Reconstruction Image By Wavelet Translation	splitting Image By Mallat Wavelet Translation then compression through FNN Vector-Classification
Origin Data Lost	NO	NO	YES	YES
Compression Rate	1.57	10	30	30
S/N (Single/Noise rate)	—	—	6	12

5 Conclusions

This paper mainly concerns a compression algorithm designed for GPRS communication system under the high congestion risk circumstance. In this paper, the basic idea of neuron network with fuzzy core is reviewed and a new structure FNN for vector classification as well as wavelet transform in compression algorithm is introduced. By using the new structure, we can appropriately adjust input and output, increase robustness, stability and working speed of the network, and improve the efficiency of the decompression image, comparing with other typical algorithms under the same compression rate. As shown by experimental results, the algorithm designed by this method is capable of controlling the size of compressing result as you want, with higher image quality than other algorithm. It proves itself that it is a more practical and more effective method than traditional methods, and it is especially better at video data transmitting under high congestion risk circumstance.

References

[1] Habib I (1995) Special feature topic on neural computing in high speed networks. IEEE Communication Mag[J],36:53-56

[2] Moh W Melody, Chen Minjia, Chu Nuiming (1995) Traffic prediction and dynamic bandwidth allocation over ATM: a neural network approach. Computer Communications [J]. 18(8): 563-571

[3] Habib I, Tarraf A, Saadawi T (1997) A neural network controller for congestion control in ATM multiplexers. Computer Networks and ISDN Systems [M]. 29:325-334

[4] Mars Fan P (1997) Access flow control scheme for ATM networks using neural-network-traffic prediction. IEEE ProcComm[J], 144(5): 295-300

[5] O.Castillo and P.Melin (2002) Hybrid Intelligent Systems for time series Prediction Using Nerual Networks, Fuzzy logic, and Fractal Theory. IEEE TANSCTIONS ON NEURAL NETWORKS [J]. 13(6): 1395-1408

[6] Y. Q. Zhang and A. Kandel (1998) Compensatory neurofuzzy systems with fast learning algorithms,, IEEE Trans. Neural Networks, vol. 9: 83–105.

[7] P J Burt, R J Kokzynski (1993) Enhanced image capture through fusion[A]. 1993 IEEE 4th Inter. Conf. on Computer Vision[C]. Berlin: IEEE, ICCV: 173-182.

[8] QU Tian-shu, DAI Yi-song, WANG Shu-xun (2002) Adaptive Wavelet Thresholding Denoising method based on SURE Estimation. ACTA ELECTRONICA SINICA. 30(2): 266-268
[9] XIE Rong-sheng, SUN Feng, HAO Yan ling (2002) Multi-wavelet Transform and Its Application in Signal Filtering. ACTA ELECTRONICA SINICA. 30(3) :419-421
[10] TANG Yan, MO Yu-long (2000) Image Coding of Tree-structured Using 2D Wavelet Transform [J]. Journal of Shanghai University (natural science), Feb.6(1): 71-74
[11] Mars Fan P (1997) Access flow control scheme for ATM networks using neural-network-traffic prediction. IEEE rocComm, 144(5): 295- 300
[12] Yi Zhong-ke, Zhu Wei-le, Gu De-ren (1997) Image Progressive Transm ission and Loss-less Coding Using Fractal Image Coding[J]. Journal of UEST of China. 26(5): 473-476

Biography

J. L. Su received the B.S. and M.S. degrees, with honors, in electronic engineering from Huaqiao University, P. R. China, in 2000 and 2004, respectively. He is currently a Ph.D. candidate in Computer Engineering and Science School of Shanghai University, P. R. China. He has published several papers in the areas of digital signal procession, image processing, artificial intelligence system, image coding, robot controlled and telecommunication. His recent research project is in the field of robot remote-control systems.

A Novel Approach for Fuzzy Connected Image Segmentation

Xiaona Zhang[1], Yunjie Zhang, Weina Wang[2], and Yi Li[3]

[1] Department of Mathematics, Dalian Maritime University
Dalian 116026, P.R. China
xiaona82@newmail.dlmu.edu.cn
[2] Jinlin Institute of Chemical Technology
Jinlin 132022, P.R. China
[3] Department of Computer, Information Engineering College, Heilongjiang Institute of Science and Technology,
Harbin 150027, P.R. China

Abstract. Image segmentation is the problem of finding the homogeneous regions (segments) in an image. The fuzzy connectedness has been applied in image segmentation, and segmented the object accurately. In recent years, how to find the reference seeds automatically for multiple objects image segmentation and speed up the process of large images segmentation as important issues to us. In this work we present a novel *TABU* search-based approach to choose the reference seeds adaptively and use a vertex set expanding method for fuzzy object extraction in image segmentation. This proposed algorithm would be more practical and with a lower computational complexity than others. The results obtained on real image confirm the validity of the proposed approach.

Keywords: fuzzy connectedness, image segmentation, vertex set expanding, *TABU* search.

1 Introduction

There has been significant interest in fuzzy-connectedness-based approaches to image segmentation in the past few years [1, 2]. The common theme underlying these approaches is the idea of "hanging-togetherness" of image elements in an object by assigning a strength of connectedness to every possible path between every possible pair of image elements.

J. K. Udupa and S. Samarasekera were the first to introduce the general approach of segmenting images by concept of fuzzy connectedness [1]. They presented a theory of fuzzy objects for n-dimensional digital spaces based on a notion of fuzzy connectedness of image elements, and led to a powerful image segmentation method based on dynamic programming whose effectiveness has been demonstrated on thousands of images in a variety of applications [2].These years, people modified and extend previously published algorithm. They developed it to Scale-Based and Tensor Scale-Based fuzzy connected image segmentation algorithm[3, 4]. And usually to solve some practical problem,especially in medical image[5].

B.-Y. Cao (Ed.): Fuzzy Information and Engineering (ICFIE), ASC 40, pp. 89–97, 2007.
springerlink.com

[6] proposed an extension to the definition of fuzzy objects, instead of defining an object on its own based on the strength of connectedness, all co-objects of importance in the image are also considered and the objects are let to compete among themselves in having spels as their members. In this competition, every pair of spels in the image will have a strength of connectedness in each object. The object in which the strength is the highest will claim membership of the spels. This approach to fuzzy object definition using relative strength of connectedness eliminates the need for a threshold of strength of connectedness that was part of the previous definition. Based on it, [7] presented relative fuzzy connectedness among multiple objects, called *k MRFOE* algorithm, for multiple relative fuzzy object extraction. It defined objects independent of the reference elements chosen as long as they are not in the fuzzy boundary between objects, and it is effectively in segmentation. But this method also has a disadvantage, in it the reference spels are specified manually, this means the algorithm can't accomplish segmentation independently and need by dint of other equipments, which may adds the complexity to the image processing, and this may become impractical when there are many important objects.

To solve this problem this paper utilizes the main idea of *TABU* search [8] to choose the reference spels adaptive. We choose one seed and find the object by the fuzzy object extraction method, after that use the *TABU* list store this object, then find the next seed, such-and-such complete segment process. In order to lower the complexity, we present Vertex Sets expanded fuzzy object extraction algorithm (*k VSEFOE* for short) for the fuzzy connected image segmentation in stead of the fuzzy object extract base on dynamic programming algorithm (*k FOE* for short) [6] in every iterative. *k VSEFOE* Algorithm uses the threshold preprocess the image scene, and then uses the Vertex Sets expanded method obtains the object. It is faster and simpler than others. Some medical image examples are presented to illustrate the visually and effectively of multiple objects image segmentation. In addition, we also test the algorithm in some images in which there are many important objects. Experiment results show that the proposed method outperforms the existing ones.

The remainder of this paper is organized as follows. Section 2 gives a brief overview of fuzzy connectedness and relative fuzzy connectedness. In section 3, a novel image segmentation method based on fuzzy connectedness is proposed, as well as computational complexity of the algorithm is analyzed by using the complexity theory. In section 4, experimental results of some images are reported. Finally, concluding remark is stated in section 5.

2 Preliminaries

2.1 A Framework for Fuzzy Connectedness, Relative Fuzzy Connectedness

A minimal set of terminology and definitions are presented to provide the preliminaries of the fuzzy connectedness of formulation employed. Here follow the terminology of Udupa et al.[1, 3]

Here states some known definitions from the theory of fuzzy subsets. [9, 10, 11]. Let X be any reference set. A fuzzy subset A of X is a set of ordered pairs. $A = \{(x, \mu_A(x)) | x \in X\}$, where $\mu_A : X \rightarrow [0, 1]$ is the membership function of A in X (μ is subscripted by the fuzzy subset under consideration). For any fuzzy relation ρ in X, ρ is called a similitude relation in X if it is reflexive, symmetric, and transitive. Here ρ is said to be reflexive if, $\forall x \in X$, $\mu_\rho(x, x) = 1$; symmetric if, $\forall x, y \in X$, $\mu_\rho(x, y) = \mu_\rho(y, x)$; Let $X = Z^n$, the set of n-tuples of integers. The pair (Z^n, α), where α is a fuzzy spel adjacency, will be referred to as a fuzzy digital space. In this fuzzy digital space, any scalar function $f : C \rightarrow [L, H]$ from a finite subset C of Z^n to a finite subset of the integers defines a scene (C, f) over (Z^n, α), in a scene S, approximate k in S is called an affinity. This is a "local" phenomenon. It is a fuzzy adjacency relation and the closer the spels are, the more adjacent they are to each other. For any spels c, $d \in S$, $\mu_k(c, d)$ indicates the local hanging-togetherness in the scene for two spels that are nearby. Here k is a function of:

$$\mu_k(c, d) = \mu_\alpha(c, d)\sqrt{\mu_\psi(c, d)\mu_\phi(c, d)} \tag{1}$$

Affinity is a local fuzzy relation in S, only reflects the local information about spels, for this presents a global fuzzy relation-fuzzy connectedness. Let k be any fuzzy spel affinity in S. A sequence $\langle c^{(1)}, c^{(2)}, \ldots, c^{(m)} \rangle$ of $m \geq 2$ spels in C stand for a path in S from a spel $c \in C$ to a spel $d \in C$, denoted by P such that $c^{(1)} = c$ and $c^{(m)} = d$. A strength of connectedness determined by k is assigned as along the path. Fuzzy k-connectedness in S is a fuzzy relation that assigns to every pair (c, d) of spels a value which is the largest of the strength of connectedness of all possible paths in S from c to d. Here use k for the fuzzy spel affinity and K for the fuzzy k-connectedness in S.

[2] gave a definition of relative connectedness. It indeed presents a fuzzy relation with the interest objects and background object. Let a 2D image composed of $m \geq 2$ regions corresponding to m objects O_1, $O_2, \ldots$, O_m is the background. Suppose we have known affinity relation for each object that assigns to every pair of nearby spels in the image a value based on the nearness of spels in space and in intensity (or in features derived from intensities). Each affinity represents local hanging-togetherness of spels in the corresponding object. Suppose we have a mechanism of combining these affinities into a single affinity relation k. So that k retains as much as possible each individual object specific flavor of the individual affinities. To every path connecting every pair of spels, the strength of connectedness is assigned which is simply the smallest pairwise affinity of spels (as per k) along the path. Let $o_1, o_2, ..., o_m$ be the reference spels selected in objects $O_1, O_2, ..., O_m$, respectively. Then any spel c is considered to belong to that object with respect to whose reference spel o has the highest strength of connectedness. This relative strength of connectedness is modulated by the affinities of the individual objects, the natural mechanism for partitioning spels into regions based on how the spels hang together among themselves relative to others.

2.2 Fuzzy $k\theta$ -Object

The image segmentation is an object extraction processing, [2] gave the object definition in detail, here describe it shortly. Let θ be any number in [0, 1]. Here defined a binary relation that $\mu_k(c, d) \geq \theta$. It is an equivalence relation in C. For any spel $o \in c$, let $O_\theta(o)$ be an equivalence class of k_θ in C that contains o. A fuzzy $k\theta$ -Object $O_\theta(o)$ of S containing o is a fuzzy subset of C defined as that if $c \in O_\theta(o)$ then $\mu_{O_\theta(o)} = \eta(f(c))$ and otherwise it is zero. Here η is an object function with range [0, 1].

3 Algorithm

3.1 Original Algorithm

The original algorithm called *k MRFOE*, for multiple relative fuzzy object extraction. It combined relative fuzzy connectedness and dynamic algorithm extract the multiple objects in an image. For this algorithm the reference spels were specified manually first. And then use the main of *k FOE* algorithm extract the objects in turn. Here the *k FOE* algorithm is the fuzzy object extraction algorithm [12]. It uses dynamic programming through computes the local affinity and the global fuzzy relation-fuzzy connectedness to find the best path from o to each spel in S. Some different here is use a function combining the affinities for each object to a single affinity. See [12] for further details on this, here reproduce the main algorithm.

Algorithm

Step 1: Find the references set S by special manually.
Step 2: Choose a reference spel o from S as seed, then use *k FOE* algorithm extract the object.
Step 3: If S not empty returns to step 2.
Step 5: Output the results.

3.2 Proposed Algorithm

This section first presents a new algorithm *k VSEFOE* algorithm, it use the threshold to precessing the image scene and then use the Vertex sets expanded algorithm find the object. Then base on it combine the *TABU* algrothm presents *k MVSEFOE* algorithm, for multiple vertex sets expanded fuzzy object extraction.

k VSEFOE Algorithm

In section 2.2, defined the fuzzy $k\theta$-Object, in *k VSEFOE* algorithm this function is used in the first step of algorithm to be a preprocessing. The principle of it is base on its definition. Let A be the adjacency matrix makes up of the fuzzy affinity k in a scene S, for any spel c and d in S, if $\mu_k(c, d) \geq \theta$, let $A_{(c,d)} = 1$, otherwise it is zero. Here $A_{(c,d)}$ is the value of $\mu_k(c, d)$ in A.

Here presents a new definition- *Vertex sets expanded* algorithm. That is also an important method to find the connected object. For an adjacency matrix which has been preprocessed, choose an element as the seed and label its neighbors which have affinity with it (here the neighbor is the element adjacent with it. In this paper, the fuzzy digital space is a Euclidean space, so the elements are 4-adjacent), then the labeled elements expanded as the same method. And so on, until travel all elements. At the end of this method gets a spels set, this set is a spanning tree [12]. We know it is a connected graph, so the object gotten by this way is a connected object.

k MVSEFOE Algorithm

This section we present a new algorithm *k MVSEFOE* algorithm, for multiple vertex sets expanded fuzzy object extraction. *TABU* search (TS) is a general technique proposed by Glover for obtaining approximate solutions to combinatorial optimization problem. *TABU* search avoids being trapped at local optimum by allowing the temporal acceptance of the recent migrations of the current solution to avoid visiting repeated solutions. Here we use the main idea of the TS after choose one seed and extract the object, create a TS list to deposit the spels which have been chosen. It make sure the spels will not be chosen in the next iterative and come true to choose the reference spels automatically. In every iterative we use the *k VSEFOE* algorithm extract objects. The main idea of this algorithm will be shown as follow. For a given scene S, and the threshold computed by condition function in 2.2.

Algorithm

Step 1: Create a TS list to keep the selected spels.
Step 2: In S, through compare all spels affinities with the threshold in turn to find the reference.
Step 3: Use *k VSEFOE* algorithm to extracts objects remove the chosen spels to the TS list.
Step 4: If S not empty returns to step 2.
Step 5: Output the result.

3.3 Algorithm Complexity Analysis

Algorithm complexity [14] is an important criterion in image processing. Especially in the real life there are many large images which make the processing work very hard. So the less complexity directly influences the algorithm feasibility. This paper presents a new algorithm which is simpler and faster than the original algorithm in [12]. This section from the process of them to prove our algorithm are simply, and use complexity theory to analysis them.

The main idea of the original algorithm use the dynamic programming through compute the local affinity and the global fuzzy relative connectedness to find the best path from o to each spel in S. Because the strength for connectedness between all possible pairs image element has to be determined, it made the workload horrendous, and at the same time added the complexity in processing. Here analysis them for detail. If there are n spels in the image:

(i) In the *k FOE* algorithm, it at most uses $n-1$ iterative, and in every iterative, the time used to compare is $O(n)$. That means the time used in every iterative is $O(n)$. So the time complexity in this algorithm is $O(n^2)$. While in *k MRFOE* algorithm we ignore the complexity of how to choose the reference spels, only consider the algorithm, if there are m important objects, that means added m iterative. So the complexity of the *k MRFOE* algorithm is $O(mn^2)$.

(ii) In the new algorithm *k MVSEFOE* algorithm, it at most uses $n-1$ iterative and in every iterative because of the spels space is an Euclidean space, the neighboring spels are 4-adjacent, so it can find the neighbor by at most $[\log 4(n-1)]+1$ steps. So the time complexity of this algorithm is $O(n\{[\log 4(n-1)]+1\})$. In multiple image segmentation, if there are m important objects, its complexity is $O(mn\{[\log 4(n-1)]+1\})$. Obviously the complexity are smaller than the *k MRFOE* algorithm.

4 Experimental and Evaluation

4.1 Experimental Result

The proposed algorithms had been tested in many experiment, include the image segmentation for separating a foreground object from a background object, multiple objects segmentation. In this section, we present some results of them to illustrate the performance of the algorithm. And give a quantitative evaluation for the new proposed algorithm and Relative fuzzy connectedness of multiple objects algorithm.

The first example shown in Fig.1 is a quantitative comparison between *k MVSEFOE* algorithm and *k MRFOE* algorithm it can prove that the latter algorithm is better. The second example in Fig. 2 is a problem about recognized follicle in Ovary Ultrasonic. Fig.3 comes from MRI of the head of a multiple sclerosis patient, comes from a CT slice from a protondensity-weighted 3D scene, and segments the two parts of the ventricle delineated as two separate objects pertains to CT. Fig.4 shows one 2D slice from a 3D scene of a patient's head. Our aim is to differentiate the different tissue segments-bones, soft tissue. The fifth and sixth examples are shown in fig.5-fig.6, there are eighty and twenty-one important objects respectively, and we use the *k MVSEFOE* algorithm process exactly.

4.2 Evaluation

It has been analyzed the complexity of two algorithms in 3.3, and demonstrated the new algorithms have less complexity than the algorithm. The examples in 4.1 illuminate the *k MRFOE* algorithm and *k MVSEFOE* algorithm are practicable. Here list a table about the time which cost by the two new algorithms in the examples in 4.1. In table R stands for the time which cost by the *k MRVSEFOE* connected algorithm(use the *k VSEFOE* algorithm to improve *k MRFOE*

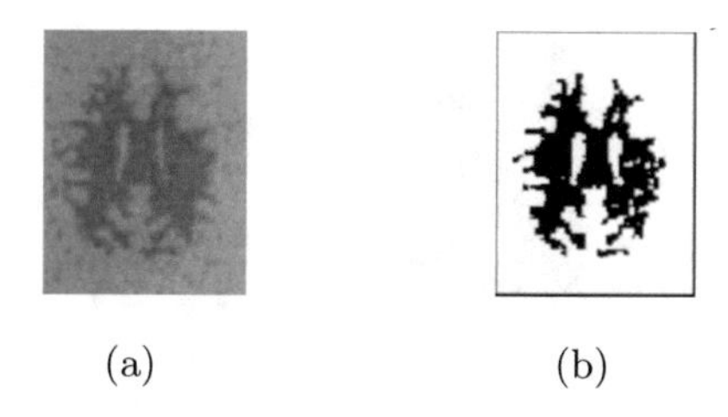

(a) (b)

Fig. 1. (a) is one of the original binary scenes from the scene was simulated, (b) is the result gotten by *k VSEFOE* algorithm

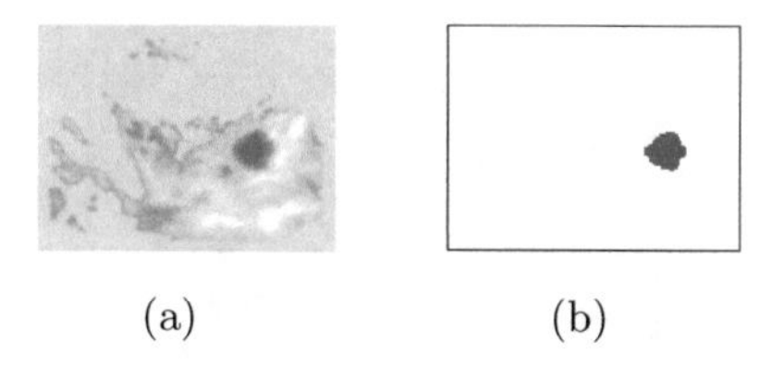

(a) (b)

Fig. 2. (a) is a scene about Ovary Ultrasonic, (b) is the segmented follicle region result obtained for the slice in (a) using the *k VSEFOE* algorithm

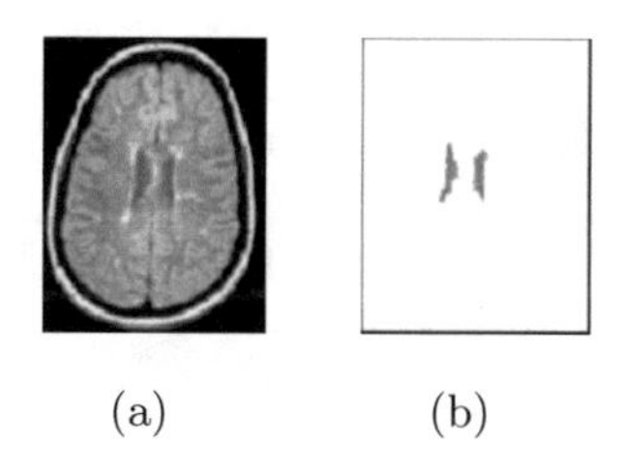

(a) (b)

Fig. 3. (a) A proton-density-weighted 2D slice of a 3D MRI scene of a patient's head. (b) The two parts of the ventricle delineated as two separate objects segmented by the *k MVSEFOE* algorithm.

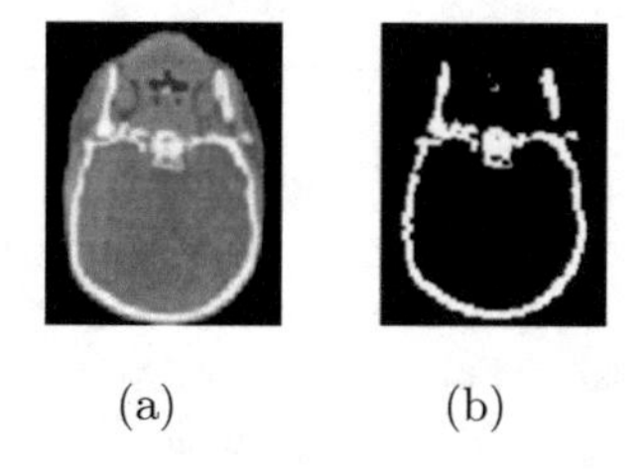

(a) (b)

Fig. 4. (a) A CT slice of a patient's head. Four objects-bone, soft tissue, from the 3D scene shown in (b), segmented by the *k MVSEFOE* algorithm.

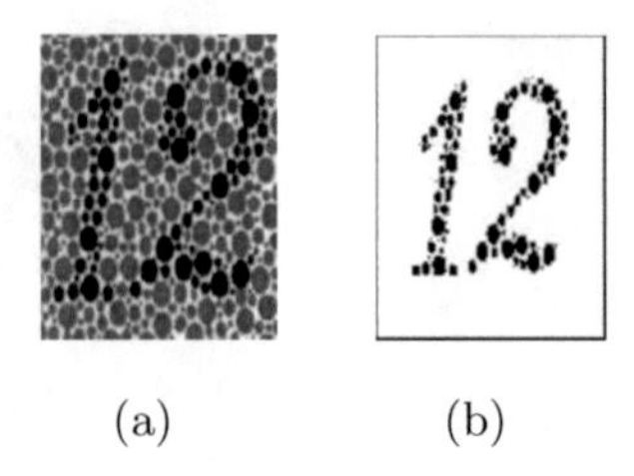

(a) (b)

Fig. 5. (a) is a 'color blindness test' image, we should segment the black part from the background. Here the foreground is number of 12, and this number make up of eighty black dots. (b) is the result segment by *k MVSEFOE* algorithm.

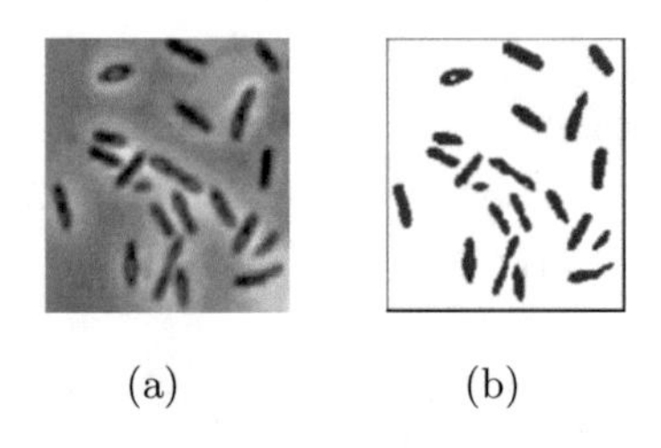

(a) (b)

Fig. 6. (a) is a medical image, it has twenty-one bacteria, (b) is the segment result of use the *k MVSEFOE* algorithm

Table 1. The time cost by the two new algorithms in the examples

Example	Size	V	R
Fig.1	100×79	12.5810_s	3.5607_m
Fig.2	160×160	15.5341_s	4.1534_m
Fig.3	130×100	20.5098_m	3.9021_m
Fig.4	100×80	17.8300_m	3.3353_m
Fig.5	190×190	30.5341_m	\
Fig.6	178×178	28.6371_m	\

algorithm), V stands for *k MVSEFOE* algorithm, m stands for minute, s stands for second. Table shows the comparison of segmentation results from various approaches. In the table we didn't consider the time cost in finding the reference spels manually in R, so in superficial V cost more time than R. But we also can find their discrepant is acceptable, so in practical the algorithm is feasibility.

5 Conculsion

For the *k MRFOE* algorithm has an inherent disadvantage that it should choose the reference spels manually, this make the algorithm can't applied widely. For this problem we present a new method, it use the main idea of the TS list to store the chosen object in multiple image segmentation. It can choose reference spels automatically and then use the k *VSEFOE* algorithm find the connected region speed up the algorithm. We have test the algorithm in some two objects segmentation and multiple objects segmentation examples. All experiments prove that this algorithm compares with original algorithms are simpler and more effectively, especially in some large image processing they would make the segmentation more feasible.

References

1. I. Bloch,(1993) Fuzzy Connectivity and Mathematical Morphology, in: Pattern Recognit.Lett.14, pp. 483-488.
2. J. K. Udupa and S. Samarasekera,(1996) Fuzzy connectedness and object definition: Theory, algorithms, and applications in image segmentation, in: Graphical Models Image Process, pp. 246-261.
3. P.K.Saha, J.K.Udupa, and D.Odhner,(2000) Scale-Based fuzzy connected image segmentation: Theory, Algorithms, and Validation, in: Compute Vision and Image Understanding 77, pp. 145-147.
4. Ying Zhuge, Jayaram K.Udupa, and Punam K.Saha,(2002) Vectorical scale-based fuzzy connected image segmentation, in: Medical Imaging 4684.
5. Yongxin Zhou, Jing Bai, (2005)Organ Segmentation Using Atlas Registration and Fuzzy Connectedness, in: Engineering in Medicine and Biology 27th Conference.
6. J. K. Udupa and S. Samarasekera,(2002) Relative fuzzy connectedness and object definition: theory, algorithm, and applications in image segmentation, in: Pattern Analysis and Machine Intelligence.
7. P.K.Saha and J.K.Udupa,(2001) Relative Fuzzy Connectedness among Multiple Objects: Theory, Algorithms, and Applications in image segmentation, in: Computer vision an image understanding.
8. Jinsing xie and wenxun xing,(1999) Modern Optimization Computation, in: China Tsinghua university press,pp.53-62.
9. A, Kaufmann, (1975)Introduction to the Theory of Fuzzy Subsets, in: Vol.1, Academic Press, New York.
10. Rosenfeld,(1992) Fuzzy digital topology, in: Information Control 40, pp.76-87.
11. A. Rosenfeld, (1991)The fuzzy geometry of image subsets, in: Pattern Recognit.Lett.2, pp. 311-317.
12. S. Dellepiane and F. Fontana,(1995) Extraction of intensity connectedness for image processing, in: Pattern Recognit. Lett.16, pp.313-324.
13. Jianhong Yin, Kaiya Wu,(2004) Graph Theory and Its Algorithm, in: China University of Technology Press, pp. 1-26, 91-98.
14. Michael Sipser, (2002)Introduction to the Theory of Computation, in: China Machine Press.

Knowledge Mass and Automatic Reasoning System in Similarity Logic $\mathbb{C}_{\mathbb{Q}}$

Yalin Zheng[1,2], Huaqiang Yuan[1], Ning Wang[1], Guang Yang[1], and Yongcheng Bai[3]

[1] Research Center of Intelligent Information Technology, Department of Computer Science and Technology, Faculty of Software, Dongguan University of Science and Technology, Dongguan 523808, China
zheng-yl@mail.tsinghua.edu.cn

[2] State Key Laboratory for Intelligent Technology and Systems, Department of Automation, Faculty of Information Science and Technology, Tsinghua University, Beijing 100084, China
zheng-yl@mail.tsinghua.edu.cn

[3] Institute of Information Science, Faculty of Mathematics and Information Science, Shaanxi University of Science and Technology, Hanzhong 723001, China
qiaozhuangx@eyou.com

Abstract. This paper introduces and investigate the $\mathbb{Q}-$knowledge mass, $\mathbb{Q}-$knowledge universe, $\mathbb{Q}-$knowledge base and $\mathbb{Q}-$automatic reasoning system bases on the similarity logic $\mathbb{C}_{\mathbb{Q}}$. We also present and investigate the $\mathbb{Q}-$true level k knowledge circle, extended $\mathbb{Q}-$ knowledge base, extended $\mathbb{Q}-$automatic reasoning system and the level (k,j) perfection of extended $\mathbb{Q}-$knowledge base $\mathbb{K}^V$ in this paper.

Keywords: Approximate Reasoning, Formula Mass, Knowledge Mass, $\mathbb{Q}-$logic, $\mathbb{Q}-$automatic Reasoning System, Extended $\mathbb{Q}-$automatic Reasoning System, Level (k,j) Perfection of Extended Knowledge Base $\mathbb{K}^V$.

1 Introduction and Preliminary

For some applications of approximate reasoning, it is too restrict of using the equivalence $\mathbb{R}$ in formula set $F(S)$. If the approximate reasoning matching scheme is designed with the similarity $\mathbb{Q}$ in $F(S)$, it may be more reasonable to reflect the nature of approximate reasoning and meet the application expectations in many aspects. The difference is that the intersections of any components of family

$$F(S)/\mathbb{Q} = \{\langle A\rangle_{\mathbb{Q}} | A \in F(S)\}$$

which is made of corresponding formula mass

$$\langle A\rangle_{\mathbb{Q}} = \{B | B \in F(S), A\mathbb{Q}B\}$$

are unnecessarily to be empty in the situation of more general similarity $\mathbb{Q}$.

In classic propositional logic $\mathbb{C} = (\hat{\mathbb{C}}, \tilde{\mathbb{C}})$, for any similarity relation $\mathbb{Q}$ in formula set $F(S)$, which satisfies

B.-Y. Cao (Ed.): Fuzzy Information and Engineering (ICFIE), ASC 40, pp. 98–112, 2007.
springerlink.com

$$\langle A\rangle_{\mathbb{Q}} \to \langle B\rangle_{\mathbb{Q}} \subseteq \langle A \to B\rangle_{\mathbb{Q}}$$

for each implication $A \to B \in \mathscr{H}$
where

$$\langle A\rangle_{\mathbb{Q}} \to \langle B\rangle_{\mathbb{Q}} = \{A^* \to B^* | A^* \in \langle A\rangle_{\mathbb{Q}}, B^* \in \langle B\rangle_{\mathbb{Q}}\}$$

$(F(S), \mathbb{Q})$ is called $\mathbb{Q}-$*space*;

$$\mathbb{C}_{\mathbb{Q}} = (\hat{\mathbb{C}}, \tilde{\mathbb{C}}, \mathbb{Q})$$

is called $\mathbb{Q}-$*logic*.

In $\mathbb{Q}-$logic $\mathbb{C}_{\mathbb{Q}}$, for each formula $A \in F(S)$, $\langle A\rangle_{\mathbb{Q}}$ is called $\mathbb{Q}-$*formula mass* with the kernel of A, or just $\mathbb{Q}-$*mass* in short. Specially, for every theorem $A \to B \in \emptyset^{\vdash}$, $\langle A \to B\rangle_{\mathbb{Q}}$ is called $\mathbb{Q}-$*knowledge mass* with the kernel of knowledge $A \to B$.

Consider the simple approximate reasoning model in $\mathbb{Q}-$logic $\mathbb{C}_{\mathbb{Q}}$

$$\begin{array}{l} A \quad \to B \\ A^* \\ \hline \qquad\quad B^* \end{array} \tag{1}$$

where $A \to B \in \emptyset^{\vdash}$ is called *knowledge*; $A^* \in F(S)$ is called *input*; $B^* \in F(S)$ is called *output* or *approximate reasoning conclusion*. If input A^* is contained in $\mathbb{Q}-$formula mass $\langle A\rangle_{\mathbb{Q}}$ with the kernel of A, the antecedent of knowledge $A \to B$, which is

$$A^* \in \langle A\rangle_{\mathbb{Q}}$$

It means that input A^* can *activate* the knowledge $A \to B$. Regard

$$B^* \in \langle B\rangle_{\mathbb{Q}}$$

as the approximate reasoning conclusion, which is the *type V* $\mathbb{Q}-$*solution* of the simple approximate reasoning with regard to input A^* under knowledge $A \to B$.

If input A^* is not contained in $\mathbb{Q}-$formula mass $\langle A\rangle_{\mathbb{Q}}$ with the kernel of A, the antecedent of knowledge $A \to B$, that is

$$A^* \notin \langle A\rangle_{\mathbb{Q}}$$

It means that input A^* can not *activate* knowledge $A \to B$. And A^* is not considered in this situation, which means there is no type V $\mathbb{Q}-$solution with regard to input A^* under knowledge $A \to B$.

This algorithm is named *type V* $\mathbb{Q}-$ *algorithm* of simple approximate reasoning.

Obviously, the solution domain, response domain and examination domain of the simple approximate reasoning type V $\mathbb{Q}-$algorithm with regard to input A^* under knowledge $A \to B$ are

$$\begin{array}{ll} Dom^{(V)}(A \to B) & = \langle A\rangle_{\mathbb{Q}} \\ Codom^{(V)}(A \to B) & = \langle B\rangle_{\mathbb{Q}} \\ Exa^{(V)}(A \to B) & = \langle A \to B\rangle_{\mathbb{Q}} \end{array}$$

Consider multiple approximate reasoning model in $\mathbb{Q}-$logic $\mathbb{C}_{\mathbb{Q}}$

$$\begin{array}{l} A_1 \rightarrow B_1 \\ A_2 \rightarrow B_2 \\ \cdots \cdots \cdots \\ A_n \rightarrow B_n \\ A^* \\ \hline \qquad B^* \end{array} \tag{2}$$

where $K = \{A_i \rightarrow B_i | i = 1, 2, \cdots, n\} \subseteq \emptyset^{\vdash}$is called *knowledge base*; $A^* \in F(S)$ is called *input*; $B^* \in F(S)$ is called *output* or *approximate reasoning conclusion.*

If there exists an $i \in \{1, 2, \cdots, n\}$ which satisfies

$$A^* \in \langle A_i \rangle_{\mathbb{Q}}$$

we say input A^* *can activate* knowledge $A_i \rightarrow B_i$ in knowledge base K.

If there exists an $i \in \{1, 2, \cdots, n\}$ which satisfies

$$A^* \notin \langle A_i \rangle_{\mathbb{Q}}$$

we say input A^* *cannot activate* knowledge $A_i \rightarrow B_i$in knowledge base K.

Suppose there exists a non-empty subset E of $\{1, 2, \cdots, n\}$, for each $i \in E$, input A^* can activate knowledge $A_i \rightarrow B_i$ in knowledge base K; while for any $j \in \{1, 2, \cdots, n\} - E$, input A^* can not activate knowledge $A_j \rightarrow B_j$ in knowledge base K. Then let

$$B_i^* \in \langle B_i \rangle_{\mathbb{Q}}$$

and

$$B^* = \bigoplus_{i \in E} B_i^*$$

which satisfy

$$B^* \in \bigcup_{i \in E} \langle B_i \rangle_{\mathbb{Q}}$$

where $\bigoplus_{i \in E} B_i^*$ is a certain aggregation of $\{B_i^* | i \in E\}$. B^*, the conclusion of approximate reasoning, is called multiple approximate reasoning *type V* $\mathbb{Q}$*−solution* with regard to input A^* under knowledge base $K = \{A_i \rightarrow B_i | i = 1, 2, \cdots, n\}$.

Suppose input A^* can not activate any knowledge $A_i \rightarrow B_i$ in knowledge base $K = \{A_i \rightarrow B_i | i = 1, 2, \cdots, n\}$, which means for each $i \in \{1, 2, \cdots, n\}$ we have

$$A^* \notin \langle A_i \rangle_{\mathbb{Q}}$$

Then, we say input A^* can not activate knowledge base $K = \{A_i \rightarrow B_i | i = 1, 2, \cdots, n\}$. In this case, we do not consider input A^*, that is, there dose not exist multiple approximate reasoning *type V* $\mathbb{Q}$*−solution* with regard to input A^* under knowledge base $K = \{A_i \rightarrow B_i | i = 1, 2, \cdots, n\}$.

This algorithm is called *type V* $\mathbb{Q}$*−algorithm*of multiple approximate reasoning.

Obviously,the solution domain, response domain and examination domain of the multiple approximate reasoning type V $\mathbb{Q}$−algorithm with regard to input A^* under knowledge base $K = \{A_i \rightarrow B_i | i = 1, 2, \cdots, n\}$ are given by

$$\begin{aligned} Dom^{(V)}(K) &= \bigcup_{i=1}^{n} \langle A_i \rangle_{\mathbb{Q}} \\ Codom^{(V)}(K) &= \bigcup_{i=1}^{n} \langle B_i \rangle_{\mathbb{Q}} \\ Exa^{(V)}(K) &= \bigcup_{i=1}^{n} \langle A_i \to B_i \rangle_{\mathbb{Q}} \end{aligned}$$

Endow the formula set $F(S)$ in classic propositional logic $\mathbb{C} = (\hat{\mathbb{C}}, \tilde{\mathbb{C}})$ with a similarity relation $\mathbb{Q}$, where for each implication $A \to B \in \mathscr{H}$ we have

$$\langle A \rangle_{\mathbb{Q}} \to \langle B \rangle_{\mathbb{Q}} = \langle A \to B \rangle_{\mathbb{Q}}$$

Then, $(F(S), \mathbb{Q})$ is called *regular* $\mathbb{Q}-$*space*, and $\mathbb{C}_{\mathbb{Q}} = (\hat{\mathbb{C}}, \tilde{\mathbb{C}}, \mathbb{Q})$ is called *regular* $\mathbb{Q}-$*logic*.

In $\mathbb{Q}-$logic $\mathbb{C}_{\mathbb{Q}}$, the type V approximate knowledge closure of knowledge base

$$K = \{A_i \to B_i | i = 1, 2, \cdots, n\}$$

is

$$K^{\textcircled{b}} = \bigcup_{A \to B \in K} \langle A \to B \rangle_{\mathbb{Q}}$$

For each knowledge $A_i \to B_i$ in knowledge base K, $\mathbb{Q}-$knowledge mass $\langle A_i \to B_i \rangle_{\mathbb{Q}}$ with the kernel of $A_i \to B_i$ in $\mathbb{Q}-$space $(F(S), \mathbb{Q})$ is also called *type V approximate knowledge mass* with the kernel of $A_i \to B_i$.

So we say in $\mathbb{Q}-$logic $\mathbb{C}_{\mathbb{Q}}$, the type V approximate knowledge closure $K^{\textcircled{b}}$ of knowledge base

$$K = \{A_i \to B_i | i = 1, 2, \cdots, n\}$$

is the union of finite number of type V knowledge mass $\langle A_i \to B_i \rangle_{\mathbb{Q}} (i = 1, 2, \cdots, n)$, that is

$$K^{\textcircled{b}} = \bigcup_{i=1}^{n} \langle A_i \to B_i \rangle_{\mathbb{Q}}$$

In $\mathbb{Q}-$logic $\mathbb{C}_{\mathbb{Q}}$, knowledge base

$$K = \{A_i \to B_i | i = 1, 2, \cdots, n\}$$

is called *type V complete*, if type V approximate knowledge closure $K^{\textcircled{b}}$ of K covers the theorem set $\emptyset^{\vdash}$, which means

$$\emptyset^{\vdash} \subseteq K^{\textcircled{b}}$$

Knowledge base K is called *type V perfect*, if type V approximate knowledge closure $K^{\textcircled{b}}$ of K is coincident with theorem set $\emptyset^{\vdash}$, that is

$$K^{\textcircled{b}} = \emptyset^{\vdash}$$

In $\mathbb{Q}-$logic $\mathbb{C}_{\mathbb{Q}}$, knowledge base

$$K = \{A_i \to B_i | i = 1, 2, \cdots, n\}$$

is type V complete, if and only if each theorem in $\emptyset^{\vdash}$ is contained in type V knowledge mass with the kernel of a knowledge in K.

In regular $\mathbb{Q}-$logic $\mathbb{C}_{\mathbb{Q}}$, knowledge base

$$K = \{A_i \to B_i | i = 1, 2, \cdots, n\}$$

is type V complete, if and only if the antecedent of each theorem $A \to B$ in $\emptyset^{\vdash}$ can activate a certain knowledge in knowledge base K and the consequence is the multiple approximate reasoning type V $\mathbb{Q}-$solution with regard to input A under knowledge base K.

In regular $\mathbb{Q}-$logic $\mathbb{C}_{\mathbb{Q}}$, knowledge base

$$K = \{A_i \to B_i | i = 1, 2, \cdots, n\}$$

is type V perfect, if and only if K is type V complete, and $A^* \to B^*$ is always a theorem in logic $\mathbb{C}$ for each input A^* which can active the knowledge in knowledge base K according to type V $\mathbb{Q}-$algorithm, that is

$$A^* \to B^* \in \emptyset^{\vdash}$$

where B^* is multiple approximate reasoning type V $\mathbb{Q}-$solution with regard to input A^* under knowledge base K.

2 Type V Knowledge Mass, Type V Knowledge Universe, Type V Knowledge Base and Type V $\mathbb{Q}-$Automatic Reasoning System

Type V knowledge universe Ω^V based on regular $\mathbb{Q}-$logic $\mathbb{C}_{\mathbb{Q}}$ is a dynamic expanding multilayered structure in formula set $F(S)$, which is built as follows.

$\Omega^{(0)} = \emptyset^{\vdash}$ is called *type V level 0 knowledge universe*, or *type V initial knowledge universe*, which is generated from the whole theorems of logic $\mathbb{C}$. Each component $A \to B$ of the knowledge universe $\Omega^{(0)} = \emptyset^{\vdash}$ is named *type V level 0 knowledge* or *type V initial knowledge*.

$\Omega^{(1)} = \bigcup\limits_{A \to B \in \Omega^{(0)}} \langle A \to B \rangle_{\mathbb{Q}}$ is called *type V level 1 knowledge universe*, each component $A^* \to B^*$ of which is named *type V level 1 knowledge*; for each $A \to B \in \Omega^{(0)}$, $\langle A \to B \rangle_{\mathbb{Q}}$ is called type V level 1 knowledge mass with the kernel of type V level 1 knowledge $A \to B$.

$\Omega^{(2)} = \bigcup\limits_{A \to B \in \Omega^{(1)}} \langle A \to B \rangle_{\mathbb{Q}}$ is called *type V level 2 knowledge universe*, each component $A^* \to B^*$ of which is named *type V level 2 knowledge*; for each $A \to B \in \Omega^{(1)}$, $\langle A \to B \rangle_{\mathbb{Q}}$ is called type V level 2 knowledge mass with the kernel of type V level 2 knowledge $A \to B$.

$\Omega^{(3)} = \bigcup\limits_{A \to B \in \Omega^{(2)}} \langle A \to B \rangle_{\mathbb{Q}}$ is called *type V level 3 knowledge universe*, each component $A^* \to B^*$ of which is named *type V level 3 knowledge*; for each $A \to B \in$

$\Omega^{(2)}$, $\langle A \to B\rangle_\mathbb{Q}$ is called type V level 3 knowledge mass with the kernel of type V level 3 knowledge $A \to B$.

$\cdots\ \cdots\ \cdots\ \cdots\ \cdots\ \cdots$

$\Omega^{(k)} = \bigcup_{A\to B\in\Omega^{(k-1)}} \langle A \to B\rangle_\mathbb{Q}$ is called *type V level k knowledge universe*, each component $A^* \to B^*$ of which is named *type V level k knowledge*; for each $A \to B \in \Omega^{(k-1)}$, $\langle A \to B\rangle_\mathbb{Q}$ is called type V level k knowledge mass with the kernel of type V level $k-1$ knowledge $A \to B$.

$\Omega^{(k+1)} = \bigcup_{A\to B\in\Omega^{(k)}} \langle A \to B\rangle_\mathbb{Q}$ is called *type V level k+1 knowledge universe*, each component $A^* \to B^*$ of which is named *type V level k+1 knowledge*; for each $A \to B \in \Omega^{(k)}$, $\langle A \to B\rangle_\mathbb{Q}$ is called type V level $k+1$ knowledge mass with the kernel of type V level k knowledge $A \to B$.

$\cdots\ \cdots\ \cdots\ \cdots\ \cdots\ \cdots$

$$\begin{aligned}
\Omega^V &= \Omega^{(\infty)} \\
&= \bigcup_{k=0}^{\infty} \Omega^{(k)} \\
&= \Omega^{(0)} \cup \bigcup_{k=1}^{\infty} \bigcup_{A\to B\in\Omega^{(k-1)}} \langle A \to B\rangle_\mathbb{Q} \\
&= \Omega^{(0)} \cup \left(\bigcup_{A\to B\in\Omega^{(0)}} \langle A \to B\rangle_\mathbb{Q} \right) \cup \\
&\quad \left(\bigcup_{A\to B\in\Omega^{(1)}} \langle A \to B\rangle_\mathbb{Q} \right) \cup \left(\bigcup_{A\to B\in\Omega^{(2)}} \langle A \to B\rangle_\mathbb{Q} \right) \\
&\quad \cup \cdots \cup \left(\bigcup_{A\to B\in\Omega^{(k-1)}} \langle A \to B\rangle_\mathbb{Q} \right) \cup \\
&\quad \left(\bigcup_{A\to B\in\Omega^{(k)}} \langle A \to B\rangle_\mathbb{Q} \right) \cup \cdots
\end{aligned}$$

From the above process, we know

$$\Omega^{(0)} \subseteq \Omega^{(1)} \subseteq \Omega^{(2)} \subseteq \Omega^{(3)} \subseteq \cdots \subseteq \Omega^{(k)} \subseteq \Omega^{(k+1)} \subseteq \cdots \subseteq \Omega^{(\infty)} = \Omega^V$$

For any $i, j = 0, 1, 2, \cdots$, a knowledge of level i is always a knowledge of level j when $i \leqslant j$, otherwise the relation dose not existence. It implies that a lower level knowledge is always a higher level knowledge, while a higher level knowledge needn't a lower level knowledge. The lower the level is , the more precise the knowledge is; The higher the level is , the less precise the knowledge is. Level 0 knowledge, the theorem, is the most precise knowledge, which is contained in all level's knowledge universes and form the kernel of the whole knowledge universe.

When unnecessary to give the specific levels, each formula $A \to B$ in Ω^V is called a *type V knowledge*, and the corresponding formula mass $\langle A \to B\rangle_\mathbb{Q}$ is called a type V knowledge mass with the kernel of knowledge $A \to B$. For each $k = 0, 1, 2, \cdots$,

$\Omega^{(k)}$ is called *current knowledge universe* for short. Specially, $\Omega^{(0)}$ is called the *initial knowledge universe*.

The above process is called the *dilation* or *expanse* of type V knowledge universe Ω^V.

Since we have constructed hierarchy type V knowledge universe Ω^V based on regular $\mathbb{Q}-$logic $\mathbb{C}_{\mathbb{Q}}$, it provides a *knowledge* platform for constructing the automatically function expanding type V knowledge base as well as the type V $\mathbb{Q}-$automatic reasoning system.

The regular $\mathbb{Q}-$logic $\mathbb{C}_{\mathbb{Q}}$ *type V knowledge base* K^V is a dynamic, expanding structure in type V knowledge universe Ω^V, which can automatically expand its functions. It can be constructed as the follow methods.

$K^{\langle 0\rangle} \in 2^{(\emptyset^{\vdash})}$, here, $2^{(X)}$ means all the non-empty finite subsets of set X. Specially, initial knowledge base $K^{\langle 0\rangle}$ may contain only one knowledge $A \to B$.

$K^{\langle 1\rangle} = K^{\langle 0\rangle} \bigcup \{A^* \to B^*\}$, where B^* is multiple approximate reasoning type V $\mathbb{Q}-$solution with regard to input A^* in current knowledge $K^{\langle 0\rangle}$.

$K^{\langle 2\rangle} = K^{\langle 1\rangle} \bigcup \{A^* \to B^*\}$, where B^* is multiple approximate reasoning type V $\mathbb{Q}-$solution with regard to input A^* in current knowledge $K^{\langle 1\rangle}$.

$\cdots\ \cdots\ \cdots\ \cdots\ \cdots\ \cdots$

$K^{\langle k\rangle} = K^{\langle k-1\rangle} \bigcup \{A^* \to B^*\}$, where B^* is multiple approximate reasoning type V $\mathbb{Q}-$solution with regard to input A^* in current knowledge $K^{\langle k-1\rangle}$.

$K^{\langle k+1\rangle} = K^{\langle k\rangle} \bigcup \{A^* \to B^*\}$, where B^* is multiple approximate reasoning type V $\mathbb{Q}-$solution with regard to input A^* in current knowledge $K^{\langle k\rangle}$.

$\cdots\ \cdots\ \cdots\ \cdots\ \cdots\ \cdots$

$K^V = K^{\langle\infty\rangle} = \bigcup_{k=0}^{\infty} K^{\langle k\rangle}$.

For each $k = 0, 1, 2, \cdots$, $K^{\langle k\rangle}$ is called *type V level 0 knowledge base* originating from initial knowledge base $K^{\langle 0\rangle}$.

From the construction process, we know

$K^{\langle 0\rangle} \subseteq K^{\langle 1\rangle} \subseteq K^{\langle 2\rangle} \subseteq K^{\langle 3\rangle} \subseteq \cdots \subseteq K^{\langle k\rangle} \subseteq K^{\langle k+1\rangle} \subseteq \cdots \subseteq K^{\langle\infty\rangle} = K^V$

Those process is called *dilation* or *expanse* of type V knowledge base K^V.

The above description illustrates a dynamic, automatically function expanding and continuously dilating knowledge base. It also described the resolving process of the multiple approximate reasoning type V $\mathbb{Q}-$algorithm which is concomitant of the automatically expanding of the knowledge base. In this way, we get a logic model of automatic reasoning system in the regular $\mathbb{Q}-$logic $\mathbb{C}_{\mathbb{Q}}$, which is called *type V $\mathbb{Q}-$automatic reasoning system.* It can be written as

$$\mathcal{E}^{\circ} = (\mathbb{C}_{\mathbb{Q}}, \Omega^V, K^V)$$

Theorem 2.1. In a type V $\mathbb{Q}-$automatic reasoning system $\mathcal{E}^{\circ} = (\mathbb{C}_{\mathbb{Q}}, \Omega^V, K^V)$, for each $k = 0, 1, 2, \cdots$,

$$K^{\langle k\rangle} \subseteq \Omega^{(k)}$$

that is, the knowledge in type V level k knowledge base $K^{\langle k\rangle}$ is always a knowledge in type V level k knowledge base $\Omega^{(k)}$.

Proof. Obviously $K^{\langle 0\rangle} \subseteq \Omega^{(0)}$.
Assume $K^{\langle i\rangle} \subseteq \Omega^{(i)}$. Then we prove $K^{\langle i+1\rangle} \subseteq \Omega^{(i+1)}$.

Because $K^{\langle i+1\rangle} = K^{\langle i\rangle} \bigcup\{A^* \to B^*\}$, $\Omega^{(i)} \subseteq \Omega^{(i+1)}$, considering mathematics induction, we assume $K^{\langle i\rangle} \subseteq \Omega^{(i)}$. It implies $K^{\langle i\rangle} \subseteq \Omega^{(i+1)}$, therefore we only need to prove $A^* \to B^* \in \Omega^{(i+1)}$.

In fact, because B^* is the type V $\mathbb{Q}-$solution of multiple approximate reasoning with regard to A^* under current knowledge base $K^{\langle i\rangle}$, there exists $A \to B \in K^{\langle i\rangle}$ which satisfies

$$A^* \in \langle A\rangle_{\mathbb{Q}}, B^* \in \langle B\rangle_{\mathbb{Q}}$$

therefore

$$A^* \to B^* \in \langle A\rangle_{\mathbb{Q}} \to \langle B\rangle_{\mathbb{Q}}$$

But from the construction of $\mathbb{Q}-$logic $\mathbb{C}_{\mathbb{Q}}$ we know

$$\langle A\rangle_{\mathbb{Q}} \to \langle B\rangle_{\mathbb{Q}} = \langle A \to B\rangle_{\mathbb{Q}}$$

therefore

$$A^* \to B^* \in \langle A \to B\rangle_{\mathbb{Q}}$$

Because $A \to B \in K^{\langle i\rangle}$ and $K^{\langle i\rangle} \subseteq \Omega^{(i)}$, we know $A \to B \in \Omega^{(i)}$. Considering the construction of type V level $i+1$ knowledge universe

$$\Omega^{(i+1)} = \bigcup_{A \to B \in \Omega^{(i)}} \langle A \to B\rangle_{\mathbb{Q}}$$

we have

$$A^* \to B^* \in \Omega^{(i+1)}$$

Now we have proven $K^{\langle i+1\rangle} \subseteq \Omega^{(i+1)}$.

So, according to mathematics induction, for each $k = 0, 1, 2, \cdots$

$$K^{\langle k\rangle} \subseteq \Omega^{(k)}$$ □

3 Type V True Level k Knowledge Circle, Extended Type V Knowledge Base and Extended Type V $\mathbb{Q}-$Automatic Reasoning System

In regular $\mathbb{Q}-$logic $\mathbb{C}_{\mathbb{Q}}$ type V knowledge universe Ω^V, type V knowledge $A \to B$ is called *type V true level k*. Or, we say its *absolute level* is level k, if $A \to B$ is type V level k knowledge and is not type V level $k-1$ knowledge, where $k = 1, 2, 3, \cdots$.

The knowledge in $\Omega^{(1)} - \Omega^{(0)}$ is of type V true level 1.

The knowledge in $\Omega^{(2)} - \Omega^{(1)}$ is of type V true level 2.

Generally, the knowledge in $\Omega^{(k)} - \Omega^{(k-1)}$ is of type V true level k, where $k = 1, 2, 3, \cdots$.

We appoint the knowledge in $\Omega^{(0)}$ is of type V true level 0.

Therefore, type V knowledge universe Ω^V can be regarded as the union of layers of *type V true level k knowledge circles*

$$\Omega^{(k)} - \Omega^{(k-1)}, k = 0, 1, 2, \cdots$$

where type V true level 0 knowledge circles is $\Omega^{(0)}$.

In fact, type V knowledge universe Ω^V can be expressed as the direct union decomposition of type V level k knowledge circles.

$$\Omega^V = \Omega^{(0)} \cup (\bigcup_{k=1}^{\infty} (\Omega^{(k)} - \Omega^{(k-1)}))$$
$$(\Omega^{(k)} - \Omega^{(k-1)}) \cap (\Omega^{(j)} - \Omega^{(j-1)}) = \emptyset,$$
$$k, j = 1, 2, 3, \cdots, k \neq j$$
$$(\Omega^{(k)} - \Omega^{(k-1)}) \cap \Omega^{(0)} = \emptyset, \; k = 1, 2, 3, \cdots$$

The absolute level of the knowledge in Ω^V is denoted by $f(A \to B)$. Notice that $\Omega^{(0)} \subseteq \Omega^{(1)} \subseteq \Omega^{(2)} \subseteq \cdots \subseteq \Omega^{(k)} \subseteq \Omega^{(k+1)} \subseteq \cdots \subseteq \Omega^{(\infty)} = \Omega^V$ It implies

$$f(A \to B) = min\{k | A \to B \in \Omega^{(k)}\}$$

The type V knowledge base K^V constructed in last section is very simple and easy to use. But it is not suitable for the confirmation of new knowledge level because it can not express the absolute level of the new knowledge in current knowledge base. In this section, we will design an extended type V knowledge base so that the absolute level of the new knowledge in current knowledge base can be confirmed to a certain extent.

An *extended type V knowledge base* $\mathbb{K}^V$ of regular $\mathbb{Q}-$logic $\mathbb{C}_{\mathbb{Q}}$ is a dynamic, automatically function expanding knowledge base, which is constructed as follows.

$\mathbb{K}_{(0,0)} \in 2^{(\emptyset^{\vdash})}$ is called *initial knowledge base* of $\mathbb{K}^V$. Specially, initial knowledge base $\mathbb{K}_{(0,0)}$ may contain only one knowledge $A \to B$.

Assume

$$\mathbb{K}_{(0,1)} = \mathbb{K}_{(0,0)} \bigcup \{A^*_{(0,1)} \to B^*_{(0,1)}\}$$
$$\mathbb{K}_{(0,2)} = \mathbb{K}_{(0,1)} \bigcup \{A^*_{(0,2)} \to B^*_{(0,2)}\}$$
$$\cdots \quad \cdots \quad \cdots \quad \cdots \quad \cdots$$
$$\mathbb{K}_{(0,0_s)} = \mathbb{K}_{(0,0_s-1)} \bigcup \{A^*_{(0,0_s)} \to B^*_{(0,0_s)}\}$$

and

$$A^*_{(0,1)} \to B^*_{(0,1)}$$
$$A^*_{(0,2)} \to B^*_{(0,2)}$$
$$\cdots \quad \cdots \quad \cdots \quad \cdots$$
$$A^*_{(0,0_s)} \to B^*_{(0,0_s)}$$

are all type V level 0 knowledge, where $B^*_{(0,i)}$ is multiple approximate reasoning type V $\mathbb{Q}-$solution with regard to the ith input $A^*_{(0,i)}$ under knowledge base $\mathbb{K}_{(0,i-1)}$. While $B^*_{(0,0_s+1)}$ which is the multiple approximate reasoning type V $\mathbb{Q}-$solution about the $(0_s+1)th$ input $A^*_{(0,0_s+1)}$ under knowledge base $\mathbb{K}_{(0,0_s)}$ and satisfies $A^*_{(0,0_s+1)} \to B^*_{(0,0_s+1)}$ is not type V level 0 knowledge. Let

$$\mathbb{K}_{(1,0)} = \mathbb{K}_{(0,0_s)} \bigcup \{A^*_{(0,0_s+1)} \to B^*_{(0,0_s+1)}\}$$

Suppose

$$\mathbb{K}_{(1,1)} = \mathbb{K}_{(1,0)} \bigcup \{A^*_{(1,1)} \to B^*_{(1,1)}\}$$
$$\mathbb{K}_{(1,2)} = \mathbb{K}_{(1,1)} \bigcup \{A^*_{(1,2)} \to B^*_{(1,2)}\}$$
$$\cdots \quad \cdots \quad \cdots \quad \cdots \quad \cdots$$
$$\mathbb{K}_{(1,1_s)} = \mathbb{K}_{(1,1_s-1)} \bigcup \{A^*_{(1,1_s)} \to B^*_{(1,1_s)}\}$$

and

$$\begin{array}{l} A^*_{(1,1)} \to B^*_{(1,1)} \\ A^*_{(1,2)} \to B^*_{(1,2)} \\ \cdots\ \cdots\ \cdots\ \cdots \\ A^*_{(1,1_s)} \to B^*_{(1,1_s)} \end{array}$$

are all type V level 1 knowledge, where $B^*_{(1,i)}$ is multiple approximate reasoning type V $\mathbb{Q}-$solution with regard to the ith input $A^*_{(1,i)}$ under knowledge base $\mathbb{K}_{(1,i-1)}$. While $B^*_{(1,1_s+1)}$ which is the multiple approximate reasoning type V $\mathbb{Q}-$solution with regard to the $(1_s+1)th$ input $A^*_{(1,1_s+1)}$ under knowledge base $\mathbb{K}_{(1,1_s)}$ and satisfies $A^*_{(1,1_s+1)} \to B^*_{(1,1_s+1)}$ is not the type V level 1 knowledge. Let

$$\mathbb{K}_{(2,0)} = \mathbb{K}_{(1,1_s)} \bigcup \{A^*_{(1,1_s+1)} \to B^*_{(1,1_s+1)}\}$$

$$\cdots\ \cdots\ \cdots\ \cdots\ \cdots$$

Suppose

$$\begin{array}{l} \mathbb{K}_{(k,0)} = \mathbb{K}_{(k-1,(k-1)_s)} \\ \quad \bigcup \{A^*_{(k-1,(k-1)_s+1)} \to B^*_{(k-1,(k-1)_s+1)}\} \\ \mathbb{K}_{(k,1)} = \mathbb{K}_{(k,0)} \bigcup \{A^*_{(k,1)} \to B^*_{(k,1)}\} \\ \mathbb{K}_{(k,2)} = \mathbb{K}_{(k,1)} \bigcup \{A^*_{(k,2)} \to B^*_{(k,2)}\} \\ \cdots\ \cdots\ \cdots\ \cdots\ \cdots \\ \mathbb{K}_{(k,k_s)} = \mathbb{K}_{(k,k_s-1)} \bigcup \{A^*_{(k,k_s)} \to B^*_{(k,k_s)}\} \end{array}$$

and

$$\begin{array}{l} A^*_{(k,1)} \to B^*_{(k,1)} \\ A^*_{(k,2)} \to B^*_{(k,2)} \\ \cdots\ \cdots\ \cdots\ \cdots \\ A^*_{(k,k_s)} \to B^*_{(k,k_s)} \end{array}$$

are all type V level k knowledge, where $B^*_{(k,i)}$ is multiple approximate reasoning type V $\mathbb{Q}-$solution with regard to the ith input $A^*_{(k,i)}$ under knowledge base $\mathbb{K}_{(k,i-1)}$. While $B^*_{(k,k_s+1)}$ which is the multiple approximate reasoning type V $\mathbb{Q}-$solution about the $(k_s+1)th$ input $A^*_{(k,k_s+1)}$ under knowledge base $\mathbb{K}_{(k,k_s)}$ and satisfied $A^*_{(k,k_s+1)} \to B^*_{(k,k_s+1)}$ is not the type V level k knowledge. Let

$$\mathbb{K}_{(k+1,0)} = \mathbb{K}_{(k,k_s)} \bigcup \{A^*_{(k,k_s+1)} \to B^*_{(k,k_s+1)}\}$$

$$\cdots\ \cdots\ \cdots\ \cdots\ \cdots$$

Finally, let

$$\mathbb{K}^V = \bigcup_{k=0}^{\infty} \mathbb{K}_{(k,k_s)}$$

From the above process, we know

$$\begin{array}{l} \mathbb{K}_{(0,0)} \subseteq \mathbb{K}_{(0,1)} \subseteq \cdots \subseteq \mathbb{K}_{(0,0_s)} \subseteq \\ \mathbb{K}_{(1,0)} \subseteq \mathbb{K}_{(1,1)} \subseteq \cdots \subseteq \mathbb{K}_{(1,1_s)} \subseteq \\ \cdots\ \cdots\ \cdots\ \cdots\ \cdots \\ \mathbb{K}_{(k,0)} \subseteq \mathbb{K}_{(k,1)} \subseteq \cdots \subseteq \mathbb{K}_{(k,k_s)} \subseteq \\ \cdots\ \cdots\ \cdots\ \cdots\ \cdots \end{array}$$

The above process is called the *dilation* or *expanse* of extended type V knowledge base K^V.

The triad form

$$\mathcal{E}^{\oplus} = (\mathbb{C}_{\mathbb{Q}}, \Omega^V, \mathbb{K}^V)$$

is called an *extended type V* $\mathbb{Q}$*−automatic reasoning system.* It is a dynamic multiple approximate reasoning system which can automatically obtain new knowledge and expand the knowledge base.

Theorem 3.1. In an extended type V automatic reasoning system $\mathcal{E}^{\oplus} = (\mathbb{C}_{\mathbb{Q}}, \Omega^V, \mathbb{K}^V)$ of regular $\mathbb{Q}$−logic $\mathbb{C}_{\mathbb{Q}}$, for each $k = 1, 2, 3, \cdots$, we have

$$\mathbb{K}_{(k,0)} \subseteq \mathbb{K}_{(k,1)} \subseteq \cdots \subseteq \mathbb{K}_{(k,k_s)} \subseteq \Omega^{(k)}$$

which means the knowledge in *type V level (k,j) knowledge base* $\mathbb{K}_{(k,j)}$ is also in type V level k knowledge universe $\Omega^{(k)}$, where $j = 0, 1, \cdots, k_s$

Proof. Only need to prove that

$$A^*_{(k-1,(k-1)_s+1)} \to B^*_{(k-1,(k-1)_s+1)}$$

is a type V level k knowledge.

In fact, because $A^*_{(k-1,(k-1)_s)}$ can activate at least one type V level $k-1$ knowledge in current knowledge base $\mathbb{K}_{(k-1,(k-1)_s+1)}$, there must be a type V $\mathbb{Q}$−solution of multiple approximate reasoning with regard to input $A^*_{(k-1,(k-1)_s)}$ under current knowledge base $\mathbb{K}_{(k-1,(k-1)_s+1)}$. Name the solution as $B^*_{(k-1,(k-1)_s+1)}$, we have

$$A \to B \in \mathbb{K}_{(k-1,(k-1)_s)}$$

which satisfies

$$A^*_{(k-1,(k-1)_s+1)} \in \langle A \rangle_{\mathbb{Q}}$$
$$B^*_{(k-1,(k-1)_s+1)} \in \langle B \rangle_{\mathbb{Q}}$$

They imply

$$A^*_{(k-1,(k-1)_s+1)} \to B^*_{(k-1,(k-1)_s+1)} \in \langle A \rangle_{\mathbb{Q}} \to \langle B \rangle_{\mathbb{Q}}$$

But from the construction of regular $\mathbb{Q}$−logic $\mathbb{C}_{\mathbb{Q}}$, we know

$$\langle A \rangle_{\mathbb{Q}} \to \langle B \rangle_{\mathbb{Q}} = \langle A \to B \rangle_{\mathbb{Q}}$$

therefore

$$A^*_{(k-1,(k-1)_s+1)} \to B^*_{(k-1,(k-1)_s+1)} \in \langle A \to B \rangle_{\mathbb{Q}}$$

which means $A^*_{(k-1,(k-1)_s+1)} \to B^*_{(k-1,(k-1)_s+1)}$ is contained in type V level k knowledge mass$\langle A \to B \rangle_{\mathbb{Q}}$ with the kernel of type V level $k-1$ knowledge $A \to B$. Besides, from the construction of type V level k knowledge universe

$$\Omega^{(k)} = \bigcup_{A \to B \in \Omega^{(k-1)}} \langle A \to B \rangle_{\mathbb{Q}}$$

we know

$$\langle A \rightarrow B \rangle_{\mathbb{Q}} \subseteq \Omega^{(k)}$$

therefore

$$A^*_{(k-1,(k-1)_s+1)} \rightarrow B^*_{(k-1,(k-1)_s+1)} \in \Omega^{(k)}$$

So, it proves that

$$A^*_{(k-1,(k-1)_s+1)} \rightarrow B^*_{(k-1,(k-1)_s+1)}$$

is a type V level k knowledge. □

Theorem 3.2. In an extended type V automatic reasoning system $\mathcal{E}^{\oplus} = (\mathbb{C}_{\mathbb{Q}}, \Omega^V, \mathbb{K}^V)$ of regular $\mathbb{Q}-$logic $\mathbb{C}_{\mathbb{Q}}$, for each $k = 0, 1, 2, 3, \cdots$ and $j = 0, 1, 2, 3, \cdots, k_s$,

$$\mathbb{K}_{(k,j)} \bigcap (\Omega^{(k)} - \Omega^{(k-1)}) \neq \emptyset$$

which implies there exists at least a true level k knowledge in type V level (k, j) knowledge base $\mathbb{K}_{(k,j)}$.

Proof. Obviously

$$A^*_{(k-1,(k-1)_s+1)} \rightarrow B^*_{(k-1,(k-1)_s+1)}$$

is a type V level k knowledge, not a type V level $k-1$ knowledge. So it is a type V true level k knowledge. Besides, for each $j = 0, 1, 2, 3, \cdots, k_s$, this knowledge is contained in type V level (k, j) knowledge base $\mathbb{K}_{(k,j)}$. □

4 The Type V Level (k,j) Perfection of Extended Type V Knowledge Base $\mathbb{K}^V$

In regular $\mathbb{Q}-$logic $\mathbb{C}_{\mathbb{Q}}$, an extended type V knowledge base $\mathbb{K}^V$ is *type V level (k,j) perfect*, if type V level (k,j) knowledge base $\mathbb{K}_{(k,j)}$ in type V level k knowledge universe $\Omega^{(k)}$ is ⓑ$-$*dense*, given by

$$(\mathbb{K}_{(k,j)})^{ⓑ} = \Omega^{(k)}$$

If $\mathbb{K}^V$ is type V level (k, j) perfect for each $j = 0, 1, 2, \cdots, k_s$, the extended type V knowledge base $\mathbb{K}^V$ is *type V level k perfect*. The extended type V knowledge base $\mathbb{K}^V$ is *type V perfect* if $\mathbb{K}^V$ is type V level k perfect for each $k = 0, 1, 2, \cdots$.

Theorem 4.1. In an extended type V expert system $\mathcal{E}^{\oplus} = (\mathbb{C}_{\mathbb{Q}}, \Omega^V, \mathbb{K}^V)$ of regular $\mathbb{Q}-$logic $\mathbb{C}_{\mathbb{Q}}$, all the knowledge of level k can be obtained by multiple approximate reasoning type V $\mathbb{Q}-$algorithm based on knowledge base $\mathbb{K}_{(k,j)}$, if $\mathbb{K}^V$ is type V level (k, j) perfect.

Proof. For any type V level k knowledge $A \rightarrow B$ in type V level k knowledge universe $\Omega^{(k)}$, because $\mathbb{K}^V$ is type V level (k, j) perfect, we have

$$(\mathbb{K}_{(k,j)})^{ⓑ} = \Omega^{(k)}$$

Therefore,

$$A \to B \in (\mathbb{K}_{(k,j)})^{\textcircled{b}}$$

Obviously,

$$\begin{aligned}
\mathbb{K}_{(0,0)} &= \{A_i \to B_i | i = 1, 2, \cdots, n\} \\
\mathbb{K}_{(0,0_s)} &= \mathbb{K}_{(0,0)} \bigcup \{A_i \to B_i | i = n+1, n+2, \cdots, n+0_s\} \\
&= \{A_i \to B_i | i = 1, 2, \cdots, n+0_s\} \\
\mathbb{K}_{(1,1_s)} &= \mathbb{K}_{(0,0_s)} \bigcup \{A_i \to B_i | i = n+0_s+1, n+0_s+2, \\
&\quad \cdots, n+0_s+1_s, n+0_s+1_s+1\} \\
&= \{A_i \to B_i | i = 1, 2, \cdots, n+0_s+1_s+1\} \\
\cdots & \cdots \cdots \cdots \cdots \cdots \\
\mathbb{K}_{(k-1,(k-1)_s)} &= \mathbb{K}_{(k-2,(k-2)_s)} \bigcup \{A_i \to B_i | \\
&\quad i = n+0_s+1_s+\cdots+(k-2)_s+1, \\
&\quad n+0_s+1_s+\cdots+(k-2)_s+2, \\
&\quad \cdots, n+0_s+1_s+\cdots+(k-2)_s+(k-1)_s, \\
&\quad n+0_s+1_s+\cdots+(k-2)_s+(k-1)_s+1\} \\
&= \{A_i \to B_i | i = 1, 2, \cdots, \\
&\quad n+0_s+1_s+\cdots+(k-2)_s+(k-1)_s+(k-1)\} \\
\mathbb{K}_{(k,j)} &= \mathbb{K}_{(k-1,(k-1)_s)} \bigcup \{A_i \to B_i | \\
&\quad i = n+0_s+1_s+\cdots+(k-1)_s+1, \\
&\quad n+0_s+1_s+\cdots+(k-1)_s+2, \\
&\quad \cdots, n+0_s+1_s+\cdots+(k-1)_s+j, \\
&\quad n+0_s+1_s+\cdots+(k-1)_s+j+1\} \\
&= \{A_i \to B_i | i = 1, 2, \cdots, \\
&\quad n+0_s+1_s+\cdots+(k-1)_s+k+j\}
\end{aligned}$$

$$(\mathbb{K}_{(k,j)})^{\textcircled{b}} = \bigcup_{i=1}^{n+0_s+1_s+\cdots+(k-1)_s+k+j} \langle A_i \to B_i \rangle_{\mathbb{Q}}$$

So, there exists at least an

$$i \in \{1, 2, \cdots, n+0_s+1_s+\cdots+(k-1)_s+k+j\}$$

which satisfies

$$A \to B \in \langle A_i \to B_i \rangle_{\mathbb{Q}}$$

It means $A \to B$ is always contained in a type V knowledge mass $\langle A_i \to B_i \rangle_{\mathbb{Q}}$ with the kernel of a certain knowledge $A_i \to B_i$. From the construction of regular $\mathbb{Q}-$logic $\mathbb{C}_{\mathbb{Q}}$, we know

$$\langle A_i \to B_i \rangle_{\mathbb{Q}} = \langle A_i \rangle_{\mathbb{Q}} \to \langle B_i \rangle_{\mathbb{Q}}$$

Therefore,

$$A \to B \in \langle A_i \rangle_{\mathbb{Q}} \to \langle B_i \rangle_{\mathbb{Q}}$$

that is

$$A \in \langle A_i \rangle_{\mathbb{Q}},\ B \in \langle B_i \rangle_{\mathbb{Q}}$$

It implies that input A activates knowledge $A_i \to B_i$ according to type V $\mathbb{Q}-$algorithm, where B can be taken as the type V $\mathbb{Q}-$solution of multiple approximate reasoning with regard to input A in current knowledge base $\mathbb{K}_{(k,j)}$. And the current knowledge base can be expanded as

$$\begin{aligned}\mathbb{K}_{(k,j+1)} &= \mathbb{K}_{(k,j)} \bigcup \{A \to B\} \\ &= \{A_i \to B_i | i = 1, 2, \cdots, n + 0_s + 1_s \\ &\quad + \cdots + (k-1)_s + k + j\} \\ &\quad \bigcup \{A \to B\} \\ &= \{A_i \to B_i | i = 1, 2, \cdots, n + 0_s + 1_s \\ &\quad + \cdots + (k-1)_s + k + j + 1\}\end{aligned}$$

□

5 Conclusion

We do research on the approximate reasoning in logic frame. Based on the similarity relation $\mathbb{Q}$ in formula set $F(S)$, we illustrate the matching scheme and corresponding algorithms for approximate reasoning. We also discuss the type V completeness and type V perfection of knowledge base in $\mathbb{Q}-$logic $\mathbb{C}_{\mathbb{Q}}$. Meanwhile, we introduces and investigate the $\mathbb{Q}-$knowledge mass, $\mathbb{Q}-$knowledge universe, $\mathbb{Q}-$knowledge base and $\mathbb{Q}-$automatic reasoning system bases on the similarity logic $\mathbb{C}_{\mathbb{Q}}$. We also present and investigate the $\mathbb{Q}-$true level k knowledge circle, extended $\mathbb{Q}-$ knowledge base, extended $\mathbb{Q}-$automatic reasoning system and the level (k,j) perfection of extended $\mathbb{Q}-$knowledge base $\mathbb{K}^V$ in this paper.

The construction of $\mathbb{Q}-$logic $\mathbb{C}_{\mathbb{Q}}$ not only ensures the approximate reasoning conclusion B^* is $\mathbb{Q}-$similar to the consequence of knowledge $A \to B$ when A^* is $\mathbb{Q}-$similar to the antecedent of knowledge $A \to B$(It satisfies the general logical requirements of approximate reasoning), but also ensures that the obtained approximate knowledge $A^* \to B^*$ is $\mathbb{Q}-$similar to the original knowledge $A \to B$(It is the more advanced and rigid logic requirement for approximate reasoning). It reflects the nature of approximate reasoning and satisfy more application expectations if we use similarity relation $\mathbb{Q}$ instead of equivalence relation $\mathbb{R}$ to design approximate reasoning matching scheme and corresponding algorithms. We hope that the ideas, expressions and methods of this paper could become one of the logical standards of approximate reasoning.

References

1. Costas PP, Nikos IK (1993) A comparative assessment of measures of similiarity of fuzzy values. Fuzzy Sets and Systems, 56:171-174
2. Costas PP (1991) Value approximation of fuzzy Systems. Fuzzy Sets and Systems, 39: 111-115
3. Wang GJ (1998) Fuzzy continuous input-output controllers are universal approximators. Fuzzy Sets and Systems, 97:95-99
4. Wang GJ (1999) On the logic foundation of fuzzy reasoning. Information Science, 117: 47-88
5. Wang GJ, Wang H (2001) Non-fuzzy versions of fuzzy reasoning in classical logics. Information Sciences, 138:211-236
6. Wang GJ, Leung Y (2003) Intergrated semantics and logic metric spaces. Fuzzy Sets and Systems, 136:71-91

7. Wang GJ, He YY (2000) Instuitionistic fuzzy sets and L-fuzzy sets. Fuzzy Sets and Systems, 110:271-274
8. Meng GW (1993) Lowen's compactness in L-fuzzy topological spaces. Fuzzy Sets and Systems, 53:329-333
9. Meng GW (1995) On countably strong fuzzy compact sets in L-fuzzy topological spaces. Fuzzy Sets and Systems, 72:119-123
10. Bai SZ (1997) Q-convergence of ideals in fuzzy lattices and its applications. Fuzzy Sets and Systems, 92:357-363
11. Bai SZ (1997) Q-convergence of nets and week separation axiom in fuzzy lattices. Fuzzy Sets and Systems, 88:379-386
12. Dug HH, Seok YH (1994) A note on the value similarity of fuzzy systems variables. Fuzzy Sets and Systems, 66:383-386
13. Hyung LK, Song YS, Lee KM (1994) Similarity measure between fuzzy sets and between elements. Fuzzy Systems and Sets, 62:291-293
14. Chen SM, Yeh MS, Hsiao PY (1995) A comparison of similarity measures of fuzzy values. Fuzzy Sets and Systems, 72:79-89
15. Chen SM (1995) Measures of smilarity between vague sets. Fuzzy Sets and Systems, 74:217-223
16. Sudkamp T (1993) Similarity, interpolation, and fuzzy rule construction. Fuzzy Sets and Systems, 58:73-86
17. Kickert WJM, Mamdami EH (1978) Analysis of a fuzzy logic controller. Fuzzy Sets and Systems, 1:29-44
18. Czogala E, Leski J (2001) On eqivalence of approximate reasoning results using different interpretations of fuzzy if-then rules. Fuzzy Sets and Systems, 117:279-296
19. Zadeh LA (2003) Computing with words and perceptions – a paradigm shift in computing and decision ananlysis. Proceedings of International Conference on Fuzzy Information Processing, Tsinghua University Press, Springer Verlag
20. Ying MS (1994) A logic for approximate reasoning. The Journal of Symbolic Logic 59: 830-837
21. Ying MS (1992) Fuzzy reasoning under approximate match. Science Bulletin 37:1244-1245
22. Ying MS (1992) Compactnees,the Lowenheim-Skolem property and the direct product of lattices of truth values. Zeitschrift fur Mathematische Logik und Grundlagen der Mathematik 38:521-524
23. Ying MS (2003) Reasoning about probabilistic sequential programs in a probabilistic logic. Acta Informatica 39:315-389
24. Zheng YL (2003) Stratified construction of fuzzy propositional logic. Proceedings of International Conference on Fuzzy Information Processing, Tsinghua University Press, Springer Verlag, 1-2:169-174
25. Zheng YL, Zhang CS, Yi X (2004) Mamdaniean logic. Proceedings of IEEE International Conference on Fuzzy Systems, Budapest, Hungary, 1-3:629-634

Image Segmentation by Multi-level Thresholding Based on C-Means Clustering Algorithms and Fuzzy Entropy

Feng Zhao and Jiulun Fan

Department of Information and Control, Xi'an Institute of Posts and Telecommunications, Xi'an, 710061, China
fengz1119@sina.com, jiulunf@xiyou.edu.cn

Abstract. C-means clustering algorithms (hard C-means clustering algorithm and fuzzy C-means clustering algorithm) are common thresholding segmentation techniques. They can be easily applied to image segmentation tasks with multi-feature and multi-threshold, but the results are very sensitive to noise. In this paper, an image segmentation method combining the C-means clustering algorithms and fuzzy entropy is presented. Firstly, pre-segmentation by multi-level thresholding was made by one of the C-means clustering algorithms; then further processing was done using fuzzy entropy. Experimental results show that the proposed method can overcome the weakness of the C-means clustering algorithms and behave well in segmenting images of low signal-to-noise ratio.

Keywords: Hard C-means clustering algorithm, Fuzzy C-means clustering algorithm, Fuzzy entropy, Post-processing, Image segmentation.

1 Introduction

Image segmentation is an important part in image processing and computer vision. It partitions an image into several regions according to image features. There are a lot of segmentation methods [1, 2], such as edge detection, region growth, histogram thresholding, clustering, and neural network, etc.

Clustering is the process of classifying elements of a data set into clusters according to some similarity criterion. Elements in the same cluster have some similar properties, while elements belonging to different clusters have some dissimilar properties. There are two well-known clustering methods [3, 4, 5, 6, 7]: hard C-means (HCM) and fuzzy C-means (FCM) clustering. The HCM assigns a sample to only one cluster. The FCM allows a fuzzy classification in which an element or data can have partial membership in more than one cluster. Most of segmentation methods only can be used to bi-level image segmentation. Extending these methods to multi-level segmentation faces the problems such as the expense of large amounts of computation and high complexity of

B.-Y. Cao (Ed.): Fuzzy Information and Engineering (ICFIE), ASC 40, pp. 113–121, 2007.
springerlink.com

algorithm, especially when more than one feature is used. The C-means clustering algorithms can be easily applied to segmentation problems with multi-threshold and multi-feature, but their drawback is that they are very sensitive to noise.

Fuzzy entropy [8] indicates the degree of fuzziness of a fuzzy set. It is an important concept in fuzzy set theory, and has been successfully applied to pattern recognition, image processing, classifier design, and neural network structure, etc. Wang et al. (2003) adopted fuzzy entropy as a post processing for segmentation results obtained using cost function. It utilized the information of regions, and achieved better segmentation results than only using cost function. Inspired by Wang et al. (2003), we propose an image segmentation method combining the C-means clustering algorithms and fuzzy entropy. Firstly, pre-segmentation to image is made by one of the C-means clustering algorithms; secondly, further processing using fuzzy entropy is done. Our method can obtain better segmentation results to images with noise.

2 Image Segmentation Based on C-Means Clustering Algorithms

2.1 C-Means Clustering Algorithms

For our convenience of description, the following is the idea of the HCM. The introduction of the FCM can be seen in Liu (1992) and Bezdek (1981).

Let $X = \{X_1, X_2, \ldots, X_N\} \subset R^S$ be a finite set of data, where S is the dimension of the set. C is the number of clusters, and d_{ij} is the Euclidean distance between the sample X_i and the centroid V_j of a specific cluster j, which can be defined as:

$$d_{ij} = \|X_i - V_j\|^2 = \sum_{l=1}^{S} (X_{il} - V_{jl})^2 \tag{1}$$

where $V_j \subset R^S (1 \leq j \leq C)$, and $V = [V_1, V_2, \ldots, V_C]$ is a $S \times C$ matrix. The HCM classifies X_i to a single cluster exclusively. Using $\mu_j(X_i)$ to denote the membership of data point X_i belonging to cluster j , we have

$$\mu_{ij} = \mu_j(X_i) == \begin{cases} 1, \, X_i \in cluster \; j \\ 0, \, X_i \notin cluster \; j \end{cases} \tag{2}$$

where $U = [\mu_{ij}]$ is the $N \times C$ matrix.

The HCM is based on minimization of the following objective function:

$$J(U,V) = \sum_{i=1}^{N} \sum_{j=1}^{C} \mu_{ij} d_{ij}^2 \tag{3}$$

where it qualify:

$$\begin{cases} \sum_{j=1}^{C} \mu_{ij} = 1, & 1 \le i \le N, \\ \mu_{ij} \in \{0,1\}, & 1 \le i \le N, 1 \le j \le C, \\ 0 < \sum_{i=1}^{N} \mu_{ij} < N, 1 \le j \le C \end{cases} \tag{4}$$

This minimization problem can be solved using the following algorithm:

Initialization: Set ε between 0 and 1 as a termination criterion. Fix the number C of clusters. Randomly choose C different samples from the data set as initial centroids $V(0)$. Let $k = 0$.

Step 1: Compute $U(k)$ according to the following formula:

$$\mu_{ij}(k) = \begin{cases} 1, d_{ij} = \min_{(1 \le r \le C)} d_{ir}(k), \\ 0, otherwise \end{cases} \tag{5}$$

Step 2: Compute $V_j(k)$ according to the following formula:

$$\forall j \; V_j(k+1) = \sum_{i=1}^{N} (\mu_{ij}(k) \cdot X_i) / \sum_{i=1}^{N} \mu_{ij}(k) \tag{6}$$

Step 3: If $\sum_{i=1}^{C} \|X_j(k+1) - V_j(k)\|^2 < \varepsilon$, stop; otherwise, let $k = k+1$, then go to **Step 1**.

Classification criterion: $X_i \in cluster\ j$, iff $\mu_{ij} = \mu_j(X_i) = 1$.

2.2 C-Means Clustering Algorithms for Image Segmentation

An image is defined as a 2D light-intensity function, which contains $M \times N$ pixels. The gray level of a pixel with coordinate (x, y) is denoted as $f(x, y)$. The local average gray level within the k-neighborhood of a pixel with coordinate (x, y) is denoted by $g(x, y)$. $g(x, y)$ can be defined as follows:

$$g(x,y) = \frac{1}{k^2} \sum_{m=-k/2}^{k/2} \sum_{n=-k/2}^{k/2} f(x+m, y+n) \tag{7}$$

For an image, Let c_{ij} be the total number of occurrence frequency of the pair (i, j) which represents pixel (x, y) with $f(x, y) = i$ and $g(x, y) = j$, $0 \le c_{ij} \le M \times N$, then the joint probability mass function p_{ij} is given by

$$p_{ij} = \frac{c_{ij}}{M \times N} \tag{8}$$

where $0 \le i, j \le L-1$, $\sum_{i=0}^{L-1} \sum_{j=0}^{L-1} p_{ij} = 1$. $\{p_{ij}\}$is the 2D histogram of the image.

For the convenience of description, we only state the case of segmenting three level images with noise, that is $C = 3$. Generally speaking, gray level images always suffer from noise. Segmentation method based on the 2D histogram can restrain the effect of noise, because it not only considers the gray level distribution, but also the spatial correlation between the pixels in the image. It performs much better than the same method based on 1D histogram. In this paper, we adopt the C-means clustering algorithms based on two image features (the gray level and local average gray level), therefore, the dimension S is 2.

The C-means clustering algorithms are effective segmentation methods. They can be easily applied to image segmentation with multi-feature and multi-threshold, but the segmentation results are sensitive to noise. We present a post process using fuzzy entropy to overcome this weakness.

3 Perform Post Process Using Fuzzy Entropy

3.1 Fuzzy Entropy and Membership Function

Fuzzy entropy indicates the degree of fuzziness of a fuzzy set. Based on the Shannon function, De Luca and Termini (1972) defined the entropy of a fuzzy set A as:

$$e(A) = \frac{1}{n \ln 2} \sum_{i=1}^{n} S(\mu_A(x_i)) \tag{9}$$

where $S(\mu_A(x_i)) = -\mu_A(x_i)\ln(\mu_A(x_i)) - (1 - \mu_A(x_i))\ln(1 - \mu_A(x_i))$.

The fuzzy entropy should satisfy the following properties [8]:

P1: $e(A) = 0$ iff A is crisp set.

P2: $e(A)$ is maximum iff $\mu_A(x_i) = 0.5$ $(\forall x_i \in A)$.

P3: $e(A) \geq e(A^*)$, where A^* is a sharpened version of A.

P4: $e(A) \geq e(A^c)$, where A^c is the complement of A.

3.2 Perform Post Process Using Fuzzy Entropy

The segmentation results of images with noise, which are achieved by the C-means clustering algorithms, have some wrong classified pixels. In this paper, in order to obtain better results, we use fuzzy entropy to reclassify those pixels. Firstly, construct three membership functions, which indicate the relationship between a pixel and its belonging region (black, gray and white region); secondly, compute three fuzzy entropies for each pixel; Lastly, classify the pixel to the class which has the minimum fuzzy entropy. The detail is as follows:

Let $Q = [q(x, y)]_{M \times N}$ be the segmentation result obtained by one of the C-means clustering algorithms. We select a $n \times n$ (n is odd) window $W_n(x, y)$ whose center is (x, y). When n=3, $W_n(x, y)$ can be defined as:

$$W_n(x, y) = \begin{bmatrix} q(x-1, y-1) & q(x, y-1) & q(x+1, y-1) \\ q(x-1, y) & q(x, y) & q(x+1, y) \\ q(x-1, y+1) & q(x, y+1) & q(x+1, y+1) \end{bmatrix} \tag{10}$$

We assume that the gray levels of Q are 0, 128 and 255, $m_0 = 0$ (black region), $m_1 = 128$(gray region), $m_2 = 255$(white region). Three fuzzy sets I_{m_0}, I_{m_1} and I_{m_2} are defined in $W_n(x, y)$, and their membership functions are

$$\mu_{m_0}(q(x+k, y+l)) = [1 + |q(x+k, y+l) - m_0|/\lambda]^{-1} \tag{11}$$

$$\mu_{m_1}(q(x+k, y+l)) = [1 + |q(x+k, y+l) - m_1|/\lambda]^{-1} \tag{12}$$

$$\mu_{m_2}(q(x+k, y+l)) = [1 + |q(x+k, y+l) - m_2|/\lambda]^{-1} \tag{13}$$

where $k = -1, 0, 1; l = -1, 0, 1$. (11), (12) and (13) respectively indicates the membership relation about the pixels in the window $W_n(x, y)$ for the black, gray and white region. m_0, m_1 and m_2 are the target values of the three region. Thus, the smaller the absolute difference between the gray level of a pixel and its corresponding target value is, the larger the membership value the pixel has, and vice versa. λ is a constant value such that $\mu_{m_i}(q(x+k, y+l))(i = 0, 1, 2)$. In this paper, $\lambda = 255$. The fuzzy entropy of the three fuzzy set are expressed as:

$$e(I_{m_0}) = -\frac{1}{(3 \times 3)\ln 2} \sum_{k=-1}^{1} \sum_{l=-1}^{1} S(\mu_{m_0}(q(x+k, y+l))) \tag{14}$$

$$e(I_{m_1}) = -\frac{1}{(3 \times 3)\ln 2} \sum_{k=-1}^{1} \sum_{l=-1}^{1} S(\mu_{m_1}(q(x+k, y+l))) \tag{15}$$

$$e(I_{m_2}) = -\frac{1}{(3 \times 3)\ln 2} \sum_{k=-1}^{1} \sum_{l=-1}^{1} S(\mu_{m_2}(q(x+k, y+l))) \tag{16}$$

From the definition of the membership function, the larger $\mu_{m_i}(q(x+k, y+l))$ is, the larger the relationship between a pixel with coordinate $(x+k, y+l)$ and its belonging m_i region is, hence, better segmentation result can be obtained. From the definition of the fuzzy entropy $e(I_{m_i})$, the larger $\mu_{m_i}(q(x+k, y+l))$ is, the smaller $e(I_{m_i})$ is. Therefore, the smaller the fuzzy entropy is, the better the segmentation effect is.

For each pixel, calculate $e(I_{m_0})$, $e(I_{m_1})$ and $e(I_{m_2})$, and find the minimum among the three fuzzy entropies. If $e(I_{m_0})$ is the minimum, set $q(x, y) = m_0$; If $e(I_{m_1})$ is the minimum, set $q(x, y) = m_1$; If $e(I_{m_2})$ is the minimum, set $q(x, y) = m_2$. Actually, this process utilize the region information of the pixel in the window. Suppose the window places at the region of m_0, that is, the gray levels of most pixels of the widow are m_0, so $e(I_{m_0})$ must be the smallest. Thus, reclassifying the pixel with coordinate (x, y) to m_0 is reasonable.

4 Experimental Results

In this paper, we evaluate the performance of our algorithm using three images with noise: House, Peppers and Tungsten. Firstly, a multi-level pre-segmentation to image is made by one of the C-means clustering algorithms; secondly, further

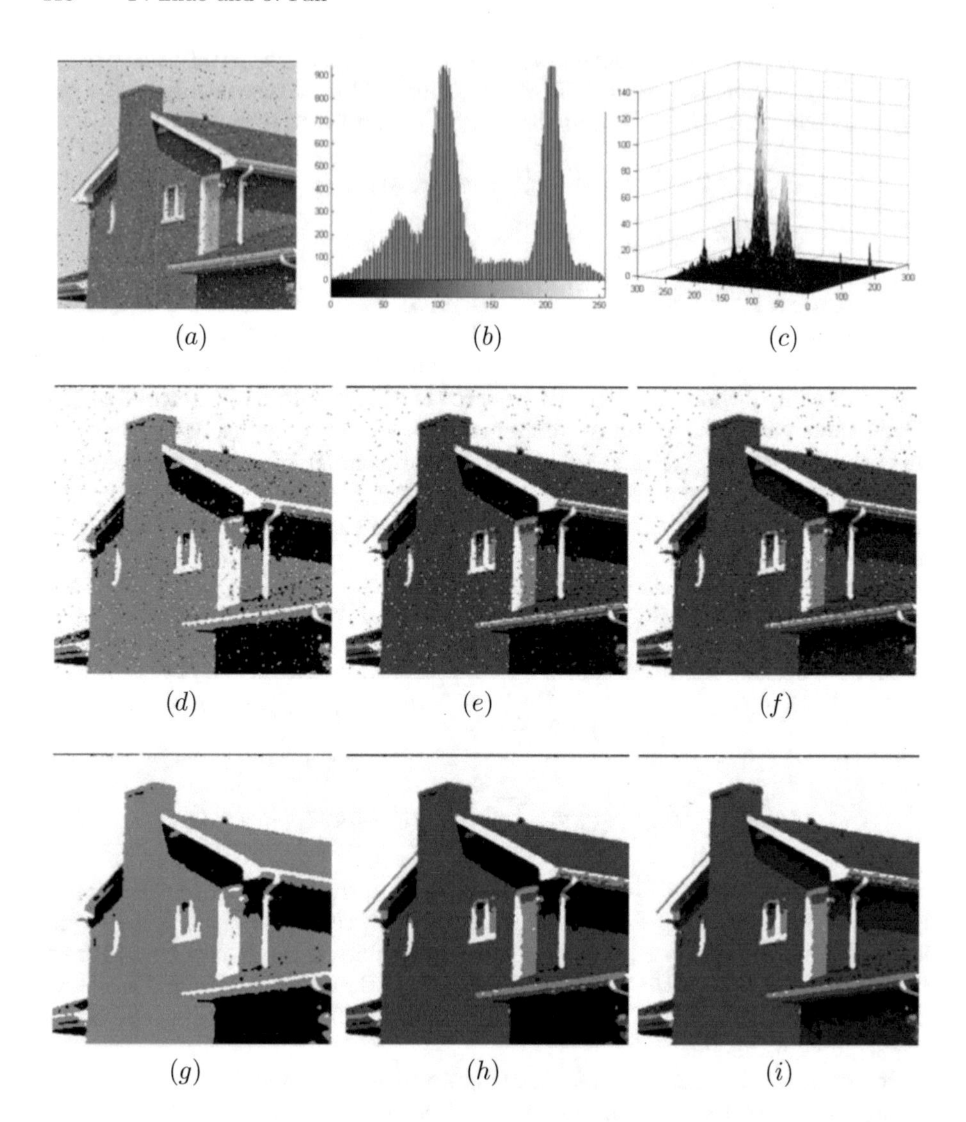

Fig. 1. Results of House image: (a) original image; (b) 1D histogram; (c) 2D histogram; (d) 3 level result of HCM; (e) 4 level result of HCM; (f) 5 level result of HCM; (g) 3 level result of our method; (h) 4 level result of our method;(i) 5 level result of our method

processing using fuzzy entropy was done. The results of HCM and FCM multi-level segmentation are almost the same. Limited by paper length, we only list the results of HCM combining fuzzy entropy in Fig1-3.

It can be seen that the segmentation results of the HCM and FCM are not satisfying, and still have a lot of wrong classified pixels. Our post processing method reclassifies these wrong classified pixels and obtains better visual effects.

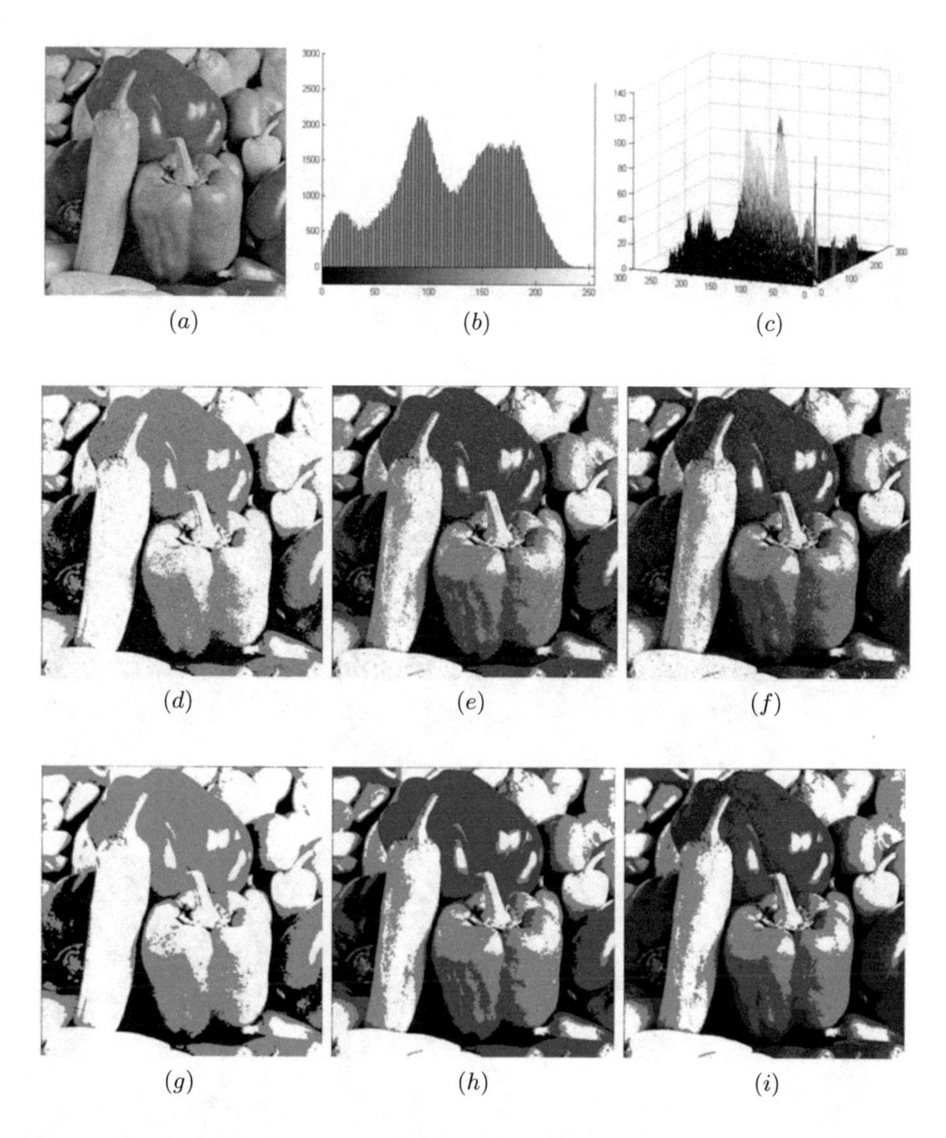

Fig. 2. Results of Peppers image: (a) original image; (b) 1D histogram; (c) 2D histogram; (d) 3 level result of HCM; (e) 4 level result of HCM; (f) 5 level result of HCM; (g) 3 level result of our method; (h) 4 level result of our method;(i) 5 level result of our method

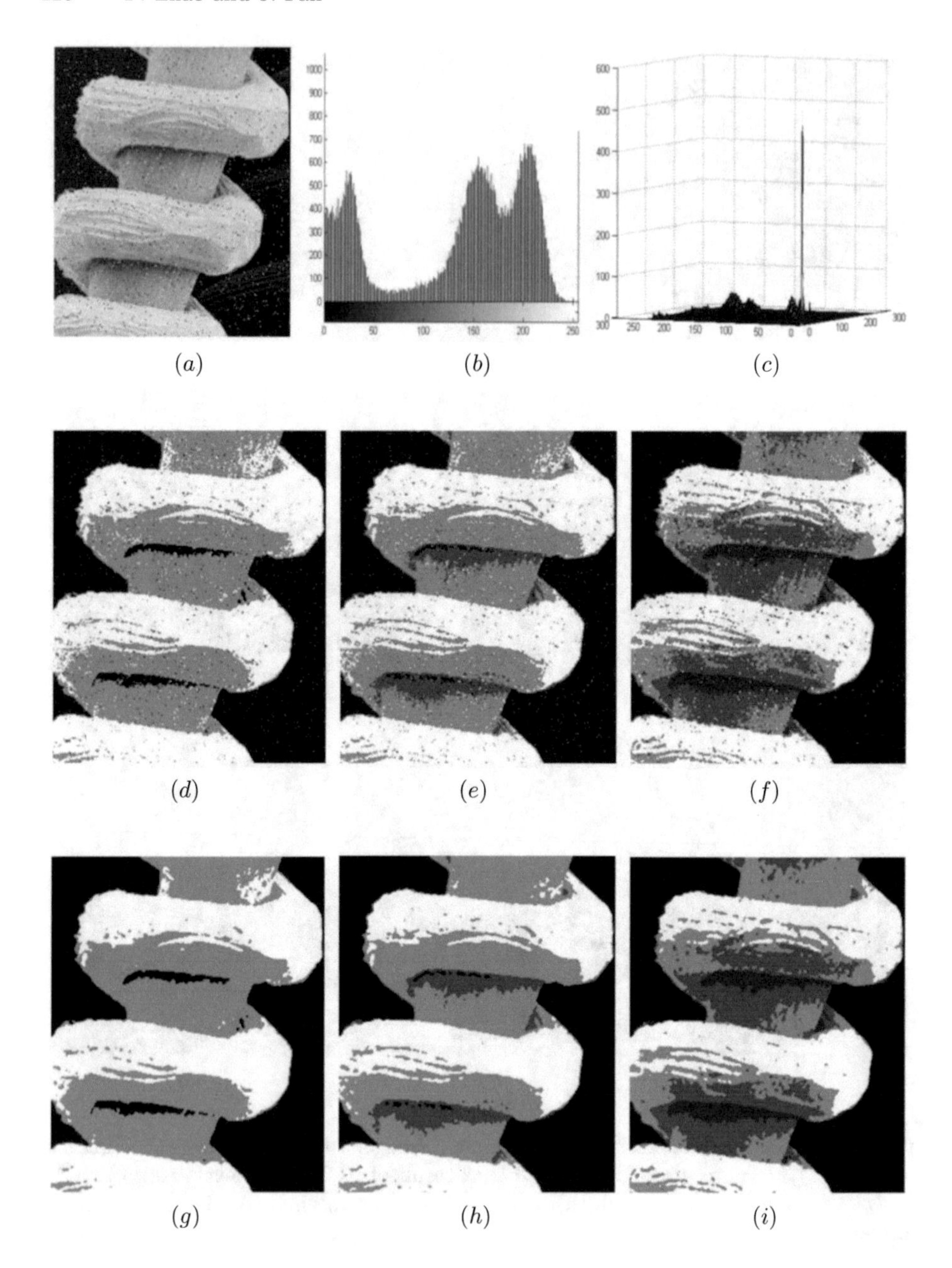

Fig. 3. Results of Tungsten image: (a) original image; (b) 1D histogram; (c) 2D histogram; (d) 3 level result of HCM; (e) 4 level result of HCM; (f) 5 level result of HCM; (g) 3 level result of our method; (h) 4 level result of our method;(i) 5 level result of our method

5 Conclusion

Image segmentation methods combining C-means clustering algorithms and fuzzy entropy are proposed. They not only possess the advantage of C-means clustering algorithms, but also overcome their sensitivity to noise using fuzzy entropy. No matter whether the pre-segmentation method is HCM or FCM, the final results are almost the same. So we suggest adopting the method combining the HCM and fuzzy entropy in the real time situations, because of less performing time of the HCM. Moreover, the idea of this paper can also be extended to other segmentation methods, and these studies will be presented in other papers.

References

1. N. R. Pal, S. K. Pal (1993) A review on image segmentation techniques, Pattern Recognition, 26(9):1277-1294.
2. X.J. Fu, L.P. Wang (2003) Data dimensionality reduction with application to simplifying RBF network structure and improving classification performance, IEEE Trans. System, Man, Cybern, Part B: Cybernetics, 33:399-409.
3. J. Kittler, J. Illingworth (1985) On threshold selection using clustering criteria, IEEE Transaction on Systems, Man and Cybernetics, 15(5):652-655.
4. J. Z. Liu, Y. Q. Tu (1992) Thresholding of images using an efficient C-mean clustering algorithm, Journal of Electronics (in Chinese), 14(4):424-427.
5. J. Z. Liu (1992) A fuzzy clustering method for image segmentation based on two-dimensional histogram, Acta Electronica Sinica (in Chinese), 20(9):40-46.
6. X. P. Zhang, J. L. Fan, J. H. Pei (2002) Texture segmentation based on spatial information and fuzzy clustering, Fuzzy Systems and Mathematics (in Chinese), 16:306-310.
7. W. Z. Zhen, J. L.Fan, W.X. Xie (2003) A new fuzzy clustering algorithm for image segmentation based on two-dimensional histogram, Computer Engineering and Applications (in Chinese), 15:86-88.
8. J. L. Fan (1999) Fuzzy entropy theory (in Chinese), Northwest University Press, Xi'an
9. B. P. Wang, J. L. Fan, W. X. Xie , C.M. Wu (2003) An image segmentation method based on cost function and fuzzy Entropy, Acta Photonica Sinica (in Chinese), 32(12):1502-1405.
10. J. C. Bezdek (1981) Pattern recognition with fuzzy objective function algorithms, Plenum Press, New York
11. A. Deluca and S.Termini (1972) A definition of a non-probobilistic entropy in the setting of fuzzy set theory. Information Control, 20(4):301-312.

Part II

Fuzzy Sets and Systems

Extension Principle of Interval-Valued Fuzzy Set

Wenyi Zeng, Yibin Zhao, and Hongxing Li

[1] School of Mathematical Sciences, Beijing Normal University, Beijing, 100875, P.R. China
{zengwy,lhxqx}@bnu.edu.cn
[2] Department of Basic Course, Institute Of Disaster Prevention Science And Technology, Yianjiao East of Beijing,101601, P.R. China
zhaoyibin5362@126.com

Abstract. In this paper, we introduce maximal and minimal extension principles of interval-valued fuzzy set and an axiomatic definition of generalized extension principle of interval-valued fuzzy set and use concepts of cut set of interval valued fuzzy set and interval-valued nested sets to explain their construction procedure in detail. These conclusions can be applied in some fields such as fuzzy algebra, fuzzy analysis and so on.

Keywords: Interval-valued fuzzy set; extension principle; reasonable extension operator.

1 Introduction

Since fuzzy set was introduced by Zadeh[19] in 1965, some new approaches and theories treating imprecision and uncertainty have been proposed. Some of theories, like intuitionistic fuzzy set proposed by Atanassov[2, 3] and $L-$fuzzy set introduced by Goguen[11] where L is an arbitrary lattice, are extensions of the classic fuzzy set. Another, well-known generalization of an ordinary fuzzy set is interval-valued fuzzy set. Interval valued fuzzy set was first introduced by Zadeh[20], after that, some authors investigated the topic and obtained some meaningful conclusions. For example, Biswas[4] and Li[13] investigated interval-valued fuzzy subgroup, Mondal[15] studied interval-valued fuzzy topology. Aimed at the important topic of approximate reasoning, Arnould[1] investigated interval-valued fuzzy backward reasoning, Chen[7, 8] investigated a method to deal with bidirectional approximate reasoning of interval-valued fuzzy sets, Gorzalczany[12] presented a method for interval-valued fuzzy reasoning based on compatibility measure and investigated some properties, Yuan[18] used normal form based on interval-valued fuzzy set to deal with approximate reasoning, Bustince[5, 6] and Cornelis[9] discussed interval-valued fuzzy relation and implication and applied them to approximate reasoning, Turksen[16, 17] proposed the definitions of interval-valued fuzzy set for the representation of combined concepts based on normal forms and investigated approximate analogical reasoning, Zeng[21, 22] proposed the concepts of cut set of interval-valued

B.-Y. Cao (Ed.): Fuzzy Information and Engineering (ICFIE), ASC 40, pp. 125–137, 2007.
springerlink.com

fuzzy set and interval-valued nested sets and investigated decomposition theorems and representation theorems of interval-valued fuzzy set and so on. These works show the importance of interval-valued fuzzy set. Moreover, some authors pointed out that there is a strong connection between intuitionistic fuzzy sets and interval-valued fuzzy sets, for more details we refer the readers to [10].

Just like that decomposition theorems and representation theorems of fuzzy set played an important role in fuzzy set theory, which helped people to develop many branches such as fuzzy algebra, fuzzy measure and integral, fuzzy analysis and so on. Extension principle introduced by Zadeh[19] is also an important tool in fuzzy set theory, it enables to extend any mapping $f: X \to Y$, where X and Y are universal sets, to a mapping $f^*: \mathcal{F}(X) \to \mathcal{F}(Y)$, where $\mathcal{F}(X)$ and $\mathcal{F}(Y)$ denote the set of all fuzzy subsets in X and Y, respectively, and

$$f^*(A)(y) = \bigvee_{y=f(x)} A(x)$$

for $A \in \mathcal{F}(X)$ and $y \in Y$. Luo[14] used the tools of nested set of fuzzy set, cylindrical extension and projection to explain the membership function of extension principle proposed by Zadeh and proposed three extension principles of fuzzy set. Zhai [23] introduced the concept of generalized extension principle and investigated minimal extension principle. In this paper, we want to study extension principle of interval-valued fuzzy set, propose maximal and minimal extension principles of interval-valued fuzzy set and an axiomatic definition of generalized extension principle of interval-valued fuzzy set and use concepts of cut set of interval-valued fuzzy set and interval valued nested sets to explain their construction procedure in detail.

The rest of our work is organized as follows. In the section 2, we recall some basic notions of interval-valued fuzzy set. In the section 3, we propose maximal extension principle of interval-valued fuzzy set and explain its construction procedure using cut set of interval valued fuzzy set and interval-valued nested set. In the section 4, we propose minimal extension principle of interval-valued fuzzy set and explain its construction procedure using cut set of interval-valued fuzzy set and interval-valued nested set. In the section 5, we introduce an axiomatic definition of generalized extension principle for interval-valued fuzzy set and give an application example. The final is conclusion.

2 Preliminaries

Throughout this paper, we write X to denote the discourse set, IVFSs(X) stands for the set of all interval-valued fuzzy subsets in X. A expresses an interval-valued fuzzy set, the operation "c" is the complement of interval-valued fuzzy set or fuzzy set in X, $\varnothing$ stands for the empty set.

Let $[I]$ be the set of all closed subintervals of the interval $[0,1]$. Specially, for an arbitrary element $a \in [0,1]$, we think a is the same as $[a,a]$, namely, $a = [a,a]$. Then according to Zadeh's extension principle[19], for any $\overline{a} = [a^-, a^+], \overline{b} = [b^-, b^+] \in [I]$, we can popularize some operators such as $\bigvee, \bigwedge$ and c to $[I]$ and have $\overline{a}\bigvee\overline{b}=[a^-\bigvee b^-,a^+\bigvee b^+]$, $\overline{a}\bigwedge\overline{b}=[a^-\bigwedge b^-,a^+\bigwedge b^+]$, $\overline{a}^c=[1-a^+,1-a^-]$, $\bigvee_{t\in W}\overline{a_t}=[\bigvee_{t\in W}a_t^-,\bigvee_{t\in W}a_t^+]$, $\bigwedge_{t\in W}\overline{a_t}=[\bigwedge_{t\in W}a_t^-,\bigwedge_{t\in W}a_t^+]$. Furthermore, we have, $\overline{a}=\overline{b}\Longleftrightarrow a^-=b^-,a^+=b^+$, $\overline{a}\leqslant\overline{b}\Longleftrightarrow a^-\leqslant b^-,a^+\leqslant b^+$ and $\overline{a}<\overline{b}\Longleftrightarrow\overline{a}\leqslant\overline{b}$ and $\overline{a}\neq\overline{b}$, then there exist a minimal element $0=[0,0]$ and a maximal element $1=[1,1]$ in $[I]$.

We call a mapping, $A:X\longrightarrow[I]$ an interval-valued fuzzy set in X. For every $A\in$IVFSs(X) and $x\in X$, then $A(x)=[A^-(x),A^+(x)]$ is the degree of membership of an element x to interval-valued fuzzy set A, thus, fuzzy sets $A^-:X\to[0,1]$ and $A^+:X\to[0,1]$ are called low and upper fuzzy set of interval-valued fuzzy set A, respectively. For simplicity, we denote $A=[A^-,A^+]$, $\mathcal{P}(X)$ stands for the set of all crisp sets in X.

For $A\in$IVFSs(X), $\lambda=[\lambda_1,\lambda_2]\in[I]$, Zeng[21] introduced the concept of cut set of interval valued fuzzy set as follows.

$$\begin{aligned}
A_\lambda^{(1,1)}&=A_{[\lambda_1,\lambda_2]}^{(1,1)}=\{x\in X|A^-(x)\geqslant\lambda_1,A^+(x)\geqslant\lambda_2\}\\
A_\lambda^{(1,2)}&=A_{[\lambda_1,\lambda_2]}^{(1,2)}=\{x\in X|A^-(x)\geqslant\lambda_1,A^+(x)>\lambda_2\}\\
A_\lambda^{(2,1)}&=A_{[\lambda_1,\lambda_2]}^{(2,1)}=\{x\in X|A^-(x)>\lambda_1,A^+(x)\geqslant\lambda_2\}\\
A_\lambda^{(2,2)}&=A_{[\lambda_1,\lambda_2]}^{(2,2)}=\{x\in X|A^-(x)>\lambda_1,A^+(x)>\lambda_2\}\\
A_\lambda^{(3,3)}&=A_{[\lambda_1,\lambda_2]}^{(3,3)}=\{x\in X|A^-(x)\leqslant\lambda_1\quad\text{or}\quad A^+(x)\leqslant\lambda_2\}\\
A_\lambda^{(3,4)}&=A_{[\lambda_1,\lambda_2]}^{(3,4)}=\{x\in X|A^-(x)\leqslant\lambda_1\quad\text{or}\quad A^+(x)<\lambda_2\}\\
A_\lambda^{(4,3)}&=A_{[\lambda_1,\lambda_2]}^{(4,3)}=\{x\in X|A^-(x)<\lambda_1\quad\text{or}\quad A^+(x)\leqslant\lambda_2\}\\
A_\lambda^{(4,4)}&=A_{[\lambda_1,\lambda_2]}^{(4,4)}=\{x\in X|A^-(x)<\lambda_1\quad\text{or}\quad A^+(x)<\lambda_2\}
\end{aligned}$$

where $A_{[\lambda_1,\lambda_2]}^{(i,j)}$ is called the $(i,j)th$ (λ_1,λ_2)-(double value) cut set of interval-valued fuzzy set A. Specially, if $\lambda=\lambda_1=\lambda_2$, $A_\lambda^{(i,j)}=A_{[\lambda,\lambda]}^{(i,j)}$ is called the $(i,j)th$ λ-(single value) cut set of interval-valued fuzzy set A.

Obviously, we have some properties in the following.

Property 1

$$A_{[\lambda_1,\lambda_2]}^{(2,2)}\subseteq A_{[\lambda_1,\lambda_2]}^{(1,2)}\subseteq A_{[\lambda_1,\lambda_2]}^{(1,1)},\quad A_{[\lambda_1,\lambda_2]}^{(2,2)}\subseteq A_{[\lambda_1,\lambda_2]}^{(2,1)}\subseteq A_{[\lambda_1,\lambda_2]}^{(1,1)}$$

$$A_{[\lambda_1,\lambda_2]}^{(4,4)}\subseteq A_{[\lambda_1,\lambda_2]}^{(3,4)}\subseteq A_{[\lambda_1,\lambda_2]}^{(3,3)},\quad A_{[\lambda_1,\lambda_2]}^{(4,4)}\subseteq A_{[\lambda_1,\lambda_2]}^{(4,3)}\subseteq A_{[\lambda_1,\lambda_2]}^{(3,3)}$$

Property 2. For $\lambda_1=[\lambda_1^1,\lambda_1^2],\lambda_2=[\lambda_2^1,\lambda_2^2]\in[I]$ and $\lambda_1^1<\lambda_2^1,\lambda_1^2<\lambda_2^2$, then $A_{\lambda_1}^{(2,2)}\supseteq A_{\lambda_2}^{(1,1)},A_{\lambda_1}^{(3,3)}\subseteq A_{\lambda_2}^{(4,4)}$.

Property 3. For $\lambda=[\lambda_1,\lambda_2]$, then

$$\left(A_\lambda^{(1,1)}\right)^c=A_\lambda^{(4,4)},\quad\left(A_\lambda^{(1,2)}\right)^c=A_\lambda^{(4,3)}$$
$$\left(A_\lambda^{(2,1)}\right)^c=A_\lambda^{(3,4)},\quad\left(A_\lambda^{(2,2)}\right)^c=A_\lambda^{(3,3)}$$

Definition 1[22]. Given a mapping

$$H : [I] \to \mathcal{P}(X) \quad \lambda = [\lambda_1, \lambda_2] \in [I] \to H(\lambda) \in \mathcal{P}(X)$$

if mapping H satisfies the following properties:

$$\begin{array}{l}(1)\ \forall \lambda_1 = [\lambda_1^1, \lambda_1^2], \lambda_2 = [\lambda_2^1, \lambda_2^2] \in [I], \lambda_1^i \leqslant \lambda_2^i, i = 1, 2 \Rightarrow H(\lambda_1) \supseteq H(\lambda_2) \\ (2)\ \bigcap_{t \in W} H(\lambda_t) \subseteq \bigcap \{H(\lambda) | \lambda < \bigvee_{t \in W} \lambda_t\}\end{array}$$

then H is called 1-1 type of interval-valued nested set on X.

Definition 2 [22]. Given a mapping

$$J : [I] \to \mathcal{P}(X) \quad \lambda = [\lambda_1, \lambda_2] \in [I] \to J(\lambda) \in \mathcal{P}(X)$$

if mapping J satisfies the following properties:

$$\begin{array}{l}(1)\ \forall \lambda_1 = [\lambda_1^1, \lambda_1^2], \lambda_2 = [\lambda_2^1, \lambda_2^2] \in [I], \lambda_1^i \leqslant \lambda_2^i, i = 1, 2 \Rightarrow J(\lambda_1) \subseteq J(\lambda_2) \\ (2)\ \bigcup_{t \in W} J(\lambda_t) \supseteq \bigcup \{J(\lambda) | \lambda < \bigvee_{t \in W} \lambda_t\}\end{array}$$

then J is called 1-2 type of interval-valued nested set on X.

Obviously, mappings H and J are generalizations of cut set of interval-valued fuzzy set. Then, we can do some operations such as $\bigcup, \bigcap$ and c on 1-1 and 1-2 types of interval-valued nested sets, respectively. For example,

$$\begin{array}{ll} H_1 \bigcup H_2 : & \Big(H_1 \bigcup H_2\Big)(\lambda) = H_1(\lambda) \bigcup H_2(\lambda) \\ H_1 \bigcap H_2 : & \Big(H_1 \bigcap H_2\Big)(\lambda) = H_1(\lambda) \bigcap H_2(\lambda) \\ \bigcup_{\gamma \in \Gamma} H_\gamma : & \Big(\bigcup_{\gamma \in \Gamma} H_\gamma\Big)(\lambda) = \bigcup_{\gamma \in \Gamma} H_\gamma(\lambda) \\ \bigcap_{\gamma \in \Gamma} H_\gamma : & \Big(\bigcap_{\gamma \in \Gamma} H_\gamma\Big)(\lambda) = \bigcap_{\gamma \in \Gamma} H_\gamma(\lambda) \\ H^c : & H^c(\lambda) = \Big(H(\lambda^c)\Big)^c \end{array}$$

In fact, interval-valued nested set J can also be defined according to above mentioned method.

Definition 3 [22]. Let $[\lambda_1, \lambda_2] \in [I]$ and $A \in \text{IVFSs}(X)$, we order $[\lambda_1, \lambda_2] \cdot A$, $[\lambda_1, \lambda_2] * A \in \text{IVFSs}(X)$, and their membership functions are defined as follows.

$$\begin{array}{l}([\lambda_1, \lambda_2] \cdot A)(x) = [\lambda_1 \wedge A^-(x), \lambda_2 \wedge A^+(x)] \\ ([\lambda_1, \lambda_2] * A)(x) = [\lambda_1 \vee A^-(x), \lambda_2 \vee A^+(x)]\end{array}$$

If $A, B \in$IVFSs(X), then the following operations can be found in Zeng[21].
$A \subseteq B$ iff $\forall x \in X, A^-(x) \leq B^-(x)$ and $A^+(x) \leq B^+(x)$,
$A = B$ iff $\forall x \in X, A^-(x) = B^-(x)$ and $A^+(x) = B^+(x)$,
$(A)^c = [(A^+)^c, (A^-)^c]$.

3 Maximal Extension Principle of Interval-Valued Fuzzy Set

In this section, based on 1-1 type of interval-valued nested set on X and representation theorem of interval-valued fuzzy set, we propose three maximal extension principles of interval-valued fuzzy set.

Definition 4 (Maximal extension principle). Given a mapping f: $X \to Y$, $x \to y = f(x)$, then f can induce two corresponding functions T_f and T_f^{-1} defined as follows.

$$T_f: \ \text{IVFSs}(X) \to \text{IVFSs}F(Y)$$
$$A \to T_f(A)(y) = \begin{cases} \bigvee\limits_{y=f(x)} A(x), & \{x|y=f(x), x \in X\} \neq \varnothing \\ 0, & \{x|y=f(x), x \in X\} = \varnothing \end{cases}$$
$$T_f^{-1}: \text{IVFSs}(Y) \to \text{IVFSs}(X)$$
$$B \to T_f^{-1}(B)(x) = B(f(x))$$

In the following, we give its construction procedure of maximal extension principle in terms of 1-1 type of interval-valued nested set.

Supposed $f: X \to Y$, $x \to y = f(x)$, $A \in$ IVFSs(X), for $[\lambda_1, \lambda_2] \in [I]$, then according to Zadeh's extension principle[19], we have $T_f(A^{(1,1)}_{[\lambda_1,\lambda_2]}) \in \mathcal{P}(Y)$, and

$$[\lambda_1, \lambda_2] \leq [\lambda_3, \lambda_4] \Rightarrow A^{(1,1)}_{[\lambda_1,\lambda_2]} \supseteq A^{(1,1)}_{[\lambda_3,\lambda_4]} \Rightarrow T_f(A^{(1,1)}_{[\lambda_1,\lambda_2]}) \supseteq T_f(A^{(1,1)}_{[\lambda_3,\lambda_4]})$$

Hence, $\{T_f(A^{(1,1)}_{[\lambda_1,\lambda_2]}) \mid [\lambda_1, \lambda_2] \in [I]\}$ is 1-1 type of interval-valued nested set on Y, known by representation theorem of interval-valued fuzzy set in Zeng[22], we have $T_f(A) = \bigcup\limits_{[\lambda_1,\lambda_2]\in[I]} [\lambda_1, \lambda_2] \cdot T_f(A^{(1,1)}_{[\lambda_1,\lambda_2]}) \in$IVFSs$(Y)$, and the following properties hold.

(1) $(T_f(A))^{(2,2)}_{[\lambda_1,\lambda_2]} \subseteq T_f(A^{(1,1)}_{[\lambda_1,\lambda_2]}) \subseteq (T_f(A))^{(1,1)}_{[\lambda_1,\lambda_2]}$,

(2) $(T_f(A))^{(1,1)}_{[\lambda_1,\lambda_2]} = \bigcap\limits_{\alpha_1<\lambda_1,\alpha_2<\lambda_2} T_f(A^{(1,1)}_{[\alpha_1,\alpha_2]})$, $\lambda_1 \cdot \lambda_2 \neq 0$;

(3) $(T_f(A))^{(1,1)}_{[\lambda_1,\lambda_2]} = \bigcup\limits_{\alpha_1>\lambda_1,\alpha_2>\lambda_2} T_f(A^{(1,1)}_{[\alpha_1,\alpha_2]})$, $\lambda_1 \neq 1$ and $\lambda_2 \neq 1$.

Similarly, for $B \in$ IVFSs(Y), we can get $T_f^{-1}(B) = \bigcup\limits_{[\lambda_1,\lambda_2]\in[I]} [\lambda_1, \lambda_2] \cdot T_f^{-1}(B^{(1,1)}_{[\lambda_1,\lambda_2]}) \in$IVFSs$(X)$.

Definition 5 (Maximal extension principle 1). Given a mapping f: $X \to Y$, $x \to y = f(x)$, then f can induce two corresponding functions T_f and T_f^{-1} defined as follows.

$$T_f: \ \text{IVFSs}(X) \to \text{IVFSs}(Y)$$
$$A \to T_f(A) = \bigcup_{[\lambda_1,\lambda_2]\in[I]} [\lambda_1,\lambda_2] \cdot T_f(A^{(1,1)}_{[\lambda_1,\lambda_2]})$$

$$T_f^{-1}: \text{IVFSs}(Y) \to \text{IVFSs}(X)$$
$$B \to T_f^{-1}(B) = \bigcup_{[\lambda_1,\lambda_2]\in[I]} [\lambda_1,\lambda_2] \cdot T_f^{-1}(B^{(1,1)}_{[\lambda_1,\lambda_2]})$$

In the following, we give Theorem 1 to show the equivalence between Definition 4 and Definition 5.

Theorem 1(Maximal extension principle 1). Given a mapping $f{:}X \to Y$, $x \to y = f(x)$,

1) If $A \in$IVFSs(X), then we have

$$\Big(\bigcup_{[\lambda_1,\lambda_2]\in[I]} [\lambda_1,\lambda_2] \cdot T_f(A^{(1,1)}_{[\lambda_1,\lambda_2]})\Big)(y) = \begin{cases} \bigvee_{y=f(x)} A(x), & \{x|y=f(x), x\in X\} \neq \varnothing \\ 0, & \{x|y=f(x), x\in X\} = \varnothing \end{cases}$$

2) If $B \in$IVFSs(Y), then we have

$$\Big(\bigcup_{[\lambda_1,\lambda_2]\in[I]} [\lambda_1,\lambda_2] \cdot T_f^{-1}(B^{(1,1)}_{[\lambda_1,\lambda_2]})\Big)(x) = B(f(x))$$

Proof. 1) If $\{x|y = f(x), x \in X\} = \varnothing$, it is obvious. Now we consider that $\{x|y = f(x), x \in X\}$ is not empty set, then we have

$$\Big(\bigcup_{[\lambda_1,\lambda_2]\in[I]} [\lambda_1,\lambda_2] \cdot T_f(A^{(1,1)}_{[\lambda_1,\lambda_2]})\Big)(y) = \bigvee_{[\lambda_1,\lambda_2]\in[I]} \Big([\lambda_1,\lambda_2] \bigwedge T_f(A^{(1,1)}_{[\lambda_1,\lambda_2]})(y)\Big)$$
$$= \bigvee_{[\lambda_1,\lambda_2]\in[I]} \Big([\lambda_1,\lambda_2] \bigwedge (\bigvee_{y=f(x)} A^{(1,1)}_{[\lambda_1,\lambda_2]}(x))\Big) = \bigvee_{y=f(x)} \Big(\bigvee_{[\lambda_1,\lambda_2]\in[I]} ([\lambda_1,\lambda_2] \bigwedge A^{(1,1)}_{[\lambda_1,\lambda_2]}(x))\Big)$$
$$= \bigvee_{y=f(x)} A(x)$$

2)

$$\Big(\bigcup_{[\lambda_1,\lambda_2]\in[I]} [\lambda_1,\lambda_2] \cdot T_f^{-1}(B^{(1,1)}_{[\lambda_1,\lambda_2]})\Big)(x) = \bigvee_{[\lambda_1,\lambda_2]\in[I]} \Big([\lambda_1,\lambda_2] \bigwedge T_f^{-1}(B^{(1,1)}_{[\lambda_1,\lambda_2]})(x)\Big)$$
$$= \bigvee_{[\lambda_1,\lambda_2]\in[I]} \Big([\lambda_1,\lambda_2] \bigwedge B^{(1,1)}_{[\lambda_1,\lambda_2]}(f(x))\Big)$$
$$= B(f(x))$$

Hence, we complete the proof of Theorem 1.

Known by decomposition theorems of interval-valued fuzzy set in Zeng[21], similarly we can give another two maximal extension principles of interval-valued fuzzy set in the following.

Theorem 2(Maximal extension principle 2). Given a mapping f: $X \to Y$, $x \to y = f(x)$,

1) If $A \in$IVFSs(X), then we have $T_f(A) = \bigcup_{[\lambda_1,\lambda_2]\in[I]} [\lambda_1,\lambda_2] \cdot T_f(A^{(1,2)}_{[\lambda_1,\lambda_2]})$;

2) If $B \in$IVFSs(Y), then we have $T_f^{-1}(B) = \bigcup_{[\lambda_1,\lambda_2]\in[I]} [\lambda_1,\lambda_2] \cdot T_f^{-1}(B^{(1,2)}_{[\lambda_1,\lambda_2]})$.

Theorem 3(Maximal extension principle 3). Given a mapping f: $X \to Y$, $x \to y = f(x)$,

1) If $A \in$IVFSs(X), then we have $T_f(A) = \bigcup_{[\lambda_1,\lambda_2]\in[I]} [\lambda_1,\lambda_2] \cdot T_f(H_A(\lambda_1,\lambda_2))$,

where $H_A[\lambda_1,\lambda_2]$ satisfies the property: $A^{(1,2)}_{[\lambda_1,\lambda_2]} \subseteq H_A[\lambda_1,\lambda_2] \subseteq A^{(1,1)}_{[\lambda_1,\lambda_2]}$;

2) If $B \in$IVFSs(Y), then we have $T_f^{-1}(B) = \bigcup_{[\lambda_1,\lambda_2]\in[I]} [\lambda_1,\lambda_2] \cdot T_f^{-1}(H_B(\lambda_1,\lambda_2))$,

where $H_B[\lambda_1,\lambda_2]$ satisfies the property: $B^{(1,2)}_{[\lambda_1,\lambda_2]} \subseteq H_B[\lambda_1,\lambda_2] \subseteq B^{(1,1)}_{[\lambda_1,\lambda_2]}$.

4 Minimal Extension Principle of Interval-Valued Fuzzy Set

Maximal extension principle implies a kind of optimistic estimation to the problems in decision-making. Based on this view of point, we consider another kind of pessimistic estimation to the problems in decision-making and propose minimal extension principle of interval-valued fuzzy set in this section.

Definition 6(Minimal extension principle)**.** Given a mapping f: $X \to Y$, $x \to y = f(x)$, then f can induce two corresponding functions S_f and S_f^{-1} defined as follows.

$$
\begin{aligned}
S_f: \ & \text{IVFSs}(X) \to \text{IVFSs}(Y) \\
& A \to S_f(A)(y) = \begin{cases} \bigwedge_{y=f(x)} A(x), & \{x|y=f(x), x \in X\} \neq \varnothing \\ 0, & \{x|y=f(x), x \in X\} = \varnothing \end{cases} \\
S_f^{-1}: \ & \text{IVFSs}(Y) \to \text{IVFSs}(X) \\
& B \to S_f^{-1}(B)(x) = B(f(x))
\end{aligned}
$$

In the following, we give its construction procedure of minimal extension principle in terms of 1-2 type of interval-valued nested set.

Supposed $f: X \to Y$, $x \to y = f(x)$, $A \in$ IVFSs(X), for $[\lambda_1,\lambda_2] \in [I]$, then according to Zadeh's extension principle[19], we have $S_f(A^{(4,4)}_{[\lambda_1,\lambda_2]}) \in \mathcal{P}(Y)$, and

$$[\lambda_1,\lambda_2] \leq [\lambda_3,\lambda_4] \Rightarrow A^{(4,4)}_{[\lambda_1,\lambda_2]} \subseteq A^{(4,4)}_{[\lambda_3,\lambda_4]} \Rightarrow S_f(A^{(4,4)}_{[\lambda_1,\lambda_2]}) \subseteq S_f(A^{(4,4)}_{[\lambda_3,\lambda_4]})$$

Hence, $\{S_f(A^{(4,4)}_{[\lambda_1,\lambda_2]}) \mid [\lambda_1,\lambda_2] \in [I]\}$ is 1-2 type of interval-valued nested set on Y, known by the representation theorem of interval-valued fuzzy set in

Zeng[22], we have $S_f(A) = \Big(\bigcap_{[\lambda_1,\lambda_2]\in[I]} [\lambda_1,\lambda_2] * S_f(A^{(4,4)}_{[\lambda_1,\lambda_2]^c})\Big)^c \in$IVFSs($Y$), and the following properties hold.

1) $(S_f(A))^{(4,4)}_{[\lambda_1,\lambda_2]} \subseteq S_f(A^{(4,4)}_{[\lambda_1,\lambda_2]}) \subseteq (S_f(A))^{(3,3)}_{[\lambda_1,\lambda_2]}$;
2) $(S_f(A))^{(4,4)}_{[\lambda_1,\lambda_2]} = \bigcap_{\alpha_1<\lambda_1,\alpha_2<\lambda_2} S_f(A^{(4,4)}_{[\alpha_1,\alpha_2]}),\ \lambda_1\cdot\lambda_2\neq 0$;
3) $(S_f(A))^{(3,3)}_{[\lambda_1,\lambda_2]} = \bigcup_{\alpha_1>\lambda_1,\alpha_2>\lambda_2} S_f(A^{(4,4)}_{[\alpha_1,\alpha_2]}),\ \lambda_1\neq 1 \text{ and } \lambda_2\neq 1$.

Similarly, for $B\in$IVFSs(Y), we can get $S_f^{-1}(B) = \Big(\bigcap_{[\lambda_1,\lambda_2]\in[I]} [\lambda_1,\lambda_2] * S_f^{-1}(B^{(4,4)}_{[\lambda_1,\lambda_2]^c})\Big)^c \in$IVFSs($X$).

Definition 7 (Minimal extension principle 1)**.** Given a mapping f: $X\to Y$, $x\to y=f(x)$, then f can induce two corresponding functions S_f and S_f^{-1} defined as follows.

$$\begin{aligned} S_f:\ &\text{IVFSs}(X)\to\text{IVFSs}(Y)\\ &A\to S_f(A)=\Big(\bigcap_{[\lambda_1,\lambda_2]\in[I]}[\lambda_1,\lambda_2]*S_f(A^{(4,4)}_{[\lambda_1,\lambda_2]^c})\Big)^c\in\text{IVFSs}(Y)\\ S_f^{-1}:\ &\text{IVFSs}(Y)\to\text{IVFSs}(X)\\ &B\to S_f^{-1}(B)=\Big(\bigcap_{[\lambda_1,\lambda_2]\in[I]}[\lambda_1,\lambda_2]*S_f^{-1}(B^{(4,4)}_{[\lambda_1,\lambda_2]^c})\Big)^c\in\text{IVFSs}(X)\end{aligned}$$

In the following, we give Theorem 4 to show the equivalence between Definition 6 and Definition 7.

Theorem 4 (Minimal extension principle 1). Given a mapping f: $X\to Y$, $x\to y=f(x)$,

1) If $A\in$ IVFSs(X), then we have

$$\Big(\bigcap_{[\lambda_1,\lambda_2]\in[I]}[\lambda_1,\lambda_2]*S_f(A^{(4,4)}_{[\lambda_1,\lambda_2]^c})\Big)^c(y)=\begin{cases}\bigwedge_{y=f(x)}A(x), & \{x|y=f(x),x\in X\}\neq\varnothing\\ 0, & \{x|y=f(x),x\in X\}=\varnothing\end{cases}$$

2) If $B\in$IVFSs(Y), then we have

$$\Big(\bigcap_{[\lambda_1,\lambda_2]\in[I]}[\lambda_1,\lambda_2]*S_f^{-1}(B^{(4,4)}_{[\lambda_1,\lambda_2]^c})\Big)^c(x)=B(f(x))$$

Proof. 1) If $\{x|y=f(x),x\in X\}=\varnothing$, it is obvious. Now we consider that $\{x|y=f(x),x\in X\}$ is not empty set, then we firstly want to prove that the following expression holds.

$$\bigwedge_{[\lambda_1,\lambda_2]\in[I]}\Big([\lambda_1,\lambda_2]\bigvee(\bigvee_{y=f(x)}A^{(4,4)}_{[\lambda_1,\lambda_2]^c}(x))\Big)=\bigvee_{y=f(x)}\Big(\bigwedge_{[\lambda_1,\lambda_2]\in[I]}([\lambda_1,\lambda_2]\bigvee A^{(4,4)}_{[\lambda_1,\lambda_2]^c}(x))\Big)$$

For every $[\lambda_1, \lambda_2] \in [I]$, we have

$$[\lambda_1, \lambda_2] \bigvee \Big(\bigvee_{y=f(x)} A^{(4,4)}_{[\lambda_1,\lambda_2]^c}(x)\Big) \geq [\lambda_1, \lambda_2] \bigvee A^{(4,4)}_{[\lambda_1,\lambda_2]^c}(x)$$

then we can get

$$\bigwedge_{[\lambda_1,\lambda_2]\in[I]} \Big([\lambda_1, \lambda_2] \bigvee \big(\bigvee_{y=f(x)} A^{(4,4)}_{[\lambda_1,\lambda_2]^c}(x)\big)\Big) \geq \bigwedge_{[\lambda_1,\lambda_2]\in[I]} \big([\lambda_1, \lambda_2] \bigvee A^{(4,4)}_{[\lambda_1,\lambda_2]^c}(x)\big)$$

Therefore, we have

$$\bigwedge_{[\lambda_1,\lambda_2]\in[I]} \Big([\lambda_1, \lambda_2] \bigvee \big(\bigvee_{y=f(x)} A^{(4,4)}_{[\lambda_1,\lambda_2]^c}(x)\big)\Big) \geq \bigvee_{y=f(x)} \Big(\bigwedge_{[\lambda_1,\lambda_2]\in[I]} \big([\lambda_1, \lambda_2] \bigvee A^{(4,4)}_{[\lambda_1,\lambda_2]^c}(x)\big)\Big)$$

Using proof by contradiction, if the equality in the above expression does not hold, then there exists $[\alpha_1, \alpha_2] \in [I]$ such that

$$\bigwedge_{[\lambda_1,\lambda_2]\in[I]} \Big([\lambda_1, \lambda_2] \bigvee \big(\bigvee_{y=f(x)} A^{(4,4)}_{[\lambda_1,\lambda_2]^c}(x)\big)\Big) > [\alpha_1, \alpha_2] > \bigvee_{y=f(x)} \Big(\bigwedge_{[\lambda_1,\lambda_2]\in[I]} \big([\lambda_1, \lambda_2] \bigvee A^{(4,4)}_{[\lambda_1,\lambda_2]^c}(x)\big)\Big)$$

Considering that $\bigvee_{y=f(x)} A^{(4,4)}_{[\lambda_1,\lambda_2]^c}(x) = 0$ or 1, we think about two kinds of cases, respectively.

(i) If $\bigvee_{y=f(x)} A^{(4,4)}_{[\lambda_1,\lambda_2]^c}(x) = 0$, then we have

$$\begin{aligned}
[\alpha_1, \alpha_2] &< \bigwedge_{[\lambda_1,\lambda_2]\in[I]} \Big([\lambda_1, \lambda_2] \bigvee \big(\bigvee_{y=f(x)} A^{(4,4)}_{[\lambda_1,\lambda_2]^c}(x)\big)\Big) \\
&\leq [\alpha_1, \alpha_2] \bigvee \big(\bigvee_{y=f(x)} A^{(4,4)}_{[\alpha_1,\alpha_2]^c}(x)\big) \\
&= [\alpha_1, \alpha_2]
\end{aligned}$$

Obviously, it is a contradiction.

(ii) If $\bigvee_{y=f(x)} A^{(4,4)}_{[\lambda_1,\lambda_2]^c}(x) = 1$, then there exists $x_k \in X$ such that $y = f(x_k)$ and $A^{(4,4)}_{[\lambda_1,\lambda_2]^c}(x_k) = 1$, thus, we have

$$\begin{aligned}
[\alpha_1, \alpha_2] &> \bigvee_{y=f(x)} \Big(\bigwedge_{[\lambda_1,\lambda_2]\in[I]} \big([\lambda_1, \lambda_2] \bigvee A^{(4,4)}_{[\lambda_1,\lambda_2]^c}(x)\big)\Big) \\
&\geq \bigwedge_{[\lambda_1,\lambda_2]\in[I]} \big([\lambda_1, \lambda_2] \bigvee A^{(4,4)}_{[\lambda_1,\lambda_2]^c}(x_k)\big) \\
&= 1
\end{aligned}$$

Obviously, it also is a contradiction.

Hence, we have

$$\bigwedge_{[\lambda_1,\lambda_2]\in[I]}\Big([\lambda_1,\lambda_2]\bigvee\big(\bigvee_{y=f(x)} A^{(4,4)}_{[\lambda_1,\lambda_2]^c}(x)\big)\Big)=\bigvee_{y=f(x)}\Big(\bigwedge_{[\lambda_1,\lambda_2]\in[I]}\big([\lambda_1,\lambda_2]\bigvee A^{(4,4)}_{[\lambda_1,\lambda_2]^c}(x)\big)\Big)$$

Further, we have

$$\begin{aligned}
&\Big(\bigcap_{[\lambda_1,\lambda_2]\in[I]}[\lambda_1,\lambda_2]*S_f(A^{(4,4)}_{[\lambda_1,\lambda_2]^c})\Big)^c(y)\\
&=\left(\bigwedge\nolimits_{[\lambda_1,\lambda_2]\in[I]}\Big([\lambda_1,\lambda_2]\bigvee S_f(A^{(4,4)}_{[\lambda_1,\lambda_2]^c})(y)\Big)\right)^c\\
&=\left(\bigvee_{y=f(x)}\Big(\bigwedge_{[\lambda_1,\lambda_2]\in[I]}\big([\lambda_1,\lambda_2]\bigvee A^{(4,4)}_{[\lambda_1,\lambda_2]^c}(x)\big)\Big)\right)^c\\
&=\bigwedge\nolimits_{y=f(x)}\Big(\bigwedge\nolimits_{[\lambda_1,\lambda_2]\in[I]}\big([\lambda_1,\lambda_2]\bigvee A^{(4,4)}_{[\lambda_1,\lambda_2]^c}(x)\big)\Big)^c\\
&=\bigwedge_{y=f(x)}\Big(\bigcap_{[\lambda_1,\lambda_2]\in[I]}\big([\lambda_1,\lambda_2]*A^{(4,4)}_{[\lambda_1,\lambda_2]^c}\big)\Big)^c(x)=\bigwedge_{y=f(x)}A(x)
\end{aligned}$$

2)

$$\begin{aligned}
&\Big(\bigcap_{[\lambda_1,\lambda_2]\in[I]}[\lambda_1,\lambda_2]*S_f^{-1}(B^{(4,4)}_{[\lambda_1,\lambda_2]^c})\Big)^c(x)=\Big(\bigwedge_{[\lambda_1,\lambda_2]\in[I]}\big([\lambda_1,\lambda_2]\bigvee S_f^{-1}(B^{(4,4)}_{[\lambda_1,\lambda_2]^c})(x)\big)\Big)^c\\
&=\Big(\bigwedge_{[\lambda_1,\lambda_2]\in[I]}\big([\lambda_1,\lambda_2]\bigvee B^{(4,4)}_{[\lambda_1,\lambda_2]^c}(f(x))\big)\Big)^c=\Big(\bigcap_{[\lambda_1,\lambda_2]\in[I]}\big([\lambda_1,\lambda_2]*B^{(4,4)}_{[\lambda_1,\lambda_2]^c}\big)\Big)^c(f(x))\\
&=B(f(x))
\end{aligned}$$

Hence, we complete the proof of Theorem 4.

Known by decomposition theorems of interval-valued fuzzy set in Zeng[21], similarly we can give another two minimal extension principles of interval-valued fuzzy set in the following.

Theorem 5(Minimal extension principle 2). Given a mapping f: $X \to Y$, $x \to y = f(x)$,

1) If $A \in$IVFSs(X), then we have $S_f(A) = \Big(\bigcap_{[\lambda_1,\lambda_2]\in[I]}[\lambda_1,\lambda_2]*S_f(A^{(4,3)}_{[\lambda_1,\lambda_2]^c})\Big)^c$;

2) If $B \in$IVFSs(Y), then we have $S_f^{-1}(B)=\Big(\bigcap_{[\lambda_1,\lambda_2]\in[I]}[\lambda_1,\lambda_2]*S_f^{-1}(B^{(4,3)}_{[\lambda_1,\lambda_2]^c})\Big)^c$.

Theorem 6(Minimal extension principle 3). Give a mapping f: $X \to Y$, $x \to y = f(x)$,

1) If $A \in$IVFSs(X), then we have $S_f(A) =\Big(\bigcap_{[\lambda_1,\lambda_2]\in[I]}[\lambda_1,\lambda_2]*S_f(H_A[\lambda_1,\lambda_2]^c)\Big)^c$,

where $H_A[\lambda_1,\lambda_2]$ satisfies the property: $A^{(4,4)}_{[\lambda_1,\lambda_2]} \subseteq H_A[\lambda_1,\lambda_2] \subseteq A^{(4,3)}_{[\lambda_1,\lambda_2]}$;

2) If $B \in$IVFSs(Y), then we have $S_f^{-1}(B) = \Big(\bigcap_{[\lambda_1,\lambda_2]\in[I]} [\lambda_1,\lambda_2] * S_f^{-1}(H_B [\lambda_1,\lambda_2]^c) \Big)^c$, where $H_B[\lambda_1,\lambda_2]$ satisfies the property: $B^{(4,4)}_{[\lambda_1,\lambda_2]} \subseteq H_B[\lambda_1,\lambda_2] \subseteq B^{(4,3)}_{[\lambda_1,\lambda_2]}$.

5 Generalized Extension Principle of Interval-Valued Fuzzy Set

Definition 4 and Definition 6 show that the main work of extension principle of interval-valued fuzzy set is how to define their membership functions of $T_f(A), T_f^{-1}(B), S_f(A)$ and $S_f^{-1}(B)$. Obviously, there exists a strong connection between $A(x)$ and $T_f(A)(y)$ or $A(x)$ and $S_f(A)(y)$. Based on this view of point, we introduce an axiomatic definition of generalized extension principle of interval-valued fuzzy set in the following.

Given two mappings $f : X \to Y, x \to y = f(x)$ and $p : \mathcal{P}(X \times [I]) \to [I]$, we construct mappings f_p and f_p^{-1} from IVFSs(X) to IVFSs(Y) and from IVFSs(Y) to IVFSs(X), respectively.

$$\begin{aligned} f_p : \text{IVFSs}(X) &\to \text{IVFSs}(Y) \\ A &\to f_p(A) \\ f_p^{-1} : \text{IVFSs}(Y) &\to \text{IVFSs}(X) \\ B &\to f_p^{-1}(B) \end{aligned}$$

Their membership functions are defined as follows, respectively.

$$f(A)(y) = \begin{cases} p\Big(\{(x, A(x))|x \in X, y = f(x)\}\Big), & \{(x, A(x))|x \in X, y=f(x)\} \neq \varnothing \\ 0, & \{(x, A(x))|x \in X, y=f(x)\} = \varnothing \end{cases}$$

$$f_p^{-1}(B)(x) = B(f(x))$$

Definition 8 (Generalized extension principle). If mappings f_p : IVFSs$(X) \to$ IVFSs(Y) and f_p^{-1} : IVFSs$(Y) \to$ IVFSs(X) satisfy properties in the following.

(P1) If $A \in \mathcal{P}(X)$, then $f_p(A) \in \mathcal{P}(Y)$. Namely, $y \in f_p(A) \Longleftrightarrow f_p(A)(y) = 1 \Longleftrightarrow$ there exists $x \in X, y = f(x)$ and $A(x) = 1$;

(P2) For every $y \in Y$ and $A \in$ IVFSs(X), we have $\bigwedge_{y=f(x)} A(x) \leq f_p(A)(y) \leq \bigvee_{y=f(x)} A(x)$;

(P3) For $A, B \in$IVFSs(X) and $A \subseteq B$, then we have $f_p(A) \subseteq f_p(B)$;

(P4) For every $B \in$IVFSs(Y), we have $f_p^{-1}(B)(x) = B(f(x))$.

Then f_p and f_p^{-1} are called reasonable extension mappings of interval-valued fuzzy set, and p is called reasonable extension operator.

Thus, we can construct many kinds of reasonable extension mappings of interval-valued fuzzy set based on the operations such as union and intersection of interval-valued fuzzy set and some ordinary arithmetical or geometrical mean operator. Therefore, it will become more effective to use interval-valued fuzzy sets to solve practical problems in real life. Of course, maximal and minimal extension principles of interval-valued fuzzy set are two special cases of reasonable extension mapping of interval-valued fuzzy set.

For example, let $X = \{x_1, x_2, \cdots, x_n\}$ and $D_y = \Big\{(x, A(X))|y = f(x), x \in X\Big\} = \Big\{(x_1, [A^-(x_1), A^+(x_1)]), (x_2, [A^-(x_2), A^+(x_2)]), \cdots, (x_n, [A^-(x_n), A^+(x_n)])\Big\}$.

For every $y \in Y$ and $p : \mathcal{P}(X \times [I]) \to [I]$, $p(D_y) = \frac{1}{n}\sum_{i=1}^{n} A(x_i) = [\frac{1}{n}\sum_{i=1}^{n} A^-(x_i), \frac{1}{n}\sum_{i=1}^{n} A^+(x_i)]$, then we can get a mapping f_p compounded by f and p.

$$\begin{aligned} f_p : \mathrm{IVFSs}(X) &\to \mathrm{IVFSs}(Y) \\ A &\to f_p(A) \end{aligned}$$

and membership function of interval-valued fuzzy set $f_p(A)$, i.e. $f_p(A)(y) = p(D_y) = \frac{1}{n}\sum_{i=1}^{n} A(x_i) = [\frac{1}{n}\sum_{i=1}^{n} A^-(x_i), \frac{1}{n}\sum_{i=1}^{n} A^+(x_i)]$.

Obviously, f_p satisfies the above axiomatic definition, therefore, f_p is reasonable extension mapping of interval valued fuzzy set.

6 Conclusion

In this paper, we introduce maximal and minimal extension principles of interval-valued fuzzy set and an axiomatic definition of generalized extension principle of interval-valued fuzzy set and use concepts of cut set of interval-valued fuzzy set and interval-valued nested sets to explain their construction procedure in detail. These conclusions can be applied in some fields such as fuzzy algebra, fuzzy analysis and so on.

Acknowledgements

This work described here is partially Supported by the grants from the National Natural Science Foundation of China (NSFC NO.60474023), Science and Technology Key Project Fund of Ministry of Education(NO.03184), and the Major State Basic Research Development Program of China (NO.2002CB312200).

References

[1] Arnould, T., Tano, S.(1995)Interval-valued fuzzy backward reasoning. IEEE Trans, Fuzzy Syst.3:425-437

[2] Atanassov, K.(1986)Intuitionistic fuzzy sets. Fuzzy Sets and Systems 20:87-96

[3] Atanassov, K.(1999)Intuitionistic Fuzzy Sets: Theory and Applications. Physica-Verlag, Heidelberg, New York

[4] Biswas, R.(1994)Rosenfeld's fuzzy subgroups with interval-valued membership functions. Fuzzy Sets and Systems 63:87-90

[5] Bustince, H.(2000)Indicator of inclusion grade for interval-valued fuzzy sets: Application to approximate reasoning based on interval-valued fuzzy sets. Internat. J. Approxi. Reasoning 23:137-209

[6] Bustince, H., Burillo, P.(2000)Mathematical analysis of interval-valued fuzzy relations: Application to approximate reasoning. Fuzzy Sets and Systems, 113:205-219

[7] Chen, S.M., Hsiao, W.H.(2000)Bidirectional approximate reasoning for rule-based systems using interval-valued fuzzy sets. Fuzzy Sets and Systems 113:185-203

[8] Chen, S.M., Hsiao, W.H., Jong, W.T.(1997)Bidirectional approximate reasoning based on interval-valued fuzzy sets, Fuzzy Sets and Systems 91:339-353

[9] Cornelis, C., Deschrijver, G., Kerre, E.E.(2004)Implication in intuitionistic fuzzy sets and interval-valued fuzzy set theory: construction, classification, application. Internat. J. Approxi. Reasoning 35: 35-95

[10] Deschrijver, G., Kerre, E.E.(2003)On the relationship between some extensions of fuzzy set theory, Fuzzy Sets and Systems 133:227-235

[11] Goguen, J.A.(1967)$L-$fuzzy sets. J. Math. Anal. Appl.18:145-174

[12] Gorzalczany, M.B.(1987)A method of inference in approximate reasoning based on interval-valued fuzzy sets. Fuzzy Sets and Systems 21:1-17

[13] Li, X.P., Wang, G.J.(2000)The S_H-interval-valued fuzzy subgroup. Fuzzy Sets and Systems 112: 319-325

[14] Luo, C.Z.(1989)Intruduction to Fuzzy Subsets. Beijing Normal University Press, Beijing, (in Chinese)

[15] Mondal, T.K., Samanta, S.K.(1999)Topology of interval-valued fuzzy sets. Indian J. Pure Appl. Math.30:23-38

[16] Turksen, I.B.(1986)Interval-valued fuzzy sets based on normal forms. Fuzzy Sets and Systems 20:191-210

[17] Turksen, I.B.(1992)Interval-valued fuzzy sets and compensatory AND. Fuzzy Sets and Systems 51:295-307

[18] Yuan, B., Pan, Y., Wu, W.M.(1995)On normal form based on interval-valued fuzzy sets and their applications to approximate reasoning. Internat. J. General Systems 23:241-254

[19] Zadeh, L.A.(1965)Fuzzy sets. Infor. and Contr. 8:338-353

[20] Zadeh, L.A.(1975)The concept of a linguistic variable and its application to approximate reasoning, Part (I), (II), (III). Infor. Sci.8:199-249; 301-357; 9: 43-80

[21] Zeng, W.Y.(2005)Shi, Y., Note on interval-valued fuzzy set.Lecture Notes in Artificial Intelligence 3316:20-25

[22] Zeng, W.Y., Shi, Y., Li H.X.(2006)Representation theorem of interval-valued fuzzy set.Internat. J. Uncert. Fuzzi. Knowl. Systems 3 in press

[23] Zhai, J.R., He, Q.(1995)Generalized extension principle.Fuzzy System and Mathematics9:16-21(in Chinese)

On Vague Subring and Its Structure

Qingjun Ren, Dexue Zhang, and Zhenming Ma

Department of Mathematics , Linyi Normal University, Linyi, 276005 P.R. China
qjr288@126.com

Abstract. Vague ring and vague ideal based on vague binary operation are defined, and some properties of them are got. At last, we give the relationships between vague ring and classical ring.

Keywords: vague ring, vague binary operation, fuzzy equality, vague ideal.

1 Introduction

In 1965, Zadeh [1] introduced the notion of fuzzy set. Then in 1971, Resonfeld [2] defined a fuzzy subgroup of a given group. Since then, the concept of fuzzy set was applied to many other algebraic systems, such as ring, module, and so on. Mordeson and his colleagues have published a monograph [8], which includes plenty of pretty results. In fuzzy setting, fuzzy relation and fuzzy function are important tools, and play a vital role in fuzzy control. In 1999, M. Demirci [3] modified the fuzzy equality defined in [4], which stood for the equal degree of two elements in classical set. And he gave new fuzzy functions based on fuzzy equality. In 2001, he applied the fuzzy equality and fuzzy function to algebraic system and defined vague group [5]. This kind of fuzzy group is a nice generalization of crisp group because of a close relationship with crisp one. In 2005, he succeeded to apply fuzzy equality to lattice theory and established the lattice of vague lattice [6, 7], and gave a concept of complete vague lattice .

In section 2 we give some facts on vague group. In section 3, the concept of vague ring and its substructures are introduced, and some good properties are got. In section 4, the concept of vague homomorphism is defined and several results as the classical one are got.

2 Preliminaries

In section, the concepts and results come from reference [5].

Definition 1. *For two nonempty sets X and Y, let E_X and E_Y be fuzzy equality on X and Y, respectively. Then a fuzzy strong fuzzy function from X to Y w.r.t. fuzzy equalities E_X and E_Y, donated by $\tilde{\cdot} : X \to Y$, iff the membership function $\tilde{\cdot} : X \times Y \to [0,1]$ of $\tilde{\cdot}$ holds.*

(F.1)$\forall x \in X, \exists y \in Y$ *such that* $\tilde{\cdot}\,(x,y) = 1$.

(F.2)$\forall x_1, x_2 \in X, y_1, y_2 \in Y$, $\tilde{\cdot}(x_1,y_1) \wedge \tilde{\cdot}(x_2,y_2) \wedge E_X(x_1,x_2) \leq E_Y(y_1,y_2)$.

B.-Y. Cao (Ed.): Fuzzy Information and Engineering (ICFIE), ASC 40, pp. 138–143, 2007.
springerlink.com

Definition 2. *(i) A strong fuzzy function* $\tilde{\cdot}: X \times X \to X$ *w.r.t. a fuzzy equality* $E_{X\times X}$ *on* $X \times X$ *and a fuzzy equality* E_X *on* X *is said to be a vague binary operation on* X *w.r.t.* $E_{X\times X}$ *and* E_X.
(ii) A vague binary operation $\tilde{\cdot}$ *on* X *w.r.t.* $E_{X\times X}$ *and* E_X *is said to be transitive of first order iff*
(T.1) $\forall a, b, c, d \in X, \tilde{\cdot}(a, b, c) \wedge E_X(c, d) \leq \tilde{\cdot}(a, b, d)$.
(iii) A vague binary operation $\tilde{\cdot}$ *on* X *w.r.t.* $E_{X\times X}$ *and* E_X *is said to be transitive of second order iff*
(T.2) $\forall a, b, c, d \in X, \tilde{\cdot}(a, b, c) \wedge E_X(b, d) \leq \tilde{\cdot}(a, d, c)$.
(iv) A vague binary operation $\tilde{\cdot}$ *on* X *w.r.t.*$E_{X\times X}$ *and* E_X *is said to be transitive of third order iff*
(T.3) $\forall a, b, c, d \in X, \tilde{\cdot}(a, b, c) \wedge E_X(a, d) \leq \tilde{\cdot}(d, b, c)$.

Definition 3. *Let* $\tilde{\cdot}$ *be a vague binary operation on* X *w.r.t.* $E_{X\times X}$ *and* E_X. *Then* X *together with* $\tilde{\cdot}$, *denoted by* $(X, \tilde{\cdot})$, *is called a vague group iff*
(VG.1) $\forall a, b, c, q, m, w \in X$,

$$\tilde{\cdot}(b, c, d) \wedge \tilde{\cdot}(a, d, m) \wedge \tilde{\cdot}(a, b, q) \wedge \tilde{\cdot}(q, c, w) \leq E_X(m, w).$$

(VG.2) There exists an (two sided) identity element $e \in X$ *such that for each* $a \in X$, $\tilde{\cdot}(a, e, a) \wedge \tilde{\cdot}(e, a, e) = 1$.
(VG3) For each $a \in X$, *there exists an (two sided) inverse element* $a^{-1} \in X$ *such that* $\tilde{\cdot}(a, a^{-1}, e) \wedge \tilde{\cdot}(a^{-1}, a, e) = 1$.

Theorem 1. *(Vague Cancellation Law) Let (X) w.r.t.* $E_{X\times X}$ *and* E_X *on* X *be a vague group. Then* $\forall\ a, b, c, u \in X$,

$$(VCL.1)\ \tilde{\cdot}(a, b, u) \wedge \tilde{\cdot}(a, c, u) \leq E_X(b, c).$$

$$(VCL.2)\ \tilde{\cdot}(b, a, u) \wedge \tilde{\cdot}(c, a, u) \leq E_X(b, c).$$

3 Vague Ring and Its Substructures

Definition 4. *Let* X *be a nonempty set, and* $\tilde{+}$, $\tilde{\cdot}$ *be vague binary operation w.r.t. a fuzzy equality* $E_{X\times X}$ *and a fuzzy equality* E_X *on* X. *Then* $(X, \tilde{+}, \tilde{\cdot})$ *is called a vague ring iff*
(VR.1) $(X, \tilde{+})$ *is a commutative vague group w.r.t. a fuzzy equality* $E_{X\times X}$ *and a fuzzy equality* E_X *on* X.
(VR.2) $(X, \tilde{\cdot})$ *is a vague semi-group w.r.t. a fuzzy equality* $E_{X\times X}$ *and a fuzzy equality* E_X *on* X.
(VR.3) for $\forall\ a, b, m, m_1, n, n_1, l, l_1, k, k_1 \in X$,
(VRDL) $\tilde{+}(a, b, m) \wedge \tilde{\cdot}(m, c, n) \wedge \tilde{\cdot}(a, c, l) \wedge \tilde{\cdot}(b, c, k) \wedge \tilde{+}(l, k, p) \leq E_X(n, p)$,
(VLDL) $\tilde{+}(a, b, m) \wedge \tilde{\cdot}(c, m, n_1) \wedge \tilde{\cdot}(c, a, l_1) \wedge \tilde{\cdot}(c, b, k_1) \wedge \tilde{+}(l_1, k_1, p_1) \leq E_X(n_1, p_1)$
If in addition
(VR.4) If X *contains an* e *such that for all* $a \in X, \tilde{\cdot}(e, a, a) \wedge \tilde{\cdot}(a, e, a) = 1$, *we call* $(X, \tilde{+}, \tilde{\cdot})$ *is a vague ring with identity* e *w.r.t. a fuzzy equality* $E_{X\times X}$ *and a fuzzy equality* E_X *on* X.

(VR.5) If $\tilde{\cdot}(a,b,m) \wedge \tilde{\cdot}(b,a,n) \leq E_X(m,n)$ *is satisfied for all* $a,b,m,n \in X$, *then we call* $(X,\tilde{+},\tilde{\cdot})$ *is commutative vague ring w.r.t. a fuzzy equality* $E_{X\times X}$ *and a fuzzy equality* E_X *on* X.

In the following part of this paper we say $(X,\tilde{+},\tilde{\cdot})$ is a vague ring for short.

Theorem 2. *Let* $(X,\tilde{+},\tilde{\cdot})$ *be vague ring. Then* $\tilde{\cdot}(a,0,0)=1$ *and* $\tilde{\cdot}(0,a,0)=1$ *for all* $a \in X$.

Proof. (i) for all $a \in X, \exists m,t \in X$ such that $\tilde{\cdot}(a,0,m)=1$, since $(X,\tilde{+})$ is a vague group, thus we can have $\tilde{+}(0,0,0)=1$ and $\tilde{+}(m,n,t)=1$,

$$1 = (\tilde{\cdot}(a,0,m) \wedge \tilde{+}(0,0,0) \wedge \tilde{\cdot}(a,0,m) \wedge \tilde{\cdot}(a,0,m) \wedge \tilde{+}(m,m,t)) \leq E_X(m,t)$$

from (VRDL), i.e. $m=t$, thus we can have $\tilde{+}(m,m,m)=1$. And $(X,\tilde{+})$ is a vague group, so $\tilde{+}(m,0,m)=1$ and $\tilde{+}(m,-m,0)=1$, thus

$$1 = (\tilde{+}(m,m,m) \wedge \tilde{+}(m,0,m) \wedge \tilde{+}(m,-m,0)) \leq E_X(m,0)$$

from (VG.1). i.e. $m=0$, thus $\tilde{\cdot}(a,0,0)=1$.

It is the same to $\tilde{\cdot}(0,a,0)=1$, so we omit it.

Theorem 3. *Let* $(X,\tilde{+},\tilde{\cdot})$ *be vague ring and* $\tilde{+}$ *be transitive on the first order. Then*

(i) $\tilde{\cdot}(-a,b,m) \wedge \tilde{\cdot}(a,-b,n) \leq E_X(m,n), \forall a,b,m,n \in X$.
(ii) $\tilde{\cdot}(-a,-b,m) \wedge \tilde{\cdot}(a,b,n) \leq E_X(m,n), \forall a,b,m,n \in X$.

Proof. (i) There exist $s,k \in X$, such that $\tilde{\cdot}(0,b,0)=1, \tilde{+}(a,-a,0)=1, \tilde{\cdot}(a,b,s)=1$, thus

$$\tilde{\cdot}(0,b,0) \wedge \tilde{+}(a,-a,0) \wedge \tilde{\cdot}(a,b,s) \wedge \tilde{\cdot}(-a,b,m) \wedge \tilde{+}(s,m,k) \leq E_X(k,0),$$

from (VDRL).

$$i.e. \tilde{\cdot}(-a,b,m) \leq E_X(k,0). \quad (1)$$

It is similar that there exists a $l \in X$ such that

$$\tilde{\cdot}(a,-b,n) \leq E_X(0,l) \quad (2)$$

and $\tilde{+}(s,n,l)=1$, thus we have

$$\tilde{\cdot}(-a,b,m) \wedge \tilde{\cdot}(a,-b,n) \leq E_X(0,l) \wedge E_X(k,0) \leq E_X(k,l).$$

Since $\tilde{+}$ be transitive on the first order, thus

$$1 = \tilde{+}(s,n,l)) \wedge E_X(k,l) \leq \tilde{+}(s,n,k),$$

i.e. $E_X(k,l) \leq \tilde{+}(s,n,k)$. We can have

$$(1 = \tilde{+}(s,m,k)) \wedge \tilde{+}(s,n,k) \leq E_X(m,n)$$

from the (VCL.1). i.e. $\tilde{+}(s,n,k) \leq E_X(m,n)$.
Thus $\tilde{\cdot}(-a,b,m) \wedge \tilde{\cdot}(a,-b,n) \leq E_X(k,l) \leq \tilde{+}(s,n,k) \leq E_X(m,n)$.
i.e. $\tilde{\cdot}(-a,b,m) \wedge \tilde{\cdot}(a,-b,n) \leq E_X(m,n)$.

The proof of (ii) is similar to (i), so we omit it.

Definition 5. *(i) Let* $(X, \tilde{+}, \tilde{\cdot})$ *be a vague ring. Then a nonempty and crisp subset* Y *of* X *is called a vague sub-ring of* X *if* $(Y, \tilde{+}, \tilde{\cdot})$ *itself is a vague sub-ring.*

(ii) Let $(X, \tilde{+}, \tilde{\cdot})$ *be a vague ring. Then a vague sub-ring* Y *of* X *is called a vague left [right] ideal if* $\forall x \in X, a \in I, \exists b \in I$ *such that*

$$\tilde{\cdot}(x, a, b) = 1 \ [\tilde{\cdot}(a, x, b) = 1],$$

and a vague ideal if it is both vague left ideal and vague right ideal.

Theorem 4. *Let* $(x, \tilde{+}, \tilde{\cdot})$ *be a vague ring. Then a nonempty and crisp subset* Y *of* X *is a vague sub-ring of* X *iff*

$$\forall\ a, b \in Y, c \in X, \tilde{+}(a, -b, c) = 1 \Rightarrow c \in Y,$$

and

$$\forall\ a, b \in Y, c \in X, \tilde{\cdot}(a, b, c) = 1 \Rightarrow c \in Y.$$

Corollary 1. *Let* $(X, \tilde{+}, \tilde{\cdot})$ *be a vague ring. Then a nonempty and crisp subset* I *of* X *is a vague left [right] ideal of* X *iff*

$$\forall\ a, b \in Y, c \in X, \tilde{+}(a, -b, c) = 1 \Rightarrow c \in Y,$$

$$\textit{and}\ \forall\ a \in I, b, c \in X, \tilde{\cdot}(b, a, c) = 1 \Rightarrow c \in I[\ \tilde{\cdot}(a, b, c) = 1 \Rightarrow c \in I].$$

And I *is a vague ideal of* X *iff it is both vague left and right ideal.*

Theorem 5. *Let* $(X, \tilde{+}, \tilde{\cdot})$ *be a vague ring, and a fixed* $a \in X$*. we suppose the set* $J = \{x \in X | \tilde{\cdot}(x, a, 0) = 1\}$*, then* J *is left ideal; suppose* $I = \{x \in X | \tilde{\cdot}(a, x, 0) = 1\}$*, then* I *is a right ideal.*

Proof. For $\forall\ x_1, x_2 \in J, \exists\ x \in X$ such that $\tilde{+}(x_1, -x_2, x) = 1$. Since $\tilde{\cdot}$ is a vague binary operation on X,$\exists m \in X$ such that $\tilde{\cdot}(x, a, m) = 1$, because $x_1, x_2 \in J$, we have $\tilde{\cdot}(x_1, a, 0) = 1$, and $\tilde{\cdot}(x_2, a, 0) = 1$. Since $\tilde{+}$ is a vague binary operation on X, $\tilde{+}(0, 0, 0) = 1$, thus we can have

$$1 = (\tilde{+}(0, 0, 0) \wedge \tilde{+}(x_1, -x_2, x) \wedge \tilde{\cdot}(x, a, m) \wedge \tilde{\cdot}(x_1, a, 0) \wedge \tilde{\cdot}(x_2, a, 0)) \leq E_X(0, m),$$

from (VRDL), i.e. $m = 0$. So $\tilde{\cdot}(x, a, 0) = 1$, it is to say $x \in J$. For $\forall x_1, x_2 \in J$, $\exists x \in X$ such that $\tilde{\cdot}(x_1, x_2, x) = 1$. Since $\tilde{\cdot}$ is a vague binary operation on X, $\exists m \in X$ such that $\tilde{\cdot}(x, a, m) = 1$, because $x_1, x_2 \in J, \tilde{\cdot}(x_1, a, 0) = 1, \tilde{\cdot}(x_2, a, 0) = 1$, and $\tilde{\cdot}(x_1, 0, 0) = 1$, so we can have

$$1 = (\tilde{\cdot}(x_1, x_2, x) \wedge \tilde{\cdot}(x, a, m) \wedge \tilde{\cdot}(x_2, a, 0) \wedge \tilde{\cdot}(x_1, 0, 0)) \leq E_X(0, m)$$

from the associate law of $\tilde{\cdot}$, i.e. $m = 0$. So $\tilde{\cdot}(x, a, 0) = 1$, thus $x \in J$, and J is a vague left ideal I; It is the same for vague right ideal I.

Theorem 6. *Let* $(X, \tilde{+}, \tilde{\cdot})$ *be a vague ring, and* I *be a vague ideal. Suppose* $X : I = \{x \in X | \tilde{\cdot}(s, x, t) = 1 \Rightarrow t \in I, \textit{for all}\ s \in X\}$*, then* $X : I$ *is a vague ideal which contains* I*.*

Proof. Suppose $x_1, x_2 \in X : I$, so there exist $t_1, t_2 \in X$, such that $\widetilde{\cdot}(s, x_1, t_1) = 1 \Rightarrow t_1 \in I$ and $\widetilde{\cdot}(s, x_2, t_2) = 1 \Rightarrow t_2 \in I$ for all $s \in X$. There exists $m \in X$ such that $\widetilde{+}(x_1, -x_2, m) = 1$, for all $s \in X, \exists n, t \in X$ such that $\widetilde{\cdot}(s, m, n) = 1$, and $\widetilde{+}(t_1, -t_2, t) = 1$, so we can have

$$1 = (\widetilde{\cdot}(s, x_1, t_1) \wedge \widetilde{\cdot}(s, x_2, t_2) \wedge \widetilde{+}(x_1, -x_2, m) \wedge \widetilde{\cdot}(s, m, n) \wedge \widetilde{+}(t_1, -t_2, t)) \leq E_X(t, n)$$

from (VRDL) i.e. $t = n$, thus $\widetilde{\cdot}(s, m, t) = 1$. I is a vague ideal, so $t \in I$ from the definition 3.2, thus $m \in X : I$; For $x_1 \in X : I, x \in X$, there exist $t_1, v \in X$ such that for all $s \in X$ $\widetilde{\cdot}(s, x_1, t_1) = 1 \Rightarrow t_1 \in I$, and $\widetilde{\cdot}(x, x_1, v) = 1$, for all $s \in X$, $\exists n, t \in X$, such that $\widetilde{\cdot}(s, v, t) = 1$.$\widetilde{\cdot}$ is a vague operation, so there exists a $u, n \in X$ such that $\widetilde{\cdot}(s, x, u) = 1$ and $\widetilde{\cdot}(u, x_1, n) = 1$,thus $(1 = \widetilde{\cdot}(x, x_1, v) \wedge \widetilde{\cdot}(s, v, t) \wedge \widetilde{\cdot}(s, x, u) \wedge \widetilde{\cdot}(u, x_1, n)) \leq E_X(t, n)$ i.e. $t = n$, because $x_1 \in X : I$, so $n \in I$, i.e. $t \in I$. Thus $X : I$ is a vague ideal. It is obvious that $I \subseteq X : I$.

Theorem 7. *Let* $(X, \widetilde{+}, \widetilde{\cdot})$ *be a vague ring and* $\{H_i : i \in I\}$ *be a nonempty family of all vague ideal of* X. *If* $\{H_i : i \in I\} \neq \emptyset$, *then* $\cap_{i \in I} H_i$ *is a vague ideal of* X.

4 The Homomorphism of Vague Ring

Definition 6. *Let* $(X, \widetilde{+}, \widetilde{\cdot})$ *and* $(Y, \widetilde{\oplus}, \widetilde{\circ})$ *be vague rings. A function (in classical sense)* $\omega : X \to Y$ *is called a vague homomorphism iff*

$$\widetilde{+}(a, b, c) \leq \widetilde{\oplus}(\omega(a), \omega(b), \omega(c)) and \widetilde{\cdot}(a, b, c) \leq \widetilde{\circ}(\omega(a), \omega(b), \omega(c)), \forall a, b, c \in X.$$

Definition 7. *Let* $(X, \widetilde{+}, \widetilde{\cdot})$ *and* $(Y, \widetilde{\oplus}, \widetilde{\circ})$ *be vague rings, and* $\omega : X \to Y$ *be a vague homomorphism. Then the crisp set* $\{a \in X : \omega(a) = 0\}$ *is called a vague kernel of* ω , *denoted by* $V ker \omega$.

Definition 8. *Let* X *and* Y *be nonempty sets, and a function* $\omega : X \to Y$ *be a vague injective w.r.t. fuzzy equalities* E_X *and* E_Y *iff*

$$E_Y(\omega(a), \omega(a)) \leq E_X(a, b). \forall a, b \in X.$$

It can be noted that a vague injective function is obviously injective in the crisp sense.

Theorem 8. *Let* $(X, \widetilde{+}, \widetilde{\cdot})$ *and* $(Y, \widetilde{\oplus}, \widetilde{\circ})$ *be vague rings, and* $\omega : X \to Y$ *be a vague homomorphism. If* ω *is a vague injective, then* $V ker \omega = \{0_X\}$.

Proof. Suppose $V\text{ker}\omega \neq \{0_X\}$, then there exists a $0_X \neq a \in X$, such that $\omega(a) = 0_Y$, since ω is a vague injective, thus

$$1 = E_Y(\omega(a), 0_Y) = E_Y(0_Y, 0_Y) \leq E_X(a, 0_X),$$

i.e. $0_X = a$. This is a contradiction.

Theorem 9. *Let* $(X, \widetilde{+}, \widetilde{\cdot})$ *and* $(Y, \widetilde{\oplus}, \widetilde{\circ})$ *be vague rings, and* $\omega : X \to Y$ *be a vague homomorphism. Then*

(i) $Vker\omega$ *is a vague ideal of* X.

(ii) For a vague ideal I *of* X, *and* ω *is a surjective, then* $\omega(I)$ *is a vague ideal of* Y.

(iii) For a vague ideal J *of* Y, $\omega^{-1}(I)$ *is vague ideal of* X.

Theorem 10. *Let* $(X, \widetilde{+}, \widetilde{\cdot})$ *and* $(Y, \widetilde{\oplus}, \widetilde{\circ})$ *be vague rings,* $\omega : X \to Y$ *be a vague homomorphism and* $\widetilde{\oplus}$ *is satisfied (T.2). Then*

$$E_Y(\omega(a), \omega(b)) \leq E_Y(\omega(c), 0_Y)$$

References

1. L.A.Zadeh (1965) Fuzzy sets. Information and Control, 8(3):338-353
2. A. Rosenfeld (1971) Fuzzy groups. J. Math. Anal. Appl., l35: 512-517
3. M.Demirci (1999) Fuzzy functions and their fundamental properties. Fuzzy Sets and Systems, 106(2):239-246
4. M.Sasaki (1993) Fuzzy function. Fuzzy Sets and Systems, 55(3) :295-301
5. M.Demirci (1999) Vague Groups. Journal of Mathematical Analysis and Applications, 230(1):142-156
6. M.Demirci (2005) A theory of vague lattices based on many-valued equivalence relations-I: general representation results. Fuzzy Sets and Systems, 151(3):437-472
7. M. Demirci (2005) A theory of vague lattices based on many-valued equivalence relations-II: complete lattices. Fuzzy Sets and Systems, 151(3):473-489
8. J. Mordeson (1998) Fuzzy Commutative Algebra. London, World Scientific

A Common Generalization of Smooth Group and Vague Group

Zhenming Ma, Wei Yang, and Qingjun Ren

Department of Mathematics, Linyi Normal University
Linyi 276005, China
{dmgywto,wyoeng,qjr288}@126.com

Abstract. The concepts of fuzzy equality and fuzzy function of Demirci (1999) are generalized. Furthermore, new fuzzy group is obtained, and the relationships with Dmeirci's fuzzy groups (1999, 2001, 2004) are studied.

Keywords: J-fuzzy relation, I-fuzzy equality, J-fuzzy function, GF-fuzzy group.

1 Introduction

The concept of fuzzy equality, which is also called the equality relation, the fuzzy equivalence relation and the similarity relation, and fuzzy functions based on fuzzy equalities have a significant concern in various fields [1-3], they had been successfully applied in category theoretical frame and fuzzy control [1, 3]. And the concept of fuzzy equality proves a useful mathematical tool to avoid Poincare-like paradoxes [3]. M. Demirci introduced the fuzzy equality to fuzzy algebraic research, where the binary operations [2, 5, 6] are fuzzily defined. The concepts of vague group, smooth group and perfect $\wedge$-vague group were given, and the elementary properties of them are studied.

In this paper, adopting a improved definition of fuzzy equality in stead of that presented in [4], which is nothing but a generalization of the fuzzy equality by the concept of tip of fuzzy set, and reformulating the definition of fuzzy function in [4], and the relationships of the fuzzy functions in [2, 4, 5] are investigated. Therefore, a new fuzzy group based on the generalized fuzzy equality and fuzzy function is defined and the properties of it are studied. At last, the relationships with the fuzzy group in [2, 5, 6] are got.

2 Common Generalization of Smooth Group and Vague Group

In this section, we generalize the definition of fuzzy equality and fuzzy function. Furthermore, we obtain the common generalization of smooth group and vague group.

B.-Y. Cao (Ed.): Fuzzy Information and Engineering (ICFIE), ASC 40, pp. 144–148, 2007.
springerlink.com

Definition 1. *Let X be a nonempty set, we call the mapping $\mu : X \to [0,1]$ a fuzzy set of set.*

We call fuzzy set f of set $X \times Y$ a fuzzy relation from X into Y. Especially, when $X = Y$ is satisfied, we call f a fuzzy relation on X. Supposed $r, s, t \in [0,1]$, $s = r \leq t$ and $(r,t] = I = J = (s,t]$. If $r = t$ is satisfied, we call I or J singleton. If $r = t = 1$ is satisfied, we call I or J strong singleton.

Definition 2. *Let ρ be fuzzy relation from X into Y, ρ is a J-fuzzy relation from X into Y if and only if $\forall (x,y) \in X \times Y$, $\rho(x,y) \in J$. Especially, J-fuzzy relation on X if and only if $\forall (x,y) \in X \times X$, $\rho(x,y) \in J$.*

Definition 3. *Let X be non-empty set, I-fuzzy relation T_X on X is a I-fuzzy equality on X if and only if*

(T_1) $T_X(x,y) = t \Leftrightarrow x = y$;

(T_2) $T_X(x,y) = T_X(y,x) = t$;

(T_3) $T_X(x,y) \wedge T_X(y,z) \leq T_X(x,z), \forall x,y,z \in X$

$T_X(x,y)$ shows the degree of I-fuzzy equality for $\forall x, y \in X$. And we have the following theorem from the definition 3.

Theorem 1. *Let X be non-empty set, T_X is a I-fuzzy equality,*

(1) If $t = 1$, then I-fuzzy equality is fuzzy equality in [4];

(2) If I is a strong singleton, then I-fuzzy equality is equality in usual sense.

Definition 4. *Let T_X and T_Y be I-fuzzy equalities on X and Y, respectively.*

(E_1) J-fuzzy relation ρ is extensive based on I-fuzzy equality T_X if and only if

$$\rho(x,y) \wedge T_X(x,x') \leq \rho(x,y)$$

(E_2) J-fuzzy relation ρ is extensive based on I-fuzzy equality T_Y if and only if

$$\rho(x,y) \wedge T_Y(y,y') \leq \rho(x,y')$$

(E_3) J-fuzzy relation ρ is extensive based on I-fuzzy equalities and if and only if (E_1) and (E_2) are both satisfied.

Definition 5. *Let T_X and T_Y be I-fuzzy equalities on X and Y, respectively.*

(F_1)$\forall x \in X, \exists y \in Y$, such that $\rho(x,y) \in J$;

(F_2)$\forall x, y \in X, \forall z, w \in Y, \mu_\rho(x,z) \wedge \mu_\rho(y,w) \wedge T_X(x,y) \leq T_Y(z,w)$

Definition 6. *Let T_X and T_Y be I-fuzzy equalities on X and Y, respectively. ρ is a J-fuzzy function based on I-fuzzy equalities T_X and T_Y on X and Y,*

(1) ρ is a J-strong singleton fuzzy function if and only if J is a strong singleton.

(2) ρ is a strong singleton fuzzy function if and only if I and J are both strong singletons.

Theorem 2. *Let X and Y be non-empty sets, ρ is a J-fuzzy function based on I-fuzzy equalities T_X and T_Y on X and Y,*

(1)If $t = 1$, then J-fuzzy function is fuzzy function;

(2)if ρ is a strong singleton J-fuzzy function, then ρ is a strong fuzzy function in [1];

(3)if ρ is a strong singleton fuzzy function, then ρ is a function in usual sense.

Proof. The proof of (1) is the direct result of the definition 3 and 5. And the same to (2) and (3) by the definition 6.

Definition 7. *Let X be non-empty set, J-fuzzy function f based on I-fuzzy equalities $T_{X\times X}$ and T_X is J-fuzzy operation if and only if is a J-fuzzy function f based on I-fuzzy equalities $T_{X\times X}$ and T_X.*

Definition 8. *Let X be non-empty set, $\bullet$ based on I-fuzzy equalities $T_{X\times X}$ and T_X is a J-fuzzy operation, $\bullet$ has the property $(*)$ if and only if*

$$\forall a,b,c,d \in X, \mu_\bullet(a,b,c) \wedge T_X(c,d) > r \Rightarrow \mu_\bullet(a,b,d) > s$$

Definition 9. *Let X be non-empty set, $\bullet$ based on I-fuzzy equalities $T_{X\times X}$ and T_X is a J-fuzzy operation,*

(G_1) X together with $\bullet$, denoted by $(X,\bullet)$, is called a GF-fuzzy semi-group if and only if the membership function $\mu_\bullet : X \times X \times X \to [0,1]$ of the J-fuzzy operation $\bullet$ satisfies

$$\mu_\bullet(a,b,c) \wedge \mu_\bullet(a,d,m) \wedge \mu_\bullet(a,b,q) \wedge \mu_\bullet(q,c,w) > s \Rightarrow T_X(m,w) > r$$

(G_2) A GF-fuzzy semi-group $(X,\bullet)$ is a GF-fuzzy monoid if and only if $\forall a \in X, \exists e \in X$ such that

$$\mu_\bullet(a,b,c) \wedge \mu_\bullet(a,d,m) > s$$

(G_3) A GF-fuzzy monoid $(X,\bullet)$ is a GF-fuzzy group if and only if $\forall a \in X, \exists b \in X$ such that

$$\mu_\bullet(a,b,e) \wedge \mu_\bullet(b,a,e) > s$$

We call e the identity of GF-fuzzy group, We call b inverse of a, denoted by a^{-1}. The set of all the GF-fuzzy groups is denoted by UFG. The identity of GF-fuzzy group is not unique, but we still have the following result:

Theorem 3. *Let $(X,\bullet)$ be GF-fuzzy group, e_1 and e_2 are the identities of the GF-fuzzy group, then $T_X(e_1,e_2) > r$.*

Theorem 4. *Let $(X,\bullet)$ be GF-fuzzy group,*

(1) $\forall a \in X, T_X((a^{-1})^{-1}, a) > r$;

(2) $T_X((a,b) > r \Rightarrow T_X((a^{-1})^{-1}, (b^{-1})^{-1}) > r$

Proof. For each $a \in X$ by (G_2) and (G_3), $\mu_\bullet(a_{-1}, (a^{-1})^{-1}, e) \wedge \mu_\bullet(a,e,a) \wedge \mu_\bullet(a,(a^{-1}),e) \wedge \mu_\bullet(e,(a^{-1})^{-1},(a^{-1})^{-1})) > r$, so we can have $T_X((a^{-1})^{-1}, a) > r$, immediately from $(G2)$. The proof of (2) is easy obtained from (1).

3 The Relationships Among Fuzzy Groups Based on Fuzzy Equalities

In this section, we consider the relationships among GF-fuzzy group, smooth group, vague group and perfect $\wedge$-vague group. It is easy to find the following result:

Theorem 5. *Let $(X, \bullet)$ be GF-fuzzy group,*

(1) If $t = 1$, and $\bullet$ has the property $(*)$, then GF-fuzzy group is smooth group in [6];

(2) If J is a strong singleton, then GF-fuzzy group is vague group in [2];

(3) If J is a strong singleton and $\bullet$ has the property $(*)$, then GF-fuzzy group is a perfect $\wedge$-vague group in [5];

(4) If $\bullet$ is a fuzzy operation with strong singleton, then GF-fuzzy group is a classical group.

The proof is easy, so we omit it.

Theorem 6. *(1) A perfect $\wedge$-vague group is also a smooth group;*

(2) A smooth group whose J is a strong singleton is a perfect $\wedge$-vague group;

(3) Smooth group and vague group are GF-fuzzy groups;

(4) Vague group with the property $(*)$ is a perfect $\wedge$-vague group in [5];

Proof. (1) It is the direct result of the definition of smooth group; (2) we can have the operation of smooth group is a fuzzy operation of perfect $\wedge$-vague group from the theorem 2(2), thus smooth group whose J is a strong singleton is a perfect $\wedge$-vague group; The proofs of (3) and (4) are obvious, so we omit it.

The set of all GF-fuzzy groups with the property $(*)$ denotes as USG, the set of all GF-fuzzy groups whose J is singleton denotes as UVG and the set of all GF-fuzzy groups with the property $(*)$ and J is singleton denotes as UPG. By theorem 5, if we fix $t \in [0, 1]$, USG, UVG and UPG are denoted by GSG, GVG and GPG, respectively. So GSG $\subset$USG, GVG $\subset$UVG and GPG $\subset$UPG. Then we can obtain the following result:

Theorem 7. *(1) There exists a bijective from GSG to SG, and SG is the set of all smooth groups.*

(2) There exists a bijective from GVG to VG, and VG is the set of all vague groups.

(3) There exists a bijective from GPG to PG, and PG is the set of all perfect $\wedge$-vague groups.

Proof. For each $(X, \bullet) \in$ GSG, we construct smooth group as follow

$$E_X(x, y) = \frac{T_X(x, y)}{t}$$

$$E_{X \times X}((x, z), (y, w)) = \frac{T_{X \times X}((x, z), (y, w))}{t}$$

and

$$\mu_{\bullet t}(a, b, c) = \frac{\mu_{\bullet}(a, b, c)}{t}$$

then it is easy to obtain $(X, \bullet_t)$ is a smooth group. Because t is a constant, so there exists a bijective. It is the same to (2) and (3). It is obvious that the mappings in theorem 5 are surjective, if we replace the sets GSG, GVG and GPG with the sets USG, UVG and UPG, respectively.

4 Conclusion

In this paper, we have examined relationships between smooth group, vague group and $\wedge$-perfect vague group. GF-fuzzy group is a common generalization of smooth group and vague group; $\wedge$-perfect vague group is a common specialization of smooth group and vague group. Some further results need to develop in order to find out which way of fuzzifying the concept of group is the best one. The question is left for future investigation.

Acknowledgements

This work is supported by a grant from Natural Science Foundation of Shandong (2004ZX13).

References

1. U. Cerruti and U. Hohle (1986) An approach to uncertainty using algebras over a monoidal closed category. Rend. Circ. Mat. Palermo Suppl. 12 (2): 47–63
2. M. Demirci (1999) Vague groups. J. Math. Anal. and App 230: 142–156
3. U. Hohle and L. N. Stout (1991) Foundations of fuzzy sets. Fuzzy Sets and Systems 40: 257–296
4. M. Demirci (1999) Fuzzy functions and their fundamental properties. Fuzzy Sets and Systems 106: 239–246
5. M. Demirci (2004) Fuzzy group, fuzzy functions and fuzzy equivalence relation. Fuzzy Sets and Systems 144: 441-458
6. M. Demirci (2001) Smooth Group. Fuzzy Sets and Systems 117: 239–246
7. A. Rosenfeld (1971) Fuzzy groups. J. Math. Anal. App l35: 512-517

Sequences of Fuzzy-Valued Choquet Integrable Functions

Rui-Sheng Wang[1] and Ming-Hu Ha[2]

[1] School of Information, Renmin University of China, Beijing 100872, China
wrs@ruc.edu.cn
[2] College of Mathematics & Computers, Hebei University, Baoding 071002, Hebei
mhha@hbu.edu.cn

Abstract. This paper deals with the properties of sequences of fuzzy-valued Choquet (for short, (C)-) integrable functions. Firstly, we introduce the concept of uniform (C)-integrabiliy and other new concepts like uniform absolute continuity and uniform boundedness for sequences of fuzzy-valued (C)-integrable functions and then discuss the relations among them. We also present several convergence theorems for sequences of fuzzy-valued (C)-integrable functions by using uniform (C)-integrability.

Keywords: uniform (C)-integrability; uniform absolute continuity; uniform boundedness; convergence theorem.

1 Introduction

Since the concepts of fuzzy measures and fuzzy integrals of real-valued functions were first developed by [1] and later extended by [4, 2, 3], integrals of set-valued functions and integrals of fuzzy-valued functions on fuzzy measure space have been also researched in different ways [5, 10]. Uniform integrability plays an important role in probability theory [6]. Ha et al. [4, 7] introduced uniform integrability of sequences of integrable functions into fuzzy measure theory respectively according to $(H), (S)$ fuzzy integrals. Choquet integral is another important kind of non-additive integrals which was first developed by Choquet [8] and has been researched by many other authors [9]. Recently, fuzzy-valued Choquet integral and its theory were also established by authors [11, 10] and has been applied in [12] for multiple criteria decision support. In this paper, we introduce the concept of uniform (C)-integrability and other new concepts like uniform absolute continuity and uniform boundedness for sequences of fuzzy-valued (C)-integrable functions. We will also discuss the relations among them and present some convergence theorems for sequences of fuzzy-valued (C)-integrable functions by using uniform (C)-integrability.

2 Preliminaries and Propositions

In this paper, $(\mathbb{X}, \mathscr{F}, \mu)$ is a fuzzy measure space. $\mathscr{F}$ is an algebra of subsets of $\mathbb{X}$. $\mu :\to [0, \infty]$ is a fuzzy measure defined on $\mathscr{F}$. $\mathbb{F}$ is the set of all real-valued

B.-Y. Cao (Ed.): Fuzzy Information and Engineering (ICFIE), ASC 40, pp. 149–158, 2007.
springerlink.com

non-negative measurable functions. $R^+ = [0, \infty]$ and $\mathcal{F}(R^+)$ is all the fuzzy sets defined on R^+.

The set of all interval numbers on R^+ is denoted by $I(R^+) = \{[a,b] \mid [a,b] \subset R^+\}$. Interval numbers have the following operation properties: for any $[a^-, a^+], [b^-, b^+] \in I(R^+)$,

(1) $[a^-, a^+] + [b^-, b^+] = [a^- + b^-, a^+ + b^+]$;

(2) $[a^-, a^+] \cdot [b^-, b^+] = [a^- \cdot b^-, a^+ \cdot b^+]$;

(3) $k \cdot [a^-, a^+] = [k \cdot a^-, k \cdot a^+]$ for any $k \in R^+$.

Definition 2.1([5]). A fuzzy set $\widetilde{a}$ in $\mathcal{F}(R^+)$ is called a *fuzzy number* iff it satisfies the following conditions:

(1) $\widetilde{a}$ is regular, i.e. there exists $x \in R^+$ such that $\widetilde{a}(x) = 1$;

(2) For any $\lambda \in (0,1]$, $a_\lambda = \{x \in R^+ \mid \widetilde{a}(x) \geq \lambda\}$ is a closed interval, denoted by $[a_\lambda^-, a_\lambda^+]$.

$\widetilde{R}^+$ denotes the set of all fuzzy numbers on R^+. Fuzzy number has the following properties: for any $\widetilde{a}, \widetilde{b} \in \widetilde{R}^+$,

(1) $(\widetilde{a} + \widetilde{b})_\lambda = a_\lambda + b_\lambda$;

(2) $(\widetilde{a} \cdot \widetilde{b})_\lambda = a_\lambda \cdot b_\lambda$;

(3) $(k \cdot \widetilde{a})_\lambda = k \cdot a_\lambda$ for any $k \in R^+$.

In this paper, we will use the following ranking approach of fuzzy number ([5]): $\widetilde{a} \leq \widetilde{b}$ iff for any $\lambda \in (0,1]$, $a_\lambda \leq b_\lambda$, i.e. $a_\lambda^- \leq b_\lambda^-, a_\lambda^+ \leq b_\lambda^+$.

Definition 2.2 ([5]). A sequence of fuzzy numbers $\{\widetilde{a}_n\}$ is said to be convergent iff for any $\lambda \in (0,1]$, $\{a_{n,\lambda}\}$ is a convergent sequence of interval numbers, i.e. $\{a_{n,\lambda}^-\}, \{a_{n,\lambda}^+\}$ are convergent real number sequences. The limit of $\{\widetilde{a}_n\}$ is defined as $\widetilde{a} = \bigcup_{\lambda \in [0,1]} \lambda[\lim\limits_{n\to\infty} a_{n,\lambda}^-, \lim\limits_{n\to\infty} a_{n,\lambda}^+]$, and we denote it by $\lim\limits_{n\to\infty} \widetilde{a}_n = \widetilde{a}$ or $\widetilde{a}_n \to \widetilde{a}$. Obviously, $\widetilde{a}$ is a fuzzy number too.

Let $\widetilde{f}$ be a fuzzy-valued function from $\mathbb{X}$ to $\widetilde{R}^+$. For any $\lambda \in [0,1]$, $(\widetilde{f}(x))_\lambda = [(\widetilde{f}(x))_\lambda^-, (\widetilde{f}(x))_\lambda^+]$. We give new notions for brevity: $f_\lambda^-(x) = (\widetilde{f}(x))_\lambda^-, f_\lambda^+(x) = (\widetilde{f}(x))_\lambda^+$. Obviously, f_λ^-, f_λ^+ are nonnegative functions on $\mathbb{X}$. In the following sections, f_n, f denote nonnegative real-valued measurable functions and $\widetilde{f}_n, \widetilde{f}$ denote fuzzy-valued functions from $\mathbb{X}$ to $\widetilde{R}^+$ unless we give special notions.

Definition 2.3 ([5]). A fuzzy-valued function $\widetilde{f}$ is said to be measurable iff f_λ^-, f_λ^+ are measurable functions for each $\lambda \in (0,1]$. $\widetilde{\mathbb{F}}$ denotes the set of all fuzzy-valued measurable functions from $\mathbb{X}$ to $\widetilde{R}^+$.

Choquet integral is a kind of monotone, nonadditive and nonlinear integrals owing to the nonadditivity of μ. More properties of Choquet integral can be found in [9,3]. Fuzzy-valued Choquet integral is a generalization of Choquet integral to fuzzy-valued function and has been applied in multiple criteria decision support [12].

Definition 2.4 ([9]). Let $f \in \mathbb{F}, A \in \mathscr{F}$. The Choquet (for short, (C)) integral of f with respect to μ on A is defined by

$$(C)\int_A f d\mu = \int_0^\infty \mu(A \cap F_\alpha) d\alpha,$$

where $F_\alpha = \{x \mid f(x) \geq \alpha\}, \alpha \in [0, \infty)$. If $(C)\int_A f d\mu < \infty$, we call f (C)-integrable. $L_1(\mu)$ is the set of all (C)-integrable functions. In addition, $(C)\int f d\mu$ implies $(C)\int_{\mathbb{X}} f d\mu$.

Definition 2.5 ([10, 11]). Let $\widetilde{f} \in \widetilde{\mathbb{F}}$ and $A \in \mathscr{F}$. The Choquet integral of $\widetilde{f}$ with respect to μ on A is defined as

$$\begin{aligned}(C)\int_A \widetilde{f} d\mu &= \bigcup_{\lambda \in [0,1]} \lambda[(C)\int_A f_\lambda^- d\mu, (C)\int_A f_\lambda^+ d\mu] \\ &= \bigcup_{\lambda \in [0,1]} \lambda[\int_0^\infty \mu(A \cap F_{\lambda,\alpha}^-) d\alpha, \int_0^\infty \mu(A \cap F_{\lambda,\alpha}^+) d\alpha]\end{aligned}$$

where $F_{\lambda,\alpha}^- = \{x \mid f_\lambda^-(x) \geq \alpha\}, F_{\lambda,\alpha}^+ = \{x \mid f_\lambda^+(x) \geq \alpha\}$ for any $\alpha \geq 0$. $\widetilde{f}$ is called (C)-integrable on A iff f_λ^+ is (C)-integrable on A for all $\lambda \in (0, 1]$. At this time, $(C)\int_A \widetilde{f} d\mu < \infty$. $\widetilde{L}_1(\mu)$ is the set of all fuzzy-valued (C)-integrable functions. In addition, $(C)\int \widetilde{f} d\mu$ means $(C)\int_{\mathbb{X}} \widetilde{f} d\mu$.

3 Uniform Integrability, Uniform Absolute Continuity and Uniform Boundedness

In this section, we will introduce the concepts of uniform (C)-integrability, uniform absolute continuity and uniform boundedness for sequences of fuzzy-valued Choquet integral and discuss the relations between them.

Definition 3.1. Let $\{\widetilde{f}_n\} \subset \widetilde{L}_1(\mu)$. $\{\widetilde{f}_n\}$ is said to be *uniformly* (C)*-integrable* iff for any given $\varepsilon > 0$, there exists $k_0 > 0$ such that

$$(C)\int_{\{x \mid \widetilde{f}_n(x) \geq k\}} \widetilde{f}_n d\mu < \varepsilon, n = 1, 2, \cdots$$

whenever $k \geq k_0$.

Definition 3.2. Let $\{\widetilde{f}_n\} \subset \widetilde{L}_1(\mu)$. The (C) integral sequence of $\{\widetilde{f}_n\}$ is said to be *uniformly absolutely continuous* iff for any given $\varepsilon > 0$, there exists $\delta > 0$ such that $(C)\int_A \widetilde{f}_n d\mu < \varepsilon$, $n = 1, 2, \cdots$ whenever $A \in \mathscr{F}$ and $\mu(A) < \delta$.

Definition 3.3. Let $\{\widetilde{f}_n\} \subset \widetilde{L}_1(\mu)$. The (C)-integral sequence of $\{\widetilde{f}_n\}$ is said to be *uniformly bounded* iff $\sup_{n \geq 1} (C)\int \widetilde{f}_n d\mu < \infty$.

Theorem 3.1. If $\widetilde{f} \in \widetilde{L}_1(\mu)$, then for any given $\varepsilon > 0$, there exists $\delta > 0$ such that $(C)\int_A \widetilde{f}d\mu < \varepsilon$ whenever $A \in \mathscr{F}$ and $\mu(A) < \delta$. (This property is referred as the absolute continuity of fuzzy-valued (C) integral)

Proof. If the conclusion does not hold, then there exists ε_0 such that for any $n \in Z^+$, there exists $A_n \in \mathscr{F}$ and $\mu(A_n) < \frac{1}{n}$ such that $(C)\int_{A_n} \widetilde{f}d\mu \geq \varepsilon_0$, i.e.

$$(C)\int_{A_n} \widetilde{f}d\mu = \bigcup_{\lambda\in[0,1]} \lambda[(C)\int_{A_n} f_\lambda^- d\mu, (C)\int_{A_n} f_\lambda^+ d\mu] \geq \varepsilon_0.$$

It is easy to see that $\left((C)\int_{A_n} \widetilde{f}d\mu\right)_\lambda = [(C)\int_{A_n} f_\lambda^- d\mu, (C)\int_{A_n} f_\lambda^+ d\mu]$. From the ranking method of interval numbers, we have $(C)\int_{A_n} f_\lambda^- d\mu \geq \varepsilon_0$ and $(C)\int_{A_n} f_\lambda^+ d\mu \geq \varepsilon_0$. On the other hand, $(C)\int_{A_n} f_\lambda^- d\mu = \int_0^\infty \mu(A_n \cap F_{\lambda,\alpha}^-)d\alpha \leq \int_0^\infty \mu(F_{\lambda,\alpha}^-)d\alpha < \infty$. When $n \to \infty$, $\mu(A_n) \to 0$, so $\mu(A_n \cap F_{\lambda,\alpha}^-) \to 0$. From the bounded convergence theorem of classical integrable function sequence, we have $(C)\int_{A_n} f_\lambda^- d\mu \to 0$. Similarly, $(C)\int_{A_n} f_\lambda^+ d\mu \to 0$. It follows from the decomposition theorem of fuzzy sets that $(C)\int_{A_n} \widetilde{f}d\mu \to 0$. This contradicts with $(C)\int_{A_n} \widetilde{f}d\mu \geq \varepsilon_0$. So the conclusion is correct.

Theorem 3.2. Let $\{\widetilde{f}_n\} \subset \widetilde{L}_1(\mu)$. If the (C)-integral sequence of $\{\widetilde{f}_n\}$ is both uniformly absolutely continuous and uniformly bounded, then $\{\widetilde{f}_n\}$ is uniformly (C)-integrable.

Proof. Since the (C)-integral sequence of $\{\widetilde{f}_n\}$ is absolutely continuous, then for any given $\varepsilon > 0$, there exists $\delta > 0$ such that

$$(C)\int_A \widetilde{f}_n d\mu < \varepsilon,\ n = 1, 2, \cdots,$$

whenever $A \in \mathscr{F}$ and $\mu(A) < \delta$. For any $k > 0$, $(C)\int_{\{x|\widetilde{f}_n(x)\geq k\}} \widetilde{f}_n d\mu \geq (C)\int_{\{x|\widetilde{f}_n(x)\geq k\}} k d\mu = k\mu(\{x \mid \widetilde{f}_n(x) \geq k\})$. So

$$\begin{aligned}\mu(\{x \mid \widetilde{f}_n(x) \geq k\}) &\leq \frac{1}{k}(C)\int_{\{x|\widetilde{f}_n(x)\geq k\}} \widetilde{f}_n d\mu \\ &\leq \frac{1}{k}(C)\int \widetilde{f}_n d\mu \leq \frac{1}{k}\sup_{n\geq 1}(C)\int \widetilde{f}_n d\mu.\end{aligned}$$

Let $k > 0$ be sufficiently large such that $\frac{1}{k}\sup_{n\geq 1}(C)\int \widetilde{f}_n d\mu < \delta$. So $\mu(\{x \mid \widetilde{f}_n(x) \geq k\}) < \delta$. With the absolute continuity of the (C)-integral sequence of $\{\widetilde{f}_n\}$, we have

$$(C)\int_{\{x|\widetilde{f}_n(x)\geq k\}} \widetilde{f}_n d\mu < \varepsilon.$$

Therefore, $\{\widetilde{f}_n\}$ is uniformly (C)-integrable according to Definition 3.1.‖

Theorem 3.3. Let $\{\widetilde{f}_n\} \subset \widetilde{L}_1(\mu)$ and μ be finite. If $\{\widetilde{f}_n\}$ is uniformly (C)-integrable, then the (C)-integral sequence of $\{\widetilde{f}_n\}$ is both uniformly absolutely continuous and uniformly bounded.

Proof. Since $\{\widetilde{f}_n\}$ is uniformly (C)-integrable, then for any given $\varepsilon > 0$, there exists $k_0 > 0$ such that

$$(C)\int_{\{x|\widetilde{f}_n(x)\geq k_0\}} \widetilde{f}_n d\mu < \frac{\varepsilon}{2},\ n = 1, 2, \cdots .$$

It follows that

$$\bigcup_{\lambda\in[0,1]} \lambda[(C)\int_{\{x|\widetilde{f}_n(x)\geq k_0\}} f^-_{n,\lambda}d\mu, (C)\int_{\{x|\widetilde{f}_n(x)\geq k_0\}} f^+_{n,\lambda}d\mu] < \bigcup_{\lambda\in[0,1]} \lambda[\frac{\varepsilon}{2}, \frac{\varepsilon}{2}] = \frac{\varepsilon}{2}.$$

From the method for ranking fuzzy numbers, we have

$$[(C)\int_{\{x|\widetilde{f}_n(x)\geq k_0\}} f^-_{n,\lambda}d\mu, (C)\int_{\{x|\widetilde{f}_n(x)\geq k_0\}} f^+_{n,\lambda}d\mu] < [\frac{\varepsilon}{2}, \frac{\varepsilon}{2}],$$

i.e. $(C)\int_{\{x|\widetilde{f}_n(x)\geq k_0\}} f^-_{n,\lambda}d\mu < \frac{\varepsilon}{2}$ and $(C)\int_{\{x|\widetilde{f}_n(x)\geq k_0\}} f^+_{n,\lambda}d\mu < \frac{\varepsilon}{2}$. So

$$\int_0^\infty \mu(\{x \mid \widetilde{f}_n(x) \geq k_0\} \cap \{x \mid f^-_{n,\lambda}(x) \geq \alpha\})d\alpha.$$
$$= \int_0^{k_0} \mu(\{x \mid \widetilde{f}_n(x) \geq k_0\})d\alpha + \int_{k_0}^\infty \mu(\{x \mid f^-_{n,\lambda}(x) \geq \alpha\})d\alpha < \frac{\varepsilon}{2}.$$

This implies that $\int_{k_0}^\infty \mu(\{x \mid f^-_{n,\lambda}(x) \geq \alpha\})d\alpha < \frac{\varepsilon}{2}$. Similarly, we also have

$$\int_{k_0}^\infty \mu(\{x \mid f^+_{n,\lambda}(x) \geq \alpha\})d\alpha < \frac{\varepsilon}{2}.$$

Let $\delta = \frac{\varepsilon}{2k_0} > 0$. When $A \in \mathscr{F}$ and $\mu(A) < \delta$, we have

$$\begin{aligned}(C)\int_A \widetilde{f}_n d\mu &= \bigcup_{\lambda\in[0,1]} \lambda[\int_0^\infty \mu(A \cap F^-_{n,\lambda,\alpha})d\alpha, \int_0^\infty \mu(A \cap F^-_{n,\lambda,\alpha})d\alpha] \\ &\leq \bigcup_{\lambda\in[0,1]} \lambda[\int_0^{k_0} \mu(A)d\alpha, \int_0^{k_0} \mu(A)d\alpha] + \bigcup_{\lambda\in[0,1]} \lambda[\frac{\varepsilon}{2}, \frac{\varepsilon}{2}] \\ &< \delta k_0 + \frac{\varepsilon}{2} = \varepsilon.\end{aligned}$$

Therefore, according to Definition 3.2, the (C)-integral sequence of $\{\widetilde{f}_n\}$ is uniformly absolutely continuous. A similar proof process can lead to

$$\begin{aligned}(C)\int \widetilde{f}_n d\mu &\leq \bigcup_{\lambda\in[0,1]} \lambda[\int_0^{k_0} \mu(\mathbb{X})d\alpha, \int_0^{k_0} \mu(\mathbb{X})d\alpha] + \bigcup_{\lambda\in[0,1]} \lambda[\frac{\varepsilon}{2}, \frac{\varepsilon}{2}] \\ &< \mu(\mathbb{X})k_0 + \frac{\varepsilon}{2}\end{aligned}$$

So $\sup_{n\geq 1}(C)\int \widetilde{f}_n d\mu < \infty$ holds. Therefore, according to Definition 3.3, the (C)-integral sequence of $\{\widetilde{f}_n\}$ is uniformly bounded.||

Corollary 3.1. Let $\{\widetilde{f}_n\} \subset \widetilde{L}_1(\mu)$ and μ be finite. The following conditions are equivalent:

(1) $\{\widetilde{f}_n\}$ is uniformly (C)-integrable.

(2) The (C)-integral sequence of $\{\widetilde{f}_n\}$ is both uniformly absolutely continuous and uniformly bounded.

Proof. It follows immediately from Theorem 3.2 and Theorem 3.3.||

Theorem 3.4. Let $\{\widetilde{f}_n\} \subset \widetilde{L}_1(\mu)$. If there exists $\widetilde{g} \in \widetilde{L}_1(\mu)$ such that for any $A \in \mathscr{F}$,

$$(C)\int_A \widetilde{f}_n d\mu \leq (C)\int_A \widetilde{g} d\mu$$

holds, then $\{\widetilde{f}_n\}$ is uniformly (C)-integrable.

Proof. It is obvious that

$$\sup_{n\geq 1}(C)\int \widetilde{f}_n d\mu \leq (C)\int \widetilde{g} d\mu < \infty,$$

so the (C)-integral sequence of $\{\widetilde{f}_n\}$ is uniformly bounded. In addition, from Theorem 3.1 (the absolute continuity of fuzzy-valued Choquet integral), for any given $\varepsilon > 0$, there exists $\delta > 0$ such that

$$(C)\int_A \widetilde{f} d\mu < \varepsilon$$

whenever $A \in \mathscr{F}$ and $\mu(A) < \delta$. It follows that

$$(C)\int_A \widetilde{f}_n d\mu \leq (C)\int_A \widetilde{g} d\mu < \varepsilon, n = 1, 2, \cdots$$

whenever $A \in \mathscr{F}$ and $\mu(A) < \delta$. This implies that the (C)-integral sequence of $\{f_n\}$ is uniformly absolutely continuous. Therefore, according to Theorem 3.2, $\{\widetilde{f}_n\}$ is uniformly (C)-integrable.||

Corollary 3.2. Let $\{\widetilde{f}_n\} \subset \widetilde{L}_1(\mu)$ and μ be null-additive. If there exists $\widetilde{g} \in \widetilde{L}_1(\mu)$ such that $\widetilde{f}_n \leq \widetilde{g}, a.e., n = 1, 2, \cdots$, then $\{\widetilde{f}_n\}$ is uniformly (C)-integrable.

Proof. If μ is null-additive and $\widetilde{f}_n \leq \widetilde{g}, a.e., n = 1, 2, \cdots$, then $(C)\int_A \widetilde{f}_n d\mu \leq (C)\int_A \widetilde{g} d\mu$. Therefore, according to Theorem 3.4, $\{\widetilde{f}_n\}$ is uniformly (C)-integrable.||

In the following theorem, we will give another sufficient condition for a sequence of fuzzy-valued (C)-integrable functions to be uniformly integrable.

Theorem 3.5. Let $\{\widetilde{f}_n\} \subset \widetilde{L}_1(\mu)$ and $\widetilde{f} \in \widetilde{L}_1(\mu)$. If $(C)\int_A \widetilde{f}_n d\mu$ converges to $(C)\int_A \widetilde{f} d\mu$ for any $A \in \mathscr{F}$, then $\{\widetilde{f}_n\}$ is uniformly (C)-integrable.

Proof. From the absolute continuity of fuzzy-valued (C)-integrals, for any $\varepsilon > 0$, there exists $\delta > 0$ such that $(C)\int_A \widetilde{f} d\mu < \frac{\varepsilon}{2}$ for any $A \in \mathscr{F}$ and $\mu(A) < \delta$. With

$$(C)\int_A \widetilde{f} d\mu = \bigcup_{\lambda\in[0,1]} \lambda[(C)\int_A f_\lambda^- d\mu, (C)\int_A f_\lambda^+ d\mu] < \bigcup_{\lambda\in[0,1]} \lambda[\frac{\varepsilon}{2}, \frac{\varepsilon}{2}]$$

we have $(C)\int_A f_\lambda^+ d\mu < \frac{\varepsilon}{2}$, $(C)\int_A f_\lambda^- d\mu < \frac{\varepsilon}{2}$. On the other hand, since $(C)\int_A \widetilde{f}_n d\mu$ converges to $(C)\int_A \widetilde{f} d\mu$ for any $A \in \mathscr{F}$, it is easily obtained from the convergence of fuzzy number sequence that

$$(C)\int_A f_{n,\lambda}^+ d\mu \to (C)\int_A f_\lambda^+ d\mu,$$

$$(C)\int_A f_{n,\lambda}^- d\mu \to (C)\int_A f_\lambda^- d\mu.$$

Therefore, there exists some positive integer N such that $(C)\int_A f_{n,\lambda}^+ d\mu < \varepsilon$ and $(C)\int_A f_{n,\lambda}^- d\mu < \varepsilon$ whenever $n > N$. As for $\widetilde{f}_1, \widetilde{f}_2, \cdots, \widetilde{f}_N$, with the absolute continuity of fuzzy-valued Choquet integral, there exist $\delta_1, \delta_2, \cdots, \delta_N$ such that $(C)\int_A f_{i,\lambda}^- d\mu < \varepsilon$ whenever $\mu(A) < \delta_i$, $i = 1, 2, \cdots, N$. Let $\delta = \min\{\delta_1, \delta_2, \cdots, \delta_N\}$, we have $(C)\int_A \widetilde{f}_n d\mu < \varepsilon$, $n = 1, 2, \cdots$ and $\mu(A) < \delta$, i.e. $\{\widetilde{f}_n\}$ is uniformly absolutely continuous. Now we only need to prove the uniform boundedness of $\{\widetilde{f}_n\}$. $(C)\int_A f_{n,\lambda}^+ d\mu$ is convergent, so there exists $M_1 > 0$ such that $(C)\int_A f_{n,\lambda}^- d\mu < M_1$. Similarly, there exists $M_2 > 0$ such that $(C)\int_A f_{n,\lambda}^+ d\mu < M_2$. Let $M = \max\{M_1, M_2\}$. Then $(C)\int_A \widetilde{f} d\mu < M$. Hence, $\{\widetilde{f}_n\}$ is uniformly bounded. Therefore, $\{\widetilde{f}_n\}$ is uniformly (C)-integrable.

4 Some Convergence Theorems

Although several researchers have considered the convergence of fuzzy-valued (C)-integrals [10, 11], there are still some areas that need to be investigated. For example, they all did not show that under what conditions the convergence in fuzzy measure can imply the convergence in (C)-mean, though the reverse conclusion has already been proved. In this section, we solve it by using the uniform (C)-integrability introduced in Section 3.

Definition 4.1. Let $A \in \mathscr{F}$, $\{\widetilde{f}_n\} \subset \widetilde{\mathbb{F}}$ and $\widetilde{f} \in \widetilde{\mathbb{F}}$. $\{\widetilde{f}_n\}$ is said to be convergent in fuzzy measure to $\widetilde{f}$ on A iff $\lim_{n\to\infty} \mu(A \cap \{x \mid |\widetilde{f}_n(x) - \widetilde{f}(x)| \geq \delta\}) = 0$ for any $\delta > 0$. We denote it by $\widetilde{f}_n \xrightarrow{\mu} \widetilde{f}$.

Definition 4.2. Let $A \in \mathscr{F}$, $\{\widetilde{f}_n\} \subset \widetilde{\mathrm{F}}$. $\{\widetilde{f}_n\}$ is said to be fundamentally convergent in fuzzy measure on A iff $\lim_{m,n\to\infty} \mu(A \cap \{x \mid |\widetilde{f}_m(x) - \widetilde{f}_n(x)| \geq \delta\}) = 0$ for any $\delta > 0$.

Definition 4.3. Let $\{\widetilde{f}_n\} \subset \widetilde{\mathrm{F}}$, $\widetilde{f} \in \widetilde{\mathrm{F}}$. $\{\widetilde{f}_n\}$ is said to be convergent in (C)-mean to $\widetilde{f}$ iff $\lim_{n\to\infty} (C) \int |\widetilde{f}_n - \widetilde{f}| d\mu = 0$. We denote it by $\widetilde{f}_n \xrightarrow{m} \widetilde{f}$.

Definition 4.4. Let $\{\widetilde{f}_n\} \subset \widetilde{\mathrm{F}}$. $\{\widetilde{f}_n\}$ is said to be fundamentally convergent in (C)-mean iff $\lim_{m,n\to\infty} (C) \int |\widetilde{f}_m - \widetilde{f}_n| d\mu = 0$.

Lemma 4.1. Let $(\mathbb{X}, \mathscr{F}, \mu)$ be a fuzzy measure space. If μ is subadditive and $\widetilde{f}$, $\widetilde{g}$ are two fuzzy-valued measurable functions with disjoint support, i.e. $\{x|\widetilde{f}(x) > 0\} \cap \{x|\widetilde{g}(x) > 0\} = \emptyset$, then $(C) \int_{\mathbb{X}} (\widetilde{f} + \widetilde{g}) d\mu \leq (C) \int_{\mathbb{X}} \widetilde{f} d\mu + (C) \int_{\mathbb{X}} \widetilde{g} d\mu$.

Proof. It is obvious from the definition of fuzzy-valued Choquet integrals and the subadditivity of real-valued Choquet integral [3].

Theorem 4.1. Let $\{\widetilde{f}_n\} \subset \widetilde{L}_1(\mu)$, $\widetilde{f} \in \widetilde{L}_1(\mu)$ and μ be subadditive. If (1) $\{\widetilde{f}_n\}$ converges to $\widetilde{f}$ in fuzzy measure μ; (2) $\{\widetilde{f}_n\}$ is uniformly (C)-integrable, then $\{\widetilde{f}_n\}$ converges to $\widetilde{f}$ in (C)-mean.

Proof. We first prove that $|\widetilde{f}_n - \widetilde{f}|$ be uniformly (C)-integrable. $|\widetilde{f}_n - \widetilde{f}| \leq \widetilde{f}_n + \widetilde{f} \leq 2\widetilde{f}_n I_{\{x|\widetilde{f}_n \geq \widetilde{f}\}} + 2\widetilde{f} I_{\{x|\widetilde{f}_n < \widetilde{f}\}}$. Since μ is subadditive, we have

$$\begin{aligned}(C) \int_A |\widetilde{f}_n - \widetilde{f}| d\mu &\leq (C) \int_A (2\widetilde{f}_n I_{\{x|\widetilde{f}_n \geq \widetilde{f}\}} + 2\widetilde{f} I_{\{x|\widetilde{f}_n < \widetilde{f}\}}) d\mu \\ &\leq 2(C) \int_{A\cap\{x|\widetilde{f}_n \geq \widetilde{f}\}} \widetilde{f}_n d\mu + 2(C) \int_{A\cap\{x|\widetilde{f}_n < \widetilde{f}\}} \widetilde{f} d\mu\end{aligned}$$

for any $A \in \mathscr{F}$. It is easy to see that $|\widetilde{f}_n - \widetilde{f}|$ is both uniformly absolutely continuous and uniformly bounded from the uniform (C)-integrability of $\{\widetilde{f}_n\}$ and the absolute continuity of fuzzy-valued (C)-integral. Therefore, $|\widetilde{f}_n - \widetilde{f}|$ is uniformly (C)-integrable, then for any given $\varepsilon > 0$, there exists $k_0 > 0$ such that

$$(C) \int_{\{x||\widetilde{f}_n(x) - \widetilde{f}(x)| \geq k_0\}} |\widetilde{f}_n - \widetilde{f}| d\mu < \varepsilon, n = 1, 2, \cdots.$$

It follows that

$$\begin{aligned}&\bigcup_{\lambda\in[0,1]} \lambda[(C) \int_{\{x||\widetilde{f}_n(x) - \widetilde{f}(x)| \geq k_0\}} |\widetilde{f}_n - \widetilde{f}|_\lambda^- d\mu, (C) \int_{\{x||\widetilde{f}_n(x) - \widetilde{f}(x)| \geq k_0\}} |\widetilde{f}_n - \widetilde{f}|_\lambda^+ d\mu] \\ &< \bigcup_{\lambda\in[0,1]} \lambda[\frac{\varepsilon}{2}, \frac{\varepsilon}{2}] = \frac{\varepsilon}{2}.\end{aligned}$$

Therefore,

$$\int_0^\infty \mu(\{x \mid |\widetilde{f}_n(x) - \widetilde{f}(x)| \geq k_0\} \cap \{x \mid |\widetilde{f}_n(x) - \widetilde{f}(x)|_\lambda^- \geq \alpha\})d\alpha$$
$$= \int_0^{k_0} \mu(\{x \mid |\widetilde{f}_n(x) - \widetilde{f}(x)| \geq k_0\})d\alpha + \int_{k_0}^\infty \mu(\{x \mid |\widetilde{f}_n(x) - \widetilde{f}(x)|_\lambda^- \geq \alpha\})d\alpha < \varepsilon.$$

This implies that $\int_{k_0}^\infty \mu(\{x \mid |\widetilde{f}_n(x) - \widetilde{f}(x)| \geq \alpha\})d\alpha < \varepsilon$. $\{\widetilde{f}_n\}$ is convergent to $\widetilde{f}$ in fuzzy measure μ, so for any given $\alpha \in (0, \infty)$, there exists n_0 such that when $n > n_0$, we have

$$\mu(\{x \mid |\widetilde{f}_n(x) - \widetilde{f}(x)| \geq \alpha\}) < \varepsilon.$$

It follows that

$$(C)\int |\widetilde{f}_n(x) - \widetilde{f}(x)|d\mu = \int_0^\infty \mu(\{x \mid |\widetilde{f}_n(x) - \widetilde{f}(x)| \geq \alpha\})d\alpha$$
$$= \int_0^{k_0} \mu(\{x \mid |\widetilde{f}_n(x) - \widetilde{f}(x)| \geq \alpha\})d\alpha + \int_{k_0}^\infty \mu(\{x \mid |\widetilde{f}_n(x) - \widetilde{f}(x)| \geq \alpha\})d\alpha < \varepsilon$$
$$< \varepsilon k_0 + \varepsilon = (k_0 + 1)\varepsilon.$$

Therefore $(C)\int |\widetilde{f}_n - \widetilde{f}|d\mu \to 0$ when $n \to \infty$. It implies that $\{\widetilde{f}_n\}$ is convergent in (C)-mean to f.‖

Theorem 4.2. Let $\{\widetilde{f}_n\} \subset \widetilde{L}_1(\mu)$ and μ be subadditive. If (1) $\{\widetilde{f}_n\}$ fundamentally converges to $\widetilde{f}$ in fuzzy measure μ; (2) $\{\widetilde{f}_n\}$ is uniformly (C)-integrable, then $\{\widetilde{f}_n\}$ is fundamentally convergent in (C) mean

Proof. The proof process is similar to that of Theorem 4.1.

5 Conclusions

Fuzzy-valued Choquet integral is a generalization to real-valued integral and has been applied in multiple decision criteria support. But the properties of sequences of fuzzy-valued (C)-integrable functions have been considered very little. This paper introduces some new concepts like uniform (C)-integrability, uniform continuity and uniform bounded for sequences of fuzzy-valued (C)-integrable functions. The relations among them are also discussed. Besides, the paper makes some complements to convergence theorems of sequences of fuzzy-valued (C)-integrable functions. According to these results, some future work can be done. For example, we can discuss the properties of $\widetilde{L}_1(\mu)$ by defining the distance of $\widetilde{f}_1, \widetilde{f}_2 \in \widetilde{L}_1(\mu)$ as $(C)\int |\widetilde{f}_1 - \widetilde{f}_2|d\mu$.

Acknowledgement

This research work is supported by Research Foundation of Renmin University of China (No.06XNB054) and National Natural Science Foundation of China (60573069; 60473045).

References

1. M.Sugeno (1974) Theory of fuzzy integrals and its applications. Ph.D. Thesis, Tokyo Inst. of Tech, Tokyo.
2. Z.Wang and G.Klir (1992) Fuzzy Measure Theory. Plenum, New York.
3. D.Denneberg (1994) Nonadditive Measure and Integral. Kluwer Academic Publishers, Boston.
4. M.H. Ha and C.X. Wu (1998) Fuzzy Measure and Fuzzy Integral Theory. Science Press, Beijing.
5. G.Zhang (1998) Fuzzy-valued Measure Theory. Tsinghua University Press, Beijing.
6. J.A. Yan (1998) Measure Theory. Science Press, Beijing.
7. M.H. Ha, X.Z. Wang, L.Z. Yang and Y. Li (2003) Sequences of (S) fuzzy integrable functions. Fuzzy Sets and Systems 138:507-522.
8. G.Choquet (1954) Theory of capacities. Ann Inst Fourier 5:131-295.
9. Z.Y. Wang, G.J.Klir, and W. Wang (1996) Monotone set functions defined by Choquet integrals. Fuzzy Sets and Systems 81:241-250.
10. G.J. Wang and X.P. Li (1999) On the convergence of the fuzzy valued functional defined by μ-integrable fuzzy valued functions. Fuzzy Sets and Systems 107: 219-226.
11. R.S. Wang, and M.H. Ha (2006) On Choquet integrals of fuzzy-valued functions. The Journal of Fuzzy Mathematics 14:89-102.
12. P. Meyera, M. Roubensb (2006) On the use of the Choquet integral with fuzzy numbers in multiple criteria decision support Fuzzy Sets and Systems 157:927-938.

Generalized Root of Theories in Propositional Fuzzy Logical Systems*

Jiancheng Zhang[1,**] and Shuili Chen[2]

[1] Department of Mathematics, Quanzhou Normal University, Fujian 362000, China
zjcqz@126.com
[2] Department of Mathematics, Jimei University, Xiamen 361021, China
shuilicheng@vip.sina.com

Abstract. In order to study the logic foundation of fuzzy reasoning and the construction of the sets $D(\Gamma)$, this paper first put forth a definition of generalized root of theoretic in Łukasiewicz propositional fuzzy logic system, *Gödel* propositional fuzzy logic system, product propositional fuzzy logic system,and the R_0-propositional fuzzy logic system. Then it goes on to prove that $D(\Gamma)$ is completely determined by its generalized root whenever theory Γ has a generalized root, thereafter it puts forward the construction of $D(\Gamma)$. Finally, it argues that every finite theory Γ has a generalized root, which can be expressed in some specific form, and that there does exist certain relationship between different theory Γ.

Keywords: Propositional fuzzy logic system; Theory; Generalized root; Deduction theorem.

1 Introduction

Fuzzy logic is the theoretical foundation of fuzzy control. With fuzzy control being widely applied and different kinds of fuzzy reasoning being put forward, more and more attention has been drawn to the logic foundation of fuzzy control. One of the main deduction rules in propositional logic systems is modus ponens (MP,for short) with the form as follows

$$\text{from } A \to B \text{ and } A \text{ infer } B. \cdots\cdots\cdots\cdots\cdots\cdots (1)$$

In common sense reasoning, if A implies B and we know a statement A^* which is close to A is true, then A^* implies certain statement B^* which is close to B, i.e., we have the following deduction form (briefly, GMP)

$$\text{from } A \to B \text{ and } A^* \text{ extract } B^*. \cdots\cdots\cdots\cdots\cdots\cdots (2)$$

the deductive form (2) is much more general and useful than the deduction form (1) because there is a certain leeway between A^* and A so a lot of different statement $A^{*'}$s are permitted in (2) and make it meaningful.

* The project was supported by the Natural Science Foundation of Fujian Province of China (No. 2006J0221 and A0510023).

** Corresponding author.

B.-Y. Cao (Ed.): Fuzzy Information and Engineering (ICFIE), ASC 40, pp. 159–169, 2007.
springerlink.com

More generally, consider the following inference model (briefly,CGMP):

$$\text{from } A_i \to B \text{ and } A_i^* \ i=1,2,\cdots,n \text{ extract } B^*. \cdots\cdots\cdots\cdots (3)$$

where the major premises $A_i \to B$, $i=1,\cdots,n$, and the minor premise A_i^*, $i=1,\cdots,n$, are given and the conclusion B^* is to be collectively determined by the pairs $\{A_i \to B, A_i^*\}$, $i=1,\cdots,n$.

Zaden first solved the probled of GMP in 1973 by applying the composition Rule of Inference (CRI) and the conclusion of GMP is as follows:

$$B^*(y) = \sup[A^*(x) \wedge R(A(x), B(y))], y \in Y. \cdots\cdots\cdots\cdots (4)$$

where $R : [0,1]^2 \to [0,1]$ is a certain implication operator such as Zadeh's operator R_Z, *Łukasiewicz's* operator $R_{Łu}$, *Gödel's* operator R_G,etc, and the proposition A, A^* and B, B^* appeared in (2) were supposed to be fuzzy subsets on universes X and Y, respectively. The well-known Triple I method to solve the problem of GMP proposed in [6], i.e, define B^* to be the smallest fuzzy subset of Y such as $\forall x \in X$, $\forall y \in Y$, $(A(x) \to B(y)) \to (A^*(x) \to B^*(y)) = 1$.

It has been proved in [6] that the Triple I conclusion B^* of GMP always exists and $B^*(y) = \sup[A^*(x) \wedge (A(x) \to B(y))]$, $y \in Y$.

Form the syntax point of view, the Triple I method of GMP and CGMP in classical propositional logic is to solve the generalized root of theories. In order to study Fuzzy reasoning's form and Fuzzy reasoning's logic foundation in Łukasiewicz propositional logic system, *Gödel* propositional logic system, product propositional logic system and the R_0-propositional logic system, the concept of generalized root of theorem is proposed and the corresponding property is discussed.

In addition, to establish some logic system, or to meet the need of reasoning, we often choose a subset Γ of well-formed formulas, which can reflect some essential properties, as the axioms of the logic system, and therefore we can deduce the so-called Γ-conclusions through some reasonable inference rules [1,2,11]. Then there arises the question: What's the essential difference between the sets $D(\Gamma)$ of all conclusions of two different theories Γ? Based on the generalized deduction theorem and completeness theorems, the present paper discusses the generalized roots of theories and the relationship between the sets $D(\Gamma)$ of two different theories in Łukasiewicz propositional fuzzy logic system, *Gödel* propositional fuzzy logic system, product propositional fuzzy logic system and the R_0-propositional fuzzy logic system. It then proves that if a theory Γ has root, $D(\Gamma)$ is completely determined by the root, and that every finite theory has a generalized root and the construction of such generalized roots can also be established. Finally the sufficient and necessary conditions for $D(\Gamma)$ being equal or included are obtained.

2 Preliminaries

2.1 Logic Systems: *Łuk*, *Göd*, *Π* and *L**

It is well known that different implication operators and Valuation lattices determine different logic systems(see [1]). Here are two kinds of valuation lattices

$L = [0,1]$ and $L = \{0, \frac{1}{n-1}, \cdots, \frac{n-2}{n-1}, 1\}$ and four popularly used implication operators and the corresponding t-norms defined as follows:

$$R_L(x,y) = (1 - x + y) \wedge 1;\ x * y = (x + y - 1) \vee 0,\ x, y \in L,$$
$$R_G(x,y) = \begin{cases} 1,\ x \leq y; \\ y,\ x > y, \end{cases} \quad x * y = x \wedge y, x, y \in L,$$
$$R_\Pi(x,y) = \begin{cases} 1,\ x \leq y; \\ \frac{y}{x},\ x > y, \end{cases} \quad x * y = x \cdot y, x, y \in L.;$$
$$R_0(x,y) = \begin{cases} 1, & x \leq y, \\ (1-x) \vee y, & x > y, \end{cases} \quad x * y = \begin{cases} x \wedge y, & x + y > 1, \\ 0, x + y \leq 1, \end{cases};\quad x, y \in L.$$

These four implication operators R_L, R_G, R_Π and R_0, are called *Łukasiewicz* implication operator, *Gödel* implication operator, product implication operator and R_0 implication operator respectively.

If we fix a t-norm $*$ above we then fix a propositional calculus: $*$ is taken for the truth function of the strong conjunction &, the residuum R becomes the truth function of the implication operator and R(.,0) is the truth function of the negation. In more detains, we have the following.

Definition 1. *The propositional calculus $PC(*)$ given by a t-norm $*$ has the set S of propositional variables $p_1, p_2, \cdots$ and connectives $\neg, \&, \rightarrow$. The set $F(s)$ of well-formed formulas in $PC(*)$ is defined inductively as follows: each propositional variable is a formula; if A, B are formulas, then $\neg A, A\&B$ and $A \rightarrow B$ are all formulas.*

Definition 2. *The formal deductive systems of $PC(*)$ given by $*$ corresponding to R_L, R_G, R_Π and R_0 are called Łukasiewicz fuzzy logic system Łuk, Gödel fuzzy logic system Göd, product propositional fuzzy logic system Π and the R_0-propositional fuzzy logic system L^*, respectively.*

Note that logic system is called the n-valued logic system if the lattice $L = \{0, \frac{1}{n-1}, \cdots, \frac{n-2}{n-1}, 1\}$ of valuation. Logic system is called the fuzzy logic if the lattice $L = [0,1]$ of valuation.

The inference rule of each logic system above is modus ponens (MP,for short): from A and $A \rightarrow B$ infer B. For the choice of axiom schemes of these logic systems we refer to [1, 2, 11]

Remark 1. (i) One can further define connectives $\wedge$ and $\vee$ in these four logic systems, but the definition of $\wedge$ and $\vee$ in L^* are different from those of $\wedge$ and $\vee$ in the other three logic systems: in $L^*, A \vee B$ is an abbreviation of $\neg((((A \rightarrow (A \rightarrow B)) \rightarrow A) \rightarrow A) \rightarrow \neg((A \rightarrow B) \rightarrow B)), A \wedge B$ is that of $\neg(\neg A \vee \neg B)$. while in *Łuk*, *Göd* and Π, $\wedge$ and $\vee$ can be defined uniformly as follows [2], $A \wedge B$ is an abbreviation of $A\&(A \rightarrow B), A \vee B$ is that of $((A \rightarrow B) \rightarrow B) \wedge ((B \rightarrow A) \rightarrow A)$.

(ii)Define on L a unary operator and two binary operators as follows:

$\neg x = R(x,0), x\&y = x * y, x \rightarrow y = R(x,y).x, y \in L$, where $(*, R)$ is an adjoint pair on L.

From Remark 1 (ii) the valuation lattice L becomes an algebra of type $(\neg, \&, \rightarrow)$. Define in the above-mentioned logic systems

$A^n := \underbrace{A\&A\&\cdots\&A}_{n}$,

and in the corresponding algebra L

$a^{(n)} := \underbrace{a * a * \cdots * a}_{n}$,

where $*$ is the t-norm adjoint to the implication operator $\rightarrow$ defined on L.

Remark 2. It is easy to verify that $a^{(n)} = a$ for every $n \in N$,in *Göd*; in L^*, $a^{(n)} = a^{(2)}$ for every $n \in N, n \geq 2$; in *Łuk*, $a^{(n)} = (na - (n-1)) \vee 0$ and in Π, $a^{(n)} = a^n$ for every $n \in N$.

Definition 3. *(i) A homomorphism* $v : F(S) \rightarrow L$ *of type* $(\neg, \&, \rightarrow)$ *from* $F(S)$ *into the valuation* L, *i.e.,* $v(\neg A) = \neg v(A), v(A\&B) = v(A) * v(B), v(A \rightarrow B) = v(A) \rightarrow v(B)$, *is called an R-valuation of* $F(S)$. *The set of all R-valuations will denoted by* Ω_R.

(ii) A formula $A \in F(S)$ *is called a tautology w.r.t.* R *if* $\forall v \in \Omega_R, v(A) = 1$ *holds.*

It is not difficult to verify in the above-mentioned four logic systems that $v(A \vee B) = max\{v(A), v(B)\}$, and $v(A \wedge B) = min\{v(A), v(B)\}$ for every valuation $v \in \Omega_R$. Moreover,one can check in *Łuk* and L^* that $A\&B$ and $\neg(A \rightarrow \neg B)$ are logically equivalent.

Definition 4. *(i) A subset* Γ *of* $F(s)$ *is called a theory.*

(ii)Let Γ *be a theory,* $A \in F(s)$. *A deduction of* A *form* Γ, *in symbols,* $\Gamma \vdash A$, *is a finite sequence of formulas* $A_1, \cdots, A_n = A$ *such that for each* $1 \leq i \leq n$, A_i *is an axiom, or* $A_i \in \Gamma$, *or there are* $j, k \in \{1, \cdots, i-1\}$ *such that* A_i *follows from* A_j *and* A_k *by MP. Equivalently, we say that* A *is a conclusion of* Γ *(or* Γ*-conclusion). The set of all conclusions of* Γ *is denoted by* $D(\Gamma)$. *By a proof of* A *we shall henceforth mean a deduction of* A *form the empty set. We shall also write* $\vdash A$ *in place of* $\phi \vdash A$ *and call* A *a theorem.*

(iii) Let $A, B \in F(s)$. *If* $\vdash A \rightarrow B$ *and* $\vdash B \rightarrow A$ *hold, then* A *and* B *are called provably equivalent and denoted* $A \sim B$.

Theorem 1. *Completeness theorem holds in every above-mentioned logic system, i.e.,* $\forall A \in F(s)$,

A is a theorem in Łuk, Göd, Π *and* L^* *iff A is a tautology in Łuk, Göd,* Π *and* L^* *respectively.*

Theorem 1 points out that semantics and syntax in these four logic systems are in perfect harmony.

2.2 Generalized Deduction Theorems in *Łuk*, *Göd*, Π and L^*

Theorem 2 (1,2). *Suppose that* Γ *is a theory,* $A, B \in F(s)$*,then*

(i) in Łuk, $\Gamma \cup \{A\} \vdash B$ *iff* $\exists n \in N, s.t. \Gamma \vdash A^n \rightarrow B$.

(ii) in Göd, deduction theorem holds, i.e., $\Gamma \cup \{A\} \vdash B$ *iff* $\Gamma \vdash A \rightarrow B$.

(iii) in Π, $\Gamma \cup \{A\} \vdash B$ *iff* $\exists n \in N, s.t. \Gamma \vdash A^n \rightarrow B$.

Theorem 3 (11). *Suppose that Γ is a theory, $A, B \in F(s)$, then in L^* holds the generalized deduction theorem: $\Gamma \cup \{A\} \vdash B$ iff $\Gamma \vdash A^2 \to B$.*

It is easy for reader to check the following Lemma 1.

Lemma 1. *Let Γ be a theory, $A \in F(s)$. If $\Gamma \vdash A$, then there exists a finite subset of Γ, say, $\{A_1, \cdots, A_m\}$ such that $\{A_1, \cdots, A_m\} \vdash A$.*

It is easy to verify that $A\&B \to C$ and $A \to (B \to C)$ are provably equivalent, hence by the definition of deduction and the generalized deduction theorem. It is easy for reader to check the following Lemma 2.

Lemma 2. *Suppose that $\Gamma = \{A_1, \cdots, A_m\}$ is a finite theory, $A \in F(s)$. Then*

(i) in Łuk and Π, $\Gamma \vdash A$ iff $\exists n_1, \cdots, n_m \in N$ such that $\vdash A_1^{n_1}\&\cdots\&A_m^{n_m} \to A$.

(ii) in Göd, $\Gamma \vdash A$ iff $\vdash A_1\&\cdots\&A_m \to A$.

(iii) in L^*, $\Gamma \vdash A$ iff $\vdash A_1^2\&\cdots\&A_m^2 \to A$.

It is easy for the reader to check the following Lemma 3, by Lemma 2.2.8 of [2] and [7].

Lemma 3. *In Łuk, Göd, Π and L^*, the following conclusions hold:*

(i) If $\vdash A, \vdash B$, then $\vdash A\&B$.

(ii) If $\vdash A \to B, \vdash C \to D$, then $\vdash A\&C \to B\&D$.

(iii) If $\vdash A \to B$, then $\vdash A\&C \to B\&C$.

3 Basic Definitions and Properties

Definition 5. *Suppose that Γ is a theory, $A \in F(S)$. If $A \in D(\Gamma)$ and for every $B \in D(\Gamma)$ there exist a $n \in N$ such that $\vdash A^n \to B$, then A is called the generalized root of Γ.*

Definition 6. *Suppose that $\Gamma_k \subset F(s), k = 1, 2, \cdots, m$. Then members of $\bigcap_{k=1}^{m} D(\Gamma_k)$ are called common conclusions of $\Gamma_1, \cdots, \Gamma_m$. Suppose that $A \in F(S)$, and A is said to be a generalized common root of $\Gamma_1, \cdots, \Gamma_m$, if $A \in \bigcap_{k=1}^{m} D(\Gamma_k)$ and $\forall B \in \bigcap_{k=1}^{m} D(\Gamma_k)$ there exist a $n \in N$ such that $\vdash A^n \to B$.*

The following propositions hold in the four logic systems in concern.

Proposition 1. *Suppose that Γ is a theory, $A \in F(S)$. If A is a generalized roots of Γ, then $D(\Gamma) = D(A)$, where $D(A)$ is an abbreviation for $D(\{A\})$.*

Proof. Let A be a generalized root of Γ. Then for every $B \in D(\Gamma)$, it follows from Definition 5 that there exist a $n \in N$ such that $\vdash A^n \to B$ holds. Since $\{A\} \vdash A^n$, it follows from the inference rule MP that $\{A\} \vdash B$. This means that $B \in D(A)$ and so $D(\Gamma) \subseteq D(A)$. For the converse, for every $B \in D(A)$, that is, $\{A\} \vdash B$. It follows from the generalized deduction theorem that there exists $n \in N$ such that $\vdash A^n \to B$. Together with the result $\Gamma \vdash A^n$ which follows from the assumption $\Gamma \vdash A$, we have that $\Gamma \vdash B$ by MP and so $D(A) \subseteq D(\Gamma)$.

Proposition 2. *Suppose that $A \in F(s)$, then $D(A) = \{B \in F(s) \mid \exists n \in N, \vdash A^n \to B\}$.*

Proof. Let $V = \{B \in F(s) \mid \exists n \in N, \vdash A^n \to B\}$. For every $C \in D(A)$, it follows from the generalized deduction theorem that there exists $n \in N$ such that $\vdash A^n \to C$. we have that $C \in V$ and so $D(A) \subseteq V$. For the converse, for every $B \in V$, that is, $\exists n \in N$ such that $\vdash A^n \to B$. It is easy to prove that $\{A\} \vdash A^n$ by Lemma 3. It follows from the inference rule MP that $\{A\} \vdash B$ holds, i.e., $B \in D(A)$. Hence $V \subseteq D(A)$ and so $D(A) = V$.

Remark 3. Proposition 2 shows that the elements of $D(\Gamma)$ are composed of A^n ($\forall n \in N$), and B ($\vdash A^n \to B, \forall n \in N$) , where A is a generalized root of Γ.

Theorem 4. *Every finite theory has a generalized root, and $A_1\&\cdots\&A_s$ is a generalized root of $\Gamma = \{A_1, \cdots, A_s\}$.*

Proof. It is only necessary to prove that $A_1\&\cdots\&A_s$ is a generalized root of Γ. By Lemma 3, $A_1\&\cdots\&A_s \in D(\Gamma)$. $\forall B \in D(\Gamma)$, It follows from Lemma 2 that there exists $n_1, \cdots, n_s \in N$ such that $\vdash A_1^{n_1}\&\cdots\&A_s^{n_s} \to B$. Let $n = max\{n_1, \cdots, n_s\}$. It is easy to prove that $\vdash (A_1\&\cdots\&A_s)^n \to A_1^{n_1}\&\cdots\&A_s^{n_s}$. Hence by Hypothetical syllogism, we get $\vdash (A_1\&\cdots\&A_s)^n \to B$. Thus $A_1\&\cdots\&A_s$ is a generalized root of Γ.

Proposition 3. *Suppose that Γ is a theory, $A \in F(S)$. If $A \in D(\Gamma)$, then A is a generalized roots of Γ iff for every $B \in \Gamma$, there exist a $n \in N$ such that $\vdash A^n \to B$.*

Proof. The necessity is trivial. For the sufficiency, $\forall C \in D(\Gamma)$, then it follows from Lemma 1 and 2 that there exists a finite string of formulas $A_1, \cdots, A_s \in \Gamma$ and $n_1, \cdots, n_s \in N$ such that $\vdash A_1^{n_1}\&\cdots\&A_s^{n_s} \to C$ holds. From the assumption there exists a $k_i \in N$ such that $\vdash A^{k_i} \to A_i, i = 1, 2, \cdots, s$. By Lemma 3, $\vdash A^{n_1k_1+\cdots+n_sk_s} \to A_1^{n_1}\&\cdots\&A_s^{n_s}$. Let $n = n_1k_1 + \cdots + n_sk_s$. Hence by Hypothetical syllogism, we get $\vdash A^n \to C$. This shows that A is a generalized roots of Γ.

Proposition 4. *Suppose that Γ_1 and Γ_2 be two theory, then $D(\Gamma_1) = D(\Gamma_2)$ iff for every $A \in \Gamma_1, \Gamma_2 \vdash A$ holds,and $\forall B \in \Gamma_2, \Gamma_1 \vdash B$ holds.*

Proof. The necessity is trivial. For the sufficiency, $\forall C \in D(\Gamma_1)$, then it follows from Lemma 1 and 2 that there exists a finite string of formulas $A_1, \cdots, A_s \in \Gamma_1$ and $n_1, \cdots, n_s \in N$ such that $\vdash A_1^{n_1}\&\cdots\&A_s^{n_s} \to C$. From the assumption, $\Gamma_2 \vdash A_i, i = 1, 2, \cdots, s$. It is easy to prove that $\Gamma_2 \vdash A_1^{n_1}\&\cdots\&A_s^{n_s}$. By the inference rule MP , we get $\Gamma_2 \vdash C$ i.e., $C \in D(\Gamma_2)$, and so $D(\Gamma_1) \subseteq D(\Gamma_2)$. For the same reason, we can get $D(\Gamma_2) \subseteq D(\Gamma_1)$.

Proposition 5. *Suppose that $\Gamma_k \subseteq F(S)$ and A_k is a the generalized root of $\Gamma_k, k = 1, \cdots, m$, then $A_1\&\cdots\&A_m$ is a generalized root of $\Gamma = \bigcup_{k=1}^{m} \Gamma_k$.*

Proof. It is easy to prove that $\Gamma \vdash A_i, i = 1, \cdots, m$ by assumption. Hence by Lemma 3, we get $\Gamma \vdash A_1 \& \cdots \& A_m$, i.e., $A_1 \& \cdots \& A_m \in D(\Gamma)$. $\forall B \in D(\Gamma)$, then there exist $B_1, \cdots, B_n \in \Gamma$ and $m_1, \cdots, m_n \in N$ such that $\vdash B_1^{m_1} \& \cdots \& B_n^{m_n} \to B$ holds. For every $B_i, i = 1, \cdots, n$, there exist $j_i \in \{1, \cdots, m\}$ and $s_i \in N$ such that $\vdash A_{j_i}^{s_i} \to B_i$ holds. $\vdash A_{j_1}^{m_1 s_1} \& \cdots \& A_{j_n}^{m_n s_n} \to B_1^{m_1} \& \cdots \& B_n^{m_n}$ by Lemma 3. Clearly, there exists $l_1, \cdots, l_m \in N$ such that $\vdash A_1^{l_1} \& \cdots \& A_m^{l_m} \to A_{j_1}^{m_1 s_1} \& \cdots, \& A_{j_n}^{m_n s_n}$. Let $k = max\{l_1, \cdots, l_m\}$. We get $\vdash (A_1 \& \cdots \& A_s)^k \to B_1^{m_1} \& \cdots \& B_n^{m_n}$. By Hypothetical syllogism $\vdash (A_1 \& \cdots \& A_s)^k \to B$ is yielede. This shows that $A_1 \& \cdots \& A_m$ is a generalized root of Γ.

Proposition 6. *Suppose that* A, B *are generalized root of* Γ_1 *and* Γ_2 *respectively. Then* $D(\Gamma_1) \subseteq D(\Gamma_2)$ *iff* $\exists n \in N, s.t. \vdash B^n \to A$.

Proof. The necessity part is obvious from the Proposition 2. For the sufficiency, $\forall C \in D(\Gamma_1)$, it follows from defined of generalized root that $\exists m \in N$ such that $\vdash A^m \to C$, $\vdash B^{mn} \to A^m$ by Lemma 3, thus, $\vdash B^{mn} \to C$ by Hypothetical syllogism, and so $C \in D(\Gamma_2)$, thus $D(\Gamma_1) \subseteq D(\Gamma_2)$.

Corollary 1. *Suppose that* $\Gamma_1 = \{A_1, \cdots, A_s\}, \Gamma_2 = \{B_1, \cdots, B_t\}$, *then* $D(\Gamma_1) \subseteq D(\Gamma_2)$ *iff* $\exists n \in N, s.t., \vdash (B_1 \& \cdots \& B_t)^n \to A_1 \& \cdots \& A_s$.

Proof. By Theorem 4 and Proposition 6, it is easy to check the Corollary above.

Proposition 7. *Suppose that* A, B *are generalized root of* Γ_1 *and* Γ_2 *respectively. If* $A \sim B$, *then* $D(\Gamma_1) = D(\Gamma_2)$.

Proof. By proposition 6 and Lemma 3, it is easy to check the proposition above. The proof is left to reader.

4 Results in the n-Valued Łukasiewicz Logic System

It has been proved that the semantics and syntax of the n-valued *Łukasiewicz* logic system is in perfect harmony, i.e. the standard completeness theorem hold. It is easy to prove the following Proposition 8 and 9.

Proposition 8. *Suppose that* $m, n \in N$. *If* $m \geq n-1$ *then for every* $a \in L$, $a^{(m)} = a^{(n-1)}$ *holds in the n-valued Łukasiewicz logic system.*

Proposition 9. *Suppose that* $A \in F(s)$. *If* $m \geq n-1$ *then* $A^m \sim A^{n-1}$ *holds in the n-valued Łukasiewicz logic system* L_n.

Theorem 5. *Suppose that* Γ *is a theory,* $A \in F(s)$. *If* A *is a generalized root of* Γ, *then* $D(\Gamma) = \{B \in F(s) \mid \vdash A^{n-1} \to B\}$.

Proof. $\forall C \in D(\Gamma)$, it follows from the assumption and Generalized deduction theorem that there exists $m \in N$ such that $\vdash A^m \to C$. We easy to prove that $\vdash A^{n-1} \to C$ by Proposition 9. For the converse, let $\vdash A^{n-1} \to C$ and

$C \in F(s)$. $\{A\} \vdash C$ by Generalized deduction theorem, thus $C \in D(\Gamma)$, and then $D(\Gamma) = \{B \in F(s) \mid \vdash A^{n-1} \to B\}$.

Theorem 6. *Suppose that A, B are generalized root of Γ_1 and Γ_2 respectively, then $D(\Gamma_1) = D(\Gamma_2)$ iff $A^{n-1} \sim B^{n-1}$.*

Proof. Suppose that $D(\Gamma_2) = D(\Gamma_1)$, $A^{n-1} \in D(\Gamma_2)$ by Lemma 3, $\exists m \in N$ such that $\vdash B^m \to A^{n-1}$ by defined of generalized root. By proposition 9, we easy to prove that $\vdash B^{n-1} \to A^{n-1}$. For the same reason, we can get $\vdash A^{n-1} \to B^{n-1}$, thus $A^{n-1} \sim B^{n-1}$. For the converse, $\forall C \in D(\Gamma_1)$, it follows from Generalized deduction theorem and Proposition 9 that $\vdash A^{n-1} \to C$. $\vdash B^{n-1} \to C$ by the assumption and Hypothetical syllogism. Hence $C \in D(\Gamma_2)$ by Generalized deduction theorem, and then $D(\Gamma_1) \subseteq D(\Gamma_2)$. For the same reason, we can get $D(\Gamma_2) \subseteq D(\Gamma_1)$,thus $D(\Gamma_2) = D(\Gamma_1)$.

Corollary 2. *Suppose that $\Gamma_1 = \{A_1, \cdots, A_s\}$, $\Gamma_2 = \{B_1, \cdots, B_t\}$, then $D(\Gamma_1) = D(\Gamma_2)$ iff $(A_1 \& \cdots \& A_s)^{n-1} \sim (B_1 \& \cdots \& B_t)^{n-1}$*

Proposition 10. *Suppose that A_i is a generalized root of $\Gamma_i, i = 1, 2, \cdots, k$, then $\bigvee_{i=1}^{k} A_i^{n-1}$ is a generalized common root of $\Gamma_1, \cdots \Gamma_k$.*

Proof. Since $\vdash A_t^{n-1} \to \bigvee_{i=1}^{k} A_i^{n-1}, t = 1, \cdots, k$ and $A_t^{n-1} \in D(\Gamma_t)$ it follows from the inference rule MP that $\bigvee_{i=1}^{k} A_i^{n-1} \in D(\Gamma_t), i = 1, \cdots, k$. $\forall B \in \bigcap_{i=1}^{k} D(\Gamma_i)$, it is easy to check $\vdash A_i^{n-1} \to B, i = 1, \cdots, k$, thus $\vdash (A_1^{n-1} \to B) \wedge \cdots \wedge (A_k^{n-1} \to B)$. Since $(A_1^{n-1} \to B) \wedge \cdots \wedge (A_k^{n-1} \to B) \sim (A_1^{n-1} \vee \cdots \vee A_k^{n-1} \to B)$, hence $\vdash A_1^{n-1} \vee \cdots \vee A_k^{n-1} \to B$ holds, then $A_1^{n-1} \vee \cdots \vee A_k^{n-1}$ is a generalized common root of $\Gamma_1, \cdots, \Gamma_k$.

5 Results in the *Gödel* Fuzzy Logic System

Proposition 11. *Suppose that A is a generalized roots of Γ, then $D(\Gamma) = \{B \in F(s) \mid \vdash A \to B\}$.*

Proof. By defined of generalized root and the deduction theorem , it is easy to check the proposition above.

Proposition 12. *Suppose that A, B are generalized roots of Γ_1 and Γ_2 respectively, then $D(\Gamma_1) \subseteq D(\Gamma_2)$ iff $\vdash B \to A$.*

Proof. Let $D(\Gamma_1) \subseteq D(\Gamma_2)$. Since $A \in D(\Gamma_2)$, then $\vdash B \to A$ by Proposition 11. For the converse, $\forall C \in D(\Gamma_1), \vdash A \to C$ by Proposition 11, it follows from the assumption and Hypothetical syllogism that $\vdash B \to C$, hence $C \in D(\Gamma_2)$ by Proposition 11,thus $D(\Gamma_1) \subseteq D(\Gamma_2)$.

The following Corollary is also obvious.

Corollary 3. *Suppose that* $\Gamma_1 = \{A_1, \cdots, A_s\}, \Gamma_2 = \{B_1, \cdots, B_t\}$, *then* $D(\Gamma_1) = D(\Gamma_2)$ *iff* $A_1 \& \cdots \& A_s \sim B_1 \& \cdots \& B_t$.

Theorem 7. *Suppose that* $\Gamma_k \subseteq F(S)$ *and* A_k *is a generalized root of* $\Gamma_k, k = 1, 2, \cdots, m$, *then* $\bigvee_{i=1}^{m} A_i$ *is a generalized common root of* $\Gamma_1, \cdots, \Gamma_m$.

Proof. Since $\vdash A_k \to \bigvee_{i=1}^{m} A_i, k = 1, \cdots, m$, $\bigvee_{i=1}^{m} A_i \in \bigcap_{k=1}^{m} D(\Gamma_k)$ by proposition 11. $\forall B \in \bigcap_{k=1}^{m} D(\Gamma_k), \vdash A_i \to B, i = 1, \cdots, m$ by proposition 11. Since $\vdash (A_1 \to B) \wedge \cdots \wedge (A_m \to B)$ and $(A_1 \to B) \wedge \cdots \wedge (A_m \to B) \sim \bigvee_{i=1}^{m} A_i \to B$, thus $\vdash \bigvee_{i=1}^{m} A_i \to B$ holds, and then $\bigvee_{i=1}^{m} A_i$ is a generalized common root of $\Gamma_1, \cdots, \Gamma_m$.

Theorem 8. *Suppose that* $\Gamma_k \subseteq F(S)$ *and* A_k *is a generalized root of* $\Gamma_k, k = 1, 2, \cdots, m$, *then* $\Gamma_1, \cdots, \Gamma_m$ *have common conclusions which non-theorem iff* $\bigvee_{i=1}^{m} A_i$ *is non-theorem.*

Proof. Suppose that A is common conclusion of $\Gamma_1, \cdots, \Gamma_m$ which non-theorem. $\vdash A_i \to A, i = 1, \cdots, m$ by proposition 11, it easy to check that $\vdash \bigvee_{i=1}^{m} A_i \to A$. It follows from the completeness theorem that $\bigvee_{i=1}^{m} A_i$ is no a tautology, and then $\bigvee_{i=1}^{m} A_i$ is non-theorem. For the converse, $\forall B \in F(s)$, since $\vdash \bigvee_{i=1}^{m} A_i \to (B \to \bigvee_{i=1}^{m} A_i)$ and $\vdash A_i \to \bigvee_{i=1}^{m} A_i, i = 1, \cdots, m$ holds, hence $\vdash A_t \to (B \to \bigvee_{i=1}^{m} A_i)$ $t = 1, \cdots m$ holds. By proposition 11, $B \to \bigvee_{i=1}^{m} A_i$ is a common conclusion of $\Gamma_1, \cdots, \Gamma_m$. Obvious, if B is a theorem then $B \to \bigvee_{i=1}^{m} A_i$ is non-theorem. Hereby, the proof comes to an end.

6 Results in the R_o – Fuzzy Logic System

Proposition 13. *Suppose that* A *is a generalized roots of* Γ, *then* $D(\Gamma) = \{B \in F(s) \mid \vdash A^2 \to B\}$.

Proof. The proof is analogous to that Proposition 2 and so is omitted.

Lemma 4 (5). *Suppose that* $A \in F(s)$. *Then* $A^n \sim A^2, n = 2, 3, \cdots$.

Theorem 9. *Suppose that A, B are generalized root of Γ_1 and Γ_2 respectively, then*
(i) $D(\Gamma_1) \subseteq D(\Gamma_2)$ iff $\vdash B^2 \to A^2$.
(ii) $D(\Gamma_1) = D(\Gamma_2)$ iff $A^2 \sim B^2$.

Proof. (i) Assume $D(\Gamma_1) \subseteq D(\Gamma_2)$, hence $\Gamma_2 \vdash A$, $\vdash B^2 \to A$ by Proposition 13. By Lemma 3, we get that $\vdash B^4 \to A^2$. then $\vdash B^2 \to A^2$ by Lemma 4. For the converse, $\forall C \in D(\Gamma_1)$, $\vdash A^2 \to C$ by Proposition 13, $\vdash B^2 \to C$ by Hypothetical syllogism and the assumption, hence $C \in D(\Gamma_2)$ by Proposition 13, thus $D(\Gamma_1) \subseteq D(\Gamma_2)$.
(ii)It is easy to prove that item (ii) by item (i).

The following corollary is also obvious.

Corollary 4. *Suppose that $\Gamma_1 = \{A_1, \cdots, A_s\}, \Gamma_2 = \{B_1, \cdots, B_t\}$, then $D(\Gamma_1) = D(\Gamma_2)$ iff $(A_1 \& \cdots \& A_s)^2 \sim (B_1 \& \cdots \& B_t)^2$.*

Theorem 10. *Suppose that $\Gamma_k \subseteq F(S)$ and A_k is a generalized root of $\Gamma_k, k = 1, 2, \cdots, m$, then $\bigvee_{i=1}^{m} A_i^2$ is a generalized common root of $\Gamma_1, \cdots, \Gamma_m$.*

Proof. By axioms, $\vdash A_i^2 \to \bigvee_{i=1}^{m} A_i^2, i = 1, \cdots, m$. By the generalized deduction theorem, we can get $\Gamma_i \vdash A^2 \vee \cdots \vee A_m^2, i = 1, \cdots, m$, i.e., $A^2 \vee \cdots \vee A_m^2 \in \bigcap_{k=1}^{m} D(\Gamma_k)$. $\forall B \in \bigcap_{k=1}^{m} D(\Gamma_k)$, by Proposition 13, $\vdash A_i^2 \to B, i = 1, \cdots, k$. Hence $\vdash (A_1^2 \to B) \wedge \cdots \wedge (A_m^2 \to B)$, it easy to check that $\vdash A_1^2 \vee \cdots \vee A_m^2 \to B$, then $A_1^2 \vee \cdots \vee A_m^2$ is a generalized common root of $\Gamma_1, \cdots, \Gamma_m$.

Theorem 11. *Suppose that $\Gamma_k \subseteq F(S)$ and A_k is a generalized root of $\Gamma_k, k = 1, 2, \cdots, m$, then $\Gamma_1, \cdots, \Gamma_m$ have common conclusions which non-theorem iff $\bigvee_{i=1}^{m} A_i^2$ is non-theorem.*

Proof. The proof is analogous to that Theorem 5.7 and so is omitted.

7 Conclusion Remarks

The present paper, based on the generalized deduction theorem and completeness theorem, gives the definition of generalized root of theory Γ in propositional logic systems. It is proved that $D(\Gamma)$ is completely determined by its generalized root whenever Γ has a generalized root. In Łukasiewicz propositional fuzzy logic systems,the n-valued Łukasiewicz propositional logic systems, *Gödel* propositional fuzzy logic system, product propositional fuzzy logic system and the R_0-propositional fuzzy logic systems, every finite theory has a generalized root. Gives the conditions for the sets $D(\Gamma)$ of two different theories being equal and included.

References

1. S.Gottwald (2001) A Treatise on Many-Valued Logics and Computation, Research Studies Press. Baldock
2. P.Hájek (1998) Metamathematics of Fuzzy Logic, Kluwer Academic Publishers, Dordrecht
3. A.G. Hamilton (1978) Logic for Mathematicians, Combridge Univ. P ress,London
4. G.J.Wang (2003) Introduction to Mathematical Logic and Resolution Principle, Science in China Press. Beijing, in Chinese
5. G.J.Wang (2000) Theory of Non-classical Mathematical Logic and Approximate Reasoning, science in China Press, Beijing, in Chinese
6. G.J.Wang (2001) Non-fuzzy versions of fuzzy reasoning in classical logics, Information Sciences, 138:211-236
7. H.B. Wu (2003)Basic R_o -algebra and basic logic system L^*, Process in Mathematics, 32: 565-576
8. G.J.Wang (1997) A formal deductive system for fuzzy propositional calculus, chinese sci. Bull 42:1521-1526
9. J.Pavelka (1979) On fuzzy logic II-Enriched residuated lattices and semantics of propositional calculi, Zeitschr. f. Math. Logik und Grundlagen d. Math. 25: 119-134
10. G.J.Wang (1997) A formal deductive system for fuzzy propositional calculus, chinese science Bulletin 42: 1521-1525
11. D.W.Pei(2001) The operation $\otimes$ and dedutive theorem in the formal deductive system L^*,Fuzzy syste.Math. 15(1):34-39

Countable Dense Subsets and Countable Nested Sets

Wenyi Zeng, Hongxing Li, and Chengzhong Luo

School of Mathematical Sciences, Beijing Normal University, Beijing, 100875, P.R. China
{zengwy,lhxqx}@bnu.edu.cn

Abstract. Fuzzy set theory becomes more popular in recent years duo to its simplicity in modeling and effectiveness in control problems. In this paper, we introduce the concepts of countable dense subset and countable nested sets, investigate countable decomposition theorems and countable representation theorems in detail. These conclusions can be applied in the general situation to develop some branches in fuzzy set theory.

Keywords: Fuzzy set, countable dense subset, countable nested set, decomposition theorem, representation theorem.

1 Introduction

Fuzzy set theory proposed by Zadeh[20] has been successfully applied in many different fields. In 1983, Luo[8] introduced the concept of nested sets, investigated the ordinary union decomposition theorems and representation theorem for fuzzy set and pointed out that any fuzzy set could be described by some nested sets. In other words, $A, B \in \mathcal{F}(X)$, $A = B$ if and only if $A_\lambda = B_\lambda$ for every cut level $\lambda \in [0, 1]$. Obviously, the condition is quite strong. The ordinary intersection decomposition theorems for fuzzy set was presented by Gu and Zhou[4] in 1995. Ralescu[10, 11, 12] investigated representation theorems for fuzzy set. Recently, Huang and Shi[5] presented union decomposition theorems and intersection decomposition theorems for fuzzy set. These works have a significant contribution to the fundamental study of fuzzy set theory.

The essence of the representation theorem is to do with the possibility of constructing a fuzzy set from a family of sets. From the development of fuzzy set theory, it is easy to see that decomposition theorems and representation theorems played an important role, they helped people to develop many branches such as fuzzy algebra, fuzzy measure and integral, fuzzy analysis, fuzzy decision making and so on.

In this paper, we introduce the concepts of countable dense subset of $(0, 1)$ and countable nested sets, investigate countable decomposition theorems and countable representation theorems in detail and point out that for fuzzy sets A, B, $A = B$ if and only if $A_\alpha = B_\alpha$ for every $\alpha \in Q \subseteq (0, 1)$. These conclusions

B.-Y. Cao (Ed.): Fuzzy Information and Engineering (ICFIE), ASC 40, pp. 170–180, 2007.
springerlink.com

can be applied in the general situation to develop some branches in fuzzy set theory. Therefore, our work is meaningful in theory.

Our work is organized as follows. In the section 2, we introduce the concept of countable dense subset of $(0,1)$ and propose countable decomposition theorems. In the section 3, we introduce the concept of countable nested set on X and propose countable representation theorems of fuzzy set. The final is conclusion.

2 Countable Decomposition Theorems

In this section, we introduce the concept of countable dense subset of $(0,1)$, discuss some properties of cut set of fuzzy set and propose three decomposition theorems for $\lambda \in [0,1]$ and six countable decomposition theorems. For simplicity, we write X to denote the discourse set, X is an nonempty set, A is a fuzzy set and $A(x)$ is its membership function, $\mathcal{F}(X)$ and $\mathcal{P}(X)$ stand for the set of all fuzzy and crisp sets in X, respectively.

Definition 1[8]. For $A \in \mathcal{F}(X)$, $\lambda \in [0,1]$,

$$A_\lambda^1 = \{x \in X | A(x) \geqslant \lambda\}, \;\; A_\lambda^2 = \{x \in X | A(x) > \lambda\}$$

$$A_\lambda^3 = \{x \in X | A(x) \leqslant \lambda\}, \;\; A_\lambda^4 = \{x \in X | A(x) < \lambda\}$$

is called the $i-$th cut set of fuzzy set A, $i = 1,2,3,4$.

Remark 1. Known by Luo[1], A_λ^2 is called strong $\lambda-$cut set of fuzzy set A. Obviously, we have the following properties.

Property

1) $A_\lambda^1 \supseteq A_\lambda^2$, $A_\lambda^3 \supseteq A_\lambda^4$
2) $\left(A_\lambda^3\right)^c = A_\lambda^2$, $\left(A_\lambda^4\right)^c = A_\lambda^1$
3) $\lambda_1 < \lambda_2 \Longrightarrow \begin{cases} A_{\lambda_1}^1 \supseteq A_{\lambda_1}^2 \supseteq A_{\lambda_2}^1 \supseteq A_{\lambda_2}^2 \\ A_{\lambda_1}^4 \subseteq A_{\lambda_1}^3 \subseteq A_{\lambda_2}^4 \subseteq A_{\lambda_2}^3 \end{cases}$
4) $(A \cup B)_\lambda^i = A_\lambda^i \cup B_\lambda^i, (A \cap B)_\lambda^i = A_\lambda^i \cap B_\lambda^i$, $i = 1,2,3,4, \forall \lambda \in [0,1]$.

Definition 2. For $\lambda \in [0,1], A \in \mathcal{F}(X)$, we order $\lambda \cdot A \in \mathcal{F}(X), \lambda * A \in \mathcal{F}(X)$ and their membership functions are defined as follows.

$$(\lambda \cdot A)(x) = \lambda \wedge A(x), \qquad (\lambda * A)(x) = \lambda \vee A(x)$$

Theorem 1[8]. $A = \bigcup\limits_{\lambda \in [0,1]} \lambda \cdot A_\lambda^1 = \bigcup\limits_{\lambda \in [0,1]} \lambda \cdot A_\lambda^2$

Theorem 2[8]. Supposed H be a mapping from $[0,1]$ to $\mathcal{P}(X)$, $H : [0,1] \to \mathcal{P}(X), \lambda \to H(\lambda)$ and satisfies $A_\lambda^2 \subseteq H(\lambda) \subseteq A_\lambda^1$, then

$$A = \bigcup_{\lambda \in [0,1]} \lambda \cdot H(\lambda)$$

and we have the following properties.

$$1)\ \lambda_1 < \lambda_2 \Longrightarrow H(\lambda_1) \supseteq H(\lambda_2)$$

$$2)\ \begin{cases} A^1_\lambda = \bigcap\limits_{\alpha<\lambda} H(\alpha),\ \lambda \neq 0 \\ A^2_\lambda = \bigcup\limits_{\alpha>\lambda} H(\alpha),\ \lambda \neq 1 \end{cases}$$

Theorem 3. $A = \Big(\bigcap\limits_{\lambda\in[0,1]} \lambda * A^3_{\lambda^c} \Big)^c$, where $\lambda^c = 1 - \lambda$.

Proof

(1) $A(x) \neq 1$ case:

$$\begin{aligned}
&\Big(\bigcap_{\lambda\in[0,1]} \lambda * A^3_{\lambda^c} \Big)^c (x) = 1 - \Big(\bigcap_{\lambda\in[0,1]} \lambda * A^3_{\lambda^c} \Big)(x) \\
&= 1 - \bigwedge_{\lambda\in[0,1]} \Big(\lambda * A^3_{\lambda^c}\Big)(x) = 1 - \bigwedge_{\lambda\in[0,1]} \Big(\lambda \vee (A^3_{\lambda^c}(x))\Big) \\
&= 1 - \bigwedge_{\lambda\in[0,1]} \{\lambda | A^3_{\lambda^c}(x) = 0\} = 1 - \bigwedge_{\lambda\in[0,1]} \{\lambda | x \notin A^3_{\lambda^c}\} \\
&= 1 - \bigwedge_{\lambda\in[0,1]} \{\lambda | A(x) > \lambda^c\} = 1 - \bigwedge_{\lambda\in[0,1]} \{\lambda | A(x) > 1 - \lambda\} \\
&= 1 - \bigwedge_{\lambda\in[0,1]} \{\lambda | \lambda > 1 - A(x)\} = 1 - (1 - A(x)) \\
&= A(x)
\end{aligned}$$

(2) $A(x) = 1$ case:

$\forall\lambda \in (0,1], A(x) > \lambda^c$, we have $A(x) > \lambda^c \Longrightarrow A^3_{\lambda^c}(x) = 0$, and $A^3_{0^c}(x) = A^3_1(x) = 1$, then we have

$$\begin{aligned}
1 &\geqslant \Big(\bigcap_{\lambda\in[0,1]} \lambda * A^3_{\lambda^c} \Big)^c (x) = 1 - \Big(\bigcap_{\lambda\in[0,1]} \lambda * A^3_{\lambda^c} \Big)(x) \\
&= 1 - \bigwedge_{\lambda\in[0,1]} \Big(\lambda * A^3_{\lambda^c}\Big)(x) = 1 - \bigwedge_{\lambda\in[0,1]} \Big(\lambda \vee (A^3_{\lambda^c}(x))\Big) \\
&\geqslant 1 - \bigwedge_{\lambda\in(0,1]} \Big(\lambda \vee (A^3_{\lambda^c}(x))\Big) = 1 - \bigwedge_{\lambda\in(0,1]} (\lambda \vee 0) \\
&= 1 - \bigwedge_{\lambda\in(0,1]} \lambda = 1 - 0 \\
&= 1
\end{aligned}$$

Therefore,

$$\Big(\bigcap_{\lambda\in[0,1]} \lambda * A^3_{\lambda^c} \Big)^c (x) = 1 = A(x)$$

Hence, we complete the proof of Theorem 3.

Theorem 4. $A = \left(\bigcap_{\lambda \in [0,1]} \lambda * A^4_{\lambda^c} \right)^c$, where $\lambda^c = 1 - \lambda$.

Proof

$$\begin{aligned}
&\left(\bigcap_{\lambda \in [0,1]} \lambda * A^4_{\lambda^c} \right)^c (x) = 1 - \left(\bigcap_{\lambda \in [0,1]} \lambda * A^4_{\lambda^c} \right)(x) \\
&= 1 - \bigwedge_{\lambda \in [0,1]} \left(\lambda * A^4_{\lambda^c} \right)(x) = 1 - \bigwedge_{\lambda \in [0,1]} \left(\lambda \vee A^4_{\lambda^c}(x) \right) \\
&= 1 - \bigwedge_{\lambda \in [0,1]} \{\lambda | A^4_{\lambda^c}(x) = 0\} = 1 - \bigwedge_{\lambda \in [0,1]} \{\lambda | x \notin A^4_{\lambda^c}\} \\
&= 1 - \bigwedge_{\lambda \in [0,1]} \{\lambda | A(x) \geqslant \lambda^c\} = 1 - \bigwedge_{\lambda \in [0,1]} \{\lambda | A(x) \geqslant 1 - \lambda\} \\
&= 1 - \bigwedge_{\lambda \in [0,1]} \{\lambda | \lambda \geqslant 1 - A(x)\} = 1 - (1 - A(x)) \\
&= A(x)
\end{aligned}$$

Hence, we complete the proof of Theorem 4.

Theorem 5. Supposed J be a mapping from $[0,1]$ to $\mathcal{P}(X)$, $J : [0,1] \to \mathcal{P}(X), \lambda \to J(\lambda)$ and satisfies $A^4_\lambda \subseteq J(\lambda) \subseteq A^3_\lambda$, then

$$A = \left(\bigcap_{\lambda \in [0,1]} \lambda * J(\lambda^c) \right)^c$$

and we have the following properties.

$$1)\ \lambda_1 < \lambda_2 \Longrightarrow J(\lambda_1) \subseteq J(\lambda_2)$$

$$2) \begin{cases} A^3_\lambda = \bigcap_{\alpha > \lambda} J(\alpha), \ \lambda \neq 1 \\ A^4_\lambda = \bigcup_{\alpha < \lambda} J(\alpha), \ \lambda \neq 0 \end{cases}$$

Proof. First, for every $\lambda \in [0,1]$, we have

$$\begin{aligned}
&A^4_\lambda \subseteq J(\lambda) \subseteq A^3_\lambda \Longrightarrow A^4_{\lambda^c} \subseteq J(\lambda^c) \subseteq A^3_{\lambda^c} \\
&\Longrightarrow \lambda * A^4_{\lambda^c} \subseteq \lambda * J(\lambda^c) \subseteq \lambda * A^3_{\lambda^c} \\
&\Longrightarrow \bigcap_{\lambda \in [0,1]} \lambda * A^4_{\lambda^c} \subseteq \bigcap_{\lambda \in [0,1]} \lambda * J(\lambda^c) \subseteq \bigcap_{\lambda \in [0,1]} \lambda * A^3_{\lambda^c} \\
&\Longrightarrow A = \left(\bigcap_{\lambda \in [0,1]} \lambda * A^4_{\lambda^c} \right)^c \supseteq \left(\bigcap_{\lambda \in [0,1]} \lambda * J(\lambda^c) \right)^c \supseteq \left(\bigcap_{\lambda \in [0,1]} \lambda * A^3_{\lambda^c} \right)^c = A
\end{aligned}$$

Therefore, $A = \left(\bigcap_{\lambda \in [0,1]} \lambda * J(\lambda^c) \right)^c$.

And known by the properties of cut set,

(1) For $\lambda_1 < \lambda_2$, we have $A^4_{\lambda_1} \subseteq A^3_{\lambda_1} \subseteq A^4_{\lambda_2} \subseteq A^3_{\lambda_2}$, so, we can get $A^4_{\lambda_1} \subseteq J(\lambda_1) \subseteq A^3_{\lambda_1} \subseteq A^4_{\lambda_2} \subseteq J(\lambda_2) \subseteq A^3_{\lambda_2}$, therefore, $J(\lambda_1) \subseteq J(\lambda_2)$.

(2) For $\forall \alpha > \lambda$, we have

$$\begin{aligned} & A_\lambda^3 \subseteq A_\alpha^4 \subseteq J(\alpha) \subseteq A_\alpha^3 \\ \Longrightarrow & A_\lambda^3 \subseteq \bigcap_{\alpha>\lambda} J(\alpha) \subseteq \bigcap_{\alpha>\lambda} A_\alpha^3 = A^3_{\bigwedge\limits_{\alpha>\lambda} \alpha} = A_\lambda^3 \end{aligned}$$

Hence, $A_\lambda^3 = \bigcap\limits_{\alpha>\lambda} J(\alpha)$.

Similarly, we have

$$\begin{aligned} & A_\lambda^4 \supseteq A_\alpha^3 \supseteq J(\alpha) \supseteq A_\alpha^4 \quad (\forall \alpha < \lambda) \\ \Longrightarrow & A_\lambda^4 \supseteq \bigcup_{\alpha<\lambda} J(\alpha) \supseteq \bigcup_{\alpha<\lambda} A_\alpha^4 = A^4_{\bigvee\limits_{\alpha<\lambda} \alpha} = A_\lambda^4 \end{aligned}$$

Therefore, $A_\lambda^4 = \bigcup\limits_{\alpha<\lambda} J(\alpha)$.

Hence, we complete the proof of Theorem 5.

Definition 3. If $Q = \{\alpha_k | k = 1, 2, \cdots\} \subseteq (0, 1)$ and Q is dense in $(0, 1)$, $1 - \alpha_k \in Q$, then we call Q countable dense subset of $(0, 1)$.

Lemma 1. Suppose Q be countable dense subset of $(0, 1)$, then we have

1) For $\lambda_1, \lambda_2 \in (0, 1)$ and $\lambda_1 < \lambda_2$, then there exists $\alpha \in Q$ such that $\lambda_1 < \alpha < \lambda_2$;

2) For $\lambda \in (0, 1)$, then we have $\lambda = \bigvee\{\alpha | \alpha < \lambda, \alpha \in Q\} = \bigwedge\{\alpha | \alpha > \lambda, \alpha \in Q\}$.

Proof. 1) Known by the definition 3, it is obvious.

2) Obviously, $\bigvee\{\alpha | \alpha < \lambda, \alpha \in Q\} \leq \lambda$. If the equality does not hold, then there exists $\gamma \in Q$ such that $\bigvee\{\alpha | \alpha < \lambda, \alpha \in Q\} < \gamma < \lambda$, it is a contradiction. Therefore, $\bigvee\{\alpha | \alpha < \lambda, \alpha \in Q\} = \lambda$.

With the same reason, we have $\bigwedge\{\alpha | \alpha > \lambda, \alpha \in Q\} = \lambda$.

Corollary 1. $A = \bigcup\limits_{\alpha \in Q} \alpha \cdot A_\alpha^1 = \bigcup\limits_{\alpha \in Q} \alpha \cdot A_\alpha^2$.

Corollary 2. Let H be defined as above, then we have $A = \bigcup\limits_{\alpha \in Q} \alpha \cdot H(\alpha)$, and

$$1)\ \alpha_1 < \alpha_2 \Longrightarrow H(\alpha_1) \supseteq H(\alpha_2)$$

$$2) \begin{cases} A_\lambda^1 = \bigcap\limits_{\alpha<\lambda} H(\alpha), \ \lambda \neq 0, \ \alpha \in Q \\ A_\lambda^2 = \bigcup\limits_{\alpha>\lambda} H(\alpha), \ \lambda \neq 1, \ \alpha \in Q \end{cases}$$

Corollary 3. $A = \left(\bigcap_{\alpha \in Q} \alpha * A^3_{\alpha^c} \right)^c = \left(\bigcap_{\alpha \in Q} \alpha * A^4_{\alpha^c} \right)^c.$

Corollary 4. Let J be defined as above, then we have $A = \left(\bigcap_{\alpha \in Q} \alpha * J(\alpha^c) \right)^c$,

and

$$1)\ \alpha_1 < \alpha_2 \Longrightarrow J(\alpha_1) \subseteq J(\alpha_2)$$

$$2) \begin{cases} A^3_\lambda = \bigcap_{\alpha > \lambda} J(\alpha),\ \lambda \neq 1,\ \alpha \in Q \\ A^3_\lambda = \bigcup_{\alpha < \lambda} J(\alpha),\ \lambda \neq 0,\ \alpha \in Q \end{cases}$$

Note 1. These corollaries are also called countable decomposition theorems.

3 Countable Representation Theorems

In this section, we introduce the concept of countable nested sets on X and propose countable representation theorems of fuzzy set.

Definition 4[8]. Given a mapping $H : [0,1] \longrightarrow \mathcal{P}(X), \lambda \rightarrow H(\lambda)$ and mapping H satisfies the property: $\lambda_1 < \lambda_2 \Longrightarrow H(\lambda_1) \supseteq H(\lambda_2)$, then H is called 1−th type of nested set on X. The set of all 1−th type of nested set on X is denoted by $\mathcal{U}^1(X)$. The 1−th type of nested set was also called nested set in Luo[8].

Theorem 6[8]. Given a mapping $S : \mathcal{U}^1(X) \longrightarrow \mathcal{F}(X), H \longrightarrow S(H) = \bigcup_{\lambda \in [0,1]} \lambda \cdot H(\lambda)$, then S is a homomorphism surjection from $(\mathcal{U}^1(X), \cup, \cap, c)$ to $(\mathcal{F}(X), \cup, \cap, c)$ and the following properties hold.

$$S(H)^2_\lambda \subseteq H(\lambda) \subseteq S(H)^1_\lambda, \quad \lambda \in [0,1]$$

$$\begin{cases} S(H)^1_\lambda = \bigcap_{\alpha < \lambda} H(\alpha),\ \lambda \in (0,1] \\ S(H)^2_\lambda = \bigcup_{\alpha > \lambda} H(\alpha),\ \lambda \in [0,1) \end{cases}$$

Definition 5. Let Q be countable dense subset of $(0,1)$, and for a given mapping

$$\begin{aligned} H :\quad Q &\longrightarrow \quad \mathcal{P}(X) \\ \alpha \in Q &\longrightarrow H(\alpha) \in \mathcal{P}(X) \end{aligned}$$

If mapping H satisfies the property: $\alpha_1 < \alpha_2 \Rightarrow H(\alpha_1) \supseteq H(\alpha_2)$, then H is called 1−th type of countable nested set for Q on X. The set of all 1−th type of countable nested set for Q on X is denoted by $\mathcal{U}^1(X, Q)$.

In the following, we introduce some operations such as union, intersection and "c" on 1−th countable nested set $\mathcal{U}^1(X, Q)$, $\forall \alpha \in Q$, we have:

$$\begin{aligned}
H_1 \bigcup H_2 :& \ \Big(H_1 \bigcup H_2\Big)(\alpha) \triangleq H_1(\alpha) \bigcup H_2(\alpha) \\
H_1 \bigcap H_2 :& \ \Big(H_1 \bigcap H_2\Big)(\alpha) \triangleq H_1(\alpha) \bigcap H_2(\alpha) \\
\bigcup_{\gamma \in \Gamma} H_\gamma :& \quad \Big(\bigcup_{\gamma \in \Gamma} H_\gamma\Big)(\alpha) = \bigcup_{\gamma \in \Gamma} H_\gamma(\alpha) \\
\bigcap_{\gamma \in \Gamma} H_\gamma :& \quad \Big(\bigcap_{\gamma \in \Gamma} H_\gamma\Big)(\alpha) = \bigcap_{\gamma \in \Gamma} H_\gamma(\alpha) \\
H^c :& \qquad H^c(\alpha) = \Big(H(1-\alpha)\Big)^c
\end{aligned}$$

Then, known by Theorem 2, we have the following theorem.

Theorem 7(Countable representation theorem I). Given a mapping $S : \mathcal{U}^1(X, Q) \longrightarrow \mathcal{F}(X), H \longrightarrow S(H) = \bigcup_{\alpha \in Q} \alpha H(\alpha)$, then S is a homomorphism surjection from $(\mathcal{U}^1(X, Q), \cup, \cap, c)$ to $(\mathcal{F}(X), \cup, \cap, c)$ and the following properties hold.

$$S(H)^2_\alpha \subseteq H(\alpha) \subseteq S(H)^1_\alpha, \ \forall \alpha \in Q$$

$$\begin{cases} S(H)^1_\lambda = \bigcap_{\alpha < \lambda} H(\alpha), \ \lambda \in (0, 1], \alpha \in Q \\ S(H)^2_\lambda = \bigcup_{\alpha > \lambda} H(\alpha), \ \lambda \in [0, 1), \alpha \in Q \end{cases}$$

Definition 6. Given a mapping $J : [0, 1] \longrightarrow \mathcal{P}(X), \lambda \rightarrow J(\lambda)$ and mapping J satisfies the property: $\lambda_1 < \lambda_2 \Longrightarrow J(\lambda_1) \subseteq J(\lambda_2)$, then J is called 2−th type of nested set on X. The set of all 2−th type of nested set on X is denoted by $\mathcal{U}^2(X)$.

Here, we introduce some operations such as union, intersection and c on nested set $\mathcal{U}^2(X)$, $\forall \lambda \in [0, 1]$.

$$\begin{aligned}
J_1 \cup J_2 :& \ \Big(J_1 \cup J_2\Big)(\lambda) \triangleq J_1(\lambda) \cup J_2(\lambda) \\
J_1 \cap J_2 :& \ \Big(J_1 \cap J_2\Big)(\lambda) \triangleq J_1(\lambda) \cap J_2(\lambda) \\
\bigcup_{\gamma \in \Gamma} J_\gamma :& \ \Big(\bigcup_{\gamma \in \Gamma} J_\gamma\Big)(\lambda) \triangleq \bigcup_{\gamma \in \Gamma} \Big(J_\gamma(\lambda)\Big) \\
\bigcap_{\gamma \in \Gamma} J_\gamma :& \ \Big(\bigcap_{\gamma \in \Gamma} J_\gamma\Big)(\lambda) \triangleq \bigcap_{\gamma \in \Gamma} \Big(J_\gamma(\lambda)\Big) \\
J^c \ :& \ J^c(\lambda) \triangleq \Big(J(1-\lambda)\Big)^c = \Big(J(\lambda^c)\Big)^c
\end{aligned}$$

Theorem 8. Given a mapping $T : \mathcal{U}^2(X) \longrightarrow \mathcal{F}(X), J \longrightarrow T(J) = \Big(\bigcap_{\lambda \in [0,1]} \lambda * J(\lambda^c)\Big)^c$, then T is a homomorphism surjection from $(\mathcal{U}^2(X), \cup, \cap, c)$ to $(\mathcal{F}(X), \cup, \cap, c)$ and the following properties hold.

$$T(J)^4_\lambda \subseteq J(\lambda) \subseteq T(J)^3_\lambda, \quad \lambda \in [0,1]$$
$$\begin{cases} T(J)^3_\lambda = \bigcap\limits_{\alpha>\lambda} J(\alpha), \ \lambda \neq 1 \\ T(J)^4_\lambda = \bigcup\limits_{\alpha<\lambda} J(\alpha), \ \lambda \neq 0 \end{cases}$$

Proof. At first, we consider an given fuzzy set A, let $J(\lambda) = A^3_{\lambda^c} \in \mathcal{U}^2(X)$, where $\lambda^c = 1 - \lambda$, then, known by Theorem 4, we have $A = T(J)$. It shows that mapping T is surjection. In the following, we want to prove that all properties hold.

1) $\forall \lambda \in [0,1]$, $\Big(T(J)\Big)^4_{\lambda^c} \subseteq J(\lambda^c) \subseteq \Big(T(J)\Big)^3_{\lambda^c}$

For every element $x \in J(\lambda^c)$, then we have $J(\lambda^c)(x) = 1$. And $\forall \alpha < \lambda$, $J(\alpha^c)(x) = 1$,

$$\begin{aligned} T(J)(x) &= \Big(\bigcap_{\alpha\in[0,1]} \alpha * J(\alpha^c)\Big)^c(x) \\ &= 1 - \Big(\bigcap_{\alpha\in[0,1]} \alpha * J(\alpha^c)\Big)(x) = 1 - \Big(\bigwedge_{\alpha\in[0,1]} (\alpha \vee J(\alpha^c)(x))\Big) \\ &= 1 - \Big(\bigwedge_{\alpha\in[\lambda,1]} (\alpha \vee J(\alpha^c)(x))\Big) \leqslant 1 - \Big(\bigwedge_{\alpha\in[\lambda,1]} \alpha\Big) \\ &= 1 - \lambda = \lambda^c \end{aligned}$$

Thus, $x \in T(J)^3_{\lambda^c}$, namely, $J(\lambda^c) \subseteq T(J)^3_{\lambda^c}$.

On the other word, if $x \notin J(\lambda^c)$, then we have $J(\lambda^c)(x) = 0$. And

$$\begin{aligned} T(J)(x) &= \Big(\bigcap_{\alpha\in[0,1]} \alpha * J(\alpha^c)\Big)^c(x) \\ &= 1 - \Big(\bigcap_{\alpha\in[0,1]} \alpha * J(\alpha^c)\Big)(x) = 1 - \Big(\bigwedge_{\alpha\in[0,1]} (\alpha \vee J(\alpha^c)(x))\Big) \\ &\geqslant 1 - \Big(\lambda \vee J(\lambda^c)(x)\Big) = 1 - \lambda \\ &= \lambda^c \end{aligned}$$

Thus, $x \notin T(J)^4_{\lambda^c}$, namely, $T(J)^4_{\lambda^c} \subseteq J(\lambda^c)$.

Therefore, for every $\lambda \in [0,1]$, we have $T(J)^4_\lambda \subseteq J(\lambda) \subseteq T(J)^3_\lambda$.

Known by Theorem 6, we can get two properties in the following.

2) $\Big(T(J)\Big)^3_\lambda = \bigcap\limits_{\alpha>\lambda} J(\alpha), \ \lambda \neq 1$

3) $\Big(T(J)\Big)^4_\lambda = \bigcup\limits_{\alpha<\lambda} J(\alpha), \ \lambda \neq 0$

4) $T\Big(\bigcup\limits_{t\in\Gamma} J_t\Big) = \bigcup\limits_{t\in\Gamma} T(J_t)$

$$\begin{aligned}\Big[T\Big(\bigcup_{t\in\Gamma} J_t\Big)\Big]_\lambda^4 &= \bigcup_{\alpha<\lambda}\Big(\bigcup_{t\in\Gamma} J_t\Big)(\alpha)\\ &= \bigcup_{\alpha<\lambda}\Big(\bigcup_{t\in\Gamma} J_t(\alpha)\Big) = \bigcup_{t\in\Gamma}\Big(\bigcup_{\alpha<\lambda} J_t(\alpha)\Big)\\ &= \Big[\bigcup_{t\in\Gamma} T\Big(J_t\Big)\Big]_\lambda^4\end{aligned}$$

Known by Theorem 5, we have

$$T(\bigcup_{t\in\Gamma} J_t) = \bigcup_{t\in\Gamma} T(J_t)$$

5) $T\Big(\bigcap_{t\in\Gamma} J_t\Big) = \bigcap_{t\in\Gamma} T(J_t)$

$$\begin{aligned}\Big[T\Big(\bigcap_{t\in\Gamma} J_t\Big)\Big]_\lambda^3 &= \bigcap_{\alpha>\lambda}\Big(\bigcap_{t\in\Gamma} J_t\Big)(\alpha)\\ &= \bigcap_{\alpha>\lambda}\Big(\bigcap_{t\in\Gamma} J_t(\alpha)\Big) = \bigcap_{t\in\Gamma}\Big(\bigcap_{\alpha>\lambda} J_t(\alpha)\Big)\\ &= \Big[\bigcap_{t\in\Gamma} T\Big(J_t\Big)\Big]_\lambda^3\end{aligned}$$

Known by Theorem 4, we have

$$T(\bigcap_{t\in\Gamma} J_t) = \bigcap_{t\in\Gamma} T(J_t)$$

6) $T(J^c) = \Big(T(J)\Big)^c$

$$\begin{aligned}\Big[T(J^c)\Big]_\lambda^3 &= \bigcap_{\alpha>\lambda} J^c(\alpha)\\ &= \bigcap_{\alpha>\lambda}\Big(J(\alpha^c)\Big)^c = \Big(\bigcup_{\alpha>\lambda} J(\alpha^c)\Big)^c\\ &= \Big(\bigcup_{1-\alpha<1-\lambda} J(\alpha^c)\Big)^c = \Big(\Big[T(J)\Big]_{1-\lambda}^4\Big)^c\\ &= \Big[\Big(T(J)\Big)^c\Big]_\lambda^3\end{aligned}$$

Known by Theorem 4, we have

$$T(J^c) = \Big(T(J)\Big)^c$$

Hence, we complete the proof of Theorem 8.

Definition 6. Let Q be countable dense subset of $(0,1)$, and for a given mapping

$$\begin{aligned}J:\quad Q &\longrightarrow \quad \mathcal{P}(X)\\ \alpha\in Q &\longrightarrow J(\alpha)\in\mathcal{P}(X)\end{aligned}$$

If mapping J satisfies the property: $\alpha_1 < \alpha_2 \Rightarrow J(\alpha_1) \subseteq J(\alpha_2)$, then J is called 2−th type of countable nested set for Q on X. The set of all 2−th type of countable nested set for Q on X is denoted by $\mathcal{U}^2(X,Q)$.

Note 2. 1−th and 2−th types of countable nested set are called countable nested sets.

Similarly, we are able to introduce some operations such as union, intersection and c on 2−th countable nested set $\mathcal{U}^2(X,Q)$, $\forall \alpha \in Q$. And we have the following theorem.

Theorem 9(Countable representation theorem II). Given a mapping T : $\mathcal{U}^2(X,Q) \longrightarrow \mathcal{F}(X), J \longrightarrow T(J) = \Big(\bigcap_{\alpha\in Q} \alpha * J(\alpha^c)\Big)^c$, then T is a homomorphism surjection from $(\mathcal{U}^2(X,Q), \cup, \cap, c)$ to $(\mathcal{F}(X), \cup, \cap, c)$ and the following properties hold.

$$T(J)^4_\alpha \subseteq J(\alpha) \subseteq T(J)^3_\alpha, \quad \alpha \in Q$$

$$\begin{cases} T(J)^3_\lambda = \bigcap\limits_{\alpha>\lambda} J(\alpha), \ \lambda \in [0,1), \alpha \in Q \\ T(J)^4_\lambda = \bigcup\limits_{\alpha<\lambda} J(\alpha), \ \lambda \in (0,1], \alpha \in Q \end{cases}$$

Obviously, if the binary relation " $\sim$" satisfies the following properties: 1) $H \sim H$; 2) $H_1 \sim H_2 \Longrightarrow H_2 \sim H_1$; 3) $H_1 \sim H_2, H_2 \sim H_3 \Longrightarrow H_1 \sim H_3$, the binary relation " $\sim$" is called an equivalent relation. Therefore, we can get its equivalent classification of nested sets $\mathcal{U}^i(X), i = 1,2$, $\langle H \rangle = \{H_1 | H_1 \sim H\}$.

Theorem 10. $\mathcal{U}^i(X)/_\sim \simeq \mathcal{F}(X), \ i = 1,2.$

Remark 2. Theorem 7 and Theorem 9 show that every fuzzy set is an equivalent classification of nested sets $\mathcal{U}^i(X)$ on $X, i = 1,2$.

Remark 3. Our proposed nested set on X and representation theorem in this paper will be able to be extensively applied in fuzzy algebra.

4 Conclusion

In this paper, we introduce the concepts of countable dense set and countable nested sets and investigate countable decomposition theorems and countable representation theorems in detail. These conclusions can be applied in the general situation to develop some branches in fuzzy set theory and enrich fuzzy set theory.

Acknowledgements

This work described here is partially Supported by National Natural Science Foundation of China (NSFC NO.60474023), Science and Technology Key Project Fund of Ministry of Education(NO.03184), and the Major State Basic Research Development Program of China (NO.2002CB312200).

References

[1] Dubois, D., Prade, H.(1978)Operations on fuzzy numbers. Internat. J. Systems Sci. 9:613-626

[2] Fang, J.X., Huang, H.(2004)On the level convergence of a sequence of fuzzy numbers. Fuzzy Sets and Systems 147:417-435

[3] Goetschel, R., Voxman, W.(1986)Elementary fuzzy calculus. Fuzzy Sets and Systems 18:31-42

[4] Gu, W.X., Zhou, J.(1995)A new resolution theory of fuzzy sets. J. Northeast Normal Univ. 2:6-7(in Chinese)

[5] Huang, Y.P., Shi, K.Q.(2000)The $\alpha-$embedded and A's fuzzy decomposition theorems. Fuzzy Sets and Systems 110: 299-306

[6] Klir, G.J., Yuan, B.(1995)Fuzzy Sets and Fuzzy Logic: Theory and Applications, Prentice-Hall International. Inc., Upper Saddle River, NJ, USA

[7] Li, D.F.(1998)Properties of b-vex fuzzy mappings and applications to fuzzy optimization. Fuzzy Sets and Systems 94: 253-260

[8] Luo, C.Z.(1983)Fuzzy sets and nested sets. J. Fuzzy Math. 4:113-126(in Chinese)

[9] Luo, C.Z.(1989)Introduction to fuzzy subsets. Beijing Normal University Press, Beijing(in Chinese)

[10] Negoita, C.V., Ralescu, D.R.(1975)Applications of fuzzy sets to systems analysis. Wiley, New York,

[11] Ralescu, D.R.(1979)A survey of the representation of fuzzy concepts and its applications in: M.M. Gupta etal..Eds., Advances in Fuzzy Sets Theory and Applications, North-Holland, Amsterdam:77-91

[12] Ralescu, D.R.(1992)A generalization of the representation theorem. Fuzzy Sets and Systems 51:309-311

[13] Swamy, U.M., Raju, D.V.(1991)Algebraic fuzzy systems. Fuzzy Sets and Systems 41:187-194

[14] Turksen, I.B.(1991)Measurement of membership function and their acquisition. Fuzzy Sets and Systems 40:5-38

[15] Wang, P.Z.(1983)Fuzzy Sets and Its Application. Shanghai Science and Technology Press, Shanghai(in Chinese)

[16] Wu, C.X., Wang, G.X.(2002)Convergence of sequence of fuzzy numbers and fixed point theorems for increasing fuzzy mappings and application. Fuzzy Sets and Systems 130:383-390

[17] Yager, R.R., Filev, D.P.(1994)Essentials of Fuzzy Modeling and Control. Wiley, NY, USA

[18] Yager, R.R., Zadeh, L.A.(Eds.)(1992)An introduction to Fuzzy Logic Applications in Intelligent Systems. Kluwer Academic Publishers, Boston, IL, USA

[19] Yao, J.F., Yao, J.S.(2001)Fuzzy decision making for medical diagnosis based on fuzzy number and compositional rule of inference, Fuzzy Sets and Systems 120:351-366

[20] Zadeh, L.A.(1965)Fuzzy sets.Infor. and Contr.8:338–353

The Construction of Power Ring

BingxueYao[1] and Yubin Zhong[2]

[1] School of Mathematics Sciences, Liaocheng University, Liaocheng 252059, China
yaobingxue@lcu.edu.cn

[2] School of Mathematics and Information Sciences, Guangzhou University, Guangzhou 510006, China
Zhong_yb@163.com

Abstract. In order to construct nonregular power rings, the concept of regular semiideal with respect to a subring is introduced. Then several constructive theorems of HX ring and power ring are established and some examples of non-regular power rings are constructed which contain some nonregular HX rings.

Keywords: HX Rings, Power Ring, Regular Power Ring, Nontrivial HX Ring, Nontrivial Power Ring.

1 Introduction

With the successful upgrade of algebraic structure of group, many researchers considered the upgrade of algebraic structure of some other algebraic systems, in which the ring was considered first. In 1988, Professor Li proposed the concept of HX ring[1] and derived some of their properties. Then Professor Zhong[2] gave the structures of HX-ring on a class of ring and cited an example of nontrivial HX ring[3], thus proving the existence of nontrivial HX ring. In this paper, we establish a series of constructive theorems of power ring and cite some examples of nontrivial HX ring and nontrivial power ring. Unless otherwise statement, $(R,+,\cdot)$ will always stand for an associative ring with zero element 0.

2 Concepts of HX Ring and Power Ring

Let $P_0(R)=P(R)\setminus\{\Phi\}$, $A,B\in P_0(R)$, we define the following operations[1]:

$$A+B=\{a+b\mid a\in A,b\in B\}, \tag{2.1}$$

$$A\cdot B=\{a\cdot b\mid a\in A,b\in B\}. \tag{2.2}$$

For convenience, we write ab, AB, $a+B$, aB and Ab instead of $a\cdot b$, $A\cdot B$, $\{a\}+B$, $\{a\}\cdot B$ and $A\cdot\{b\}$, respectively.

B.-Y. Cao (Ed.): Fuzzy Information and Engineering (ICFIE), ASC 40, pp. 181–187, 2007.
springerlink.com

Definition 2.1[1]**.** Let $\Re$ be a nonempty subset of $P_0(R)$ such that $\Re$ forms a ring for the operations (2.1) and (2.2). Then $\Re$ is called an HX ring on R whose zero element is denoted by Q and the negative element of $A \in \Re$ is denoted by $-A$.

Example 2.2. Let S be a subring of R. Then $\{\{s\} \mid s \in S\}$ is an HX ring on R.. If A is a nonempty subset of R such that $A + A = AA = A$, then $\{A\}$ is also an HX ring. These HX rings are trivial HX rings.

Example 2.3[3] **.** Let C^0 be the set of all nonzero complex numbers. The operations "$\oplus$" and "$\otimes$" of C^0 are defined as follows. $a \oplus b = ab,\ a \otimes b = |a|^{|\ln|b||}$, $\forall a, b \in C^0$. Then $(C^0, \oplus, \otimes)$ forms a ring. Let $I = (1, +\infty)$ and $H = \{1, -1, i, -i\}$. Then $\Re = \{a \oplus I \mid a \in H\}$ is an HX ring on $(C^0, \oplus, \otimes)$.

Remark. A quotient ring of a given ring R may not be an HX ring, because the multiplication of the quotient ring is not coincident with the operation (2.2), although its addition is coincident with the operation (2.1).

Example 2.4[4]**.** Let Z denote the ordinary integral number ring and let $I = (10)$ be an ideal of Z. Then the quotient ring $(Z/I, +, \circ)$ of Z oncerning I satisfies: $(z_1 + I) + (z_2 + I) = (z_1 + z_2) + I,\ (z_1 + I) \circ (z_2 + I) = z_1 z_2 + I,\ \forall z_1, z_2 \in Z$. Let $z_1 = 2, z_2 = 4$. Then

$$(z_1 + I)(z_2 + I) = \{8 + 20m + 40n + 100mn \mid m, n \in Z\},$$
$$z_1 z_2 + I = \{8 + 10m \mid m \in Z\}.$$

Clearly, $(z_1 + I) \circ (z_2 + I) = z_1 z_2 + I \neq (z_1 + I)(z_2 + I)$, since $18 \in z_1 z_2 + I$ and $18 \notin (z_1 + I)(z_2 + I)$. This means that $(Z/I, +, \circ)$ is not an HX ring. In fact, if I is an ideal of R, then for all $r_1, r_2 \in R$ we have $(r_1 + I)(r_2 + I) \subseteq r_1 r_2 + I$. However, in general,"$\subseteq$" can not be replaced by"=".

Definition 2.5[5]**.** Let $\Re$ be a nonempty subset of $P_0(R)$ consisting two operations, addition"+" and multiplication"$\circ$", such that $A + B = \{a + b \mid a \in A, b \in B\}$ and $A \circ B \supseteq AB$ for all $A, B \in \Re$. If $(\Re, +, \circ)$ forms a ring, then $(\Re, +, \circ)$ is called a power ring on R whose zero element is denoted by Q.

Clearly, (1) if $(\Re, +, \circ)$ is a power ring on R, then $(\Re, +)$ is a power group on $(R, +)$; (2) all quotient rings of every subring of R and all HX rings are also power rings on R.

Definition 2.6. All quotient rings of every subring of R and all trivial HX rings are called trivial power rings on R.

Let $(\Re,+,\circ)$ be a power ring on R. Then $\Re^* = \bigcup\{A \mid A \in \Re\}$ is called the basis element set of $(\Re,+,\circ)$.

Definition 2.7. Let S be a nonempty subset of R such that $S+S \subseteq S$ and $SS \subseteq S$. Then S is called a subsemiring of R.

3 Construction of HX Ring and Power Ring

Upgrade of algebraic structure of ring is more difficult than that of group, because it is not easy to construct some general examples. In this section, we will first establish several constructive theorems of HX ring and power ring, then give some examples of nontrivial HX ring and nontrivial power ring.

Theorem 3.1. Let H be a subring of R and let I be a nonempty subset of R such that $I+I=I$. If $(h_1+I)(h_2+I)=h_1h_2+I$ holds for all $h_1,h_2 \in H$, then $H/I=\{h+I \mid h \in H\}$ is an HX ring on R and $H/_{H \cap \underline{I}} \cong H/I$, where $\underline{I}=\{x \in R \mid x+I=I\}$.

Example 3.2. Let $Z^{2\times 2}$ be the ordinary matrix ring of order 2 over integral set and let

$$H=\left\{\begin{pmatrix} m & n \\ 0 & 0 \end{pmatrix} \mid m,n \in Z\right\},\ I=\left\{\begin{pmatrix} 0 & m \\ 0 & n \end{pmatrix} \mid m,n \in Z\right\}.$$

Then H is a subring of $Z^{2\times 2}$ and $I+I=II=I$.

Let $h_1=\begin{pmatrix} m_1 & n_1 \\ 0 & 0 \end{pmatrix} \in H$ and $h_2=\begin{pmatrix} m_2 & n_2 \\ 0 & 0 \end{pmatrix} \in H$. Then

$$h_1+I=\left\{\begin{pmatrix} m_1 & m \\ 0 & n \end{pmatrix} \mid m,n \in Z\right\},\ h_2+I=\left\{\begin{pmatrix} m_2 & m' \\ 0 & n' \end{pmatrix} \mid m',n' \in Z\right\}.$$

So $(h_1+I)(h_2+I)=\left\{\begin{pmatrix} m_1m_2 & m_1m'+mn' \\ 0 & nn' \end{pmatrix} \mid m,n,m',n' \in Z\right\}$

$$=\begin{pmatrix} m_1m_2 & m_1n_2 \\ 0 & 0 \end{pmatrix}+\left\{\begin{pmatrix} 0 & m_1(m'-n_2)+mn' \\ 0 & nn' \end{pmatrix} \mid m,m,m',n' \in Z\right\}.$$

Considering $I=\left\{\begin{pmatrix} 0 & m_1(m'-n_2)+mn' \\ 0 & nn' \end{pmatrix} \mid m,n \in Z, m'=n_2, n'=1\right\}$

$\subseteq\left\{\begin{pmatrix} 0 & m_1(m'-n_2)+mn' \\ 0 & nn' \end{pmatrix} \mid m,n,m',n' \in Z\right\}\subseteq I$, we have that

$$I=\left\{\begin{pmatrix} 0 & m_1(m'-n_2)+mn' \\ 0 & nn' \end{pmatrix} \mid m,n,m',n' \in Z\right\}.$$

Hence

$$(h_1+I)(h_2+I)=\begin{pmatrix} m_1m_2 & m_1n_2 \\ 0 & 0 \end{pmatrix}+I=\begin{pmatrix} m_1 & n_1 \\ 0 & 0 \end{pmatrix}\begin{pmatrix} m_2 & n_2 \\ 0 & 0 \end{pmatrix}+I$$

$$=h_1h_2+I.$$

From Theorem 3.1 we see that $H/I=\{h+I \mid h \in H\}$ is a regular HX ring on $Z^{2\times 2}$.

Theorem 3.2. Let H be a subring of R and let I be a nonempty subset of R such that $I+I=II=I$. If $HI=IH=\{0\}$. Then $H/I=\{h+I \mid h \in H\}$ is an HX ring on R and $H/_{H\cap \underline{I}} \cong H/I$, where $\underline{I}=\{x \in R \mid x+I=I\}$.

Example 3.3. Let $R^{2\times 2}$ be the ordinary matrix ring of order 2 over real number field and let $H=\left\{\begin{pmatrix} a & 0 \\ 0 & 0 \end{pmatrix} \mid a \in R\right\}$, $I=\left\{\begin{pmatrix} 0 & 0 \\ 0 & b \end{pmatrix} \mid b \in R, b>0\right\}$. Then H is a subring of $R^{2\times 2}$ and $I+I=II=I$. Clearly, $IH=HI=\left\{\begin{pmatrix} 0 & 0 \\ 0 & 0 \end{pmatrix}\right\}$, so H/I is an HX ring on $R^{2\times 2}$ from Theorem 3.2 and it is a nonregular HX ring.

Theorem 3.4. Let I be a nonempty subset of R such that $I+I=II=I$. Then

(1) $H=\{x \in R \mid xI=Ix=\{0\}\}$ is a subring of R,

(2) $H_1/I = \{h + I \mid h \in H_1\}$ is an HX ring on R, where H_1 is an any subring of H.

Definition 3.5[1]. Let H be a subring of R and let I be a subsemiring (subring) of R such that $HI \bigcup IH \subseteq I$. Then I is called a regular semiideal (regular ideal) with respect to H. In particular, I is called a regular semiidesl (regular ideal) of H if I satisfies condition $I \subseteq H$.

Theorem 3.6. Let H be a subring of R and let $I \in P_0(R)$ be a regular semiideal with respect to H .Then

(1) $\forall h_1, h_2 \in H, h_1 + I \subseteq h_2 + I$, if and only if $h_1 - h_2 \in I$.

(2) $(H/I, +, \circ)$ is a regular power ring on R such that $H/_{H \bigcap \underline{I}} \cong H/I$, where

$H/I = \{h + I \mid h \in H\}$, $(h_1 + I) \circ (h_1 + I) = h_1 h_2 + I, \forall h_1, h_2 \in H$.

Corollary 3.7. Let H be a subring of R and let $I \in P_0(R)$ be a regular ideal with respect to H. Then $(H/I, +, \circ)$ is a uniform power ring on R.

Remark. The power ring $(H/I, +, \circ)$ in Corollary 3.7 is not the quotient ring of H, because $I \subseteq H$ does not hold.

Example 3.8. Let $Z^{2\times 2}$ be the ordinary matrix ring of order 2 over integral set and let

$$H = \left\{ \begin{pmatrix} 0 & m \\ 0 & 0 \end{pmatrix} \mid m \in Z \right\}, I = \left\{ \begin{pmatrix} 0 & 2m \\ 0 & 2n \end{pmatrix} \mid m, n \in Z, n \geq 0 \right\}.$$

Then H is a subring of $Z^{2\times 2}$ and I is a regular semiideal of $Z^{2\times 2}$ with respect to H. So $(H/I, +, \circ)$ is a regular power ring on $Z^{2\times 2}$ from Theorem 3.6. Considering

$II = \left\{ \begin{pmatrix} 0 & 4m \\ 0 & 4n \end{pmatrix} \mid m, n \in Z, n \geq 0 \right\}, \begin{pmatrix} 0 & 0 \\ 0 & 2 \end{pmatrix} \in I, \begin{pmatrix} 0 & 0 \\ 0 & 2 \end{pmatrix} \notin II$, we have that $II \neq I$.

Hence H/I is not an HX ring.

Definition 3.9. Let H be a subring of R and let $I \in P_0(R)$ be a regular semiideal with respect to H. Then $(H/I,+,\circ)$ is called the regular quasi-quotient ring of H concerning I.

From Theorem 3.6 we can obtain the follows.

Theorem 3.10. Let $(\mathfrak{R},+,\circ)$ be a regular power ring on R. Then

(1) Q is a regular semiideal with respect to $\overline{\mathfrak{R}}*$.
(2) $(\mathfrak{R},+,\circ)$ is the same as the regular quasi-quotien ring of $\overline{\mathfrak{R}}*$ concerning Q.

If $I \in P_0(R)$ is a regular semiideal with respect to H, then the power ring

$(H/I,+,\circ)$ is a regular power ring. To construct more general power ring, we introduce the following concept.

Definition 3.11. Let H be a subring of R and let I be a subsemiring (subring) of R such that $I+I=I$ and $HI \cup IH \subseteq I \cup \{0\}$. Tthen I is called a semiideal (ideal) with respect to H. In particular, if I satisfies condition $I \subseteq H$, then I is called a semiidesl (ideal) of H.

Theorem 3.12. Let H be a subring of R and let $I \in P_0(R)$ be a semiideal with respect to H. Then $(H/I,+,\circ)$ is a power ring on R such that $H/_{H \cap \underline{I}} \cong H/I$,

where $\underline{I} = \{x \in R \mid x+I=I\}, H/I = \{h+I \mid h \in H\}$, $(h_1+I)\circ(h_2+I) =$

$h_1h_2+I, \forall h_1,h_2 \in H$.

Remark. The conclusion of Theorem 3.12 is similar to that of Theorem 3.6, but there is a essential distinction, for the power ring in Theorem 3.6 must be a regular power ring while the power ring in Theorem 3.12 may not be regular.

Definition 3.13. Let H be a subring of R and let $I \in P_0(R)$ be a semiideal with respect to H. Then $(H/I,+,\circ)$ is called the quasi-quotient ring of H concerning I.

Example 3.14. Let $R^{2\times2}$ be the ordinary matrix ring of order 3 over real number field and let

$$H=\left\{\begin{pmatrix} a & 0 & 0 \\ 0 & 0 & 0 \\ 0 & 0 & 0 \end{pmatrix} \mid a\in R\right\},\quad I=\left\{\begin{pmatrix} 0 & 0 & 0 \\ 0 & b & 0 \\ 0 & 0 & 2m \end{pmatrix} \mid b\in R, b>0, m\in Z\right\}.$$

Then H is a subring of $R^{3\times 3}$. Obviously, we have $I+I=I, II\subseteq I$ and $HI=IH=\left\{\begin{pmatrix} 0 & 0 & 0 \\ 0 & 0 & 0 \\ 0 & 0 & 0 \end{pmatrix}\right\}$. So I is an semiideal with respect to H. Owing to that $\begin{pmatrix} 0 & 0 & 0 \\ 0 & 0 & 0 \\ 0 & 0 & 0 \end{pmatrix} \notin I$, $(H/I,+,\circ)$ is a nonregular power ring on $R^{3\times 3}$ from Theorem 3.12. Moreover, from that $\begin{pmatrix} 0 & 0 & 0 \\ 0 & 1 & 0 \\ 0 & 0 & 2 \end{pmatrix} \in I$ and $\begin{pmatrix} 0 & 0 & 0 \\ 0 & 1 & 0 \\ 0 & 0 & 2 \end{pmatrix} \notin II$ we have $II\neq I$. It means that H/I is not an HX ring.

Acknowledgement

This subject is supported by Ministry of Education, the People's Republic of China (No.206089).

References

1. Hongxing Li (1988) HX Ring. BUSEFAL 34: 3-8
2. Yubin Zhong (2000) The Existence of HX-Ring. Applied Mathematics-A Journal of Chinese University 15(2): 134-138
3. Yubin Zhong (1995) The structure of HX-ring on a class of ring. Fuzzy Systems and Mathematics 9(4): 73-77
4. Bingxue Yao (2001) Isomorphism Theorems of Regular Power Rings. Italian Journal of Pure and Applied Mathematics 9: 91-96
5. Bingxue Yao, Hongxing Li (2000) Power Ring. Fuzzy Systems and Mathematics 14(2): 15-19

On Fuzzy Ideals in BCH-Algebras

Feng-xiao Wang

Department of Mathematics, Shaanxi University of Technology
Hanzhong 723000
wangfx@snut.edu.cn

Abstract. The aim of this paper is to introduce the notions of fuzzy quasi-associative ideal and other fuzzy ideals of BCH-algebras and to investigate their properties. We give several characterizations of quasi-associative fuzzy ideals and closed fuzzy ideals. The relations among various fuzzy ideals are discussed as well. Finally, we characterize quasi-associative BCH-algebras via fuzzy ideals.

Keywords: BCH-algebra; Ideal; Quasi-associative; Fuzzy ideal.

1 Introduction

BCK/BCI-algebras are two important classes of logical algebras introduced by Iséki in 1966. Since then, a great deal of literature has been produced on the theory of BCK/BCI-algebras. Hu and Li[1] introduced the concept of BCH-algebras based upon BCK/BCI-algebras, and subsequently gave examples of proper BCH-algebras. For the general development of the BCK/BCI/BCH-algebras the ideal theory plays an important role. The concept of fuzzy sets was introduced by L.A.Zadeh[2]. Since then these ideals have been applied to other algebraic structures such as semigroups, groups, rings, ideals, modules,etc. In 1991, Xi [3,4] applied the concept of fuzzy sets to BCK-algebras. From then on Jun, Meng et al,etc [5-9] applied the concept to the ideals theory of BCK/BCI-algebras. The notions of fuzzy ideal of BCH-algebras were introduced by Hu and He [10] and were extensively investigated by many researchers[11,12].

In this paper, we define and study an interesting class of fuzzy ideals in BCH-algebras, called a class of quasi-associative fuzzy ideals. This class happens to be a proper subclass of a class of closed fuzzy ideals defined and studied by Y. B. Jun [7]. Moreover, we characterize closed fuzzy ideals in BCH-algebras. Finally, we characterize quasi-associative BCH-algebras via fuzzy ideals.

2 Preliminaries

An algebra $(X, *, 0)$ of type (2,0) is called a BCH-algebra[1], if it satisfies the following axioms:

(1) $x * x = 0$;
(2) $x * y = y * x = 0 \Longrightarrow x = y$;
(3) $(x * y) * z = (x * z) * y$.

B.-Y. Cao (Ed.): Fuzzy Information and Engineering (ICFIE), ASC 40, pp. 188–193, 2007.
springerlink.com

for all $x, y, z \in X$. In a BCH-algebra X, we can define a partial ordering $\leq$ by putting $x \leq y$ if and only if $x * y = 0$.

In a BCH-algebra, the following hold:

(4) $x * 0 = x$.

(5) $0 * (0 * (x * y)) = (0 * y) * (0 * x)$.

(6) $(0 * y) * (0 * x) = 0 * (y * x)$.

(7) $0 * (0 * (0 * x)) = 0 * x$.

Definition 1. *A BCH-algebra $(X, *, 0)$ is called quasi-associative, if for all $x, y, z \in X$, $(x * y) * z \leq x * (y * z)$.*

It is shown that the following are equivalent in BCH-algebras:

1. X is quasi-associative.
2. $0 * x = 0 * (0 * x)$.
3. $0 * x \leq x$.
4. $0 * (x * y) = 0 * (y * x)$.
5. $(0 * x) * y = 0 * (x * y)$.

A subset I of a BCH-algebra X is called an ideal, if for any $x, y \in X$:(I_1) $0 \in A$, $(I_2) x * y \in I$ and $y \in I$ imply $x \in I$.

An ideal I is called quasi-associative [13], if for each $x \in I$, $0 * x = 0 * (0 * x)$. Moreover, we have

Theorem 1. *In a BCH-algebra X, the following are equivalent:*

(1) I is quasi-associative ideal.

(2) $0 * x \leq x$.

(3) $0 * (x * y) \leq 0 * (y * x)$ for all $x, y \in I$.

Definition 2. *Let S be a set. A fuzzy set in S is a function $\mu \to [0, 1]$.*

Definition 3. *Let μ be a fuzzy set in a set S. For $t \in [0, 1]$, the set $\mu_t = \{s \in S | \mu(s) \geq t\}$ is called a level subset of μ.*

Definition 4. *Let X be a BCH-algebra. A fuzzy subset μ in X is said to be a fuzzy ideal, if it satisfies*

1. $\mu(0) \geq \mu(x)$,

2. $\mu(x) \geq min\{\mu(x * y), \mu_A(y)\}$, for any $x, y \in X$.

Definition 5. *Let$(X; *, 0)$ be a BCH-algebra. A fuzzy set μ of X is said to be a closed fuzzy ideal, if, for all $x \in X$, $\mu(0 * x) \geq \mu(x)$.*

If μ is a fuzzy ideal of X and $x \leq y$, then it follows that $\mu_A(x) \geq \mu_A(y)$.

Throughout this paper,$X = (X; *, 0)$ will denote a BCH-algebra.The set of all fuzzy subsets of X is called a fuzzy power set and denoted by $F(X)$.

3 Fuzzy Quasi-associate Ideals of BCH-Algebras

The following theorem gives characterizations of closed fuzzy ideals in BCH-algebras.

Theorem 2. *Let μ be a fuzzy ideal of a BCH-algebra X. Then for all $x, y \in X$, the following are equivalent:*
(a). $\mu_A(0 * x) \geq \mu_A(0 * (0 * x))$;
(b). $\mu_A(0 * x) \geq \mu_A(x)$;
(c). $\mu_A(0 * (x * y)) \geq \mu_A(0 * (y * x))$.

Proof. Since $0 * (0 * x) \leq x$, then $\mu_A(0 * (0 * x)) \geq \mu_A(x)$. Therefore $\mu_A(0 * x) \geq \mu_A(0 * (0 * x)) \geq \mu_A(x)$, which gives (b). Replacing x by $0 * y$ in (b), we get (a).This shows (a)and (b) are equivalent.
From (a), $\mu_A(0*(y*z)) \geq \mu_A(0*(0*(y*z)) = \mu_A((0*z)*(0*y)) = \mu_A(0*(z*y))$, or $\mu_A(0 * (y * z)) \geq \mu_A(0 * (z * y))$, which is (c). Put $y = 0$ in (c)and get (a). □

Now, we define quasi-associative fuzzy ideals as:

Definition 6. *A fuzzy ideal μ in X is called quasi-associative, if for all $x, y, z \in X$,*

$$\mu_A((x * y) * z) \geq \mu_A(x * (y * z)). \tag{1}$$

Clearly, if X is quasi-associative, then every fuzzy ideal in X is quasi-associative. If we put $x = y = 0$ in (1), we get (a) of theorem 2. Thus

Theorem 3. *Every quasi-associative fuzzy ideal in a BCH-algebra is a closed fuzzy ideal, but the converse is not true.*

Proof. Let μ be a quasi-associative fuzzy ideal of X. Putting $x = y = 0$ in (1), we get (a) of theorem 2. Thus μ is a closed fuzzy ideal.
To show the last half part, we consider the following example.

Example 2.5 Let $X = \{0, 1, 2, 3\}$ be a proper BCH-algebra with Cayley table given by

*	0	1	2	3
0	0	0	0	0
1	1	0	3	3
2	2	0	0	2
3	3	0	0	0

Define $\mu : X \to [0, 1]$ by $\mu(0) = \mu(1) = t_0$, $\mu(2) = \mu(3) = t_1$. Where $t_0 \geq t_1$ and $t_0, t_1 \in [0, 1]$. By routine calculation give that μ is a closed fuzzy ideal. But

$$t_1 = \mu((1 * 0) * 2) = \mu(3) \leq \mu(1 * (0 * 2)) = \mu(1) = t_0.$$

Hence μ is not a quasi-associative fuzzy ideal.

Theorem 4. *A fuzzy ideal μ of a BCH-algebra X is quasi-associative, if for every $t \in [0, 1]$, the level ideal $\mu_t = \{x \in X | \mu_A(x) \geq t\}$ is quasi-associative, where $\mu_t \neq \emptyset$.*

Proof. Suppose that a fuzzy ideal A is not quasi-associative, then there exist $x_0, y_0, z_0 \in A$such that

$$\mu_A((x_0 * y_0) * z_0) < \mu_A(x_0 * (y_0 * z_0)).$$

Let $t_0 = [\mu_A((x_0 * y_0) * z_0) + \mu_A(x_0 * (y_0 * z_0))]/2$. Then

$$\mu_A((x_0 * y_0) * z_0) < t_0;$$

and

$$\mu_A(x_0 * (y_0 * z_0)) > t_0.$$

This shows that $x_0 * (y_0 * z_0) \in \mu_{t_0}$. Since μ_t is a quasi-associative ideal of X, therefore $(x_0 * y_0) * z_0 \leq x_0 * (y_0 * z_0)$ gives that $(x_0 * y_0) * z0 \in \mu_{t_0}$ and hence $\mu_A((x_0 * y_0) * z_0) \geq t_0$ contradicts (1). This completes the proof. □

Corollary 1. *Let μ be a quasi-associative fuzzy ideal of a BCH-algebra X. For any $x_0 \in X$, $A_{x_0} = \{x \in X|\mu(x) \geq \mu(x_0)\}$ is quasi-associative.*

Proof. Let μ be a quasi-associative fuzzy ideal of X. Then for any $x_0 \in X$, $A_{x_0} = \{x \in X|\mu(x) \geq \mu(x_0)\}$ is quasi-associative of X with putting $t = \mu(x_0)$ by Theorem 4. □

Corollary 2. *Let μ be a fuzzy ideal of a BCH-algebra X. If ideal $I = \{x \in X|\mu(x) = \mu(0)\}$ is quasi-associative, then μ is quasi-associative.*

Definition 7. *Let $(X; *, 0)$ be a BCH-algebra. A fuzzy set μ of X is called a fuzzy subalgebra of X, if and only if for any $x, y \in X$, $\mu(x * y) \geq min(\mu(x), \mu(y))$.*

Theorem 5. *Let μ be a quasi-associative fuzzy ideal in a BCH-algebra X. Then for any $x, y \in X$, $\mu(x * y) \geq min\{\mu(x), \mu(y)\}$, that is, f is a fuzzy subalgebra in X.*

Proof. Let μ be a quasi-associative fuzzy ideal. Since $((x * y) * x) * (0 * y) = 0$, it follows that $\mu(x * y) \geq \mu(0 * y) \wedge \mu(x)$. By Definition 6, we have $\mu(0 * x) \geq \mu(0 * (0 * x))$. Since $0 * (0 * x) \leq x$, then $\mu(0 * (0 * x)) \geq \mu(x)$. This shows that $\mu(0 * x)) \geq \mu(x)$. We have that $\mu(x * y) \geq min\{\mu(x), \mu(y)\}$. Hence μ is a fuzzy subalgebra in X, ending the proof. □

Let A be a fuzzy ideal in X with membership function μ_A. Then $f(A)$ is a fuzzy ideal in Y with membership function $\mu_{f(A)}$ defined as:

$$\mu_{f(A)}(y) = \sup\{\mu_A(x)|x \in f^{-1}(y), y \in Y\}.$$

and $\mu_{f(A)}(y) = 0$ if $f^{-1}(y) = \emptyset$.

Definition 8. *A fuzzy set A with membership function μ_A in a BCH-algebra X has sup property, if for any subset T of X, there exists $t_0 \in T$ such that $\mu_A(t_0) = \sup\{\mu_A(t)|t \in T\}$.*

We call A a sup fuzzy ideal. Using this property, we prove the following:

Theorem 6. *Assume that X and Y be two BCH-algebras. Let $f : X \to Y$ be an onto homomorphism and A a sup quasi-associative fuzzy ideal of X. Then $f(A)$ is a quasi-associative fuzzy ideal of Y.*

Proof. Let $y_1, y_2, y_3 \in Y$ and $x_1 = f^{-1}(y_1)$, $x_2 = f^{-1}(y_2)$, $x_3 = f^{-1}(y_3)$ such that

$$\mu_A((x_1 * x_2) * x_3) = \sup\{\mu_A(t) | t \in f^{-1}((y_1 * y_2) * y_3)\},$$

and

$$\mu_A(x_1 * (x_2) * x_3) = \sup\{\mu_A(t) | t \in f^{-1}(y_1 * (y_2) * y_3)\}.$$

$$\begin{aligned} \mu_{f(A)}((y_1 * y_2) * y_3) &= \sup\{\mu_A(t) | t \in f^{-1}((y_1 * y_2) * y_3)\} \\ &= \mu_A((x_1 * x_2) * x_3) \\ &\geq \mu_A(x_1 * (x_2) * x_3) \\ &= \sup\{\mu_A(t) | t \in f^{-1}(y_1 * (y_2) * y_3)\} \\ &= \mu_{f(A)}(y_1 * (y_2) * y_3). \end{aligned}$$

Thus

$$\mu_{f(A)}((y_1 * y_2) * y_3) \geq \mu_{f(A)}(y_1 * (y_2) * y_3).$$

It follows that $f(A)$ is a quasi-associative fuzzy ideal of Y completing the proof. □

Let $f : X \to Y$ be an onto mapping. Let B be a fuzzy set in Y with membership function μ_B. Then $f^{-1}(B)$ is a fuzzy set in X with membership function $\mu_{f^{-1}(B)}$ given by $\mu_{f^{-1}(B)}(x) = \mu_B f(x)$, for all $x, y \in X$.

Theorem 7. *Let $f : X \to Y$ be an onto BCH-homomorphism. If a fuzzy ideal B of Y with membership function μ_B is quasi-associative, then the fuzzy ideal $f^{-1}(B)$ with membership function $\mu_{f^{-1}(B)}$ is quasi-associative.*

Proof. Since f is an onto BCH-homomorphism, therefore for $x, y, z \in X$, we have

$$\begin{aligned} \mu_{f^{-1}(B)}((x * y) * z) &= \mu_B f((x * y) * z) \\ &= ((f(x) * f(y)) * f(z)) \\ &\geq (f(x) * (f(y) * f(z)) \\ &= \mu_B f(x * (y * z)) \\ &= \mu_{f^{-1}(B)}(x * (y * z)). \end{aligned}$$

This gives that the fuzzy ideal $f^{-1}(B)$ is a quasi-associative in X. □

References

1. Hu Q.P and Li X(1985) On Proper BCH-algebras. *Math. Japonica*, **30**,659-661.
2. Zadeh L.A(1965) Fuzzy sets.*Information and Control*, **8**,338-353.
3. Hu B. Q and He J.J(2003) Fuzzy ideal in BCH-algebras,*Proceedings of the 22nd International Conference of the North American Fuzzy Information Processing Society-NAFIPS*, 254-259.
4. Hu B.Q., He J.J and Liu M(2004) Fuzzy Ideals and Fuzzy H-ideals of BCH-algebras. *Fuzzy Systems and Mathematics.***18**,9-15.

5. Ougen X (1991) Fuzzy BCK-algebras, *Math. Japonica*,**36**, No. 5, 512-517.
6. Ahmad B (1993) Fuzzy BCI-algebras,*J. Fuzzy Math.*, Vol. 1, No. 2,445-452.
7. Jun Y. B (1993) Closed Fuzzy ideals in BCI-algebras,*Math. Japonica*,**38**, NO. 1, 199-202.
8. Liu Y.J and Meng J (2001) Fuzzy ideals in BCI-algebras,*Fuzzy Sets and Systems*, **123**, 227-237.
9. Meng J and Guo X(2005) On fuzzy ideals in BCK/BCI-algebras,*Fuzzy Sets and Systems*, **149**, 509-525.
10. Xi C(1990) On a class of BCI-algebras, *Math. Japonica*, **35**, No. 1, 13-17.
11. Hoo C. S(1987) BCI-algebras with condition (S),*Math. Japonica*, **32**, 749-756.
12. Wang F.X(2006) Quasi-associative fuzzy ideals in BCH-algebras. *Fuzzy Systems and Mathematics* (to appear)

Minimization of Lattice Automata

Liao Zekai[1] and Shu Lan[2]

[1,2] Department of Applied Mathematics, University of Electronic Science and Technology of China, Chengdu, Sichuan, China
ANDYLIAO1982@163.COM, SHUL@UESTC.EDU.CN

Abstract. Theories of minimization of fuzzy automata have been developed by several authors, and most of which applied methods of algebraic theory, more specifically, the equivalence relation, the quotient space and equivalence class, congruence and homomorphism, in studying fuzzy automata. In this paper, we also apply the algebraic theory in the study of lattice automata, and obtain some results similar to the ones of fuzzy automata. In this paper, concepts of refining equivalence and refining congruence are defined, the quotient lattice automaton with respect to refining congruence is formulated, the concept of lattice automaton is reviewed, the equivalence of a lattice automaton and its quotient automaton is proved, the minimal property of quotient automaton is shown, and the minimization algorithm of lattice automata is proposed. The main idea of this paper is that by putting forward the concepts of refining equivalence and refining congruence, we can derive the quotient lattice automaton with respect to refining congruence, and via showing that the quotient lattice automaton is not only equivalent to the lattice automaton but also a minimal automaton, we obtain the minimization of a lattice automaton, and thus get the minimization algorithm of lattice automata.

Keywords: Lattice Automata, Refining Equivalence, Refining Congruence, Quotient Lattice Automata, Minimization.

1 Introduction

Fuzzy sets theory was introduced by Zadeh [7] in 1965. Fuzzy finite automata was first proposed by Wee [8] in 1967. The definition of lattice automata, which is the generalization of fuzzy automata, was given by Mordeson and Malik [9]. Some authors, such as Malik et al. [3], Basak and Gupta [2], Cheng and Mo [4] and Tatjana Petkovic [1], had contributed in the field of algebraic theory of fuzzy automata, more specifically, the minimization algorithm of fuzzy automata.

The idea of this paper is mainly from [1] and [2]. We slightly modify the definition of congruence defined in [1] and give the definition of refining congruence. We don't limit ourselves to concerning only fuzzy automata, but extended to lattice automata in a natural way.

2 Lattice Automata

The definition of a lattice automaton has been given by Mordeson and Malik [9].

Definition 2.1. A lattice automaton (LA) is a 6-tuple $A=(Q,\Sigma,L,\delta,I,F)$, where Q is a finite set of states, Σ is a finite set of input symbols, $L=(L,\wedge,\vee,0,1)$ is a complete distributive lattice and called the weighting space, where 0 and 1 are the

B.-Y. Cao (Ed.): Fuzzy Information and Engineering (ICFIE), ASC 40, pp. 194–205, 2007.
springerlink.com

minimal and maximal elements of the lattice, δ is a weighting function such as $\delta: Q\times\Sigma\times Q\rightarrow L$, and is called a state transition function. For any $\sigma\in\Sigma$, $p,q\in Q$, the value $\delta(p,\sigma,q)$ of $(p,\sigma,q)\in Q\times\Sigma\times Q$ represents the weight of transition from state p to state q when the input symbol is σ. I is an initial distribution function, where $I: Q\rightarrow L$. F is a final distribution function, where $F: Q\rightarrow L$. We denote $F_0 \triangleq \text{supp}F=\{q\in Q\mid F(q)>0\}$, and call F_0 the support set of F. For any $\sigma\in\Sigma$, $M(\sigma)$ denotes the transition matrix of input symbol σ, whose entries are defined as follows $M(\sigma)_{ij}=\delta(q_i,\sigma,q_j)$.

Let Σ^* denote the set of all words of finite length over Σ, and for any word $x\in\Sigma^*$, $|x|$ denotes its length. The empty word is denoted by ε.

We extend the transition function δ from $Q\times\Sigma\times Q$ to $Q\times\Sigma^*\times Q$ as follows, and still denote it as δ, if there is no confusion.

$$\delta(p,\varepsilon,q)=\begin{cases}1, & p=q\\ 0, & p\neq q\end{cases},$$

$$\delta(p,\sigma x,q)=\vee_{r\in F_0}(\delta(p,\sigma,r)\wedge\delta(r,x,q)) \text{ for all } p,q\in Q,\ \sigma\in\Sigma,$$

$x\in\Sigma^*$.

3 Refining Congruence and Quotient Automata

We slightly change the definition of congruence in [1] and get the definition of refining congruence. To that end, we give the following definition first.

Definition 3.1. A relation ρ on the state set Q of an LA $A=(Q,\Sigma,L,\delta,I,F)$ is called a refining equivalence on an LA A if it satisfies the following condition:

$$(p,p')\in\rho\Leftrightarrow(F(p)=F(p')=0) \text{ or } (F(p)>0 \text{ and } F(p')>0)$$

for any $p,p'\in Q$.

Definition 3.2. A refining equivalence θ on an LA $A=(Q,\Sigma,L,\delta,I,F)$ is called a refining congruence if it satisfies the following condition:

$$p\theta p'\Leftrightarrow \vee_{q'\theta q}\delta(p,\sigma,q')=\vee_{q'\theta q}\delta(p',\sigma,q') \quad (1)$$

for any $q\in F_0$, $\sigma\in\Sigma$, $p,p',q'\in Q$.

Lemma 3.3. Let θ be a refining congruence on an LA $A=(Q,\Sigma,L,\delta,I,F)$. Then for any $q\in F_0$, $x\in\Sigma^*$, $p,p',q'\in Q$,

$$p\theta p' \Leftrightarrow \underset{q'\theta q}{\vee} \delta(p,x,q') = \underset{q'\theta q}{\vee} \delta(p',x,q').$$

Proof. We prove it by induction on the length of x.

For $|x|=1$, the claim holds from (1).

Assume that it holds for any word of length $k \in \mathbb{N}$. Now we show that it is also true for any word of length $k+1$.

For any $q \in F_0$, $x \in \Sigma^*$ ($|x|=k$), $p, p', q' \in Q$, if $p\theta p'$, then

$$\underset{q'\theta q}{\vee} \delta(p,\sigma x,q')$$

$$= \underset{q'\theta q}{\vee} \underset{r\in F_0}{\vee} (\delta(p,\sigma,r) \wedge \delta(r,x,q'))$$

$$= \underset{r\in F_0}{\vee} (\delta(p,\sigma,r) \wedge \underset{q'\theta q}{\vee} \delta(r,x,q'))$$

$$= \underset{r'/\theta\in F_0/\theta}{\vee} \underset{r\in r'/\theta}{\vee} (\delta(p,\sigma,r) \wedge \underset{q'\theta q}{\vee} \delta(r',x,q'))$$

$$= \underset{r'/\theta\in F_0/\theta}{\vee} (\underset{r\theta r'}{\vee} \delta(p,\sigma,r) \wedge \underset{q'\theta q}{\vee} \delta(r',x,q'))$$

$$= \underset{r'/\theta\in F_0/\theta}{\vee} (\underset{r\theta r'}{\vee} \delta(p',\sigma,r) \wedge \underset{q'\theta q}{\vee} \delta(r',x,q'))$$

$$= \underset{r'/\theta\in F_0/\theta}{\vee} \underset{r\in r'/\theta}{\vee} (\delta(p',\sigma,r) \wedge \underset{q'\theta q}{\vee} \delta(r',x,q'))$$

$$= \underset{r'/\theta\in F_0/\theta}{\vee} \underset{r\in r'/\theta}{\vee} (\delta(p',\sigma,r) \wedge \underset{q'\theta q}{\vee} \delta(r,x,q'))$$

$$= \underset{r\in F_0}{\vee} (\delta(p',\sigma,r) \wedge \underset{q'\theta q}{\vee} \delta(r,x,q'))$$

$$= \underset{q'\theta q}{\vee} \underset{r\in F_0}{\vee} (\delta(p',\sigma,r) \wedge \delta(r,x,q'))$$

$$= \underset{q'\theta q}{\vee} \delta(p',\sigma x,q').$$

From the proof above, it is easy to show that, from $\underset{q'\theta q}{\vee} \delta(p,x,q') = \underset{q'\theta q}{\vee} \delta(p',x,q')$, we get $p\theta p'$, so $p\theta p' \Leftrightarrow \underset{q'\theta q}{\vee} \delta(p,x,q') = \underset{q'\theta q}{\vee} \delta(p',x,q')$. □

Definition 3.4. Let θ be a refining congruence on an LA $A=(Q,\Sigma,L,\delta,I,F)$. The quotient lattice automaton A/θ determined by θ is defined as follows:

$$A/\theta=(Q/\theta,\Sigma,L,\delta_\theta,I_\theta,F_\theta)$$

where

$$Q/\theta=\{q/\theta \mid q\in Q\},$$

$$q/\theta=\{q' \mid q'\in Q, q'\theta q\},$$

$$\delta_\theta(p/\theta,\sigma,q/\theta)=\vee_{p'\theta p}\vee_{q'\theta q}\delta(p',\sigma,q') \tag{2}$$

$$I_\theta(q/\theta)=\vee_{q'\theta q} I(q'),$$

$$F_\theta(q/\theta)=\vee_{q'\theta q} F(q'),$$

$$F_{\theta_0}=\mathrm{supp}F_\theta=\{q/\theta\in Q/\theta \mid F_\theta(q/\theta)>0\},$$

for any $p,q\in Q$, $\sigma\in\Sigma$. These mappings are well-defined.

Lemma 3.5. Let θ be a refining congruence on an LA $A=(Q,\Sigma,L,\delta,I,F)$. Then for any $x\in\Sigma^*$, $p,q\in Q$, $\delta_\theta(p/\theta,x,q/\theta)=\vee_{p'\theta p}\vee_{q'\theta q}\delta(p',x,q')$.

Proof. We prove it by induction on the length of x.

For $|x|=1$, the claim holds from (2)

Assume that it holds for any word of length $k\in\mathbb{N}$. Now we show that it is also true for any word of length $k+1$.

For any $\sigma\in\Sigma$, $x\in\Sigma^*$, $|x|=k$,

$$\delta_\theta(p/\theta,\sigma x,q/\theta)$$

$$=\vee_{r/\theta\in F_0/\theta}(\delta_\theta(p/\theta,\sigma,r/\theta)\wedge\delta_\theta(r/\theta,x,q/\theta))$$

$$=\vee_{r/\theta\in F_0/\theta}(\vee_{p'\theta q}\vee_{r'\theta r}\delta(p',\sigma,r')\wedge\vee_{r'\theta r}\vee_{q'\theta q}\delta(r',x,q'))$$

$$= \vee_{r/\theta \in F_0/\theta} (\vee_{r'\theta r} \vee_{p'\theta q} \delta(p', \sigma, r') \wedge \vee_{r'\theta r} \vee_{q'\theta q} \delta(r', x, q'))$$

$$= \vee_{r/\theta \in F_0/\theta} \vee_{r'\theta r} (\vee_{p'\theta q} \delta(p', \sigma, r') \wedge \vee_{q'\theta q} \delta(r', x, q'))$$

$$= \vee_{r' \in F_0} (\vee_{p'\theta q} \delta(p', \sigma, r') \wedge \vee_{q'\theta q} \delta(r', x, q'))$$

$$= \vee_{r' \in F_0} \vee_{p'\theta q} (\delta(p', \sigma, r') \wedge \vee_{q'\theta q} \delta(r', x, q'))$$

$$= \vee_{r' \in F_0} \vee_{p'\theta q} \vee_{q'\theta q} (\delta(p', \sigma, r') \wedge \delta(r', x, q'))$$

$$= \vee_{p'\theta q} \vee_{q'\theta q} (\vee_{r' \in F_0} (\delta(p', \sigma, r') \wedge \delta(r', x, q')))$$

$$= \vee_{p'\theta q} \vee_{q'\theta q} \delta(p', \sigma x, q').$$ □

4 Equivalent and Minimal Automata

In order to get the minimization of lattice automata, first we give the definition of equivalent and minimal automaton.

Definition 4.1. An LA $A = (Q, \Sigma, L, \delta, I, F)$, the weight of the word x, denoted $\omega(x)$, is defined as follows: $\omega : \Sigma^* \to L$ for any $x \in \Sigma^*$, $\omega(x) = \vee_{p \in Q} \vee_{q \in Q} (I(p) \wedge \delta(p, x, q) \wedge F(q))$.

Definition 4.2. Let $A_i = (Q_i, \Sigma, L, \delta_i, I_i, F_i)$ be an LA, and $\omega_i(x)$ the weight of word x, $i = 1, 2$. A_1 and A_2 are said to be equivalent, written $A_1 \equiv A_2$, if for any $x \in \Sigma^*$, $\omega_1(x) = \omega_2(x)$.

Theorem 4.3. Let $A = (Q, \Sigma, L, \delta, I, F)$ be an LA. Let θ be a refining congruence on A. Then the quotient lattice automaton A/θ is equivalent to A, i.e. $A \equiv A/\theta$.

Proof. For any $x \in \Sigma^*$,

$$\omega(x)$$

$$= \vee_{p \in Q} \vee_{q \in Q} (I(p) \wedge \delta(p, x, q) \wedge F(q))$$

$$= \vee_{p \in Q} \vee_{q'/\theta \in Q/\theta} \vee_{q \in q'/\theta} (I(p) \wedge \delta(p, x, q) \wedge F(q))$$

$$= \bigvee_{p\in Q}\bigvee_{q'/\theta\in Q/\theta}(I(p)\wedge(\bigvee_{q\in q'/\theta}\delta(p,x,q))\wedge(\bigvee_{q\in q'/\theta}F(q)))$$

$$= \bigvee_{q'/\theta\in Q/\theta}\bigvee_{p'/\theta\in Q/\theta}\bigvee_{p\in p'/\theta}(I(p)\wedge(\bigvee_{q\theta q'}\delta(p,x,q))\wedge F_\theta(q'/\theta))$$

$$= \bigvee_{q'/\theta\in Q/\theta}\bigvee_{p'/\theta\in Q/\theta}((\bigvee_{p\theta p'}I(p))\wedge(\bigvee_{p\theta p'}\bigvee_{q\theta q'}\delta(p,x,q))\wedge F_\theta(q'/\theta))$$

$$= \bigvee_{q'/\theta\in Q/\theta}\bigvee_{p'/\theta\in Q/\theta}(I_\theta(p'/\theta)\wedge\delta_\theta(p'/\theta,x,q'/\theta)\wedge F_\theta(q'/\theta))$$

$$= \omega_\theta(x) .$$ □

Definition 4.4. Let $A=(Q,\Sigma,L,\delta,I,F)$ be an LA. Let θ be a refining congruence on A. A is said to be minimal if for all $p,p'\in Q$, $p\theta p'$ implies $p=p'$.

Theorem 4.5. Let $A=(Q,\Sigma,L,\delta,I,F)$ be an LA. Let θ be a refining congruence on A. Then the quotient lattice automaton A/θ is minimal.

Proof. Let π be a refining congruence on A/θ. For any p/θ, $p'/\theta\in Q/\theta$, suppose $p/\theta\,\pi\,p'/\theta$, now we prove that $p/\theta=p'/\theta$.

$\because p/\theta\ \pi\ p'/\theta$,

$\therefore$ For $\forall q/\theta\in F_{\theta_0},\sigma\in\Sigma,q'/\theta\in Q/\theta$,

$$\bigvee_{q'/\theta\pi q/\theta}\delta_\theta(p/\theta,\sigma,q'/\theta)=\bigvee_{q'/\theta\pi q/\theta}\delta_\theta(p'/\theta,\sigma,q'/\theta),$$

that is

$$\bigvee_{q'/\theta\pi q/\theta}\bigvee_{p_1\theta p}\bigvee_{q''\theta q'}\delta(p_1,\sigma,q'')=\bigvee_{q'/\theta\pi q/\theta}\bigvee_{p_2\theta p'}\bigvee_{q''\theta q'}\delta(p_2,\sigma,q'') \tag{3}$$

We now prove that $q'\in F_0$.

$\because q'/\theta\,\pi\,q/\theta$ and $q/\theta\in F_{\theta_0}$,

$\therefore F(q'/\theta)>0$, i.e. $\bigvee_{q''\theta q'}F(q'')>0$,

$\therefore \exists q_0 \theta q'$, s.t. $F(q_0) > 0$,

$\therefore F(q') > 0$, i.e. $q' \in F_0$.

So (3) can be rewritten as follows:

$$\mathop{\vee}_{q'/\theta\pi q/\theta} \mathop{\vee}_{p_1\theta p} \mathop{\vee}_{q''\theta q'} \delta(p,\sigma,q'') = \mathop{\vee}_{q'/\theta\pi q/\theta} \mathop{\vee}_{p_2\theta p'} \mathop{\vee}_{q''\theta q'} \delta(p',\sigma,q'').$$

$$\therefore \mathop{\vee}_{q'/\theta\pi q/\theta} \mathop{\vee}_{q''\theta q'} \delta(p,\sigma,q'') = \mathop{\vee}_{q'/\theta\pi q/\theta} \mathop{\vee}_{q''\theta q'} \delta(p',\sigma,q''),$$

$$\therefore \mathop{\vee}_{q''\theta q'} \delta(p,\sigma,q'') = \mathop{\vee}_{q''\theta q'} \delta(p',\sigma,q''),$$

$\therefore p\theta p'$, i.e. $p/\theta = p'/\theta$. □

5 Minimization Algorithm

By theorem 4.3 and 4.5, we know that quotient lattice automaton A/θ is the minimization of lattice automaton A. So, the procedure to obtain the minimization of an LA A is in fact the procedure to find the refining congruence on A.

Theorem 5.1. Let θ_1 be a refining equivalence and θ a refining congruence. For $k \in \mathbb{N}$, we set $\theta_{k+1} = \{(p,p') \in \theta_k \mid \mathop{\vee}_{q'\theta_k q} \delta(p,\sigma,q') = \mathop{\vee}_{q'\theta_k q} \delta(p',\sigma,q')$ for $\forall \sigma \in \Sigma, q \in F_0\}$. Then

(1) $\theta_1 \supseteq \cdots \supseteq \theta_k \supseteq \theta_{k+1} \supseteq \cdots \supseteq \theta$;

(2) if $\theta_k = \theta_{k+1}$ for some $k \in \mathbb{N}$, then $\theta_k = \theta_{k+m}$ for every $m \in \mathbb{N}$;

(3) if $\theta_k = \theta_{k+1}$ for some $k \in \mathbb{N}$, then $\theta_k = \theta$.

Proof. (1) It is obvious that $\theta_k \supseteq \theta_{k+1}$ for any $k \in \mathbb{N}$. It remains to show that $\theta \subseteq \theta_k$ for every $k \in \mathbb{N}$. We prove it by induction on k.

This holds for $k = 1$, for a refining congruence is of course a refining equivalence. Assume that $\theta \subseteq \theta_k$ for some $k \in \mathbb{N}$. Now we prove that $\theta \subseteq \theta_{k+1}$.

Set $(p,p') \in \theta$, then $(p,p') \in \theta_k$. Since θ is a refining congruence, we have

$\mathop{\vee}_{q'\theta q} \delta(p,\sigma,q') = \mathop{\vee}_{q'\theta q} \delta(p',\sigma,q')$, for any $q \in F_0$, $\sigma \in \Sigma$, $q' \in Q$.

From $\theta \subseteq \theta_k$, we get $q/\theta_k = \bigcup_{i=1}^{m} q_i/\theta$ for some $m \in \mathbb{N}$ and $q_i \in F_0$ ($i = 1, \cdots, m$). This implies that $\underset{q'\theta_k q}{\vee} \delta(p,\sigma,q') = \overset{m}{\underset{i=1}{\vee}} \underset{q'\theta q_i}{\vee} \delta(p,\sigma,q') = \overset{m}{\underset{i=1}{\vee}} \underset{q'\theta q_i}{\vee} \delta(p',\sigma,q') = \underset{q'\theta_k q}{\vee} \delta(p',\sigma,q')$, thus $(p, p') \in \theta_{k+1}$.

(2) We prove it by induction on m.

For $m = 1$, it is trivial.

Assume that $\theta_k = \theta_{k+m}$ for some $m \in \mathbb{N}$. Now we prove that $\theta_k = \theta_{k+m+1}$.

From (1), we have $\theta_{k+m+1} \subseteq \theta_k$. So it remains to show that $\theta_k \subseteq \theta_{k+m+1}$.

Consider $(p, p') \in \theta_k$, since $\theta_k = \theta_{k+1}$, it follows that for any $\sigma \in \Sigma$, $q \in F_0$, $\underset{q'\theta_k q}{\vee} \delta(p,\sigma,q') = \underset{q'\theta_k q}{\vee} \delta(p',\sigma,q')$.

By using $\theta_k = \theta_{k+m+1}$, we obtain $(p, p') \in \theta_{k+m}$ and

$\underset{q'\theta_{k+m} q}{\vee} \delta(p,\sigma,q') = \underset{q'\theta_{k+m} q}{\vee} \delta(p',\sigma,q')$, for any $\sigma \in \Sigma$, $q \in F_0$.

Thus $(p, p') \in \theta_{k+m+1}$. Hence $\theta_k \subseteq \theta_{k+m+1}$. So $\theta_k = \theta_{k+m+1}$.

(3) Assume that $(p, p') \in \theta_k$, since $\theta_k = \theta_{k+1}$, we get $(p, p') \in \theta_{k+1}$, it follows that for any $\sigma \in \Sigma$, $q \in F_0$, $\underset{q'\theta_k q}{\vee} \delta(p,\sigma,q') = \underset{q'\theta_k q}{\vee} \delta(p',\sigma,q')$.

On the other hand, from $\underset{q'\theta_k q}{\vee} \delta(p,\sigma,q') = \underset{q'\theta_k q}{\vee} \delta(p',\sigma,q')$, it follows that $(p, p') \in \theta_{k+1}$, since $\theta_k = \theta_{k+1}$, hence $(p, p') \in \theta_k$. That is to say, θ_k satisfies condition (1), thus θ_k is a refining congruence. So $\theta_k = \theta_{k+1} = \cdots = \theta$.

Algorithm 5.2. For a given LA $A=(Q,\Sigma,L,\delta,I,F)$, to compute the refining congruence θ, we proceed along the following steps.

Step 1: $L=\{(p,q)\in\theta \mid p<q\}$; $L_{\text{old}}=L$;

Step 2: take $\sigma\in\Sigma$;

Step 3: for every $p\in Q$

for every $q\in F_0$, $Max(\sigma)_{p,q}=M(\sigma)_{p,q}$;

for every $(q,q')\in L$,

$Max(\sigma)_{p,q}=Max(\sigma)_{p,q'}=Max(\sigma)_{p,q}\vee Max(\sigma)_{p,q'}$;

Step 4: for every $(p,p')\in L$,

if $Max(\sigma)_{p,q}\neq Max(\sigma)_{p',q}$ for some $q\in F_0$,

then remove (p,p') from L;

Step 5: if Σ is not exhausted, take new $\sigma\in\Sigma$ and go to Step 3;

Step 6: if $L\neq L_{\text{old}}$ then $L_{\text{old}}=L$ and go to Step 2;

Step 7: $\theta=L\cup L^{-1}\cup E_Q$.

Remark: Some notations above are worth mentioning. $p<q$ means p appears before q in the state set Q. $M(\sigma)_{p,q}=\delta(p,\sigma,q)$, while $M(\sigma)$ is the transition matrix. If $(p,q)\in L$, then $(q,p)\in L^{-1}$. $E_Q=\{(q,q)\mid q\in Q\}$.

6 An Example

To illustrate the proceedings of the minimization algorithm of lattice automata, we now give an example.

Consider an LA $A=(Q,\Sigma,[0,1],\delta,I,F)$, where $Q=\{1,2,\cdots,9\}$, $\Sigma=\{\sigma_0,\sigma_1\}$, I=(0.3 0.7 0.8 0.2 1 0.4 0.5 0.9), $F=(0\,0\,0\,0\,0\,1\,1\,1\,1)$ and

$$M(\sigma_0)=\begin{bmatrix} 1 & 0.3 & 0.5 & 0.8 & 0.7 & 0.9 & 0.7 & 0.9 & 0.3 \\ 0.3 & 1 & 0.2 & 0.4 & 0.8 & 0.9 & 0.5 & 0.4 & 0.9 \\ 0.5 & 0.2 & 1 & 0.3 & 0.8 & 0.1 & 0.9 & 0.9 & 0.7 \\ 0.8 & 0.4 & 0.3 & 1 & 0.2 & 0.1 & 0.5 & 0.3 & 0.2 \\ 0.7 & 0.8 & 0.8 & 0.2 & 1 & 0.5 & 0.4 & 0.5 & 0.5 \\ 0.9 & 0.9 & 0.1 & 0.1 & 0.5 & 1 & 0.8 & 0.7 & 0.4 \\ 0.7 & 0.5 & 0.9 & 0.5 & 0.4 & 0.8 & 1 & 0.6 & 0.7 \\ 0.9 & 0.4 & 0.9 & 0.3 & 0.5 & 0.7 & 0.6 & 1 & 0.2 \\ 0.3 & 0.9 & 0.7 & 0.2 & 0.5 & 0.4 & 0.7 & 0.2 & 1 \end{bmatrix},$$

$$M(\sigma_1)=\begin{bmatrix} 1 & 0.6 & 0.7 & 0.2 & 0.5 & 0.8 & 0.7 & 0.2 & 0.4 \\ 0.6 & 1 & 0.2 & 0.5 & 0.3 & 0.4 & 0.8 & 0.1 & 0.4 \\ 0.7 & 0.2 & 1 & 0.3 & 0.5 & 0.8 & 0.1 & 0.8 & 0.8 \\ 0.2 & 0.5 & 0.3 & 1 & 0.7 & 0.5 & 0.6 & 0.6 & 0.5 \\ 0.5 & 0.3 & 0.5 & 0.7 & 1 & 0.6 & 0.2 & 0.5 & 0.6 \\ 0.8 & 0.4 & 0.8 & 0.5 & 0.6 & 1 & 0.3 & 0.2 & 0.1 \\ 0.7 & 0.8 & 0.1 & 0.6 & 0.2 & 0.3 & 1 & 0.1 & 0.2 \\ 0.2 & 0.1 & 0.8 & 0.6 & 0.5 & 0.2 & 0.1 & 1 & 0.7 \\ 0.4 & 0.4 & 0.8 & 0.5 & 0.6 & 0.1 & 0.2 & 0.7 & 1 \end{bmatrix}.$$

Step 1: $L=\{(1,2),(1,3),(1,4),(1,5),(2,3),(2,4),\ (2,5),(3,4),(3,5),$ $(4,5),(6,7),(6,8),(6,9),(7,8),(7,9),(8,9)\}$.

Step 2: take σ_0.

Step 3: results in

$$Max(\sigma_0)=\begin{bmatrix} 1 & 0.3 & 0.5 & 0.8 & 0.7 & 0.9 & 0.9 & 0.9 & 0.9 \\ 0.3 & 1 & 0.2 & 0.4 & 0.8 & 0.9 & 0.9 & 0.9 & 0.9 \\ 0.5 & 0.2 & 1 & 0.3 & 0.8 & 0.9 & 0.9 & 0.9 & 0.9 \\ 0.8 & 0.4 & 0.3 & 1 & 0.2 & 0.5 & 0.5 & 0.5 & 0.5 \\ 0.7 & 0.8 & 0.8 & 0.2 & 1 & 0.5 & 0.5 & 0.5 & 0.5 \\ 0.9 & 0.9 & 0.1 & 0.1 & 0.5 & 1 & 1 & 1 & 1 \\ 0.7 & 0.5 & 0.9 & 0.5 & 0.4 & 1 & 1 & 1 & 1 \\ 0.9 & 0.4 & 0.9 & 0.3 & 0.5 & 1 & 1 & 1 & 1 \\ 0.3 & 0.9 & 0.7 & 0.2 & 0.5 & 1 & 1 & 1 & 1 \end{bmatrix}.$$

Step 5: results in

$$L=\{(1,2),(1,3),(2,3),(4,5),(6,7),(6,8),(6,9),(7,8),(7,9),(8,9)\}.$$

Step 6: take σ_1, go to Step 3.

From Step 3 to Step 6, there are no changes in L and Σ exhausted.

Step 7: $L\neq L_{\text{old}}$, thus $L_{\text{old}}=L$ and go to Step 2.

From Step 2 to Step 7, there are no changes in L, i.e. $L_{\text{old}}=L$.

Step 8: $\theta=L\cup L^{-1}\cup E_Q$, while

$$L=\{(1,2),(1,3),(2,3),(4,5),(6,7),(6,8),(6,9),(7,8),(7,9),(8,9)\}.$$

Hence, the minimization of this LA, which is its quotient lattice automaton, is $A/\theta=(\{p,q,r\},\Sigma,[0,1],\delta_\theta,I_\theta,F_\theta)$, where $p=\{1,2,3\}$, $q=\{4,5\}$, $r=\{6,7,8,9\}$, $I_\theta=(0.8\ \ 1\ \ 0.9)$, $F_\theta=(0\ 0\ 1)$, and

$$M_\theta(\sigma_0)=\begin{bmatrix} 1 & 0.8 & 0.9 \\ 0.8 & 1 & 0.5 \\ 0.9 & 0.5 & 1 \end{bmatrix},\ M_\theta(\sigma_1)=\begin{bmatrix} 1 & 0.5 & 0.8 \\ 0.5 & 1 & 0.6 \\ 0.8 & 0.6 & 1 \end{bmatrix}.$$

7 Conclusions

We extended the work in [1] by expanding the minimization algorithm from fuzzy automata to lattice automata. Concepts of refining equivalence and refining congruence are formulated, the quotient lattice automaton with respect to refining congruence is defined, the equivalence of a lattice automaton and its quotient automaton is proved, the minimal property of quotient automaton is shown, and the minimization algorithm of lattice automata is proposed.

Acknowledgements

This work described here is partially supported by the grants from the National Natural Science Foundation of China (No. 10671030) and the Fostering Plan for Young and Middle Age Leading Researchers in UESTC (No. Y02018023601033).

References

1. Tatjana Petkovic (2006) Congruences and homomorphisms of fuzzy automata. Fuzzy Sets and Systems 157:444-458
2. Basak NC, Gupta A (2002) On quotient machines of a fuzzy automaton and the minimal machine. Fuzzy Sets and Systems 125:223-229
3. Malik DS, Mordeson JN, Sen MK (1999) Minimization of fuzzy finite state machines. J. Inform. Sci. 113:323-330
4. Cheng W, Mo Z (2004) Minimization algorithm of fuzzy finite automata. Fuzzy Sets and Systems 141:439-448
5. Mizumoto M, Toyoda J, Tanaka K (1969) Some considerations on fuzzy finite automata. J. Comput. System Sci. 3:409-422
6. Malik DS, Mordeson JN, Sen MK (1997) Products of fuzzy finite state machines. Fuzzy Sets and Systems, 92:95-102
7. Zadeh LA (1965) Fuzzy sets. J. Inform. and Control 8:338-353
8. Wee WG (1967) On generalizations of adaptive algorithm and application of the fuzzy sets concept to pattern classification. Ph.D. thesis, Purdue University
9. Mordeson JN, Malik DS (2002) Fuzzy Automata and Languages: Theory and Applications. Chapman&Hall/CRC, Boca Raton, London

Fixed Points in $\mathcal{M}$-Fuzzy Metric Spaces

Jin Han Park[1,⋆], Jong Seo Park[2], and Young Chel Kwun[3,⋆⋆]

[1] Division of Mathematical Sciences, Pukyong National University, Pusan 608-737, South Korea
jihpark@pknu.ac.kr
[2] Department of Mathematics Education, Chinju National University of Education, Chinju 660-756, South Korea
parkjs@cue.ac.kr
[3] Department of Mathematics, Dong-A University, Pusan 604-714, South Korea
yckwun@dau.ac.kr

Abstract. In this paper, we give some common fixed point theorems for five mappings satisfying some conditions in $\mathcal{M}$-fuzzy metric spaces.

Keywords: $\mathcal{M}$-fuzzy metric space, Fixed point, Compatible mapping of type $(*)$.

1 Introduction

The theory of fuzzy sets proposed by Zadeh [14] has showed successful applications in various fields and laid the foundation of fuzzy mathematics. Especially, Deng [3], Erceg [4], Kaleva and Seikkala [8] and Kramosil and Michalek [9] introduced the concepts of fuzzy metric spaces in different ways. George and Veeramani [5] modified the concept of fuzzy metric spaces due to Kramosil and Michalek and defined the Hausdorff topology of fuzzy metric spaces. Recently, many authors [1, 7, 10, 11] have also studied the fixed point theory in these fuzzy metric spaces. Sedghi et al. [13] introduced the concept of $\mathcal{M}$-fuzzy metric spaces which is a generalization of fuzzy metric spaces due to George and Veeramani and proved common fixed point theorems for two mappings under the conditions of weak compatible and R-weakly commuting mappings in complete $\mathcal{M}$-fuzzy metric spaces. In this paper, we show that every D-metric and fuzzy metric induce a $\mathcal{M}$-fuzzy metric, respectively, introduce the concept of compatible mapping of type $(*)$ in $\mathcal{M}$-fuzzy metric spaces and give common fixed point theorems for five mappings satisfying some conditions.

Definition 1. [2] Let X be a nonempty set. A generalized metric (or D-metric) on X is a function $D : X^3 \to \mathbf{R}^+$ satisfying the following conditions: for all $x, y, z, a \in X$,

(D-1) $D(x, y, z) \geq 0$,

(D-2) $D(x, y, z) = 0$ if and only if $x = y = z$,

⋆ This work was supported by the Korea Research Foundation Grant funded by the Korean Government(MOEHRD) (KRF-2006- 521-C00017).

⋆⋆ Corresponding author.

B.-Y. Cao (Ed.): Fuzzy Information and Engineering (ICFIE), ASC 40, pp. 206–215, 2007.
springerlink.com

(D-3) $D(x, y, z) = D(p\{x, y, z\})$ (symmetry), where p is a permutation function,
(D-4) $D(x, y, z) \leq D(x, y, a) + D(a, z, z)$.
The pair (X, D) is called a generalized metric (or D-metric) space.

Immediate examples of D-metric are
(a) $D(x, y, z) = \max\{d(x, y), d(y, z), d(z, x)\}$,
(b) $D(x, y, z) = d(x, y) + d(y, z) + d(z, x)$, where d is the ordinary metric on X.

Definition 2. [12] A binary operation $* : [0, 1] \times [0, 1] \to [0, 1]$ is called a continuous t-norm if $([0, 1], *)$ is an Abelian topological monoid with unit 1 such that $a * b \leq c * d$ whenever $a \leq c$ and $b \leq d$ for all $a, b, c, d \in [0, 1]$.

Typical examples of continuous t-norm are $a * b = ab$ and $a * b = \min\{a, b\}$.

Definition 3. [13] The 3-tuple $(X, \mathcal{M}, *)$ is called a $\mathcal{M}$-fuzzy metric space if X is an arbitrary set, $*$ is a continuous t-norm, and $\mathcal{M}$ is a fuzzy set on $X^3 \times (0, \infty)$ satisfying the following conditions: for all $x, y, z, a \in X$ and $t, s > 0$,
(FM-1) $\mathcal{M}(x, y, z, t) > 0$,
(FM-2) $\mathcal{M}(x, y, z, t) = 1$ if and only if $x = y = z$,
(FM-3) $\mathcal{M}(x, y, z, t) = \mathcal{M}(p\{x, y, z\}, t)$ (symmetry), where p is a permutation function,
(FM-4) $\mathcal{M}(x, y, a, t) * \mathcal{M}(a, z, z, s) \leq \mathcal{M}(x, y, z, t + s)$,
(FM-5) $\mathcal{M}(x, y, z, \cdot) : (0, \infty) \to [0, 1]$ is continuous.

In the following examples, we know that both D-metric and fuzzy metric induce a $\mathcal{M}$-fuzzy metric.

Example 1. Let (X, D) be a D-metric space. Define $a * b = a.b$ for all $a, b \in [0, 1]$ and for all $x, y, z \in X$ and $t > 0$,

$$\mathcal{M}(x, y, z, t) = \frac{t}{t + D(x, y, z)}. \tag{1}$$

Then $(X, \mathcal{M}, *)$ is a $\mathcal{M}$-fuzzy metric space.

Example 2. Let $(X, M, *)$ is a fuzzy metric space. If we define $\mathcal{M} : X^3 \times (0, \infty) \longrightarrow [0, 1]$ by

$$\mathcal{M}(x, y, z, t) = M(x, y, t) * M(y, z, t) * M(z, x, t)$$

for all $x, y, z \in X$, then $(X, \mathcal{M}, *)$ is a $\mathcal{M}$-fuzzy metric space.

Lemma 1. [13] *Let $(X, \mathcal{M}, *)$ be a $\mathcal{M}$-fuzzy metric space. For any $x, y, z \in X$ and $t > 0$, we have*
(a) $\mathcal{M}(x, x, y, t) = \mathcal{M}(x, y, y, t)$.
(b) $\mathcal{M}(x, y, z, \cdot)$ *is nondecreasing.*

Definition 4. [2] Let $(X, \mathcal{M}, *)$ be a $\mathcal{M}$-fuzzy metric space and $\{x_n\}$ be a sequence in X.

(a) $\{x_n\}$ is said to be convergent to a point $x \in X$ (denoted by $\lim_{n\to\infty} x_n = x$) if $\lim_{n\to\infty} \mathcal{M}(x, x, x_n, t) = 1$ for all $t > 0$.

(b) $\{x_n\}$ is called a Cauchy sequence if $\lim_{n\to\infty} \mathcal{M}(x_{n+p}, x_{n+p}, x_n, t) = 1$ for all $t > 0$ and $p > 0$.

(c) A $\mathcal{M}$-fuzzy metric in which every Cauchy sequence is convergent is said to be complete.

Remark 1. Since $*$ is continuous, it follows from (MF-4) that the limit of sequence is uniquely determined.

Let $(X, \mathcal{M}, *)$ be a $\mathcal{M}$-fuzzy metric space with the following condition:

(FM-6) $\lim_{t\to\infty} \mathcal{M}(x, y, z, t) = 1$ for all $x, y, z \in X$ and $t > 0$.

Lemma 2. *Let $\{x_n\}$ be a sequence in a $\mathcal{M}$-fuzzy metric space $(X, \mathcal{M}, *)$ with the condition* (FM-6)*. If there exists a number $k \in (0, 1)$ such that*

$$\mathcal{M}(x_{n+2}, x_{n+1}, x_{n+1}, kt) \geq \mathcal{M}(x_{n+1}, x_n, x_n, t) \tag{2}$$

for all $t > 0$ and $n = 1, 2, \cdots$, then $\{x_n\}$ is a Cauchy sequence in X.

Proof. By the simple induction with the condition (2), we have, for all $t > 0$ and $n = 1, 2, \cdots$,

$$\mathcal{M}(x_{n+2}, x_{n+1}, x_{n+1}, t) \geq \mathcal{M}(x_2, x_1, x_1, \frac{t}{k^n}). \tag{3}$$

Thus, by (FM-6) and (3), for any positive integer p and $t > 0$, we have

$$\mathcal{M}(x_n, x_{n+p}, x_{n+p}, t) \geq \mathcal{M}(x_1, x_2, x_2, \frac{t}{pk^{n-1}}) * \cdots * \mathcal{M}(x_1, x_2, x_2, \frac{t}{pk^{n+p-2}}).$$

Therefore, by (FM-6), we have

$$\lim_{n\to\infty} \mathcal{M}(x_n, x_{n+p}, x_{n+p}, t) \geq 1 * \cdots * 1 \geq 1,$$

which implies that $\{x_n\}$ is a Cauchy sequence in X. □

Lemma 3. *Let $(X, \mathcal{M}, *)$ be a $\mathcal{M}$-fuzzy metric space with the condition* (FM-6)*. If, for all $x, y \in X$ and for a number $k \in (0, 1)$,*

$$\mathcal{M}(x, y, z, kt) \geq \mathcal{M}(x, y, z, t),$$

then $x = y = z$.

Proof. By hypothesis and (FM-6), we have

$$1 \geq \mathcal{M}(x, y, z, t) \geq \mathcal{M}(x, y, z, \frac{t}{k}) \geq \cdots \geq \mathcal{M}(x, y, z, \frac{t}{k^n}) \to 1 \text{ as } n \to \infty.$$

Hence, by (FM-2), $x = y = z$. □

2 Compatible Mappings of Type (*)

In this section, we introduce the concept of compatible mappings of type (*) and give some properties of these mappings for our main results.

Definition 5. [13] Let A and B be mappings from a $\mathcal{M}$-fuzzy metric space $(X, \mathcal{M}, *)$ into itself. The mappings are said to be compatible if

$$\lim_{n\to\infty} \mathcal{M}(ABx_n, BAx_n, BAx_n, t) = 1$$

for all $t > 0$, whenever $\{x_n\}$ is a sequence in X such that $\lim_{n\to\infty} Ax_n = \lim_{n\to\infty} Bx_n = z$ for some $z \in X$.

Definition 6. Let A and B be mappings from a $\mathcal{M}$-fuzzy metric space $(X, \mathcal{M}, *)$ into itself. The mappings are said to be compatible of type (*) if

$$\lim_{n\to\infty} \mathcal{M}(ABx_n, BBx_n, BBx_n, t) = 1 \text{ and } \lim_{n\to\infty} \mathcal{M}(BAx_n, AAx_n, AAx_n, t) = 1$$

for all $t > 0$, whenever $\{x_n\}$ is a sequence in X such that $\lim_{n\to\infty} Ax_n = \lim_{n\to\infty} Bx_n = z$ for some $z \in X$.

Proposition 1. *Let $(X, \mathcal{M}, *)$ be a $\mathcal{M}$-fuzzy metric space and A and B be continuous mappings from X into itself. Then A and B are compatible if and only if they are compatible of type (*).*

Proof. Let $\{x_n\}$ be a sequence in X such that $\lim_{n\to\infty} Ax_n = \lim_{n\to\infty} Bx_n = z$ for some $z \in X$. Since A is continuous, we have

$$\lim_{n\to\infty} AAx_n = \lim_{n\to\infty} ABx_n = Az.$$

Further, since A and B are compatible, we get

$$\lim_{n\to\infty} \mathcal{M}(ABx_n, BAx_n, BAx_n, t) = 1$$

for all $t > 0$. Thus, from the inequality

$$\begin{aligned}&\mathcal{M}(AAx_n, BAx_n, BAx_n, t)\\&\geq \mathcal{M}(AAx_n, AAx_n, ABx_n, \frac{t}{2}) * \mathcal{M}(ABx_n, BAx_n, BAx_n, \frac{t}{2}),\end{aligned}$$

it follows that $\mathcal{M}(AAx_n, BAx_n, BAx_n, t) = 1$. Similarly, we also obtain $\mathcal{M}(BBx_n, BBx_n, ABx_n, t) = 1$. Hence, A and B are compatible of type (*).

Conversely, let $\{x_n\}$ be a sequence in X such that $\lim_{n\to\infty} Ax_n = z$ and $\lim_{n\to\infty} Bx_n = z$ for some $z \in X$. Then, since B is continuous, we have

$$\lim_{n\to\infty} BAx_n = \lim_{n\to\infty} BBx_n = Bz.$$

Since A and B are compatible of type (*), we get

$$\lim_{n\to\infty} \mathcal{M}(ABx_n, BBx_n, BBx_n, \frac{t}{2}) = \lim_{n\to\infty} \mathcal{M}(BAx_n, AAx_n, AAx_n, \frac{t}{2}) = 1.$$

for all $t > 0$. Hence, from the inequality

$$\begin{aligned} &\mathcal{M}(ABx_n, BAx_n, BAx_n, t) \\ &\geq \mathcal{M}(ABx_n, BBx_n, BBx_n, \frac{t}{2}) * \mathcal{M}(BBx_n, BBx_n, BAx_n, \frac{t}{2}), \end{aligned}$$

it follows that $\lim_{n\to\infty} \mathcal{M}(ABx_n, BAx_n, BAx_n, t) \geq 1 * \cdots * 1 \geq 1$ and so $\lim_{n\to\infty} \mathcal{M}(ABx_n, BAx_n, BAx_n, t) = 1$. Hence, A and B are compatible. □

Proposition 2. *Let $(X, \mathcal{M}, *)$ be a $\mathcal{M}$-fuzzy metric space and A and B be mappings from X into itself. If A and B are compatible of type $(*)$ and $Az = Bz$ for some $z \in X$, then $ABz = BBz = BAz = AAz$.*

Proof. Let $\{x_n\}$ be a sequence in X defined by $x_n = z$ for some $z \in X$ and $n = 1, 2, \cdots$ and $Az = Bz$. Then we have

$$\lim_{n\to\infty} Ax_n = \lim_{n\to\infty} Bx_n = Az.$$

Since A and B are compatible of type $(*)$, we get

$$\mathcal{M}(ABz, BBz, BBz, t) = \lim_{n\to\infty} \mathcal{M}(ABx_n, BBx_n, BBx_n, t) = 1$$

and hence $ABz = BBz$. Similarly, we have $BAz = AAz$. But, $Az = Bz$ implies $BBz = BAz$. Therefore, we obtain $ABz = BBz = BAz = AAz$. □

Proposition 3. *Let $(X, \mathcal{M}, *)$ be a $\mathcal{M}$-fuzzy metric space and A and B be mappings from X into itself. If A and B are compatible of type $(*)$ and $\{x_n\}$ is a sequence in X such that $\lim_{n\to\infty} Ax_n = \lim_{n\to\infty} Bx_n = z$ for some $z \in X$, then*
(a) *$\lim_{n\to\infty} BAx_n = Az$ if A is continuous at z.*
(b) *$ABz = BAz$ and $Az = Bz$ if A and B are continuous at z.*

Proof. (a) Since A is continuous at z and $\lim_{n\to\infty} Ax_n = z$, $\lim_{n\to\infty} AAx_n = Az$. Since A and B is compatible of type $(*)$, for all $t > 0$, we have

$$\lim_{n\to\infty} \mathcal{M}(BAx_n, AAx_n, AAx_n, t) = 1$$

and thus from (FM-4) we get

$$\begin{aligned} &\lim_{n\to\infty} \mathcal{M}(BAx_n, Az, Az, t) \\ &\geq \lim_{n\to\infty} \mathcal{M}(BAx_n, AAx_n, AAx_n, \frac{t}{2}) * \lim_{n\to\infty} \mathcal{M}(AAx_n, Az, Az, \frac{t}{2}) \geq 1, \end{aligned}$$

i.e., $\lim_{n\to\infty} \mathcal{M}(BAx_n, Az, Az, t) = 1$. Hence we have $\lim_{n\to\infty} BAx_n = Az$.

(b) Since $\lim_{n\to\infty} Ax_n = \lim_{n\to\infty} Bx_n = z$ and A and B are continuous at z, by (a), we have

$$\lim_{n\to\infty} ABx_n = Az \text{ and } \lim_{n\to\infty} BAx_n = Bz.$$

Thus we obtain $Az = Bz$ by the uniqueness of the limit and so by Proposition 2, we have $BAz = ABz$. □

3 Common Fixed Point Theorems

In this section, we prove some common fixed point theorems for mappings satisfying some conditions.

Theorem 1. *Let $(X, \mathcal{M}, *)$ be a complete $\mathcal{M}$-fuzzy metric space with $t * t \geq t$ for all $t \in [0,1]$ and the condition* (FM-6). *Let A, B, S, T and P be mappings from X into itself such that*

(1.a) *$P(X) \subset AB(X)$ and $P(X) \subset ST(X)$,*

(1.b) *there exists a number $k \in (0,1)$ such that*

$$\begin{aligned}&\mathcal{M}(Px, Py, Py, kt)\\ &\quad\geq \mathcal{M}(ABx, Px, Px, t) * \mathcal{M}(STy, Py, Py, t) * \mathcal{M}(STy, Px, Px, \alpha t)\\ &\quad\ * \mathcal{M}(ABx, Py, Py, (2-\alpha)t) * \mathcal{M}(ABx, STy, STy, t)\end{aligned}$$

for all $x, y \in X$, $\alpha \in (0,2)$ and $t > 0$,

(1.c) *$PB = BP$, $PT = TP$, $AB = BA$ and $ST = TS$,*

(1.d) *A and B are continuous,*

(1.e) *the pair $\{P, AB\}$ are compatible of type $(*)$,*

(1.f) *$\mathcal{M}(x, STx, STx, t) \geq \mathcal{M}(x, ABxABx, t)$ for all $x \in X$ and $t > 0$.*

Then A, B, S, T and P have a common fixed point in X.

Proof. Since $P(X) \subset AB(X)$, for ant $x_0 \in X$, we can choose a point $x_1 \in X$ such that $Px_0 = ABx_1$. Since $P(X) \subset ST(X)$, for this point x_1, we can choose a point $x_2 \in X$ such that $Px_1 = STx_2$. Thus by induction, we can define a sequence $\{y_n\}$ in X as follows:

$$y_{2n} = Px_{2n} = ABx_{2n+1} \text{ and } y_{2n+1} = Px_{2n+1} = STx_{2n+2} \tag{4}$$

for $n = 1, 2, \cdots$. By (1.b), for all $t > 0$ and $\alpha = 1 - q$ with $q \in (0,1)$, we have

$$\begin{aligned}&\mathcal{M}(y_{2n+1}, y_{2n_2}, y_{2n+2}, kt) = \mathcal{M}(Px_{2n+1}, Px_{2n+2}, Px_{2n+2}, kt)\\ &\geq \mathcal{M}(y_{2n}, y_{2n+1}, y_{2n+1}, t) * \mathcal{M}(y_{2n+1}, y_{2n+2}, y_{2n+2}, t)\\ &\quad * \mathcal{M}(y_{2n+1}, y_{2n+1}, y_{2n+1}, t) * \mathcal{M}(y_{2n}, y_{2n+2}, y_{2n+2}, (1+q)t)\\ &\quad * \mathcal{M}(y_{2n}, y_{2n+1}, y_{2n+1}, t)\\ &\geq \mathcal{M}(y_{2n}, y_{2n+1}, y_{2n+1}, t) * \mathcal{M}(y_{2n+1}, y_{2n+2}, y_{2n+2}, t)\\ &\quad * \mathcal{M}(y_{2n}, y_{2n+1}, y_{2n+1}, qt).\end{aligned}$$

Since $*$ is continuous and $\mathcal{M}(x, y, z, \cdot)$ is continuous, letting $q \to 1$ in above equation, we get

$$\begin{aligned}&\mathcal{M}(y_{2n+1}, y_{2n+2}, y_{2n+2}, kt)\\ &\quad\geq \mathcal{M}(y_{2n}, y_{2n+1}, y_{2n+1}, t) * \mathcal{M}(y_{2n+1}, y_{2n+2}, y_{2n+2}, t).\end{aligned} \tag{5}$$

Similarly, we have

$$\begin{aligned}&\mathcal{M}(y_{2n+2}, y_{2n+3}, y_{2n+3}, kt)\\&\quad\geq \mathcal{M}(y_{2n+1}, y_{2n+2}, y_{2n+2}, t) * \mathcal{M}(y_{2n+2}, y_{2n+3}, y_{2n+3}, t).\end{aligned} \tag{6}$$

Thus, from (5) and (6), it follows that

$$\mathcal{M}(y_{n+1}, y_{n+2}, y_{n+2}, kt) \geq \mathcal{M}(y_n, y_{n+1}, y_{n+1}, t) * \mathcal{M}(y_{n+1}, y_{n+2}, y_{n+2}, t)$$

for $n = 1, 2, \cdots$ and then for positive integers n and p,

$$\mathcal{M}(y_{n+1}, y_{n+2}, y_{n+2}, kt) \geq \mathcal{M}(y_n, y_{n+1}, y_{n+1}, t) * \mathcal{M}(y_{n+1}, y_{n+2}, y_{n+2}, \frac{t}{k^p}).$$

Thus, since $\mathcal{M}(y_{n+1}, y_{n+2}, y_{n+2}, \frac{t}{k^p}) \to 1$ as $p \to \infty$, we have

$$\mathcal{M}(y_{n+1}, y_{n+2}, y_{n+2}, kt) \geq \mathcal{M}(y_n, y_{n+1}, y_{n+1}, t).$$

By Lemma 2, $\{y_n\}$ is a Cauchy sequence in X and since X is complete, $\{y_n\}$ converges to a point $z \in X$. Since $\{Px_n\}$, $\{ABx_{2n+1}\}$ and $\{STx_{2n+2}\}$ are subsequences of $\{y_n\}$, they also converge to the point z. Since A, B are continuous and the pair $\{P, AB\}$ is compatible of type $(*)$, by Proposition 3 (a), we have

$$\lim_{n\to\infty} PABx_{2n+1} = ABz \quad \text{and} \quad \lim_{n\to\infty} (AB)^2 x_{2n+1} = ABz.$$

By (1.b) with $\alpha = 1$, we get

$$\begin{aligned}&\mathcal{M}(PABx_{2n+1}, Px_{2n+2}, Px_{2n+2}, kt) \geq\\&\mathcal{M}((AB)^2x_{2n+1}, PABx_{2n+2}, PABx_{2n+2}, t) * \mathcal{M}(STx_{2n+2}, Px_{2n+2}, Px_{2n+2}, t)\\&* \mathcal{M}(STx_{2n+2}, PABx_{2n+1}, PABx_{2n+1}, t) * \mathcal{M}((AB)^2x_{2n+1}, Px_{2n+2}, Px_{2n+2}, t)\\&* \mathcal{M}((AB)^2x_{2n+1}, STx_{2n+2}, STx_{2n+2}, t),\end{aligned}$$

which implies that

$$\begin{aligned}\mathcal{M}(ABz, z, z, kt) &= \lim_{n\to\infty} \mathcal{M}(PABx_{2n+1}, Px_{2n+2}, Px_{2n+2}, kt)\\&\geq 1 * 1 * \mathcal{M}(z, z, ABz, t) * \mathcal{M}(ABz, z, z, t) * \mathcal{M}(ABz, z, z, t)\\&\geq \mathcal{M}(ABz, z, z, t).\end{aligned}$$

Hence, by Lemma 3, we have $ABz = z$. By (1.f), since $\mathcal{M}(z, z, STz, t) \geq \mathcal{M}(z, z, ABz, t) = 1$ for all $t > 0$, we get $STz = z$. Again, by (1.b) with $\alpha = 1$, we have

$$\begin{aligned}&\mathcal{M}(PABx_{2n+1}, Pz, Pz, kt)\\&\geq \mathcal{M}((AB)^2x_{2n+1}, PABx_{2n+1}, PABx_{2n+1}, t) * \mathcal{M}(STz, Pz, Pz, t)\\&\ * \mathcal{M}(STz, PABx_{2n+1}, PABx_{2n+1}, t) * \mathcal{M}((AB)^2x_{2n+1}, Pz, Pz, t)\\&\ * \mathcal{M}((AB)^2x_{2n+1}, STz, STz, t).\end{aligned}$$

This implies that

$$\begin{aligned}\mathcal{M}(ABz, Pz, Pz, kt) &= \lim_{n\to\infty} \mathcal{M}(PABx_{2n+1}, Pz, Pz, kt) \\ &\geq 1 * 1 * 1 * \mathcal{M}(ABz, Pz, P, t) * 1 \\ &\geq \mathcal{M}(ABz, Pz, Pz, t)\end{aligned}$$

and so, by Lemma 3, $ABz = Pz$.

Now, we show that $Bz = z$. In fact, by (1.b) with $\alpha = 1$ and (1.c), we get

$$\begin{aligned}&\mathcal{M}(Bz, z, z, kt) = \mathcal{M}(BPz, Pz, Pz, kt) = \mathcal{M}(PBz, Pz, Pz, kt) \\ &\quad\geq \mathcal{M}(ABBz, PBz, PBz, t) * \mathcal{M}(STz, Pz, Pz, t) * \mathcal{M}(STz, PBz, PBz, t) \\ &\quad\quad * \mathcal{M}(ABBz, Pz, Pz, t) * \mathcal{M}(ABBz, STz, STz, t) \\ &\quad= 1 * 1 * \mathcal{M}(z, Bz, Bz, t) * \mathcal{M}(Bz, z, z, t) * \mathcal{M}(Bz, z, z, t) \\ &\quad\geq \mathcal{M}(z, Bz, Bz, t),\end{aligned}$$

which implies that $Bz = z$. Since $ABz = z$, we have $Az = z$.

Next, we show that $Tz = z$. Indeed, by (1.b) with $\alpha = 1$ and (1.c), we get

$$\begin{aligned}&\mathcal{M}(Tz, z, z, kt) = \mathcal{M}(TPz, Pz, Pz, kt) = \mathcal{M}(Pz, Pz, TPz, kt) \\ &\quad= 1 * 1 * \mathcal{M}(Tz, z, z, t) * \mathcal{M}(z, Tz, Tz, t) * \mathcal{M}(z, Tz, Tz, t) \\ &\quad\geq \mathcal{M}(Tz, z, z, t),\end{aligned}$$

which implies that $Tz = z$. Since $STz = z$, we have $Sz = STz = z$. Therefore, by combining the above results, we obtain $Az = Bz = Sz = Tz = Pz = z$, that is, z is the common fixed point of A, B, S, T and P.

Finally, the uniqueness of the fixed point of A, B, S, T and P follows easily from (1.b). □

From Theorem 1 with $B = T = I_X$ (the identity mapping on X), we have the following.

Corollary 1. *Let $(X, \mathcal{M}, *)$ be a complete $\mathcal{M}$-fuzzy metric space with $t * t > t$ for all $t \in [0, 1]$ and the condition* (FM-6). *Let A, B and P be mappings from X into itself such that*

(1.g) *$P(X) \subset A(X)$ and $P(X) \subset S(X)$,*

(1.h) *there exists a number $k \in (0, 1)$ such that*

$$\begin{aligned}\mathcal{M}(Px, Py, Py, kt) \geq\ & \mathcal{M}(Ax, Px, Px, t) * \mathcal{M}(Sy, Py, Py, t) * \mathcal{M}(Ax, Sy, Sy, t) \\ & * \mathcal{M}(Ax, Py, Py, (2 - \alpha)t) * \mathcal{M}(Sy, Px, Px, \alpha t)\end{aligned}$$

for all $x, y \in X$, $\alpha \in (0, 2)$ and $t > 0$,

(1.i) *A is continuous,*

(1.j) *the pair $\{P, A\}$ are compatible of type $(*)$,*

(1.k) *$\mathcal{M}(x, Sx, Sx, t) \geq \mathcal{M}(x, AxAx, t)$ for all $x \in X$ and $t > 0$.*

Then A, S and P have a common fixed point in X.

From Theorem 1 with $A = B = S = T = I_X$ (the identity mapping on X), we have the following.

Corollary 2. *Let* $(X, \mathcal{M}, *)$ *be a complete* $\mathcal{M}$*-fuzzy metric space with* $t * t \geq t$ *for all* $t \in [0,1]$ *and the condition* (FM-6)*. Let* P *be a mapping from* X *into itself such that there exists a number* $k \in (0,1)$ *such that*

$$\mathcal{M}(Px, Py, Py, kt) \geq \mathcal{M}(x, Px, Px, t) * \mathcal{M}(y, Py, Py, t) * \mathcal{M}(y, Px, Px, \alpha t) \\ *\mathcal{M}(x, Py, Py, (2-\alpha)t) * \mathcal{M}(x, y, y, t)$$

for all $x, y \in X$, $\alpha \in (0,2)$ *and* $t > 0$*. Then* A*,* S *and* P *have a common fixed point in* X*.*

Corollary 3. *Let* $(X, \mathcal{M}, *)$ *be a complete* $\mathcal{M}$*-fuzzy metric space with* $t * t \geq t$ *for all* $t \in [0,1]$ *and the condition* (FM-6)*. Let* P *be a mapping from* X *into itself such that there exists a number* $k \in (0,1)$ *such that*

$$\mathcal{M}(Px, Py, Py, kt) \geq \mathcal{M}(x, y, y, t)$$

for all $x, y \in X$ *and* $t > 0$*. Then* P *has a fixed point in* X*.*

Remark 2. Corollary 3 is an extension of Banach contraction theorem [6] in fuzzy metric spaces to a contractive mapping on complete $\mathcal{M}$-fuzzy metric spaces.

By using Theorem 1, we have the following:

Theorem 2. *Let* $(X, \mathcal{M}, *)$ *be a complete* $\mathcal{M}$*-fuzzy metric space with* $t*t \geq t$ *for all* $t \in [0,1]$ *and the condition* (FM-6)*. Let* A, B, S, T *and* $\{P_\alpha\}_{\alpha \in \Lambda}$ *be mappings from* X *into itself such that the conditions* (1.d) *and* (1.f) *hold and*
(1.l) $\cup_{\alpha \in \Lambda} P_\alpha(X) \subset AB(X)$ *and* $\cup_{\alpha \in \Lambda} P_\alpha(X) \subset ST(X)$,
(1.m) *there exists a number* $k \in (0,1)$ *such that*

$$\mathcal{M}(P_\alpha x, P_\alpha y, P_\alpha y, kt) \\ \geq \mathcal{M}(ABx, P_\alpha x, P_\alpha x, t) * \mathcal{M}(STy, P_\alpha y, P_\alpha y, t) * \mathcal{M}(STy, P_\alpha x, P_\alpha x, \alpha t) \\ *\mathcal{M}(ABx, P_\alpha y, P_\alpha y, (2-\alpha)t) * \mathcal{M}(ABx, STy, STy, t)$$

for all $x, y \in X$, $\alpha \in (0,2)$, $\alpha \in \lambda$ *and* $t > 0$,
(1.n) $P_\alpha B = BP_\alpha$, $P_\alpha T = TP_\alpha$, $AB = BA$ *and* $ST = TS$ *for all* $\alpha \in \Lambda$,
(1.o) *the pair* $\{P_\alpha, AB\}$ *is compatible of type* $(*)$*.*
Then A, B, S, T *and* $\{P_\alpha\}_{\alpha \in \Lambda}$ *have a unique common fixed point in* X*.*

References

1. Chang SS, Cho YJ, Lee BS, Jung JS, Kang SM (1997) Coincidence point and minimization theorems in fuzzy metric spaces. Fuzzy Sets and Systems 88: 119–128.
2. Dhage BC (1992) Generalized metric spaces and mappings with fixed point. Bull Calcutta Math Soc 84: 329–336.
3. Deng ZK (1982) Fuzzy pseudo-metric spaces. J Math Anal Appl 86: 74–95.
4. Erceg MA (1979) Metric spaces in fuzzy set theory. J Math Anal Appl 69: 205–230.
5. George A, Veeramani P (1994) On some results in fuzzy metric spaces. Fuzzy Sets and Systems 64: 395–399.

6. Grabiec M (1988) Fixed point in fuzzy metric spaces. Fuzzy Sets and Systems 27: 385–389.
7. Jung JS, Cho YJ, Kim JK (1994) Minimization theorems for fixed point theorems in fuzzy metric spaces and applications. Fuzzy Sets and Systems 61: 199–207.
8. Kaleva O, Seikkala S (1984) On fuzzy metric spaces. Fuzzy Sets and Systems 12: 215–229.
9. Kramosil O, Michalek J (1975) Fuzzy metric and statistical metric spaces. Kybernetica 11: 336–344.
10. Mishra SN, Sharma N, Singh SL (1994) Common fixed points of maps on fuzzy metric spaces. Internat J Math Math Sci 17: 253–258.
11. Park JH, Park JS, Kwun YC (2006) A common fixed point theorem in the intuitionistic fuzzy metric spaces. In: Jiao L et. al. (eds) Advances in Natural Computation Data Mining. Xidian University, Xian, 293-300.
12. Schweizer B, Sklar A (1960) Statistical metric spaces. Pacific J Math 10: 313–334.
13. Sedghi S, Shobe N, Park JH, A common fixed point theorem in $\mathcal{M}$-fuzzy metric spaces, submitted.
14. Zadeh LA, (1965) Fuzzy sets. Inform and Control 8: 338–353.

Common Fixed Points of Maps on Intuitionistic Fuzzy Metric Spaces

Jin Han Park[1,*], Jong Seo Park[2], and Young Chel Kwun[3,**]

[1] Division of Mathematical Sciences, Pukyong National University, Pusan 608-737, South Korea
jihpark@pknu.ac.kr
[2] Department of Mathematics Education, Chinju National University of Education, Chinju 660-756, South Korea
parkjs@cue.ac.kr
[3] Department of Mathematics, Dong-A University, Pusan 604-714, South Korea
yckwun@dau.ac.kr

Abstract. The purpose of this paper is to obtain the common fixed point theorems for asymptotically commuting maps on intuitionistic fuzzy metric spaces defined by Park, Park and Kwun [10].

Keywords: Common fixed point, Intuitionistic fuzzy metric space, Asymptotically commuting map.

1 Introduction

There have been several attempts to formulate fixed point theorems in fuzzy mathematics since the fixed point theory is one of the preeminent basic tools to handle various physical formulations [1, 2, 3, 17]. However, it appears that Kramosil and Michalek's research [7] of fuzzy metric spaces paves a way for very soothing machinery to develop fixed point theorems especially for contractive type maps. Grabiec [3] obtained the fuzzy version of Banach contraction principle and Mishra et al. [8] proved the common fixed points of maps on fuzzy metric spaces [7]. Recently, Park [9] introduced and studied the intuitionistic fuzzy metric spaces. Park, Kwun and Park [13] proved a fixed point theorem in the intuitionistic fuzzy metric spaces. Also, Park, Park and Kwun [10] modified the definition of intuitionistic fuzzy metric spaces due to Park [9] and studied a common fixed point theorem in the intuitionistic fuzzy metric spaces.

In this paper, we are to obtain the common fixed points of maps on intuitionistic fuzzy metric spaces defined by Park, Park and Kwun [10]. Our research is an extension of Mishra, Sharma and Singh's result [8].

* This work was supported by the Korea Research Foundation Grant funded by the Korean Government(MOEHRD) (KRF-2006- 521-C00017).

** Corresponding author.

B.-Y. Cao (Ed.): Fuzzy Information and Engineering (ICFIE), ASC 40, pp. 216–225, 2007.
springerlink.com

2 Preliminaries

Now, we will give some definitions, properties and notation of the intuitionistic fuzzy metric space following by Schweizer and Sklar [14], Grabiec [3] and Park et al. [10].

Definition 1. [14] A operation $* : [0,1] \times [0,1] \to [0,1]$ is continuous t-norm if $*$ is satisfying the following conditions:
(a) $*$ is commutative and associative,
(b) $*$ is continuous,
(c) $a * 1 = a$ for all $a \in [0,1]$,
(d) $a * b \leq c * d$ whenever $a \leq c$ and $b \leq d$ $(a, b, c, d \in [0,1])$.

Definition 2. [14] A binary operation $\diamond : [0,1] \times [0,1] \to [0,1]$ is continuous t-conorm if $\diamond$ is satisfying the following conditions:
(a) $\diamond$ is commutative and associative,
(b) $\diamond$ is continuous,
(c) $a \diamond 0 = a$ for all $a \in [0, 1]$,
(d) $a \diamond b \geq c \diamond d$ whenever $a \leq c$ and $b \leq d$ $(a, b, c, d \in [0,1])$.

Remark 1. [9] The following conditions are satisfied:
(a) For any $r_1, r_2 \in (0,1)$ with $r_1 > r_2$, there exist $r_3, r_4 \in (0,1)$ such that $r_1 * r_3 \geq r_2$ and $r_4 \diamond r_2 \leq r_1$.
(b) For any $r_5 \in (0,1)$, there exist $r_6, r_7 \in (0,1)$ such that $r_6 * r_6 \geq r_5$ and $r_7 \diamond r_7 \leq r_5$.

Definition 3. [10] The 5-tuple $(X, M, N, *, \diamond)$ is said to be an intuitionistic fuzzy metric space if X is an arbitrary set, $*$ is a continuous $t-$norm, $\diamond$ is a continuous t-conorm and M, N are fuzzy sets on $X^2 \times (0, \infty)$ satisfying the following conditions; for all $x, y, z \in X$,
(a) $M(x, y, t) > 0$,
(b) $M(x, y, t) = 1 \Leftrightarrow x = y$,
(c) $M(x, y, t) = M(y, x, t)$,
(d) $M(x, y, t) * M(y, z, s) \leq M(x, z, t+s)$,
(e) $M(x, y, \cdot) : (0, \infty) \to (0, 1]$ is continuous,
(f) $N(x, y, t) > 0$,
(g) $N(x, y, t) = 0 \Leftrightarrow x = y$,
(h) $N(x, y, t) = N(y, x, t)$,
(i) $N(x, y, t) \diamond N(y, z, s) \geq N(x, z, t+s)$,
(j) $N(x, y, \cdot) : (0, \infty) \to (0, 1]$ is continuous.
Then (M, N) is called an intuitionistic fuzzy metric on X. The functions $M(x, y, t)$ and $N(x, y, t)$ denote the degree of nearness and the degree of non-nearness between x and y with respect to t, respectively.

Remark 2. [13, 10] In an intuitionistic fuzzy metric space X, $M(x, y, \cdot)$ is non-decreasing and $N(x, y, \cdot)$ is nonincreasing for all $x, y \in X$.

In all that follows stands **N** for the set of natural numbers and X stands for an intuitionistic fuzzy metric space $(X, M, N, *, \diamond)$ with the following properties:

$$\lim_{t\to\infty} M(x,y,t) = 1, \qquad \lim_{t\to\infty} N(x,y,t) = 0 \quad \text{for all } x,y \in X.$$

Lemma 1. [3, 11] *Let $\{x_n\}$ be a sequence in an intuitionistic fuzzy metric space X. If there exists a positive number $0 < k < 1$ such that*

$$M(x_{n+2}, x_{n+1}, kt) \geq M(x_{n+1}, x_n, t),$$
$$N(x_{n+2}, x_{n+1}, kt) \leq N(x_{n+1}, x_n, t), \quad t > 0,\ n \in \mathbf{N}.$$

Then $\{x_n\}$ is a Cauchy sequence.

Lemma 2. [3] *If x, y are any two points in an intuitionistic fuzzy metric space X and k is a positive number with $k < 1$, and*

$$M(x,y,kt) \geq M(x,y,t), \quad N(x,y,kt) \leq N(x,y,t),$$

then $x = y$.

3 Result

In this section, we will prove the fixed point theorems extended from Mishra, Sharma and Singh's result [8] in the intuitionistic fuzzy metric spaces.

Definition 4. Self maps P and S of an intuitionistic fuzzy metric space X will be called *asymptotically commuting* if and only if for all $t > 0$,

$$\lim_{n\to\infty} M(PSx_n, SPx_n, t) = 1, \ \lim_{n\to\infty} N(PSx_n, SPx_n, t) = 0,$$

where $\{x_n\}$ is a sequence in X such that $\lim_{n\to\infty} Px_n = \lim_{n\to\infty} Sx_n = z$ for some $z \in X$.

For an equivalent formulation in a metric space refer to Jungck [5, 6] and Trivari and Singh [16]. Following Jungck's nomenclature, asymptotically commuting maps may also be called compatible maps. Such maps are more general than commuting and weakly commuting maps [15] both.

Lemma 3. *If Q and T are asymptotically commuting maps on intuitionistic fuzzy metric space X and $Qx_n \to z$, $Tx_n \to z$ for some $z \in X(\{x_n\} \subset X)$, then $QTx_n \to Tz$ provided T is continuous at $z \in X$.*

Proof. Since T is continuous at $z \in X$, $TQx_n \to Tz$ and $TTx_n \to Tz$. By the asymptotic commutativity of Q and T,

$$\lim_{n\to\infty} M(QTx_n, TQx_n, t) = 1, \ \lim_{n\to\infty} N(QTx_n, TQx_n, t) = 0.$$

Therefore

$$M(Tz, QTx_n, t) \geq M(Tz, TQx_n, \tfrac{t}{2}) * M(TQx_n, QTx_n, \tfrac{t}{2}),$$
$$N(Tz, QTx_n, t) \leq N(Tz, TQx_n, \tfrac{t}{2}) \diamond N(TQx_n, QTx_n, \tfrac{t}{2})$$

yields, by Remark 1,

$$\lim_{n\to\infty} M(Tz, QTx_n, t) \geq 1 * 1 = 1,$$
$$\lim_{n\to\infty} N(Tz, QTx_n, t) \leq 0 \diamond 0 = 0.$$

Hence

$$\lim_{n\to\infty} QTx_n = Tz.$$

Theorem 1. *Let X be a complete intuitionistic fuzzy metric space and $P, Q : X \to X$. If there exist continuous maps $S, T : X \to X$ and a constant $k \in (0,1)$ such that*

(1.1) $ST = TS$,

(1.2) $\{P, S\}$ and $\{Q, T\}$ are asymptotically commuting pairs,

(1.3) $PT(X) \cup QS(X) \subset ST(X)$,

(1.4) $M(Px, Qy, kt) \geq M(Sx, Ty, t) * M(Px, Sx, t) * M(Qy, Ty, t) * M(Px, Ty, \alpha t) * M(Qy, Sx, (2-\alpha)t)$,

(1.5) $N(Px, Qy, kt) \leq N(Sx, Ty, t) \diamond (Px, Sx, t) \diamond N(Qy, Ty, t) \diamond N(Px, Ty, \alpha t) \diamond N(Qy, Sx, (2-\alpha)t)$

for all $x, y \in X$, $t > 0$ and $\alpha \in (0,2)$. Then P, Q, S and T have a unique common fixed point.

Proof. Let x_0 be given in X. We construct a sequence $\{x_n\}$ as follows:

$$PTx_{2n} = STx_{2n+1},\ QSx_{2n+1} = STx_{2n+2},\ n = 0, 1, 2, \cdots.$$

We can do this since (1.3) holds.

Let $z_n = STx_n$. Then for $\alpha = 1 - q$, $q \in (0,1)$, by (1.4) and (1.5),

$$\begin{aligned}
&M(z_{2n+1}, z_{2n+2}, kt)\\
&= M(PTx_{2n}, QSx_{2n+1}, kt)\\
&\geq M(STx_{2n}, TSx_{2n+1}, t) * M(PTx_{2n}, STx_{2n}, t) * M(QSx_{2n+1}, TSx_{2n+1}, t)\\
&\qquad * M(PTx_{2n}, TSx_{2n+1}, (1-q)t) * M(QSx_{2n+1}, STx_{2n}, (1+q)t)\\
&= M(z_{2n}, z_{2n+1}, t) * M(z_{2n+1}, z_{2n+2}, t) * M(z_{2n+2}, z_{2n}, (1+q)t)\\
&\geq M(z_{2n}, z_{2n+1}, t) * M(z_{2n+1}, z_{2n+2}, t) * M(z_{2n}, z_{2n+1}, qt)
\end{aligned}$$

and

$$\begin{aligned}
&N(z_{2n+1}, z_{2n+2}, kt)\\
&= N(PTx_{2n}, QSx_{2n+1}, kt)\\
&\leq N(STx_{2n}, TSx_{2n+1}, t) \diamond N(PTx_{2n}, STx_{2n}, t) \diamond N(QSx_{2n+1}, TSx_{2n+1}, t)\\
&\qquad \diamond N(PTx_{2n}, TSx_{2n+1}, (1-q)t) \diamond N(QSx_{2n+1}, STx_{2n}, (1+q)t)\\
&= N(z_{2n}, z_{2n+1}, t) \diamond N(z_{2n+1}, z_{2n+2}, t) \diamond N(z_{2n+2}, z_{2n}, (1+q)t)\\
&\leq N(z_{2n}, z_{2n+1}, t) \diamond N(z_{2n+1}, z_{2n+2}, t) \diamond N(z_{2n}, z_{2n+1}, qt) \diamond N(z_{2n+1}, z_{2n+2}, t)\\
&= N(z_{2n}, z_{2n+1}, t) \diamond N(z_{2n+1}, z_{2n+2}, t) \diamond N(z_{2n}, z_{2n+1}, qt).
\end{aligned}$$

Since $*$ and $\diamond$ are continuous and $M(x,y,\cdot), N(x,y,\cdot)$ are continuous, making $q \to 1$ gives

$$M(z_{2n+1}, z_{2n+2}, kt) \geq M(z_{2n}, z_{2n+1}, t) * M(z_{2n+1}, z_{2n+2}, t),$$
$$N(z_{2n+1}, z_{2n+2}, kt) \leq N(z_{2n}, z_{2n+1}, t) \diamond N(z_{2n+1}, z_{2n+2}, t).$$

Similarly, by (1.4) and (1.5), for $q' \in (0,1)$,

$$\begin{aligned}
&M(z_{2n+3}, z_{2n+2}, kt)\\
&= M(PTx_{2n+2}, QSx_{2n+1}, kt)\\
&\geq M(z_{2n+2}, z_{2n+1}, t) * M(z_{2n+3}, z_{2n+2}, t) * M(z_{2n+2}, z_{2n+1}, t)\\
&\qquad * M(z_{2n+3}, z_{2n+1}, (1+q')t) * M(z_{2n+2}, z_{2n+2}, (1-q')t)\\
&= M(z_{2n+2}, z_{2n+1}, t) * M(z_{2n+3}, z_{2n+2}, t) * M(z_{2n+3}, z_{2n+1}, (1+q')t)\\
&\geq M(z_{2n+2}, z_{2n+1}, t) * M(z_{2n+3}, z_{2n+2}, t)\\
&\qquad * M(z_{2n+1}, z_{2n+2}, q't) * M(z_{2n+2}, z_{2n+3}, t)\\
&= M(z_{2n+2}, z_{2n+1}, t) * M(z_{2n+3}, z_{2n+2}, t) * M(z_{2n+1}, z_{2n+2}, q't).
\end{aligned}$$

and

$$\begin{aligned}
&N(z_{2n+3}, z_{2n+2}, kt)\\
&= N(PTx_{2n+2}, QSx_{2n+1}, kt)\\
&\leq N(z_{2n+2}, z_{2n+1}, t) \diamond N(z_{2n+3}, z_{2n+2}, t) \diamond N(z_{2n+2}, z_{2n+1}, t)\\
&\qquad \diamond N(z_{2n+3}, z_{2n+1}, (1+q')t) \diamond N(z_{2n+2}, z_{2n+2}, (1-q')t)\\
&= N(z_{2n+2}, z_{2n+1}, t) \diamond N(z_{2n+3}, z_{2n+2}, t) \diamond N(z_{2n+3}, z_{2n+1}, (1+q')t)\\
&\leq N(z_{2n+2}, z_{2n+1}, t) \diamond N(z_{2n+3}, z_{2n+2}, t)\\
&\qquad \diamond N(z_{2n+1}, z_{2n+2}, q't) \diamond N(z_{2n+2}, z_{2n+3}, t)\\
&= N(z_{2n+2}, z_{2n+1}, t) \diamond N(z_{2n+3}, z_{2n+2}, t) \diamond N(z_{2n+1}, z_{2n+2}, q't).
\end{aligned}$$

Since $*$ and $\diamond$ are continuous and $M(x,y,\cdot\,), N(x,y,\cdot\,)$ are continuous, making $q' \to 1$ gives

$$M(z_{2n+3}, z_{2n+2}, kt) \geq M(z_{2n+1}, z_{2n+2}, t) * M(z_{2n+2}, z_{2n+3}, t),$$

$$N(z_{2n+3}, z_{2n+2}, kt) \leq N(z_{2n+1}, z_{2n+2}, t) \diamond N(z_{2n+2}, z_{2n+3}, t).$$

Hence, putting $2n+1 = m \in N$,

$$M(z_{m+1}, z_{m+2}, kt) \geq M(z_m, z_{m+1}, t) * M(z_{m+1}, z_{m+2}, t),$$

$$N(z_{m+1}, z_{m+2}, kt) \leq N(z_m, z_{m+1}, t) \diamond N(z_{m+1}, z_{m+2}, t).$$

Consequently, for $m, p \in N$

$$M(z_{m+1}, z_{m+2}, kt) \geq M(z_m, z_{m+1}, t) * M(z_{m+1}, z_{m+2}, \frac{t}{k^p}),$$

and

$$N(z_{m+1}, z_{m+2}, kt) \leq N(z_m, z_{m+1}, t) \diamond N(z_{m+1}, z_{m+2}, \frac{t}{k^p}).$$

Since, by our hypothesis,

$$\lim_{p\to\infty} M(z_{m+1}, z_{m+2}, \frac{t}{k^p}) = 1, \quad \lim_{p\to\infty} N(z_{m+1}, z_{m+2}, \frac{t}{k^p}) = 0,$$

we have

$$M(z_{m+1}, z_{m+2}, kt) \geq M(z_m, z_{m+1}, t),$$

and

$$N(z_{m+1}, z_{m+2}, kt) \leq N(z_m, z_{m+1}, t).$$

By Lemma 1, $\{z_n\}$ is a Cauchy sequence and it has a limit in X. That is, $\lim_{n\to\infty} z_n = z$. $\{PTx_{2n}\}$ and $\{QSx_{2n+1}\}$ being the subsequences of $\{STx_n\}$ converges to z.

Let $y_n = Tx_n$ and $w_n = Sx_n$, $n \in N$. Then $Py_{2n} \to z$, $Sy_{2n} \to z$, $Tw_{2n+1} \to z$ and $Qw_{2n+1} \to z$. So, for $t > 0$, from (1.2),

$$M(PSy_{2n}, SPy_{2n}, t) \to 1, \ \ M(QTw_{2n+1}, TQw_{2n+1}, t) \to 1,$$
$$N(PSy_{2n}, SPy_{2n}, t) \to 0, \ \ N(QTw_{2n+1}, TQw_{2n+1}, t) \to 0.$$

Moreover, by continuity of T and Lemma 3,

$$TTw_{2n+1} \to Tz, \ TQw_{2n+1} \to Tz \text{ and } QTw_{2n+1} \to Tz.$$

By (1.4) and (1.5) with $\alpha = 1$,

$$\begin{aligned}
&M(Py_{2n}, QTw_{2n+1}, kt)\\
&\geq M(Sy_{2n}, TTw_{2n+1}, t) * M(Py_{2n}, Sy_{2n}, t) * M(QTw_{2n+1}, TTw_{2n+1}, t)\\
&\qquad * M(Py_{2n}, TTw_{2n+1}, t) * M(QTw_{2n+1}, Sy_{2n}, t)\\
&N(Py_{2n}, QTw_{2n+1}, kt)\\
&\leq N(Sy_{2n}, TTw_{2n+1}, t) \diamond N(Py_{2n}, Sy_{2n}, t) \diamond N(QTw_{2n+1}, TTw_{2n+1}, t)\\
&\qquad \diamond N(Py_{2n}, TTw_{2n+1}, t) \diamond N(QTw_{2n+1}, Sy_{2n}, t).
\end{aligned}$$

Therefore by Remark 1,

$$M(z, Tz, kt) \geq M(z, Tz, t), \quad N(z, Tz, kt) \leq N(z, Tz, t).$$

By Lemma 2, $Tz = z$. Similarly, we have $Sz = z$. Again by (1.4) and (1.5) with $\alpha = 1$,

$$M(Py_{2n}, Qz, kt) \geq M(Sy_{2n}, Tz, t) * M(Py_{2n}, Sy_{2n}, t) * M(Qz, Tz, t)$$
$$* M(Py_{2n}, Tz, t) * M(Qz, Sy_{2n}, t)$$
$$N(Py_{2n}, Qz, kt) \leq N(Sy_{2n}, Tz, t) \diamond N(Py_{2n}, Sy_{2n}, t) \diamond N(Qz, Tz, t)$$
$$\diamond N(Py_{2n}, Tz, t) \diamond N(Qz, Sy_{2n}, t).$$

Thus, as $n \to \infty$,

$$\begin{aligned} M(z, Qz, kt) &\geq M(z, Tz, t) * M(z, z, t) * M(Qz, Tz, t) \\ &\quad * M(z, Tz, t) * M(Qz, z, t) \\ &\geq M(Qz, z, t) \end{aligned}$$

and

$$\begin{aligned} N(z, Qz, kt) &\leq N(z, Tz, t) \diamond N(z, z, t) \diamond N(Qz, Tz, t) \\ &\quad \diamond N(z, Tz, t) \diamond N(Qz, z, t) \\ &\leq N(Qz, z, t). \end{aligned}$$

Therefore

$$M(z, Qz, kt) \geq M(z, Qz, t), \;\; N(z, Qz, kt) \leq N(z, Qz, t).$$

By Lemma 2, $Qz = z$. Similarly, we have $Pz = z$. For two fixed points x, y of S, T, P and Q, we have, from (1.4) and (1.5),

$$\begin{aligned} M(x, y, kt) &= M(Px, Qy, kt) \\ &\geq M(Sx, Ty, t) * M(Px, Sx, t) * M(Qy, Ty, t) \\ &\quad * M(Px, Ty, \alpha t) * M(Qy, Sx, (2-\alpha)t) \\ &= M(x, y, t) * M(x, x, t) * M(y, y, t) \\ &\quad * M(x, y, \alpha t) * M(x, y, (2-\alpha)t) \\ &\geq M(x, y, t) \text{ with } \alpha = 1 \end{aligned}$$

and

$$\begin{aligned} N(x, y, kt) &= N(Px, Qy, kt) \\ &\leq N(Sx, Ty, t) \diamond N(Px, Sx, t) \diamond N(Qy, Ty, t) \\ &\quad \diamond N(Px, Ty, \alpha t) \diamond N(Qy, Sx, (2-\alpha)t) \\ &= N(x, y, t) \diamond N(x, x, t) \diamond N(y, y, t) \\ &\quad \diamond N(x, y, \alpha t) \diamond N(x, y, (2-\alpha)t) \\ &\leq N(x, y, t) \text{ with } \alpha = 1. \end{aligned}$$

Therefore, by Lemma 2, $x = y$.

From Theorem 1 with $S = T = I_X$ (the identity mapping on X), we have the following.

Corollary 1. *Let X be a complete intuitionistic fuzzy metric space and $P, Q : X \to X$. If there exists a constant $k \in (0,1)$ such that*

$$\begin{aligned} M(Px, Qy, kt) &\geq M(x, y, t) * M(x, Px, t) * M(y, Px, \alpha t) \\ &\quad * M(x, Qy, (2-\alpha)t) * M(y, Qy, t), \\ N(Px, Qy, kt) &\leq N(x, y, t) \diamond N(x, Px, t) \diamond N(y, Px, \alpha t) \\ &\quad \diamond N(x, Qy, (2-\alpha)t) \diamond N(y, Qy, t) \end{aligned} \tag{1}$$

for all $x, y \in X$, $t > 0$ and $\alpha \in (0,2)$, then P and Q have a unique common fixed point.

From Theorem 1 with $S = T$ and $P = Q$, we have the following.

Corollary 2. *Let X be a complete intuitionistic fuzzy metric space and P, S asymptotically commuting maps on X such that $P(X) \subset S(X)$. If S is continuous and there exists a constant $k \in (0,1)$ such that*

$$\begin{aligned} M(Px, Py, kt) &\geq M(Sx, Sy, t) * M(Sx, Px, t) * M(Sy, Py, t) \\ &\quad * M(Sy, Px, \alpha t) * M(Sx, Py, (2-\alpha)t), \end{aligned} \tag{2}$$

$$\begin{aligned} N(Px, Py, kt) &\leq N(Sx, Sy, t) \diamond N(Sx, Px, t) \diamond N(Sy, Py, t) \\ &\quad \diamond N(Sy, Px, \alpha t) \diamond N(Sx, Py, (2-\alpha)t) \end{aligned} \tag{3}$$

for all $x, y \in X$, $t > 0$ and $\alpha \in (0,2)$, then P and S have a unique common fixed point.

Corollary 3. (Intuitionistic Fuzzy Banach Contraction Theorem) *Let X be a complete intuitionistic fuzzy metric space and $P : X \to X$ such that*

$$M(Px, Py, kt) \geq M(x, y, t), \quad N(Px, Py, kt) \leq N(x, y, t) \tag{4}$$

for all $x, y \in X$, $k \in (0,1)$. Then P has a unique fixed point.

Proof. It follows from Corollary 1 since (1) with $P = Q$ includes (4).

Theorem 2. *Let X be a complete intuitionistic fuzzy metric space and P, Q two maps on the product $X \times X$ with values in X. If there exists a constant $k \in (0,1)$ such that*

$$\begin{aligned} & M(P(x,y), Q(u,v), kt) \\ &\quad \geq M(P(x,y), x, t) * M(Q(u,v), u, t) * M(x, u, t) \\ &\quad * M(y, v, t) * M(P(x,y), u, \alpha t) * M(Q(u,v), x, (2-\alpha)t), \end{aligned} \tag{5}$$

$$\begin{aligned} & N(P(x,y), Q(u,v), kt) \\ &\quad \leq N(P(x,y), x, t) \diamond N(Q(u,v), u, t) \diamond N(x, u, t) \\ &\quad \diamond N(y, v, t) \diamond N(P(x,y), u, \alpha t) \diamond N(Q(u,v), x, (2-\alpha)t) \end{aligned} \tag{6}$$

for all $x, y, u, v \in X$, $t > 0$ and $\alpha \in (0,2)$, then there exists exactly one point $w \in X$ such that

$$P(w, w) = w = Q(w, w).$$

Proof. From (5), (6) and Remark 1,

$$M(P(x,y),Q(u,y),kt) \geq M(P(x,y),x,t) * M(Q(u,y),u,t) * M(x,u,t)$$
$$*M(P(x,y),u,\alpha t) * M(Q(u,y),x,(2-\alpha)t)$$

and

$$N(P(x,y),Q(u,y),kt) \leq N(P(x,y),x,t) \diamond N(Q(u,y),u,t) \diamond N(x,u,t)$$
$$\diamond N(P(x,y),u,\alpha t) \diamond N(Q(u,y),x,(2-\alpha)t)$$

for all $x, y, u \in X$. Therefore, by Corollary 1, for each $y \in X$, there exists one and only one $z(y) \in X$ such that

$$P(z(y),y) = z(y) = Q(z(y),y). \tag{7}$$

For any $y, y' \in X$, by (5) and (6) with $\alpha = 1$, we get

$$\begin{aligned}
&M(z(y),z(y'),kt) = M(P(z(y),y),Q(z(y'),y),kt) \\
&\geq M(P(z(y),y),z(y),t) * M(Q(z(y'),y),z(y'),t) * M(z(y),z(y'),t) \\
&\qquad *M(y,y',t) * M(P(z(y),y),z(y'),t) * M(Q(z(y'),y),z(y),t) \\
&= M(z(y),z(y),t) * M(z(y'),z(y'),t) * M(z(y),z(y'),t) \\
&\qquad *M(y,y',t) * M(z(y),z(y'),t) * M(z(y'),z(y),t) \\
&= M(z(y),z(y'),t) * M(y,y',t)
\end{aligned}$$

and

$$\begin{aligned}
&N(z(y),z(y'),kt) = N(P(z(y),y),Q(z(y'),y),kt) \\
&\leq N(P(z(y),y),z(y),t) \diamond N(Q(z(y'),y),z(y'),t) \diamond N(z(y),z(y'),t) \\
&\qquad \diamond N(y,y',t) \diamond N(P(z(y),y),z(y'),t) \diamond N(Q(z(y'),y),z(y),t) \\
&= N(z(y),z(y),t) \diamond N(z(y'),z(y'),t) \diamond N(z(y),z(y'),t) \\
&\qquad \diamond N(y,y',t) \diamond N(z(y),z(y'),t) \diamond N(z(y'),z(y),t) \\
&= N(z(y),z(y'),t) \diamond N(y,y',t).
\end{aligned}$$

Therefore

$$M(z(y),z(y'),kt) \geq M(y,y',t) * M(z(y),z(y'),\frac{t}{k^n}),$$
$$N(z(y),z(y'),kt) \leq N(y,y',t) \diamond N(z(y),z(y'),\frac{t}{k^n}).$$

Consequently,

$$\lim_{n\to\infty} M(z(y),z(y'),kt) \geq M(y,y',t),$$
$$\lim_{n\to\infty} N(z(y),z(y'),kt) \leq N(y,y',t).$$

Thus Corollary 3 yields that the map $z(\cdot) \in X$ into itself has exactly one fixed point $w \in X$, that is, there exist $w \in X$ such that $z(w) = w$. Hence, by (7),

$$P(w,w) = z(w) = w = Q(w,w).$$

Remark 3. In an intuitionistic fuzzy metric space X, if $x, y \in X$ are such that $P(x, y) = x$ and $Q(x, y) = y$, then it can be seen that $x = y$ using (5) and (6).

References

1. Badard R (1984) Fixed point theorems for fuzzy numbers. Fuzzy Sets and Systems 13: 291–302.
2. Butnariu D (1982) Fixed points for fuzzy mappings. Fuzzy Sets and Systems 7: 191–207.
3. Grabiec M (1988) Fixed point in fuzzy metric spaces. Fuzzy Sets and Systems 27: 385–389.
4. Jungck G (1976) Commutating maps and fixed points. Amer Math Monthly 83: 261–263.
5. Jungck G (1986) Compatible mappings and common fixed points. Internat J Math Math Sci 9: 771–779.
6. Jungck G (1988) Common fixed points for commuting and compatible maps on compacta. Proc Amer Math Soc 108: 977–983.
7. Kramosil I, Michalek J (1975) Fuzzy metric and statistical metric Kybernetica 11: 336–344.
8. Mishra SN, Sharma N, Singh SL (1994) Common fixed points of maps on fuzzy metric spaces. Internat J Math Math Sci 17: 253–258.
9. Park JH (2004) Intuitionistic fuzzy metric spaces. Chaos, Solitons & Fractals 22: 1039–1046.
10. Park JH, Park JS, Kwun YC (2006) A common fixed point theorem in the intuitionistic fuzzy metric spaces. In: Jiao L et. al. (eds) Advances in Natural Computation Data Mining. Xidian University, Xian, 293–300.
11. Park JS, Kim SY (1999) A fixed point Theorem in a fuzzy metric space. Far East J Math Sci 1: 927–934.
12. Park JS, Kwun YC, Some fixed point theorems in the intuitionistic fuzzy metric spaces, to appear.
13. Park JS, Kwun YC, Park JH (2005) A fixed point theorem in the intuitionistic fuzzy metric spaces. Far East J Math Sci 16: 137–149.
14. Schweizer B, Sklar A (1983) Probabilistic metric spaces. North-Holland, New York, Oxford.
15. Sessa S (1982) On a weak commutativity condition of mappings in fixed point considerations. Publ Inst Math (Beograd) 32(46): 149–153.
16. Tivari BML, Singh SL (1986) A note on recent generalizations of Jungck contraction principle. J UPGC Acad Soc 3: 13–18.
17. Weiss MD (1975) Fixed points, separation and induced topologies for fuzzy sets. J Math Anal Appl 50: 142–150.
18. Zadeh LA (1965) Fuzzy sets. Inform and Control 8: 338–353.

Generalized Fuzzy B-Algebras

A. Zarandi Baghini and A. Borumand Saeid

Dept. of Mathematics, Islamic Azad university, Kerman Branch, Kerman, Iran
{arsham,zarandi}@iauk.ac.ir

Abstract. By two reletions belonging to ($\in$) and quasi-coincidence (q) between fuzzy points and fuzzy sets, we define the concept of (α, β)-fuzzy subalgebras where α, β are any two of $\{\in, q, \in \vee q, \in \wedge q\}$ with $\alpha \neq \in \wedge q$. We state and prove some theorems in (α, β)-fuzzy B-algebras.

Mathematics Subject Classification: 03G25, 03B05, 03B52, 06F35.

Keywords: B-algebra, (α, β)-fuzzy subalgebra, fuzzy point.

1 Introduction

Y. Imai and K. Iseki [4] introduced two classes of abstract algebras: BCK-algebras and BCI-algebras. It is known that the class of BCK-algebras is a proper subclass of the class of BCI-algebras. In [8], J. Neggers and H. S. Kim introduced the notion of d-algebras, which is generalization of BCK-algebras and investigated relation between d-algebras and BCK-algebras. Also they introduced the notion of B-algebras [7], which is a generalization of BCK-algebra. In 1980, P. M. Pu and Y. M. Liu [8], introduced the idea of quasi-coincidence of a fuzzy point with a fuzzy set, which is used to generate some different types of fuzzy subgroups, called (α, β)-fuzzy subgroups, introduced by Bhakat and Das [2]. In particular, $\{\in, \in \vee q\}$-fuzzy subgroup is an important and useful generalization of Rosenfeld's fuzzy subgroup. In this note we introduced the notion of (α, β)-fuzzy B-algebras. We state and prove some theorems discussed in (α, β)-fuzzy Bc-subalgebras and level subalgebras.

2 Preliminary

Definition 2.1. [7] A B-algebra is a non-empty set X with a consonant 0 and a binary operation $*$ satisfying the following axioms:

(I) $x * x = 0$,
(II) $x * 0 = x$,
(III) $(x * y) * z = x * (z * (0 * y))$,

for all $x, y, z \in X$.

Example 2.2. [5] Let $X = \{0, 1, 2, 3\}$ be a set with the following table:

B.-Y. Cao (Ed.): Fuzzy Information and Engineering (ICFIE), ASC 40, pp. 226–233, 2007.
springerlink.com

$*$	0	1	2	3
0	0	3	2	1
1	1	0	3	2
2	2	1	0	3
3	3	2	1	0

Then $(X, *, 0)$ is a B-algebra. But $(X, *, 0)$ is not a BCK-algebra, since $0*1 \neq 0$.

Theorem 2.3. [7] If X is a B-algebra, then $x * y = x * (0 * (0 * y))$, for all $x, y \in X$;

A non-empty subset I of a B-algebra X is called a subalgebra of X if $x*y \in I$ for any $x, y \in I$.

A mapping $f : X \longrightarrow Y$ of B-algebras is called a B-homomorphism if $f(x * y) = f(x) * f(y)$ for all $x, y \in X$.

We now review some fuzzy logic concept (see [10]).

We now review some fuzzy logic concepts (see [2] and [10]).

Let X be a set. A fuzzy set A on X is characterized by a membership function $\mu_A : X \longrightarrow [0, 1]$.

Let $f : X \longrightarrow Y$ be a function and B a fuzzy set of Y with membership function μ_B. The inverse image of B, denoted by $f^{-1}(B)$, is the fuzzy set of X with membership function $\mu_{f^{-1}(B)}$ defined by $\mu_{f^{-1}(B)}(x) = \mu_B(f(x))$ for all $x \in X$.

Conversely, let A be a fuzzy set of X with membership function μ_A. Then the image of A, denoted by $f(A)$, is the fuzzy set of Y such that

$$\mu_{f(A)}(y) = \begin{cases} \sup\limits_{x \in f^{-1}(y)} \mu_A(x) & \text{if } f^{-1}(y) \neq \emptyset \\ 0 & otherwise \end{cases}$$

A fuzzy set μ of a set X of the form

$$\mu(y) := \begin{cases} t & \text{if } y = x, \\ 0 & \text{otherwise} \end{cases}$$

where $t \in (0, 1]$ is called a fuzzy point with support x and value t and is denoted by x_t.

Consider a fuzzy point x_t, a fuzzy set μ on a set X and $\alpha \in \{\in, q, \in \vee q, \in \wedge q\}$, we define $x_t \alpha \mu$ as follow:

(i) $x_t \in \mu$ (resp. $x_t q \mu$) means that $\mu(x) \geq t$ (resp. $\mu(x) + t > 1$) and in this case we said that x_t belong to (resp. quasi-coincident with) fuzzy set μ.

(ii) $x_t \in \vee q \mu$ (resp. $x_t \in \wedge q \mu$) means that $x_t \in \mu$ or $x_t q \mu$ (resp. $x_t \in \mu$ and $x_t q \mu$).

Definition 2.1. (1) *Let μ be a fuzzy set of a B-algebra X. Then μ is called a fuzzy B-algebra (subalgebra) of X if*

$$\mu(x * y) \geq \min\{\mu(x), \mu(y)\}$$

for all $x, y \in X$.

Example 2.2. (1) *Let* $X = \{0, 1, 2, 3\}$ *be a set with the following table:*

$*$	0	1	2	3
0	0	1	2	3
1	1	0	1	1
2	2	2	0	2
3	3	3	3	0

Then $(X, *, 0)$ *is a* B*-algebra. Define a fuzzy set* $\mu : X \to [0, 1]$ *on* X*, by* $\mu(0) = \mu(1) = t_0$ *and* $\mu(2) = \mu(3) = t_1$*, for* $t_0, t_1 \in [0, 1]$ *and* $t_0 > t_1$*. Then* μ *is a fuzzy* B*-algebra of* X*.*

Definition 2.3. (2) *Let* μ *be a fuzzy set of* X*. Then the upper level set* $U(\mu; \lambda)$ *of* X *is defined as following :*

$$U(\mu; \lambda) = \{x \in X \mid \mu(x) \geq \lambda\}.$$

Definition 2.4. *Let* $f : X \longrightarrow Y$ *be a function. A fuzzy set* μ *of* X *is said to be* f*-invariant, if* $f(x) = f(y)$ *implies that* $\mu(x) = \mu(y)$*, for all* $x, y \in X$*.*

3 (α, β)-Fuzzy B-Algebras

From now on X is a B-algebra and α, $\beta \in \{\in, q, \in \vee q, \in \wedge q\}$ unless otherwise specified. By $x_t\overline{\alpha}\mu$ we mean that $x_t\alpha\mu$ does not hold.

Theorem 3.1. *Let* μ *be a fuzzy set of* X*. Then* μ *is a fuzzy* B*-algebra if and only if*

$$x_{t_1}, y_{t_2} \in \mu \Rightarrow (x * y)_{\min(t_1, t_2)} \in \mu, \tag{1}$$

for all $x, y \in X$ *and* $t_1, t_2 \in [0, 1]$*.*

Note that if μ is a fuzzy set of X defined by $\mu(x) \leq 0.5$ for all $x \in X$, then the set $\{x_t \mid x_t \in \wedge q\mu\}$ is empty.

Definition 3.2. *A fuzzy set* μ *of* X *is said to be an* (α, β)*-fuzzy subalgebra of* X*, where* $\alpha \neq \in \wedge q$*, if it satisfies the following condition:*

$$x_{t_1}\alpha\mu, y_{t_2}\alpha\mu \Rightarrow (x * y)_{\min(t_1, t_2)}\beta\mu$$

for all t_1*,* $t_2 \in (0, 1]$

Proposition 3.3. μ *is an* $(\in, \in)$*-fuzzy subalgebra of* X *if and only if for all* $t \in [0, 1]$*, the nonempty level set* $U(\mu; t)$ *is a subalgebra of* X*.*

Example 3.4. *Let* $X = \{0, 1, 2, 3\}$ *be a set with the following table:*

$*$	0	1	2	3
0	0	1	2	3
1	1	0	3	2
2	2	3	0	1
3	3	2	1	0

Then $(X, *, 0)$ is a B-algebra. Let μ be a fuzzy set in X defined $\mu(0) = 0.6$, $\mu(1) = 0.7$ and $\mu(2) = \mu(3) = 0.3$. Then μ is an $(\in, \in \vee q)$-fuzzy subalgebra of X. But

(1) μ is not an $(\in, \in)$-fuzzy subalgebra of X since $1_{0.62} \in \mu$ and $1_{0.66} \in \mu$, but $(1 * 1)_{\min(0.62,0.66)} = 0_{0.62} \overline{\in} \mu$.

(2) μ is not a $(q, \in \vee q)$-fuzzy subalgebra of X since $1_{0.41} q \mu$ and $2_{0.77} q \mu$, but $(1 * 2)_{\min(0.41,0.77)} = 3_{0.41} \overline{\in \vee q} \mu$.

(3) μ is not an $(\in \vee q, \in \vee q)$-fuzzy subalgebra of X since $1_{0.5} \in \vee q \mu$ and $3_{0.8} \in \vee q \mu$, but $(1 * 3)_{\min(0.5,0.8)} = 2_{0.5} \overline{\in \vee q} \mu$.

Theorem 3.5. *Let μ be a fuzzy set. Then the following diagram shows the relationship between (α, β)-fuzzy subalgebras of X, where α, β are one of $\in$ and q.*

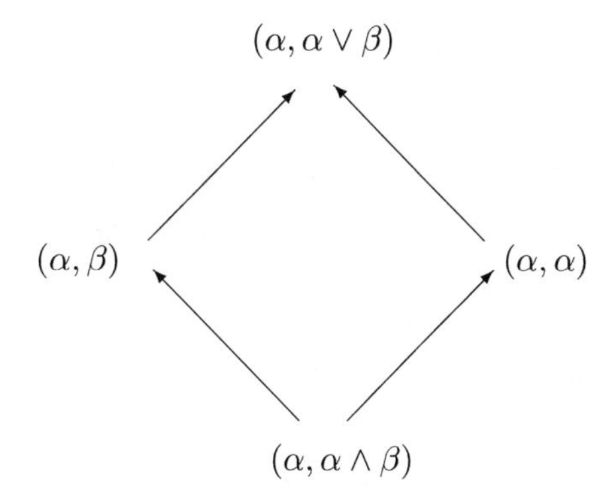

and also we have

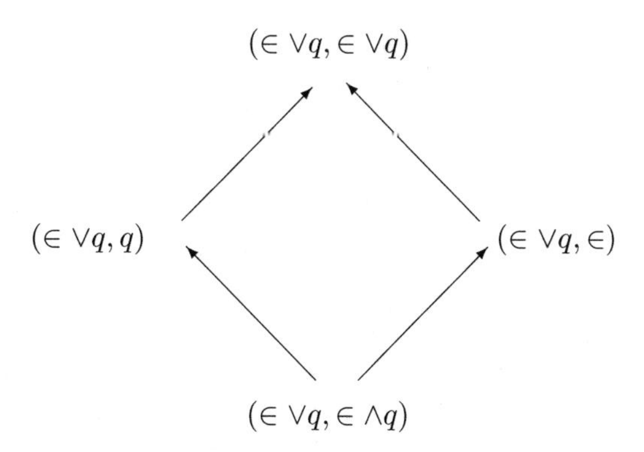

Proposition 3.6. *If μ is a nonzero (α, β)-fuzzy subalgebra of X, then $\mu(0) > 0$.*

For a fuzzy set μ in X, we denote the support μ by, $X_0 := \{x \in X \mid \mu(x) > 0\}$.

Proposition 3.7. *If μ is a nonzero $(\in, \in \vee q)$-fuzzy subalgebra of X, then the set X_0 is a subalgebra of X.*

Proposition 3.8. *If μ is a nonzero $(q, \in \vee q)$-fuzzy subalgebra of X, then the set X_0 is a subalgebra of X.*

Theorem 3.9. *Let μ be a nonempty (α, β)-fuzzy subalgebra, where $\alpha, \beta \in \{\in, q, \in \vee q, \in \wedge q\}$ and $\alpha \neq \in \wedge q$. Then X_0 is a subalgebra of X.*

Theorem 3.10. *Any non-zero (q, q)-fuzzy subalgebra of X is constant on X_0.*

Theorem 3.11. *μ is a non-zero (q, q)-fuzzy subalgebra if and only if there exists subalgebra S of X such that*

$$\mu(x) = \begin{cases} t & \text{if } x \in S \\ 0 & \text{otherwise} \end{cases}$$

for some $t \in (0, 1]$.

Theorem 3.12. *μ is a non-zero (q, q)-fuzzy subalgebra of X if and only if $U(\mu; \mu(0)) = X_0$ and for all $t \in [0, 1]$, the nonempty level set $U(\mu; t)$ is a subalgebra of X.*

The following example shows that the condition " $U(\mu; \mu(0)) = X_0$" is necessary.

Example 3.13. *Let $X = \{0, 1, 2, 3\}$ be B-algebra in Example 3.3. Define fuzzy set μ on X by*

$$\mu(0) = 0.6, \quad \mu(1) = \mu(2) = \mu(3) = 0.3$$

Then $X_0 = X$, $U(\mu; \mu(0)) = \{0\} \neq X_0$ and also

$$U(\mu; t) = \begin{cases} X & \text{if } 0 \leq t \leq 0.3 \\ \{0\} & \text{if } 0.3 < t \leq 0.6 \\ \emptyset & \text{if } t > 0.6 \end{cases}$$

is a subalgebra of X, while by Theorem 3.11, μ is not a (q, q)-fuzzy subalgebra.

Theorem 3.14. *Every (q, q)-fuzzy subalgebra is an $(\in, \in)$-fuzzy subalgebra.*

Note that in Example 3.13 μ is an $(\in, \in)$-fuzzy subalgebra, while it is not a (q, q)-fuzzy subalgebra. So the converse of the above theorem is not true in general.

Theorem 3.15. *If μ is a non-zero fuzzy set of X. Then there exists subalgebra S of X such that $\mu = \chi_S$ if and only if μ is an (α, β)-fuzzy subalgebra of X, where (α, β) is one of the following forms:*

(i) $(\in, q)$, *(ii)* $(\in, \in \wedge q)$,
(iii) $(q, \in)$, *(iv)* $(q, \in \wedge q)$,
(v) $(\in \vee q, q)$, *(vi)* $(\in \vee q, \in \wedge q)$,
(vii) $(\in \vee q, \in)$.

Theorem 3.16. *Let S be a subalgebra of X and let μ be a fuzzy set of X such that*

(a) $\mu(x) = 0$ *for all* $x \in X \backslash S$,
(b) $\mu(x) \geq 0.5$ *for all* $x \in S$.

Then μ is a $(q, \in \vee q)$-fuzzy subalgebra of X.

Theorem 3.17. *Let μ be a $(q, \in \vee q)$-fuzzy subalgebra of X such that μ is not constant on the set X_0. Then there exists $x \in X$ such that $\mu(x) \geq 0.5$. Moreover, $\mu(x) \geq 0.5$ for all $x \in X_0$.*

Theorem 3.18. *Let μ be a non-zero fuzzy set of X. Then μ is a $(q, \in \vee q)$-fuzzy subalgebra of X if and only if there exists subalgebra S of X such that*

$$\mu(x) = \begin{cases} a & \text{if } x \in S \\ 0 & \text{otherwise} \end{cases} \quad \text{or} \quad \mu(x) = \begin{cases} \geq 0.5 & \text{if } x \in S \\ 0 & \text{otherwise} \end{cases}$$

for some $a \in (0, 1]$

Theorem 3.19. *Let μ be a non-zero $(q, \in \vee q)$-fuzzy subalgebra of X. Then the nonempty level set $U(\mu; t)$ is a subalgebra of X, for all $t \in [0, 0.5]$.*

Theorem 3.20. *Let μ be a non-zero fuzzy set of X, $U(\mu; 0.5) = X_0$ and the nonempty level set $U(\mu; t)$ is a subalgebra of X, for all $t \in [0, 1]$. Then μ is a $(q, \in \vee q)$-fuzzy subalgebra of X.*

Theorem 3.21. *A fuzzy set μ of X is an $(\in, \in \vee q)$-fuzzy subalgebra of X if and only if $\mu(x * y) \geq \min(\mu(x), \mu(y), 0.5)$, for all $x, y \in X$.*

Theorem 3.22. *Let μ be an $(\in, \in \vee q)$-fuzzy subalgebra of X.*
(i) If there exists $x \in X$ such that $\mu(x) \geq 0.5$, then $\mu(0) \geq 0.5$.
(ii) If $\mu(0) < 0.5$, then μ is an $(\in, \in)$-fuzzy subalgebra of X.

Lemma 3.23. *Let μ be a non-zero $(\in, \in \vee q)$ fuzzy subalgebra of X. Let $x, y \in X$ such that $\mu(x) < \mu(y)$. Then*

$$\mu(x * y) = \begin{cases} \mu(x) & \text{if } \mu(y) < 0.5 \text{ or } \mu(x) < 0.5 \leq \mu(y) \\ \geq 0.5 & \text{if } \mu(x) \geq 0.5 \end{cases}$$

Theorem 3.24. *Let μ be an $(\in, \in \vee q)$-fuzzy subalgebra of X. Then for all $t \in [0, 0.5]$, the nonempty level set $U(\mu; t)$ is a subalgebra of X. Conversely, if the nonempty level set μ is a subalgebra of X, for all $t \in [0, 1]$, then μ is an $(\in, \in \vee q)$-fuzzy subalgebra of X.*

Theorem 3.25. *Let S be a subset of X. The characteristic function χ_S of S is an $(\in, \in \vee q)$-fuzzy subalgebra of X if and only if S is a subalgebra of X.*

Lemma 3.26. *Let $f : X \to Y$ be a B-homomorphism and G be a fuzzy set of Y with membership function μ_G. Then $x_t \alpha \mu_{f^{-1}(G)} \Leftrightarrow f(x)_t \alpha \mu_G$, for all $\alpha \in \{\in, q, \in \vee q, \in \wedge q\}$.*

Theorem 3.27. *Let $f : X \to Y$ be a B-homomorphism and G be a fuzzy set of Y with membership function μ_G.*
(i) If G is an (α, β)-fuzzy subalgebra of Y, then $f^{-1}(G)$ is an (α, β)-fuzzy subalgebra of X,
(ii) Let f be epimorphism. If $f^{-1}(G)$ is an (α, β)-fuzzy subalgebra of X, then G is an (α, β)-fuzzy subalgebra of Y.

Theorem 3.28. *Let $f : X \to Y$ be a B-homomorphism and H be an $(\in, \in \vee q)$-fuzzy subalgebra of X with membership function μ_H. If μ_H is an f-invariant, then $f(H)$ is an $(\in, \in \vee q)$-fuzzy subalgebra of Y.*

Lemma 3.29. *Let $f : X \to Y$ be a B-homomorphism.*
(i) If S is a subalgebra of X, then $f(S)$ is a subalgebra of Y,
(ii) If S' is a subalgebra of Y, then $f^{-1}(S')$ is a subalgebra of X.

Theorem 3.30. *Let $f : X \to Y$ be a B-homomorphism. If H is a non-zero (q, q)-fuzzy subalgebra of X with membership function μ_H, then $f(H)$ is a non-zero (q, q)-fuzzy subalgebra of Y.*

Theorem 3.31. *Let $f : X \to Y$ be a B-homomorphism. If H is an (α, β)-fuzzy subalgebra of X with membership function μ_H, then $f(H)$ is an (α, β)-fuzzy subalgebra of Y, where (α, β) is one of the following form*

(i) $(\in, q)$, *(ii)* $(\in, \in \wedge q)$,
(iii) $(q, \in)$, *(iv)* $(q, \in \wedge q)$,
(v) $(\in \vee q, q)$, *(vi)* $(\in \vee q, \in \wedge q)$,
(vii) $(\in \vee q, \in)$, *(viii)* $(q, \in \vee q)$.

Proof. The proof is similar to the proof of Theorem 3.30, by using of Theorems 3.15 and 3.18.

Theorem 3.32. *Let $f : X \to Y$ be a B-homomorphism and H be an $(\in, \in)$-fuzzy subalgebra of X with membership function μ_H. If μ_H is an f-invariant, then $f(H)$ is an $(\in, \in)$-fuzzy subalgebra of Y.*

Theorem 3.33. *Let $\{\mu_i \mid i \in \Lambda\}$ be a family of $(\in, \in \vee q)$-fuzzy subalgebra of X. Then $\mu := \bigcap_{i \in \Lambda} \mu_i$ is an $(\in, \in \vee q)$-fuzzy subalgebra of X.*

Theorem 3.34. *Let $\{\mu_i \mid i \in \Lambda\}$ be a family of $(\in, \in)$-fuzzy subalgebra of X. Then $\mu := \bigcap_{i \in \Lambda} \mu_i$ is an $(\in, \in)$-fuzzy subalgebra of X.*

Theorem 3.35. *Let $\{\mu_i \mid i \in \Lambda\}$ be a family of (α, β)-fuzzy subalgebra of X. Then $\mu := \bigcap_{i \in \Lambda} \mu_i$ is an (α, β)-fuzzy subalgebra of X, where (α, β) is one of the following form*
(i) $(\in, q)$, *(ii)* $(\in, \wedge q)$,
(iii) $(q, \in)$, *(iv)* $(q, \in \wedge q)$,
(v) $(\in \vee q, q)$, *(vi)* $(\in \vee q, \in \wedge q)$,
(vii) $(\in \vee q, \in)$, *(viii)* $(q, \in \vee q)$,
(ix) (q, q).

References

1. S. K Bhakat and P. Das, *$(\in, \in \vee q)$-fuzzy subgroups*, Fuzzy Sets and Systems 80 (1996), 359-368.
2. A. Borumand Saeid, *Fuzzy topological B-algebras*, International Journal of Fuzzy Systems, (to appear).
3. Y. Imai and K. Iseki, *On axiom systems of propositional calculi*, XIV Proc. Japan Academy, 42 (1966), 19-22.
4. Y. B. Jun, E. H. Roh, Chinju and H. S. Kim, *On Fuzzy B-algebras*, Czechoslovak Math. J. 52 (2002) 375-384.
5. J. Meng and Y.B. Jun, *BCK-algebras*, Kyung Moonsa, Seoul, Korea, (1994).
6. P. M. Pu and Y. M. Liu, *Fuzzy Topology I, Neighborhood structure of a fuzzy point and Moore-Smith convergence*, J. Math. Anal. Appl. 76 (1980), 571-599.
7. J. Neggers and H. S. Kim, *On B-algebras*, Math. Vensik 54 (2002), 21-29.
8. ———, *On d-algebras*, Math. Slovaca 49 (1999), 19-26.
9. A. Rosenfeld, *Fuzzy Groups*, J. Math. Anal. Appl. 35 (1971), 512-517.
10. L. A. Zadeh, *Fuzzy Sets*, Inform. Control, 8 (1965), 338-353.
11. ———, *The concept of a linguistic variable and its application to approximate reasoning. I*, Information Sci. 8 (1975), 199-249.

Fuzzy Set Theory Applied to QS-Algebras

Arsham Borumand Saeid

Dept. of Mathematics, Islamic Azad university, Kerman Branch, Kerman, Iran
arsham@iauk.ac.ir

Abstract. In this paper the notion of fuzzy QS-subalgebra and fuzzy topological QS-algebras are introduced. We state and prove some theorem in fuzzy QS-subalgebras and level subalgebras. Finally the Foster's results on homomorphic images and inverse images in fuzzy topological QS-algebras are studied.

Keywords: Fuzzy QS-algebra, fuzzy QS-subalgebras, level subalgebras, fuzzy topological QS-algebras.

1 Introduction

In 1966, Y. Imai and K. Iseki [5] introduced two classes of abstract algebras: BCK-algebras and BCI-algebras. It is known that the class of BCK-algebras is a proper subclass of the class of BCI-algebras. In [3] Q. P. Hu and X. Li introduced a wide class of abstract algebras: BCH-algebra. They shown that the class of BCI-algebras is a proper subclass of the class of BCH-algebras. J. Neggers, S. S. Ahn and H. S. Kim introduced the notion of Q-algebras [9], which is a generalization of $BCH/BCI/BCK$-algebras. In [1], S. S. Ahn and H. S. Kim introduced the notion of QS-algebras which is a generalization of Q-algebras.

The concept of a fuzzy set, which was introduced in [11] by L. A. Zadeh. Provides a natural framework for generalizing many of the concepts of general mathematics and topology. D. H. Foster (cf. [2]) combined the structure of a fuzzy topological spaces with that of a fuzzy group, introduced by A. Rosenfeld (cf. [10]), to formulated the elements of a theory of fuzzy topological groups.

In the present paper, we introduced the concept of fuzzy QS-subalgebras and fuzzy topological QS-algebras and study this structure. We state and prove some theorem discussed in fuzzy QS-subalgebras and level subalgebras. Finally some of Fosters results on homomorphic images and inverse images in fuzzy topological QS-algebras are studied.

2 Preliminary

Definition 2.1. [1] A QS-algebra is a non-empty set X with a consonant 0 and a binary operation $*$ satisfying the following axioms:

(I) $x * x = 0$,

(II) $x * 0 = x$,

B.-Y. Cao (Ed.): Fuzzy Information and Engineering (ICFIE), ASC 40, pp. 234–242, 2007.
springerlink.com

(III) $(x * y) * z = (x * z) * y$,
(IV) $(x * y) * (x * z) = z * y$,
for all $x, y, z \in X$.

In X we can define a binary relation by $x \leq y$ if and only if $x * y = 0$.

Example 2.2. [1] Let $\mathcal{Z}$ be the set of all integers and let $n\mathcal{Z}= \{nz \mid z \in \mathcal{Z}\}$. Then $(\mathcal{Z}; -, 0)$ and $(n\mathcal{Z};-, 0)$ are both QS-algebras, where "-" is the usual subtraction of integers. Also $(\mathcal{R}; -, 0)$ and $(\mathcal{C}; -, 0)$ are QS-algebras where $\mathcal{R}$ is the set of all real numbers, $\mathcal{C}$ is the set of all complex numbers and "-" is the usual subtraction of real (complex) numbers.

Definition 2.3. [1] Let X be a QS-algebra. Then for any $x, y$and z in X, the following hold:
(a) $x \leq y$ implies $z * y \leq z * x$,
(b) $x \leq y$ and $y \leq z$ imply $x \leq z$,
(c) $x * y \leq z$ implies $(x * z \leq y$,
(d) $(x * z) * (y * z) \leq x * y$,
(e) $x \leq y$ implies $x * z \leq y * z$,
(f) $0 * (0 * (0 * x)) = 0 * x$.

Definition 2.4. A non-empty subset S of a QS-algebra X is called a subalgebra of X if $x * y \in S$ for any $x, y \in S$.

A mapping $f : X \longrightarrow Y$ of QS-algebras is called a QS-homomorphism if $f(x * y) = f(x) * f(y)$ for all $x, y \in X$.

We now review some fuzzy logic concept (see [11]).

Let X be a set. A fuzzy set A in X is characterized by a membership function $\mu_A : X \longrightarrow [0, 1]$. Let f be a mapping from the set X to the set Y and let B be a fuzzy set in Y with membership function μ_B.

The inverse image of B, denoted $f^{-1}(B)$, is the fuzzy set in X with membership function $\mu_{f^{-1}(B)}$ defined by $\mu_{f^{-1}(B)}(x) = \mu_B(f(x))$ for all $x \in X$.

Conversely, let A be a fuzzy set in X with membership function μ_A Then the image of A, denoted by $f(A)$, is the fuzzy set in Y such that:

$$\mu_{f(A)}(y) = \begin{cases} \sup\limits_{z \in f^{-1}(y)} \mu_A(z) & \text{if } f^{-1}(y) = \{x : f(x) = y\} \neq \emptyset, \\ 0 & \text{otherwise} \end{cases}$$

A fuzzy set A in the QS-algebra X with the membership function μ_A is said to be have the sup property if for any subset $T \subseteq X$ there exists $x_0 \in T$ such that

$$\mu_A(x_0) = \sup_{t \in T} \mu_A(t)$$

A fuzzy topology on a set X is a family τ of fuzzy sets in X which satisfies the following condition :

(i) For $c \in [0,1], k_c \in \tau$, where k_c has a constant membership function,
(ii) If $A, B \in \tau$, then $A \cap B \in \tau$,
(iii) If $A_j \in \tau$ for all $j \in J$, then $\bigcup_{j \in J} A_j \in \tau$.

The pair (X, τ) is called a fuzzy topological space and members of τ are called open fuzzy sets.

Let A be a fuzzy set in X and τ a fuzzy topology on X. Then the induced fuzzy topology on A is the family of fuzzy subsets of A which are the intersection with A of τ-open fuzzy sets in X. The induced fuzzy topology is denoted by τ_A, and the pair (A, τ_A) is called a fuzzy subspace of (X, τ).

Let (X, τ) and (Y, υ) be two fuzzy topological space. A mapping f of (X, τ) into (Y, υ) is fuzzy continuous if for each open fuzzy set U in υ the inverse image $f^{-1}(U)$ is in τ.

Conversely, f is fuzzy open if for each fuzzy set V in τ, the image $f(V)$ is in υ.

Let (A, τ_A) and (B, υ_B) be fuzzy subspace of fuzzy topological spaces (X, τ) and (Y, υ) respectively, and let f be a mapping from (X, τ) to (Y, υ).

Then f is a mapping of (A, τ_A) into (B, υ_B) if $f(A) \subseteq B$. Furthermore f is relatively fuzzy continuous if for each open fuzzy set $V^{'}$ in υ_B the intersection $f^{-1}(V^{'}) \cap A$ is in τ_A. Conversely, f is relatively fuzzy open if for each open fuzzy set $U^{'}$, the image $f(U^{'})$ is in υ_B.

Lemma 2.5. [2] Let $(A, \tau_A), (B, \upsilon_B)$ be fuzzy subspace of fuzzy topological space (X, τ), (Y, υ) respectively, and let f be a fuzzy continuous mapping of (X, τ) into (Y, υ) such that $f(A) \subset B$. Then f is a relatively fuzzy continuous mapping of (A, τ_A) into (B, υ_B).

3 Fuzzy QS-Subalgebra

From now on X is a QS-algebra, unless otherwise is stated.

Definition 3.1. Let μ be a fuzzy set in a QS-algebra. Then μ is called a fuzzy QS-subalgebra (algebra) of X if

$$\mu(x * y) \geq \min\{\mu(x), \mu(y)\}$$

for all $x, y \in X$.

Example 3.2. Let $X = \{0, 1, 2, 3\}$ be a set with the following table:

*	0	1	2
0	0	0	0
1	1	0	0
2	2	0	0

Then $(X, *, 0)$ is a QS-algebra, but not a $BCH/BCI/BCK$-algebra.

Define a fuzzy set $\mu : X \rightarrow [0,1]$ by $\mu(0) = 0.7 > 0.1 = \mu(x)$ for all $x \in \{1,2\}$. Then μ is a fuzzy QS-subalgebra of X.

Lemma 3.3. If A is a fuzzy QS-subalgebra of X, then for all $x \in X$

$$\mu_A(0) \geq \mu_A(x).$$

Proof. For all $x \in X$, we have $x * x = 0$ hence $\mu_A(0) = \mu_A(x * x) \geq min\{\mu_A(x), \mu_A(x)\} = \mu_A(x)$.

Proposition 3.4. Let A be a fuzzy QS-subalgebra of X, and let $n \in \mathcal{N}$. Then

(i) $\mu_A(\prod^n x * x) \geq \mu_A(x)$, for any odd number n,

(ii) $\mu_A(\prod^n x * x) = \mu_A(x)$, for any even number n.

Theorem 3.5. Let A be a fuzzy QS-subalgebra of X. If there exists a sequence $\{x_n\}$ in X, such that

$$\lim_{n \to \infty} \mu_A(x_n) = 1$$

Then $\mu_A(0) = 1$.

Proof. By above lemma we have $\mu_A(0) \geq \mu_A(x)$, for all $x \in X$, thus $\mu_A(0) \geq \mu_A(x_n)$, for every positive integer n. Consider

$$1 \geq \mu_A(0) \geq \lim_{n \to \infty} \mu_A(x_n) = 1.$$

Hence $\mu_A(0) = 1$.

Theorem 3.6. Let A_1 and A_2 are fuzzy QS-subalgebras of X. Then $A_1 \cap A_2$ is an i-v fuzzy QS-subalgebras of X.

Proof. Let $x, y \in A_1 \cap A_2$. Then $x, y \in A_1$ and A_2, since A_1 and A_2 are fuzzy QS-subalgebras of X by above theorem we have:

$$\begin{aligned}
\mu_{A_1 \cap A_2}(x * y) &= min\{\mu_{A_1}(x * y), \mu_{A_2}(x * y)\} \\
&\geq min\{min(\mu_{A_1}(x), \mu_{A_1}(y)), min(\mu_{A_2}(x), \mu_{A_2}(y))\} \\
&= min\{\mu_{A_1 \cap A_2}(x), \mu_{A_1 \cap A_2}(y)\}
\end{aligned}$$

Which Proves theorem.

Corollary 3.7. Let $\{A_i | i \in \Lambda\}$ be a family of fuzzy QS-subalgebras of X. Then $\bigcap_{i \in \Lambda} A_i$ is also an fuzzy QS-subalgebras of X.

Definition 3.8. Let A be a fuzzy set in X and $\lambda \in [0,1]$. Then the level QS-subalgebra $U(A;\lambda)$ of A and strong level QS-subalgebra $U(A;>,\lambda)$ of X are defined as following:

$$U(A;\lambda) := \{x \in X \mid \mu_A(x) \geq \lambda\},$$

$$U(A;>,\lambda) := \{x \in X \mid \mu_A(x) > \lambda\}.$$

Theorem 3.9. Let A be a fuzzy QS-subalgebra of X with the least upper bound λ_0. Then the following condition are equivalent :

(i) A is an fuzzy QS-subalgebra of X.

(ii) For all $\lambda \in Im(\mu_A)$, the nonempty level subset $U(A;\lambda)$ of A is a QS-subalgebra of X.

(iii) For all $\lambda \in Im(\mu_A) \setminus \lambda_0$, the nonempty strong level subset $U(A;>,\lambda)$ of A is a QS-subalgebra of X.

(iv) For all $\lambda \in [0,1]$, the nonempty strong level subset $U(A;>,\lambda)$ of A is a QS-subalgebra of X.

(v) For all $\lambda \in [0,1]$, the nonempty level subset $U(A;\lambda)$ of A is a QS-subalgebra of X.

Theorem 3.10. Each QS-subalgebra of X is a level QS-subalgebra of a fuzzy QS-subalgebra of X.

Proof. Let Y be a QS-subalgebra of X, and A be an fuzzy set on X defined by

$$\mu_A(x) = \begin{cases} \alpha & \text{if } x \in Y \\ 0 & \text{Otherwise} \end{cases}$$

where $\alpha \in [0,1]$. It is clear that $U(A;\alpha) = Y$. Let $x, y \in X$. We consider the following cases:

case 1) If $x, y \in Y$, then $x * y \in Y$ therefore

$$\mu_A(x * y) = \alpha = min\{\alpha, \alpha\} = min\{\mu_A(x), \mu_A(y)\}.$$

case 2) If $x, y \notin Y$, then $\mu_A(x) = 0 = \mu_A(y)$ and so

$$\mu_A(x * y) \geq 0 = min\{0, 0\} = min\{\mu_A(x), \mu_A(y)\}.$$

case 3) If $x \in Y$ and $y \notin Y$, then $\mu_A(x) = \alpha$ and $\mu_A(y) = 0$. Thus

$$\mu_A(x * y) \geq 0 = min\{\alpha, 0\} = min\{\mu_A(x), \mu_A(y)\}.$$

case 4) If $y \in Y$ and $x \notin Y$, then by the same argument as in case 3, we can conclude that $\mu_A(x * y) \geq min\{\mu_A(x), \mu_A(y)\}$.

Therefore A is a fuzzy QS-subalgebra of X.

We in the next theorem generalize the above lemma.

Theorem 3.11. Let X be a QS-algebra. Then for any chain of subalgebras

$$A_0 \subset A_1 \subset \ldots \subset A_r = X$$

there exists a fuzzy subalgebra μ of X whose level subalgebras are exactly the subalgebras of this chain.

Proof. Consider a set of numbers

$$t_0 > t_1 > \ldots > t_r$$

where each t_i be in $[0,1]$. Define $\mu : X \to [0,1]$ by $\mu(A_i \backslash A_{i-1}) = t_i$ for all $0 < i \leq r$, and $\mu(A_0) = t_0$.

We prove that μ is a fuzzy subalgebra of X. Let $x, y \in X$, we consider the following cases:

Case 1) Let $x, y \in A_i \backslash A_{i-1}$, then $\mu(x) = t_i = \mu(y)$. Since A_i is a subalgebra thus $x * y \in A_i$, so $x * y \in A_i \backslash A_{i-1}$ or $x * y \in A_{i-1}$, and in each of them we have

$$\mu(x * y) \geq t_i = min\{\mu(x), \mu(y)\}.$$

Case 2) Let $x \in A_i \backslash A_{i-1}$, $y \in A_j \backslash A_{j-1}$, where $i < j$. Then $\mu(x) = t_i$ $\mu(y) = t_j$, since $A - j \subseteq A_i$ and A_i is a subalgebra of X, then $x * y \in A_i$. Hence

$$\mu(x * y) \geq t_j = min\{\mu(x), \mu(y)\}.$$

It is clear that $Im(\mu) = \{t_0, t_1, \ldots, t_r\}$, therefore the level subalgebras of μ are given by the chain of subalgebras

$$\mu_{t_0} \subset \mu_{t_1} \ldots \subset \mu_{t_r} = X.$$

We have $\mu_{t_0} = \{x \in X \mid \mu(x) \geq t_0\} = A_0$. it is clear that $A_i \subseteq \mu_{t_i}$. Let $x \in \mu_{t_i}$ then $\mu(x) \geq t_i$ then $x \notin A_j$ for $j > i$. So $\mu(x) \in \{t_0, t_1, \ldots, t_i\}$, thus $x \in A_k$ for $k \leq i$, since $A_k \subseteq A_i$ we get that $x \in A_i$. Hence $A_i = \mu_{t_i}$ for $0 \leq i \leq r$.

Theorem 3.12. Let X be a QS-algebra. Then two level subalgebras μ_{t_1}, μ_{t_2} (where $t_1 < t_2$) of μ are equal if and only if there is no $x \in X$ such that $t_1 \leq \mu(x) < t_2$.

Proof. In contrary let $\mu_{t_1} = \mu_{t_2}$ where $t_1 < t_2$ and there exists $x \in X$ such that $t_1 \leq \mu(x) < t_2$. Then μ_{t_2} is a proper subset of μ_{t_1}, which is a contradiction.

Conversely, suppose that there is no $x \in X$ such that $t_1 \leq \mu(x) < t_2$. Since $t_1 < t_2$ then $\mu_{t_1} \subseteq \mu_{t_2}$. If $x \in \mu_{t_1}$, then $\mu(x) \geq t_1$ by hypotheses we get that $\mu(x) \geq t_2$. Therefore $x \in \mu_{t_2}$, then $\mu_{t_2} \subseteq \mu_{t_1}$. Hence $\mu_{t_1} = \mu_{t_2}$.

Theorem 3.13. Let Y be a subset of X and A be a fuzzy set on X which is given in the proof of Theorem 3.10. If A is a fuzzy QS-subalgebra of X, then Y is a QS-subalgebra of X.

Proof. Let A be a fuzzy QS-subalgebra of X, and $x, y \in Y$. Then $\mu_A(x) = \alpha = \mu_A(y)$, thus

$$\mu_A(x * y) \geq min\{\mu_A(x), \mu_A(y)\} = min\{\alpha, \alpha\} = \alpha.$$

which implies that $x * y \in Y$.

Theorem 3.14. If A is a fuzzy QS-subalgebra of X, then the set

$$X_{\mu_A} := \{x \in X \mid \mu_A(x) = \mu_A(0)\}$$

is a QS-subalgebra of X.

Proof. Let $x, y \in X_{\mu_A}$. Then $\mu_A(x) = \mu_A(0) = \mu_A(y)$, and so

$$\mu_A(x * y) \geq min\{\mu_A(x), \mu_A(y)\} = min\{\mu_A(0), \mu_A(0)\} = \mu_A(0).$$

by Lemma 3.3, we get that $\mu_A(x * y) = \mu_A(0)$ which means that $x * y \in X_{\mu_A}$.

Theorem 3.15. Let N be a fuzzy sub set of X. Let N be a fuzzy set defined by μ_N as:

$$\mu_N(x) = \begin{cases} \alpha & \text{if } x \in N \\ \beta & \text{Otherwise} \end{cases}$$

for all $\alpha, \beta \in [0, 1]$ with $\alpha \geq \beta$. Then N is a fuzzy QS-subalgebra if and only if N is a QS-subalgebra of X. Moreover, in this case $X_{\mu_N} = N$.

Proof. Let N be an fuzzy QS-subalgebra. Let $x, y \in X$ be such that $x, y \in N$. Then

$$\mu_N(x * y) \geq min\{\mu_N(x), \mu_N(y)\} = min\{\alpha, \alpha\} = \alpha$$

and so $x * y \in N$.

Conversely, suppose that N is a QS-subalgebra of X, let $x, y \in X$.
(i) If $x, y \in N$ then $x * y \in N$, thus

$$\mu_N(x * y) = \alpha = min\{\mu_N(x), \mu_N(y)\}$$

(ii) If $x \notin N$ or $y \notin N$, then

$$\mu_N(x * y) \geq \beta = min\{\mu_N(x), \mu_N(y)\}$$

This shows that N is a fuzzy QS-subalgebra.

Moreover, we have

$$X_{\mu_N} := \{x \in X \mid \mu_N(x) = \mu_N(0)\} = \{x \in X \mid \mu_N(x) = \alpha\} = N.$$

4 Fuzzy Topological QS-Algebra

Proposition 4.1. Let f be a QS-homomorphism from X into Y and G be a fuzzy QS-algebra of Y with the membership function μ_G. Then the inverse image $f^{-1}(G)$ of G is a fuzzy QS-algebra of X.

Proof. Let $x, y \in X$. Then

$$\begin{aligned}\mu_{f^{-1}(G)}(x * y) &= \mu_G(f(x * y)) \\ &= \mu_G(f(x) * f(y)) \\ &\geq \min\{\mu_G(f(x)), \mu_G(f(y))\} \\ &= \min\{\mu_{f^{-1}(G)}(x), \mu_{f^{-1}(G)}(y)\}.\end{aligned}$$

Proposition 4.2. Let f be a QS-homomorphism from X onto Y and D be a fuzzy QS-algebra of X with the sup property. Then the image $f(D)$ of D is a fuzzy QS-algebra of Y.

Proof. Let $a, b \in Y$, let $x_0 \in f^{-1}(a), y_0 \in f^{-1}(b)$ such that

$$\mu_D(x_0) = \sup_{t \in f^{-1}(a)} \mu_D(t), \qquad \mu_D(y_0) = \sup_{t \in f^{-1}(b)} \mu_D(t)$$

Then by the definition of $\mu_{f(D)}$, we have

$$\begin{aligned}\mu_{f(D)}(x * y) &= \sup_{t \in f^{-1}(a*b)} \mu_D(t) \\ &\geq \mu_D(x_0 * y_0) \\ &\geq \min\{\mu_D(x_0), \mu_D(y_0) \\ &= \min\{\sup_{t \in f^{-1}(a)} \mu_D(t), \sup_{t \in f^{-1}(b)} \mu_D(t)\} \\ &= \min\{\mu_{f(D)}(a), \mu_{f(D)}(b)\}.\end{aligned}$$

For any QS-algebra X and any element $a \in X$ we denote by R_a the right translation of X defined by $R_a(x) = x * a$ for all $x \in X$. It is clear that $R_0(x) = 0 = R_x(x)$ for all $x \in X$.

Definition 4.3. Let τ a fuzzy topology on X and D be a fuzzy QS-algebra of X with induced topology τ_D. Then D is called a fuzzy topological QS-algebra of X if for each $a \in X$ the mapping $R_a : (D, \tau_D) \to (D, \tau_D)$ is relatively fuzzy continuous.

Theorem 4.4. Let X and Y be two QS-algebras, $f : X \to Y$ be a QS-homomorphism. Let τ and υ be the fuzzy topologies on X and Y respectively, such that $\tau = f^{-1}(\upsilon)$. Let G be a fuzzy topological QS-algebra of Y with membership function μ_G. Then $f^{-1}(G)$ is a fuzzy topological QS-algebra of X with membership function $\mu_{f^{-1}(G)}$.

Theorem 4.5. Given QS-algebras X and Y and a QS-homomorphism f from X onto Y, let τ be the fuzzy topology on X and υ be the fuzzy topology on Y such that $f(\tau) = \upsilon$. Let D be a fuzzy topological QS-algebra of X. If the membership function μ_D of D is a f-invariant, then $f(D)$ is a fuzzy topological QS-algebra of Y.

References

1. S. S. Ahn and H. S. Kim (1999) *On QS-algebras*, J. of the Chungcheong Math. Soc. 12: 1-7.
2. D. H. Foster (1979) *Fuzzy topological groups*, J. Math.Anal. Appl. 67: 549-564.
3. Q. P. Hu and X. Li (1983) *On BCH-algebras*, Mathematics Seminar Notes 11: 313-320.
4. Q. P. Hu and X. Li (1985) *On proper BCH-algebras*, Math. Japanica, 30: 659-661.
5. Y. Imai and K. Iseki (1966) *On axiom systems of propositional calculi*, XIV Proc. Japan Academy, 42: 19-22.
6. K. Iseki and S. Tanaka (1978) *An introduction to the theory of BCK-algebras*, Math. Japanica, 23: 1-26.
7. K. Iseki (1980) *On BCI-algebras*, Mathematics Seminar Notes, 8: 125-130.
8. Y.B. Jun and E. H. Roh (1993) *On the $BCI - G$ part of BCI-algebras*, Math. Japanica, 38: 697-702.
9. J. Neggers, S. S. Ahn and H. S. Kim (2001) *On Q-algebras*, Int. J. Math. & Math. Sci. 27: 749-757.
10. A Rosenfeld (1971) *Fuzzy Groups*, J. Math. Anal. Appl. 35: 512-517.
11. L. A. Zadeh (1965) *Fuzzy Sets*, Inform. Control, 8: 338-353.

L-Topological Dynamical System and Its Asymptotic Properties

Leilei Chu[1] and Hongying Zhang[2]

[1] Faculty of Science, Xi'an Jiaotong University, Xi'an, 710049, P.R. China
chull@mail.xjtu.edu.cn

[2] Faculty of Science, Xi'an Jiaotong University, Xi'an, 710049, P.R. China
hyzhang@sohu.com

Abstract. With the development of the theory of fuzzy topological space, more and more research has been focused on fuzzy topological dynamical systems. In this paper the fuzzy topological dynamical system on L-topological space is defined axiomatically in terms of an order-homomorphism mapping. This definition includes as special cases crisp single and multivalued dynamical systems on crisp topological space. Some algebraic structures of the L-topological dynamical system are summarized. It is demonstrated that orbit of system is associated with the ones of the molecules which could be used to derive the orbit of system. Meanwhile asymptotic properties of the system are investigated using convergence theory on L-topological space, and its conjugate system is discussed as well.

Keywords: Lattice; fuzzy topology; L-topological space; topological dynamical system.

1 Introduction

Let S be a non empty set and $f : S \to f(S) \subseteq S$, then a dynamical system on S is defined in terms of mapping f with composition of operation. If S is a set with crisp topology, we will have a crisp topological dynamical system. However, if S is a set with fuzzy topology, we will get a fuzzy topological dynamical system. Fuzzy sets were first discussed in 1965 by Zadeh in his expository paper[1], and also fuzzy dynamical systems were first discussed briefly in his important paper[2], and have since frequently been mentioned, mainly in an optimization or control context.

The first systematic treatment of abstract fuzzy dynamical systems was in 1973 by Nazaroff[3], who fuzzified Halkin's crisp topological polysystems to obtain fuzzy topological polysystems. These were further investigated by Kloeden[4]. The purpose of his paper is to define abstract fuzzy dynamical systems to explicitly exhibit the time dependent dynamical character of the systems. To this end, a fuzzy dynamical system on an underlying complete, locally compact metric state space X is defined axiomatically in terms of a fuzzy attainability set mapping $F : X \times T \to C$, where T is time set and C is the collection of all non empty compact fuzzy subsets of X with Hausdorff metric. Analogously with crisp attainability set mappings, the domain of F can be extended from $X \times T$

B.-Y. Cao (Ed.): Fuzzy Information and Engineering (ICFIE), ASC 40, pp. 243–251, 2007.
springerlink.com

to $C \times T$, hence a fuzzy topological dynamical system on the fuzzy metric space C is obtained. But this system, strictly speaking, is an extended type of crisp generalized semi-dynamical system.

However, fuzzy topological dynamical systems and fuzzy topological entropy based on the theory of fuzzy topological spaces were not studied much, which was not like the case of crisp topological dynamical systems. The study of fuzzy topological dynamical systems depends on the development of the theory of fuzzy topological spaces, but there exist essential difficulties with two basic concepts, convergence and compactness. Although Chang used the term of fuzzy topology in [5], in fact the kind of topologies being defined are not fuzzy topologies , but rather topologies made up of fuzzy subsets,i.e. L-topology. The kind of topologies being defined by Goguen[1] are also the case. Further studies were done until recent ten years(see [6,7,8]).

In this paper, the concept of L-topological dynamical system on L-topological space is defined axiomatically. In section 2, the concept of F-lattice dynamical system is introduced and the properties and algebraic structures of the system and subsystem are discussed. In section 3, asymptotic properties of L-topological dynamical system are investigated, so is the topological conjugate system. In section 4 the conclusion is drawn and some problems worth further research are put forward.

2 Dynamical Systems on *F*-Lattice

The lattice L is called an F-lattice if L is a completely distributive lattice with order reversing involution "$\prime$". Let $(L, \leq)$ be an F-lattice and let us consider the case $\otimes = \wedge$. Then $(L, \leq, \wedge)$ is always a *cqm*$-$lattice, so $L \in |\mathbf{CQMIL}|$ with $\otimes = \wedge$. Let X be a non empty crisp set and let L be an F-lattice with the largest element " 1" and the smallest element " 0". Obviously L^X is an F-lattice and $A \in L^X$ is called an LF-set. The element $a \in L$ is called a molecule if $a \neq 0$ and the following condition holds:

$$a = x \vee y \Rightarrow a = x \text{ or } a = y, \ \forall x, y \in L.$$

The set of all molecules in L is denoted by $M(L)$.

If $x \in X$ and $\lambda \in L$, then x_λ will denote the L-fuzzy set of L^X defined by

$$x_\lambda(t) = \begin{cases} \lambda, \ t = x, \\ 0, \ t \neq x, \end{cases}$$

for each $t \in X$. When $\lambda \neq 0, x_\lambda$ is an L-fuzzy point of L^X. Obviously, if $\lambda \in M(L)$, then $x_\lambda \in M(L^X)$.

Definition 1. *Let L be F-lattice and $\varphi \in \mathbf{SET}(X, Y)$. A mapping $\varphi_L^{\rightarrow} : L^X \rightarrow L^Y$ is called an order-homomorphism if the following conditions hold:*

(1) $\varphi_L^{\rightarrow}(\bigvee a_i) = \bigvee \varphi_L^{\rightarrow}(a_i)$;

(2) $(\varphi_L^{\rightarrow})^{\vdash}(b') = ((\varphi_L^{\rightarrow})^{\vdash}(b))'$.

The symbols $\eta(x_\lambda)$ and $\eta^-(x_\lambda)$ denote the collections of all R-neighborhoods and CR-neighborhoods of x_λ respectively.

A molecule net in L^X is a mapping $S : D \to M(L^X)$, denoted by $S = \{S(n) \mid n \in D\}$, where D is a directed set. If $A \in L^X$ and for each $n \in D, S(n) \in A$, then S is called a molecule net in A.

Definition 2. *A L-fuzzy point $x_\lambda \in L^X$ is called limit point of a molecule net $S = \{S(n) \mid n \in D\}$ (or S converges to x_λ, denoted by $S \to x_\lambda$), if for each $A \in \eta(x_\lambda)$ we have eventually $S(n) \not\leq A$. x_λ is called a cluster point of S, denoted by $S\infty x_\lambda$, if for each $A \in \eta(x_\lambda)$ we have frequently $S(n) \not\leq A$.*

2.1 Systems

Let L be an F-lattice. The concept of L-topological dynamical system will be introduced in this section, and the algebraic structure of its orbits will be given.

Definition 3. *Let X be a non empty set and let G be topological additive group of $\mathbf{R}$ or $\mathbf{Z}$ and a mapping $\psi : G \times L^X \to L^X$. (L^X, ψ) is referred to as dynamical system on L^X if for each $A \in L^X$ the following statements hold:*

(1) $\psi(0, A) = A$;

(2) $\psi(s + t, A) = \psi(s, \psi(t, A)), \forall s, t \in G$.

If X is equipped with L-topology δ, then (L^X, ψ)(or ψ) is referred to as L-topological dynamical system, LTDS for short.

Next note $f = \varphi_L^{\rightarrow}$, so $f^{\vdash} = (\varphi_L^{\rightarrow})^{\vdash} = \varphi_L^{\leftarrow}$, and suppose f is order-homomorphism.

Suppose that $f : L \to L_1$ and $g : L_1 \to L_2$ are order-homomorphisms, then $g \circ f : L \to L_2$ is also an order-homomorphism[9].

Theorem 1. *Let $G = \mathbf{Z}_+$ and $f : L^X \to L^X$ be an order-homomorphism. Then (L^X, f) is a dynamical system on L^X and is called a semi-dynamical system. If (X, δ) is an L-topological space and f is continuous on $M(L^X)$, then (L^X, f) is a continuous LTDS on L^X.*

Now the semi-dynamical system (L^X, f) will be discussed.

Definition 4. *Let $A \in L^X$. A map $orb_+ : L^X \to (L^X)^{\mathbf{Z}_+}$ is called orbij map, and*

$$orb_+(A) = (A, f(A), f^2(A), \cdots) \in (L^X)^{\mathbf{Z}_+}$$

is called positive semi-orbit of LF-set A on space L^X, and orbit for short.

If $f : L^X \to L^X$ is an order-homomorphism, then for each $m \in \mathbf{N}, f^m : M(L^X) \to M(L^X)$ is an order-homomorphism as well, so the following definition is proper.

Definition 5. *Let $e \in M(L^X)$. Then*

$$orb_+(e) = (e, f(e), f^2(e), \cdots) \in (M(L^X))^{\mathbf{Z}_+}$$

is called orbit of molecule e on space L^X.

Let $A \in L^X$. Then $(A] = \{B \in L^X \mid B \leq A\}$ is an ideal generated by A. Note $\pi(A) = \{x_\lambda \in M(L^X) \mid x_\lambda \leq A\}$ as the set of molecules in $(A]$. Note

$$\bigvee_{x_\lambda \in \pi(A)} orb_+(x_\lambda) = (\bigvee_{x_\lambda \in \pi(A)} x_\lambda, \bigvee_{x_\lambda \in \pi(A)} f(x_\lambda), \bigvee_{x_\lambda \in \pi(A)} f^2(x_\lambda), \cdots),$$

which is called orbit set of molecules in $(A]$.

Theorem 2. *Let $A \in L^X$. Then $orb_+(A)$ is the orbit set of the molecules in ideal $(A]$, namely*

$$orb_+(A) = \bigvee_{x_\lambda \in \pi(A)} orb_+(x_\lambda).$$

Definition 6. *Let $A \in L^X$. If there exists $m \in \mathbf{N}$ such that $f^m(A) = A$, then $orb_+(A)$ is called periodic orbits. The period of $orb_+(A)$(or A) is the smallest natural number m such that $f^m(A) = A$.*

It is obvious that for each $A \in L^X$, there exists period when L and X are finite sets.

Theorem 3. *A number $m \in \mathbf{N}$ is the period of $A \in L^X$ iff*

$$\sup\{f^m(x_\lambda) \mid x_\lambda \in \pi(A)\} = \sup \pi(A);$$

moreover $n \in \mathbf{N}$ is the period of molecule $x_\lambda \in M(L^X)$ iff for each $t \in X$ we have

$$f^n(x_\lambda)(t) = \begin{cases} \lambda, & t = x \\ 0, & t \neq x. \end{cases}$$

Theorem 4. *Let $\varphi \in \mathbf{SET}(X,Y)$. If $\varphi_L^{\rightarrow} : L^X \to L^Y$ preserves arbitrary joins and is one-to-one surjection, then*

$$\varphi_L^{\leftarrow}(B) = A \Leftrightarrow \varphi_L^{\rightarrow}(A) = B, \ \forall A \in L^X, B \in L^Y; \tag{1}$$

if $\varphi_L^{\rightarrow}$ also satisfies the condition (2) of Definition 1, then $\varphi_L^{\leftarrow}$ is an order-homomorphism.

Proof. Firstly as $\varphi_L^{\rightarrow}$ preserves arbitrary joins, we get that $(\varphi_L^{\rightarrow})^{\vdash}$ exists uniquely and $(\varphi_L^{\rightarrow})^{\vdash} = \varphi_L^{\leftarrow}$. Suppose $\varphi_L^{\leftarrow}(B) = A.$, then we have $\varphi_L^{\rightarrow}(A) = (\varphi_L^{\rightarrow} \circ \varphi_L^{\leftarrow})(B) \leq B$. As $\varphi_L^{\rightarrow}$ is surjection, there exists $C \in L^X$ such that $\varphi_L^{\rightarrow}(C) = B$, so $B = \varphi_L^{\rightarrow}(C) \leq \varphi_L^{\rightarrow}(\varphi_L^{\leftarrow}\varphi_L^{\rightarrow}(C)) = \varphi_L^{\rightarrow}(\varphi_L^{\leftarrow}(B)) = \varphi_L^{\rightarrow}(A) \leq B$, that is $\varphi_L^{\rightarrow}(A) = B$. On the other side, suppose $\varphi_L^{\rightarrow}(A) = B$, then $B = \varphi_L^{\rightarrow}(A) \leq \varphi_L^{\rightarrow}(\varphi_L^{\leftarrow}\varphi_L^{\rightarrow}(A)) = \varphi_L^{\rightarrow}\varphi_L^{\leftarrow}(B) \leq B$, that is $\varphi_L^{\rightarrow}(A) = \varphi_L^{\rightarrow}(\varphi_L^{\leftarrow}(B))$. Because $\varphi_L^{\rightarrow}$ is ont-to-one, we have $\varphi_L^{\leftarrow}(B) = A$, hence (1) holds.

Secondly, it is obvious that $\varphi_L^{\leftarrow}$ preserves arbitrary joins. Suppose $\varphi_L^{\rightarrow}(A') = C$. By (1) $\varphi_L^{\leftarrow}(C) = A'$ or $A = (\varphi_L^{\leftarrow}(C))' = \varphi_L^{\leftarrow}(C')$, hence $\varphi_L^{\rightarrow}(A) = \varphi_L^{\rightarrow}\varphi_L^{\leftarrow}(C') = C' = (\varphi_L^{\rightarrow}(A'))'$, and it follows that $\varphi_L^{\leftarrow}$ is order-homomorphism. This completes the proof.

If $f : L^X \to L^X$ is an order-homomorphism and one-to-one surjection, then by Theorem 2.4, $f^{\vdash} : L^X \to L^X$ exists and is also an order-homomorphism. Similarly, the orbit generated by $f^{\vdash}$ is denoted by orb$_{\vdash}$(A) or orb$_{\vdash}$(e), which is called adjoint orbit of f and has the above properties.

2.2 Subsystems

Suppose $Y \subset X$ and $Y \neq \emptyset, A \in L^X$. Define $A \mid Y \in L^Y$ by $(A \mid Y)(y) = A(y)$, $\forall y \in Y$. $A \mid Y$ is called restriction of A on Y and $(Y, \delta \mid Y)$ is called subspace of (X, δ). Suppose $A \in L^Y$. $A^* \in L^X$ is defined as extended mapping of A in X by

$$A^*(x) = \begin{cases} A(x), \; x \in Y, \\ 0, \qquad x \bar{\in} Y, \; x \in X. \end{cases} \tag{2}$$

Suppose $f : L^X \to L_1^Z$ is order-homomorphism, then $f \mid Y : L^Y \to L_1^Z$ is defined as restriction of f on Y as follows:

$$(f \mid Y)(A) = f(A^*), \; \forall A \in L^Y. \tag{3}$$

The following propositions can be proved easily.

Proposition 1. *Let $\{A_t \mid t \in T\} \subset L^X$ and $A \in L^X$. Then we have*
(1) $(\bigvee_{t \in T} A_t) \mid Y = \bigvee_{t \in T} (A_t \mid Y)$,
(2) $(\bigwedge_{t \in T} A_t) \mid Y = \bigwedge_{t \in T} (A_t \mid Y)$,
(3) $A' \mid Y = (A \mid Y)'$.

Proposition 2. *Let $\{A_t \mid t \in T\} \subset L^X$. Then*

$$(\bigvee_{t \in T} A_t)^* = \bigvee_{t \in T} A_t^*, \; (\bigwedge_{t \in T} A_t)^* = \bigwedge_{t \in T} A_t^*.$$

Suppose $f : L^X \to L^X$ is order-homomorphism. We say L^Y is invariant about $f \mid Y$ if $f \mid Y : L^Y \to L^Y$.

Theorem 5. *Let $f : L^X \to L^X$ be order-homomorphism and let Y be a non empty subset of X such that L^Y is invariant about $f \mid Y$. Then $f \mid Y : L^Y \to L^Y$ is also an order-homomorphism.*

Proposition 3. *Let $f : L^X \to L^X$ be order-homomorphism and let L^Y be invariant about $f \mid Y$. Then for arbitrary $x_\lambda \in M(L^Y)$, we have $(f \mid Y)(x_\lambda) = f(x_\lambda)$.*

Proposition 4. *Assume that (X, δ) is LTS, (L^X, f) is LTDS and Y is a non empty subset of X such that L^Y is invariant about $f \mid Y$. Then the dynamical system $(L^Y, f \mid Y)$ on subspace $(L^Y, \delta \mid Y)$ is called subsystem of (L^X, f).*

Suppose that A^* is the extension of $A \in L^Y$. Note

$$orb_+(A^*) \mid Y = (A^* \mid Y, f(A^*) \mid Y, f^2(A^*) \mid Y, \cdots).$$

Theorem 6. *Suppose $A \in L^Y$ and L^Y is invariant about $f \mid Y$. Suppose $orb_+(A)$ is the orbit of system $(L^Y, f \mid Y)$ in L^Y. Then we have*

$$orb_+(A) = orb_+(A^*) \mid Y.$$

Proof. It is evident that we only need to show $f^m(A^*) \mid Y = (f \mid Y)^m(A)$ for all $m \in \mathbf{Z}_+$ and $A \in L^Y$. At first, we are to prove that

$$(f \mid Y)^m(x_\lambda) = f^m(x_\lambda^*) \mid Y = f^m(x_\lambda),\ \forall x_\lambda \in M(L^Y). \tag{4}$$

Since $f : L^X \to L^X$ is order-homomorphism and L^Y is invariant about $f \mid Y$, then $f \mid Y : L^Y \to L^Y$ is order-homomorphism by Theorem 5, hence $f \mid Y : M(L^Y) \to M(L^Y)$. Let $m = 0$. Then (4) holds and $f : M(L^Y) \to M(L^Y)$ by Proposition 3. Let $m = k - 1$, (4) holds as well. So for each $x_\lambda \in M(L^Y)$ we have

$$\begin{aligned}(f \mid Y)^k(x_\lambda) &= (f \mid Y)^{k-1}((f \mid Y)(x_\lambda)) = (f \mid Y)^{k-1}(f(x_\lambda)) \\ &= f^{k-1}(f(x_\lambda)) = f^k(x_\lambda).\end{aligned}$$

This means that (4) holds when $m \in \mathbf{Z}_+$. Secondly, by (4) we get

$$\begin{aligned}f^m(A^*) \mid Y = f^m(\bigvee_{\pi(A^*)} x_\lambda^*) \mid Y &= \bigvee_{\pi(A)} f^m(x_\lambda) = \bigvee_{\pi(A)} (f \mid Y)^m(x_\lambda) \\ &= (f \mid Y)^m(\bigvee_{\pi(A)} x_\lambda) = (f \mid Y)^m(A).\end{aligned}$$

This completes the proof.

3 Asymptotic Properties

Let L be an F-lattice, $(X, \delta) \in |L\text{-}\mathbf{TOP}|$ and $\varphi \in L\text{-}\mathbf{TOP}\ (X, X)$. Note $f = \varphi_L^{\rightarrow}$. Suppose that $f : L^X \to L^X$ is continuous order-homomorphism on (X, δ), $G = \mathbf{Z}_+$, and (L^X, f) is discrete semi-dynamical system. Now we discuss asymptotic properties of f on $M(L^X)$.

3.1 Recurrence

Definition 7. *$x_\lambda \in M(L^X)$ is called fuzzy recurrent point of f, or briefly, recurrent point of f, if in $\mathbf{Z}_+$ there exists monotone increasing sequence $\{n_i \mid i \in \mathbf{N}\}, n_i \to +\infty (i \to +\infty)(n_i \nearrow +\infty$ for short), such that $f^{n_i}(x_\lambda) \to x_\lambda$ when $i \to +\infty$. Namely, for arbitrary $P \in \eta(x_\lambda)$, there exists $k \in \mathbf{N}$, when $i \geq k, P \in \eta(f^{n_i}(x_\lambda))$.*

Clearly if $x_\lambda \in M(L^X)$ is period point then it is recurrent point. The sets of all fixed points, all period points and all recurrent points of f in $M(L^X)$ are denoted by $Fix(f), Per(f)$ and $Rec(f)$ respectively.

Let $E \subseteq M(L^X)$ be ordinary set and note $f(E) = \{f(x_\lambda) \mid x_\lambda \in E\}$. Set E is said to be invariant about f, if $f(E) \subseteq E$, particularly E is called invariant set of f if $f(E) = E$. It is obvious that $Fix(f)$ is invariant set of f. Let $A \in L^X$. LF-set A is said to be invariant about f, if $f(A) \leq A$, particularly A is called invariant LF-set, if $f(A) = A$.

Proposition 5. *The following statements hold:*

(1) Let $E \subset M(L^X)$. Then $f(E) \subset E$ iff for any $x_\lambda \in E$ we have $orb_+(x_\lambda) \in E^G$.

(2) Let $A = \bigvee\{y_\mu \mid y_\mu \in E\}$. Then $f(A) \leq A$ iff for arbitrary $x_\lambda \leq A, m \in G$, we have $orb_+(x_\lambda)(m) \leq A$.

Theorem 7. *The following statements hold:*

(1) $Rec(f)$ is invariant about f, i.e. $f(Rec(f)) \subseteq Rec(f)$.

(2) Note $A = \bigvee\{x_\lambda \mid x_\lambda \in Rec(f)\}$, then A is an invariant LF-set about f.

Definition 8. *$x_\lambda \in M(L^X)$ is called fuzzy non wandering point of f, if for any R-neighborhood $P \in \eta(x_\lambda)$, there exist $y_\mu \in M(L^X)$ and natural number k such that $P \in \eta(y_\mu)$ and $P \in \eta(f^k(y_\mu))$. The set of all fuzzy non wandering points is called non wandering set, denoted by $\Omega(f)$.*

Proposition 6. *$\Omega(f) = \{x_\lambda \in M(L^X) \mid \forall P \in \eta(x_\lambda), \exists y_\mu \in M(L^X), k \in \mathbf{N}$, s.t. $P \in \eta(y_\mu)$ and $(f^\vdash)^k(P) \in \eta(y_\mu)\}$, where $(f^\vdash)^k = f^\vdash \circ (f^\vdash)^{k-1}$.*

Theorem 8. *The following conclusions hold:*

(1) Non wandering set $\Omega(f)$ is invariant set about f.

(2) Let $A = \bigvee\{x_\lambda \in M(L^X) \mid x_\lambda \in \Omega(f)\}$. Then LF-set A is an invariant closed set about f.

Proof. At first to verify (1). Let f be continuous order-homomorphism. For any $x_\lambda \in \Omega(f)$, we have $f(x_\lambda) \in M(L^X)$ and for any $Q \in \eta(f(x_\lambda)), f^\vdash(Q) \in \eta(x_\lambda)$. At the same time there exist $y_\mu \in M(L^X)$ and $k \in \mathbf{N}$ such that

$$\mu \not\leq (f^\vdash(Q) \bigvee (f^\vdash)^k(f^\vdash(Q)))(y).$$

From $\mu \in M(L)$, it follows that $\mu \not\leq f^\vdash(Q)(y)$ and $\mu \not\leq (f^\vdash)^k(f^\vdash(Q))(y)$. Since $y_\mu \not\leq f^\vdash(Q) \Leftrightarrow f(y_\mu) \not\leq Q$, and $y_\mu \not\leq (f^\vdash)^k(f^\vdash(Q)) \Leftrightarrow f(y_\mu) \not\leq (f^\vdash)^k(Q)$, $Q \in \eta(f(y_\mu))$ and $(f^\vdash)^k(Q) \in \eta(f(y_\mu))$, i.e.$f(x_\lambda) \subset \Omega(f)$. By the arbitrariness of $x_\lambda \in \Omega(f)$, we know $f(\Omega(f)) \subseteq \Omega(f)$.

Secondly to verify (2). Similar to the proof of conclusion (2) of Theorem 7, we get $f(A) \leq A$. Now we are to prove A is closed set. It is only need to show that the closure $A^- \leq A$. For any $x_\lambda \leq A^-$, there exists molecule net S such that $S \leq A$ and $S \to x_\lambda$. So for any CR-neighborhood $P \in \eta(x_\lambda)$, there exists $n_0 > 0$, when $n > n_0$, $P \in \eta(S(n))$. By $S \leq A$, i.e. S is in A, then there exist $y_\mu \in M(L^X)$ and $k \in \mathbf{N}$ such that $P \in \eta(y_\mu)$ and $(f^\vdash)^k(P) \in \eta(y_\mu)$. Because P is an arbitrary CR-neighborhood of x_λ, by Proposition 6 we have $x_\lambda \leq A$, so $A^- = \bigvee\{x_\lambda \in M(L^X) \mid \exists$ molecule net $S \leq A, S \to x_\lambda\} \leq A$. This completes the proof of this theorem.

3.2 ω-Limit Sets

Definition 9. *Suppose $x_\lambda \in M(L^X)$. $y_\mu \in M(L^X)$ is called ω-limit point of x_λ, if there exists strictly monotone increasing sequence $n_i \nearrow +\infty (i \to +\infty)$, such* *that* $f^{n_i}(x_\lambda) \to y_\mu$. *The set of all* ω*-limit points of* x_λ *is denoted by* $L_\omega(x_\lambda)$, *namely*

$$L_\omega(x_\lambda) = \{y_\mu \in M(L^X) \mid \exists n_i \nearrow +\infty, f^{n_i}(x_\lambda) \to y_\mu(i \to +\infty)\}.$$

Note $L_\omega(f) = \bigcup\limits_{x_\lambda \in M(L^X)} L_\omega(x_\lambda)$, *which is called* ω*-limit set of* f.

Theorem 9. $\bigvee\{y_\mu \mid y_\mu \in L_\omega(x_\lambda)\} = \bigwedge\limits_{n\geq 0}\{\bigvee\limits_{k\geq n} f^k(x_\lambda)\}^-$.

Theorem 10. *Suppose* $A = \bigvee\{y_\mu \mid y_\mu \in L_\omega(x_\lambda)\}$. *Then* A *and* $L_\omega(x_\lambda)$ *are invariant sets about* f.

Theorem 11. *All of the following sets are invariant, and they have relation*

$$Fix(f) \subseteq Per(f) \subseteq Rec(f) \subseteq L_\omega(f) \subseteq \Omega(f).$$

Proof. It is only need to show $L_\omega(f) \subseteq \Omega(f)$, and the others are direct consequences of the definitions. For any $y_\mu \in L_\omega(f)$, there exists $x_\lambda \in M(L^X)$ such that $y_\mu \in L_\omega(x_\lambda)$, namely, there exists $n_i \nearrow +\infty$, such that $f^{n_i}(x_\lambda) \to y_\mu$. For any $P \in \eta(y_\mu)$, there exists $m \in \mathbf{N}$, when $n_i \geq m, P \in \eta(f^{n_i}(x_\lambda))$. Let $k = n_i - m > 0$, and note $e = f^m(x_\lambda) \in M(L^X)$. Then $P \in \eta(e)$ and $P \in \eta(f^k(e))$, hence $y_\mu \in \Omega(f)$.

3.3 Topological Conjugacy

Let $h \in$**POSET**(L^X, L^Y) be a one-to-one surjection from L^X to L^Y such that L^X has arbitrary joins (or $\vee$) and h preserves them. Then by Theorem 4 there is a unique right adjoint operator $h^\vdash$ such that

$$h^\vdash(B) = A \Longleftrightarrow h(A) = B,\ \forall A \in L^X,\ B \in L^Y. \tag{5}$$

Particularly, if L is an F-lattice and $h : L^X \to L^Y$ is an order-homomorphism and one-to-one surjection, then $h^\vdash : L^X \leftarrow L^Y$ is also an order-homomorphism.

Definition 10. *Let* $h : L^X \to L^Y$ *be order-homomorphism. Then it is referred to as homeomorphism mapping if it is a continuous and one-to-one surjection and* $h^\vdash$ *is also continuous.*

Proposition 7. *Let* $h : L^X \to L^Y$ *be order-homomorphism. Then it is homeomorphism mapping iff it is one-to-one surjection and continuous open(closed) mapping.*

Let (L^X, f) and (L^Y, g) be LTDS.

Definition 11. *The dynamical systems* (L^X, f) *and* (L^Y, g) *are said to be topological conjugate, if there exists homeomorphism mapping* $h : L^X \to L^Y$ *such that* $h \circ f = g \circ h$, *where* h *is called topological conjugate mapping.*

$L^f_\omega(x_\lambda)$ and $L^g_\omega(y_\mu)$ are used to denote ω-limit sets of f at $x_\lambda \in M(L^X)$ and g at $y_\mu \in M(L^Y)$ respectively, similarly $orb^f(x_\lambda)$ and $orb^g(y_\mu)$ are defined.

Theorem 12. *Suppose that (L^X, f) and (L^Y, g) are topological conjugate, and $h : L^X \to L^Y$ is topological conjugate mapping. Then for any $e \in M(L^Y)$ there exists $x_\lambda \in M(L^X)$ such that $e = h(x_\lambda)$, and the following conclusions hold:*
(1) $orb^g(e) = h(orb^f(x_\lambda))$;
(2) if $f^m(x_\lambda) = x_\lambda$, then $g^m(e) = e$;
(3) $Rec(g) = \{h(t_\xi) \mid t_\xi \in Rec(f)\}$;
(4) $L^g_\omega(e) = \{h(t_\xi) \mid t_\xi \in L^f_\omega(x_\lambda)\}$;
(5) if f and g are bijections, then $\Omega(g) = \{h(x_\lambda) \mid x_\lambda \in \Omega(f)\}$.

We get immediately that

Theorem 13. *On the conditions of Theorem 12, both of the following statements hold:*
*(1) $orb^g = h \circ orb^f$; (2) $\bigvee\{e \mid e \in A\} = \bigvee\{\sigma \mid \sigma \in h(B)\}$, where $(A, B) \in \{(Rec(g), h(Rec(f))), (L^g_\omega, h(L^f_\omega)),$
$(\Omega(g), h(\Omega(f)))\}$.*

4 Conclusion

Based on the theory of L-topological space, L-topological dynamical system and subsystem are defined in this paper. Similar to crisp dynamical system, the concepts of fuzzy recurrent point, non wandering point and ω-limit point are introduced. Furthermore, asymptotic and topological conjugate properties are discussed. There are also many problems worth researching about asymptotic, stability, topological entropy and nonlinear properties in LTDS. At the same time, it is worth investigating the L-fuzzy topological dynamical system based on L-fuzzy topological space as well.

References

1. Zadeh LA (1965) Fuzzy sets and systems, in: Proc. Symp. on Systems Theory, Polytechnic Institute Press, Brooklyn, NY
2. Zadeh LA (1965) Fuzzy sets, Information and Control, 8: 338–353
3. Nazaroff GJ (1973) Fuzzy topological polysystems, J. Math. Anal. Appl., 41: 478–485
4. Kloeden PE (1982) Fuzzy dynamical systems, Fuzzy Sets and Systems, 7: 275-296
5. Chang CL (1968) Fuzzy topological spaces, J. Math. Anal. Appl. 24: 191–201
6. U. Höhle, S. E. Rodabaugh(Eds.)(1999) Mathematics of Fuzzy Sets: Logic, Topology, and Measure Theory, The Handbooks of Fuzzy Sets Series, Volume 3, Kluwer Academic Publisher, Dordrecht
7. Liu Yinming, Luo Maokang (1998) Fuzzy Topology, World Scientific, Singapore
8. Rodabaugh SE, Klement EP(Eds.) (2003) Topological And Algebraic Structures In Fuzzy Sets: A Handbook of Recent Developments in the Mathematics of Fuzzy Sets, Kluwer Academic Publishers, Dordrecht
9. Wang Guojun (1984) Order-homomorphisms on fuzzes, Fuzzy Sets and Systems, 12: 281–288

Integrating Probability and Quotient Space Theory: Quotient Probability

Hong-bin Fang

Department of Mathmatics, Anhui University, Hefei 230039, P.R. China
Department of Key Laboratory of Intelligent Computing & Signal Processing of Ministry of Education, Institute of Artifical Intelligence, Anhui University Hefei 230039, P.R. China
fhb1396666@sohu.com

Abstract. In this paper, quotient probability is presented in order to integrate quotient space theory and probability theory . Quotient probability can be updated basing on quotient spaces fusion caused by different equivalence relations. Distance between different quotient probabilities basing on corresponding grain-sized quotient spaces can reflect different granularity relation, namely the distance value is zero under the condition of consistency information for different equivalence relations, or is positive under the condition of inconsistency information for different equivalence relations.

Keywords: Granular Computing, Quotient Space Theory, Probability Theory.

1 Introduction

Granular computing emerges as a new multidisciplinary study and has received much attention in recent years[1,2,3,4]. Many methods and models of granular computing have been proposed and studied, especially rough set theory[5] and fuzzy set theory[6] as well as quotient space theory[7] are in importance . Zadeh primarily discussed the representation of granular by rough set, namely he used the "language", "word" to represent the thinking and reasoning of mankind. Zadeh L.A. brought forward three main conceptions about man cognition: granulation, organization, causality in [8,9,10]. Granular computing based on rough set theory primarily discussed the representation and characterization of granular, and the relations between granular and concept. Granular computing based on quotient space theory describes granularity by the equivalent classes, and is a kind of theory studying the relations of interconversion and interdependency at different grain-size worlds considering topology structure among elements on the space.

In other hand, probability theory has been widely used in uncertain problems, so some new theories emerged recently with respect to combining fuzzy set theory or rough set theory with probability theory[11,12,13]. In this paper, a new concept, namely quotient probability, is presented in order to integrate quotient space theory and probability theory, as the same time its corresponding properties are indicated.

The remainder of this paper is as follows. In section 2, the definition and some results of quotient probability are introduced. In section 3, conclusion is given.

B.-Y. Cao (Ed.): Fuzzy Information and Engineering (ICFIE), ASC 40, pp. 252–259, 2007.
springerlink.com

2 Quotient Probability and Some Results

Assume (X, T) is a semi-order space, i.e., there exists a semi-order relation T(<) among part of elements on X satisfying (1) If x<y and y<z, then x<z; (2) If x<y and y<x, then x=y. X is a finite set. X_1 and X_2 are quotient sets of X. There exist semi-order structures $[T]_1$ and$[T]_2$ (namely acyclic directed graph (abbr. DAG))on the quotient sets X_1 and X_2 respectively, namely (X_1,T_1) and (X_2,T_2) are both semi-order spaces.

Definition 1 (consistency information)**.** The information of the semi-order spaces (X_1,T_1) and (X_2,T_2) is consistency if the following condition is satisfied:

1) X_3 -- the least upper bound space of X_1 and X_2, namely $X_3=\{a_i \cap b_j / \forall a_i \in X_1, \forall b_j \in X_2\}$;
2) $\forall x, y \in X_3, x=a_1 \cap b_1, y=a_2 \cap b_2, a_1, a_2 \in X_1, b_1, b_2 \in X_2$, $x<y \Leftrightarrow$ $a_1<a_2, b_1<b_2$.

Definition 2 (inconsistency information)**.** The information of the semi-order spaces (X_1,T_1) and (X_2,T_2) is inconsistency if the following condition is satisfied:

1) X_3 -- the least upper bound space of X_1 and X_2, namely $X_3=\{a_i \cap b_j / \forall a_i \in X_1, \forall b_j \in X_2\}$;
2) $\exists x, y \in X_3, x=a_1 \cap b_1, y \in a_2 \cap b_2, a_1, a_2 \in X_1, b_1, b_2 \in X_2$, $x<y \Leftrightarrow a_1<a_2, b_1>b_2$ or $a_1>a_2, b_1<b_2$.

When an original sample space X is given , P_r is its probability distribution as well as $\mathcal{F}$ is its σ –field. The original space (X, P_r,$\mathcal{F}$, T) is translated into new space ($[X]_R$, P_R , $\mathcal{F}_R$, $[T]_R$) according to equivalence relation R , in which T is semi-order structure on X , $[X]_R$ is quotient set of X , $\mathcal{F}_R$ and P_R are σ -field and probability of $[X]_R$ respectively, and $[T]_R$ is quotient semi-order structure of T. Here the representation of P_R is as follows:

$\forall A \in \mathcal{F}_R$, A=$\bigcup_{i\in I} a_i = \bigcup_{i \in I}\left(\bigcup_{j\in J} x_{ji}\right) \in \mathcal{F}$ and $a_i = \bigcup_{j\in J} x_{ji}$, $a_i \in [X]_R$, where I and J are all index set, $x_{ji} \in$ X, P_R (A)= $P_r\left(\bigcup_{i\in I}\left(\bigcup_{j\in J} x_{ji}\right)\right)$ (1)

Definition 3. Probability distribution P_R of Space($[X]_R$, P_R , $\mathcal{F}_R$, $[T]_R$) is called quotient probability of $[X]_R$, satisfying the following conditions:

1) $\forall x \in [X]_R, x \neq \varnothing$, x is an atom element of $[X]_R$ (namely $x \neq \phi$ and if $\forall c \subset x, c=\phi$);

2) $\forall x \in [X]_R, x \neq \varnothing,\ P_R(x) > 0$, namely all nonempty elememts of $[X]_R$ consist of support set of probability distribution P_R .

All nonempty elememts of $[X]_R$ constitute an exclusive and complete events with respect to P_R by definition 3.

2.1 Probability Update and Spaces Fusion

Assume that original space (X, P_r, $\mathcal{F}$, T)is translated into two new spaces ($[X]_1$, P_1 , $\mathcal{F}_1$, $[T]_1$) and ($[X]_2$, P_2 , $\mathcal{F}_2$, $[T]_2$) according to equivalence relations $R_1$and R_2 respectively, as the same time P_1 and P_2 are corresponding quotient probability on its space respectively. Now, we can assume nonempty elements set of $[X]_1$ is $\{\gamma_1', \gamma_2', \ldots, \gamma_n'\}, \forall \gamma_i', \gamma_l', \gamma_i' \cap \gamma_l' = \varnothing, \sum_{i=1}^{n} \gamma_i' = [X]_1$, according to quotient probability P_1 , $\{\gamma_1', \gamma_2', \ldots, \gamma_n'\}$ is support set and complete and exclusive events of P_1 , namely

$$P_1(\gamma_i') = \beta_i', \beta_i' > 0 \ \text{, i=1,2,…,n ;} \ \sum_{i=1}^{n} \beta_i' = 1 \tag{2}$$

as the same time, we can assume nonempty elements set of $[X]_2$ is $\{\gamma_1'', \gamma_2'', \ldots, \gamma_m''\}, \forall \gamma_i'', \gamma_l'', \gamma_i'' \cap \gamma_l'' = \varnothing, \sum_{j=1}^{m} \gamma_j'' = [X]_2$ according to quotient probability P_2 , $\{\gamma_1'', \gamma_2'', \ldots, \gamma_m''\}$ is support set and complete and exclusive events of P_2, namely

$$P_2(\gamma_j'') = \beta_j'', \beta_j'' > 0 \ \ j=1,2,\ldots,m; \ \sum_{j=1}^{m} \beta_j'' = 1 \tag{3}$$

$$\text{Let } \{\gamma_k^*\} = \{\gamma_i' \cap \gamma_j'' \,|\, \forall \gamma_i' \in \{\gamma_1', \gamma_2', \ldots, \gamma_n'\}, \forall \gamma_j'' \in \{\gamma_1'', \gamma_2'', \ldots, \gamma_m''\}\} \tag{4}$$

so $\{\gamma_k^*\}$ is the fusion of $[X]_1$ and $[X]_2$ caused by equivalence relations R_1、 R_2 respectively . Now, we denote ($[X]^*$, P^* , $\mathcal{F}^*$, $[T]^*$) is the fusion space of ($[X]_1$, P_1 , $\mathcal{F}_1$, $[T]_1$) and ($[X]_2$, P_2 , $\mathcal{F}_2$, $[T]_2$) ; here $[X]^*$= $\{\gamma_k^*\}$, P^* and $\mathcal{F}^*$ are probability and σ –field of $[X]^*$ respectively; $[T]^*$ is the fusion topology of $[T]_1$ and $[T]_2$, in which the fusion topology method is accordance with literature [14] considering consistency information and inconsistency information.

In literature [15], topology structure of a problem in different quotient spaces can be combined into a result under the condition of consistency information as well as the result is optimization basing on topology structure of fusion quotient space.

P^* of the space ($[X]^*$, P^* , $\mathcal{F}^*$, $[T]^*$) is put forward as follows:

1) Under the condition of consistency information with respect to ($[X]_1$, P_1 , $\mathcal{F}_1$, $[T]_1$) and ($[X]_2$, P_2 , $\mathcal{F}_2$, $[T]_2$) basing on definition 1 , $\forall A \in \mathcal{F}^*$,

$A=\bigcup_{i\in I} a_i = \bigcup_{i\in I}\left(\bigcup_{j\in J} x_{ji}\right) \in \mathcal{F}$ and $a_i \in [X]^*$, where I and J are all index set, $x_{ji} \in X$, then $P^*(A)= P_r\left(\bigcup_{i\in I}\left(\bigcup_{j\in J} x_{ji}\right)\right)$;

2) Under the condition of inconsistency information with respect to ($[X]_1$, P_1, $\mathcal{F}_1$, $[T]_1$) and ($[X]_2$, P_2, $\mathcal{F}_2$, $[T]_2$) basing on definition 2, $\forall \gamma_k^* \in [X]^*, \gamma_k^* \neq \varnothing$,

$$P^*(\gamma_k^*) = P_1 \otimes P_2 = \frac{\sum_{\gamma_i' \cap \gamma_j'' = \gamma_k^*} P_1(\gamma_i') P_2(\gamma_j'')}{\sum_{\gamma_i' \cap \gamma_j'' \neq \varnothing} P_1(\gamma_i') P_2(\gamma_j'')} \tag{5}$$

Proposition 1. *1)* Under the condition of consistency information , P^* of the space ($[X]^*$, P^* , $\mathcal{F}^*$, $[T]^*$) is quotient probability of $[X]^*$ if X is the support set of P_r with respect to the original space (X, P_r, $\mathcal{F}$, T) ;

2) Under the condition of inconsistency information, P^* of the space ($[X]^*$, P^* , $\mathcal{F}^*$, $[T]^*$) is quotient probability of $[X]^*$ basing on formula (5).

Proof. *1)* According to the condition of consistency information and the definition of support set , it is correct;

2) $\forall \gamma_k^* \in [X]^*, \gamma_k^* \neq \varnothing$,the following is obtained in light of formula (4):

$$\exists \gamma_s' \in \{\gamma_1', \gamma_2', \ldots, \gamma_n'\},\ \exists \gamma_l'' \in \{\gamma_1'', \gamma_2'', \ldots, \gamma_m''\},\ \gamma_k^* = \gamma_s' \cap \gamma_l'' \tag{6}$$

$P_1(\gamma_s') > 0, P_2(\gamma_l'') > 0$ considering formula (2) and (3), so that formula (7) is true basing on formula (5), namely:

$$P^*(\gamma_k^*) = P_1 \otimes P_2 = \frac{\sum_{\gamma_i' \cap \gamma_j'' = \gamma_k^*} P_1(\gamma_i') P_2(\gamma_j'')}{\sum_{\gamma_i' \cap \gamma_j'' \neq \varnothing} P_1(\gamma_i') P_2(\gamma_j'')} = \frac{P_1(\gamma_s') P_2(\gamma_l'')}{\sum_{\gamma_i' \cap \gamma_j'' \neq \varnothing} P_1(\gamma_i') P_2(\gamma_j'')} > 0 \tag{7}$$

The other hand , $\forall \gamma_k^*, \gamma_t^* \in [X]^*, \gamma_k^*, \gamma_t^* \neq \varnothing$, formula (8) is obtained under the condition of formula (2) 、(3) and (4) , namely

$$\gamma_k^* \cap \gamma_t^* = \varnothing \tag{8}$$

Basing on formula (5), Formula (9) is true as follows:

$$\sum_k P^*(\gamma_k^*) = \sum_k \frac{\sum_{\gamma_i' \cap \gamma_j'' = \gamma_k^*} P_1(\gamma_i') P_2(\gamma_j'')}{\sum_{\gamma_i' \cap \gamma_j'' \neq \varnothing} P_1(\gamma_i') P_2(\gamma_j'')} = \frac{\sum_k \sum_{\gamma_i' \cap \gamma_j'' = \gamma_k^*} P_1(\gamma_i') P_2(\gamma_j'')}{\sum_{\gamma_i' \cap \gamma_j'' \neq \varnothing} P_1(\gamma_i') P_2(\gamma_j'')} = 1 \tag{9}$$

In conclusion, $\left\{\gamma_k^*\right\}=\left\{\gamma_i^{'}\cap\gamma_j^{''}\middle|\forall\gamma_i^{'}\in\left\{\gamma_1^{'},\gamma_2^{'},\ldots,\gamma_n^{'}\right\},\forall\gamma_j^{''}\in\left\{\gamma_1^{''},\gamma_2^{''},\ldots,\gamma_m^{''}\right\}\right\}$ is atom events set of $[X]^*$ and support set of P^* .

Basing on definition 3 and results above, 2) part of Proposition 1 is proved.

2.2 Distance Between Two Different Quotient Probabilities

P_{r_1}, P_{r_2} are both probability distributions of sample space X as well as arbitrary complete and exclusive events set of X is presented as following:

$$\{\gamma_1,\gamma_2,\ldots,\gamma_N\},\forall\gamma_i,\gamma_j,\gamma_i\cap\gamma_j=\varnothing,\sum_{i=1}^{v}\gamma_i=X$$

$$\forall\gamma_i\in\{\gamma_1,\gamma_2,\ldots,\gamma_N\},P_{r_1}(\gamma_i)>0,P_{r_2}(\gamma_i)>0$$

Definition 4. The distance between P_{r_1} and P_{r_2} about $\{\gamma_1,\gamma_2,\ldots,\gamma_N\}$, a complete and exclusive events set of X, is defined as follows:

$$D\left(P_{r_1},P_{r_2}\right)=\max_{i=1}^{N}\frac{P_{r_1}(\gamma_i)}{P_{r_2}(\gamma_i)}-\min_{i=1}^{N}\frac{P_{r_1}(\gamma_i)}{P_{r_2}(\gamma_i)} \tag{10}$$

Definition 5 [15]. R_1, R_2 are equivalence relations on problem space (X, P_r, $\mathcal{F}$, T), $R_1 \lhd R_2$ *if* $\forall x,y\in X$, $xR_2y\Rightarrow xR_1y$;here xRy represents element x ,y equivalent with respect to equivalence relation R.

Proposition 2. Under the condition of consistency information, the original space(X, P_r, $\mathcal{F}$, T) translate into two new spaces($[X]_1$, P_1 , $\mathcal{F}_1$, $[T]_1$) and ($[X]_2$, P_2 , $\mathcal{F}_2$, $[T]_2$) according to equivalence relations R_1、R_2 respectively, as well as $R_1<R_2$, then given an arbitrary complete and exclusive events set of $[X]_1$, according to formula (10), $D(P_1, P_2)=0$.

Proof. $\because$ Under the condition of consistency information, equivalence relations R_1、R_2 satisfy $R_1<R_2$ as well as $P_1$and P_2 are defined by formula (1) on $[X]_1$and $[X]_2$ respectively.

$\therefore$ an arbitrary complete and exclusive events set of $[X]_1$, $\left\{\gamma_1^{'},\gamma_2^{'},\ldots,\gamma_n^{'}\right\}$, presents the following :

$\forall\gamma_s^{'}\in\left\{\gamma_1^{'},\gamma_2^{'},\ldots,\gamma_n^{'}\right\}$, $\gamma_s^{'}=\sum_{k=1}^{h}\gamma_{j_k}^{''}$, $\left\{\gamma_{j_k}^{''},k=1,2,\ldots,h\right\}$ is part of all atom events of $[X]_2$.

$$\therefore P_1\left(\gamma_s^{'}\right)=\sum_{k=1}^{h}P_2\left(\gamma_{j_k}^{''}\right)=P_2\left(\sum_{k=1}^{h}\gamma_{j_k}^{''}\right)=P_2\left(\gamma_s^{'}\right)$$

$\therefore$ According to formula (10), $D(P_1,P_2)=0$.

Corollary 1. original space (X, P_r, $\mathcal{F}$, T)translate into two new spaces($[X]_1$, P_1, $\mathcal{F}_1$, $[T]_1$) and ($[X]_2$, P_2, $\mathcal{F}_2$, $[T]_2$) according to equivalence relations R_1, R_2 respectively. Under the condition of consistency information, the space ($[X]_3$, P_3, $\mathcal{F}_3$, $[T]_3$) corresponding to equivalence relation R_3 is the fusion of spaces ($[X]_1$, P_1, $\mathcal{F}_1$, $[T]_1$) and ($[X]_2$, P_2, $\mathcal{F}_2$, $[T]_2$). According to formula (10), $D(P_1,P_3)=0$ for an arbitrary complete and exclusive events set of $[X]_1$, as the same time $D(P_2,P_3)=0$ for an arbitrary complete and exclusive events set of $[X]_2$.

Proof. $\because$ the space ($[X]_3$, P_3, $\mathcal{F}_3$, $[T]_3$)corresponding to equivalence relation R_3 is the fusion of spaces ($[X]_1$, P_1, $\mathcal{F}_1$, $[T]_1$) and ($[X]_2$, P_2, $\mathcal{F}_2$, $[T]_2$) under the condition of consistency information.

$\therefore$ $R_1<R_3$ and $R_2<R_3$

$\therefore$ Corollary 1 is right in view of proposition 2.

Proposition 3. Original space (X, P_r, $\mathcal{F}$, T) is translated into two new spaces($[X]_1$, P_1, $\mathcal{F}_1$, $[T]_1$) and ($[X]_2$, P_2, $\mathcal{F}_2$, $[T]_2$) according to equivalence relations R_1、R_2 respectively under the condition of inconsistency information . Space ($[X]^*$, P^*, $\mathcal{F}^*$, $[T]^*$) is the fusion of spaces ($[X]_1$, P_1, $\mathcal{F}_1$, $[T]_1$) and ($[X]_2$, P_2, $\mathcal{F}_2$, $[T]_2$) , then according to formula (10) , $D(P_1, P^*)\neq 0$ and $D(P_2, P^*)\neq 0$.

Proof. If according to formula (10) , $D(P_1, P^*)=0$, for an arbitrary complete and exclusive events set $\{\gamma_1',\gamma_2',\ldots,\gamma_n'\}$ of $[X]_1$ and $\forall\gamma_s'\in\{\gamma_1',\gamma_2',\ldots,\gamma_n'\}$, $P_1(\gamma_s')=P^*(\gamma_s')$; otherwise, at least $\exists\gamma_g'\in\{\gamma_1',\gamma_2',\ldots,\gamma_n'\}$, $P_1(\gamma_g')<P^*(\gamma_g')$.

$$\because \sum_{k=1}^{n}P_1(\gamma_k')=1,\sum_{k=1}^{n}P^*(\gamma_k')=1 \text{ by formula (1)}$$

$$\therefore \text{at least } \exists\gamma_t'\in\{\gamma_1',\gamma_2',\ldots,\gamma_n'\},\ P_1(\gamma_t')>P^*(\gamma_t')$$

$$\therefore D(P_1, P^*)=\max_{i=1}^{n}\frac{P_1(\gamma_i')}{P^*(\gamma_i')}-\min_{i=1}^{n}\frac{P_1(\gamma_i')}{P^*(\gamma_i')}>0\ , \quad \text{contradiction.}$$

$\therefore \forall\gamma_s'\in\{\gamma_1',\gamma_2',\ldots,\gamma_n'\}$, $P_1(\gamma_s')=P^*(\gamma_s')$, if $D(P_1, P^*)=0$ according to formula (10).

Similarly, for an arbitrary complete and exclusive events set $\{\gamma_1'',\gamma_2'',\ldots,\gamma_m''\}$ of $[X]_2$, $\forall\gamma_p''\in\{\gamma_1'',\gamma_2'',\ldots,\gamma_m''\}$, $P_2(\gamma_p'')=P^*(\gamma_p'')$, if $D(P_2, P^*)=0$ according to formula (10).

$\because$ Under the condition of inconsistency information, ($[X]^*$, P^*, $\mathcal{F}^*$, $[T]^*$) is the fusion of spaces ($[X]_1$, P_1, $\mathcal{F}_1$, $[T]_1$) and ($[X]_2$, P_2, $\mathcal{F}_2$, $[T]_2$) ; P^* is obtained by formula (5), and P^* is nonlinear compromise between P_1 and P_2 , so it is contradicted with the results above unless P_1 and $P_2$are the same uniform distribution. .

$\therefore$ According to formula (10) , $D(P_1, P^*) \neq 0$ and $D(P_2 , P^*) \neq 0$.

3 Conclusion

In the paper, quotient probability is presented in order to integrate quotient space theory and probability theory . Some meaning outcomes about quotient probability are given concerning granularity relation. Distance between different quotient probabilities basing on corresponding grain-sized quotient spaces equals zero under the condition of consistency information, or is positive under the condition of inconsistency information.

Acknowledgement

This work was supported by the National Natural Science Foundation of China Grant (No. 60575023),and Outstanding Person Team Construction in Anhui University.

References

[1] Inuiguchi, M., Hirano, S. and Tsumoto, S.(2003) (Eds.) Rough Set Theory and Granular Computing, Springer, Berlin

[2] Lin, T.Y. (2003) Granular computing, LNCS 2639, Springer, Berlin, 16-24

[3] Lin, T.Y., Yao, Y.Y. and Zadeh L.A. (2002) (Eds.) Data Mining, Rough Sets and Granular Computing, Physica-Verlag, Heidelberg

[4] Liu, Q. and Jiang, S. L. (2002) Reasoning about information granules based on rough logic, International Conference on Rough Sets and Current Trends in Computing, 139-143

[5] Pawlak Z (1991) Rough Sets Theoretical Aspects of Reasoning About Data. Dordrecht: Kluwer Academic Publishers

[6] Zadeh L.A. (1965) Fuzzy sets, In Control 8, 338-353

[7] Zhang L, Zhang B (2003) Theory of fuzzy quotient space (Methods of fuzzy granular computing), Journal of Software,14(4):770-776(in Chinese)

[8] Zadeh L.A. (1996) fuzzy logic=computing with words. IEEE Transactions on Fuzzy Systems,4(1):103~111

[9] Zadeh L.A. (1997)Towards a theory of fuzzy information granulation and its centrality in human reasoning and fuzzy logic. Fuzzy Sets and Systems, 19(1):111~127

[10] Zadeh L.A. (1998) Some reflections on soft computing, granular computing and their roles in the conception, design and utilization of information/intelligent systems. Soft Computing,2(1):23~25

[11] Beaubouef T., Petry F.E., Arora G.(1999) Information measure for rough and fuzzy sets and application to uncertainty in relational database. In: Pal, S.K., Skowron, A. (Eds.) Rough Fuzzy Hybridization: A New Trend in Decision-making. Springer, Singapore, pp. 200–214

[12] Y. Feng (2001) Sums of independent fuzzy random variables, Fuzzy Sets and Systems 123, 11–18

[13] Y. Feng (2002) On the convergence of fuzzy martingale, Fuzzy Sets and Systems 130, 67–73
[14] Hongbin Fang, Ling Zhang (2005) Semi-Order Structure Fusion Model of Quotient Spaces under the Condition of Inconsistency Information. IEEE International Conference on Granular Computing, Beijing
[15] Bo Zhang and Ling Zhang (1992) "Theory and Application of Problem Solving," North-Holland, Elsevier Science Publishers B.V

ω-Convergence Theory of Filters in $L\omega$-Spaces

Shui-Li Chen[1], Guo-Long Chen[2], and Jian-Cheng Zhang[3]

[1] School of Science, Jimei University, Xiamen, Fujian 361021, P.R. China
sgzx@jmu.edu.cn
[2] College of Mathematics and Computer Science, Fuzhou University, Fuzhou, Fujian 350002, P.R. China
cgl@fzu.edu.cn
[3] Department of Mathematics, Quanzhou Normal University, Quanzhou, Fujian 362000, P.R. China
zjcqz@126.com

Abstract. In this paper, an ω-convergence theory of filters in an $L\omega$-space is established. By means of the ω-convergence theory, some important characterizations with respective to the ω-closed sets, ωT_2 separation and (ω_1, ω_2)-continuous mappings are obtained. Moreover, the mutual relationships among ω-convergence of molecular nets, ω-convergence of ideals and ω-convergence of filters are given.

Keywords: Fuzzy lattice; $L\omega$-space; Filter; Molecular net; Ideal; Fuzzy mapping; ω-convergence.

1 Introduction

The Moore-Smith convergence theory in fuzzy topology was first introduced by Pu and Liu [21]. Wang [22,23] and Lowen [20] further extended its scope by establishing the equivalence of a fuzzy net respectively. Since then, many convergence theories, such as θ-convergence theory, δ-convergence theory, U-convergence theory, σ-convergence theory, R-convergence theory, P-convergence theory, S-convergence theory, N-convergence theory, and so on [1-13, 24], were presented by means of multifarious closure operators. In order to unify various convergence theories, we established a generalized Moore-Smith convergence theory which called the ω-convergence theory with respective to molecular nets and ideals in an $L\omega$-space [14-17]. In this paper, we shall further enrich and consummate the ω-convergence theory and give its some applications.

2 Preliminaries

Throughout the paper, L denotes a fuzzy lattice while M denotes the set consisting of all molecules [22,23], i. e., nonzero $\vee$-irreducible elements in L. 0 and 1 are the least and greatest element of L respectively. For each non-empty set X, L^X will be the family of all L-sets defined on X and with value in L, $M^*(L^X)$ will stand for the set of all molecules in L^X, and the constant set taking on the constant values 1 and 0 at each x in X will be denoted by 1_X and 0_X respectively.

B.-Y. Cao (Ed.): Fuzzy Information and Engineering (ICFIE), ASC 40, pp. 260–268, 2007.
springerlink.com

Definition 2.1. (Wang [23]). Let L be a complete lattice, $e\in L$, $B\subset L$. B is called a minimal family of e if $B\neq\varnothing$ and (1) sup $B=e$; (2) $\forall A\subset L$, sup $A\geq e$ implies that $\forall x\in B$, there exists $y\in A$ such that $y\geq x$.

According to Hutton [19], in a completely distributive lattice, each element $e\in L$ has a greatest minimal family which will be denoted by $\beta(e)$. For each $e\in M$, $\beta^*(e)=\beta(e)\cap M$ is a minimal family of e and is said to be the standard minimal family of e.

Definition 2.2. (Chen [14,15]). Let X be a non-empty crisp set.

(i) An operator ω: $L^X\to L^X$ is said to be an ω-operator if (1) $\omega(1_X)=1_X$; (2) $\forall A$, $B\in L^X$ and $A\leq B$, $\omega(A)\leq\omega(B)$; (3) $\forall A\in L^X$, $A\leq\omega(A)$.
(ii) An L-set $A\in L^X$ is called an ω-set if $\omega(A)=A$.
(iii) Put $\Omega=\{A\in L^X \mid \omega(A)=A\}$, and call the pair (L^X,Ω) an $L\omega$-space.

Definition 2.3. (Chen [14,15]). Let (L^X,Ω) be an $L\omega$-space, $A\in L^X$ and $x_\alpha\in M^*(L^X)$. If there exists a $Q\in\Omega$ such that $x_\alpha\not\leq Q$ and $P\leq Q$, then call P an ωR-neighborhood of x_α. The collection of all ωR-neighborhoods of x_α is denoted by $\omega\eta(x_\alpha)$.

Definition 2.4. (Chen [14,15]). Let (L^X,Ω) be an $L\omega$-space, $A\in L^X$ and $x_\alpha\in M^*(L^X)$. If $A\not\leq P$ for each $P\in\omega\eta(x_\alpha)$, then x_α is called an ω-adherence point of A, and the union of all ω-adherence points of A is called the ω-closure of A, and denoted by ωcl(A). If $A=\omega$cl(A), then call A an ω-closed set. If A is an ω-closed set, then call A' an ω-open set. If $P=\omega$cl(P) and $x_\alpha\not\leq P$, then P is said to be an ω-closed R-neighborhood (briefly, ωCR-neighborhood) of x_α, and the collection of all ωCR-neighborhoods of x_α is denoted by $\omega\eta^-(x_\alpha)$.

Definition 2.5. (Chen [14,15]). Let (L^X,Ω) be an $L\omega$-space, $A\in L^X$ and ωint$(A)=\vee\{B\in L^X \mid B\leq A$ and B is an ω-open set in $L^X\}$. We call ωint(A) the ω-interior of A. Obviously, A is ω-open if and only if $A=\omega$int(A).

Definition 2.6. (Chen [16]). Let N be a molecular net in L^X and $x_\alpha\in M^*(L^X)$. Then x_α is said to be an ω-limit (ω-cluster) point of N, or N ω-converges (ω-accumulates) to x_α, in symbols, $N\to_\omega x_\alpha$ ($N\propto_\omega x_\alpha$), if N is eventually (frequently) not in P for each $P\in\omega\eta^-(x_\alpha)$. The union of all ω-limit (ω-cluster) points of N will be denoted by ω-lim N (ω-ad N).

Definition 2.7. (Chen[16]). Let I be an ideal in L^X and $x_\alpha\in M^*(L^X)$. Then x_α is said to be an ω-limit (ω-cluster) point of I, or I ω-converges (ω-accumulates) to x_α, in symbols, $I\to_\omega x_\alpha$ ($I\propto_\omega x_\alpha$), if $\omega\eta^-(x_\alpha)\subset I$ ($P\vee B\neq 1_X$ for each $P\in\omega\eta^-(x_\alpha)$ and each $B\in I$). The union of all ω-limit (ω-cluster) points of I will be denoted by ω-lim$I$ (ω-ad$I$).

Proposition 2.1. (Wang [23]). *Let L be a completely distributive lattice. Then each element of L is the union of some $\vee$-irreducible elements.*

3 ω-Convergence of Filters

In this section, we shall present the concepts of ωQ-neighborhoods (resp. ωOQ-neighborhoods) of a molecule and ω-convergence of a filter in an $L\omega$-space, and discuss their properties.

Definition 3.1. Let (L^X, Ω) be an $L\omega$-space, $B \in L^X$ and $x_\alpha \in M^*(L^X)$. If there is an ω-open set G such that $x_\alpha \not\leq G'$ and $G \leq B$, then we say that B (resp. G) is an ωQ-neighborhood (resp. ωOQ-neighborhood) of x_α, and the collection of all ωQ-neighborhoods (resp. ωOQ-neighborhoods) of x_α is denoted by $\omega\mu\ (x_\alpha)$ (resp. $\omega\mu^o(x_\alpha)$).

Evidently, every ωQ-neighborhood (resp. ωOQ-neighborhood) of x_α is a Q-neighborhood (resp. open Q-neighborhood) of x_α [21] when ω is the fuzzy closure operator, and B (resp. G) is an ωQ-neighborhood (resp. ωOQ-neighborhood) of x_α if and only if B' (resp. G') is an ωR-neighborhood (resp. ωCR-neighborhood) of x_α.

Definition 3.2. Let (L^X, Ω) be an $L\omega$-space, $x_\alpha \in M^*(L^X)$ and let $\mathcal{F}$ be a filter in L^X. Then x_α is said to be an ω-limit (ω-cluster) point of $\mathcal{F}$, or $\mathcal{F}$ ω-converges (ω-accumulates) to x_α, in symbols, $\mathcal{F} \rightarrow_\omega x_\alpha$ ($\mathcal{F} \propto_\omega x_\alpha$), if $\omega\mu^o(x_\alpha) \subset \mathcal{F}$ ($F \wedge G \neq 0_X$ for each $G \in \omega\mu^o(x_\alpha)$ and each $F \in \mathcal{F}$). The union of all ω-limit (ω-cluster) points of $\mathcal{F}$ will be denoted by ω-lim$\mathcal{F}$ (ω-ad$\mathcal{F}$).

Theorem 3.1. *Let (L^X, Ω) be an $L\omega$-space, $e \in M^*(L^X)$ and let $\mathcal{F}$ be a filter in L^X. Then:*

(1) *$\mathcal{F} \rightarrow_\omega e$ if and only if $\mathcal{F} \rightarrow_\omega b$ for each $b \in \beta^*(e)$;*

(2) *$\mathcal{F} \propto_\omega e$ if and only if $\mathcal{F} \propto_\omega b$ for each $b \in \beta^*(e)$;*

(3) *$\mathcal{F} \propto_\omega e$ if and only if $e \leq \omega\mathrm{cl}(F)$ for each $F \in \mathcal{F}$.*

Proof. (1) Suppose that $\mathcal{F} \rightarrow_\omega e$, $b \in \beta^*(e)$ and $G \in \omega\mu^o(b)$. Then $G \in \omega\mu^o(e)$ because of the fact that $b \not\leq G'$ and $b \leq e$, and hence $G \in \mathcal{F}$ by $\mathcal{F} \rightarrow_\omega e$. Conversely, if e is not an ω-limit point of $\mathcal{F}$, then there exists an $G \in \omega\mu^o(e)$ such that $G \notin \mathcal{F}$. Since $e = \beta^*(e)$, there is a $b \in \beta^*(e)$ with $G \in \omega\mu^o(b)$. This means that b is not an ω-limit point of $\mathcal{F}$. Hence, the sufficiency is proved.

(2) Similar to the proof of (1).

(3) Let $\mathcal{F} \propto_\omega e$. Then $F \wedge G \neq 0_X$ for each $G \in \omega\mu^o(x_\alpha)$ and each $F \in \mathcal{F}$ by Definition 3.2, equivalently, $F \not\leq G'$ for each $G' \in \omega\eta^-(x_\alpha)$ and each $F \in \mathcal{F}$. Therefore, $e \leq \omega\mathrm{cl}(F)$ for each $F \in \mathcal{F}$. Conversely, if $e \leq \omega\mathrm{cl}(F)$ for each $F \in \mathcal{F}$, then $F \not\leq G'$ for each $G' \in \omega\eta^-(x_\alpha)$ by Definition 2.4, in other words, $F \wedge G \neq 0_X$ for each $G \in \omega\mu^o(x_\alpha)$. Consequently, $\mathcal{F} \propto_\omega e$ by arbitrariness of F.

Proposition 3.1. *Let (L^X, Ω) be an $L\omega$-space, $b, d \in M^*(L^X)$ and let $\mathcal{F}$ be a filter in L^X. Then:*

(1) *if $\mathcal{F} \rightarrow_\omega d$ and $b \leq d$, then $\mathcal{F} \rightarrow_\omega b$;*

(2) *if $\mathcal{F} \propto_\omega d$ and $b \leq d$, then $\mathcal{F} \propto_\omega b$.*

Proof. The proof is straightforward.

Theorem 3.2. *Let* (L^X, Ω) *be an* $L\omega$*-space,* $e\in M^*(L^X)$ *and let* F *be a filter in* L^X*. Then*:
(1) $F\to_\omega e$ *if and only if* $e\leq\omega\text{-lim}F$;
(2) $F\propto_\omega e$ *if and only if* $e\leq\omega\text{-ad}F$;
(3) $\omega\text{-lim}F\leq\omega\text{-ad}F$.

Proof. The proof is straightforward, we only the proof of (1). If $F\to_\omega e$, then $e\leq\omega$-lim$F$ by the definition of ω-lim$F$. Conversely, if $e\leq\omega$-lim$F$, then for each $b\in\beta^*(e)$, there exists an ω-limit point d of F with $b\leq d$ by virtue of the fact that $e=\sup\beta^*(e)$ and the definition of ω-lim$F$. Consequently, $F\to_\omega e$ according to Proposition 3.1 and Theorem 3.1.

Proposition 3.2. *Suppose that* F_1 *and* F_2 *are two filters in an* $L\omega$*-space* (L^X, Ω) *which* F_2 *is finer than* F_1 *(i.e.,* $F_1\subset F_2$*) and* $x_\alpha\in M^*(L^X)$*. Then*:
(1) *if* x_α *is an* ω*-limit point of* F_1*, then* x_α *is also an* ω*-limit point of* F_2;
(2) *if* x_α *is an* ω*-cluster point of* F_2*, then* x_α *is also an* ω*-cluster point of* F_1.

Proof. (1) If x_α is an ω-limit point of F_1, then $\mu^\rho(x_\alpha)\subset F_1$. Since $F_1\subset F_2$, $\mu^\rho(x_\alpha)\subset F_2$. Therefore, x_α is an ω-limit point of F_2.

(2) Let F_2 ω-accumulates to x_α. Then for each $G\in\omega\mu^\rho(e)$ and each $F\in F_2$ we have $F\wedge G\neq 0_X$, specially, for each $F\in F_1$, $F\wedge G\neq 0_X$ by virtue of $F_1\subset F_2$. Hence, F_1 ω-accumulates to x_α.

Theorem 3.3. *Let* F *be a filter in an* $L\omega$*-space* (L^X, Ω) *and* $x_\alpha\in M^*(L^X)$*. Then* x_α *is an* ω*-cluster point of* F *if and only if there exists a filter* F^* *which is finer than* F *such that* x_α *is an* ω*-limit point of* F^*.

Proof. Assume that x_α is an ω-cluster point of F. By Definition 3.2(ii), $F\wedge G\neq 0_X$ for each $G\in\omega\mu^\rho(e)$ and each $F\in F$. Write $F^*=\{H\in L^X \mid F\wedge G\leq H$ for each $G\in\omega\mu^\rho(x_\alpha)$ and each $F\in F\}$, then F^* is a filter which is finer than F, and $G\in F^*$ for each $G\in\omega\mu^\rho(x_\alpha)$. This implies that x_α is an ω-limit point of F^*.

Conversely, suppose that F^* is a filter which is finer than F, and F^* ω-converges to x_α. According to Definition 3.2, for each $G\in\omega\mu^\rho(x_\alpha)$ we have $G\in F^*$, hence $F\wedge G\in F^*$ for each $F\in F^*$ and each $G\in\omega\mu^\rho(x_\alpha)$, and hence $F\wedge G\neq 0_X$ by the definition of filter. Therefore, x_α is an ω-cluster point of F.

Theorem 3.4. *Let* (L^X, Ω) *be an* $L\omega$*-space,* $A\in L^X$ *and* $x_\alpha\in M^*(L^X)$*. Then the following conditions are equivalent*:
(1) x_α *is an* ω*-adherence point of* A;
(2) *there exists a filter* F *with* $A\in F$ *such that* x_α *is an* ω*-limit point of* F;
(3) *there exists a filter* F *with* $A\in F$ *such that* x_α *is an* ω*-cluster point of* F.

Proof. The proof is straightforward.

Theorem 3.5. *Let* (L^X, Ω) *be an* $L\omega$*-space and* $A \in L^X$. *Then the following conditions are equivalent*:

(1) *A is an* ω*-closed set*;
(2) *for each filter* F *containing A as an element in* L^X, ω-lim$F \leq A$;
(3) *for each filter* F *containing A as an element in* L^X, ω-ad$F \leq A$.

Proof. (1)⇒(2): Suppose that A is an ω-closed set, F is a filter containing A as an element, and $x_\alpha \in M^*(L^X)$. If $x_\alpha \leq \omega$-lim$F$, then $x_\alpha \leq \omega$cl$(A)=A$ in line with Theorem 3.4. Therefore, ω-lim$F \leq A$.

(2)⇒(3): It follows from (2) and Theorem 3.2(3).

(3)⇒(1): Assume that ω-ad$F \leq A$ for each filter containing A as an element and $x_\alpha \leq \omega$cl(A). Then by Theorem 3.4 we know that $x_\alpha \leq \omega$-ad$F \leq A$. This means that ωcl(A) $\leq A$, i.e., A is ω-closed.

Theorem 3.6. *If* F *be a filter in an* $L\omega$*-space* (L^X, Ω), *then* ω-lim$F$ *and* ω-ad$F$ *are both* ω*-closed sets in* (L^X, Ω).

Proof. For each $x_\alpha \in M^*(L^X)$, if $x_\alpha \leq \omega$cl$(\omega$-lim$F)$, then ω-lim$F \nleq G$ for each $G' \in \omega\mu^\rho(x_\alpha)$,. With reference to Proposition 2.1 we can choose a molecule $b \leq \omega$-lim$F$ such that $b \nleq G$, thus $G' \in \omega\mu^\rho(b)$. Consequently, $G' \in F$ by $b \leq \omega$-lim$F$. This shows that $x_\alpha \leq \omega$-lim$F$, and thus ωcl$(\omega$-lim$F) \leq \omega$-lim$F$. On the other hand, ω-lim$F \leq \omega$cl$(\omega$-lim$F)$ by Theorem 2.5 in [14].

Similarly, we can easily verify that ω-ad$F$ is an ω-closed set in (L^X, Ω).

Definition 3.3. Let (L^X, Ω) be an $L\omega$-space, $x_\alpha \in M^*(L^X)$ and let F_0 be a filter base in L^X. Then x_α is said to be an ω-limit (ω-cluster) point of F_0, or F_0 ω-converges (ω-accumulates) to x_α, in symbols, $F_0 \rightarrow_\omega x_\alpha$ ($F_0 \propto_\omega x_\alpha$), if $F \rightarrow_\omega x_\alpha$ where F is the filter generated by F_0, i.e.,

$$F=\{F \in L^X \mid \text{there exists a } H \in F_0 \text{ such that } H \leq F\}.$$

The union of all ω-limit (ω-cluster) points of F_0 will be denoted by ω-limF_0 (ω-adF_0).

Theorem 3.7. *Let* F_0 *be a filter base in* (L^X, Ω) *and* $x_\alpha \in M^*(L^X)$. *Then*:

(1) $F_0 \rightarrow_\omega x_\alpha$ *if and only if every* ωOQ*-neighborhood of* x_α *contains a member of* F_0;
(2) $F_0 \propto_\omega x_\alpha$ *if and only if every* ωOQ*-neighborhood of* x_α *intersects all member of* F_0.

Proof. It follows straightforward from Definition 3.2 and Definition 3.3.

4 Relationships Among ω-Convergence of Filters, ω-Convergence of Nets and ω-Convergence of Ideals

In the section, we shall respectively discuss the relationships between ω-convergence of filters and ω-convergence of nets, and between ω-convergence of filters and ω-convergence of ideals.

Definition 4.1. Let $\mathcal{F}$ be a filter in an $L\omega$-space (L^X, Ω) and $D(\mathcal{F})=\{(F, G) \mid F\in \mathcal{F}$ and $F\not\leq G\}$. In $D(F)$, we define a relation "$\leq$" as follows:

$$(F_1, G_1)\leq (F_2, G_2) \text{ iff } F_2\leq F_1,\ G_1\leq G_2.$$

Then $D(\mathcal{F})$ is a directed set as per the relation "$\leq$". Put

$$N(\mathcal{F})=\{N(F, G)\in M^*(L^X) \mid N(F, G) \leq F, N(F, G)\not\leq G \text{ and } (F, G)\in D(\Omega)\}.$$

It is clear that $N(\mathcal{F})$ is a molecular net in L^X, we say that $N(\mathcal{F})$ is the molecular net induced by $\mathcal{F}$.

Theorem 4.1. *Let $\mathcal{F}$ be a filter in an $L\omega$-space (L^X, Ω) and $x_\alpha\in M^*(L^X)$. Then $\mathcal{F}$ ω-accumulates to x_α if and only if $N(\mathcal{F})$ ω-accumulates to x_α.*

Proof. Suppose that $\mathcal{F}$ ω-accumulates to x_α. Then $F\not\leq G$ for each $G'\in \omega\mu^\circ(x_\alpha)$ and each $F\in \mathcal{F}$ and hence $(F, G)\in D(\mathcal{F})$. As per Definition 4.1, for each $(P, Q)\in D(\mathcal{F})$ with $(P, Q) \geq (F, G)$ we have $N(P,Q)\not\leq Q$ and $G\leq Q$, thus $N(P, Q)\not\leq G$. This shows that $N(\mathcal{F})$ is frequently not in G. Therefore $N(\mathcal{F})$ ω-accumulates to x_α.

Conversely, if $N(\mathcal{F}) \propto_\omega x_\alpha$, then for each $G'\in \omega\mu^\circ(x_\alpha)$ and each $(F, G)\in D(\mathcal{F})$, there exists a $(P, Q)\in D(\mathcal{F})$ with $(P, Q) \geq (F, G)$ such that $N(P, Q)\not\leq G$. Since $N(P, Q)\leq P\leq F$, $F\not\leq G$. In the sequel, $\mathcal{F}$ ω-accumulates to x_α.

Definition 4.2. Let $N=\{N(n), n\in D\}$ be a molecular net in an $L\omega$-space (L^X, Ω), $F_m = \vee\{N(n), n\geq m\}$ $(m\in D)$ and $\mathcal{F}(N)=\{F\in L \mid$ there exists $m\in D$ with $F\geq F_m\}$. Then $\mathcal{F}(N)$ is a filter in (L^X, Ω) and is called the filter induced by N.

Theorem 4.2. *Let $N=\{N(n), n\in D\}$ be a molecular net in an $L\omega$-space (L^X, Ω) and $x_\alpha\in M^*(L^X)$. Then N ω-accumulates to x_α if and only if $\mathcal{F}(N)$ ω-accumulates to x_α.*

Proof. Assume that N ω-accumulates to x_α. Then for each $G'\in \omega\mu^\circ(x_\alpha)$ and each $m\in D$, there exists $n\in D$ satisfying $n\geq m$ and $N(n)\not\leq G$, and hence $F_m\not\leq G$. According to the definition of $\mathcal{F}(N)$ we know that for each $F\in \mathcal{F}(N)$, there is $m\in D$ with $F_m\leq F$, and so $F\not\leq G$. Therefore, x_α is an ω-cluster point of $\mathcal{F}(N)$.

Conversely, suppose that $\mathcal{F}(N)$ ω-accumulates to x_α. Then $F\not\leq G$ for each $G'\in \omega\mu^\circ(x_\alpha)$ and each $F\in \mathcal{F}(N)$, specially, $F_m\not\leq G$ for each $m\in D$. By the definition of F_m, we can choose $n\in D$ satisfying $n\geq m$ and $N(n)\not\leq G$. Hence $N\propto_\sigma x_\alpha$.

It follows that the ω-convergence of molecular nets and filters are closed relative to each other. The following theorems further bring to light the close relation.

Theorem 4.3. *Let $N=\{N(n), n\in D\}$ be a molecular net in an $L\omega$-space (L^X, Ω) and $x_\alpha\in M^*(L^X)$. Then ω-ad$N=\wedge\{\omega\text{cl}(F_m) \mid m\in D\}$.*

Proof. Suppose that $x_\alpha\in \omega$-ad N, $G'\in \omega\mu^\circ(x_\alpha)$ and $m\in D$. Then there exists $n\in D$ such that $n\geq m$ and $N(n)\not\leq G$. Since $N(n)\leq F_m$, we know that $F_m\not\leq G$, i.e., $x_\alpha\leq \omega\text{cl}(F_m)$. Conversely, if $x_\alpha\not\leq \omega$-ad N, then there exists $G'\in \omega\mu^\circ(x_\alpha)$ and $m\in D$ satisfying $N(n)\leq G$

wherever $n \geq m$. Hence $F_m \leq G$ by the definition of F_m. This shows that $x_\alpha \nleq \omega\mathrm{cl}(F_m)$, and hence $x_\alpha \nleq \wedge\{\omega\mathrm{cl}(F_m) \mid m \in D\}$.

Theorem 4.4. *Let N be a molecular net in* (L^X, Ω). *Then* $\omega\text{-ad}\mathcal{F}(N) = \wedge\{\omega\mathrm{cl}(F) \mid F \in \mathcal{F}\}$

Definition 4.3. Let $\mathcal{F}$ be a filter and I an ideal in (L^X, Ω). Then $I(\mathcal{F}) = \{F' \mid F \in \mathcal{F}\}$ is an ideal and $\mathcal{F}(I) = \{B' \mid B \in I\}$ is a filter in (L^X, Ω). We call $I(\mathcal{F})$ ($\mathcal{F}(I)$) the ideal (filter) induced by $\mathcal{F}$ (I).

Theorem 4.5. *Let* $\mathcal{F}$ *be a filter and I an ideal in* (L^X, Ω), *and let* $x_\alpha \in M^*(L^X)$. *Then*:

(1) $\mathcal{F}$ *ω-converges to* x_α *if and only if* $I(\mathcal{F})$ *ω-converges to* x_α;
(2) $\mathcal{F}$ *ω-accumulates to* x_α *if and only if* $I(\mathcal{F})$ *ω-accumulates to* x_α.
(3) *I ω-converges to* x_α *if and only if* $\mathcal{F}(I)$ *ω-converges to* x_α;
(4) *I ω-accumulates to* x_α *if and only if* $\mathcal{F}(I)$ *ω-accumulates to* x_α.

Proof. The proof is straightforward, and is omitted.

5 Some Applications of ω-Convergence of Filters

In this section, we shall give some characterizations of (ω_1, ω_2)-continuity of L-fuzzy mapping and ω-separations by means of ω-convergence theory of filters.

Definition 5.1. (Chen [15,17]) Let f be an L-fuzzy mapping from an $L\omega_1$-space (L^X, Ω_1) into an $L\omega_2$-space (L^Y, Ω_2). Then:

(i) f is called (ω_1, ω_2)-continuous if $f^{-1}(B) \in \omega_1 O(L^X)$ for each $B \in \omega_2 O(L^Y)$;
(ii) f is called (ω_1, ω_2)-continuous at $x_\alpha \in M^*(L^X)$ if $f^{-1}(B) \in \omega_1\eta^-(x_\alpha)$ for each $B \in \omega_2\eta^-(f(x_\alpha))$.

Theorem 5.1. (Chen [15,17]) *Let f be an L-fuzzy mapping from an* $L\omega_1$*-space* (L^X, Ω_1) *into an* $L\omega_2$*-space* (L^Y, Ω_2). *Then f is* (ω_1, ω_2)*-continuous if and only if for every* $x_\alpha \in M^*(L^X)$, *f is* (ω_1, ω_2)*-continuous at* x_α.

Theorem 5.2. *Let f be an L-fuzzy mapping from an* $L\omega_1$*-space* (L^X, Ω_1) *into an* $L\omega_2$*-space* (L^Y, Ω_2), *and* $x_\alpha \in M^*(L^X)$. *Then f is* (ω_1, ω_2)*-continuous at* x_α *if and only if* $f^{-1}(B) \in \omega_1\mu^\circ(x_\alpha)$ *for each* $B \in \omega_2\mu^\circ(f(x_\alpha))$.

Proof. Since $B \in \omega_2\mu^\circ(f(x_\alpha))$ if and only if $B' \in \omega_2\eta^-(f(x_\alpha))$, and $f^{-1}(B) \in \omega_1\mu^\circ(x_\alpha)$ if and only if $f^{-1}(B') \in \omega_1\eta^-(x_\alpha)$, the proof is obvious by Definition 5.1(ii).

Theorem 5.3. *Let f be an L-fuzzy mapping from an* $L\omega_1$*-space* (L^X, Ω_1) *into an* $L\omega_2$*-space* (L^Y, Ω_2). *Then f is* (ω_1, ω_2)*-continuous if and only if for every filter base* $\mathcal{F}_0$ *in*

(L^X, Ω_1) and the filter base $f(F_0)=\{f(F) \mid F\in F_0\}$ in (L^Y, Ω_2) we have $f(\omega_1\text{-lim}F_0)\leq \omega_2\text{-lim}f(F_0)$.

Proof. Suppose that f is (ω_1,ω_2)-continuous, F_0 is an filter base in (L^X, Ω_1) and $f(F_0)=\{f(F) \mid F\in F_0\}$ is the filter base in (L^Y, Ω_2). Then for each molecule $y_\alpha\leq f(\omega_1\text{-lim}F_0)$, there exists a molecule $x_\alpha\leq\omega\text{-lim}F_0$ with $y_\alpha=f(x_\alpha)$. We affirm that $y_\alpha\leq\omega_2\text{-lim}f(F_0)$. In fact, for each $B\in \omega_2\eta^-(f(x_\alpha))$ we have $f^{-1}(B)\in \omega_1\eta^-(x_\alpha)$ being the continuity of f, in other words, for each $B'\in \omega_2\mu^0(f(x_\alpha))$ we have $f^{-1}(B')\in \omega_1\mu^0(x_\alpha)$. Hence, there exists a member $F\in F_0$ such that $F\leq f^{-1}(B')$, i.e., $f(F)\leq B'$ according to Theorem 3.7. This means that $y_\alpha\leq\omega_2\text{-lim}f(F_0)$.

Conversely, assume that the condition of the theorem holds. If f is not (ω_1,ω_2)-continuous, then there exists an ω_2-closed set B in L^Y such that $f^{-1}(B)\neq\omega_1\text{cl}(f^{-1}(B))$, i.e., there is a molecule $x_\alpha\leq\omega_2\text{cl}(f^{-1}(B))$ with $x_\alpha\not\leq f^{-1}(B)$ by Proposition 2.1. Hence, we can choose a filter base F_0 in (L^X, Ω_1) which contains $f^{-1}(B)$ as member such that x_α is an ω-limit point of F_0 in the light of Theorem 3.7, and thus $f(x_\alpha)\leq f(\omega_1\text{-lim}F_0)\leq\omega_2\text{-lim}f(F_0)$ according to the assumption. Since B is ω_2-closed, $f(x_\alpha)\leq B$ according to Theorem 3.5. This contradicts $x_\alpha\not\leq f^{-1}(B)$. Therefore, f is (ω_1,ω_2)-continuous.

Theorem 5.4. *An $L\omega$-space (L^X, Ω) is an ωT_2 space[18] if and only if every filter in L^X has at most one ω-limit point.*

Proof. The proof is straightforward, and is omitted.

Acknowledgements

This paper is supported by the National Natural Science Foundation of China (No. 10471083) and the Natural Science Foundation of Fujian Province of China (No. A0510023).

References

1. S. L. Chen (1992) Moore-Smith θ-convergence theory on completely distributive lattices, Proc. Fuzzy Math. and Systems, Hunan Science and Technology Press, Changsha, China, pp. 34-36.
2. S. L. Chen (1993) θ-convergence theory of ideals on completely distributive lattices, J. Math. 13(2): 47-52.
3. S. L. Chen (2004) σ-convergence theory and its applications in fuzzy lattices, Information Sciences 165: 45-58.
4. S. L. Chen, S. T. Chen(2000) A new extension of fuzzy convergence, Fuzzy Sets and Systems 109: 1999-1204.
5. S. L. Chen (1995) Urysohn-convergence of fuzzy ideals, J. Fuzzy Math. 3(3): 555-558.
6. S. L. Chen, J. S. Cheng (1997) θ-convergence of nets of L-fuzzy sets and its applications, Fuzzy Sets and Systems 86: 235-240

7. S. L. Chen, W. N. Zhou (1999) *SU*-Separation axioms in topological molecular lattices, J. Fuzzy Math. 7(4): 987-995.
8. S. L. Chen, J. R. Wu (1999) σ-convergence theory of ideals in fuzzy lattices, J. Wuhan Urban Construction Institute 16(4): 43-47
9. S. L. Chen, X. G. Wang (2000) $S\theta$-convergence theory in *L*-fuzzy topology, J. Fuzzy Math. 8(2): 501-516
10. S. L. Chen, J. R. Wu (2000) *SR*-convergence theory in fuzzy lattices, Information Sciences125: 233-247.
11. S.L. Chen, J.S. Cheng (1993) Theory of *R*-convergence of ideals on fuzzy lattices and its applications, Proceeding of the Fifth IFSA World Congress, Vol. I, Korea, pp. 269-273.
12. S. L. Chen (1995) *U*-convergence and *L*-fuzzy *U*-sets, Information Sciences 87 (4): 205-213.
13. S. L. Chen (2000) σ-convergence of molecular nets in fuzzy lattices, Fuzzy Systems and Math. 14(1): 13-18.
14. S. L. Chen and J. S. Cheng (2005) On $L\omega$-spaces, Proc. Eleventh IFSA Word Congress, Tsinghua University Press, Vol. I: pp. 257-261
15. S. L. Chen and C. Q. Dong (2002) *L*-order-preserving operator spaces, Fuzzy Systems and Math. 16(ZJ): 36-41
16. S. L. Chen (2002) Moore-Smith convergence theory on *L*-fuzzy order-preserving operator spaces, J. Jimei Univ. 7(3): 271-277
17. S. L. Chen (2006) On *L*-fuzzy order-preserving operator ω-spaces, J. Fuzzy Math. 14(2): 481-498.
18. Z. X. Huang and S. L. Chen (2005) ω-separation axioms in $L\omega$-space, J. Math. 25(4): 383-388.
19. B. Hutton (1977) Uniformities on fuzzy topological spaces, J. Math. Anal. Appl. 58: 559-571.
20. R. Lowen (1983) Relation between filter and net convergence in fuzzy topological spaces, Fuzzy Math. 4: 41-52.
21. P. M. Pu and Y. M. Liu (1980) Fuzzy topology I: Neighborhood structure of a fuzzy point and Moore-Smith convergence, J. Math. Anal. Appl. 76: 57-599.
22. G. J. Wang (1983) A new fuzzy compactness defined by fuzzy nets, J. Math. Anal. Appl. 94: 1-23.
23. G. J. Wang (1992) Theory of topological molecular lattices, Fuzzy Sets and Systems 47: 351-376.
24. K. Wu (2000) Convergences of fuzzy sets based on decomposition theory and fuzzy polynomial function, Fuzzy Sets and Systems 109(2): 173-186.

On the Perturbation of Fuzzy Matrix Equations with $\vee$-T Composition

Yifen Lai and Shuili Chen

School of Sciences, Jimei University, Xiamen 361021,
China phlaiyf@sohu.com, shuilichen@vip.sina.com

Abstract. In this paper, the algorithms for the solution of fuzzy matrix equations with $\vee$-T composition are simplified. By the fuzzy solution-invariant matrix, the perturbation issues of fuzzy matrix equations are considered.

Keywords: Fuzzy matrix equation, Solution, Perturbation.

The perturbation method is an important mathematical tool used to solve engineering and scientific problems. The perturbation theory of fuzzy matrix equations has been widely applied to fields of fuzzy control, fuzzy inference, fuzzy logic, etc. [2] studied the perturbation issues of fuzzy matrix equation $A \circ X=B$ with max-min composition. [5] studied the perturbation issues of fuzzy matrix equation $A \circ X=B$ with $\vee$-• composition. They gave out some beneficial results. Here, we discuss the perturbation issues of fuzzy matrix equation $A \circ_T X=B$ with $\vee$-T composition, where T is a t-norm with continuity and strict monotonicity, and generalize the results in [5].

1 Preliminaries

In this paper, the letter I is used to denote the real unit interval [0,1].

Definition 1.1 [3]. A t-norm is a binary operation T on I satisfying the following conditions:

① $aTb=bTa$; (commutative law)
② $(aTb)Tc=aT(bTc)$; (associative law)
③ if $b \le c$, then $aTb \le aTc$; (monotonicity)
④ $aT1=a$, (boundary condition)

Where a,b and c are arbitrary elements in I.

It is clear that $aT0=0$.

From the point of view of function, t-norms are binary functions. In this paper, t-norms on I are always continuous. Moreover, t-norms are strictly monotone, that is, for any $a,b,c \in I$, if $b<c$, then $aTb<aTc$. For instance, the operations $P:(a,b) \mapsto ab$, $Q:(a,b) \mapsto \dfrac{ab}{1+(1-a)(1-b)}$ on I are t-norms with continuity and strict monotonicity.

B.-Y. Cao (Ed.): Fuzzy Information and Engineering (ICFIE), ASC 40, pp. 269–279, 2007.
springerlink.com

Lemma 1.1. If $a,b\in I$, $a\neq 0$ and $a\geq b$, then the equation

$$aTx=b \tag{1}$$

is always solvable. Moreover, the solution of Eq.(1) is unique.

Proof. Since the function $y=f(x)=aTx$ is continuous on I, $f(0)=0$, $f(1)=a$ and $f(0)\neq f(1)$, it follows from intermediate value theorem of continuous function that for any $b\in[0,a]$, there exists a $x_0\in I$ such that $f(x_0)=b$. Therefore, for any $a,b\in I$, if $a\neq 0$ and $a\geq b$, then the equation $aTx=b$ is always solvable.

Below,we show that the solution of Eq.(1) is unique.

Assume that $x_1<x_2$.If x_1 and x_2 are solution of Eq.(1),then $aTx_1=aTx_2=b$. This contradicts the fact that T is strictly monotone and hence proves the uniqueness part of this Lemma.

Definition 1.2. [1]. Let T be a t-norm on I.A binary operation α_T on I,defined by

$$a\alpha_T b=\vee\{x\in I \mid aTx\leq b\},\ \forall a,b\in I,$$

is called the generalized inverse operation of T.

Definition 1.3. [3]. A t-norm T on I is said to be infinitely $\vee$-distributive if $a,b_i\in I$, $i\in K$, where K is any nonempty index set $\Rightarrow aT(\vee_{i\in K} b_i)=\vee_{i\in K}(aT\ b_i)$.

Throughout this paper, unless otherwise stated, T stands for any given infinitely $\vee$-distributive t-norm on I. Clearly, $aT(a\alpha_T b)\leq b$. When T is the (algebraic) product P,

$$a\alpha_P b=\begin{cases}1, & \text{if } a\leq b\\ b/a, & \text{if } a>b\end{cases}.$$

Corollary 1.1. If $a,b\in I$, $a\neq 0$ and $a\geq b$, then $x=a\alpha_T b$ is a solution of Eq.(1).

Proof. Suppose x_0 is a solution of Eq.(1). Then $aTx_0=b$ and hence $x_0\leq a\alpha_T b$. On the other hand, $aT(a\alpha_T b)\leq b=aTx_0$ and hence $a\alpha_T b\leq x_0$. Therefore we have $x_0=a\alpha_T b$. The proof is complete.

Corollary 1.2. If $a,b\in I$, $a\neq 0$ and $a\geq b$, then $a\alpha_T b>c \Leftrightarrow aTc<b$.

Proof. By Lemma 1.1 and Corollary 1.1, we know that $x=a\alpha_T b$ is the unique solution of Eq.(1), i.e., $aT(a\alpha_T b)=b$. If $a\alpha_T b>c$, then by strict monotonicity of T, $aTc<aT(a\alpha_T b)=b$. Conversely, if $aTc<b$, then $aTc<aT(a\alpha_T b)$. Thus, it follows from strict monotonicity of T that $c<a\alpha_T b$. This completes the proof.

Lemma 1.2. [1]. Let $a,b,c\in I$. Then $a\leq b\Rightarrow a\alpha_T c\geq b\alpha_T c$.

Lemma 1.3. If $a,b,c\in I$ and $a>b\geq c$, then $a\alpha_T c<b\alpha_T c$.

Proof. Since $a>b$, by Lemma 1.2, $a\alpha_T c \leq b\alpha_T c$. If $b=0$, it is clear that $a\neq 0$, $c=0$, $a\alpha_T c=0$ and $b\alpha_T c=1$. Thus $a\alpha_T c<b\alpha_T c$. If $b\neq 0$, suppose that $a\alpha_T c=b\alpha_T c \stackrel{\wedge}{=} d$. Since $a,b\geq c$, $a\neq 0$ and $b\neq 0$, by Corollary 1.1, $aTd=bTd=c$, and hence $a=b$. This contradicts the condition. Therefore, $a\alpha_T c<b\alpha_T c$.

Lemma 1.4. Let $a,b\in I$. Then $(a\alpha_T b)\alpha_T b\geq a$.

Proof. Let $x=a\alpha_T b$. Then $aTx\leq b$. According to the definition of α_T, $x\alpha_T b\geq a$. Hence, $(a\alpha_T b)\alpha_T b\geq a$.

Definition 1.4. Let T be a t-norm on I. Let $A=[a_{ij}]_{\mathrm{m\times n}}$, $B=[b_{jk}]_{\mathrm{n\times r}}$ be matrices over I. The operation $A\circ_T B$, defined by

$$A\circ_T B=[p_{ik}]_{\mathrm{m\times r}},\ p_{ik}=\bigvee_{j=1}^{n}(a_{ij}Tb_{jk})$$

is called the $\vee$-T composition of A and B. If T is the (algebraic) product P, then $A\circ_T B$ is referred to as the $\vee$-• composition of A and B and written as $A\circ B$.

Let

$$A\circ_T X=B \tag{2}$$

be a fuzzy matrix equation, Where $A=(a_{ij})_{\mathrm{m\times n}}$, $a_{ij}\in I$, X is a $n\times 1$ matrix and B is a $m\times 1$ fuzzy matrix. Similar to definition in [5], perturbation elements in matrix A can be defined as follows:

Definition 1.5. In a fuzzy matrix equation (2), if $\forall\ \varepsilon>0$, $a_{ij}-\varepsilon\geq 0$, $a_{ij}+\varepsilon\leq 1$, when a_{ij} perturbs in $[a_{ij}-\varepsilon, a_{ij}+\varepsilon]$, the solution set of the Eq.(2) varies, then a_{ij} is called an element without perturation in A.

Definition 1.6. In a fuzzy matrix equation (2), if a_{ij} perturbs within $(a'_{ij}, 1]$ $(a_{ij}\geq a'_{ij})$, the solution set of the equation is invariable, then a_{ij} is called an upper perturbation element in A; if a_{ij} perturbs within $[0, a'_{ij})$ $(a_{ij}\leq a'_{ij})$, the solution set of the equation is invariable, then a_{ij} is called a lower perturbation element in A; if a_{ij} perturbs within I, the solution set of the equation is invariable, then a_{ij} is called full perturbation element in A.

Similarly, we can define upper-closed perturbation elements and lower-closed perturbation elements. Obviously, a full perturbation element in A must be an upper one and a lower one in A.

2 Fuzzy Solution-Invariant Matrix

Lemma 2.1. [4,Theorem 2.5(1)] Eq.(2) is solvable if and only if $\overline{X}=A^T\circ_{\alpha_T}B$ is a solution of Eq.(2), where

$$A^T \circ_{\alpha_T} B = (\mathop{\wedge}_{k=1}^{m}(a_{k1}\alpha_T b_k), \mathop{\wedge}_{k=1}^{m}(a_{k2}\alpha_T b_k), \cdots, \mathop{\wedge}_{k=1}^{m}(a_{kn}\alpha_T b_k))^T .$$

When this is the case, $\overline{X}$ is the largest solution of Eq.(2).

Definition 2.1. If Eq.(2) is solvable, the following notation is used to denote the set of solution of Eq.(2):

$$S(A,B)=\{X_0 \mid A \circ_T X_0=B\}.$$

In what follows, we always assume that $S(A,B)\neq\Phi$, i.e., Eq.(2) is solvable.

Theorem 2.1. Given a fuzzy matrix equation (2), define

$$A^{(0)} = (a_{ij}^{(0)}),\ a_{ij}^{(0)} = \begin{cases} 0, & if\ a_{ij} < b_i \\ a_{ij}, & if\ a_{ij} \geq b_i \end{cases},\ i=1,2,\ldots,m,\ j=1,2,\ldots,n.$$

Then $S(A,B)=S(A^{(0)},B)$.

Proof. Given a fuzzy matrix equation $A \circ_T X=B$. Suppose $\overline{X} = (\overline{x}_1, \overline{x}_2, \cdots, \overline{x}_n)^T$ is the largest solution of $A \circ_T X=B$.

For any given $i\in\{1,2,\ldots,m\}$,write

$$J_1=\{j \mid a_{ij}<b_i\},$$
$$J_2=\{j \mid a_{ij}\geq b_i\}.$$

Suppose $X=(x_1,x_2,\ldots,x_n)^T\in S(A,B)$. Then for any i,

$$b_i = \mathop{\vee}_{j=1}^{n}(a_{ij}Tx_j) = (\mathop{\vee}_{j\in J_1}(a_{ij}Tx_j)) \vee (\mathop{\vee}_{j\in J_2}(a_{ij}Tx_j)).$$

Since $\mathop{\vee}_{j\in J_1}(a_{ij}Tx_j) \leq \mathop{\vee}_{j\in J_1} a_{ij} < b_i$, we deduce that $b_i = \mathop{\vee}_{j\in J_2}(a_{ij}Tx_j) = \mathop{\vee}_{j=1}^{n}(a_{ij}^{(0)}Tx_j)$ and so $X\in S(A^{(0)},B)$.

Conversely, for any given $j\in\{1,2,\ldots,n\}$, write

$$I_1=\{i \mid a_{ij}<b_i\},$$
$$I_2=\{i \mid a_{ij}\geq b_i\}.$$

Then $\overline{x}_j = \mathop{\wedge}_{i=1}^{m}(a_{ij}\alpha_T b_i) = (\mathop{\wedge}_{i\in I_1}(a_{ij}\alpha_T b_i)) \wedge (\mathop{\wedge}_{i\in I_2}(a_{ij}\alpha_T b_i))$. If $i\in I_1$, then $a_{ij}\alpha_T b_i=1$.Since $0\alpha_T b_i=1$, we have $\overline{x}_j = \mathop{\wedge}_{i\in I_2}(a_{ij}\alpha_T b_i) = \mathop{\wedge}_{i=1}^{m}(a_{ij}^{(0)}\alpha_T b_i)$. Hence, $A^{(0)} \circ_T X=B$ has the same larget solution as $A \circ_T X=B$.

Suppose $X=(x_1,x_2,\ldots,x_n)^T \in S(A^{(0)},B)$. Then from $b_i = \mathop{\vee}_{j=1}^{n}(a_{ij}^{(0)}Tx_j) \leq \mathop{\vee}_{j=1}^{n}(a_{ij}Tx_j) \leq \mathop{\vee}_{j=1}^{n}(a_{ij}T\overline{x}_j) = b_i$ deduce that $b_i = \mathop{\vee}_{j=1}^{n}(a_{ij}Tx_j)$ and hence $X\in S(A,B)$.

Therefore, by the preceding proof, $S(A,B)=S(A^{(0)},B)$.

Definition 2.2. Given a fuzzy matrix equation (2), define

$$A^{(1)}=(a_{ij}^{(1)}),\ a_{ij}^{(1)}=\begin{cases}a_{ij}, & \text{if } a_{ij}\geq b_i \text{ and } a_{ij}T\overline{x}_j=b_i\\ 0, & \text{elsewhere}\end{cases},$$

where $\overline{X}=(\overline{x}_1,\overline{x}_2,\cdots,\overline{x}_n)^T$ is the largest solution of Eq.(2). $A^{(1)}$ is called simple matrix of A.

Theorem 2.2. Given a fuzzy matrix equation $A\circ_T X=B$, we have $S(A,B)=S(A^{(1)},B)$.

Proof. Given a fuzzy matrix equation $A\circ_T X=B$. Suppose $\overline{X}=(\overline{x}_1,\overline{x}_2,\cdots,\overline{x}_n)^T$ is the largest solution of $A\circ_T X=B$.

For any given $i\in\{1,2,\dots,m\}$,write

$$J_1=\{j\mid a_{ij}<b_i\},$$
$$J_2=\{j\mid a_{ij}\geq b_i,\ a_{ij}T\overline{x}_j=b_i\},$$
$$J_3=\{j\mid a_{ij}\geq b_i,\ a_{ij}T\overline{x}_j<b_i\},$$
$$J_4=\{j\mid a_{ij}\geq b_i,\ a_{ij}T\overline{x}_j>b_i\}.$$

We claim that for any $i\in\{1,2,\dots,m\}$, $J_4=\Phi$. Otherwise, there exists an i such that $J_4\neq\Phi$. Assume that $j\in J_4$. Then $a_{ij}T\overline{x}_j>b_i$, and hence $\bigvee_{j=1}^{n}(a_{ij}T\overline{x}_j)>b_i$, contradiction.

Suppose $X=(x_1,x_2,\dots,x_n)^T\in S(A,B)$. Then for any i,

$$b_i=\bigvee_{j=1}^{n}(a_{ij}Tx_j)=(\bigvee_{j\in J_1}(a_{ij}Tx_j))\vee(\bigvee_{j\in J_2}(a_{ij}Tx_j))\vee(\bigvee_{j\in J_3}(a_{ij}Tx_j)).$$

Since $\bigvee_{j\in J_1}(a_{ij}Tx_j)\leq\bigvee_{j\in J_1}a_{ij}<b_i$ and $\bigvee_{j\in J_3}(a_{ij}Tx_j)\leq\bigvee_{j\in J_3}(a_{ij}T\overline{x}_j)<b_i$, we have $b_i=\bigvee_{j\in J_2}(a_{ij}Tx_j)=\bigvee_{j=1}^{n}(a_{ij}^{(1)}Tx_j)$. Therefore, $X\in S(A^{(1)},B)$.

Conversely, for any given $j\in\{1,2,\dots,n\}$, write

$$I_1=\{i\mid a_{ij}<b_i\},$$
$$I_2=\{i\mid a_{ij}\geq b_i,\ a_{ij}T\overline{x}_j=b_i\},$$
$$I_3=\{i\mid a_{ij}\geq b_i,\ a_{ij}T\overline{x}_j<b_i\},$$
$$I_4=\{i\mid a_{ij}\geq b_i,\ a_{ij}T\overline{x}_j>b_i\}.$$

We claim that for any $j\in\{1,2,\dots,n\}$, $I_4=\Phi$. Otherwise, there exists a j such that $I_4\neq\Phi$. This implies that $\bigvee_{j=1}^{n}(a_{ij}T\overline{x}_j)>b_i$ for $i\in I_4$. It contradicts the fact. Therefore,

$$\overline{x}_j=\bigwedge_{i=1}^{m}(a_{ij}\alpha_T b_i)=(\bigwedge_{i\in I_1}(a_{ij}\alpha_T b_i))\wedge(\bigwedge_{i\in I_2}(a_{ij}\alpha_T b_i))\wedge(\bigwedge_{i\in I_3}(a_{ij}\alpha_T b_i)).$$

If $i \in I_1$, then $a_{ij}\alpha_T b_i=1$. Thus, $\wedge_{i\in I_1}(a_{ij}\alpha_T b_i)=1$. If $a_{ij}=0$, it is clear that $I_3=\Phi$. If $a_{ij}\neq 0$, then, for $i\in I_3$, applying Corollary 1.2, we have $a_{ij}\alpha_T b_i > \overline{x}_j$, and hence $\wedge_{i\in I_3}(a_{ij}\alpha_T b_i) > \overline{x}_j$. Therefore, $\overline{x}_j = \wedge_{i\in I_2}(a_{ij}\alpha_T b_i) = \wedge_{i=1}^{m}(a_{ij}^{(1)}\alpha_T b_i)$, that is, $A^{(1)}\circ_T X=B$ has the same larget solution as $A\circ_T X=B$.

Suppose $X=(x_1,x_2,\ldots,x_n)^T \in S(A^{(1)},B)$. Then from $b_i = \vee_{j=1}^{n}(a_{ij}^{(1)}Tx_j) \leq \vee_{j=1}^{n}(a_{ij}Tx_j) \leq \vee_{j=1}^{n}(a_{ij}T\overline{x}_j) = b_i$ deduce that $b_i = \vee_{j=1}^{n}(a_{ij}Tx_j)$, and hence $X \in S(A,B)$.

Therefore, by the preceding proof, $S(A,B)=S(A^{(1)},B)$.

Definition 2.3. Let $A\circ_T X=B$ and $A'\circ_T X=B$ be fuzzy matrix equations. If $S(A,B)= S(A',B)$, then A' is referred to as a fuzzy solution-invariant matrix of A.

3 Fuzzy Perturbation Issues

Theorem 3.1. Given a fuzzy matrix equation (2), if $a_{i_0 j_0} \neq 0$ and $a_{i_0 j_0}^{(0)} = a_{i_0 j_0}^{(1)} = 0$, then $a_{i_0 j_0}$ is a lower perturbation element in A, and it can perturb within $[0, \overline{x}_{j_0}\alpha_T b_{i_0})$.

Proof. If $a_{i_0 j_0} \neq 0$ and $a_{i_0 j_0}^{(0)} = a_{i_0 j_0}^{(1)} = 0$, then, by the definition of $A^{(0)}$ and $A^{(1)}$, we know that $a_{i_0 j_0} < b_{i_0}$ and hence $a_{i_0 j_0}\alpha_T b_{i_0} = 1$. $\forall a'_{i_0 j_0} \in [0, \overline{x}_{j_0}\alpha_T b_{i_0})$, we replace $a_{i_0 j_0}$ in A by $a'_{i_0 j_0}$ (Note: $a_{i_0 j_0} < b_{i_0} \leq \overline{x}_{j_0}\alpha_T b_{i_0}$), keep the other elements in A unchanged and get a new fuzzy matrix $A' = (a'_{ij})_{m\times n}$. Below, we show that the matrix equation $A'\circ_T X=B$ is solvable.

Since $a_{i_0 j_0}T\overline{x}_{j_0} \leq a_{i_0 j_0} < b_{i_0}$ and $a'_{i_0 j_0}T\overline{x}_{j_0} < (\overline{x}_{j_0}\alpha_T b_{i_0})T\overline{x}_{j_0} \leq b_{i_0}$,we see that

$$
\begin{aligned}
b_{i_0} &= \vee_{j=1}^{n}(a_{i_0 j}T\overline{x}_j) \\
&= (\vee_{j\neq j_0}(a_{i_0 j}T\overline{x}_j)) \vee (a_{i_0 j_0}T\overline{x}_{j_0}) \\
&= \vee_{j\neq j_0}(a_{i_0 j}T\overline{x}_j) \\
&= (\vee_{j\neq j_0}(a_{i_0 j}T\overline{x}_j)) \vee (a'_{i_0 j_0}T\overline{x}_{j_0})
\end{aligned}
$$

$$= (\vee_{j \neq j_0} (a'_{i_0 j} T \overline{x}_j)) \vee (a'_{i_0 j_0} T \overline{x}_{j_0})$$

$$= \vee_{j=1}^{n} (a'_{i_0 j} T \overline{x}_j).$$

Since only one element $a_{i_0 j_0}$ in A is changed, it is clear that if $i \neq i_0$, then $a'_{ij} = a_{ij}$, and hence $b_i = \vee_{j=1}^{n} (a_{ij} T \overline{x}_j) = \vee_{j=1}^{n} (a'_{ij} T \overline{x}_j)$. Therefore, $\overline{X} = (\overline{x}_1, \overline{x}_2, \cdots, \overline{x}_n)^T$ is a solution of equation $A' \circ_T X = B$.

Suppose $\overline{X}' = (\overline{x}'_1, \overline{x}'_2, \cdots, \overline{x}'_n)^T$ is the largest solution of $A' \circ_T X = B$.

If $a'_{i_0 j_0} \in [0, b_{i_0})$, then $(a'_{i_0 j_0})^{(0)} = 0 = a^{(0)}_{i_0 j_0}$. Since only one element $a_{i_0 j_0}$ in A is changed, it is clear that $(a'_{ij})^{(0)} = a^{(0)}_{ij}$ if $i \neq i_0$ or if $j \neq j_0$. Therefore, $(A')^{(0)} = A^{(0)}$. It follows from Theorem 2.1 that $S(A', B) = S((A')^{(0)}, B) = S(A^{(0)}, B) = S(A, B)$.

If $a'_{i_0 j_0} \in [b_{i_0}, \overline{x}_{j_0} \alpha_T b_{i_0})$, then, applying Lemma 1.2, 1.3 and 1.4, we obtain $1 = b_{i_0} \alpha_T b_{i_0} \geq a'_{i_0 j_0} \alpha_T b_{i_0} > (\overline{x}_{j_0} \alpha_T b_{i_0}) \alpha_T b_{i_0} \geq \overline{x}_{j_0}$, i.e., $a'_{i_0 j_0} \alpha_T b_{i_0} \in (\overline{x}_{j_0}, 1]$. Thus,

$$\overline{x}_{j_0} = \wedge_{i=1}^{m} (a_{ij_0} \alpha_T b_i)$$

$$= (\wedge_{i \neq i_0} (a_{ij_0} \alpha_T b_i)) \wedge (a_{i_0 j_0} \alpha_T b_{i_0})$$

$$= \wedge_{i \neq i_0} (a_{ij_0} \alpha_T b_i)$$

$$= (\wedge_{i \neq i_0} (a_{ij_0} \alpha_T b_i)) \wedge (a'_{i_0 j_0} \alpha_T b_{i_0})$$

$$= (\wedge_{i \neq i_0} (a'_{ij_0} \alpha_T b_i)) \wedge (a'_{i_0 j_0} \alpha_T b_{i_0})$$

$$= \wedge_{i=1}^{m} (a'_{ij_0} \alpha_T b_i) = \overline{x}'_{j_0}.$$

Since only $a_{i_0 j_0}$ in A is changed, it is clear that if $j \neq j_0$,t hen $\overline{x}_j = \wedge_{i=1}^{m} (a_{ij} \alpha_T b_i)$ is invariable. Therefore, $\overline{X} = \overline{X}'$. Since

$$a'_{i_0 j_0} T \overline{x}'_{j_0} = a'_{i_0 j_0} T \overline{x}_{j_0} = \overline{x}_{j_0} T a'_{i_0 j_0} < \overline{x}_{j_0} T (\overline{x}_{j_0} \alpha_T b_{i_0}) \leq b_{i_0},$$

by Definition 2.2, $(a'_{i_0 j_0})^{(1)} = 0$, and so $(a'_{i_0 j_0})^{(1)} = a^{(1)}_{i_0 j_0}$. Moreover, if $i \neq i_0$ or if $j \neq j_0$, then $a^{(1)}_{ij} = (a'_{ij})^{(1)}$. Therefore, $(A')^{(1)} = A^{(1)}$. Applying Theorem 2.2, we obtain $S(A', B) = S(A, B)$.

Therefore, by the preceding proof, $a_{i_0 j_0}$ can perturb within $[0, \overline{x}_{j_0} \alpha_T b_{i_0})$.

A noteworthy special case of Theorem 3.1 is the

Corollary 3.1. Given a fuzzy matrix equation $A \circ X=B$, if $a_{i_0 j_0} \neq 0$ and $a^{(0)}_{i_0 j_0} = a^{(1)}_{i_0 j_0} = 0$, then $a_{i_0 j_0}$ is a lower perturbation element in A. If $\overline{x}_{j_0} \leq b_{i_0}$, then it can perturb within $[0, 1)$; if $\overline{x}_{j_0} > b_{i_0}$, then it can perturb within $[0, b_{i_0} / \overline{x}_{j_0})$.

Proof. When T is the (algebraic) product P,

$$\overline{x}_{j_0} \alpha_P b_{i_0} = \begin{cases} 1, & \text{if } \overline{x}_{j_0} \leq b_{i_0} \\ b_{i_0} / \overline{x}_{j_0}, & \text{if } \overline{x}_{j_0} > b_{i_0} \end{cases}.$$

By Theorem 3.1, the Corollary evidently hold ture.

Note: Under conditions of Corollary 3.1, [5] gives the result (see Theorem 3.1 of [5]) as follows: $a_{i_0 j_0}$ is a lower perturbation element in A and it can perturb within $[0, b_{i_0})$. Obviously, Corollary 3.1 is a generalization of Theorem 3.1 of [5].

Theorem 3.2. Given a fuzzy matrix equation (2), if $a^{(0)}_{i_0 j_0} \neq 0$ and $a^{(1)}_{i_0 j_0} = 0$, then $a_{i_0 j_0}$ is a lower perturbation element in A, and it can perturb within $[0, \overline{x}_{j_0} \alpha_T b_{i_0})$.

Proof. If $a^{(0)}_{i_0 j_0} \neq 0$ and $a^{(1)}_{i_0 j_0} = 0$, then $a_{i_0 j_0} \geq b_{i_0}$ and $a_{i_0 j_0} T \overline{x}_{j_0} < b_{i_0}$, and so $a_{i_0 j_0} \neq 0$. Hence, $a_{i_0 j_0} \alpha_T b_{i_0} > \overline{x}_{j_0}$ by Corollary 1.2. $\forall a'_{i_0 j_0} \in [0, \overline{x}_{j_0} \alpha_T \ b_{i_0})$, we replace $a_{i_0 j_0}$ in A by $a'_{i_0 j_0}$, keep the other elements in A unchanged and get a new fuzzy matrix $A' = (a'_{ij})_{m \times n}$. Below, we show that the matrix equation $A' \circ_T X=B$ is solvable.

Since $a_{i_0 j_0} T \overline{x}_{j_0} < b_{i_0}$ and $a'_{i_0 j_0} T \overline{x}_{j_0} < (\overline{x}_{j_0} \alpha_T b_{i_0}) T \overline{x}_{j_0} \leq b_{i_0}$, we see that

$$\begin{aligned} b_{i_0} &= \bigvee_{j=1}^{n} (a_{i_0 j} T \overline{x}_j) \\ &= (\bigvee_{j \neq j_0} (a_{i_0 j} T \overline{x}_j)) \vee (a_{i_0 j_0} T \overline{x}_{j_0}) \\ &= \bigvee_{j \neq j_0} (a_{i_0 j} T \overline{x}_j) \\ &= (\bigvee_{j \neq j_0} (a_{i_0 j} T \overline{x}_j)) \vee (a'_{i_0 j_0} T \overline{x}_{j_0}) \\ &= (\bigvee_{j \neq j_0} (a'_{i_0 j} T \overline{x}_j)) \vee (a'_{i_0 j_0} T \overline{x}_{j_0}) \\ &= \bigvee_{j=1}^{n} (a'_{i_0 j} T \overline{x}_j). \end{aligned}$$

Since only one element $a_{i_0 j_0}$ in A is changed, it is clear that if $i \neq i_0$, then $a'_{ij} = a_{ij}$, and hence $b_i = \bigvee_{j=1}^{n}(a_{ij} T \bar{x}_j) = \bigvee_{j=1}^{n}(a'_{ij} T \bar{x}_j)$. Therefore, $\bar{X} = (\bar{x}_1, \bar{x}_2, \cdots, \bar{x}_n)^T$ is a solution of equation $A' \circ_T X = B$.

Suppose $\bar{X}' = (\bar{x}'_1, \bar{x}'_2, \cdots, \bar{x}'_n)^T$ is the largest solution of $A' \circ_T X = B$.

If $a'_{i_0 j_0} \in [0, b_{i_0})$, then $(a'_{i_0 j_0})^{(1)} = 0 = a^{(1)}_{i_0 j_0}$ and $a'_{i_0 j_0} \alpha_T b_{i_0} = 1$. Thus,

$$\begin{aligned}
\bar{x}_{j_0} &= \bigwedge_{i=1}^{m}(a_{ij_0} \alpha_T b_i) \\
&= (\bigwedge_{i \neq i_0}(a_{ij_0} \alpha_T b_i)) \wedge (a_{i_0 j_0} \alpha_T b_{i_0}) \\
&= \bigwedge_{i \neq i_0}(a_{ij_0} \alpha_T b_i) \\
&= (\bigwedge_{i \neq i_0}(a_{ij_0} \alpha_T b_i)) \wedge (a'_{i_0 j_0} \alpha_T b_{i_0}) \\
&= (\bigwedge_{i \neq i_0}(a'_{ij_0} \alpha_T b_i)) \wedge (a'_{i_0 j_0} \alpha_T b_{i_0}) \\
&= \bigwedge_{i=1}^{m}(a'_{ij_0} \alpha_T b_i) = \bar{x}'_{j_0}.
\end{aligned}$$

Since only one element $a_{i_0 j_0}$ is changed, it is clear that if $j \neq j_0$, then $\bar{x}_j = \bigwedge_{i=1}^{m}(a_{ij} \alpha_T b_i)$ is invariable. Therefore, $\bar{X} = \bar{X}'$. Moreover, if $i \neq i_0$ or if $j \neq j_0$, then $a^{(1)}_{ij} = (a'_{ij})^{(1)}$. Therefore, $(A')^{(1)} = A^{(1)}$. Applying Theorem 2.2, we obtain $S(A',B)=S(A,B)$.

If $a'_{i_0 j_0} \in [b_{i_0}, \bar{x}_{j_0} \alpha_T b_{i_0})$, then, applying Lemma 1.2, 1.3 and 1.4, we obtain $1 = b_{i_0} \alpha_T b_{i_0} \geq a'_{i_0 j_0} \alpha_T b_{i_0} > (\bar{x}_{j_0} \alpha_T b_{i_0}) \alpha_T b_{i_0} \geq \bar{x}_{j_0}$, i.e., $a'_{i_0 j_0} \alpha_T b_{i_0} \in (\bar{x}_{j_0}, 1]$. Thus,

$$\begin{aligned}
\bar{x}_{j_0} &= \bigwedge_{i=1}^{m}(a_{ij_0} \alpha_T b_i) \\
&= (\bigwedge_{i \neq i_0}(a_{ij_0} \alpha_T b_i)) \wedge (a_{i_0 j_0} \alpha_T b_{i_0}) \\
&= \bigwedge_{i \neq i_0}(a_{ij_0} \alpha_T b_i) \\
&= (\bigwedge_{i \neq i_0}(a_{ij_0} \alpha_T b_i)) \wedge (a'_{i_0 j_0} \alpha_T b_{i_0}) \\
&= (\bigwedge_{i \neq i_0}(a'_{ij_0} \alpha_T b_i)) \wedge (a'_{i_0 j_0} \alpha_T b_{i_0}) \\
&= \bigwedge_{i=1}^{m}(a'_{ij_0} \alpha_T b_i) = \bar{x}'_{j_0}.
\end{aligned}$$

Since only one element $a_{i_0j_0}$ is changed, it is clear that if $j \neq j_0$, then $\bar{x}_j = \bigwedge_{i=1}^{m}(a_{ij}\alpha_T b_i)$ is invariable. Therefore, $\bar{X} = \bar{X}'$. Since $a'_{i_0j_0} \geq b_{i_0}$ and $a'_{i_0j_0}T\bar{x}'_{j_0} = a'_{i_0j_0}T\bar{x}_{j_0} < (\bar{x}_{j_0}\alpha_T b_{i_0})T\bar{x}_{j_0} \leq b_{i_0}$, $(a'_{i_0j_0})^{(1)} = 0 = a^{(1)}_{i_0j_0}$. Moreover, if $i \neq i_0$ or if $j \neq j_0$, then $a^{(1)}_{ij} = (a'_{ij})^{(1)}$. Therefore, $(A')^{(1)}=A^{(1)}$. Applying Theorem 2.2, we obtain $S(A',B)=S(A,B)$.

Therefore, by the preceding proof, $a_{i_0j_0}$ can perturb within $[0, \bar{x}_{j_0}\alpha_T b_{i_0})$.

A noteworthy special case of Theorem 3.2 is the

Corollary 3.2. Given a fuzzy matrix equation $A \circ X=B$, if $a^{(0)}_{i_0j_0} \neq 0$ and $a^{(1)}_{i_0j_0} = 0$, i.e., $a_{i_0j_0} \geq b_{i_0}$ and $a_{i_0j_0}T\bar{x}_{j_0} < b_{i_0}$, then $a_{i_0j_0}$ is a lower perturbation element in A. If $\bar{x}_{j_0} \leq b_{i_0}$, then it can perturb within $[0, 1)$; if $\bar{x}_{j_0} > b_{i_0}$, then it can perturb within $[0, b_{i_0} / \bar{x}_{j_0})$.

Note: Under conditions of Corollary 3.2, [5] gives the result (see Theorem 3.2 of [5]) as follows: If $a_{i_0j_0} = b_{i_0}$, then $a_{i_0j_0}$ can perturb within $[0, b_{i_0}]$; if $a_{i_0j_0} > b_{i_0}$, then $a_{i_0j_0}$ can perturb within $[b_{i_0}, b_{i_0} / \bar{x}_{j_0})$. Obviously, Corollary 3.2 is a generalization of Theorem 3.2 of [5].

Theorem 3.3. Given a fuzzy matrix equation (2), $a_{i_0j_0}$ can not perturb within $(\bar{x}_{j_0}\alpha_T b_{i_0}, 1]$.

Proof. Assume that $a_{i_0j_0}$ can perturb within $(\bar{x}_{j_0}\alpha_T b_{i_0}, 1]$. $\forall a'_{i_0j_0} \in (\bar{x}_{j_0}\alpha_T b_{i_0}, 1]$, we replace $a_{i_0j_0}$ in A by $a'_{i_0j_0}$, keep the other elements in A unchanged and get a new fuzzy matrix $A' = (a'_{ij})_{m \times n}$. If the equation $A' \circ_T X=B$ has no solution, then the proposition evidently hold ture. If the equation $A' \circ_T X=B$ is solvable, suppose $\bar{X}' = (\bar{x}'_1, \bar{x}'_2, \cdots, \bar{x}'_n)^T$ is its largest solution. Since $a'_{i_0j_0} > \bar{x}_{j_0}\alpha_T b_{i_0}$, according to the definition of α_T, $\bar{x}_{j_0}Ta'_{i_0j_0} > b_{i_0}$, $a'_{i_0j_0}\alpha_T b_{i_0} < \bar{x}_{j_0}$, and hence, $\bar{x}'_{j_0} \leq a'_{i_0j_0}\alpha_T b_{i_0} < \bar{x}_{j_0}$. Therefore, $\bar{X}' \neq \bar{X}$. This completes the proof.

Acknowledgement

The project was supported by the National Natural Science Foundation of China (10471083) and the Natural Science Foundation of Fujian Province of China (A0510023).

References

1. Li SL, Yu YD, Wang ZD (1997) *T*-congruence *L*-relations on groups and rings. Fuzzy Sets and Systems 92: 365-381
2. Tang FC (2000) Perturbation techniques for fuzzy matrix equations. Fuzzy Sets and Systems 109: 363-369
3. Wang ZD (1994) On *L*-subsets and *TL*-subalgebras. Fuzzy Sets and Systems 65: 59-69
4. Wang ZD (2004) *T*-type regular *L*-relations on a complete Brouwerian lattice. Fuzzy Sets and Systems 145: 313-322
5. Zhang CY, Dang PA (2002) On the perturbation of fuzzy matrix equations with max-product composition. Fuzzy Systems and Mathematics 16(4): 53-56

Normal Distribution Fuzzy Sets

Zehua Lv[1], Chuanbo Chen[2], and Wenhai Li[1]

[1] College of Computer Science & Technology, Huazhong University of Science and Technology, Wuhan 430074, China
lzhhust@163.com, lwhaymail@21cn.com

[2] School of Software Engineering, Huazhong University of Science and Technology, Wuhan 430074, China
chuanboc@163.com

Abstract. Owing to the theory of intuitionistic fuzzy sets for fuzzy sets generalization extends the membership degree from a single value in [0, 1] to a subinterval in [0, 1], it brings forward two questions: 1. Whether all the values in the subinterval have the same probabilities as the membership degree or not? 2. If the probabilities in the subinterval are different, which kind of distribution will they be? In this paper, a method for expressing an intuitionistic fuzzy set by a series of normal distribution functions has been presented according to the investigation of vote model. The theory of normal distribution fuzzy sets is established. This theory solve the problems existing in intuitionistic fuzzy sets, the probability distribution of membership degree in [0, 1] can be clearly recognized. The notion of inclusion, union, intersection, and complement extending to such sets and the properties of normal distribution fuzzy sets are discussed in detail. The relationship among fuzzy sets, intuitionistic fuzzy sets and normal distribution fuzzy sets is specified.

Keywords: Fuzzy sets, intuitionistic fuzzy sets, normal distribution function.

1 Introduction

Since Zadeh introduced fuzzy sets in 1965 [1], many new approaches and theories treating imprecision and uncertainty have been proposed. Some of these theories are extensions of fuzzy sets; others try to handle imprecision and uncertainty in different or better ways. Such as intuitionistic fuzzy sets [2], two-fold fuzzy sets [19], L-fuzzy sets [20], L-intuitionistic fuzzy sets [24], interval-valued fuzzy sets [19, 25], interval valued intuitionistic fuzzy sets [24], vague sets [3], etc. Many scholars [21, 22, 23] have discussed the relationship of these theories from different point of views.

The theory of intuitionistic fuzzy sets proposed by Atanassov is characterized by two functions expressing the degree of belongingness and the degree of no belongingness, respectively. This idea is a natural generalization of fuzzy set. It seems to be useful in many general soft sciences, such as social science, linguistic, psychology, economics, etc. In 1993 Gau and Buehere proposed vague set [3], Bustince and Burillo [4] pointed out the notion of vague sets coinciding with that of intuitionistic fuzzy sets proposed by Atanassov. Other well-known

B.-Y. Cao (Ed.): Fuzzy Information and Engineering (ICFIE), ASC 40, pp. 280–289, 2007.
springerlink.com

generalization of an ordinary fuzzy set is the so-called interval-valued fuzzy set. Generally, the idea of interval-valued fuzzy set was attributed to Gorzalczany [5] and Turksen [6], but actually they appear earlier in the papers [7, 8, 9]. Sometimes, these approaches are even mathematically equivalent, the only difference is that they have arisen on different ground and they have different semantics.

At the same time, the properties and applications of fuzzy sets generalization are also investigated widely. Szmidt and Kacprzyk [10] proposed a no-probabilistic type of entropy measure for intuitionistic fuzzy sets. De et al. [11] studied Sanchez's approach for medical diagnosis and extended this concept with the notion of intuitionistic fuzzy set theory. Turanli and Coker [12] introduced several types of fuzzy connectedness in intuitionistic fuzzy topological spaces. Szmidt and Kacprzyk [13] and Weiqiong Wang [27] discussed distances between intuitionistic fuzzy sets. Bustince [14] presented different theorems to build intuitionistic fuzzy relations on a set with predetermined properties. Moreover, the similarity measure between intuitionistic fuzzy sets is another hotspot, which is very important in pattern recognition research. Shyi-Ming Cheng [26] proposed two similarity measures for measuring the degree of similarity between vague sets. Li and Cheng [15] studied similarity measures of intuitionistic fuzzy sets and applied to pattern recognitions. Liang [16] and Mitchell [17] introduced other kinds of similarity measures to avoid the drawbacks proposed by Li and Cheng. Huang Yang [18] proposed another kind of similarity measures of intuitionistic of fuzzy sets based on Hausdorff distance, and many examples verified that this kind of similarity measure is very simple and suited well to be used into linguistic variables. Przemyslaw Grzegorzewski [28] proposed a new method of measuring distances between intuitionistic fuzzy sets based on the Hausdorff metric.

In this paper, according to reviewing the evolution of fuzzy theory, we present the brand new theory of normal distribution fuzzy sets. The rest of this paper is organized as follows: In section , the conception of normal distribution fuzzy sets is given; in section , the notions of inclusion, union, intersection and complement are extended to normal distribution fuzzy sets, and the properties are established; in section , the relationship among fuzzy sets, intuitionistic fuzzy sets and normal distribution fuzzy sets is discussed; conclusion is drawn in section .

2 Normal Distribution Fuzzy Sets

According to reviewing the evolutive process of fuzzy theory, we find that intuitionistic fuzzy set is the most typical theory among those fuzzy set generalizations. Therefore, we use the theory of intuitionistic fuzzy sets as the representative to discuss the evolutive process of fuzzy theory in this paper. The main difference between fuzzy sets and classical crisp sets is that the fuzzy sets have not clear boundaries, we can not distinguish an object subjecting to a fuzzy set or not, but we know the membership degree. Intuitionistic fuzzy sets are more vague, the characteristic is that not only whether an object belonging to an intuitionistic fuzzy set is not known, but also the membership degree is not precise, we only know the membership degree is in a subinterval of [0,1]. For example, assume an intuitionistic fuzzy set $B = \{(x_1, 0.2, 0.6)\}$, we merely know that the degree of x_1 belonging to B is in the subinterval $[0.2, 0.4]$. In other words, the inevitability of 1 belonging to B is 0.2 and the possibility is 0.4. Therefore, we will naturally consider the following two questions: 1. whether all the values in the subinterval have the same probability as the membership degree of x_1?2. If the values' probabilities as membership in the subinterval are not the same, which kind of distribution will they be?

Despite of great differences existing between intuitionistic fuzzy sets and fuzzy sets, intuitionistic fuzzy sets are generalization of fuzzy sets in substance. Therefore, we can use some kinds of method to translate an intuitionistic fuzzy set into a fuzzy set.

Firstly, let us take example for the vote model, assume an intuitionistic fuzzy set $\{(x, 0.3, 0.3)\}$, it can be interpreted as "the vote for a resolution is 3 in favor 3 against and 4 abstentions", how to draw a conclusion to this vote? The general viewpoint is that the finally result is 0.5, or take an average of 0.3 and 0.7. Therefore, we can take $[t(x) + 1 - \gamma(x)]/2$ as the degree of x subjecting to the set. Another interpretation is: in the vote model, we evaluate the assenters 1, the objectors 0 and the neutrals 0.5, and then standardize the result by dividing the total number of voters. For example, if 3 voters sustain, 3 voters object and 4 voters disclaim, the result should be $(3 \times 1 + 4 \times 0.5 + 3 \times 0)/10 = 0.5$, if 4 voters sustain, 2 voters object and 4 voters disclaim, the result is$(4 \times 1 + 4 \times 0.5 + 2 \times 0)/10 = 0.6$.

Secondly, let us consider a more detailed problem in intuitionistic fuzzy set. An intuitionistic set A is characterized by a truth-membership function $t_A(x_i)$ and a false-membership of x_i derived from the evidence for x_i, $\gamma_A(x_i)$ is a lower bound on the degree of membership of x_i derived from the evidence against x_i. These lower bounds can produce a subinterval of $[0, 1]$, denoted by $[t_A(x_i),\ 1-\gamma_A(x_i)]$. In other words, the membership degree of x_i belonging to A is in the interval $[t_A(x_i), 1 - \gamma_A(x_i)]$, we want to know which value is equal to the membership degree of x_i best of all subjecting to A in the interval. The general idea is that the median value, $\mu = [t_A(x_i) + 1 - \gamma_A(x_i)]/2$, has the largest probability, the probabilities of other values in the subinterval satisfy this rule: With the distance away from μ becomes bigger, the probability becomes smaller. Therefore, we naturally think that the probability of the values in the subinterval accord with normal distribution, and then we can use a normal distribution function substituting $[t_A(x_i),\ 1 - \gamma_A(x_i)]$ to express the membership degree of x_i.

In the rest of this section, we firstly introduce a method to express the membership degrees of elements in an intuitionistic fuzzy set by a series of normal distribution functions. Let A be an intuitionistic fuzzy set in the universe $X = \{x_1,\ x_2,\ \cdots,\ x_n\}$, $A = \{< x_i,\ t_A(x_i),\ \gamma_A(x_i) > | x_i \in X\}$. Then, we use the set A_N to express it as follows:

$$A_N = \{< x_i,\ \varphi_i(z) > | x_i \in X\} \tag{1}$$

Here, $\varphi_i(z)$ is a normal distribution function and $\varphi_i(z) \sim N(\mu_i,\ \sigma_i)$.

$$\mu_i = E(\varphi_i(z)) = [t_A(x_i) + 1 - \gamma_A(x_i)]/2 \tag{2}$$

$$\sigma_i = D(\varphi_i(z)) = \frac{1}{\sqrt{2\pi}(1 - \|I_A(x_i)\|/2)} \tag{3}$$

Here $\|I_A(x_i)\| = 1 - \gamma_A(x_i) - t_A(x_i)$. $E(\varphi_i(z))$ denotes the expectation of the normal distribution function $\varphi_i(z)$, σ_i denotes the variance of the normal distribution function $\varphi_i(z)$. With the methods mentioned above, we can express the membership subinterval by a normal distribution function. Next, we will give the definition of normal distribution fuzzy sets.

Definition 1. *(Normal Distribution Sets, NFSs) Let X be a space of objects, with the generic element of X denoted by x_i, a normal distribution fuzzy set A_N in X is characterized by a set of normal distribution functions $\{\varphi_i(z)\}$, for each $\varphi_i(z)$, it satisfy the conditions listed below*

$$\mu_i = E(\varphi_i(z)) \in [0,\ 1], \quad \sigma_i = D(\varphi_i(z)) \geq 1\Big/\sqrt{2\pi} \tag{4}$$

If X is discrete, a normal distribution fuzzy set A_N can be written as $A_N = \sum_{i=1}^{n} \varphi_i(z)/x_i$; if X is continuous, a normal distribution fuzzy set A_N can be written as $A_N = \int \varphi_i(z)/x_i$

In this paper, we mainly discuss the normal distribution fuzzy sets in the discrete universe. We can express it by Fig.1.

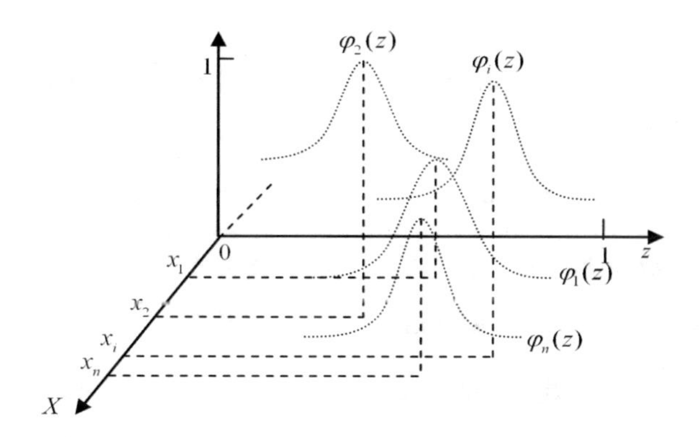

Fig. 1. A normal distribution fuzzy set

Because the normal distribution function is decided by its expectation and variance, we can denote the normal distribution fuzzy sets A_N in X as follows

$$A_N = \{(x_i,\ \mu_i,\ \sigma_i)|x_i \in X\} \tag{5}$$

The null set ϕ_N is defined as $\phi_N = \{(x_i, 0, 1/\sqrt{2\pi})|x_i \in X\}$, the universal set is defined as $\phi_N = \{(x_i, 1, 1/\sqrt{2\pi})|x_i \in X\}$.

Definition 2. *The complement of a normal distribution fuzzy set is denoted by* $\bar{A}_N$ *and defined by*

$$\bar{A}_N = \sum_{i=1}^{n} \varphi_i'(z)/x_i \tag{6}$$

Here $\mu_i' = E(\varphi_i'(z)) = 1 - E(\varphi_i(z))$,$\sigma_i' = D(\varphi_i'(z)) = D(\varphi_i(z))$,$\bar{A}_N$ *can also be denoted as follows*

$$\bar{A}_N = \{(x_i,\ 1 - \mu_i,\ \sigma_i)|x_i \in X\} \tag{7}$$

Definition 3. *(Containment)A normal distribution fuzzy set* $A_N = \{(x_i,\ \mu_i^A,\ \sigma_i^A)|x_i \in X\}$ *is contained in the other normal distribution fuzzy set* $B_N = \{(x_i,\ \mu_i^B,\ \sigma_i^B)|x_i \in X\}$, $A_N \subseteq B_N$ *if and only if* $\forall\ i$, $\mu_i^A \leq \mu_i^B$, $\sigma_i^A \geq \sigma_i^B$.

Definition 4. *(Union)*A_N *and* B_N *are two normal distribution fuzzy sets in* X,$A_N = \{(x_i,\ \mu_i^A,\ \sigma_i^A)|x_i \in X\}$, $B_N = \{(x_i,\ \mu_i^B,\ \sigma_i^B)|x_i \in X\}$, *the union is a normal distribution fuzzy set, written as* $C_N = A_N \cup B_N$,C_N *is defined by*

$$\mu_i^C = \max(\mu_i^A,\ \mu_i^B), \quad \sigma_i^C = \min(\sigma_i^A,\ \sigma_i^B) \tag{8}$$

Definition 5. *(Intersection)*A_N *and* B_N *are two normal distribution fuzzy sets in* X,$A_N = \{(x_i,\ \mu_i^A,\ \sigma_i^A)|x_i \in X\}$, $B_N = \{(x_i,\ \mu_i^B,\ \sigma_i^B)|x_i \in X\}$, *the union is a normal distribution fuzzy set, written as* $C_N = A_N \cap B_N$,C_N *is defined by*

$$\mu_i^c = \min(\mu_i^A,\ \mu_i^B), \quad \sigma_i^c = \max(\sigma_i^A,\ \sigma_i^B) \tag{9}$$

3 Some Properties of Union, Complementation and Intersection

With the operations of union, intersection and complementation defined in section, it is easy to extend many of the basic identities which hold for ordinary sets to normal distribution fuzzy sets. The main properties are listed as follows.

Property 1. Commutativity: Assume $A_N, B_N \in NFSs$

$$A_N \cup B_N = B_N \cup A_N, \quad A_N \cap B_N = B_N \cap A_N \tag{10}$$

Proof. Follows quickly from the commutativity of min and max.

Property 2. Associativity: Assume $A_N, B_N, C_N \in NFSs$

$$A_N \cup (B_N \cup C_N) = (A_N \cup B_N) \cup C_N \tag{11}$$

$$A_N \cap (B_N \cap C_N) = (A_N \cap B_N) \cap C_N \tag{12}$$

Proof. Follows quickly from the associability of min and max.

Property 3. Involution: Assume $A_N \in NFSs$, $\bar{\bar{A}}_N$=A_N

Proof. Suppose $A_N = \{(x_i,\ \mu_i^A,\ \sigma_i^A)|x_i \in X\}$
from definition2, we have $\overline{A}_N$=$\{(x_i,\ 1-\mu_i^A,\ \sigma_i^A)|x_i \in X\}$
$\overline{\overline{A}}_N$=$\{(x_i,\ 1-(1-\mu_i^A),\ \sigma_i^A)|x_i \in X\}$
$= \{(x_i,\ \mu_i^A,\ \sigma_i^A)|x_i \in X\} = A$

Property 4. Idempotence: Assume $A_N \in NFSs$

$$A_N \cup A_N = A_N, \quad A_N \cap A_N = A_N \tag{13}$$

Proof. Follows the idempotence of min and max.

Property 5. Distributive: Assume $A_N, B_N, C_N \in NFSs$

$$A_N \cup (B_N \cap C_N) = (A_N \cup B_N) \cap (A_N \cup C_N) \tag{14}$$

$$A_N \cap (B_N \cup C_N) = (A_N \cap B_N) \cup (A_N \cap C_N) \tag{15}$$

Proof. Follows the distributivity of min and max.

Property 6. Assume $A_N \in NFSs$, $A_N \cap \phi_N = \phi_N$,$A_N \cup \phi_N = A_N$,$A_N \cap X_N = A_N$,$A_N \cup X_N = X_N$, where $\phi_N = \{(x_i,\ 0,\ 1/\sqrt{2\pi})|x_i \in X\}$,$X_N = \{(x_i,\ 1,\ 1/\sqrt{2\pi})|x_i \in X\}$

Proof. Suppose $A_N = \{(x_i,\ \mu_i^A,\ \sigma_i^A)|x_i \in X\}$,let $B_N = A_N \cap \phi$.Then, $\mu_i^B = \min(0,\ \mu_i^A) = 0$, $\sigma_i^B = \max(+\infty,\ \sigma_i^A) = +\infty$,obviously $B_N = \phi_N$.Let $C_N = A_N \cup X_N$, then $\mu_i^c = \max(1,\ \mu_i^A) = 1$, $\sigma_i^c = \min(1/\sqrt{2\pi},\ \mu_i^A) = 1/\sqrt{2\pi}$. Then, we get $C_N = X_N$. Other equations can be proved similarly.

Property 7. Absorption: Assume $A_N, B_N \in NFSs$, $A_N \cup (A_N \cap B_N) = A_N$, $A_N \cap (A_N \cup B_N) = A_N$

Proof. $\mu_i^{A\cap B} = \min(\mu_i^A,\ \mu_i^B)$, $\mu_i^{A\cup(A\cap B)} = \max(\mu_i^A,\ \min(\mu_i^A,\ \mu_i^B)) = \mu_i^A$, $\sigma_i^{A\cap B} = \max(\sigma_i^A,\ \sigma_i^B)$, $\sigma_i^{A\cup(A\cap B)} = \min(\sigma_i^A,\ \max(\sigma_i^A,\ \sigma_i^B)) = \sigma_i^A$, then, $A_N \cup (A_N \cap B_N) = A_N$. $A_N \cap (A_N \cup B_N) = A_N$can be proved similarly.

Property 8. DeMorgan's: Assume $A_N, B_N \in NFSs$, $\overline{A_N \cup B_N} = \bar{A}_N \cap \bar{B}_N$,$\overline{A_N \cap B_N} = \bar{A}_N \cup \bar{B}_N$

Proof. $\mu_i^{A\cup B} = \max(\mu_i^A,\ \mu_i^B)$, $\mu_i^{\overline{A\cup B}} = 1-\max(\mu_i^A,\ \mu_i^B)$, $\mu_i^{\overline{A}} = 1-\mu_i^A$, $\mu_i^{\overline{B}} = 1-\mu_i^B$, $\mu_i^{\bar{A}\cap\bar{B}} = \min(\mu_i^{\bar{A}},\mu_i^{\bar{B}}) =, \min(1-\mu_i^A, 1-\mu_i^B) = 1-\max(\mu_i^A,\mu_i^B)$. Then, $\overline{A_N \cup B_N} = \bar{A}_N \cap \bar{B}_N$. $\overline{A_N \cap B_N} = \bar{A}_N \cup \bar{B}_N$ can be proved similarly.

4 The Relationship Among Fuzzy Sets, Intuitionistic Fuzzy Sets and Normal Distribution Fuzzy Sets

Normal distribution fuzzy sets are the extension of intuitionistic fuzzy sets. Therefore, we mainly discuss the relationship among fuzzy sets, intuitionistic fuzzy sets and normal distribution fuzzy sets in this section.

Lemma 1. *Intuitionistic fuzzy sets and fuzzy sets are the $\alpha-cut$ of normal distribution sets.*

Let A_N is a normal distribution fuzzy set in the universe of X, $A_N = \sum_{i=1}^{n} \varphi_i(z)/x_i$, $\alpha \in [0,\ 1]$, then the $\alpha - cut$ of A_N is defined as follows: Suppose $\varphi_i(z_{i1}) = \varphi_i(z_{i2}) = \alpha$ and $z_{i1} \leq z_{i2}$ the $\alpha - cut$ of A_N denoted as $A_N^{\alpha} = \{(x_i, z_{i1}, 1 - z_{i2}) | x_i \in X\}$ for convenience we use the form of vague sets to express A_N^{α}

$$A_N^{\alpha} = \sum_{i=1}^{n} [z_{i1},\ z_{i2}]/x_i, \qquad x_i \in X \tag{16}$$

We can show the transformation of element x_i in Fig.2.

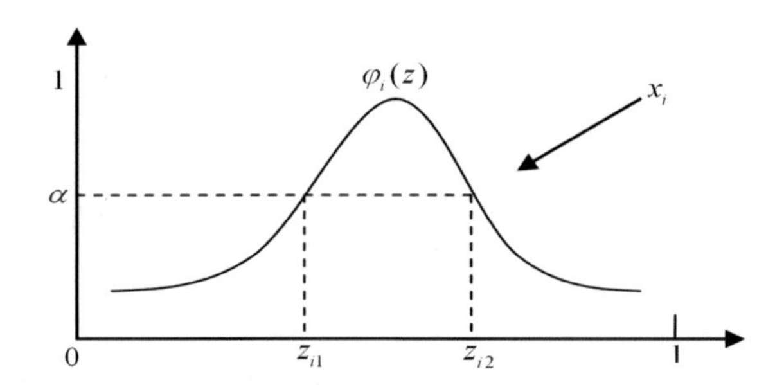

Fig. 2. Transformation form $NDFs$ to IFs

If $z_{i1} \leq 0$, the interval $[z_{i1},\ z_{i2}]$ is substituted by $[0,\ z_{i2}]$, if $z_{i2} \geq 1$, the interval $[z_{i1},\ z_{i2}]$ is substituted by $[z_{i1},\ 1]$.

While transforming a normal distribution fuzzy set to an intuitionistic fuzzy set, we can use the same level α for each element or use different level α for different element. Therefore, an intuitionistic fuzzy set is decided by three parameters, μ_i, σ_i and α_i . We can denote an intuitionistic fuzzy set A as follows:

$$A = A_N^{\alpha} = \{(x_i,\ \mu_i,\ \sigma_i,\ \alpha_i) | x_i \in X\} \tag{17}$$

If for each α_i, $z_{i1} = z_{i2}$, the interval $[z_{i1},\ z_{i2}]$ degenerates to one point, denotes μ_i ,and it satisfy $\varphi(\mu_i) = \alpha_i$. Then, the $\alpha - cut$ of the normal distribution fuzzy set becomes a fuzzy set.

If the level $\alpha_i = \varphi(\mu_i) = 1$, it satisfy both $\sigma_i = 1/\sqrt{2\pi}$ and $\varphi_i(z) = \exp(-\pi(z - \mu_i)^2)$. Here we call the element x_i normal element, if each element in A_N^{α} is normal element, we call the set A_N^{α} normal fuzzy set. Then, the normal fuzzy set can denote as follows

$$A_N^1 = \{(x_i,\ \mu_i,\ 1\big/\sqrt{2\pi},\ 1) | x_i \in X\} \tag{18}$$

If $\alpha_i = \varphi_i(\mu_i) < 1$, we call element x_i non-normal element, if there is one non-normal element in A_N^{α} , A_N^{α} is a non-normal fuzzy set, denoted as follows:

$$A_N = \{(x_i,\ \mu_i,\ \sigma_i,\ \varphi_i(\mu_i)) | x_i \in X\} \tag{19}$$

The normal element x_i and non-normal element x_j are shown in Fig.3.

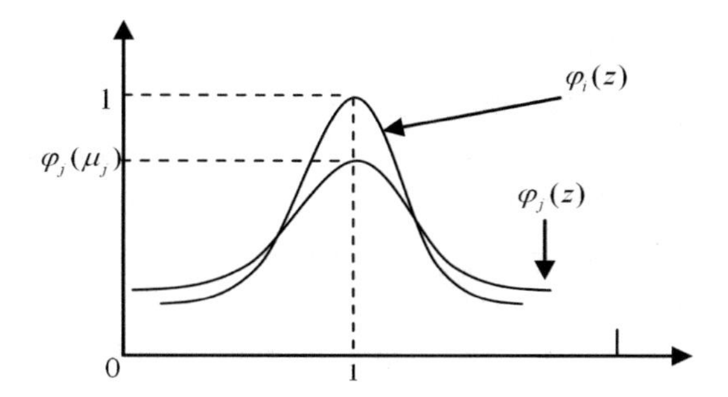

Fig. 3. Normal element x_i and non-normal element x_j

Example 1. Let $X=\{x_1,\ x_2,\ x_3,\ x_4, x_5\}$, $X_N = \{(x_1, 1, 1/\sqrt{2\pi}), (x_2, 1, 1/\sqrt{2\pi}), (x_3, 1, 1/\sqrt{2\pi}),\ (x_4, 1, 1/\sqrt{2\pi}),\ (x_5, 1, 1/\sqrt{2\pi})\}$, $\phi_N=\{(x_1, 0, +\infty),(x_2, 0, +\infty), (x_3, 0, +\infty), (x_4, 0, +\infty), (x_5, 0, +\infty)\}$, $A_N=\{(x_1, 0.2, 2),(x_2, 0.5, 1), (x_3, 0.7, 0.8), (x_4, 0.8, 0.5), (x_5, 0.9, 1/\sqrt{2\pi})\}$, A_N is shown in Fig.4, in order to compare the difference, we draw the five normal distribution function curves under one coordinate system. Note that these curves are only abridged general views, which are used to show the relationship among these functions.

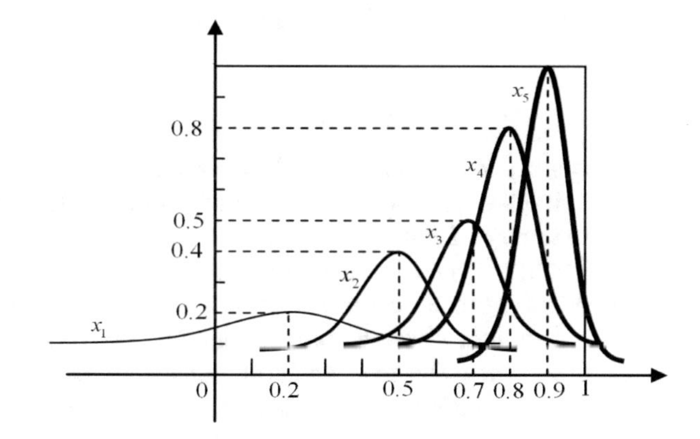

Fig. 4. Abridged general view of A_N

We now discuss the $\alpha - cut$ of A_N for different level α , we firstly use the fixed level, when $\alpha = 0.4$.

For element x_1: the corresponding function is $\varphi(z) = \frac{1}{2\sqrt{2\pi}} \exp(-\frac{(x-0.2)^2}{8})$, therefore, $0.4 = \frac{1}{2\sqrt{2\pi}} \exp(-\frac{(x-0.2)^2}{8})$. This equation has no roots, so the element x_1 is not in $A_N^{0.4}$.

For element x_2 : the corresponding function is $\varphi(z) = \frac{1}{\sqrt{2\pi}} \exp(-\frac{(z-0.5)^2}{2})$, therefore, $0.4 = \frac{1}{\sqrt{2\pi}} \exp(-\frac{(z-0.5)^2}{2})$. The roots are $z_1 = z_2 = 0.5$, then the membership interval of x_2 is [0.5,0.5] under the level 0.4.

For the element x_3: the corresponding function is $\varphi(z)=\frac{1}{0.8\times\sqrt{2\pi}}\exp(-\frac{(z-0.7)^2}{2\times 0.8^2})$, therefore, $0.4 = \frac{1}{0.8\times\sqrt{2\pi}}\exp(-\frac{(z-0.7)^2}{2\times 0.8^2})$. The roots are $z_1 = 0.17$, $z_2 = 1.23$, because $z_2 > 1$, the membership interval of element x_3 is [0.17,1].

For the element x_4: the corresponding function is $\varphi(z)=\frac{1}{0.5\times\sqrt{2\pi}}\exp(-\frac{(x-0.8)^2}{2\times 0.5^2})$, therefore, $0.4 = \frac{1}{0.5\times\sqrt{2\pi}}\exp(-\frac{(x-0.8)^2}{2\times 0.5^2})$. The roots are $z_1 = 0.21$, $z_2 = 1.39$, then the membership interval of x_4 is [0.211] under this level.

For the element x_5 : the corresponding function is $\varphi(z) = \exp(-\frac{(z-0.9)^2}{2\times(1/\sqrt{2\pi})^2})$, therefore, $0.4 = \exp(-\frac{(z-0.9)^2}{2\times(1/\sqrt{2\pi})^2})$, the roots are $z_1 = 0.36$,$z_2 = 1.44$, and the membership interval of element x_5 is [0.361] under the level α 0.4. So the $\alpha - cut$ of A_N under the level of $\alpha 0.4$ is

$$A_N^{0.4} = \{(x_2, [0.5, 0.5]), (x_3, [0.17, 1]), (x_4, [0.21, 1]), (x_5, [0.36, 1])\} \tag{20}$$

Let $\alpha = 1$, $A_N^{\alpha} = 0.95/x_5$, the element x_5 is a normal element, this $\alpha - cut$ of A_N is a normal fuzzy set.

We can also set different level for different element, if $\alpha_1 = 0.2$, $\alpha_2 = 0.4$, $\alpha_3 = 0.5$, $\alpha_4 = 0.8$, $\alpha_5 = 1$, we can get a non-normal fuzzy set

$$A_N^{\alpha i} = 0.2/x_1 + 0.5/x_2 + 0.7/x_3 + 0.8/x_4 + 0.9/x_5 \tag{21}$$

5 Conclusion

Intuitionistic fuzzy sets are generalization of fuzzy sets, they extend the membership degree from a single value in [0, 1] to a subinterval in [0, 1]. In this paper, we have presented the theory of normal distribution fuzzy sets according to the investigation of vote mode, which is an extension of intuitionistic fuzzy sets. This theory can solve the problem well existing in intuitionistic fuzzy sets. The notions of inclusion, union, intersection, and complement have been extended to normal distribution fuzzy sets. The relationship among fuzzy sets, intuitionistic fuzzy sets and normal distribution fuzzy sets is discussed. In a word, the normal distribution fuzzy sets are generalization of intuitionistic fuzzy sets, which proposes a new method to deal with fuzzy information.

References

1. L.A. Zadeh(1965) Fuzzy sets, Inform. Control 8 338-353
2. K. Atanassov(1986)Intuitionistic fuzzy sets, Fuzzy Sets and Systems 20: 87-96
3. Gau, W.L., Buehere, D.J.(1993) Vague sets. IEEE Trans. Systems Man Cybernet. 23 (2): 610-614
4. P. Burillo, H. Bustince(1996) Vague sets are intuitionistic fuzzy sets, Fuzzy Sets and Systems 79: 403-405
5. B. Gorzalczany(1983) Approximate inference with interval-valued fuzzy sets-an outline, in: Proc. Polish Symp. on Interval and Fuzzy Mathematics, Poznan,pp. 89-95

6. B. Turksen (1986) Interval valued fuzzy sets based on normal forms, Fuzzy Sets and Systems 20: 191-210
7. I. Grattan-Guiness(1975)Fuzzy membership mapped onto interval and many-valued quantities, Z. Math. Logic Grundladen Math. 22:149-160
8. K.U. Jakn(1975)Intervall-wertige Mengen, Math. Nach. 68:115-132
9. R. Sambuc(1975)Fonctions phi-:oues, Application a l'Aide au diagnostic en pathologie thyroidienne, Ph.D. thesis, University of Marseille, France.
10. E. Szmidt, J. Kacprzyk(2001) Entropy for intuitionistic fuzzy sets, Fuzzy Sets and Systems 118: 467-477
11. S.K. De, R. Biswas, A.R. Roy(2001) An application of intuitionistic fuzzy sets in medical diagnosis, Fuzzy Sets and Systems117:209-213
12. N. Turanli, D. Coker(2000) Fuzzy connectedness in intuitionistic fuzzy topological spaces, Fuzzy Sets and Systems 116:369-375
13. E. Szmidt, J. Kacprzyk(2001) Distances between intuitionistic fuzzy sets, Fuzzy Sets and Systems 114:505 - 518
14. H. Bustince(2000) Construction of intuitionistic fuzzy relations with predetermined properties, Fuzzy Sets and Systems 109:379-403
15. Li, D., Cheng(2002) new similarity measures of intuitionistic fuzzy sets and application to pattern tecognition. Pattern Recognition Lett. 23:221-225
16. Liang, Z., Shi, P.(2003)Similarity measures on intuitionistic fuzzy stes. Pattern Recognition Lett. 24: 2687-2693
17. Mitchell, H.B.(2003)On the Dengfeng-Chuntian similarity measure and its application to pattern recognition. Pattern recognition Lett., 24:3101-3104
18. Wen-Liang Hung, Miin-Shen Yang(2004)Similarity measures of intuitionistic fuzzy sets based on Hausdorff distance. Pattern Recognition Lett., 25:1603-1611
19. D.Dubois, W. Ostaniewica, H. Prade(2000) Fuzzy sets: history and basic notions, in: D.Dubois, H. Prade (Eds.), Fundamentals of Fuzzy Sets, Kluwer Academic Publishers, DOrdrecht.
20. J.Goguen.(1967)L-fuzzy sets, J. .Math. Anal. Appl. 18:145-174
21. Glad Deschrijver, Etienne E. Kerre.(2003) On the relationship between some extensions of fuzzy set theory. Fuzzy Sets and Systems 133: 227-235
22. Guo-jun Wang, Ying-yu He(2000)Intuitionistic fuzzy sets and L-fuzzy sets. Fuzzy Sets and Systems 110:271-274
23. Ying-jie Lei, Bao-shu Wang.(2004) On the equivalent mapping between extensions of fuzzy set theory. Systems Engineering and Electronics.26(10)
24. K.T. Atanassov(1999) Intuitionistic Fuzzy Sets, Physica-Verlag, Heidelberg, New York.
25. M.B. Gorza lczany (1987)A method of inference in approximate reasoning based on interval valued fuzzy sets, Fuzzy Sets and Systems 21:1-17
26. Shyi-Chen(1995)Measures of similarity between vague sets. Fuzzy Sets and Systems 74:217-223
27. Weiqiong Wang, Xiaolong Xin(2005) Distance measure between intuitionistic fuzzy sets. Pattern Recognition Letters. 26:2063-2069
28. Przemyslaw Grzegorzewski (2004)Distances between intuitionistic fuzzy sets and/or interval-valued fuzzy sets based on Hausdorff metric. Fuzzy Sets and Systems 148:319-328

Probabilistic Fuzzy Hypernear-Rings

Xueling Ma[1], Jianming Zhan[1], and Yang Xu[2]

[1] Department of Mathematics, Hubei Institute for Nationalities,Enshi, Hubei Province, 445000, P.R. China
zhanjianming@hotmail.com
[2] Department of Applied Mathematics, Southwest Jiaotong University,Chengdu, Sichuan 610031, P.R. China
xuyang@home.swjtu.edu.cn

Abstract. In this paper, using the concept of T-fuzzy hyperideals of hypernear-rings, we define a probabilistic version of hypernear-rings using random sets and show that fuzzy hyperideals defined in triangular norms are consequences of probabilistic hyperideals under certain conditions.

Keywords: T-fuzzy hyperideal; Hypernear-ring; Probability space; Random set.

1 Introduction

The study of algebraic hyperstructure is a well established branch of classical algebraic theory. Hyperstructure theory was born in 1934 [1] when Marty defined hypergroups, began to analyse their properties and applied them to groups, rational functions and algebraic functions. Now, they are widely studied from the theoretical point of view and from their applications to many subjects of pure and applied mathematics nowadays, for example, polygroups which are certain subclasses of hypergroups are used to study color algebra. A comprehensive review of the theory of hyperstructures appears in [2-4].

After the introduction of fuzzy sets by Zadeh [5], the theory was developed in many directions. The study of fuzzy algebraic structures started with the introduction of the concept of the fuzzy subgroupoid (the subgroup) of a groupoid (a group) in the pioneering paper of Rosenfeld. Since then many researchers were engaged in extending the concepts and results of abstract algebra to the broader framework of the fuzzy setting. In 1979, Anthony and Sherwood [6] redefined the fuzzy subgroups using the statistical triangular norms. Moreover, Schweizer and Sklar [7] generalized the ordinary triangular inequality in a metric space to the general probabilistic metric space. In [8], we introduced the concept of T-fuzzy hyperideals of hypernear-rings with respect to triangular norms and obtain some related results. Other important results were obtained in [9-12].

In this paper, using the concept of T-fuzzy hyperideals of hypernear-rings, we define a probabilistic version of hypernear-rings using random sets and show that fuzzy hyperideals defined in triangular norms are consequences of probabilistic hyperideals under certain conditions.

B.-Y. Cao (Ed.): Fuzzy Information and Engineering (ICFIE), ASC 40, pp. 290–295, 2007.
springerlink.com

2 Preliminaries

A *hyperstructure* is a non-empty set H together with a mapping "$\circ$" : $H \times H \to \rho(H)$, where $\rho(H)$ is the set of all the non-empty subsets of H. If $x \in H$ and $A, B \in \rho(H)$, then by $A \circ B$, $A \circ x$ and $x \circ B$, we mean that $A \circ B = \bigcup_{a \in A, b \in B} a \circ b$, $A \circ x = A \circ \{x\}$ and $x \circ B = \{x\} \circ B$, respectively.

Now,we call a hyperstructure $(H, \circ)$ a *canonical hypergroup* [13] if the following axioms are satisfied:

(i) for every $x, y, z \in H, x \circ (y \circ z) = (x \circ y) \circ z$;

(ii) for every $x, y \in H, x \circ y = y \circ x$;

(iii) there exists a $0 \in H$ such that $0 \circ x = x$, for all $x \in H$;

(iv) for every $x \in H$, there exists a unique element $x' \in H$ such that $0 \in x \circ x'$. (we call the element x' the opposite of x).

(v) for every x, y, z, elements of H, from $x \in y \circ z$, it follows, $y \in x \circ z'$.

Definition 1. ([14]) *A hypernear-ring is an algebraic structure $(R, +, \cdot)$ which satisfies the following axioms:*

(1) $(R, +)$ is a quasicanonical hypergroup , i.e., in $(R, +)$ the following hold:

(a) $x + (y + z) = (x + y) + z$ for all $x, y, z \in R$;

(b) There exists $0 \in R$ such that $x + 0 = 0 + x = x$ for all $x \in R$;

(c) For any $x \in R$, there exists one only one $x' \in R$ such that $0 \in x + x'$ (we shall write $-x$ for x' and we call it the opposite of x)

(d) $z \in x + y$ implies $y \in -x + z$ and $x \in z - y$.

If $x \in R$ and A, B are subsets of R, then by $A + B, A + x$ and $x + B$ we mean $A + B = \bigcup_{a \in A, b \in B} a + b, A + x = A + \{x\}, x + B = \{x\} + B$ respectively.

(2) With respect to the multiplication, $(R, \cdot)$ is a semigroup having a bilaterally absorbing element 0, i.e., $x \cdot 0 = 0 \cdot x = 0$ for all $x \in R$.

(3) The multiplication is distributive with respect to the hyperoperation + on the left side, i.e., $x \cdot (y + z) = x \cdot y + x \cdot z$ for all $x, y, z \in R$.

A quasicanonical subhypergroup $A \subsetneq R$ is called *normal* if for all $x \in R$ we have $x + A - x \subseteq A$.

A normal quasicanonical subhypergroup A of the hypergroup $(R, +)$ is

(i) a *left hyperideal* of R if $x \cdot a \in A$ for all $x \in R$ and $a \in A$;

(ii) a *right hyperideal* of R if $(x + A) \cdot y - x \cdot y \subseteq A$ for all $x, y \in R$;

(iii) a *hyperideal* of R if $(x + A) \cdot y - x \cdot y \bigcup z \cdot A \subseteq A$ for all $x, y, z \in R$.

Note that for all $x, y \in R$, we have $-(-x) = x, 0 = -0$, 0 is unique, and $-(x + y) = -y - x$.

Definition 2. ([15]) *Let $(R, +, \cdot)$ be a hypernear-ring and μ a fuzzy subset of R. We say that μ is a fuzzy subhypernear-ring of R if it satisfies the inequality:*

(FH1) $min\{\mu(x), \mu(y)\} \leq \inf_{z \in x+y} \mu(z)$ for all $x, y \in R$;

(FH2) $\mu(x) \leq \mu(-x)$ for all $x \in R$;

(FH3) $min\{\mu(x), \mu(y)\} \leq \mu(x \cdot y)$ for all $x, y \in R$.

Furthermore μ is called a fuzzy hyperideal of R if μ is a fuzzy subhypernear-ring of R and

(FH4) $\mu(y) \leq \inf_{z\in x+y-x} \mu(z)$ *for all* $x, y \in R$;
(FH5) $\mu(y) \leq \mu(x \cdot y)$ *for all* $x, y \in R$;
(FH6) $\mu(i) \leq \inf_{z\in(x+i)\cdot y - x\cdot y} \mu(z)$ *for all* $x, y, i \in R$.

Definition 3. ([7]) *A triangular norm , said also "t-norm", is a function* $T : [0,1] \times [0,1] \to [0,1]$ *satisfying for every* $x, y, z \in [0,1]$:
(i) $T(x,y) = T(y,x)$;
(ii) $T(x,y) \leq T(x,z)$ *if* $y \leq z$;
(iii) $T(x, T(y,z)) = T(T(x,y), z)$;
(iv) $T(x,1) = x$.

Obviously, the function "min" defined on $[0,1] \times [0,1] \to [0,1]$ is a t-norm. Other t-norms which are frequently encountered in the study of probabilistic spaces are T^m and $Prod$ defined by $T^m(a,b) = max\{a+b-1, 0\}$, $Prod(a,b) = ab$ for every $a, b \in [0,1]$.

3 Probabilistic Fuzzy Hypernear-Rings

In what follows, let R be a hypernear-ring unless otherwise specified.

Definition 4. ([8]) *Let* $(R, +, \cdot)$ *be a hypernear-ring and* μ *a fuzzy subset of* R. *We say that* μ *is a fuzzy subhypernear-ring of* R *with respect to* T *(briefly, a* T*-fuzzy subhypernear-ring of* R*) if it satisfies the inequality:*
(TH1) $T(\mu(x), \mu(y)) \leq \inf_{z\in x+y} \mu(z)$ *for all* $x, y \in R$;
(FH2) $\mu(x) \leq \mu(-x)$ *for all* $x \in R$;
(TH3) $T(\mu(x), \mu(y)) \leq \mu(x \cdot y)$ *for all* $x, y \in R$.
Furthermore μ *is called a* T*-fuzzy hyperideal of* R *if* μ *is a* T*-fuzzy subhypernear-ring of* R *and*
(FH4) $\mu(y) \leq \inf_{z\in x+y-x} \mu(z)$ *for all* $x, y \in R$;
(FH5) $\mu(y) \leq \mu(x \cdot y)$ *for all* $x, y \in R$;
(FH6) $\mu(i) \leq \inf_{z\in(x+i)\cdot y - x\cdot y} \mu(z)$ *for all* $x, y, i \in R$.

Example 1. Let $R = \{0, a, b\}$ be a set with a hyperoperation "+" and a binary operation " $\cdot$ " as follows:

$+$	0	a	b
0	$\{0\}$	$\{a\}$	$\{b\}$
a	$\{a\}$	$\{0,a,b\}$	$\{a,b\}$
b	$\{b\}$	$\{a,b\}$	$\{0,a,b\}$

$\cdot$	0	a	b
0	0	0	0
a	0	a	b
b	0	a	b

Then $(R, +, \cdot)$ is a hypernear-ring. Define a fuzzy set $\mu : R \to [0,1]$ by $\mu(a) = \mu(b) = 0.5$ and $\mu(0) = 1$, and a t-norm T by $T(x,y) = \frac{xy}{2-(x+y-xy)}$. Routine calculations give that μ is a T-fuzzy hyperideal of R.

Definition 5. *Let* μ *be a fuzzy set. For every* $\alpha \in [0,1]$, *the set* $\mu_\alpha = \{x \in R | \mu(x) \geq \alpha\}$ *is called the level subset of* μ.

The concept of level subset is very important in the relationship between fuzzy sets and crisp sets.

Theorem 1. *A fuzzy set μ is a fuzzy hyperideal of M if and only if non-empty subset μ_α of μ is a hyperideal of R.*

In the theory of probability we start from $(\Omega, \mathbb{A}, P)$ where Ω is a set of elementary events and $\mathbb{A}$, σ-algebra of subsets of ω called events. A probability on $\mathbb{A}$ is defined as a countable additive and positive function P such that $P(\Omega) = 1$.

The following definition is an extract from [16-18].

Given an universe U , for each arbitrary $u \in U$, let $\dot{u} = \{A|u \in A$ and $A \subseteq U\}$.

For each A in the power set of U , let $\dot{A} = \{\dot{u}|u \in A\}$. An ordered pair $(\rho(U), \mathbb{B})$ is said to be a hyper-measure structure on U if $\mathbb{B}$ is a σ-field in $\rho(U)$ and satisfies the condition: $\dot{U} \subseteq \mathbb{B}$.

Definition 6. *Given a probability space $(\Omega, \mathbb{A}, P)$ and an hyper-measurable structure $(\rho(R), \mathbb{B})$ on U , a random set on U is defined to be a mapping $R_a : \Omega \to \rho(U)$, that is $\mathbb{A} - \mathbb{B}$ measurable, that is , $\forall C \in \mathbb{B}, R_a^{-1}(C) = \{\omega|\omega \in \Omega$ and $R_a(\omega) \in C\} \in \mathbb{A}$.*

Definition 7. *Let $(\Omega, \mathbb{A}, P)$ be a probability space , and $R_a : \Omega \to \rho(U)$ be a random set , where $\rho(R)$ is the set of all subsets of R. If for any $\omega \in \Omega$, $R_a(\omega)$ is a hyperideal of R, then the falling shadow S of the random R_a, i.e., $S(x) = P(\{\omega|x \in R_a(\omega)\})$ is called a π-fuzzy hyperideal of R.*

Based on the concept of a falling shadow, we establish a theoretical approach of the fuzzy hyperideals.

Theorem 2. *Let S be a π-fuzzy hyperideal of R, then for all $x, y, i \in R$, we have*
(i) $\inf_{z \in x-y} S(z) \geq T^m(S(x), S(y))$;
(ii) $S(x \cdot y) \geq T^m(S(x), S(y))$;
(iii) $\inf_{z \in x+y-x} S(z) \geq S(y)$;
(iv) $S(xy) \geq S(y)$;
(v) $\inf_{z \in (x+i) \cdot y - x \cdot y} S(z) \geq S(i)$.

Proof. We only show that the conclusion (i)holds and the others are similar. Since $R_a(\omega)$ is a hyperideal of R. For any $x \in R_a(\omega)$ and $y \in R_a(\omega)$, then $x - y \in R_a(\omega)$. So for every $z \in x - y$, we have $\{\omega|z \in R_a(\omega)\} \supseteq \{\omega|x \in R_a(\omega)\} \cap \{\omega|y \in R_a(\omega)\}$.

Then $S(z) = P(\omega|z \in R_a(\omega))$
$\geq P(\{\omega|x \in R_a(\omega)\} \cap \{\omega|y \in R_a(\omega)\})$
$\geq P(\omega|x \in R_a(\omega)) + P(\omega|y \in R_a(\omega)) - P(\omega|x \in R_a(\omega)$ or $y \in R_a(\omega))$
$\geq S(x) + S(y) - 1$
Hence $\inf_{z \in x-y} S(z) \geq T^m(S(x), S(y))$.

Theorem 3. *(i) Let $\mathbb{R}$ denote the set of all hyperideals of R. Let $R_x = \{A|A \in \mathbb{R}, x \in A\}$ for each $x \in R$. Let $(\mathbb{R}, \sigma)$ be a measurable space, where σ is a σ-algebra that contains $\{R_x|x \in R\}$ and P a probability measure on $(\mathbb{R}, \sigma)$. Define $\mu : R \to [0,1]$ as follows: $\mu(x) = P(R_x)$ for all $x \in R$. Then μ is a T^m-fuzzy hyperideal of R;*

(ii) Suppose that there exists $\mathbb{A} \in \sigma$ such that $\mathbb{A}$ is a chain with respect to set inclusion relation and $P(\mathbb{A}) = 1$. Then μ is a fuzzy hyperideal of R.

Proof. (i) We only show that the condition (i) of Definition 4 holds and the others are similar. Let $x, y \in R$, then for all $R_z \subseteq R_x \cup R_y$ for all $z \in x - y$, and so

$\mu(z) = P(R_z) \geq P(R_x \cup R_y) \geq max\{P(R_x) + P(R_y) - 1, 0\} = T^m(\mu(x), \mu(y))$.

(ii) We only show that the condition (i) of Definition 2 and the others are similar. Since P is a probability measure and $P(\mathbb{A}) = 1$, we have $P(R_x \cap \mathbb{A}) = P(R_x)$ for all $x \in R$. Therefore, for every $z \in x - y$, we have

$\mu(z) = P(R_z) \geq P(R_x \cup R_y) = P(R_x \cap \mathbb{A}) \cap P(R_y \cap \mathbb{A})$.

Since A with the set inclusion forms a chain, it follows that either $R_x \cap \mathbb{A} \subseteq R_y \cap \mathbb{A}$ or $R_y \cap \mathbb{A} \subseteq R_x \cap \mathbb{A}$. Therefore

$\mu(z) \geq min\{P(R_x \cap \mathbb{A}), P(R_y \cap \mathbb{A})\} = min\{\mu(x), \mu(y)\}$,

and so $\inf_{z \in x - y} \mu(z) \geq min\{\mu(x), \mu(y)\}$.

In the study of a unified treatment of uncertainty modeled by means of combining probability and fuzzy set theory, Goodman [19] pointed out the equivalence of a fuzzy set and a class of random sets. Let $(\Omega, \sigma, P) = ([0,1], \sigma, m)$, where σ is a Borel field on $[0,1]$ and m the usual Lebesgue measure. Then

$$R_a : [0,1] \to \rho(R), \alpha \mapsto \mu_\alpha$$

is a measurable function. This notion was first investigated by Goodman in [19], also see [16-17].

Theorem 4. *Let μ be a fuzzy hyperideal of R. Then there exists a probability space $(\Omega, \mathbb{A}, P)$ such that for some $A \in \mathbb{A}, \mu(x) = P(A)$.*

Proof Suppose $\Omega = \mathbb{R}$, the set of all hyperideals of R. We know that $([0,1], \sigma, m)$, is a probability pace, where σ is a Borel field on $[0,1]$ and m the usual Lebesgue measure on the measurable space $([0,1], \sigma)$. Let $R_a : [0,1] \to \rho(R)$ given by $\alpha \mapsto \mu_\alpha$ is a random set. Define $\mathbb{A} = \{A|A \in \mathbb{R}, R_a^{-1}(A) \in \sigma\}$ and $P = m \circ R_a^{-1}$. It is easy to see that $(\mathbb{R}, \mathbb{A}, P)$ is a probability space. If we put $R_x = \{A|A \in \mathbb{R}, x \in A\}$, then for $x \in R$, we have $\mu_\alpha \in R_x$ for all $\alpha \in [0, \mu(x)]$ and μ_s for all $s \in (\mu(x), 1]$. So $R_a^{-1}(R_x) = [0, \mu(x)]$ are hence $R_x \in \mathbb{A}$. Now we get $P(R_x) = m \circ R_a^{-1}(R_x) = m([0, \mu(x)]) = \mu(x)$.

4 Conclusions

In this paper, we showed that fuzzy hyperideals defined in triangular norms are consequences of probabilistic hyperideals under certain conditions. Our future work will focus on characterizing probabilistic rough sets by a fuzzy set which is determined by the rough membership function. One possible application may be in measuring the uncertainty for inconsistent information system.

Acknowledgement. This work is supported by the National Natural Science Foundation of China (60474022) and the Key Science Foundation of Education Committee of Hubei Province, China (2004Z002; D200529001).

References

1. Marty F(1934) Sur une generalization de la notion de groupe. 8th Congress Math Scandianaves Stockholm
2. Corsini P, Leoreanu V(2003) Applications of hyperstructure theory. Advances in Mathematics, Kluwer Academic Publishers
3. Davvaz B(2003) A brief survey of the theory of H_v-structures , Algebraic hyperstructures and applications (Proceedings of the Eighth International Congress AHA Samothraki, 2002), 39–70, Edited by T. Vougiouklis, Printed by Spanidis Press
4. Vougiouklis T(1994) Hyperstructures and their representations. Hadronic Press Inc, Palm Harber, USA
5. Zadeh LA(1965) Fuzzy sets. Inform and Control 8: 338-353
6. Anthony JM, Sherwood H(1979) Fuzzy group defined. J Math Anal Appl 69: 124-130
7. Schweizer B, Sklar A(1960) Statistical metric space. Pacific J Math 10:313-334
8. Zhan JM(2006) On properties of fuzzy hyperideals in hypernear-rings with t-norms. J Appl Math Computing 20: 255-277
9. Davvaz B(1999) Interval-values fuzzy subhypergroups. J Appl Math Computing 6:197-202
10. Davvaz B(2000) Strong regularity and fuzzy strong regularity in semihypergroups. J Appl Math Computing 7:205-213
11. Zhan JM, Dudek WA(2006) Interval valued intuitionistic (S, T)-fuzzy H_v-submodules. Acta Math Sinica, English Series 22: 963-970
12. Zhan JM, Tan ZS(2006) Approximations in hyperquasigroups. J Appl Math Computing 21:485-494
13. Krasner M(1983) A class of hyperring and hyperfields. Int J Math Math Sci 2: 307-312
14. Dasic V(1990) Hypernear-rings. Proc 4th Int Congress AHA, Xanthi Greece World Scientific, 75-85
15. Davvaz B(1999) On hypernear-rings and fuzzy hyperideals. J Fuzzy Math 7: 745-753
16. Li QD, Lee ES(1995) On random α-cuts. J Math Anal Appl 190:546-558
17. Tan SK, Wang PZ, Lee ES(1993) Fuzzy set operations based on the theory of falling shadows. J Math Anal Appl 174:242-255
18. Yuan XH, Lee ES(1997) A fuzzy algebraic system based on the theory of falling shadows. J Math Anal Appl 208:243-251
19. Goodman IR(1982) Fuzzy sets as equivalence classes of random sets. Recent Development in Fuzzy set and Possibility Theory, Pergamona New York

Isomorphic Fuzzy Sets and Fuzzy Approximation Space

Zhang Chengyi[1] and Fu Haiyan[1,2]

[1] Department of Computer Science, Hainan Normal University, Haikou, Hainan, 571158, China
chengyizh@hainnu.edu.cn
[2] School of Mathematics and System Science, Shandong University, Jinan, Shandong, 250100, China
yanzi78@163.com

Abstract. This paper discussed the relation between fuzzy sets and rough sets by theory of the isomorphism and homomorphism of fuzzy sets, and two main conclusions are reached. Firstly, for any group of fuzzy sets which are isomorphic for each other in X, an approximation space could be defined uniquely. Secondly, for any group of fuzzy equivalence relations which are isomorphic, similarly a group of fuzzy approximation spaces which are isomorphic could be defined. Moreover, the construction of fuzzy approximation space is given. Finally, we illustrate an concrete example to show the construction of the fuzzy approximation space, and discussed their related properties.

Keywords: fuzzy sets, rough sets, fuzzy equivalent relation, approximation space, granule computing.

1 Introduction

Among the major fields which underlie granular computing there are three theories that stand out in importance. They are: fuzzy set theory(FS), rough set theory (RS)and quotient space theory(QS). Zadeh primarily discussed the representation of granular by FS. Namely he used the "language", "word" to represent the thinking and reasoning of mankind. In Zadeh L.A [4-6], Zadeh put forward that there are three basic concepts that underlie human cognition: granulation, organization and causation. Informally, granulation involves decomposition of whole into parts; organization involves integration of parts into whole; and causation involves association of causes with effects. Moreover, the concept of fuzzy information granulation and "computing with words" are proposed.

Granular computing based on RS theory primarily discussed the representation and characterization of granular, and the relations between granular and concept. Pawlak thought that the intelligence (knowledge)of person was a classified ability, and proposed that a concept could be represented by a subset in universe X. Giving an equivalent relation in U, it means giving a knowledge base (X,R).A concept x can be represented by the knowledge in (X,R) (the union of sets in knowledge base). For some sets which can not be represented by the knowledge in (X,R) , he used R-lower and R-upper approximation to approximating. Now, RS theory has already extensive application into some fields, especially into data mining and acquire the success Pawlak Z [1], Dubois. D [3], Yao Y.Y [8].

Granular computing based on QS theory, although also describes granular by the equivalent classes and describes concept by granular, the point of discussion is

B.-Y. Cao (Ed.): Fuzzy Information and Engineering (ICFIE), ASC 40, pp. 296–306, 2007.
springerlink.com

different to RS theory. QS theory puts greater emphasis on the transform relation between the different granular spaces. Great differences lie in that the universe of QS is a topology space but the universe of RS is a simple point set. Thus RS is just an example of QS. In Zhang B [10], Zhang Bo and Zhang Ling established a whole set of theories and calculating methods for this granular space model, and achieved remarkable accomplishment in some fields, for example, heuristic search and path programming .

Intensive study of the relations of these three kinds of theories and their coordination will have much significance for the development of granular computing theory and its application. This paper discussed the relation between FS and RS by theory of the isomorphism and homomorphism of fuzzy sets. We proved the following two main conclusions. Firstly, for any group of fuzzy sets which are isomorphic for each other in X, an approximation space could be defined uniquely. Secondly, for any group of fuzzy equivalence relations which are isomorphic, similarly a group of fuzzy approximation spaces which are isomorphic could be defined.

Moreover, the construction of fuzzy approximation space is given. Finally, we discussed their related properties. Their results preliminary have established connection between FS and RS.

2 Isomorphism and Homomorphism of Fuzzy Sets

Definition 2.1. Let X be an universe. Mapping $\underset{\sim}{A}$: X→[0,1] is called a fuzzy subset in X, $x \in$ X, $\underset{\sim}{A}(x)$ is called membership degree of x about $\underset{\sim}{A}$, $\underset{\sim}{A}(\bullet)$ is called the membership function of $\underset{\sim}{A}$. Given two fuzzy sets $\underset{\sim}{A}, \underset{\sim}{B}$, the operations of fuzzy sets are defined as following :

$$(\underset{\sim}{A} \cup \underset{\sim}{B})(x) = \max \{ \underset{\sim}{A}(x), \underset{\sim}{B}(x) \},$$

$$(\underset{\sim}{A} \cap \underset{\sim}{B})(x) = \min \{ \underset{\sim}{A}(x), \underset{\sim}{B}(x) \},$$

$$\underset{\sim}{A}^{c}(x) = 1 - \underset{\sim}{A}(x), \forall x \in \mathrm{X}.$$

($\underset{\sim}{A}^{c}$ denoted the complement of $\underset{\sim}{A}$)

The set of whole fuzzy sets in X is denoted as F(X).

Definition 2.2. Let $f: \underset{\sim}{A} \rightarrow f(\underset{\sim}{A})$ be a mapping from F(X) to itself. Then

1. f is called an isomorphism on F(X) and we call $\underset{\sim}{A}$ and $f(\underset{\sim}{A})$ are isomorphic, denoted as $\underset{\sim}{A} \cong f(\underset{\sim}{A})$ if the following two conditions are satisfied:

$$\forall \underset{\sim}{A} \in \mathrm{F(X)}, \forall x, y \in X, \ \underset{\sim}{A}(x) < \underset{\sim}{A}(y) \Leftrightarrow f(\underset{\sim}{A})(x) < f(\underset{\sim}{A})(y), \tag{1}$$

$$\underset{\sim}{A}(x) = \underset{\sim}{A}(y) \Leftrightarrow f(\underset{\sim}{A})(x) = f(\underset{\sim}{A})(y), \tag{2}$$

2. f is called a homormorphism on F(X) and we call $\underset{\sim}{A}$ and f($\underset{\sim}{A}$) are homormorphic, denoted as $\underset{\sim}{A}$~f($\underset{\sim}{A}$) if the following two conditions are satisfied:

$\forall \underset{\sim}{A} \in F(X), \forall x, y \in X,$

$$\underset{\sim}{A}(x) \leq \underset{\sim}{A}(y) \Rightarrow f(\underset{\sim}{A})(x) \leq f(\underset{\sim}{A})(y), \tag{3}$$

$$f(\underset{\sim}{A})(x) < f(\underset{\sim}{A})(y) \Rightarrow \underset{\sim}{A}(x) < \underset{\sim}{A}(y). \tag{4}$$

Note1: easily illustrate an example: $\underset{\sim}{A} \cong f(\underset{\sim}{A}) \Rightarrow \underset{\sim}{A} \sim f(\underset{\sim}{A})$, but the opposite result is false.

Definition 2.3. Suppose $\underset{\sim}{A} \in F(X)$, $A_\lambda = \{x \mid \underset{\sim}{A}(x) \geq \lambda\}$ is the λ-cut set of $\underset{\sim}{A}$, then

$$H(\underset{\sim}{A}) = \{A_\lambda \mid \lambda \in \mathrm{Im}(\underset{\sim}{A})\} \tag{5}$$

is called the irreducible nested set of $\underset{\sim}{A}$.

Theorem 2.1. Let $\underset{\sim}{A}, \underset{\sim}{B} \in F(X)$, $H(\underset{\sim}{A}) = \{A_\lambda \mid \lambda \in \mathrm{Im}(\underset{\sim}{A})\}$ and $H(B) = \{B_\mu \mid \mu \in \mathrm{Im}(\underset{\sim}{B})\}$ the irreducible nested sets of $\underset{\sim}{A}$ and $\underset{\sim}{B}$ respectively. Then $\underset{\sim}{A} \cong \underset{\sim}{B}$ if and only if $H(\underset{\sim}{A}) = H(\underset{\sim}{B})$.

Proof (Sufficiency). Suppose that $H(\underset{\sim}{A}) = H(\underset{\sim}{B})$, $\forall x, y \in X$, if $\underset{\sim}{A}(x) < \underset{\sim}{A}(y)$, then $\exists \lambda_1, \lambda_2 \in \mathrm{Im}(\underset{\sim}{A})$ such that $y \in A_{\lambda_1}$, $x \in A_{\lambda_2} \setminus A_{\lambda_1}$, $A_{\lambda_1} \subset A_{\lambda_2}$, Then $\exists \mu_1, \mu_2 \in \mathrm{Im}(\underset{\sim}{B})$, $\mu_1 > \mu_2$ satisfy $A_{\lambda_1} = B_{\mu_1}$, $A_{\lambda_2} = B_{\mu_2}$, Thus $\underset{\sim}{B}(x) < \mu_1 \leq \underset{\sim}{B}(y)$, i.e. $\forall x, y \in X$, $\underset{\sim}{A}(x) < \underset{\sim}{A}(y) \Rightarrow \underset{\sim}{B}(x) < \underset{\sim}{B}(y)$. Similarly, we can prove that $\underset{\sim}{B}(x) < \underset{\sim}{B}(y) \Rightarrow \underset{\sim}{A}(x) < \underset{\sim}{A}(y)$.

Suppose $\underset{\sim}{A}(x) = \underset{\sim}{A}(y)$, then $\exists A_\lambda \in H(\underset{\sim}{A})$, such that $x, y \in A_\lambda$. If $\underset{\sim}{B}(x) \neq \underset{\sim}{B}(y)$, we assuming $\underset{\sim}{B}(x) < \underset{\sim}{B}(y)$, then, we get $\underset{\sim}{A}(x) < \underset{\sim}{A}(y)$ from (1), it is a contradiction. Thus, we have $\forall x, y \in X, \underset{\sim}{A}(x) = \underset{\sim}{A}(y) \Leftrightarrow \underset{\sim}{B}(x) = \underset{\sim}{B}(y)$, hence $\underset{\sim}{A} \cong \underset{\sim}{B}$.

(**Necessity**) Suppose $\underset{\sim}{A} \cong \underset{\sim}{B}$, then $\forall x, y \in X$, $\underset{\sim}{A}(x) < \underset{\sim}{A}(y) \Leftrightarrow \underset{\sim}{B}(x) < \underset{\sim}{B}(y)$, $\underset{\sim}{A}(x) = \underset{\sim}{A}(y) \Leftrightarrow \underset{\sim}{B}(x) = \underset{\sim}{B}(y)$. Let $\underset{\sim}{A}(x) = \lambda_1$, $\underset{\sim}{A}(y) = \lambda_2$ and $\underset{\sim}{B}(x) = \mu_1$, $\underset{\sim}{B}(y) = \mu_2$, we only need prove $A_{\lambda_1} = B_{\mu_1}$.

$\forall z \in A_{\lambda_1}$, since $\underset{\sim}{A}(z) \geq \lambda_1 = \underset{\sim}{A}(x)$, then $\underset{\sim}{B}(z) \geq \underset{\sim}{B}(x) = \mu_1$, thus $z \in B_{\mu_1}$, it implies that $A_{\lambda_1} \subseteq B_{\mu_1}$. Similarly, we can prove $B_{\mu_1} \subseteq A_{\lambda_1}$. Hence $\forall \lambda \in \mathrm{Im}(\underset{\sim}{A})$, there existing $\mu \in \mathrm{Im}(\underset{\sim}{B})$, such that $A_\lambda = B_\mu$. Conversely, we have also $\forall \mu \in \mathrm{Im}(\underset{\sim}{B})$, there existing $\lambda \in \mathrm{Im}(\underset{\sim}{A})$, such that $B_\mu = A_\lambda$. Thus $H(\underset{\sim}{A}) = H(\underset{\sim}{B})$.

It is clear that the isomorphic relation of fuzzy sets is an equivalent relation on $F(X)$. Let $\underset{\sim}{A} \in F(X)$,the isomorphic equivalent class contained $\underset{\sim}{A}$ is denoted $[\underset{\sim}{A}]$ and $H(\underset{\sim}{A})$ is called the irreducible nested set of $[\underset{\sim}{A}]$.The following two statements are equivalent: (1) an isomorphic equivalent class $[\underset{\sim}{A}]$ given in X; (2) an irreducible nested set given in X.

Theorem 2.2. Let $\underset{\sim}{A}, \underset{\sim}{B} \in F(X)$, $H(\underset{\sim}{A})$ and $H(\underset{\sim}{B})$ the irreducible nested sets of $\underset{\sim}{A}$ and $\underset{\sim}{B}$ respectively. Then $\underset{\sim}{A} \sim \underset{\sim}{B}$ if and only if $H(\underset{\sim}{B}) \subseteq H(\underset{\sim}{A})$.

The proof of the theorem of 2.2 is similar to that of Theorem of 2.1.

3 Fuzzy Sets and Approximation Spaces

Let X denote a finite and non-empty set called the universe, and let $R \subseteq U \times U$ denote an equivalence relation on X, i.e., R is a reflexive, symmetric and transitive relation. If two elements x, y in X belong to the same equivalence class, i.e., xRy, we say that they are indistinguishable. The pair $W=(X,R)$ is called an approximation space. The equivalence relation R partitions the set X into disjoint subsets. It defines the quotient set X/R consisting of equivalence classes of R. The equivalence class $[x]_R$ containing x plays dual roles. It is a subset of X if considered in relation to the universe, and an element of X/R if considered in relation to the quotient set. The empty set $\varnothing$ and equivalent classes are called the elementary sets. The union of one or more elementary sets is called a composed set. The family of all composed sets is denoted by Com(W). It is a subalgebra of the Boolean algebra 2^X formed by the power set of X.

$\forall A \subseteq X$,let $R\_(A)=\{x \in X | [x]_R \subseteq A\}$, $R^-(A)=\{x \in X | [x]_R \cap A \neq \varnothing\}$,then the pair $(R\_(A),R^-(A))$ is called a rough set, denoted as $A=(R\_(A),R^-(A))$,and $R\_(A)$, $R^-(A)$ are called the lower and upper approximations of A respectively. $Bn_R(A) = R^-(A) - R\_(A)$ is called as the R-boundary of A.

Following, we will discuss the relation between fuzzy sets and approximation spaces.

Given $\underset{\sim}{A} \in F(X)$, let A_λ and $A_{[\lambda]}=\{x \mid \underset{\sim}{A}(x) > \lambda\}$ be the λ -cut and λ -strong cut of $\underset{\sim}{A}$ respectively and $A[\lambda] = A_\lambda \setminus A_{[\lambda]} = \{x \mid \underset{\sim}{A}(x) = \lambda\}$. Thus, we can define a relation $R_{\underset{\sim}{A}}$ on X: $\forall x, y \in X$, $x R_{\underset{\sim}{A}} y$ if and only if $\underset{\sim}{A}(x) = \underset{\sim}{A}(y)$, then $R_{\underset{\sim}{A}}$ is an equivalent relation on X and $\{A[\lambda] \mid \lambda \in \mathrm{Im}(R_{\underset{\sim}{A}})\}$ are equivalent classes of X under $R_{\underset{\sim}{A}}$. We can get an approximation space $(X, R_{\underset{\sim}{A}})$.

Theorem 3.1. Let $\underset{\sim}{A}, \underset{\sim}{B} \in F(X)$, $\underset{\sim}{A} \cong \underset{\sim}{B}$. Then $(X, R_{\underset{\sim}{A}})=(X, R_{\underset{\sim}{B}})$.

Proof. Let $\underset{\sim}{A}, \underset{\sim}{B} \in$ F(X), $\underset{\sim}{A} \cong \underset{\sim}{B}$ .If $x\mathrm{R}_{\underset{\sim}{A}}y$, then $\underset{\sim}{A}(x) = \underset{\sim}{A}(y)$, $x, y \in A[\lambda]$. Since $H(\underset{\sim}{A})$=$H(\underset{\sim}{B})$, $\forall \mathrm{A}_{\lambda} \in \mathrm{H}(\underset{\sim}{A})$, $\exists \mathrm{B}_{\mu} \in \mathrm{H}(\underset{\sim}{\mathrm{B}})$,such that A_{λ}=B_{μ} .Thus

$\mathrm{A}_{\lambda_1} \subseteq \mathrm{A}_{\lambda_2} \Leftrightarrow \mathrm{B}_{\mu_1} \subseteq \mathrm{B}_{\mu_2}$ if A_{λ_1}=B_{μ_1}, A_{λ_2}=B_{μ_2} .

From $A_{[\lambda]} = \bigcup_{a>\lambda} A_{\lambda}$, we have $\mathrm{A}[\lambda]$=$\mathrm{A}_{\lambda} \setminus \mathrm{A}_{[\lambda]} =$ $\mathrm{B}_{\mu} \setminus \mathrm{B}_{[\mu]} = B[\mu]$, so $x\mathrm{R}_{\underset{\sim}{\mathrm{B}}}y$. Similarly, $x\mathrm{R}_{\underset{\sim}{\mathrm{B}}}y \Rightarrow x\mathrm{R}_{\underset{\sim}{A}}y$. Then $\mathrm{X}/\mathrm{R}_{\underset{\sim}{A}}$=$\mathrm{X}/\mathrm{R}_{\underset{\sim}{\mathrm{B}}}$.

Definition 3.1. Assume R_1 and R_2 are two equivalent relations on X, (X, R_1),(X, R_2) are the approximation spaces defined by R_1 and R_2.We call X/R_1 is a refinement of X/R_2 if $\forall a \in X / R_1$,$\exists b \in \mathrm{X}/\mathrm{R}_2$,such that $a \subseteq b$. Moreover, (X, R_2) is called a quotient approximation space of (X, R_1).

Theorem 3.2. Let $\underset{\sim}{A}, \underset{\sim}{B} \in$ F(X), $\underset{\sim}{A} \sim \underset{\sim}{\mathrm{B}}$ and $\mathrm{R}_{\underset{\sim}{A}}$, $\mathrm{R}_{\underset{\sim}{\mathrm{B}}}$ are equivalent relations efined by $\underset{\sim}{A}$ and $\underset{\sim}{\mathrm{B}}$ .Then $\mathrm{X}/\mathrm{R}_{\underset{\sim}{A}}$ is a refinement of $\mathrm{X}/\mathrm{R}_{\underset{\sim}{\mathrm{B}}}$ and (X, $\mathrm{R}_{\underset{\sim}{\mathrm{B}}}$) is a quotient approximation space of (X, $\mathrm{R}_{\underset{\sim}{A}}$).

The proof of the theorem of 3.2 is similar to that of theorem of 3.1.

Theorem 3.3. Let $\underset{\sim}{A} \in$ F(X), $\mathrm{R}_{\underset{\sim}{A}}$ is the equivalent relation on X defined by $\underset{\sim}{A}$ and (X, $\mathrm{R}_{\underset{\sim}{A}}$)is an approximation space defined by $\mathrm{R}_{\underset{\sim}{A}}$. Then a normalized isosceles distance function $\bar{d}(.,.)$ on X / R_{A} can be defined such that (X / R_{A} , $\bar{d}$) is a distance space , denoted as the quotient distance space defined by $\mathrm{R}_{\underset{\sim}{A}}$.

Proof. Let $\underset{\sim}{A} \in$ F(X), $\forall x, y \in X$, define $d(x, y) = |\underset{\sim}{A}(x) - \underset{\sim}{A}(y)|$, then we have

(1)$1 \geq d(x, y) \geq 0$, (2) $d(x, y)$=$d(y, x)$,

(3) $\forall x, y, z \in X$,$d(x, z) \leq d(x, y) + d(y, z)$.

Thus $d(.,.)$ is a distance function. Since $1 \geq d(x, y) \geq 0$, it is called a normalized isosceles distance .

Suppose that $\mathrm{R}_{\underset{\sim}{A}}$ is the equivalent relations on X defined by $\underset{\sim}{A}$ and(X, $\mathrm{R}_{\underset{\sim}{A}}$)is the approximation space defined by $\mathrm{R}_{\underset{\sim}{A}}$.

$\forall a, b \in X / R_{\mathrm{A}}$, Let $\bar{d}(a, b) =| \underset{\sim}{A}(x) - \underset{\sim}{A}(y)|$ for $\forall x \in a \in X / R_{\mathrm{A}}$, $\forall y \in b \in X / R_{\mathrm{A}}$, then $\forall y \in b \in X / R_{\mathrm{A}}$, $x_1, x_2 \in a \in X / R_{\mathrm{A}}$ from $d(x_1, x_2)$=0 implies $d(x_1, y)$ $\leq d(x_1, x_2)$+$d(x_2, y) \leq d(x_2, y)$. Similarly, $d(x_1, y) \geq d(x_2, y)$, so $d(x_1, y)$=$d(x_2, y)$. Thus, we have $\forall y_1, y_2 \in b \in X / R_{\mathrm{A}}$, $x_1, x_2 \in a \in X / R_{\mathrm{A}}$, $d(x_1, y_1) = d(x_2, y_1) =$ $d(x_2, y_2)$, it expresses that $\bar{d}(.,.)$ is well-defined.

Moreover, $\bar{d}(.,.)$ has following properties:

1. $\bar{d}(a,b) \geq 0$ and from $\bar{d}(a,b)=0$ can imply that $a=b$;
2. $\bar{d}(a,b)=\bar{d}(b,a)$;
3. $\forall a,b,c \in X/R_A, \forall x \in a, \forall y \in b, \forall z \in c$,

$$\bar{d}(a,c)=|\underset{\sim}{A}(x)-\underset{\sim}{A}(z)| \leq |\underset{\sim}{A}(x)-\underset{\sim}{A}(y)|+|\underset{\sim}{A}(y)-\underset{\sim}{A}(z)| = \bar{d}(a,b)+\bar{d}(b,c).$$

Thus, $\bar{d}(.,.)$ is a normalized isosceles distance function on X/R_A and $(X/R_A, \bar{d})$ is a quotient distance space.

Theorm 3.4. Let $(X/R, d)$ be a quotient distance space with a normalized isosceles distance $\bar{d}$ .It can define a group of fuzzy sets which are isomorphic for each other.

Proof. Let $(X/R, d)$ be a quotient space with a normalized isosceles distance $\bar{d}$. Suppose $a \in X/R$, for any $b \in X/R$,we can calculate $\bar{d}(a,b)$ respectively. To line up all $\bar{d}(a,b)$ from big to small, and denote as $u_1 = 1$, u_2, …,u_n. Suppose that $\bar{d}(a,b_{ij}) = u_i$ ($b_{ij} \in X/R$, $j=1,2,\cdots,k$).Let $a_i = \bigcup_{j=1}^{k} b_{ij}$ ($i=1,2,\cdots,n$),we can construct an irreducible nested sets as following:

$$A_1 = a_1, A_{k+1} = A_k \cup a_k, k=1,2,\cdots,n$$

By theorem2.1, this irreducible nested sets $\{A_k | k=1,2,\cdots,n\}$ can define an isomorphic equivalent class $[\underset{\sim}{A}]$, moreover, can define a group of fuzzy sets which are isomorphic for each other.

Theorm 3.5. Let $\underset{\sim}{A} \in F(X)$, $R_{\underset{\sim}{A}}$ be the equivalent relation on X defined by $\underset{\sim}{A}$ and $(X, R_{\underset{\sim}{A}})$ be the approximation space defined by $R_{\underset{\sim}{A}}$. If $\ker(\underset{\sim}{A}) = \{x \mid \underset{\sim}{A}(x)=1\} \neq \varnothing$, then the isomorphic equivalent class on F(X) defined by the quotient distance space $(X/R_{\underset{\sim}{A}}, \bar{d})$ just is $[\underset{\sim}{A}]$.(distance function $\bar{d}$ defined as above)

Proof is obvious.

4 Fuzzy Relations and Fuzzy Approximation Spaces

Definition 4.1. Let $\underset{\sim}{R} \in F(X \times X)$. $\underset{\sim}{R}$ is called a fuzzy equivalent relation on X if the following conditions are satisfied:

(1) $\forall x \in X$, $\underset{\sim}{R}(x,x)=1$;

(2) $\forall x,y \in X$, $\underset{\sim}{R}(x,y)=\underset{\sim}{R}(y,x)$; (3) $\forall x,y,z \in X$, $\underset{\sim}{R}(x,z) \geq \sup_y(\min(\underset{\sim}{R}(x,y), \underset{\sim}{R}(y,z)))$.

Lemma 4.1 [15]. Let $R_\lambda = \{(x,y) | \underset{\sim}{R}(x,y) \geq \lambda\}, 0 \leq \lambda \leq 1$. Then R_λ is an equivalent relation on X ,called as the λ-cut relation of $\underset{\sim}{R}$.

Let $\underset{\sim}{R}$ be a fuzzy equivalent relation on X and $R_\alpha(\alpha \in \mathrm{Im}(\underset{\sim}{R}))$ be the α-cuts of $\underset{\sim}{R}$. Then $\{(X, R_\alpha) | \alpha \in \mathrm{Im}(\underset{\sim}{R})\}$ is a group of approximations space on X defined by R_α.The following properties are obvious:

(1) $0 \leq \alpha_2 \leq \alpha_1 \leq 1 \Leftrightarrow R_{\alpha_1} \subseteq R_{\alpha_2} \Leftrightarrow X / R_{\alpha_2}$ is quotient approximation space of X / R_{α_1}.

(2) Let $0 \leq \alpha_2 \leq \alpha_1 \leq 1$. $\forall x \in X$,if $x \in a_{\alpha_1} \in X / R_{\alpha_1}$ and $x \in a_{\alpha_2} \in X / R_{\alpha_2}$,then $a_{\alpha_1} \subseteq a_{\alpha_2}$.

Definition 4.2. Let $\underset{\sim}{R}$ be a fuzzy equivalent relation on X and $(X, R_\alpha) | \alpha \in \mathrm{Im}(\underset{\sim}{R})\}$ be a group of quotient approximation spaces defined by the λ-cut relation R_λ of $\underset{\sim}{R}$. Suppose that $[a]_\alpha(\alpha \in \mathrm{Im}(\underset{\sim}{R}))$ is the equivalent class closed to x in R_α. Then

(1) $X / \underset{\sim}{R} = \bigcup_{\alpha \in \mathrm{Im}(\underset{\sim}{R})} \alpha(X / R_\alpha)$ is called as the fuzzy approximation space in X under fuzzy equivalent relation $\underset{\sim}{R}$.

(2) $[a]_{\underset{\sim}{R}} = \bigcup_{\alpha \in \mathrm{Im}(\underset{\sim}{R})} \alpha[a]_{R_\alpha}$ is the fuzzy equivalent class closed to x in $X / \underset{\sim}{R}$.

Example 4.1. The classification of some environment units. Suppose each unit involve four factors: air, moisture, soil and crop. Polluted status of environment unit is weighed by content of contamination in 4 elements. Now suppose we have 5 environment units, polluted data as follows:

$$X = \{a_1, a_2, \cdots, a_5\}$$

$a_1=(5,5,3,2)$, $a_2=(2,3,4,5)$, $a_3=(5,5,2,3)$, $a_4=(1,5,3,1)$, $a_5=(2,4,5,1)$, give the clustering analysis of X.

First suppose c=0.1 , and establishing the fuzzy equivalent relation matrix $\underset{\sim}{R}=(r_{ij})_{5\times5}$ basing on the absolute value subtrahend method:

$r_{ij}=1-0.1\sum_{k=1}^{4}|a_{ik}-a_{jk}|$,

then $$\underset{\sim}{R} = \begin{pmatrix} 1 & 0.4 & 0.8 & 0.5 & 0.5 \\ 0.4 & 1 & 0.4 & 0.4 & 0.4 \\ 0.8 & 0.4 & 1 & 0.5 & 0.5 \\ 0.5 & 0.4 & 0.5 & 1 & 0.6 \\ 0.5 & 0.4 & 0.5 & 0.6 & 1 \end{pmatrix}$$

We get by computing:

$X/R_1=\{(a_1),(a_2),(a_3),(a_4),(a_5)\}$; $X/R_{0.8}=\{(a_1,a_3),(a_2),(a_4),(a_5)\}$;
$X/R_{0.6}=\{(a_1,a_3),(a_2),(a_4,a_5)\}$; $X/R_{0.5}=\{(a_1,a_3,a_4,a_5),(a_2)\}$;
$X/R_{0.4}=\{(a_1,a_2,a_3,a_4,a_5)\}$.

Then $X/\underset{\sim}{R}=X/R_1\cup 0.8\circ(X/R_{0.8})\cup 0.6\circ(X/R_{0.6})\cup 0.5\circ(X/R_{0.5})\cup 0.4\circ(X/R_{0.4})$

$=\{[a_1]_{\underset{\sim}{R}},[a_2]_{\underset{\sim}{R}},[a_3]_{\underset{\sim}{R}},[a_4]_{\underset{\sim}{R}},[a_5]_{\underset{\sim}{R}}\}$.

Where $[a_1]_{\underset{\sim}{R}}=\{a_1\}\cup 0.8\circ\{a_1,a_3\}\cup 0.6\circ\{a_1,a_3\}\cup$

$0.5\circ\{a_1,a_3,a_4,a_5\}\cup 0.4\circ\{a_1,a_2,a_3,a_4,a_5\}$,

$[a_1]_{\underset{\sim}{R}}(a_1)=1$, $[a_1]_{\underset{\sim}{R}}(a_2)=0.4$, $[a_1]_{\underset{\sim}{R}}(a_3)=0.8$, $[a_1]_{\underset{\sim}{R}}(a_4)=[a_1]_{\underset{\sim}{R}}(a_5)=0.5$.

$[a_2]_{\underset{\sim}{R}}=\{a_2\}\cup 0.4\circ\{a_1,a_2,a_3,a_4,a_5\}$,

$[a_2]_{\underset{\sim}{R}}(a_2)=1$, $[a_2]_{\underset{\sim}{R}}(a_1)=[a_2]_{\underset{\sim}{R}}(a_3)=[a_2]_{\underset{\sim}{R}}(a_4)=[a_2]_{\underset{\sim}{R}}(a_5)=0.4$.

$[a_3]_{\underset{\sim}{R}}=\{a_3\}\cup 0.8\circ\{a_1,a_3\}\cup 0.6\circ\{a_1,a_3\}\cup$

$0.5\circ\{a_1,a_3,a_4,a_5\}\cup 0.4\circ\{a_1,a_2,a_3,a_4,a_5\}$,

$[a_3]_{\underset{\sim}{R}}(a_3)=1$, $[a_3]_{\underset{\sim}{R}}(a_2)=0.4$, $[a_3]_{\underset{\sim}{R}}(a_1)=0.8$, $[a_3]_{\underset{\sim}{R}}(a_4)=[a_3]_{\underset{\sim}{R}}(a_5)=0.5$.

$[a_4]_{\underset{\sim}{R}}=\{a_4\}\cup 0.6\circ\{a_4,a_5\}\cup 0.5\circ\{a_1,a_3,a_4,a_5\}\cup 0.4\circ\{a_1,a_2,a_3,a_4,a_5\}$,

$[a_4]_{\underset{\sim}{R}}(a_4)=1$, $[a_4]_{\underset{\sim}{R}}(a_1)=[a_4]_{\underset{\sim}{R}}(a_3)=0.5$, $[a_4]_{\underset{\sim}{R}}(a_5)=0.6$, $[a_4]_{\underset{\sim}{R}}(a_2)=0.4$.

$[a_5]_{\underset{\sim}{R}}=\{a_5\}\cup 0.6\circ\{a_4,a_5\}\cup 0.5\circ\{a_1,a_3,a_4,a_5\}\cup 0.4\circ\{a_1,a_2,a_3,a_4,a_5\}$,

$[a_5]_{\underset{\sim}{R}}(a_5)=1$, $[a_5]_{\underset{\sim}{R}}(a_4)=0.6$, $[a_5]_{\underset{\sim}{R}}(a_1)=[a_5]_{\underset{\sim}{R}}(a_3)=0.5$, $[a_5]_{\underset{\sim}{R}}(a_2)=0.4$.

We may choose different threshold values to determine different cluster basing on the requirement of diverse clustering, which could be applied to pattern recognition, artificial intelligent and so on.

Theorem 4.2. Let $X/\underset{\sim}{R}$ be the fuzzy approximation space under fuzzy equivalent relation $\underset{\sim}{R}$. Then

(1) $[a]_{\underset{\sim}{R}}(x)=\underset{\alpha\in \mathrm{Im}(\underset{\sim}{R})}{\vee}(\alpha\wedge[a]_{R_\alpha}(x))=\vee\{\alpha\mid x\in[a]_{R_\alpha},\alpha\in \mathrm{Im}(\underset{\sim}{R})\}$;

(2) $X/\underset{\sim}{R}=\{[a]_{\underset{\sim}{R}}\mid a\in U\}$.

Proof. We prove only (2).

$$X/\underset{\sim}{R}=\underset{\alpha\in \mathrm{Im}(\underset{\sim}{R})}{\cup}\alpha(X/R_\alpha)=\underset{\alpha\in \mathrm{Im}(\underset{\sim}{R})}{\cup}(\alpha(\{[a]_{R_\alpha}\mid[a]_{R_\alpha}\in X/R_\alpha\})$$

$$=\underset{\alpha\in \mathrm{Im}(\underset{\sim}{R})}{\cup}\{\alpha[a]_{R_\alpha}\mid[a]_{R_\alpha}\in X/R_\alpha\}=\{\underset{\alpha\in \mathrm{Im}(\underset{\sim}{R})}{\cup}(\alpha[a]_{R_\alpha})\mid[a]_{R_\alpha}\in X/R_\alpha,\ \alpha\in \mathrm{Im}(\underset{\sim}{R})\}$$

$$=\{[a]_{\underset{\sim}{R}}\mid[a]_{\underset{\sim}{R}}=\underset{\alpha\in \mathrm{Im}(\underset{\sim}{R})}{\cup}\alpha[a]_{R_\alpha}\}.$$

Similarly to Theorem 4.2 we have

Definition 4.2. Let $X/\underset{\sim}{R}, X/\underset{\sim}{Q}$ be two fuzzy approximation spaces defined by fuzzy equivalent relations $\underset{\sim}{R}, \underset{\sim}{Q}$ on X respectively. If there exists a bijection $\varphi: \mathrm{Im}(\underset{\sim}{R}) \to \mathrm{Im}(\underset{\sim}{Q})$ such that

$\forall \lambda \in \mathrm{Im}(\underset{\sim}{R})$, $X/R_\lambda = X/Q_{\varphi(\lambda)}$, then we call the fuzzy approximation space $X/\underset{\sim}{R}$ is isomorphic to $X/\underset{\sim}{Q}$, denoted as $X/\underset{\sim}{R} \cong X/\underset{\sim}{Q}$.

Theorem 4.3. Let $[\underset{\sim}{T}]$ be a group of fuzzy isomorphic equivalent relations on X, then $\forall \underset{\sim}{R}, \underset{\sim}{Q} \in [\underset{\sim}{T}], X/\underset{\sim}{R} \cong X/\underset{\sim}{Q}$.

Example 4.2. For the data above Example 4.1, suppose c=0.05, and establishing the fuzzy equivalent relation matrix $\underset{\sim}{Q} = (q_{ij})_{5\times 5}$ basing the method of absolute value subtrahend, $q_{ij} = 1 - 0.05\sum_{k=1}^{4} |a_{ik} - a_{jk}|$, then

$$\underset{\sim}{Q} = \begin{pmatrix} 1 & 0.7 & 0.9 & 0.75 & 0.75 \\ 0.7 & 1 & 0.7 & 0.7 & 0.7 \\ 0.9 & 0.7 & 1 & 0.75 & 0.75 \\ 0.75 & 0.7 & 0.75 & 1 & 0.8 \\ 0.75 & 0.7 & 0.75 & 0.8 & 1 \end{pmatrix}$$

we get by computing:

$X/Q_1 = \{(a_1),(a_2),(a_3),(a_4),(a_5)\}$; $X/Q_{0.9} = \{(a_1,a_3),(a_2),(a_4),(a_5)\}$;
$X/Q_{0.8} = \{(a_1,a_3),(a_2),(a_4,a_5)\}$; $X/Q_{0.75} = \{(a_1,a_3,a_4,a_5),(a_2)\}$;
$X/Q_{0.7} = \{(a_1,a_2,a_3,a_4,a_5)\}$.

Then

$$X/\underset{\sim}{Q} = X/Q_1 \cup 0.9\circ(X/Q_{0.9}) \cup 0.8\circ(X/Q_{0.8}) \cup 0.75\circ(X/Q_{0.75}) \cup 0.7\circ(X/Q_{0.7})$$
$$= \{[a_1]_{\underset{\sim}{Q}}, [a_2]_{\underset{\sim}{Q}}, [a_3]_{\underset{\sim}{Q}}, [a_4]_{\underset{\sim}{Q}}, [a_5]_{\underset{\sim}{Q}}\}.$$

Where

$[a_1]_{\underset{\sim}{Q}} = \{a_1\} \cup 0.9\circ\{a_1,a_3\} \cup 0.8\circ\{a_1,a_3\} \cup 0.75\circ\{a_1,a_3,a_4,a_5\} \cup 0.7\circ\{a_1,a_2,a_3,a_4,a_5\}$,

$[a_1]_{\underset{\sim}{Q}}(a_1) = 1$, $[a_1]_{\underset{\sim}{Q}}(a_2) = 0.7$, $[a_1]_{\underset{\sim}{Q}}(a_3) = 0.9$, $[a_1]_{\underset{\sim}{Q}}(a_4) = [a_1]_{\underset{\sim}{Q}}(a_5) = 0.75$.

$[a_2]_{\underset{\sim}{Q}} = \{a_2\} \cup 0.7\circ\{a_1,a_2,a_3,a_4,a_5\}$,

$[a_2]_{\underset{\sim}{Q}}(a_2) = 1$, $[a_2]_{\underset{\sim}{Q}}(a_1) = [a_2]_{\underset{\sim}{Q}}(a_3) = [a_2]_{\underset{\sim}{Q}}(a_4) = [a_2]_{\underset{\sim}{Q}}(a_5) = 0.7$.

$[a_3]_{\underset{\sim}{Q}} = \{a_3\} \cup 0.9\circ\{a_1,a_3\} \cup 0.8\circ\{a_1,a_3\} \cup$

$0.75\circ\{a_1,a_3,a_4,a_5\} \cup 0.7\circ\{a_1,a_2,a_3,a_4,a_5\}$, $[a_3]_{\underset{\sim}{Q}}(a_3) = 1$, $[a_3]_{\underset{\sim}{Q}}(a_1) = 0.9$,

$[a_3]_{\tilde{Q}}(a_2)=0.7$, $[a_3]_{\tilde{Q}}(a_4)=[a_3]_{\tilde{Q}}(a_5)=0.75$.

$[a_4]_{\tilde{Q}}=\{a_4\}\cup 0.8\circ\{a_4,a_5\}\cup 0.75\circ\{a_1,a_3,a_4,a_5\}\cup\ 0.7\circ\{a_1,a_2,a_3,a_4,a_5\}$,

$[a_4]_{\tilde{Q}}(a_4)=1$, $[a_4]_{\tilde{Q}}(a_1)=[a_4]_{\tilde{Q}}(a_3)=0.75$, $[a_4]_{\tilde{Q}}(a_5)=0.8$, $[a_4]_{\tilde{Q}}(a_2)=0.7$.

$[a_5]_{\tilde{Q}}=\{a_5\}\cup 0.8\circ\{a_4,a_5\}\cup 0.75\circ\{a_1,a_3,a_4,a_5\}\cup\ 0.7\circ\{a_1,a_2,a_3,a_4,a_5\}$,

$[a_5]_{\tilde{Q}}(a_5)=1$, $[a_5]_{\tilde{Q}}(a_4)=0.8$, $[a_5]_{\tilde{Q}}(a_1)=[a_5]_{\tilde{Q}}(a_3)=0.75$, $[a_5]_{\tilde{Q}}(a_2)=0.7$.

It is clear that $X/\tilde{R}$ in Example 4.1 is isomorphic to $X/\tilde{Q}$.

Theorem 4.4. Let $\tilde{R}$ and $\tilde{Q}$ be two fuzzy equivalent relations on X. If $\tilde{R}\sim\tilde{Q}$, then $X/\tilde{Q}$ is a fuzzy quotient approximation space on $X/\tilde{R}$.

Theorem 4.5. Let $\tilde{R}$ be a fuzzy equivalent relation on X and (X, R_α) be an approximation space defined by R_α. $\forall$ a ,b$\in$ X / R_α, let d (a ,b)=1- $\tilde{R}$ (x ,y), $\forall$ x$\in$ a, y$\in$ b. Then d (.,.) is a distance function on X / R_α.

Theorem 4.6. Let $[\tilde{T}]$ be a group of fuzzy equivalent relations which are isomorphic for each other on X. Then, $[\tilde{T}]$ can define uniquely a fuzzy approximation space on X.

Definition 4.3. Let $\{(X, R_\alpha) \mid \alpha\in \mathrm{Im}(\tilde{R})\}$ be a group of approximation spaces satisfied that:

$0\le \alpha_2\le\alpha_1\le 1 \Leftrightarrow$ X / R_{α_2} is a quotient approximation space on X / R_{α_1}. Then, $\{(X, R_\alpha) \mid \alpha\in \mathrm{Im}(\tilde{R})\}$ is called a nested quotient approximation spaces on X.

Theorem 4.7. Let $\{(X, R_\alpha) \mid \alpha\in \mathrm{Im}(\tilde{R})\}$ be a nested quotient approximation spaces on X . Then,

$\{(X, R_\alpha) \mid \alpha\in \mathrm{Im}(\tilde{R})\}$ can define uniquely a fuzzy approximation space on X. Moreover, it can define a group of fuzzy equivalent relations on X which are isomorphic for each other.

Proof. By Definition4.3 and the method in Theorem3.4 and Theorem3.5,we can complete the proof of this theorem.

5 Conclusion

We discussed the relation between FS and RS by theory of the isomorphism and homomorphism of fuzzy sets. We proved the following two main conclusions. Firstly, for any group of fuzzy sets which are isomorphic for each other in X, an approximation space could be defined uniquely. Secondly, for any group of fuzzy equivalence relations which are isomorphic, similarly a group of fuzzy approximation spaces which are isomorphic could be defined. Moreover, the construction of fuzzy approximation space is given. These results preliminary have established connection between fuzzy sets and quotient spaces. It is beneficial to the development of granular computing and its applications.

Acknowledgements

I am grateful to the anonymous referees for their valuable comments and suggestions. This work supported by the National Natural Science Foundation of China (NSFC) under Projects 60364001 and 70461001 and National Natural Science Foundation of Hainan under Projects 80401.

References

1. Pawlak Z(1991). " Rough Sets Theoretical Aspects of Reasoning About Data". *Dordrecht*, Kluwer Academic Publishers.
2. Dubois, D, Prade H(1990). "Rough fuzzy sets and fuzzy rough sets", *International Journal of General Systems* 2:191-209.
3. Dubois, D. and Prade, H(1992)."Putting rough sets and fuzzy sets together," *Intelligent Decision Support: Handbook of Applications and Advances of the Rough Sets Theory*, 203-222Slowinski, R., Ed., Kluwer Academic Publishers, Boston.
4. Zadeh L.A(1996). "fuzzy logic=computing with words", *IEEE Transactions on Fuzzy Systems* 1:103-111.
5. Zadeh L.A(1997). "Towards a theory of fuzzy information granulation and its centrality in human reasoning and fuzzy logic". *Fuzzy Sets and Systems* 1:111-127.
6. Zadeh L.A(1998). "Some reflections on soft computing, granular computing and their roles in the conception, design and utilization of information/intelligent systems", *Soft Computing* 1:23-25.
7. Yao Y.Y(2000). "Granular computing: basic issues and possible solutions", *Wang PP, ed. Proceedings of the 5th Joint Conference on Information Sciences* 186-189, Association for Intelligent Machinery.
8. Yao Y.Y, Li X(1996). "Comparison of rough-set and interval-set models for uncertain reasoning", *Fundamental Informatics*1: 289-298.
9. Yao, Y.Y(1995). "On combining rough and fuzzy sets," *Proceedings of the CSC'95 Workshop on Rough Sets and Database Mining*, Lin, T.Y. (Ed.),San Jose State University.
10. Zhang B, Zhang L(1992).*Theory and Applications of Problem Solving*. Elsevier Science Publishers B.V.

Part III

Soft Computing

Particle Swarm Optimization Algorithm Design for Fuzzy Neural Network

Ming Ma[1] and Li-Biao Zhang[2]

[1] Information Manage Center, Beihua University, Jilin 132013, China
mam@mail.edu.cn
[2] College of Computer Science and Technology, Jilin University, Changchun 130012 , China

Abstract. Designing a fuzzy neural network can be considered as solving an optimization problem. A pruning algorithm for solving the optimization problem is presented based on particle swarm optimization with division of work. It can optimize the fuzzy rules automatically, and determine the structure of the fuzzy neural network based on the effective rules obtained through fuzzy rules searching. Numerical simulations show the effectiveness of the proposed algorithm.

Keywords: particle swarm optimization, fuzzy neural network, fuzzy rule.

1 Introduction

We have encountered some problems in the course of setting up fuzzy neural network, for example it is difficult to obtain fuzzy rules for more complex system. Some small fuzzy system conducted employing expert experience, but in application most of the fuzzy rules can't be obtained directly. Therefore, designing a fuzzy neural network can be considered as solving an optimization problem. This has promoted research on how to identify an optimal and efficient fuzzy neural network structure for a given problem.

Particle Swarm Optimization (PSO) is an optimization algorithm proposed by Kennedy and Eberhart in 1995 [1][2]. It is easy to be understood and realized, and it has been applied in many optimization problems [3][4][5]. The PSO is more effective than traditional algorithms in most cases. The application of PSO in the above optimization problem could be very promising. In this paper, a Particle Swarm Optimization with Division of Work is proposed, and on this basis a pruning algorithm is proposed to obtain fuzzy rules. Satisfactory results through experiments are obtained.

The rest of this paper is organized as follows: The weighted fuzzy neural network is introduced in Section 2. The proposed PSO with Division of Work is introduced in Section 3. The proposed pruning algorithm is described in Section 4. The simulation and experimental results are presented in Section 5. Finally, concluding remarks are given in Section 6.

2 Fuzzy Neural Network Architecture

Different fuzzy inference mechanism can be distinguished by the consequents of the fuzzy if-then rules[6], such as Mamdani and Takagi-Sugeno inference system.

B.-Y. Cao (Ed.): Fuzzy Information and Engineering (ICFIE), ASC 40, pp. 309–314, 2007.
springerlink.com

Mamdani inference mechanism is adopted in this paper. The weighted fuzzy neural network (WFNN) is an adaptive network based on improving fuzzy weighted reasoning method. In improving fuzzy weighted reasoning method, we try to list all possible fuzzy rules. The fuzzy if –then rules have the form:

$$\underset{j=1}{\overset{m}{OR}} \; (IF \; (\underset{i=1}{\overset{n}{AND}} \; (x_i \text{ is } A_{ij})) \text{ THEN } Y \text{ is } w_{j1}/z_1, w_{j2}/z_2, \ldots, w_{jl}/z_l), \tag{1}$$

where m is the number of fuzzy rules, n is the number of input variables , l is the number of consequent fuzzy set, z_i(i=1,…,l)is a constant, w_{ji} (j=1,…,m, i=1,…,l)is weight parameter, and A_{ij}(j=1,…,m) is the antecedent fuzzy set of the ith rule. Then the output value is calculated as following.

$$z_0 = \frac{\sum_{i=1}^{l} (f(\sum_{j=1}^{m} u_j w_{ji}) \times z_i)}{\sum_{i=1}^{l} (f(\sum_{j=1}^{m} u_j w_{ji}))}, \tag{2}$$

where u_j is the extent to which a rule is activated. Each u_j is calculated as $u_j=u_{A1j}(x_1)$ AND $u_{A2j}(x_2)$…AND $u_{Anj}(x_n)$, and $u_{Aij}(x_i)$ is the membership function value of x_i in the A_{ij} fuzzy set.

The weighted fuzzy neural network is a seven layers feedforward network, as shown in Fig 1.

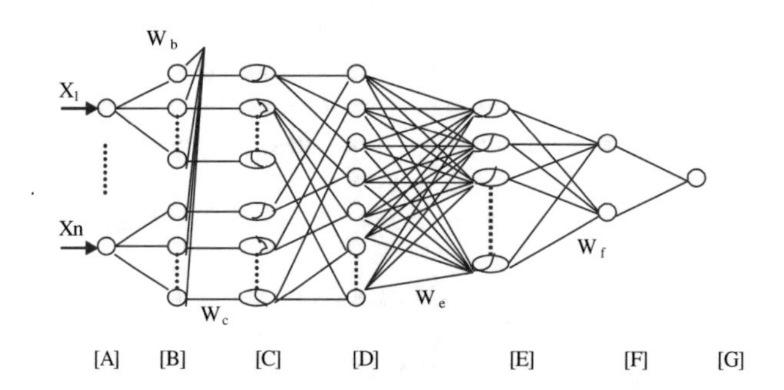

Fig. 1. The weighted fuzzy neural network architecture. The layer of A is an input layer, and all the x_i (i=1…n) are input signals; the layer of B and C perform fuzzification of the input data, and w_b,w_c are parameters of membership functions ;the layer of D and E perform the fuzzy inference, all the w_e are weighted parameters, and each w_{ei} represents the importance of the corresponding fuzzy rule;the layer of F and G perform defuzzification of the output data, and all the w_f are weighted parameters.

3 Particle Swarm Optimization with Division of Work

PSO originated from the research of food hunting behaviors of birds. Each swarm of PSO can be considered as a point in the solution space. If the scale of swarm is N, then the position of the i-th (i=1,2…N) particle is expressed as Xi . The “best” position passed by the particle is expressed as pBest [i] . The speed is expressed with Vi . The index of the position of the "best" particle of the swarm is expressed with g. Therefore, swarm i will update its own speed and position according to the following equations [1][2]:

$$\begin{gathered} Vi=w*Vi+c1*rand()*(pBest[i]-Xi)+c2*Rand()*(pBest[g]-Xi), \\ Xi=Xi+Vi, \end{gathered} \tag{3}$$

where c1 and c2 are two positive constants, rand() and Rand() are two random numbers within the range [0,1],and w is the inertia weight.

The theorems and the related prove for the relationship between parameters setting and particles convergence are presented in reference [7]. It is pointed out that there are reasonable parameters w、c1 and c2 with which the particle will converge onto a line segment whose endpoints are pBest[i] and pBest[g] respectively in certain assumed conditions. Although there is certain difference between assumed conditions and actual situations, it is rather significant reference for improving the PSO. On this basis a Particle Swarm Optimization with Division of Work is proposed. In the strategy of work-dividing, the whole swarm is divided into two subgroups: P_Near, and P_Far. The subgroup P_Near is responsible for finding new individual best and swarm best, and the subgroup P_Far is responsible for exploring new searching areas, as shown in follows.

P_Near: suppose that Xj(t) =pBest[j]= pBest[g], we compute the distance between each particle and pBest[g] ,all the particles which the corresponding distance is smaller than a fixed value form P_Near from the whole swarm. In P_Near each particle update its own speed and position according to the equation (3), and choose proper parameters w, c1 and c2 and satisfy $w > \frac{c_1 + c_2}{2} - 1$, so the particle's searching areas are considered as the most hopeful to find new individual best and swarm best.

To Xj(t) we adopt the ideas of EP and seeks new swarm best near the optimum, as shown in following.

$$Xj'(t) = Xj(t) + \sigma N(0,1) \tag{4}$$

where σ is the mutation step size, and N (0,1) is a random number within the range [0,1].

We choose the best one as pBest [j] in the particle Xj'(t) and the particle Xj(t) .

P_Far:all the particles which are relatively far with pBest[g] form P_far from the whole swarm. In P_far each particle update its own speed and position according to the equation (3), and choose proper parameters w, c1 and c2 and satisfy $w < \frac{c_1 + c_2}{2} - 1$. The particles diverge in P_Far and they act as a role for exploring new searching areas during the operation of the algorithm. These particles can keep the diversity of the population and avoid the "premature" phenomenon to taking place.

In the execution of the algorithm, element of two subgroup change constantly too, and particle of P-Far might become members of the P-Near in the next. It mean at this moment that there are new searching areas that are opened up.

4 Pruning Algorithm

In the proposed algorithm, we define a real-valued vector as a particle to represent a solution., the vector has the form:

$$X=(x_1, x_2 \ldots x_m, x_{m+1}, \ldots x_n) , \tag{5}$$

where m is the number of all the connections between D layer and E layer, n is the number of all the parameters, the x_i (i=1,…,m) represent the weighted parameters of the connection between D layer and E layer, and the x_i (i=m+1,…,n) represent the other connection parameters or the parameters of membership functions. As illustrated in Fig 1, each x_i (i=1,…,m) represents the corresponding w_{ei}, and each x_i (i=m+1,…,n) represents the other parameter. According to the improving fuzzy weighted reasoning method, we let x_i (i=1,…,m)>0 , and let x_i be a random number within the range (0,1) when x_i <0 in the execution of the algorithm.

The following functions have been used for evaluation of PSO.

$$f_2(X) = \frac{1}{\frac{1}{2}\sum_K (O - T)^2} , \tag{6}$$

where K is the number of sample ,T is the teacher signal , and O is the output of network.

As illustrated in Figure 1, each node in D layer is connected with each node in E layer, and each connection between D layer and E layer represents a fuzzy rule, by this token, the fuzzy rules are redundant. The extraction of fuzzy rules is how to cut some connections between D layer and E layer. In the weighted fuzzy neural network, the value of each w_e represents a corresponding connection between D layer and E layer, when the value is very small, that is to say, the connection is redundant. Because each x_i (i=1,…,m) represents the corresponding w_e , we can confirm the structure of the weighted fuzzy neural network by computing the value of each x_i (i=1,…,m) . Now we describe our method in following.

The pruning algorithm based on PSO is formed of two phases. In the first phase, each particle updates its own speed and position according to (3) with the corresponding parameters. If the fixed precision is achieved, then go to the second phase. In the second phase, we adopt the following step to cut the redundant connection between D layer and E layer.

Step 1: we let all connections from a node of D layer in Fig 1 be a group, and then each group is composed of corresponding x_is.

Step 2: calculate the sum of all the x_i's value in each group, and calculate the proportion of each group to all the groups. The proportion represents the importance of fuzzy rules in the corresponding group. If it's value is smaller than a small value, that is to say the importance of those fuzzy rules is little enough, then those fuzzy rules are disabled.

Step 3: in each group calculate the proportion of each x_i's value to the sum. The proportion represents the importance of the corresponding fuzzy rule in the group.

If it's value is much bigger than the others in the group, the corresponding fuzzy rule is enabled, the others fuzzy rules are disabled; if it's value is smaller than a small value, the corresponding fuzzy rule is disabled.

Initialize the particle swarm again, and loop the process until it achieves the termination condition, we will obtain suitable fuzzy rules in the end.

5 Numerical Simulations

Box and Jenkins's gas furnace [8] is a famous example of system identification. The well-known Box-Jenkins data set consists of 296 input-output observations, where the input u(t) is the rate of gas flow into a furnace and the output y(t) is the CO_2 concentration in the outlet gases. Since the process is dynamical, there are different values of the variables that can be considered as candidates to affect the present output y(t).In our case we have considered y(t- 1) and u(t- 3) as inputs to predict y(t).

The fuzzy sets of the input consist {small, middle, big}, the fuzzy sets of the output consist {small, middle, big}, and use the Gaussian membership function for each fuzzy subset. In the proposed pruning algorithm, the parameters of the PSO are these: in the subgroup P_Near learning rate c1=c2=1.5,and inertia weight is 1.2; in the subgroup P_Far learning rate c1=c2=2, and inertia weight is taken from 0.8 to 0.4 with a linear decreasing rate. The population of particles was set to 40, and stopping condition: 3000 generations. Before the execution of the algorithm the weighted fuzzy neural network has 27 fuzzy rules. The comparative results between the model in this paper and other models are summarized in Table 1.

Table 1. Comparative results

Model	inputs	rules	error
Tong [9]	y(t-1),u(t-4)	19	0.469
Pedrycz [9]	y(t-1),u(t-4)	81	0.320
Sugeno [9]	y(t-1),u(t-3), u(t-4)	6	0.190
In this paper	y(t-1),u(t-3)	9	0.0935

From the Table 1, we can see that we obtain a better performance with a number of rules smaller in comparison with other methods known in the literature.

6 Conclusion

Based on the PSO with Division of Work and weighted fuzzy neural network, the proposed pruning algorithm can solve the fuzzy neural network optimization problem. The numerical experiments indicate the effectiveness of the algorithm.

Acknowledgement

This work was supported by the National Natural Science Foundation of China under Grant No. 60433020.

References

1. Eberhart, R., Kennedy, J(1995) A New Optimizer Using Particle Swarm Theory. In:Proc. Sixth International Symposium on Micro Machine and Human Science(Nagoya, Japan). IEEE Service Center,Piscataway,NJ : 39-43
2. Kennedy, J., Eberhart, R(1995) Particle Swarm Optimization.In: IEEE International Conference on Neural Networks (Perth, Australia). IEEE Service Center, Piscataway, NJ , Volume IV: 1942-1948
3. Van den Bergh F, Engelbrecht A P(2000) Cooperative Learning in Neural Networks using Particle Swarm Optimizers. South Africam Computer Journal, (26): 84-90.
4. Parsopoulos, K.E., Vrahatis, M. N(2001) Particle Swarm Optimizer in Noisy and Continuously Changing Environments. In: Artificial Intelligence and Soft Computing. IASTED/ACTA Press,Anaheim,CA,USA: 289-294
5. Li-Biao Zhang, Chun-Guang Zhou, Yan-Chun Liang(2003) Solving Multi Objective Optimization Problems Using Particle Swarm Optimization. The 2003 Congress on Evolutionary Computation, Canberra, Australia,Volume IV. IEEE Service Center, Piscataway, NJ: 2400-2405
6. Ludmila, I. K(2000) How Good Are Fuzzy If-Then Classifiers?. IEEE Trans on Systems, Man and Cybernetics Part B: Cybernetics 30(4) :501-509
7. Quan-Sheng Dou,Chun-Guang Zhou,Ming Ma(2005) Two Improvement Strategies for Particle Swarm Optimization. Journal of Computer Research and Development, 42(5):897-904(in Chinese)
8. G E P Box, G M. Jenkins(1970) Time Series Analysis, Forecasting, and Control. Ssn Francisco, CA: Holden Day.
9. Ji-Chang Lo , Chien-Hsing Yang(1999) A heuristic error-feedback learning algorithm for fuzzy modeling . IEEE Trans on Systems , Man and Cybernetics , 29(6) : 686 -691.

A Method of Intuitionistic Fuzzy Reasoning Based on Inclusion Degree and Similarity Measure

Lin Lin[1] and Xue-hai Yuan[2]

[1] Department of Science, Shenyang Institute of Aeronautical Engineering, Shenyang, 110136, P.R. China
authorllin@yahoo.com.cn

[2] Department of Mathematics, Liaoning Normal University, Dalian, 116029, P.R. China
yuanxuehai@yahoo.com.cn

Abstract. In this paper, a new method of intuitionistic fuzzy reasoning is proposed. First, the definition of inclusion degree based on fuzzy implication operators and the similarity measure between intuitionistic fuzzy sets(*IFSs*) are introduced. Then, the similarity measure of *IFSs* is applied to intuitionistic fuzzy reasoning. Finally, the process of reasoning is illustrated with an example.

Keywords: Intuitionistic fuzzy sets, Intuitionistic fuzzy reasoning, Inclusion degree, Similarity measure.

1 Introduction

In 1965, the definition of fuzzy sets was introduced by Zadeh in the paper[1]. As its generalizations, the notion of intuitionistic fuzzy set(*IFS*) was introduced by Atanassov[2] in 1986. Since the *IFS* processes the information of membership and non-membership, it provides more choices for the attribute description of an object and has stronger ability to express uncertainty. These *IFSs* have been widely studied and applied in a variety of areas such as logic programming[3], decision making problems[4] and in medical diagnostics[5].

In many applications, the approximate reasoning of *IFSs* is very important. At present, lots of methods for intuitionistic fuzzy reasoning in literatures are based on the composition of triangular norms. In 2004, the method of intuitionistic fuzzy compositional rule of inference(ICRI) was proposed by Deschrijver[6].

In fuzzy sets theory, the methods of approximate reasoning based on similarity measure are usually used. Recently, many similarity measures have been proposed by distance between *IFSs*[7, 8]. However, these similarity measures may not be effective in some cases.

In this study, we present a new method to calculate the degree of similarity between *IFSs* based on inclusion degree. Numerical results show that the proposed

B.-Y. Cao (Ed.): Fuzzy Information and Engineering (ICFIE), ASC 40, pp. 315–322, 2007.
springerlink.com

similarity measure is more reasonable than existing methods. The organization of this paper is as follows: In Section 2, the inclusion degree between *IFSs* based on fuzzy implication operators is proposed. Section 3 uses the proposed inclusion degree to generate a new similarity measure to calculate the degree of similarity between *IFSs*, and compares this similarity measure with exiting methods. A method of intuitionistic fuzzy reasoning on the basis of similarity measure is introduced in Section 4 and in the same section, the process of reasoning is illustrated with an example. Conclusions are made in Section 5.

2 Inclusion Degree Based on Fuzzy Implication Operators

The notion of intuitionistic fuzzy set is introduced by Atanassov in 1986.

Definition 1. *Let $X \neq \emptyset$ be a given set. An intuitionistic fuzzy set in X is an expression A given by*

$$A = \{< x, \mu_A(x), \nu_A(x) > | x \in X\}$$

where $\mu_A : X \to [0,1]$, $\nu_A : X \to [0,1]$ with the condition $0 \leq \mu_A(x)+\nu_A(x) \leq 1$, for all x in X.

The numbers $\mu_A(x)$ and $\nu_A(x)$ denote, respectively, the membership degree and the nonmembership degree of the element x in A.

For convenience of notation, we abbreviate "intuitionistic fuzzy set" to *IFS* and represent *IFSs(X)* as the set of all the *IFSs* in X.

The operations of *IFS* are defined as follows, for every $A, B \in$ *IFSs*(X),
$A = B$ if and only if $A \subseteq B$ and $B \subseteq A$.
$A \subseteq B$ if and only $\mu_A(x) \leq \mu_B(x)$,$\nu_A(x) \geq \nu_B(x)$;
$A \cap B = \{< x, \min(\mu_A(x), \mu_B(x)), \max(\nu_A(x), \nu_B(x)) >\}$;
$A \cup B = \{< x, \max(\mu_A(x), \mu_B(x)), \min(\nu_A(x), \nu_B(x)) >\}$.

Let $X = \{x_1, x_2, \cdots, x_n\}$. The definition of inclusion degree was proposed in the paper[9].

Definition 2. *Denote σ :IFS(X)×IFS(X)→ $[0,1]$, $\sigma(A,B)$ is said to be the degree of IFS B including IFS A, and σ is called the inclusion degree function, if $\sigma(A,B)$ satisfies the properties (D1)-(D3).*
(D1) $A \subseteq B \Rightarrow \sigma(A,B) = 1$;
(D2) $\sigma(X, \emptyset) = 0$;
(D3) $\sigma(C,A) \leq \min\{\sigma(B,A), \sigma(C,B)\}$ *if* $A \subseteq B \subseteq C$.

Theorem 1. *For all $A, B \in$IFSs(X). Let I be an implication operator. If I satisfies the property*

$$\forall a, b \in [0,1], \quad a \leq b \Rightarrow I(a,b) = 1.$$

Denote:

$$\sigma(A,B) = \frac{1}{n}\sum_{i=1}^{n}(\lambda I(\mu_A(x_i),\mu_B(x_i)) + (1-\lambda)I(1-\nu_A(x_i),1-\nu_B(x_i))), \quad (1)$$

$\lambda \in [0,1]$.
Then, σ *:IFSs(X)×IFSs(X)→* $[0,1]$ *is an inclusion degree function.*

Proof. Obviously, $\sigma(A,B)$ satisfies (D1) and (D2) of Definition 2. In the following, $\sigma(A,B)$ will be proved to satisfy (D3).

For I is a fuzzy implication operator, $\forall a,b,c \in [0,1]$,

$$b \le c \Rightarrow I(a,b) \le I(a,c),$$

$$a \le c \Rightarrow I(a,b) \ge I(c,b).$$

For any $C = \{< x, \mu_C(x), \nu_C(x) > | x \in X\}$, $A \subseteq B \subseteq C$.

We have $\mu_A(x_i) \le \mu_B(x_i) \le \mu_C(x_i)$ and $\nu_A(x_i) \ge \nu_B(x_i) \ge \nu_C(x_i)$.

It is easy to follow that

$$I(\mu_C(x_i),\mu_A(x_i)) \le I(\mu_C(x_i),\mu_B(x_i)),$$

$$I(1-\nu_C(x_i),1-\nu_A(x_i)) \le I(1-\nu_C(x_i),1-\nu_B(x_i))$$

Then

$$\lambda I(\mu_C(x_i),\mu_A(x_i)) + (1-\lambda)I(1-\nu_C(x_i),1-\nu_A(x_i))$$
$$\le \lambda I(\mu_C(x_i),\mu_B(x_i)) + (1-\lambda)I(1-\nu_C(x_i),1-\nu_B(x_i))$$

So we have $\sigma(C,A) \le \sigma(C,B)$.

In the similar way, we can prove that $\sigma(C,A) \le \sigma(B,A)$ if $A \subseteq B \subseteq C$. Therefore $\sigma(A,B)$ satisfies (D3) in Definition 2. So we have finished the proof of the theorem.

3 Similarity Measure Between *IFSs* Based on Inclusion Degree

In the study of the similarity between *IFSs*, Li and Cheng[8] introduced the following definition.

Definition 3. *For all* $A,B,C \in$ *IFS(X). Denote* δ : *IFS(X)× IFS(X)→* $[0,1]$. $\delta(A,B)$ *is said to be the degree of similarity between* A *and* B, δ *is called the similarity measure function, if* $\delta(A,B)$ *satisfies the properties (L1)-(L4).*
(L1) $\delta(A,B) = \delta(B,A)$;
(L2) $\delta(A,A) = 1$;
(L3) $\delta(X,\emptyset) = 0$;
(L4) $\delta(A,C) \le \min\{\delta(A,B),\delta(B,C)\}$ *if* $A \subseteq B \subseteq C$.

Theorem 2. *Let σ be an inclusion degree function on IFS(X). For all $A, B \in$ IFS(X), denote:*

$$\delta(A,B) = \sigma(A \cup B, A \cap B) \tag{2}$$

Then $\delta(A,B)$ is the degree of similarity between A and B.

Proof. Obviously, $\delta(A,B)$ satisfies (L1)-(L3) of Definition 3. In the following, $\delta(A,B)$ will be proved to satisfy (L4).

From (2), we have $\delta(A,C) = \sigma(A \cup C, A \cap C)$, $\delta(A,B) = \sigma(A \cup B, A \cap B)$.

For $A \subseteq B \subseteq C$, $A \cup C \supseteq A \cup B, A \cap C = A \cap B$.

Then $\sigma(A \cup C, A \cap C) \leq \sigma(A \cup B, A \cap B)$, $\delta(A,C) \leq \delta(A,B)$.

In the similar way, we can prove that $\delta(A,C) \leq \delta(B,C)$ if $A \subseteq B \subseteq C$. Therefore $\delta(A,B)$ satisfies (L4) in Definition 3. So we have finished the proof of the theorem.

Now, let's compare the method with exiting methods.

In the paper[7], the degree of similarity between the two *IFSs* A and B can be calculated as follows:

$$\delta_C(A,B) = 1 - \frac{1}{n}\sum_{i=1}^{n} \frac{\mid (\mu_A(x_i) - \mu_B(x_i)) - (\nu_A(x_i) - \nu_B(x_i)) \mid}{2} \tag{3}$$

In the paper[8], the degree of similarity is calculated as

$$\delta_L(A,B) = 1 - \sqrt[p]{\frac{1}{n}\sum_{i=1}^{n} [\frac{\mid (\mu_A(x_i) - \mu_B(x_i)) - (\nu_A(x_i) - \nu_B(x_i)) \mid}{2}]^p} \tag{4}$$

To overcome the drawbacks of $\delta_C(A,B)$ and $\delta_L(A,B)$, Hong and Kim[10] proposed the following similarity measures between *IFSs*.

$$\delta_H(A,B) = 1 - \frac{1}{n}\sum_{i=1}^{n} \frac{\mid \mu_A(x_i) - \mu_B(x_i) \mid + \mid \nu_A(x_i) - \nu_B(x_i) \mid}{2} \tag{5}$$

But the above methods also have drawbacks. For example, Let $X = (x_1, \cdots, x_n)$, $A, B_1, B_2, B_3 \in$ *IFS*(X), where,

$$A = \{(x_i, 0, 0) \mid x_i \in X\}, B_1 = \{(x_i, 0.2, 0.8) \mid x_i \in X\},$$

$$B_2 = \{(x_i, 0.4, 0.6) \mid x_i \in X\}, B_3 = \{(x_i, 0.5, 0.5) \mid x_i \in X\}.$$

By (5), we have

$$\delta_H(A, B_1) = \delta_H(A, B_2) = \delta_H(A, B_3) = 0.5.$$

Obviously, the method is not effective in this case.

Let $I(x,y) = \begin{cases} 1 & x \leq y \\ 1 - x + y & x > y \end{cases}$. By (1) and (2), we have

$$\delta(A, B_1) = 0.2 + 0.6\lambda, \quad \delta(A, B_2) = 0.4 + 0.2\lambda, \quad \delta(A, B_3) = 0.5.$$

Because different people has different opinion, measureing the similarity of the two *IFSs* based on the $\lambda \in [0,1]$ is more reasonable to express different people's opinion. For example, let $\lambda = 0.1$, then we have

$$\delta(A, B_1) = 0.26, \quad \delta(A, B_2) = 0.42, \quad \delta(A, B_3) = 0.5.$$

4 Intuitionistic Fuzzy Reasoning Based on Similarity Measure

Let $X = \{x_1, x_2, \cdots, x_n\}$, $Y = \{y_1, y_2, \cdots, y_n\}$.

Definition 4. *Let an intuitionistic fuzzy set on* X *be*

$$A = \sum_{i=1}^{n} (\mu_A(x_i), \nu_A(x_i))/x_i, x_i \in X.$$

For all $a \in [0,1]$, *the quantitative product of IFS* A *and* a *is defined as*

$$aA = \sum_{i=1}^{n} (a\mu_A(x_i), 1 - a(1 - \nu_A(x_i)))/x_i. \tag{6}$$

Especially, $aA = \sum_{i=1}^{n} (0,1)/x_i$ if $a = 0$; $1A = A$ if $a = 1$.

There are two models of intuitionistic fuzzy reasoning. One is Intuitionistic Fuzzy Modus Ponens (IFMP, for short). It can be expressed as:

$$\text{(M} - 3.1) \qquad \begin{array}{l} A \longrightarrow B \\ A^* \\ \hline \qquad\quad B^* \end{array}$$

The other is Intuitionistic Fuzzy Modus Tollens (IFMT, for short). It can be expressed as:

$$\text{(M} - 3.2) \qquad \begin{array}{r} A \longrightarrow B \\ B^* \\ \hline A^* \qquad\quad \end{array}$$

In this paper, we only deal with the IFMP problem.

4.1 Single Rule Case of IFMP

In (M-3.1), let $\delta(A^*, A) = \delta$. Then the conclusion is

$$B^* = \delta B = \sum_{i=1}^{n} (\delta\mu_B(y_i), 1 - \delta(1 - \nu_B(y_i)))/y_i. \tag{7}$$

Obviously, when $A^* = A$, $\delta = 1$ and $B^* = B$.

4.2 Multi-rules Case of IFMP

The model of multi-rules reasoning is

$$(\mathrm{M}-3.3)\qquad \begin{array}{l} \text{IF } x \text{ is } A_1 \text{ THEN } y \text{ is } B_1 \\ \text{IF } x \text{ is } A_2 \text{ THEN } y \text{ is } B_2 \\ \quad\vdots \\ \text{IF } x \text{ is } A_p \text{ THEN } y \text{ is } B_p \\ \quad x \text{ is } A^* \\ \hline \qquad\qquad y \text{ is } B^* \end{array}$$

where p is the quantity of rules.

Let $\delta(A^*, A_i) = \delta_i$. $i = 1, 2, \cdots, p$. By (7), we can get the conclusion B_i^* of each rule,

$$B_i^* = \sum_{j=1}^{n} (\delta_i \mu_{B_i}(y_j), 1 - \delta_i(1 - \nu_{B_i}(y_j)))/y_j. \tag{8}$$

$i = 1, 2, \cdots, p$.

Then we have

$$B^* = B_1^* \vee_{L^*} B_2^* \vee_{L^*} \cdots \vee_{L^*} B_p^*. \tag{9}$$

4.3 Numerical Example

Suppose we have a system:

$$\begin{array}{ll} \text{Rules}: & \text{IF } x \text{ is } A_1 \text{ THEN } y \text{ is } B_1 \\ & \text{IF } x \text{ is } A_2 \text{ THEN } y \text{ is } B_2 \\ & \text{IF } x \text{ is } A_3 \text{ THEN } y \text{ is } B_3 \\ & \text{IF } x \text{ is } A_4 \text{ THEN } y \text{ is } B_4 \\ & \text{IF } x \text{ is } A_5 \text{ THEN } y \text{ is } B_5 \\ \text{Input}: & \quad x \text{ is } A_0 \\ \hline \text{Output}: & \qquad\qquad y \text{ is } B_0 \end{array}$$

where x, y are the names of objects, $A_0, A_1, \cdots, A_5$ are intuitionistic fuzzy sets on X, $B_0, B_1, \cdots, B_5$ are intuitionistic fuzzy sets on Y, and it is known

$$A_0 = \{(0.2, 0.6), (0.4, 0.4), (0.6, 0.2), (1, 0), (0.6, 0.2), (0.4, 0.4), (0.2, 0.6)\};$$

$$A_1 = \{(0, 1), (0, 0.8), (0.2, 0.6), (0.4, 0.4), (0.6, 0.2), (1, 0), (0.6, 0.2)\};$$

$$A_2 = \{(0, 0.8), (0.2, 0.6), (0.4, 0.4), (0.6, 0.2), (1, 0), (0.6, 0.2), (0.4, 0.4)\};$$

$$A_3 = \{(0.2, 0.6), (0.4, 0.4), (0.6, 0.2), (1, 0), (0.6, 0.2), (0.4, 0.4), (0.2, 0.6)\};$$

$$A_4 = \{(0.4, 0.4), (0.6, 0.2), (1, 0), (0.6, 0.2), (0.4, 0.4), (0.2, 0.6), (0, 0.8)\};$$

$$A_5 = \{(0.6, 0.2), (1, 0), (0.6, 0.2), (0.4, 0.4), (0.2, 0.6), (0, 0.8), (0, 1)\}.$$

$B_1 = A_5$; $B_2 = A_4$; $B_3 = A_3$; $B_4 = A_2$; $B_5 = A_1$.

Let $\lambda = 0.1$, $I(x,y) = \begin{cases} 1 & x \leq y \\ 1 - x + y & x > y \end{cases}$. By (2), we have

$$\delta_1 = \delta(A_1, A_0) = \sigma(A_1 \cup A_0, A_1 \cap A_0) = 0.654.$$

$$\delta_2 = \delta(A_2, A_0) = \sigma(A_2 \cup A_0, A_2 \cap A_0) = 0.769.$$

$$\delta_3 = \delta(A_3, A_0) = \sigma(A_3 \cup A_0, A_3 \cap A_0) = 1.$$

$$\delta_4 = \delta(A_4, A_0) = \sigma(A_4 \cup A_0, A_4 \cap A_0) = 0.94.$$

$$\delta_5 = \delta(A_5, A_0) = \sigma(A_5 \cup A_0, A_5 \cap A_0) = 0.654.$$

By (8), the results of each rule are

$B_1^* = \{(0.392, 0.131), (0.654, 0), (0.392, 0.131), (0.262, 0.262), (0.131, 0.392), (0, 0.523), (0, 0.654)\}$.

$B_2^* = \{(0.308, 0.308), (0.461, 0.154), (0.769, 0), (0.461, 0.154), (0.308, 0.308), (0.154, 0.461), (0, 0.615)\}$.

$B_3^* = \{(0.2, 0.6), (0.4, 0.4), (0.6, 0.2), (1, 0), (0.6, 0.2), (0.4, 0.4), (0.2, 0.6)\}$.

$B_4^* = \{(0, 0.635), (0.159, 0.476), (0.317, 0.317), (0.476, 0.159), (0.794, 0), (0.476, 0.159), (0.317, 0.317)\}$.

$B_5^* = \{(0, 0.654), (0, 0.523), (0.131, 0.392), (0.262, 0.262), (0.392, 0.131), (0.654, 0), (0.392, 0.131)\}$.

By (9), the conclusion of reasoning is

$$\begin{aligned} B^* &= B_1^* \vee_{L^*} B_2^* \vee_{L^*} B_3^* \vee_{L^*} B_4^* \vee_{L^*} B_5^* \\ &= \{(0.392, 0.138), (0.654, 0), (0.769, 0), (1, 0), \\ &\qquad (0.794, 0), (0.654, 0), (0.392, 0.131)\} \end{aligned}$$

5 Conclusion

In this paper, we proposed a new method of intuitionistic fuzzy reasoning. The definition of inclusion degree and the similarity measure between intuitionistic fuzzy sets based on fuzzy implication operators are introduced. Here, we only discussed the IFMP problem.

Acknowledgement

The authors are grateful for three anonymous referees for their helpful comments on an earlier version of this paper. The research was supported by the foundation of Shenyang Institute of Areonautical Engineering under Grant Number: 06YB24.

References

1. L. A. Zadeh (1965) Fuzzy sets. Inform. and Control 8: 338-353.
2. K. T. Atanassov (1986) Intuitionistic fuzzy sets. Fuzzy Sets and Systems 20: 87-96.
3. K. T. Atanassov, G. Gargov (1990) Intuitionistic fuzzy logic. Comptes Rendus de L'Academe Bulgare des Sciences 43: 9-12.
4. D. F. Li(2005) Multiattribute decision-making medols and methods using intuitionistic fuzzy set. Journal of Computer and System Sciences 70: 73-85.
5. S. K. De, R. Biswas, A. R. Roy (2001) An application of intuitionistic fuzzy sets in medical diagnosis. Fuzzy Sets and Systems 117: 209-213.
6. C. Cornelis, G. Deschrijver, E. E. Kerre (2004) Implication in intuitionistic fuzzy and interval-valued fuzzy set theory: construction, classification, application. International Journal of Approximate Reasoning 35: 55-95.
7. S. M. Chen (1995) Measure of similarity between vague sets. Fuzzy Sets and Systems 74(2): 217-223.
8. D. F. Li, C. Cheng (2002) New similarity measure of intuitionistic fuzzy sets and application to patten recognition. Pattern Recognition Letters 23: 221-225.
9. H. Bustine (2000) Indicator of inclusion grade for interval-valued fuzzy sets.Application for approximate reasoning based on interval-valued fuzzy sets. International Journal of Approximate Reasoning 23(3): 137-209.
10. D. H. Hong, C.Kim (1999) A note on similarity measures between vague sets and between elements. Information Sciences 115: 83-96.

A Discrete Particle Swarm Optimization Algorithm for the Multiobjective Permutation Flowshop Sequencing Problem

Guo Wenzhong[1], Chen Guolong[1,2,⋆], Huang Min[1], and Chen Shuili[3]

[1] College of Mathematics and Computer Science, Fuzhou University, Fuzhou 350002, China
{guowenzhong,cgl}@fzu.edu.cn, huangm1059@163.com
[2] School of Computer Science, National University of Defense Technology, Changsha 410073, China
[3] Department of Mathematics, School of Sciences, Jimei University, Xiamen 361021, China
shuilichen@vip.sina.com

Abstract. The application of Particle Swarm Optimization (PSO) on combinatorial optimization problems is limited, and it is due to the continuous nature of PSO. In order to solve the Multiobjective Permutation Flowshop Sequencing Problem (MPFSP), a Discrete Particle Swarm Optimization (DPSO) algorithm is proposed. To obtain a well approximation of true Pareto front, the phenotype sharing function of the objective space is applied in the definition of fitness function. The effectiveness of the proposed DPSO has been analyzed using 5 problems with the objectives of minimizing the makespan and the total tardiness, and 20 benchmarks problems with the objectives of minimizing the makespan and the total flowtime. The result shows that the proposed DPSO can reach a good approximation of true Pareto front.

Keywords: Multiobjective Permutation Flow Shop Sequencing, Discrete Particle Swarm, Makespan, Total Tardiness, Total Flowtime.

1 Introduction

Permutation Flowshop Sequencing Problem (PFSP) is one of the most well-known problems in the area of scheduling. It deals with the sequencing x of n processing jobs, where each job is to be processed with an identical sequence of machines. Most of research in PFSP is concerned with determining the sequence based on a single specific criterion. However, real-life scheduling decisions involve the consideration of more than one objective at a time. In such Multi-objective Permutation Flowshop Sequencing Problems (MPFSP), there is a set of trade-off solutions that are superior to the rest of solutions. These solutions are known as non-dominated solutions or Pareto-optimal solutions [1]. Attempts have been made to consider the MPFSP lately. Ishibichi and Murata et.al [2,3] proposed a Multi-Objective Genetic Local Search (MOGLS) algorithm for flow

⋆ Corresponding author. Email: cgl@fzu.edu.cn.

B.-Y. Cao (Ed.): Fuzzy Information and Engineering (ICFIE), ASC 40, pp. 323–331, 2007.
springerlink.com

shop scheduling. Bagchi [4] presented a genetic algorithm (called ENGA) for multiobjective scheduling problem. Chang et al. [5] proposed a gradual priority weighting approach based on a genetic algorithm (GPWGA). And Pasupathy et al. [6] proposed a Pareto genetic algorithm with an archive of non-dominated solutions subjected to a local search (PGA-ALS).

Particle Swarm Optimization (PSO) is a relatively popular Swarm Intelligence method originally developed by Kennedy and Eberhart [7]. The main algorithm is relatively simple and it's straightforward for implementation. The success of the PSO algorithm as a single-objective optimizer has motivated researchers to extend it to other areas [10]. But the PSO mainly succeeds when dealing with continuous search spaces. The major drawback of successfully applying a PSO algorithm to combinatorial problems is due to its continuous nature. In order to solve the single objective PFSP, Tasgetiren et al. recently presented a PSO algorithm in [8, 9]. But the applications of PSO in multiobjective combinatorial problems are very rare.

In this paper, a Discrete Particle Swarm Optimization (DPSO) algorithm is proposed to solve the Multiobjective Permutation Flowshop Sequencing Problem. The proposed DPSO is first applied into 5 small problems generated by Ishibichi and Murata [2] with the objectives of minimizing the makespan and the total tardiness, and the result is compared with the true Pareto front generated by enumeration. In addition, the DPSO is compared with four existing algorithms on another 20 benchmark problems proposed by Taillard [12] with the objectives of minimizing the makespan and total flowtime. These algorithms are PGA-ALS (2004), MOGLS (1998), ENGA (1999) and GPWGA (2002).

2 The Discrete Particle Swarm Optimization Algorithm for the MPFSP

In the PSO algorithm, each particle moves around the search space with a velocity, which is constantly updated by the particle's own experience and the experience of the particle's neighbors or the experience of the whole swarm. The basic concepts of PSO are described as follows.

Assume that the search space is n-dimensional. A current position of the i-th particle in the swarm is denoted as $X_i^t = (x_{i1}^t, x_{i2}^t, \ldots, x_{in}^t)$ and the current velocity of the i-th particle is represented as $V_i^t = (v_{i1}^t, v_{i2}^t, \ldots, v_{in}^t)$. A current personal best position is represented as $P_i^t = (p_{i1}^t, p_{i2}^t, \ldots, p_{in}^t)$, and a current global best position is denoted as $G_i^t = (g_1^t, g_2^t, \ldots, g_n^t)$. In addition, w is the inertia weight, which is a parameter to control the impact of previous velocities on the current velocity. c_1 and c_2 are acceleration coefficients for the personal and global best. The current velocity of the j-th dimension of the i-th particle can be updated as follows:

$$v_{ij}^t = w^t v_{ij}^{t-1} + c_1 rand_1 (p_{ij}^{t-1} - x_{ij}^{t-1}) + c_2 rand_2 (g_j^{t-1} - x_{ij}^{t-1}) \quad (1)$$

It's obvious that the standard PSO equation cannot be used to generate a discrete job permutation since its continuous nature.

2.1 Discrete PSO for the Single Objective PFSP

Pan et al. [9] proposed a newly designed method for updating the position of the particle based on discrete job permutation. In their resent work (http://logistics.iem.yzu.edu.tw/teachers/Ycliang/Ycliang.htm), the position of the i-th particle at iteration t can be updated as follows:

$$X_i^t = c_2 \oplus F_3(c_1 \oplus F_2(w \oplus F_1(X_i^{t-1}), P_i^{t-1}), G_i^{t-1}) \tag{2}$$

The update equation consists of three components as follows, where r_1, r_2 and r_3 are uniform random numbers generated between 0 and 1.

(1) λ_i^t represents the velocity of the particle.

$$\lambda_i^t = w \oplus F_1(X_i^{t-1}) = \begin{cases} F_1(X_i^{t-1}), & r_1 < w \\ X_i^{t-1}, & else \end{cases} \tag{3}$$

where F_1 indicates the insert (mutation) operator with the probability of w.

(2) δ_i^t is the "*cognition*" part of the particle for the private thinking of the particle itself.

$$\delta_i^t = c_1 \oplus F_2(\lambda_i^t, P_i^{t-1}) = \begin{cases} F_2(\lambda_i^t, P_i^{t-1}), & r_2 < c_1 \\ \lambda_i^t, & else \end{cases} \tag{4}$$

where F_2 represents the crossover operator with the probability of c_1.

(3) X_i^t is the "*social*" part of the particle representing the collaboration among particles.

$$X_i^t = c_2 \oplus F_3(\delta_i^t, G_i^{t-1}) = \begin{cases} F_3(\delta_i^t, G_i^{t-1}), & r_3 < c_2 \\ \delta_i^t, & else \end{cases} \tag{5}$$

where F_3 represents the crossover operator with the probability of c_2.

2.2 The Proposed DPSO for the MPFSP

Quite different from the Single-objective Optimization Problem, there is a set of Pareto optimal solutions which are non-dominated by any other in Multiobjective Optimization Problem (MOP). In such multiobjective permutation flowshop scheduling problems, it's common to obtain a good approximation set of the true Pareto front. In general, there are three main goals to achieve [10]: 1) Maximize the number of Pareto optimal solutions found, 2) Minimize the distance of the produced Pareto front with respect to the true Pareto front, 3) Maximize the spread of solutions found.

2.2.1 The Update of the Particle's Position

Inspired by the basic idea of the above discrete PSO for the single objective PFSP, the proposed DPSO updates the position of the particle by equation (2). Differently, the two-point order crossover and the arbitrary two-element change mutation are adopted in the algorithm. In addition, the selection of global best for guiding flying of swarm is rather different, which is described in 2.2.2. The crossover and mutation operators are illustrated in Fig.1.

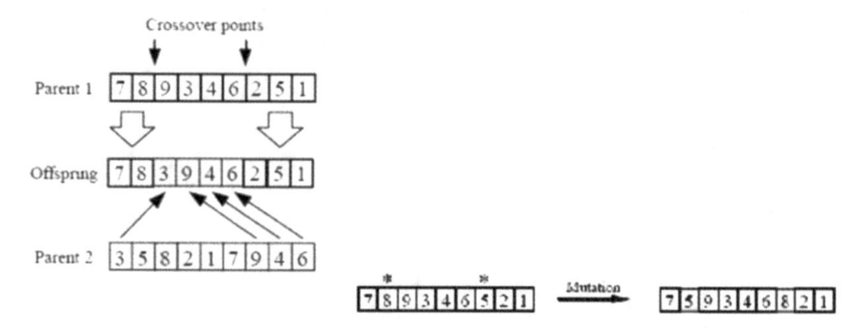

Fig. 1. The crossover and the mutation operators

2.2.2 Selection of Global Best for Guiding Flying of Swarm

In the MPFSP, there is a set of Pareto optimal solutions that can be selected to lead the flying of the swarm. How to select the leader and thus to converge to the true Pareto front and keep diversity of swarm are the main issues when applying PSO in multiobjective optimization (MOP). In other words, it should promote the swarm flying towards the true Pareto front and distributing along the front as uniform as possible. The issue can be well solved by applying the sharing function in the definition of fitness function, that is, the particle is evaluated by both the Pareto dominance and the neighborhood density. Therefore the Pareto optimal solutions with lower neighborhood density can be preferred and are selected as the leader.

There are two classes of sharing in general: the genotypes sharing of decision spaces and the phenotype sharing of the objective spaces. As different solutions in decision spaces may map to a same value in objective spaces, phenotype sharing can reflect the neighborhood density of the Pareto front more exactly and it's more suitable for the MOP than genotypes sharing. In the previous work, a phenotype sharing based fitness function has been defined [11]. The fitness function is adopted in the proposed DPSO algorithm to find the global best for guiding flying. In particular, when there are several global best solutions with respect to the fitness value, a random one is selected as the leader of flying among all the global bests.

2.2.3 The Main Procedure

The algorithm keeps a fixed size of swarm. The main algorithm is the following.

1) Initialize parameters
2) Initialize swarms.

3) Evaluate each of the particles by the fitness function in the initial swarm.
4) While the end condition has not been met do
   a) Find the personal best.
   b) Find the global best.
   c) Update position by the equations (2).
   d) Evaluate the swarm with the fitness function.

3 Experimental Results

The 5 small problems generated by Ishibichi and Murata [2, 3] with the objectives of minimizing the makespan and the total tardiness are considered, with each problem reflecting a different type of Pareto front. In addition, 20 benchmark problems given by Taillard [12] are solved, with the objectives of minimizing the makespan and total flowtime.

3.1 The 5 Small Problems with the Objectives of Minimizing the Makespan and the Total Tardiness

The first five m-machine n-job permutation flowshop scheduling problems are generated in the same manner as Ishibichi and Murata described in [2, 3]. In order to evaluate the approximation of the true Pareto front, an enumeration method is applied to compute all the non-dominated solutions of a given problem. As it is unpractical to enumerate all solutions when the problem size is large, only the problems of small size are considered. All the five test problems have 2 machines (i.e., $m = 2$) and 8 jobs (i.e., $n = 8$), with 2 objectives to minimize the makespan and the total tardiness.

In the following experiment, the size of swarm is set as 20 and the maximum time of evolution is set to 500. And the parameters of DPSO are set as follows $w = 0.5, c_1 + c_2 = 1.0$, where c_1 decreases linearly as the evolution performs. And the parameters for the fitness function are set as $\alpha = 1, \beta = 1, \sigma_s^1 = 1, \sigma_s^2 = 1$.

Although the test problems are rather smaller when compared with other literatures, each problem is selected to reflect different types of true Pareto front distribution. The true Pareto front distributions and the Pareto optimal solutions obtained by the proposed DPSO of a random run are showed in the Fig. 2.

For each test problem, the Pareto optimal solutions obtained by the proposed DPSO are compared with the true Pareto front generated by the enumeration method and the ratio (denoted as P_{true} Ratio) of obtained number of true Pareto optimal solutions is calculated to evaluate the approximation of the true Pareto front. It's obvious that a higher value of P_{true} Ratio implies a better approximation of the true Pareto front. In order to reduce the influence of random in the algorithm, each problem is performed with 20 independent runs and an average P_{true} Ratio is obtained. The results of P_{true} Ratio of the 5 test problems are given in Table 1.

As it's implied in Table 1, the DPSO obtains a good result for all the 5 problems. The average P_{true} Ratio of all the 5 problems with all 20 runs is 0.94.

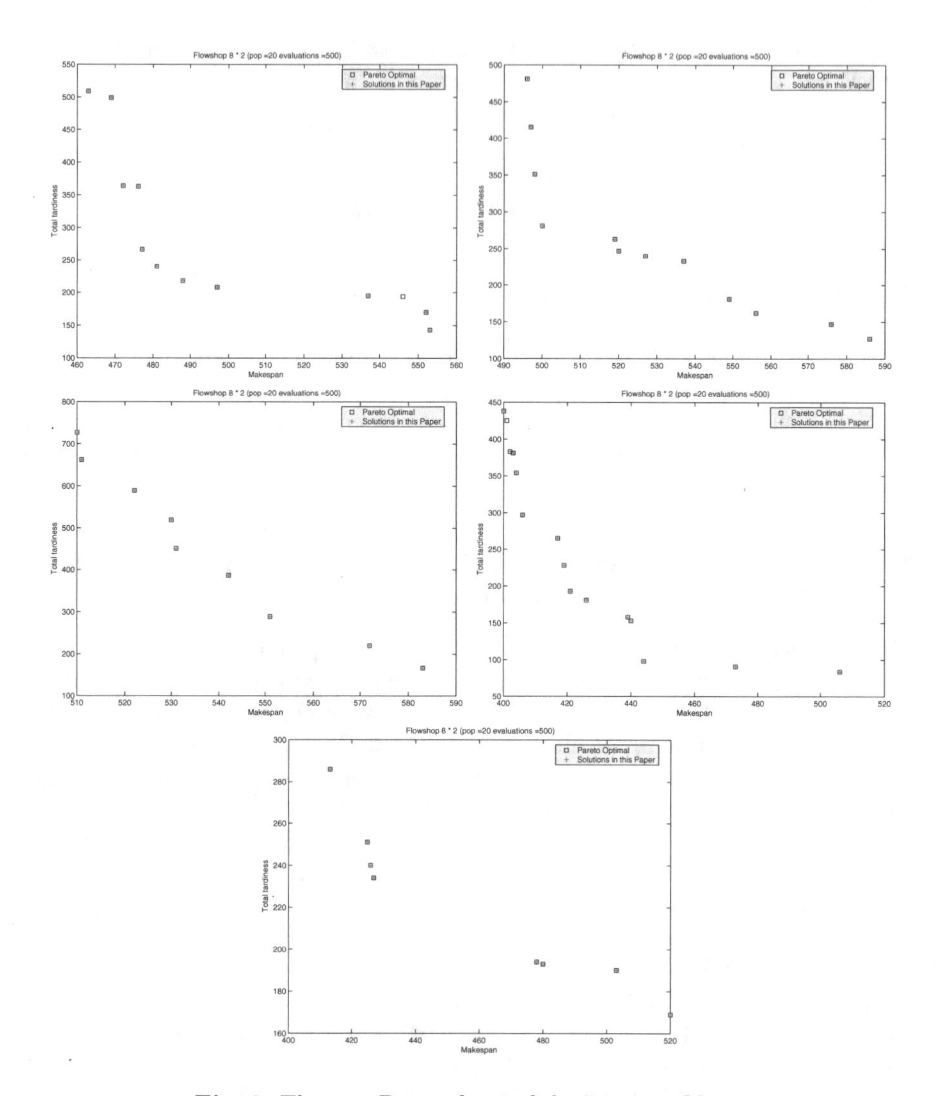

Fig. 2. The true Pareto front of the 5 test problems

It means that the DPSO can get 94% of true Pareto optimal solutions in a single run on average, which implies the proposed DPSO can obtain a good approximation of the true Pareto front. For the problem 4 in particular, the DPSO algorithm reaches all the true Pareto optimal in all the 20 runs. It may conclude that the proposed DPSO is very suitable for the problems with a uniform distribution of true Pareto front. Although the DPSO performs worst in the problem 1, where the true Pareto front is most disconnected, it still gets an average P_{true} Ratio of 0.85, which means the proposed DPSO can also reach a good approximation of the true Pareto front.

Table 1. The results of P_{true} Ratio of the 5 test problems

No.	Run Time																				Arg.[a]
	1	2	3	4	5	6	7	8	9	10	11	12	13	14	15	16	17	18	19	20	
1	0.83	0.83	0.83	0.92	0.83	0.92	0.75	0.67	0.92	0.92	0.75	0.67	0.83	1.00	0.92	0.92	0.83	0.83	0.75	1.00	0.85
2	0.92	1.00	1.00	1.00	1.00	0.83	1.00	0.92	1.00	1.00	1.00	1.00	1.00	1.00	1.00	1.00	1.00	1.00	1.00	1.00	0.98
3	0.93	0.93	0.87	0.87	1.00	1.00	0.93	0.93	0.93	1.00	1.00	0.80	1.00	1.00	0.93	0.93	0.93	0.80	0.80	0.93	0.93
4	1.00	1.00	1.00	1.00	1.00	1.00	1.00	1.00	1.00	1.00	1.00	1.00	1.00	1.00	1.00	1.00	1.00	1.00	1.00	1.00	1.00
5	1.00	1.00	1.00	1.00	1.00	0.88	1.00	1.00	1.00	1.00	1.00	1.00	1.00	1.00	0.63	1.00	1.00	1.00	0.88	1.00	0.97

[a] The average P_{true} Ratio of the 20 runs.

3.2 The 20 Problems of Taillard with the Objectives of Minimizing the Makespan and the Total Flowtime

The 20 problems generated by Taillard [12] are available in OR-Library (http:// people.brunel.ac.uk/ mastjjb/jeb/info.html). The number of jobs considered is 50, and the number of machines is 5 and 10 respectively. And the proposed DPSO is compared with another 4 existing algorithms namely, PGA-ALS, MOGLS, ENGA and GPWGA.

In the following experiment, the size of swarm is set as 50 and the maximum time of evolution is set to 5000. The parameters of DPSO are set as follows: $c_1 + c_2 = 1.0$, c_1 and w decreases linearly as the evolution performs. And the parameters for the fitness function are set as $\alpha = 1, \beta = 1, \sigma_s^1 = 1, \sigma_s^2 = 1$.

Table 2. Net set of non-dominated solutions obtained from various multiobjective flow shop scheduling algorithms for the problem size (50×5)

Problem 1		Problem 2		Problem 3		Problem 4		Problem 5		Problem 6		Problem 7		Problem 8		Problem 9		Problem 10	
MS	TFT	MS	TFT	MS	TFT	MS	TFT	MS	TFT	MS	TFT	MS	TFT	MS	TFT	MS	TFT	MS	TFT
2735	66215^5	2901	69375^1	2644	65837^1	2776	71751^5	2891	70892^1	2832	70602^5	2745	68553^5	2705	65391^1	2571	65818^5	2789	73132^5
2738	66193^5	2894	69400^1	2700	65078^1	2778	71700^5	2904	70495^1	2833	70417^5	2746	68436^1	2758	65291^1	2575	65202^5	2791	73055^5
2741	66126^5	2859	70122^1	2656	65712^1	2785	70217^1	2864	71102^1	2844	69220^5	2771	67950^1	2701	65045^1	2578	65200^5	2792	71678^5
2746	66107^5	2843	70894^1	2656	65712^1	2784	70239^1	2863	77928^4	2845	69097^5	2769	67298^1	2686	69606^1	2580	65199^5	2793	71502^5
2752	66046^5	2838	71668^1	2655	65720^1	2783	70246^1			2846	69079^5	2782	67218^1	2692	65436^1	2589	65075^5	2794	71447^5
2729	68036^1	2839	71647^1	2622	66517^1	2865	69754^1			2849	69078^5	2774	67237^1			2590	64903^5	2796	71420^5
2731	67028^1	2956	69162^1	2643	66082^1	2863	69875^1			2948	67887^1	2795	67209^1			2607	64893^5	2843	71405^5
2765	66024^2	2945	69241^1	2649	65787^1	2862	69922^1			2874	68869^1	2767	67371^1			2613	64866^5	2846	71392^5
2763	66032^2	2941	69355^1	2657	65292^1	2857	69974^1			2839	69754^1	2768	67361^1			2566	66711^1	2903	71365^5
2799	65963^2	2915	69356^1	2685	65136^1	2782	70311^1			2841	69409^1	2732	72026^1			2567	66686^1	2782	76605^4
2770	65979^2	2890	69481^1	2664	65232^1	2807	70096^1			2857	68969^1	2734	71691^1			2568	66673^1	2783	76499^4
2724	71531^4	2879	69566^1	2630	66135^1	2805	70105^1			2882	68606^2	2762	68089^1			2565	66954^1	2787	76147^4
		2885	69510^1	2631	66110^1	2789	70157^1			2886	68549^2	2741	69521^1			2564	67187^1		
		2873	69986^2	2644	65837^1	2791	70148^1			2887	67912^2	2736	70277^1			2561	68851^1		
		2867	70039^2	2621	70731^4	2786	70212^1					2747	68155^1			2617	64602^2		
						2774	72022^1					2746	68436^1			2664	64234^2		
						2768	72510^1									2633	64496^2		
						2772	72353^1												
						2771	72422^1												
						2769	72483^1												
						2761	72764^1												
						2759	73009^1												
						2785	70217^1												

[a] (1) On every solution, the superscript indicates the algorithm that yielded the corresponding solution: 1 is PGA-ALS; 2 is MOGLS; 3 is ENGA; 4 is GPWGA; and 5 is DPSO (2) MS decontes the makespan and TFT denotes the total flow time that are obtained by a sequence.

Table 3. Net set of non-dominated solutions obtained from various multiobjective flow shop scheduling algorithms for the problem size (50 × 10)

Problem 1		Problem 2		Problem 3		Problem 4		Problem 5	
MS	TFT	MS	TFT	MS	TFT	MS	TFT	MS	TFT
3117	90640^5	3057	86342^1	2973	84128^5	3110	91767^1	3063	91137^5
3139	90407^5	3114	86112^1	2981	84044^5	3218	89711^1	3066	91060^5
3298	90075^1	3154	85999^1	2986	83938^5	3180	90073^1	3067	90889^5
3249	90099^1	3142	86049^1	3021	83380^1	3153	90105^1	3073	89745^1
3137	90408^1	3181	85959^1	3020	83437^1	3151	90172^1	3070	89903^1
3135	90448^1	3108	86196^1	2944	85042^1	3150	90230^1	3126	89149^1
3148	90364^1	3038	86464^1	2935	85248^1	3142	90299^1	3129	89086^1
3052	93013^1	3058	86326^1	2930	85607^1	3139	90447^1	3127	89101^1
3047	93511^1	3049	86393^1	2929	85615^1	3136	90463^1	3133	89075^1
3070	92508^1	3046	86461^1	2920	86647^1	3135	90572^1	3120	89270^1
3059	92666^1	3059	86321^1	2922	86471^1	3134	91227^1		
3063	92602^1	3077	86319^1	2921	86600^1	3133	91268^1		
3075	92124^1	3018	86488^1	2924	86452^1	3131	91307^1		
3074	92493^1	3013	86627^1	2928	86410^1	3129	91462^1		
3156	90254^1	3014	86533^1	2932	85573^1	3250	89471^1		
3152	90305^1	3017	86511^1	2933	85559^1	3219	89612^1		
3237	90158^1	3002	86659^1	2942	85115^1	3183	89987^2		
3209	90165^1	2982	87849^1	2943	85112^1	3209	89811^2		
3197	90207^1	2976	88635^1	2941	85132^1				
3111	91149^1	2984	87080^1	2913	90304^1				
3076	91757^1	3109	86129^1	2983	84017^1				
3099	91236^1	3109	86129^1	2989	83695^1				
3097	91256^1	3091	86217^1	3002	83630^1				
3087	91688^1	3087	86273^1	2998	83668^1				
		2975	90226^1	3045	83257^1				
		3080	86276^2	3044	83279^1				
				3022	83355^1				
				3036	83305^1				
				3030	83349^1				
				3037	83301^1				
				2959	84739^1				
				2945	84790^1				
				2962	84298^1				

Problem 6		Problem 7		Problem 8		Problem 9		Problem 10	
MS	TFT	MS	TFT	MS	TFT	MS	TFT	MS	TFT
3110	90278^5	3157	95120^5	3143	89928^1	3141	88164^1	3869	128130^1
3114	90014^5	3159	93262^5	3144	89903^1	3250	87946^1	3855	128334^1
3116	89812^5	3162	92941^5	3139	89935^1	3334	87796^1	3854	138496^1
3118	89733^5	3229	91684^1	3153	89773^1	3275	87900^1	3924	127901^1
3120	89666^5	3226	91751^1	3075	94951^1	3226	87951^1	3901	128105^1
3123	89653^5	3242	91554^1	3075	94951^1	3179	88048^1	3899	128122^1
3135	89522^5	3250	91468^1	3084	92945^1	3198	88023^1	3907	128037^1
3216	89014^1	3246	91482^1	3094	91881^1	3205	87994^1	3974	127305^1
3194	89089^1	3313	91467^1	3089	91932^1	3315	87810^1	3985	127179^1
3230	89009^1	3317	91450^1	3087	91958^1	3369	87767^1	4014	127112^1
3093	90379^1	3372	91350^1	3230	89094^1	3286	87837^1	4061	127110^1
3075	90910^1	3401	91277^1	3301	89039^1	3301	87818^1	4064	127076^1
3080	90841^1	3352	91375^1	3181	89465^1	3124	88336^1	3997	127145^1
3065	91181^1	3391	91304^1	3180	89501^1	3065	88667^1	4004	127144^1
3071	91116^1	3324	91422^1	3174	89659^1	3068	88491^1	3983	127214^1
3068	91122^1	3165	92095^1	3178	89629^1	3067	88583^1	3929	127474^1
3168	89128^2	3191	91940^1	3225	89316^1	2995	90282^1	3957	127358^1
3166	89168^2			3227	89293^1	3007	90181^1	3926	127724^1
3129	89636^2			3104	90428^1	3005	90187^1	3925	127746^1
3173	89111^2			3112	90423^1	3020	90106^1		
3162	89261^2			3103	90487^1	3014	90108^1		
				3102	90642^1	3082	88399^1		
				3118	90063^1	3178	88146^1		
				3117	90339^1	3031	88709^1		
				3150	89838^1	3025	88713^2		

Pasupathy et al. have provided the net set of Pareto optimal solutions obtained by PGA-ALS, MOGLS, ENGA and GPWGA [6]. The Pareto optimal solutions obtained by DPSO are combined with them, and a new net set of Pareto optimal solutions of the 5 algorithms is yielded. The result of the 20 problems for the given size ($n \times m$) is showed in Table 2 and Table 3. As it showed in the tables, the proposed DPSO can produce many numbers of the non-dominated solutions.

4 Conclusions

This paper deals with the application of PSO to Multiobjective Permutation Flowshop Scheduling Problem (MPFSP). The continuous nature of the stand PSO has limited its application in such combinational problems. In order to solve the discrete job permutation, a Discrete Particle Swarm Optimization algorithm (DPSO) is proposed. In the terms of leading the swarm flying to a uniform distribution of Pareto optimal solutions, a fitness function considering both the Pareto dominance and the phenotype neighborhood density is applied to select the leader of flying. The proposed DPSO is first verified with the true Pareto front produced by enumeration on 5 different problem instances, where each problem reflects a different type of Pareto front. And it's compared with 4 algorithms (PGA-ALS, MOGLS, ENGA and GPWGA) on 20 benchmark problems. The results show that the proposed DPSO can obtain a good approximation of the true Pareto front in MPFSP with a small problem size. Future research will include improvement of the performance of the PSO algorithm especially in the adjustment of the parameters of it. In addition, the study of applying PSO to the anomaly detection system will be continued.

Acknowledgement

Supported by the National Natural Science Foundation of China under Grant No.60673161, the Key Project of Chinese Ministry of Education under Grant No.206073, Fujian Provincial Natural Science Foundation of China under Grant No.A0610012.

References

1. Srinivas N, Deb K (1995) Multiobjective optimization using nondominated sorting in genetic algorithms. Evolutionary Computation 2(3): 221–248
2. Ishibuchi H, Murata T (1998) A multi-objective genetic local search algorithm and its application to flowshop scheduling. IEEE Trans Syst Man Cybern C 28:392–403
3. Ishibuchi H, Yoshida T, Murata T (2003) Balance between genetic search and local search in memetic algorithms for multiobjective permutation flowshop scheduling. IEEE Trans Evol Comput 7: 204–223
4. Bagchi TP (1999) Multiobjective scheduling by genetic algorithms.Kluwer, Boston
5. Chang P-C, Hsieh J-C, Lin SG (2002) The development of gradual priority weighting approach for the multi-objective flowshop scheduling problem. Int J Prod Econ 79: 171–183
6. Pasupathy T, Rajendran C, Suresh RK (2006) A multi-objective genetic algorithm for scheduling in flow shops to minimize the makespan and total flow time of jobs. The International Journal of Advanced Manufacturing Technology 27(7-8): 804–815
7. Kennedy, J, Eberhart RC (1995) Particle swarm optimization. Proc IEEE Int'l Conf On Neural Networks: 1942–1948
8. Tasgetiren MF, Sevkli M, Liang YC, Gencyilmaz G (2004) Particle swarm optimization algorithm for permutation flowshop sequencing problem. Proceedings of the 4th International Workshop on Ant Colony Optimization and Swarm Intelligence: 382–390
9. Pan QK, Tasgetiren MF, Liang YC (2005) A discrete particle swarm optimization algorithm for the no-wait flowshop scheduling problem with makespan criterion. Proceedings of the International Workshop on UK Planning and Scheduling Special Interest Group: 34–43
10. Reyes-Sierra M, Coello Coello CA (2006) Multi-Objective Particle Swarm Optimizers: A Survey of the State-of-the-Art. Technical Report EVOCINV-01-2006. Evolutionary Computation Group at CINVESTAV-IPN, Mexico
11. Chen G-L Guo W-Z, Tu X-Z, Chen H-W (2005) An Improved Genetic Algorithm for Multi-objective Optimization. ISICA'2005: Progress in Intelligent Computation and Its Applications: 204–210
12. Taillard E (1993) Benchmarks for basic scheduling problems. Eur J Oper Res 64:278-285

An Optimization Method for Fuzzy *c*-Means Algorithm Based on Grid and Density

Zou Kaiqi, Deng guannan, and Kong Xiaoyan

College of Information Engineering, Dalian University, University Key Lab of Information Sciences and Engineering, Dalian University, Dalian 116622, China
zoukq@vip.sina.com

Abstract. In this paper an optimization method for fuzzy c-means algorithm based on grid and density is proposed in order to solve the problem of initializing fuzzy c-means algorithm. Grid and density are needed to extract approximate clustering center from sample space. Then, an optimization method for fuzzy c-means algorithm is proposed by using amount of approximate clustering centers to initialize classification number, and using approximate clustering centers to initialize initial clustering centers. Experiment shows that this method can improve clustering result and shorten clustering time validly.

Keywords: Fuzzy c-means algorithm; Grid; Density.

1 Introduction

Fuzzy c-means algorithm is the most widespread clustering algorithm. But research shows that fuzzy c-means algorithm strongly depends on initial parameters' state. In order to improve clustering result, people begin to research initialization methods. Now there are mainly two modes of initializations. First is initializing partition matrix. Second is initializing initial clustering centers. But most algorithms of this type need to know the classification number c before analyzing, but it is difficult. Solution is treating classification number c as initial parameter to initialize. So In this paper we propose an initialization method for fuzzy c-means algorithm based on grid and density. Grid and density are two common clustering algorithms, which could solve large scale and high dimensional clustering problem with great efficiency and could find any clustering sharp. Experiment shows that this method can improve clustering result and shorten clustering time validly.

2 Grid and Density

Definition 1. Sample space S could be separated into $\xi_1, \xi_2, \cdots, \xi_n$ intervals if inputting parameters $\xi_1, \xi_2, \cdots, \xi_n$. So the whole sample space turns into finite disjoint rectangle-like units, and each of these units could be described as $U_i = \{u_{i1}, u_{i2}, \cdots, u_{in}\}$, where $u_{ij} = [l_{ij}, h_{ij})$. U_i is called grid unit.

B.-Y. Cao (Ed.): Fuzzy Information and Engineering (ICFIE), ASC 40, pp. 332–336, 2007.
springerlink.com

Definition 2. Density of grid unit U_i is:

$$D(U_i) = \text{Amount of samples in grid unit / Total amount of samples}$$

Definition 3. Grid unit U_i is a dense unit, only needs $D(U_i) > \tau$, where density threshold value τ is an inputting parameter.

Definition 4. $U_i = \{u_{i1}, u_{i2}, \cdots, u_{id}\}$ is a grid unit. If U_i is a dense unit, then the geometric centre p of U_i is called condensation point.

3 Initialization Method for Fuzzy c-Means Algorithm Based on Grid and Density

In this paper we will initialize not only clustering centers but also classification number. We will use the amount of clustering centers to initialize classification number. So how to extract approximate clustering center is the most important part in this paper.

3.1 Extract Approximate Clustering Center Algorithm Based on Grid and Density

This algorithm separate sample space into grid unit, then find out sample's condensation points through analyzing densities of grid units. According to analyzing condensation points we can get approximate location of clustering centre. Its' processes are shown as follows:

Partition grid

Before running this algorithm, sample space must be separated into grid unit. There are many strategies to partition grid, the simplest method is partitioning grid directly through inputting parameters $\xi_1, \xi_2, \cdots, \xi_k$. This method is applicable to lower dimensional data. What's more, to high dimensional data we can use some particular methods such as adaptive grid partition method, bisecting partitioning grid method, etc.

Searching condensation points

According to the definition of condensation point, the distribution of condensation points is also denseness if the distribution of samples is denseness. So the distribution of condensation points reflects the distribution of clustering samples. Thus, we can change extracting condensation points from sample set to condensation point set.

Searching condensation points need adapt to the grid partition method. If partitioning grid directly we can get condensation points directly according to its definition. If bisecting partitioning grid, the densities of grid units will be computed with the process of partitioning grid. So when finishing partitioning grid, the condensation points can be extracted directly.

Extract approximate clustering center

Clustering center has the characteristic that data points are dense, so the condensation point P which has the maximal density must be an approximate clustering center. We add point P into approximate clustering center set. Then, we get rid of the point P and all of the points which could be in the same class with P, so the class that P belongs to would get rid of from condensation points set. Repeat the steps above we will get every approximate clustering point. Detail is shown as follows:

Step 1. Select the condensation point P which has the maximal density and add it into the approximate clustering center set.

Step 2. Search the condensation points which could be in the same class with point P, and then delete these points and point P from condensation point set.

According to the distribution of samples we know that the adjacent classes maybe intersect each other. In order to solve this problem, according to the distribution of samples that its density decreases from clustering center to its boundary, we judge whether the adjacent condensation points belong to the same class through comparing the changes of their densities. If the density begins to increase, it indicates that we have reached the boundary of this class.

Step 3. If there aren't usable condensation points in the condensation point set, then the algorithm finishes, else repeat step 1.

3.2 Initialization

Apply the extract approximate clustering centre algorithm based on grid and density to the given sample set. When algorithm finishes, the points in the approximate clustering center set are the approximation clustering canters of the clustering samples, and the amount of the approximation clustering centers is the classification number. So if initialize fuzzy c-means algorithm only need use the amount of the approximation clustering centers initialize classification number c and use the approximate clustering center initialize initial clustering center $P^{(0)}$.

4 Experiments

This paper uses a set of data produced by Matlab for clustering experiment. These data contain eight classes, and each class contains 50 data. After inputting parameters 0.05, 0.05, the sample space is separated into 400 grid units as shown in Fig.1. Where, dots denote samples.

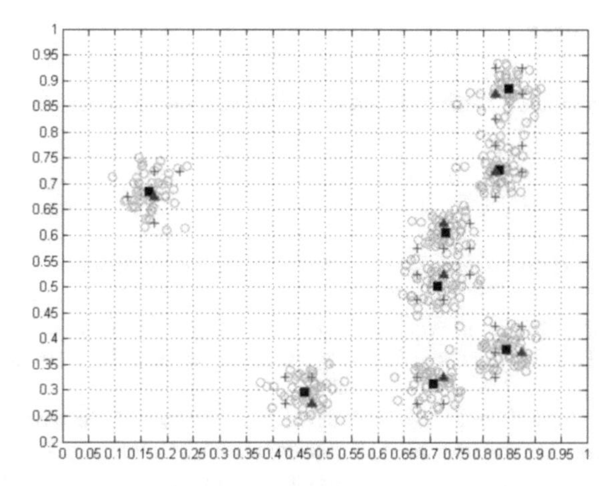

Fig. 1. Comparing with fuzzy c-means algorithm

Input density threshold value $\tau = 0.01$, and extract condensation points. Here we get 38 condensation points as shown in Fig.1. where, the solid dots denote condensation points.

We start searching approximate clustering centers from the condensation point P which has the maximal density, and get these approximate clustering centers:

(0.8250, 0.7250), (0.7250, 0.3250), (0.7250, 0.5250), (0.7250, 0.6250)

(0.8750, 0.3750), (0.4750, 0.2750), (0.8250, 0.8750), (0.1750, 0.6750)

As shown in Fig.1. Where, the solid triangles denote approximate clustering centers.

In order to prove validity of this algorithm, this paper uses the program of fuzzy c-means algorithm provided by Matlab to classify these data, and gets these clustering centers:

(0.8330, 0.7280), (0.7067, 0.3130), (0.7135, 0.5033), (0.7288, 0.6067)

(0.8457, 0.3796), (0.4613, 0.2977), (0.8500, 0.8857), (0.1668, 0.6845)

The detail is also shown in Fig.1. Where, solid squares denote clustering centers.

From Fig.1 we can see that the approximate clustering centers are extracted through grid and density near to the real clustering centers. So if we use the approximate clustering centers to initialize fuzzy c-means algorithm, we would get a good clustering result in shorter time.

5 Conclusions

In this paper we proposed an initialization method for fuzzy c-means algorithm based on grid and density in order to solve the problem of initializing fuzzy c-means algorithm. Experiment proves that this method can initialize classification number and initial clustering center validly.

Acknowledgement

This item is supported by National Nature Science Foundation of China (60573072).

References

1. Bezdek J C(1981) Pattern Recognition with Fuzzy Objective Function Algorithms. Plenum, New York
2. Chen Zhuo, Meng Qingchun, Wei Zhengang et al(2005) A Fast Clustering Algorithm Based on Grid and Density Condensation Point. Journal of Harbin Institute of Technology 37:1654-1657
3. Deng Guannan, Zou Kaiqi(2006) Optimizing the structure of clustering neural networks based on grid and density. Journal of Harbin Engineering University 27:55-58
4. Hu Yang, Chen Gang(2003) An Effective Cluster Analysis Algorithm Based on Grid and Intensity . Computer Applications 23:64-67

Obstacle Recognition and Collision Avoidance of a Fish Robot Based on Fuzzy Neural Networks

Seung Y. Na[1], Daejung Shin[2], Jin Y. Kim[1], Seong-Joon Baek[1], and So Hee Min[1]

[1] Dept. of Electronics and Computer Engineering, Chonnam National University,
300 Yongbong-dong, Buk-gu, Gwangju, South Korea 500-757
{syna,beyondi,tozero,minsh}@chonnam.ac.kr
[2] BK21, Chonnam National University,
300 Yongbong-dong, Buk-gu, Gwangju, South Korea 500-757
djshin71ha@hotmail.com

Abstract. Detection and recognition of obstacles are the most important concerns for fish robots to avoid collision for path planning as well as natural and smooth movements. The more information about obstacle shapes we obtain, the better control of fish robots we can apply. The method employing only simple distance measuring sensors without cameras is proposed. We use three fixed IR sensors and one IR sensor, which is mounted on a motor shaft to scan a certain range of foreground from the head of a fish robot. The fish robot's ability to recognize the features of an obstacle is improved to avoid collision based on the fuzzy neural networks. Evident features such as obstacles' sizes and angles are obtained from the scanned data by a simple distance sensor through neural network training algorithms. Experimental results show the successful path control of the fish robot without hitting on obstacles.

Keywords: Fish Robot, Obstacle Recognition, Collision Avoidance.

1 Distance Scanning System for Fish Robots

Several types of small scale fish robots and submersibles have been constructed in our lab. They have been tested in a tank, whose dimensions are 120 □ 120 □ 180cm, for collision avoidance, maneuverability, control performance, posture maintenance, path design, and data communication. Depth control with strain gauges, acceleration sensors, illumination, and the control of motors for fins are processed based on the MSP430F149 by TI. User commands, sensor data and images are transmitted by Bluetooth modules between robots and a host notebook PC while fish robots are operated within a depth of 10cm. RF modules are used when the depth is greater than 10cm. They are operated in autonomous and manual modes in calm water. Manual operations are made by remote control commands in various ranges of Bluetooth protocol depending on antenna configurations.

The use of simpler IR type distance sensors is proposed rather than using a camera or a sonar module for fish robot's eyes. A distance measuring sensor for a general purpose is mounted on a motor shaft to scan a certain range of the foreground from the head of a fish robot. Simple plane obstacles with various sizes and approaching angles are considered in our experiments. Sonar sensors are not used to make the robot structure simple and compact. All circuits, sensors, motors and a processor card are contained in a chassis whose dimensions are 30 x 10 x 15cm.

B.-Y. Cao (Ed.): Fuzzy Information and Engineering (ICFIE), ASC 40, pp. 337–344, 2007.
springerlink.com

Successive direction changes of the propulsion are necessary to avoid collision with obstacles. Direction changes of the fish robot are determined on the basis of the location, sizes, and approaching angles to an obstacle. The set of information is obtained by a distance sensor that scans the foreground from the head of a fish robot, in addition to the three fixed sensors at the front, left and right sides of the body, respectively.

In this paper, basic rules about the location, sizes, and approaching angles to obstacles have been obtained from distance measurements. Simplified obstacles are assumed to be combinations of planes with several different angles between the planes. From the raw data of the sensor output voltage, a sufficient amount of information can be deduced for obstacle shape recognition. Proper direction changes to avoid collision are determined based on the rules of obstacle shapes. Estimated values of the location, sizes, and approaching angles to an obstacle are calculated using a general ANN algorithm. Experimental results employing the proposed method show the successful path trajectories of the fish robot without collision.

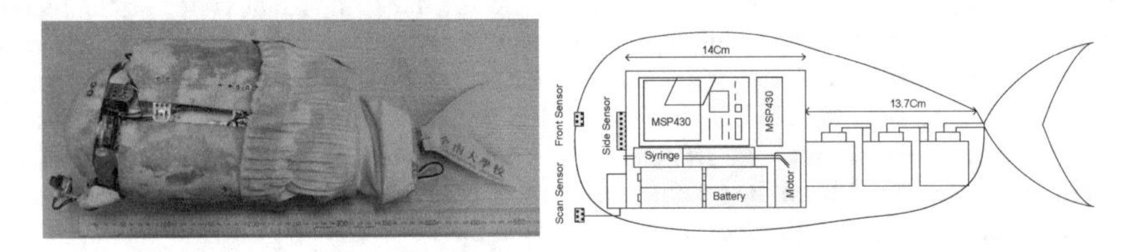

Fig. 1. Fish robot and distance sensors

2 Distance Scanning System for Fish Robots

Sonar sensors are used in the water as well as in the air to detect obstacles or to measure distances in a variety of applications. The use of ultrasonic sensors, however, is not appropriate at short range or in congested situations. Fish robots in this study are concerned mostly about obstacles that exist close to. Therefore, the fish robot in our experiments uses IR distance sensors to measure distances from the robot to obstacles in water. Due to structural limits, a resolution level depends on distance and is also affected by noise. The underwater environment causes a much higher level of noise to IR distance sensors; moreover, the range of underwater measurement gets more shortened than that of the measurement in the air. Since data can be obtained only for one point at one time, it makes the system insufficient for shape recognition. Therefore, to deal with these disadvantages of IR sensors for shape recognition, a proper scanning method should be used. In this paper we use a scanning IR sensor with a servo motor at the center, in addition to three fixed IR sensors at the front and both sides, to get the pattern information of the obstacles in water. Also, a potentiometer is used to measure the rotation angle of the motor.

The fixed sensors, which are located at the front and both sides of the fish robot's body, are used to detect the existence of obstacles. They are also used to get distance data for a quick measurement or collision avoidance in the emergencies that no accurate object recognition is available. The shape measurement IR sensor system is shown at the front of the fish robot in Fig. 1.

The detection of obstacles or the outputs of distance measurements by IR sensors are not sufficient for the preparation of collision avoidance. Typical examples, as shown in Fig. 2, have the same distance measurements, but the direction changes which are necessary for collision avoidance are quite different from each other.

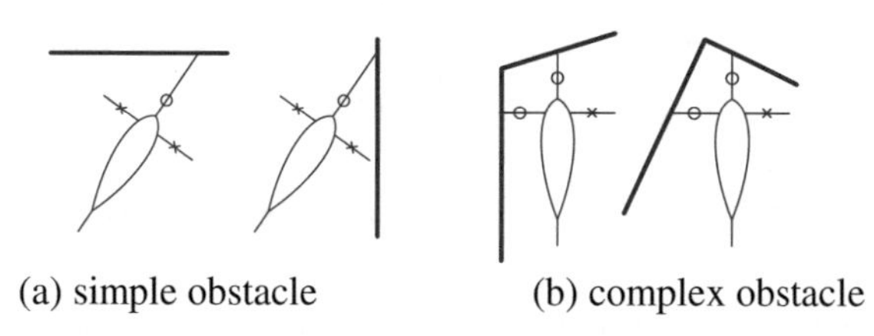

(a) simple obstacle (b) complex obstacle

Fig. 2. The same measurements for different situations

To get more information that is necessary to figure out the shapes and positions of obstacles, the scanning IR sensor at the front center begins to scan a certain range immediately when one of the three fixed sensors detects an obstacle. When the fixed sensor at the center detects an object, the scanning sensor scans symmetrically. When a sensor at the left or right side detects an object, the scanning sensor scans the corresponding side range. If more than one fixed sensors detects an object, the scanning sensor scans a wider range. The magnitudes of range are determined according to the rules, which are derived from the number of obstacle detections and the obstacle shapes by scanning results.

A basic set of positions of a fish robot relative to simplified obstacles and the corresponding operations of a fish robot for natural swimming are described in Table 1 and Fig. 3. The front, left and right fixed IR sensors are represented as F, L, and R, respectively. Detection of an obstacle for each sensor is depicted as "1" while "0" means no detection for the corresponding sensor.

Since the obstacles have random shapes, a set of proper models for obstacle shapes are necessary. The simplest model of obstacles that is adopted in this study consists of piecewise planes. Fundament rules that are necessary to avoid obstacle collision for

Table 1. Positions of a fish robot with obstacles and its operations for cruise

F	L	R	Position	Operation
0	0	0	S1	Swim forward
0	0	1	S2	Swim forward without colliding with the right obstacle
0	1	0	S3	Swim forward without colliding with the left obstacle
0	1	1	S4	Swim forward and keep middle position
1	0	0	S5	When $\theta_F \geq 90^o$, turn left and try not to collide with right obstacle When $\theta_F < 90^o$, turn right and try not to collide with left obstacle
1	0	1	S6	Turn left and try not to collide with the right obstacle
1	1	0	S7	Turn right and try not to collide with the left obstacle
1	1	1	S8	When $L \geq R$, turn left and try not to collide with the right obstacle When $L < R$, turn right and try not to collide with the left obstacle

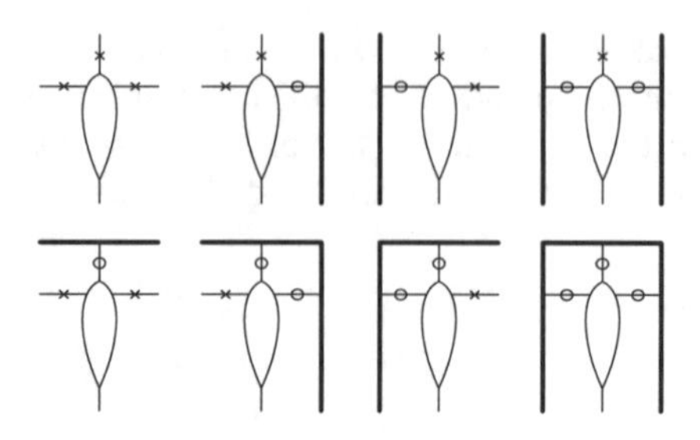

Fig. 3. Positions of a fish robot relative to simplified obstacles

natural swimming are derived for a set of plane angles of 180°, 150°, 120°, and 90°. Also, a set of angles of 30 °, 60°, 90°, 120°, and 150° are assumed to be the approaching angles of a fish robot to an obstacle. Possible combinations of angles are summarized in Fig. 4.

Due to restricted computation resources of a fish robot, the simplified models of shapes and angles are necessary for fast actions to avoid collision. When the plane angle is less than 90° it is very hard to escape without collision. Thus the case is not included for the rules of forward swimming. By the same logic, the case of plane angles that are greater than 180° is excluded for the rules since it is very easy to escape without collision.

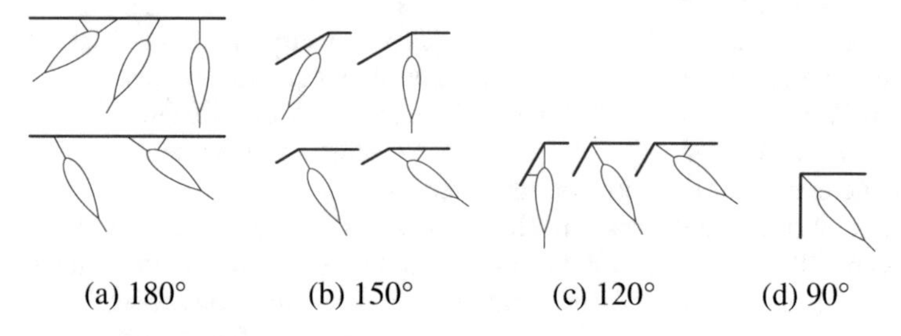

Fig. 4. Simplified obstacle shapes and approaching angles

Fig. 5 represents the angles of the front obstacles while the fish robot swims along the left wall. When the front obstacle is detected, the front range is scanned since the left wall is detected beforehand. The turning angle to the right is determined by the scanning results of obstacle shapes.

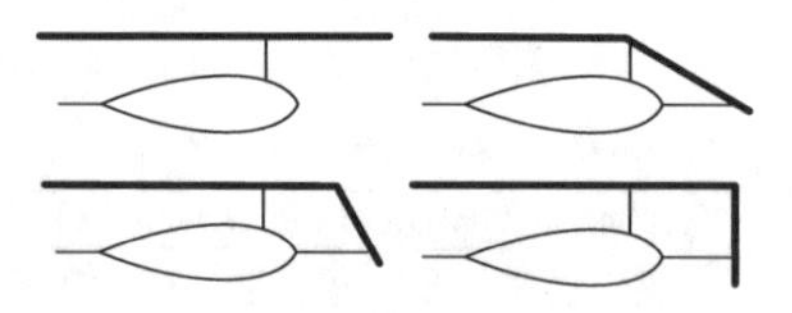

Fig. 5. Obstacle angles while cruising

3 Obstacle Recognition System

Since the obstacles are assumed to be near, the recognition routines should not take much time. Since the scanning time is proportional to the scanning range, we try to restrict the scanning angle as much as possible. Also the scanning direction is determined by the detection of an obstacle by the three fixed IR sensors.

3.1 Scanning Obstacles

Fig. 6 shows the basic scanned results of the simplified obstacle models and several approaching angles of a fish robot. The case 1 is when the approaching angle is less

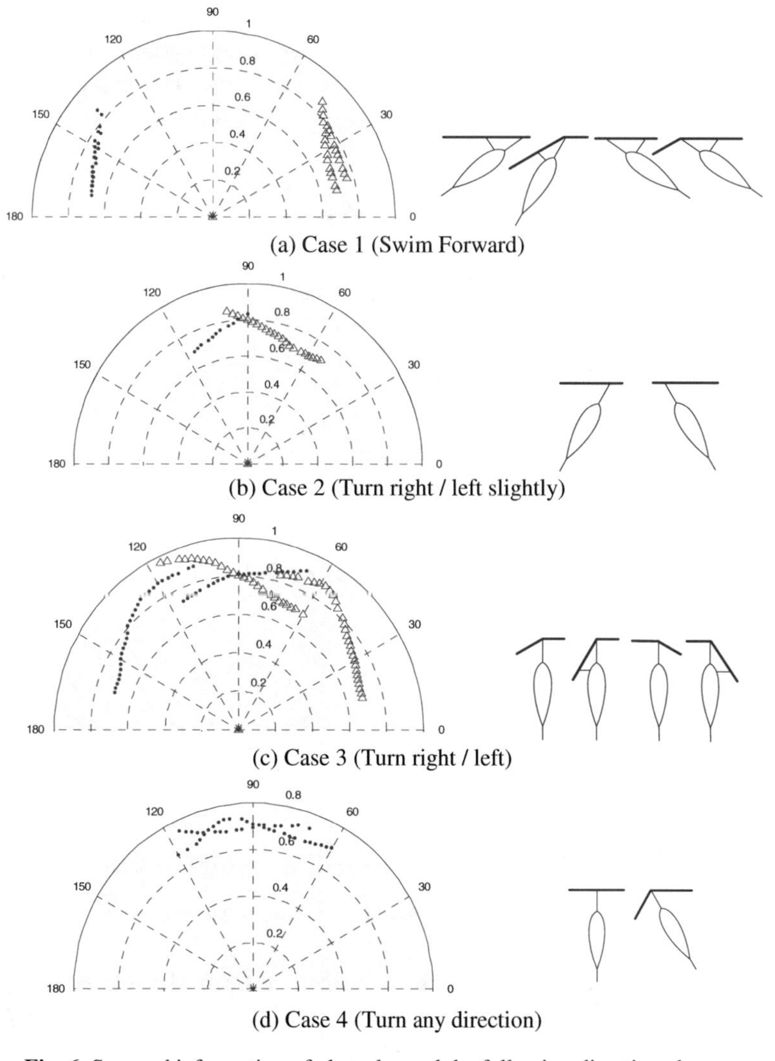

(a) Case 1 (Swim Forward)

(b) Case 2 (Turn right / left slightly)

(c) Case 3 (Turn right / left)

(d) Case 4 (Turn any direction)

Fig. 6. Scanned information of obstacles and the following direction changes

than 30° with respect to the closest obstacle. The scanned results of the obstacles of 180° and 150° are represented as dots for left side ones and as triangles for right side ones. The case 2 is for flat obstacles when right or left turn is necessary after obstacle recognition. The case 4 is when either right or left turn is necessary for symmetrical obstacles. From the scanned results, the obstacle angles that correspond to the simplified piecewise obstacle shapes are recognized. Depending on this obstacle angle, the amount of turning that a fish robot makes is determined to avoid collision.

3.2 Recognition of Shapes and Estimation of Approaching Angles

Two fuzzy inference systems are proposed: one is to recognize the angles of two planes in the simplified models of obstacles, and the other is to estimate the approaching angles of a fish robot to an obstacle. The scanned distance measurements in Fig. 6 are the basic input data for the inference. Sugeno type fuzzy systems are implemented by a subtractive clustering method using genfis2 in Matlab. Training has been made by using *anfis* of an adaptive neuro-fuzzy inference system.

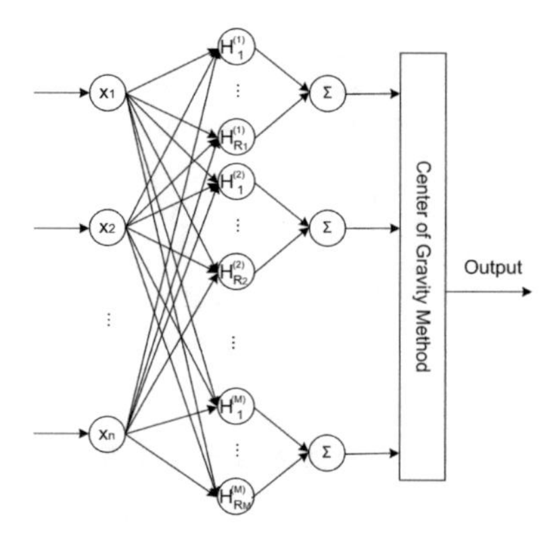

Fig. 7. Network architecture of a fuzzy inference system

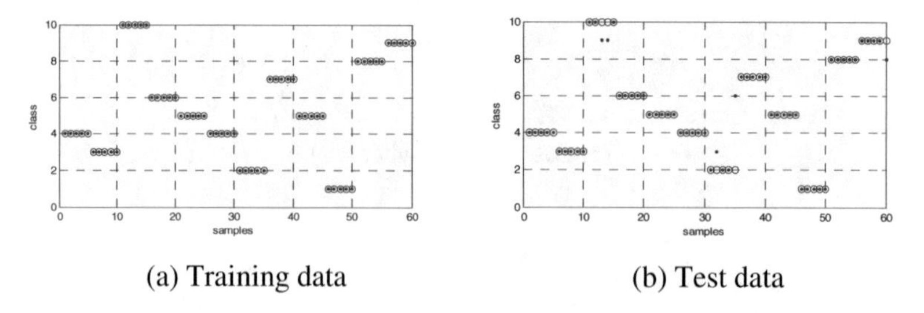

(a) Training data (b) Test data

Fig. 8. Estimation results of inference systems

First, the outputs of the inferences are angles between obstacle planes and approaching angles of a fish robot, respectively. Second, the estimation results by the inference engines are used for the classification of the relative situations of the fish robot against the obstacles.

Assuming the two right side cases in Fig. 6(a) as one and two left side cases in Fig. 6(a) as one in terms of necessary turning direction changes after obstacle angle recognitions, there are ten classes to identify for proper direction changes. For the 120 scanned distance data for twelve cases in Fig. 6, 60 data sets are used for training and the other 60 sets are used for tests to classify the ten classes for proper direction changes. The success rate for experimental classification was 91.2%(55 out of 60 sets) and the success rate for the necessary direction changes was 98.3%(59 out of 60 sets).

3.3 Experiments

Appropriate turning direction and magnitude for successive path changes are determined according to the results of recognition inference systems for the angles between two planes of obstacles and for the approaching angles. The approaching angles and the obstacle shape angles between the consisting planes are estimated first. Then the results are used for the classification of the basic ten classes in consideration for proper direction changes to avoid collision. Angles between the basic sets of angles are solved using interpolation. The results of inferences are implemented on the fish robot of Fig. 1 to show the effectiveness of the estimation algorithms. Fig. 9 shows swimming trajectories of the fish robot for several angles of an obstacle with different approaching angles. For obstacle angles that are greater than 90°, direction changes to avoid collision by IR distance sensors are successfully made by using only forward movement.

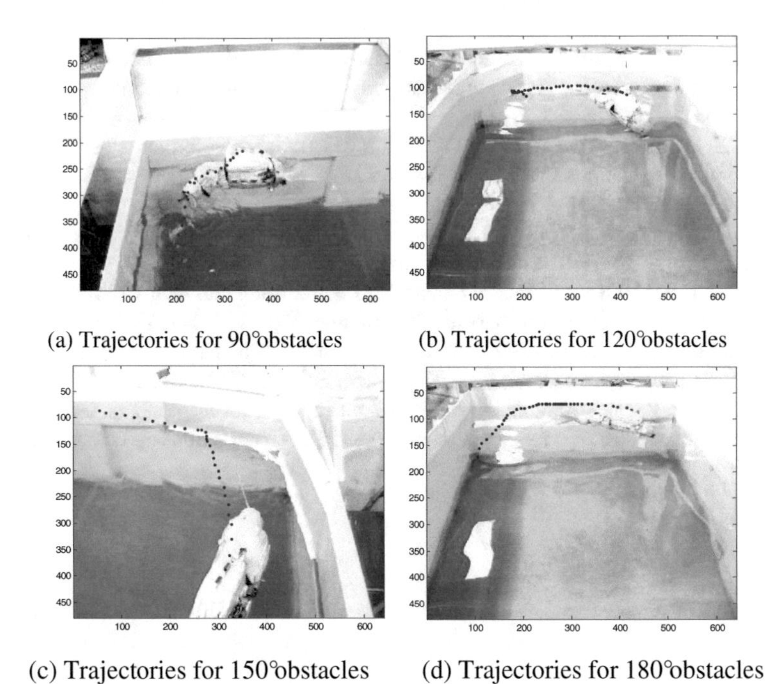

(a) Trajectories for 90°obstacles (b) Trajectories for 120°obstacles

(c) Trajectories for 150°obstacles (d) Trajectories for 180°obstacles

Fig. 9. Swimming trajectories for several obstacle and approaching angles

4 Conclusions

Detection and recognition of obstacles to avoid collision, regardless of the approaching angles for fish robots, are considered fundamental to generate natural movement. Simple IR distance sensors are used instead of cameras or sonar systems. Based on the simplified models of obstacles, proper sets of obstacle angles and approaching angles are chosen to establish the basic rules for inference algorithms. Arbitrary angles for obstacle shapes and approach lines are solved by these inference results using interpolation of the basic training sets. Successful experimental results in an aquarium show the effectiveness of the implementation on the fish robot.

Acknowledgments

This work was supported by the Second BK21, CNU and the NURI-CEIA, CNU.

References

1. Yu, J., Tan, M., Wang, S., Chen, E. (2004) Development of a Biomimetic Robotic Fish and Its Control Algorithm. IEEE Transactions on Systems, Man, and Cybernetics—Part B: Cybernetics 34: 1798-1810
2. Liu, J., Hu, H. (2005) Mimicry of Sharp Turning Behaviours in a Robotic Fish. Proceedings of the 2005 IEEE International Conference on Robotics and Automation, Barcelona, Spain, April : 3329-3334
3. Antonelli, G., Chiaverini, S., Finotello, R., Schiavon, R. (2001) Real-time path planning and obstacle avoidance for RAIS: an autonomous underwater vehicle. IEEE Journal of Oceanic Engineering 26(2) : 216-227
4. Petillot, Y., Ruiz, T., Tena, I., Lane, D. M. (2001) Underwater vehicle obstacle avoidance and path planning using a multi-beam forward looking sonar. IEEE Journal of Oceanic Engineering 26(2) : 240-251
5. Panella, M., Gallo, A. S. (2005) An Input-Output Clustering Approach to the Synthesis of ANFIS Networks. IEEE Transactions on Fuzzy Systems 13(1) : 69-81
6. Mar, J., Lin, F. (2001) An ANFIS Controller for the Car-Following Collision Prevention System. IEEE Transactions on Vehicular Technology 50(4) : 1106-1113
7. Frayman, Y., Wang, L. P. (1998) Data mining using dynamically constructed recurrent fuzzy neural networks. Lecture Notes in Computer Science 1394 : 122-131
8. Wai, R., Chen, P. (2004) Intelligent tracking control for robot manipulator including actuator dynamics via TSK-type fuzzy neural network. IEEE Transactions on Fuzzy Systems 12 : 552-560

Generalization of Soft Set Theory: From Crisp to Fuzzy Case

Xibei Yang[1], Dongjun Yu[1], Jingyu Yang[1], and Chen Wu[1,2]

[1] School of Computer Science and Technology, Nanjing University of Science and Technology, Nanjing, Jiangsu, 210094, P.R. China
yangxibei@hotmail.com

[2] College of Information Science and Technology, Drexel University, Philadelphia, PA 19104, USA
wuchenzj1@gmail.com

Abstract. The traditional soft set is a mapping from parameter to the crisp subset of universe. However, the situation may be more complex in real world because the fuzzy characters of parameters. In this paper, the traditional soft set theory is expanded to be a fuzzy one, the fuzzy membership is used to describe parameter-approximate elements of fuzzy soft set. Furthermore, basic fuzzy logic operators are used to define generalized operators on fuzzy soft set and then the DeMorgan's laws are proved. Finally, the parametrization reduction of fuzzy soft set is defined, a decision-making problem is analyzed to indicate the validity of the fuzzy soft set.

Keywords: Soft set, Fuzzy soft set, Tabular representation, Difference function, Reduction.

1 Introduction

Soft set theory[1] was firstly proposed by Russia researcher, Molodtsov in 1999. It is a general mathematical tool for dealing with uncertain, fuzzy, not clearly defined objects. What the soft set theory different with the traditional tool for dealing with uncertainties, such as theory of probability, theory of fuzzy set[2] and theory of rough set[3, 4], is that it is free from the inadequacy of the parametrization tools of those theories[1]. In [1], Molodtsov successfully applied the soft theory into several directions, such as smoothness of functions, game theory, operations research, Riemann-integration, Perron integration, probability, theory of measurement and so on. Based on Molodtsov's work, Maji[5] and his colleagues defined some operators on soft set and then they presented subset, complement of soft set, some immediate examples are also given for explaining their theoretical fruits. Furthermore, Maji[6] presented an application of soft set in the decision-making problem that is based on the knowledge reduction of rough set, it represents that the soft set theory has a tremendous potential in the field of intelligent information processing. However, from a further investigation of the decision-making problem, [7] pointed out that the knowledge reduction of rough set could not be applied in the soft set theory because the attributes

B.-Y. Cao (Ed.): Fuzzy Information and Engineering (ICFIE), ASC 40, pp. 345–354, 2007.
springerlink.com

reduction in rough set theory is designed to find a minimal attributes set that retains the classification ability of the indiscernibility relation. The idea of employing the attributes reduction in rough set theory to reduce the number of parameters to compute the optimal objects seems meaningless. Therefore, the parametrization reduction [7] of soft set was proposed. The reduction of parameters in soft set theory is designed to offer minimal subset of the conditional parameters set to keep the optimal choice objects.

It is clear that above theoretical and practical discussions are all based on the standard soft set theory. However, in the practical applications of soft sets, the situation can be more complex. For instance, in the decision-making problem[6] that was analyzed by Maji, words "expensive, beautiful, cheap, modern" and so on are looked as a set of parameters. According to Molodtsov's soft set theory, for every parameter, there is a mapping into one of the crisp subset of the universe(set of houses). Nevertheless, it is well known that all those words are fuzzy concepts in real world from the viewpoint of Zadeh's fuzzy set theory[2]. Or simply, for any one house in the universe, it is unreasonable to say that the house is absolutely beautiful or not. What we can say is that how degree of beautiful the house is. From what have been discussed above, it is clear that soft set needs to be expanded to improve its potential ability in practical engineering applications. This is the topic of our paper, the soft set theory is expanded to be a fuzzy one and then some immediate outcomes could be discussed.

This paper has been organized as follows. In section 2, basic notions about soft set is reviewed and then the fuzzy soft set theory is strictly defined. Section 3 mainly focuses on binary operators on fuzzy soft set those are based on the fuzzy logic operators. Therefore, two immediate properties about operators on fuzzy soft set are proved. In section 4, the reduction of fuzzy soft set is defined, then a decision-making problem is analyzed by fuzzy soft set. Conclusion is summarized in section 5.

2 Theory of Fuzzy Soft Set

2.1 Soft Set Theory

Molodtsov[1] defined the soft set in the following way. Let U be an initial universe set and E be a set of parameters.

Definition 1. *A pair $< F, E >$ is called a soft set (over U) if and only if F is a mapping of E into the set of all subsets of the set U.*

In other words, the soft set is not a kind of set, but a parameterized family of subsets of the set U[1, 5]. Let us denote $P(U)$ the power set on U and then F is a mapping such as $F : E \rightarrow P(U)$. For $\forall \epsilon \in E$, $F(\epsilon)$ may be considered as the set of ϵ-approximate elements of the soft set $< F, E >$.

Example 1. Suppose the following,

- U is the set the houses under consideration and $U = \{h_1, h_2, h_3, h_4, h_5, h_6\}$.

- E is the set of parameters and $E = \{e_1, e_2, e_3, e_4, e_5\}$={beautiful, wooden, cheap, in the green surroundings,in good repair}.

Suppose that $F(e_1) = \{h_2, h_4\}$, $F(e_2) = \{h_1, h_3\}$, $F(e_3) = \{h_3, h_4, h_5\}$, $F(e_4) = \{h_1, h_3, h_5\}$, $F(e_5) = \{h_1\}$, then, the soft set $< F, E >$ can be looked as a collection of approximation as below: $< F, E >$={beautiful houses = $\{h_2, h_4\}$; wooden houses = $\{h_1, h_3\}$; cheap houses = $\{h_3, h_4, h_5\}$; in the green surroundings = $\{h_1, h_3, h_5\}$; in good repair = $\{h_1\}$}.

Tabular representation[8] is a useful tool to show information system in rough set theory[3, 4]. On the other hand, it also can be used to represent soft set[6]. The tabular representation of the soft set in Example 1 can be seen in Table 1. For $\forall \epsilon_j \in E$, if $h_i \in F(\epsilon)$, then we denote $h_{ij} = 1$, otherwise $h_{ij} = 0$. From the above example, it is clear that the mapping only classify the objects into two simple classes(Yes or No), i.e. $h_{ij} \in \{0, 1\}$. However, from the viewpoint of fuzzy set theory, it is more reasonable to use membership to represent the degree of the object who holds the parameter. That is to say, for $\forall \epsilon_j \in E$ and $\forall h_i \in U$, $h_{ij} \in [0, 1]$ other than $\{0, 1\}$ is more suitable for representation of real world. Therefore, in Table 1, h_{i1} could be looked as a membership to indicate the beautiful degree of house h_i is.

Table 1. Tabular representation of a soft set

U	e_1	e_2	e_3	e_4	e_5
h_1	0	1	0	1	1
h_2	1	0	0	0	0
h_3	0	1	1	1	0
h_4	1	0	1	0	0
h_5	0	0	1	1	0
h_6	0	0	0	0	0

2.2 Fuzzy Soft Set Theory

Definition 2. *Let U be an initial universe and E be a set of parameters, $\mathcal{F}(U)$ is the fuzzy power set on U, a pair $< \widetilde{F}, E >$ is called a fuzzy soft set (over U) where $\widetilde{F}$ is a mapping given by*

$$\widetilde{F} : E \to \mathcal{F}(U). \tag{1}$$

From the above definition, it is clear that fuzzy soft set is a generalization of the standard soft set. In general, for $\forall \epsilon \in E$, $\widetilde{F}(\epsilon)$ is a fuzzy subset of U and it is called fuzzy value set of parameter ϵ. If for $\forall \epsilon \in E$, $\widetilde{F}(\epsilon)$ is a crisp subset of U, then $< \widetilde{F}, E >$ is degenerated to be the standard soft set. In order to distinguish with the standard soft set, let us denote $\mu_{\widetilde{F}(\epsilon)}(x)$ by the degree that object x holds parameter ϵ where $x \in U$ and $\epsilon \in E$. In other words, $\widetilde{F}(\epsilon)$ can be written as a fuzzy set such that $\widetilde{F}(\epsilon) = \Sigma_{x \in U} \mu_{\widetilde{F}(\epsilon)}(x)/x$.

Example 2. Table 2 is a tabular representation of fuzzy soft set $< \widetilde{F}, E >$ where universe and set of parameters are all same to Example 1. For instance, we can see that $\mu_{\widetilde{F}(e_1)}(h_1) = 0.9$, this means that house h_1 has a degree of 0.9 to be regarded as a beautiful one.

From the above example, it is clear that different objects may have different memberships of parameters, this is reasonable in our decision-making. Suppose that, Mr. X is interested to buy a house which is very cheap. On the other hand, Mr. Y is like to choose a house which is not very expansive. The standard soft set has not the ability to distinguish with these two kind of fuzzy characteristic, however, the fuzzy soft set can do this work. The detailed illustration will be in section 4.

Table 2. Tabular representation of fuzzy soft set

U	e_1	e_2	e_3	e_4	e_5
h_1	0.9	0.7	0.3	0.8	0.7
h_2	0.8	0.8	0.8	0.9	0.1
h_3	0.6	0.2	0.7	0.7	0.9
h_4	0.8	0.1	0.7	0.8	0.2
h_5	0.8	0.3	0.9	0.2	0.5
h_6	0.8	0.8	0.2	0.7	0.8

Definition 3. *The class of all fuzzy value sets of fuzzy soft set $< \widetilde{F}, E >$ is called fuzzy value-class of the fuzzy soft set and is denoted by $C_{<\widetilde{F},E>}$, then we have*

$$C_{<\widetilde{F},E>} = \{\epsilon \in E : \widetilde{F}(\epsilon)\}. \tag{2}$$

Example 3. For the fuzzy soft set that is represented in Table 2, $C_{<\widetilde{F},E>}$ can be written as follows:

$$\widetilde{F}(e_1) = 0.9/h_1 + 0.8/h_2 + 0.6/h_3 + 0.8/h_4 + 0.8/h_5 + 0.8/h_6$$
$$\widetilde{F}(e_2) = 0.7/h_1 + 0.8/h_2 + 0.2/h_3 + 0.1/h_4 + 0.3/h_5 + 0.8/h_6$$
$$\widetilde{F}(e_3) = 0.3/h_1 + 0.8/h_2 + 0.7/h_3 + 0.7/h_4 + 0.9/h_5 + 0.2/h_6$$
$$\widetilde{F}(e_4) = 0.8/h_1 + 0.9/h_2 + 0.7/h_3 + 0.8/h_4 + 0.2/h_5 + 0.7/h_6$$
$$\widetilde{F}(e_5) = 0.7/h_1 + 0.1/h_2 + 0.9/h_3 + 0.2/h_4 + 0.5/h_5 + 0.8/h_6$$

Definition 4. *Let $< \widetilde{F}, E_1 >$ and $< \widetilde{G}, E_2 >$ be two fuzzy soft sets over a common universe U, we say that $< \widetilde{F}, E_1 >$ is a fuzzy soft subset of $< \widetilde{G}, E_2 >$ if and only if $E_1 \subset E_2$ and for $\forall \epsilon \in E_1$, $\widetilde{F}(\epsilon) = \widetilde{G}(\epsilon)$ holds, it can be represented as $< \widetilde{F}, E_1 > \widetilde{\subset} < \widetilde{G}, E_2 >$.*

Example 4. For the fuzzy soft set that is represented in Table 2, suppose that $E_1 = \{e_1, e_2, e_3\}$, $E_2 = E$, then we have $< \widetilde{F}, E_1 > \widetilde{\subset} < \widetilde{F}, E_2 >$.

Definition 5. *Let* $<\widetilde{F}, E_1>$ *and* $<\widetilde{G}, E_2>$ *be two fuzzy soft sets,* $<\widetilde{F}, E_1>$, $<\widetilde{G}, E_2>$ *are said to be fuzzy soft equal if and only if* $<\widetilde{F}, E_1>$ *is a fuzzy soft subset of* $<\widetilde{G}, E_2>$ *and* $<\widetilde{G}, E_2>$ *is a fuzzy soft subset of* $<\widetilde{F}, E_1>$*, it can be represented as* $<\widetilde{F}, E_1> \widetilde{=} <\widetilde{G}, E_2>$.

In the standard soft set theory, Maji considered two special soft sets[5], named as Null Soft Set and Absolute Soft Set, respectively. As far as the fuzzy soft set $<\widetilde{F}, E>$ on U is concerned, if for $\forall \epsilon \in E$, $\forall x \in U$, we have $\mu_{\widetilde{F}(\epsilon)}(x) = 0$, then $<\widetilde{F}, E>$ is called a Null fuzzy soft set; on the other hand, if $\forall \epsilon \in E$, $\forall x \in U$, we have $\mu_{\widetilde{F}(\epsilon)}(x) = 1$, then $<\widetilde{F}, E>$ is called an Absolute fuzzy soft set.

3 Operators on Fuzzy Soft Set

3.1 Fuzzy Logic Operators

First of all, let us review several fuzzy logic operators founded in [9, 10].

- A negation is a function $N : [0, 1] \to [0, 1]$, such that $N(0) = 1$ and $N(1) = 0$. An usual representation of the negation is $N(x) = 1 - x$ and it is called as standard negation.
- A triangular norm, or shortly $T-$Norm is a continuous, non-decreasing function $T(x, y) : [0, 1]^2 \to [0, 1]$ such that $T(x, 1) = x$. Clearly a T-Norm stands for a conjunction. Usual representations of T-Norm are:
  - the standard min operator: $T(x, y) = min(x, y)$;
  - the algebraic product: $T(x, y) = xy$;
  - the Lukasiewicz $T-$Norm (also called the bold intersection): $T(x, y) = max(x + y - 1, 0)$.
- A triangular conorm, or shortly $T-$Conorm is a continuous, non-decreasing function $S(x, y) : [0, 1]^2 \to [0, 1]$ such that $S(0, y) = y$. Clearly a T-Conorm stands for a disjunction. Usual representations of T-Conorm are:
  - the standard max operator: $S(x, y) = max(x, y)$;
  - the algebraic product:$S(x, y) = xy$;
  - the Lukasiewicz $T-$Conorm (also called the bold sum): $S(x, y) = min(x + y, 1)$.

Given a negator N, a $T-$Norm T and $T-$Conorm S are dual with respect to N if and only if De Morgan laws are satisfied, i.e. $S(N(x), N(y)) = N(T(x, y))$, $T(N(x), N(y)) = N(S(x, y))$.

3.2 Operators on Fuzzy Soft Set

Definition 6. *The negator of a fuzzy soft set* $<\widetilde{F}, E>$ *is denoted by* $N(\widetilde{F}, E)$ *and it is defined by* $N(\widetilde{F}, E) = (\widetilde{F}^N, \sim E)$*, where* $\widetilde{F}^N :\sim E \to \mathcal{F}(U)$ *is a mapping given by*

$$\begin{aligned} \widetilde{F}^N(\sim \epsilon) &= \Sigma_{x \in U} \mu_{F(\sim \epsilon)}(x)/x \\ &= \Sigma_{x \in U} N(\mu_{F(\epsilon)}(x))/x \end{aligned} \quad (3)$$

in which for $\forall \epsilon \in E$.

Example 5. For the fuzzy soft set that is represented in Table 2, suppose that $N(x) = 1 - x$, then we have:

$$\widetilde{F}(\sim e_1) = 0.1/h_1 + 0.2/h_2 + 0.4/h_3 + 0.2/h_4 + 0.2/h_5 + 0.2/h_6$$
$$\widetilde{F}(\sim e_2) = 0.3/h_1 + 0.2/h_2 + 0.8/h_3 + 0.9/h_4 + 0.7/h_5 + 0.2/h_6$$
$$\widetilde{F}(\sim e_3) = 0.7/h_1 + 0.2/h_2 + 0.3/h_3 + 0.3/h_4 + 0.1/h_5 + 0.8/h_6$$
$$\widetilde{F}(\sim e_4) = 0.2/h_1 + 0.1/h_2 + 0.3/h_3 + 0.2/h_4 + 0.8/h_5 + 0.3/h_6$$
$$\widetilde{F}(\sim e_5) = 0.3/h_1 + 0.9/h_2 + 0.1/h_3 + 0.8/h_4 + 0.5/h_5 + 0.2/h_6$$

Definition 7. *The T operator on two fuzzy soft set. If $< \widetilde{F}, E_1 >$ and $< \widetilde{G}, E_2 >$ are two fuzzy soft sets, then $< \widetilde{F}, E_1 > T < \widetilde{G}, E_2 >$ is defined by $< \widetilde{F}, E_1 > T < \widetilde{G}, E_2 >=< \widetilde{H}, E_1 \times E_2 >$ where*

$$\widetilde{H}(\alpha, \beta) = \Sigma_{x \in U} T(\mu_{\widetilde{F}(\alpha)}(x), \mu_{\widetilde{G}(\beta)}(x))/x \tag{4}$$

in which for $\forall(\alpha, \beta) \in E_1 \times E_2$.

Example 6. For the fuzzy soft set that is represented in Table 2, $U = \{h_1, h_2, h_3, h_4, h_5, h_6\}$, suppose that $E_1 = \{e_1, e_2, e_3\}$, $E_2 = \{e_4, e_5\}$, $T-$Norm is regarded as standard min operator, then we have

$$\widetilde{H}(e_1, e_4) = 0.8/h_1 + 0.8/h_2 + 0.6/h_3 + 0.8/h_4 + 0.2/h_5 + 0.7/h_6$$
$$\widetilde{H}(e_1, e_5) = 0.7/h_1 + 0.1/h_2 + 0.6/h_3 + 0.2/h_4 + 0.5/h_5 + 0.8/h_6$$
$$\widetilde{H}(e_2, e_4) = 0.7/h_1 + 0.8/h_2 + 0.2/h_3 + 0.1/h_4 + 0.2/h_5 + 0.7/h_6$$
$$\widetilde{H}(e_2, e_5) = 0.7/h_1 + 0.1/h_2 + 0.2/h_3 + 0.1/h_4 + 0.3/h_5 + 0.8/h_6$$
$$\widetilde{H}(e_3, e_4) = 0.3/h_1 + 0.8/h_2 + 0.7/h_3 + 0.7/h_4 + 0.2/h_5 + 0.2/h_6$$
$$\widetilde{H}(e_3, e_5) = 0.3/h_1 + 0.1/h_2 + 0.7/h_3 + 0.2/h_4 + 0.5/h_5 + 0.2/h_6$$

Definition 8. *The S operator on two fuzzy soft set. If $< \widetilde{F}, E_1 >$ and $< \widetilde{G}, E_2 >$ are two fuzzy soft sets, then $< \widetilde{F}, E_1 > S < \widetilde{G}, E_2 >$ is defined by $< \widetilde{F}, E_1 > S < \widetilde{G}, E_2 >=< \widetilde{O}, E_1 \times E_2 >$ where*

$$\widetilde{O}(\alpha, \beta) = \Sigma_{x \in U} S(\mu_{\widetilde{F}(\alpha)}(x), \mu_{\widetilde{G}(\beta)}(x))/x \tag{5}$$

in which for $\forall(\alpha, \beta) \in E_1 \times E_2$.

Example 7. Following Example 3.2, $T-$Conorm is regarded as the standard max operator, then we have

$$\widetilde{O}(e_1, e_4) = 0.9/h_1 + 0.9/h_2 + 0.7/h_3 + 0.8/h_4 + 0.8/h_5 + 0.8/h_6$$
$$\widetilde{O}(e_1, e_5) = 0.9/h_1 + 0.8/h_2 + 0.9/h_3 + 0.8/h_4 + 0.8/h_5 + 0.8/h_6$$
$$\widetilde{O}(e_2, e_4) = 0.8/h_1 + 0.9/h_2 + 0.7/h_3 + 0.8/h_4 + 0.3/h_5 + 0.8/h_6$$
$$\widetilde{O}(e_2, e_5) = 0.7/h_1 + 0.8/h_2 + 0.9/h_3 + 0.2/h_4 + 0.5/h_5 + 0.8/h_6$$
$$\widetilde{O}(e_3, e_4) = 0.8/h_1 + 0.9/h_2 + 0.7/h_3 + 0.8/h_4 + 0.9/h_5 + 0.7/h_6$$
$$\widetilde{O}(e_3, e_5) = 0.7/h_1 + 0.8/h_2 + 0.9/h_3 + 0.7/h_4 + 0.9/h_5 + 0.8/h_6$$

Theorem 1. *Let* $<\widetilde{F}, E_1>$ *and* $<\widetilde{G}, E_2>$ *are two fuzzy soft set, then*

$$N\{<\widetilde{F}, E_1> T <\widetilde{G}, E_2>\} = \{N <\widetilde{F}, E_1>\} S \{N <\widetilde{G}, E_2>\}; \tag{6}$$
$$N\{<\widetilde{F}, E_1> S <\widetilde{G}, E_2>\} = \{N <\widetilde{F}, E_1>\} T \{N <\widetilde{G}, E_2>\}. \tag{7}$$

Proof. Suppose that

$$\begin{aligned}\{N <\widetilde{F}, E_1>\} S \{N <\widetilde{G}, E_2>\} &= <\widetilde{F}^N, \sim E_1> S <\widetilde{G}^N, \sim E_2> \\ &= <\widetilde{J}, \sim E_1 \times \sim E_2> \\ &= <\widetilde{J}, \sim (E_1 \times E_2)>\end{aligned}$$

Then, for $\forall \alpha, \beta \in (E_1 \times E_2)$, we have

$$\begin{aligned}\widetilde{J}(\sim \alpha, \sim \beta) &= \Sigma_{x \in U} S(\mu_{\widetilde{F}(\sim\alpha)}(x), \mu_{\widetilde{G}(\sim\beta)}(x))/x \\ &= \Sigma_{x \in U} S(N(\mu_{\widetilde{F}(\alpha)}(x)), N(\mu_{\widetilde{G}(\beta)}(x)))/x \\ &= \Sigma_{x \in U} N(T(\mu_{\widetilde{F}(\alpha)}(x), \mu_{\widetilde{G}(\beta)}(x)))/x\end{aligned}$$

Therefore, we have $N\{<\widetilde{F}, E_1> T <\widetilde{G}, E_2>\} = \{N <\widetilde{F}, E_1>\} S \{N <\widetilde{G}, E_2>\}$.

Suppose that

$$\begin{aligned}\{N <\widetilde{F}, E_1>\} T \{N <\widetilde{G}, E_2>\} &= <\widetilde{F}^N, \sim E_1> T <\widetilde{G}^N, \sim E_2> \\ &= <\widetilde{K}, \sim E_1 \times \sim E_2> \\ &= <\widetilde{K}, \sim (E_1 \times E_2)>\end{aligned}$$

Then, for $\forall \alpha, \beta \in (E_1 \times E_2)$, we have

$$\begin{aligned}\widetilde{K}(\sim \alpha, \sim \beta) &= \Sigma_{x \in U} T(\mu_{\widetilde{F}(\sim\alpha)}(x), \mu_{\widetilde{G}(\sim\beta)}(x))/x \\ &= \Sigma_{x \in U} T(N(\mu_{\widetilde{F}(\alpha)}(x)), N(\mu_{\widetilde{G}(\beta)}(x)))/x \\ &= \Sigma_{x \in U} N(S(\mu_{\widetilde{F}(\alpha)}(x), \mu_{\widetilde{G}(\beta)}(x)))/x\end{aligned}$$

Therefore, we have $N\{<\widetilde{F}, E_1> S <\widetilde{G}, E_2>\} = \{N <\widetilde{F}, E_1>\} T \{N <\widetilde{G}, E_2>\}$

4 Application of Fuzzy Soft Set

Maji's example of decision-making is based on the standard soft set theory. Let $U = \{h_1, h_2, h_3, h_4, h_5, h_6\}$ be a set of six houses, $E =$ {expensive; beautiful; wooden; cheap; in the green surroundings; modern; in good repair; in bad repair} be a set of parameters. Suppose that Mr. X wants to buy a house on the basis of his choice parameters "beautiful, wooden, cheap, in the green surrounding, in good repair". The problem is to select the house which qualifies with all(or with maximum number of) parameters those chosen by Mr.X. Therefore, Maji defined the choice value[6] to measure the important of the house. The choice value of an object $h_i \in U$ is c_i, given by $c_i = \sum_j h_{ij}$ where $h_{ij} \in \{0, 1\}$.

In this section, we will use the fuzzy soft set to help Mr.Y select a suitable house because his need is different with Mr. X's. Let $U = \{h_1, h_2, h_3, h_4, h_5, h_6\}$ be set of six houses, E_1 ={beautiful; wooden; cheap; in the green surroundings; in good repair} be a set of parameters, this fuzzy soft set has been showed in Table 2. Suppose that Mr. Y wants to buy a house that is the most beautiful one, in other words, the degree of beauty of the house is 1.0. Similarly, the degrees of wooden, cheap and in good repair are 0.8, the degree of green surrounding is 0.6, which could be written in a fuzzy set $\mathbf{F} = 1.0/e_1 + 0.8/e_2 + 0.8/e_3 + 0.6/e_4 + 0.8/e_5$. In order to find which house is the most suitable for Mr.Y and which parameter is the key one for his choice, let us define the difference function as follows.

Definition 9. *Let* $< \widetilde{F}, E >$ *be a fuzzy soft set,* $\mathbf{F} = \sum_{\epsilon \in E} \mu_{\mathbf{F}}(\epsilon)/\epsilon$ *be the target, then for* $\forall x \in U$*, the difference function between* x *and* $\mathbf{F}$ *is defined as*

$$d(x) = \sum_{\epsilon \in E} |\mu_{\mathbf{F}}(\epsilon) - \mu_{\widetilde{F}(\epsilon)}(x)| \tag{8}$$

According to the above definition, we can get Table 3 in which every house has a difference value with Mr.Y's exception. Let us denote by M_E the collection of objects in U which has the minimal difference value. In Table 3, we have $M_E = \{h_6\}$.

Table 3. Fuzzy soft set with difference function

U	e_1	e_2	e_3	e_4	e_5	$d(x)$
h_1	0.9	0.7	0.3	0.8	0.7	1.0
h_2	0.8	0.8	0.8	0.9	0.1	1.2
h_3	0.6	0.2	0.7	0.7	0.9	1.3
h_4	0.8	0.1	0.7	0.8	0.2	1.8
h_5	0.8	0.3	0.9	0.2	0.5	1.5
h_6	0.8	0.8	0.2	0.7	0.8	0.9

Definition 10. *Let* $< \widetilde{F}, E >$ *be a fuzzy soft set, for* $\forall A \subset E$*, if* $M_{E-A} = M_E$*, then* A *is called a dispensable set in* E*, otherwise* A *is called an indispensable set in* E*.*

The parameter set E is called independent if for $\forall A \subset E$, A is indispensable in E, otherwise E is dependent.

Definition 11. *Let* $< \widetilde{F}, E >$ *be a fuzzy soft set,* $B \subseteq E$ *is called a reduction of* E *if* B *is independent and* $M_B = M_E$*.*

For the fuzzy soft set in Table 3, if we delete e_1, e_3, e_4, then $M_{e_2,e_5} = \{h_6\}$, that is to say, $\{e_1, e_3, e_4\}$ is a dispensable set in E. However, set $\{e_2, e_5\}$ is dependent because if we delete e_2 from $\{e_2, e_5\}$, $M_{e_5} = \{h_6\}$; on the other hand, if we delete e_5 from $\{e_2, e_5\}$, then $M_{e_2} = \{h_2, h_6\}$. From what have been discussed above, fuzzy soft set in Table 2 has a parameter reduction e_5, e_5 is the key parameter in Mr. Y's selection of house. From the reduction of fuzzy soft set, we get the Table 4, it is clear that h_6 should be the best choice for Mr. Y.

Table 4. Reduction of fuzzy soft set

U	h_1	h_2	h_3	h_4	h_5	h_6
e_5	0.7	0.1	0.9	0.2	0.5	0.8
$d(x)$	0.1	0.7	0.1	0.6	0.3	0.0

5 Conclusion

The soft set theory proposed by Molodtsov offers a general mathematical tool for dealing with uncertain or vague objects. In the present paper, the standard soft set theory is expanded to be a fuzzy one because the fuzzy character of parameters in real world is taken into consideration. Moreover, not only fuzzy logic operators are used to define operators on fuzzy soft set, but also a decision-making problem is analyzed by fuzzy soft set. To sum up, the fuzzy soft theory may play a more important role in intelligent information processing.

Acknowledgement

This work was supported in part by the Natural Science Foundation of China (No.60472060, No.60572034 and No.60632050) and Natural Science Foundation of Jiangsu Province (No.BK2006081).

References

1. Molodtsov D (1999) Soft Set Theory–First Results. Computers and Mathematics with Applications 37:19–31
2. Zadeh L A (1965) Fuzzy Set. Information and Control 8:338–353
3. Pawlak Z (1982) Rough sets. International Journal of Computer and Information Sciences 11:341–356
4. Pawlak Z (2002) Rough sets and intelligent data analysis. International Journal of Information Sciences 147:1–12
5. Maji P K, Biswas R, Roy A R (2003) Soft Set Theory. Computers and Mathematics with Applications 45:555–562
6. Maji P K, Roy A R (2002) An Application of Soft Sets in A Decision Making Problem. Computers and Mathematics with Applications 44:1077–1083
7. Denggang C, Tsang E C C, Yeung D S, Xizhao W (2005) The Parameterization Reduction of Soft Sets and its Applications. Computers and Mathematics with Applications 49:757–763

8. Yao Y Y (1998) Relational Interpretations of Neighbourhood Operators and Rough Set Approximation Operators. International Journal of Information Sciences 111:239–259
9. Radzikowska A M, Kerre E E (2002) A Comparative Study of Fuzzy Rough Sets. Fuzzy Sets and Systems 126:137–155
10. Morsi N N, Yakout M M Axiomatics For Ruzzy Rough Set. Fuzzy Sets and Systems 100:327–342

A New QPSO Based BP Neural Network for Face Detection

Shouyi Li, Ronggui Wang, Weiwei Hu, and Jianqing Sun

School of Computer and Information, Heifei University of Technology,
Hefei, Anhui 230009, China
lishouyi710@yahoo.com.cn, wangrgui@mail.hf.ah.cn,
hendry8426@yahoo.com.cn, sunjianq83@yahoo.com.cn

Abstract. Quantum-behaved Particle Swarm Optimization (QPSO) algorithm is a quantum-inspired version of the particle swarm optimization (PSO) algorithm, which outperforms traditional PSOs in search ability as well as having less parameter to control. In this paper, QPSO technique is introduced into BP neural network to instead adopting gradient descent method in BP learning algorithm. Due to the characteristic of the QPSO algorithm, the problems of traditional BPNN can be avoided, such as easily converging to local minimum. Then we adopt the new learning algorithm to training a neural network for face detection, and the experiment results testify its efficiency in this.

Keywords: Quantum Particle Swarm Optimization, machine learning, BPNN, face detection.

1 Introduction

Particle Swarm Optimization (PSO) is a population-based evolutionary search technique which is originally introduced by Kennedy and Eberhart in 1995[3]. It is inspired by the emergent motion of a flock of birds searching for food. Compared with other stochastic optimization techniques like genetic algorithms (GA), PSO has fewer complicated operations and fewer defining parameters. However, the classical PSOs are not global-convergence guaranteed algorithms [1]. Recently, a global convergence-guaranteed PSO Quantum-behaved Particle Swarm Optimization (QPSO) was proposed in [4]. The idea is to permit all particles to have quantum behavior instead of the classical Newtonian random motion assumed in the particle swarm optimization algorithm (PSO). It outperforms traditional PSOs in search ability as well as having less parameter to control [6].

In this paper, we introduce the QPSO algorithm into the BP neural network, and proposed an improved BPNN learning algorithm. Due to the characters of QPSO algorithm, the improved method can solve the problems of traditional BPNN well, such as easily converging to local minimum. Then a neural network for face detection was trained using the new algorithm. And the experiment results show that the new learning algorithm has better performance in face detection.

The rest of this paper is organized as follows. In Section 2, a brief introduction of PSO and QPSO is given. In Section 3, a modified BPNN learning algorithm based on QPSO is introduced. And face detection using the new learning algorithm is described

B.-Y. Cao (Ed.): Fuzzy Information and Engineering (ICFIE), ASC 40, pp. 355–363, 2007.
springerlink.com

in Section 4, followed by experiment results in section 5. Finally the paper is concluded in Section 6.

2 Quantum Particle Swarm Optimization

2.1 PSO Algorithm

PSO is a population-based optimization algorithm. The swarm refers to a number of potential solutions to the optimization problem, initially having a population of random solutions. Each potential solution is represented by a particle which is given a random velocity and flies through the problem space. Every particle remembers its previous best position called *pbest* and corresponding fitness. The swarm remember another value called *gbest*, which is the best solution discovered by all particles.

In PSO algorithm, the velocity vector and position vector of the particles in D-dimensional space are adjusted as follows [9]:

$$v_{ij}(t+1) = w * v_{ij}(t) + c_1 * r_1(p_{ij} - x_{ij}) + c_2 * r_2(p_{gj} - p_{ij}) \quad (1)$$

$$x_{ij}(t+1) = x_{ij}(t+1) + v_{ij}(t+1) \quad (2)$$

where parameter w is the inertia factor, c_1 and c_2 are positive constants, r_1 and r_2 are two random functions in the interval [0,1]. $X_i = (x_{i1}, x_{i2}, ..., x_{iD})$ represents the position vector of the ith particle. $V_i = (v_{i1}, v_{i2}, ..., v_{iD})$ represents its velocity vector. $P_i = (p_{i1}, p_{i2}, ..., p_{iD})$ represents the best previous position(the position give best fitness value) of the ith particle called pbest. $P_g = (p_{g1}, p_{g2}, ..., p_{gD})$ is the position of the best particle among all the particles in the population and called gbest. In [8], to guarantee PSO algorithm to convergence, each particle in the PSO system converges to its local point $P = (p_1, p_2, ..., p_D)$ given by

$$P_d = (c_1 * r_1 * p_{id} + c_2 * r_2 * g_d) / (r_1 + r_2) \quad d=1,2,...D \quad (3)$$

2.2 QPSO Algorithm

In QPSO algorithm, particles have quantum behavior. The dynamic behavior of the particle is widely divergent from that of the particle in traditional PSO sys-tems in that the exact values of position and velocity cannot be determined simul- taneously. In [4] the state of the particle is depicted by wave function $\Psi(\bar{x}, t)$ whose square value describes the probability of the particles appearing in position $\bar{x}$. And Delta potential well was employed with the center on point $\vec{p} = (p_1, p_2, \ldots, p_D)$ to constraint the quantum particles in PSO in order that the particle can converge to their local P without explosion. The probability density function and distribution function are got through solving the Schrŏdinger equation, Employed Monte Carlo method, then the position of the ith particle can be obtained:

$$x_{ij}(t) = p_j \pm \frac{L_{ij}}{2} \text{In}(1/u) \quad (4)$$

where p_j is described as equation (3), u is a random number uniformly distributed between 0 and 1. The value of L is given by [5]:

$$L_{ij}(t+1) = 2 * \beta * \left| mbest_j - x(t)_{ij} \right| \tag{5}$$

$$mbest_j = \frac{1}{M} \sum_{i=1}^{M} p_{ij} \qquad j = 1, 2, \ldots, D \tag{6}$$

$$mbest = (mbest_1, mbest_2, \ldots, mbest_D) \tag{7}$$

where β is Contraction-Expansion coefficient, M is the size of the population. Then the position can be rewritten as follows

$$x_{ij}(t+1) = p_j \pm \beta * \left| mbest_j - x(t)_{ij} \right| * In(1/u) \tag{8}$$

In addition, an adaptive parameter control method in [6] is used to enhance QPSO in convergence speed. In this paper, we modified the method properly as follows:

$$\Delta F = (F_i - F_{gbest}) / F_{gbest} \tag{9}$$

where the error function ΔF is used to identify how close the particle is to the global best position, gbest. The smaller the value of the error function for a certain particle, the closer to the gbest the particle is. F_i is the fitness of the ith particle; F_{gbest} is the fitness of gbest. Then β is given by

$$\beta = 0.2 + \frac{1}{In(\Delta F + 1.0) + 0.75} \tag{10}$$

The equation can guarantee particles far away from the gbest can be given smaller value of β, whereas those close to the gbest can be given larger value of β.

3 Improved BPNN Learning Algorithm Based on QPSO

Back Propagation neural network (BPNN) is one kind of neural networks with most wide application. It is based on gradient descent method which minimizes the sum of the squared errors between the actual and the desired output values. Because of adopting the gradient method, the problems including slowly learning convergent velocity and easily converging to local minimum can not be avoided. QPSO is a global search algorithm, which is superior to classical PSO in performance and has fewer parameters to control. Therefore better performance can be obtained if QPSO replace the gradient descent method in BPNN learning algorithm. In this section, we'll describe a modified BPNN learning algorithm based on QPSO.

In the traditional BP network, the process is: considering the network has three layers

$$H_j = f\left(\sum_{i}^{n} z_i * w_{ij}\right) \tag{11}$$

$$Y_k = f\left(\sum_{i}^{m} H_j * v_{jk}\right) \tag{12}$$

$$E(\vec{w}) = \frac{1}{N}\sum_{p=1}^{N}\sum_{j=1}^{M}(D_{pj} - Y_{pj})^2 \tag{13}$$

where z_i (i=0,1,2....l) is input of the neural network, H_j (j=0,1,2.....m) is hidden node, Y_k (k=0,1,2.....m) is output of the network, w_{ij} is weight value between input layer and hidden layer, v_{jk} is weight between hidden layer and output layer, the size of the training set is N, M is the number of output neuron, $E(\vec{w})$ is the average error measure of the entire individual when the weight vector is $\vec{w}$ in the network, D_{pj} is the jth expectation output value of the pth input training individual, Y_{pj} is the corresponding practice output value, and f is the sigmoid function. Then weight w in the network changes according gradient descent method.

$$\Delta w\,(t+1) = -\eta \frac{\partial E}{\partial w}(t) + \alpha * \Delta w\,(t) \tag{14}$$

$$w(t+1) = w(t) + \Delta w\,(t+1) \tag{15}$$

where η and α are positive constants.

Instead of using gradient descent method, QPSO algorithm is adopted according to the reasons described above. Particle $X_i = (x_{i1}, x_{i2}, ..., x_{iD})$ represents all weight references of the network. And the weights alter through the change of the particle's state in QPSO algorithm. The fitness function is $E(X_i)$ (13). Then the improved learning algorithm is described in detail as follows:

```
    initialize population: random Xi
do
        for i=1 to population size M
          compute E(Xi) using equation (13)
        if E(Xi)<E(pi) then pi=xi
          pg=min(pi)
          find out mbest using equation (7)
          for d=1 to dimension D
        r1=rand(0,1) , r1=rand(0,1)
        p=(r1*pid+r2*pgd)/(r1+r2)
        u=rand(0,1)
        if rand(0,1)>0.5
              xid=p-beta*abs(mbestd-xid)*In(1/u)
        else xid=p+beta*abs(mbestd-xid)*In(1/u)
until termination criterion is met
```

4 Face Detection Based on the Improved Learning Algorithm

4.1 Preprocessing

The samples comes from MIT CBCL face database in which each sample is 19×19 pixels. Before designing and training a network for face detection using the samples, a preprocessing procedure needs to be done. It includes the following steps:

(1) illumination correction: A minimal brightness plane is subtracted from the original image, allowing reduction of light and heavy shadows.

(2) histogram normalization: histogram normalization is performed over the images in order to normalize the contrast of the gray images.

(3) masking: that is to mask the corner pixels because the values of these pixels are highly unstable. Figure #1 shows the three processes.

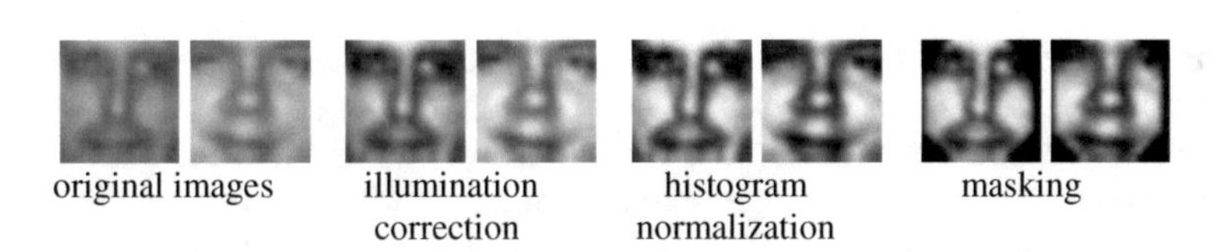

Fig. 1. preprocessing procedure

4.2 Designing the Network

Although each sample has 283 dimensions from originally 361(19×19) dimensions after the 3ird step above, the search space is still very large if a network is designed using the new algorithm. In order to reduce computation cost, we adopt principal component analyses (PCA) to reduce the dimension of the input training sample from 283 to 50. Therefore a three-layer network is designed. The input layer has 50 neurons, respectively receiving each value of the 50 dimensions of the project of input training sample. The hidden layer has 8 neurons, and output layer has 1 neuron. So the network has 408 weights in total. All of these weights are represented by a particle which has 408 dimensions in the new learning algorithm. And the size of the population is set to be 50 in this experiment.

4.3 Training

During the training phase, each sample(19*19) is represented in a vector, which is then projected on the feature subspace learned by PCA after the preprocessing procedure. The vector is the input of the network. The desired output is set to be 1 for samples that represent face and 0 for samples that are non-face. And bootstrap method [2] is used to find representative images which are non-faces, avoiding the problem of using a huge training set for non-faces. After each training iteration, the *gbest*, *pbest* and corresponding fitness are stored and used for next training iteration. Finally the weights are given by *gbest*. By this method, another 3000 representative non-face images were obtained and added into the training set as negative samples.

4.4 Detection

During the process of detection, the test window (19*19) is represented in a vector, which then is also projected on the feature subspace learned by PCA after the pre-processing procedure. Then the project is used as the input of the network. The test-window with an output value higher than a threshold is considered as a face candidate. To detect faces of variable sizes and locations in the input image, image pyramid method [2] was used here. The idea is that repeatedly scaling the input image down by a factor to different size, the input image at each size is scanned exhaustively by the test window. Finally the problem of overlapping detection is solved by using the method in [7].

5 Experiment Results

To train the network, 2400 face samples and 2400 non-face samples extracted randomly from the training set are used as the originally training set. In order to show the comparison between the new learning algorithm (QPSO) and BP, the training results on the originally training set using the two learning algorithms are shown in Graph 1. Abscissa n means training iteration times; E is learning error (12). The parameters η and α in BP algorithm are set to be 0.2 and 0.5 respectively. And the only parameter beta in QPSO algorithm changes adaptively according equation (9). From the graph, the convergent velocity of BP learning algorithm becomes slow gradually, ultimately converging to local minimum. However the QPSO algorithm can avoid the problem and can be easier convergence to the global minimum. Therefore the conclusion that the new learning algorithm has better performance than traditional BPNN can be got.

After the network has been trained using the new algorithm, face detection is performed, which is described in section 4. In the experiment, two sets of images are

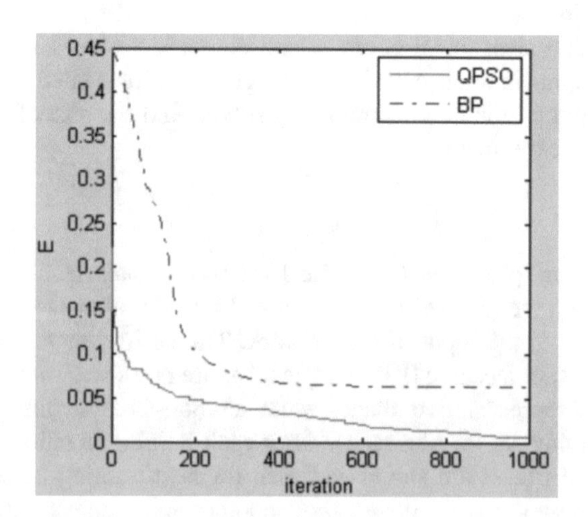

Graph 1. Error curves of BP and QPSO learning algorithms

Table 1. Detection results

Test1 60 images				Test2 40 images			
BP		QPSO based BP		BP		QPSO based BP	
Detec-Tion Rate	False positi-ves	Detec-Tion Rate	False positi-ves	Detec-Tion Rate	False positi-ves	Detec-Tion Rate	False positi-ves
88.3%	109	89.6%	106	91.6%	94	92.4%	93

used to evaluate the performance of the system. The test set 1 contains 40 images which have better quality, and each image has fewer faces. Test set 2 consists of 60 images from CMU and MIT face test set. Then the threshold is set to be 0.5 as the judgment of face and non-face. The detection results of using BP and QPSO based BPare listed in Table 1. From the detection result in the table, it is confirmed that the proposed QPSO learning algorithm works better in face detection. Finally, some detection examples are given in Figure #2. From these examples, we can see that the detections are relatively accurate and robust, even to the low image quality.

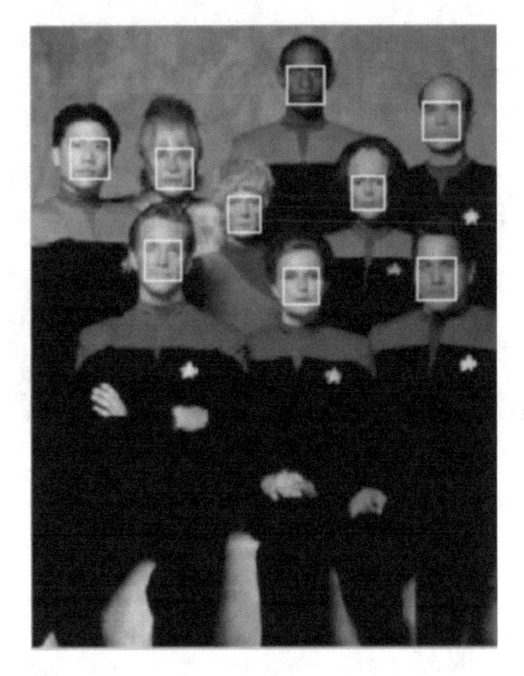

Fig. 2. Face detection examples

Fig. 2. (*continued*)

6 Conclusion

In this paper, QPSO algorithm is introduced into BP neural network learning algorithm. And the superiority is obtained by using the modified BPNN learning algorithm based on QPSO. Then a neural network for face detection was trained using the new learning algorithm. And the experiment result shows the method is effective in face detection.

Acknowledgements

This paper is supported by the National Natural Science Foundation of China (Grant No. 60575023), and Ph. D. Programs Foundation of Ministry of Education of China (Grant No.20050359012).

References

1. F. Van den Bergh (2001) An Analysis of Particle Swarm Optimizers. PhD Thesis. University of Pretoria.
2. H. Rowley, S. Baluja, and T. Kanade (1998) Neural network-based face de- tection. IEEE Transactions on Pattern Analysis and Machine Intelligence 20: 23-38.
3. J. Kennedy, R. C. Eberhart (1995) Particle Swarm Optimization. IEEE Int'l Conference on Neural Networks, 1942-1948.
4. Jun Sun, Bin Feng, and Wenbo Xu (2004) Particle swarm optimization with particles having quantum behavior. Congress on Evolutionary Computation 1: 325-331.
5. Jun Sun, Wenbo Xu, and Bin Feng (2004) A Global Search Strategy of Quantum-behaved Particle Swarm Optimization. IEEE Conference on Cybernetics and Intelligent Systems 1: 111-116.
6. Jun Sun, Wenbo Xu, and Bin Feng (2005) Adaptive Parameter Control for Quantum-behaved Particle Swarm Optimization on Individual Level. IEEE International Conference on Systems, Man and Cybernetics 4: 3049-3054.
7. Linlin Huang, Akinobu Shimizu, and Hidefuni Kobatake (2002) Face Detection using a modified Radial Basis Function Neural Network. IEEE International Conference on Pattern Recognition 2: 342-345.
8. M. Clerc and J. Kennedy (2002) The particle swarm: explosion, stability, and convergence in a multi-dimensional complex space. IEEE Transactions on Evolutionary Computation 6: 58-73.
9. Y. Shi, R. C. Eberhart (1998) A Modified Particle Swarm optimizer. IEEE International Conference on Evolutionary Computation, 69-73.

ECC-Based Fuzzy Clustering Algorithm

Yi Ouyang[1], Yun Ling[2], and AnDing Zhu[3]

[1] College of Computer and Information Engineering, Zhejiang Gongshang University
oyy@mail.hzic.edu.cn
[2] College of Computer and Information Engineering, Zhejiang Gongshang University
yling@mail.hzic.edu.cn
[3] College of Computer and Information Engineering, Zhejiang Gongshang University
jason_hi@163.com

Abstract. Recently various clustering approaches have been developed for web pages clustering optimization. Traditional methods take the vector model as their free-text analytical basis. However these algorithms cannot perform well on these problems which involving many Ecommerce information objectives. A novel approach based on the ECC vector space model FCM clustering algorithm is proposed to deal with these problems in this paper. By introducing Ecommerce concept (ECC) model, the Automatic Constructing Concept algorithm is proposed at first. Through the ACC algorithm and fields keywords table, the Ecommerce concept objects are established automatically. The ECC-Based Fuzzy Clustering (EFCM) is used to divide web pages into the different concept subsets. The experiment has compared it with Kmeans, Kmedoid, and Gath-Geva clustering algorithm, and results demonstrate the validity of the new algorithm. According to classification performance, the EFCM algorithm shows that it can be a clustering method for the Ecommerce semantic information searching in Internet.

Keywords: Unsupervised Clustering Methods, Ecommerce Concept, Kmeans, Fuzzy C-means.

1 Introduction

The growth of the Internet has been seen an information explosion. The traditional information searching has become inefficient and most results of information retrieval are irrelative with the users' demands. It is in urgent need of some efficient methods for information retrieval.

The web pages clustering is the Key technology in semantic searching system, and it can be used in different applied environment. By web pages clustering, information retrieval system can not only search the relevant articles of the web pages through automatically, but can attract users to browse more useful contents effectively. It is convenient for user to find a favorite topic and meanwhile it can improve systematic clicking rate. Through clustering the semantic similar web pages together, we can organize and index the information of web pages to reduce the quantity of searching the candidate sets. It also can reduce the systematic calculation when performing a retrieval task to improve the response speed searched.

B.-Y. Cao (Ed.): Fuzzy Information and Engineering (ICFIE), ASC 40, pp. 364–372, 2007.
springerlink.com

Fuzzy C-means (FCM) is a data clustering technique in which each data point belongs to a cluster to some degree that is specified by a membership grade. This technique was originally introduced by Jim Bezdek in 1981[4] as an improvement on earlier clustering methods. It provides a method that shows how to group data points into a specific cluster. In this paper, we present the hierarchical models for ECommerce Concept (ECC) analyzing at first, then Kmeans[1], Kmedoid, Gath-Geva[2] and ECC-Based EFCM clustering algorithms are used to cluster the web pages respectively. Finally, we compared the ECC-Based fuzzy clustering with other methods.

2 The Hierarchical Model of ECC

The Ecommerce Web pages have same similarity. Such as, there are try to introduce their company or products, so most of web pages have title, introduction, list table, figure as well as advertisement. The purpose of our research is searching Ecommerce commodity more efficient. The hierarchical model of ECC have four layers, Keywords layer, Concept layer, Document layer and Fields layers, and the Concept layer is composed by a set of keywords which have same lexical link in WordNet[11].

According to [3] that list 62 kinds of commercial object, we search 8,500 piece of web pages from Internet at first, and use the statistic method to get the business activity object that people concern about. In order to get the eigenvalue of the web pages, the formula of the keywords frequency extracted from web pages as follows:

$$kfreq_{i,j} = \log \frac{\sum\limits_{i=1}^{n} |N_{i,j}|}{\sum\limits_{i=1}^{n} |D_i|} \tag{1}$$

where, $|N_{i,j}|$ means the number of amount j-th keywords in i-th web page, $|D_i|$ means the number of total words in i-th web page. Through formula 2.1, we can get the fields words list when they meet with the $kfreq_{i,j} \geq threshold$.

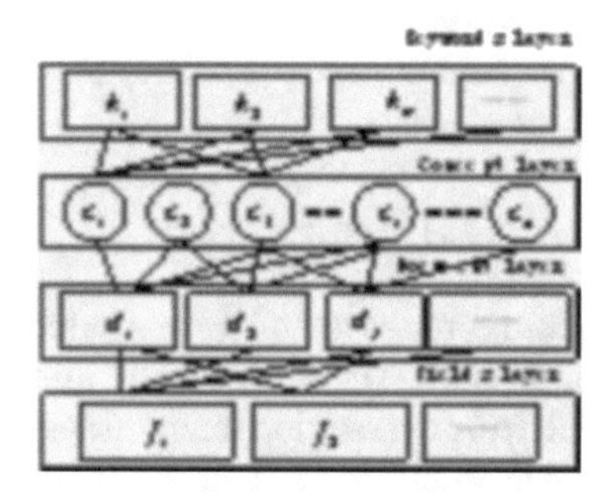

Fig. 1. The structure of semantic concept layers

An intermediate concept model, which lies between keywords structure and web pages, is the hierarchical structure. This model represents that the Ecommerce fields objects composed by several WebPages set, and each WebPages set composed by several relative concept. The concept object is composed by several relative keywords, and the relationship between words can extracted from WordNet[11]. Fig.1. shows this hierarchical structure structure.

3 ECC-Based Vector Space Model

The simplification from web page to a hierarchy allows the adoption of faster algorithms to solve queries. So, we define the Concept Vector Model to be used in the web pages clustering.

Concept space is an n-dimensional space composed of n independent concept axes. Each concept vector, c_i, represents one concept, and has a magnitude of $|c_i|$. In concept space, a document vector, T is represented by the sum of n-dimensional concept vectors,c_i

Documents generally contain various concepts, and we must determine those concepts if we are to comprehend the means of a document. In accordance with the accepted view in the linguistics literature that lexical chains provide a good representation of discourse structures and topicality of segments [6]. Hirst and St-Onge adapted the Rogets-based relations to WordNet-based ones, extra-strong, strong and medium-strong relations [7]. Two words are related in a medium-strong fashion if there exists an allowable path between words, and a path is allowable if it contains no more than five links and conforms to one of the eight patterns described in Hirst and St-Onge. Consequently, numerous kinds of relations can be used in composing lexical chains. The large number of possible word relations means that, if the proposed method is used to index a massive number of documents, there would be a large number of parameters and hence the indexing and retrieval computations would take a long time. Therefore, in the present work on the clustering of lexical items, there are only four kinds of relations—-identity, synonymy, hypernymy(hyponymy), and meronymy be considered.

Definition 1. *Let N be the total number of WebPages in the system and n_j be the number of web pages in which the index term k_i appears. Let $freq_{i,j}$ be the raw frequency of term k_i in the webpage d_j. Then, the normalized frequency $f_{i,j}$ of term k_i in web page d_j is given by*

$$f_{i,j} = \frac{freq_{i,j}}{\max[freq_{t,j}]} \tag{2}$$

where the maximum is computed over all terms which are mentioned in the text of the web page d_j. If the term k_i does not appear in the web page d_j then $f_{i,j} = 0$. Further, let idf_i,inverse document frequency for k_i, be given by $idf_i = \log \frac{N}{n_i}$. The tf-idf term-weighting schemes use weights which are given by $w_{i,j} = f_{i,j} \times \log \frac{N}{n_i}$. Let $w_{i,j}$ be the term-weighting of term k_i in the web page d_j.

Definition 2. *Let* $K = \{k_1, k_2, ..., k_n\}$ *be the set of nouns in a document, and* $R = \{identity, synonym, hyponym, meronym\}$ *be the set of lexical relations. Let* $C = \{c_1, c_2, ..., c_n\}$ *be the set of concept object in a document. Concept object* c_j *is composed of several keywords* k_i*, and each* k_i *and* c_j *have a weight that represents their respective degrees of semantic importance within a document. The type of lexical relation decides the weight value* $r_{i,j}$*.*

The way of constructing a concept object from a given document is described in the following algorithm. Based on the concept object definition, we first proposed the Automatic Construct Concept(ACC) algorithm which use the WordNet as the lexical relation analyzing tools.

Algorithm 1 (Automatic Construct Concept algorithm)
Input: Input fields keywords words table T
Output: A Fields Concept object-hash table H
Procedure:
Step1: Select a fields keywords table T $=\{w_1, w_2, ..., w_n\}$ and load them into the concept hash table processing identity relation;
Step2: initialize the hash table H;
Step3: while $(T \notin \Phi)$
choose a new keyword from T
initialize the current concept pointer c_k
for $w' \in T$ do
if w_i does not display in H do
Link the word w_j to concept c_k;
remove word w_i from T;
end if
if hasrelation(w',$c_i\{k_1, k_2, ...k_n\}$)defined in WordNet do
Get the lexical relation $r_{i,j}$ from WordNet;
Link word w' to concept c_i and order by $r_{i,j}$;
Remove word w' from T;
end if
end for
k=k+1;
end while Through ACC algorithm, the relevance keywords is clustered into a concept object as Fig.2.

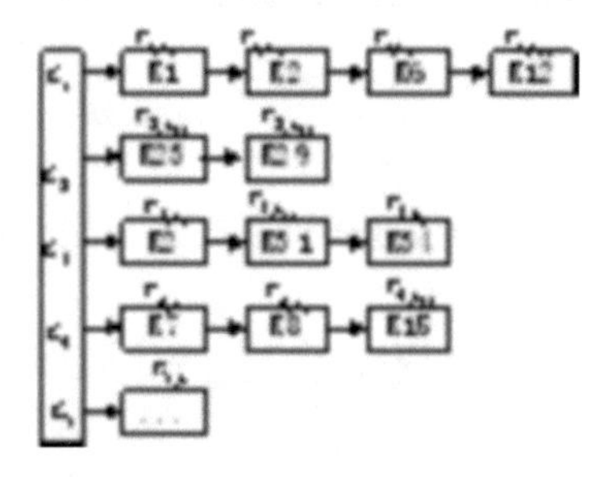

Fig. 2. Keywords cluster into concept object

Definition 3. *Let the weight $\tau_{i,j}$ associated with a pair $[c_i, d_j]$ is positive and non-binary. Further, the index terms in the query are also weighted. Let $\tau_{i,q}$ be the weight associated with the pair $[c_i, q]$. The value of $\tau_{i,j}$ can get by compare the occurred frequency of keywords set of concept object c_i in document j.*

$$\tau_{i,j} = \frac{\sum\limits_{k_i \in C_i} |r'_{i,j}|}{\sum\limits_{k_i \in C_i} |r_{i,j}|} \tag{3}$$

where $\sum\limits_{k_i \in C_i} |r_{i,j}|$ is the amount of the relation among keywords occurring in document j , and $\sum\limits_{k_i \in C_i} |r_{i,j}|$is the amount of the relation keywords in concept object c_i .Then a web page can be represents as a vector of concept object $D_j = \{\tau_{1,j}, \tau_{2,j}, ...\tau_{n,j}\}$,where n means the total number of concept object, and c means the concept object identity.

4 ECC-Based Fuzzy Clustering Algorithm

In this paper, we choose the FCM and GG algorithm as the basic algorithm. Using the $\lambda$and β as control variable, the cluster covariance matrix is used in conjunction with an "exponential" distance and Euclidean distance, and the clusters are not constrained in volume.

Fuzzy clustering methods have been developed following the general fuzzy set theory strategies by Zadeh[5]. The difference between the traditional hard clustering and fuzzy clustering can be stated as follows.

Fuzzy c-Means method is one of the most widely used methods in fuzzy clustering. It is proposed by Dunn[10] and generalized by Bezdek[4]. The aim of FCM is to find cluster centers (centroids) that minimize a dissimilarity function. Usually, membership functions are defined based on a distance function, such that membership degrees express proximities of entities to cluster centers. By choosing a suitable distances function, different cluster shapes can be identified. However, these approaches typically fail to explicitly describe how the fuzzy cluster structure relates to the data from which it is derived.

To accommodate the introduction of fuzzy partitioning, the membership matrix(U) is randomly initialized according to Equation4

$$\sum_{i=1}^{c} u_{ij} = 1, \forall j = 1, ..., n \tag{4}$$

The dissimilarity function which is used in FCM is given Equation5

$$J(U, c_1, c_2, ..., c_c) = \sum_{i=1}^{c}\sum_{j=1}^{n} u_{i,j}^{m} \|x_j - c_i\|^2 \tag{5}$$

Where u_{ij} is between 0 and 1, c_i is the centroid of cluster i;$\|x_j - c_i\|^2$ is the Euclidean distance between i-th centroid and j-th concept data point, $m \in [1, \infty]$

is a weighting exponent. To reach a minimum of dissimilarity function there are two conditions. These are given in Equation 6 and Equation 7

$$c_i = \frac{\sum_{j=1}^{n} u_{i,j}^m x_j}{\sum_{j=1}^{n} u_{ij}^m} \tag{6}$$

$$u_{i,j} = \frac{1}{\sum_{k=1}^{c} \left(\frac{d_{i,j}}{d_{kj}}\right)^{2/(m-1)}} \tag{7}$$

Detailed algorithm of fuzzy c-means proposed by Bezdek[5]. This algorithm determines the following steps. Give a set of data X specify c, choose a weighting exponent m>1 and a termination tolerance ε >0. Initialize the partition matrix with a more robust method.

Repeat for l=1,2,...

Step1 Calculate the cluster centers:

$$c_i^{(l)} = \frac{\sum_{j=1}^{n} u_{i,j}^{(l-1)m} x_j}{\sum_{j=1}^{n} u_{i,j}^m}$$

Step2 Compute the distance measure $D_{i,j}^2$.The distance to the prototype is calculated based the fuzzy covariance matrices of the cluster

$$F_i^{(l)} = \frac{\sum\limits_{j=1}^{N} (u_{i,j}^{(l-1)})^m (x_j - c_i^{(l)})(x_j - c_i^{(l)})^T}{\sum\limits_{j=1}^{N} (u_{i,j}^{(l-1)})^m}, 1 \leq i \leq c$$

The distance function is chosen as

$$D_{i,j}^2(x_j, c_i) = \lambda \frac{\sqrt{2\pi F_i^{(l)}}}{a_i} \exp(\frac{1}{2}(x_j - c_i^{(l)})^T F_i^{-1}(x_j - c_i^{(l)})) + \beta(x_k - c_i)^T (x_k - c_i)$$

With the a priori probability $a_i = \frac{1}{N} \sum_{j=1}^{N} u_{ij}$,and $\lambda + \beta = 1$

Step3 Update the partition matrix

$$u_{i,j}^{(l)} = \frac{1}{\sum\limits_{k=1}^{c} (D_{i,j}(x_j, c_i)/D_{k,j}(x_j, c_k))^{2/(m-1)}}$$

where 1≤i≤c,1≤k≤N.

Until $\left\|U^{(l)} - U^{(l-1)}\right\| < \varepsilon$

5 Validation

Cluster validity refers to fuzzy partitions suitable for all data. The clustering algorithm always tries to find the best fit for the fixed number of clusters and the parameterized cluster shapes. However this does not mean that the best fit is meaningful. Either the number of clusters might be wrong or the cluster shapes might not correspond to the groups in the data, if the data cannot be grouped into a meaningful way. Different scalar validity measures have been proposed in the literature, such as Partition Coefficient(PC),Classification Entropy(CE)[4],Partition Index(SC),Separation Index(S)[8],Xie and Beni's Index(XB)[9].

6 Experimental Results and Analysis

We have collected 8,000 commercial web pages, and has divided them into 30 kinds at first. All the data was grouped into 30 classes by two ways . One way was used for clustering based on TF-IDF model, and another way was used for clustering based on ECC model. The tests were conducted on currently available algorithms (kmeans,kmedoid ,Gath-Geva and ECC-based EFCM algorithm. When λ=0.3, β= 0.7, we compared the EFCM algorithm with other clustering algorithms. The results of those three algorithms and ECC-based clustering algorithm as follows:

As seen from Fig .3.,Fig .4. ,Fig .5. and Fig .6., the ECC-based clustering method are more high than the TF-IDF clustering method. And for ECC-based EFCM clustering algorithm, since the concept relations and control factors have influenced and adjusted the distance value, the result is much improved. The average precision of each method among 30 categories, the ECC-Based EFCM algorithm is better than other methods.

The semantic relation among commercial concept object is not too close, but both of them play important role in real business activities and have intimate relations. For example, the information interactions of computer mainly depend on monitor as the output device, while monitor and keyboard are the output and input devices respectively with similar functions. ECC-based algorithm has greatly increased their semantic similarity to computer, which suits the actual important

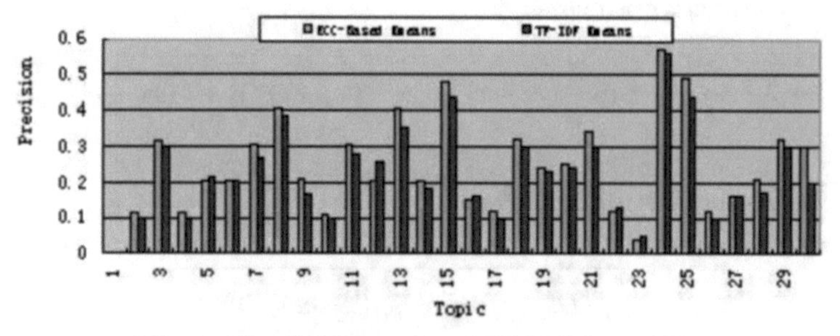

Fig. 3. The ECC-Based and TF-IDF-Based Kmeans

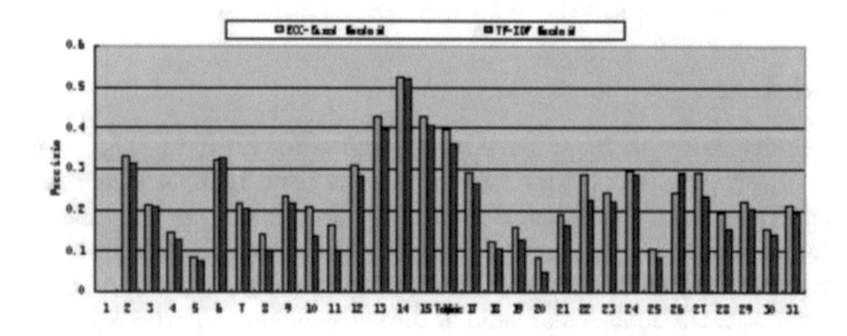

Fig. 4. The ECC-Based and TF-IDF-Based Kmedoid

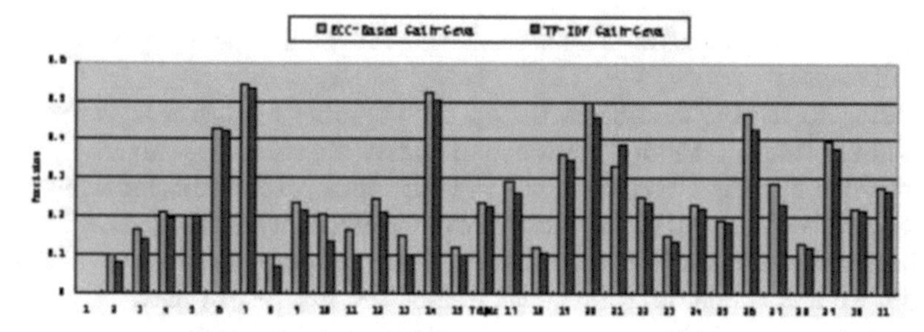

Fig. 5. The ECC-Based and TF-IDF-Based Gath-Geva

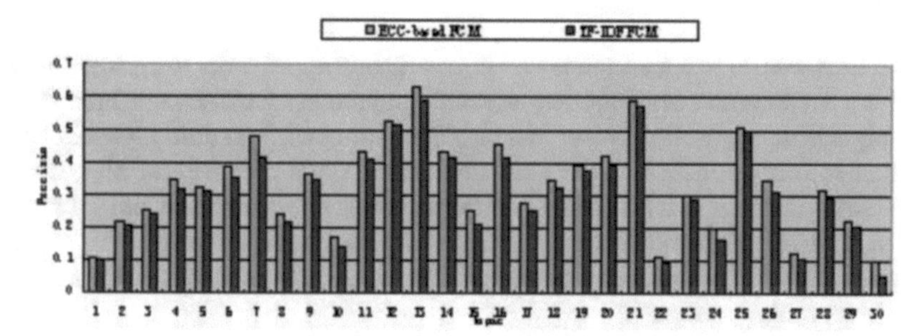

Fig. 6. The ECC-Based and TF-IDF-Based EFCM

Table 1. The numerical values of validity measures

CN	*PC*	*CE*	*SC*	*S*	*XB*
5	0.7739	0.3121	0.6935	0.0064	14.5271
10	0.7743	0.3468	0.6591	0.0062	8.8294
15	0.755	0.3548	0.6356	0.0061	7.3455
20	0.7422	0.4511	0.6243	0.0067	6.4216
25	0.7291	0.5679	0.6155	0.0060	6.0929
30	0.7539	0.6055	0.5126	0.0050	6.3459
35	0.7075	0.3246	0.4581	0.0054	5.1688
40	0.6833	0.3387	0.4365	0.0052	5.2679
45	0.6957	0.2872	0.4434	0.0050	3.4507
50	0.6554	0.4987	0.4242	0.0045	2.3316

relations between those concepts in business activities. Through the structure of semantic concept layers , we could get the semantic relevant web pages from the Fields Concept object-hash table H, while in actuality they do have relations. According to performance validation method, we can conclude that ECC-based EFCM clustering algorithm from the aspect of business application could more properly clustering precision. The indexes show in Table 1,where CN means the number of classes.

According to the PC definition, the optimal number of cluster is at the maximum value.In Table 1., there are more informative diagrams are shown: SC and S hardly decreases at the CN= 30 point. The XB index reaches this local minimum at CN = 35. Considering that SC and S are more useful, when comparing different clustering methods with the same class, we chose the optimal number of clusters to 30. So, we can conclude from Table 1. that the results are better than other classified results, when the web page are divided into 30 classes.

7 Conclusion

In this study, we use fuzzy c-means and hard kmeans algorithms to cluster the web page data. In Ecommerce products searching systems, ECC-Based Fuzzy clustering algorithm gives the better results than kmeans, kmedoid,GG algorithm according to our application. Another important feature of EFCM algorithm is membership function and an object can belong to several classes at the same time but with different degrees. This is a useful feature for Ecommerce products system. At a result, EFCM methods can be important supportive tool for the Ecommerce semantic searching system.

References

1. J. B. MacQueen (1967) Some Methods for classification and Analysis of Multivariate Observations Proceedings of 5-th Berkeley Symposium on Mathematical Statistics and Probability, Berkeley, University of California Press,1:281-297
2. I. Gath and A.B. Geva (1989) Unsupervised optimal fuzzy clustering. IEEE Transactions on Pattern Analysis and Machine Intelligence 7:773-781
3. Ling-Yun, Ouyang-Yi,Li-Biwei (2006) The ECommerce Information Model Driven Semantic Searching Alogrithm. 2006 Inernational Symposium on Distributed Computing and Applications to Business, Engineering and Science 2:840-844
4. J. C. Bezdek (1981) Pattern Recognition with Fuzzy Objective Function Algorithms. Plenum Press, New York 21-29
5. L. Zadeh (1965) Fuzzy sets. Information and Control 8,338–351.
6. Morris, J., and Hirst, G (1991) Lexical cohesion computed by thesaural relations as an indicator of the structure of text. Computational Linguistics, 17(1), 21-43
7. Hirst, G., and St-Onge, D (1998) Lexical chains as representations of context for the detection and correction of malapropisms. In Christiane Fellbaum (Ed.), WordNet: An electronic lexical database. Cambridge, MA: The MIT Press
8. A.M. Bensaid, L.O. Hall, J.C. Bezdek, L.P. Clarke, M.L. Silbiger, J.A. Arrington, and R.F. Murtagh(1996) Validity-guided (Re)Clustering with applications to imige segmentation. IEEE Transactions on Fuzzy Systems, 4:112-123
9. X. L. Xie and G. A. Beni(1991) Validity measure for fuzzy clustering. IEEE Trans. PAMI,3(8):841-846
10. Dunn, J.C (1974) Some recent investigations of a new fuzzy partition algorithm and its application to pattern classification problems. J.Cybernetics 4, 1-15
11. Christiane Fellbaum (Ed.)(1998) WordNet: An Electronic Lexical Database. MIT Press, 45-64

A New Algorithm for Attribute Reduction Based on Discernibility Matrix

Guan Lihe

Institute of information and calculation science, Chongqing Jiaotong University, Chongqing, 400074, P.R. China
guanlihe@cquc.edu.cn

Abstract. Attribute reduction is one of the key problems in the theoretical research of Rough Set, and many algorithms have been proposed and studied about it. These methods may be divided into two strategies (addition strategy and deletion strategy), which are based on adopting different heuristics or fitness functions for attribute selection. In this paper, we propose a new algorithm based on frequency of attribute appearing in the discernibility matrix. It takes the core as foundation, and joins the most high frequency attribute in the discernibility matrix, until the discernibility matrix is empty. In order to find optimum Pawlak reduction of decision table, this paper adds the converse eliminate action until it cannot delete. The time complexity of the algorithm in this paper is $O(mn^2)$,and the testing indicates that the performance of the method proposed in this paper is faster than that of those algorithms.

Keywords: Rough Sets,Pawlak Reduction,Discernibility Matrix,Attribute Frequency.

1 Introduction

The theory of rough sets was provided by Poland scientist Pawlak in 1982, which has been applied in data analysis, data mining and knowledge discovery. A fundamental notion supporting such applications is the concept of reduction [1], and it is one of main contents of rough set research. A reduction is a subset of attributes that is jointly sufficient and individually necessary for preserving the same information under consideration as provided by the entire set of attributes. It has been proved that finding a reduction with the minimal number of attributes is NP-hard [2]. Researchers' efforts on reduction construction algorithms are therefore mainly focus on designing search strategies and heuristics for finding a good reduction efficiently.

A review of existing reduction construction algorithms shows that most of them may be divided into two strategies (addition strategy and deletion strategy), which are based on adopting different heuristics or fitness functions for attribute selection. By considering the properties of reductions, we observe that the deletion strategy always results in a reduction [3,4]. On the other hand, algorithms based on a straightforward application of the addition strategy only produce a superset of a reduction [5,6,7,8]. These algorithms can not obtain a minimal (optimal) reduction efficiently.

B.-Y. Cao (Ed.): Fuzzy Information and Engineering (ICFIE), ASC 40, pp. 373–381, 2007.
springerlink.com

In this paper, we propose a new algorithm based on frequency of attribute appearing in the discernibility matrix. It takes the core as foundation, and joins the most high frequency attribute in the discernibility matrix, until the discernibility matrix is empty. In order to find optimum Pawlak reduction of decision table, this paper adds the converse eliminate action until it cannot delete. The testing indicates that the performance of the method proposed in this paper is faster than that of the CEBARKCC and CEBARKNC[12].

The structure of the rest is organized as follows. Section 2 briefly reviews necessary concepts of decision tables and discernibility matrices. In section 3, a new algorithm of attributes reduction is proposed based on frequency of attribute appearing in the discernibility matrix. Section 4 presents a test of algorithm introduced in section 3. Finally, some concluding remarks are shown in section 5.

2 Basic Concepts

In this section, some concepts, notations, and results related to the problem of reduction construction will be recalled briefly.

2.1 Decision Tables and Pawlak Reduction

In rough set theory [1], Pawlak has defined two sorts of information systems without and with decision attributes, denoted as $\langle U, C\rangle$ and $\langle U, C \cup D\rangle$ respectively. Here U is the universe of objects, while C and D are the condition attributes set and the decision attributes set respectively. Their reductions are called reduction and relative reduction respectively. In this paper, we mainly discuss the relative reduction of the information systems with decision attributes.

Definition 1. (Decision Table) A decision table is called a decision information system S, which is the tuple $S = (U, A = C \cup D, \{V_a|a \in A\}, \{I_a|a \in A\})$,where U is a finite nonempty set of objects, $A = C \cup D$ is a finite nonempty set of attributes, C stands for a set of condition attributes that describe the features of objects, and D is a set of decision attributes. Usually,D contains only one decision attribute that indicates the labels of objects, i.e. $D = \{d\}$,V_a is a nonempty set of values for an attribute $a \in A$, and $I_a : U \rightarrow V_a$ is an information function, such that for an object $x \in U$, an attribute $a \in A$,and a value $v \in V_a$,$I_a(x) = v$ means the object x has the value v on attribute a.

Definition 2. (Discernibility Relation) With any subset , a discernibility relation is defined by $E_B = \{(x, y) \in U^2|\forall a \in A, I_a(x) = I_a(y)\}$.Readily understood,it is an equivalence relation, which partitions U into disjoint subsets. Such a partition of the universe is denoted by $U/E_B = \{[x]_B|x \in U\}$, $[x]_B$is an equivalence class of x concerning B.

Definition 3. ($POS_P(Q)$) $POS_P(Q) = \cup_{X \in U/E}\underline{P}(X)$ is the P positive region of Q, where P and Q are both attribute subsets of an information system, $\underline{P}(X) = \{Y|Y \in U/E_P \wedge Y \subseteq X\}$ is the lower approximation of X to P.

Definition 4. (Consistent Decision Table) A decision table $S = (U, C \cup D, \{V_a\}, \{I_a\})$ is consistent if and only if $POS_C(D) = U$, otherwise, inconsistent.

By Pawlak, a reduction must satisfy an independent condition, that is, a reduction $\emptyset \neq R \subseteq C$,R is independent if$\forall r \in R, E_{R-\{r\}} \neq E_R$for information system without decision attributes; or if $\forall r \in R, POS_{R-\{r\}}(D) \neq POS_R(D)$ for decision tables.

Definition 5. (Pawlak Reduction) Given a decision table $S = (U, C \cup D, \{V_a\}, \{I_a\})$, a subset $R \subseteq C$ is called a Pawlak reduction, if R satisfies the flowing two conditions:

(i)$POS_R(D) = POS_C(D)$;

(ii)$\forall r \in R, POS_{R-\{r\}}(D) \neq POS_R(D)$.

Given an information system, there may be exist more than one reduction. The intersection of all reductions is defined as the core.

Definition 6. Given an information system, the intersection of all the reductions is called the set of core attributes, denoted as $CORE_D(C)$.

2.2 Discernibility Matrices

A decision table $S = (U, C \cup D, \{V_a\}, \{I_a\})$ is a sort of data representation. Discernibility matrix is the other sort of data representation, which stores all the differences between any two objects of the universe regarding the entire attribute set. Since Skowron [10] putted forward discernibility matrix, many researchers have studied about it. Especially, Wang [11] finds some errors in some former results, and studies the problem of calculating the attribute core of a decision table based on the discernibility matrix.

In the following, we will give the discernibility matrices definitions of consistent decision table and inconsistent decision table.

Definition 7. Given a consistent decision table $S = (U, C \cup D, \{V_a\}, \{I_a\}), U = \{x_1, x_2, \cdots, x_n\}$, its discernibility matrix, denoted by M_{con}, is a matrix; each matrix element $m_{ij} \in M_{con}$ is defined by

$$m_{ij} = \{a \in C | (x_i, x_j \in U) \wedge I_a(x_i) \neq I_a(x_j) \wedge \mathrm{d}(x_i) \neq \mathrm{d}(x_j)\}.$$

The physical meaning of m_{ij} is that objects x_i and x_j are distinguished by any of the elements in m_{ij}.

Definition 8. Given an inconsistent decision table $S = (U, C \cup D, \{V_a\}, \{I_a\})$, $U = \{x_1, x_2, \cdots, x_n\}, D = \{d\}$, its discernibility matrix, denoted by M_{incon}, is a matrix; each matrix element $m_{ij} \in M_{incon}$ is defined by

$$m_{ij} = \{a \in C | (x_i, x_j \in U) \wedge I_a(x_i) \neq I_a(x_j) \wedge \mathrm{d}(x_i) \neq \mathrm{d}(x_j) \wedge w(x_i, x_j)\},$$

where $w(x_i, x_j) \equiv x_i \in POS_C(\mathrm{D}) \vee x_j \in POS_C(\mathrm{D})$.

We may easily draw the following proposition.

Proposition 1. $CORE_D(C) = \cup_{m_{ij} \in M \wedge |m_{ij}|=1} m_{ij}$, where $M = M_{con} \cdot or \cdot M_{incon}$.

Example 1. Given two decision tables DT_1, DT_2 corresponding to Table1 (consistent decision table) and Table2 (inconsistent decision table) respectively, where a, b, c are condition attributes, d is decision attribute.$M_{con}(DT_1)$ and $M_{incon}(DT_2)$ are the discernibility matrices of DT_1 and DT_2, respectively. These discernibility matrices are displayed as follows.

Table 1. Decision table DT_1

U_1	a	b	c	d
$x1$	0	0	2	0
$x2$	1	1	2	1
$x3$	1	0	1	0
$x4$	0	0	3	1

Table 2. Decision table DT_2

U_2	a	b	c	d
$y1$	0	2	1	0
$y2$	0	0	3	1
$y3$	3	1	2	1
$y4$	3	0	3	0
$y5$	0	2	1	1

$$M_{con}(DT_1) = \begin{bmatrix} \emptyset & a,b & \emptyset & c \\ & \emptyset & b,c & \emptyset \\ & & \emptyset & a,b,c \\ & & & \emptyset \end{bmatrix} \quad M_{incon}(DT_2) = \begin{bmatrix} \emptyset & b,c & a,b,c & \emptyset & \emptyset \\ & \emptyset & \emptyset & a & \emptyset \\ & & \emptyset & b,c & \emptyset \\ & & & \emptyset & a,b,c \\ & & & & \emptyset \end{bmatrix}$$

By proposition1, it is easy to gain the two decision tables' core attribute sets $CORE_{DT_1}(C) = \{c\}$ and $CORE_{DT_2}(C) = \{a\}$.

According to the definition of the discernibility matrices, the independence of reduction can be redefined as follows [9].

Definition 9. (The Independence of Pawlak Reductions) Let $\emptyset \neq R \subseteq C, r \in R$.Define set $M^* = \{\alpha | \alpha \cap (R - \{r\}) = \emptyset, \emptyset \neq \alpha \in M\}$,$r$ is dispensable in R if $M^* = \emptyset$, otherwise r is indispensable in R. R is independent if each $r \in R$ is indispensable in R, otherwise R is dependent.

Now, another definition of Pawlak reduction may be given as follows.

Definition 10. (Pawlak Reduction) $\emptyset \neq R \subseteq C$ is a Pawlak reduction of decision table $S = (U, C \cup D)$, if $\{\alpha | \alpha \cap R = \emptyset, \emptyset \neq \alpha \in M\} = \emptyset$ and R is independent.

3 Algorithms

Because finding a reduction with the minimal number of attributes is NP-hard, at present, many reduction algorithms are based upon heuristics, in which how to define the significance of attributes is the key problem. Actually, the reduction strategy depends on the definition of the attribute significance. There are many definitions about the attributes significance, for example, information gain, roughness of attributes, etc.

In this paper,we take the frequency of attribute appearing in the discernibility matrix as elicitation, and then do reduction for decision information table. In order to find optimum reduction, we add the converse eliminate action until it cannot delete. Because of no using of the basic concept of rough set, only using simple calculation, the algorithm has improved efficiency.

3.1 Algorithm Principle

Proposition 2. Given a decision information table $S = (U, C \cup D, \{V_a\}, \{I_a\})$ and its discernibility matrix M, a subset $R \subseteq C$ is a relative-reduction if and only if R satisfies two conditions:

(1) for all no empty $m \in M, m \cap R \neq \emptyset$;

(2) for $\forall a \in R$, there exists at least one $m \in M$ such that $m \cap \{R - \{a\}\} = \emptyset$.

According to definition9, we can prove the proposition easily and immediately.

According the proposition2, the intersection of a reduction and every item of the discernibility matrix is not empty. If it is empty, object i and j are not discerned by the reduction. So the result is contradicted with the existing theory that the reduction can discern all objects.

we may design algorithm as follows: firstly, suppose that candidate reduction set is empty, namely $R = \emptyset$ .Secondly, we check the intersection of every item of no empty m in the discernibility matrix M and candidate reduction set R. If it is empty, we randomly select a attribute from m into R, or else overleap the item. Thirdly, we repeat the process, until every item of the discernibility matrix is checked. In this time, R is a reduction.

If the intersection of certain item in the discernibility matrix and existing candidate is empty, and the item is made of many attributes,in this condition, how to know which attribute is precedence checked? When the discernibility matrix is created, the frequency of attribute appeared in the discernibility matrix is recorded. The frequency evaluates the significance of attribute, and is took as standard of attribute precedence. The calculation method for the attribute frequency is calculated very simply. The frequency of attribute is defined as follows: $\delta(a) = |\{m \in M | a \in m\}|$.

3.2 Algorithm Describing

Algorithm1
Import: a decision table $S = (U, C \cup D, \{V_a\}, \{I_a\})$,where $C = \{a_1, a_2, \cdots, a_n\}$, $D = \{d\}$.
Output: reduction R.

(1)$R = \emptyset$;

(2) Calculating the discernibility matrix M of S, and constructing the discernibility matrix element set $M^* = \{m | m \in M \wedge m \neq \emptyset\}$. Deleting the iterative elements and dealing with the absorptive law in M^*.

(3) Calculating core attributes set C^*,and let $R = R \cup C^*$;

(4) Calculating $M^* = \{m | m \in M^* \wedge m \cap R = \emptyset\}$;

(5) If $M^* \neq \emptyset$ then continuing, else output;

(6) For all $a \in C - R$, Calculating $\delta(a) = |\{m \in M^* | a \in m\}|$;

(7) Selecting an attribute from $C - R$, whose frequency function value $\delta(a)$ is maximal, if there are not less than two attributes whose values are the same and maximal, then random attribute will be selected. i.e., $R = R \cup \{a | \forall a^* \in C - R \wedge \delta(a) \geq \delta(a^*)\}$, go to (4).

Unfortunately, the result of the algorithm1 is not a Pawlak reduction, which is only a superset of Pawlak reduction. Here is a counterexample.

Example 2. Given discernibility matrix $M^0 = \{ab, ac, ad, be, cf, dg\}$,where $a, b, c, d, e, f, g \in C$,and M^0 has been deleted the iterative elements and dealted with the absorptive law. Find $\delta(x), \delta(a) = 3, \delta(b) = \delta(c) = \delta(d) = 2, \delta(e) = \delta(f) = \delta(g) = 1$, According to Algorithm1, attribute 'a' should be chosen as the first attribute in the reduction. By the algorithm, M^0 is changed to $M^1 = \{be, cf, dg\}, \delta(b) = \delta(c) = \delta(d) = \delta(e) = \delta(f) = \delta(g) = 1$. 'b' should be chosen as the second attribute in the reduction. Finally, we obtain a reduction $\{a, b, c, d\}$. This reduction is not independent. In fact, its proper subset $\{b, c, d\}$ is also a reduction. By Definition9, we can verify that $\{b, c, d\}$ is independent. Since $\{a, b, c, d\} - \{a\} = \{b, c, d\}, \{a, b, c, d\}$ is not a Pawlak reduction.

If we analyze the process of the algorithm1 carefully, we can find that algorithm1 satisfies only the first of the proposition2. Therefore, in order to obtain Pawlak reduction, we must modify the algorithm1, and then algorithm2 is given as follows.

Algorithm 2. ABFAADM (Algorithm based on frequency of attribute appearing in the discernibility matrix)
Import: a decision table $S = (U, C \cup D, \{V_a\}, \{I_a\})$,where $C = \{a_1, a_2, \cdots, a_n\}$, $D = \{d\}$.
Output: reduction R.

(1)Find the reduction R and the set of core attribute C^* of decision table S by algorithm1;

(2) For all $r \in R - C^*$, let $M^" = \{\alpha | \alpha \cap (R - \{r\}) = \emptyset, \emptyset \neq \alpha \in M\}$,if $M^" = \emptyset$,then deleting the attribute 'r' from R, i.e.,$R = R - \{r\}$;

(3)Output: R.

Apparently, the result of the algorithm2 is a Pawlak reduction.

3.3 Complexity Analysis of the Algorithm

Let$|U| = n$ and $|C| = m$.There are $n(n-1)/2$items in discernibility matrix M, the cost of calculating discernibility matrix is $O(mn(n-1)/2)$. The time complexity of calculating C^* is $O(n(n-1)/2)$. Therefore, the time complexity of algorithm1 is $O(mn^2)$. In the second step of algorithm2 the complexity of validating independent of R is $O((mn(n-1)/2)$. Therefore, the time complexity of algorithm2 is $O(mn^2)$ too.In fact, there are some empty elements in M, and the number of containing attributes most of elements in M is far less than m . Therefore, the efficiency of this algorithm will be faster.

4 Experimental Analysis

Yu and Wang[12] developed two algorithms CEBARKCC and CEBARKNC basing on conditional entropy, whose time complexity are $O(mn^2 + n^3)$ and $O(m^2n + mn^3)$. Algorithm ABFAADM of this paper is based on frequency of attribute appearing in the discernibility matrix, whose time complexity is $O(mn^2)$ and less than that of the above two algorithms. CEBARKCC and ABFAADM are heuristic algorithms with computing core. Algorithm CEBARKNC does not compute core.

In our experiments, the Bact94database and some databases from UCI repository are used. We test each database on a P4 1.5G PC of 256M memory with the three algorithms. The results of some experiments are shown in Table 3 and 4.

In Table 3, $S\#$ is the sequential number of a database, $N\#$ is the number of samples in the database, $C_0\#$ is the cardinality of the core of the database, $B\#$ is the reduction of it, and RT is the CPU time (by second) of the reducing process.

In Table 4, GR is the number of generated rules, RR is the recognition rate (%), and Set1 is B,C,D,E,F,G,I,L,Q,T,U,V,W,Y,Z,A1,B1,C1,D1,J1,K1,N1,C2,D4,Y4.

The 8 databases used here are: 1-Protein Localization Sites Database; 2-Heart Disease Database; 3-Australian Credit Approval Database(1); 4-BUPA Liver Disorders Database; 5-Pima Indians Diabetes Database(1); 6-HSV Database; 7-Pima Indians Diabetes Database (2); 8-Bact94database.

Table 3. Comparison of the CPU time of the algorithms

$S\#$	N#	$C_0\#$	CEBARKCC		CEBARKNC		ABFAADM	
			B#	RT	B#	RT	B#	RT
1	306	0	3	1.5	3	0.8	3	0.5
2	150	0	2	6.9	2	0.35	2	0.29
3	345	1	3	30	3	2.4	3	2.2
4	315	2	4	1.1	4	0.6	4	0.4
5	384	3	4	0.7	4	1.4	4	0.41
6	92	5	5	0.04	5	0.15	5	0.01
7	738	5	5	1.1	5	4.1	5	0.9
8	10459	25	25	221	25	9621	25	103

Table 4. Comparison of the reduction results of the algorithms

S#	CEBARKCC			CEBARKNC			ABFAADM		
	Reduciton	GR	RR	Reduciton	GR	RR	Reduciton	GR	RR
1	{A,B,F}	106	100	{A,B,F}	106	100	{A,B,F}	106	100
2	{H,E}	200	72	{H,E}	200	72	{H,E}	200	72
3	{B,C,M}	550	80	{B,M,N}	397	80	{B,C,M}	550	80
4	{B,C,D,E}	167	96	{B,C,D,E}	167	96	{B,C,D,E}	167	96
5	{B,D,F,G}	189	77	{B,E,F,G}	185	77	{B,D,F,G}	189	77
6	{A,B,C,D,G}	67	77	{A,B,C,D,G}	67	77	{A,B,C,D,G}	67	77
7	{A,B,C,D,H}	397	70	{A,B,C,D,H}	397	70	{A,B,C,D,H}	397	70
8	Set1	2224	82	{Set1}	2224	82	{Set1}	2224	82

From the experiment results, we can find the following conclusions:

(1) From table3, Algorithm ABFAADM is faster than algorithm CEBARKCC and CEBARKNC;

(2)From table4, we can find that algorithm ABFAADM developed in this paper is feasible to reduction, and they can get the minimal reduction in most cases. The recognition rate depends on the quantity of conflict samples in the database.

5 Conclusions

In this paper, an effective heuristic attribute reduction algorithm is developed according to the frequency of attribute appearing in the discernibility matrix. After gaining superset of reduction, in order to find optimum Pawlak reduction of decision tables, this paper adds the converse eliminate action until it cannot delete. The time complexity of algorithm in this paper is $O(mn^2)$, which is less than that of CEBARKCC and CEBARKNC[12]. The experiment results indicate that the performance of the method proposed in this paper is faster than that of those algorithms and feasible to reduction. What's more, algorithm ABFAADM can get not only the Pawlak reduction but also the minimal reductions in most cases.

Acknowledgements

This paper is partially supported by Scientific Research Foundation for the High Level Scholars of Chongqing Jiaotong University (No.2004-2-19).

References

1. Pawlak Z(1991)Rough Sets: Theoretical Aspects of Reasoning about Data.Volume9 of System Theory, Knowledge Engineering and Problem Solving. Kluwer Academic Publishers, Dordrecht,The Netherlands.
2. Wong S,Ziarko W(1985)On optimal decision rules in decision tables, Bulletin of the Polish Academy of Sciences and Mathematics,693-696.

3. Hu XH(2001)Using rough sets theory and database operations to construct a good ensemble of classiers for data mining applications, Proceedings of ICDM'01, 233-240.
4. Ziarko W(2002)Rough set approaches for discovering rules and attribute dependencies, in: Klosgen, W. and Zytkow, J.M. (eds.), Handbook of Data Mining and Knowledge Discovery,328-339,Oxford.
5. Hu XH,Cercone N(1995)Learning in relational databases: a rough set approach, Computation Intelligence: An International Journal,**11(2)**:323-338.
6. Jenson R,Shen Q(2001)A rough set-aided system for sorting WWW bookmarks, in: Zhong, N. et al. (eds), Web Intelligence: Research and Development,95-105.
7. Miao DQ,Wang J(1999)An information representation of the concepts and operations in rough set theory, Journal of Software, **10**:113-116.
8. Shen Q,Chouchoulas A(2000)A modular approach to generating fuzzy rules with reduced attributes for the monitoring of complex systems, Engineering Applications of Artificial Intelligence, **13(3)**:263-278.
9. Wang J,Wang J(2001)Reduction algorithms based on discernibility matrix: the ordered attributes method, Journal of Computer Science and Technology,**16(6)**: 489-504.
10. Skowron A,Rauszer C(1992)The discernibility matrices and functions in information systems, In Slowinski, R. (Ed.): Intelligent Decision Support-Handbook of Applications and Advances of the Rough Sets Theory. Kluwer Academic Publisher,Dordrecht,331-362.
11. Wang GY(2002)Attribute Core of Decision Table, In: J. J. Alpigini, J. F. Peters,A. Skowron, N. Zhong (Eds.): Rough Sets and Current Trends in Computing, Springer-Verlag(LNAI2475):213-217.
12. Yu H, Wang GY, Yang DC, Wu Y(2002)Knowledge Reduction Algorithms Based on Rough Set and Conditional Information Entropy, in Data Mining and Knowledge Discovery: Theory, Tools, and Technology IV, Belur V. Dasarathy, Editor,Proceedings of SPIE,**4730**:422-431.

Optimal Computing Budget Allocation Based Compound Genetic Algorithm for Large Scale Job Shop Scheduling

Wang Yong-Ming[1], Xiao Nan-Feng[1], and Yin Hong-Li[2]

[1] School of Computer Science and Engineering, South China University of Technology, Guangzhou 510640, China
yymm_wang@163.com
[2] School of Computer Science and Information, Yunnan Normal University, Kunming 650092,China
Hongli_yin@163.com

Abstract. Job shop scheduling, especially the large scale job shop scheduling problem has earned a reputation for being difficult to solve. Genetic algorithms have demonstrated considerable success in providing efficient solutions to many non-polynomial hard optimization problems. But unsuitable parameters may cause poor solution or even no solution for a specific scheduling problem when evolution generation is limited. Many researchers have used various values of genetic parameters by their experience, but when problem is large and complex, they cannot tell which parameters are good enough to be selected since the trial and error method need unaffordable time-consuming computing. This paper attempts to firstly find the fittest control parameters, namely, number of generations, probability of crossover, probability of mutation, for a given job shop problem with a fraction of time. And then those parameters are used in the genetic algorithm for further more search operation to find optimal solution. For large-scale problem, this compound genetic algorithm can get optimal solution efficiently and effectively; avoid wasting time caused by unfitted parameters. The results are validated based on some benchmarks in job shop scheduling problems.

Keywords: Control Parameters, Genetic Algorithm (GA), Optimal Computing Budget Allocation, Large Scale Job Shop Scheduling.

1 Introduction

Many optimization problems in the industrial engineering world and, in particular, in manufacturing systems are very complex in nature and difficult to solve by conventional optimization techniques. The use of formal approaches in industry is frequently limited owing to the complexity of the models or algorithms. Since the 1960s, there has been an increasing interest in imitating living beings to solve difficult optimization problems. Simulating the natural evolutionary process of human beings results in stochastic optimization techniques called evolutionary algorithms, which can often outperform conventional optimization methods when applied to difficult real-world problems [1]. There are currently three main avenues of this research: genetic algorithms, evolutionary programming, and evolution strategies. Genetic algorithms are perhaps the most widely known types of evolutionary algorithm today.

B.-Y. Cao (Ed.): Fuzzy Information and Engineering (ICFIE), ASC 40, pp. 382–396, 2007.
springerlink.com

The performance of a specialized genetic algorithm is often sensitive to problem instances. As a common practice, researchers tune their algorithm to its best performance on the test set reported. This causes problem when the algorithms are to be used in industry applications where the algorithm configured for today's problem may perform poorly for tomorrow's instances [2]. It is not practical to conduct massive experiments in a regular basis for the selection of a more effective algorithm setting. Furthermore, when the size of problem is large, we cannot afford to conduct the needed massive experiments. This problem-sensitive nature is most evident in algorithms for job shop scheduling. Clearly, given a set of parameters, one can find a parameter setting for a particular algorithm that is the best than all other parameters under consideration. The ability to make this decision effectively has great practical importance. For example, job shop scheduling problems in a manufacturing plant may vary in a frequent basis due to its highly dynamic nature. Due-date tightness, seasonality of order volumes and mixes, machine or tooling status, and changing operating policies may all alter the structure of the scheduling problem to be solved. As a result, the best algorithm or algorithm setting for solving a preconceived static problem may perform poorly in a day-to-day basis.

2 Job Shop Scheduling Problem Formulation

Scheduling allocates resources over time in order to perform a number of tasks. Typically, resources are limited and therefore tasks are assigned to resources in a temporal order. The jobs are input to the scheduling engine of a production planning system. It determines the sequences and periods for processing jobs on its dedicated machines and determines the reduction of the work-in-process inventory produced by increasing the throughput of jobs. Jobs often follow technological constraints that define a certain type of shop floor. Typically, the objectives are the reduction of makespan of an entire production program, the minimization of mean tardiness, the maximization of machine load or some weighted average of many similar criteria [3]. Within the great variety of production scheduling problems, the general job shop problem (JSP) is probably the one most studied during the last decade. It has earned a reputation for being difficult to solve. It illustrates at least some of the demands required by a wide range of real-world problems.

The classical JSP can be stated as follows: there are m different machines and n different jobs to be scheduled. Each job is composed of a set of operations and the operation order on machines is pre-specified. Each operation is characterized by the required machine and the fixed processing time.

The problem is to determine the operation sequences on the machines in order to minimize the makespan, i.e., the time required to complete all jobs.

The JSP with makespan objective can be formulated as follows:

$$\min \quad \max_{1\le k\le m}\{\max_{1\le i\le n}\{c_{ik}\}\} \tag{1}$$

$$\text{s.t.} \quad c_{ik}\text{-}t_{ik} = M(1\text{-}a_{ihk}) \ge c_{ih}, \quad i=1,2,\cdots,n, \quad h,k=1,2,\cdots,m \tag{2}$$

$$c_{jk}\text{-}c_{ik} = M(1\text{-}x_{ijk}) \geq p_{jk}, \quad i, j = 1,2,\cdots,n, \quad k = 1,2,\cdots,m \tag{3}$$

$$c_{ik} \geq 0, \quad i = 1,2,\cdots,n, \quad k = 1,2,\cdots,m \tag{4}$$

$$x_{ijk} = 0 \quad or \quad 1, \quad i, j = 1,2,\cdots,n, \quad k = 1,2,\cdots,m \tag{5}$$

Where c_{jk} is the completion time of job j on machine k, t_{jk} is the processing time of job j on machine k, M is a big positive number, a_{ihk} is an indicator coefficient defined as follows:

$$a_{ihk} = \begin{cases} 1, & \text{if processing on machine } h \text{ precedes that on machine } k \text{ for job } i \\ 0, & \text{others} \end{cases}$$

and x_{ijk} is an indicator variable defined as follows:

$$x_{ijk} = \begin{cases} 1, & \text{if job } i \text{ precedes job } j \text{ on machine } k \\ 0, & \text{others} \end{cases}$$

The objective is to minimize makespan. Constraint (2) ensures that the processing sequence of operations for each job corresponds to the prescribed order. Constraint (3) ensures that each machine can process only one job at a time.

3 Heuristic and Search Techniques for Job Shop Scheduling Problem

Various types of scheduling problem are solved in different job shop environments. Varieties of algorithms are employed to obtain optimal or near optimal schedules. Traditionally, the automatic generation of scheduling plans for job shops has been addressed using optimization and approximation approaches [4,5]. Optimization algorithms include enumerative procedures and mathematical programming techniques. The approximation techniques used are branch–bound, priority rule based, heuristics, local search algorithms, and evolutionary programs.

Optimization algorithms provide satisfactory or optimal results if the problems to be solved are not large. Most scheduling problems are NP-hard, that is, the computational requirement grows exponentially as a function of the problem size, and this degrades the performance of conventional optimization techniques and hence optimization algorithms are ruled out in practice. The quality of solutions using branch and bound algorithms depends on the bounds; a good bound requires a substantial amount of computation. Local search-based heuristics are known to produce excellent results in short run times, but they are susceptible to being stuck in local entrapments (local minima). Evolutionary programs, which belong to random search processes, are regarded as better than those approach mentioned above in the sense that they guarantee near optimal solutions in actual cases. By changing the evaluation parameters of the genetic search process, the solutions can also be obtained for other objectives and can be more flexible. These are useful considerations for any difficult optimization problems.

The above discussion indicates that there are many avenues to explore in the area of heuristic search algorithms for job shop scheduling. There is much literature available in this area; but very few authors have attempted to evaluate the optimal control parameters for GA, and this is very crucial especially for large-scale scheduling problems. In this paper, an attempt has been made to establish a algorithm to identify optimal GA control parameters from a group of candidate parameters combination, when GA is used to solve large scale NP-hard job shop scheduling problem. Therefore, given a specific large-scale job shop scheduling problem, we presented a compound genetic algorithm to resolve it in two steps: firstly, identify the fittest GA control parameters from candidate parameters use optimal computing budget allocation based comparison algorithm with a fraction of time; secondly, genetic algorithm with the fittest parameters do further endeavor to find optimal solution.

4 Genetic Algorithm for Job Shop Scheduling

The majority of job shop scheduling problem is NP-hard, i.e., no method can guarantee a global optimum in polynomial time. Exact search techniques can only be applied to small-scale problems. The quality of the conventional method is often unsatisfied. Meta-heuristics, such as Simulated Annealing Algorithm, Genetic Algorithm and Tabu Search Technique are general and can achieve a satisfied quality, but their performance depends a great deal on the parameters used.

Among all the meta-heuristics, GA may be the one most widely used. Based on the mechanics of artificial selection and genetics, a GA combines the concept of survival of the fittest among the solutions with a structured yet randomized information exchange and offspring creation. A GA repeats evaluation, selection, crossover and mutation after initialization until the stopping condition is satisfied. GA is naturally parallel, exhibits implicit parallelism [6], because it does not evaluate and improve a single solution but analyses and modifies a set of solutions simultaneously. Even if, with random initialization, a selection operator can select some relatively better solutions as seeds, a crossover operator can generate new solutions, which hopefully retain good features from the parents, and the mutation operator can enhance the diversity and provide a chance to escape from local optima. GA is an iterative learning process with a certain learning ability, and thus it is regarded as an aspect of computational intelligence. However, there still exist many weaknesses, such as premature convergence, parameter dependence, and a hard to determine stopping criterion.

4.1 The Encoding Representation and Fitness Value

Because of the existence of the precedence constraints of operations, JSP is not as easy as traveling salesmen problem to find out a nature representation. There is not a good representation for the precedence constraints with a system of inequalities [7]. Therefore, a penalty approach is not easily applied to handle such kinds of constraints. Orvosh and Davis [8] have shown that for many combinatorial optimization problems, it is relatively easy to repair an infeasible or illegal chromosome and the repair strategy did indeed surpass other strategies such as rejecting strategy or penalizing strategy. Most GA/JSP researchers prefer to take repairing strategy to handle the infeasibility and illegality. Consequently, one very important issue in building a genetic algorithm for the job shop problem is to devise an appropriate representation of solutions together with problem-specific genetic operations in order that all chromosomes generated in either initial phase or evolutionary process will produce feasible schedules. This is a crucial phase that conditions all the subsequent steps of genetic algorithms. During the last few years, there are nine representations for the job shop scheduling problem have been proposed [9].

In this paper, operation-based representation method was used. This representation encodes a schedule as a sequence of operations and each gene stands for one operation. One natural way to name each operation is using a natural number, like the permutation representation for traveling salesman problem. Unfortunately because of the existence of the precedence constraints, not all possible permutations of these numbers define feasible schedules. Gen et al. proposed an alternative: they named all operations for a job with the same symbol and then interpreted it according to the order of occurrence in the given chromosome [10, 11]. For a n-job and m-machine problem, a chromosome contains $(n \bullet m)$ genes. Each job appears in the chromosome exactly m times and each repeating (each gene) does not indicate a concrete operation of a job but refers to a unique operation which is context-dependent. It is easy to see that any permutation of the chromosome always yields a feasible schedule. A schedule is decoded from a chromosome with the following decoding procedure: (1) firstly translate the chromosome to a list of ordered operations, (2) then generate the schedule by a one-pass heuristic based on the list. The first operation in the list is scheduled first, then the second operation, and so on. Each operation under treatment is allocated in the best available processing time for the corresponding machine the operation requires. The process is repeated until all operations are scheduled. A schedule generated by the procedure can be guaranteed to be an active schedule [12].

The scheduling objective is minimizing the makespan $C_{max}(X_i)$. But in GA, we transform it to the fitness function $f(X_i)$, the chromosome with maximal $f(X_i)$ is the fittest.

$$f(X_i) = 1 - C_{\max}(X_i) / \sum_{i=1}^{P_s} C_{\max}(X_i) \tag{6}$$

4.2 The Selection, Crossover and Mutation Operator

In the selection step of GA, if a real number r randomly generated between 0 and 1 satisfies then individual X_i is selected.

$$\sum_{j=1}^{i-1} f(X_j) / \sum_{j=1}^{P_s} f(X_j) \le r \le \sum_{j=1}^{i} f(X_j) / \sum_{j=1}^{P_s} f(X_j) \tag{7}$$

The crossover operator can introduce new individuals by recombining the current population. The literature provides many crossover operators [13,14], such as partially mapped crossover (PMX), ordinal mapped crossover (OMX), and edge crossover (EX) and so on, they use a single-point or a two-point crossover. It has been shown that there is not much statistically significant difference in the solution quality for different crossover operators for job shop scheduling problem [15], so only PMX is used here. In particular, PMX firstly chooses two crossover points and exchanges the subsection of the parents between the two points, and then fills up the chromosomes by partial mapping. For example, if two parents are (2 6 4 3 5 7 1) and (4 5 2 1 6 7 3), the two crossover points are 3 and 5. Then the children will be (2 3 4| 1 6| 7 5) and (4 1 2| 3 5 | 7 6).

The purpose of mutation is to introduce new genetic material or to recreate good genes to maintain diversity characteristic in population. Three operators widely used for Operation-based encoding are SWAP, inverse (INV) and insert (INS) [16]. Since there is not much statistically significant difference in solution quality among them [15], only SWAP is applied here, i.e., two distinct elements are randomly selected and swapped.

4.3 Genetic Algorithm Scheme Designing

Therefore the genetic algorithm used in optimal parameters evaluation and testing is designed and described as the following, and when the fittest parameters is identified, the same genetic algorithm model is used to search solution for a specific large-scale job shop scheduling.

The scheme of genetic search process

Step 0 Given the parameters required, such as population size P_s, crossover probability P_c and mutation probability P_m, etc. Set k=1 and randomly generate an initial population

$P(k) = \{X_1(k), X_2(k), \cdots, X_{P_s}(k)\}$.

Step 1 Evaluate all the individuals in $P(k)$, and determine the best one X^*.

Step 2 Output X^* and its performance if k reaches the maximum generation N_g, otherwise, continue the steps.

Step 3 Set l=0.

Step 4 Select two parents from $P(k)$, which are denoted by X_1 and X_2.

Step 5 If a real number r which is randomly generated between 0 and 1 is less than P_c, then perform a crossover for X_1 and X_2 to generate two new individual X_1' and X_2'; otherwise let $X_1'=X_1$ and $X_2'=X_2$.

Step 6 If a real number r which is randomly generated between 0 and 1 is less than P_m, then perform mutation for X_1' to generate individual X_1''; else let $X_1''=X_1'$. Similarly, X_2'' is generated. Then, put X_1'' and X_2'' into $P(k+1)$ and let $l=l+1$.

Step 7 If $l<P_s/2$ then go to step 4; otherwise, go to step8.

Step 8 Evaluate all the individuals in $P(k+1)$ and use the best one to update X^* if it is better than X^*, then let $k=k+1$ and go to step 2.

Thus, when P_s, P_c, P_m and N_g are given, the above GA can be applied for job shop scheduling problem to test performance of some GA control parameters or obtain a satisfactory or maybe optimal solution using fittest parameters. If it is used for parameters performance testing and comparison, in order to guarantee the fair comparison, the product of P_s and N_g must be fixed on N_e. The results of a single run are not enough, since the outcome of genetic algorithm is uncertain and inconsistent. A typical experimental analysis is to carry out many independent runs and present the average, standard deviations [17]. It is very time-consuming to determine the optimal parameters, and it is not easy to tell how many runs are required. In the next section, a methodology based on optimal computing budget allocation for automatically determining the fittest parameters for a specific large scale job shop scheduling problem will be presented.

5 Optimal Computing Budget Allocation Algorithm for GA Parameters Selection

In some sense, developing efficient techniques that identify which algorithm or which control parameter of the same algorithm performs well under which specific conditions is at least as important as developing new efficient algorithms. When GA is used to big size job shop scheduling problem, the identifying of control parameters especially the crossover probability and mutation probability is very crucial. In the following, we will use the optimal computing budget allocation based algorithm performance comparison method to select the fittest parameters of GA for a specific large scale scheduling problem.

5.1 Problem Statement

Suppose we wish to compare the performance of genetic algorithms with different population size (P_s), crossover probability (P_c) and mutation probability (P_m). There are a total of k different algorithm-parameter combinations. For convenience we will call these combinations k different algorithms indexed by i, where $i= 1,2,\cdots,k$. Our objective is to find an algorithm (or more accurately, an algorithm with a particular parameter setting), which performs the best over a particular problem instance as well as a specified range of variations for that instance. We define a best algorithm as one that provides the best-expected performance for the current problem instances. Denote $h(\boldsymbol{\theta}_i)$ as the outcome of applying algorithm with parameters vector $\boldsymbol{\theta}_i$. $h_j(\boldsymbol{\theta}_i)$ is a random variable characterized by uncertain outcome of a genetic algorithm for a problem instances. Specifically

$$h(\theta_i) = E_j[h_j(\theta_i)]$$

Thus a best algorithm i^* can be chosen based on the expected performance measure $E_j[h_j(\boldsymbol{\theta}_i)]$, i.e., $B^*=$ arg min$_i$ $E_j[h_j(\boldsymbol{\theta}_i)]$.For most real-life problems, neither the closed-form expression of $h_j(\boldsymbol{\theta}_i)$ nor that of $E_j[h_j(\boldsymbol{\theta}_i)]$ exists. To estimate $E_j[h_j(\boldsymbol{\theta}_i)]$. One may take a specific problem instance, and then repeated the genetic algorithm with parameter vector $\boldsymbol{\theta}_i$ for n times. Thus $E_j[h_j(\boldsymbol{\theta}_i)]$ is approximated by the value

$$\hat{E}_j[h_j(\theta_i)] = \frac{1}{n}\sum_{j=1}^{n} h_j(\theta_i) \tag{8}$$

If the variance is finite, as the strong law of large numbers dictates, the following property holds with probability 1:

$$\frac{1}{n}\sum_{j=1}^{n} h_j(\theta_i) \rightarrow E_j[h_j(\theta_i)], \quad \text{as} \quad n \rightarrow \infty.$$

Since it is not possible to conduct an infinite number of repetitions, the best algorithm must be chosen without knowing the exact value of the performance measure. The main difficulty is that with traditional methods the estimate

$\frac{1}{n}\sum_{j=1}^{n} h_j(\theta_i)$ converges slowly. In general, the rate of convergence for such a value estimate is at best $O(1/\sqrt{n})$ [18]. The large required for a good approximation implies that each algorithm must be repeated a large number of times, which need a long and unaffordable computer time. In this section we present a new approach for algorithm comparison using the optimal-computing-budget- allocation performance comparison method. Given a specified confidence interval, our method seeks to identify the best genetic algorithm control parameter vector among a group of candidate vector using a fraction of the computing effort required for traditional methods.

5.2 The Method for Algorithm Comparison and Selection

Suppose we select an algorithm using the following criterion

$$B \equiv \arg\min_i \hat{E}_j[h_j(\theta_i)] \equiv \left(\frac{1}{n}\sum_{j=1}^{n} h_j(\theta_i)\right). \tag{9}$$

Given the fact that we can only conduct a finite number of repetitions n, $\hat{E}_j[h_j(\theta_i)]$ is an approximation to the true expected performance $E_j[h_j(\theta_i)]$. An algorithm B with the smallest value of $\hat{E}_j[h_j(\theta_i)]$ is not necessarily the true best algorithm.

Let us define correct selection (CS) as the event that the selected algorithm B is actually the best algorithm. The confidence probability $P\{CS\}=P\{$*the current top-ranking algorithm B is actually the best algorithm*$\}$. Therefore the purpose is to make the probability $P\{CS\}$ sufficiently high, although the value of $\hat{E}_j[h_j(\theta_i)]$ may converge slowly.

C. H. Chen [19] developed an estimation technique to quantify the confidence level for ordinal comparison when the number of designs is large. In addition to the confidence probability, this approach also provides sensitivity information for each algorithm. The sensitivity information is useful if incremental computing effort is to be allocated during the comparison.

Theorem 1: let $\tilde{J}_i, i \in \{1,2,\cdots,B-1,B,B+1,\cdots,k\}$, denote the random variable whose probability distribution is the posterior distribution of the expected performance for algorithm I under a Bayesian model. For a minimization problem

$$P\{\text{CS}\} \geq \prod_{i=1,i\neq B}^{k} P\{\tilde{J}_B < \tilde{J}_i\} \equiv \textit{Approximate Probability of Correct Selection (APCS)}$$

$$= \prod_{i=1,i\neq B}^{k} \phi\left(\frac{\frac{1}{n}\sum_{j=1}^{n} h_j(\theta_i) - \frac{1}{n}\sum_{j=1}^{n} h_j(\theta_B)}{\sqrt{\frac{\sigma_i^2}{n} + \frac{\sigma_B^2}{n}}}\right) \tag{10}$$

If the variance is unknown, σ_i^2 can be replaced by the following

$$S_i^2 = \frac{1}{n-1}\sum_{j=1}^{n}\left\{h_j(\theta_i) - \left[\frac{1}{n}\sum_{j=1}^{n} h_j(\theta_i)\right]\right\}^2 \tag{11}$$

Where, ϕ is the standard normal cumulative distribution. Numerical testing in [19] shows that *APCS* provides a good approximation to $P\{\text{CS}\}$. We will therefore use *APCS* to approximate $P\{\text{CS}\}$ in this paper. Intuitively *APCS* provides a convenient stopping criterion for the process of genetic algorithm comparison. As the number of testing repetitions n increases, the variance σ_i^2/n decreases and more confidence can be given to the repetition mean. Using the *APCS* measure and the basic property of $\tilde{J}_i$, we design an iterative genetic algorithm comparison steps as follows.

Consider a set of genetic algorithms (GA with different control parameters) for comparison. Allocate a small number of testing repetitions for each genetic algorithm, and then rank the algorithms according to their estimated relative performance. Select the best (highest ranked) algorithm. Compute the approximate probability of correct selection (*APCS*) for the current comparison. If the current selection reaches the desired level of confidence, stop; otherwise, allocate more number of testing repeat times to genetic algorithms. Continue until the desired level of confidence is reached. But when allocating test samples to algorithms, we use optimal computing budget allocation to decision the number of testing repeat times in next iteration. This is based on the following idea. Intuitively, to ensure a high P(CS) or *APCS*, a large portion of the computing budget should be allocated to those designs which are potentially good or critical in order to reduce the estimator variance. On the other hand, limited computational effort should be expanded on non-critical designs that have little effect on identifying the good designs. Based on such a motivation, we use an optimal computing budget allocation technique to choose the optimal number of repeat times for all the designs to maximize the efficiency of genetic algorithm selection with a given computing budget or confidence level.

5.3 Optimal Computing Budget Allocation in Algorithm Comparison

Suppose we could find the allocation of testing times to all genetic algorithms, which minimizes total computation cost while obtaining the desired confidence level. Then we can optimally decide which algorithm will receive how many computing budgets in each iteration step of the experiment. Let N_i be the number of testing samples of algorithm i. the total computation cost can be approximated by $N_1+N_2+\cdots+N_k$. The goal is to choose N_i for all i such that the total computation cost is minimized, subject to the restriction that the confidence level defined by *APCS* is greater than some satisfactory level. This optimization problem in its simplest form can be stated as follows:

$$\min_{N_1, N_2, \cdots, N_k} \{N_1 + N_2 + \cdots + N_k\} \quad \text{s.t. } APCS \geq P^* \tag{12}$$

where P^* is a user-defined confidence level requirement.

Let us define $EPCS(N_1,N_2,\cdots N_{s-1}, N_s+T,N_{s+1},\cdots, N_k)$ as an estimated $P\{CS\}$ if additional T testing repeat times are performed on algorithm s. *EPCS* is computed using the statistical information after $N_1,N_2,\cdots, N_k$ testing are completed for genetic algorithms $1,\cdots,k$, respectively.

To minimize the computing of reaching a desired confidence level, in each iteration step we conduct further test to the algorithm that has a maximum promising index (PI) defined as follows:

$$PI(s) \equiv EPCS(N_1,N_2,\cdots,N_{s-1},N_s+T,N_{s+1},\cdots,N_k) - APCS(N_1,N_2,\cdots,N_{s-1},N_s,N_{s+1},\cdots,N_k)$$

C. H. Chen [20] suggests a simple and effective way to estimate the *EPCS* as the following:

If $s \neq B$

$$EPCS(N_1,N_2,\cdots,N_{s-1},N_s+T,N_{s+1},\cdots,N_k) = P\{\tilde{J}_B < \hat{J}_s\} \bullet \prod_{i=1,i\neq B,i\neq s}^{k} P\{\tilde{J}_B < \tilde{J}_i\} \quad (13)$$

where

$$\tilde{J}_i \sim N\left(\frac{1}{N_i}\sum_{j=1}^{N_i} h_j(\theta_i), \frac{\sigma_i^2}{N_i}\right) \qquad \hat{J}_i \sim N\left(\frac{1}{N_s}\sum_{j=1}^{N_s} h_j(\theta_s), \frac{\sigma_s^2}{N_s+T}\right)$$

and if $s=B$

$$EPCS(N_1,N_2,\cdots,N_{B-1},N_B+T,N_{B+1},\cdots,N_k) = \prod_{i=1,i\neq B}^{k} P\{\hat{J}_B < \tilde{J}_i\} \quad (14)$$

where

$$\tilde{J}_i \sim N\left(\frac{1}{N_i}\sum_{j=1}^{N_i} h_j(\theta_i), \frac{\sigma_i^2}{N_i}\right) \qquad \hat{J}_B \sim N\left(\frac{1}{N_B}\sum_{j=1}^{N_B} h_j(\theta_B), \frac{\sigma_B^2}{N_B+T}\right)$$

Thus, *EPCS* provides sensitivity information about how *APCS* will change if additional T times testing repeats are performed on algorithm. Under the above framework, promising refers to high improvement of the overall comparison confidence level. Since we intend to minimize the total number of testing samples, we select and test a subset of most promising algorithms in each iteration, then repeat the process until *APCS* achieves the desired level P^*.

Therefore the GA control parameters selection algorithm based on optimal computing budget allocation is as the following: We first test all algorithms with n_0 times respectively. A promising index (*PI*) is then calculated for all algorithms and a subset of m algorithms are selected for further testing based on their *PI*'s. The procedure is summarized as follows.

Genetic algorithm comparison approach with optimal computing budget allocation

Step 0. Perform n_0 time for all genetic algorithms respectively. Setting: l=0, $N_1^l = N_2^l = \cdots = N_k^l = n_0$.

Step 1. If $APCS(N_1^l, N_2^l, \cdots, N_k^l) \geq P^*$, stop; otherwise, go to step 2.

Step 2. Calculate $PI(s)$ for all algorithms $s = 1,2,\cdots, k$.

Step 3. Find the set $S(m) \equiv \{s: Pi(s)$ is among the highest $m\}$.

Step 4. Perform additional T times repeat testing for genetic algorithm $i, i \in S(m)$.

Setting: $N_i^{l+1} \leftarrow N_i^l + T$, for $i \in S(m)$; and $N_i^{l+1} \leftarrow N_i^l$, for $i \notin S(m)$; $l \leftarrow l+1$, go to step 1.

In the above algorithm, n_0 cannot be too small; otherwise the estimates of the mean and the variance may be very poor, resulting in terminating the comparison too early. A good choice for n_0 is between 10 and 20 [21]. A large T can result in wasting computation time to obtain an unnecessarily high confidence level. On the other hand, if T is small, we need to compute *APCS* and *PI* many times. Since the cost of computing *APCS* is much cheaper than runs of the algorithm testing, it is advisable to select a smaller, in genetic algorithm comparison, T is set to 5. The selection of m is trivial. m is the number of promising designs for further testing. A large m is definitely a bad choice. Consider an extreme case that $m=k$. This means that we simulate all designs every time, which is equivalent to no use of optimal computing budget allocation. However, we don't want m is too small because small m means that we need to perform Steps 2 and 3 many times. A good choice for m is any number between $k/20$ to $k/10$.

6 Optimal-Computing-Budget-Allocation Based Compound Genetic Algorithm

Suppose there is a large scale job shop scheduling problem to solve using genetic algorithm, researcher often set parameters such as population size P_s, crossover probability P_c, mutation probability P_m, by their experience. They cannot tell the performance of the algorithm until many practical runs. But as we all know that the majority of production scheduling problem is NP problem. We even can't afford the price of a single run for those time-consuming problems, not to say a large number times needed for obtaining algorithm performance measurement. In this paper, an optimal computing budget allocation based compound genetic algorithm is used to handle the large scale job shop scheduling problems. Firstly the fittest combination of P_s, P_c, and P_m for a given problem is selected using a reasonable computational time. Because of the using of optimal computing budget allocation method, we can obtain a pre-specified correct selection confident level with much less time. Then the genetic algorithm with the fittest parameters is applied to conduct more searches to find the optimal solution of NP scheduling problems. The procedure of the compound genetic algorithm is as the following.

Compound genetic algorithm for large scale job shop scheduling problem

Step 1. For a given large size scheduling problem, we design genetic algorithm according to the particular need. Set the candidate parameter combination of P_s, P_c, and P_m for the fittest parameters selection. Set the evaluation times N_e=600 for one time genetic search testing.

Step 2. Specify a correct selection confidence level P^* or the overall running times N for all candidate genetic algorithms, $N=N_1+N_2+\cdots+N_k$.

Step 3. Perform n_0 time for all genetic algorithms (all combination of P_s, P_c, and P_m) respectively. Setting: l=0, $N_1^l = N_2^l = \cdots = N_k^l = n_0$.

Step 4. Calculate the average makespan value and the variance according to the related formula in section 3.

Step 5. If $APCS\ (N_1^l, N_2^l, \cdots, N_k^l) \geq P^*$, or $N_1^l + N_2^l + \cdots + N_k^l \geq N$ go to step 9; otherwise, go to step 6.

Step 6. Calculate $PI(s)$ for all genetic algorithms $s = 1,2,\cdots, k$.

Step 7. Find the set $S\ (m) \equiv \{s: Pi(s)$ is among the highest $m\}$.

Step 8. Perform additional T times repeat testing for genetic algorithm $i, i \in S(m)$.

Setting: $N_i^{l+1} \leftarrow N_i^l + T$, for $i \in S(m)$; and $N_i^{l+1} \leftarrow N_i^l$, for $i \notin S(m)$; $l \leftarrow l+1$, go to step 4.

Step 9. Output the selected fittest parameter combination of P_s, P_c, and P_m, and use the genetic algorithm controlled by these parameters to do further search to find optimal solution for the scheduling problem.

7 Computational Experiments and Analysis

In this paper, the large scale job shop problem benchmarks named SWV2, SWV4, SWV6, SWV9, SWV10, SWV12, SWV15, respectively, which are given by Storer [22], only SWV2 is listed in Table 1. SWV4 is 20-job-10-machine problem; SWV6, SWV9 and SWV10 are 20-job-15-machine problem; SWV12 and SWV15 are 50-job-10-machine problem. These hard problems can be downloaded from http://mscmga.ms.ic.ac.uk.

For every benchmark, we set 4 choices {10, 20, 30, 50} for population size, 4 choices {0.5, 0.6, 0.7,0.8, 0.9} for crossover probability and 4 choices {0.01, 0.05, 0.1, 0.2} for mutation probability. Thus, there are a total of 64 parameter combinations.

Table 1. SWV2-Hard Problem Given by Storer in 1992

Job	Operations sequence (machine, processing time)										Total proc. time
J1	(4,62)	(1,60)	(3,64)	(2,12)	(0,39)	(5,2)	(7,64)	(6,87)	(9,21)	(8,60)	471
J2	(2,66)	(1,71)	(3,23)	(4,75)	(0,78)	(7,74)	(6,35)	(9,24)	(8,23)	(5,50)	519
J3	(1,5)	(3,92)	(4,6)	(0,69)	(2,80)	(7,13)	(5,17)	(9,89)	(6,80)	(8,47)	498
J4	(0,82)	(3,84)	(1,24)	(2,47)	(4,93)	(7,85)	(5,34)	(6,73)	(8,28)	(9,91)	641
J5	(4,55)	(0,57)	(3,63)	(2,24)	(1,40)	(7,30)	(6,37)	(5,99)	(8,88)	(9,41,)	534
J6	(1,75)	(2,47)	(3,68)	(0,7)	(4,78)	(7,80)	(6,2)	(9,23)	(8,49)	(5,50)	479
J7	(0,91)	(4,25)	(2,10)	(1,21)	(3,94)	(8,6)	(7,59)	(5,84)	(9,75)	(6,70)	535
J8	(2,85)	(1,31)	(0,94)	(4,94)	(3,11)	(5,21)	(9,7)	(6,61)	(8,50)	(7,93)	547
J9	(1,27)	(0,77)	(4,13)	(2,30)	(3,2)	(5,88)	(7,4)	(9,39)	(6,53)	(8,54)	387
J10	(1,34)	(2,12)	(3,31)	(0,24)	(4,24)	(7,16)	(5,6)	(9,88)	(8,81)	(6,11)	327

Table 2. Result of SWV2 When The Parameters Selection Computing Is Ran Out

(a)

P_s	10				20			
P_m	0.01	0.05	0.1	0.2	0.01	0.05	0.1	0.2
P_c=0.5	956.9	971.2	969.5	975.6	949.7	964.2	959.0	955.1
P_c=0.6	974.5	973.0	973.1	979.0	958.6	953.9	968.1	949.0
P_c=0.7	972.9	977.1	980.6	974.3	978.9	982.1	980.2	967.9
P_c=0.8	970.3	966.5	976.8	980.1	987.3	975.8	986.0	973.8
P_c=0.9	973.1	966.8	981.0	985.7	983.1	953.9	976.3	982.4

(b)

P_s	30				50			
P_m	0.01	0.05	0.1	0.2	0.01	0.05	0.1	0.2
P_c=0.5	949.6	952.1	955.9	956.5	948.7	954.2	950.9	951.5
P_c=0.6	955.4	950.3	961.3	949.0	952.2	945.6	961.8	949.3
P_c=0.7	969.2	971.7	966.0	963.4	968.9	971.2	972.0	959.7
P_c=0.8	953.0	956.6	978.6	978.1	953.3	941.0	968.7	968.3
P_c=0.9	971.3	968.6	974.3	977.5	971.6	939.3	973.6	974.2

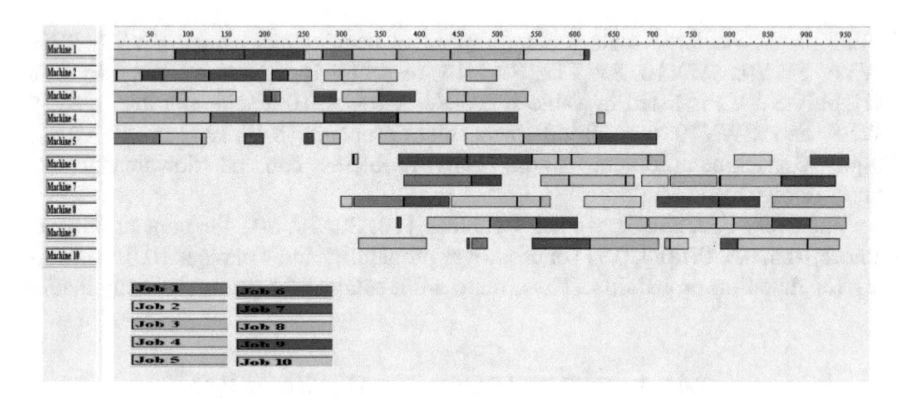

Fig. 1. The Gantt Chart of SWV2's Scheduling Results

We set n_0=10, T=5 and total times of testing repeat constraint N=3000. and when the fittest parameters are identified, we further run the selected genetic algorithm for 2000 generation to find optimal scheduling solution.

Table 2 are the results of benchmark problem SWV2 in optimal genetic control parameters selection phase when the total3000 times of testing repeat is ran out. According these outcomes we can tell which parameters combination is optimal. As to SWV2, obviously P_s=50, P_c=0.8, and P_m=0.05 is the optimal parameters combination, and the correct selection confidence level is 95.3%. Fig. 1 is Gantt Chart of SWV2's scheduling results. From Table 3, we can see that large scale job shop problem converge to optimal solution very slowly. The best thing we can do is using a fraction of computing time to selection the fittest parameters firstly, then we can put our limited computing resource on the most efficient algorithm. This is very significant for find solution of some NP-hard combinational problem, especially large scale job shop scheduling problems. By avoiding blind search, algorithm may find optimal solution more possibly. Otherwise, it is clear that a large population size, a high crossover probability and a low mutation probability are suggested for the job shop scheduling problems on a large scale.

Table 3. Results of 7 Benchmark Problems Using Compound Genetic Algorithm

Problem name	(Jobs, Machines)	P^*	P_s	P_c	P_m	Mean makespan	Makspan after further *2K* times iteration
SWV2	(10,10)	95.3%	30	0.8	0.01	1207.3	941
SWV4	(20,10)	93.4%	30	0.8	0.1	1471.2	1452
SWV6	(20,10)	94.0%	50	0.95	0.05	1682.6	1586
SWV9	(20,15)	92.8%	50	0.95	0.05	1650.4	1610
SWV10	(20,15)	92.6%	50	0.95	0.05	1703.5	1642
SWV12	(50,10)	92.1%	50	0.95	0.01	2998.8	2977
SWV15	(50,10)	92.0%	50	0.95	0.01	2901.6	2891

8 Conclusions

In job shop scheduling problem, as the number of jobs and number of machines rise, the problem's complexity increase exponentially. So the most of large scale JSP is NP-hard problem. In this paper, we presented a compound genetic algorithm to resolve these difficult problems in two steps: firstly, identify the fittest GA control parameters from a group of candidate parameters use optimal computing budget allocation based comparison algorithm with a fraction of time; secondly, genetic algorithm with the fittest parameters do further endeavor to find optimal solution. After carrying out experiments on some job shop scheduling benchmarks with various sets of parameter values, the results obtained are presented. The fittest control parameters selected in experiments will produce an efficient solution and will lead to the optimum solution faster and avoid entrapments in local minima. Future work includes applying the optimal computing budget allocation method to comparison other parameter dependent heuristic algorithm, such as annealing simulation algorithm and artificial neutral network and so on; and using the compound genetic algorithm for solution of other large scale combinatorial optimization problems.

References

1. M. Gen, R. Cheng (1997) Genetic Algorithm and Engineering Design. John Wiley, New York
2. Back, T. (1992) The interaction of mutation rate, selection, and self-adaptation within genetic algorithm. Parallel Problem Solving from Nature 2: 85-94

3. D. C. Mattfeld (1996) Evolutionary Search and the Job Shop: Investigations on Genetic Algorithm for Production Scheduling. Springer-Verlag, Heidelberg, Germany
4. J. Blazewicz, K. Ecker, G. Schmidt and J. Weglarz (1993) Scheduling in Computer and Manufacturing Systems. Springer-Verlag, Berlin,Heidelberg
5. P. Brucker (1995) Scheduling Algorithms. Springer-Verlag, Berlin, Heidelberg
6. M.C. Portmann (1997) Scheduling methodologies: optimization and compusearch approaches. The Planning and Scheduling of Production Systems, Chap.9: 271–300
7. Ponnambalam SG, Jawahar N, Kumar BS (2003) Estimation of optimum genetic control parameters for job shop scheduling. Int J Adv Manuf Technol 19: 224-234
8. D. Orvosh, L. Davis (1994) Using a genetic algorithm to optimize problems with feasibility constraints. Proc. of the First IEEE Conf. on Evolutionary Computation, IEEE Press, Florida: 548-552
9. Runwei Cheng (1996) A tutorial survey of job-shop scheduling problems using genetic algorithms-I: representation. Computers ind. Engng 30(4): 986-995
10. M. Gen, Y. Tsujimura and E. Kubota (1994) Solving job-shop scheduling problem using genetic algorithms. Proc. of the 16th Int. Conf. on Computer and Industrial Engineering, Ashikaga, Japan: 576-579
11. A. Kubota (1995) Study on optimal scheduling for manufacturing system by genetic algorithms. Master's thesis, Ashikaga Institute of Technology, Ashikaga, Japan
12. K. Baker (1974) Introduction to Sequencing and Scheduling. New York, John Wiley and Sons
13. De Jong KA (1975) An analysis of the behavior of a class of genetic adaptive systems. Dissertation, University of Michigan: 76-81
14. Grefenstette JJ (1999) Optimization of control parameters for genetic algorithms. IEEE Trans Sys Man Cybern 16(1): 122-128
15. Wang L, Zheng DZ (2003) An effective hybrid heuristic for flow shop scheduling. Int J Adv Manuf Technol 21(1): 38-44
16. Goldberg DE (1989) Genetic algorithms in search, optimization and machine learning. Addison-Wesley
17. Eiben AE, Schoenauer M (2002) Evolutionary computing. Informat Process Lett 82: 1-6
18. H. J. Kushner and D. S. Clark (1978) Stochastic Approximation for Constrained and Unconstrained Systems. New York, Springer-Verlag: 26-27
19. C. H. Chen (1996) A lower bound for the correct subset-selection probability and its application to discrete event system simulations. IEEE Trans. Automat. Contr. 41: 1227–1231
20. H. C. Chen, C. H. Chen, L. Dai, and E. Yucesan (1997) New development of optimal computing budget allocation for discrete event simulation. Proc. Winter Simulation Conf. Dec: 334–341
21. Chen CH, Lin J, Yucesan E, Chick SE (2000) Simulation budget allocation for further enhancing the efficiency of ordinal optimization. Discr Event Dynam Sys 10: 251-270
22. Storer R H, Wu SD, Vaccari R (1992) New search spaces for sequencing problems with applications to job-shop scheduling. Management Science 38(10): 1495-1509

Part IV

Fuzzy Engineering

Multi-Immune-Agent Based Power Quality Monitoring Method for Electrified Railway

Hongsheng Su and Youpeng Zhang

School of Automatic and Electrical Engineering
Lanzhou Jiaotong University, Lanzhou 730070, P.R. China
shsen@163.com, zhangyp@mail.lzjtu.cn

Abstract. In view of the fact that existing power quality monitoring system can't satisfy increasing demands such as wide-area large-scale far-reaching distributions of monitoring points of electrified railway as well as uncertainties of environmental varieties such as loads change in point of common coupling (PCC), traditional power quality monitoring methods for electrified railway are difficult to adapt it well. Based on it, combining autonomous decentralized systems (ADS), artificial immune systems (AIS) and multi-agent (MA) technology, in this paper a novel power quality monitoring method for electrified railway called cooperative multi-immune-agent system is proposed. It not only can be used to monitor the distributed process state in real time and diagnose faults in advance, but also can improve the operational condition of power system and enhance the efficiency of power supply, and is a new approach for distributed artificial intelligence (DAI). The study shows cooperative multi-immune-agent system is a quite incompact self-organizing system, more flexible and robust and updated locally, quite suitable to solve indeterminate problems under dynamic environment.

Keywords: Electrified Railway, Multi-Immune-Agent, Power Quality Monitoring.

1 Introduction

With the fast advancement of distributed artificial intelligence (DAI), agent as well as multi-agent is considered as a ubiquitous guidance idea for the complicated advanced problem solving [1]. In addition, as scale expansion, systems become more and more complicated not to be effectively controlled, control the whole system only using a central controller becomes extremely difficult, fault tolerance capabilities of system become extensively weak. To resolve the problem, a concept called autonomous decentralized systems (ADS) has been reported [2], whose architecture is dynamic can't be designed and confirmed beforehand. All autonomous main bodies commonly construct autonomous fields and by which behavior of each autonomous main body is restricted. Meantime, with the developments of gene project and computer networks technology, artificial neural networks (ANN), fuzzy systems (FS), DNA and artificial immune systems (AIS), are attracting more and more researchers spreading the whole globe due to their unique information processing abilities. And now immune system has been broadly seen as a technology measure exceeding any conventional ways [3]. After immune theory based immune algorithm is successfully applied to diverse fields, enormous achievements have already been achieved [4].

B.-Y. Cao (Ed.): Fuzzy Information and Engineering (ICFIE), ASC 40, pp. 399–410, 2007.
springerlink.com

Considering existing power quality monitoring system can't meet requirements such as wide-area large-scale distribution of monitoring points of electrified railway as well as uncertainties of environment alteration such as loads change in point of common coupling (PCC), traditional power quality monitoring means for electrified railway is difficult to suit it well. To change the situation, multi-agent based state monitoring and control system has already been developed recently in [5-6]. Distributed systems have been wide-area large-scale applications such as those related to public utilities and power station facilities, and exhibit excellent monitoring and control effects. Consistent with this trend, a multi-immune-agent based system is studied and developed in this paper, which can automatically control equipments based on data from a network of field immune agents (ImAs) as well as cooperative methods among them. In this research, to adapt to dynamic environment, ImA is used to act as field intelligence terminals. To cooperate with the existing systems, ADS is served as communication networks, In addition, multi-immune-agent based cooperative mechanism is also studied so as to improve the serviceability.

2 Immune Agent Model

Enlightened by nature-inspired, several kinds of organism algorithms are simulated, e.g., ANN, FS, DNA and AIS, and etc. In accordance with immune theory and multi-agent (MA) technology, in [7] an immune agent model and its many characteristics are researched, e.g., self-balance ability to disturbance, self-defining and non-self defining process, and knowledge processing mechanism based on learning-memory-forgetting of immune systems and so forth. Moreover, Immune agent is more intelligent to fully meet complicated decentralized advanced problem solving and processing demands. When immune agents monitor and protect any other systems, it also safeguards itself. The essence of distribution and decentralization lies in that there is no way to inflict impact to it from the outside, and can't implement central control in inside either.

Clone and negative selection are two important theoretical doctrines. Clone selection explains why antibodies can recognize antigens and yield immune memory etc, why formation and evolvement of antibodies develop towards existing antigens. Negative selection considers that mass antibodies are randomly produced in interior organism in advance, those who can produce destruction role to self antibodies will be cleared for ever, otherwise, disease of immune system itself will be aroused, only remained antibodies can used to monitor all outside antigens. The two processes are applied to turn out effective antibodies library and keep self-stability. Design of ImAs possesses the above two characteristics, therefore, whose structure is flexible and technique is constantly evolutionary.

3 Immune Supervision Networks Model

Due to high decentralization of subsystems, flexibility and fault tolerance capabilities of ADS are dramatically improved. Non-operational states of anyone subsystem such as fault, maintenance and update can be seen as normal state in term of the overall system, there is no any discrimination between normal and abnormal state of system. Fault, expansion and dissipation are also seen as a normal state. Another important characteristic of ADS is its dynamic structure that can't be designed and confirmed in advance. All characteristics of ADS, like AIS, are very suitable to construct distributed autonomous networks. Base on Jerne' idiotypic networks model[8], combining ADS and MA approach, a formation networks model for states monitoring and fault diagnosis beforehand for electrified railway is proposed as shown in Fig.1, where n agents constitute a set ImA={ImA_1,ImA_2,…,ImAn},under initial states, each ImA is located at node of problem solving, say, traction substations or PCC, so as to monitor the states and diagnose fault beforehand . Every ImA determines fitness of self-survival according to ambient environment. Consequently, evolution of the overall ImA systems is also promoted. Each ImA gains fielded knowledge under initial states, and can recognize and defend unexpected events of the governed range. Secondary response lets each ImA learn and remember reference experience coming from foreign ImA to enrich and update home knowledge base, meantime, self-immune and defense function are boosted up. Moreover, ImA transforms own accumulating experiment into vaccinia, and broadcasts it to foreign ImA by stimulating format over ADS date field to complement and amend fielded knowledge.

4 Architecture and Algorithm of Immune Agent

Internal structure of ImA shows a hierarchy recursive agent model as shown Fig.1, i.e., local function of each agent is implemented by interior subagent, e.g., sensor agent for identifying antigens, knowledge base agent for running knowledge and information processing agent as well as reasoning agent, and etc. Local agent may be further compartmentalized according to actual functional requirements. Thus, ImA owns powerful function and risk differentiation abilities, called super agent. In a multi-immune-agent system, immune agents must have communication interface (communication agent) to monitor and correspond. Multi-immune-agent can communicate and cooperate through ADS data field, the work principle of single ImA is described below [7].

Step 1. According to prior experience as well as fielded experts knowledge, home knowledge library is established, after negative selection, antibodies pattern base $\boldsymbol{X}$ is set beforehand below.

$$\boldsymbol{X}=(x_1, x_2,\ldots,x_n) \tag{1}$$

Step 2. To accept stimulatory vaccines coming from foreign ImAs, home agent updates native knowledge base described by

$$\boldsymbol{X'}=(x_1',x_2', \ldots, x_n') \tag{2}$$

Step 3. When $t=k$, antigen sensor agent collects local range information, after filtering and data fusion m patterns are formed by

$$\boldsymbol{Y}(k)=(y_1(k), y_2(k), \ldots , y_m(k)) \tag{3}$$

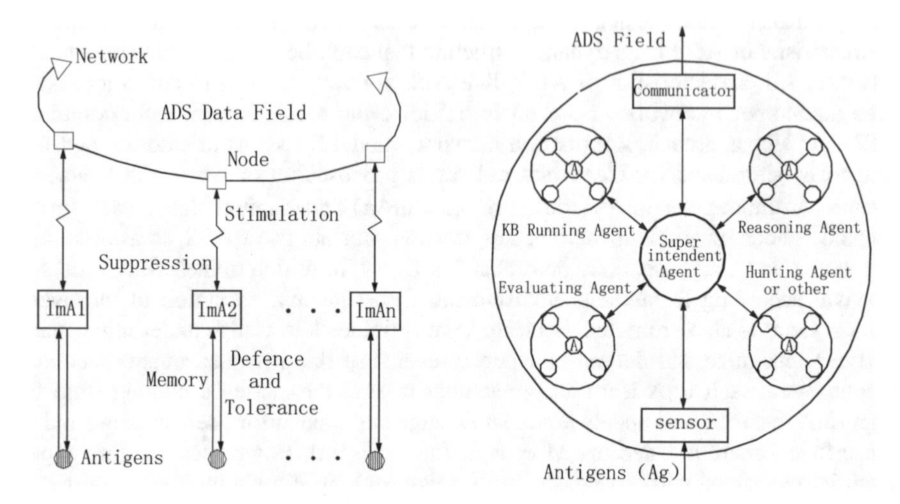

Fig. 1. Supervision network model (*left*) and hierarchical agent architecture (*right*)

Step 4. According to the affinity calculation results between the antibody (Ab) and the antigen (Ag), if $\exists y_i(t) \notin \boldsymbol{X}'$, $y_i(k) \in \boldsymbol{Y}(k)$, then judging $y_i(k)$ has larger probability leading to the abnormities, whereas seeking solving schemes from foreign ImAs, keep waiting and doing immune tolerance, then turning to step9; If $\exists y_i(t) \in \boldsymbol{X}'$, $y_i(k) \in \boldsymbol{Y}(k)$, then corresponding solving scheme exists in home knowledge base, therefore, secondary immune response is carried through.

Step 5. Controller agent sends out orders to execution unit to eliminate antigens and keep immune tolerance.

Step 6. Let the pattern as vaccine, and broadcast it to foreign ImAs over ADS data fields, as experience to instruct them dealing with resemble problems.

Step 7. While $t=k+1$, systems enter into next time circle.

Step 8. According to evolution strategy, negative selection are carried through, local pattern library is updated.

Step 9. If other ImAs send back reference vaccines, and turn to step5. If not, initial immune response should be done, learning, memory and update local knowledge base for secondary immune response.

5 Cooperative Multi-Immune-Agent Based Power Quality Monitoring Systems

ImAs based distributed control technology not only can evaluate diverse information received from other immune agents, but also cooperates with those ImAs to enhance-serviceability. In conventional systems, the host CPU controls the ways in which conflicts are handled, and conversely, here conflicts are resolved by having each ImA cooperate with other functional ImAs by using priority levels, conditions and evaluation points. Regarding distributed systems, multi-agent systems can improve both the cooperation and the effectiveness of communication. To enhance serviceability the conflicts that occur when the data processed by different ImAs is in consistent or irregular need to be resolved, these conflicts are resolved by having the ImA cooperate with other ImAs, using immune-terminal parameters (abnormal information $\boldsymbol{X}$, assistance information $\boldsymbol{Y}$). The designed ImA that can compare the $\boldsymbol{X}$ with $\boldsymbol{Y}$, an adequate ImA or several ImAs with capability to assist, will support malfunctioning information ImA based on cooperative assistance algorithm, ImA can negotiate and cooperate with each other to resolve the conflicts that occur when the data processed by different ImAs is in consistent or irregular.

5.1 Immune-Terminal Parameter

The structure of abnormal information is expressed using $\boldsymbol{X}$ $(x1,...,xn)$ as seen in Fig.2, for instance, x_1 means sign of ImA that appears abnormal, and x_2 specifies the degree of abnormity, this means the degree of exceeding limit of reactive power or harmonic or imbalance voltage of traction substation as well as temporality and space properties of data. The content of $\boldsymbol{X}$ is introduced in detail below.

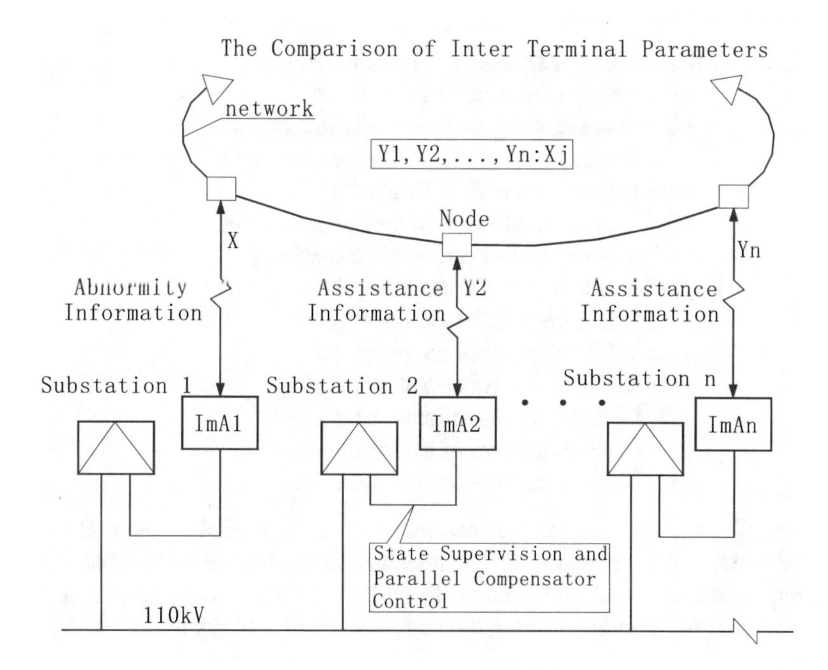

Fig. 2. The main functions of the proposed method

X=($x1$, $x2$, …, $x13$): Information number.
$x1$: ImA sign that yields abnormity.
$x2$: Degree of emergency.
$x3$: Loads important degree of the governed substation
$x4$: Level of traction substation harmonic.
$x5$ Level of voltage imbalance of substation.
$x6$: Level of reactive power of substation.
$x7$: Status of compensators that belong to ImA or substation.
$x8$: Evaluation point of parameter $x2$.
$x9$: Evaluation point of parameter $x3$.
$x10$: Evaluation point of parameter $x4$.
$x11$: Evaluation point of parameter $x5$.
$x12$: Evaluation point of parameter $x6$.
$x13$: Evaluation point of parameter $x7$.

Assistance information Y($y1$,…,$y13$) is also illustrated below. For example, $y2$ is the maximum margin degree to boundaries of its own power quality. The ImA with the maximum value of Y has the responsibility to assist ImA with its substation power quality abnormity until the level is restored to within the specified range. The algorithm for detail comparison to calculate ***X*, *Y*** is shown section 5.2. Also, the evaluation point parameters $y8$ to $y13$ are assigned according to the factors influencing the overall system like X values.

Y=($y1$,$y2$,…,$y13$): Information number.
$y1$: ImA sign that supplies assistance.
$y2$: Degree of margin of the assistance ImAs.
$y3$ Loads capacity of the assistance substation.
$y4$: Level of harmonic of substation.
$y5$: Level of unbalanced voltage of substation.
$y6$: Level of reactive power of substation.
$y7$: Condition of compensators belong to ImA.
$y8$: Evaluation point of parameter $y2$.
$y9$: Evaluation point of parameter $y3$.
$y10$: Evaluation point of parameter $y4$.
$y11$: Evaluation point of parameter $y5$.
$y12$: Evaluation point of parameter $y6$.
$y13$: Evaluation point of parameter $y7$.

The difference between the conventional architecture and the proposed method is that immune agent is added on MA architecture to enhance the following functions: Transfer, Negotiation and cooperation, in the case of supervising power quality for electrified railway, ImA inters from the margin of PCC capacity as well as other connected substations and the abnormal degree of a abnormal ImA, and cooperates with related agents to make abnormal substation accord with power quality requirements by choosing solitary or co-operative assistance of ImAs. In a word, the role of ImA in DAI can be summarized as follows:

Transfer: Assistance obligation of other ImAs to the abnormal ImA is transferred in accordance with the result of ***X,Y*** comparisons.

Negotiation: In case of that a conflict exists among ImAs, it is resolved by negotiation based on ***X, Y*** conditions such as margin of substation power quality, evaluation points etc.

Cooperation: Several ImAs with perfect power quality cooperate to control own input/output of parallel compensators or let out margin power quality to save a abnormal ImA based on the result of ***X,Y*** comparisons.

5.2 Control Algorithm

To resolve the conflicts, we define three priority levels that enable the ImAs to prioritize a problem. They are described in Table 1. The control methods and the ImA judgment algorithms of ***X*** and *Y* are shown in Fig.3.

Table 1. Priority levels of problems in ImAs

priority	event	event definition
1	$\triangle p, \triangle h, \triangle v < c$	normal
2	$\triangle p, \triangle h, \triangle v > c$	minor problem
3	$\triangle p, \triangle h, \triangle v >> c$	critical problem

p: reactive power, *h*: harmonic, *v*: voltage imbalance, *c*: constant value.

1. *Normal event*: [$\triangle p$, $\triangle h$ or $\triangle v<c$](c is a constant)

Under normal conditions, reactive power and harmonic and voltage imbalance can be adequately balanced by its own parallel compensators.

2. *Minor problem event*: [$\triangle p$, $\triangle h$ or $\triangle v>c$]

Under minor problem conditions, the ImA immediately calculates $x2$, which is a urgent degree of exceeding limitation of voltage, harmonic and imbalance voltage, using a conversion function. Then it sets $x2$ and the other information $x3$ to $x13$, and in a circular table and broadcasts it over the network at certain intervals in order to obtain assistance information *Y* from other ImAs.

3. *Key problem event*:[$\triangle p$, $\triangle h$ or$\triangle v>> c$]

Under critical problem conditions, though these control patterns do not theoretically exist in the changes in $\triangle p$, $\triangle h$ and$\triangle v$, these are cases in which many conflicts or critical problems occur simultaneously. When this happens, the ImA sets $x2$=0 and the other information $x3$ to $x13$ in a circular table and broadcasts it immediately, $x2$ is set zero because the $x2$ should always be less than the $y2$ in order to get the assistance ***Y*** of normally operating ImAs in supplying parallel compensators to the substation of a malfunctioning ImA.

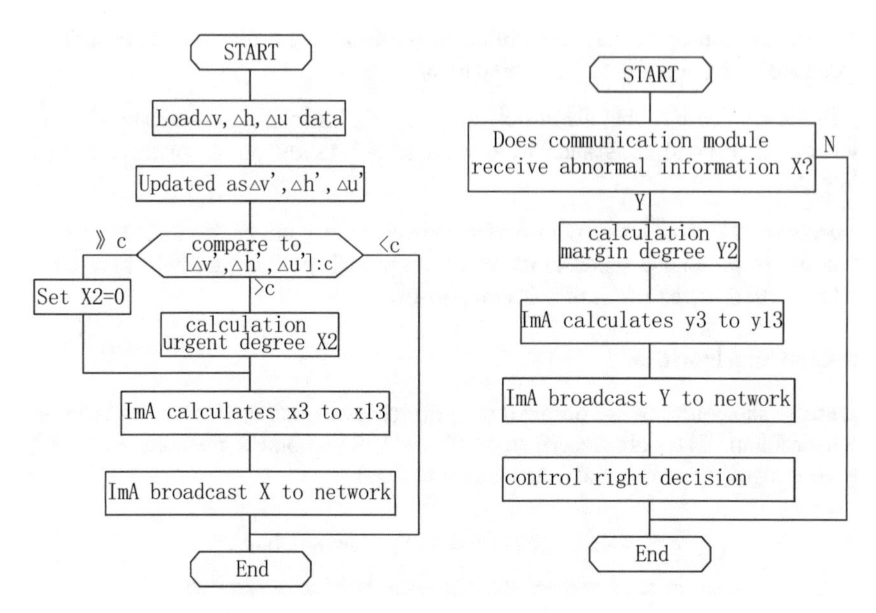

Fig. 3. ImA judgment algorithm of ***X*** (*left*) and ***Y***.(***right***)

When an ImA receives ***X*** from other ImAs, it calculates *y*2 and the average mean value of *y*3 to *y*13 and it transfers ***Y***(*y*1 to *y*13) to the network at certain intervals. Firstly, ImA compares *X*j with *Y*2 to *Yn*, and compares degrees of emergency/margin [Max (*y*k2-*x*j2>0)], where *y*k2 expresses the agent *k* that it provides help, *x*j2 expresses the agent *j* that it needs help. If one of deviations are +, then ImA calculates evaluation points based on (*y*3 to *y*13), and compares the result of evaluations. And then, if its own evaluation points are largest, ImA has a solitary assistance control mode. If not, return to ImA comparing function. If [Max (*y*k2-*x*j2<0)] all of deviations are -, then ImAs have a cooperative assistance control.

5.3 Control Decision Algorithm

Each ImA decides control parameters according to the result of comparison ***X*** with ***Y***. If an ImA has a right of control, ImA predicts a trend of over/underflow of power quality in a malfunctioning ImA. Then, if it is overflow-trended, back-up parallel compensators are full-open on according to an interval of sending ***X***. on the other hand, it is underflow-trended, back-up parallel compensators are full-closed on according to an interval of sending ***X***. back-up equipments mean parallel compensators, series compensators, voltage regulators and reactive compensators and so on. Also, value of control parameters (*p*1,*p*2,… ,*pn*) is expressed below.

*p*1=1(parallel compensator: full-open)
*p*1=0(parallel compensator: full-close)
*p*2=1(series compensator: full-open)
*p*2=1(series compensator: full-close)
*p*3=1(voltage ajustor: full-open)
*p*3=0(voltage ajustor: full-close)

Thus, control parameter calculation function decides an output value of control parameter based on the result of comparison ***X*** with ***Y*** via ImA.

5.4 Resolving Conflicts

The communication modules exchange data among ImAs so as to prevent inconsistent assessment of the priority levels. Conflicting judgments are identified by the discrimination function of the ImAs, and a process to ensure one order of priority levels is run by the ImAs in accordance with the priority levels listed in Table 2. However, the ImAs for distributed systems use an asynchronous communication method, and each ImA is independent, so it is impossible for one ImA to supervise all the conditions of the system on a real time basis. Each ImA therefore negotiates with the other ImAs to achieve adequate control, but conflict among ImAs will still occur, resulting in inconsistent overall solutions. An example conflict that occurs when the data is inconsistent or irregular is the following. If a priority level determined by one ImA is greater than those of the other ImAs in the group, the CP(control parameters) values determined by the ImA is available. If, on the other hand, the priority levels of some other ImAs are greater, the CP values determined by the ImA are used to determine the threshold level. Here conflicts resolution method, evaluation points are assigned to each priority level according to the conditions listed in Table 2. if the priority levels of all ImAs are the same, these points are added to evaluation points, and the cumulative criteria are evaluated. These evaluation points are assigned to according to factors affecting the overall system. They are used to the best CPs and to resolve conflicts among ImAs. Thus, each ImA compare the conditions (i.e., normal, minor, or critical) and negotiates and cooperates with the other ImAs to select the suitable CPs to avoid further failures. Each ImA instructs its control mechanism to maintain level of power quality within an allowed range.

Table 2. Definition of evaluation point

priority	conditions	evaluation point
1	importance of substation	5
2	Level of harmonic	4
3	Level of imbalance voltage	3
4	Level of reactive power	2
5	Compensators status	1

6 Experiment Evaluation

A local power system of certain electrified railway is shown in Fig. 4, a PCC *A* (220kV), four traction substations respectively called *B,C,D* and *E*, a large scale aluminum plant *F* and other small users *G*. Where electrified engines are random harmonic sources, electrolytic aluminum plant is a stable harmonic source, other users are small random harmonic loads.

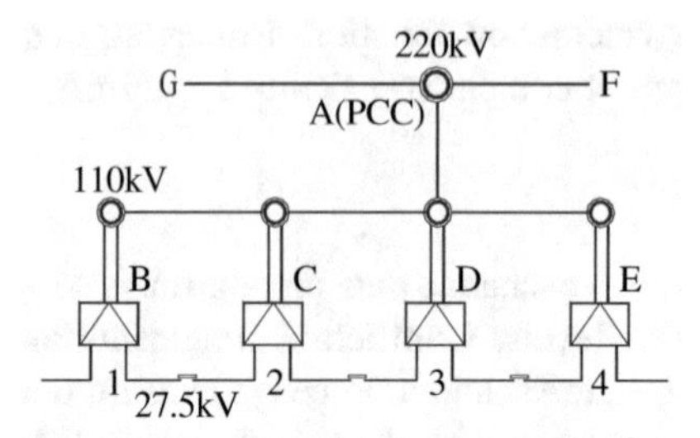

Fig. 4. Local electrical power system of electrified railway

Assume that under initialization states, all nodes *A,B,C,D* and *E* install ImAs respectively named *a,b,c,d* and *e* to monitoring power quality of individual node (traction substation), for convenience analysis, taking harmonic as an example, monitoring networks model in Fig .2 is applied, every ImA gains its field knowledge under initialized conditions.

6.1 Evaluation

To illustrate the working principle of multi-immune-agent systems well, let us consider the following three examples.

1)Immune agent *b* senses harmonic information of the governed range (antigens), after matching with self-patterns (antibodies) finding that harmonic capacities exceed limitation, but home knowledge base exists scheme solving such as parallel compensator switching and so on, then corresponding measure is adopted to drop harmonic capacities. This belongs to normal event.

2)Immune agent *b* senses harmonic information of the governed scope, after patterns matching finding that harmonic capacities appear seriously abnormal, namely, home knowledge library has not corresponding strategies (antibodies) to solve harmonic exceeding limit problems, it has to seek scheme solving from other agents through ADS data fields and do immune tolerance. If foreign ImA, say, PCC agent *a*, has gained control right by comparison of the results and sends back information that at the moment harmonic capacities of PCC *A* are enough lower to be able to accommodate harmonic capacities of agent *b*, so problems shall be solved. And inversely, not solved, home ImA must cooperate with all agents, say, dynamic coordinating harmonic limit values of agent *c, d* and *e* so as to ensure allowed harmonic capacities in PCC *A* of power system. And the achieved experience as vaccinia is sent to other Immune agents over ADS, meanwhile, native knowledge base is also updated. This belongs to minor problem event.

3)Assume that ImA *a* senses that harmonic information of PCC become lower due to technology improvement of electrolytic aluminum plant *F*, it may inform this news to other agents and cooperate with them to coordinate each knowledge base, i.e. harmonic limit value of agent *b,c,d,*and *e* may be enlarge to allow trains with larger harmonic capacities to pass. This shows their cooperation mechanism under dynamic environment is extremely effective. That is to say, knowledge library is constantly evolved according to environment, and not pre-fixed well at the very start.

6.2 Solution to Conflicts

If one time, agent *b* and agent *c* sense harmonic information at the same time, and they broadcast information *Xb* and *Xc* to network, thus, parallel conflicts occur, if their priority levels *x2* are not equal, ImA can determine which one to save first in accordance with priority level, otherwise, if their priority levels are equal, especially anti-priority levels, (i.e., overflow vs underflow) occur simultaneously, ImA can not decide which one should be saved first. This deadlock between ImAs is called "infinite loop". To resolve these deadlocks, decisions were made based on following control flow course, in the first step, each ImA detects that there are any malfunctioning substations or ImAs. If ImA detects that several abnormal ImAs occur simultaneously, each ImA compares evaluation points ($x2$ to $x13$) and select the largest evaluation points of an ImA. Then ordinary ImAs select the cooperative mode and calculate control parameters according to prediction of over/underflow of power quality levels in substation. The critical points of saving method for simultaneous abnormal ImAs are illustrated as follows:

- Save abnormal ImA according to the degrees of urgency level.
- Decide the ImA to save according to large number of evaluation points in turn.
- Ordinary ImA calculates control *P* in cooperative mode until abnormal ImAs are dismissed.

7 Conclusions

Cooperative ImAs-based monitoring networks model for electrified railway is a tight cohesion and quite incompact self-organizing systems, a high decentralized and distributed autonomous networks model. Meanwhile, ADS is also applied to act as detecting networks model, all kinds of heterogeneous data and networks can be integrated into a wide-area large-scale complicated system not to influence individual primary work, whereas, system robustness and fault tolerance abilities are boosted up. All kinds of existing expert systems such as distribution power systems reconstruction and recovery as well as substation automation etc can be combined into a larger system to realize more complicated function.

References

1. Jennings NR.(1998) A roadmap of agent research and development. Auto Agents & Multi-Agent System 1: 7-38
2. Mori K.(1984) A proposal of the autonomous decentralized system concept. IEICE-C 104: 303-310
3. De Castro LN, Von Zuben, FJ (1999) Artificial immune systems: Part I-Basic Theory and Applications. Technical Report-RT DCA 1, pp.89
4. Wei W, Zhan, G(2002) Artificial immune system and its applications in control system. Control Theory and Appl.19:157-166

5. Dong HY, Bai JS, Xue JY (2003) Approach for substation fault diagnosis based on multi-agent. Advanced Technology of Electrical Engineering and Energy 22:21-24
6. Zhao, LP, Xie SF, Li QZ (2004) Study on multi-agent based power quality monitoring and control system for electrified railway. Power System Technology 28: 69-73
7. Su HS, Li QZ (2005) Study on substation fault diagnosis approach based on multi-immune-agent. In: Guan ZC (eds) Proceedings of the XIVth International Symposium on High Voltage Engineering, Tsinghua University Press, Beijing, pp.427
8. Jerne NK (1974) Towards a network theory of the immune system. Ann Immunology 125: 373-389

BF Hot Metal Silicon Content Prediction Using Unsupervised Fuzzy Clustering

Luo Shi-hua[1,2] and Zeng Jiu-sun[2]

[1] Department of mathematics, Ningbo University, Ningbo, 315211, China
luoshihua@nbu.edu.cn
[2] Institute of System Optimum Technique, Zhejiang University, Hangzhou 310027, China
superman100f@yahoo.com.cn

Abstract. Since a similar series of events that lead to a similar result would be clustered together. The switches from one stationary state to another, which are usually vague and not focused on any particular time point, are naturally treated by means of fuzzy clustering. A unsupervised fuzzy clustering approach is established in this paper for predicting silicon content in molten iron which collected online from No.7 BF at Handan Iron and Steel Group Co.. This new approach consists of five steps: step 1 rearranges the silicon content ([Si]) time-series into sliding widows after confirming the best value for the dimension of the temporal patterns; step 2 Clusters the temporal patterns into an optimal number of fuzzy sets, find the degree of membership of each temporal pattern in each cluster; step 3 groups similar temporal patterns together into clusters, which may represent the different states of the dynamic system, by an unsupervised fuzzy clustering procedure; step 4 Fits a prediction model (AR) to each cluster; step 5 predicts the future samples of [Si] by a fuzzy mixture of the above prediction models weighted by the degree of membership of the latest temporal pattern in each of the corresponding clusters. The rate of hit shot of [Si] is 86% in [Si] $\pm 0.1\%$ range using such new algorithm which only using the last [Si] time series.

Keywords: Silicon Content Prediction, Unsupervised Fuzzy Clustering.

1 Introduction

The main purpose of a blast furnace (BF) is to chemically reduce and physically convert iron oxides into liquid iron called "hot metal". Such process is highly complicated; whose operating mechanism is characteristic of nonlinearity, time lag, high dimension, big noise and distribution parameter etc [1]. It has not come true to realize automation of BF ironmaking process in metallurgical technology from the eighteenth of the twentieth century after trying methods of classical cybernetics and modern cybernetics, because of its complexity and no appropriate mathematical models of BF ironmaking process. The quality and the quantity of the different input material as well and many environment factors all influence the quality of the molten iron. Not only is silicon content in molten iron an important quality variable, it also reflects the internal state of the high-temperature lower region of the BF [2], uniformity in silicon content and its accurate and advance prediction can greatly help to stabilize BF operations. In past years, efforts have been made to build up effective model to predict

B.-Y. Cao (Ed.): Fuzzy Information and Engineering (ICFIE), ASC 40, pp. 411–418, 2007.
springerlink.com

silicon content in molten iron [3-7]. But designing a predictive controller, which can forecast accurately silicon content in molten iron ([Si]) is still a puzzle. We have designed a new fuzzy algorithm that has good performance [8], but such algorithm needs more information which some time can not be obtained integrally.

In this paper, an unsupervised fuzzy clustering is established to predict [Si] which only using the last time series of [Si]. Clustering is one of the common methods used for finding a structure in given data in particular for finding structure related to time. Applying the clustering methods to a set of continuously sliding windows of [Si] is useful for grouping similar temporal patterns which are dispersed along the signal. While most of the common system identification methods (like ARMA or NN) train their prediction model by using the previous sample-window [9], we suggest training a separate prediction model for each cluster, using all of its accrued temporal patterns. By this means, on the one hand we can capture similar rare events and fit them into their own model, while, on the other hand, models fitted onto similar common events will be unbiased by these rare events. The switches from one stationary state to another, which are usually vague and not focused on any particular time point, can be naturally treated by means of fuzzy clustering. An adaptive selection of the number of clusters can overcome the general non-stationary nature of the signals [10].

"Hard" hierarchical clustering methods are very well known methods for recursively partitioning sets to subsets [11]. In here, we will use an algorithm that were recently presented for a natural "soft" top-down hierarchical partition, which fully follows the fuzzy sets rules by means of unsupervised fuzzy clustering.

2 General Scheme and Principle of the Methods

The clustering procedure can be applied directly to the continuous overlapping windows of the sampled raw data. In the clustering phase of the algorithm, similar temporal note that the procedure can be partially applied and patterns from the past are grouped together into clusters, while in the next stage a separate model is fitted for each cluster of similar patterns. This prediction method combines unsupervised learning in the clustering phase and supervised learning in the modeling phase.

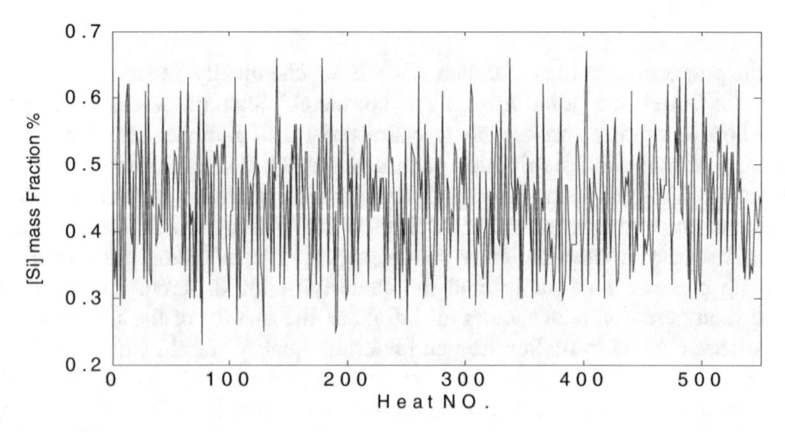

Fig. 1. Time series graph of the [Si] in No.7 BF at Handan Iron and Steel Group Co

Time series data of the [Si] in hot metal which gain from the Intelligent Control Expert System on 2000m3 BF Handan Iron & Steel Co. in China are taken as sample space with the capacity of 550 tap numbers and sampling time interval of 2h. The observed values of such samples are shown as Fig.1:

2.1 Established Temporal Patterns

If we have L samples of a time series of [Si], $[Si]_n \in X$, $n=1,2,\ldots,L$,we want to predict its d-th sample ahead $[Si]_{L+d}$. Fist, we must construct the N-dimensional data set for clustering from the following P temporal patterns (column vectors):

$$X_i = ([Si]_i, [Si]_{i+1}, \ldots, [Si]_{i+N-1}) \in R^N, \; i = 1, 2, \ldots, P, \tag{1}$$

Where $P=L-N-d+2$. In here, how to confirm the best value for the dimension of the temporal patterns (N) is an open problem.

In the ironmaking process, solid raw material such as iron ore, coke and limestone are charged to the top of the blast furnace, and preheated air is blown into the furnace from the bottom through the tuyeres. The time cost from solid raw material to molten hot metal is about 6h and the time sample [Si] is about interval of 2h. We can hypothesize that feeble influence is being between $[Si]_{i-4}$ and $[Si]_i$ [12]. By calculating the sequence correlation, we can proof that our hypothesis is legitimacy [13].

2.2 Fuzzy Clustering by State Recognition

By 2.1, we have gained some temporal patterns. Then, we Cluster the temporal patterns into an optimal number of fuzzy sets, K, i.e., find the degree of membership, $0 \le u_{j,i} \le 1$, of each temporal pattern X_i, $i=1,2,\ldots,P$, in each cluster, $j=1,2,\ldots,K$, such that :

$$\sum_{j=1}^{k} u_{j,i} = 1, \quad i = 1, 2, \cdots, P. \tag{2}$$

Because in practical application, BF operators describe the range of [Si] as five classes, normally, they consider that the lowest 25% [Si] to be low, the next 60%to be on the appropriate values and the next 15% as high values (Note that if better operating achieved, the percentage may be as20%, 70%, 10%). So the value of K is 3×3×3=27 (Sorting of them as (L, L, L), (L, L, M), (L, L, H), (L, M, L), (L, M, M), (L, M, H) etc) and table 1 gives its fuzzy status.

Table 1. Fuzzy status of [Si]

Fuzzy status	Low (L)	Moderate (M)	High (H)
Range of [Si]	[0.2, 0.37]	[0.37, 0.53]	[0.53, 0.7]

Then, assign each status a fuzzy membership function μ. The shape of each membership function is triangular, one vertex lies at the center of the status and has membership value unity; the other two vertices lie at centers of the two neighboring status respectively [14]. For the ith temporal patterns such as (1), let:

$$u_{j,i} = \mu_j([Si]_i) \cdot \mu_j([Si]_{i+1}) \cdot \mu_j([Si]_{i+2}), \ (j=1,2,\ldots,K) \tag{3}$$

2.3 Fit a Prediction Model to Each Fuzzy Cluster

In this part, we fit a prediction model to each fuzzy cluster, Using only similar temporal patterns for the training of each model simplifies the learning Time Series task and enables unbiased prediction. First, we must find the set of its "maximal members", i.e., the set of temporal patterns that have the maximal degree of membership in the j-th cluster, $A_j=\{X_i \ /i \in J_j\}$, $j=1,2,\ldots,K$, and a column vector of the corresponding predictions, $b_j=\{[Si]_{I+N-1+d} \ /i \in J_j\}$, $j=1,2,\ldots,K$, as a learning set, where

$$J_j = \left\{ i \,|\, u_{j,i} = \max_{k \in \{1,2,\cdots,K\}} (u_{k,i}),\ i=1,2,\cdots,P-1 \right\}. \tag{4}$$

J_j is the set of indices of the temporal patterns that have the maximal degree of membership in the j-th cluster. Being such processes, we can see that each temporal pattern is a member in one and only one, set of maximal members.

Second, we fit a prediction model (AR) of order 3, $b_j = c_j \cdot A_j$, to each cluster, $j=1,2,\ldots,K$, each of the cluster can estimate 3-dimensional coefficients (row) vectors c_j by $c_j = b_j \cdot pinv(A_j)$, "pinv" means generalized inverse of a matrix [15].

2.4 Predicting by a Combination of the Models

Now, we can predict the d-th sample ahead $[Si]_{L+d}$ by a fuzzy combination of all the prediction models, which have gained in the previous 2.3. Using the degree of membership of the last pattern $u_{j,P}$ in all the clusters, $j=1,2,\ldots,K$. For weighting all of the models' prediction results, the lastly prediction expressions is established:

$$[Si]_{L+d} = \sum_{j=1}^{K} u_{j,P} \cdot c_j \cdot X_P \tag{5}$$

3 Simulation of Predictive Algorithm

Using above model to predict the value of sample data that are shown as Fig.1, the data set contains 550 data pairs, In this paper, we only predict the 1-th sample ahead

$[Si]_{L+1}$, using 3D temporal patterns (N=3). The Fig.2 shows the original status of the 3D temporal patterns. We can see some infrequence evens on the edge of the figure.

The first 500 points of the data pairs were used as training data, and the final 50 points were used as test data. Two cases were simulated: (1) 250 training data (from 251 to 500) were used to confirm c_j; (2) 500 training data (from 1 to500) were used. Figs. 3 and 4 show the result of each case. Comparing Figs.3 and 4, we can find that we obtain an evidently improved prediction. It indicates that if similar temporal patterns for training are all too small, the fitting model of each cluster may be has big error, so the prediction results cannot be well. We also have simulated another tow cases which using 700 and 900 training data, we only obtain slightly improved prediction. It indicates that too big training data may be bringing more noise to model.

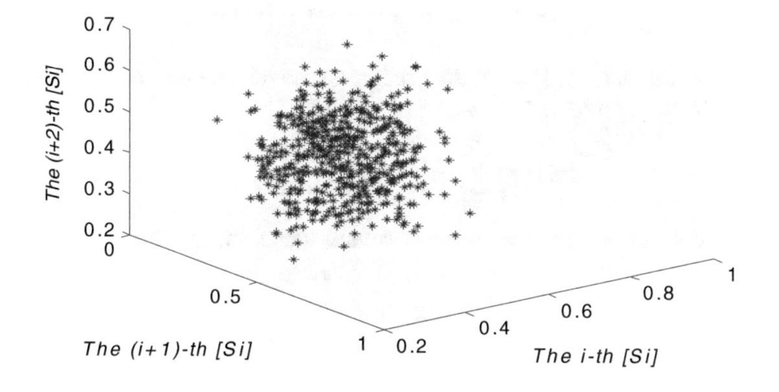

Fig. 2. The original status of the 3D temporal patterns

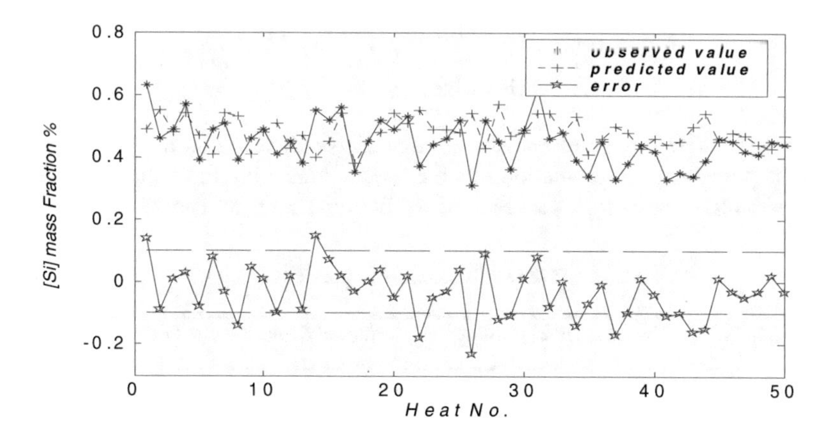

Fig. 3. The predictive result of case (1)

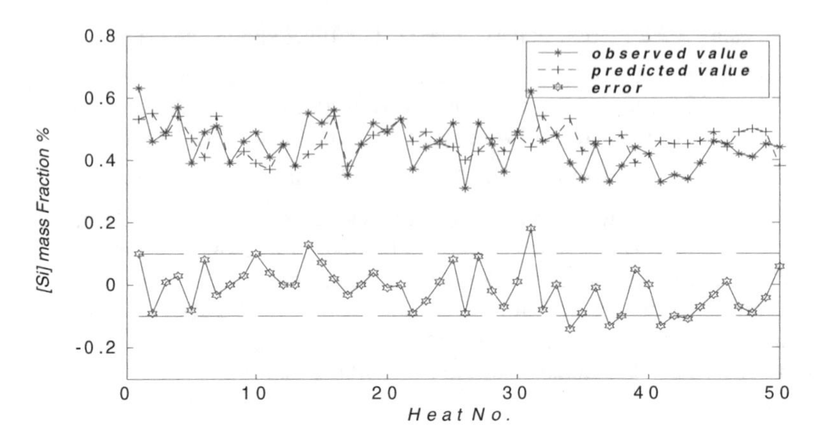

Fig. 4. The predictive result of case (2)

To evaluate the performance of the prediction model, some important criteria used in practice are considered as follows:

$$\mathrm{Perr} = \sum_{j=1}^{N_p} (x'_j - x_j)^2 / \sum_{j=1}^{N_p} x_j^2 \ , \tag{6}$$

where x_j' is the predicted value, x_j the observed value and N_p the total predicted tap numbers.

$$J = \frac{1}{N_p} (\sum_{j=1}^{N_p} H_j) \times 100\% \ , \tag{7}$$

where

$$H_j = \begin{cases} 1 & \text{where} \left| x'_j - x_j \right| \le 0.1 \\ 0 & \text{else.} \end{cases} \tag{8}$$

Generally, if H_j equals to one we say that the prediction hits the target, and J denotes the percentage of prediction hits the target. Analyzing the predicted values and observed values, we can get the result of prediction, shown as Table 2:

Table 2. The result of prediction

Group No.	The number of training	Standard deviation	Perr	J
Case (1)	250	0.087	0.0385	72%
Case (2)	500	0.089	0.0261	86%

With these criteria, the percentage of prediction hitting the target is 84% and *P*err is in the magnitude of 10^{-2} in case (2), which is helpful for operator to make right decision to operate blast furnace.

4 Conclusions and Discussion

The main conclusions in this paper are: (1) A method for dynamic system identification by unsupervised fuzzy clustering has established, which algorithm have two main advantages. First, it finds and groups similar events (even some infrequence evens) in all the "history" of the time series, and avoids the use of non-relevant information that can bias the prediction results. Second, it uses only this minimal required number of similar temporal patterns of [Si], improves the robustness and reduces the computation time of any prediction algorithm that is used. (2) Finds some skills to confirm the best value for the dimension of the temporal patters, the value of K and each $u_{j,i}$. (3) The essence of such method is a hierarchical-fuzzy algorithm, so it tries to exploit the advantages of hierarchical clustering while overcoming its disadvantages by mean of fuzzy clustering. The hierarchical partition helps to fathom the inner structure of the data, and thus enables multiscale compression and reconstruction of the data. (4) The new algorithm was applied to predict [Si] only using the last [Si] time series, and good performance is shown due to the high percentage of prediction hitting the target.

Future research on this project should focus on the following issues: (1) The essence of criteria for temporal pattern classification is the operators experience, how to establish more effective criteria should be further investigated. (2) How to decide the number of training data to get the best result in reasonable computing time. (3) Prediction of other important output variables, such as the hot metal temperature and the slag basicity.

Acknowledgements

This work described is partially supported by the grants from the Ministry of Science and Technology Foundation of China (No. 2005EC000166) and Ningbo Natural Science Foundation (No.2006A610032).

Refences

1. Biswas AK (1984) Principles of blast furnace ironmaking. SBA Publication, Calcutta
2. Chen J (2001) A Predictive System for Blast Furnaces by Integrating a Neural Network with Qualitative Analysis. Engineering application of artificial intelligence 14: 77-85
3. Gao CH, Zhou ZM, Shao ZJ (2005) Chaotic Local-region Linear Prediction of Silicon Content in Hot Metal of Blast Furnace. Acta Metall. Sin 41: 433-436
4. Yao B., Yang TJ, Ning XJ (2000) An Improved Artificial Neural Network Model for Predicting Silicon Content of Blast Furnace Hot Metal. Journal of University of Science and Technology Beijing 7: 269-272

5. Miyano T, Kimoto S, Shibuta H. et. al. (2000) Time Series Analysis and Prediction on Complex Dynamical Behavior Observed in a Blast Furnace. Physica D 135: 305-330
6. Jose AC, Hiroshi N, Jun-ichiro Y (2002) Three-dimensional Multiphase Mathematical Modeling of the Blast Furnace Based on the Multifluid Model. ISIJ International 42: 44-52
7. Juan J, Javier M, Jesus SA, et. al.(2004) Blast Furnace Hot Metal Temperature Prediction Through Neural Networks-based Models. ISIJ International 44: 573-580
8. Luo SH, Liu XG, Zhao M (2005) Prediction for silicon content in molten iron using a combined fuzzy-associative-rules bank. Lecture Notes in Artificial Intelligence 3614: 667-675
9. Gath I, Geva AB (1989) Unsupervised Optimal Fuzzy Clustering. IEEE Trans. on Pattern Anal. Machine Intell. 7: 773-781
10. Geva AB (1998) Feature Extraction and State Recognition in Biomedical Signals with Hierarchical Unsupervised Fuzzy Clustering Methods. Medical & Biological Engineering & Computing 36: 608-614
11. Amir BG (1999) Non-stationary Time-Series prediction Using Fuzzy Clustering. Fuzzy Information Processing Society 99: 413-417
12. Liu XG, Liu F (2003) Optimization and Intelligent Control System of Blast Furnace Ironmaking Process. Metallurgical industry Press, Beijing
13. Chen XR (2000) Probability Theory and Mathematical Statistics. Science Press, Beijing
14. Wang LX, Mendel JM (1992) Generating Fuzzy Rules by Learning from Examples. IEEE Trans. on Systems, Man and Cybern. 22: 1414-1427
15. Hanselman D (2002) Mastering Matlab 6. Tsinghua University Press, Beijing

Speech Emotion Pattern Recognition Agent in Mobile Communication Environment Using Fuzzy-SVM

Youn-Ho Cho, Kyu-Sik Park, and Ro Jin Pak

Dankook University Division of Information and Computer Science San 8, Hannam-Dong, Yongsan-Ku, Seoul Korea, 140-714
{adminmaster,kspark,rjpak}@dankook.ac.kr

Abstract. In this paper, we propose a speech emotion recognition agent in mobile communication environment. The agent can recognize five emotional states - neutral, happiness, sadness, anger, and annoyance from the speech captured by a cellular-phone in real time. In general, the speech through the mobile network contains both speaker environmental noise and network noise, thus it can causes serious performance degradation due to the distortion in emotional features of the query speech. In order to minimize the effect of these noises and so improve the system performance, we adopt a simple MA (Moving Average) filter which has relatively simple structure and low computational complexity. Then a SFS (Sequential Forward Selection) feature optimization method is implemented to further improve and stabilize the system performance. For a practical application to call center problem, we created another emotional engine that distinguish two emotional states - "agitation" which includes anger, happiness and annoyance, and "calm" which includes neutral and sadness state. Two pattern classification methods, k-NN and Fuzzy-SVM, is compared for emotional state classifications. The experimental results indicate that the proposed method provides very stable and successful emotional classification performance as 72.5% over five emotional states and 86.5% over two emotional states.

Keywords: Fuzzy-SVM, Mobile communication, Speech Emotion Recognition, MA filtering, SFS.

1 Introduction

Recently, the problem of speech emotion recognition has gained increased attention, because the human-computer interaction (HCI) grows its importance in ubiquitous communication services. Besides human facial expressions, speech has proven as one of the most promising media for the automatic recognition of human emotions.

Most of speech emotion classification has two stages of a pattern recognition problem: feature extraction and classification based on the selected feature. Depending on the various combinations of these stages, several strategies are employed in the literature. Dellaert et al. [1] used 17 features and compared three classifiers: maximum likelihood Bayes classification, kernel regression and k-NN (Nearest Neighbor). They reached 60% - 65% accuracy with four

B.-Y. Cao (Ed.): Fuzzy Information and Engineering (ICFIE), ASC 40, pp. 419–430, 2007.
springerlink.com

emotion categories. Scherer [2] extracted 16 features by the jack-knifing procedure and achieved an overall accuracy 40.4% for fourteen emotional states. As a first direct attempt for emotion recognition in call center application, Petrushin [3] suggested neural network classifier with pitch, first and second formant, energy and the speaking rate as emotional features and they distinguish between two emotional states - agitation and calm with the accuracy of 77%. Yacoub et. al. [4] focused on distinguishing anger versus neutral speech for call center applications and they achieved a maximum accuracy of 94%. However these papers did not report any analysis and results with an environmental and network noise effect in call center application. Other good works on general speech emotional classification can be found in [5-7].

Although many combinations of emotional features and classifiers have been evaluated in those works, little attention has been paid on speech emotion recognition in mobile service environment. Previous methods are tend to fail when the query speech signal contains background noises and network errors as in mobile environment.

In contrast to previous works, this paper focuses on the following issues of the emotion recognition problem. Firstly, the proposed system accepts query sound captured by a cellular phone in real-time using INTEL Dialogic D/4PCI board and an agent can recognize five emotional states - neutral, happiness, sadness, anger, and annoyance. For a practical application, another emotional engine was implemented to distinguish two emotional states - "agitation" which includes anger, happiness and annoyance, and "calm" which includes neutral and sadness state - which is salient to call center application. This kind of system allows automatic call routing of angry customer and provide a feedback to an operator or a supervisor for monitoring purposes. Secondly, in order to minimize the effect of mobile noises and so improve the system performance, we adopt a simple MA (Moving Average) filter which has relatively simple structure and low computational complexity. Thirdly, a SFS (Sequential Forward Selection) feature optimization method is implemented to further improve and stabilize the system performance. Aim to our implementation we choose Fuzzy-SVM[8] as our classification algorithm and compared the performance with simple k-NN[9] algorithm.

This paper is organized as follows. Section 2 describes proposed mobile-based emotion recognition system. Section 3 explains the methods of noise-robust feature extraction and optimization. Section 4 compares experimental results of the proposed system. Finally, a conclusion is given in section 5.

2 Proposed System

The proposed system is illustrated in Fig. 1. The system consist 3 stages - speech signal acquisition, emotional classification, and reporting for the resulting emotional state. From the figure, a query speech signal is picked up by the single microphone of the cellular phone and then transmitted to the emotion recognition server. The transmitted signal is acquired in real-time by using INTEL dialogic D4PCI-U board in 8 kHz sampling rate, 16 bit, MONO. Then the queried speech is classified and the classification result is reported and summarized.

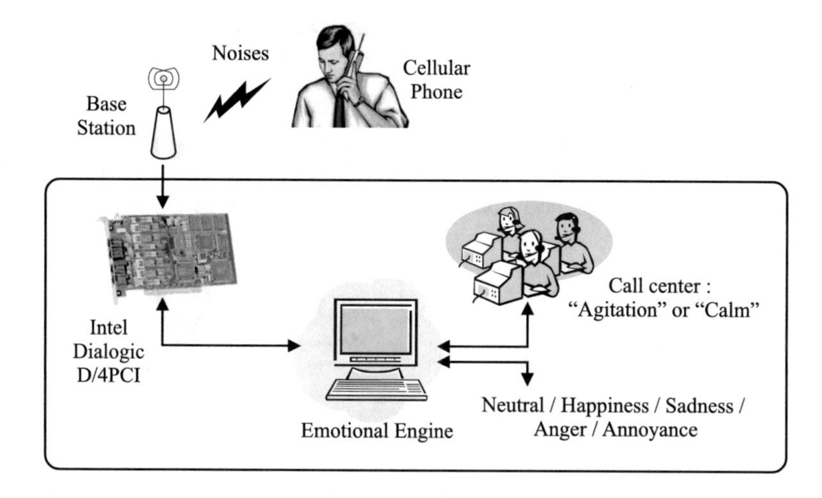

Fig. 1. Proposed mobile-based speech emotion recognition agent

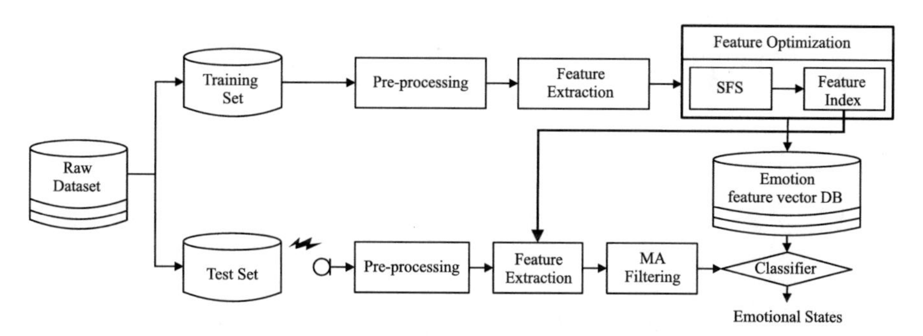

Fig. 2. Proposed emotional engine

Fig. 2 shows the block diagram of the proposed emotional engine. The proposed algorithm consists of 2 phases - training and testing.

Raw speech data set is randomly divided into two sets of data set - training set and test set. At the training phase, feature set as defined in section 3.2 is extracted from preprocessed speeches in training set. Then a SFS feature optimization is applied to select only those features that are most significant in the system classification performance with respect to k-NN and SVM classifier. These feature set is used to build up emotional feature vector DB. At the testing phase, a test input speech is picked up by a microphone and preprocessed as before, and the same indexed feature set from the SFS optimization stage is extracted. Then the 5th order MA (MA(5)) filter is applied to feature vector to minimize the effect of background noise. Finally the k-NN, Fuzzy-SVM pattern classifier is tested for the classification of queried speech and the performance of the system is evaluated.

3 Emotional Feature Extractions and Optimization

3.1 Pre-processing

Speech pre-processing stage is required to build up the robust emotional features and it consists of a BPF (Band Pass Filtering), 20 ms speech segmentation, hamming windowing, and end-point detection procedure as shown in fig. 3.

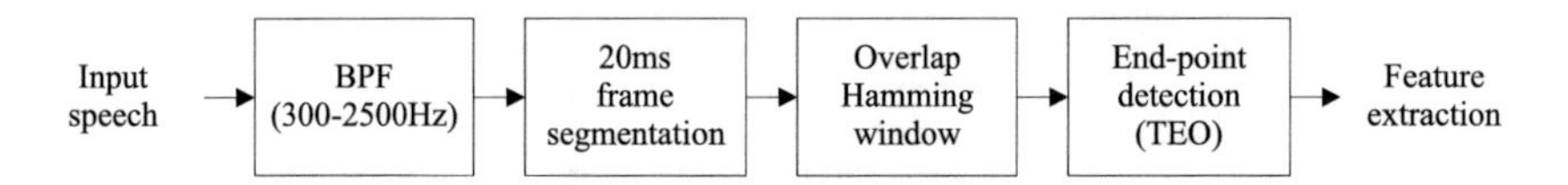

Fig. 3. Speech pre-processing procedure

The input speech signal is first bandpass filtered with a passband 300 Hz - 2500 Hz. The reason for this BPF is clear if we compare the spectrogram of the original clean speech and the speech acquired from the cellular-phone as in fig. 4. As seen from the figure, the speech through the mobile network undergoes the two main distortions - bandwidth reduction to 300 Hz-2500 Hz and mobile noise. Here the BPF allows extracting only the meaningful signal portion of the noisy speech.

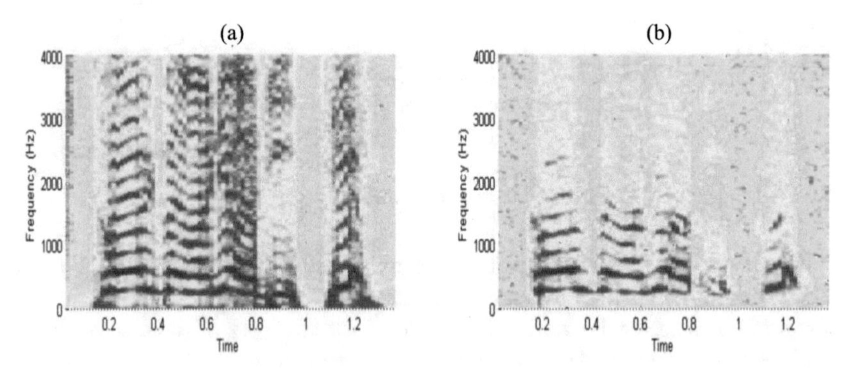

Fig. 4. Spectrogram comparison of (a) clean speech and (b) speech through mobile network

At the sampling rate of 8 kHz, the filtered signal is divided into 20ms frames with 50% overlapped hamming window at the two adjacent frames. Then the endpoint detection algorithm is applied to distinguish voiced and nonvoiced segment from the input speech so that the emotional features are extracted only from the voiced portion of speech. In this paper, we used a Li. Gus [10] endpoint detection algorithm. This algorithm is based on TEO (Teager Energy Operator) and energy entropy, and it is known very effective when the speech is in high background noise.

3.2 Emotional Feature Extraction

Two types of emotional features are computed from every 20 ms frame - one is the prosodic features such as pitch and energy, and the other is speech phoneme feature such as Mel Frequency Cepstral Coefficients (MFCC). The means and standard deviations of these original features and their delta values are computed over each frame to form a total of 32 dimensional feature vector. Followings are short descriptions of the features used in this paper.

A few pitch detection algorithms including HPS[11], AMDF[12], SHR[13] are tested and compared in noisy speech environment. Among them, SHR algorithm shows good robust characteristics against mobile noise and so the SHR is used in this paper. SHR (Subharmonic-to-Harmonic Ratio) transforms the speech signal into the FFT domain, and then decides two candidate pitches by detecting the peak magnitude of the log spectrum. The final pitch is determined by comparing SHR to a certain threshold. If SHR is less than a threshold value, then is chosen as the final pith. Otherwise is selected. Here DA( ) is difference function [11].

$$SHR = 0.5\frac{DA(log f_1) - DA(log f_2)}{DA(log f_1) + DA(log f_2)} \tag{1}$$

Short time energy of the speech signal provides a convenient representation that reflects amplitude variations. The short time energy is defined as

$$E_n = \sum_{m=-\infty}^{\infty} x^2(m) \cdot h(n-m) \tag{2}$$

MFCC is the most widely used feature in speech recognition. It captures short-term perceptual features of human auditory system. In this paper, 6th order MFCC is used.

3.3 SFS Feature Optimization

Not all the 32-dimensional features in the previous section are used for emotional state classification purpose. Some features are highly correlated among themselves and some feature dimension reduction can be achieved using the feature redundancy. In order to reduce the computational burden and so speed up the search process, while maintaining a system performance, an efficient feature dimension reduction and selection method is desired. In ref. [14], a sequential forward selection (SFS) method is used to meet these needs. In this paper, we adopt the same SFS method for feature selection to reduce dimensionality of the features and to enhance the classification accuracy. Firstly, the best single feature is selected and then one feature is added at a time which in combination with the previously selected features to maximize the classification accuracy [14]. This process continues until all 32 dimensional features are selected. After completing the process, we pick up the best feature lines that maximize the classification accuracy. We note that SFS method described here allows not only choosing the best features, but also it helps to stabilize the system performance.

3.4 MA (Moving Average) Filter to Minimize Noise Effect

In general, the speech through the mobile network contains mobile noises such as speaker environmental noise and network noise, thus it can causes serious performance degradation due to the distortion in emotional features of the query speech. In order to minimize the effect of these noises and so improve the system performance, we adopt a simple MA (Moving Average) filter that has relatively simple structure and low computational complexity.

MA filter is essentially a lowpass filter that can smooth out any spikes in the time sequence. As will be seen, this simple filtering operation results in further significant improvements. The key idea is to apply MA filter to the feature sequences on feature domain to smooth out spikes due to the mobile noises. Even though in clean speech the spikes might contain important information about the speech utterance, in noisy speech these spikes are more likely to be caused by noise. In order to apply MA filter, we represent feature data by a $T \times D$ dimension matrix FM_{td} as follow.

$$FM_{td} = \begin{vmatrix} x_{1,1} & x_{1,2} & \dots & x_{1,D} \\ x_{2,1} & x_{2,2} & \dots & x_{2,D} \\ \dots & \dots & \dots & \dots \\ x_{T,1} & x_{T,2} & \dots & x_{T,D} \end{vmatrix}_{T \times D} \tag{3}$$

Here *t=1,2,...T* represents the number of frames in time sequence and is the dimension of the feature space. In other words, each row of represents a feature vector and each column represents a time sequence. Then the features in column order are normalized to zero mean and unit variance in order to avoid numerical problems caused by small variances of the feature values. This normalized time sequence of features is further processed by MA filter as follows.

$$\hat{x}_{td} = \frac{1}{M} \sum_{i=0}^{M} x_{(t-i),d}, x_{(t-i),d} = feature\ coefficients \tag{4}$$

In this paper, the filter order M=5 was found to yield the best results for our experiments. An example is illustrated in figure 5, which shows 1st MFCC coefficient change over the frame index. Fig. 5 (a) is the feature sequence that was extracted from clean speech and fig 5(b) show the feature sequence from the noisy speech captured by cellular phone. The resulting features after MA(5) filtering operation to fig. 5(b) is shown in fig 5(c). As we compare the figures, the lowpass MA filter smoothes out the feature sequence toward temporal similitude, further minimize the differences between the clean (fig 5(a)) and the noisy (fig 5(b)) feature sequences. Again, as seen from the fig. 5, the simple feature post-processing step has a good positive influence on minimizing noise effect. We note that MA(2) filter is also applied to the features when we setup the emotional DB as shown in figure 2 and the reason for this is to minimize the differences in feature coefficients between the query and DB due to the MA filtering operation.

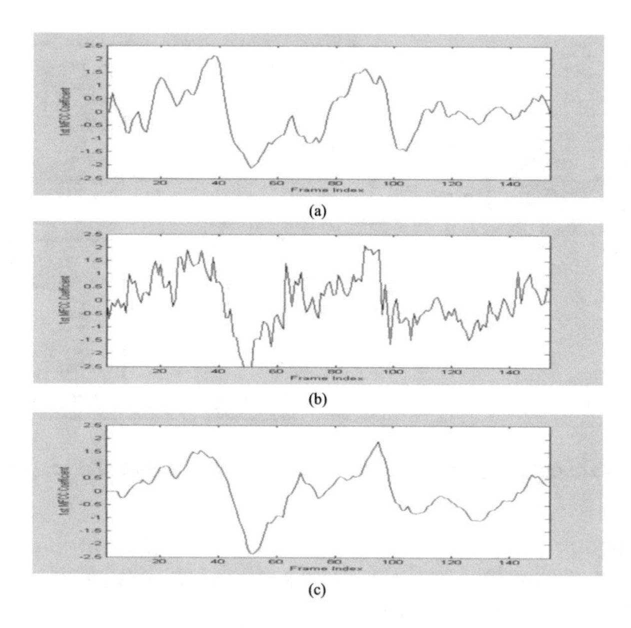

Fig. 5. Alleviation of feature distortion using MA filter: (a) 1st MFCC coefficient over the frame index for the clean speech, (b) feature sequences from the noisy speech captured by cellular phone, (c) resulting features sequences by applying the MA(5) filter to (b)

4 Experimental Results

4.1 Speech Database and Experimental Setup

In our experiment, we used the database from the Yonsei University [15]. The original data set includes short utterances by fifteen semi-professional actor and actress and it consists of total of 5400 utterances across 5 emotional states, i.e., normal, joy, sadness, anger and annoyance in 8 kHz, 16 bit recording. Among them, 2000 utterances (1000 utterances for each emotional state) were chosen as training set to define an optimal feature set by SFS optimization stage and to setup the emotional feature vector DB. Another 200 utterances are used as the test data set for system evaluation. Fig. 6 shows a block diagram of experimental setup. In order to compare the system performance, two sets of experiment have been performed.

One is the system (called proposed system) with a proposed method as in fig. 2 and the other is the system (called reference system) without any noise reduction technique and the feature optimization method. Here the proposed system works as follows. Firstly, a test speech is acquired through actual cellular phone via INTEL dialogic D4PCI-U board in 8 kHz sampling rate, 16 bit, MONO. Secondly, emotional features are extracted according to the feature index from the SFS feature optimization stage and then MA(5) filter is applied to feature

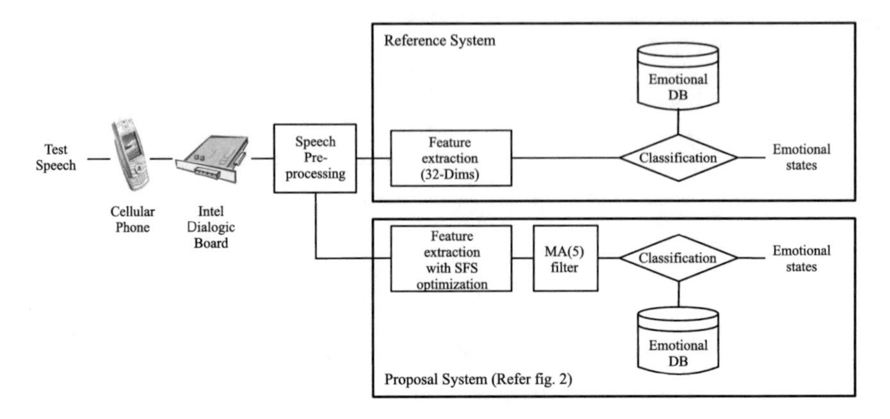

Fig. 6. Experimental setup for performance comparison of the proposed system

vector in order to minimize the effect of mobile noises. Finally the k-NN, Fuzzy-SVM pattern classifier is tested for the classification of queried speech and the performance of the system is evaluated.

4.2 SFS Feature Optimization Experiment

Fig. 7 describes the SFS feature optimization procedure to select best feature coefficients with k-NN and Fuzzy-SVM classifiers. It displays the average classification accuracy across five emotional states at each feature dimensions with k-NN/Fuzzy-SVM classifiers. From the figure, we see that the classification performance increases with the increase of features up to certain number of features, while it remains almost constant after that. Thus based on the observation of this boundary, we can select first 10 features up to the boundary and ignore the rest of them. In this way, we can determine the number of best feature sets for each classifier. For both k-NN/Fuzzy-SVM classifiers, we observed that MFCC feature is most significant in the classification result. We note that the SFS feature optimization procedures shown in fig. 7 is performed with respect to the clean speech instead of noisy speech. The reason for this is to prevent the feature variations due to the random mobile noise characteristics, and it is only used to choose the stable feature set for usage with the proposed mobile emotion recognition system. As we intuitively know, the less number of feature set is always desirable.

4.3 Classification Results for Five Emotional States

Table 1 compares the classification results for five emotional states between the proposed system and the reference system under the experimental setup as shown in fig. 6. We note that the classification accuracy for table 1 is measured with the noisy query, so this accuracy is a little bit lower than the ones shown in fig. 7.

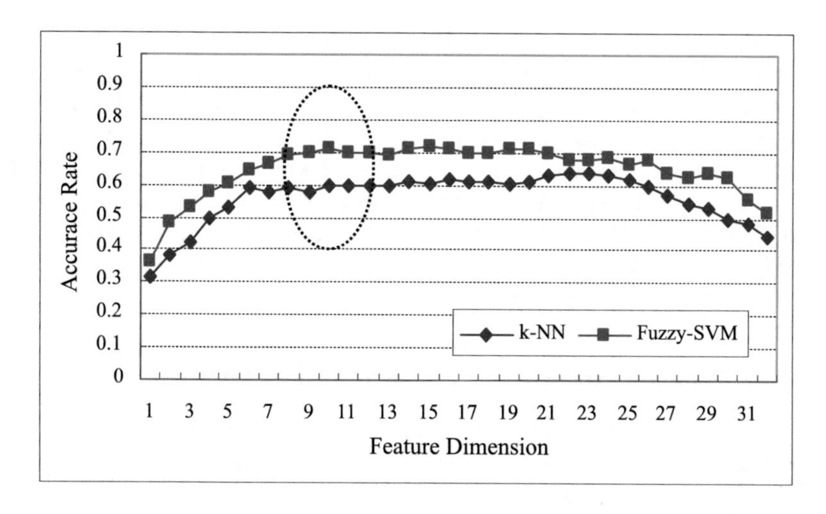

Fig. 7. SFS feature optimization procedure with k-NN/Fuzzy-SVM classifier

Table 1. Average classification results for five emotional states

	k-NN (Feature Dim.)	Fuzzy-SVM (Feature Dim.)
Reference system	44.5% (32)	51.5% (32)
Proposed system	63.5% (10)	72.5% (10)

Table 2. Classification accuracy in a form of confusion matrix (Fuzzy-SVM)

Query	Neutral	Happiness	Sadness	Anger	Annoyance
Neutral	37(4)	0(0)	1(0)	2(0)	0(0)
Happiness	3(11)	24(23)	4(4)	2(0)	7(2)
Sadness	4(14)	2(1)	25(24)	4(0)	5(1)
Anger	2(16)	4(2)	2(5)	25(3)	7(4)
Annoyance	1(5)	4(8)	0(14)	1(0)	34(13)
Average Classification Accuracy 72.5% (51.5%)					

As seen on the table 1, the proposed method achieves more than 20% higher accuracy than the reference system even with less number of feature dimensions.

Table 2 shows Fuzzy-SVM classification performance of the proposed system in a form of a confusion matrix. As a comparison purpose, the classification result of the reference system is included in the table and they are shown in parenthesis. The numbers of correct classification results lie in the diagonal of the confusion matrix.

From the table 2, we see much improvement of classification performance with proposed MA filtering and a SFS feature optimization. It actually can achieves more than 20% improvement even with fewer number of feature set. The proposed method works fairly well over the emotional states of neutral and annoyance while the average number of correct classification is little lower in happiness, sadness and anger state. Happiness and anger state is often misclassified as annoyance state. From the confusion matrix, we observe that the proposed method is very efficient especially to the anger and annoyance states where these were quite susceptible to the mobile noise.

4.4 Classification Results with Two Emotional States

Table 3 compares the classification results for two emotional states between the proposed system and the reference system under the experimental setup as shown in Fig. 6. As seen from table 3, we also see much improvement of classification performance in proposed system over the reference system. It actually can achieves more than 8% - 14.5% improvement even with fewer number of feature set derived from SFS feature optimization method.

Table 3. Performance comparison of proposed system for two emional states

	Reference System		Proposed System	
	k-NN	Fuzzy-SVM	English	Fuzzy-SVM
Classification Accuracy	67.5%	72%	73.5%	86.5 %

Table 4 compares the classification performance between the proposed system and reference system in a form of a confusion matrix. Here (a) and (b) shows the classification result with k-NN and Fuzzy-SVM classifier respectively for the reference system. On the other hand, (c) and (d) describes the results for the proposed system. The numbers of correct classification result lies in the diagonal term and misclassification statistics are listed in off -diagonal terms.

Table 4. Comparison of classification results in a form of confusion matrix. (a) reference system with k-NN, (b) reference system with Fuzzy-SVM, (c) proposed system with k-NN, (d) proposed system with Fuzzy-SVM.

	Reference System				Proposed System			
	(a) k-NN		(b) Fuzzy-SVM		(a) k-NN		(b) Fuzzy-SVM	
Query	Agitation	Calm	Agitation	Calm	Agitation	Calm	Agitation	Calm
Agitation	73	27	46	54	68	32	80	20
Calm	38	62	2	98	21	79	7	93
Average Accuracy	67.5%		72%		73.5%		86.5%	

As seen on table 4, in reference system, k-NN classifier shows the misclassification rate 27% in "Calm" state and 38% in "Agitation" state, while Fuzzy-SVM classifier shows the misclassification rate 54% in "Calm" state, but the "Agitation" state is easily recognizable with only 2% error. For the case of the proposed system, in k-NN classifier, the misclassification rate 32% for "Calm" state and 21% for "Agitation" state, so we see some improvements in classification of anger category. On the other hand for the Fuzzy-SVM classifier, we see a significant improvement over the "Calm" category where the misclassification rate decreases from 54% to 20%. From the table 4, we observe that the "Calm" state is quite susceptible to the noises from mobile network and environment, while the "Agitation" state is quite easily recognizable even under the mobile noise conditions.

5 Conclusion

In this paper, we propose a speech emotion recognition agent in mobile communication environment. The agent can recognize five emotional states - neutral, happiness, sadness, anger, and annoyance from the speech captured by a cellular-phone in real time. For a practical application to call center problem, we created another emotional engine that distinguish two emotional states - "agitation" which includes anger, happiness and annoyance, and "calm" which includes neutral and sadness state. In order to alleviate noise effect due to the mobile network and environment, we adopt a simple MA filter on the feature domain. SFS feature optimization is also utilized to improve and stabilize the system performance. The proposed system has been tested and compared with cellular phones in the real mobile environment and it shows a very stable and successful emotional classification performance as 72.5% over five emotional states and 86.5% over two emotional states. Future work will involve the development of new emotional features and further analysis of the system for practical implementation.

Acknowledgment

This work was supported by grant No. R01-2004-000-10122-0 from the Basic Research Program of the Korea Science & Engineering Foundation.

References

1. F. Dellaert, T. Polzin, and A. Waibel(1996) Recognizing emotion in Speech, In Proc. International Conf. on Spoken Language Processing, pp 1970-1973
2. Petrushin, V(1996) Adding the affective dimension: A new look in speech analysis and synthesis, In Proc. International Conf. on Spoken Language Processing, pp 1808-1811
3. K.R. Scherer(1999) Emotion recognition in speech signal: recognition and application to call centers, in Proc. Artificial Neural Networks in Engineering, pp 7-10, Nov.

4. S. Yacoub, S. Simske, X. Lin, and J. Burns(2003) Recognition of emotions in interactive voice response system, in Eurospeech 2003 Proc.
5. V. Kostov and S. Fukuda(2000) Emotion in user interface, Voice Interaction system, IEEE Intl Conf. on systems, Man, Cybernetics Representation, no. 2, pp. 798-803
6. T. M Oriyama and Oazwa(1999) Emotion recognition and synthesis system on speech, IEEE Intl. Conference on Multimedia Computing and Systems, pp. 840-844
7. C. M. Lee, S. Narayanan, and R. Pieraccini(2002) Classifying emotions in human-machine spoken dialogs, ICME'02
8. Shigeo Abe and Takuya Inoue(2002) Fuzzy Support Vector Machines for Multi-class Problems, ESANN'2002 proceedings, Bruges (Belgium), 24-26 April.
9. Anil K. Jin, Robert P.W. Duin, Jianchang Mai(2000) Statistical Pattern Recognition: A Review, IEEE Trans. Pattern Analysis and Machine Intelligence, vol. 22, no. 1, Jan.
10. Li. Gu and S. A. Zahorian(2002) A new robust algorithm for isolated word endpoint detection, ICASSP2002, Orlando, USA.
11. Noll, M.(1969) Pitch determination of human speech by the harmonic product spectrum, the harmonic sum spectrum, and a maximum likelihood estimate, In Proceedings of the Symposium on Computer Processing Communications, pp. 779-797
12. M.J. Ross, H.L. Shaer, A. Cohen, R. Freudberg, and H. J. Manley(1974) Average magnitude difference function pitch extractor, ASSP-22:353-362, Oct.
13. Xuejing Sun(2000) A pitch determination algorithm based on subharmonic-to harmonic ratio, ICSLP, pp. 676-679
14. M. Liu and C. Wan(2001) A study on content-based classification retrieval of audio database, Proc. of the International Database Engineering and Applications Symposium, pp. 339 - 345
15. Kang, Bong-Seok(2000) A text-independent emotion recognition algorithm using speech signal, MS Thesis, Yonsei University

Efficient Wavelet Based Blind Source Separation Algorithm for Dependent Sources

Rui Li[1,2] and FaSong Wang[3]

[1] School of Sciences, Henan University of Technology, ZhengZhou 450052, China
[2] Department of Mathematics, Zhenghou University, ZhengZhou 450001, China
slxlr@haut.edu.cn
[3] China Electronics Technology Group Corporation the 27th Research Institute, Zhengzhou 450005, China
fasongwang@126.com

Abstract. The purpose of this paper is to develop novel Blind Source Separation (BSS) algorithms from linear mixtures of them, which enable to separate and extract (Blind Signal Extraction (BSE)) dependent source signals. Most of the proposed algorithms for solving BSS problem rely on independence or at least uncorrelation assumption of source signals. However, in practice, the latent sources are usually dependent to some extent. On the other hand, there is a large variety of applications that require considering sources that usually behave light or strong dependence. The proposed algorithm is developed based on the wavelet coefficient representations using continuous wavelet transformation(CWT) which only requires slight differences in the CWT coefficient of the considered signals in the same scale. Moreover the proposed algorithm can extract the desired signals in the overcomplete conditions. Simulation results show that the proposed algorithm is able to extract the dependent signals and yield ideal performance.

Keywords: Blind Source Separation, Statistically Dependent, Continuous Wavelet Transformation, Independent Component Analysis, Overcomplete.

1 Introduction

Blind source separation(BSS) has attracted considerable attention in the signal processing and neural network fields, since it has rapidly growing applications in various subjects, such as speech processing, image enhancement, and biomedical signal processing [1, 2].

The classical problem of BSS consists of recovering source signals from several observed mixtures of them. The are obtained from a set of sensors, each receiving a different combination of source signals. The problem is called **blind** because no information is available about the mixtures and sources. This challenging research problem was studied comprehensively in the last decades especially for the case of independent sources which leads to the so called independent component analysis(ICA) following a precise mathematical framework of ICA was

B.-Y. Cao (Ed.): Fuzzy Information and Engineering (ICFIE), ASC 40, pp. 431–441, 2007.
springerlink.com

constructed by Comon in [3] and many algorithms were developed by researchers using the concept of contrast functions mainly based on minimization of mutual information (MMI) [3-5].

Most of the known and efficient algorithms assume that primary sources are statistically independent, stationary and non-Gaussian, however in many applications the sources are not completely independent. Due to these limitations, poor performance is often obtained when dealing with real sources, such as audio signals. Some authors [6-8] have proposed different approaches which take advantage of the nonstationarity of such sources in order to achieve better performance than the classical methods, but they still require their independence or uncorrelation.

In this paper we focus on the separation of dependent sources and propose an algorithm based on continuous Wavelet Transformation(CWT) and the method called Time Frequency Ratio Of Mixtures(TIFROM) [9]. Using WT is a desirable alternative to existing temporal and multi-spectral decorrleation methods, not least because efficient wavelet based CWT representations reduce the computational cost of covariance estimation, and are free from the cross-term issues associated with STFDs. Unlike classical BSS methods, this algorithm sets no conditions on the stationarity, independence or non-Gaussianity of the sources, the separation condition is that the sources have slight differences in the CWT coefficient of the considered signals. Its other attractive feature is that it can be used in underdetermined mixtures.

This paper is organized as follows: Section 2 introduces briefly the BSS model and the BSS indeterminacies; Then in section 3, we describe the CWT based BSS algorithm for dependent sources in detail; In section 4 and section 5, CWT based general BSS algorithm and the procedure of the proposed BSS algorithm are given; Simulations illustrating the good performance of the proposed method are given in section 6; Finally, section 7 concludes the paper.

2 BSS Model and Indeterminacies

For our purposes, the problem of BSS can be formulated as follows:

$$\mathbf{x}(t) = \mathbf{A}\mathbf{s}(t) + \mathbf{n}(t), \tag{1}$$

where $\mathbf{s}(t) = [s_1(t), s_2(t), \cdots, s_n(t)]^T$ is the unknown source vector, and the source signals $s_i(t)$ are dependent. Matrix $\mathbf{A} \in R^{m \times n}$ is an unknown nonsingular mixing matrix. The observed mixtures $\mathbf{x}(t) = [x_1(t), x_2(t), \cdots, x_m(t)]^T$ are sometimes called as sensor outputs and $\mathbf{n}(t) = [n_1(t), n_2(t), \cdots, n_m(t)]^T$ is a vector of additive noise.

The following assumptions for the model to be identified are needed: 1) The sources $s_1, \cdots, s_n$, are zero mean, i.e. $E(s_i) = 0$, and unit variance, i.e. $E(s_i^2) = 1$; 2) Here we consider that $\mathbf{n}(t) = 0$; 3) $\mathbf{A}$ has full column rank, i.e. rank($\mathbf{A}$)$= n$ and $a_{ij} \neq 0, \forall i, j$.

In BSS, complete identification of the mixing matrix is impossible because an exchange of a fixed scalar factor between a source signal and corresponding column of $\mathbf{A}$ does not affect the observations [3]. So, we can only get the separation signals $\mathbf{y}(t)$ satisfied

$$\mathbf{y}(t) = \mathbf{Q\Lambda s}(t) = (\lambda_1 s_{k_1}, \cdots, \lambda_n s_{k_n})$$

where $\mathbf{\Lambda} = diag\{\lambda_1, \lambda_2, \cdots, \lambda_n\}$ is a nonsingular diagonal matrix and $\mathbf{Q}$ is a permutation matrix. That is $y_i(t) = \lambda_i s_{k_i}(t)$,$1 \leq i \leq n$ and $(k_1, k_2, \cdots, k_N)$ is a permutation of $(1, 2, \cdots, n)$.

So there are two indeterminacies in BSS: (1) scaling ambiguity; (2) permutation ambiguity. But this does not affect the application of BSS, because the main information of the signals is included in the waveform of them.

3 CWT Based BSS Algorithm for Dependent Sources

As we use the linear instantaneous mixing model, the linear time-frequency representation(TFR) must be consider. Mathematically, TFR corresponds to a joint function $T_x(t,f)$ of time t and frequency f. We shall call $T_x(t,f)$ a "TFR" of the signal $x(t)$ [10]. All linear TFRs satisfy the superposition or linearity principle which states that if $x(t)$ is a linear combination of some signal components, then the TFR of $x(t)$ is the same linear combination of the TFRs of each of the signal components:

$$x(t) = c_1 x_1(t) + \cdots + c_n x_n(t) \Longrightarrow T_x(t,f) = c_1 T_{x_1}(t,f) + \cdots + c_n T_{x_n}(t,f)$$

Linearity is a desirable property in any application involving multi-component signals (e.g., speech). This property is well correspond to the linear BSS problem. Two linear TFRS of basic importance are the short-time Fourier transform (STFT) and the continuous wavelet transform(CWT).

A few TF BSS methods have previously been reported. The first one [11] provided limited performance. [12,13] give good separation even for underdetermined mixtures with another type of methods, which require the sources to be sparse e.g. in the TF domain. A new BSS method called TIFROM which uses TF information was given in [14] which use the STFT as the TFR.

3.1 Problem Statement and Why Use WT

We here consider the following noiseless linear instantaneous mixture of n real-valued sources:

$$\begin{cases} x_1(t) = a_{11}s_1(t) + a_{12}s_2(t) + \cdots + a_{1n}s_n(t) \\ x_2(t) = a_{21}s_1(t) + a_{22}s_2(t) + \cdots + a_{2n}s_n(t) \\ \quad\vdots \qquad\quad \vdots \qquad\quad \vdots \qquad\quad \vdots \qquad \vdots \\ x_m(t) = a_{m1}s_1(t) + a_{m2}s_2(t) + \cdots + a_{mn}s_n(t) \end{cases} \tag{2}$$

where the coefficients a_{ij} of the mixing matrix $\mathbf{A}$ are real, constant and different from zero as discussed above. For simplicity, we first see the two sources condition:

$$x_1(t) = a_{11}s_1(t) + a_{12}s_2(t),$$
$$x_2(t) = a_{21}s_1(t) + a_{22}s_2(t).$$

BSS may be seen as a method for finding an estimate of $\hat{\mathbf{A}}^{-1} = \mathbf{Q}\mathbf{\Lambda}\mathbf{A}^{-1}$, where $\mathbf{Q}$ and $\mathbf{\Lambda}$ defined as before. Then we focus on:

$$\hat{\mathbf{A}}^{-1} = \begin{bmatrix} 1 & 1 \\ 1/c_1 & 1/c_2 \end{bmatrix}^{-1} \tag{3}$$

where

$$c_1 = \frac{a_{11}}{a_{21}}, \quad c_2 = \frac{a_{12}}{a_{22}}. \tag{4}$$

are called cancelling coefficient values which make it possible to cancel $s_1(t)$ or $s_2(t)$ and therefore to extract $s_2(t)$ or $s_1(t)$. By applying (3), we can get the output vector:

$$\mathbf{y}(t) = \hat{\mathbf{A}}^{-1}\mathbf{x}(t) = [a_{11}s_1(t), a_{12}s_2(t)]^T. \tag{5}$$

A simple time domain BSS method may then be derived when $x_1(t)$ or $x_2(t)$ contain only the contribution of one source. If we find one of these locations, the above cancelling coefficient values ci are easy to compute. As discussed in [13], this time t_i may hardly be determined in practical situations, since both sources are simultaneously active. As a result, this approach based on temporal analysis is therefore restricted to the very special case when each source occurs alone in large enough time intervals. Similar BSS methods, which also consider temporally sparse sources have already been reported in [14].

In order to get the ideal separation result, we use CWT information to cancel a source signal contributions from a set of linear instantaneous mixtures of these sources as done in [13].

The idea of the CWT is to project a signal $x(t)$ on a family of zero-mean functions (the wavelets) deduced from an elementary function (the mother wavelet) by translations and dilations:

$$X_i^{\Psi}(t,f) = \int_{t'} x(t')\Psi_{t,a}^*(t')dt', \tag{6}$$

where $\Psi_{t,a}(t') = |a|^{-1/2}\Psi(\frac{t'-t}{a})$, $a = f_0/f$ and parameter f_0 equals to the center frequency of Ψ. The variable a corresponds now to a scale factor, in the sense that taking $|a| > 1$ dilates the wavelet Ψ and taking $|a| < 1$ compresses Ψ. By definition, the CWT is more a time-scale than a TF representation. However, for wavelets which are well localized around a non-zero frequency ν_0 at scale $a = 1$, a TF interpretation is possible thanks to the formal identification $\nu = \nu_0/a$.

The basic difference between the CWT and the STFT is as follows: when the scale factor a is changed, the duration and the bandwidth of the wavelet are both changed but its shape remains the same. And in contrast to the STFT, which uses a single analysis window, the CWT uses short windows at high frequencies and long windows at low frequencies. This partially overcomes the resolution limitation of the STFT: the bandwidth B is proportional to ν, or B/ν=Q(constant),

We call it a constant Q analysis. The CWT can also be seen as a filter bank analysis composed of band-pass filters with constant relative bandwidth. So unlike using STFT in [13], we here consider the CWT.

3.2 How to Use CWT and the Basic Algorithm

We now show how CWT may be used to identify the cancelling coefficient values (4). First, define the complex ratio:

$$\alpha(t_j, f_k) = \frac{X_1(t_j, f_k)}{X_2(t_j, f_k)}, \tag{7}$$

which is computed for each TF window. Taking into account Equ. (2),(6) and the linear TFR of CWT, this ratio may be written as:

$$\alpha(t_j, f_k) = \frac{a_{11}S_1(t_j, f_k) + a_{12}S_2(t_j, f_k)}{a_{21}S_1(t_j, f_k) + a_{22}S_2(t_j, f_k)}, \tag{8}$$

Therefore, if one source occurs alone in the TF window (t_j, f_k), then $\alpha(t_j, f_k)$ is equal to the cancelling coefficient value, which makes it possible to cancel this source and thus to extract the other one. This situation when sources only disappear in some areas of the TF plane is much more frequent than the case when they disappear at all frequencies during a whole time period [13].

In order to get ideal separation result, we request the following assumptions [13]:

Assumption 1: The power of each source is non-negligible at least at some times t.

Assumption 2: For each source s_i, there exist some adjacent TF windows (t_j, f_k) centered on time t_j and frequency f_k where only s_i occurs, i.e. where: $S_l(t_j, f_k) \ll S_i(t_j, f_k), \forall l \neq i$.

Assumption 3: When several sources occur in a given set of adjacent TF windows they should vary so that (t_j, f_k) does not take the same value in all these windows. Especially, 1) at least one of the sources must take significantly different TF values in these windows so that the variance of the ratio is non-negligible and 2) the sources should not vary proportionally.

Thus, if only source $s_i(t)$ is present in several time-adjacent windows (t_j, f_k), then $\alpha(t_j, f_k)$ equals to c_i over these successive windows. To exploit this phenomenon, we compute the sample variance of the ratio $\alpha(t, f)$ on series T_p of M short half-overlapping time windows corresponding to adjacent t_j and apply this approach to each frequency f_k. We resp. define the sample mean and variance of $\alpha(t, f)$ on T_p and f_k by [13]

$$\bar{\alpha}(T_p, f_k) = \frac{1}{M} \sum_{j=1}^{M} \alpha(t_j, f_k), \tag{9}$$

$$var[\alpha(T_p, f_k)] = \frac{1}{M} \sum_{j=1}^{M} |\alpha(t_j, f_k) - \bar{\alpha}(T_p, f_k)|. \tag{10}$$

If e.g. $S_1(t_j, f_k)$ for these M windows, then (8) shows that $\alpha(t_j, f_k)$ is constant over them, so that its variance $var[\alpha(T_p, f_k)]$ is equal to zero. Conversely, under Assumption 3, if both $S_1(t_j, f_k)$ and $S_2(t_j, f_k)$ are different from zero, then $var[\alpha(T_p, f_k)]$ is significantly different from zero. So, by searching for the lowest value of $var[\alpha(T_p, f_k)]$ in all the available series of windows (T_p, f_k), we directly find a TF domain (T_p, f_k) with only one source. The corresponding value c_i which cancels this source is then estimated by the mean $\bar{\alpha}(T_p, f_k)$. We find the second cancelling coefficient value c_i by searching the next lowest value of $var[\alpha(T_p, f_k)]$ vs. (T_p, f_k) associated to a significantly different value of $\bar{\alpha}(T_p, f_k)$ using a threshold set to the minimum difference that we request between the two values in (3). We thus obtain estimates of the two cancelling coefficient values defined in (3). The separated signals are then derived from these values by using (i) successive source cancellations or (ii) the global matrix inversion.

4 CWT Based General BSS Algorithm

We now show how the above method may be extended to the case when n mixtures of n source signals are available.

4.1 $m = 2$ Mixtures and n Source Signals

As in [13], for clarity, we first consider an special situation $m = 2, n > 2$: we consider 2 observed mixtures, but they now contain more than 2 source signals. The observations then become:

$$\begin{cases} x_1(t) & = a_{11}s_1(t) \ + a_{12}s_2(t) \ + \cdots + a_{1n}s_n(t) \\ x_2(t) & = a_{21}s_1(t) \ + a_{22}s_2(t) \ + \cdots + a_{2n}s_n(t) \end{cases} \tag{11}$$

The complex ratio $\alpha(t, f)$ defined in (7) becomes:

$$\alpha(t, f) = \frac{\sum_{s=1}^{n} a_{1s}S_s(t, f)}{\sum_{s=1}^{n} a_{2s}S_s(t, f)}, \tag{12}$$

Under the same assumptions as above, consider a TF window (t_j, f_k) where only source s_i occurs. The complex ratio then becomes:

$$\alpha(t_j, f_k) = \frac{a_{1i}}{a_{2i}}. \tag{13}$$

It is easily shown that applying to the vector $\mathbf{x}(t)$ any 2×2 partial inverse matrix

$$\hat{\mathbf{A}}^{-1} = \begin{bmatrix} 1 & 1 \\ 1/c_{i2} & 1/c_{j2} \end{bmatrix}^{-1} \tag{14}$$

where $c_{i2} = a_{1i}/a_{2i}$ and $c_{j2} = a_{1j}/a_{2j}$ provides two different outputs with resp. cancellation of s_j and s_i. The BSS method defined in Subsection 3.2 is therefore straightforwardly extended to the current case, but then leads to a partial separation, i.e. to the cancellation of only one of the existing sources in each output signal. This is of high practical interest in signal enhancement applications anyway, as this method gives an efficient solution for removing the contribution of an undesirable source.

4.2 General Case: n Mixtures and n Source Signals

As suggested above, the areas (T_p, f_k) where a given source appears alone in observations are the same for all mixtures. This phenomenon is called the *coherence of the TF maps.* Because of this coherence, single-source areas may be detected for most mixing matrices by analyzing the variance of the ratio

$$\alpha(t,f) = \frac{X_i(t,f)}{X_j(t,f)}$$

associated to only one arbitrary pair of observations: this variance is low in and only in single-source areas under Assumption 3. Now, we suppose that n observed mixtures of n sources are available.

We first perform a single variance analysis with two observations. This yields all the TF areas where only one source occurs for all the observations. Then we adapt the approach of Subsection 4.1 to each pair of observations (x_1, x_j). We thus compute the mean of the ratio $X_1(t,f)/X_j(t,f)$ in the area given by the variance analysis where only s_i exists, which yields the value $c_{ij} = a_{1i}/a_{ji}$. Using these values, we then build a matrix $\hat{\mathbf{A}}^{-1}$ which achieves global inversion up to a scale factor, i.e:

$$\hat{\mathbf{A}} = \begin{bmatrix} 1 & \cdots & 1 \\ 1/c_{12} & \cdots & 1/c_{n2} \\ \vdots & \vdots & \vdots \\ 1/c_{1n} & \cdots & 1/c_{nn} \end{bmatrix}, \tag{15}$$

which yields:

$$\mathbf{y}(t) = \hat{\mathbf{A}}^{-1}\mathbf{x}(t) = [a_{11}s_1(t), \cdots, a_{1n}s_n(t)]^T.$$

This efficient global algorithm leads to a complete BSS in one step when no sources are initially hidden, i.e. when all the c_{ij} may be derived directly from the observations.

However, when $\hat{\mathbf{A}}$ misses the coefficients c_{ij} associated to the hidden sources and then cannot be used directly to separate all sources by means of its inverse for the non-square. Suppose there are P visible sources (a source is said *visible* in the TF domain if there exist at least one TF area where it occurs alone [15]), we can derive $P \times P$ square sub-matrices from $\hat{\mathbf{A}}$ by keeping its first line and $P-1$ arbitrary other lines. If the inverse of any such sub-matrix is multiplied by the vector containing the mixed signal $s_i(t)$ which correspond to the P lines kept from $\hat{\mathbf{A}}$, we get a vector of P *recombined signals.* Each such signal with index j only consists of contributions from the visible sources with the same index j and from all $n-P$ initially hidden sources. We repeat this procedure by keeping different sets of lines from $\hat{\mathbf{A}}$ and deriving corresponding sub-matrices and recombined signals. Then the initial BSS problem reduced to P independent BSS sub-problems and each only involving $n-P+1$ linearly mixtures of the same $n-P+1$ sources. Moreover, additional sources may now be visible in these new mixtures, because they are no more hidden by other sources, which have been cancelled when deriving these new mixtures, at last get the final separation. The above algorithm can easily be extended to the underdetermined case [13].

5 Procedure of the BSS Algorithm

The procedure of the algorithm gives as follows:

Step 1: Calculate the CWT $X_i(t_j, f_k)$ of each mixtures $x_i(t), i = 1, \cdots, n$;

Step 2: Calculate the ratio $\alpha(t_j, f_k) = X_1(t_j, f_k)/X_i(t_j, f_k)$ for a single couple (X_1, X_i), with $i \neq 1$;

Step 3: Select the number M of successive time windows included in the series of windows T_p and then calculate $var[\alpha(T_p, f_k)]$ for all the available series of windows (T_p, f_k);

Step 4: Sort the values $var[\alpha(T_p, f_k)]$ in ascending order. Only use the TF areas (T_p, f_k) associated to the first values in this ordered list. The first and subsequent areas in this beginning of the list are successively used as follows. Each considered area is kept, as the jth area, only if all distances between (i) the column of values $c_{i,j}$ with $i = 1, \cdots, n$ corresponding to this TF area and (ii) the columns of values corresponding to the previously kept TF areas in this procedure are higher than a user-defined threshold. This yields P columns of values, with $P \leq n$, where P is the number of initially visible sources. These columns are gathered in the matrix defined as

$$\hat{\mathbf{A}} = \begin{bmatrix} 1 & \cdots & 1 \\ 1/c_{12} & \cdots & 1/c_{P2} \\ \vdots & \vdots & \vdots \\ 1/c_{1n} & \cdots & 1/c_{Pn} \end{bmatrix}; \tag{16}$$

Step 5: 1) If $P = n$, use the inverse of this square matrix $\hat{\mathbf{A}}$ to achieve a global separation.

2) if $P < n$, then derive the $n-P+1$ square sub-matrices with size $P \times P$:

$$\hat{\mathbf{A}} = \begin{bmatrix} 1 & \cdots & 1 \\ 1/c_{12} & \cdots & 1/c_{P2} \\ \vdots & \vdots & \vdots \\ 1/c_{1,P-1} & \cdots & 1/c_{P,P-1} \\ 1/c_{1,q} & \cdots & 1/c_{P,q} \end{bmatrix} \tag{17}$$

with $q = P, \cdots, n$ and calculate the recombined signals

$$\mathbf{y}^q(t) = \hat{\mathbf{A}}_q^{-1}[x_1(t), \cdots, x_{P-1}(t), x_q(t)]^T.$$

Then create P new independent subsystems $\mathbf{x}^i(t), i = 1, \cdots, P$. Each of them consists of $N - p + 1$ linearly independent mixtures of the same $N - p + 1$ sources. At last, recursively apply the whole process independently to each of the P subsystems until complete source separation.

6 Simulation Results

In order to confirm the validity of the proposed wavelet transformation based BSS algorithm, simulations are given below. We simulate it on computer using four source signals which have different waveforms. The source correlation values are shown in Table 1.

Table 1. The Correlation Values Between Source Signals

	source 1	source 2	source 3	source 4
source 1	1	0.6027	0.3369	0.4113
source 2	0.6027	1	0.4375	0.4074
source 3	0.3369	0.4375	1	0.5376
source 4	0.4113	0.4074	0.5376	1

So the sources have high dependent, most of the classic BSS algorithm may fail separate the original source signals, but the proposed wavelet based BSS algorithm can separate the desired signals properly. Next, for comparison we execute the mixed signals with different BSS algorithms: JADE Algorithm [16], FPICA algorithm(Gaussian) [17], SOBI algorithm [18], EVD24 algorithm [19] and SANG algorithm [20]. At the same convergent conditions, the proposed algorithm was compared along the criteria statistical whose performance was measured using a performance index called cross-talking error index E defined as

$$E = \sum_{i=1}^{N}\left(\sum_{j=1}^{N}\frac{|p_{ij}|}{max_k|p_{ik}|} - 1\right) + \sum_{j=1}^{N}\left(\sum_{i=1}^{N}\frac{|p_{ij}|}{max_k|p_{kj}|} - 1\right),$$

where $\mathbf{P}(p_{ij}$ is the entries of $\mathbf{P})$ is the performance matrix. The separation results of the four different sources are shown in Table 2 for various BSS algorithms(averaged over 100 Monte Carlo simulation).

Table 2. The results of the separation are shown for various BSS algorithms

Algorithm	JADE	FPICA	SOBI	EVD24	SANG	WT-BSS
E	0.4368	0.3700	0.7844	0.3752	0.4064	0.1105

The waveforms of source signals, mixed signals and the separated signals are given from up to down are shown in Fig.1(the first 512 observations are given).

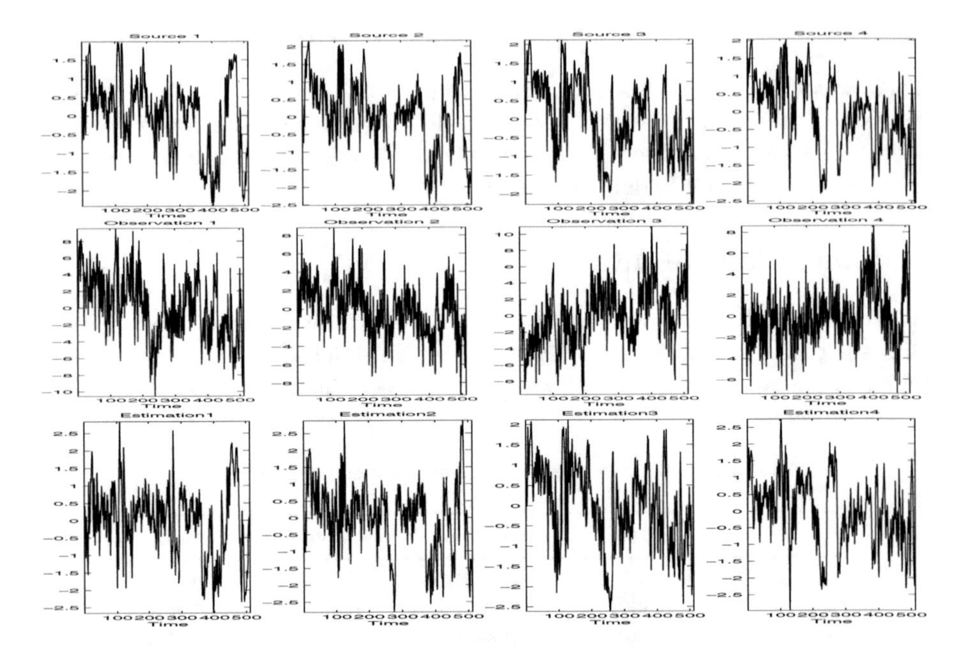

Fig. 1. The Source Signals, Observed Signals and Experiment Results Showing the Separation of Dependent Sources Using the Proposed WT Based BSS Algorithm

7 Conclusion

Most of the proposed algorithms for solving BSS problem rely on independence or at least uncorrelation assumption of source signals. However, in practice, the latent sources are usually dependent to some extent. On the other hand, there is a large variety of applications that require considering sources that usually behave light or strong dependence. In this paper, we develop novel BSS algorithms which can separate and extract dependent source signals from linear mixtures of them. The proposed algorithm is given based on the wavelet coefficient using CWT which only requires slight differences in the CWT coefficient of the considered signals. Moreover the proposed algorithm can extract the desired signals in the overcomplete conditions or even in the noisy conditions. Simulation results show that the proposed algorithm is able to extract the dependent signals and yield ideal performance but some other classic BSS algorithms give poor separation result.

Acknowledgments

This work is partially supported by the Science Foundation of Henan University of Technology under Grant No.06XJC032 and National Natural Science Foundation of China under Grant No.60472062. The authors also gratefully acknowledge the helpful comments and suggestions of the reviewers, which have improved the presentation.

References

1. Cichocki A, Amari S (2002) Adaptive Blind Signal and Adaptive Blind Signal and Image Processing. John Wiley&Sons, New York
2. Hyvarinen A, Karhunen J, Oja E (2001) Independent component analysis. John Wiley&Sons, New York
3. Comon, P (1994) Independent Component Analysis: A New Concept? Signal Processing 36: 287-314
4. Amari S (1998) Natural Gradient Works Efficiently in Learning. Neural Computation 10: 251-276
5. Zhang XD, Zhu XL, Bao Z (2003) Grading Learning for Blind Source Separation. Science in China(Series F) 46: 31-44
6. Hyvarinen A (2001) Blind source separation by nonstationarity of variance: a cumulant-based approach. IEEE Trans. Neural Networks 12: 1471-1474
7. Parra L, Spence C (2000) Convolutive blind separation of nonstationary sources. IEEE Trans. Speech Audio Processing 8: 320-327
8. Pham DT, Cardoso JF (2000) Blind separation of instantaneous mixtures of nonstationary sources. IEEE Trans. Signal Processing 49: 1837-1848
9. Abrard F, Deville Y (2005) A time-frequency blind signal separation method applicable to underdetermined mixtures of dependent sources. Signal Processing 85: 1389-1403
10. Hlawatsch F, Boudreaux-Bartels GF (1992) Linear and quadratic time-frequency signal representations. IEEE Signal Processing Mag. 9: 21-67
11. Belouchrani A, Amin MG (1998) Blind source separation based on time-frequency signal representations. IEEE Trans. Signal Processing 46: 2888-2897
12. Bofill P, Zibulevsky M (2001) Underdetermined blind source separation using sparse representation. Signal Processing 81: 2353-2362.
13. Abrard F, Deville Y (2005) A time-frequency blind signal separation method applicable to underdetermined mixtures of dependent sources. Signal Processing 85:1389-1403
14. Lee TW, Lewicki M, Girolami M, Sejnowski TJ (1999) Blind source separation of more sources than mixtures using overcomplete representations. IEEE Signal Processing Lett. 6: 87-90
15. Puigt M, Deville Y (2005) Time-frequency ration-based blind separation methods for attenuated and time-delayed sources. Mechanical Systems and Signal Processing 19: 1348-1379
16. Cardoso JF (1999) High-order contrasts for independent component analysis. Neural Computation 11: 157-192
17. Hyvarinen A, Oja E (1998) A fast fixed-point algorithm for independent component analysis. Neural Computation 10 :1483-1492
18. Belouchrani A, Abed-Meraim K, Cardoso JF, Moulines E (1998) A blind source separation technique using second order statistics. IEEE Trans. on Signal Processing 45: 434-444
19. Georgiev P, Cichocki A (2002) Robust blind source separation utilizing second and fourth order statistics. Proceedings of International Conference on Artificial Neural Network(ICANN2002), Madrid, August 27-30, Lecture Notes in Computer Science 2415: 1162-1167
20. Amari S, Chen T, Cichocki A (2000) Nonholonomic orthogonal learning algorithms for blind source separation. Neural Computation 12: 1463-1484.

An Incentive Mechanism for Peer-to-Peer File Sharing

Hongtao Liu[1,2,3], Yun Bai[3,4], and Yuhui Qiu[1,3]

[1] Computer and Information Science Faculty of Southwest University, Chongqing, China
[2] Computer and Modern Education Technology Department of Chongqing Education College, China
[3] Semantic Grid Laboratory of Southwest University, Chongqing, China
[4] School of Electronic and Information Engineering of Southwest University
lht01@swu.edu.cn

Abstract. With the growth of P2P networks, it is an important problem to encourage peers in P2P to share its resource. All the rational and self-interested peers will want to maximize those resources' utility. This paper models the behavior of peers in P2P network; we make use of the storage space as the key factor, and restrict the peers' resource getting ability by counting the peers' communication capability and the average size of files in the peers. In this mechanism, peers increase the motivation of sharing resources with others. Finally, it can increase the utility of P2P network.

Keyword: P2P, Incentive mechanisms, Storage space, Utility function.

1 Introduction

Increasing personal computers connected by the Internet exciting opportunities for online communication and business. P2P applications which aim to exploiting the cooperative paradigm of information exchange to greatly increase the accessibility of information to a large number of users. P2P allow users to share networks resources and services with peers to the benefit of peers' behavior. In P2P networks the resources and services include the exchange of information and content files, processing cycles, cache storage, and disk storage for data files. Normal P2P networks do not provide service incentive for users consequently users can obtain services without themselves contributing any information or service to a P2P community. This leads to the "Free-riding" [1]. Free-riding is an example of strategy where rational users free ride and consume a resource but do not produce at the same level as their consumption and "Tragedy of the Commons"[2,3] problems, in which the majority of information requests are directed towards a small number of P2P pees willing to share their resources. P2P network are dynamically changing and self-organizing, distributed resource-sharing networks. There is no central authority or peer to mandate or coordinate the resources that each peer should contribute. In the same time the voluntary peers contribute their resources are not unpredictable. So that P2P systems must be designed to take user incentives and self-interest into consideration [4]. In this paper we propose a Mechanism to differentiation in a P2P network based on the Behavior of each peer, thereby encouraging peers to share their resources.

B.-Y. Cao (Ed.): Fuzzy Information and Engineering (ICFIE), ASC 40, pp. 442–447, 2007.
springerlink.com

The rest of this paper is organized as follows: in Section 2, we introduce some relation work of incentive Mechanism P2P service. In Section 3 we develop a mechanism to incentive peers in P2P networks to contribution resources. In Section 4 we demonstrate the incentive Mechanism in the experiments. Finally, in Section 5 we present the conclusions and future work.

2 Relation Work

We now briefly present some relation work. Golle et al. [5] proposed a simple game theoretic model of agent behavior in centralized P2P systems and shown that our model predicts free riding in the original Napster mechanism. a reputation based mechanism is already used by the KaZaA [6] file sharing system; in that system, it's called the *participation level*. Quantifying a user's reputation and prevention of faked reputations, however, are thorny problems. In [7] present a static and a dynamic pricing incentive mechanism to motivate each peer to behave rationally while still achieving good overall system performance. In this paper study the behavior of the peers under two pricing mechanisms and evaluate the impact of free riding using simulations. In[8]introduce a resource distribution incentive mechanism between all information sharing nodes. The mechanism is driven by a distributed algorithm which has linear time complexity and guarantees Pareto-optimal resource allocation. Besides giving incentive, the mechanism distributes resources in a way that increases the aggregate utility of the whole network, then it model the whole resource request and distribution process as a competition game between the competing nodes. It shows that this game has a Nash equilibrium and is collusion-proof. To realize the game, It propose a protocol in which all competing nodes interact with the information providing node to reach Nash equilibrium in a dynamic and efficient manner. The mechanism such as the auditing described by Ngan et al.[9] presents architectures for fair sharing of storage resources that are robust against collusions among nodes. It requiring nodes to publish auditable records of their usage can give nodes economic incentives to report their usage truthfully.

3 Incentive Mechanism

We first introduce relative parameters used in this paper for our incentive mechanism P2P network.

N : a set representing all peers in the P2P system with $|N| = \mathrm{N}$, n_i represent peer i

μ: $(u_1, u_2, ...u_N)$ u_i represents the maximal upload bandwidth of peer i, where $i \in N$

D: $(d_1, d_2, ..d_N)$: where d_i represents the *maximal* download bandwidth of peer i, where $i \in N$

U_i : represents the utility function of peer i
F_i : represents the average number files of peer i share in P2P
$\mathrm{G_i}$: represents local copies of files the peer want to get in the network
$\mathrm{C_i}$: represents reciprocal space or files the peer contributes for remote peers

The other peers can get the $\mathrm{C_i}$ that must share the resources by the upload bandwidth of peer $i : \mu_i$,and the same way ,the other's resources peer can share by download bandwidth of peer $i : d_i$

So the relationship among, $\mathrm{G_i}$ and $\mathrm{C_i}$ are

$$C_i = \Delta t \mu_i \quad \text{and} \quad G_i = \Delta t d_i \tag{1}$$

when we can get share resources we contribute our resources at the same time

$$\frac{C_i}{\mu_i} = \Delta t = \frac{G_i}{d_i} \Rightarrow \tag{2}$$

So we get $C_i = \dfrac{\mu_i}{d_i} G_i = \phi_i G_i$

ϕ_i lie on the peer's communication capability $\phi_i = \dfrac{\mu_i}{d_i}$ defines an important character of the P2P system .The higher values of ϕ_i. increase the efficiency of finding a node with free space to accept storage requests and the lower cost of effective capacity in the P2P system. Because peers join in and leave the network is random, the probability function peers access the resource

$$P_i = p_i(\phi) = 1 - e^{-\phi_i F_i} \tag{3}$$

We can find more smaller file's size the higher probability peer can share it in spite of peer get the files or contribute the files. But if peers want get files in high probability or the files' size bigger than the average size of system with the same probability, peers must increase ϕ_i . On the other way peers are encouraged to division those files into segments. Peers in P2P are rational users if peers have not motivation to share file in P2P peers will disconnect to the network at the same time peers will care about utility of peers' resource. We use compute the peers' cost and the profit to modeling the peers' utility function.

Peers' cost: $G_i + C_i$ and the expectation of peers' can get resources: $P_i G_i$.

Because the peers request the resources are not always be accepted. So the peers' utility function

$$U_i = \frac{P_i G_i}{G_i + C_i}$$

$$= \frac{P_i G_i}{G_i + \phi_i G_i} = \frac{P_i}{1+\phi_i} = \frac{1-e^{-\phi_i F_i}}{1+\phi_i} \tag{4}$$

Although the peers have the possibly to get all resources that those wanted but those must prepare enough storage space otherwise peers will take a risk to lose some import resources.

If the maximize of utility function we can get so that the formula will be existed

$$\frac{\partial U_i}{\partial \phi_i} = \frac{F_i e^{-\phi_i F_i}(1+\phi_i) - (1-e^{-\phi_i F_i})}{(1+\phi_i)^2} = \frac{F_i e^{-\phi_i F_i} + \phi_i F_i e^{-\phi_i F_i} + e^{-\phi_i F_i} - 1}{(1+\phi_i)^2} = 0 \tag{5}$$

Because $\phi_i \succ 0$ so we can make sure that $(1+\phi_i)^2$ will be a positive real figure.

Then the formula(6) is exist.

$$e^{-\phi_i F_i}(F_i + \phi_i + 1) = 1 \tag{6}$$

It can be find $e^{-\phi_i F_i} = \frac{1}{(F_i + \phi_i + 1)}$. From the we can find if the U_i has the maximize value the formula must come into existence. So the peers' utility function

$$U_i = \frac{1 - \frac{1}{F_i + \phi_i + 1}}{1+\phi_i} = \frac{F_i + \phi_i}{(1+\phi_i)(F_i + \phi_i + 1)} \tag{7}$$

From formula (7) we can get that peers' utility function are decide by ϕ_i and F_i.

4 Experiments

The motivation of rational peers connects to P2P systems are want get files what they want. So if the utility is receivable (make the most utility of resources) or peers not get the files peers will not disconnect to P2P system.

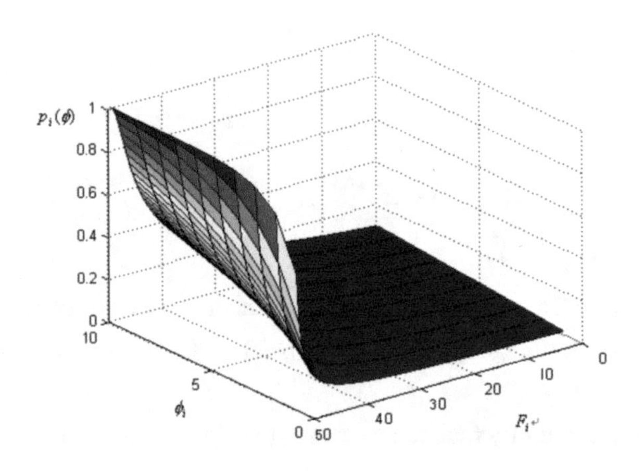

Fig. 1. Probability Function $p_i(\phi)$

From Figure1, we can get the result that $p_i(\phi)$ is decided by peer's communicate ability and average file's size in peer i. The relation ϕ_i, F_i and U_i is explained by figure 2.

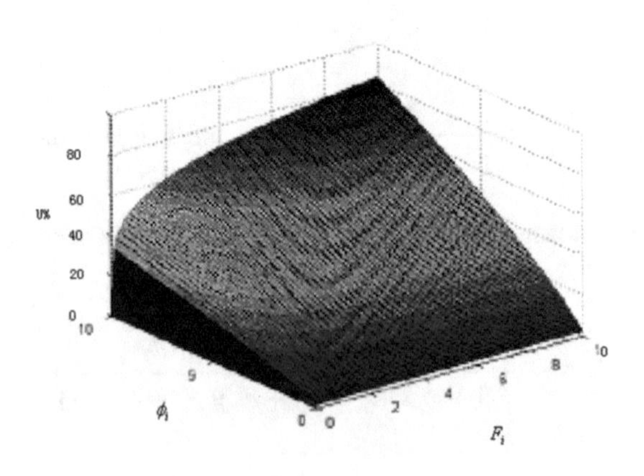

Fig. 2. The relation between ϕ_i, F_i and U_i

We can see that if the ϕ_i , and F_i are less than 3 they have almost the same affect to the utility function U_i , but if ϕ_i , and F_i not in the scope, ϕ_i have much more effect than F_i , and the result we also can get from the formula (7).

If the node is positive to share their resource the utility it will have high probability to get the resource so that get higher and higher utility. At the same time if the node have get what it want and don't want get more resource utility will have a drop in the U_i to node.

5 Conclusions and Future Work

We have presented an incentive mechanism for P2P networks. The files share among peers based on each peer's communication capability and peers behavior utility function. So that rational and self-interested peers will like to join the network. Every peer' utility function is relation with its communication capability and space of its contribution that will decide the probability they access files what they want. They will not be discriminated and the behavior of division those share files into segments is encouraged. Because it can increase the probability they access files.

The future work if some peers with similar C_i , ϕ_i and F_i we will propose an incentive to cluster together and reveal their preferences to others peers.

References

[1] Michal Feldman (2004) "Free Riding and Whitewashing in Peer to Peer Systems", SIGCOMM'04 Workshop, Aug. Portlad, oregon USA:153-160

[2] C. Kenyon, G. Cheliotis, R. Buyya. (2004) "10 Lessons from Finance for Commercial Sharing of IT Resources". In: Peer-to-Peer Computing: The Evolution of a Disruptive Technology. IRM Press

[3] G,Hardin . (1968) "The Tragedy of the Commons", Science, 162:1243-1248

[4] J. Feigenbaum and S. Shenker.(2002.) "Distributed Algorithmic Mechanism Design: Recent Results and Future Directions". In Proc. 6th Int'l Workshop on Discrete Algorithms and Methods for Mobile Computing and Communications, Atlanta, GA, Sept:1-13

[5] P. Golle, K. Leyton-Brown, I. Mironov, and M. Lillibridge. (2001)"Incentives for Sharing in Peer-to-Peer Networks", In Proc. 3rd ACM Conf. on Electronic Commerce, Tampa, FL, Oct:327-338

[6] John Borland. (2002)"U.S. liability looms over Kazaa." CNET News.Com, November 25

[7] Bin Yu and Munindar P. Singh.(2003) "Incentive Mechanisms for Peer-to-Peer Systems", Proceedings of the 2nd International Workshop on Agents and Peer-to-Peer Computing (AP2PC), Melbourne, July, Springer: 526-538

[8] Richard T. B. Ma. (2004)"A Game Theoretic Approach to Provide Incentive and Service Differentiation in P2P Networks ", SIGMETRICS/Performance'04, June 12–16:478-497

[9] T.-W. J. Ngan, D. S.Wallach, and P. Druschel. (2003) "Enforcing Fair Sharing of Peer-to-Peer Resources". In Proc. 2nd Int'l Workshop on Peer-to-Peer Systems, Berkeley, CA, Feb: 361-367

Application of the BP Neural Network in the Checking and Controlling Emission of Vehicle Engines

Lee Jun[1,2] and Qin Datong[2]

[1] School of mechantronics and automotive engineerinng, Chongqing Jiaotong University, Chongqing 400074, China
[2] Chongqing University, Chongqing, China
cqleejun@sina.com.cn

Abstract. This paper discusses the application of the BP Neural Network in checking and controlling the emission of vehicle engines based on the principle of the neural network. The mathematical model of the BP Neural Network is established. This checking technology is a new method for testing and experiment the vehicle emission and it is very important in studying and researching the automobile emission.

Keywords: BP neural network; engine emission; examination.

1 Introduction

The ingredients of the vehicle emissions are very complex. Besides including serious of the NO_X, CO_X, HC, there are many poison particles-PM. Since there are many unarranged test-points and unmeasured factors with the equipments, the emission of the automobiles can't be tested and controlled in the checking the engine emissions and the real-time controlling automobile. While the condition of the engine operation is a non-linear process, the conventional mathematical model must be simplified and treated in researching the engine emission. Therefore, the checking is not accurate and the precision is not better. If we calculate by the experienced formula, there are only a few of the main factors usually in computation and the error of the result is greater by the experienced formula.[1~4] Artificial neural network is a method simulated the construction of the organism neural network and the mode of thinking. It is a mathematical model beyond to the researching things and the non-linear problems of the multi-dimension space can be resolved. [5~7]The technology of the checking and the controlling emission based on the BP neural network can resolve the arbitrary complex and non-linear problems in the neural network, and there are advantages in the study by itself and the greater fault tolerance. This method is a good means of studying and researching in the checking and controlling automobile emission.[8]

2 Model of the BP Neural Network

Back Propagation Network (BP) is a multi-layer feed-forward neural network consisted of some hidden layers. The theory of the multi-layer BP Neural Network was first

B.-Y. Cao (Ed.): Fuzzy Information and Engineering (ICFIE), ASC 40, pp. 448–454, 2007.
springerlink.com

presented in 1974 by Werbos. Then Mr. Rumelhart and others advanced and studied the BP Neural Network learning calculation in 1985 and the imagine of the multi-layer network had been realized. In this neural network there are not only input points and output points, but also the hidden joint-points.[9,10] The results are outputted through computation and analysis. If the number of the input points is M in the neural network and the number of the output points is L, this neural network can be regarded as the greater non-linear map from M-dimension Euclidean space to L-dimension Euclidean space. The technology of the step seeking is employed in the BP learning calculation so that the mean square error is minimum between the real outputs and expected outputs. The process of the neural network learning and calculation is a progress of the revision the weights from the error direction to the back propagation.[11~13]

2.1 Construction of the BP Neural Network

The BP Neural Network construction is a three layers neural network construction, including the input layer, hidden layer and output layer.[2,14,15] In this BP Neural Network in the checking and controlling emission of the automobile engines, the numbers of the input layers are 5 (including rotation rate, inlet temperature, inlet pressure, throttle opening and compression ratio). The hidden layers are 12 and the output layers are 5. The output layers consist of the HC, CO, NO, NO_2, PM. This neural network construction is shown in Fig. 1.

2.2 Equation of the BP Neural Network

The output in the unit of the ith of the hidden or output layers is Oi :

$$O_i = f\left(\sum_{j=1}^{n} \omega_{ij} * x_j + g_i\right)$$

Where, the ωij is the relative weights between the jth connected point in the forward layer and the ith connected point in this layer. The xj is the output of the jth connected point in the forward layer. The gi is the threshold of the ith connected point in the forward layer. The n is the sum in the forward layer. The f(x) is S function (Sigmoid Function).

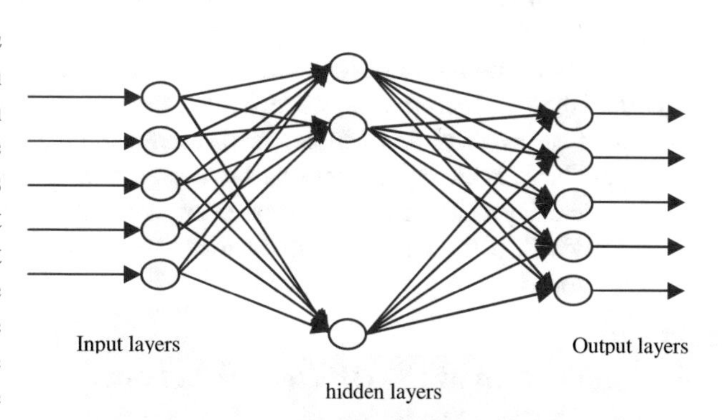

Fig. 1. BP Neural Network Construction

The formula is:

$$f(x) = (1+e^{-x})^{-1}$$

(1) The adjusted formula of the weights is:

$$\omega(n-lager)[j][i] = \omega(n-lager) + \alpha*bfb(n-lager)[j]*bfa(n-lager)[i] + \beta*\Delta\omega(n-lager)[j][i]$$

Where, the α, $\beta$are the kinetic energy parameters; The bfb(n – lager) [j] is the expected output in the hidden layer or the output layer; The bfa(n – lager) [i] is the real output in the hidden layer or the output layer.

(2) The formula of the threshold in the output layer is:

$$g[3][j] = g[3][j] + bfb[3][j] - bfa[3][j]*bfa[3][j]*(1 - bfa[3][j])$$

Where, the bfb[3][j] is the expected output in the output layer; the bfa[3][j] is the real output in the output layer.

(3) The formula of the adjusted threshold in the hidden layer:

$$g[2][j] = g[2][j] + bfa[2][j]*(1 - bfa[2][j])*sum$$

$$sum = sum + bfb[2][k]*\omega[2][k][j]$$

Where, the bfa[2][j] is the output in the hidden layer; The g[2][j] is the threshold of the output.

3 Process of the Neural Network Leaning

(1) First, choice the primary data of the weights and the thresholds;
(2) Second, input the trained samples;
(3) Third, compute the output values from some of the input values and the threshold values. If the sum of the mean square error of the sample is smaller than the required value, the learning train stops. Otherwise, it reverses the 4th step.
(4) According to the error between the outputted values in calculation and the real outputted values of this sample, the threshold and the weights are adjusted from the output lager to the input layer step-by-step backward propagation, until the ahead of the input layer. Then these processes are repeated. The threshold and the weights are revised. If the results is satisfactory and required, the computation is stopped. The procedure is shown in figure 2.

4 Application of the BP Neural Network in Checking and Controlling Emission

According to the method above them, we have experimented and checked the CAMRY AZ-FE engine emission in practice. The kinetic energy parameters(α=0.5, β=0.5)are

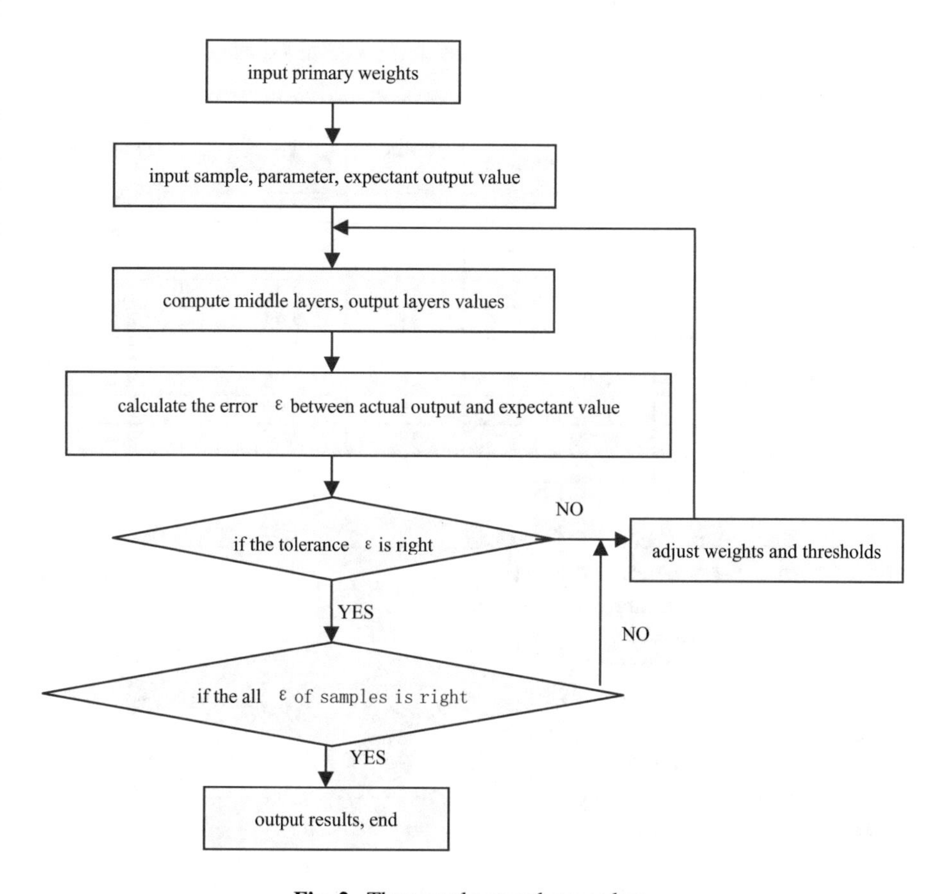

Fig. 2. The neural network procedure

chosen as the neural network parameters. After 8154 times, the result is satisfactory and precise. The computation and examination is more accurate.

Measuring and checking the CAMRY AZ-FE engine emission, we have received about 200~300 groups data of the HC, CO, SO_2, NO_2, PM in various of the speed, temperature and pressure of the inlet gas, differential inlet valve, and compression ratio. Through learning and training, the system for checking and controlling the engine have been taken and applied in measure and control the engine in practice. For example, we have taken the CO data about 250 groups. Except the great faults, there are 232 groups in leaning and training the system, the control and checking engine system have been set up. Then inspecting and comparing with the simulation system in MATLAB, the practical system fault is in allowing value and is smaller. The results are following table 1and figure 3.

From the results, we know there is smaller tolerance in comparing between the calculation results and practical results within the permitted exact condition. Then these

Table 1. Results about CO data of the practical system and simulation system

number	CO value in practice %	CO value in simulation %	number	CO value in practice %	CO value in simulation %
1	5.54	5.32	11	2.17	2.16
2	4.26	4.27	12	2.19	2.35
3	3.28	3.27	13	2.23	2.28
4	3.09	3.12	14	2.25	2.25
5	1.8	1.7	15	2.28	2.28
6	1.2	1.2	16	2.42	2.31
7	1.14	0.9	17	3.35	3.12
8	1.02	1.05	18	3.88	3.37
9	1.06	1.09	19	3.78	3.46
10	2.13	1.86	20	4.58	4.50

practical data can been supplied to the neural network system in training and leaning continually, so as that the system can approach to the practical condition and increase the precision. Then figure 4 is HC emission in experiment and simulation.

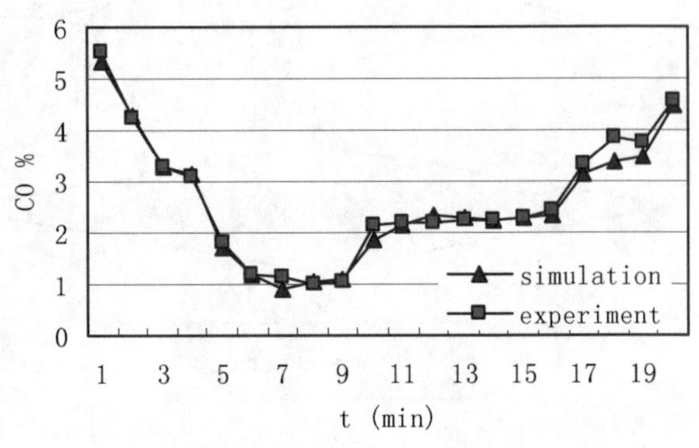

Fig. 3. CO emission in experiment and simulation

The BP neural network system can also realize the fuzzy control in vehicle engine in practice. To control the engine emission, we can take the system to calculate the HC, CO, SO_2, NO_2, PM of the engine emission, and to compare with the expectant values, and to compute the tolerance and ratio of error. Then these data can be input-signal in fuzzy control, and the engine can be controlled in the fuzzy neural network control. In reality, the engine control have been only the data of the engine revolution speed, inlet temperature and pressure, throttle valve opening angle, and so on, without checking the HC, CO, SO_2, NO_2, PM of the engine emission in practice. Through

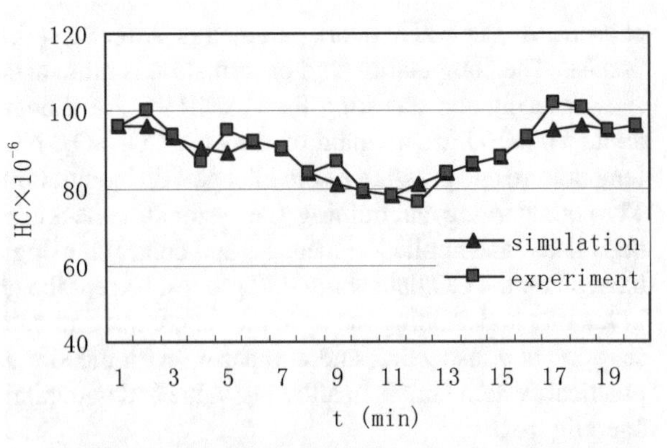

Fig. 4. HC emission in experiment and simulation

analyzing and computing in the neural network system, the HC, CO, SO_2, NO_2, PM of the engine emission can be predicted in really, so the system may reduce the difficulty in checking the engine emission, upgrade the fuzzy control precision and real-time control.

5 Result

The BP Neural Network is applied in checking engine emission and it can solve better the non-linear problems of automobile engine emission. This method can realize and accomplish the soft-measurement to automobile engine and can solve the problems of lacking of the special instruments and the unarranged test-points, so that it can complicate the real-time testing and controlling the automobile engine emission.

Acknowledgements

The work was supported by Natural Science Foundation Project of Chongqing CSTC, China (No. 2006BB3410).

References

1. Gao Jun, Zhu Meilin, Wu Limin(2001) Artificial Neural Network of Torque and Smoke Density for Vehicle Diesel EngineJournal Huazhong University of Science & Technology 11: 98~100
2. Wei Shaoyuan, Zhang Lei(2001) Applied Researches on BP Neural Net Engaged into Automobile Trouble Diagnosis. Journal of Liaoning Institute of Technology 1: 12~14
3. Yuan Quan, Zhang Jianyang, He Yong, Chen Kakao(1998) Application of Artificial Neural network to fault Diagnostic of gasoline engine. Journal of Zhejiang Agricultural University 4: 369~371
4. Chris M A. Virtual Sensing: A Neural Network-based Intelligent Performance and Emission Prediction System for On-Board Diagnostics and Engine Control. SAE 980516
5. Li Wei, Jia Zhixin, Meng Huirong(2002) Reliability model of fault data of vehicles transmission system by neural network method. Machinery 1: 4~5
6. S. A. Grossberg (1988)Nonlinear Neural Networks. Principles, and Architectures. Neural Network,1:47~61
7. Hertz, John (1991)Introduction to neural computation. Addition-Wesley Publishing Company
8. Ayeb M. SI Engine Modeling Using Neural Networks. SAE 980790
9. Li Xiaoying, Hou Zhiying (2002)Application of the Soft-measurement based on Neural Network in Automobile Engine Emissions. Computer and Communications 3:46~48
10. Guo Shiyong, Wang Yu(1997)BP Neural Network Applied to Auto. Failure Diagnosis. Journal of Xinjiang Institute of Technology. 2: 121~125
11. Wang Wencheng(1998)Neural Network and It Applies in Vehicle Engineering. Beijing Institute and Technology Publishing

12. Wei Shaoyuan, Zhang Lei(2001)Study on Auto Fault Diagnosis Expert System Based on Neural Network. Journal of Highway and Transmission Research and development 2: 78~81
13. Daniel Svozil, Vladimir Kvasnicka, Jiri Posichal (1997) Introduction to Multi-layer Feed-forward Neural Networks. Chemometrics and intelligent laboratory systems 39:43~62
14. A. de Lucas, A. Duran, M. Carmona, M. Lapuerta(2001)Modeling Diesel Particulate Emission with Neural Networks. Fuel 80:539~548
15. H. C. Krijnsen, J. C. M. van Leeuwen, R. Bakker, C. M. van den Bleek, H. P. A. Calis(2001)Potimum NO_X Abatement in Diesel Exhaust Using Inferential Feed-forward Reductant Control. Fuel 80:1001~1008

Oil and Gas Pipeline Limited Charge Optimum Maintenance Decision-Making Analysis Based on Fuzzy-Gray-Element Theory and Fuzzy Analytical Hierarchy Process*

Peng Zhang, Xingyu Peng, and Xinqian Li

SouthWest Petroleum University, Chengdu 610500, China
zp_swpi@sina.com

Abstract. This papers analyses the choice, combination and optimization of maintenance measures in the maintenance of oil and gas pipelines, initially proposes 4 factors system that influence effect of maintenance on pipelines, which includes person factor, fieldwork difficult degree, fieldwork management and environment condition, establishes hierarchy model of maintenance measures, introduces fuzzy analytical hierarchy process based on triangular fuzzy number, and compares the effects of different maintenance measures in order to determine the weight under single factor condition and composite matter-element of maintenance measures. On the basis of the above analysis, the paper proposes 4 methods of maintenance projects fund distribution based on different weight of efficiency factor of maintenance measures under the condition of limited charge. Using the fuzzy-gray-element theory combining fuzzy mathematics, gray theory and matter-element analysis, the paper analyses maintenance measure fund distribution based on different weight of efficiency factor of maintenance measures with fuzzy-gray-element correlation decision-making analysis, and conclusively determines optimization of maintenance project according to criteria of maximized correlative degree.

Keywords: Fuzzy-gray-element; fuzzy analytical hierarchy process; maintenance decision-making.

1 Foreword

Transportation of oil and gas pipelines is a relatively safer transporting method than of road and railroad, the failure probability of which is smaller. But in view of serious condition of the service pipelines, the failure accidents of pipelines are inevitable with the decadence of pipelines. Considering the characteristics of flammability, explosion and poison of transporting media, failure of pipelines will do serious harm to the safety of human and environment and produce great loss to economy [1]. As a result, reliability analysis and risk assessment of oil and gas pipelines is very important. Risk assessment of oil and gas pipelines is able to rank pipes with different risk grades in order to distinguish high-risk pipe and thus to operate maintenance to reduce the risk to the controlling level [2]. But, which maintenance measure and degree is appropriate for the company to apply to the high-risk pipes still needs optimum maintenance

B.-Y. Cao (Ed.): Fuzzy Information and Engineering (ICFIE), ASC 40, pp. 455–463, 2007.
springerlink.com

decision-making analysis [3]. Thus, this paper apply fuzzy-gray-element theory and fuzzy analytical hierarchy process to the analysis of maintenance measure for oil and gas pipelines, and the operators can carry out optimization maintenance measures on pipelines according to fuzzy- gray-element correlation decision-making theory.

2 Theoretical Basis of Fuzzy-Gray-Element

2.1 Concept of Fuzzy-Gray-Element

Fuzzy theory proposes to use fuzzy set and membership degree to signify the quantification of meanings in order to solve uncertainty and fuzziness in the reality. Traditional mathematics is on the basis of binary logic. In the contrary, uncertain problems in the reality may be solved by fuzzy theory [4].

Professor Junlong Deng of China proposed fuzzy gray system theory [5] in 1982. Gray system means a kind of system that contains both known and unknown information. The symbol of gray system is the relationship among gray number, gray element and gray relation. Gray element means uncompleted element. If decision-making model contains gray element, it belongs to gray decision-making models. Because of multiplicity of maintenance measures for oil and gas pipelines and uncertainty of maintenance cost, maintenance measures decision-making of pipelines in this paper is gray decision-making problem.

The basis of fuzzy-gray-element theory is matter-element model. Matter-element analysis is a theory established by professor Wen Cai to research the rules and methods of incompatibility in the reality. A matter-element can be described by "material, character and value", namely (M,C,V). In practical decision-making problems, different characters of the material relating to it must be taken into consideration, and a matter-element is described as $(M,(C_1,C_2,...C_n),(V_1,V_2...,V_n))$.

2.2 Correlation Analysis

Correlation analysis is the shortened from correlation degree analysis, the essence of which is a quantificational and comparative analytical process of dynamic process development trend and an analytical method for measuring the correlation degree between factors. Correlation degree is defined as the quantification of relation degree between different methods and the standard method. It is described as $K_i(i=1,2,\cdots,m)$, namely the correlation degree of the material i.

The basis of correlation analysis is the determination of correlation coefficient, and gray system theory proposes the calculation method of it as followed:

$$\tilde{\otimes}_f(\xi_{ij})=\frac{\Delta_{min}+\rho\Delta_{max}}{\Delta_{oij}+\rho\Delta_{max}}, i=1,2,\cdots m; j=1,2,\cdots n \tag{1}$$

ρ is the discerning coefficient whose area is $[0.1,0.5]$, and we usually take $\rho=0.5$. Δ_{oij} is the absolute value of whitening fuzzy gray value of standard method and the recommending method after the change of data.

$$\Delta o_{ij} = \left|\tilde{\otimes}_{oj} - \tilde{\otimes}_{fij}\right|, i = 1,2,\cdots m; j = 1,2,\cdots n \tag{2}$$

Where, $\Delta_{max}, \Delta_{min}$ signify the maximal and minimal value of the above absolute values.

After the determination of correlation coefficient, different arithmetic operators including $M(\bullet,+)$、 $M(\wedge,\vee)$、 $M(\wedge,\oplus)$ can be used to calculate correlation degree. For example, when $M(\wedge,\vee)$ is applied, the calculation formula is as followed:

$$K_i = \bigvee_{j=1}^{n} \left(a_j \wedge \tilde{\otimes}_f(\xi_{ij})\right), i = 1,2,\cdots m \tag{3}$$

Where, a_j signifies every weight. According to different weight of maintenance measures, in this paper hierarchy analysis is applied to determine every weight of maintenance measure.

3 Fuzzy Hierarchy Analysis Method Assurance the Maintenance Measure Weight

3.1 The Analytic Hierarchy Process Brief Introduction

AHP[6] (The Analytic Hierarchy Process) was putted forward by the professor Satty.T.L of American Pittsburgh university, in the mid 70's. It was a kind of system analysis method. According to the property and the general index mark of the system, AHP resolve system constitute factor, and according to of each factor is mutually connection, belong to a relation, be divided into a different layer combination, thus constitute a multilayer of system analysis structure model. According to the model and relation system factor, the important weight of the bottom factor compared to the topic factor is calculated.

3.2 Maintenance Measure Support Layer Model Establishment

The choice that aims at maintenance measure of high risk pipeline is decided by the concrete influence factor of the high risk pipeline segment, it to should be adopt of the maintenance measure is totally different, the maintenance measure layer model that this paper build up only for the sake of general method, don't aim at the concrete pipeline segment, so this paper doesn't put forward concrete maintenance measure, but with generally of replace with the measure.

According to maintenance measure effect to calculate weight of maintenance measure, then establishing AHP model. Establish the general index mark is “the maintenance measure effect"(ME-measure efficiency) this is the model first floor. The maintenance measure effect (the ME-measure efficiency) includes *PF*-person factor, *FD*-fieldwork difficult degree, *FM*-fieldwork management, *EC*-environment condition, which is the second floor of the model. In order to simplify model, these index signs no longer subdivide, the third layer is each concrete maintenance measure. The valuation layer model of maintenance measure effect is gained (see figure 1).

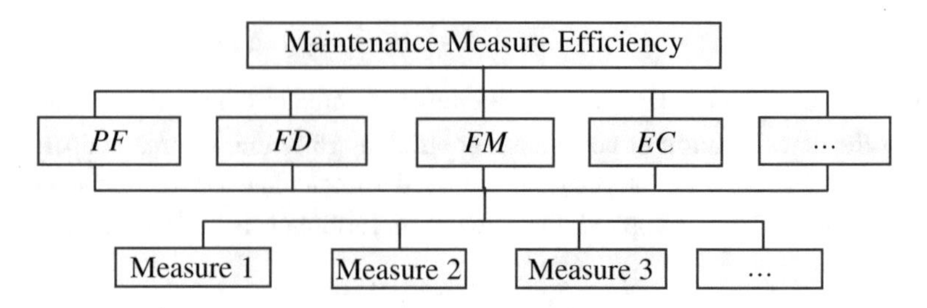

Fig. 1. Analysis Model of Maintenance

3.3 Establishment Triangle Fuzzy Number Complementary Judgment Matrix

The function that judgment matrix is at up one layer some factor of under stipulation condition carry on a comparison to the opposite importance of same layer factor, it can reflect one factor is relevant of all layer factors of hobby degree. This paper use triangle fuzzy number to build up complementary judgment matrix, which reflect the fuzzy and the uncertainty of the comparison factor.

Assuming the triangle fuzzy number complementary judging matrix is $\tilde{A}=(a_{ij})_{n\times n}$, which element $a_{ij}=(a_{lij},a_{mil},a_{uij})$ and $a_{ji}=(a_{lji},a_{mji},a_{uji})$ is complementary to each other, that is $a_{lij}+a_{uji}=a_{mij}+a_{mji}=a_{uij}+a_{lji}=1$ and $a_{lij}>a_{mij}>a_{uij}$, $i,j\in n$.

For four influence factors of maintenance measure: person factor, fieldwork difficult degree, fieldwork management and environment condition. Mutuality between arbitrary two maintenance measures is shown by a triangle fuzzy number (see table 1).

Table 1. Triangle Fuzzy Number of Comparative Language Variables

Comparative Language Variable	Triangle Fuzzy Number
Measure i inferior to measure j	(0,0,025)
Measure i relatively inferior to measure j	(0,0.25,05)
Measure i as well as measure j	(0.5,0.5,0.5)
Measure i relatively superior to measure j	(0.5,0.75,1)
Measure i superior to measure j	(0.75,1,1)

3.4 Calculation Triangle Fuzzy Number Weight Vector of Maintenance Factor

For the calculation of weight of complementary judgment matrix of maintenance factor, the currently common calculation method includes: The characteristic vector method, logarithms minimum two multiplications and minimum deviation method. For triangle fuzzy number complementary judgment matrix [7] in this paper, the each factor weight of it is gained by the formula (4)

$$\tilde{w}_i=\frac{\sum_{i=1}^{n}a_{ij}}{\sum_{i=1}^{n}\sum_{j=1}^{n}a_{ij}}=\left[\frac{\sum_{i=1}^{n}a_{lij}}{\sum_{i=1}^{n}\sum_{j=1}^{n}a_{uij}},\frac{\sum_{i=1}^{n}a_{mij}}{\sum_{i=1}^{n}\sum_{j=1}^{n}a_{mij}},\frac{\sum_{i=1}^{n}a_{uij}}{\sum_{i=1}^{n}\sum_{j=1}^{n}a_{lij}}\right],\quad i,j=1,2,\cdots,n \tag{4}$$

3.5 Calculation Weight of Maintenance Factor

The triangle fuzzy weight vector is not complete state of the index factor weight of needed. By a kind of possibility method, triangle fuzzy weight vector is handled. Assuming $\tilde{w}_i = (w_{il}, w_{im}, w_{iu})$ and $\tilde{w}_j = (w_{jl}, w_{jm}, w_{ju})$, $i, j = 1,2,\cdots,n$, thus the calculation formation of possibility[8] for $p_{ij}(\tilde{w}_i > \tilde{w}_j)$ is as followed:

$$p_{ij}(\tilde{w}_i > \tilde{w}_j) = 0.5max\left[1 - max\left(\frac{\tilde{w}_{jm} - \tilde{w}_{il}}{\tilde{w}_{im} - \tilde{w}_{il} + \tilde{w}_{jm} - \tilde{w}_{jl}}, 0\right), 0\right] + 0.5max\left[1 - max\left(\frac{\tilde{w}_{ju} - \tilde{w}_{im}}{\tilde{w}_{iu} - \tilde{w}_{im} + \tilde{w}_{ju} - \tilde{w}_{jm}}, 0\right), 0\right] \quad (5)$$

The gain of $p_{ij}(\tilde{w}_i > \tilde{w}_j)$ can be used to establish the possibility matrix $p = (p_{ij})_{n \times n}$. So for the sequence problem of triangle fuzzy number is converted into solving sequence vector of the possibility matrix, and finally the formula of calculation weight as follows (see Table 2):

$$w_i = \frac{1}{n}\left(\sum_{j=1}^{n} p_{ij} + 1 - \frac{n}{2}\right), \quad i \in N \quad (6)$$

Composite weight matter-element is gained as follows:

$$R_a = \begin{bmatrix} & S_1 & S_2 & \cdots & S_n \\ w_j & w_1 & w_2 & \cdots & w_n \end{bmatrix} \quad (7)$$

For weight of various maintenance measures, based on which to distribute fund could be considered as the optimum project under the finite charge (see Table 1).

Table 2. Weight of Maintenance Measures Based on various efficient factors

In / Wei / Meas	Index of Assessment				Rank of weight ω_j
	$p_{PF,i}$	$p_{FD,i}$	$p_{FM,i}$	$p_{EC,i}$	
	0.112	0.527	0.298	0.063	
MeasureM_1	0.250	0.329	0.008	0.402	0.229
MeasureM_2	0.250	0.605	0.008	0.008	0.350
MeasureM_3	0.250	0.033	0.582	0.582	0.255
MeasureM_4	0.250	0.033	0.402	0.008	0.166

4 The Fuzzy-Grey-Element Correlative Decision-Making Model

4.1 The Establishment of the Maintenance Project Decision Model

By applying the fuzzy-grey-element theory and the fuzzy analytical hierarchy process based on the triangular fuzzy numbers, which were both introduced in the paper before, we build up the fuzzy- grey-element correlative decision-making analysis model of oil and gas pipelines, its main step is as follows:

(1) The establishment of the maintenance project fuzzy-grey-element. We assume that d_i means the number i maintenance project, M_j means the number j maintenance measure index sign, $\tilde{\otimes}_{fij}$ means the maintenance funds whiten fuzzy grey magnitude of the number j maintenance measure in the number i maintenance project, and $\tilde{\otimes}_f R_{m \times n}$ means the composite fuzzy-grey-element.

$$\tilde{\otimes}_f R_{mn} = \begin{bmatrix} & d_1 & d_2 & d_3 & d_4 \\ M_1 & \tilde{\otimes}_{f11} & \tilde{\otimes}_{f21} & \tilde{\otimes}_{f31} & \tilde{\otimes}_{f41} \\ M_2 & \tilde{\otimes}_{f12} & \tilde{\otimes}_{f22} & \tilde{\otimes}_{f32} & \tilde{\otimes}_{f42} \\ M_3 & \tilde{\otimes}_{f13} & \tilde{\otimes}_{f23} & \tilde{\otimes}_{f33} & \tilde{\otimes}_{f43} \\ M_4 & \tilde{\otimes}_{f14} & \tilde{\otimes}_{f24} & \tilde{\otimes}_{f34} & \tilde{\otimes}_{f44} \end{bmatrix} \tag{8}$$

Note: The article provides 4 kinds of maintenance projects: d_1, d_2, d_3, d_4, and 4 kinds of maintenance measures: M_1, M_2, M_3, M_4.

(2) The establishment of the maintenance project correlative coefficient composite fuzzy gray-element. Based on the correlative coefficient calculation formula (1) in gray theory, we get the whiten fuzzy gray correlative coefficient value of each maintenance measure index sign in each maintenance project, thus we build up decision-making correlative coefficient composite fuzzy-gray-element, as follows:

$$\tilde{\otimes}_f R_{\varsigma} = \begin{bmatrix} & d_1 & d_2 & d_3 & d_4 \\ M_1 & \tilde{\otimes}_f(\xi_{11}) & \tilde{\otimes}_f(\xi_{21}) & \tilde{\otimes}_f(\xi_{31}) & \tilde{\otimes}_f(\xi_{41}) \\ M_2 & \tilde{\otimes}_f(\xi_{12}) & \tilde{\otimes}_f(\xi_{22}) & \tilde{\otimes}_f(\xi_{32}) & \tilde{\otimes}_f(\xi_{42}) \\ M_3 & \tilde{\otimes}_f(\xi_{13}) & \tilde{\otimes}_f(\xi_{23}) & \tilde{\otimes}_f(\xi_{33}) & \tilde{\otimes}_f(\xi_{43}) \\ M_4 & \tilde{\otimes}_f(\xi_{14}) & \tilde{\otimes}_f(\xi_{24}) & \tilde{\otimes}_f(\xi_{34}) & \tilde{\otimes}_f(\xi_{44}) \end{bmatrix} \tag{9}$$

(3) The establishment of the maintenance project correlative degree composite fuzzy-gray-element $\tilde{\otimes}_f R_k$. Based on the correlative coefficient value of each maintenance measure in each project in $\tilde{\otimes}_f R_k$, and combined with the weight determined by fuzzy analytic hierarchy process based on the triangular fuzzy numbers, using degree of association computation principle, we can get the degree of association values between 4 kinds of maintenance projects and optimum maintenance association projects which were based on maintenance measure weight, thus we build up maintenance project correlative coefficient composite fuzzy-gray-element. After contrasting and analyzing various different operators, such as $M(\bullet,+)$, $M(\wedge,\vee)$, $M(\wedge,\oplus)$, the paper adopts $M(\bullet,+)$ operator to determine the value of correlative degree.

$$\tilde{\otimes}_f R_k = R_a * \tilde{\otimes}_f R_{\xi} = \begin{bmatrix} & d_1 & d_2 \quad d_3 & d_4 \\ k_i & k_1 = \sum_{j-1}^{4} w_j \tilde{\otimes}_f(\xi_{1j}) & \cdots \quad \cdots & k_4 = \sum_{j=1}^{4} w_j \tilde{\otimes}_f(\xi_{4j}) \end{bmatrix} \tag{10}$$

In above formula, "$*$"represents the $M(\bullet,+)$ operator, means adding after multiplying, what's more, the maintenance measure weight composite matter-element is as follows:

$$R_a = \begin{bmatrix} & S_1 & S_2 & S_3 & S_4 \\ w_j & w_1 & w_2 & w_3 & w_4 \end{bmatrix} \tag{11}$$

(4) Correlative sequencing determines the optimum maintenance project. Sequencing the degrees of association $\tilde{\otimes}_f R_k$ which is in the maintenance project correlative coefficient composite fuzzy-gray-element by their value, by using the max correlative degree methods, max degree of association is determined, $k^* = \max\{k_1,k_2,k_3.k_4\}$, which corresponds to the optimum maintenance project of pipeline.

4.2 Determination of Optimum Project Based on Fuzzy-Gray-Element Theory

In view of complicity of maintenance condition, influence of multiple elements of maintenance efficiency and unpredictable factors of maintenance operation in the practical projects which cause uncertainty in the maintenance work, the pipeline company can not only consider reference to single factor to determine optimum maintenance project for distributing maintenance fund with limited charge, which can not get the real optimum combination, namely the optimum maintenance measure only meeting the condition with single element is unable to reach the optimum measure combination as whole.

After the consideration of fund distributing methods of maintenance measures under 4 different conditions, fuzzy-gray-element theory is applied to analyze correlative degree based on the rules of the index, which the larger is the better, and final optimum maintenance project is determined according to principle of maximal correlative degree. The process is as followed:

(1) Experts propose 4 maintenance projects of fund distribution, in which fund is distributed based on the weight of singer factor.
(2) Establishment of fuzzy-gray-element $\tilde{\otimes}_f R_{mn}$ of maintenance project.

$$\tilde{\otimes}_f R_{mn} = \begin{bmatrix} & d_1 & d_2 & d_3 & d_4 \\ M_1 & 25 & 32.9 & 0.8 & 40.2 \\ M_2 & 25 & 60.5 & 0.8 & 0.8 \\ M_3 & 25 & 3.3 & 58.2 & 58.2 \\ M_4 & 25 & 3.3 & 40.2 & 0.8 \end{bmatrix}$$

(3) Establishment of compound fuzzy-ray-element $\tilde{\otimes}_f R_\varsigma$ of correlative coefficient of maintenance project.

Because the larger the fund of maintenance measure is, the better it is, the standard value takes the maximization signified by $\tilde{\otimes}_{fomax}$, namely

$$\tilde{\otimes}_{fomax} = max\{\tilde{\otimes}_{fij}\} i = 1,2,\cdots m;\ j = 1,2,\cdots n$$

$$f_{oj} = \begin{bmatrix} & f_{o1} & f_{o2} & f_{o3} & f_{o4} \\ f_o & 40.2 & 60.5 & 58.2 & 40.2 \end{bmatrix}$$

(4) Establishment of composite fuzzy-gray-element $\tilde{\otimes}_f R_\varsigma$ of correlative coefficient of maintenance project.

$$\tilde{\otimes}_f R_\varsigma = \begin{bmatrix} & d_1 & d_2 & d_3 & d_4 \\ M_1 & 0.663 & 0.803 & 0.432 & 1.000 \\ M_2 & 0.457 & 1.000 & 0.333 & 0.333 \\ M_3 & 0.473 & 0.352 & 1.000 & 1.000 \\ M_4 & 0.663 & 0.447 & 1.000 & 0.431 \end{bmatrix}$$

(5) Establishment of composite fuzzy-gray-element $\tilde{\otimes}_f R_\varsigma$ of correlation of maintenance project.

$$\tilde{\otimes}_f R_k = R_a * \tilde{\otimes}_f R_\xi = \begin{bmatrix} & d_1 & d_2 & d_3 & d_4 \\ k_i & k_1 = 0.542 & k_2 = 0.698 & k_3 = 0.636 & k_4 = 0.672 \end{bmatrix}$$

Where, $R_a = \begin{bmatrix} & M_1 & M_2 & M_3 & M_4 \\ w_j & 0.229 & 0.350 & 0.255 & 0.166 \end{bmatrix}$

(6) Determination of optimum maintenance project of relation rank

According the rule of maximal correlative degree, the maximal correlative degree is the $k^* = \max\{k_1, k_2, k_3. k_4\} = k_3 = 0.698$, so among the 4 maintenance projects, Project 2 (based on weight of fieldwork difficulty degree): the fund of maintenance measure M_1 is 329000 RMB, the fund of maintenance measure M_2 is 605000 RMB, the fund of maintenance measure M_3 is33000 RMB, the fund of maintenance measure M_4 is 33000 RMB, is the optimum maintenance project with the limited charge of 1 million.

5 Conclusion

Optimum integrity maintenance decision-making of oil and gas pipeline based on the assessment of quantity risk of pipeline distinguishes the high risk pipes with unacceptable risk level according to the criteria of risk acceptance, proposes a serial of measures for resistance to risk in order to reduce the risk level of dangerous pipes through the main perilous factors influencing the high risk pipes, and analyze the optimum project among the proposed maintenance measures based on single weight, thus, to select the optimization maintenance measure combination with limited charge, to reduce the risk of high risk pipes economically and reasonably, and to make the maintenance decision-making system systematic, accurate and predicable.

Maintenance measure of oil and gas pipelines is a dynamic process, which is not supposed to be the last choice of the measure or the process decided by single expert or manager. It is important to avoid individual subject and blindness and carry out investigation of maintenance project operation and information feedback after the determination of optimum maintenance project in order to adjust project in time and achieve reasonable operation and realization of maintenance project.

Acknowledgements

This work described here is partially supported by the grants from the National Natural Science Foundation of China (NSFC No. 50678154), the Specialized Research Fund for the Doctoral Program of Higher Education (SRFDPHE No. 20060615003), Risk Innovation Fund of Petrochina Company Limited (RIFPCL No. 050511-5-2), and Postgraduate Innovation Fund of Southwest Petroleum University (2007-10).

References

[1] Zhang Peng, Chen Liqiong, Hou Xiangqin (2005) Utility Function Method of Optimal Decision Risk-Based for Oil/Gas Pipeline Integrity Maintenance, Papers of 2005 China International Oil & Gas Pipeline Technology (Integrity) Conference, Sept. 15-17, Shanghai, PRC. China: 163-174

[2] Peng Zhang, Xingyu Peng, Liqiong Chen, Ling Qin and Yamei Hu (2006) Study on Risk Mitigation Extent Model of Maintenance Measure to Long-distance Oil/gas External Corrosion Pipeline, In: Jia Xisheng, Huang Hongzhong, Wang Wenbin, NEW CENTURY and NEW MAINTENANCE, Proceeding of the First International Conference on Maintenance Engineering, October 15-18, 2006, Beijing: Science Press: 1085-1093

[3] Zhang Pen, Cheng Li-qiong, Hou Xiang-qin (2004) Optimum Integrity Maintnance Measure Based on Risk for Oil and Gas Pipelines (1), Natural Gas Industry.24(9): 159-162

[4] XiaoCiyuan (2004) Engineering Fuzzy System, Beijing: Science Press

[5] Zhang Xiao-yu, Li Wei (1997) Analysis of Fuzzy Gray Element and Application[M], Beijing: Petroleum Industry Press

[6] T L Satty (1980) The Analytic Hierarchy Process. New York: McGraw-Hill

[7] L A Zadeh (1965) Fuzzy Sets. Information and Control 8(3): 338-353

[8] Qu Ze-shui (1999) Study on the Relation between Two Classes of Scales in AHP. Systems Engineering-theory & Practice 19(7): 98-101

The Fuzzy Seepage Theory Based on Fuzzy Structuring Element Method

Fu Yi, Bing Liang, and Ying-he Guo

Liaoning Technical University, 123000, Fu-xin, Liao-ning, China
Yifu9716@163.com

Abstract. Fuzzy structuring element is a new method in solving analysis express problem of fuzzy number and fuzzy value function. The paper starts with basic conception of fuzzy structuring element, introduces the method of expressing fuzzy number by fuzzy structuring element, and puts up with analysis expression of fuzzy value function, discuss the existence of fuzzy qualified differential equation and gives the express method of it's solution. By analyzing the uncertainty of seepage rate in landfill, leads-in the conception of fuzzy seepage rate, establishes fuzzy seepage model of landfill gas migrate, obtains analysis expression of one-dimension ideal fuzzy seepage model by using of fuzzy structuring element method. This gives a new mathematical method in studying seepage problem in complex geological circumstance.

Keywords: Fuzzy Seepage Theory, Fuzzy Structuring Element, landfill gas seepage.

1 Introduction

In the exploitation of underground fluid resource (petroleum, natural gas, underground water etc.) and hazard resistant (environmental protection, outburst, nuclear waste storage etc.) and other correlated engineering field, seepage mechanics is widely used and given important basic theory and technical method. The seepage rate is considered as a definitive value in classical seepage theory. In fact, because of material's heterogeneous and anisotropic and lacking advanced testing means at present, seepage rate difficult to be confirmed accurately. Uncertainty and immeasurability of seepage rate make it a fatal difficulty for studying seepage theory by classical mathematical physics.

Landfill gas migration is a typical seepage problem. Massmann J.W.[1] gave semi-analysis and numerical model of landfill gas migration in the consideration of gas pressure and density variation lesser; Anderson and Callinan[2] established the landfill gas cross-migration model by diffusion theory and Darcy's law; Jia-jun Chen[3] established the landfill gas migration model in the consideration of medium moisture content variation. The common characteristic for these models is considering internal landfill as isotropic porous medium and seepage rate as constant. This disagrees with the fact of inner landfill heterogeneous and anisotropic. In fact, the inner landfill

B.-Y. Cao (Ed.): Fuzzy Information and Engineering (ICFIE), ASC 40, pp. 464–471, 2007.
springerlink.com

seepage rate is not definite. According to the characteristic of uncertain seepage rate, Si-cong Guo[4] studied fuzzy numerical solution of stope gas seepage for irregular pore medium. The author of the paper considered seepage rate as a fuzzy value function, established fuzzy seepage model of landfill gas, and then proved its fuzzy solution expressible by using of structuring element method in fuzzy analysis. By with the author gave the fuzzy seepage equation's analysis expression of one-dimensional ideal condition, and obtained fuzzy distribution of gas pressure in inner landfill. This not only achieved a new method for studying gas migration law in heterogeneous medium, but also provided theory foundation for designing and management of landfill gas control system.

2 Fuzzy Structuring Element and Fuzzy Value Function

2.1 Fuzzy Structuring Element

Definition 1: Assume E is a fuzzy set in real number field, membership function is E(x), $x \in R$, if E(x) meet the following properties:

1) $E(0)=1, \quad E(1+0)=E(-1-0)=0$;

2) E(x) is monotone increasing left-continuous function in interval [-1, 0), and monotone decreasing right-continuous function in interval (0, 1];

3) E(x)=0, when $-\infty < x < -1$ or $1 < x < +\infty$;

So call E is fuzzy structuring element in R.

Definition 2: Assume E is fuzzy structuring element in R, if meet:

1) $\forall x \in (-1,1), E(x) > 0$;

2) E(x) is continuous and strict monotone increasing in interval [-1, 0), and continuous and strict monotone decreasing in interval (0, 1];

So call E is canonical fuzzy structuring element.

From the fuzzy structuring element definition we can know, fuzzy structuring element is a normal convex fuzzy set in R, a idiosyncratic fuzzy number. The following is a typical fuzzy structuring element:

Assume fuzzy set E with membership function:

$$E(x)=\begin{cases} 1+x & x \in [-1,0] \\ 1-x & x \in (0,-1] \\ 0 & other \end{cases}$$

Call it as triangle structuring element, Fig.1 is membership function expressing by graph:

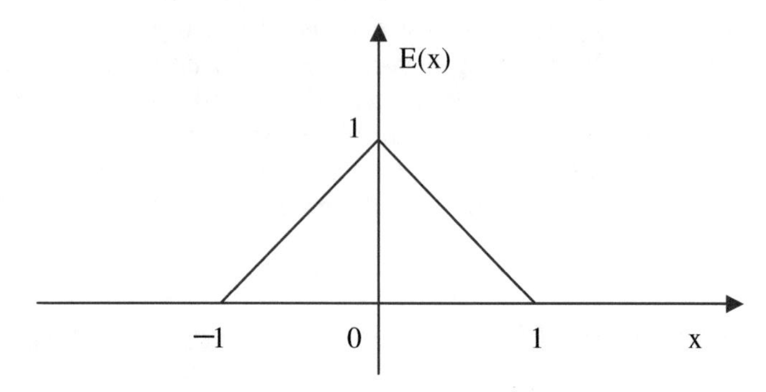

Fig. 1. Triangle Structuring element

2.2 Analysis Expression of Fuzzy Value Function

Theorem 1: For a given canonical fuzzy structuring element E and random bounded closed fuzzy number A, always exist a monotone bounded function in interval [-1, 1], make $A = f(E)$.

For two-dimension real space $X \times Y$, E is a canonical fuzzy structuring element in Y, assume $g(x, y)$ is binary function in $X \times Y$, and for random $x \in X$, $g(x, y)$ is always a monotone bounded function about Y in interval [-1, 1], call $\tilde{f}_E(x) = g(x, E)$ is a fuzzy value function generated from fuzzy structuring element E in X, abridged as $\tilde{f}(x)$.

Theorem 2: For a given canonical fuzzy structuring element E and random bounded fuzzy value function $\tilde{f}(x)$, always exist a binary function $g(x, y)$ in $X \times Y$, and for random $x \in X$, $g(x, y)$ is always a monotone bounded function about Y in interval [-1, 1], make $\tilde{f}_E(x) = g(x, E)$.

Theorem 3: For a random finite fuzzy number A in R, always exist a finite real number α, β, make $A = \alpha + \beta E$.

We can see theorem proving mentioned above in reference [5], theorem 1 and theorem 2 respectively give expressive method of fuzzy number and fuzzy value by fuzzy structuring element, theorem 3 explicate that any finite fuzzy number can be obtained by dilate and translate transformation from fuzzy structuring element. This can solve the problem of fuzzy seepage rate expressed in fuzzy seepage theory.

3 Establish of Fuzzy Seepage Model

3.1 Classical Landfill Gas Seepage Model

Landfill gas migrate speed is larger than percolating water, so we approximately consider it as single-phase flow when establish landfill gas seepage mathematics model. According to the dynamics theory of porous media flow, we assume mini-deformation and percolating water migrate is more stable than landfill gas, and consider landfill gas as ideal gas. Gas pressure varies slightly. The basic equation of landfill gas seepage is:

$$S_S \frac{\partial p}{\partial t} = \nabla \cdot \left| \frac{\bar{k}}{\mu} (\nabla p + \bar{\rho} g \nabla z) \right| + q \tag{1}$$

Where: $S_S = \frac{S_G n_e W_m}{\bar{\rho} RT}$, S_G ——rate of aeration; n_e ——effective drainage porosity; W_m ——mole mass, kg/mol ; $\bar{\rho}$ ——average gas density, kg/m^3 ; R ——ideal gas constant, $m^3 \cdot Pa/(mol \cdot K)$; T ——temperature, K; p ——pressure, Pa; t ——time, d; $\bar{k}$ ——tensor of seepage rate, m^2; μ ——gas coefficient of viscosity, Pa·s; g ——gravity acceleration, m/s^2; z ——vertical position, m; q ——source-sink item, l/d。

Intact landfill gas seepage model need essential solve condition, next initial condition and boundary condition are considered in the model:

1) Initial condition:

$$p\big|_{t=0} = p_0 \tag{2}$$

2) Boundary condition:

$$p\big|_{\Gamma_1} = p_1 \tag{3}$$

$$-\vec{n} \cdot \left| \frac{\bar{k}}{\mu} (\nabla p + \bar{\rho} g \nabla z) \right|_{\Gamma_2} = q_2 \tag{4}$$

Where: Γ_1 ——the first boundary condition(known gas pressure boundary); Γ_2——the second boundary condition(known flux boundary). Equation (1)(2)(3)(4) composite landfill gas seepage mathematical model.

3.2 Fuzzy Qualified Differential Equation

Definition 3: Assume $\tilde{G}(t)$ is a given fuzzy value function, from the theory of fuzzy value function, $\forall \lambda \in (0,1]$, λ -cut-set of $\tilde{G}(t)$ is a interval value, $G_\lambda(t) = [g_\lambda^-(t), g_\lambda^+(t)]$, call common function $g_\lambda^-(t)$ and $g_\lambda^+(t)$ are λ-qualified function. All the λ-qualified function of $\tilde{G}(t)$ notate as C_λ.

Definition 4: Assume $\tilde{G}(t)$ is a known fuzzy value function, $g(t)$ is a λ-qualified function of $\tilde{G}(t)$, $\mu_{\tilde{G}(t)}(g(t)) = \lambda$. For a given differential equation $F(t; x; g(t)) = 0$ with parameter $g(t)$, assume equation has solution at initial condition and its solution is $x(t, g(t))$, call $F(t; \tilde{x}; \tilde{G}(t)) = 0$ is fuzzy qualified differential equation, define its solution as:

$$\tilde{x}(t) = \bigcup_{g(t) \in C_\lambda} \lambda \cdot x(t, g(t)) \tag{5}$$

It is apparent of fuzzy qualified differential equation and its solution, $\tilde{G}(t)$ is fuzzy variation parameter in differential equation, $g(t)$ is common non-fuzzy parameter, the degree of membership of $\tilde{G}(t)$ is λ, and then, the degree of membership between $x(t, g(t))$, the solution of common differential equation $F(t; x; g(t)) = 0$ with parameter $g(t)$ and $\tilde{x}(t) = \bigcup_{g(t) \in C_\lambda} \lambda \cdot x(t, g(t))$, the solution of fuzzy differential equation $F(t; \tilde{x}; \tilde{G}(t)) = 0$ with parameter $\tilde{G}(t)$ is λ.

Substitute certain parameter known by people in common differential equation with fuzzy parameter, then we get fuzzy qualified differential equation. Whereas, If we substitute fuzzy parameter in fuzzy qualified differential equation with definite parameter, we will get common differential equation. We call the equation as describe equation of fuzzy qualified differential equation, and describe parameter for corresponding definite parameter.

3.3 Fuzzy Seepage Equation

Based on the definition of fuzzy qualified differential equation, we lead-in the conception of fuzzy seepage rate, take the seepage rate of random site (x, y, z) in landfill for a fuzzy number, notated as $\tilde{k} = \tilde{k}(x, y, z)$, so seepage rate in all zone is a fuzzy function. And then, fuzzy seepage equation of landfill gas migrate mathematical model is:

$$S_S \frac{\partial p}{\partial t} = \nabla \cdot \left| \frac{\tilde{k}}{\mu} (\nabla p + \bar{\rho} g \nabla z) \right| + q \tag{6}$$

$$p\big|_{t=0} = p_0 \tag{7}$$

$$p\big|_{\Gamma_1} = p_1 \tag{8}$$

$$-\vec{n} \cdot \left| \frac{\tilde{k}}{\mu} (\nabla p + \bar{\rho} g \nabla z) \right|_{\Gamma_2} = q_2 \tag{9}$$

4 Fuzzy Solution of Fuzzy Seepage Model

4.1 Analysis Solution of One-Dimension Ideal Seepage Model

The model of landfill gas seepage is a nonlinear partial differential equation, it has analysis solution only in simple condition. The paper emphasis on analysis solution of fuzzy seepage, so study one-dimension seepage problem for solves conveniently. Initial pressure known as p_x, left side pressure changes from p_0 to p_1 suddenly at $t = 0$, right side is no aeration boundary. The mathematical model of such problem is:

$$\begin{cases} S_S \dfrac{\partial p}{\partial t} = \dfrac{\partial}{\partial x} \left| \dfrac{k}{\mu} \dfrac{\partial p}{\partial x} \right| & 0 \le x \le L \\ p\big|_{t=0} = p_x & 0 \le x \le L \\ p\big|_{x=0} = p_1, & \left.\dfrac{\partial p}{\partial x}\right|_{x=L} = 0 \end{cases} \tag{10}$$

According to the mathematical physics, we first lead-in intermediate variable transfer above model to homogeneous, then make the model sine transform and solve, and inverse sine transform, at last we obtain the solution of this problem:

$$p = p_x + (p_1 - p_0) F(x, t, k) \tag{11}$$

$$F(x,t,k)=1-\frac{4}{\pi}\sum_{n=0}^{\infty}\frac{1}{2n+1}\cdot\sin\frac{2n+1}{2}\pi\frac{x}{L}\cdot e^{-\frac{1}{4}(2n+1)^2\pi^2\frac{tk}{L^2 S_S \mu}} \tag{12}$$

4.2 Expressible Problem of Fuzzy Differential Equation Solution

If equation only has one fuzzy parameter, then call it as one-parameter fuzzy differential equation. For such equation, if describe parameter of correspondence describe equation and solution of the equation are monotonic, the solution of this fuzzy differential equation can be expressed. Assume $F(t;\tilde{x};\tilde{G}(t))=0$ is one-parameter fuzzy qualified differential equation, correspondence describe equation is $F(t;x;g(t))=0$, if common differential equation with parameter $g(t)$ has solution in given solve condition D_0 and the solution is $f(t)=f_s(g(t),D_0)$, so the solution of fuzzy qualified differential equation is:

$$\tilde{x}(t,D_0)=f_s(g(t),D_0)\Big|_{g(t)=\tilde{G}(t)}=f_s(\tilde{G}(t),D_0) \tag{13}$$

For fuzzy seepage model of landfill gas migrate, equation (10) is it's describe equation of one-dimension ideal fuzzy seepage, the solution of above describe equation is monotone decreasing. For fuzzy seepage model with only one-parameter fuzzy seepage rate $\tilde{k}$, its solution can be expressed.

4.3 Analysis Expression of Fuzzy Seepage Model

For describe parameter fuzzy seepage rate $\tilde{k}$, it generate from fuzzy structuring element. Assume $\tilde{k}=\alpha(x)+\beta(x)E$, where $\alpha(x)$ is translation function, it describe regularities of distribution of seepage in landfill; $\beta(x)$ is dilation function, it give the distribution of fuzzy degree; E is a canonical fuzzy structuring element. Thus the fuzzy solution of landfill gas one-dimension ideal fuzzy seepage model is:

$$\tilde{p}=p_x+(p_1-p_0)F(x,t,\tilde{k}) \tag{14}$$

$$
\begin{aligned}
F(x,t,\tilde{k}) &= 1-\frac{4}{\pi}\sum_{n=0}^{\infty}\frac{1}{2n+1}\cdot\sin\frac{2n+1}{2}\pi\frac{x}{L}\cdot e^{-\frac{1}{4}(2n+1)^2\pi^2\frac{t}{L^2 S_S\mu}(\alpha+\beta E)} \\
&= 1-\frac{4}{\pi}\sum_{n=0}^{\infty}\frac{1}{2n+1}\cdot\sin\frac{2n+1}{2}\pi\frac{x}{L}\cdot e^{-\frac{1}{4}(2n+1)^2\pi^2\frac{t\alpha}{L^2 S_S\mu}}\cdot e^{-\frac{1}{4}(2n+1)^2\pi^2\frac{t\beta E}{L^2 S_S\mu}}
\end{aligned}
\tag{15}
$$

5 Conclusions

The paper starts with basic conception of fuzzy structuring element, introduces the method of expressing fuzzy number by fuzzy structuring element, and puts up with analysis expression of fuzzy value function, discuss the existence of fuzzy qualified differential equation and gives the expressive method of it's solution. By analyzing the uncertainty of seepage rate in landfill, leads-in the conception of fuzzy seepage rate, establishes fuzzy seepage model of landfill gas migrate, obtains analysis expression of one-dimension ideal fuzzy seepage model by using of the method of fuzzy structuring element. This gives a new mathematical method in studying fuzzy theory and a new method for the designing of landfill gas control system. At the same time, fuzzy seepage theory can be used in underground water seepage in complex geological circumstance. It also has important theory significance and applicable value in exploitation of coal seam gas and underground petroleum pollution.

References

1. Massann J.W (1989) Applying groundwater models in vapor extraction system design [J].Journal of Environmental Engineering. 129-149
2. Anderson D.R. and Calllnan J.P (1971) Gas generation and movement in landfills [J]. Proceeding of National Industrial Solid Waste Management Conference, University of Houston, USA。
3. Jia-jun CHEN, Hong-qi WANG and jin-sheng WANG (1999) Numerical model of landfill gas migration [J]. Environmental science.93~96
4. Si-cong GUO (2004) Study on fuzzy numerical solution of stope gas seepage flow of non-uniform pore medium [J]. Science technology and engineering.99~102
5. Si-cong GUO (2004) Principle of fuzzy mathematical analysis based on structured element [M]. Shenyang: North-east university press.

Ecological Safety Comprehensive Evaluation on Mineral-Resource Enterprises Based on AHP

Zheng Yunhong[1] and Li Kai[2]

School Of Business Administration, Northeastern University,
Shenyang, P.R. China, 110004
yhzheng@mail.com.cn

Abstract. This paper confirmed AHP as the evaluation basic method, set up a universal evaluation index system of mineral-resource enterprises ecological safety. This index system consisted of twenty indexes, it contained three aspect of environment pressure and quality and resources environment protection ability, so the scientificity and integrity of index system were ensured. Based on the characteristic, level and relevant standard of every evaluation index, evaluation criterion and the relevant grade were established. An application example was giving showing the feasibility and effectiveness of the proposed approach. After empirical analysis, pointed out that evaluating the ecological environment safety of mine-enterprise is not only necessary but also the base of building ecological safety alarming mechanism. Meanwhile, it is the decision base of executing the sustainable development strategy in mine-enterprise.

Keywords: mineral-resource enterprise, ecological safety, index system, AHP.

1 Introduction

Ecological safety evaluation is mainly used on the evaluation about national or regional ecological environment condition [1]. It is almost not used on evaluating these enterprises whose productive activities causing serious influence on the ecological environment. In fact, it is necessary to evaluate the ecological safety of mineral-resource enterprises (MRE for short) [2,3,4], bring this work into production safety management will change the enterprises' present condition of "first pollution and wasting resources, then administer", decrease enterprises' cost of production and increase enterprises' benefits of economy and society. It is also an important basic work of preparing for the foundation of alarming mechanism in enterprises. Because of these above, after confirming the intension of the ecological safety evaluation about MRE, considering these enterprises' features, this article set up the sustainable ecological safety index system, using for reference of the experiences of the region ecological safety evaluation, made out a method of comprehensive evaluation. And on the basis of it, a demonstration analyses be finished.

B.-Y. Cao (Ed.): Fuzzy Information and Engineering (ICFIE), ASC 40, pp. 472–480, 2007.
springerlink.com

2 Method of Ecological Safety Evaluation for MRE

2.1 Meaning of Ecological Safety Evaluation for MRE

Ecological safety for MRE is that in the course of developing mine resources, preventing enterprises from threatening the environment of themselves and surrounding areas, caused by their unsuitable guiding ideology, technology and ways. It includes three meanings:

- Poor productive environmental quality, especially poisonous chemicals that threaten worker's health;
- Release of industrial wastes threaten surrounding environment;
- Disorder exploitation for resource threatens the sustainable development of enterprises.

This evaluation is guided by the thought of sustainable development and makes a ration description of the safety quality of mine ecological environmental system. It is to evaluate the status of MRE and the subsequently bad result caused by it after mine ecological environment or natural resources is influenced by more than one threatening factors, in order to offer scientific proofs for mine developmental strategy on the basis of realizing and maintaining mine ecological environmental safety. It contains that according to the results of evaluation, providing decision-maker with reports and applying the results to mine strategy.

The content of ecological safety evaluation for MRE includes two parts:

- The condition and perspective of mine ecological safety, evaluating department should provide the decision-maker with relative information about it;
- Key factors, order and scale that can influence miner ecological safety, ascertaining main influential factors, scale and degree [1,5].

2.2 Index System of Ecological Safety Evaluation

Although mining industry has no sustainable development features, but the exploitation speed and the degree of destructing environment could be controlled. The precondition is that the behaviors of government, enterprise and person should be guided by the sustainable development strategy. Thus, the evaluation should start from the view of sustainable development, according to the general indicator system for regional sustainable development ecological safety evaluation, combine the productive features of MRE, regard resources and ecological environment as the center, make use of Delphi's method, finish the selection of index and decide on evaluation index system from three aspects of resources ecological environmental pressure, resources environment quality and resources environmental protection ability. It can be used as a general index system of ecological safety evaluation for MRE, the enterprises can make relative regulation in accord with their own production feature (Table1) [1,5,6,7,8].

Table 1. The ecological safety evaluation index system of MRE（O）

A_i	A_{ij}		A_{ijk}	
Resources ecological environmental pressure A_1	Land pressure	A_{11}	Resources output per excavation unit	A_{111}
			Land area mined per ton	A_{112}
	Water pressure	A_{12}	Quantity of water used	A_{121}
			Burthen of industrial waste water on per unit water resource	A_{122}
	Pollutant burthen	A_{13}	Burthen of the three industrial wastes on per unit land area	A_{131}
			Burthen of dressing powder on per unit land area	A_{132}
Resources environment quality A_2	Resources quantity	A_{21}	Loss rate of the unit ore	A_{211}
			Dilution of the unit ore	A_{212}
			Dressing recovery rate of the unit ore	A_{213}
	Ecological environment quantity	A_{22}	Coverage of forest network of mine area	A_{221}
			Air quantity index of mine area	A_{222}
			Occurrence rate of geological disaster	A_{223}
			Loss rate of geological disaster	A_{224}
Resources environmental protection ability A_3	Input ability	A_{31}	Total production value per capita	A_{311}
			Input-intensity of ecology constructing	A_{312}
			Input-intensity on pollution control	A_{313}
	Technology ability	A_{32}	Treatment rate in three industrial wastes	A_{321}
			Comprehensive utilization rate of solid waste	A_{322}
			Proportion of R&D expenditure in per ten thousand total production value	A_{323}
			Quality of staffs in the mine	A_{324}

Where

O denotes mine resources ecological safety target level index;
A_i represents system level index;
A_{ij} stands for state level index;
A_{ijk} is variable level index.

2.3 Method of Evaluating the Ecological Safety of MRE

The evaluation on ecological safety usually applies the way of multiple indices, landscape ecology, evaluation of the ecological safety capacity, analytic hierarchy process (AHP for short), and so on. Among these methods, AHP is the most suitable way to evaluate the ecological safety of MRE, because the systematic diversity of sustainable development demands that the weighting applied during comprehensive evaluation should be determined through the combination of qualitative and quantitative. While the way AHP determines the weighting accords better with the demands for systematic sustainable development [1].

The evaluating method consists of the following major steps [1,8,9].

(1) Confirm evaluation object, analyses the correlation of the ecological environment elements and find out the correlation of them.

(2) Creation of an evaluating hierarchical structure of ecological safety for MRE under the precondition of sustainable development. The interrelated elements that

influence ecological safety can be fallen into different hierarchy, then constructing a tree-like hierarchical structure.

(3) Building judgment matrix. The element's value of judgment matrix reflected that how people understand the significance (or priorities, prejudice, strength. etc) of each influence factor in the hiberarchy structure of safety evaluation to the exploitation of mine resources. Then the experts carry on judging grade to the elements of each level by means of the pairwise comparisons of importance, making use of the judgment matrix eigenvectors confirm the contribution degree of the lower lever index to the upper level index, thus we can get the ranking result of the importance for variables level index to goal level index.

(4) Confirming the evaluation index weighting. According to the principle of AHP, the index weighting vectors of each level can be ensured. If the evaluation indices in first level be represented as A_i(i=1 to 3), then the corresponding weighting vectors is $W=(w_1,w_2,w_3)$; If the evaluation indices in second level be represented in turn as A_{1j}(j=1 to 3), A_{2j}(j=1 to 2), A_{3j}(j=1 to 2), then the corresponding weighting vectors are $W_1=(w_{11},w_{12},w_{13})$, $W_2=(w_{21},w_{22})$, $W_3=(w_{31},w_{32})$; If the evaluation indices in third level be represented in turn as A_{11k} (k=1 to 2), A_{12k} (k=1 to 2), A_{13k} (k=1 to 2), A_{21k} (k=1 to 3), A_{22k} (k=1 to 4), A_{31k} (k=1 to 3), A_{32k} (k=1 to 4), then the corresponding weighting vectors are $W_{11}=(w_{111},w_{112})$, $W_{12}=(w_{121},w_{122})$, $W_{13}=(w_{131},w_{132})$, $W_{21}=(w_{211},w_{212},w_{213})$, $W_{22}=(w_{221},w_{222},w_{223},w_{224})$, $W_{31}=(w_{311},w_{312},w_{313})$, $W_{32}=(w_{321},w_{322},w_{323},w_{324})$. The combination weighting (W_{ci}) that the evaluation indices in third level about evaluation object should be confirmed, we assume combination weighting be represented as following:

$$W_{ci} = \left(w_{c1}, w_{c2}, w_{c3}, \cdots\cdots w_{c20}\right)$$

where

ci=the number of the indices in third level

(5) Level ranking and consistency inspection. Carrying on the single structure taxis of each level first, then calculating the opposite eigenvectors of judgment matrix eigenvalue for each level. After eigenvector be standardized, it become the relative importance ranking weighting of the element of certain level for the corresponding element of upper level. Then do the consistency inspection to judgment matrix and calculate the consistency rate CR. When $CR \leqslant 0.1$, the single structure taxis of each level is thought as a result of satisfied consistency, otherwise the element value of judgment matrix need to be adjusted. Secondly, carrying on the total taxis of the hierarchy and inspecting the hierarchies' consistency. According to the sequence from the highest level to the lowest level, calculate the relative importance taxis weighting of

all elements in the same level, get the relative weighting of each element that influence ecological environment safety in the process of exploiting mine resources, in the meantime, the consistency inspection for the total taxis of hierarchy can be done. When $CR \leqslant 0.1$ (CR is the stochastic consistency rate for the total taxis of the hierarchy, the total taxis of the hierarchy be thought as a result of satisfied consistency, otherwise the element value of judgment matrix need to be adjusted.

(6) Confirm evaluation criteria. According to the difficult in the process of getting the relative datum, the relative criterion about international, national and regional will be chosen.

(7) Comprehensive evaluation. The ecological safety evaluation for MRE should first be based on safety. We begin from the unsafe perspective, denoted by unsafe index, then calculating the safety index according to the unsafe index. Because there is only one evaluation object, the original datum be standardized in terms of the below method, at the same time, building the evaluation model to confirm the unsafe index value of each indexes.

In order to express clearly, we assume X_i is evaluation index value of the index i in third hiberarchy (where i=1 to n represent the number of index), $P(C_i)$ is unsafe index value of the index i (where C_i is the number of index, i=1 to n, $0 \leq P(C_i) \leq 1$), XS_i is the criteria value for evaluation index i.

a. For the index that is safer with a larger value:

Suppose the safety value to be the criteria value.

$$\text{If } X_i \geq XS_i, \text{then } P(C_i)=0\text{; If } X_i < XS_i, \text{then } P(C_i)=1-X_i/XS_i \times 100\%$$

Suppose the unsafe value to be the criteria value.

$$\text{If } X_i \leq XS_i, \text{then } P(C_i)=1\text{; If } X_i > XS_i, \text{then } P(C_i)=XS_i/X_i \times 100\%$$

b. For the index that is safer with a smaller value:

Suppose the safe value to be the criteria value.

$$\text{If } X_i \leq XS_i, \text{then } P(C_i)=0\text{; If } X_i > XS_i, \text{then } P(C_i)=1-XS_i/X_i \times 100\%$$

Suppose the unsafe value to be the criteria value.

$$\text{If } X_i \geq XS_i, \text{then } P(C_i)=1\text{; If } X_i < XS_i, \text{then } P(C_i)=X_i/XS_i \times 100\%$$

c. Translate the unsafe value into safe value, with $Q(C_i)$ indicating safe value, then $Q(C_i)=1-P(C_i)$

(8) Analysis of the result. Confirm the evaluating graded and corresponding standard firstly, then present out the comment collection for ecological safety of MRE

$$V=\{v_1, v_2, v_3, v_4\}$$

Where v_j is evaluation result, j is evaluation grading, j=1 to 4. This paper defines ecological safety degree of a mine in four levels(table 2) [1,8,9].

Table 2. Evaluation rank and corresponding value

Degree of safety	judgment	value
v_1	dangerous	$0.00\le V<0.25$
v_2	relatively unsafe	$0.25\le V<0.50$
v_3	relatively safe	$0.50\le V<0.75$
v_4	safe	$0.75\le V\le 1.00$

According to the following formulae, the ecological safety index value of evaluation indices for MRE can be calculated. Then we can confirm the ecological safety condition of MRE by table 2.

for each index

$$Q(O-C_i)=W_{ci}\times Q(C_i) \tag{1}$$

the general ecological safety index value of MRE is:

$$Q(C_i)=\sum_{1}^{n}W_{ci}\times Q(C_i) \tag{2}$$

(W_{ci} is the weighting of index i, i=1 to n)

3 Example for the Applying of Comprehensive Evaluation on Ecological Safety of MRE

In order to illustrate the ecological safety evaluation of MRE more thoroughly, this thesis takes Z gold mine company as an example to give a comprehensive evaluation on its ecological safety according to the upper index system and evaluation model. The evaluating process as following:

(1) Building the evaluating index system. We can use the general ecological safety evaluation index system for MRE as table1.

(2) Confirming weighting of the evaluation index. After certain experts' pairwise comparison and judgment of the relative importance of each level, gathering the evaluation results come from the experts. Then, with the method of square root we can get the relative importance judgment matrix of each evaluation index and the relative weightings of the elements of each level (table 3).

Table 3. Relative weightings of the elements of each level

Level	Index vector	Weighting vector
First level	A_i (i=1 to 3)	W=(0.333,0.333,0.333)
Second level	A_{1j} (j=1 to 3)	W_1=(0.349,0.484,0.168)
	A_{2J}(J=1 to2)	W_2=(0.333,0.667)
	A_{3j}(j=1 to 2)	W_3=(0.333,0.667)
Third level	A_{11k}(k=1 to 2)	W_{11}=(0.667,0.333)
	A_{12k}(k=1 to 2)	W_{12}=(0.333,0.667)
	A_{13k}(k=1 to 2)	W_{13}=(0.667,0.333)
	A_{21k}(k=1 to 3)	W_{21}=(0.493,0.196,0.311)
	A_{22k}(k=1 to 4)	W_{22}=(0.109,0.189,0.351,0.351)
	A_{31k}(k=1 to 3)	W_{31}=(0.163,0.540,0.297)
	A_{32k}(k=1 to 4)	W_{32}=(0.413,0.292,0.108,0.187)

The combination weighting (Wci) that the evaluation indices in third level about evaluation object:

Wc_i =(0.078,0.039,0.054,0.108,0.037,0.019,0.055,0.022,0.035,0.024,0.042,0.078,0.078,0.018,0.060,0.031,0.092,0.065,0.024,0.042)

(3) Consistency inspection to judgment matrix. The consistency rate can be calculated as follows:

$$CR = CI/RI$$

Where

CI is the stochastic consistency index, RI is average stochastic consistency index, if $CR \leqslant 0.1$, then the consistency of the comparisons judgment matrix can be accepted. Through inspection, the consistency rate of judgment matrix about evaluation index in every level is: $CR \leqslant 0.1$.

(4) Confirm evaluation criterion and calculate the safety index value. The determining of ecological safety evaluating criterion of MRE is a significantly explorative work. There are several ways to determine it, with the consideration of the peculiarities of MRE and the trait of each index, we choose criterion as follows:

- Based on the international average level.
- Based on the national average level.
- For the index that lacks statistics date but counts much important in the index system, we can ask the specialists to determine it.

The ecological safety index value of evaluation indices can be calculated with formula (1), and the general ecological safety index value of MRE can be confirmed with formula (2). The evaluating result is presented in table 4.

Table 4. The result of ecological safety evaluation for Z mine company

index	Safety index value $Q(C_i)$		weight (W_{Ci})	Safety degree (score) $W_{ci}\times Q(C_i)$		Safety degree (score) $\sum W_{Ci}\times Q(C_i)$		Safety degree (score) $\sum W_{Ci}\times Q(C_i)$	
	1998	2002		1998	2002	1998	2002	1998	2002
Gold output per excavating unit (kg/t)	0.748	0.924	0.078	0.058	0.072	0.097	0.111	0.179	0.232
Land area mined per ton (m^2/t)	1.000	1.000	0.039	0.039	0.039				
Quantity of water used (t)	0.746	1.000	0.054	0.040	0.054	0.040	0.065		
Burthen of industrial waste water on the unit water resource %	0.000	0.101	0.108	0.000	0.011				
Burthen of the three industrial wastes on the unit land area (kg/m^2)	0.643	0.997	0.037	0.024	0.037	0.043	0.056		
Burthen of dressing powder on the unit land area (kg/m^2)	1.000	1.000	0.019	0.019	0.019				
Loss rate of the unit ore %	1.000	1.000	0.055	0.055	0.055	0.109	0.112	0.304	0.333
Dilution of the unit ore %	0.879	1.000	0.022	0.019	0.022				
Dressing recovery rate of the unit ore %	1.000	1.000	0.035	0.035	0.035				
Coverage of forest network of mine area %	0.000	0.977	0.024	0.000	0.023	0.195	0.221		
Air quantity index of mine area	0.930	1.000	0.042	0.039	0.042				
Occurrence rate of geological disaster %	1.000	1.000	0.078	0.078	0.078				
Loss rate of geological disaster %	1.000	1.000	0.078	0.078	0.078				
Total production value per capita (yuan/person)	0.739	0.885	0.018	0.013	0.016	0.053	0.061	0.193	0.247
Input-intensity of ecology constructing %	0.166	0.237	0.060	0.009	0.014				
Input-intensity on pollution control %	1.000	1.000	0.031	0.031	0.031				
Treatment rate in three industrial wastes %	0.600	0.920	0.092	0.056	0.085	0.140	0.186		
Comprehensive utilization rate of solid waste %	0.773	1.000	0.065	0.050	0.065				
Proportion of R&D expenditure in per ten thousand total production value %	0.021	0.030	0.024	0.001	0.001				
Quality of staffs in the mine %	0.814	0.841	0.042	0.034	0.035				
The general ecological environment safety degree of Z mine company								0.677	0.812

(5) Comprehensive evaluation

The ecological safety condition of MRE can be graded in the way of comprehensive grading, with ecological safety index value indicating ecological environmental safety condition, the safety degree of the evaluating object can be confirmed by table 2. The larger the safety index value is, the higher the ecological safety degree of MRE is.

Through the comprehensive evaluation on the ecological safety condition of Z mine company in 1998 and 2002, we get the result that, the general ecological safety index value is 0.677 in 1998, which is in a condition of relatively safe; The general ecological safety index value is 0.812 in 2002, which is in the condition of safe. From this we can see the significant improvement of Z company on ecological safety condition from 1998 to 2002.

4 Conclusion

With the declination of the quality of resource, the disturbance and destroying to the ecological environment will increase with the same excavating quantity. If the mine-enterprises don't protect the environment and treat the position in the process of exploring, the natural capital input in mine resource exploration will increase. From the perspective of mine-enterprises, they can acquire short-term profit without investing in environmental protection but for the long-term profit in sustainable development, the loss will be severe.

From the analysis above we can determine that, ecological safety evaluation as a method of measuring the safety of ecological system, can be applied not only in the eco-environmental quality evaluation in a country or a region, but in the component of social-economical system such as mine-enterprise, as an evaluation on its ecological safety. Through this work, the continual usage of irrenewable mine resource and the self-sustainable development of MRE can be realized. Thus, it is absolutely necessary to bring ecological safety evaluation into the managing process of MRE, moreover, this work is the important basis for the foundation of environmental safety pro-warning system in MRE and simultaneously, it is an important decision base of executing the sustainable development strategy in MRE.

References

[1] Liu Y, Liu YZ,(2004) Ecological security evaluation for land resource. Resources Science 5, 69-75, (in Chinese)
[2] Bybochkin,A.M., (1995) Ecological safety in prospecting and development of deposits of solid mineral resources. Bezopasnost Truda v Promyshlennosti 2, 2-6.
[3] Lzumi, Shigelcazu,(2004) Environmental safety issues for semiconductors (Research on scarce materials recycling). Thin Solid Films 8, 7-12.
[4] Peter Calow,(1998) Ecological risk assessment: risk for what? how do we decide?. Ecotoxicology and Environmental Safety 40, 15-18.
[5] Sun XG., (2003) Research on the measurement of sustainable development. Dalian:Dongbei University of Finance & Economics Press,100-134, 152-160. (in Chinese)
[6] Friedhelm Korte, Frederick Conlston,(1995) From single-substance evaluation to ecological process concept: the dilemma of processing Gold with Cyanide. Ecotoxicology and Environmental Safety 32, 96-101.
[7] K.N.Trubetskoi; Yu.P.Galchenko, (2004) Methodological Basis of ecological safety Standards for the Technogenic Impact of Mineral Resource Exploitation. Russian journal of ecology 2, 65-70.
[8] Tran LT, Knight CG, O'Nell RV.,(2002) Fuzzy decision analysis for integrat environmental vulnerability assessment of the Mid-Atlantic region. Environment Management 29, 845-859.
[9] E.Cagno, A.Digiulio, P.Trucco,(2000) Risk and causes-of-risk assessment for an effective industrial safety management. International. Journal of Reliability, Quality and Safety Engineering 7, 113-128.

A Note on the Optimal Makespan of a Parallel Machine Scheduling Problem

Yumei Li[1], Yundong Gu[2], Kaibiao Sun[3], and Hongxing Li[3]

[1] School of Information Engineering, Beijing Technology and Business University, Beijing 100037, P.R. China
{brandaliyumei}@hotmail.com

[2] School of mathematics and physics, North China Electric Power University, Beijing 102206, P.R. China
{guyund}@126.com

[3] School of Mathematical Sciences, Beijing Normal University, Beijing 100875, P.R. China
{fuzzyskb}@126.com, {lhxqx}@bnu.edu.cn

Abstract. For the parallel machine scheduling problem under consideration, the authors in two literatures of 1961 and 2002 respectively gave the proofs for the optimal makespan under Level Algorithm. But, some errors in their proofs are found by us with three counterexamples, and no one has given the correct proof until now. In this paper, a new algorithm is proposed. And the new algorithm is more convenient and easier for theoretical analysis than Level Algorithm does. Then, it is showed that the result schedule obtained by using the new algorithm is consistent with that by Level Algorithm in the sense that they can give the same result schedule. Finally, by using the proposed new algorithm, the proof for the optimal makespan is accomplished.

Keywords: Scheduling problem, parallel machine, algorithm, result schedule, optimal makespan.

1 Introduction

In this paper, a parallel machine scheduling problem is considered. Assume that a set of jobs $j \in N = \{1, 2, \cdots, n\}$ with unit processing times $p_j = 1$ has to be processed on m parallel identical machines $1, 2, \cdots, m$. Each machine can process at most one job at a time and each job can be processed on any of the machines. Precedence relations $i \to j$ among the jobs may be given, that is, job j can not start before job i is completed, where i is called a predecessor of j and j is called a successor of i. Furthermore, job i is called a direct predecessor of j and reversely j is called a direct successor of i if there does not exist job k such that $i \to k$ and $k \to j$. We restrict our considerations to intree precedence relations, i.e., each job has at most one direct successor. The problem is to find a feasible schedule minimizing the makespan, i.e. minimizing the completion time of the last job.

For the parallel machine scheduling problem, Level Algorithm is first proposed and a proof for the optimal makespan is given at the same time in [3]. Afterwards,

B.-Y. Cao (Ed.): Fuzzy Information and Engineering (ICFIE), ASC 40, pp. 481–493, 2007.
springerlink.com

another proof for the optimal makespan under Level Algorithm is presented in [7]. By giving three counterexamples, some errors in the proofs for the optimal makespan are found by us in both [3] and [7], although the conclusions in the two literatures are correct. Additionally, in [2], [4], [5], [6] and [8], the authors respectively showed that an optimal schedule can be obtained for the problem by using Level Algorithm, but they did not give any corresponding proof for the optimal makespan. Until now, the proof for the optimal makespan keeps open. In this paper, a new algorithm is proposed for the problem. It is showed that the result schedule obtained by using the new algorithm is consistent with that by Level Algorithm in the sense that they give the same result schedule. But, the new algorithm is more convenient and easier for theoretical analysis. This makes us be able to provide a correct proof for the optimal makespan.

2 Level Algorithm

The whole intree precedence relations can be represented by a graph consisting of n nodes representing jobs and directed arcs representing the precedence relations restrictions. In an intree graph, the single job with no successors is called *root* and is located at *level* 1. The jobs directly preceding the root are at level 2; the jobs directly preceding the jobs at level 2 are at level 3, and so on(see Figure 1). Let $l_{\max}$ be the highest level of the whole intree.

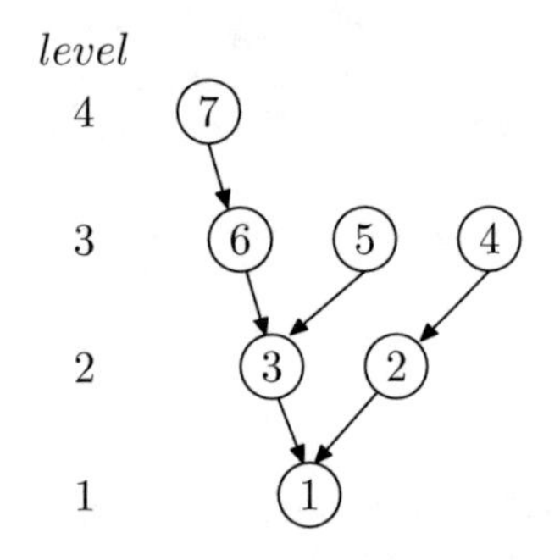

Fig. 1. An intree graph with $l_{\max}$=4

Denote the parallel machine scheduling problem under consideration by $P|p_j = 1, intree|C_{\max}$, where $C_{\max} = \max\{C_j|j \in N\}$, C_j is the completion time of job j in the schedule. Assume that the schedule starts from time $t = 0$. Nodes that are completed are considered to be removed from the graph, and call intree precedence relations composed of the remaining jobs at some time t as the current intree at time t. Level Algorithm proposed in [3] for the problem $P|p_j = 1, intree|C_{\max}$ is summarized in [1] as following:

Algorithm 0: Level Algorithm
Begin
calculate levels of the jobs;
t:=0;
Repeat
construct list L_t consisting of all the jobs without predecessors at time t;
–all these jobs either have no predecessors
–or their predecessors have been scheduled in time interval $[0, t-1]$
Order L_t in nonincreasing order of job levels;
Schedule m jobs (if any) to processors at time t from the beginning of list L_t;
Remove the scheduled jobs from the graph and from the list;
$t := t + 1$;
Until all jobs have been scheduled;
end

By Level Algorithm, among the jobs which are **released** simultaneously at any time t, the jobs at the higher levels are scheduled first by Level Algorithm, where **released at time t** means that the jobs either have no predecessors or their predecessors have been scheduled before time t.

3 The Errors in Proofs for the Optimal Makespan

The number of jobs at level l is denoted by $N(l)$, and the highest level in the whole intree is also denoted by $l_{\max}$. Obviously, the total number of jobs with $level \geq l_{\max} + 1 - r$ is

$$N(l_{\max}) + N(l_{\max} - 1) + \cdots + N(l_{\max} - r + 1) = \sum_{k=1}^{r} N(l_{\max} + 1 - k)$$

where $r = 1, 2, \cdots, l_{\max}$.

By using Level Algorithm for the problem $P|p_j = 1, intree|C_{\max}$, the following two cases are proved in [3] from different aspects. In the case that

$$\max_r \{\frac{\sum_{k=1}^{r} N(l_{\max} + 1 - k)}{r}\} \leq m,$$

they proved that $C_{max} = l_{max}$. In another case that

$$\max_r \{\frac{\sum_{k=1}^{r} N(l_{\max} + 1 - k)}{r}\} > m,$$

they proved that $C_{\max} = l_{\max} + c$, where c is related with $l_{\max}$, m and the number of jobs in each level of the whole intree. Their proofs about $C_{\max} = l_{\max} + c$ are composed of the proofs of $C_{\max} \geq l_{\max} + c$ and $C_{\max} \leq l_{\max} + c$. Although their conclusions are correct, there are some errors in several key steps for proving $C_{\max} \leq l_{\max} + c$ in their argumentation.

3.1 The Error in the Proof of $C_{\max} \leq l_{\max} + c$ in [3]

For an intree with $l_{\max}$ levels, it is clear that $C_{\max} \geq l_{\max}$. In [3], $l_{\max} + c$ is assumed to be the prescribed time to complete all jobs in and let m be a positive integer such that

$$m - 1 \leq \max_r \{\frac{\sum_{k=1}^{r} N(l_{\max} + 1 - k)}{r + c}\} \leq m. \tag{1}$$

Let $P(r)$ be the set of jobs with $level \geq l_{\max} + 1 - k, k = 1, \cdots, r$.

A node j is called a starting node in a graph if there does not exist a node i such that $i \rightarrow j$. When the number of the starting nodes in the intree graph is larger than m from the beginning of the schedule, an integer r' is defined[3] by the following conditions:

1. Using Level Algorithm, remove m starting nodes at a time, at time units $t = 1, \cdots, c'$, until the number of starting nodes is less than m. Remove all the starting nodes in the current graph at $t = c' + 1$, where c' is a non-negative integer.
2. All nodes in $P(r)$ are removed after $c' + 1$ units of time.
3. r' is the largest of such integers. In the proof of $C_{max} \leq l_{max} + c$ in [3], the following inequality is used:

$$\sum_{k=1}^{r'} N(l_{\max} + 1 - k) \geq (c' + \lambda)m, 0 < \lambda < 1.$$

But, we find that the inequality dose not hold. Furthermore, the following counterexample is constructed by us to show that $\sum_{k=1}^{r'} N(l_{\max} + 1 - k) < (c' + \lambda)m$.

Example 1. Consider the following instance of $P|p_j = 1, intree|C_{\max}$ with 16 jobs. The jobs are subject to the precedence relation constraints depicted in Figure 2. Schedule the jobs by using Level Algorithm within $l_{\max} + c$ units of time, where $c = 1$.

Since $c = 1$, by inequality (1), we have $m = 3$. From the result schedule(see Table 1), it follows that $c' = 4$. Also from the definition of r', we have $r' = 4$, i.e. the jobs with $level \geq l_{\max} + 1 - r' = 6 + 1 - 4 = 3$ are all scheduled till time $c' + 1 = 5$ in the result schedule. Thus, $\sum_{k=1}^{4} N(l_{\max} + 1 - k) = 5 + 2 + 2 + 2 = 11$, and $(c' + \lambda)m = (4 + \lambda)3 = 12 + 3\lambda, 0 < \lambda < 1$. Obviously, we have $11 < 12 + 3\lambda, 0 < \lambda < 1$, i.e. $\sum_{k=1}^{r'} N(l_{\max} + 1 - k) < (c' + \lambda)m$ holds. □

Table 1. A schedule under level Algorithm for Example 1

14	11	5	2			
15	12	9	7	4		
16	13	10	8	6	3	1

0 1 2 3 4 6 5 7

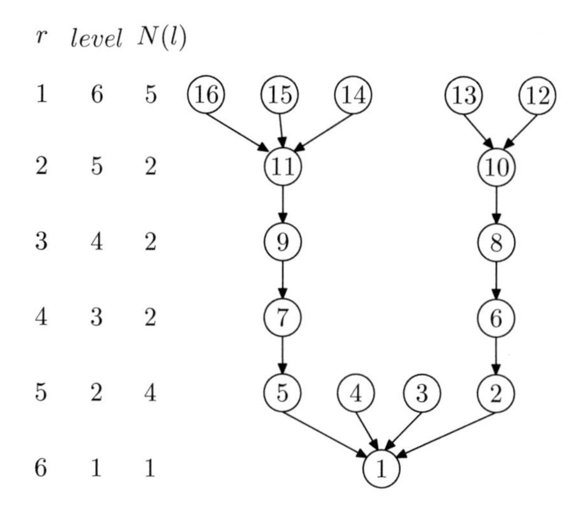

Fig. 2. An intree graph of 16 jobs

3.2 The Errors in the Proof of $C_{\max} \leq l_{\max} + c$ in [7]

First, assume that m machines are given in [7]. Then, in the case that

$$\max_r\{\frac{\sum_{k=1}^{r} N(l_{\max}+1-k)}{r}\} > m,$$

the smallest positive integer c is chosen such that

$$\max_r\{\frac{\sum_{k=1}^{r} N(l_{\max}+1-k)}{r+c}\} \leq m < \max_r\{\frac{\sum_{k=1}^{r} N(l_{\max}+1-k)}{r+c-1}\}.$$

Consider the value of

$$\frac{\sum_{k=1}^{r} N(l_{\max}+1-k)}{r}$$

as a function of r and denote

$$\rho(r) = \frac{\sum_{k=1}^{r} N(l_{\max}+1-k)}{r}, r = 1, 2, \cdots, l_{\max}.$$

Let $r^{''}$ be the largest integer such that $\rho(r) \geq m$. That is, $\rho(r) < m$ is hold for any r with $r^{''} + 1 \leq r \leq l_{\max}$.

Assume that $t' + 1$ is a time satisfying the following conditions:

1. By using Level Algorithm, all jobs at *level* $\geq l_{\max} + 1 - r^{''}$ are completed at time $t' + 1$.

2. In the time interval $[0, t']$, m machines are all used to process jobs.

3. In the time interval $[t', t' + 1]$, some machines are idle and some machines are used to process the remaining jobs at *level* $= l_{\max} + 1 - r^{''}$ or the jobs at lower levels.

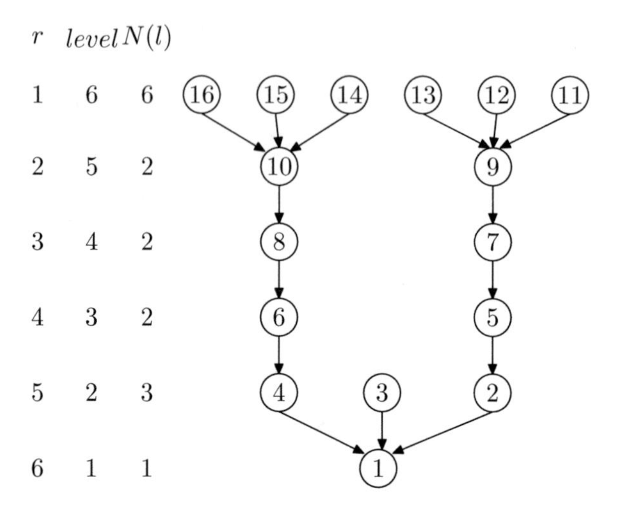

Fig. 3. An intree graph of 16 jobs

Table 2. An intree graph of 12 jobs

14	11	3				
15	12	9	7	5	2	
16	13	10	8	6	4	1

0 1 2 3 4 5 6 7

In the proof of $C_{max} \leq l_{max} + c$ in [7], the following inequality is used:

$$\sum_{k=1}^{r''} N(l_{\max}+1-k) \geq m(t'+\lambda), 0 < \lambda \leq 1. \tag{2}$$

However, we find that the above mentioned t' dose not exist in some concrete instances. Furthermore, the inequality (2) may not hold in some instances even if there exists the t'.

Example 2. Consider the following instance of $P|p_j = 1, intree|C_{\max}$ with 16 jobs. The jobs are subject to the precedence constraints depicted in Figure 3. Given 3 machines, schedule the jobs by using Level Algorithm.

First, we compute the $\rho(r)$ values:
$\rho(1) = \frac{6}{1} = 6, \rho(2) = \frac{6+2}{2} = 4, \rho(3) = \frac{6+2+2}{3} = 3.33, \rho(4) = \frac{6+2+2+2}{4} = 3,$
$\rho(5) = \frac{6+2+2+2+3}{5} = 3, \rho(6) = \frac{6+2+2+2+3+1}{6} = 2.66.$

From the $\rho(r)$ values and the definition of r'', we have $r'' = 5$. Thus, $l_{\max} + 1 - r'' = 6+1-5 = 2$. From the result schedule(see Table 2), we can see that in time interval $[0, 3]$ all machines are used to process jobs. In time interval $[3, 4]$ one machine is idle, but the jobs processed on the other two machines are neither the remaining jobs at level 2 nor the jobs at lower levels. Therefore, in this example, the t' satisfying the above three conditions does not exist. □

Example 3. Consider the following instance of $P|p_j = 1, intree|C_{\max}$ with 12 jobs. The jobs are subject to the precedence constraints depicted in Figure 4. Given 3 machines, schedule the jobs by using Level Algorithm.

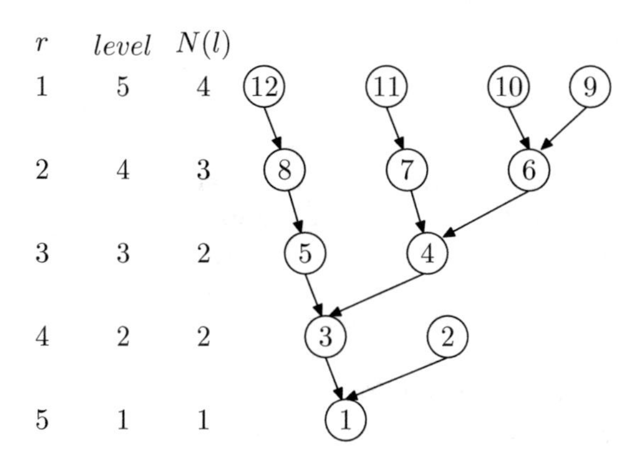

Fig. 4. An intree graph of 12 jobs

Table 3. A schedule under Level Algorithm for Example 3

10	7	2			
11	8	5			
12	9	6	4	3	1

0 1 2 3 4 5 6

First, we compute the $\rho(r)$ values:

$\rho(1) = \frac{4}{1} = 4, \rho(2) = \frac{4+3}{2} = 3.5, \rho(3) = \frac{4+3+2}{3} = 3, \rho(4) = \frac{4+3+2+2}{4} = 2.75,$

$\rho(5) = \frac{4+3+2+2+1}{5} = 2.4.$

From the $\rho(r)$ values and the definition of r'', it follows that $r'' = 3$.

From the result schedule(see Table 3) and the definition of t', we have $t' = 3$. Thus, $\sum_{k=1}^{r''} N(l_{\max}+1-k) = 4+3+2 = 9$ and $m(t'+\lambda) = 3(3+\lambda) = 9+3\lambda, 0 < \lambda \leq 1$. Obviously, in this example, the inequality (2) dose not hold. □

4 A New Algorithm and the Proof for the Optimal Makespan of $P|p_j = 1, intree|C_{\max}$

In this Section, we consider the problem $P|p_j = 1, intree|C_{\max}$ in the case that

$$\max_r\{\frac{\sum_{k=1}^{r} N(l_{\max}+1-k)}{r}\} > m.$$

Assume that m machines are given. Let c be the smallest positive integer such that

$$\max_r\{\frac{\sum_{k=1}^{r} N(l_{\max}+1-k)}{r+c}\} \leq m < \max_r\{\frac{\sum_{k=1}^{r} N(l_{\max}+1-k)}{r+c-1}\}.$$

Let l_i, $s(i)$, $PS(i)$ and $DS(i)$ denote the level, the starting time, the set of direct predecessors and the set of direct successor of job i respectively, $i \in N$. J_l^* is the set of jobs at level l such that for any $i \in J_l^*, i$ is a direct successor of some jobs at level $l+1$, $l < l_{\max}$. J_l is composed of the level l jobs with no predecessors and some jobs eliminated from J_l^* by the following Algorithm at some time. $|J_l|$ denotes the number of jobs in J_l, and $\overline{N}(t)$ counts the number of jobs assigned at time t. In addition, let δ_i denote a dummy start time for job $i \in N$ for the convenience of constructing the following new algorithm.

Algorithm 1: A New Algorithm for $P|p_j = 1, intree|C_{\max}$

1. For $i = 1$ to n Do $\delta_i = 0$;
2. For $l = 1$ to $l_{\max}$ Do $J_l^* = \emptyset$ and $J_l = \{i|l_i = l \text{ and } i \text{ has no predecessors}, i \in N\}$;
3. $t = 0; l = l_{\max}$;
4. For $t = 0$ to n Do $\overline{N}(t) = 0$;
5. **While** $l \geq 1$ **Do Begin**
6. **While** $J_l^* \neq \emptyset$ **Do Begin**
7. Take any job i from J_l^*;
8. If $\overline{N}(\delta_i) = m$ do $J_l = J_l \bigcup \{i\}$
9. Else if $\overline{N}(\delta_i) < m$
10. Let $s(i) = \delta_i$; If $DS(i) \neq \emptyset$, for $j \in DS(i)$, do $\delta_j = \max\{\delta_j, s(i)+1\}$;
11. For $j \in DS(i)$, $PS(j) = PS(j) \setminus \{i\}$, and if $PS(j) = \emptyset$, let $J_{l-1}^* = J_{l-1}^* \cup \{j\}$;
12. $J_l^* = J_l^* \setminus \{i\}$
    End
13. $t = \min\{t|\overline{N}(t) < m\}$;
14. **While** $J_l \neq \emptyset$ **Do Begin**
15. If $|J_l| \leq m - \overline{N}(t)$ Do Begin
16. $s(i) = t$ for all $i \in J_l$; $\overline{N}(t) = \overline{N}(t) + |J_l|$;
17. For any $i \in J_l$, if $DS(i) \neq \emptyset$, for $j \in DS(i)$, do $\delta_j = \max\{\delta_j, t+1\}$;
18. For $j \in DS(i)$, $PS(j) = PS(j) \setminus \{i\}$, and if $PS(j) = \emptyset$, let $J_{l-1}^* = J_{l-1}^* \cup \{j\}$;
19. $J_l = \emptyset$

EndIf
20. If $|J_l| > m - \overline{N}(t)$ Do Begin
21. take $m - \overline{N}(t)$ jobs $i_1, i_2, \cdots, i_{m-\overline{N}(t)}$ from J_l such that $s(i) = t$ for $i \in \{i_1, i_2, \cdots, i_{m-\overline{N}(t)}\}$;
22. For any $i \in \{i_1, i_2, \cdots, i_{m-\overline{N}(t)}\}$, if $DS(i) \neq \emptyset$, for $j \in DS(i)$, do $\delta_j = \max\{\delta_j, t+1\}$;
23. For $j \in DS(i)$, $PS(j) = PS(j) \setminus \{i\}$, and if $PS(j) = \emptyset$, let $J^*_{l-1} = J^*_{l-1} \cup \{j\}$;
24. $J_l = J_l \setminus \{i_1, i_2, \cdots, i_{m-\overline{N}(t)}\}$;
25. $t = t + 1$
EndIf
End
26. $l = l - 1$
End

In Algorithm 1, the jobs are scheduled level by level from $l_{\max}$ downwards to 1. The jobs of the same level are sorted into two subsets for which different measures are taking to schedule the jobs. Moreover, when schedule a job, the direct successor (if any) of the job is dealt with accordingly in terms of the intree precedence relations and the time progress, see Steps 10, 11, 17, 18, 22, 23. Just because of these operations, it is more convenient and easier to do theoretical analysis, especially to prove the optimal makespan of the problem $P|p_j = 1, intree|C_{\max}$.

Lemma 1. In the process of scheduling jobs by using Algorithm 1, if there are no idle machines in some interval $[k, k+1]$, then there are no idle machines in any interval $[\overline{k}, \overline{k}+1]$ with $\overline{k} < k$.

Proof: By contradiction, assume that there are no idle machines in some interval $[k, k+1]$ but there exist idle machines in an interval $[\overline{k}, \overline{k}+1]$ with $\overline{k} < k$. Without loss of generality, assume that $[\overline{k}, \overline{k}+1]$ is the first interval that there are idle machines backwards from $[k, k+1]$, i.e. there are idle machines in $[\overline{k}, \overline{k}+1]$ but there are no idle machines from time $\overline{k}+2$ to $k+1$. Thus, the number of jobs scheduled in $[\overline{k}, \overline{k}+1]$ is less than that in $[\overline{k}+1, \overline{k}+2]$. Since a job has at most one direct successor in an intree, there exists some job j scheduled in $[\overline{k}+1, \overline{k}+2]$ such that j is not the direct successor of any job i scheduled in $[\overline{k}, \overline{k}+1]$. Thus, this job j can be scheduled at least in interval $[\overline{k}, \overline{k}+1]$ according to Steps 10 and 13 of Algorithm 1, and this is a contradiction. □

Remark 1. In fact, according to the proving process of Lemma 1, when we schedule the jobs by using Algorithm 1, for any two adjacent time intervals $[k-1, k]$ and $[k, k+1]$, the jobs scheduled in time interval $[k, k+1]$ are not more than that in time interval $[k-1, k]$ at any time. Thus, if there are idle machines at some time point k, then there must be idle machines at the time point $k+1$. So, Step 25 in Algorithm 1 is reasonable.

Remark 2. In the process of scheduling jobs by using Algorithm 1, if $\overline{N}(\delta_i) = m$ for any $i \in J^*_l$, then for any nonnegative integer $k < \delta_i$ there is $\overline{N}(k) = m$ by

Lemma 1. That means, there are no idle machines before $\delta_i + 1$. Thus, this job i can be put into J_l and the schedule of job i in J_l will not violate the precedence relations, because, its predecessors are scheduled before δ_i but the schedule of the jobs in J_l can begin not earlier than time $\delta_i + 1$ according to Step 13 of Algorithm 1. Therefore, Steps 8 and 13 are indeed reasonable in Algorithm 1.

Theorem 1. A result schedule produced by Algorithm 1 is one of the result schedules produced by Level Algorithm.

Proof: We only need prove this Theorem in the case $N(l_{\max}) < m$. Otherwise, if $N(l_{\max}) \geq m$, then Algorithm 1 always schedule the jobs in the set $J_{l_{\max}}$ until some time t at which the number of the unscheduled $l_{\max}$ level jobs is not more than m. On the other hand, at each integer time k with $0 \leq k \leq t$, Level Algorithm always order the $l_{\max}$ level jobs in the initial part of the released job list L_k and thus the $l_{\max}$ jobs can always be scheduled first. Moreover, Level Algorithm can order the $l_{\max}$ jobs in its Step 4 referring to the job's scheduling sequence obtained by Algorithm 1, i.e., order the jobs which are scheduled at time t by Algorithm 1 in the initial part of list L_t at the same time t, and thus, Level Algorithm and Algorithm 1 can schedule the same jobs in each time interval from time 0 to t. Therefore, we can eliminate the jobs which are scheduled before time t and this can not influence our proof.

In the following, assume that a result schedule S has already been obtained by using Algorithm 1. Then by analyzing the scheduling process of Level Algorithm, we will show that Level Algorithm can produce the same result schedule by induction.

As for Algorithm 1, first, the $N(l_{\max})$ jobs of level $l_{\max}$ are put into the set $J_{l_{\max}}$ and then scheduled in time interval $[0, 1]$. If in the time interval [0,1], there are not idle machines in the result schedule S, then there must exist a level $l_1 \leq l_{\max}$ such that there are jobs in $J_{l_1} \neq \emptyset$ being scheduled in [0,1] according to Algorithm 1. Moreover, if also the jobs of J_{l_1} are not enough for occupying the machines in [0,1], then there must exist another level $l_2 \leq l_1$ such that there are jobs in $J_{l_2} \neq \emptyset$ being scheduled in [0,1]. Until some $l_r \geq 1$ such that all machines in [0,1] are occupied. Furthermore, we claim that all jobs in $J_{l_1}, J_{l_2}, \cdots, J_{l_r}$ are also the leaf jobs in the current intree at time $t = 0$. Otherwise, without loss of generality, suppose that there is certain job $i \in J_{l_k}, l_1 \geq l_k \geq l_r$ such that i has predecessors, then it must be first put into $J^*_{l_k}$ according to Algorithm 1. Afterwards, it is in J_{l_k}, so this means that it is eliminated from $J^*_{l_k}$, and further means that all machines in interval $[\delta_i, \delta_i + 1]$ are occupied such that i can't be scheduled at time $t = \delta_i$. It's a contradiction that there are idle machines in [0,1] when Algorithm 1 is scheduling the jobs of level l_r according to Remark 2. Moreover, from Step 13 of Algorithm 1, it is easy to see that l_1 is the first level lower than $l_{\max}$ such that in l_1 there are leaf jobs of the current intree at time $t = 0$, l_2 is the first level lower than l_1 such that in l_2 there are leaf jobs of the current intree at time $t = 0$, and so on.

Now, we consider the schedule of Level Algorithm in time interval $[0, 1]$. At time $t = 0$, Level Algorithm orders all leaf jobs in the current intree level by level beginning from $l_{\max}$ downwards and then a released job list L_0 is obtained. In L_0, the $l_{\max}$ level jobs are ordered in the initial part, and the leaf jobs of level l_1 as above mentioned in the schedule of Algorithm 1 are ordered closely, then the leaf jobs of level $l_2, \cdots$, the leaf jobs of level $l_r \cdots$, and so on. Level Algorithm will take m jobs from the released job list to schedule on the machines in time interval $[0, 1]$. From the order of the jobs in L_0, it will first take the $l_{\max}$ level jobs and schedule them, then the jobs of level l_1, $l_2, \cdots, l_r \cdots$, and so on until all machines are occupied in interval [0,1]. Moreover, it is clear that the jobs which are taken by Level Algorithm in level l_1, $l_2, \cdots, l_r \cdots$ can be the same with those scheduled by Algorithm 1 in interval [0,1]. Therefore, Level Algorithm can schedule the same jobs with those by Algorithm 1 in [0,1].

Assuming that the two algorithms have scheduled the same jobs in each time interval from [0,1] to some interval $[k-1, k]$, we will prove that in the time interval $[k, k+1]$ Level Algorithm can schedule the same jobs with those by Algorithm 1 in the following two cases.

(1) If there are idle machines in $[k-1, k]$ in the result schedule produced by the two algorithms, then the jobs which are scheduled in $[k, k+1]$ are all the direct successors (if any) of the jobs scheduled in $[k-1, k]$ according to the two algorithms. But, by the induction, in time interval $[k-1, k]$ the same jobs are scheduled by the two algorithms, so, the same direct successors of the jobs scheduled in $[k-1, k]$ can be scheduled in $[k, k+1]$ by the two algorithms. Therefore, Level Algorithm can schedule the same jobs with those by Algorithm 1 in time interval $[k, k+1]$.
(2) If there are no idle machines in $[k-1, k]$ in the result schedule produced by the two algorithms, we also can prove the fact that Level Algorithm can schedule the same jobs with those by Algorithm 1 in time interval $[k, k+1]$ in the following. Without loss of generality, assume that the highest level of the jobs scheduled in $[k-1, k]$ by Algorithm 1 is some level l_{k_1}, that means, at time $t = k-1$ Algorithm 1 is scheduling the jobs of level l_{k_1}. Two subcases need to be considered below.

(a) If at time $t = k$, all jobs of level l_{k_1} are all scheduled, then Algorithm 1 reaches level $l_{k_1} - 1$. Similar to the proof in time interval [0,1], we can prove that Level Algorithm can schedule the same jobs with those by Algorithm 1 in time interval $[k, k+1]$.

(b) If at time $t = k$, some l_{k_1} level jobs are not scheduled completely by Algorithm 1, then Algorithm 1 is still scheduling the jobs of level l_{k_1}. In $[k, k+1]$, Algorithm 1 first schedule the jobs in $J^*_{l_{k_1}}$ (if any), then the jobs in $J_{l_{k_1}}$. If the remaining jobs of level l_{k_1} at time k can't occupy all machines in time interval $[k, k+1]$, some lower level's leaf jobs (if any) of the current intree at time $t = k$ will be scheduled in this time interval. Also similar to the proof process in [0,1], we can prove that Level Algorithm can schedule the same jobs in time interval $[k, k+1]$. If the remaining jobs of level l_{k_1} at time k can occupy all machines in time interval $[k, k+1]$, then Level Algorithm can also schedule these jobs which are scheduled in $[k, k+1]$ by Algorithm 1 because Level Algorithm can order these jobs in the initial part of list L_k.

In a word, Level Algorithm can schedule the same jobs in time interval $[k, k+1]$. □

In fact, the difference between Algorithm 1 and Level Algorithm is that, Level Algorithm schedule the jobs mainly in terms of the time process assisted with the job levels; Algorithm 1 schedules the jobs mainly in terms of the job levels assisted with the time progress. Since the result schedule produced by Level Algorithm is optimal ([2], [4], [5], [6], [8]), the makespan of an instance is of course unique for the problem $P|p_j = 1, intree|C_{max}$.

Theorem 2. For the problem $P|p_j = 1, intree|C_{max}$, in the case that

$$\max_r\{\frac{\sum_{k=1}^{r} N(l_{\max}+1-k)}{r}\} > m,$$

all jobs can be scheduled within $l_{\max}+c$ units of time by using Algorithm 1, where $c \geq 1$ is the smallest positive integer such that

$$\max_r\{\frac{\sum_{k=1}^{r} N(l_{\max}+1-k)}{r+c}\} \leq m < \max_r\{\frac{\sum_{k=1}^{r} N(l_{\max}+1-k)}{r+c-1}\}.$$

Proof: Without loss of generality, assume that the schedule starts from time$t = 0$.

Note that $c \geq 1$ is the smallest positive integer such that

$$\max_r\{\frac{\sum_{k=1}^{r} N(l_{\max}+1-k)}{r+c}\} \leq m.$$

When $r = 1$, $N(l_{\max})/(1+c) \leq m$, i.e. $N(l_{\max}) \leq (1+c)m$. So, the jobs at level $l_{\max}$ can be scheduled within $1+c$ units of time according to Algorithm 1.

When $r = 2$,

$$\frac{N(l_{\max}) + N(l_{\max}-1)}{2+c} \leq m,$$

i.e. $N(l_{\max}) + N(l_{\max}-1) \leq (2+c)m$.

After finishing the schedule of the jobs at level $l_{\max}$, we begin to schedule the jobs at level $l_{\max} - 1$ now. At first, schedule the jobs in $J^*_{l_{\max}-1}$, i.e. schedule the direct successors of the level $l_{\max}$ jobs according to Algorithm 1. For any job $j \in J^*_{l_{\max}-1}$, either it is scheduled at time δ_j or it is eliminated to the set $J_{l_{\max}-1}$. However, despite any of the two cases occurring for all jobs $j \in J^*_{l_{\max}-1}$, after dealing with the jobs in $J^*_{l_{\max}-1}$, all the remaining idle machines from time $t = 0$ to time $t = c+2$ can be used to process all the jobs in $J_{l_{\max}-1}$ according to Step 13 of Algorithm 1. Since $N(l_{\max}) + N(l_{\max}-1) \leq (2+c)m$, and $\delta_j \leq (2+c)$ for $j \in J^*_{l_{\max}-1}$ according to Algorithm 1 and the above schedule of the jobs at level $l_{\max}$, all jobs at level $l_{\max} - 1$ can be scheduled within $c+2$ units of time.

By analogy, when $r = l_{max}$, all the jobs in the whole intree can be scheduled within $l_{\max} + c$ units of time by using Algorithm 1. □

By Theorem 2 and the correct proof of $C_{\max} \geq l_{\max} + c$ in [3] and [7], there is $C_{\max} = l_{\max} + c$. Thus, the proof for the optimal makespan $C_{\max} = l_{\max} + c$ of the scheduling problem $P|p_j = 1, intree|C_{\max}$ under Algorithm 1 is accomplished.

5 Conclusion

In [3] and [7], some errors in the proofs for the optimal makespan of problem $P|p_j = 1, intree|C_{\max}$ are presented with three counterexamples by us. Then, a new algorithm, more convenient for theoretical analysis than Level Algorithm does, is proposed. And, it is proved that the result schedule obtained by using the new algorithm is consistent with that by Level Algorithm in the sense that they can give the same result schedule. Finally, by using the new algorithm, the proof for the optimal makespan of $P|p_j = 1, intree|C_{\max}$ is accomplished.

In the future, we can expect that, some new algorithms with lower computational complexity, may be similarly constructed for some other important scheduling problems such as $Pm|p_j = 1, intree|\sum C_j$ and $Fm|p_j = 1, intree|\sum C_j$ etc.

Acknowledgements

Thanks to the support by National Natural Science Foundation of China (Grant No. 60474023), Science and Technology Key Project Fund of Ministry of Education(Grant No. 03184), and the Major State Basic Research Development Program of China (Grant No. 2002CB312200).

References

1. Blazewicz J, Ecrer K, Schmidt G and Weglarz J (1993) Scheduling in Computer and Manufacturing Systems. Springer-Verlag, Berlin, 124
2. Brucker P, Garey MR, Johnson DS (1977) Scheduling Equal-Length Tasks under Treelike Precedence Constraints to Minimize Maximum Lateness. Mathematics of Operations Research 2(3): 275-284
3. Hu TC (1961) Parallel Sequencing and Assembly Line Problems. Operations Research 9: 841-848
4. Leung JY-T (2004) Handbooks of Scheduling•Algorithms, Models, and Performance Analysis. CRC Press Company, Boca Raton London, New York, Washington, DC, page3-1–page3-4
5. Monma CL (1982) Linear-Time Algorithms for Scheduling on Parallel Processors, Operations Research 30: 116-124.
6. Tanaev VS, Gordon VS and Shafransky YM (1994) Scheduling Theory•Single-Stage Systems Kluwer Academic Publishers, Dordrecht/Boston/London, 122-126
7. Tang HY and Zhao CL (2002) Introduction to scheduling. Beijing: Science Press, 49-52
8. Yue MY (2001) Introduction to Combinatorial Optimization. Zhejiang: Zhejiang Science and Technology Press, 70-72

Part V

Fuzzy Operation Research and Management

Advances in Fuzzy Geometric Programming

Bing-yuan Cao[1,2] and Ji-hui Yang[2]

[1] School of Mathematics and Information Science, Guangzhou University, Guangdong, ZIP 510006, P.R. China
caobingy@163.com
[2] Department of Mathematics, Shantou University Guangdong, ZIP 515063, P.R. China
yangjihui@163.com

Abstract. Fuzzy geometric programming emerges as a new branch subject of optimization theory and fuzzy mathematics has received much attention in recent years. Its conceptual framework is presented by B.Y.Cao in IFSA 1987. At present, dozens of researchers from China, India, Cuba, England, Canada, Belgian and China Taiwan and etc. have joined the team of studying. This theory has been applied in power system, environmental engineering, economic management and so on with the range of its application expecting to be further expanded.

Keywords: Geometric programming; Fuzzy geometric programming; Optimal solution: Application of fuzzy geometric programming.

1 Introduction

Geometric programming (GP) is an important branch of Operations Research. Established in 1961, its founders include Zener from Electric Corporation Research Laboratories, Duffin from Carnegie Mellon University and Peterson from North Carolina State University[1]. The general form of GP is

$$
\begin{array}{lll}
(GP) & \min & g_0(x) \\
 & \text{s.t.} & g_i(x) \leqslant 1 (1 \leqslant i \leqslant p'), \\
 & & g_i(x) \geqslant 1 \ (p'+1 \leqslant i \leqslant p), \\
 & & x > 0
\end{array}
$$

it is a reversed posynomial GP, where all of $g_i(x) (0 \leqslant i \leqslant p)$ are posynomials function of variable x with $g_i(x) = \sum\limits_{k=1}^{J_i} v_{ik}(x)$, here

$$
v_{ik}(x) = \begin{cases} c_{ik} \prod\limits_{l=1}^{m} x_l^{\gamma_{ikl}}, & (1 \leqslant k \leqslant J_i; 0 \leqslant i \leqslant p'), \\ c_{ik} \prod\limits_{l=1}^{m} x_l^{-\gamma_{ikl}}, & (1 \leqslant k \leqslant J_i; p'+1 \leqslant i \leqslant p). \end{cases} \tag{1}
$$

is a monomial function of variable x, and coefficients $c_{ik} > 0$,, variable $x = (x_1, x_2, \cdots, x_n)^T > 0$, exponents γ_{ikl} $(1 \leq k \leq J_i, 0 \leq i \leq p, 1 \leq l \leq n)$ are arbitrary real number.

B.-Y. Cao (Ed.): Fuzzy Information and Engineering (ICFIE), ASC 40, pp. 497–502, 2007.
springerlink.com

Applications of GP can be found in many field such as mechanical engineering, civil engineering, structural design and optimization, chemical engineering, optimal control, decision making, network flows, theory of inventory, balance of machinery, analog circuitry, design theory, communication system, transportation, fiscal and monetary, management science, electrical engineering, electronic engineering, environmental engineering, nuclear engineering, and technical economical analysis. Until now, its application scope is continuously being expanded [2].

However, in reality, in order to deal with a large number of problems, constraint conditions need softening in classical GP, coefficients, exponents, even variables in it are regarded as fuzzy parameters or fuzzy variables in objective functions as well as in constraint conditions.

2 Fuzzy Geometric Programming

In economic management, we often meet with a problem as follows.

Suppose that we manufacture a case for transporting cotton. The case is Vm^3 in volume and has a bottom, but no top. The bottom and two sides are made from Cm^2 of a special flexible material with negligible cost. The material for the other two sides cost more than A yuan/m^2 (yuan means RMB) and it costs about k yuan to transport the case. What is the cost at least to ship one case of cotton? Such a problem can be posed for solution by a GP. Since a classical GP cannot give a good account for the problem, or obtain a practical solution to it, at IFSA 1987 conference, B.Y.Cao proposed the fuzzy geometric programming (FGP) theory for the first time [3], which is effective to solve the problem. The general form of FGP is

$$(\widetilde{GP}) \qquad \begin{aligned} &\widetilde{\min}\ \tilde{g}_0(\tilde{x}) \\ &\text{s.t. } \tilde{g}_i(\tilde{x}) \lesssim 1 (1 \leqslant i \leqslant p') \\ &\qquad \tilde{g}_i(\tilde{x}) \gtrsim 1 (p'+1 \leqslant i \leqslant p) \\ &\qquad \tilde{x} > 0 \end{aligned}$$

a fuzzy reversed posynomial GP, where all

$$\tilde{g}_i(\tilde{x}) = \sum_{k=1}^{J_i} \tilde{v}_{ik}(\tilde{x}) (0 \leqslant i \leqslant p)$$

are fuzzy posynomials function of fuzzy variable $\tilde{x}$, here

$$\tilde{v}_{ik}(\tilde{x}) = \begin{cases} \tilde{c}_{ik} \prod_{l=1}^{m} \tilde{x}_l^{\tilde{\gamma}_{ikl}}, & (1 \leqslant k \leqslant J_i; 0 \leqslant i \leqslant p'), \\ \tilde{c}_{ik} \prod_{l=1}^{m} \tilde{x}_l^{-\tilde{\gamma}_{ikl}}, & (1 \leqslant k \leqslant J_i; p'+1 \leqslant i \leqslant p) \end{cases}$$

are fuzzy monomial function of fuzzy variable $\tilde{x}$, where fuzzy coefficients $\tilde{c}_{ik} \gtrsim 0$ are all freely fixed in the closed interval $[c_{ik}^-, c_{ik}^+]$, and degree of accomplishment is determined by

$$\tilde{c}_{ik}(c_{ik}) = \begin{cases} 0, & \text{if } \ c_{ik} < c_{ik}^-, \\ \left(\frac{c_{ik}-c_{ik}^-}{c_{ik}^+-c_{ik}^-}\right)^r, & \text{if } \ c_{ik}^- \leqslant c_{ik} \leqslant c_{ik}^+, \\ 1, & \text{if } \ c_{ik} > c_{ik}^+, \end{cases} \tag{2}$$

where $0 < c_{ik}^- < c_{ik}^+, c_{ik}^-, c_{ik}^+$ is left and right endpoints in the intervals, $c_{ik}^-, c_{ik}^+, c_{ik}$ and r stand for an arbitrary real number.

Similarly, exponents $\tilde{\gamma}_{ikl}(1 \leqslant k \leqslant J_i, 0 \leqslant i \leqslant p, 1 \leqslant l \leqslant m)$ can be defined, and its degree of accomplishment is determined by

$$\tilde{\gamma}_{ikl}(\gamma_{ikl}) = \begin{cases} 0, & \text{if } \ \gamma_{ikl} < \gamma_{ikl}^-, \\ \left(\frac{\gamma_{ikl}-\gamma_{ikl}^-}{\gamma_{ikl}^+-\gamma_{ikl}^-}\right)^r, & \text{if } \ \gamma_{ikl}^- \leqslant \gamma_{ikl} \leqslant \gamma_{ikl}^+, \\ 1, & \text{if } \ \gamma_{ikl} > \gamma_{ikl}^+, \end{cases} \tag{3}$$

here $\gamma_{ikl}^- < \gamma_{ikl}^+, \gamma_{ikl}^-, \gamma_{ikl}^+$ is left and right endpoints in the intervals, and $\gamma_{ikl}, \gamma_{ikl}^-$ and γ_{ikl}^+ is an arbitrary rational number.

For each item $\widetilde{v}_{ik}(\tilde{x})$ $(1 \leqslant k \leqslant J_i; p'+1 \leqslant i \leqslant p)$ in the reversed inequality $\tilde{g}_i(\tilde{x}) \gtrsim 1$, $\tilde{x}_l$ acts as an exponent in the item by $-\tilde{\gamma}_{ikl}$ instead of by $\tilde{\gamma}_{ikl}$, where symbol '$\lesssim$', or '$\gtrsim$'denotes the fuzzifiness version of $\leqslant$, or $\geqslant$ and has the linguistic interpretation "essentially smaller than or equal", or "essentially larger than or equal", and $\widetilde{\min}$ is an extension of min operation. The membership function of objective $g_0(\tilde{x})$ is defined by

$$\tilde{A}_0(\tilde{x}) = \tilde{B}_0(\phi_0(\tilde{g}_0(\tilde{x}))) = \begin{cases} 1, & \text{if } \ \tilde{g}_0(\tilde{x}) \leqslant Z_0 \\ e^{-\frac{1}{d_0}\left(\tilde{g}_0(\tilde{x})-Z_0\right)}, & \text{if } \ Z_0 < \tilde{g}_0(\tilde{x}) \leqslant Z_0 + d_0 \end{cases} \tag{4}$$

and the membership functions of constraints defined by

$$\tilde{A}_i(\tilde{x}) = \tilde{B}_i(\phi_i(\tilde{g}_i(\tilde{x}))) = \begin{cases} 1, & \text{if } \ \tilde{g}_i(\tilde{x}) \leqslant 1 \\ e^{-\frac{1}{d_i}\left(\tilde{g}_i(\tilde{x})-1\right)}, & \text{if } \ 1 < \tilde{g}_i(\tilde{x}) \leqslant 1 + d_i \end{cases} \tag{5}$$

at $1 \leqslant i \leqslant p'$, and

$$\tilde{A}_i(\tilde{x}) = \tilde{B}_i(\phi_i(\tilde{g}_i(\tilde{x}))) = \begin{cases} 0, & \text{if } \ \tilde{g}_i(\tilde{x}) \leqslant 1 \\ e^{-\frac{1}{d_i}\left(1-\tilde{g}_i(\tilde{x})\right)}, & \text{if } \ 1 < \tilde{g}_i(\tilde{x}) \leqslant 1 + d_i \end{cases} \tag{6}$$

at $p'+1 \leqslant i \leqslant p$ hold, respectively, where Z_0 is an expectation value of the objective function $\tilde{g}_0(\tilde{x}); \tilde{g}_i(\tilde{x})(1 \leqslant i \leqslant p)$ are constraint functions; $d_i \geqslant 0(0 \leqslant i \leqslant p)$ is a flexible index of i-th fuzzy function $\tilde{g}_i(\tilde{x})$, which is subjectively chosen to be constants of admissible violations in a constraint function or an objective function, respectively.

3 Present Situation of Fuzzy Geometric Programming

In the last 20 years, FGP has been in a long-term development in its theory and application. In 2002, B. Y. Cao published the first monograph of fuzzy geometric programming as applied optimization in Vol.76 by Kluwer Academic Publisher (present Springer-Verlag) [4]. In the book he gives readers a thorough understanding of FGP.

The theory development of FGP includes:

I. *From form*
1 Fuzzy posynomial geometric programming [5];
2 Fuzzy reverse posynomial geometric programming [6];
3 Fuzzy multi-objective geometric programming [7-8];
4 Fuzzy PGP classification and its corresponding class properties [9];
5 Fuzzy fractional geometric programming [10];
6 Extension geometric programming [11];
5 Fuzzy relation geometric programming [12].
II. *From Coefficients*
1 GP with interval and fuzzy valued coefficients;
2 GP with type $(\cdot, c)$ fuzzy coefficients;
3 GP with L-R fuzzy coefficients;
4 GP with flat fuzzy coefficients.
III. *From variable*
1 GP with T-fuzzy variables;
2 GP with trapezoidal fuzzy variables.
IV. *From algorithm*
1 The primal algorithm, dual algorithm of GP;
2 Lagrange problem of GP;
3 Solving GP by soft algorithm.

The content of II, III, IV can be found in Reference [4].
A large number of applications have been discovered in FGP in a wide variety of scientific and non-scientific fields since FGP is much superior to classical GP in dealing with question in some fields including power system [13], environmental engineering [14], postal services [15], economical analysis [16], transportation [17], inventory theory [18], and etc. It seems clear that even more remain to be discovered. For these reasons, fuzzy geometric programming arguably potentially becomes a ubiquitous optimization technology the same as fuzzy linear programming, fuzzy objective programming, and fuzzy quadratic programming.

4 Future Development of Fuzzy Geometric Programming

The FGP will attract us to further research because many aspects remain untouched. In the basic field, we shall consider the following topics.

1. Fuzzy reverse GP, including a GP problem with mixed sign-terms, is much more complex than the fuzzy convex (resp. concave) GP and fuzzy PGP, which differ a lot in their properties.

2. Fuzzy fractional GP model and its method remain to be researched.

3. The discussion of the GP with fuzzy coefficients and fuzzy variables is yet to be further perfected and extended.

4. It is very interesting to solve antinomy in the realistic world by using fuzzy GP further [19].

5. Solving fuzzy relation geometric programming.

6. Decision making based on FGP.

7. It will be a creative job to combine fuzzy GP and other fuzzy programming with Data Mining before dealing with their data and classification.

8. Solving FGP through meta-heuristic algorithms, such as Genetic Algorithm, Simulated Annealing Algorithm, Artificial Neural Network, Tabu Search, Ant Algorithms, it is worth researching how to effectively apply those algorithms to solving FGP [20].

9. We have developed the concepts of extension convex functions in extendable-valued sets based on the extendable sets and its convex sets, built a matter-element model and a mathematical model in extendable GP, and discussed solution properties in an extendable mathematical model and given an algorithm to the programming which can be changed into an extendable LP by the solution to contrary theory. This model plays a practical role in solving fuzziness at the border as well as contradictory conflicts in building a fuzzy non-compatible GP model and in turning a non-compatible problem into a compatible one.

In the real world, the problems, which can be expressed once by an exponent polynomial functions in the relation under fuzzy environment, are concluded as a fuzzy GP. Such research directions as mentioned above are not limited to those fields.

Acknowledgments

This research is supported by the National Natural Science Foundation of China (No. 70271047) and by the Foundation of Guangzhou University, China.

References

1. E. L. Peterson (2001) The Origins of Geometric Programming. Annals of Operations Research, 105:15-19.
2. J. H. Yang, B. Y. Cao (2006) The origin and its application of geometric programming, Proc.of the Eighth National Conference of Operations Research Society of China. Global-Link Publishing Company, Hong Kong: 358-363.ISBN: 962-8286-09-9.
3. B. Y. Cao (1987) Solution and theory of question for a kind of fuzzy positive geometric program, in Proceedings of the 2nd IFSA Conference, Tokyo, Japan, 1: 205- 208.
4. B. Y. Cao (2002) Fuzzy Geometric Programming. Dordrecht: Kluwer Academic Publishers.
5. B. Y. Cao (1993) Fuzzy geometric programming (I), Int. J. of Fuzzy Sets and Systems, 53(2): 135-154.
6. B. Y. Cao (1996) Fuzzy geometric programming (II), J. of Fuzzy Mathematics, 4(1): 119-129.

7. R. K. Verma (1990) Fuzzy geometric programming with several objective functions. Fuzzy Sets and Systems,35(1): 115-120.
8. A. P. Burnwal, S. N. Mukherjee, D. Singh (1996) Fuzzy geometric programming with nonequivalent objectives. Ranchi University Mathematical J.,27: 53-58.
9. B. Y. Cao (1995) Classification of fuzzy posynomial geometric programming and corresponding class properties. Jounal of Fuzzy Systems and Mathematics, 9(4): 60-64.
10. B. Y. Cao (2000) Parameterized solution to a fractional geometric programming. Proc.of the sixth National Conference of Operations Research Society of China. Global-Link Publishing Company, Hong Kong: 362-366.ISBN: 962-8286-09-9.
11. B. Y. Cao (2001) Extension posynomial geometric programming. Jounal of Guangdong University of Techonology 18(1): 61-64.
12. J. H. Yang, B. Y. Cao (2005) Geometric Programming with Fuzzy Relation Equation Constraints. Proceedings of the IEEE International Conference on Fuzzy Systems, 557- 560.
13. B. Y. Cao (1999) Fuzzy geometric programming optimum seeking in power supply radius of transformer subsation. Proceedings of the Fuzz-IEEE99 Conference, Seoul, Korea, 3: 1749-1753.
14. B. Y. Cao (1995) Fuzzy geometric programming optimum seeking of scheme for waste-water disposal in power plant. Proceedings of the Fuzz-IEEE /IFES95 Conference, Yokohama, Japan, 5: 793-798.
15. M. P. Biswal (1992) Fuzzy programming technique to solve multi-objective geometric programming problems. Fuzzy Sets and Systems,51(1): 67-71.
16. Shiang-Tai Liu (2004) Fuzzy geometric programming approach to a fuzzy machining economics model. International Journal of Production Research, 42(16): 3253-3269.
17. Islam. Sahidul, R. T. Kumar (2006) A new fuzzy multi-objective programming: entropy based geometric programming and its application of transportation problems. European Journal of Operational Research, 173(2): 387-404.
18. N. K. Mandal, T. K. Roy, M. Maiti (2005) Multi-objective fuzzy inventory model with three constraints: a geometric programming approach. Fuzzy Sets and Systems, 150(1): 87-106.
19. B. Y. Cao (2004) Antinomy in posynomial geometric programming. Advances in Systems Science and Applications, USA, 4(1): 7-12.
20. M. Gen, Y. S. Yun (2006) Soft computing approach for reliability optimization: State-of-the-art survey. Reliability Engineering & System Safety, 91(9): 1008-1026.

A Method for Estimating Criteria Weights from Intuitionistic Preference Relations

Zeshui Xu

Department of Management Science and Engineering, School of Economics and Management, Tsinghua University, Beijing 100084, China
xu_zeshui@263.net

Abstract. Intuitionistic preference relation is a powerful means to express decision maker's intuitionistic preference information over criteria in the process of multi-criteria decision making. In this paper, we define the concept of consistent intuitionistic preference relation and give the equivalent interval fuzzy preference relation of an intuitionistic preference relation. Then we develop a method for estimating criteria weights from intuitionistic preference relations, and finally, we use two numerical examples to illustrate the developed method.

Keywords: Intuitionistic preference relation, consistent intuitionistic preference relation, weak transitivity, priority vector, linear programming model.

1 Introduction

In [1], Atanassov introduced the concept of intuitionistic fuzzy set, which emerges from the simultaneous consideration of the degrees of membership and non-membership with a degree of hesitancy. The intuitionistic fuzzy set has been studied and applied in a variety of areas. For example, Atanassov and Gargov [2] gave the notion of interval intuitionistic fuzzy set. De *et al.* [3] defined some operations on intuitionistic fuzzy sets. De *et al.* [4] applied the intuitionistic fuzzy sets to the field of medical diagnosis. Deschrijver and Kerre [5] established the relationships between intuitionistic fuzzy sets, L-fuzzy sets, interval-valued fuzzy sets and interval valued intuitionistic fuzzy sets. Some authors investigated the corrections [6-10] and similarity measures [11-15] of intuitionistic fuzzy sets. Deschrijver *et al.* [16] extended the notion of triangular norm and conorm to intuitionistic fuzzy set theory. Deschrijver and Kerre [17] introduced some aggregation operators on the lattice L^*, and considered some particular classes of binary aggregation operators based on t-norms on the unit interval. They also studied the properties of the implicators generated by these classes. Park [18] defined the notion of intuitionistic fuzzy metric spaces. Park and Park [19] introduced the notion of generalized intuitionistic fuzzy filters based on the notion of generalized intuitionistic fuzzy sets given by Mondal and Samanta [20], and defined the notion of Hausdorffness on generalized intuitionistic fuzzy filters. Deschrijver and Kerre [21] introduced the notion of uninorm in interval-valued fuzzy set theory. Cornelis *et al.* [22] constructed a representation theorem for Łukasiewicz implicators on the lattice L^* which serves as the underlying algebraic structure for both intuitionistic fuzzy and interval-valued fuzzy sets. Gutiérez Garcá and

B.-Y. Cao (Ed.): Fuzzy Information and Engineering (ICFIE), ASC 40, pp. 503–512, 2007.
springerlink.com

Rodabaugh [23] demonstrated two meta-mathematical propositions concerning the intuitionistic approaches to fuzzy sets and fuzzy topology, as well as the closely related interval-valued sets and interval-valued intuitionistic sets. Dudek *et al.* [24] considered the intuitionistic fuzzification of the concept of sub-hyperquasigroups in a hyperquasigroup and investigated some properties of such sub-hyperquasigroups. Xu and Yager [25] investigated the aggregation of intuitionistic information, and developed some geometric aggregation operators, such as the intuitionistic fuzzy weighted geometric operator, the intuitionistic fuzzy ordered weighted geometric operator, and the intuitionistic fuzzy hybrid geometric operator. They also gave an application of these operators to multiple attribute decision making based on intuitionistic fuzzy sets. Szmidt and Kacprzyk [26], and Xu [27] investigated the group decision making problems with intuitionistic preference relations. Herrera *et al.* [28] developed an aggregation process for combining numerical, interval valued and linguistic information. Furthermore, they proposed different extensions of this process to deal with contexts in which can appear other type of information as intuitionistic fuzzy sets or multi-granular linguistic information.

In the process of multi-criteria decision making under intuitionistic fuzzy information, intuitionistic preference relation is a powerful tool to express the decision maker' intuitionistic preference information over criteria, the priority weights derived from the intuitionistic preference relation can be used as the weights of criteria. Thus, how to estimate criteria weights from intuitionistic preference relations is an interesting and important issue, which no investigation has been devoted to. In this paper, we shall develop a method for estimating criteria weights from intuitionistic preference relations. In order to do so, the rest of this paper is organized as follows: In Section 2, we review some basic concepts. Section 3 introduces the notion of consistent intuitionistic preference relation, and gives the equivalent interval fuzzy preference relation of an intuitionistic preference relation. In Section 4, we develop a method for estimating criteria weights from intuitionistic preference relations based on linear programming models, and finally, two numerical examples are given.

2 Preliminaries

Let X be a universe of discourse. Atanassov [1] introduced the notion of intuitionistic fuzzy set A, which can be shown as follows:

$$A=\{<x_j,\mu_A(x_j),v_A(x_j)>|\, x_j\in X\} \tag{1}$$

The intuitionistic fuzzy set A assigns to each element $x_j \in X$ a membership degree $\mu_A(x_j)\in[0,1]$ and a nonmembership degree $v_A(x_j)\in[0,1]$, with the condition

$$0\le \mu_A(x_j)+v_A(x_j)\le 1,\ \forall x_j\in X \tag{2}$$

For each $x_j \in X$. The value

$$\pi_A(x_j)=1-\mu_A(x_j)-v(x_j)$$

is called the indeterminacy degree or hesitation degree of x_j to A. Especially, if

$$\pi_A(x_j)=1-\mu_A(x_j)-v_A(x_j)=0, \text{ for each } x_j \in X$$

then, the intuitionistic fuzzy set A is reduced to a common fuzzy set [29].

Consider a multi-criteria decision making problem with a finite set of n criteria, and let $X=\{x_1,x_2,...,x_n\}$ be the set of criteria. In [27], we introduced the notion of imtuitionistic preference relation as follows:

Definition 1 [27]. *An intuitionistic preference relation B on X is represented by a matrix* $B=(b_{ij})\subset X\times X$ *with* $b_{ij}=<(x_i,x_j)>,\mu(x_i,x_j),v(x_i,x_j)>$, *for all* $i,j=1,2,...,n$. *For convenience, we let* $b_{ij}=(\mu_{ij},v_{ij})$, *for all* $i,j=1,2,...,n$, *where* b_{ij} *is an intuitionistic fuzzy value, composed by the certainty degree* μ_{ij} *to which* x_i *is preferred to* x_j *and the certainty degree* v_{ij} *to which* x_i *is non-preferred to* x_j, *and* $1-\mu_{ij}-v_{ij}$ *is interpreted as the hesitation degree to which* x_i *is preferred to* x_j. *Furthermore,* μ_{ij} *and* v_{ij} *satisfy the following characteristics:*

$$0\le\mu_{ij}+v_{ij}\le1,\ \mu_{ji}=v_{ij},\ v_{ji}=\mu_{ij},\ \mu_{ii}=v_{ii}=0.5,\ \textit{for all } i,j=1,2,...,n \tag{3}$$

3 Consistent Intuitionistic Preference Relation

By Definition 1, we know that each element b_{ij} in the intuitionistic preference relation B consists of the pair (μ_{ij},v_{ij}). Consider that each pair (μ_{ij},v_{ij}) must satisfy the condition $\mu_{ij}+v_{ij}\le1$, i.e., $\mu_{ij}\le1-v_{ij}$. This condition is exactly the condition under which two real numbers form an interval [28]. As a result, we can transform the pair $b_{ij}=(\mu_{ij},v_{ij})$ into the interval number $\dot{b}_{ij}=[\mu_{ij},1-v_{ij}]$, and thus, the intuitionistic preference relation $B=(b_{ij})_{n\times n}$ is equivalent to an interval fuzzy preference relation [30,31] $\dot{B}=(\dot{b}_{ij})_{n\times n}$, where $\dot{b}_{ij}=[\dot{b}_{ij}^-,\dot{b}_{ij}^+]=[\mu_{ij},1-v_{ij}]$, for all $i,j=1,2,...,n$, and

$$\dot{b}_{ij}^-+\dot{b}_{ji}^+=\dot{b}_{ij}^++\dot{b}_{ji}^-=1,\ \dot{b}_{ij}^+\ge\dot{b}_{ij}^-\ge0,\ \dot{b}_{ii}^+=\dot{b}_{ii}^-=0.5, \text{ for all } i,j=1,2,...,n$$

For convenience, we denote by the intuitionistic preference relation $B=(b_{ij})_{n\times n}$, where $b_{ij}=[\mu_{ij},1-v_{ij}]$, for all $i,j=1,2,...,n$. Especially, if

$$\mu_{ij}+v_{ij}=1, \text{ for all } i,j=1,2,...,n$$

then the intuitionistic preference relation $B=(b_{ij})_{n\times n}$ is reduced to a fuzzy preference relation [32-38] $R=(r_{ij})_{n\times n}$, where

$$0 \le r_{ij} \le 1,\ r_{ij}+r_{ji}=1,\ r_{ii}=0.5, \text{ for all } i,j=1,2,...,n$$

Let $w=(w_1,w_2,...,w_n)^T$ be the vector of priority weights, where w_i reflects the importance degree of the criterion x_i, and

$$w_i \ge 0,\ i=1,2,...,n,\ \sum_{i=1}^{n} w_i = 1 \tag{4}$$

then, a fuzzy preference relation $R=(r_{ij})_{n\times n}$ is called a consistent fuzzy preference relation, if the following additive transitivity [33] is satisfied

$$r_{ij}=r_{ik}-r_{jk}+0.5, \text{ for all } i,j,k=1,2,...,n$$

and such a fuzzy preference relation is given by [39,40]:

$$r_{ij}=0.5(w_i-w_j+1), \text{ for all } i,j=1,2,...,n \tag{5}$$

By (5), in the following, we define the concept of consistent intuitionistic preference relation:

Definition 2. *Let* $B=(b_{ij})_{n\times n}$ *be an intuitionistic preference relation, where* $b_{ij}=[\mu_{ij},1-v_{ij}]$, *for all* $i,j=1,2,...,n$, *if there exists a vector* $w=(w_1,w_2,...,w_n)^T$, *such that*

$$\mu_{ij} \le 0.5(w_i-w_j+1) \le 1-v_{ij}, \text{ for all } i=1,2,...n-1;\ j=i+1,...,n \tag{6}$$

where w *satisfies the condition (4), then we call* B *a consistent intuitionistic preference relation; otherwise, we call* B *an inconsistent intuitionistic preference relation.*

In the next section, we shall develop a method for estimating criteria weights from intuitionistic preference relations.

4 A Method for Estimating Criteria Weights

If $B=(b_{ij})_{n\times n}$ is *a* consistent intuitionistic preference relation, then the priority vector $w=(w_1,w_2,...,w_n)^T$ of B should satisfy Eqs.(4) and (6). Thus, motivated by the idea [41], we utilize Eqs.(4) and (6) to establish the following linear programming model:

(M-1) $w_i^- = Min\, w_i$ and $w_i^+ = Max\, w_i$

$$s.t.\ 0.5(w_i - w_j + 1) \geq \mu_{ij},\quad i = 1,2,...,n-1;\ j = i+1,...,n$$

$$0.5(w_i - w_j + 1) \leq 1 - v_{ij},\quad i = 1,2,...,n-1;\ j = i+1,...,n$$

$$w_i \geq 0,\ i = 1,2,...,n,\ \sum_{i=1}^{n} w_i = 1$$

Solving the model (M-1), we can get the weight intervals $[w_i^-, w_i^+]$, $i = 1,2,...,n$. Especially, if $w_i^- = w_i^+$, for all i, then we get a unique priority vector $w = (w_1, w_2, ..., w_n)^T$ from the intuitionistic preference relation B.

If $B = (b_{ij})_{n \times n}$ is an inconsistent intuitionistic preference relation, then Eq.(6) does not always hold. In such case, we relax Eq.(5) by introducing the deviation variables d_{ij}^- and d_{ij}^+, $i = 1,2,..., n-1;\ j = i+1,..., n$:

$$\mu_{ij} - d_{ij}^- \leq 0.5(w_i - w_j + 1) \leq 1 - v_{ij} + d_{ij}^+,\ \text{for all } i = 1,2,...,n-1;\ j = i+1,...,n \quad (7)$$

where d_{ij}^- and d_{ij}^+ are both nonnegative real numbers. Obviously, the smaller the deviation variables d_{ij}^- and d_{ij}^+, the closer B to an inconsistent intuitionistic preference relation. As a result, we establish the following optimization model:

(M-2) $J_1^* = Min \sum_{i=1}^{n-1} \sum_{j=i+1}^{n} (d_{ij}^- + d_{ij}^+)$

$$s.t.\ 0.5(w_i - w_j + 1) + d_{ij}^- \geq \mu_{ij},\quad i = 1,2,...,n-1;\ j = i+1,...,n$$

$$0.5(w_i - w_j + 1) - d_{ij}^+ \leq 1 - v_{ij},\quad i = 1,2,...,n-1;\ j = i+1,...,n$$

$$w_i \geq 0,\ i = 1,2,...,n,\ \sum_{i=1}^{n} w_i = 1$$

$$d_{ij}^-, d_{ij}^+ \geq 0,\ i = 1,2,...,n-1;\ j = i+1,...,n$$

Solving this model, we can get the optimal deviation values $\dot{d}_{ij}^-$ and $\dot{d}_{ij}^+$, $i = 1,2,..., n-1;\ j = i+1,..., n$.

Based on the optimal deviation values $\dot{d}_{ij}^-$ and $\dot{d}_{ij}^+$, $i = 1,2,...,n-1;$ $j = i+1,...,n$, we further establish the following optimization model:

(M-3) $w_i^- = Min\, w_i$ and $w_i^+ = Max\, w_i$

$$s.t.\ 0.5(w_i - w_j + 1) + \dot{d}_{ij}^- \geq \mu_{ij},\quad i = 1,2,\ldots,n-1;\ j = i+1,\ldots,n$$

$$0.5(w_i - w_j + 1) - \dot{d}_{ij}^+ \leq 1 - v_{ij},\quad i = 1,2,\ldots,n-1;\ j = i+1,\ldots,n$$

$$w_i \geq 0,\ i = 1,2,\ldots,n,\ \sum_{i=1}^{n} w_i = 1$$

Solving the model (M-3), we can get the priority weight intervals $[w_i^-, w_i^+]$, $i = 1,2,\ldots, n$. Especially, if $w_i^- = w_i^+$ for all i, then we get a unique priority vector $w = (w_1, w_2, \ldots, w_n)^T$ from the intuitionistic preference relation B.

Example 1. For a multiple criteria decision making problem, there are five criteria $x_i (i = 1,2,\ldots,5)$. A decision maker compares each pair of criteria x_i and x_j, and provides his/her intuitionistic fuzzy preference value $a_{ij} = (\mu_{ij}, v_{ij})$, composed by the certainty degree μ_{ij} to which x_i is preferred to x_j and the certainty degree v_{ij} to which x_i is non-preferred to x_j degree, and then constructs the following intuitionistic preference relation:

$$A = (a_{ij})_{5\times 5} = \begin{bmatrix} (0.5,0.5) & (0.6,0.3) & (0.4,0.2) & (0.7,0.2) & (0.4,0.5) \\ (0.3,0.6) & (0.5,0.5) & (0.5,0.3) & (0.6,0.1) & (0.3,0.6) \\ (0.2,0.4) & (0.3,0.5) & (0.5,0.5) & (0.6,0.2) & (0.4,0.5) \\ (0.2,0.7) & (0.1,0.6) & (0.2,0.6) & (0.5,0.5) & (0.3,0.6) \\ (0.5,0.4) & (0.6,0.3) & (0.5,0.4) & (0.6,0.3) & (0.5,0.5) \end{bmatrix}$$

We first transform the intuitionistic preference relation A into its equivalent interval fuzzy preference relation $B = (b_{ij})_{n\times n}$ (here, $b_{ij} = [\mu_{ij}, 1 - v_{ij}]$, $i, j = 1,2,\ldots,5$):

$$B = (b_{ij})_{5\times 5} = \begin{bmatrix} [0.5,0.5] & [0.6,0.7] & [0.4,0.8] & [0.7,0.8] & [0.4,0.5] \\ [0.3,0.4] & [0.5,0.5] & [0.5,0.7] & [0.6,0.9] & [0.3,0.4] \\ [0.2,0.6] & [0.3,0.5] & [0.5,0.5] & [0.6,0.8] & [0.4,0.5] \\ [0.2,0.3] & [0.1,0.4] & [0.2,0.4] & [0.5,0.5] & [0.3,0.4] \\ [0.5,0.6] & [0.6,0.7] & [0.5,0.6] & [0.6,0.7] & [0.5,0.5] \end{bmatrix}$$

then by solving the model (M-2), we get $J_1^* = 0.075$, and the optimal deviation values are as follows:

$$\dot{d}_{12}^- = \dot{d}_{12}^- = 0,\ \dot{d}_{13}^- = \dot{d}_{13}^+ = 0,\ \dot{d}_{14}^- = 0.025,\ \dot{d}_{14}^+ = 0,\ \dot{d}_{15}^- = \dot{d}_{15}^+ = 0$$

$$\dot{d}_{23}^- = \dot{d}_{23}^+ = 0\ \ \dot{d}_{24}^- = 0.025,\ \dot{d}_{24}^+ = 0,\ \dot{d}_{25}^- = \dot{d}_{25}^+ = 0,\ \dot{d}_{34}^- = 0.025$$

$$\dot{d}_{34}^+ = 0,\ \dot{d}_{35}^- = \dot{d}_{35}^+ = 0\ \ \dot{d}_{45}^- = \dot{d}_{45}^+ = 0$$

Thus, by Definition 2, A is an inconsistent intuitionistic preference relation. Based on the optimal deviation values $\dot{d}_{ij}^-$ and $\dot{d}_{ij}^+$, $i = 1,2,3,4;\ j = i+1,\dots,5$, we solve the model (M-3) and get the following:

$$w_1^- = 0.35,\ w_1^+ = 0.35,\ w_2^- = 0.15,\ w_2^+ = 0.15,\ w_3^- = 0.15,\ w_3^+ = 0.15$$

$$w_4^- = 0,\ w_4^+ = 0,\ w_5^- = 0.35,\ w_5^+ = 0.35$$

thus, we get a unique priority vector $w = (0.35, 0.15, 0.15, 0, 0.35)^T$, i.e., the weights of the criteria $x_i (i = 1,2,\dots,5)$ are $w_1 = 0.35$, $w_2 = 0.15$, $w_3 = 0.15$, $w_4 = 0$, and $w_5 = 0.35$ respectively.

Example 2. Suppose that a decision maker provides his/her preference information over a collection of criteria x_1, x_2, x_3, x_4 with the following intuitionistic preference relation:

$$A = (a_{ij})_{4\times4} = \begin{bmatrix} (0.5, 0.5) & (0.6, 0.2) & (0.5, 0.4) & (0.7, 0.1) \\ (0.2, 0.6) & (0.5, 0.5) & (0.4, 0.3) & (0.6, 0) \\ (0.4, 0.5) & (0.3, 0.4) & (0.5, 0.5) & (0.7, 0.1) \\ (0.1, 0.7) & (0, 0.6) & (0.1, 0.7) & (0.5, 0.5) \end{bmatrix}$$

We first transform the intuitionistic preference relation A into its equivalent interval fuzzy preference relation

$$B = (b_{ij})_{4\times4} = \begin{bmatrix} [0.5, 0.5] & [0.6, 0.8] & [0.5, 0.6] & [0.7, 0.9] \\ [0.2, 0.4] & [0.5, 0.5] & [0.4, 0.7] & [0.6, 1] \\ [0.4, 0.5] & [0.3, 0.6] & [0.5, 0.5] & [0.7, 0.9] \\ [0.1, 0.3] & [0, 0.4] & [0.1, 0.3] & [0.5, 0.5] \end{bmatrix}$$

then by solving the model (M-2), we get $J_1^* = 0$, and all the optimal deviation values $\dot{d}_{ij}^-$ and $\dot{d}_{ij}^+$ ($i = 1,2,3;\ j = i+1,\dots,4$) are equal to zero. Thus, by Definition 2, A is a consistent intuitionistic preference relation, and then we solve the model (M-1) and get a unique priority vector $w = (0.4, 0.2, 0.4, 0)^T$, i.e., the weights of the criteria $x_i (i = 1,2,3,4)$ are $w_1 = 0.4$, $w_2 = 0.2$, $w_3 = 0.4$, and $w_4 = 0$ respectively.

5 Conclusions

We have introduced the notion of consistent intuitionistic preference relation and established two simple linear programming models to develop a method for estimating criteria weights from intuitionistic preference relations. The method can be applicable to multi-criteria decision making problems in many fields, such as the high technology project investment of venture capital firms, supply chain management, medical diagnosis, etc. In the future, we shall study the approach to improving the consistency of inconsistent intuitionistic preference relations.

Acknowledgements

The work was supported by the National Natural Science Foundation of China (No.70571087) and the National Science Fund for Distinguished Young Scholars of China (No.70625005).

References

1. Atanassov K (1986) Intuitionistic fuzzy sets. Fuzzy Sets and Systems 20: 87-96
2. Atanassov K, Gargov G (1989) Interval-valued intuitionistic fuzzy sets. Fuzzy Sets and Systems 31: 343-349
3. De SK, Biswas R, Roy AR (2000) Some operations on intuitionistic fuzzy sets. Fuzzy Sets and Systems 114: 477-484
4. De SK, Biswas R, Roy AR (2001) An application of intuitionistic fuzzy sets in medical diagnosis. Fuzzy Sets and Systems 117: 209-213
5. Deschrijver G, Kerre EE (2003) On the relationship between some extensions of fuzzy set theory. Fuzzy Sets and Systems 133: 227-235
6. Bustince H, Burillo P (1995) Correlation of interval-valued intuitionistic fuzzy sets. Fuzzy Sets and Systems 74: 237-244
7. Gerstenkorn T, Mańko J (1991) Correlation of intuitionistic fuzzy sets. Fuzzy Sets and Systems 44: 39-43
8. Hong DH, Hwang SY (1995) Correlation of intuitionistic fuzzy sets in probability spaces. Fuzzy Sets and Systems 75: 77-81
9. Hung WL, Wu JW (2002) Correlation of intuitionistic fuzzy sets by centroid method. Information Sciences 144: 219-225
10. Mitchell HB (2004) A correlation coefficient for intuitionistic fuzzy sets. International Journal of Intelligent Systems 19: 483-490
11. Szmidt E, Kacprzyk J (2000) Distances between intuitionistic fuzzy sets. Fuzzy Sets and Systems 114: 505-518
12. Li DF, Cheng CT (2002) New similarity measures of intuitionistic fuzzy sets and application to pattern recognitions. Pattern Recognition letters 23: 221-225
13. Liang ZZ, Shi PF (2003) Similarity measures on intuitionistic fuzzy sets. Pattern Recognition Letters 24: 2687-2693
14. Grzegorzewski P (2004) Distances between intuitionistic fuzzy sets and/or interval-valued fuzzy sets based on the Hausdorff metric. Fuzzy Sets and Systems 148: 319-328

15. Wang WQ, Xin XL (2005) Distance measure between intuitionistic fuzzy sets. Pattern Recognition Letters 26: 2063-2069
16. Deschrijver G, Cornelis C, Kerre EE (2004) On the representation of intuitionistic fuzzy t-norms and t-conorms. IEEE Transactions on Fuzzy Systems 12: 45-61
17. Deschrijver G, Kerre EE (2005) Implicators based on binary aggregation operators in interval-valued fuzzy set theory. Fuzzy Sets and Systems 153: 229-248
18. Park JH (2004) Intuitionistic fuzzy metric spaces. Chaos, Solitons and Fractals 22: 1039-1046
19. Park JH, Park JK (2004) Hausdorffness on generalized intuitionistic fuzzy filters. Information Sciences 168: 95-110
20. Mondal TK, Samanta SK (2002) Generalized intuitionistic fuzzy sets. Journal of Fuzzy Mathematics 10: 839-861
21. Deschrijver G, Kerre EE (2004) Uninorms in L^*-fuzzy set theory. Fuzzy Sets and Systems 148: 243-262
22. Cornelis C, Deschrijver G, Kerre EE (2004) Implication in intuitionistic fuzzy and interval-valued fuzzy set theory: construction, classification, application. International Journal of Approximate Reasoning 35: 55-95
23. Gutiérez Garcá J, Rodabaugh SE (2005) Order-theoretic, topological, categorical redundancies of interval-valued sets, grey sets, vague sets, interval-valued "intuitionistic" sets, "intuitionistic" fuzzy sets and topologies. Fuzzy Sets and Systems 156: 445-484
24. Dudek WA, Davvaz B, Jun YB (2005) On intuitionistic fuzzy sub-hyperquasigroups of hyperquasigroups. Information Sciences 170: 251-262
25. Xu ZS, Yager RR (2006) Some geometric aggregation operators based on intuitionistic fuzzy sets. International Journal of General Systems 35: 417-433
26. Szmidt E, Kacprzyk J (2002) Using intuitionistic fuzzy sets in group decision making. Control and Cybernetics 31: 1037-1053
27. Xu ZS (2007) Intuitionistic preference relations and their application in group decision making. Information Sciences, in press
28. Herrrera F, Martńez L, Sáchez PJ (2005) Managing non-homogeneous information in group decision making. European Journal of Operational Research 166: 115-132
29. Zadeh LA (1965) Fuzzy Sets. Information and Control 8: 338-353
30. Xu ZS (2004) On compatibility of interval fuzzy preference matrices. Fuzzy Optimization and Decision Making 3: 217-225
31. Xu ZS (2006) A C-OWA operator based approach to decision making with interval fuzzy preference relation. International Journal of Intelligent Systems 21: 1289-1298
32. Orlovsky SA (1978) Decision-making with a fuzzy preference relation. Fuzzy Sets and Systems 1: 155-167
33. Tanino T (1984) Fuzzy preference orderings in group decision making. Fuzzy Sets and Systems 12: 117-131
34. Kacprzyk J, Roubens M (1988) Non-conventional preference relations in decision-making. Springer, Berlin
35. Chiclana F, Herrera F, Herrera-Viedma E (1998) Integrating three representation models in fuzzy multipurpose decision making based on fuzzy preference relations. Fuzzy Sets and Systems 97: 33-48
36. Chiclana F, Herrera F, Herrera-Viedma E (2001) Integrating multiplicative preference relations in a multipurpose decision-making based on fuzzy preference relations. Fuzzy Sets and Systems 122: 277-291
37. Xu ZS, Da QL (2002) The uncertain OWA operator. International Journal of Intelligent Systems 17: 569-575

38. Xu ZS, Da QL (2005) A least deviation method to obtain a priority vector of a fuzzy preference relation. European Journal of Operational Research 164: 206-216
39. Chiclana F, Herrera F, Herrera-Viedma E (2002) A note on the internal consistency of various preference representations. Fuzzy Sets and Systems 131: 75-78
40. Xu ZS (2004) Uncertain multiple attribute decision making: methods and applications. Tsinghua University Press, Beijing
41. Wang YM, Yang JB, Xu DL (2005) A two-stage logarithmic goal programming method for generating weights from interval comparison matrices. Fuzzy Sets and Systems 152: 475-498

The Area Compensation Method About Fuzzy Order and Its Application

Zhang-xia Zhu[1] and Bing-yuan Cao[2]

[1] Department of Mathematics & Physics, Anhui University of Science & Technology, Huainan, 232001, P.R. China
zhxzhu@tom.com

[2] School of Mathematics & Information Sciences, Guangzhou University, Guangzhou, 510006, P. R. China
caobingy@163.com

Abstract. In this paper,we have presented a simple ranking method,which associated an intuitive geometrical representation:the area compensation,then a new fuzzy order relation between fuzzy numbers is defined. According to the new order, we can transform the fuzzy linear programming with fuzzy variables into the two-stage multiple objective linear programming,then the solving method is given.

Keywords: Interval numbers; fuzzy numbers; triangular fuzzy numbers; the area compensation; the fuzzy linear programming with the fuzzy variables.

1 Introduction

The fuzzy optimization, especially the fuzzy linear programming problems have been put forward since the 70th of 20st century, and since then they have received people's great concern, and much research achievements have been made. In this paper,we mainly discuss the fuzzy linear programming with fuzzy variables ,whose standard form is

$$\begin{aligned} \max\ & Z = c\widetilde{x} \\ \text{s.t.}\ \ & A\widetilde{x} \leq \widetilde{b}, \\ & \widetilde{x} \geq 0. \end{aligned}$$

where

$$c = (c_1, c_2, \cdots, c_n), A = (a_{ij})_{m\times n}, \widetilde{b} = (\widetilde{b}_1, \widetilde{b}_2, \cdots, \widetilde{b}_m)^T$$

and

$$\widetilde{x} = (\widetilde{x}_1, \widetilde{x}_2, \cdots, \widetilde{x}_n)^T.$$

Solving this kind of fuzzy linear programming problem is based on an order relation between fuzzy numbers.In this paper,we have presented a simple ranking method,which associated an intuitive geometrical representation:the area compensation,then a new fuzzy order relation between fuzzy numbers is defined.We have extended the initial definition related to the fuzzy numbers to the case of non-normalized convex fuzzy sets. According to the new order, we can transform

B.-Y. Cao (Ed.): Fuzzy Information and Engineering (ICFIE), ASC 40, pp. 513–522, 2007.
springerlink.com

the fuzzy linear programming with fuzzy variables into the two-stage multiple objective linear programming,then the solving method is given.

2 Basic Concept

In order to study the fuzzy linear programming problems with fuzzy variables, firstly we introduce some elementary knowledge as follows. In practical fuzzy mathematical programming problem,interval numbers,triangular fuzzy numbers are most commonly used,because they have intuitive appeal and can be easily specified by the decision maker.So we will mainly discuss the ranking method of interval numbers and triangular fuzzy numbers.

Definition 2.1. *Let $a = [\underline{a}, \overline{a}] = \{x|\underline{a} \leq x \leq \overline{a}\}$, Then closed interval a is said to be an interval number, where $\underline{a}, \overline{a} \in R, \underline{a} \leq \overline{a}$.*

From the extension principle, the properties of interval numbers can be described as follows.

Properties 2.1. *Let $a = [\underline{a}, \overline{a}]$,$b = [\underline{b}, \overline{b}]$ are two interval numbers, then*
(1)$a + b = [\underline{a} + \underline{b}, \overline{a} + \overline{b}]$;
(2)$a - b = [\underline{a} - \overline{b}, \overline{a} - \underline{b}]$;
(3) $ka = \begin{cases} [k\underline{a}, k\overline{a}], & k \geq 0, \\ [k\overline{a}, k\underline{a}], & k < 0. \end{cases}$

Definition 2.2. *Let $a = [\underline{a}, \overline{a}]$,$b = [\underline{b}, \overline{b}]$,then $a \leq b$ only and if only $\underline{a} \leq \underline{b}$,$\overline{a} \leq \overline{b}$.*

Definition 2.3. *Fuzzy set $\widetilde{A}$ is called a fuzzy number in R, if it satisfies conditions*
(1) There exists $x_0 \in R$ such that $\widetilde{A}(x_0) = 1$;
(2) For any $\lambda \in [0, 1]$, $\widetilde{A}_\lambda = \{x|\widetilde{A}(x) \geq \lambda\}$is a closed interval on R,and which is denoted by $\widetilde{A}_\lambda = [A^L_\lambda, A^R_\lambda]$.

Here we regard $F(R)$ as the set of all fuzzy numbers in R, among the which, the triangular fuzzy numbers are often seen and used, namely

Definition 2.4. *If $\widetilde{A} \in F(R)$,and its membership function A can be expressed as*

$$\widetilde{A}(x) = \begin{cases} \frac{x - A^L}{A^C - A^L}, & A^L \leq x \leq A^C, \\ \frac{x - A^R}{A^C - A^R}, & A^C \leq x \leq A^R, \\ 0, & otherwise. \end{cases}$$

Then $\widetilde{A}$ is called a triangular fuzzy number, and which is denoted by $\widetilde{A} = (A^L, A^C, A^R)$. Here A^L, A^C and A^R are called three parameter variables.

Obviously, when these variables are changed, $\widetilde{A}$ is also changed.

For the triangular fuzzy numbers, they meet the following properties that we all know.

Property 2.2. *Let*

$$\widetilde{A} = (A^L, A^C, A^R), \widetilde{B} = (B^L, B^C, B^R).$$

Then

(1) $\widetilde{A} + \widetilde{B} = (A^L + B^L, A^C + B^C, A^R + B^R)$;

(2) $\widetilde{A} - \widetilde{B} = (A^L - B^R, A^C - B^C, A^R - B^L)$;

(3) $k\widetilde{A} = \begin{cases} (kA^L, kA^C, kA^R), & k \geq 0, \\ (kA^R, kA^C, kA^L), & k < 0. \end{cases}$

3 Ranking Fuzzy Numbers

In this section,we propose a comparison method based on the area compensation determined by the membership functions of two fuzzy numbers,then a new fuzzy order relation between fuzzy numbers is defined.

Definition 3.1. *Let* $\widetilde{A}, \widetilde{B} \in F(R)$ *,for any* $\lambda \in [0,1], \widetilde{A}_\lambda = [A_\lambda^L, A_\lambda^R], \widetilde{B}_\lambda = [B_\lambda^L, B_\lambda^R]$. *We define*

$$S_L(\widetilde{A} \geq \widetilde{B}) = \int_{U(\widetilde{A},\widetilde{B})} [A_\lambda^L - B_\lambda^L] d\lambda \tag{1}$$

$$S_R(\widetilde{A} \geq \widetilde{B}) = \int_{V(\widetilde{A},\widetilde{B})} [A_\lambda^R - B_\lambda^R] d\lambda \tag{2}$$

where $U(\widetilde{A}, \widetilde{B}) = \{\lambda | 0 \leq \lambda \leq 1, A_\lambda^L \geq B_\lambda^L\}$,$V(\widetilde{A}, \widetilde{B}) = \{\lambda | 0 \leq \lambda \leq 1, A_\lambda^R \geq B_\lambda^R\}$.*And*

$$S_L(\widetilde{B} \geq \widetilde{A}) = \int_{U(\widetilde{B},\widetilde{A})} [B_\lambda^L - A_\lambda^L] d\lambda \tag{3}$$

$$S_R(\widetilde{B} \geq \widetilde{A}) = \int_{V(\widetilde{B},\widetilde{A})} [B_\lambda^R - A_\lambda^R] d\lambda \tag{4}$$

where $U(\widetilde{B}, \widetilde{A}) = \{\lambda | 0 \leq \lambda \leq 1, B_\lambda^L \geq A_\lambda^L\}$,$V(\widetilde{B}, \widetilde{A}) = \{\lambda | 0 \leq \lambda \leq 1, B_\lambda^R \geq A_\lambda^R\}$.

Obviously,$S_L(\widetilde{A} \geq \widetilde{B})$ is the area which claims that the left slope of$\widetilde{A}$ is greater to the corresponding part of $\widetilde{B}$,$S_R(\widetilde{B} \geq \widetilde{A})$ is the area which claims that the right slope of$\widetilde{B}$ is greater to the corresponding part of $\widetilde{A}$.

Property 3.1. *Let* $\widetilde{A}, \widetilde{B} \in F(R)$ *,and we denote*

$$d_L(\widetilde{A}, \widetilde{B}) = \frac{1}{2}[S_L(\widetilde{A} \geq \widetilde{B}) - S_L(\widetilde{B} \geq \widetilde{A})] \tag{5}$$

$$d_R(\widetilde{A}, \widetilde{B}) = \frac{1}{2}[S_R(\widetilde{A} \geq \widetilde{B}) - S_R(\widetilde{B} \geq \widetilde{A})] \tag{6}$$

Then

$$d_L(\widetilde{A},\widetilde{B}) = \frac{1}{2}\int_0^1 (A_\lambda^L - B_\lambda^L)d\lambda \tag{7}$$

$$d_R(\widetilde{A},\widetilde{B}) = \frac{1}{2}\int_0^1 (A_\lambda^R - B_\lambda^R)d\lambda \tag{8}$$

Proof
$d_L(\widetilde{A},\widetilde{B}) = \frac{1}{2}[S_L(\widetilde{A} \geq \widetilde{B}) - S_L(\widetilde{B} \geq \widetilde{A})] = \frac{1}{2}[\int_{U(\widetilde{A},\widetilde{B})}[A_\lambda^L - B_\lambda^L]d\lambda + \int_{U(\widetilde{B},\widetilde{A})}[A_\lambda^L - B_\lambda^L]d\lambda] = \frac{1}{2}\int_0^1(A_\lambda^L - B_\lambda^L)d\lambda$;
$d_R(\widetilde{A},\widetilde{B}) = \frac{1}{2}[S_R(\widetilde{A} \geq \widetilde{B}) - S_R(\widetilde{B} \geq \widetilde{A})] = \frac{1}{2}[\int_{U(\widetilde{A},\widetilde{B})}[A_\lambda^R - B_\lambda^R]d\lambda + \int_{U(\widetilde{B},\widetilde{A})}[A_\lambda^R - B_\lambda^R]d\lambda] = \frac{1}{2}\int_0^1(A_\lambda^R - B_\lambda^R)d\lambda$.

According to Property 3.1,we can easily get

Property 3.2. *Let*$\widetilde{A},\widetilde{B},\widetilde{C} \in F(R)$,*then*
(1) $d_L(\widetilde{A},\widetilde{A}) = 0$;
(2) $d_L(\widetilde{A},\widetilde{B}) = -d_L(\widetilde{B},\widetilde{A})$;
(3) $d_L(\widetilde{A},\widetilde{B}) + d_L(\widetilde{B},\widetilde{C}) = d_L(\widetilde{A},\widetilde{C})$;
(4) $d_L(k\widetilde{A},k\widetilde{B}) = kd_L(\widetilde{A},\widetilde{B})$.

Definition 3.2. *For two fuzzy numbers* $\widetilde{A}$*and*$\widetilde{B}$, *we define*
(1) $\widetilde{A} \leq \widetilde{B} \Leftrightarrow d_L(\widetilde{A},\widetilde{B}) \leq 0$,*and* $d_R(\widetilde{A},\widetilde{B}) \leq 0$;
(2) $\widetilde{A} \geq \widetilde{B} \Leftrightarrow d_L(\widetilde{A},\widetilde{B}) \geq 0$,*and* $d_R(\widetilde{A},\widetilde{B}) \geq 0$.

Especially,we have

Property 3.3. *Let* $\widetilde{A},\widetilde{B}$*are the fuzzy triangular numbers,then*

$$A^L + A^C \geq B^L + B^C, A^C + A^R \geq B^C + B^R \Leftrightarrow \widetilde{A} \geq \widetilde{B} \tag{9}$$

$$A^L + A^C \leq B^L + B^C, A^C + A^R \leq B^C + B^R \Leftrightarrow \widetilde{A} \leq \widetilde{B} \tag{10}$$

Proof: For any $\lambda \in [0,1]$,$A_\lambda^L = A^L + \lambda(A^C - A^L)$,$A_\lambda^R = A^R - \lambda(A^R - A^C)$, $B_\lambda^L = B^L + \lambda(B^C - B^L)$,$B_\lambda^R = B^R - \lambda(B^R - B^C)$.

According to property 3.1, we can get

$d_L(\widetilde{A},\widetilde{B}) = \frac{1}{2}\int_0^1(A_\lambda^L - B_\lambda^L)d\lambda = A^L + A^C - B^L - B^C$, $d_R(\widetilde{A},\widetilde{B}) = \frac{1}{2}\int_0^1(A_\lambda^R - B_\lambda^R)d\lambda = A^C + A^R - B^C - B^R$.

So

(1)$A^L + A^C \geq B^L + B^C, A^C + A^R \geq B^C + B^R \Leftrightarrow \widetilde{A} \geq \widetilde{B}$;

(2)$A^L + A^C \leq B^L + B^C, A^C + A^R \leq B^C + B^R \Leftrightarrow \widetilde{A} \leq \widetilde{B}$.

If we turn $\leq$and$\geq$ to$<$and $>$,then we can get the corresponding ranking about the fuzzy numbers.

Example 3.1. Let$\widetilde{A} = (1,2,3), \widetilde{B} = (2,3,4)$,then
$d_L(\widetilde{A},\widetilde{B}) = \frac{1}{2}\int_0^1 (A_\lambda^L - B_\lambda^L)d\lambda = \frac{1}{2}\int_0^1 (1+\lambda-2-\lambda)d\lambda = -\frac{1}{2} < 0,$
$d_R(\widetilde{A},\widetilde{B}) = \frac{1}{2}\int_0^1 (A_\lambda^R - B_\lambda^R)d\lambda = \frac{1}{2}\int_0^1 (3-\lambda-4+\lambda)d\lambda = -\frac{1}{2} < 0.$
So $\widetilde{A} < \widetilde{B}$.

4 Approach to the Fuzzy Linear Programming with Fuzzy Variables

Now let we consider the following fuzzy linear programming problem with fuzzy variables:

$$\begin{aligned} \max\ & Z = c\widetilde{x} \\ \text{s.t.}\ & A\widetilde{x} \leq \widetilde{b}, \\ & \widetilde{x} \geq 0. \end{aligned} \tag{11}$$

where

$$c = (c_1, c_2, \cdots, c_n), A = (a_{ij})_{m\times n}, \widetilde{b} = (\widetilde{b}_1, \widetilde{b}_2, \cdots, \widetilde{b}_m)^T$$

and

$$\widetilde{x} = (\widetilde{x}_1, \widetilde{x}_2, \cdots, \widetilde{x}_n)^T.$$

By property 2.2,model(11)can be transformed as

$$\begin{aligned} &\max z = (\sum_{j=1}^n c_j x_j^L, \sum_{j=1}^n c_j x_j^C, \sum_{j=1}^n c_j x_j^R) \\ &s.t.(\sum_{j=1}^n a_{ij} x_j^L, \sum_{j=1}^n a_{ij} x_j^C, \sum_{j=1}^n a_{ij} x_j^R) \leq (b_i^L, b_i^C, b_i^R) \\ &(x_j^L, x_j^C, x_j^R) \geq 0, i = 1, 2, \cdots m. \end{aligned} \tag{12}$$

According to property 3.3,we have

$$\max z^L = \sum_{j=1}^n c_j x_j^L$$

$$\max z^C = \sum_{j=1}^n c_j x_j^C$$

$$\max z^R = \sum_{j=1}^n c_j x_j^R$$

$$s.t. \sum_{j=1}^{n}(a_{ij}x_j^L + a_{ij}x_j^C) \leq b_i^L + b_i^C \tag{13}$$

$$\sum_{j=1}^{n}(a_{ij}x_j^C + a_{ij}x_j^R) \leq b_i^C + b_i^R$$

$$(x_j^L, x_j^C, x_j^R) \geq 0, x_j^C - x_j^L \geq 0, x_j^R - x_j^C \geq 0, i = 1, 2, \cdots m.$$

Obviously,model (13)is a multiple objective linear programming problem with three objects,and we can get the optimal solutions with the help of the methods about the multiple objective linear programming.The other hand,we know that the triangular fuzzy numbers $(\sum_{j=1}^{n} c_j x_j^L, \sum_{j=1}^{n} c_j x_j^C, \sum_{j=1}^{n} c_j x_j^R)$ can get the optimal solutions,firstly $\sum_{j=1}^{n} c_j x_j^C$ can get the optimal values,then both $\sum_{j=1}^{n} c_j x_j^L$and $\sum_{j=1}^{n} c_j x_j^R$ can get the optimal values. So (13) can be turned into two-stage multiple objective linear programming problem.

The first-stage linear programming problem

$$\max z^C = \sum_{j=1}^{n} c_j x_j^C$$

$$s.t. \sum_{j=1}^{n}(a_{ij}x_j^L + a_{ij}x_j^C) \leq b_i^L + b_i^C \tag{14}$$

$$\sum_{j=1}^{n}(a_{ij}x_j^C + a_{ij}x_j^R) \leq b_i^C + b_i^R$$

$$(x_j^L, x_j^C, x_j^R) \geq 0, x_j^C - x_j^L \geq 0, x_j^R - x_j^C \geq 0, i = 1, 2, \cdots m.$$

Denoting $(x_j^C)^*, j = 1, 2, \cdots n$ the optimal solutions of (14).

The second-stage linear programming problem

$$\max z^L = \sum_{j=1}^{n} c_j x_j^L$$

$$\max z^R = \sum_{j=1}^{n} c_j x_j^R$$

$$s.t. \sum_{j=1}^{n}(a_{ij}x_j^L + a_{ij}(x_j^C)^*) \leq b_i^L + b_i^C \tag{15}$$

$$\sum_{j=1}^{n}(a_{ij}(x_j^C)^* + a_{ij}x_j^R) \leq b_i^C + b_i^R$$

$$(x_j^L, x_j^R) \geq 0, (x_j^C)^* - x_j^L \geq 0, x_j^R - (x_j^C)^* \geq 0, i = 1, 2, \cdots m.$$

For model(15), x_j^L and x_j^R are uncorrelated variables, so it can be transformed to two linear programming problem

$$\max z^L = \sum_{j=1}^{n} c_j x_j^L$$

$$s.t. \sum_{j=1}^{n} (a_{ij} x_j^L + a_{ij}(x_j^C)^*) \leq b_i^L + b_i^C \tag{16}$$

$$x_j^L \geq 0, (x_j^C)^* - x_j^L \geq 0, i = 1, 2, \cdots m.$$

Denoting $(x_j^L)^*, j = 1, 2, \cdots n$ the optimal solutions of (16).
And

$$\max z^R = \sum_{j=1}^{n} c_j x_j^R$$

$$\sum_{j=1}^{n} (a_{ij}(x_j^C)^* + a_{ij} x_j^R) \leq b_i^C + b_i^R \tag{17}$$

$$x_j^R \geq 0, x_j^R - (x_j^C)^* \geq 0, i = 1, 2, \cdots m.$$

Denoting $(x_j^R)^*, j = 1, 2, \cdots n$ the optimal solutions of(17).
So the optimal solution of (11)is $\widetilde{x} = ((x_j^L)^*, (x_j^C)^*, (x_j^R)^*)^T, j = 1, 2, \cdots n$.

Example 4.1. Solving the following fuzzy linear programming problem with fuzzy variables

$$\begin{aligned} \max\ & Z = c\widetilde{x} \\ \text{s.t.}\ & A\widetilde{x} \leq \widetilde{b}, \\ & \widetilde{x} \geq 0. \end{aligned}$$

where $c = (2, 3), A = (a_{ij})_{2\times 2}, a_{11} = 1, a_{12} = 2, a_{21} = 3, a_{22} = 4, \widetilde{b}_1 = (1, 2, 3), \widetilde{b}_2 = (2, 3, 4)$ and $\widetilde{x} = (\widetilde{x}_1, \widetilde{x}_2)^T$.

According to the above method,we can get the first programming problem

$$\max z^C = 2x_1^C + 3x_2^C$$

$$s.t. (x_1^L + x_1^C) + 2(x_2^L + x_2^C) \leq 3$$

$$3(x_1^L + x_1^C) + 4(x_2^L + x_2^C) \leq 5$$

$$(x_1^R + x_1^C) + 2(x_2^R + x_2^C) \leq 5$$

$$3(x_1^R + x_1^C) + 4(x_2^R + x_2^C) \leq 7$$

$$(x_j^L, x_j^C, x_j^R) \geq 0, x_j^C - x_j^L \geq 0, x_j^R - x_j^C \geq 0, i, j = 1, 2.$$

and the second programming problem

$$\max z^L = 2x_1^L + 3x_2^L$$

$$s.t. (x_1^L + x_1^C) + 2(x_2^L + x_2^C) \leq 3$$

$$3(x_1^L + x_1^C) + 4(x_2^L + x_2^C) \leq 5$$
$$x_j^L \geq 0, x_j^C - x_j^L \geq 0, i, j = 1, 2.$$

and the third programming problem

$$\max z^R = 2x_1^R + 3x_2^R$$
$$(x_1^R + x_1^C) + 2(x_2^R + x_2^C) \leq 5$$
$$3(x_1^R + x_1^C) + 4(x_2^R + x_2^C) \leq 7$$
$$x_j^R \geq 0, x_j^R - x_j^C \geq 0, i, j = 1, 2.$$

For the above three programming problems, we can get the optimal solution x_j^C,$x_j^L$and x_j^R according to the method about the linear programming problem. So the optimal solution about the original problem is $\widetilde{x} = (x_j^L, x_j^C, x_j^R)^T, j = 1, 2.$

5 Extensions for the Area Compensation Ranking Method

In this section,we simply propose a comparison method based on the area compensation determined by the membership functions of two non-normalized convex fuzzy sets.

Definition 5.1. *Fuzzy set $\widetilde{A}$ is called a non-normalized convex fuzzy sets on R, if for any $\lambda \in [0, 1]$, $\widetilde{A}_\lambda$is a convex subsets of R which lower and upper limits are represented by $A_\lambda^L = inf\{x|\widetilde{A}(x) \geq \lambda\}$ and $A_\lambda^R = sup\{x|\widetilde{A}(x) \geq \lambda\}$.*
We suppose that both limits are finite.

Denoting $hgt(\widetilde{A})$ the maximum value of the membership function $\mu_{\widetilde{A}}$, then we have

Definition 5.2. *Let $\widetilde{A}, \widetilde{B}$ are two non-normalized convex fuzzy sets, we define*

$$S_L(\widetilde{A} \geq \widetilde{B}) = \int_{U(\widetilde{A},\widetilde{B})} [A_\lambda^L - B_\lambda^L] d\lambda \tag{18}$$

$$S_R(\widetilde{A} \geq \widetilde{B}) = \int_{V(\widetilde{A},\widetilde{B})} [A_\lambda^R - B_\lambda^R] d\lambda \tag{19}$$

where $U(\widetilde{A}, \widetilde{B}) = \{\lambda | 0 \leq \lambda \leq hgt(\widetilde{A}), A_\lambda^L \geq B_\lambda^L\}$,$V(\widetilde{A}, \widetilde{B}) = \{\lambda | 0 \leq \lambda \leq hgt(\widetilde{A}), A_\lambda^R \geq B_\lambda^R\}$.

$$S_L(\widetilde{B} \geq \widetilde{A}) = \int_{U(\widetilde{B},\widetilde{A})} [B_\lambda^L - A_\lambda^L] d\lambda \tag{20}$$

$$S_R(\widetilde{B} \geq \widetilde{A}) = \int_{V(\widetilde{B},\widetilde{A})} [B_\lambda^R - A_\lambda^R] d\lambda \tag{21}$$

where $U(\widetilde{B},\widetilde{A}) = \{\lambda | 0 \leq \lambda \leq hgt(\widetilde{A}), B_\lambda^L \geq A_\lambda^L\}$,$V(\widetilde{B},\widetilde{A}) = \{\lambda | 0 \leq \lambda \leq hgt(\widetilde{A}), B_\lambda^R \geq A_\lambda^R\}$.

Obviously,$S_L(\widetilde{A} \geq \widetilde{B})$ is the area which claims that the left slope of$\widetilde{A}$ is greater to the corresponding part of $\widetilde{B}$,$S_R(\widetilde{B} \geq \widetilde{A})$ is the area which claims that the right slope of$\widetilde{B}$ is greater to the corresponding part of $\widetilde{A}$.

Property 5.1. *Let $\widetilde{A}, \widetilde{B}$ are two non-normalized convex fuzzy sets,and we denote*

$$d_L(\widetilde{A},\widetilde{B}) = \frac{1}{2hgt(\widetilde{A})}[S_L(\widetilde{A} \geq \widetilde{B}) - S_L(\widetilde{B} \geq \widetilde{A})] \tag{22}$$

$$d_R(\widetilde{A},\widetilde{B}) = \frac{1}{2hgt(\widetilde{A})}[S_R(\widetilde{A} \geq \widetilde{B}) - S_R(\widetilde{B} \geq \widetilde{A})] \tag{23}$$

Then

$$d_L(\widetilde{A},\widetilde{B}) = \frac{1}{2hgt(\widetilde{A})}\int_0^{hgt(\widetilde{A})}(A_\lambda^L - B_\lambda^L)d\lambda \tag{24}$$

$$d_R(\widetilde{A},\widetilde{B}) = \frac{1}{2hgt(\widetilde{A})}\int_0^{hgt(\widetilde{A})}(A_\lambda^R - B_\lambda^R)d\lambda \tag{25}$$

Proof

$d_L(\widetilde{A},\widetilde{B}) = \frac{1}{2hgt(\widetilde{A})}[S_L(\widetilde{A} \geq \widetilde{B}) - S_L(\widetilde{B} \geq \widetilde{A})] = \frac{1}{2hgt(\widetilde{A})}[\int_{U(\widetilde{A},\widetilde{B})}[A_\lambda^L - B_\lambda^L]d\lambda + \int_{U(\widetilde{B},\widetilde{A})}[A_\lambda^L - B_\lambda^L]d\lambda] = \frac{1}{2hgt(\widetilde{A})}\int_0^{hgt(\widetilde{A})}(A_\lambda^L - B_\lambda^L)d\lambda$;

$d_R(\widetilde{A},\widetilde{B}) = \frac{1}{2hgt(\widetilde{A})}[S_R(\widetilde{A} \geq \widetilde{B}) - S_R(\widetilde{B} \geq \widetilde{A})] = \frac{1}{2hgt(\widetilde{A})}[\int_{U(\widetilde{A},\widetilde{B})}[A_\lambda^R - B_\lambda^R]d\lambda + \int_{U(\widetilde{B},\widetilde{A})}[A_\lambda^R - B_\lambda^R]d\lambda] = \frac{1}{2hgt(\widetilde{A})}\int_0^{hgt(\widetilde{A})}(A_\lambda^R - B_\lambda^R)d\lambda$.

Definition 5.3. *For two non-normalized convex fuzzy sets $\widetilde{A}$and$\widetilde{B}$, we define*

(1)$\widetilde{A} \leq \widetilde{B}$only and if only $d_L(\widetilde{A},\widetilde{B}) \leq 0$,and $d_R(\widetilde{A},\widetilde{B}) \leq 0$;

(2)$\widetilde{A} \geq \widetilde{B}$only and if only $d_L(\widetilde{A},\widetilde{B}) \geq 0$,and $d_R(\widetilde{A},\widetilde{B}) \geq 0$.

If we turn $\leq$and$\geq$ to$<$and $>$,then we can get the corresponding ranking about the fuzzy numbers.

6 Conclusion

In this paper,we have presented a simple ranking method,which associated an intuitive geometrical representation:the area compensation.We have extended the initial definition related to the fuzzy numbers to the case of non-normalized convex fuzzy sets.

With the help of the the area compensation,we define a new order relation between fuzzy numbers and apply it to the fuzzy linear programming with fuzzy variables.

Acknowledgment

The authors would like to thank the supports of National Natural Science Foundation of China (No.70271047 and No.79670012) and Youth Science Foundation of Anhui University of Science & Technology.

References

1. H.-J.Zimmerman(1985) Fuzzy Set Theory-and Its Applications.*Kluwer-Nijhoff Publishing.*
2. Cheng-zhong Luo(1989)*Fuzzy set.* Bei-jing Normal University Press.
3. Fan Li(1997)*Fuzzy information disposal system.* Bei-jing university press.
4. Yi-Jin,Qing-li Da, Nan-rong Xu(1995)*Study of the multiobjective fuzzy linear programming problem with fuzzy variables.* Journal of Systems Engineering 10(2): 24-32.
5. Xiao-zhong Li, Qing-de Zhang(1998)*Fuzzy linear programming problems with fuzzy variables and fuzzy constraints.* Journal of Liaocheng Teachers University 11(1): 7-11.
6. Jing-shing Yao,Kweimei Wu(2000)*Ranking fuzzy numbers based on decomposition principle and signed distance.* Fuzzy Sets and Systems 116: 275-288.
7. Philippe Fortemps,marc Roubens(1996) Ranking and defuzzificaton methods based on area compensation. *Fuzzy Sets and Systems*82:319-330.
8. H.R.Maleki,M.Tata,M.Mashinchi(2000)Linear programming with fuzzy variables.*Fuzzy Sets and Systems*109: 21-33.
9. Qiao Zhong,Wang Guang-yuan(1993) On solutions and distribution problems of the linear programming with fuzzy random variables coefficients.*Fuzzy Sets and Systems*58: 155-170.
10. Hua-wen Liu(2004) Comparison of fuzzy numbers based on a fuzzy distance measure. *Shandong University Transaction*39(2):31-36.
11. Xin-wang Liu and Qing-li Da(1998) The solution for fuzzy linear programming with constraint satisfactory function.*Journey of Systems Engineering,*13(3):36-40.
12. L Tran and L Duckstein(2002)Comparison of fuzzy numbers using a fuzzy distance measure. *Fuzzy Set and Systems*130: 331-341.

A Method for the Priority Vector of Fuzzy Reciprocal Matrix

Wenyi Zeng*, Yan Xing, and Hongxing Li

School of Mathematical Sciences, Beijing Normal University, Beijing 100875, China

Abstract. In this paper, we discuss some new properties of fuzzy consistent matrix, propose a novel method to transform fuzzy reciprocal matrix into fuzzy consistent matrix and calculate the corresponding priority vector. Simultaneously, some concepts including fuzzy consistent index and fuzzy consistent ratio are put forward, and an acceptable value(< 0.1) is given to adjust fuzzy reciprocal matrix. Finally, we investigate the convergence of our proposed algorithm and use two numerical examples to illustrate our proposed method reasonable.

Keywords: Fuzzy reciprocal matrix, Fuzzy consistent matrix, Priority vector.

1 Introduction

In decision analysis, pairwise comparison of alternatives is widely used[1, 2]. Generally, decision makers express their pairwise comparison information in two formats: multiplicative preference relations[2, 3] and fuzzy reciprocal relations [4, 5, 6]. The analytic hierarchy process(AHP) with multiplicative preference relations has been applied extensively in many fields, such as economic analysis, technology transfer and population forecast[3]. However, decision makers may also use fuzzy reciprocal relations to express their preference[5] due to their different culture and educational backgrounds, personal habits and vague nature of human judgement. There are some research issues between multiplicative preference relations and fuzzy reciprocal relations, as both are based on pairwise comparison. Therefore, research progress in multiplicative preference relations can benefit research in fuzzy reciprocal relations.

The consistency in preference relations(whether multiplicative or fuzzy) given by decision makers is difficult to satisfy the ranking results of the final decision [7, 8]. To date, a great deal of research has been conducted on the consistency problems of multiplicative preference relations. The issue of consistency in AHP was first addressed by Saaty, who originally developed the notions of perfect consistency and acceptable consistency. Recently, the consistency of fuzzy reciprocal matrix become an interesting topic. Generally, fuzzy reciprocal matrix proposed by experts hardly satisfies consistency. To ensure the credibility and accuracy of the priority vector of fuzzy reciprocal matrix, people usually need to adjust fuzzy reciprocal matrix. Several authors did some meaningful works, respectively. For example, Song [9] proposed an additive consistent index ρ to measure the difference between fuzzy reciprocal matrix and fuzzy

* Corresponding author: Zeng Wenyi(zengwy@bnu.edu.cn).

B.-Y. Cao (Ed.): Fuzzy Information and Engineering (ICFIE), ASC 40, pp. 523–533, 2007.
springerlink.com

consistent matrix and a kind of method to solve the priority vector of fuzzy reciprocal matrix, Sun [10] used the goal programming method to adjust consistency of fuzzy reciprocal matrix, Xu[11] based on minimum variance method, minimum difference method and eigenvector method to revise fuzzy reciprocal matrix and calculate the priority vector, Ma [12] pointed out the shortage of the algorithm proposed by Xu[11], put forward a kind of new algorithm and converted fuzzy reciprocal matrix to fuzzy consistent matrix.

In this paper, we discuss some new properties of fuzzy consistent matrix, propose a novel method to transform fuzzy reciprocal matrix into fuzzy consistent matrix and calculate the corresponding priority vector. Simultaneously, some concepts including fuzzy consistent index and fuzzy consistent ratio are put forward and an acceptable value(< 0.1) is given to adjust fuzzy reciprocal matrix. Finally, we investigate the convergence of our proposed algorithm and use two numerical examples to illustrate our proposed method reasonable.

The rest of this work is organized as follows. In section 2, some properties of fuzzy consistent matrix are given. In section 3, some concepts including fuzzy consistent index and fuzzy consistent ratio and two kinds of methods for adjusting inconsistency of fuzzy reciprocal matrix, along with theoretical proofs of convergence for our algorithm are put forward. In section 4, two numerical examples are used to illustrate our proposed method reasonable. The conclusion is in section 5.

2 Some Properties of Fuzzy Consistent Matrix

Throughout this paper, we write $X = \{x_1, x_2, \cdots, x_n\}$ to denote a finite set of alternatives, where x_i denotes the ith alternative. In a decision problem. the preference information of decision maker on X is described by a fuzzy relation(fuzzy matrix), $R = (r_{ij})_{n\times n}$, with membership function $\mu_R : X \times X \to [0,1]$, where $\mu_R(x_i, x_j) = r_{ij}$ denotes the preference degree of alternativex_i over x_j.

Definition 1[13]. Let $R = (r_{ij})_{n\times n}$, if $0 \leq r_{ij} \leq 1$, $\forall i,j = 1,\cdots,n$, then R is fuzzy matrix.

Definition 2[13]. Let $R = (r_{ij})_{n\times n}$, if R is fuzzy matrix and $r_{ij} + r_{ji} = 1$, $\forall i,j = 1,\cdots,n$, then R is fuzzy reciprocal matrix.

Property 1[14]. $\frac{1}{2} \leq \sum\limits_{j=1}^{n} r_{ij} \leq n - \frac{1}{2}, \forall i,j = 1,\cdots,n$

Property 2[14]. $\sum\limits_{i=1}^{n}\sum\limits_{j=1}^{n} r_{ij} = \frac{n^2}{2}$

Definition 3[13]. Let $R = (r_{ij})_{n\times n}$, if R is fuzzy matrix and $r_{ij} = r_{ik} - r_{jk} + \frac{1}{2}$, $\forall i,j,k = 1,\cdots,n$, then R is fuzzy consistent matrix. Several scholars[10, 11, 13, 15] have investigated fuzzy consistent matrix and obtained some meaningful results.

Property 3[13]. Let R be fuzzy consistent matrix, if we run arbitrary line and its corresponding row of R, then the new matrix is also fuzzy consistent matrix.

Property 4[13]**.** R satisfies that:

$$\lambda \geq \frac{1}{2}, r_{ij} \geq \lambda, r_{jk} \geq \lambda \Longrightarrow r_{ik} = r_{ij} + r_{jk} - 0.5 \geq \lambda$$

$$\lambda \leq \frac{1}{2}, r_{ij} \leq \lambda, r_{jk} \leq \lambda \Longrightarrow r_{ik} = r_{ij} + r_{jk} - 0.5 \leq \lambda$$

Property 5[11]**.** Let $R = (r_{ij})_{n\times n}$ be fuzzy consistent matrix, then its element r_{ij} can be stated by the following equation

$$r_{ij} = r_{i1} - r_{j1} + \frac{1}{2} = r_{1j} - r_{1i} + \frac{1}{2}, \forall i, j = 1, \cdots, n$$

.

Theorem 1[16]**.** A fuzzy reciprocal matrix $R = (r_{ij})$ is fuzzy consistent matrix if and only if its elements satisfy $r_{ij} + r_{jk} + r_{ki} = \frac{3}{2}, \forall i, j, k = 1, \cdots, n.$

Theorem 2[14]**.** A fuzzy matrix $R = (r_{ij})$ is fuzzy consistent matrix if and only if there exist $1 \times n$ non-negative and normalized vector $\omega = (\omega_1, \omega_2, \cdots, \omega_n)'$ and a positive number a such that

$$r_{ij} = a(\omega_i - \omega_j) + \frac{1}{2}, \forall i, j = 1, \cdots, n$$

Theorem 3[14]**.** Let R be fuzzy consistent matrix and $\omega = (\omega_1, \omega_2, \cdots, \omega_n)'$ be the priority vector, then they satisfy the following equation.

$$\omega_i = \frac{1}{n} - \frac{1}{2a} + \frac{1}{na}\sum_{k=1}^{n} r_{ik}, \forall i = 1, \cdots, n \tag{1}$$

Further, we have

$$\omega_i = \frac{1}{na}\sum_{k=1}^{n} r_{1k} + \frac{1}{n} - \frac{1}{a} r_{1i}, i = 1 \cdots, n$$

or

$$\begin{pmatrix} \omega_1 \\ \omega_2 \\ \vdots \\ \omega_n \end{pmatrix} = \frac{1}{na}\begin{pmatrix} 1-n & 1 & \cdots & 1 \\ 1 & 1-n & \cdots & 1 \\ \cdots & \cdots & \cdots & \cdots \\ 1 & 1 & \cdots & 1-n \end{pmatrix}\begin{pmatrix} r_{11} \\ r_{12} \\ \vdots \\ r_{1n} \end{pmatrix} + \frac{1}{n}\begin{pmatrix} 1 \\ 1 \\ \vdots \\ 1 \end{pmatrix} \tag{2}$$

Theorem 4[12] **.**Let $Q = (q_{ij})$ be fuzzy reciprocal matrix, then we can construct fuzzy matrix $R = (r_{ij})_{n\times n}$, where

$$r_{ij} = \frac{1}{n}\sum_{k=1}^{n}(q_{kj} - q_{ki} + \frac{1}{2}) \tag{3}$$

Furthermore, we are also able to adjust the fuzzy matrix R based on the following two situations, for r_{ij} ,$\forall i,j=1,\cdots,n$, then we have:
Case 1. $r_{ij}\geq 0,\forall i,j=1,\cdots,n$. Let $R^{'}=(r^{'}_{ij})_{n\times n}=R$, then $R^{'}$ is fuzzy consistent matrix;
Case 2. $\exists r_{ij}<0$, let $R^{'}=(r^{'}_{ij})_{n\times n}$, where $r^{'}_{ij}=\dfrac{r_{ij}+b}{1+2b},\forall i,j=1,\cdots,n$, $b=\max\{|r_{ts}||r_{ts}<0,t,s=1,\cdots,n\}$, then fuzzy matrix $R^{'}$ is fuzzy consistent matrix.
Property 6. If $R=(r_{ij})_{n\times n}$ is fuzzy consistent matrix , then its rank of R is 1 or 2.

Proof

$$R=(r_{ij})_{n\times n}=\begin{pmatrix} \frac{1}{2} & r_{12} & \cdots & r_{1n} \\ 1-r_{12} & \frac{1}{2} & \cdots & r_{1n}-r_{12}+\frac{1}{2} \\ \cdots & \cdots & \cdots & \cdots \\ 1-r_{1n} & r_{12}-r_{1n}+\frac{1}{2} & \cdots & \frac{1}{2} \end{pmatrix}$$
$$\xrightarrow{\text{other rows subtract the first row}}\begin{pmatrix} \frac{1}{2} & r_{12} & \cdots & r_{1n} \\ \frac{1}{2}-r_{12} & \frac{1}{2}-r_{12} & \cdots & \frac{1}{2}-r_{12} \\ \cdots & \cdots & \cdots & \cdots \\ \frac{1}{2}-r_{1n} & \frac{1}{2}-r_{1n} & \cdots & \frac{1}{2}-r_{1n} \end{pmatrix}$$

If $\frac{1}{2}-r_{1j}=0,\forall j=2,\cdots,n$, then the above matrix is changed into a matrix whose rank is 1. Otherwise, there exists an element r_{1j} such that $\frac{1}{2}-r_{1j}\neq 0$, thus, the above matrix is changed into a matrix whose rank is 2.

Remark 1. If the rank of fuzzy consistent matrix R is 1, then we can get its corresponding priority vector: $\omega=\frac{1}{n}(1,1,\cdots,1)^{'}$.

Remark 2

$$|R|=\begin{cases} \frac{1}{2}, & n=1 \\ \frac{1}{4}-r_{12}+r_{12}^{2}, & n=2 \\ 0, & n\geq 3 \end{cases}$$

where $|R|$ is the determinant of fuzzy matrix R.
Property 7. A fuzzy consistent matrix $R=(r_{ij})_{n\times n}$ have two non-zero real eigenvalues at most.
Proof

$$|R-\lambda I|=\begin{vmatrix} \frac{1}{2}-\lambda & r_{12} & \cdots & r_{1n} \\ 1-r_{12} & \frac{1}{2}-\lambda & \cdots & r_{1n}-r_{12}+\frac{1}{2} \\ \cdots & \cdots & \cdots & \cdots \\ 1-r_{1n} & r_{12}-r_{1n}+\frac{1}{2} & \cdots & \frac{1}{2}-\lambda \end{vmatrix}$$
$$=\begin{vmatrix} \frac{1}{2}-\lambda & r_{12} & \cdots & r_{1n} \\ \frac{1}{2}-r_{12}+\lambda & \frac{1}{2}-r_{12}-\lambda & \cdots & \frac{1}{2}-r_{12} \\ \vdots & \vdots & \ddots & \vdots \\ \frac{1}{2}-r_{1n}+\lambda & \frac{1}{2}-r_{1n} & \cdots & \frac{1}{2}-r_{1n}-\lambda \end{vmatrix}$$
$$=\begin{vmatrix} 1 & 0 & 0 & 0 & \cdots & 0 \\ 0 & \frac{1}{2}-\lambda & r_{12} & \cdots & & r_{1n} \\ \frac{1}{2}-r_{12} & \frac{1}{2}-r_{12}+\lambda & \frac{1}{2}-r_{12}-\lambda & \cdots & & \frac{1}{2}-r_{12} \\ \vdots & \vdots & \vdots & \ddots & & \vdots \\ \frac{1}{2}-r_{1n} & \frac{1}{2}-r_{1n}+\lambda & \frac{1}{2}-r_{1n} & \cdots & & \frac{1}{2}-r_{1n}-\lambda \end{vmatrix}=\lambda^{n-2}(\lambda-a)(\lambda-b)$$

where, $a = \frac{n}{2} - \sum_{j=1}^{n} r_{1j}, b = \sum_{j=1}^{n} r_{1j}$.

Let $|R - \lambda I| = 0$, then we have

$$\lambda_1 = \frac{n}{2} - \sum_{j=1}^{n} r_{1j}, \lambda_2 = \sum_{j=1}^{n} r_{1j}, \lambda_3 = \lambda_4 = \cdots = \lambda_n = 0$$

Known by the property 1, we have $-\frac{n-1}{2} \leq \lambda_1 \leq \frac{n-1}{2}$.

If $\lambda_1 = 0$, then we have only one non-zero real eigenvalue.

If $\lambda_1 \neq 0$, then we have two non-zero real eigenvalues.

Property 8. For fuzzy matrix R determined by Eq. (3), then $-1 < r_{ij} < 1, \forall i, j = 1, 2, \cdots, n$.

Proof. Known by property 1, then we have:

$$r_{ij} = \frac{1}{n}\sum_{k=1}^{n}(q_{kj} - q_{ki} + \frac{1}{2}) \leq \frac{1}{n}(n - \frac{1}{2} - \frac{1}{2} + \frac{1}{2}) = \frac{n - \frac{1}{2}}{n} < 1, \forall i, j = 1, \cdots, n$$

$$r_{ij} \geq \frac{1}{n}(\frac{1}{2} - n + \frac{1}{2} + \frac{1}{2}) = \frac{\frac{3}{2} - n}{n} > -1$$

Therefore, we have $-1 < r_{ij} < 1$.

Remark 3. A fuzzy matrix R determined by Eq. (3) isn't always consistent matrix.

Example 1

$$Q = \begin{pmatrix} 0.5 & 1 & 1 & 1 \\ 0 & 0.5 & 0.6 & 0.4 \\ 0 & 0.4 & 0.5 & 1 \\ 0 & 0.6 & 0 & 0.5 \end{pmatrix} \xrightarrow{(2)} R = \begin{pmatrix} 0.5 & 1 & 0.9 & 1 \\ 0 & 0.5 & 0.4 & 0.6 \\ 0.1 & 0.6 & 0.5 & 0.7 \\ -0.1 & 0.4 & 0.3 & 0.5 \end{pmatrix}$$

It is easy to see that fuzzy matrix R isn't fuzzy consistent matrix.

Property 9. Suppose Q be fuzzy reciprocal matrix, and fuzzy consistent matrix R' is converted based on theorem 6, a vector $\omega = (\omega_1, \cdots, \omega_n)'$ is the priority vector of R', then we have the following result.

$$\begin{pmatrix} \omega_1 \\ \omega_2 \\ \vdots \\ \omega_n \end{pmatrix} = \begin{pmatrix} \frac{1}{n} + \frac{1}{2a(1+2b)} - \frac{1}{na(1+2b)} \sum_{k=1}^{n} q_{k1} \\ \frac{1}{n} + \frac{1}{2a(1+2b)} - \frac{1}{na(1+2b)} \sum_{k=1}^{n} q_{k2} \\ \vdots \\ \frac{1}{n} + \frac{1}{2a(1+2b)} - \frac{1}{na(1+2b)} \sum_{k=1}^{n} q_{kn} \end{pmatrix} \quad (4)$$

where, $b = \max\{|r_{ts}| | r_{ts} \leq 0, t, s = 1, \cdots, n\}$

Proof. By Eq. (1), we have,

$$\begin{aligned}
\omega_i &= \frac{1}{n} - \frac{1}{2a} + \frac{1}{na}\sum_{l=1}^{n} r'_{il} \\
&= \frac{1}{n} - \frac{1}{2a} + \frac{1}{na}\sum_{l=1}^{n}\left(\frac{1}{n}\sum_{k=1}^{n}\frac{(q_{kl} - q_{ki} + \frac{1}{2}) + b}{1+2b}\right) \\
&= \frac{1}{n} - \frac{1}{2a} + \frac{1}{n^2 a(1+2b)} \times \frac{n^2}{2} - \frac{1}{na(1+2b)}\sum_{k=1}^{n} q_{ki} + \frac{1}{2a(1+2b)} + \frac{b}{a(1+2b)} \\
&= \frac{1}{n} + \frac{1}{2a(1+2b)} - \frac{1}{na(1+2b)}\sum_{k=1}^{n} q_{ki}, \forall i = 1, \cdots, n
\end{aligned}$$

If $b = 0$,then we have

$$\begin{pmatrix} \omega_1 \\ \omega_2 \\ \vdots \\ \omega_n \end{pmatrix} = \begin{pmatrix} \frac{1}{2a} + \frac{1}{n} - \frac{1}{na}\sum\limits_{k=1}^{n} q_{k1} \\ \frac{1}{2a} + \frac{1}{n} - \frac{1}{na}\sum\limits_{k=1}^{n} q_{k2} \\ \vdots \\ \frac{1}{2a} + \frac{1}{n} - \frac{1}{na}\sum\limits_{k=1}^{n} q_{kn} \end{pmatrix} \tag{5}$$

It is easy to find that the fuzzy matrix constructed by Q is fuzzy consistent matrix.

3 Algorithm

Let $Q = (q_{ij})_{n\times n}$ be fuzzy reciprocal matrix, known by theorem 2, we can construct fuzzy consistent matrix $R = (r_{ij})_{n\times n}$, $\omega = (\omega_1, \cdots, \omega_n)'$ is the priority vector of R, where $r_{ij} = a(\omega_i - \omega_j) + \frac{1}{2}$ and a is a positive real number.In general, we order $a = 1$ and $q_{ij} = a(\omega_i - \omega_j) + \frac{1}{2} + \epsilon_{ij}$, where ϵ_{ij} is a real number.

Theorem 5. Let $Q = R + E$ and $E = (\epsilon_{ij})_{n\times n}$, the matrix E is deviation matrix and satisfies the following properties:

(1) $E = (\epsilon_{ij})_{n\times n}$is a antisymmetric matrix; (2) Q has the same real eigenvalues as R.

Proof. (1) Known by properties of fuzzy reciprocal matrix Q, we have

$$\begin{aligned}
1 = q_{ij} + q_{ji} &= a(\omega_i - \omega_j) + \frac{1}{2} + \epsilon_{ij} + a(\omega_j - \omega_i) + \frac{1}{2} + \epsilon_{ji} \\
&= \epsilon_{ij} + \epsilon_{ji} + 1, \forall i, j = 1, 2, \cdots, n,
\end{aligned}$$

thus, $\epsilon_{ij} = -\epsilon_{ji}, \forall i, j = 1, 2, \cdots, n$.

Therefore, E is an antisymmetric matrix.

(2) Suppose that λ is a real eigenvalue of matrix Q and α is its corresponding eigenvalue vector.

$$\alpha' Q\alpha = \alpha' (R+E)\alpha = \alpha' R\alpha = \lambda$$

Therefore, $R\alpha = \lambda\alpha$, it shows that λ is a real eigenvalue of the matrix R and α is its corresponding eigenvalue vector.

Similarly, we can prove that the real eigenvalue of matrix R is also the eigenvalue of the matrix Q.

In the following, we propose two concepts such as fuzzy consistency index CI and fuzzy consistency ratio CR.

Definition 4. $CI = \frac{\lambda_{\max}-n}{n-1}$ is called fuzzy consistency index of fuzzy reciprocal matrix Q, where $\lambda_{\max}$ is the maximal real eigenvalue of matrix $A = (a_{ij})_{n\times n} = (e^{q_{ij}-\frac{1}{2}})_{n\times n}$ and n is the order of the matrix Q; $CR = \frac{CI}{RI}$ is called fuzzy consistency ratio of Q.

Theorem 6. $CI = \frac{\lambda_{max}-n}{n-1} = \frac{1}{n(n-1)} \sum\limits_{1\le i<j\le n} (e^{\epsilon_{ij}} + e^{-\epsilon_{ij}} - 2)$,and $CI \ge 0$.

Proof. Let $A = (a_{ij})_{n\times n}$,where $a_{ij} = e^{q_{ij}-\frac{1}{2}} = e^{r_{ij}-\frac{1}{2}+\epsilon_{ij}} = e^{a(\omega_i-\omega_j)+\epsilon_{ij}} = \frac{e^{a\omega_i}}{e^{a\omega_j}} \times e^{\epsilon_{ij}}$.

Obviously,$a_{ji} = \frac{e^{a\omega_j}}{e^{a\omega_i}} \times e^{\epsilon_{ji}} = \frac{e^{a\omega_j}}{e^{a\omega_i}} \times e^{-\epsilon_{ij}} = \frac{1}{a_{ij}}$

So, A is a judgement matrix.

Suppose there is a matrix $B = (b_{ij})_{n\times n}$,where $b_{ij} = e^{a(\omega_i-\omega_j)}$. Obviously, $b_{ij} \cdot b_{jk} = b_{ik}$, for every $i, j, k = 1, \cdots, n$. Thus, B is consistent matrix.

Suppose $\lambda_{\max}$ be the maximal real eigenvalue of matrix $A = (a_{ij})_{n\times n}$, $\eta = (e^{a\omega_i})_{n\times 1}$ is its corresponding eigenvalue vector, then we have $A\eta = \lambda_{max}\eta$,

$$\lambda_{max}e^{a\omega_i} = \sum_{j=1}^{n} a_{ij}e^{a\omega_j} = \sum_{j=1}^{n} e^{a\omega_i}e^{\epsilon_{ij}}$$

$$\lambda_{max} = \sum_{j=1}^{n} e^{\epsilon_{ij}}$$

$$n\lambda_{max} = \sum_{i=1}^{n}\sum_{j=1}^{n} e^{\epsilon_{ij}}$$

$$n\lambda_{max} - n = \sum_{i=1}^{n}\sum_{\substack{j=1\\ j\ne i}}^{n} e^{\epsilon_{ij}} = \sum_{1\le i<j\le n} (e^{\epsilon_{ij}} + e^{-\epsilon_{ij}})$$

$$\begin{aligned} CI &= \frac{\lambda_{max}-n}{n-1} = -1 + \frac{\lambda_{max}-1}{n-1} \\ &= -1 + \frac{1}{n(n-1)} \sum_{1\le i<j\le n} (e^{\epsilon_{ij}} + e^{-\epsilon_{ij}}) \\ &= \frac{1}{n(n-1)} \sum_{1\le i<j\le n} (e^{\epsilon_{ij}} + e^{-\epsilon_{ij}} - 2) \end{aligned}$$

Obviously, $CI \ge 0$.

Table 1. Random consistent index RI

n	1	2	3	4	5	6	7	8	9	10
RI	0	0	0.52	0.89	1.12	1.26	1.36	1.41	1.46	1.49

Remark 4. Let Q be fuzzy reciprocal matrix, if A is consistent matrix, then Q is fuzzy consistent matrix.

Algorithm 1. Let Q be fuzzy reciprocal matrix,we order $\alpha \in (0,1)$.

(1) $Q^{(0)} = Q$, known by theorem 4, and we have fuzzy consistent matrix $R^{(0)}$ and deviation matrix $E^{(0)}$.

(2) $Q^{(k)} \xrightarrow{\text{Theorem 4}} R^{(k)} \to E^{(k)}$. and $A^{(k)} = (a_{ij}^{(k)})$, where $\omega^{(k)} = (\omega_1^{(k)}, \cdots, \omega_n^{(k)})'$ is the priority vector of $R^{(k)}$.

(3) Calculating $CI^{(k)}$ and $CR^{(k)} = \frac{CI^{(k)}}{RI}$. If $CR^{(k)} < 0.1$, then go to (5); if $CR^{(k)} \geq 0.1$, then go to (4), where RI is a random consistent index proposed by Saaty and its value is shown by the following table 1.

(4) Let $A^{(k+1)} = (a_{ij}^{(k+1)})$, where $a_{ij}^{(k+1)} = (a_{ij}^{(k)})^{\alpha} (\frac{e^{a\omega_i^{(k)}}}{e^{a\omega_j^{(k)}}})^{(1-\alpha)} = \frac{e^{a\omega_i^{(k)}}}{e^{a\omega_j^{(k)}}} e^{\alpha\epsilon_{ij}^{(k)}} = e^{r_{ij}^{(k)} - \frac{1}{2} + \alpha\epsilon_{ij}^{(k)}}$, let $Q^{(k+1)} = R^{(k)} + \alpha E^{(k)}$, $k = k+1$;

(5) Calculating the rank of the matrix $R^{(k)}$.

If its rank is 1, known by remark 1, then we have the priority vector of fuzzy reciprocal matrix $R^{(k)}$, $\omega = \frac{1}{n}(1, 1, \cdots, 1)'$;

If its rank is 2, known by Eq.(1), we have the priority vector of fuzzy reciprocal matrix $R^{(k)}$, ω is the priority vector of matrix $Q^{(k)}$.

Algorithm 2. Replace (4) of the above algorithm with the following (4').

(4)$'$ Let $A^{(k+1)} = (a_{ij}^{(k+1)})$, then we can choose an element of matrix $E^{(k)}$ which is the maximal absolute value.

$$a_{st}^{(k+1)} = \begin{cases} e^{r_{ij}^{(k)} - \frac{1}{2} + \alpha\epsilon_{ij}^{(k)}}, & (s,t) = (i,j) \\ e^{r_{ji}^{(k)} - \frac{1}{2} + \alpha\epsilon_{ji}^{(k)}} & (s,t) = (j,i) \\ e^{r_{st}^{(k)} - \frac{1}{2} + \epsilon_{st}^{(k)}}, & \text{other cases} \end{cases}$$

$k = k+1$;

Theorem 7. In Algorithm 1, if $CI^{(k+1)} < CI^{(k)}$, then it shows that our algorithm is convergence.

Theorem 8. In Algorithm 2, if $CI^{(k+1)} < CI^{(k)}$, then it shows that our algorithm is convergence.

Remark 5. If $CR = 0$,then $e^{\epsilon_{ij}} + e^{-\epsilon_{ij}} - 2 = 0, \forall i,j = 1, \cdots, n$.The following statement is equivalent: $\epsilon_{ij} = 0, \forall i,j = 1, \cdots, n$.So $E = 0$.We can get a conclusion that Q is a fuzzy consistent matrix.

Remark 6. According to Algorithm 1 or 2, we can obtain the priority vector $\omega = (\omega_1, \cdots, \omega_n)'$ of matrix R. If there exist some negative elements of ω_i, then we order

$$\omega_i = \frac{1}{na}\sum_{j=1}^{n} r_{1j} - \frac{1}{a}r_{1i} + \frac{1}{n} + \varepsilon \geq 0, \ \ \forall i = 1, \cdots, n.$$

$\varepsilon \geq -\frac{1}{na}\sum\limits_{j=1}^{n} r_{1j} + \frac{1}{a}r_{1i} - \frac{1}{n}, \forall i = 1, \cdots, n$

Choose a real number ε, $\varepsilon = \max\limits_{i=1,\cdots,n}\{-\frac{1}{na}\sum_{j=1}^{n} r_{1j} + \frac{1}{a}r_{1i} - \frac{1}{n}\} > 0.$, then it ensure that $\omega_i \geq 0, \forall i = 1, \cdots, n$.

Remark 7. If $\sum\limits_{i=1}^{n} \omega_i \neq 1$, then we order, $\omega_0 = \sum\limits_{i=1}^{n} \omega_i$, and $\bar{\omega_i} = \frac{\omega_i}{\omega_0}, i = 1, \cdots, n,$ thus, we have adjusted vector $\bar{\omega} = (\bar{\omega_1}, \cdots, \bar{\omega_n})'$.

4 Numerical Examples

Example 2. $X = \{x_1, x_2, x_3\}$, we have fuzzy reciprocal matrix Q,

$$Q = \begin{pmatrix} 0.5 & 1 & 0.1 \\ 0 & 0.5 & 0.8 \\ 0.9 & 0.2 & 0.5 \end{pmatrix}$$

Known by theorem 4, then we have fuzzy consistent matrix R_1 for Q, where

$$R_1 = \begin{pmatrix} 0.5 & 0.6 & 0.5 \\ 0.4 & 0.5 & 0.4 \\ 0.5 & 0.6 & 0.5 \end{pmatrix}$$

Here, $CR = 0.1559 \geq 0.1$. In the following, we need to adjust matrix Q according to algorithm 1.

Case 1: $\alpha = 0.1$ and $a = 1$

$$Q^{(1)} = \begin{pmatrix} 0.5 & 0.64 & 0.46 \\ 0.36 & 0.5 & 0.44 \\ 0.54 & 0.56 & 0.5 \end{pmatrix} \rightarrow R_1^{(1)} = \begin{pmatrix} 0.5 & 0.6 & 0.5 \\ 0.4 & 0.5 & 0.4 \\ 0.5 & 0.6 & 0.5 \end{pmatrix} \rightarrow CR^{(1)} = 0.0015$$

The priority vector of $R_1^{(1)}$, $\omega = (0.3667, 0.2667, 0.3667)'$, it is the priority vector ω of $Q^{(1)}$.

Case 2: $\alpha = 0.3$ and $a = 1$

$$Q^{(1)} = \begin{pmatrix} 0.5 & 0.72 & 0.38 \\ 0.28 & 0.5 & 0.52 \\ 0.62 & 0.48 & 0.5 \end{pmatrix} \rightarrow R_1^{(1)} = \begin{pmatrix} 0.5 & 0.6 & 0.5 \\ 0.4 & 0.5 & 0.4 \\ 0.5 & 0.6 & 0.5 \end{pmatrix} \rightarrow CR^{(1)} = 0.0139$$

The priority vector of $R_1^{(1)}$, $\omega = (0.3667, 0.2667, 0.3667)'$, it is the priority vector ω of $Q^{(1)}$.

Example 3. Let Q be fuzzy reciprocal matrix,

$$Q = \begin{pmatrix} 0.5 & 1 & 1 & 1 \\ 0 & 0.5 & 0.6 & 0.4 \\ 0 & 0.4 & 0.5 & 1 \\ 0 & 0.6 & 0 & 0.5 \end{pmatrix} \xrightarrow{(2)} R = \begin{pmatrix} 0.5 & 1 & 0.9 & 1 \\ 0 & 0.5 & 0.4 & 0.6 \\ 0.1 & 0.6 & 0.5 & 0.7 \\ -0.1 & 0.4 & 0.3 & 0.5 \end{pmatrix}$$

$$\xrightarrow{\text{Theorem 4}} R' = \begin{pmatrix} 0.5 & 0.8472 & 0.7778 & 0.9167 \\ 0.1528 & 0.5 & 0.4306 & 0.5694 \\ 0.2222 & 0.5694 & 0.5 & 0.6389 \\ 0.0833 & 0.4306 & 0.3611 & 0.5 \end{pmatrix}$$

$$\rightarrow E = \begin{pmatrix} 0 & 0.1528 & 0.2222 & 0.0833 \\ -0.1528 & 0 & 0.1694 & -0.1694 \\ -0.2222 & -0.1694 & 0 & 0.3611 \\ -0.0833 & 0.1694 & -0.3611 & 0 \end{pmatrix}$$

Here, $CR = 0.0252 < 0.1$, $w = (0.5104\,, 0.1632\,, 0.2326\,, 0.0938)'$

5 Conclusion

In this paper, we put forward some concepts including fuzzy consistent index and fuzzy consistent ratio, discuss some new properties of fuzzy consistent matrix, propose a novel method to transform fuzzy reciprocal matrix into fuzzy consistent matrix and calculate the corresponding priority vector. Simultaneously, we propose a kind of method to adjust fuzzy reciprocal matrix according to an acceptable value(< 0.1). Finally, we investigate the convergence of our proposed algorithm and use two numerical examples to illustrate our proposed method reasonable.

Acknowledgements

This work described here is partially supported by National Natural Science Foundation of China(60474023), Doctor Foundation of Ministry of Education (20020027013),Science and Technology Key Project Fund of Ministry of Education(03184) and the Major State Basic Research Development Program of "973" of China (2002CB312200).

References

[1] J.Kacprzyk(1986) Group decision making with a fuzzy linguistic majority. Fuzzy Sets and Systems 18: 105-118
[2] T.L.Saaty(1980) The Analytic Hierarchy Process. McGraw-Hill. New York
[3] L.G.Vargas(1999) An overview of the analytic hierarchy process and its applications. European J. Oper. Res 116: 443-449
[4] F.Chiclana, F.Herrera, E.Herrera-Viedma(2001) Integrating multiplicative preference relation in a multipurpose decision making model based on fuzzy preference relations. Fuzzy Sets and Systems 122: 277-291

[5] T.Tanino(1990) On group decision making under fuzzy preferences, in: J.Kacprzyk, M.Fedrizzi(Eds.). Multiperson Decision Making Using Fuzzy Sets and Possibility Theory. Kluwer. Netherlands. pp.172-185

[6] Q.Zhang, J.C.H.Chen, Y.Q.He, J.Ma, D.N.Zhou(2003) Multiple attribute decision making: approach integrating subjective and objective information. Internat. J. Manuf. Technol Management 5: 338-361

[7] J.S.Finan, W.J.Hurley(1997) The analytic hierarchy process: Does adjusting a pairwise comparison matrix to improve the consistency ratio help?. Comput. Oper. Res 24: 749-755

[8] Z.Switalski(2001) Transitivity of fuzzy preference relation—-an empirical study. Fuzzy Sets and Systems 118: 503-508

[9] G.X.Song, D.L.Yang (2003) Methods for Identifying and Improving the Consistency of Fuzzy Judgement Matrix, Systems Engineering 21(1): 110-116(in chinese)

[10] Z.X.Sun, W.H.Qiu(2005) A method for improving the complementary and consistency of fuzzy judgment matrix. Systems Engineering 23(4): 101-104(in chinese)

[11] Z.S.Xu(2004) Uncertain multiple attribute decision making: Methods and Applications. Beijing. Tsinghua Publishing Press

[12] J.Ma, Z.P.Fan, Y.P.Jiang, J.Y.Mao, L.Ma(2006) A method for repairing the inconsistency of fuzzy preference relations. Fuzzy sets and Systems 157: 20-33

[13] M.Yao, Z.Shen(1997) Applications of fuzzy consisent matrix in soft science.Systems Engineering 15(2): 54-57(in chinese)

[14] Y.J.Lu(2002) Weight calculation method of fuzzy Analytical Hierarchy Process.Fuzzy Systems and Maticsmaths 16(2): 80-85(in chinese)

[15] J.J.Zhang(2000) Fuzzy Analytic Hierarchy Process(FAHP). Fuzzy Systems and Maticsmaths 14(2): 80-88(in chinese)

[16] E.Herrera-Viedma, F.Herrera, F.Chiclana, M.Luque(2004) Some issues on consistency of fuzzy preference relations. Europearn J. Oper. Res 154: 98-109

The Theory of Fuzzy Logic Programming

Dong-Bo Liu[1,2] and Zheng-Ding Lu[1]

[1] College of Computer Science & Technology,
Huazhong University of Science & Technology, 430000, Wuhan, China
[2] Institute of China Electronic System Engineering, 100039, Beijing, China
ldb0853@sina.com

Abatract. The complete formal specification of fuzzy Horn clause logic and its semantics interpretation is presented as well as a proof theory for fuzzy Horn clauses. We show that, the procedural interpretation for Horn clauses can be developed in much the same way for fuzzy Horn clauses. Then the theory of fuzzy logic programming is developed.

Keywords: Fuzzy Logic Programming, Fuzzy Horn Clauses, Fuzzy Proof Theory, Fuzzy Procedural Interpretation

1 Introduction

One have found that the classical binary logic contained the truth values *true* and *false* is not sufficient for human reasoning. There are plenty of states, which are fuzzy, between *true* and *false* [1,4].

In the classical logic programming, we have adopted the Horn clause subset of first order predicate logic. The Horn clause logic can be generalized to the fuzzy case based on L. A. Zadeh's fuzzy sets theory. We first regard the truth value *false* as the real number 0, *true* as the real number 1, and extend the concept of truth value to include all real numbers in the interval [0,1]. Then we generalize two important concepts in Horn clause logic to the fuzzy case. The first concept is the implication which can be regarded as transferring truth value from the condition to the conclusion. We can associate each implication with a factor f which is called implication strength. If the truth value of the condition is t, then the truth value of the conclusion is $f \otimes t$, where $\otimes$ is a soft product operator. The second concept is interpretation. We consider fuzzy interpretations as fuzzy subsets of the Herbrand base. They can be dealt with in a similar way as the classical Horn clause logic [3].

Based on the fuzzy sets theory, the complete formal specification of fuzzy Horn clause logic and its semantics interpretation is presented as well as a proof theory for fuzzy Horn clauses. We show that, the procedural interpretation for Horn clauses [2,5] can be developed in much the same way for fuzzy Horn clauses. Then the theory of fuzzy logic programming is developed in this paper.

2 Syntax and Semantics of Fuzzy Horn Clause Logic

In the first order predicate logic, the following implication rule

$$A \leftarrow B_1 \wedge B_2 \wedge \ldots \wedge B_n, \qquad n \geq 0.$$

B.-Y. Cao (Ed.): Fuzzy Information and Engineering (ICFIE), ASC 40, pp. 534–542, 2007.
springerlink.com

is called a Horn clause, where B_1, B_2, ..., B_n are $n(\geq 0)$ conditions, and A is the conclusion [2,5].

Similarly, a fuzzy Horn clause also has one conclusion and $n(\geq 0)$ conditions, the difference is that each fuzzy Horn clause has a implication strength f, where $f \in (0,1]$. The formal specification of a fuzzy Horn clause is as following:

$$A \leftarrow (f) — B_1 \wedge B_2 \wedge \ldots \wedge B_n, \quad n \geq 0.$$

Let the truth value of B_i is $t(B_i)$, i=1,2,...,n, and the truth value of all conditions is t. If n=0, we define t =1. If n>0, $t=min\{t(B_i)|i=1,2,\ldots,n\}$. The truth value of the conclusion $t(A)=f \otimes t$, where $\otimes$ is a soft product operator. To simplified the discussion, we suppose the $\otimes$ operator is the mathematical product $\times$.

Now we generalize some basic concepts of Horn clause logic.

Definition 2.1. Let Γ be a set of fuzzy Horn clauses, the Herbrand base $H(\Gamma)$ of the Γ is the set of all ground atomic formulas that can be formed with the symbols contained in Γ.

Definition 2.2. Let Γ be a set of fuzzy Horn clauses, $H(\Gamma)$ denotes the Herbrand base of Γ. The Herbrand interpretation I of Γ is defined as a mapping $H(\Gamma) \rightarrow [0,1]$.

In this case, a Herbrand interpretation is regarded as a fuzzy subset of $H(\Gamma)$. The mapping $H(\Gamma) \rightarrow [0,1]$ can be thought of as the membership function characterizing a fuzzy subset I of Γ.

Definition 2.3. For a set Γ of fuzzy Horn clauses and its interpretation I,

(a) Γ is true in I if and only if every one of its fuzzy Horn clauses is true in I.
(b) A fuzzy Horn clause C in Γ is true in I if and only if every one of its ground instances is true in I.
(c) A ground instance $A \leftarrow (f) — B_1 \wedge B_2 \wedge \ldots \wedge B_n$ of the fuzzy Horn clause C is true in I if and only if $\mu_I(A) \geq f \times min\{\mu_I(B_i)|i=1,2, \ldots, n\}$.
(d) A ground instance $A \leftarrow (f)$— is true in I if and only if $\mu_I(A) \geq f$.

Here, we define $min\phi$=1.

Definition 2.4. A Herbrand interpretation I such that a set Γ of fuzzy Horn clauses is true in I is called a Herbrand model of Γ.

Definition 2.5. For any set Γ of fuzzy Horn clauses (let its Herbrand base be $H(\Gamma)$), any $A \in H(\Gamma)$, and any $f \in (0,1]$, $\Gamma \Rightarrow \{A \leftarrow (f)—\}$ if and only if the right-hand side is true in every Herbrand model of Γ.

Note that the symbol "$\Rightarrow$" means truth in all Herbrand models rather than in all models. For a set Γ of fuzzy Horn clauses, it is clear that

$$\Gamma \Rightarrow \{A \leftarrow (f_1)—\} \text{ implies } \Gamma \Rightarrow \{A \leftarrow (f_2)—\}, \forall f_1, f_2 \in [0,1] \text{ and } f_1 \geq f_2$$

Let Γ be a set of fuzzy Horn clauses, and $M(\Gamma)$ denotes the set of Herbrand models of Γ. $\cap M(\Gamma)$ is defined in the fuzzy case by adopting Zadeh's rule [1] for intersections $\mu_{\cap S}(A) = inf\ \{\mu_I(A)|I \in S\}$, where S is a set of Herbrand interpretations and inf is the greatest lower bound.

Theorem 2.1. Let Γ be a set of fuzzy Horn clauses, and $\cap M(\Gamma)$ denotes the intersection of all Herbrand models of Γ. Therefore

$$\mu_{\cap M(\Gamma)}(A) = sup\{f \mid \Gamma \Rightarrow \{A \leftarrow (f) \text{—}\}\},$$

where *sup* is the least upper bound.

Proof. If Γ is a set of fuzzy Horn clauses, I is a Herbrand model of Γ, $A \in H(\Gamma)$, and $\Gamma \Rightarrow \{A \leftarrow (f) \text{—}\}$, then $\{A \leftarrow (f) \text{—}\}$ is true in I, and $\mu_I(A) \geq f$, by Defintion 2.3. Therefore,

$$\mu_I(A) \geq sup\{f \mid \Gamma \Rightarrow \{A \leftarrow (f) \text{—}\}\},$$

for any Herbrand model $I \in M(\Gamma)$, and

$$\mu_{\cap M(\Gamma)}(A) \geq sup\{f \mid \Gamma \Rightarrow \{A \leftarrow (f) \text{—}\}\}.$$

However, we have $\Gamma \Rightarrow \{A \leftarrow (g) \text{—}\}$, where $g = \mu_{\cap M(\Gamma)}(A)$, for any Γ and $A \in H(\Gamma)$. So the relation $\mu_{\cap M(\Gamma)}(A) > sup\{f \mid \Gamma \Rightarrow \{A \leftarrow (f) \text{—}\}\}$ is impossible. So far, the theorem has been proved.

For the binary case, fixpoint theory [5,6,7] associates each set Γ of Horn clauses with a mapping T from interpretations to interpretations, and it shows that fixpoints of T are models of Γ. Following the same way, we establish the fixpoint semantics for a set Γ of fuzzy Horn clauses [3,8,9].

Definition 2.6. Let Γ be a set of fuzzy Horn clauses, T is a mapping associated with Γ from interpretations to interpretations. For every $A \in H(\Gamma)$,

$$\mu_T(A) = sup\{f \times min\{\mu_I(B_i) \mid i = 1,2, \ldots, n\} \mid A \leftarrow (f) \text{—} B_1 \wedge B_2 \wedge \ldots \wedge B_n \text{ is a ground instance of a fuzzy Horn clause in } \Gamma\}.$$

Definition 2.7. Let Γ be a set of fuzzy Horn clauses, and $H(\Gamma)$ denotes the Herbrand base of Γ. For two interpretations $I, J \subseteq H(\Gamma)$,

$$I \subseteq J \text{ if and only if } \mu_I(A) \leq \mu_J(A),\ \forall A \in H(\Gamma).$$

As in the classical Horn clause logic, we also define $I=J$ if and only if $I \subseteq J$ and $J \subseteq I$, for the fuzzy interpretations.

Theorem 2.2. For any set Γ of fuzzy Horn clauses, the function T associated with Γ is monotone.

Proof. Let $I \subseteq H(\Gamma)$, $J \subseteq H(\Gamma)$, and $I \subseteq J$. By Definition 2.7, if

$$A \leftarrow (f) \text{—} B_1 \wedge B_2 \wedge \ldots \wedge B_n$$

is the ground instance of a fuzzy Horn clause in Γ, then

$$\mu_I(B_i) \leq \mu_J(B_i),\ i = 1,2, \ldots, n.$$

For any $A \in H(\Gamma)$,

$$\mu_I(A) \leq \mu_J(A),$$

by Definition 2.6. Therefore $T(I) \subseteq T(J)$. It is said that T is a monotone function.

Inference 2.3. For the set of fuzzy Horn clauses, the monotonicity of Γ implies that the least fixpoint $lfp(T)$ of T, namely $\cap\{I|T(I)=I\}$ exists and is equal to $\cap\{I|T(I)\subseteq I\}$.

The following theorem will associate models with fixpoints in the fuzzy case.

Theorem 2.4. For any set Γ of fuzzy Horn clauses, and any $I\subseteq H(\Gamma)$,

$$I\in M(\Gamma) \text{ if and only if } T(I)\subseteq I.$$

Proof. ($\Longrightarrow$) If $I\in M(\Gamma)$, then Γ is true in I. For any ground instance $A\leftarrow(f)—B_1 \wedge B_2 \wedge ... \wedge B_n$ of a fuzzy Horn clause in Γ, we have

$$\mu_I(A)\geq f\times min\{\mu_I(B_i)|i=1,2,...,n\}$$

by Definition 2.3. Hence

$$\mu_I(A)\geq sup\{f\times min\{\mu_I(B_i)|i=1,2,...,n\}|A\leftarrow(f)—B_1\wedge B_2\wedge ...\wedge B_n$$
$$\text{is a ground instance of a fuzzy Horn clause in } \Gamma\},$$

and $\mu_I(A)\geq \mu_{T(I)}(A)$ by Definition 2.6. Therefore $T(I)\subseteq I$.

($\Longleftarrow$) If $T(I)\subseteq I$, then $\mu_I(A)\geq \mu_{T(I)}(A)$ for any $A\in H(\Gamma)$, by Definition 2.7. In addition, for any ground instance $A\leftarrow(f)—B_1\wedge B_2\wedge ...\wedge B_n$ of a fuzzy Horn clause in Γ, we have

$$\mu_{T(I)}(A)\geq f\times min\{\mu_I(B_i)|i=1,2,...,n\}$$

by Definition 2.6. Hence

$$\mu_I(A)\geq f\times min\{\mu_I(B_i)|i=1,2,...,n\}$$

and this implies that Γ is true in I by Definition 2.3. It is said that $I\in M(\Gamma)$.

Theorem 2.4 enables us to discover properties of Herbrand models by studing fixpoints of $T(I)$.

Theorem 2.5. For any set Γ of fuzzy Horn clauses, $\cap M(\Gamma) = lfp(T)$.

Proof. We have known that $lfp(T) = \cap\{I|T(I)=I\}$ by Inference 2.3. By Theorem 2.4, $T(I)=I$ if and only if $I\in M(\Gamma)$. Hence

$$lfp(T) = \cap\{I|T(I)=I\} = \cap\{I|I\in M(\Gamma)\} = \cap M(\Gamma).$$

Theorem 2.6. For any set Γ of fuzzy Horn clauses, mapping T associated with Γ is continuous, *i.e.*,

$$\cup\{T(I_j)|j\in N\}=T(\cup\{I_j|j\in N\}).$$

for all sequences $I_1\subseteq I_2\subseteq ...$ of Herbrand interpretations.

Proof. For any atomic formula $A\in H(\Gamma)$, we have

$$\mu_{T(\cup\{I_j|j\in N\})}(A) = sup\{f\times min\{\mu_{\cup\{I_j|j\in N\}}(B_k)|k=1,2,...,n\}|A\leftarrow(f)—$$
$$B_1\wedge B_2\wedge ...\wedge B_n \text{ is a ground instance of a fuzzy Horn clause in } \Gamma\}.$$

suppose $A\leftarrow(f_\alpha)—B_{\alpha 1}\wedge B_{\alpha 2}\wedge ...\wedge B_{\alpha m}$ is the α-th ground instance of fuzzy Horn clauses in Γ having A as conclusion. The above expression can be shortened to

$$\mu_{T(\cup\{Ij|j\in N\})}(A) = sup_\alpha(f_\alpha \times min_k \mu_{\cup\{Ij|j\in N\}}(B_{\alpha k})).$$

However,

$$\mu_{\cup\{Ij|j\in N\}}(B_{\alpha k}) = sup_j \mu_{Ij}(B_{\alpha k}),$$

where j indexes the monotone sequence $I_1 \subseteq I_2 \subseteq$ ...of Herbrand interpretations, therefore

$$\mu_{T(\cup\{Ij|j\in N\})}(A) = sup_\alpha sup_j\, f_\alpha \times min_k \mu_{Ij}(B_{\alpha k}) = sup_\alpha sup_j\, \upsilon_{\alpha j}$$

Where $\upsilon_{\alpha j} = f_\alpha \times min_k \mu_{Ij}(B_{\alpha k})$. Using the same method we find

$$\mu_{\cup\{T(Ij)|j\in N\}}(A) = sup_j\; sup_\alpha \upsilon_{\alpha j}.$$

Now $sup_\alpha sup_j\, \upsilon_{\alpha j} = sup_j\; sup_\alpha \upsilon_{\alpha j}$ needs to be proved.

The set consisting of all $\upsilon_{\alpha j}$ is bounded above; therefore, it has a least upper bound, say υ. Hence, $sup_\alpha sup_j\, \upsilon_{\alpha j} \geq \upsilon$.

On the other hand, for all $\alpha \in N$, we have $sup_j\, \upsilon_{\alpha j} < \upsilon$, therefore $sup_\alpha sup_j\, \upsilon_{\alpha j} < \upsilon$. Hence, $sup_\alpha sup_j\, \upsilon_{\alpha j} = \upsilon$.

Similarly, we show that $sup_j\; sup_\alpha \upsilon_{\alpha j} = \upsilon$. Therefore

$$\cup\{T(I_j)|j\in N\}=T(\cup\{I_j|j\in N\}).$$

It is easy to prove the important property of T by the above Theorem 2.6.

Theorem 2.7. For any set Γ of fuzzy Horn clauses,

$$lfp(T) = \cup\{T^n(\phi)|n\in N\},$$

where ϕ is a special interpretation such that

$$\mu_\phi(A) = 0,\;\; \forall A\in H(\Gamma).$$

Now we present a theorem that can serve as foundation for the completeness result on the fuzzy proof theory. A completeness result for a proof method is of the form: if an assertion is true, then it can be proved according to the method.

We assume $\mu_{M(\Gamma)}(A)=\upsilon$, and want to show that A←(υ)— can be derived from Γ. By Theorem 2.5 and Theorem 2.7, we have

$$\cap M(\Gamma)= \cup\{T^n(\phi)|n\in N\}.$$

But we try to draw the stronger conclusion from $\mu_{M(\Gamma)}(A)=\upsilon$ that there exists an $n\in N$ such that $\mu_{T^n(\phi)}(A)=\upsilon$. Here is one of the methods.

Lemma 2.8. For any finite set Γ of fuzzy Horn clauses, any $A\in H(\Gamma)$, and any real number $\varepsilon > 0$,

$$\{\; \mu_{T^n(\phi)}(A)|\; n\in N \wedge \mu_{T^n(\phi)}(A) \geq\varepsilon \;\}$$

is finite.

Proof. Let $F(\Gamma)$ be the set of implications of fuzzy Horn clauses in Γ. Note that Γ is finite, therefore $F(\Gamma)$ is a finite set. Let m be the greatest element of $F(\Gamma)$ such that m<1. The real number $\mu_{T^n(\phi)}(A)$ is a product of a sequence of elements of $F(\Gamma)$. In this sequence, if r is the smallest integer such that $m<\varepsilon$, then at most r elements can be less than 1. The sequence can have any length, because 1 can occur in the sequence any number of time. So we conclude that the number of different products ($\geq\varepsilon$) of the sequences of elements of $F(\Gamma)$ is not greater than $|F(\Gamma)|^r$. It is that $\{ \mu_{T^n(\phi)}(A)| \, n\in N \wedge \mu_{T^n(\phi)}(A) \geq\varepsilon \}$ is a finite set.

Theorem 2.9. For any finite set Γ of fuzzy Horn clauses, and any $A\in H(\Gamma)$, there exists an $n\in N$ such that $\mu_{\cap M(\Gamma)}(A)= \mu_{T^n(\phi)}(A)$.

Proof. If $\upsilon=\mu_{M(\Gamma)}(A)=0$, then there exists $n=0\in$ N such that the expression holds. Suppose $\upsilon>0$, then $\cap M(\Gamma)=lfp(T)= \cup\{T^n(\phi)|n\in N\}$, by Definition 2.5 and Definition 2.7. Hence

$$\mu_{\cap M(\Gamma)}(A) = sup\{\mu_{T^n(\Phi)}(A) \mid n\in N \}= sup\{\mu_{T^n(\Phi)}(A) \mid n\in N \wedge \mu_{T^n(\Phi)}(A) \geq\varepsilon\}. \qquad (2.1)$$

for any $\varepsilon<\mu$. If we choose such an ε positive, according to $\mu>0$, then (2.1) is finite by Lemma 2.8. Hence the least upper bound is attained for an $n\in N$.

Note that the sets of fuzzy Horn clauses discussed in this section are finite, it is not a superfluous condition.

3 Fuzzy Proof Theory

In this section we describe a proof theory precisely for fuzzy Horn clauses, and justify its results using the semantics results presented in the previous section.

As in the binary case, the fuzzy proof procedure for fuzzy Horn clauses is also a search of an *and*/*or* tree. This tree determined by a set Γ of fuzzy Horn clauses and an initial atom G is defined as follows.

Definition 3.1

(a) There are two kinds of nodes: *and*-nodes and *or*-nodes.
(b) Each *or*-node is labeled by a single atomic formula.
(c) Each *and*-node is labeled by a fuzzy Horn clause in Γ and a substitution.
(d) The descendants of each *or*-node are all *and*-nodes, and the descendants of each *and*-node are all *or*-nodes.
(e) The root is an *or*-node labeled by G.
(f) For each fuzzy Horn clause C in Γ with a left-hand side unifying with the atomic formula A (with the most general substitution θ) in an *or*-node, there is an *and*-node descendant of the *or*-node labeled with C and θ. An *and*-node with no descendants is called a failure node.
(g) For each atomic formula B in the right-hand side of the fuzzy Horn clause labeling an *and*-node, there is a descendant *or*-node labeled with B. An *and*-node with no descendants is called a success node.
(h) Each node is associated with a real number which is called the value of the node. The value of a success node is the implication of its associated fuzzy Horn clause. The value of a nonterminal *and*-node is $f\times t$, where f is the implication of

the fuzzy Horn clause labeling the *and*-node, and t is the minimum of the values of its descendants. The value of a failure node is 0. The value of a nonterminal *or*-node is the maximum of the values of its descendants.

In the fuzzy case, a proof tree is a subtree of an *and/or* tree defined as follows.

Definition 3.2

(a) The root of the proof tree is the root of the *and/or* tree.
(b) An *or*-node of the proof tree which also occurs in the *and/or* tree has one descendant in the proof tree which is one of the descendants of that node in the *and/or* tree.
(c) An *and*-node in the proof tree which also occurs in the *and/or* tree has descendants in the proof tree, all of the descendants of that node in the *and/or* tree.
(d) All terminal nodes in a proof tree are success nodes.
(e) Each node of the proof tree is assigned a real number as the value of the node (with the same method as in the *and/or* tree).

In the binary case, correctness of the (SLD-resolution) proof procedure says in the most elementary form: if $A\in H(\Gamma)$ is proved, then $A\in M(\Gamma)$. We can express correctness like this: results of the proof procedure are not more true than they are in the minimal model $\cap M(\Gamma)$.

In the fuzzy case limited to finite *and/or* trees, the form of the corresponding correctness is suggested.

Theorem 3.1. For any set Γ of fuzzy Horn clauses with a finite *and/or* tree and any $A\in H(\Gamma)$, the value of the root in the *and/or* tree with A as root is not greater than $\mu_{\cap M(\Gamma)}(A)$.

Proof. Note first that the value of the root in the *and/or* tree is the maximum of the values of the roots of its constituent proof trees. It can easily be verified that the value of the root of a proof tree with A as root is not greater than $T^{n+1}(\phi)$, where n is the length of a longest path from the root to a terminal node. Here one unit of path length is from *or*-node to *or*-node along the path. It is proved by Theorem 2.9.

The following is completeness of the fuzzy proof procedure.

Theorem 3.2. For any set Γ of fuzzy Horn clauses with a finite and/or tree and any $A\in H(\Gamma)$, the value of the root in the and/or tree with A as root is at least $\mu_{\cap M(\Gamma)}(A)$.

Proof. By induction on n, we prove that $\mu \geq \mu_{T}n_{(\phi)}(A)$, for all $n\in N$. Then we conclude that

$$\mu \geq sup\{\mu_{T}n_{(\phi)}(A)|n\in N\} = \mu_{\cup\{T}n_{(\phi)|\, n\in N\}}(A) = \mu_{\cap M(\Gamma)}(A).$$

Now let's start the inductive proof of $\mu \geq \mu_{T}n_{(\phi)}(A)$.

(1) for $n = 0$, it is true.
(2) suppose it holds for $n = n_0$, then

$$\mu_{T}n0{+}1_{(\phi)}(A) = sup\{f\times min\{\mu_{T}n0_{(\phi)}(B_k)|k\in N\}|\, A\leftarrow(f)\text{—}B_1 \wedge B_2 \wedge \ldots \wedge B_n$$
$$\text{is a ground instance of a fuzzy Horn clause in } \Gamma\}.$$

The set over which the superemum is taken is finite by Lemma 2.8. Therefore the superemum must be attained for ground instance

$$A \leftarrow (f) — B_1 \wedge B_2 \wedge ... \wedge B_n$$

of a fuzzy Horn clause C:

$$A' \leftarrow (f) — B_1' \wedge B_2' \wedge ... \wedge B_n'$$

in Γ. Hence

$$\mu_{T^{n0+1}(\phi)}(A) = f \times min\{\mu_{T^{n0}(\phi)}(B_k) | k \in N\}. \quad (3.1)$$

Let us consider the *and/or* tree for Γ having A as root. One of the descendants of the root must be the fuzzy Horn clause. Because its left-hand side A' has A as ground instance, there is a most general substitution θ of A' and A. Hence one of the descendants of the root is the node (C,θ) labeled with C and θ.

Its descendants are $B_1'\theta$, $B_2'\theta$, ..., $B_k'\theta$ with values μ_1', μ_2', ..., μ_k' and having B_1, B_2, ..., B_k respectively as ground instances.

By the induction hypothesis, B_1, B_2, ... , B_k are roots of *and/or* tree having values μ_1, μ_2, ... , μ_k such that $\mu_i \geq \mu_T n0_{(\phi)}(B_i)$, i =1,2,...,k. Because $B_i'\theta$ has B_i as instance, we must have $\mu_i' \geq \mu_i$. For the value μ of the entire *and/or* tree, with A as root, we have

$$\mu = f \times min\{\mu_i' | i = 1,2,...,k\}$$

and hence

$$\mu \geq f \times min\{\mu_T n0_{(\phi)}(B_i) | i = 1,2,...,k \}.$$

We conclude

$$\mu \geq \mu_T n0{+}1_{(\phi)}(A)$$

by (3.1), which completes the induction proof.

4 Fuzzy Procedural Interpretation

In the logic programming system, each Horn clause is interpreted as a procedural [2,5]. Fuzzy Horn clauses can be interpreted procedurally in a similar way:

(a) □, an empty clause, containing no atomic fomulas, is interpreted as a halt statement.

(b) $\leftarrow (f) — A_1 \wedge A_2 \wedge ... \wedge A_n$, a clause consisting of $n(\geq 1)$ conditions and no conclusions, is interpreted as a fuzzy goal (related to the f).

(c) $A \leftarrow (f) — B_1 \wedge B_2 \wedge ... \wedge B_n$, a clause consisting of exacting one conclusion and $n(\geq 0)$ conditions, is interpreted as a fuzzy procedure. For a given question $\leftarrow (g) — A$, when $g > f$, this fuzzy procedure for A goes nowhere; when $g \leq f$, subquestions $\leftarrow (f_0) — B_1$, $\leftarrow (f_0) — B_2$, ..., and $\leftarrow (f_0) — B_n$ $(n \geq 0)$ need to be answered first, where $f_0 = g/f$. The question answering procedure is according to the sequence from left to right.

(d) A←(f)—, a clause consisting of only one conclusion, is interpreted as a fuzzy assertion. For a given question ←(g)—A, if $g \leq f$, then it is satisfied; if $g > f$, then it is not satisfied. The question is answered directly with no subquestion derivation.

A set Γ of fuzzy Horn clauses can be seen as a fuzzy logic program. It is initiated by an initial fuzzy goal. By using fuzzy procedures constantly, new fuzzy goals can be derived from old ones, so as to advance the computation procedure. Finally, it terminates with the derivation of the halt statement □ (it is derived from ←(f_1)—A and A←(f_2)—, where $f_1 \leq f_2$).

Such a fuzzy goal oriented derivation from an initial set Γ of fuzzy Horn clauses and from an initial fuzzy goal $\boldsymbol{G}_1$ ($\boldsymbol{G}_1 \in \Gamma$) is a sequence of fuzzy goals

$$\boldsymbol{G}_1, \boldsymbol{G}_2, ..., \boldsymbol{G}_n.$$

where $\boldsymbol{G}_i$ ($n \geq 1$) contains a single selected fuzzy procedure call according to the given strategy, and $\boldsymbol{G}_{i+1}$ is obtained from $\boldsymbol{G}_i$ by procedure invocation.

5 Conclusions

The theory of fuzzy logic programming is developed in this paper. The set of fuzzy Horn clauses can be seen as a non-deterministic programming language, because given a single fuzzy goal, several fuzzy procedures may have a name which matches the selected fuzzy procedure call at the same time. Each fuzzy procedure gives rise to a new fuzzy goal. A proof procedure, which sequences the generation of derivations in the search for a refutation, is a search of an and/or tree. In the proof procedure, each value of f is given according to the *min-max* rule.

References

1. Zadeh, LA (1965) Fuzzy sets. J. Information and Control 8:338-353
2. Kowalski, RA (1974) Predicate logic as programming language. In: Rosenfeld JL(ed) Information Processing 74, Proceedings of IFIP congress 74, Stockholm, North Holland, Amsterdum, pp569-574
3. Liu, DB, Li, DY (1988) Fuzzy reasoning based on f-Horn clause rules In: Grabowski J, Lescanne P, Wechler W (eds) Algebraic and Logic Programming, Proceedings of 1st Int. Workshop on Algebraic and Logic Programming, Gaussig, GDR, pp214-222
4. Li, DY, Liu, DB (1990) A fuzzy PROLOG database system. Research Studies Press, Taunton, Somerset, England, John Wiley & Sons, New York
5. van Emden, MH, Kowalski, RA (1976) The semantics of logic as a programming language. J. Assoc. Comput. Mach. 23:733-742
6. Apt, KR, van Emden, MH (1982) Contributions to the theory of logic programming. J. Assoc. Comput. Mach. 29:841-862
7. Park, D (1969) Fixpoint induction and proofs of program propreties. In: Meltzer, B, Michie, D (eds) Machine Intelligence 5, Edinburgh University Press, Edinburgh, pp59-78
8. van Emden, MH (1986) Quantitative deduction and its fixpoint theory. J. Logic Programming 1:37-53
9. Vojtas, P (2001) Fuzzy logic programming. J. Fuzzy Sets and Systems 1:361-370

Multiobjective Matrix Game with Vague Payoffs

Xiaoguang Zhou[1], Yuantao Song[2], Qiang Zhang[3], and Xuedong Gao[1]

[1] School of Economics and Management, University of Science and Technology Beijing, Beijing 100083, China
xiaoguang@manage.ustb.edu.cn
[2] College of Engineering, Graduate School of the Chinese Academy of Sciences, Beijing 100049, China
[3] School of Management and Economics, Beijing Institute and Technology, Beijing 100081, China

Abstract. There are always uncertainty, incompleteness and imprecision existing in decision making information. These are why fuzzy set theory is commonly used in decision making. Vague set can indicate the decision makers' preference information in terms of favor、against and neutral. When dealing with uncertain information, vague set indicates information more abundant than fuzzy set. According to the theories of multiobjective decision making and fuzzy game, the multiobjective two-person zero-sum matrix game, whose payoff values are vague values is researched. Firstly, the concept of vague set and the order function is introduced. Then the model of multiobjective two-person zero-sum matrix game based on vague set is described. Thirdly, two solutions of vague multiobjective two-person zero-sum matrix game are discussed: one is making the vague multiobjective game problem crisp through the order function of vague values, then turning it into single objective game; the other method is turning vague multiobjective game problem into vague single objective game problem first, then making it crisp through the order function of vague values. Finally, numerical examples are given to apply the proposed methods.

Keywords: Vague Set; Matrix Game; Order Function; Multiobjective Game.

1 Introduction

Two-person zero-sum game with fuzzy payoffs is an important problem during the research of fuzzy game theory. Campos dealt with matrix game with fuzzy payoffs and formulated a problem yielding a maximin solution by applying fuzzy mathematical programming[1]. Nishizaki and Sakawa have considered equilibrium solutions with respect to the degree of attainment of the fuzzy goal in a multiobjective bimatrix game with fuzzy payoffs[2]. Maeda has defined Nash equilibrium strategies based on possibility and necessity measures and investigated its properties[3]. Vijay and his cooperators gained the equilibrium solutions by using a suitable defuzzification function[4].

Gau and Buehrer brought out the concept of vague set in 1993[5]. In 1996, Bustince and Burillo pointed out vague set was the same as intuitionistic set[6]. Vague set can indicate the decision-makers' preference information in terms of favor、against and

B.-Y. Cao (Ed.): Fuzzy Information and Engineering (ICFIE), ASC 40, pp. 543–550, 2007.
springerlink.com

neutral. When dealing with uncertain information, vague set indicates information more abundant and flexible than fuzzy set. Therefore, vague set becomes focused on recently[7].

The rest of this paper is organized as follows. In section 2, the concept and order function of vague set are introduced. In section 3, the model of multiobjective two-person zero-sum matrix game based on vague set is described. In section 4, the solutions are proposed, and numerical examples are given to apply the proposed methods. Finally, the conclusion is given.

2 Definition and Order Function of Vague Set

Definition 1. Let U be a space of points, with a generic element of U denoted by x. A vague set in U is characterized by a truth-membership function $t_A(x)$ and a false-membership function $f_A(x)$. $t_A(x)$ and $f_A(x)$ both associate a real number in the interval [0, 1] with each point in U, where $t_A(x)+f_A(x)\leq 1$. That is

$$t_A:\ U\rightarrow[0,1],\quad f_A:\ U\rightarrow[0,1],$$

$t_A(x)$ is a lower bound on the grade of membership of x derived from the evidence for x, and $f_A(x)$ is a lower bound on the negation of x derived from the evidence against x.

Let $x\in U$, the close interval $[t_A(x), 1-f_A(x)]$ is called a vague value of vague set A at the point of x. It indicates the membership degree both for $x\in A$ and against $x\in A$ at the same time. For example, if a vague value is [0.5, 0.8], then we can see that $t_A(x)$=0.5, $f_A(x)$=0.2. It can be interpreted as "the degree of element x belongs to vague set A is 0.5, and the degree of element x does not belongs to vague set A is 0.2." In a voting model, it can be interpreted as "the vote for resolution is 5 in favor, 3 against and 2 abstentions." As thus, the meaning of vague value $[t_A(x), 1-f_A(x)]$ at the point x, is more abundant than that of fuzzy value $u(x)$.

Order function mainly resolves the degree that alternative A_i matches the demand of decision maker[8]. Part of the abstained group may be inclined to vote for the proposal, while other part is apt to against and the rest still make for abstaining during decision making. According to the voting results, Liu[9] divided the abstained part π_{A_i} into three groups: $t_{A_i}\pi_{A_i}$, $f_{A_i}\pi_{A_i}$, $(1-t_{A_i}-f_{A_i})\pi_{A_i}$, which separately indicates the proportions of the three kinds votes in abstaining part. The order function is:

$$L(E(A_i))=t_{A_i}+t_{A_i}(1-t_{A_i}-f_{A_i}). \tag{1}$$

It indicates that the more $L(E(A_i))$ is, the more that alternative A_i matches the demand of decision maker.

Example 1. If $E(A_1)=[0.5, 0.9]$, $E(A_2)=[0.6, 0.8]$, according to Equation(1), $L(E(A_1))=0.5+0.5\times(1-0.5-0.1)=0.7$, $L(E(A_2))=0.6+0.6\times(1-0.6-0.2)=0.72$, this indicates that alternative A_2 is superior to A_1.

3 Model of Multiobjective Two-Person Zero-Sum Matrix Game Based on Vague Set

Suppose there are two players P_1 and P_2 in the game, the pure strategy set is signed as $S_1=\{\alpha_1, \alpha_2, \ldots, \alpha_m\}$, $S_2=\{\beta_1, \beta_2, \ldots, \beta_n\}$, and N objectives need considering. When a game situation (α_i, β_j) is formed, that is, player P_1 selects strategy $\alpha_i \in S_1$ and player P_2 select strategy $\beta_j \in S_2$, then P_1 gains payoff value of the kth(k=1, 2, …, N) objective is $\tilde{a}_{ij}^k$, and P_2 correspondingly loses $\tilde{a}_{ij}^k$, that is payoff value is $-\tilde{a}_{ij}^k$. Although $\tilde{a}_{ij}^k+(-\tilde{a}_{ij}^k)$ is probably not equal to zero, it's still called a zero-sum game because how many P_1 gains actually is what P_2 loses. To simplify, the payoff matrix of the kth objective of player P_1 is signed as $\tilde{A}^k=(\tilde{a}_{ij}^k)m\times n$, $\tilde{a}_{ij}^k=[t_{ij}, t_{ij}^*]$, $t_{ij}^*=1-f_{ij}$, i=1, 2, …, m; j=1, 2, …, n; k=1, 2, …, N.

$$\tilde{A}^k = \begin{array}{c} \\ \alpha_1 \\ \alpha_2 \\ \vdots \\ \alpha_m \end{array} \begin{array}{c} \begin{array}{cccc} \beta_1 & \beta_2 & \cdots & \beta_n \end{array} \\ \begin{bmatrix} \tilde{a}_{11}^k & \tilde{a}_{12}^k & \cdots & \tilde{a}_{1n}^k \\ \tilde{a}_{21}^k & \tilde{a}_{22}^k & \cdots & \tilde{a}_{2n}^k \\ \vdots & \vdots & \vdots & \vdots \\ \tilde{a}_{m1}^k & \tilde{a}_{m2}^k & \cdots & \tilde{a}_{mn}^k \end{bmatrix} \end{array} \qquad (2)$$

All payoff matrixes of N objectives of Player P_1 is signed as $A=(\tilde{\alpha}_{ij})_{m\times n}$, i=1, 2, …, m, j=1, 2, …, n. That is,

$$\tilde{A} = \begin{array}{c} \\ \alpha_1 \\ \alpha_2 \\ \vdots \\ \alpha_m \end{array} \begin{array}{c} \begin{array}{cccc} \beta_1 & \beta_2 & \cdots & \beta_n \end{array} \\ \begin{bmatrix} \tilde{\alpha}_{11} & \tilde{\alpha}_{12} & \cdots & \tilde{\alpha}_{1n} \\ \tilde{\alpha}_{21} & \tilde{\alpha}_{22} & \cdots & \tilde{\alpha}_{2n} \\ \vdots & \vdots & \cdots & \vdots \\ \tilde{\alpha}_{m1} & \tilde{\alpha}_{m2} & \cdots & \tilde{\alpha}_{mn} \end{bmatrix} \end{array} \qquad (3)$$

4 Solutions of Multiobjective Two-Person Zero-Sum Matrix Game Based on Vague Set

There are several solutions to solve multiobjective two-person zero-sum whose payoff values are vague values, and two representative methods are discussed in the following content.

(1) Convert the payoff matrix $\tilde{A}^k$ of each objective into crisp payoff matrix, and transform the multiobjective game to a single objective game by selecting proper objective weight.

Suppose the players use the same weight vector $\omega = (\omega_1, \omega_2, \ldots, \omega_N)^T$, in which $\omega_k (k=1, 2, \ldots, N)$ is the weight of objective f_k, if it holds that $\omega_k \geq 0$, $\sum_{k=1}^{N} w_k = 1$.

Thus, the relative membership degree of all objectives can convert into a linear weighted sum, that is

$$b_{ij} = \sum_{k=1}^{N} w_k b_{ij}^k , \tag{4}$$

where b_{ij}^k is crisp payoff value of each objective, i=1, 2, …, m; j=1, 2, …, n. Let matrix $B=(b_{ij})_{mxn}$, then the multiobjective two-person zero-sum matrix game is converted into a single objective two-person zero-sum matrix game of payoff matrix B whose player is P_1, in which V is the expected payoff value of player P_1, that is $V=y^T Bz$. According to the solution of matrix game, the equilibrium strategy y^* and z^* of player P_1 and P_2 and the expected payoffs value v can be calculated.

Example 2. Suppose there are two companies P_1 and P_2 aiming to enhance the sales amount and market share of a product in a targeted market. Under the circumstance that the demand amount of the product in the targeted market basically is fix, the sales amount and market share of one company increases, following the decrease of the sales amount and market share of another company, but the sales amount is not certain to be proportional to the market share. The two companies are considering about the three strategies to increase the sales amount and market share: strategy α_1: advertisement; strategy α_2: reduce the price; strategy α_3: improve the package.

The above problem actually is a multiobjective two-person zero-sum matrix game. Let company 1 be player P_1, adopting strategy (α_1, α_2, α_3); company2 be player P_2, adopting strategy (β_1, β_2, β_3). Under the three strategies, the payoff matrix $\tilde{A}^1$、$\tilde{A}^2$ of targeted sales quantity f_1(million) and market share f_2 (%) are separately indicated by vague value:

$$\begin{array}{cc} & \begin{array}{ccc} \beta_1 & \beta_2 & \beta_3 \end{array} \\ \tilde{A}^1 = \begin{array}{c} \alpha_1 \\ \alpha_2 \\ \alpha_3 \end{array} & \begin{bmatrix} [0.1,0.2] & [0.2,0.4] & [0.5,0.7] \\ [0.2,0.3] & [0.3,0.5] & [0.4,0.6] \\ -[0.3,0.5] & -[0.2,0.6] & [0.1,0.2] \end{bmatrix} \end{array},$$

$$\begin{array}{cc} & \begin{array}{ccc} \beta_1 & \beta_2 & \beta_3 \end{array} \\ \tilde{A}^2 = \begin{array}{c} \alpha_1 \\ \alpha_2 \\ \alpha_3 \end{array} & \begin{bmatrix} [0.2,0.3] & [0.5,0.6] & [0.8,0.9] \\ [0.3,0.4] & [0.2,0.3] & [0.4,0.7] \\ -[0.3,0.4] & -[0.2,0.3] & [0,0.1] \end{bmatrix} \end{array}.$$

The element"-[0.3, 0.5]"in matrix $\tilde{A}^1$ indicates that, when company 1 takes steps to improve the package and company 2 takes measures to advertisement, the sales amount of company 1 decreases between 300 thousand and 500 thousand, while towards company 2, the sales amount increases between 300 thousand and 500 thousand. The minus in the Equation shows that the sales amount of company 1 decrease accompanies the increase of sales amount of company 2. Other elements are analogous with this explanation.

The element"-[0.3, 0.4]"in matrix $\tilde{A}^2$ indicates that, when company 1 takes steps to improve the package and company 2 takes measures to advertisement, the market share of company 1 decreases between 3% and 4%, while towards company 2, the market occupation proportion increases between 3% and 4%. The minus in the Equation shows that the market share of company 1 decrease accompanies the increase of market share of company 2. Other elements are analogous with this explanation.

Suppose that the two companies both agree that the weight of objective f_1 and f_2 separately are 0.4 and 0.6. According to Equation (1), $\widetilde{A}^1$ and $\widetilde{A}^2$ are defuzzified as

$$B^1 = \begin{bmatrix} 0.11 & 0.24 & 0.60 \\ 0.22 & 0.36 & 0.48 \\ -0.36 & -0.28 & 0.11 \end{bmatrix},$$

$$B^2 = \begin{bmatrix} 0.22 & 0.55 & 0.88 \\ 0.33 & 0.22 & 0.52 \\ -0.33 & -0.22 & 0 \end{bmatrix}.$$

Considering the weight of each objective, the two objectives can be linearly aggregated as:

$$B = \begin{bmatrix} 0.176 & 0.426 & 0.768 \\ 0.286 & 0.276 & 0.504 \\ -0.342 & -0.244 & 0.044 \end{bmatrix}.$$

According to the solution of classical matrix game, the equilibrium strategy y^* is (0.038, 0.962, 0), z^* is (0.038, 0.423, 0), and the expected payoff value v is 0.282. That is, when company 1 selects strategy α_1 by the probability 0.038, and selects strategy α_2 by the probability 0.962; company 2 selects strategy β_1 by the probability 0.577, and selects strategy β_2 by the probability 0.423, the two players achieve multiobjective equilibrium, and the equilibrium value is 0.282.

(2) The essence idea of the second solution of multiobjective two-person zero-sum matrix game is: all objectives can be linearly aggregated according to the weight of each objective, and defuzzifies the aggregated vague payoff matrix through order function, then the matrix game can be solved.

After acquiring the equilibrium strategy y^* and z^*, the minimum possible equilibrium value and maximum possible equilibrium value can be achieved:

$$\underline{v} = \sum_{i-1}^{m} \sum_{j=1}^{n} y_i^* \underline{\widetilde{a}}_{ij} z_j^* \tag{5}$$

$$\overline{v} = \sum_{i-1}^{m} \sum_{j=1}^{n} y_i^* \overline{\widetilde{a}}_{ij} z_j^* \tag{6}$$

Where $\underline{\tilde{a}}_{ij}$ is the lower bound of vague value of optimum strategy of vague payoff matrix $\tilde{A}$, $\bar{\tilde{a}}_{ij}$ is the upper bound of vague value of optimum strategy of vague payoff matrix $\tilde{A}$.

Example 3. Except the weight of objectives are vague value ω_1 and ω_2, ω_1=[0.3, 0.5], ω_2 =[0.5, 0.7], the others are the same as example 2.

The payoff matrix $\tilde{A}^1$, $\tilde{A}^2$ can be aggregated as $\tilde{A}$ at first.

$$\tilde{A} = \begin{bmatrix} [0.127,0.289] & [0.295,0.536] & [0.490,0.760] \\ [0.201,0.388] & [0.181,0.408] & [0.296,0.643] \\ -[0.227,0.460] & -[0.154,0.447] & [0.030,0.163] \end{bmatrix}.$$

According to Equation (1), $\tilde{A}$ is defuzzified as

$$B = \begin{bmatrix} 0.148 & 0.366 & 0.622 \\ 0.239 & 0.222 & 0.399 \\ -0.279 & -0.199 & 0.034 \end{bmatrix}.$$

According to the solution of classical matrix game, the equilibrium strategy y^* is (0.071, 0.929, 0), z^* is (0.613, 0.387, 0), and the expected payoff value v is 0.232. That is, when company 1 selects strategy α_1 by the probability 0.071, and selects strategy α_2 by the probability 0.929; company 2 selects strategy β_1 by the probability 0.613, and selects strategy β_2 by the probability 0.387, the two players achieve multiobjective equilibrium, and the equilibrium value is 0.232.

According to Equation (5) and (6), the minimum possible equilibrium value and maximum possible equilibrium value can be achieved as

$$\underline{v} = \sum_{i-1}^{m} \sum_{j=1}^{n} y_i^* \underline{a}_{ij} z_j^* = 0.193,$$

$$\bar{v} = \sum_{i-1}^{m} \sum_{j=1}^{n} y_i^* \bar{a}_{ij} z_j^* = 0.395.$$

5 Conclusion

This paper has dealt with multiobjective two-person zero-sum matrix game whose payoff values are vague values under the condition of same weight. In the future, we will study the situation that players have different weight of each object.

Acknowledgements

This work described here was partially supported by the grants from the National Natural Science Foundation of China (No. 70471063) and the Program of New Century Excellent Talents of China (No: NCET-05-0097).

References

1. Campos L (1989) Fuzzy linear programming models to solve fuzzy matrix games. Fuzzy Sets and Systems 32: 275-289
2. Nishizaki I, Sakawa M (2000) Equilibrium solutions in multiobjective bimatrix games with fuzzy payoffs and fuzzy goals. Fuzzy Sets and Systems 111: 99-116
3. Maeda Takashi (2003) On characterization of equilibrium strategy of two-person zero-sum games with fuzzy payoffs. Fuzzy Sets and Systems 139: 283–296
4. Vijay V, Chandra S, Bector C.R (2005) Matrix games with fuzzy goals and fuzzy payoffs. The international Journal of Management Science 33: 425–429
5. Gau W L, Buehrer D J (1993) Vague sets. IEEE Transactions on Systems, Man, and Cybernetics 23(2): 610-614
6. Bustince H, Burillo P (1996) Vague sets are intuitionistic fuzzy sets. Fuzzy Sets and Systems 79: 403-405
7. Zhou XG, Zhang Q (2005) Comparison and improvement on similarity measures between vague sets and between elements. Journal of Systems Engineering 20(6): 613-619 (in Chinese)
8. Zhou XG, Zhang Q (2005) Aggregation vague opinions under group decision making. Proceeding of the 4th Wuhan International Conference on E-business, Wuhan 6: 1736-1742
9. Liu HW (2004) Vague set methods of multicriteria fuzzy decision making. System Engineering Theory and Practice 5: 103-109 (in Chinese)

Fuzzy Geometric Object Modelling

Qingde Li[1] and Sizong Guo[2]

[1] Department of Computer Science, University of Hull, Hull, HU6 7RX, UK
q.li@dcs.hull.ac.uk
[2] Institute of Sciences, Liaoning Engineering Technique University, Fuxin, Liaoning 123000, China
guosizong@tom.com

Abstract. Fuzzy geometric object modelling has been one of the fundamental tasks in computer graphics. In this paper, we investigated and developed techniques for modelling fuzzy objects based on fuzzy set theory. With the proposed techniques any conventional geometric objects, such as points, lines, triangles, and curved surfaces can all be easily extended as certain kind of fuzzy sets to represent geometric objects with fuzzy boundaries. In addition, experimental results are presented to demonstrate how 3D fuzzy shapes represented in this way can be rendered efficiently and effectively in multiple passes using shader programs running entirely on programmable GPUs.

Keywords: Fuzzy geometry, fuzzy curves, fuzzy shape modelling.

1 Introduction

Geometric objects modelling has been an important task in CAGD, computer graphics, computer games, and in the creation of special effects in movies. So far, various shape modelling techniques have been proposed, such as polygonal meshes, subdivision method, spline objects and implicit geometric objects. However, many natural objects with fuzzy boundaries, such as fog, cloud, smoke, explosion, fire, fur, hair, and gas, cannot be described properly using these techniques. Currently the most popular technique used for modelling fuzzy objects has been the particle systems. However, this is a quite expensive method in terms of computational efficiency. In this paper, we investigate how to represent and display fuzzy shapes in terms of fuzzy sets. It is found that not only can the fuzzy objects be represented naturally and intuitively as fuzzy sets, but also the fuzzy objects represented in this way can be implemented easily, thanks to the availability of programmable graphics hardware.

The main idea underlying the proposed techniques is implicit geometric shape modelling. Modelling geometric objects as implicit functions was first addressed in Ricci's pioneering work [1] in 1973. However, implicitly represented objects are expensive to display. Because of this, it is much less popular than the parametrically represented shapes and mesh objects in the area of graphics object

B.-Y. Cao (Ed.): Fuzzy Information and Engineering (ICFIE), ASC 40, pp. 551–563, 2007.
springerlink.com

modelling. However, increasing attention is being received with its advantages being recognized by more and more people. For instance, implicitly modelled shapes are easy to blend. It is also much easier to detect when an object collides with an implicitly described object. In addition, implicit shapes are much less computational expensive to ray trace.

The idea of modelling geometric objects is quite simple. Imagine an object as something solid in space. Then it can be described as an ordinary set. In mathematics, a set A can be conveniently described by its membership function $\mu_A(\mathbf{P})$. Generally, a 2D or 3D geometric object can be represented as a function $F : \mathcal{R}^n \rightarrow \mathcal{R}, (n = 2, 3)$, which in general partitions the space $\mathcal{R}^n$ into three parts corresponding to $F(\mathbf{P}) < 0$, $F(\mathbf{P}) = 0$, and $F(\mathbf{P}) > 0$ respectively. Therefore, a real function F naturally associates with a geometric shape either by representing its boundary as the real roots of the equation $F(\mathbf{P}) = 0$ or as a solid defined by $\{\mathbf{P} : F(\mathbf{P}) \geq 0\}$ or by $\{\mathbf{P} : F(\mathbf{P}) \leq 0\}$. A complex solid geometric object can be considered to be constructed by blending a set of simple geometric primitives using set-theoretic operations such as union, intersection, and subtraction. When each geometric primitive is characterized as a real function, the key to this technique is to define blending operations that combine a set of functions representing simple geometric objects into a single function representing the required geometric shape. The simplest and the most natural blending operation is the one corresponding exactly to the Boolean operations $\max(x, y)$ or $\min(x, y)$. Assume that the two solid objects A and B are represented implicitly by functions $F_A(\mathbf{P})$ and $F_B(\mathbf{P})$ respectively as $A = \{\mathbf{P} : F_A(\mathbf{P}) \geq 0\}$ and $B = \{\mathbf{P} : F_B(\mathbf{P}) \geq 0\}$. Then, the union $A \cup B$, the intersection $A \cap B$, and the subtraction $A - B$ of set A and B can be represented by functions $\max\{F_A(\mathbf{P}), F_B(\mathbf{P})\}$, $-\max\{-F_A(\mathbf{P}), -F_B(\mathbf{P})\}$, and $-\max\{-F_A(\mathbf{P}), F_B(\mathbf{P})\}$ respectively.

However, general implicit functions cannot be used directly to represent fuzzy shapes, where the geometric objects do not have a distinct boundary. One obvious problem is that a general function does not always provide geometric intuitiveness for modelling geometric shapes. For example, a circle (or sphere) with radius R can be represented using $F(\mathbf{P}) = \|\mathbf{P} - \mathbf{P}_0\|^n - R^n$, where n is any positive number. Though the value of the function $F(\mathbf{P})$ can be used to interpret how far a point closes to the circle, it cannot properly describe to what degree a point belongs to the set characterized by the circle. Secondly, the value of an implicit function can vary from $-\infty$ to ∞ and the shapes described by the function can be quite sensitive with the location of $\mathbf{P}$. As a result, it can be difficult to predict the exact shape represented by an implicit function when considering some delicate issues associated with practical problems, such as shape deformation and surface detail modelling. In this paper, we propose to use fuzzy set to model geometric objects with a focus on fuzzy shape modelling. As will be seen later, fuzzy sets based shape modelling exhibits more geometric intuitiveness and offers us an easy way to model fuzzy objects, such as furs, and smoke.

So far, much of the basic concepts of conventional geometry have been extended to define their fuzzy counterparts [2, 3, 4, 5, 6] and an overview on various issues on fuzzy geometry can be found in [7]. In [8], the problem of edge representation using fuzzy sets in blurred images was investigated. More recently, some general discussions on the representation, construction and display approaches for fuzzy shapes were made in [9]. The aim of this paper is to investigate how to represent fuzzy shape using implicit functions with a focus on practical implementation using shaders.

2 Modelling Fuzzy Geometric Objects Using Smooth Unit Step Function and Implicit Functions

In this section, we consider how to convert a given implicit function into a membership function of a fuzzy set by using smooth unit step functions (SUSF).

As is commonly known that a fuzzy set on space R^n is a mapping from R^n to $[0, 1]$. As fuzzy sets are extension of conventional sets, a fuzzy shape should be a natural extension of conventional shape. Consider the problem of modelling an animal with skin fur. One natural way is to first model the body (without fur) of the animal as a solid A and then add its fuzzy skin on to the body. When the fuzzy skin is modelled as a fuzzy set, the addition of the fuzzy skin to the solid body is simply a union of fuzzy sets. Now the problem is how to create fuzzy shapes so that they can be implemented in practice efficiently. One cheap way to generate fuzzy shapes is the fuzzification of ordinary geometric objects. It is found in this research that both implicitly and explicitly defined shapes can be fuzzified by using a SUSF and displayed by using shader programs implemented in multiple passes. In this section we only discuss how to create fuzzy shapes from implicit functions and the discussion on how to create fuzzy objects from a polygonal mesh will be given in the next section.

Let A be a solid geometric shape defined implicitly by an equation $F(\mathbf{P}) \geq 0$ and let $H(x)$ be the Heaviside unit step function. That is

$$H(x) = \begin{cases} 0, & x < 0; \\ \frac{1}{2}, & x = 0; \\ 1, & x > 0. \end{cases} \tag{1}$$

Now consider the composition

$$\mu(\mathbf{P}) = H \circ F(\mathbf{P}). \tag{2}$$

It is obvious that $\mu(\mathbf{P})$ defines a fuzzy set. In fact, we have

$$A = \{\mathbf{P} : \mu(\mathbf{P}) > 0\} = \{\mathbf{P} : F(\mathbf{P}) \geq 0\}.$$

The boundary of the shape corresponds to the set

$$\partial A = \{\mathbf{P} : F(\mathbf{P}) = 0\} = \{\mathbf{P} : \mu(\mathbf{P}) = \frac{1}{2}\}.$$

To specify a really fuzzy shape, we need only replace the Heaviside unit step function with a smooth unit step function, a continuous function taking the following form:

$$\mu(x) = \begin{cases} 0, & x < a; \\ f(x), & x \in [a,b]; \\ 1, & x > b. \end{cases} \tag{3}$$

where $a < b$, and $f(x)$, with $0 \leq f(x) \leq 1$, is strictly increasing on $[a,b]$. Various such kind of functions can be defined directly (see [10]). But the SUSFs specified using the following standard SUSF are more preferable in terms of the computational efficiency and the flexibility in defining the general SUSFs.

Let

$$\begin{aligned} H_0(x) &= \begin{cases} 0, & x < 0; \\ \frac{1}{2}, & x = 0; \\ 1, & x > 0. \end{cases} \\ H_n(x) &= \frac{1}{2}(\ (1 + \frac{x}{n})H_{n-1}(x+1) \\ &\quad +(1 - \frac{x}{n})H_{n-1}(x-1)\ ), \\ &\quad n = 1, 2, 3, \cdots. \end{aligned} \tag{4}$$

It can be shown that $H_n(x)$ has the following properties:

Proposition 2.1. *For each function $H_n(x)$, we have*

(1) $H_n(x)$ is C^{n-1}-continuous for $n > 1$;
(2) $H_n(x)$ is a piecewise-polynomial function;
(3) $H_n(x)$ is monotonically increasing and takes value 1 when $x \geq n$, and 0 when $x \leq -n$;
(4) $H_n(x) + H_n(-x) = 1$, $H_n(0) = \frac{1}{2}$;
(5) $x(2H_n(x) - 1) \leq x(2H_{n-1}(x) - 1)$, $n = 1, 2, \cdots$.

For an arbitrary number $R > 0$, the SUSF with $[-R, R]$ can be defined as follows:

$$H_{n,R}(x) = H_n(nx/R). \tag{5}$$

More general SUSFs with an arbitrary rising interval $[a,b]$ can also be easily specified using the standard SUSF defined in (4). In fact, for a given rising interval $[a,b]$, the corresponding SUSF can be written out directly as:

$$H_{n,[a,b]} = H_n\left(\frac{2n(x-a)}{b-a} - n\right).$$

Now we consider how the SUSFs can be used to define fuzzy objects from implicitly specified geometric shapes.

2.1 Method 1: Fuzzification of a Solid Geometric Object

Let $F(\mathbf{P})$ be a function representing an ordinary solid geometric object with the internal part of the shape corresponding to $F(\mathbf{P}) > 0$. Then the object can be fuzzified by defining a fuzzy set with the following membership function:

$$\mu(\mathbf{P}) = H_{n,R} \circ F(\mathbf{P}). \tag{6}$$

It can be shown directly that $\mu(\mathbf{P}) = 1$ when $F(\mathbf{P}) \geq R$ and $\mu(\mathbf{P}) = 0$ when $F(\mathbf{P}) \leq -R$.

Figure 1 illustrate how an implicitly defined ellipse and a triangle can be turned into fuzzy shapes in this way.

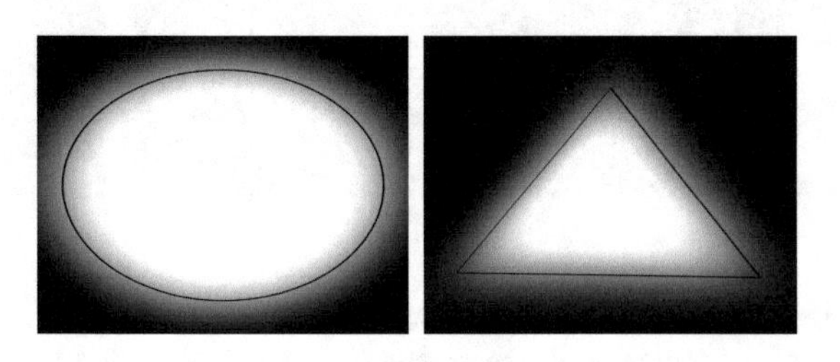

Fig. 1. A solid fuzzy ellipse and a solid fuzzy triangle obtained by blurring their boundaries

2.2 Method 2: Fuzzification of the Boundary of a Geometric Object

Fuzzy objects can also be created by blurring a given boundary object, such as a curve or a surface.

Assume that the boundary of a geometry object is represented implicitly by $F(\mathbf{P}) = 0$ using a function $F(\mathbf{P})$. Let R be the range of the fuzzy boundary. Then

$$\widetilde{\mathcal{F}}(\mathbf{P}) = H_{n,R}\left(R - \frac{2}{R}\mathcal{F}(\mathbf{P})^2\right) \tag{7}$$

represents a fuzzy object. It can be seen from the properties of SUSF $H(x)$, $\widetilde{\mathcal{F}}(\mathbf{P}) = 1$ when $F(\mathbf{P}) = 0$ and $\widetilde{\mathcal{F}}(\mathbf{P}) = 0$ when $|F(\mathbf{P})| > R$.

Fuzzy points: A fuzzy point is a fuzzy set with its base shape to be a point. Let $\mathbf{P}_0$ be a point. Then a fuzzy point located at $\mathbf{P}_0$ can be defined by defining its membership function as:

$$\widetilde{\mathcal{F}} = H_{n,R}\left(R - \frac{2}{R}\|\mathbf{P} - \mathbf{P}_0\|^2\right) \tag{8}$$

Fuzzy 2D lines: A fuzzy straight line is a fuzzy set with its core shape to be a line. Let $\mathbf{n}$ be a 2D unit vector. Then the line passing through a point $\mathbf{P}_0$ with normal $\mathbf{n}$ can be represented as

$$\mathcal{L}(\mathbf{P}) = (\mathbf{P} - \mathbf{P}_0) \cdot \mathbf{n} = 0.$$

According to equation (7), the corresponding fuzzy line can be defined by

$$\widetilde{\mathcal{F}} = H_{n,R}\left(R - \frac{2}{R}\mathcal{L}(\mathbf{P})^2\right) \tag{9}$$

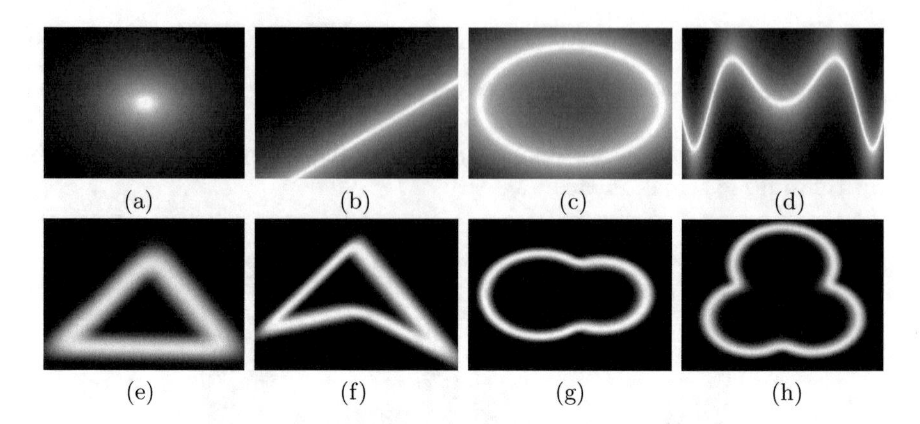

Fig. 2. (a) A fuzzy point defined by equation 8 with its centre located at $[0.5, 0.5]$. (b) A fuzzy line defined by equation 9, where the core line is specified by a point $[0.5, 0.5]$ and a normal vector $(-\frac{1}{\sqrt{2}}, \frac{1}{\sqrt{2}})$. (c), (d). 2D fuzzy curves.

Fuzzy 2D polylines and curves: In general, as long as we can represent a 2D polyline (or, in general, a curve) as an implicit function $F(\mathbf{P}) = 0$, the polyline (or the curve) can be fuzzified in the same way as the fuzzification of a point and a line. Figure 2 lists some examples of fuzzy polylines and curves described in this way.

Fuzzy planes: A fuzzy plane can be defined in a similar way as the fuzzy line. Let $\mathbf{n}$ be a 3D unit vector. Then the plane passes through a point $\mathbf{P}_0$ with normal $\mathbf{n}$ can be represented as

$$\mathcal{P}(\mathbf{P}) = (\mathbf{P} - \mathbf{P}_0) \cdot \mathbf{n} = 0,$$

and the corresponding fuzzy plane can be specified by

$$\widetilde{\mathcal{F}} = H_{n,R}\left(R - \frac{2}{R}\mathcal{P}(\mathbf{P})^2\right) \tag{10}$$

2.3 Method 3: Adding a Fuzzy Boundary to an Ordinary Solid Geometric Object

Let $F(\mathbf{P}) \geq 0$ be an implicitly described solid geometric object. Then a fuzzy boundary can be added to the object from either outside or inside of the object. For example, the following function can be introduced to create a fuzzy set with specified internal fuzzy boundary:

$$\mu(\mathbf{P}) = H_n\left(\frac{2F(\mathbf{P})^2}{R} - R\right). \tag{11}$$

As can be seen from the above equation, $\mu(\mathbf{P}) = 0$ whenever $F(\mathbf{P}) \leq 0$. $\mu(\mathbf{P}) = 1$ when $F(\mathbf{P}) \geq R$.

3 Modelling Fuzzy Geometric Objects Using Geometric Meshes and Parametrically Defined Geometric Shapes

As is commonly known that most geometric objects used in computer graphics and computer games are geometric meshes, where the shape of a geometric object is represented as a collection of well connected simple polygons. These ordinary geometric meshes can be easily turned into objects with fuzzy boundaries following the decomposition theorem about fuzzy sets. According to this theorem, any fuzzy set can be represented as the union of its α-cuts. Suppose the boundary of the base shape $\mathcal{S}$ of a fuzzy set is represented as a parametric surface $\mathbf{P} = S(u, v)$ or as a polygonal mesh. Let the fuzzy boundary of the object is $\widetilde{A}$. Then the expected fuzzy model can be represented as $\mathcal{S} \cup \widetilde{A}$. According to the α-cut theorem, we have

$$\widetilde{A} = \bigcup \lambda A_\lambda. \tag{12}$$

Suppose the fuzzy range is modelled by a number R. Let the solid geometry corresponding to the parametric surface $\mathcal{S}$ to be A_0. Then the fuzzy object can be represented as

$$\widetilde{S} = A_0 \bigcup \left(\bigcup_{\lambda=0}^{R} \lambda \mathcal{S}_\lambda \right), \tag{13}$$

where $\mathcal{S}_\lambda$ represents the geometric solid obtained by scaling $\mathcal{S}$ along its normals by $1 + \lambda$.

This modelling technique can be immediately implemented using shader programming language to run on modern programmable graphics hardware.

Fuzzy objects displayed in figures 3 are created in this way.

An interesting thing is that when the surface detail of a mesh is added using certain texture mapping techniques, various kinds of fur textures can be added to the mesh using the above fuzzy modelling technique. Figure 4 shows some fuzzy objects created in this way.

4 Fuzzy Geometric Objects Blending

When the underline geometric objects are modelled implicitly as fuzzy sets, they can be blended easily using fuzzy set operations. In this section, we discuss the issues relating to the combination of simple fuzzy geometric objects.

When two geometric objects are represented as fuzzy sets $\widetilde{A}$ and $\widetilde{B}$, we can find their union, intersection and difference as usual. One important issue needs to be addressed here is the choice of the fuzzy set operation. One typical feature of the most commonly used fuzzy objects is that their membership functions are smooth to reflect the nature of fuzziness. However, many of the traditional fuzzy sets operations are not smooth. The main problem of using non-smooth fuzzy set operations is that they may lead to non-smooth fuzzy sets. Another significant feature of fuzzy sets is that they are frequently described by a piecewise polynomial membership function. Thus algebraic or piecewise algebraic operations are

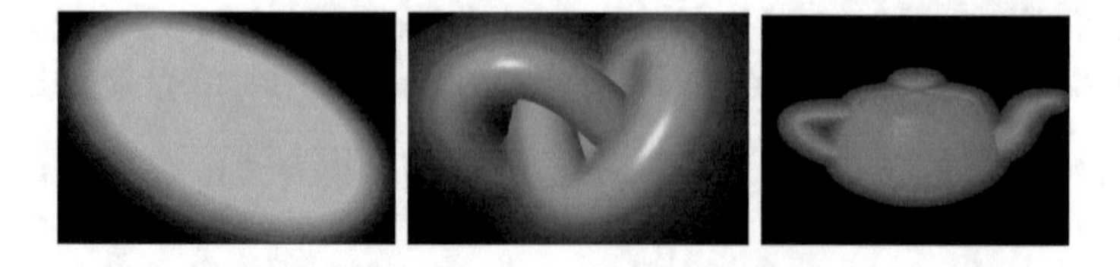

Fig. 3. 3D Soft shapes modelled as the unions of a collection of fuzzy sets obtained from some ordinary polygonal meshes

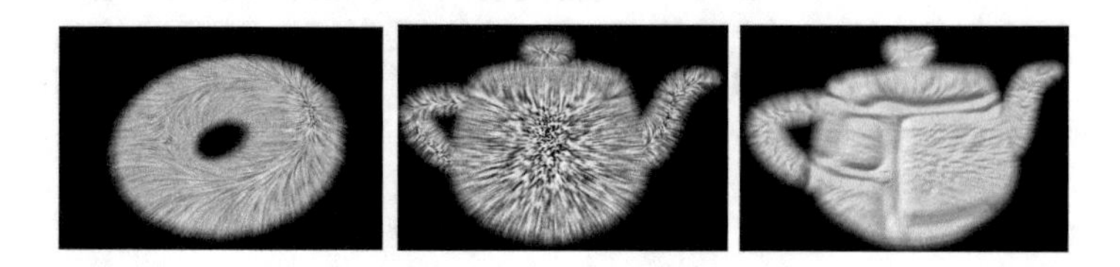

Fig. 4. 3D Soft shapes modelled as the unions of a collection of fuzzy sets obtained from some textured polygonal meshes

preferred since the fuzzy sets produced by these kinds of operations will also be described by piecewise polynomials. The fuzzy shapes presented in figure 5 are created using the well known algebraic operators.

Complex fuzzy shapes can be obtained by combining a set of simple fuzzy shapes. For instance, a 2D fuzzy polygon can be considered to be the result of a sequence of fuzzy set operations on a set of fuzzy half plane. Similarly, various fuzzy polyhedra can be created using a collection of fuzzy half spaces. More general fuzzy objects can be created by combining a set of simple fuzzy shapes like fuzzy spheres, fuzzy cylinders, and fuzzy planes.

5 Shape Preserving Fuzzy Geometric Blending

As have been pointed out that most of the commonly used fuzzy set operations can be used to combine two fuzzy shapes when they are described as fuzzy sets. However, to generate a fuzzy geometric shape with required smoothness, smooth fuzzy set operations must be used. Algebraic operator serves as a good fuzzy shape blending operation. It is simple to compute and always produce smooth membership functions as long as the fuzzy shapes involved in the operation are all smooth. The main drawback of algebraic operation is that it is not shape preserving in terms of controlling the blending range. When two shapes are combined using the algebraic operation, no matter whether it is a union, intersection, or subtraction, no parts of the newly obtained shape will be the same as those of the original shapes. The operations defined using $\min(x, y)$ are shape preserving but they are not smooth. Dombi's operator[11] could be used approximately as a shape preserving operation but it has some drawbacks. It is not piecewise algebraic and can be expensive to compute. But the main

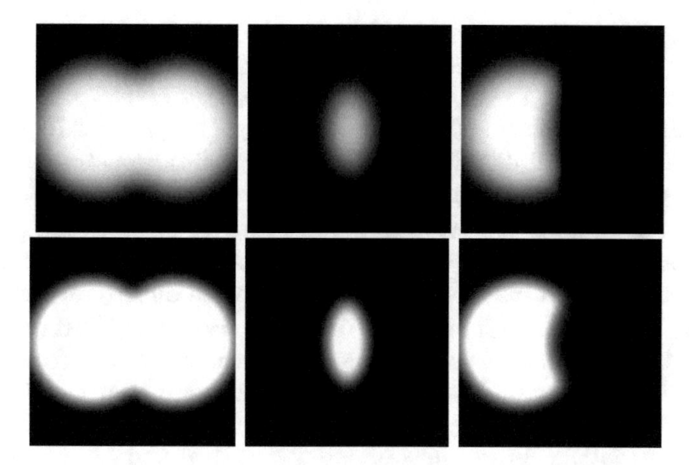

Fig. 5. Fuzzy shapes created by blending two fuzzy discs using conventional algebraic fuzzy set operation

disadvantage of Dombi's operator is its lack of intuition in terms of specifying the blending range.

The shape preserving feature of a blending operation plays a crucial role in implicit modelling when parts of geometric primitives are reconstructed from real data. In this case, one would hope that these real shapes should be kept as unchanged as possible as one could. In addition, the tessellation of blended implicit surfaces obtained from shape-preserving blending operations can be made much more effectively and efficiently as many parts of the blended shapes can be identical with those implicit shapes involved in the blending. Thus, many polygons in the meshes representing the primitive implicit surfaces can be reused to construct the polygonal mesh corresponding to the blended shapes. In terms of smoothness, a blending operation should be defined easily with any required smoothness, such as C^1 and C^2 continuity. Moreover, the operation should also be defined with clear geometric meaning and geometric intuition. It should have a simple mathematical form and be easy to construct. In the past few years, several smooth shape preserving blending operations have been proposed. In [12], Smooth blending range controllable operations were defined using scalar function. In [13], a blending technique using R-function was introduced to achieve exactly the same purpose through the introduction of control points and bounding solid. Smooth shape preserving Boolean operators were also introduced in the work of Barthe et al[14],[15]. The major limitation of all these blending operations is that they only have a C^1 or G^1 continuity and it is difficult to extend to define operators with higher order of smoothness.

In this last section, we present two constructive methods to define smooth piecewise algebraic operations for combining two fuzzy shapes with specified blending range, namely $\wedge_{n,\delta}(x,y)$ and $\widehat{\wedge}_{n,\delta}(x,y)$. The main features of these fuzzy set operations are that they can be constructed to have whatever degree of smoothness required and to approximate Zadeh's minimum functions to any required precision. In addition, the binary operator $\widehat{\wedge}_{n,\delta}(x,y)$ can be specified to preserve $\min(x,y)$ to any required extent. More precisely, $\widehat{\wedge}_{n,\delta}(x,y)$ can be defined to be identical to the ordinary binary operation $\min(x,y)$ in a subregion of $[0,1]^2$, with the area of the subregion to be $(1-3\delta)^2, \delta \in (0,1/2]$. These novel operations are all described by recursively defined functions and can be implemented cheaply and elegantly. The main idea used here is the introduction of smooth absolute functions $|x|_n$ and $\widehat{|x|}_n$, both of which are generalized from the conventional absolute function $|x|$. With smooth absolute functions, the conventional minimum function $\min(x,y)$ is extended directly to smooth ones, namely the degree n smooth minimum functions $min_{n,\delta}(x,y)$ and $\widehat{min}_{n,\delta}(x,y)$, where $\delta > 0$ is a parameter used to control the smooth fusion range along the line $y = x$. These two kinds of minimum functions are then modified further respectively to build the binary operations $\wedge_{n,\delta}(x,y)$ and $\widehat{\wedge}_{n,\delta}(x,y)$ on $[0,1]$ for fuzzy set aggregation. As will be seen later, $\wedge_{n,\delta}(x,y)$ and $\widehat{\wedge}_{n,\delta}(x,y)$ have the following properties:

1. Both become the conventional Boolean AND operation when confined on $\{0,1\}$;
2. $\wedge_{n,\delta}(1,x) = \wedge_{n,\delta}(x,1) = x$, $\widehat{\wedge}_{n,\delta}(1,x) = \widehat{\wedge}_{n,\delta}(x,1) = x$;
3. $\wedge_{n,\delta}(0,x) = \wedge_{n,\delta}(x,0) = 0$, $\widehat{\wedge}_{n,\delta}(0,x) = \widehat{\wedge}_{n,\delta}(x,0) = 0$;
4. $\wedge_{n,\delta}(x,y)$ is C^n-smooth, and $\widehat{\wedge}_{n,\delta}(x,y)$ is C^{n-1}-smooth.
5. $\wedge_{n,\delta}(x,y)$ and $\widehat{\wedge}_{n,\delta}(x,y)$ can be set to approximate $\min(x,y)$ with any specified precision by adjusting parameter δ.

The construction of these functions is briefly introduced here. The detailed discussion on these operators can be found in [16].

Definition 1. *Let $|x| : \mathbb{R} \rightarrow \mathbb{R}$ be the conventional absolute function. That is, $|x| = x$ when $x \geq 0$ and $|x| = -x$ when $x < 0$. Then we introduce the following generalized absolute functions:*

$$\begin{aligned} |x|_0 &= |x|; \\ |x|_n &= \frac{1}{2(n+1)}\big(\ (n-x)|1-x|_{n-1} \\ &\qquad\qquad +(n+x)|1+x|_{n-1} \ \big). \\ &\qquad n = 1,2,3,\cdots \end{aligned} \tag{14}$$

$|x|_n$ is called degree n upper absolute function.

It can be shown that $|x|_n$ has the following properties

Proposition 5.1. *(1) $|x|_n \geq |x|$; and $|x|_n = |x|$ when $|x| \geq n$;*
(2) $|x|_n$ is C^n-continuous;
(3) $|x|_n$ is a piecewise polynomial function.

For smooth absolute function $|x|_n$, the difference between $|x|_n$ and $|x|$ is in the range $[-n,n]$. In the following discussion, we will refer this range of difference as the span of $|x|_n$. Smooth absolute functions with an arbitrary span $[-\delta,\delta]$ $(\delta > 0)$ can be easily introduced using $|x|_n$.

Definition 2. *For $\delta > 0$ and $n > 0$, we define*

$$|x|_{n,\delta} = \frac{\delta}{n}\left|\frac{nx}{\delta}\right|_n. \tag{15}$$

It can be shown immediately that $|x|_{n,\delta} = |x|$ when $|x| \geq \delta$. Smooth functions that approximate the conventional absolute function can also be introduced using the SUSFs.

Definition 3. *Now consider the sequence of functions defined in the following way using smooth unit step functions $H_n(x)$:*

$$\widehat{|x|}_n = xSgn_n(x), \qquad n = 0, 1, 2, \cdots \tag{16}$$

where $Sgn_n(x) = 2H_n(x) - 1$. $\widehat{|x|}_n$ is called degree n lower absolute function.

As with $|x|_{n,\delta}$, $|\widehat{x}|_{n,\delta}$ can be introduced using $\widehat{|x|}_n$.

The above two smooth absolute functions can be used to define what we called the smooth minimum functions.

Definition 4. *For $\delta > 0$, let*

$$min_{n,\delta}(x, y) = \frac{1}{2}(x + y - |x - y|_{n,\delta}); \tag{17}$$

$$\widehat{min}_{n,\delta}(x, y) = \frac{1}{2}(x + y - |\widehat{x - y}|_{n,\delta}). \tag{18}$$

We call $min_{n,\delta}(x, y)$ the degree n lower smooth minimum function and $\widehat{min}_{n,\delta}$ (x, y) the degree n upper smooth minimum function, where δ is a parameter referred to as the approximation accuracy to the minimum operation $\min(x, y)$.

These two functions have first been introduced in [17] for implicit shape modelling with controllable blending range. They are then modified in [16] into $\wedge_{n,\delta}(x, y)$ and $\widehat{\wedge}_{n,\delta}(x, y)$ for approximating the Zadeh's minimum fuzzy set operations[18]. Both of these two operations can be used for blending two fuzzy sets, where the blending range can be well controlled by specifying an appropriate δ values intuitively.

To summarize, we have proposed in this paper how fuzzy set theory can be applied to model fuzzy geometric objects. The proposed techniques are simple in their mathematical representation and are practical to implement.

References

1. Ricci, A.: A constructive geometry for computer graphics. Computer Journal **16**(3) (1973) 157–160
2. Rosenfeld, A.: The diameter of a fuzzy set. Fuzzy Sets and Systems **13**(6) (1984) 241–24

3. Rosenfeld, A.: Fuzzy rectangles. Pattern Recognition Lett **11** (1990) 677–794
4. Rosenfeld, A.: Fuzzy plane geometry: triangles. Pattern Recogn. Lett. **15**(12) (1994) 1261–1264
5. Buckley, J.J., Eslami, E.: Fuzzy plane geometry i: Points and lines. Fuzzy Sets and Systems **86**(2) (1997) 179–187
6. Buckley, J.J., Eslami, E.: Fuzzy plane geometry ii: Circles and polygons. Fuzzy Sets and Systems **87**(1) (1997) 79–85
7. Rosenfeld, A.: Fuzzy geometry: an updated overview. Inf. Sci. **110**(3-4) (1998) 127–133
8. Kim, T.Y., Han, J.H.: Edge representation with fuzzy sets in blurred images. Fuzzy Sets Syst. **100**(1-3) (1998) 77–87
9. Zhang, J.L., Pham, B., Chen, Y.P.P.: Modelling and indexing fuzzy complex shapes. In: Proceedings of the IFIP TC2/WG2.6 Sixth Working Conference on Visual Database Systems, Deventer, The Netherlands, The Netherlands, Kluwer, B.V. (2002) 401–415
10. Li, Q., Griffiths, J.G., Ward, J.: Constructive implicit fitting. Comput. Aided Geom. Des. **23**(1) (2005) 17–44
11. Dombi, J.: A general class of fuzzy operators, de morgan class of fuzzy operators and fuzziness measures induced by fuzzy operators. Fuzzy Sets and Systems (2) (1982) 149–163
12. Hsu, P.C., Lee, C.: The scale method for blending operations in functionally-based constructive geometry. Computer Graphics Forum **22**(2) (2003) 143–158
13. Pasko, G.I., Pasko, A.A., Kunii, T.L.: Bounded blending for function-based shape modeling. IEEE Computer Graphics and Applications **25**(2) (2005) 36–44
14. Barthe, L., Dodgson, N.A., Sabin, M.A., Wyvill, B., Gaildrat, V.: Two-dimensional potential fields for advanced implicit modeling operators. COMPUTER GRAPHICS forum **22**(1) (2003) 23–33
15. Barthe, L., Wyvill, B., Broot, E.D.: Controllable binary CSG operators for "soft objects". International Journal of Shape Modeling **10**(2) (2004) 135–154
16. Li, Q.: Smooth piecewise algebraic fuzzy set operations. Technical report, Department of Computer Science, University of Hull (2006)
17. Li, Q.: Smooth piecewise polynomial blending operations for implicit shapes. Computer Graphics Forum (To appear)
18. Zadeh, L.: Fuzzy sets. Information and Control **8** (1965) 338–353
19. Wang, P.Z.: Fuzzy Sets and Random Set Falling Shadows. Beijing Normal Univ. Press, Beijing (1985)
20. Zimmermann, H.J.: Fuzzy Set Theory and Its Applications, 3rd ed. Kluwer Academic Publishers, Boston, MA (1996)
21. Kaufmann, A., Gupta, M.M.: Introduction to Fuzzy Arithmetic Theory and Application. Van Nostrand Reinhold, New York (1991)
22. Barthe, L., Gaildrat, V., Caubet, R.: Extrusion of 1d implicit profiles: Theory and first application. International Journal of Shape Modeling **7**(2) (2001) 179–199
23. Hoffmann, C.M.: Implicit curves and surfaces in CAGD. IEEE Computer Graphics and Appl **13** (1993) 79–88
24. Middleditch, A.E., Sears, K.H.: Blend surfaces for set-theoretic volume modelling systems. Proceedings of SIGGRAPH 85 (1985) 161–170
25. Rockwood, A.P.: The displacement method for implicit blending surfaces in solid models. ACM Transactions on Graphics **8**(4) (1989) 279–297
26. Bloomenthal, J., Wyvill, B.: Interactive techniques for implicit modeling. Computer Graphics (1990 Symposium on Interactive 3D Graphics) **24**(2) (1990)109–116

27. Pasko, A.A., Adzhiev, V., Sourin, A., Savchenko, V.V.: Function representation in geometric modeling: concepts, implementation and applications. The Visual Computer **11**(8) (1995) 429–446
28. Wickerhauser, M.V.: Adapted Wavelet Analysis from Theory to Software. A K Peters, Wellesley (1994)

A Fuzzy Portfolio Selection Methodology Under Investing Constraints

Wei Chen[1], Runtong Zhang[1], Wei-Guo Zhang[2], and Yong-Ming Cai[1]

[1] School of Economics and Management, Beijing Jiaotong University, Beijing, 100044, P.R. China
cwcw2001@163.com
[2] School of Business Administration, South China University of Technology, Guangzhou, 510641, P.R. China
cwnxyc@sina.com

Abstract. It is well-known that the financial market is affected by many non-probabilistic factors. In a fuzzy uncertain economic environment, the future states of returns and risks of risky assets cannot be predicted accurately. Based on this fact, possibilistic portfolio selection problem under investing constrains is discussed in this paper. The possibilistic mean value of the return is termed measure of investment return and the possibilistic variance of the return is termed measure of investment risk. We further present a quadratic programming model replaced Markowitz's mean-variance model when returns of assets are trapezoidal fuzzy numbers, the conventional probabilistic mean-variance model is simplified and extended. A numerical example of a possibilistic fuzzy portfolio selection problem is given to illustrate our proposed effective means and approaches.

Keywords: Portfolio Selecion, Possiblistic mean, Possiblistic variance, Investing Constrains.

1 Introduction

The mean-variance methodology for the portfolio selection problem, proposed originally by Markowitz [1,2], has played an important role in the development of modern portfolio selection theory. It combines probability with optimization techniques to model the behavior investment under uncertainty. The return is measured by the mean, and the risk is measured by the variance, of the return of a portfolio of assets. Most of existing portfolio selection models are based on probability theory. Previous researches on the mean-variance portfolio selection problem include Sharpe [3], Merton [4], Perold [5], Pang [6] Vörös [7] and Best [8].

The basic assumption for using Markowitz's mean-variance model is that the situation of assets in the future can be correctly reflected by asset data in the past, that is to say, the means, variances and covariances in future are similar to those in the past. However, it is hard to ensure this kind of assumption for real ever-changing asset markets.

B.-Y. Cao (Ed.): Fuzzy Information and Engineering (ICFIE), ASC 40, pp. 564–572, 2007.
springerlink.com

Recently, a number of researchers investigated fuzzy portfolio selection problem. Watada [9], Inuiguchi [10], Ramaswamy [11], León et al [12] and Wang [13] discussed portfolio selection using fuzzy decision theory. Tanaka and Guo [14,15] proposed two kinds of portfolio selection models based on fuzzy probabilities and exponential possibility distributions, respectively. Zhang and Nie [16] introduced the admissible efficient portfolio model under the assumption that the expected returns and risks of assets have admissible errors. Zhang and Wang [17] proposed the admissible efficient portfolio model under the assumption that there exists a risk-less asset. The closed form solution of the admissible efficient frontiers were derived from three cases: the risk-less asset can either be only lent, or borrowed, or both lent and borrowed. Carlsson [18] introduced a possibilistic approach to selecting portfolios with highest utility score under the assumption that the returns of assets are trapezoidal fuzzy numbers. Zhang and Wang [19] discussed the general weighted possibilistic portfolio selection problem when the risk-free asset can be lent and short sales are not allowed on all risky assets.

In this paper, we discuss the portfolio selection problem under investing constraints based on the possibilistic theory under the assumption that the returns of assets are fuzzy numbers. Some properties as in probability theory based on the Carlsson and Fullérs' notations are discussed in Section 2. In section 3 , we present a possibilistic portfolio model under investing constrains. Especially, when returns of assets are trapezoidal fuzzy numbers a quadratic programming model is given to replaced Markowitz's mean-variance model, the conventional probabilistic mean-variance model is simplified and extended. A numerical example of a possibilistic fuzzy portfolio selection problem is given to illustrate our proposed effective means and approaches in Secion 4. Finally, we conclude this paper in Section 5.

2 Possibilistic Mean and Variance

Let us introduce some definitions, which we need in the following section. A fuzzy number A is a fuzzy set of the real line $\mathcal{R}$ with a normal, fuzzy convex and continuous membership function of bounded support. The family of fuzzy numbers is denoted by $\mathcal{F}$. Let $A \in \mathcal{F}$ be fuzzy number with $\gamma-$ level set $[A]^\gamma = [a_1(\gamma), a_2(\gamma)](\gamma > 0)$, $\gamma \in [0,1]$. A $\gamma-$ level set of a fuzzy number A is defined by $[A]^\gamma = \{t \in \mathcal{R} | A(t) \geq \gamma\}$ if $\gamma > 0$ and $[A]^\gamma = cl\{t \in \mathcal{R} | A(t) > 0\}$ (the closure of the support of A) if $\gamma = 0$.

In 2001, Carlsson and Fullér [18] defined the notations of the possibilistic mean values and variance of a fuzzy number A as

$$M(A) = \int_0^1 \gamma(a_1(\gamma) + a_2(\gamma))d\gamma = \frac{M^L(A) + M^U(A)}{2},$$

where

$$M^L(A) = \int_0^1 \gamma a_1(\gamma)d\gamma = \frac{\int_0^1 a_1(\gamma)Pos[A \leq a_1(\gamma)]d\gamma}{\int_0^1 Pos[A \leq a_1(\gamma)]d\gamma},$$

$$M^U(A) = \int_0^1 \gamma a_2(\gamma) d\gamma = \frac{\int_0^1 a_2(\gamma) Pos[A \geq a_2(\gamma)] d\gamma}{\int_0^1 Pos[A \geq a_2(\gamma)] d\gamma},$$

where Pos denotes possibility, i.e.

$$Pos[A \leq a_1(\gamma)] = \Pi([-\infty, a_1(\gamma)]) = \gamma,$$

$$Pos[A \geq a_2(\gamma)] = \Pi([a_2(\gamma), \infty]) = \gamma.$$

Then the following lemma can directly be proved using the definition of interval-valued possibilistic mean.

Lemma 2.1. *Let $A_1, A_2, \ldots, A_n$ be n fuzzy numbers, and let λ is a real number. Then*

$$M^L(\sum_{i=1}^n A_i) = \sum_{i=1}^n M^L(A_i),$$

$$M^U(\sum_{i=1}^n A_i) = \sum_{i=1}^n M^U(A_i),$$

$$M^L(\lambda A_i) = \begin{cases} \lambda M^L(A_i) & \text{if } \lambda \geq 0, \\ \lambda M^U(A_i) & \text{if } \lambda < 0. \end{cases}$$

$$M^U(\lambda A_i) = \begin{cases} \lambda M^U(A_i) & \text{if } \lambda \geq 0, \\ \lambda M^L(A_i) & \text{if } \lambda < 0. \end{cases}$$

where the addition and multiplication by a scalar of fuzzy numbers are defined by the sup-min extension principle [20].

The following theorem obviously holds by Lemma 2.1.

Theorem 2.1. *Let $A_1, A_2, \ldots, A_n$ be n fuzzy numbers, and let $\lambda_0, \lambda_1, \lambda_2, \ldots, \lambda_n$ be $n+1$ real numbers. Then*

$$M(\sum_{i=1}^n \lambda_i A_i + \lambda_0) = \lambda_0 + \sum_{i=1}^n \lambda_i M(A_i).$$

The possibilistic variance of fuzzy number A with $[A]^\gamma = [a_1(\gamma), a_2(\gamma)]$ and the possibilistic covariance between fuzzy numbers A and B with $[B]^\gamma = [b_1(\gamma), b_2(\gamma)] (\gamma \in [0,1])$ were, defined as [20]

$$Var(A) = \frac{1}{2} \int_0^1 \gamma (a_2(\gamma) - a_1(\gamma))^2 d\gamma$$

and

$$Cov(A, B) = \frac{1}{2} \int_0^1 \gamma [(a_2(\gamma) - a_1(\gamma))(b_2(\gamma) - b_1(\gamma))] d\gamma,$$

respectively.

The following conclusions are given in [18].

Lemma 2.2. *Let A and B be two fuzzy numbers. Then*

$$Var(\lambda A + \mu B) = \lambda^2 Var(A) + \mu^2 Var(B) + 2|\lambda\mu| Cov(A, B).$$

Theorem 2.2 obviously holds by Lemma 2.2.

Theorem 2.2. *Let $A_1, A_2, \ldots, A_n$ be n fuzzy numbers and let $\lambda_1, \lambda_2, \ldots, \lambda_n$ be n real numbers. Then*

$$Var(\sum_{i=1}^{n} \lambda_i A_i) = \sum_{i=1}^{n} \lambda_i^2 Var(A_i) + 2 \sum_{i<j=1}^{n} |\lambda_i \lambda_j| Cov(A_i, A_j).$$

Let $b_{ij} = Cov(A_i, A_j), i, j = 1, \ldots, n$. Then the matrix

$$\mathbf{Cov} = (b_{ij})_{n \times n}$$

is called as the possibilistic covariance matrix of fuzzy vector $(A_1, A_2, \ldots, A_n)$.

The following theorem shows that **Cov** is nonnegative definite.

Theorem 2.3. *Let $A_1, \ldots, A_n$ be n fuzzy numbers. Then the possibilistic covariance matrix of fuzzy vector $(A_1, A_2, \ldots, A_n)$,* **Cov**, *is nonnegative definite matrix.*

Proof. It follows by the definition of possibilistic covariance that

$$Cov(A_i, A_j) = Cov(A_j, A_i), i, j = 1, \ldots, n.$$

Therefore, **Cov** is a real symmetric matrix.

In particular,

$$b_{ii} = Cov(A_i, A_i) = Var(A_i), i = 1, 2, \ldots, n.$$

Let $[A_i]^\gamma = [a_{i1}(\gamma), a_{i2}(\gamma)], i = 1, \ldots, n$, then for any $t_i \in \mathcal{R}, i = 1, \ldots, n$, it follows that

$$\begin{aligned}
\sum_{i,j=1}^{n} b_{ij} t_i t_j &= 2 \sum_{i,j=1}^{n} \int_0^1 \gamma [a_{i2}(\gamma) - a_{i1}(\gamma)][a_{j2}(\gamma) - a_{j1}(\gamma)] t_i t_j d\gamma \\
&= 2 \int_0^1 \gamma \sum_{i=1}^{n} [a_{i2}(\gamma) - a_{i1}(\gamma)] t_i \sum_{j=1}^{n} [a_{j2}(\gamma) - a_{j1}(\gamma)] t_j d\gamma \\
&= 2 \int_0^1 \gamma [\sum_{i=1}^{n} t_i (a_{i2}(\gamma) - a_{i1}(\gamma)]^2 d\gamma \geq 0.
\end{aligned}$$

Thus, **Cov** is a nonnegative definite matrix.

This concludes the proof of the theorem.

3 A Fuzzy Portfolio Selection Model Under Investing Constraints

It is well-known that the financial market is affected by many non-probabilistic factors. In a fuzzy uncertain economic environment, the future states of returns and risks of risky assets cannot be predicted accurately. Based on this fact, a portfolio selection problem under the assumption that $r_i, i = 1, \ldots, n$ are considered as n fuzzy numbers is useful to describe an uncertain investment environment with vagueness and ambiguity. The return associated with the portfolio $\mathbf{x} = (x_1, x_2, \ldots, x_n)'$ is

$$r = \sum_{i=1}^{n} x_i r_i = \mathbf{r}'\mathbf{x}.$$

From theorem 2.1, the possibilistic mean value of r are given by

$$M(r) = \sum_{i=1}^{n} M(x_i r_i) = \sum_{i=1}^{n} x_i M(r_i).$$

From theorem 2.3, the possibilistic variance of r are given by

$$Var(r) = \sum_{i=1}^{n} x_i^2 Var(r_i) + 2 \sum_{i>j=1}^{n} x_i x_j Cov(r_i, r_j).$$

In order to describe conventiently, we introduce the following notations:

$$\begin{aligned} \mathbf{M} &= (M(r_1), M(r_2), \ldots, M(r_n))', \\ \mathbf{Cov} &= (Cov(r_i, r_j))'_{n\times n}, \\ \mathbf{F} &= (1, 1, \ldots, 1)'. \end{aligned}$$

Thus, the possibilistic mean value and variance of r can be rewritten as

$$M(r) = \mathbf{M}'\mathbf{x},$$

$$Var(r) = \mathbf{x}'\mathbf{Cov}\mathbf{x}.$$

Analogous to Markowitz's mean-variance methodology for the portfolio selection problem, the possibilistic mean value correspond to the return, while the possibilistic variance correspond to the risk. From this point of view, the possibilistic mean-variance model of portfolio selection can be formulated as

$$\begin{aligned} \min \quad & \mathbf{x}'\mathbf{Cov}\mathbf{x} \\ s.t. \quad & \mathbf{M}'\mathbf{x} \geq \mu, \\ & \mathbf{F}'\mathbf{x} = 1, \\ & \mathbf{x} \geq 0. \end{aligned} \tag{1}$$

From [17], the optimal solution of (1), $\mathbf{x}^*$, is called as the possibilistic efficient portfolio. The possibilistic efficient portfolios for all possible μ construct the possibilistic efficient frontier. Solving (1) for all possible μ, the possibilistic efficient frontier is derived explicitly.

In the following section, we assume that $r_i (i = 1, \ldots, n)$ are trapezoidal numbers.

A fuzzy number A is called trapezoidal with tolerance interval $[a, b]$, left width α and right width β if its membership function has the following form:

$$A(s) = \begin{cases} 1 - \frac{a-s}{\alpha} & \text{if } a - \alpha \leq s \leq a, \\ 1 & \text{if } a \leq s \leq b, \\ 1 - \frac{s-b}{\beta} & \text{if } b \leq s \leq b + \beta, \\ 0 & \text{otherwise.} \end{cases}$$

and we use the notation $A = (a, b, \alpha, \beta)$. It can easily be shown that

$$[A]^\gamma = [a - (1-\gamma)\alpha, b + (1-\gamma)\beta], \forall \gamma \in [0, 1].$$

Here, we assume that $r_i (i = 1, \ldots, n)$ are trapezoidal numbers with tolerance interval $[a_i, b_i]$, left width $\alpha_i > 0$ and right width $\beta_i > 0$. A $\gamma-$ level sets of r_i is computed by

$$[r_i]^\gamma = [a_i - (1-\gamma)\alpha_i, b_i + (1-\gamma)\beta_i],$$

for all $\gamma \in [0.1], i = 1, \ldots, n$.

It is easy to get that

$$\begin{aligned} M(r_i) &= \int_0^1 \gamma(a_i - (1-\gamma)\alpha_i + b_i + (1-\gamma)\beta_i) d\gamma \\ &= \frac{a_i + b_i}{2} + \frac{\beta_i - \alpha_i}{6}, \end{aligned}$$

$$\begin{aligned} Var(r_i) &= \frac{(b_i - a_i)^2}{4} + \frac{(b_i - a_i)(\alpha_i + \beta_i)}{6} + \frac{(\alpha_i + \beta_i)^2}{24} \\ &= [\frac{b_i - a_i}{2} + \frac{\alpha_i + \beta_i}{6}]^2 + \frac{(\alpha_i + \beta_i)^2}{72}. \end{aligned}$$

By Theorem 2.1, the possibilistic means of the return associated with the portfolio $(x_1, x_2, \ldots, x_n)$ is given by

$$M(\sum_{i=1}^n x_i r_i) = \sum_{i=1}^n \frac{1}{2}[(a_i + b_i) + \frac{\beta_i - \alpha_i}{3}] x_i$$

By Theorem 2.2, the possibilistic variance of r is given by

$$Var(\sum_{i=1}^n x_i r_i) = (\sum_{i=1}^n \frac{1}{2}([b_i - a_i + \frac{1}{3}(\alpha_i + \beta_i)] x_i)^2 + \frac{1}{72}[\sum_{i=1}^n (\alpha_i + \beta_i) x_i]^2.$$

Thus, the model (1) is equal to the following model:

$$\begin{aligned} \min \quad & (\sum_{i=1}^n \frac{1}{2}[b_i - a_i + \frac{1}{3}(\alpha_i + \beta_i)] x_i)^2 + \frac{1}{72}[\sum_{i=1}^n (\alpha_i + \beta_i) x_i]^2 \\ s.t. \quad & \sum_{i=1}^n \frac{1}{2}[(a_i + b_i) + \frac{\beta_i - \alpha_i}{3}] x_i \geq \mu, \\ & \mathbf{F}'\mathbf{x} = 1, \\ & \mathbf{x} \geq 0. \end{aligned} \tag{2}$$

Noted that quantity constraints are also very important for constructing portfolios in reality. The quantity of each asset i that is included in the portfolio is limited within a given interval. Specifically, a minimum l_i and a maximum u_i for each asset i are given, and we impose that either $x_i = 0$ or $l_i \leq x_i \leq u_i$. So, the possibilistic portfolio model with quantity constrains can be formulated as follows:

$$\begin{aligned} \min \quad & (\sum_{i=1}^{n} \frac{1}{2}[b_i - a_i + \frac{1}{3}(\alpha_i + \beta_i)]x_i)^2 + \frac{1}{72}[\sum_{i=1}^{n}(\alpha_i + \beta_i)x_i]^2 \\ s.t. \quad & \sum_{i=1}^{n} \frac{1}{2}[(a_i + b_i) + \frac{\beta_i - \alpha_i}{3}]x_i \geq \mu \\ & \mathbf{F}'\mathbf{x} = 1 \\ & \mathbf{l} \leq \mathbf{x} \leq \mathbf{u} \end{aligned} \quad (3)$$

where $\mathbf{l} = (l_1, l_2, \ldots, l_n)'$, $\mathbf{u} = (u_1, u_2, \ldots, u_n)'$.

4 Numerical Example

In order to illustrate our proposed effective means and approaches , we consider a real portfolio selection example. We select five stocks from Shenzhen Stock Exchange. Original data come from every week's closed prices from April 2002 to January 2004. Based on the historical data and the corporations' financial reports and the future information, we can estimate their returns with the following possibility distributions:

Table 1. Security Possibilitic Return

Stock	a_i	b_i	α_i	β_i
1	0.073	0.093	0.054	0.087
2	0.085	0.115	0.075	0.102
3	0.108	0.138	0.096	0.123
4	0.128	0.168	0.126	0.162
5	0.158	0.208	0.168	0.213

According to model (2), we can compute out the coefficients of the possibilistic efficient portfolio selection model as follows.

We try to solve our example by Matlab language, after computing model (2) and (3) we summarize the numerical results in Tables 3 – 4.

From Table 3 and Table 4, we conclude that, for the same expected return, the portfolio without quantity constrains has more investment risk than that with quantity constrains, it is because that investor must pay more additional costs under quantity constrains.

In addition, comparing Table 3 and Table 4, it is clear that in the case of $u = 0.05$ the posibilistic portfolio without quantity constrains only has two

Table 2. The coefficients of the possibilistic efficient portfolio selection model

Stock	$[(a_i + b_i) + \frac{\beta_i - \alpha_i}{3}]$	$[b_i - a_i + \frac{1}{3}(\alpha_i + \beta_i)]$	$\alpha_i + \beta_i$
1	0.1770	0.0670	0.1410
2	0.2090	0.0890	0.1770
3	0.2550	0.1030	0.2190
4	0.3080	0.1360	0.2880
5	0.3810	0.1770	0.3880

Table 3. Investment proportion with quantity constranins, l=[0 0 0 0 0], u=[1 1 1 1 1]

Expected return (μ)	1	2	3	4	5	Portfolio risk
0.05	0.7160	0.2840	0.0000	0.0000	0.0000	0.0017
0.10	0.4816	0.3949	0.1097	0.0138	0.0000	0.0020

Table 4. Investment proportion with quantity constranins, l=[0.08 0.05 0.15 0.12 0.05], u=[0.50 0.45 0.35 0.85 0.65]

Expected return (μ)	1	2	3	4	5	Portfolio risk
0.05	0.4539	0.2261	0.1500	0.1200	0.0500	0.0026
0.10	0.6300	0.0200	0.1000	0.0500	0.2000	0.0029

security while that with quantity constrains has five securities, at same time, in the case of $u = 0.10$ the posibilistic portfolio without quantity constrains has four securities while that with quantity constrains has five securities. It means that portfolio with quantity constrains tends to take more distributive investment than that without quantity constrains.

5 Conclusions

Since the returns of risky assets are in a fuzzy uncertain economic environment and vary from time to time, the fuzzy number is a power tool used to describe an uncertain environment with vagueness and ambiguity. In this paper, We discuss the portfolio selection problem with investing constraints based on the possibilistic theory under the assumption that the returns of assets are fuzzy numbers. The possibilistic mean value of the return is termed measure of investment return and the possibilistic variance of the return is termed measure of investment risk. We further present a quadratic programming model replaced Markowitz's mean-variance model when returns of assets are trapezoidal fuzzy numbers, the conventional probabilistic mean-variance model is simplified and extended. A numerical example of a possibilistic fuzzy portfolio selection problem is given to illustrate our proposed efective means and approaches.

References

1. Markowitz H (1952) Portfolio selection. Journal of Finance 7:77-91.
2. Markowitz H (1959) Portfolio selection: Efficient Diversification of Investments. Wiley, New York.
3. Sharpe WF(1970) Portfolio theory and capital markets. McGraw-Hill, New York.
4. Merton RC (1972) An analytic derivation of the efficient frontier. Journal of Finance and Quantitative Analysis, September 1851-1872.
5. Perold AF (1984) Large-scale portfolio optimization. Management Science 30:1143-1160.
6. Pang JS (1980) A new efficient algorithm for a class of portfolio selection problems. Operational Research 28: 754-767.
7. Vörös J (1986) Portfolio analysis-An analytic derivation of the efficient portfolio frontier. European Journal of Operational Research 203: 294-300.
8. Best MJ, Hlouskova J (2000) The efficient frontier for bounded assets. Math.Meth.Oper.Res 52:195-212.
9. Watada J (1997) Fuzzy portfolio selection and its applications to decision making. Tatra Mountains Mathematical Publication 13:219-248.
10. Inuiguchi M, Tanino T (2000) Portfolio selection under independent possibilistic information. Fuzzy Sets and Systems 115: 83-92.
11. Ramaswamy S (1998) Portfolio selection using fuzzy decision theory. Working Paper of Bank for international Settlements 59.
12. León, Liern V, Vercher E (2002) Viability of infeasible portfolio selection problems: A fuzzy approach. European Journal of Operational Research 139: 178-189.
13. Wang SY , Zhu SS (2002) On fuzzy portfolio selection problem. Fuzzy Optimization and Decision Making 1: 361-377.
14. Tanaka H, Guo P (1999) Portfolio selection based on upper and lower exponential possibility distributions. European Journal of Operational Research 114: 115-126.
15. Tanaka H, Guo P, Türksen IB (2000) Portfolio selection based on fuzzy probabilities and possibility distributions. Fuzzy sets and systems 111: 387-397.
16. Zhang WG, Nie ZK (2004) On admissible efficient portfolio selection problem. Applied mathematics and computation 159: 357-371.
17. Zhang WG, Liu WA , Wang YL (2006) On admissible efficient portfolio selection problem: Models and algorithms. Applied mathematics and computation 176: 208-218.
18. Carlsson C, Fullér R, Majlender P (2002) A possibilistic approach to selecting portfolios with highest utility score. Fuzzy Sets and Systems 131: 13-21.
19. Zhang WG, Wang YL (2005) Portfolio selection: Possibilistic mean-variance model and possibilistic efficient frontier. Lecture Notes in Computer Science 3521: 203-213.
20. Zadeh LA (1965) Fuzzy sets. Inform. and Control 8: 338-353.

The Interaction Among Players in Fuzzy Games

Shu-jin Li[1] and Qiang Zhang[2]

[1] School of Management and Economics, Beijing Institute of Technology, Beijing 100081, China; The Central Institute for Correctional Police, Baoding 071000, China
lishugold@yahoo.com.cn
[2] School of Management and Economics, Beijing Institute of Technology, Beijing 100081, China
qiangzhang@bit.edu.cn

Abstract. In this paper, the derivative among levels is defined which characterize the interaction among participation levels in fuzzy coalition. At the same time, the concept of independence among levels based on a coalition is established. Based on these concepts and the assumption that fuzzy characteristic functions concerned are continuous, the (absolute) interaction among levels with respect to other players is defined, then the (absolute) interaction among players with respect to other players is defined and the properties of absolute interaction among players are proved as well.

Keywords: Fuzzy game, Cooperative game, Interaction,Independence.

1 Introduction

The interaction phenomena among players in crisp cooperative games has been paid more attention to in recent years. Owen [1] proposed the index of interaction, which is in terms of the situation of superadditive games. Subsequently, more researchers devote to the study of interaction phenomena among elements(players) of coalition, Murofushi and Soneda [2], Grabisch [3] and Grabisch and Rubens [4] proposed the Shapley interaction index and Banzhaf interaction index in their study of multicriteria decision making. As interaction index can not describe the independence among elements(players) in coalition effectively, Ivan Kojadinovic proposed a concept of marginal amount of interaction [5], furthermore he proposed the axiomatic characterizations of the Shapley interaction index and Banzhaf interaction index in 2005 [6]. The interaction phenomena in crisp cooperative games has been described in detail, but in fuzzy cooperative games, how the phenomena of interaction and independence should be modeled among players? For this purpose, in this paper, the concepts of -derivative and independence among levels are defined, then the notions of interaction and independence among players are established, and some properties of interaction among players are proved as well.

B.-Y. Cao (Ed.): Fuzzy Information and Engineering (ICFIE), ASC 40, pp. 573–582, 2007.
springerlink.com

2 Interaction for Crisp Cooperative Games

Let $N = \{1, 2, \cdots, n\}$ be a nonempty set of players, considering possibilities of cooperation. This kind of situation are modeled in [7] by means of games in characteristic function form. A cooperative game with player set N is defined by a set function $w : 2^N \to R$, assigning to each group of players(coalition) $S \subseteq N$, its worth $w(S)$ obtained as a result of cooperation; conventionally, it is assumed that $w(\phi) = 0$, we write 2^N to denote the power set of N, each nonempty element of 2^N is called a coalition, the set of cooperative crisp games with player set N is denoted by G^N .

For simplifying notations, we will write $W(i), S \cup i$ instead of $W(\{i\}), S \cup \{i\}$, furthermore, we write ij, ijk instead of $\{ij\}, \{ijk\}$, for example $S \cup ijk$. We denote $S \setminus T$ the set difference of S and T. Cardinality of sets $S, T, \cdots$ will be denoted by $|S|, |T|, \cdots$ or $s, t, \cdots$.

Definition 2.1. Given a game $w \in G^N$ and $S, T \subseteq 2^N$, we denote by $\delta_S w(T)$ the S-derivative of w in the presence of T, which recursively is defined by $\delta_i w(T) = w(T \cup i) - w(T \setminus i), \forall i \in N$ and $\delta_S w(T) = \delta_i(\delta_{S \setminus i} w(T)), \forall i \in S$ With convention $\delta_\phi w(T) = w(T)$. By induction on S , it is easy to prove

$$\delta_S w(T) = \delta_S w(T \setminus S) = \sum_{L \subseteq S} (-1)^l w((T \cup S) \setminus L), \forall S, T \subseteq N$$

Then we have

$$\delta_S w(T) = \sum_{L \subseteq S} (-1)^{s-l} w(T \cup L), \forall S \subseteq N, T \subseteq N \setminus S$$

Definition 2.2. Given a game $w \in G^N$, players of a coalition $S(|S| \geq 2)$ are said to be marginally mutually independent in the presence of $T(T \subseteq N \setminus S)$ when, for any $K \subseteq S, |K| \geq 2, \delta_K w(T) = 0$.

Proposition 2.1. Given a game $w \in G^N$, if the players of coalition $S(|S| \geq 2)$ are marginally mutually independent in the presence of $T(T \subseteq N \setminus S)$, then for any $K \subseteq S, w(T \cup K) = \sum_{i \in K} w(T \cup i) - (|K| - 1)w(T)$.

Proof. We can prove the proposition by induction.

It is clear the equality is true when $|K| = 0$ or 1. If $|K| = 2$, no loss of generality, we suppose $K = \{1, 2\} \subseteq S$, then for $T \subseteq N \setminus S$, from the definition 2.2, we have $\delta_{12} w(T) = 0$, i.e.

$$w(T \cup 12) - w(T \cup 1) - w(T \cup 2) + w(T) = 0$$

$$w(T \cup 12) = w(T \cup 1) + w(T \cup 2) - w(T)$$

Now we prove the equality is true for any $K \subseteq S$.
Assume the equality is true for any $K' \subseteq S, |K'| \leq |K|$. From the definition of marginally mutual independence, we have $\delta_K w(T) = 0$ for any $K \subseteq S$, that is

$$\sum_{L \subseteq K} (-1)^{|K|-|L|} w(T \cup L) = 0 (from\ definition\ 2.1)$$

then

$$w(T \cup K) = \sum_{L \subseteq K, L \neq K} (-1)^{|K|-|L|-1} w(T \cup L)$$

By induction hypothesis, for any $|L| < |K|, w(T \cup L) = \sum_{i \in L} w(T \cup i) - (|L| - 1)w(T)$, then
$w(T \cup K) = \sum_{L \subseteq K, L \neq K} (-1)^{|K|-|L|-1} (\sum_{i \in L} w(T \cup i) - (|L| - 1)w(T))$
$= \sum_{L \subseteq K, L \neq K} (-1)^{|K|-|L|-1} \sum_{i \in L} w(T \cup i) - \sum_{L \subseteq K, L \neq K} (-1)^{|K|-|L|-1} (|L| - 1)w(T) = \sum_{i \in K} w(T \cup i) - (|K| - 1)w(T)$

3 Interaction Among Players in Games with Fuzzy Coalitions

3.1 Fuzzy Coalitions and LP−Derivative

Fuzzy coalitions are introduced to n -person cooperative games by Aubin[9,10] and Butnareu[11,12], then the fuzziness in formation of coalitions can be taken into account in n -person cooperative games, In games with fuzzy coalitions, players belonging to a coalition do not always transfer their rights of decision making to the coalition completely, they take action only according to the level of participation.

Definition 3.1. A fuzzy coalition variable of player set N is a n-dimension variable $x = (x_1, x_2, \cdots, x_n)$ in $[0,1]^N$, where the ith coordinate x_i is referred to as the level variable of player i.

Let $x_i = s_i (i = 1, \cdots, n)$, then we call $s = (s_1, s_2, \cdots, s_n)$ a fuzzy coalition, if there is no fear of confusion, we also call x fuzzy coalition.

In this section, we denote $N, M, \cdots$, the sets of players; $Lx = \{x_1, x_2, \cdots, x_n\}$ the set of level variable; $s, t, \cdots$ the fuzzy coalition; $LP, LT, \cdots$ the subsets of Lx; $id(LP)$ the set of indicator of LP, i.e. $id(LP) = \{i | x_i \in LP\}$; $|N|, |M|, \cdots$ the cardinality of $N, M, \cdots$; $|LP|, |LT|, \cdots$ the cardinality of $LP, LT, \cdots$. the empty coalition in a fuzzy setting is $\mathbf{0} = e^{\phi} = (0, 0, \cdots, 0)$, and e^S, with $S \in 2^N$, denotes a crisp-like coalition, e^S is called a crisp-like coalition because it corresponds to the situation where the players within S fully cooperate(i.e. they have participation level 1) and the players outside S are not involved at all(i.e. they have participation level 0). $e^N = (1, 1, \cdots, 1)$ is called the grand coalition. We write e^i instead of $e^{\{i\}}$, the set of fuzzy coalitions is denoted by F^N or $[0,1]^N$.

Definition 3.2. A function v which associates a fuzzy coalition $t \in [0,1]^N$ with a real number $v(t)$ is called a fuzzy characteristic function. with $v(\mathbf{0}) = 0$. Here $\mathbf{0} = (0, 0, \cdots, 0)$, $v(t)$ represents the total amount of a side-payment (transferable utility) that a fuzzy coalition t could earn according to the rate of participation of every member in N, n-person cooperative games with fuzzy coalitions are represented by a pair (N, v), the set of fuzzy games with player set N is denoted by FG^N.

Definition 3.3. A function $v : [0,1]^N \to R$ is called a supermodular function on $[0,1]^N$ if $v(s \vee t) + v(s \wedge t) \geq v(s) + v(t)$ for all $s, t \in [0,1]^N$.

Where $s \vee t$ and $s \wedge t$ are those elements of $[0,1]^N$ with the ith coordinate equal to $max\{s_i, t_i\}$ and $min\{s_i, t_i\}$ respectively. The set operations $\vee$ and $\wedge$ play the same role for the fuzzy coalitions as the union and intersection for crisp coalitions. For $s, t \in [0,1]^N$ we use the notation $s < t$ iff $s_i < t_i$ for each $i \in N$.

Given fuzzy coalition variable $x = (x_1, x_2, \cdots, x_n)$, we call $x^{\#}$ the set of fuzzy coalition variable created by x, which is defined by

$$x^{\#} = \{t | t \leq x, and \;\; t_i = x_i \;\; or \;\; t_i = 0 \;\; for \;\; each \;\; i \in N\}$$

Given fuzzy coalition $x = (x_1, x_2, \cdots, x_n)$, the interaction phenomena among level variable of x can be modeled in a similar way to that of crisp cooperative game.

Definition 3.4. We call $\delta_{x_i} v(t)$ the x_i-derivative of v in the presence of $t(t \in (x - x_i e^i)^{\#})$, which is defined by $\delta_{x_i} v(t) = v(t \vee x_i e^i) - v(t)$

The x_i-derivative of v in the presence of t is a function, when $x = s$, we call $\delta_{x_i} v(t)|_{x=s}$ the x_i derivative value of v based on s. If there is no fear of confusion, we call both the derivative function and the derivative value the derivative.

Definition 3.5. We call $\delta_{x_i x_j} v(t)$ the $x_i x_j$-derivative of v in the presence of $t(t \in (x - x_i e^i - x_j e^j)^{\#})$, which is defined by

$$\delta_{x_i x_j} v(t) = v(t \vee x_i e^i \vee x_j e^j) - v(t \vee x_i e^i) - v(t \vee x_j e^j) + v(t) \tag{1}$$

Accordingly, $\delta_{x_i x_j} v(t)|_{x=s}$ is the x_i, x_j-derivative(value) of v in the presence of t based on s.

When $\delta_{x_i x_j} v(t)|_{x=s} > 0$, We say there exists positive interaction between levels x_i and x_j in the presence of t based on s; $\delta_{x_i x_j} v(t)|_{x=s} < 0$ means that there exists negative interaction between levels x_i and x_j in the presence of t based on s, and $\delta_{x_i x_j} v(t)|_{x=s} = 0$ means there is no interaction between levels x_i and x_j in the presence of t based on s.

Definition 3.6. Generally, We call $\delta_{LP} v(t)$ the LP-derivative of v in the presence of $t(t \in (x - \sum_{x_i \in LP} x_i e^i)^{\#})$, which is defined by $\delta_{LP} v(t) = \delta_{x_i}(\delta_{LP \setminus x_i} v(t))$, where LP is a subset of Lx.

By definition 3.6, it is easy to prove, for any $LP \subseteq Lx, t = \sum_{x_i \in LP} x_i e^i, LT \subseteq Lx \setminus LP$

$$\delta_{LP} v(t) = \sum_{LR \subseteq LP} (-1)^{|LP| - |LR|} v(t \bigvee_{x_i \in LR} x_i e^i) \tag{2}$$

When $x = s$, we call $\delta_{LP} v(t)|_{x=s}$ the LP-derivative of v in the presence of t based on s.

From definition 3.6, it is easy to verify that $\delta_{LP} v(t) = 0$ when $x_i = 0 (x_i \in LP)$, which means that for any $t = \sum_{x_i \in LT} x_i e^i, LT \subseteq Lx \setminus LP$, there is no simultaneous interaction among levels of LP. This can be interpreted intuitively, when a player of N does not participate in a coalition, apparently, he can not affect others, that is, there exists no simultaneous interaction between the player and others.

When $\delta_{LP}v(t)|_{x=s} > 0(< 0)$, We say there exists positive(negative) simultaneous interaction among levels of LP in the presence of T based on S; $\delta_{LP}v(t)|_{x=s} = 0$ means there is no simultaneous interaction among levels of LP in the presence of t based on s.

Proposition 3.1. Given $v \in FG^N$ is a supermodular, s, s' are two fuzzy coalitions, satisfying $s_i \leq s'_i, s_j \leq s'_j$ and for any $k \in N \setminus ij$, $s_k = s'_k$. Then for any $t(t \in (x - x_i e^i - x_j e^j)^{\#})$, $\delta_{x_i x_j} v(t)|_{x=s'} \geq \delta_{x_i x_j} v(t)|_{x=s}$.

Proof. Suppose $s = \{s_1, \cdots s_i, \cdots s_j, \cdots s_n\}, s' = \{s_1, \cdots s'_i, \cdots s'_j, \cdots s_n\}$, let $s^1 = \{s_1, \cdots s'_i, \cdots s_j, \cdots s_n\}$.

First we prove $\delta_{x_i x_j} v(t)|_{x=s^1} \geq \delta_{x_i x_j} v(t)|_{x=s}$. From the definition of $x_i x_j$ -derivative, for any $t \in (x - x_i e^i - x_j e^j)^{\#}$, we need only to prove

$$\begin{aligned} &v(t \vee s'_i e^i \vee s_j e^j) - v(t \vee s'_i e^i) - v(t \vee s_j e^j) + v(t) \\ &\geq v(t \vee s_i e^i \vee s_j e^j) - v(t \vee s_i e^i) - v(t \vee s_j e^j) + v(t) \end{aligned}$$

That is we need to prove

$$v(t \vee s'_i e^i \vee s_j e^j) - v(t \vee s'_i e^i) \geq v(t \vee s_i e^i \vee s_j e^j) - v(t \vee s_i e^i). \tag{3}$$

Let $\tau_1 = t \vee s_i e^i \vee s_j e^j, \tau_2 = t \vee s'_i e^i$, because v is a supermodular, then there is

$$v(\tau_1 \vee \tau_2) + v(\tau_1 \wedge \tau_2) \geq v(\tau_1) + v(\tau_2)$$

Inserting the expressions of τ_1 and τ_2, the inequality of (3) is easily to be obtained.

In a similar way we can prove $\delta_{x_i x_j} v(t)|_{x=s'} \geq \delta_{x_i x_j} v(t)|_{x=s^1}$. Hence $\delta_{x_i x_j} v(t)|_{x=s'} \geq \delta_{x_i x_j} v(t)|_{x=s}$.

3.2 Mutual Independence Among Levels Based on s

Given a fuzzy coalition s, LP is a subset of Lx, $t \in (x - \sum_{x_i \in LP} x_i e^i)^{\#}$, We have indicated above that $\delta_{LP}v(t)|_{x=s} = 0$ implies there does not exist simultaneous interaction among levels of LP in the presence of t based on s, but this does not imply that the levels of LP are independent based on s. In the following, a definition of marginal independence among levels based on s is proposed.

Definition 3.7. The levels of $LP(LP \subseteq Lx)$ are called marginally mutual independent in the presence of t based on fuzzy coalition s if for any $LK \subseteq LP(|LK| \geq 2)$, there is $\delta_{LK}v(t)|_{x=s} = 0$. here $t \in (x - \sum_{x_i \in LP} x_i e^i)^{\#}$.

Proposition 3.2. Given a game $v \in FG^N$ and $s \in F^N, LP \subseteq Lx$, if the levels of LP are marginally mutually independent in the presence of $t(t \in (x - \sum_{x_i \in LP} x_i e^i)^{\#})$ based on s, then for any $LK \subseteq LP$,

$$v(t \bigvee_{x_i \in LK} x_i e^i) = \sum_{x_i \in LK} v(t \vee x_i e^i) - (|LK| - 1)v(t)$$

Proof. The proof of this proposition 3.2 is similar to that of proposition 2.1 and therefore it is omitted.

3.3 Interaction Among Levels of $LP(|LP| \geq 2)$ with Respect to Players of M

The interaction phenomena among levels based on a certain fuzzy coalition can be modeled similar to that of the crisp games. however,how the interaction among levels of some players with respect to other players should be modeled completely? and how the independence among levels should be characterized when the participation levels of others change from 0 to 1? In this section, we should focus on these problems. In the following, we always assume that the characteristic function of fuzzy coalitions are continuous with respect to level variables.

Definition 3.8. The levels of $LP(|LP| \geq 2)$ are called mutually independent in $x_i = s_i(x_i \in LP)$ with respect to players $M \subseteq N \backslash id(LP)$ if $\delta_{LK} v(t)|_{x_i=s_i, i \in id(LK)} = 0$ for any $t \leq \sum_{k \in M} e^k$ and any $LK \subseteq LP$.

From definition 3.8, it is easily to be seen, If the levels of LP are mutually independent in $x_i = s_i(x_i \in LP)$ with respect to players $M \subseteq N \setminus id(LP)$, then they are mutually independent with respect to any subsets of M. On the other hand, if the levels of $LP(|LP| \geq 2)$ are mutually independent in $x_i = s_i(x_i \in LP)$ with respect to players $M \subseteq N \backslash id(LP)$, then the levels of $LP' \subseteq LP(|LP'| \geq 2)$ are mutually independent in $x_i = s_i(x_i \in LP')$ with respect to players M.

Definition 3.9. The interaction among levels of $LP(|LP| \geq 2$ with respect to players $M \subseteq N \setminus id(LP)$ is defined by

$$B(v, LP, M) = \int_{[0,1]^m} \delta_{LP} v(t) dLM \tag{4}$$

And the absolute interaction among levels of LP with respect to players $M \subseteq N \setminus id(LP)$ is defined by

$$AB(v, LP, M) = \int_{[0,1]^m} |\delta_{LP} v(t)| dLM \tag{5}$$

here $t = \sum_{k \in M} x_k e^k, dLM = \prod_{k \in M} dx_k, |M| = m$.

$B(v, LP, M)$ describe the accumulation of simultaneous interaction among levels of LP when the level of each player of M changes from 0 to 1. For $x_i = s_i(x_i \in LP)$, $B(v, LP, M)|_{x_i=s_i(x_i \in LP)}$ is a real number, which measures the accumulation of simultaneous interaction among $x_i = s_i(x_i \in LP)$ when $x_i(i \in M)$ change from 0 to 1.

$AB(v, LP, M)$ describe the accumulation of the simultaneous absolute interaction among levels of LP with respect to players of M when the level of each player of M change from 0 to 1. When $AB(v, LP, M)|_{x_i=s_i(x_i \in LP)} = 0$, it means that $\delta_{LP} v(t) = 0$ for any $t \leq \sum_{k \in M} e^k$, i.e. there is no simultaneous interaction among $x_i = s_i(x_i \in LP)$ in the presence of any $t \leq \sum_{k \in M} e^k$.

3.4 The Interaction Among Players of $P(|P| \geq 2)$

The interaction phenomena among levels of LP have been studied above, now a problem arises: how shall we characterize the mutual independence among players of $P(|P| \geq 2)$? Which index can be used to express the independence among players of $P(|P| \geq 2)$ effectively? in this section, we should give an answer to this question.

Definition 3.10. The players of $P(|P| \geq 2)$ are called mutually independent with respect to players of $M \subseteq N \setminus P$, if for any $s \leq \sum_{i\in M} e^i + \sum_{j\in P} e^j$ and any $t \in (x - \sum_{i\in P} x_i e^i)^{\#}$, $\delta_{LP_0} v(t)|_{x=s} = 0$. here $LP_0 \subseteq LP = \{x_i | i \in P\}$.

From the definition above, we can see that if the players of $P(|P| \geq 2)$ are mutually independent with respect to players of M, then they are mutually independent with respect to any subsets of M. On the other hand, if the players of $P(|P| \geq 2)$ are mutually independent with respect to players M, then the players of $P_0(P_0 \subseteq P, |P_0| \geq 2)$ are mutually independent with respect to players M.

In the following, the concepts of dull player and substitute player for fuzzy cooperative games are defined.

Definition 3.11. Given $v \in FG^N$, x is fuzzy coalition variable, player i is called a dull player of v if $v(x) = v(x - x_i e^i) + v(x_i e^i)$.

Definition 3.12. Given $v \in FG^N$, players i, j are called two substitute players if for any $a \in [0, 1]$, any $t \leq e^N - e^i - e^j, v(t \vee ae^i) = v(t \vee ae^j)$.

Definition 3.13. The interaction among players of $P(|P| \geq 2)$ with respect to players $M \subseteq N \setminus P$ is defined by

$$B(v, P, M) = \int_{[0,1]^p} B(v, LP, M) dLP \tag{6}$$

i.e. $B(v, P, M) = \int_{[0,1]^p} \int_{[0,1]^m} \delta_{LP} v(t) dLM dLP$ when $p \geq 2$.

The absolute interaction among players of $P(|P| \geq 2)$ with respect to players $M \subseteq N \setminus P$ is defined by

$$AB(v, P, M) = \int_{[0,1]^p} AB(v, LP, M) dLP \tag{7}$$

i.e. $AB(v, P, M) = \int_{[0,1]^p} \int_{[0,1]^m} |\delta_{LP} v(t)| dLM dLP$ when $p \geq 2$. Here $t = \sum_{k\in M} x_k e^k, dLM = \prod_{k\in M} dx_k, dLP = \prod_{k\in P} dx_k, |M| = m, |P| = p$. $B(v, P, M)$ is the integral of $B(v, LP, M)$ when the levels of LP change from 0 to 1. It can be regarded as a measure of the simultaneous interaction among players P with respect to players M. However, the independence among players of P with respect to players M can not be characterized by $B(v, P, M) = 0$.

$AB(v, P, M)$ describes the integral of $AB(v, LP, M)$ when each $x_i \in LP$ changes from 0 to 1. When $AB(v, P, M) = 0$, it is clear that $\delta_{LP} v(t)|_{x=s} = 0$ for any $s \leq \sum_{i\in M} e^i + \sum_{j\in P} e^j$, here $t = \sum_{k\in M} x_k e^k$. In this case , we say there is no simultaneous interaction among players of P with respect to player M.

Proposition 3.3. If i is a dull players in fuzzy cooperative game, then $AB(v, P\cup i, M) = 0$ for any $M \subseteq N \setminus (P \cup i)$.

Proof. From definition 3.13, we have

$$AB(v, P\cup i, M) = \int_{[0,1]^{p+1}} AB(v, LP\cup x_i, M)dx_i dLP$$
$$= \int_{[0,1]^{p+1}} \int_{[0,1]^m} |\delta_{LP\cup x_i} v(t)| dLM dx_i dLP,$$

By definition 3.6

$\delta_{LP\cup x_i} v(t) = \sum_{LR\subseteq LP\cup x_i} (-1)^{|LP|+1-|LR|} v(t \bigvee_{x_k\in LR} x_k e^k)$
$= \sum_{LR\subseteq LP} (-1)^{|LP|+1-|LR|}$
$v(t \bigvee_{x_k\in LR} x_k e^k) + \sum_{LR\subseteq LP} (-1)^{|LP|+1-|LR|-1} v(t \bigvee_{x_k\in LR} x_k e^k \vee x_i e^i)$
$= \sum_{LR\subseteq LP} (-1)^{|LP|+1-|LR|} v(t \bigvee_{x_k\in LR} x_k e^k) + \sum_{LR\subseteq LP} (-1)^{|LP|+1-|LR|-1}$
$(v(t \bigvee_{x_k\in LR} x_k e^k) + v(x_i e^i))(by\ definition\ 3.11)$
$= \sum_{LR\subseteq LP} (-1)^{|LP|+1-|LR|} v(t \bigvee_{x_k\in LR} x_k e^k) + \sum_{LR\subseteq LP} (-1)^{|LP|-|LR|}$
$v(t \bigvee_{x_k\in LR} x_k e^k) + \sum_{LR\subseteq LP} (-1)^{|LP|-|LR|} v(x_i e^i) = 0$

Hence $AB(v, P\cup i, M) = 0$.

Proposition 3.4. If i, j are two substitute players in fuzzy cooperative game $v \in FG^N$, then $AB(v, P\cup i, M) = AB(v, P\cup j, M)$ for any $M \subseteq N \setminus (P \cup ij)$.

Proof. From definition 3.13, we have
$AB(v, P\cup i, M) = \int_{[0,1]^{p+1}} AB(v, LP\cup x_i, M)dx_i dLP$
$= \int_{[0,1]^{p+1}} \int_{[0,1]^m} |\delta_{LP\cup x_i} v(t)| dLM dx_i dLP$

Now we prove $\delta_{LP\cup x_i} v(t) = \delta_{LP\cup x_j} v(t)$ for $t = \sum_{k\in M} x_k e^k$. By definition 3.6
$\delta_{LP\cup x_i} v(t) = \sum_{LR\subseteq LP\cup x_i} (-1)^{|LP|+1-|LR|} v(t \bigvee_{x_k\in LR} x_k e^k)$
$= \sum_{LR\subseteq LP} (-1)^{|LP|+1-|LR|} v(t \bigvee_{x_k\in LR} x_k e^k) + \sum_{LR\subseteq LP} (-1)^{|LP|-|LR|}$
$v(t \bigvee_{x_k\in LR} x_k e^k \vee x_i e^i)$
$= \sum_{LR\subseteq LP} (-1)^{|LP|+1-|LR|} v(t \bigvee_{x_k\in LR} x_k e^k) + \sum_{LR\subseteq LP} (-1)^{|LP|-|LR|}$
$v(t \bigvee_{x_k\in LR} x_k e^k \vee x_j e^j)(by\ definition\ 3.12)$
$= \sum_{LR\subseteq LP\cup x_j} (-1)^{|LP|+1-|LR|} v(t \bigvee_{x_k\in LR} x_k e^k) = \delta_{LP\cup x_j} v(t)$

then we have $AB(v, P\cup i, M) = AB(v, P\cup j, M)$.

Proposition 3.5. If i, j are two substitute players in fuzzy cooperative game $v \in FG^N$, then $AB(v, P, M\cup i) = AB(v, P, M\cup j)$ for any $M \subseteq N \setminus (P \cup ij)$.

Proof. From definition 3.13, we have

$$AB(v, P, M\cup i) = \int_{[0,1]^p} AB(v, LP, M\cup i)dLP =$$
$$= \int_{[0,1]^p} \int_{[0,1]^{m+1}} |\delta_{LP} v(t \vee x_i e^i)| dLM dx_i dLP$$

We need to prove $\delta_{LP}v(t \vee x_i e^i) = \delta_{LP}v(t \vee x_j e^j)$ for any $t = \sum_{k \in M} x_k e^k$.
By definition 3.6
$\delta_{LP}v(t \vee x_i e^i = \sum_{LR \subseteq LP}(-1)^{|LP|-|LR|}v(t \vee x_i e^i \bigvee_{x_k \in LR} x_k e^k)$
$= \sum_{LR \subseteq LP}(-1)^{|LP|-|LR|}v(t \vee x_j e^j \bigvee_{x_k \in LR} x_k e^k)(by\ \ definition\ \ 3.13)$
$= \delta_{LP}v(t \vee x_j e^j$
then $AB(v, P, M \cup i) = AB(v, P, M \cup j)$

Proposition 3.6. If i, j are two substitute players in fuzzy cooperative game $v \in FG^N$, then $AB(v, P \cup i, M \cup j) = AB(v, P \cup j, M \cup i)$ for any $M \subseteq N \backslash (P \cup ij)$.

Proof. Similar to the proof of proposition 3.4 and 3.5. it is easy to get the result.

Example. Let $v(x) = -11x_1x_2x_3 + 6x_1x_2 + 7x_2x_3 + 8x_1x_3$ be a fuzzy characteristic function, $x = (x_1, x_2, x_3)$ is fuzzy coalition variable, Given $s = (s_1, s_2, s_3) = (0.3, 0.5, 0.8)$, then $v(s) = v(0.3, 0.5, 0.8) = 4.3$.

By equality (1)(2), we have
$\delta_{x_1x_2}v(\phi)|_{x=s} = 0.9$, $\delta_{x_1x_2}v(0, 0, x_3)|_{x=s} = -1.14$, $\delta_{x_1x_3}v(\phi)|_{x=s} = 1.92$, $\delta_{x_1x_3}v(0, x_2, 0)|_{x=s} = -0.12$, $\delta_{x_2x_3}v(\phi)|_{x=s} = 2.8$, $\delta_{x_2x_3}v(x_1, 0, 0)|_{x=s} = 0.76$, $\delta_{x_1x_2x_3}v(\phi)|_{x=s} = -2.04$

By the dates above we can say, based on s, there exists the strongest positive interaction 2.8 between the levels $x_2 = 0.5$ and $x_3 = 0.8$ in the presence of $t = \phi$, and there exists the strongest negative simultaneous interaction -2.04 among levels $x_1 = 0.3, x_2 = 0.5$ and $x_3 = 0.8$, and in the presence of $t = \phi$.

By equality (4), the interaction between levels $x_1 = 0.3$ and $x_2 = 0.5$ with respect to ϕ and 3 are
$B(v, \{x_1, x_2\}, \phi)|_{x_1=0.3, x_2=0.5} = \delta_{x_1x_2}v(\phi)|_{x_1=0.3, x_2=0.5} = 0.9$
$B(v, \{x_1, x_2\}, 3)|_{x_1=0.3, x_2=0.5} = \int_0^1 \delta_{x_1x_2}v(0, 0, x_3)|_{x_1=0.3, x_2=0.5}dx_3 = 0.075$

By (5), the absolute interaction between levels $x_1 = 0.3$ and $x_2 = 0.5$ with respect to players ϕ and 3 are
$AB(v, \{x_1, x_2\}, \phi)|_{x_1=0.3, x_2=0.5} = |\delta_{x_1x_2}v(\phi)|_{x_1=0.3, x_2=0.5}| = 0.9$
$AB(v, \{x_1, x_2\}, 3)|_{x_1=0.3, x_2=0.5} = \int_0^1 |\delta_{x_1x_2}v(0, 0, x_3)|_{x_1=0.3, x_2=0.5}|dx_3$
$= \int_0^1 |-1.65x_3 + 0.9|dx_3 = \int_0^{\frac{6}{11}}(-1.65x_3 + 0.9)dx_3 + \int_{\frac{6}{11}}^1 (1.65x_3 - 0.9dx_3 = \frac{2013}{4840}$

By equality (6), the interaction between player 1 and 2 with respect to ϕ and player 3 are
$B(v, \{1, 2\}, \phi) = \int_0^1 \int_0^1 B(v, \{x_1, x_2\}, \phi)dx_1dx_2 = \int_0^1 \int_0^1 6x_1x_2dx_1dx_2 = \frac{3}{2}$
$B(v, \{1, 2\}, \{3\}) = \int_0^1 \int_0^1 B(v, \{x_1, x_2\}, \{3\})dx_1dx_2$
$= \int_0^1 \int_0^1 \int_0^1 (-11x_1x_2x_3 + 6x_1x_2dx_3dx_2dx_1 = \frac{1}{8}$

By equality (7), the absolute interaction between player 1 and 2 with respect to players ϕ and player 3 are
$AB(v, \{1, 2\}, \phi) = \int_0^1 \int_0^1 AB(v, \{x_1, x_2\}, \phi)dx_1dx_2 = \int_0^1 \int_0^1 |6x_1x_2|dx_1dx_2 = \frac{3}{2}$.
$AB(v, \{1, 2\}, \{3\}) = \int_0^1 \int_0^1 AB(v, \{x_1, x_2\}, \{3\})dx_1dx_2 = \int_0^1 \int_0^1 \int_0^1 |-11x_1x_2x_3 + 6x_1x_2|dx_3dx_2dx_1 = \frac{61}{88}$.

4 Conclusion

Inspired by the interaction among players in crisp cooperation game, we defined the LP-derivative among levels based on fuzzy coalition s in fuzzy cooperative game, furthermore, we defined the interaction among levels of LP with respect to players M, which considering the effect of each level value of $x_i (i \in M)$ to LP when x_i change from 0 to 1. At last we introduce the absolute interaction among players of P and prove the properties satisfied by the absolute interaction. All of the approaches are useful for a manager to make decision in a setting of fuzzy cooperative game.

Acknowledgements

This work described here is partially supported by the grants from the National Natural Science Foundation of China (No. 70471063), the National Innovation Base in Philosophy and Social Science, titled Management on National Defense Science and Technology and National Economy Mobilization, of the Second Phase of "985 Project" of China (No. 107008200400024), the Main Subject Project of Beijing of China (NO: xk100070534).

References

1. G.Owen (1971/72) Multilinear extensions of games, Management Sci: 1864-79.
2. T. Murofushi, S. Soneda (1993) Techniques for reading fuzzy measures (iii): interaction index, in: Nineth Fuzzy System Symposium, Saporo, Japan: 693-696.
3. M. Grabisch (1997) k-order additive discrete fuzzy measures and their representation, Fuzzy Sets and Systems 92(2): 167-189.
4. M. Grabisch, M. Roubens (1999) An axiomatic approach to the concept of interaction among players in cooperative games, Int J Game Theory 28: 547-565.
5. I. Kojadinovic (2003) Modeling interaction phenomena using fuzzy measures: on the notions of interaction and independence, Fuzzy Sets and Systems 135 (3): 317-340.
6. I. Kojadinovic (2005) An axiomatic approach to the measurement of the amount of interaction among criteria or players, Fuzzy Sets and Systems 152 (3): 417-435.
7. J.von Neumann, O. Morgenstern (1944) The Theory of Games and Economic Behavior, Princeton University Press, Princeton, NJ.
8. M.Grabisch (2000) The interaction and Mobius representations of fuzzy measures on finite spaces, k-additive measures: a survy. In M.Grabisch, T.Murofushi, and M.Sugeno, editors, Fuzzy Measures and Integrals-Theory and Applications, Physica Verlag: 70-93.
9. J.P.Aubin (1979) Mathematical Methods of Game and Economic Theory, North-Holland.
10. J.P. Aubin (1984) Cooperative fuzzy game: the static and dynamic points of view, TIMS/Studies in the Management Science, Vol.20: 407-428.
11. D.Butnaviu (1978) Fuzzy games; a description of the concept, Fuzzy Sets and Systems, Vol.1: 181-192.
12. D.Butnariu (1980) Stability and shapely value for an n-persons fuzzy game, Fuzzy Sets and Systems, Vol.4: 63-72.

Novel Method for Fuzzy Hybrid Multiple Attribute Decision Making

Congjun Rao, Jin Peng, and Wenlue Chen

College of Mathematics and Information Sciences, Huanggang Normal University, Huanggang 438000, Hubei, China
raocjun79@163.com

Abstract. In order to solve the problems of fuzzy hybrid multiple attribute decision making with precision numbers, interval numbers and fuzzy numbers, this paper presents a new decision making method based on grey relational degree, and gives the principle and process of decision making. Finally, this new method is applied to solve a practical problem of hybrid multiple attribute decision making. The practical example indicates that the new method is not only feasible and reasonable but also simple in computing.

Keywords: Fuzzy hybrid multiple attribute decision making ; Grey relational degree; Interval number; Fuzzy number; Approach degree.

1 Introduction

In the fields of natural science, social science, economy, management and military affairs, there exist many problems of fuzzy hybrid multiple attribute decision making. For example, in evaluating the alternatives of new weapon manufacturing, we mainly consider the precision of hitting the target, warhead load, mobile capability, price, reliability, and repairable characteristics. In these indexes, the precision and warhead load are denoted by precision numbers, mobile capability and price are denoted by interval numbers, reliability and repairable characteristics are denoted by fuzzy numbers or language variables by consulting experts.This is a problem of system analysis whose index values containing precision numbers, interval numbers and fuzzy numbers, and all that problems are called the problems of fuzzy hybrid multiple attribute decision making[1,2].

The methods for the fuzzy multiple attribute decision making with single index of precision numbers, interval numbers and fuzzy numbers are given respectively in [3]. In [1] , it presents a technique of order preference by similarity to ideal solution for fuzzy hybrid multiple attribute decision making, but the research on fuzzy hybrid multiple attribute decision making is not perfect yet[3,5].This paper presents a new decision making method based on grey relational degree for the problems of fuzzy multiple attribute decision making with

B.-Y. Cao (Ed.): Fuzzy Information and Engineering (ICFIE), ASC 40, pp. 583–591, 2007.
springerlink.com

precision numbers, interval numbers and fuzzy numbers, and proves that the relational degree meets the invariability of the whole parallel move, the whole similarity and the standardization. Finally, a decision making example is given to demonstrate the feasibility and superiority for our new method. Therefore, this paper provides a new and effective way to solve the problems of fuzzy hybrid multiple attribute decision making.

2 The Description for the Problem of Fuzzy Hybrid Multiple Attribute Decision Making

Let $X = \{X_1, X_2, ..., X_m\}$ be a set of alternatives for the problem of fuzzy hybrid multiple attribute decision making, and let $U = \{U_1, U_2, ..., U_n\}$ be a index set. $W = \{w_1, w_2, ..., w_n\}$ is the weight vector of indexes, where w_j satisfies

$$0 \leq w_j \leq 1, \sum_{j=1}^{n} w_j = 1$$

The evaluation value of alternative $X_i, (i = 1, 2, ..., m)$ to index U_j is denoted as a_{ij} , and we denote:

$$N_1 = \{1, 2, ..., n_1\}$$

$$N_2 = \{n_1 + 1, n_1 + 2, ..., n_2\}$$

$$N_3 = \{n_2 + 1, n_2 + 2, ..., n\}$$

When $j \in N_1$, a_{ij} are precision numbers. When $j \in N_2$, a_{ij} are interval numbers. When $j \in N_3$, a_{ij} are fuzzy numbers. All values of a_{ij} are formed a hybrid decision making matrix, it denoted as $A = (a_{ij})_{m \times n}$. We need to make evaluation and rank order for all alternatives.

3 The Principle and Method of the Decision Making

3.1 Processing Data for Indexes

Firstly, several relational definitions are given.

Definition 1. *Suppose that*

$$a = [a^L, a^U] = \{x | 0 \leq a^L \leq x \leq a^U, a^L, a^U \in R\}$$

then a is called an interval number. Especially, if $a^L = a^U$, then a is a real number. The set of all interval numbers is denoted as $\overline{R}$.

Definition 2. *Suppose that S_a={worst, worse, bad, common, good, better, best} ([3])or S_b={lowest, lower, low, common, high, higher, highest}([5]), and they are denoted as $S_a = \{s_1, s_2, \cdots, s_7\}$,or $S_b = \{s_1, s_2, \cdots, s_7\}$, then S_a and S_b are called fuzzy number set , and their corresponding interval numbers are defined as follows:*

$$S_1=[0,0.1],\ S_2=[0.1,0.25],\ S_3=[0.25,0.4],$$
$$S_4=[0.4,0.6],\ S_5=[0.6,0.75],\ S_6=[0.75,0.9],$$
$$S_7=[0.9,1],$$

where

$$s_7 \succ s_6 \succ s_5 \succ s_4 \succ s_3 \succ s_2 \succ s_1$$

the corresponding interval numbers for S_b is the same as S_a.

Definition 3. *Let $a = [a^L, a^U], b = [b^L, b^U]$ be two interval numbers, $k > 0$, then the algorithm for a, b are defined as follows:*

$$a + b = [a^L, a^U] + [b^L, b^U] = [a^L + b^L, a^U + b^U];$$
$$a \times b = [a^L \times b^L, a^U \times b^U];$$
$$a \div b = [\frac{a^L}{b^U}, \frac{a^U}{b^L}];$$
$$ka = k[a^L, a^U] = [ka^L, ka^U];$$
$$\frac{1}{a} = \frac{1}{[a^L, a^U]} = [\frac{1}{a^U}, \frac{1}{a^L}];$$

Definition 4. *Let A be a matrix, where $A = (a_{ij})_{m\times n}$, if the elements of A are all interval numbers, then A is called a grey matrix.*

In the fuzzy hybrid multiple attribute decision making, the following steps are given to process index data.

(1)Using definition 1 and definition 2 to transform all the index data(including precision numbers, interval numbers and fuzzy numbers) into interval numbers.

So all the index data are formed a grey matrix, we denoted it as $B = (b_{ij})_{m\times n}$. B is called a decision making matrix.

(2)Normalizing the decision making matrix B.

Usually, the benefit type and the cost type are two main kinds of indexes. We denote the benefit type index as I_1 , and denote the cost type index as I_2 . The following algorithms are given to normalize the matrix B ([3,4]). Suppose that the matrix B is transformed into matrix Y , where

$$Y = (y_{ij})_{m\times n}, \qquad y_{ij} = [y_{ij}^L, y_{ij}^U]$$
$$i = 1, 2, ..., m; \qquad j = 1, 2, ..., n$$

when $j \in I_1$, we have

$$y_{ij}^L = \frac{b_{ij}^L}{\sum\limits_{i=1}^{m} b_{ij}^U}, \qquad y_{ij}^U = \frac{b_{ij}^U}{\sum\limits_{i=1}^{m} b_{ij}^L} \tag{1}$$

when $j \in I_2$, we have

$$y_{ij}^L = \frac{\frac{1}{b_{ij}^U}}{\sum\limits_{i=1}^{m}(\frac{1}{b_{ij}^L})}, \qquad \frac{\frac{1}{b_{ij}^L}}{\sum\limits_{i=1}^{m}(\frac{1}{b_{ij}^U})} \tag{2}$$

3.2 The New Method Based on the Grey Relational Degree

Grey relational space is the measure space of difference information which satisfies grey relational axioms, and difference information is the numerical value which represents the difference between reference sequence and compared sequence [6].

Firstly, we give some relational definitions and conclusions as follows.

Definition 5. *Let $a = [a^L, a^U], b = [b^L, b^U]$ be two interval numbers, the distance between a and b is defined[7]:*

$$D(a,b) = \frac{\sqrt{2}}{2}\sqrt{(a^L - b^L)^2 + (a^U - b^U)^2}$$

Definition 6. *Let $Y = (y_{ij})_{m\times n}$ be a grey matrix, where*

$$y_{ij} = [y_{ij}^L, y_{ij}^U], \qquad i = 1, 2, ..., m; \qquad j = 1, 2, ..., n$$

Then $f = (f_1, f_2, ..., f_n)$ is called ideal solution, where

$$f_j = [f_j^L, f_j^U] = [\max_i(y_{ij}^L), \max_i(y_{ij}^U)] \tag{3}$$

and $g = (g_1, g_2, ..., g_n)$ is called anti-ideal solution, where

$$g_j = [g_j^L, g_j^U] = [\min_i(y_{ij}^L), \min_i(y_{ij}^U)] \tag{4}$$

Definition 7. *Suppose X is the grey relational factors set of hybrid index sequences,$X_0 \in X, X_i \in X, i = 1, 2, ..., m.X_0$ is the reference sequence, X_i are compared sequences, where*

$$x_0 = (a_{01}, a_{02}, ..., a_{0n})$$

$$x_i = (a_{i1}, a_{i2}, ..., a_{in}), \qquad i = 1, 2, ..., m$$

$a_{ij} = [a_{ij}^L, a_{ij}^U]$are all interval numbers. Then the grey relational coefficient and grey relational degree are defined:

$$r(a_{0j}, a_{ij}) = \frac{\rho D_{max}}{D_{0i}(j) + \rho D_{max}} \tag{5}$$

$$r(x_0, x_i) = \sum_{j=1}^{n} w_j r(a_{0j}, a_{ij}) \tag{6}$$

where $D_{0i}(j) = D(a_{0j}, a_{ij})$ is the distance between a_{0j} and a_{ij},$D_{\max} = \max_i \max_j D_{0i}(j)$, ρ is the distinguishing coefficient, $\rho \in (0, 1)$, and usually we set $\rho = 0.5$, w_j is the jth weight and satisfies

$$0 \leq w_j \leq 1, \qquad \sum_{j=1}^{n} w_j = 1$$

Theorem 1. *The relational degree $r(x_0, x_i)$ defined above satisfies relational four axioms.*

1) Normality

$$0 \leq r(x_0, x_i) \leq 1$$

$$r(x_0, x_i) = 0 \iff x_0, x_i \in \emptyset$$

$$r(x_0, x_i) = 1 \iff x_0 = x_i$$

2) Symmetry

$$r(x_0, x_i) = r(x_i, x_0) \iff X = \{x_0, x_i\}$$

3) Wholeness

$$x_i, x_j \in X = \{x_s \,| s = 0, 1, 2, \cdots, m, m \geq 2\}$$

and usually $r(x_i, x_j) \neq r(x_j, x_i)$

4) Closing

The smaller $D_{0i}(j)$ is, the larger $r(x_0, x_i)$ is.

Proof. Here we merely prove (1).

$0 \leq r(x_0, x_i) \leq 1$ is obvious.

$$r(x_0, x_i) = 1 \iff \forall j, r(a_{0j}, a_{ij}) = 1 \iff \forall j, D_{0i}(j) = 0$$

By the definition of $D_{0i}(j) = D(a_{0j}, a_{ij})$, we have

$$D_{0i}(j) = 0 \iff \forall j, a_{0j}^L = a_{ij}^L, a_{0j}^U = a_{ij}^U \iff a_{0j} = a_{ij}$$

so

$$r(x_0, x_i) = 1 \iff x_0 = x_i$$

and we can also get that 2), 3), 4) are obviously true.

Theorem 2. *The relational degree $r(x_0, x_i)$ keeps the invariability about the three generating operations, namely, the whole parallel move, the standardization, and the whole similarity.*

Proof. The Theorem 2 may be proved by a similar process of Ref.[6].

Next we give the new method based on the relational degree as follows.

In the decision making matrix $Y = (y_{ij})_{m\times n}$, we can construct the ideal solution $f = (f_1, f_2, ..., f_n)$ and the anti-ideal solution $g = (g_1, g_2, ..., g_n)$ respectively by using formula (3) and (4).

Let $f = (f_1, f_2, ..., f_n)$ and $g = (g_1, g_2, ..., g_n)$ be as the reference sequence respectively , and let $x_i = (a_{i1}, a_{i2}, ..., a_{in}), i = 1, 2, ..., m$ be as the compared sequences, where

$$a_{ij} = [y_{ij}^L, y_{ij}^U], j = 1, 2, ..., n$$

By using formula (5) and (6), we can calculate the grey relational degree $r_i^+ = r(x_i, f)$ between $x_i, (i = 1, 2, ..., m)$ and f, and calculate the grey relational degree $r_i^- = r(x_i, g)$ between $x_i (i = 1, 2, ..., m)$ and g.

For the grey relational degree r_i^+ and r_i^-, the greater the value of r_i^+ is, the better the alternative x_i is, and the smaller the value of r_i^- is, the better the alternative x_i is. So if the alternative x_i is the best alternative, it must be satisfied two conditions, namely, the value of r_i^+ is as great as possible and the value of r_i^- is as small as possible. Therefore, let alternative x_i be subject to f by the approach degree of u_i, and let project x_i be subject to g by the approach degree of $1 - u_i$, then the following model is established to determine the value of u_i.

$$\min\{V(\mathbf{u}) = \sum_{i=1}^{m}[(1-u_i)r(f,x_i)]^2 + \sum_{i=1}^{m}[u_i r(g,x_i)]^2\} \tag{7}$$

where $\mathbf{u} = (u_1, u_2, ..., u_m)$, it is the vector of optimal solution.

Theorem 3. *The optimal solution of model (7) can be expressed as*

$$u_i = \frac{r^2(f,x_i)}{r^2(f,x_i) + r^2(g,x_i)}, \qquad i = 1, 2, ..., m. \tag{8}$$

Proof. By

$$\frac{\partial V(\mathbf{u})}{\partial u_k} = 0,$$

we have

$$2(1-u_i)r^2(f,x_i) - 2u_i r^2(g,x_i) = 0,$$

so

$$u_i = \frac{r^2(f,x_i)}{r^2(f,x_i) + r^2(g,x_i)}, \qquad i = 1, 2, ..., m$$

We can rank order for all alternatives according to the values of u_i. The greater the value of u_i is, the better the alternative x_i is.

3.3 The Basic Steps of the New Decision Making Method

According the above definitions and conclusions, we give all steps of the new method for fuzzy hybrid multiple attribute decision making.

Step1: Using Definition 1 and Definition 2 to transform all index data into interval numbers.

Step2: Using formula (1) and (2) to process index data, the raw decision making matrix is normalized as matrix Y, where $Y = (y_{ij})_{m \times n}$.

Step3: Using formula (3) and (4) to construct the ideal solution $f = (f_1, f_2, ..., f_n)$ and the anti-ideal solution $g = (g_1, g_2, ..., g_n)$ respectively.

Step4: Using formula (5) and (6) to calculate the grey relational degree $r_i^+ = r(x_i, f)$ and $r_i^- = r(x_i, g), i = 1, 2, ..., m$.

Step5: Using formula (8) to calculate the approach degree $u_i, i = 1, 2, ..., m$.

Step6: Ranking order for all alternatives according to u_i.

4 An Application Example

In order to prove the feasibility and superiority of our new method, here we consider the following example which given in [1].

A Department of Defense of a country plans to develop a kind of tactical missile weapon equipment, the department of development provides the information of four kinds of missile types. There are six indexes are given to evaluate these four kinds of missile types,i.e. U_1:Precision of hitting the target(km);U_2:Warhead load(kg);U_3:Mobile capability (km/h);U_4:Price (10^6);U_5:Reliability;U_6:Repairable characteristics. The experts appointed by the Department of Defense examine the tactical and technical index of the four types of the missile in detail, and the results are shown in Table 1. If our aim is to get the best benefit, then which type of the missile should be chosen.

Table 1. The index data of four kinds of missile types

Types	U_1	U_2	U_3	U_4	U_5	U_6
1	2.0	500	[55,66]	[4.7,5.7]	common	higher
2	2.5	540	[30,40]	[4.2,5.2]	low	common
3	1.8	480	[50,60]	[5.0,6.0]	high	high
4	2.2	520	[35,45]	[4.5,5.5]	common	common
Weights	0.2	0.2	0.1	0.1	0.2	0.2

Table 2. Decision making matrix B

Types	U_1	U_2	U_3	U_4	U_5	U_6
x_1	[2.0,2.0]	[500,500]	[55,66]	[4.7,5.7]	[0.40,0.60]	[0.75,0.90]
x_2	[2.5,2.5]	[540,540]	[30,40]	[4.2,5.2]	[0.25,0.40]	[0.40,0.60]
x_3	[1.8,1.8]	[480,480]	[50,60]	[5.0,6.0]	[0.60,0.75]	[0.60,0.75]
x_4	[2.2,2.2]	[520,520]	[35,45]	[4.5,5.5]	[0.40,0.60]	[0.40,0.60]

Table 3. Normalized decision making matrix Y

Types	U_1	U_2	U_3	U_4	U_5	U_6
x_1	[0.235,0.235]	[0.245,0.245]	[0.261,0.388]	[0.201,0.297]	[0.170,0.364]	[0.263,0.419]
x_2	[0.294,0.294]	[0.265,0.265]	[0.142,0.235]	[0.220,0.332]	[0.106,0.242]	[0.140,0.279]
x_3	[0.212,0.212]	[0.235,0.235]	[0.237,0.353]	[0.191,0.279]	[0.255,0.455]	[0.211,0.349]
x_4	[0.259,0.259]	[0.255,0.255]	[0.166,0.265]	[0.208,0.310]	[0.170,0.364]	[0.140,0.279]

1) Using Definition 1 and Definition 2 to transform all the index data into interval numbers, we obtain the raw decision making matrix B.

2) By the normalization formula (1) and (2), the decision making matrix B is normalized as matrix Y.

3) Determining the ideal solution f and the anti-ideal solution g.

$$f = ([0.294, 0.294], [0.265, 0.265], [0.261, 0.388],$$
$$[0.220, 0.332], [0.255, 0.455], [0.263, 0.419])$$
$$g = ([0.212, 0.212], [0.235, 0.235], [0.142, 0.235],$$
$$[0.191, 0.279], [0.106, 0.242], [0.140, 0.279])$$

4) Calculating the grey relational degree $r_i^+ = r(x_i, f)$ and $r_i^- = r(x_i, g), i = 1, 2, 3, 4$.

$$r_1^+ = 0.765, \quad r_2^+ = 0.689, \quad r_3^+ = 0.720, \quad r_4^+ = 0.639,$$
$$r_1^- = 0.646, \quad r_2^- = 0.825, \quad r_3^- = 0.726, \quad r_4^- = 0.749.$$

5) Calculating the approach degree $u_i, i = 1, 2, 3, 4$.

$$u_1 = 0.584, \quad u_2 = 0.411, \quad u_3 = 0.496, \quad u_4 = 0.421$$

Therefore, we get the order of the four alternatives:

$$x_1 \succ x_3 \succ x_4 \succ x_2$$

It shows the order by the method in this paper is basically the same as the order in [1], and proves that the method is rational and feasible, and the new method is simpler in computing and programming, and it has higher precision and better properties.

5 Conclusions

This paper presents a new method based on the grey relational degree for fuzzy hybrid multiple attribute decision making. This method is specific, fine in properties and simple for computing, and it is of a great value in practice.

Acknowledgments

This work is supported by the National Natural Science Foundation of China Grant No.70671050, the Significant Project No.Z200527001, and the Key Project of Hubei Provincial Department of Education No. D200627005.

References

1. Xia Y, Wu Q (2004) A Technique of Order Preference by Similarity to Ideal Solution for Hybrid Multiple Attribute Decision Making Problems.Journal of systems engineering 6, 630-634
2. Song Y, Zhang S (2002) Hybrid Multiple Criteria Decision Making with Time Series Based on Fuzzy Pattern Recognition.Systems Engineering and Electronics 4, 1-4
3. Xu Z (2004) Uncertain Multiple Attribute Decision Making: Methods and Applications. Beijing, Tsinghua University Press,108-110,164-167
4. Fan Z, Gong X (1999) Methods of Normalizing the Decision Matrix for Multiple Attribute Decision Making Problems with Intervals.Journal of Northeastern University (Natural Science)3, 326-329
5. Li R (2002) Fuzzy Multiple Attribute Decision Making Theory and Its Applications. Beijing, Science Press,141-248
6. Deng J (2002) Foundation of Grey Theory. Wuhan, Huazhong University of Science and Technology Press, 251-258
7. Xiao X, Song Z, Li F (2005) Foundation of Grey Technology and its Application. Beijing, Science Press,26-50

Auto-weighted Horizontal Collaboration Fuzzy Clustering

Fusheng Yu, Juan Tang, Fangmin Wu, and Qiuyan Sun

School of Mathematical Sciences, Beijing Normal University, Beijing, China, 100875
yufusheng@263.net

Abstract. Horizontal Collaboration Fuzzy C-Means (HC-FCM) is such a clustering method that does clustering in a set of patterns described in some feature space by considering some external sources of clustering information which are about the same set of patterns but described in different feature spaces. Because of potential security and privacy restrictions, the external information is provided only by some partition matrices obtained by Fuzzy C-Means (FCM). HC-FCM quantifies the influence of these partition matrices by constructing a new objective function which combines these partition matrices with suitable weights. These weights are crucial in HC-FCM, but how to determine them still remains a problem. This paper first puts forward a concept of similarity measure of partition matrices, then based on this similarity measure, proposes some simple and practical methods for automatically determining the weights, and finally gives a new version of HC-FCM, named auto-weighted HC-FCM. With the work of this paper, HC-FCM becomes more practical. Some experiments are carried to show the performance and reveal the validity of our method.

Keywords: FCM, HC-FCM, Auto-weighted HC-FCM.

1 Introduction

Fuzzy C-Means (FCM) is a typical one of fuzzy clustering methods, which is very effective in single data-set clustering and analysis [3,7]. By this method, we can investigate the structure of some given data set under granules of different sizes. While in reality, due to the enormous abundance and diversity data, thorough data analysis requires a highly collaborative, orchestrated and unified approach. The independent clustering for each data set is incomplete; we should appreciate the value of external sources of information. Under this condition, Pedrycz proposed a new version of FCM, named collaborative FCM(C-FCM)[4] with horizontal collaborative fuzzy clustering and vertical collaborative fuzzy clustering being its two typical forms. The horizontal collaborative clustering method deals with such clustering of one set influenced by the same sets clustered in different feature spaces. Because of potential security and privacy restrictions, outer data offer information only with partition matrix. So in this process, the quantification of the relationships among the different partition matrices of the same data sets is crucial. Similar to FCM, C-FCM realizes clustering by means of a suitable objective function constructed with the partition matrices produced by clustering of the same data set on different feature spaces. In his paper[5], Pedrycz gave a manner to incorporate

B.-Y. Cao (Ed.): Fuzzy Information and Engineering (ICFIE), ASC 40, pp. 592–600, 2007.
springerlink.com

these partition matrices with a group weights($\alpha[ii, jj]$)) which reflect the influence of other data sets on the data set the decision-maker has, and showed the results under different weights group. From his paper, we can conclude that different weights will result in different clustering results. But he did not tell us how to give these weights. This confines the application of the method. This paper will concentrate on this problem and give some approaches to quantity these weights from partition matrixes.

This paper is organized in this way: We will present preliminary in Section 2, concentrate on the major work and show the experiment results in Section 3; and conclude this study in Section 4.

2 Prelimimary

2.1 Fuzzy C-Means (FCM)

FCM is an effective clustering means based on one fuzzy partition. It optimizes the clustering through cyclic iteration of partition matrix and ends the cycle when the objective function reaches an appropriate threshold. The corresponding objective function is as follows:

$$\text{Minimize } \mathrm{J(U,V)} = \sum_{i=1}^{c} u_{ik}{}^{2} d_{ik}{}^{2} \text{ ,subject to} \sum_{\mathrm{i=1}}^{\mathrm{c}} \mathrm{u_{ik}} = 1$$

Where $U = (u_{ij})_{cn}$ is the partition matrix, $V = (v_{ij})_{cl}$ is the matrix of centroids of clusters, d_{ik} is the distance between the *k*th patterns and the centroid of the *i*th cluster(*i*=1,2,…,*c*; k=1,2, …,*n*). By Lagrange multipliers method, we can obtain the optimal solution of the optimization problem [1,2].

2.2 Horizontal Collaboration Fuzzy C-Means

The background of Horizontal Collaboration Fuzzy C-Means(HC-FCM)[5] may be outlined as follows: given a reference data set $X[ii]$, do clustering on it by referring to some useful information about some external data sets. The reference data set and the external data sets describe the same group of patterns but in different feature spaces. In reality, the external data sets are usually owned by different departments. Because of potential security and privacy restrictions, we usually cannot have the data sets themselves. What we can get is the information of the clustering results of external data sets, which is provided by partition matrices U produced by FCM [6]. The general scheme of HCFCM is depicted in fig. 1.

Considering the external data sources, HC-FCM refines its objective function as follows:

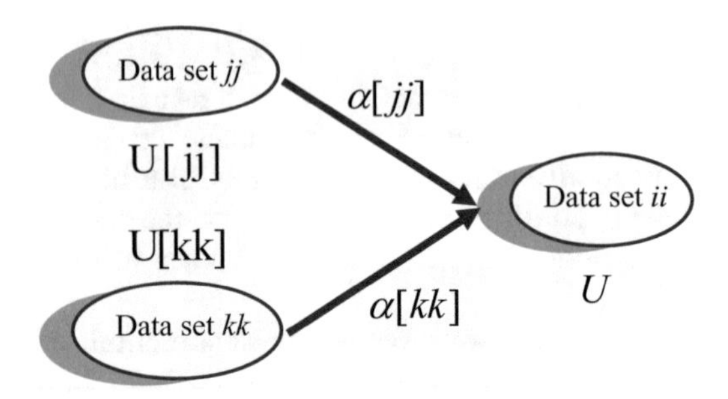

Fig. 1. General scheme of horizontal fuzzy clustering

$$\text{Minimize } Q = \sum_{i=1}^{c} u_{ik}^{2} d_{ik}^{2} + \sum_{jj=1}^{p} \alpha[jj] \sum_{i=1}^{c} (u_{ik} - u_{ik}[jj])^{2} d_{ik}^{2}$$

$$\text{subject to } \sum_{i=1}^{c} u_{ik} = 1$$

Where $U = (u_{ij})_{cn}$ is the partition matrix of reference data set X ; p denotes the number of external data sets $X[1],\ldots,X[p]$; $U[ii]$ denotes the corresponding partition matrix of $X[ii](ii = 1,2,\ldots, p)$; $X[ii]$ stands for the iith external data set; d_{ik} is the distance between the kth pattern and ith cluster in reference data set X . $\alpha[jj]$ assumes nonnegative values, the higher the value of interaction coefficient, the stronger of the effect of data set $X[jj]$ on data set X .

We may also obtain the optimal solution to the optimization problem by Lagrange multiplier method. We omit the formulas of optimal partition matrix and centroids of clusters here, readers may refer to papers [3,5].

In the objective function of HC-FCM, the first term is the same to the one of FCM. The second term reflects the total effect of external data sets on the clustering of the iith data set. In the second term, there appears a group of weights $\alpha[ii, jj]$, each of which quantifies the single effect of each data set $X[jj]$ on the reference data set $X[ii]$. So, there are two factors that may influent the clustering result of the reference data set. They are partition matrices and weights whose effects are showed in papers [3,4]. The matrices are provided by FCM clustering on the external data sets, which remain unchanged. As for the weights, we can change them. How to determine them? This is the problem we will concentrate on in this paper. We will discuss the conditions and limitations of HC-FCM, and then propose some feasible approaches to determine the weights.

3 The Determining of Weights in HC-FCM

Generally speaking, there are two kinds of methods for determining weights. One is to give weights subjectively and experientially according to some acquired knowledge or information. But in the process of horizontal collaborative fuzzy clustering, because of potential security and privacy restrictions, the knowledge or information we may take means of to determine the weights is only the partition matrices [5]. Meanwhile this kind of knowledge is somehow difficult to directly take means of to determine weights. In this circumstance, the better way is other one by which weights are automatically determined according to the considerations of partition matrices. Even though it is difficult for us to directly give the weights, it is easy to give some principle for determining the weights. For example, we may expect to think a lot of the influence of data sets that have similar clustering result to that of the reference data set, we may also expect to think a lot of the influence of data sets that have different clustering result, or expect to treat the external data sets equally without discrimination. These expectations may be used as principles for determining weights, while they are easy to be given out. In the remainder of this section, we will give some approaches to automatically calculate the weights according to these principles.

3.1 Measure of Partition Similarity

Before giving the algorithms for calculating the weights in HC-FCM, we should first give a similarity measure to measure the similarity between the clustering results of the reference data set and those of external data sets. If the similarity between two clustering results can be calculated, then the weights can be determined.

Let $X=\{x_1,x_2,\ldots x_n\}$ be the set of all the patterns to be clustered, c be the number of clusters to be generated, $\Omega(c)$ be the set of all c-partition matrices on X. $U=(u_1,u_1,\ldots u_c)^T$ is a partition matrix with $u_i,(i=1,2,\ldots,c)$ being its ith row which represents a cluster produced in clustering

Definition 1. Let $U=(u_1,u_1,\ldots u_c)^T$, $V=(v_1,v_1,\ldots v_c)^T$ be any two elements in $\Omega(c)$. A similarity measure of partition matrices is a mapping $f:\Omega(c)\times\Omega(c)\to[0,1]$ satisfying the conditions:
(C1) $f(U,U)=1$; (C2) $f(U,V)=f(V,U)$;
(C3) $f(U,U^*)=1$; where $\mathrm{U}^*=(\mathrm{u}_{\mathrm{i}_1},\mathrm{u}_{\mathrm{i}_2},\ldots,\mathrm{u}_{\mathrm{i}_c})^{\mathrm{T}}$; $\mathrm{u}_{\mathrm{i}_1},\mathrm{u}_{\mathrm{i}_2},\ldots,\mathrm{u}_{\mathrm{i}_c}$ is an all permutation of $\mathrm{u}_1,\mathrm{u}_1,\ldots,\mathrm{u}_{\mathrm{c}}$.

Example 1. Define

$$f_1(U,V)=\sum_{i=1}^{c}\max_{j=1}^{c}\left\{\sum_{k=1}^{n}(u_{ik}\wedge v_{jk})\right\}/\sum_{i=1}^{c}\min_{j=1}^{c}\left\{\sum_{k=1}^{n}(u_{ik}\vee v_{jk})\right\}$$

Example 2. Define $f_2(U,V)=\sum_{i=1}^{c}\max_{j=1}^{c}\left\{\sum_{k=1}^{n}\left(u_{ik}\wedge v_{jk}\right)\right\}/n$

it is easy to verify that $f_1(U,V)$ and $f_2(U,V)$ are all similarity measures of partition matrices.

The following example verifies the rationality of similarity measures in example 1 and 2.

$$U=\begin{pmatrix}0.2 & 06 & 0.1\\ 0.7 & 0.2 & 0.1\\ 0.1 & 0.2 & 0.8\end{pmatrix} \qquad V_1=\begin{pmatrix}0.1 & 0.2 & 0.8\\ 0.2 & 0.6 & 0.1\\ 0.7 & 0.2 & 0.1\end{pmatrix}$$

$$V_2=\begin{pmatrix}0.1 & 0.5 & 0.1\\ 0.7 & 0.2 & 0.2\\ 0.2 & 0.3 & 0.7\end{pmatrix} \qquad V_3=\begin{pmatrix}0.33 & 0.33 & 0.34\\ 0.33 & 0.34 & 0.33\\ 0.34 & 0.33 & 0.33\end{pmatrix}$$

Ostensibly, U ,V_1,V_2 have clear classification of clusters. Especially, U and V_1 are same without considering the order or rows. While, V_3 has vague classification. The results illustrate the validity of the definition of similarity measure of partition matrices.

$$f_1(U,V_1)=1 \qquad f_1(U,V_2)=0.8182 \qquad f_1(U,V_3)=0.4706$$
$$f_2(U,V_1)=1 \qquad f_2(U,V_2)=0.9 \qquad f_2(U,V_3)=0.64$$

3.2 Encouragement Approach

In this approach, we assume such an encouragement principle:

The more similar the partition matrix to the reference partition matrix, the larger the effect of the external data set.

In terms of this principle, we can choose a suitable similarity measure of partition matrices, and calculate the similarity of the external partition matrices to the reference partition matrix. The following three approaches to calculate weights are all encouragement approaches.

Suppose $r[ii]$ is the similarity of partition matrix $U[ii]$ to the reference partition matrix U $(ii=1,\cdots,p)$, which is calculated under a similarity measure. Then the weight $\alpha[ii]$ expressing the degree of influence that $U[ii]$ exerts onto $U,(ii=1,2,\ldots,p)$ may be given by the following three approaches:

$$\textit{Approach } 1\text{: } \alpha[ii] = r[ii] \Big/ \sum_{ii=1}^{p} r[ii]$$

$$\textit{Approach } 2\text{: } \alpha[\mathrm{ii}] = r[ii] \Big/ \min_{\mathrm{ii}=1}^{\mathrm{p}}\{\mathrm{r[ii]}\}$$

$$\textit{Approach } 3\text{: } \alpha[\mathrm{ii}] = r[ii] \Big/ \max_{\mathrm{ii}=1}^{\mathrm{p}}\{\mathrm{r[ii]}\}$$

Experiment 1. We consider the data set X distributed in two subspaces. In the horizontal fuzzy clustering, we get three external partition matrices V_1, V_2, V_3 to do horizontal collaboration clustering with the original partition matrix U of X. The similarity measure between partition matrices is set the one defined in example 1. By calculation, we get

$$f_1(U, V_1) = 0.7356, \quad f_1(U, V_2) = 0.9438, \quad f_1(U, V_3) = 0.9978$$

$\alpha[ii, jj]$ is set according to above three approaches.

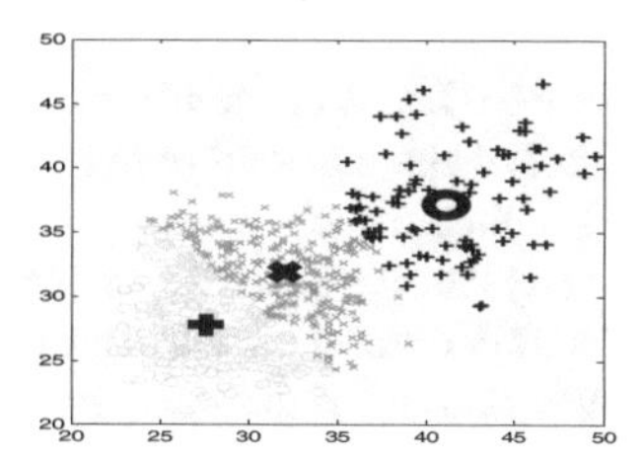

Fig. 2. Results of Clustering without collaboration

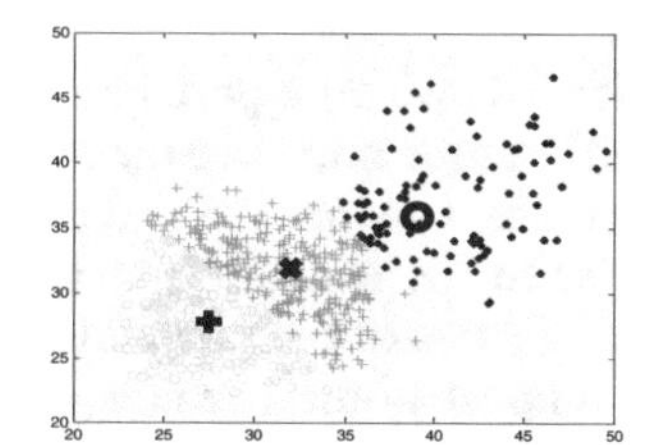

Fig. 3. Collaborative Clustering results under weight approach 1

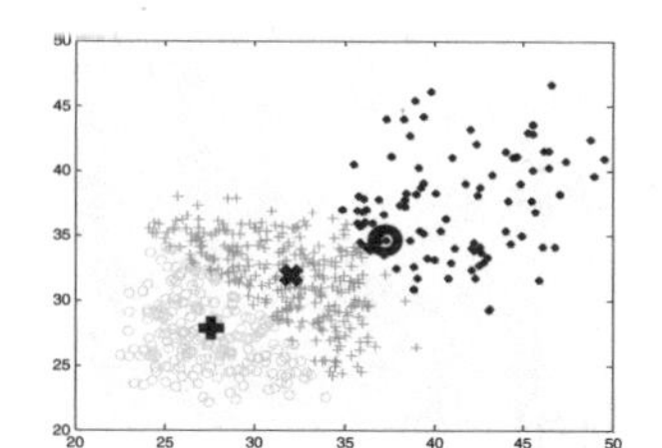

Fig. 4. Collaborative Clustering results under weight approach 2

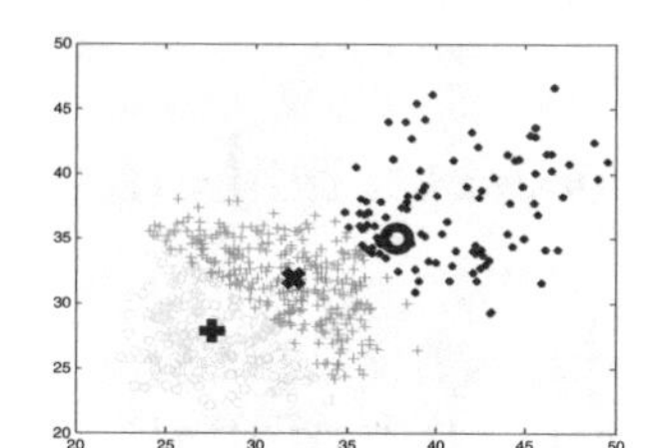

Fig. 5. Collaborative Clustering results under weight approach 3

From the above experiments, it is clear to see that the collaborative phenomenon of clustering with different weights. Theoretically, the lager weight is, the more influence clustering will get. In approach 2 every vector of weight α is no less than 1. So collaborative clustering following approach 2 will get more influence. It is apparent from above figures. (fig. 4 has more differences from fig.2 with comparing to fig. 3, 5) It can also be verified from other aspects.

Let u, v be the original partition matrix and prototype of X before collaboration, U, V be the ones after collaboration respectively. We compare the indices in table1 according to above approaches.

Table 1. Comparison of three Encouragement weight approaches

Approach	weights	‖u-U‖	‖v-V‖	Sim(u,U)
1	0.27, 0.35, 0.38	1.72034	2.42863	0.90425
2	1.00, 1.28, 1.36	2.15546	4.60223	0.88407
3	0.74, 0.95,1.00	2.07434	3.94327	0.88723

Obviously, the indices in approach 2 are all largest, so collaboration clustering following approach2 gets more influence.

3.3 Penalty Approach

In section 3.2, we set a principle with three approaches. The more similar the partition matrix to the reference one, the greater the effect of the external data set on the reference one. Obviously, the three algorithms are all encouragement approaches.

Similar to the encouragement approach, similarity between reference partition matrix and external ones should first be calculated under some similarity measure. But in contrast to the encouragement approach, weights are calculated by such a principle:

the more similar the partition matrix to the reference partition matrix, the smaller the corresponding weight.

The following approaches are for determining weights in penalty way:

$$\textit{Approach 1: } \alpha[\mathrm{ii}] = (1 - r[ii]) \Big/ \sum_{\mathrm{ii}=1}^{p} (1 - r[ii])$$

$$\textit{Approach 2: } \alpha[\mathrm{ii}] = (1 - r[ii]) \Big/ \min_{i=1}^{p} (1 - r[i])$$

$$\textit{Approach 3: } \alpha[\mathrm{ii}] = (1 - r[ii]) \Big/ \max_{i=1}^{p} (1 - r[i])$$

Experiment 2. Using the same data set X and three reference partition matrices in experiment 1, we compare collaboration clustering under the above three penalty approaches.

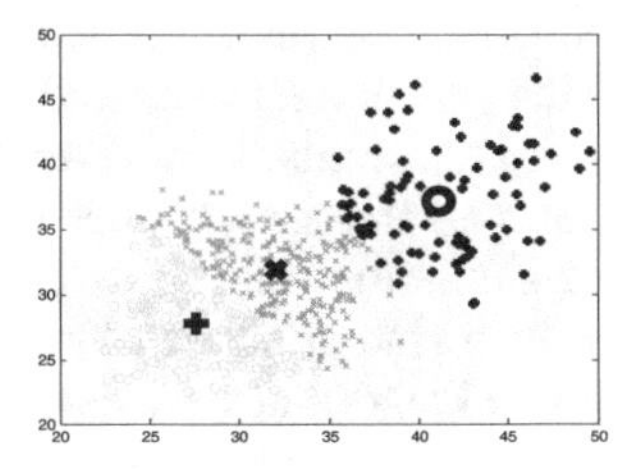

Fig. 6. Results of Clustering without collaboration

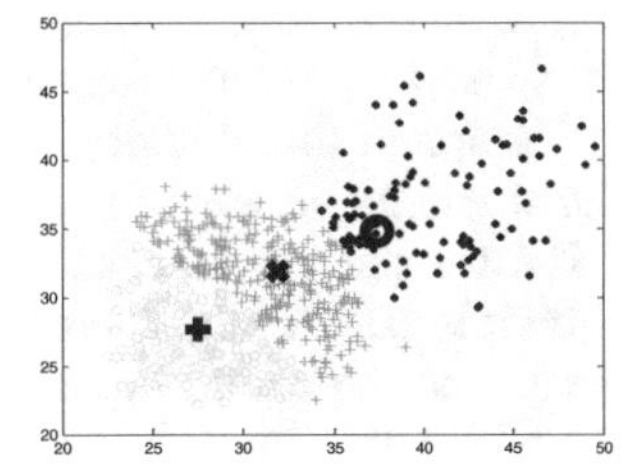

Fig. 7. Collaborative Clustering results under weight approach 1

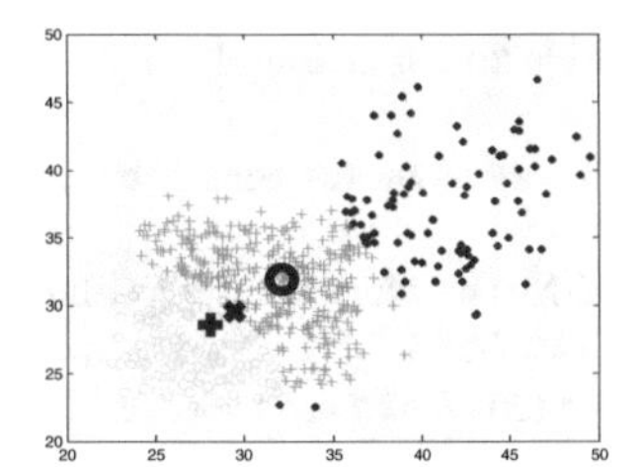

Fig. 8. Collaborative Clustering results under weight approach 2

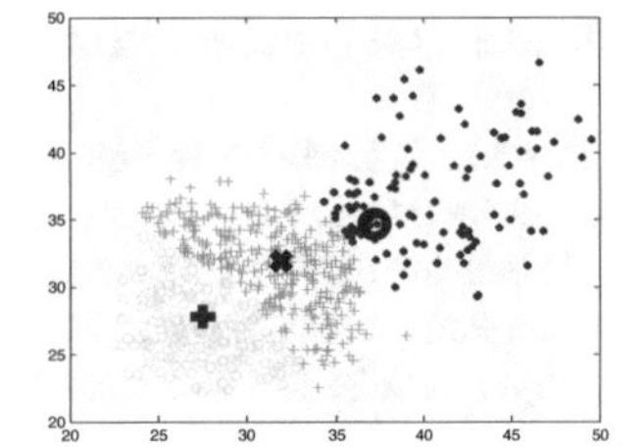

Fig. 9. Collaborative Clustering results under weight approach 3

The referential indices are listed in table 2.

Table 2. Comparison of three penalty weight approaches

Approach	weights	||u-U||	||v-V||	Sim(u,U)
1	0.82, 0.17, 0.01	3.94676	4.41178	0.79469
2	131, 28, 1	6.19599	11.00414	0.76986
3	1.00, 0.212, 0.01	4.14373	4.63051	0.78872

In table 2, we find that approach 2 of penalty approach usually puts highest weight to the reference partition matrix with the lowest similarity. While, it may make partition matrix U depart far away from the original one. Due to irrationality, approach 2 is not commended. Its irrationality can also be reflected in fig. 8. In fig. 8, one prototype escapes from its corresponding cluster.

4 Conclusions

In this paper, we discuss a problem in the horizontal collaborative fuzzy clustering proposed by W. Pedrycz. The problem is how to determine the weights appeared in the objective function. In the horizontal collaborative fuzzy clustering, because of potential security and privacy restrictions, giving weights subjectively is not feasible.

So, we concentrate on auto-weighted horizontal collaborative fuzzy clustering. We first discuss how to measure the similarity between the partition matrices and give an axiom definition of this similarity measure, then by a given similarity measure to calculate the similarity between the external partition matrices and the reference partition matrix. The weights are calculated according to the similarities and some principle. Meanwhile it is easy for the decision-makers to give such kind principle. We do some experiments and the results show that the new proposed algorithms for determining weights are suitable. Thus our work makes the horizontal collaborative fuzzy clustering algorithm more feasible and useful.

References

1. J.C. Bezdek (1981) Pattern recognition with fuzzy objective function algorithms, Plenum Press, New York
2. J.C. Bezdek, R.Ehrlich, W.Full (1984) FCM: the fuzzy C-means clustering algorithm, Computers and Geosciences 10:191-203
3. M.P. Windham (1982) Cluster validity for fuzzy C-Means clustering algorithms, IEEE Trans. On Pattern Analysis and Machine Intelligence 11:57-363
4. W. Pedrycz (2002) Collaborative fuzzy clustering, Pattern Recognition Letters 23: 1675-1686
5. W. Pedrycz (2005) Knowledge-Based Clustering: From Data to Information Granules, Wiley Inter-Science
6. W. Pedrycz (2003) Knowledge-Based Networking in Granular Worlds, In Rough-Neural Computing, edited by S.P. Pal etc., Springer, 109-138
7. X.L. Xie, G, Beni (1991) A validity measure for fuzzy clustering, IEEE Trans. On Pattern Analysis and Machine Intelligence13: 841-647

A Web-Based Fuzzy Decision Support System for Spare Parts Inventory Control

Yurong Zeng[1], Lin Wang[2], and Jinlong Zhang[2]

[1] School of Computer, Hubei University of Economics, Wuhan 430205, P.R. China
zengyurong@sohu.com
[2] School of Management, Huazhong University of Science and Technology, Wuhan 430074, P.R. China
wanglin@mail.hust.edu.cn

Abstract. This paper presents a web-based fuzzy decision support system (FDSS) for spare parts inventory control in a nuclear power plant. In this study, we integrate an effective algorithm of meta-synthesis based on Delphi, analytic hierarchy process, grey theory, and fuzzy comprehensive evaluation to identify the criticality class of spare parts, and the web-based inventory control FDSS to obtain reasonable replenishment parameters that will be helpful for the reducing of total inventory holding costs. The proposed FDSS was successful in decreasing inventories holding costs significantly by modifying the unreasonable purchase applications while maintaining the target service level.

Keywords: spare parts, inventory control, fuzzy decision support system.

1 Introduction

Inventory control of spare parts (SPs) plays an increasingly important role in modern operations management. Based on the experience of operations and maintenance, the demands for SPs are usually uncertain and excessive stocking is very expensive [1]. For example, there are about 5,000 SPs that are vital for safely production in a nuclear power plant in China. Most of them are non-standard and purchased from France with a lead-time ranging from 6 to 20 weeks. The company is usually obliged to carry inventories consisting of over 110 millions dollars of SPs ready for maintenance.

A key distinguishing feature of the SPs inventory system is the need to specify the criticality class of the SPs [2]. Usually, we identify the optimal control parameters according to the criticality class, that is associated with the setting of target service level, and other constrains condition. Therefore, identifying criticality class for a new SP in an accurate and efficient way becomes a challenge for inventory management. The best way to manage an inventory is thorough the development of better technique for identifying the criticality class (H, M or L) of a SP and management inventories from the point of view of their necessity in production and maintenance operation. But this decision procedure is rather complex. However, decision support system (DSS) which has the capability of easy storage and retrieval of the required information would be a powerful tool to support decision-makers in making decisions [3].

So, there is a need to develop a DSS to assist managers to control inventory costs effectively while achieve the required service level.

There are many papers about spare parts inventory control [4,5]. However, there is currently no integrated DSS that combines the effective criticality class evaluation and web-based decision implementation to SPs replenishment decision-making. We therefore propose an integrated framework that can implement the identifying of criticality class that will be used to specify the target service level and present optimal inventory purchase advices. Our approach is as follows: first, the SPs criticality class evaluation system (SPCCES) provides the evaluation results of criticality class. Second, the web-based replenishment DSS (WRDSS) is developed to provide some replenishment suggestions. Moreover, the integrated SPs inventory control DSS (SPICDSS) is designed to provide managers with daily purchase suggestions.

2 The Framework for SPICDSS

2.1 The General Integrated Framework

The integrated framework of SPICDSS proposed will be developed to support decision making in SP inventory control with the use of the criticality class evaluating results and other constraint conditions. Based on this framework, we develop an SPICDSS that can be used by many companies.

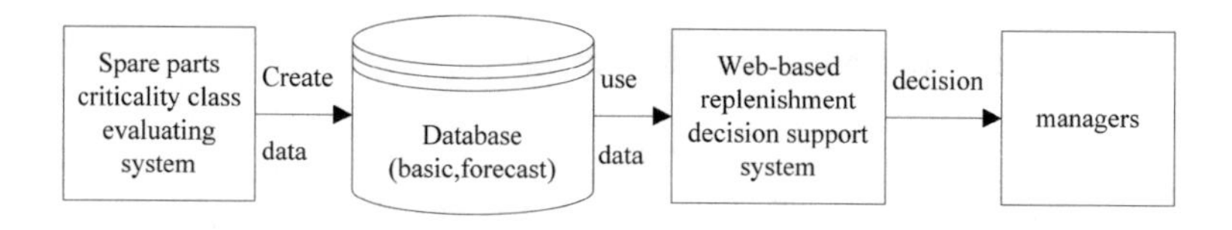

Fig.1. The general integrated framework

SPICDSS consists of two subsystems: one is SPCCES, which is the off-line training and forecasting unit; the other is WRDSS, which is the daily on-line decision unit. The former can provide classification data for the latter. The general framework is shown in Fig.1.

2.2 The Criticality Class Evaluation Methodology

Factors such as costs of spares, availability, storage considerations, probability of requirement of a SP, machine downtime costs, etc., are generally considered while managing SPs inventories. However, little work has been conducted on the criticality analysis of SPs using systematic and well-structured procedures. Simple and straightforward procedures such as ABC and analytic hierarchy process (AHP) analysis have been used in practice for confirming the criticality class of the SPs [6]. But the index considered is so simple and may results in inaccuracy result. However, the criticality of SPs must to be evaluated and this is a difficult task which is often accomplished using subjective judgments. It is a fact that criticality class evaluation is relatively difficult to most companies, and only limited information is available. The application of fuzzy and grey theory to criticality class analysis seems appropriate; as such analysis is highly subjective and related to inexact and grey information [7]. Grey theory was developed by Deng[8], based upon the concept that information is sometimes incomplete or unknown. In the following section, we will discuss the proposed criticality class evaluation model.

(1) Confirm the index set

By using group-discussing and anonymous questionnaire methods at the same time, necessary information is gathered and the 40 experts' ideas are analyzed. The following factors are selected to evaluate the criticality class of a SP.

Table 1. The index set for evaluating the criticality class

	Index name	Desription
P_1	The specificity	For standard parts the availability is usually good; For the user-specific parts: suppliers are unwilling to stock the special, low volume parts.
P_2	Predictability of demand	It is easily to divide the parts into two categories: parts with random failures and parts with a predictable wearing pattern.
P_3	Status of availability of the production facility	Alternative production facility available; alternative production facility available if suitable modifications are made in machine or process; no alternative production facility available.
P_4	Functional necessity	A SP is more crucial when it's pivotal to production or safety problem.
P_5	Lead-time	The difficulty to obtain a SP is associated with the lead-time of procurement.
P_6	Reparability character	If a SP can't be repaired or the time for repair is so long, the difficulty to manage a SP is high.
P_7	The stage of lifecycle	If a SP is in his initial or decay stage, the difficulty to obtain a SP in a short time will become higher.
P_8	Supply market	When a SP is always readily available from several suppliers, the difficulty to manage the SPs is low.

(2) Confirm the weight of each index

In fact, AHP is an effective multi-criteria decision making tool to find out the relative priorities to be assigned to different criteria and alternatives which characterize a decision. We can identify the weight of each index using AHP, i.e. $W=\{W_1, W_2,\ldots, W_n\}$.

(3) Confirm the sample matrix

A score among 1 to 10 can be given for every index according to the difficulty to obtain a SP or the analysis of its influence to production when a SP is unavailable. Supposing the number of experts is and assuming the value of index i that given by expert L is $d_{ki,}$ then the sample matrix can be expressed as:

$$D = \begin{bmatrix} d_{11} & d_{12} & \cdots & d_{1n} \\ d_{21} & d_{22} & \cdots & d_{2n} \\ \vdots & \vdots & \vdots & \vdots \\ d_{r1} & d_{r2} & \cdots & d_{rn} \end{bmatrix}$$

(4) Confirm the evaluation class

According to scientific estimate theory, we confirm the criticality class of SPs as m. That is to say, the comprehensive evaluate standard matrix is: $V=\{V_1,V_2,\ldots,V_m\}$.

(5) Confirm the evaluation grey number

①Upper end level, grey number $\otimes \in [0,\infty]$, the whitening function (WF) is:

$$f_1(d_{ki}) = \begin{cases} d_{ki}/d_1 & d_{ki} \in [0, d_1] \\ 1 & d_{ki} \in [d_1, \infty] \\ 0 & d_{ki} \in (-\infty, 0) \end{cases} \tag{1}$$

② Middle level, grey number $\otimes \in [0, d_1, 2d_1]$, the WF is:

$$f_2(d_{ki}) = \begin{cases} d_{ki}/d_1 & d_{ki} \in [0, d_1] \\ 2 - d_{ki}/d_1 & d_{ki} \in [d_1, 2d_1] \\ 0 & d_{ki} \notin (0, 2d_1] \end{cases} \tag{2}$$

③Low end level, grey number $\otimes \in (0, \ d_1, d_2)$, the WF is:

$$f_3(d_{ki}) = \begin{cases} 1 & d_{ki} \in [0, d_1] \\ \dfrac{d_2 - d_{ki}}{d_2 - d_1} & d_{ki} \in [d_1, d_2] \\ 0 & d_{ki} \notin (0, d_2] \end{cases} \tag{3}$$

(6) Calculate grey statistics

We can obtain $f_j(d_{ki})$ which represents the degree of d_{ki} belongs to index j (j=1,2,…,m)by grey theory, then n_{ij} and n_i can be calculated by Eq.4 and Eq. 5.

$$n_{ij} = \sum_{k=1}^{r} f_j(d_{ki}) \tag{4}$$

$$n_i = \sum_{j=1}^{m} n_{ij} \tag{5}$$

(7) Calculate grey evaluation value and fuzzy matrix
Then, r_{ij} can be calculated by formula $r_{ij}=n_{ij}/ n_i$, thus:

$$R = \begin{bmatrix} r_{11} & r_{12} & \dots & r_{1m} \\ r_{21} & r_{22} & \dots & r_{2m} \\ \vdots & \vdots & \vdots & \vdots \\ r_{n1} & r_{n2} & \dots & r_{nm} \end{bmatrix} \tag{6}$$

(8) Calculate fuzzy comprehensive matrix

$$B = (b_1, b_2, \dots, b_m) = (w_1, w_2, \dots, w_n) \cdot \begin{bmatrix} r_{11} & r_{12} & \dots & r_{1m} \\ r_{21} & r_{22} & \dots & r_{2m} \\ \vdots & \vdots & \vdots & \vdots \\ r_{n1} & r_{n2} & \dots & r_{nm} \end{bmatrix} \tag{7}$$

(9) Calculate the result of evaluation

Firstly, we should confirm the score of criticality class for a spare part: $C=(V_1,V_2,\dots,V_m)^T$. Thus, Z can be obtained by formula $Z=(W\cdot R)\cdot C$, that is the result of criticality class evaluation.

2.3 The Web-based Replenishment DSS (WRDSS)

To be more explicit, we assume that there is a fixed cost associated with each replenishment. Related assumptions are as follows: (1) We pay attention to the case of Poisson demand; (2) The replenishment lead-time is of constant length; (3) The entire cost is assigned to the item that triggers the replenishment; (4) Inventory costs are proportional to the average inventory level; (5) Service is defined as the fraction of demand to be satisfied directly from shelf [9]. The following notations are defined.

λ_i , poisson demand rate for item i, in pieces/year;
L_i, lead-time for item i;
h_i, inventory holding cost per item per year;
r_i,, price of item i;
k_i, fixed cost per replenishment;
S_i, order-up-to-level of item i;
s_i, must-order point of item i;

$P_{p0\leq}(x_0/z)$, probability that a Poisson variable with parameter z takes on a value less than or equal to x_0.

We know our target can be expressed by Eq. 8.

$$\min \quad EC_i = \frac{S_i - s_i + 1}{2} + (s_i - \lambda_i L_i) h_i r_i + \frac{k_i \lambda_i}{S_i - s_i} \tag{8}$$

Subject to

$$1-\frac{1}{S_i - s_i}\sum_{x_0=s_i+1}^{\infty}(x_0 - s_i)\,p_{p0\leq}(x_0|\ \lambda_i L_i) \geq p \tag{9}$$

Strictly speaking, the pair (S_i-, s_i) which minimize Eq.8 while satisfying Eq.9 should be found. A simple approach we first select the order quantity which minimize EC_i, ignoring the constraint. Then, given this value of S_i- s_i, the constraint is used to find the lowest value for s_i, the first step leads to the usual economic order quantity expression

$$Q_i^* = S_i - s_i = \sqrt{2k_i\lambda_i / h_i r_i} \tag{10}$$

In general this will not be an integer. Therefore, we test the two integer values Q_i Then, the lowest s_i which satisfies the following inequality should be used.

$$\lambda_i L_i - s_i - \lambda_i L_i p_{po\leq}(s_i - 1|\ \lambda_i L_i) + s_i p_{po\leq}(s_i|\ \lambda_i L_i) \leq Q_i(1-p) \tag{11}$$

Of course, we can also implement more complex inventory review policy and adopt different service level defines or other demand pattern assumption, and thus corresponding optimal purchasing strategies will be generated.

Fig.2 shows the basic architecture of this subsystem. Basically, this system contains the following components: (1) **User Interface**. Used by decision-makers to input some data and query related data, and to select some joint replenishment decision methods and retrieve answers WJRDSS. It also displays to the users and allows them to interact with the system to arrive at the satisfactory solutions. (2) **MB**. Used to deal with models of the many different types, which may be chosen. (3) **DB**. Used as a repository of basic data and classification data for all model programs. At the same time, it provides data support for the KB and MB needed. (4) **Security System**. Not only used for user protection purposes, but also checks for user access levels to determine the level of flexibility and interaction that the system would provide.

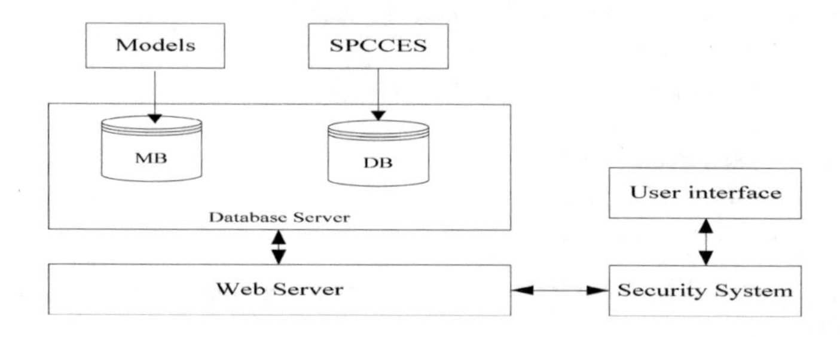

Fig. 2. The architecture of the WRDSS

According to the previous WRDSS architecture, we develop corresponding system using the browser/server mode, which contains a three-tier structure. WRDSS partial user interfaces are shown in Fig.3.

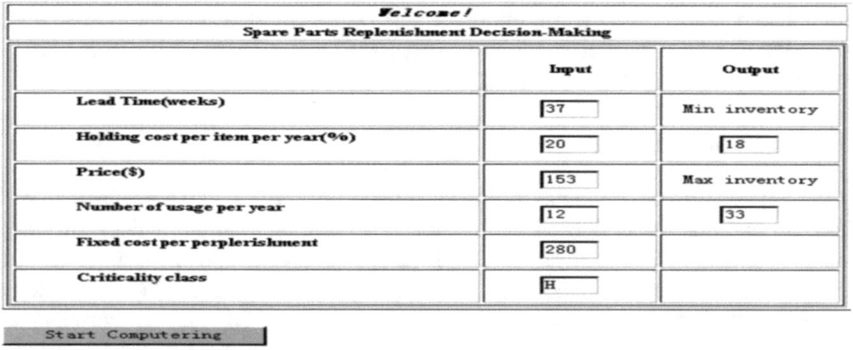

Welcome!

Spare Parts Replenishment Decision-Making

	Input	Output
Lead Time(weeks)	37	Min inventory
Holding cost per item per year(%)	20	18
Price($)	153	Max inventory
Number of usage per year	12	33
Fixed cost per perplerishment	280	
Criticality class	H	

Start Computering

Fig. 3. WRDSS partial user interface

3 The Development of SPICDSS and Application

3.1 Prototype System

Based on the general framework and the description of two subsystems, we construct an SPICDSS for users' convenience. The formation of SPICDSS is the integration of the SPCCES and WRDSS. The integration process is that the SPCCES is embedded into the WRDSS. The SPCCES is developed by Powerbuilder8.0 and adopts the client/server mode here. The advantage of this mode is that the server's function can be exerted as it can in terms of either program execution efficiency or program interaction [10]. In SPCCES, the input data are transformed into appropriate input vectors by corresponding rules and heuristic algorithms. In order to translate those results into corresponding purchasing advice, the WRDSS is constructed. In the same way, the SPCCES is considered to be an embedded system and has been a part of WRDSS. Based on some models and knowledge bases and databases, we develop WRDSS with a three-tier structure using the popular browser/server mode. When entering the WRDSS, we can click the "SPCCES" button; then the SPCCES will run in the server

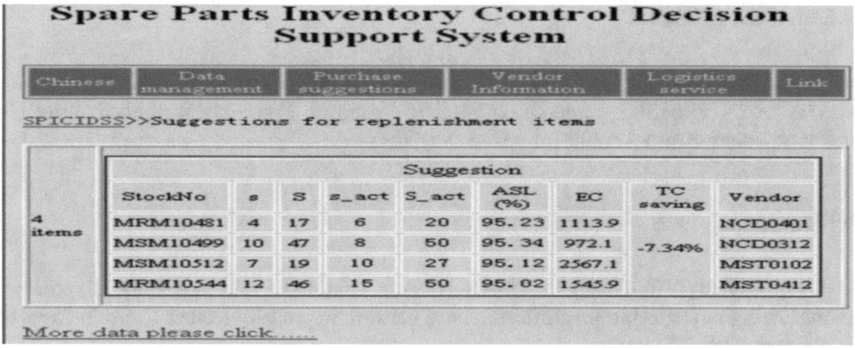

Spare Parts Inventory Control Decision Support System

Chinese | Data management | Purchase suggestions | Vendor Information | Logistics service | Link

SPICIDSS>>Suggestions for replenishment items

	Suggestion								
	StockNo	s	S	s_act	S_act	ASL (%)	EC	TC saving	Vendor
4 items	MRM10481	4	17	6	20	95. 23	1113.9	-7.34%	NCD0401
	MSM10499	10	47	8	50	95. 34	972.1		NCD0312
	MSM10512	7	19	10	27	95. 12	2567.1		MST0102
	MRM10544	12	46	15	50	95. 02	1545.9		MST0412

More data please click......

Fig. 4. The implementation of SPICDSS

site, and meanwhile the results obtained will be transmitted into the corresponding DB of the database server. Fig.4 shows an SPICDSS interface when we query the spare parts replenishment suggestions.

3.2 Application Analysis

By implementation of SPICDSS and practical application in DaYa Bay Nuclear Power Plant, we find that the DSS is integrated, user-oriented; their purchase decision suggestions are reliable by practical testing. It can help inventory managers to make scientific decisions by adopting the replenishment advices provided by SPICDSS. For example, the criticality class of a bearing is corrected from "H" to "M" and the corresponding minimum inventory setting is modified from 7 to 5 with the help of the advice of the SPICDSS. So, the total inventory holding costs of this SP can be saved clearly.

It is obvious that the performance of SPICDSS will depend on the accuracy of the proposed criticality class evaluation model of spare parts. In order to study the effectiveness of the proposed model, we must test the evaluation accuracy by reference to expertise's judgment. The test for the criticality class evaluation of 5100 SPs at a nuclear power plant shows the accuracy is about 97.3%. By deploying these models, the unreasonable target service level settings confirmed by the corresponding criticality class and unnecessary purchase applications were modified and the inventory at DaYa Bay Nuclear Power Plant consisting of over 110 billion dollars worth of SPs was reduced by 8.3%.

4 Conclusions

This study briefly introduces an integrated framework for SPs inventory control decision and implementation of the framework. Based on the proposed framework, we construct an integrated SPs inventory control DSS that can improve the overall performance of forecasting and decision process for inventory managers. In the future, we plan to make more progress in establishing more reasonable inventory models to improve the decision support ability of SPICDSS.

Acknowledgement

This research is partially supported by National Natural Science Foundation of China (No.70571025, 70672039) and the Scientific Research Foundation of Huazhong University of Science and Technology (No.06030B).

References

1. Wang L, Zeng YR, Zhang JL, Huang W, Bao YK (2006) The criticality of spare parts evaluating model using an artificial neural network approach. Lecture Notes in Computer Science, 3991:728-735
2. Chang PL, Chou YC, Huang MG (2005) A (r, r, Q) inventory model for spare parts involving equipment criticality. International Journal of Production Economics 97(1): 66-74

3. Petrovic D, Xie Y, Burnham K (2006) Fuzzy decision support system for demand forecasting with a learning mechanism. Fuzzy Sets and Systems 157:1713-1725
4. Kennedy WJ, Patterson JW (2002) An overview of recent literature on spare parts inventories. International Journal of Production Economics 75:201-215
5. Kukreja A, Schmidt CP (2005) A model for lumpy demand parts in a multi-location inventory system with transshipments. Computer & Operations Research 32 (8):2059-2075
6. Gajpal P, Ganesh L, Rajendran C (1994) Criticality analysis of spare parts using the analytic hierarchy process. International Journal of Production Economics 35:293-298
7. Hsu C, Wen YH (2000) Application of grey theory and multiobjective programming towards airline network design. European Journal of Operational Research, 127(1): 44-68
8. Deng JL (1987) A basic method of grey system. Wuhan: Huazhong University of Science and Technology Press
9. Hua ZS, Zhang B,Yang J, Tan DS(2007)A new approach of forecasting intermittent demand for spare parts inventories in the process industries. Journal of the Operational Research Society 58(1):52-61
10. Lai KK, Yu LA, Wang SY (2004) A neural network and web-based decision support system for forex forecasting and trading. Lecture Notes in Computer Science 3327:243-253

Fuzzy Ranking for Influence Factor of Injury Surveillance Quality in Hospital

Ji-hui Yang and Li-ping Li*

[1] Department of Mathematics, Shantou University
Guangdong, ZIP 515063, P.R. China
yangjihui@163.com
[2] Injury Prevention Research Center Medical College of Shantou University,
Guangdong, ZIP 515041, P.R. China
lpli@stu.edu.cn

Abstract. To scientifically evaluate the import extent of quality of injury surveillance in hospital, and to provide baseline data and scientific evidence for the evaluation of quality of injury surveillance in hospital. **Methods:** The fuzzy complementary judgement matrix about the factor sets is gotten by questionnaire to decision making experts and 0.50-0.90 scale , and then the fuzzy ranking is given for influence factor of injury surveillance quality in hospital based on the modern group decision making theory. **Results:** The ranking of import extent of quality of injury surveillance in hospital is: Factor from the filling staff>Factor from the staff of collecting and checking>Factor from the process of importing>Factor from the staff of supervising and guidance>Factor from the group of surveillance work>Factor from the injury patients>Factor from quality control and managing. **Conclusion:** The ranking method is not only important in theory to evaluate injury surveillance quality in hospital, but it also directionless to injury surveillance sites operation.

Keywords: Injury surveillance quality; Scale; Group decision making; Fuzzy complementary judgement matrix; Ranking.

1 Introduction[5][6]

Injury has been recognized as an important public health problem associated closely with health and life. According to World Health Organization, injury, communicable diseases and noncommunicable diseases have been ranked three major public health problems which we are facing at present. Injury events which often happened brought in large losses for country, family and people who were injured.With the support of the government and health departments, we built injury surveillance spots based in hospital. Through monitoring injury patients and injury circumstances and mastering injury data, we can predict the injury tendency and make some injury prevention strategies which is the effective way to reduce injury morbidity and to avoid injury happening. For

* Corresponding author.

B.-Y. Cao (Ed.): Fuzzy Information and Engineering (ICFIE), ASC 40, pp. 610–618, 2007.
springerlink.com

many years, injury surveillance system has been built in many cities of China. In the operating of injury surveillance sites, the factors of impacting hospital injury surveillance were complicated. If we can grasp these factors, it can help us to use limited fund effectively, to mobilize injury workers reasonably and to guide injury prevention scientifically. In the process of several years's practice monitoring, and after seeking suggestions of famous injury prevention specialists in China, the factors about impacting the quality hospital injury surveillance are as follows:

1.1 Factors from the Filling Staff Include

They are too busy to fill with the injury surveillance report cards timely;

The level of operation level and understanding to report card;

Responsibility and patience;

Whether they get to train about injury surveillance uniformly;

Mental condition;

Negligence without other reasons;

The affect of dialect;

Filling staff's identification.

1.2 Factors from the Injury Patients Include

If injury patients need to do examination further, the filling staff may forget to fill with the examination results;

If injury patients discharged from hospital initiatively because of cost or other reasons, some injury information may lose;

If patients in a coma were sent to hospital by police or witnesses, injury information could not gain;

The degree of understanding the surveillance questions by injury patients;

The degree of cooperation by injury patients;

Recall bias of injury patients;

Lie deliberately or civility bias of injury patients;

Injury patients may be reluctant to answer some injury questions (Such as suicide, intentional injury, sexy violate and so on);

The information proxies replied of injury patients may be not enough accurate.

1.3 Factors from the Staff of Collecting and Checking Include

Whether they collected and examined the report cards on regular time and whether they feedback the problem that they founded to filling staff;

Whether they distributed the cards timely;

Whether they conserved the delivered cards and the undelivered cards properly;

Whether they kept them secretly.

1.4 Factors from the Supervising and Guidance Include

The filling staff may fill with the cards more actively and more accurate when the supervising staff were present;

The frequency of supervising and guidance;

Whether they addressed advices and feedbacked to the principal promptly;

1.5 Factors from the Process of Inputting Include

Entry clerks' attitude;

Entry clerks' operation ability;

The burden of entry because of large data and limited time;

Rationality of entry process.

1.6 Factors from the Group of Surveillance Work Include

The frequency of holding injury training;

The effect of training and the ability of trainer;

The frequency of supervising and guidance irregularly;

Whether they feedback and solve problems they found in time

1.7 Factors from the Quality Control and Managing Include

Whether they communicated with each other and solve problems in time;

How much the surveillance money has been invested, whether the surveillance fund was workable;

They did not carry out quality control by request;

The leader of hospital did not pay much attention to injury surveillance and did not founded related systems.

Aimed at the above seven factors, we made the evaluation form of the importance of injury surveillance quality in hospital, and then we got the fuzzy evaluation of the seven factors by questionnaire to injury prevention experts, based on the modern group decision making theory, a ranking of the seven influence factors was gained. The ranking from reality, combined with group suggestions of injury experts, it had a certain guiding role for reasonable manipulating and improving the quality of injury surveillance sites.

2 Some Basic Concepts[1][2]

To narrate conveniently, suppose that $X = \{x_1, x_2, \cdots, x_n\}$ is factor set which is composed of n factors, where $I = \{1, 2, \cdots, n\}$ is the first n natural number sets.

Definition 1. *Matrix* $A = [a_{ij}]_{m\times n}(0 \leq a_{ij} \leq 1)$*is called* $n \times n$ *fuzzy matrix.*

Definition 2. *If the* $n \times n$ *fuzzy matrix* $A = [a_{ij}]_{m\times n}$ *is such that:*

$$a_{ij} + a_{ji} = 1(i \neq j), \forall i, j \in I,$$
$$a_{ii} = 0, \forall i \in I$$

then A *is called the fuzzy complementary judgement matrix.*

Definition 3. *If the* $n \times n$ *fuzzy complementary judgement matrix* $A = [a_{ij}]_{m\times n}$ *is such that:*

$$a_{ij} = a_{ik} - a_{kj} + 0.5(i \neq j), \forall i, j, k \in I,$$

then A *is called the additive complete consistency fuzzy complementary judgement matrix.*

Definition 4. *If the* $n \times n$ *fuzzy complementary judgement matrix* $A = [a_{ij}]_{m\times n}$ *is such that:*

$$a_{ik}a_{kj}a_{ji} = a_{ki}a_{jk}a_{ij} + 0.5(i \neq j), \forall i, j, k \in I,$$

then A *is called the multiplicative complete consistency fuzzy complementary judgement matrix.*

3 0.50-0.90 Scale[3]

For the given factor sets X, decision making experts compared with every two factors of X by fuzzy linguistics firstly, and then constructed the fuzzy judgement matrix based on some scale. We took 0.50-0.90 scale for example to explain the process of construction.

Meaning of 0.1-0.9 Scale

Scale Value a_{ij}	Fuzzy Linguistics
0.5	a_i is equally important than a_j
0.6	a_i is moderately important to a_j
0.7	a_i is strongly important than a_j
0.8	a_i is very strongly important than a_j
0.9	a_i is extremely important than a_j

Note: The intermediate values of two adjacent judgements in the table are 0.55, 0.65, 0.75, and 0.85. It is easy to know that the judgement matrix constructed by 0.50-0.90 scale is the fuzzy complementary judgement matrix.

4 Ranking Based on the Fuzzy Complementary Judgement Matrix[4]

Without loss of generality, we assume that $X = \{x_1, x_2, \cdots x_7\}$ is the factor set, where

x_1 denotes the factor from the filling staff.
x_2 denotes the factor from the injury patients.
x_3 denotes the factor from the staff of collecting and checking.
x_4 denotes the factor from the staff of supervising and guidance.
x_5 denotes the factor from the process of importing.
x_6 denotes the factor from the group of surveillance work.
x_7 denotes the factor from quality control and managing.

Simultaneity, we invited three decision making experts from Jinan University, Shantou University and the Second Affiliated Hospital of Shantou University Medical College, respectively. Based on 0.50-0.90 scale, the method of pairwise comparison factor used by three decision making experts, and then the three fuzzy complementary judgement matrix can be gotten as follows:

$$A^1 = \begin{bmatrix} 0.50 & 0.60 & 0.60 & 0.60 & 0.80 & 0.70 & 0.80 \\ 0.60 & 0.55 & 0.50 & 0.80 & 0.80 & 0.85 & 0.90 \\ 0.60 & 0.55 & 0.50 & 0.40 & 0.50 & 0.70 & 0.50 \\ 0.60 & 0.80 & 0.40 & 0.50 & 0.70 & 0.60 & 0.50 \\ 0.80 & 0.80 & 0.50 & 0.70 & 0.50 & 0.60 & 0.50 \\ 0.70 & 0.85 & 0.70 & 0.60 & 0.60 & 0.50 & 0.50 \\ 0.80 & 0.90 & 0.50 & 0.50 & 0.50 & 0.50 & 0.50 \end{bmatrix}.$$

$$A^2 = \begin{bmatrix} 0.50 & 0.50 & 0.70 & 0.80 & 0.80 & 0.60 & 0.20 \\ 0.50 & 0.50 & 0.80 & 0.80 & 0.70 & 0.80 & 0.50 \\ 0.70 & 0.80 & 0.50 & 0.70 & 0.70 & 0.60 & 0.30 \\ 0.80 & 0.80 & 0.70 & 0.50 & 0.40 & 0.20 & 0.10 \\ 0.80 & 0.70 & 0.70 & 0.40 & 0.50 & 0.30 & 0.10 \\ 0.60 & 0.80 & 0.60 & 0.20 & 0.30 & 0.50 & 0.30 \\ 0.20 & 0.50 & 0.30 & 0.10 & 0.10 & 0.30 & 0.50 \end{bmatrix}.$$

$$A^3 = \begin{bmatrix} 0.50 & 0.70 & 0.60 & 0.55 & 0.70 & 0.65 & 0.50 \\ 0.70 & 0.50 & 0.30 & 0.20 & 0.40 & 0.25 & 0.15 \\ 0.60 & 0.30 & 0.50 & 0.40 & 0.65 & 0.55 & 0.35 \\ 0.55 & 0.20 & 0.40 & 0.50 & 0.70 & 0.65 & 0.50 \\ 0.70 & 0.40 & 0.65 & 0.70 & 0.50 & 0.40 & 0.15 \\ 0.65 & 0.25 & 0.55 & 0.65 & 0.40 & 0.50 & 0.35 \\ 0.50 & 0.15 & 0.35 & 0.50 & 0.15 & 0.35 & 0.50 \end{bmatrix}.$$

Generally speaking, the above three fuzzy complementary judgement matrixes from decision making experts are not consistent. It can not be used before modified it into fuzzy consistency complementary judgement matrix. We therefore firstly introduce the following formula:

$$b_i^p = \sum_{k=1}^{7} a_{ik}^p, p = 1, 2, 3; i = 1, 2, \cdots, 7,$$

we can compute the ranking index of importance of every matrix by the above formula. The results are

$$b^1 = (4.60, 5.00, 3.75, 4.10, 4.40, 4.45, 4.20)$$
$$b^2 = (4.10, 4.60, 4.30, 3.50, 3.50, 3.30, 2.00)$$
$$b^3 = (4.20, 2.50, 3.35, 3.50, 3.30, 3.35, 2.50)$$

Secondly, we introduce the following formula:

$$c_{ij}^p = \frac{b_i^p - b_j^p}{12} + 0.5, p = 1, 2, 3; i, j = 1, 2, \cdots, 7,$$

we can get fuzzy consistency complementary judgement matrixes of every matrix as follows:

$$C^1 = \begin{bmatrix} 0.50 & 0.47 & 0.57 & 0.54 & 0.52 & 0.51 & 0.53 \\ 0.53 & 0.50 & 0.60 & 0.58 & 0.55 & 0.55 & 0.57 \\ 0.43 & 0.40 & 0.50 & 0.47 & 0.45 & 0.44 & 0.46 \\ 0.46 & 0.42 & 0.53 & 0.50 & 0.48 & 0.47 & 0.49 \\ 0.48 & 0.45 & 0.55 & 0.52 & 0.50 & 0.50 & 0.52 \\ 0.49 & 0.45 & 0.56 & 0.53 & 0.50 & 0.50 & 0.52 \\ 0.47 & 0.43 & 0.54 & 0.51 & 0.48 & 0.48 & 0.50 \end{bmatrix}.$$

$$C^2 = \begin{bmatrix} 0.50 & 0.46 & 0.48 & 0.55 & 0.55 & 0.57 & 0.68 \\ 0.54 & 0.50 & 0.53 & 0.59 & 0.59 & 0.61 & 0.72 \\ 0.52 & 0.47 & 0.50 & 0.57 & 0.57 & 0.58 & 0.69 \\ 0.45 & 0.41 & 0.43 & 0.50 & 0.50 & 0.52 & 0.63 \\ 0.45 & 0.41 & 0.43 & 0.50 & 0.50 & 0.52 & 0.63 \\ 0.43 & 0.39 & 0.42 & 0.48 & 0.48 & 0.50 & 0.61 \\ 0.32 & 0.28 & 0.31 & 0.37 & 0.37 & 0.39 & 0.50 \end{bmatrix}.$$

$$C^3 = \begin{bmatrix} 0.50 & 0.64 & 0.57 & 0.56 & 0.58 & 0.57 & 0.64 \\ 0.36 & 0.50 & 0.43 & 0.42 & 0.43 & 0.43 & 0.50 \\ 0.43 & 0.57 & 0.50 & 0.49 & 0.50 & 0.50 & 0.57 \\ 0.44 & 0.58 & 0.51 & 0.50 & 0.52 & 0.51 & 0.60 \\ 0.42 & 0.57 & 0.50 & 0.48 & 0.50 & 0.50 & 0.57 \\ 0.43 & 0.57 & 0.50 & 0.49 & 0.50 & 0.50 & 0.57 \\ 0.36 & 0.50 & 0.43 & 0.40 & 0.43 & 0.43 & 0.50 \end{bmatrix}.$$

We take the weight vector of the three decision making experts as: $w = (\frac{1}{3}, \frac{1}{3}, \frac{1}{3})$. Then we integrate fuzzy consistency complementary judgement matrixes C^1, C^2, C^3 and w by the following simple weighted average operator:

$$c_{ij} = \frac{c_{ij}^1 + c_{ij}^2 + c_{ij}^3}{3}, \; i, j = 1, 2, \cdots, 7.$$

According to the above operator, we can get the group fuzzy consistency complementary judgement matrix:

$$C = \begin{bmatrix} 0.50 & 0.52 & 0.54 & 0.55 & 0.55 & 0.55 & 0.62 \\ 0.48 & 0.50 & 0.52 & 0.53 & 0.52 & 0.53 & 0.60 \\ 0.46 & 0.48 & 0.50 & 0.51 & 0.51 & 0.51 & 0.57 \\ 0.45 & 0.47 & 0.49 & 0.50 & 0.50 & 0.50 & 0.57 \\ 0.45 & 0.48 & 0.49 & 0.50 & 0.50 & 0.51 & 0.57 \\ 0.45 & 0.47 & 0.49 & 0.50 & 0.49 & 0.50 & 0.57 \\ 0.38 & 0.40 & 0.43 & 0.43 & 0.43 & 0.43 & 0.50 \end{bmatrix}.$$

Finally with the ranking formula of fuzzy consistency complementary judgement matrix:

$$\omega_i = \frac{1}{42}\sum_{j=1}^{7}(c_{ij} + 2.50), i = 1, 2, \cdots, 7,$$

we can get the ranking vector of factor sets:

$$\omega = (0.1507, 0.1417, 0.1438, 0.1423, 0.1428, 0.1421, 0.1310).$$

The results of ranking are:

$$x_1 \succ x_3 \succ x_5 \succ x_4 \succ x_6 \succ x_2 \succ x_7,$$

where the symbol " $\succ$ " means " is preferred or superior to".

5 Conclusion

Base on 0.50-0.90 scale and fuzzy judgement which is given by decision making experts, the decision making experts' judgement is expressed as fuzzy complementary judgement matrix, and then a ranking of the factor sets by modern group decision making theory. Judged from the result of ranking, factor from the filling staff is the most important, because filling staff locate at "entrance position" of the whole injury surveillance process, integrity and exact filling data are the precondition of data coordinate and analysis at later surveillance. Therefore, to guard with "entrance gate" of the injury surveillance process is very important, filling staff must select the person who is experienced, earnest and liable. Filling staff's work is the base of the whole injury surveillance quality. We can also make out by the result of ranking: the affect factors of quality control and managing are "the least important", the reason is, if the affect factors 1, 3, 4, 5, 6 are dealt with very well , it can explain that quality and managing have well done, quality and managing rely on frontal factors directly. We can also make out from the result of ranking that affect factor from injury patients has little effect to the whole injury quality surveillance, we can also find that from practice, to inquire injury patients and their relations, as long as inquiry methods are appreciate and explain the purpose of the work, we can easily gain the correlative data of injury patients, and the quality of data is very good. In a word, the result of ranking is according with the result which sample from injury surveillance sites, and the theory may directionless to injury surveillance sites operation in practice.

Acknowledgments

I am grateful to the experts of injury prevention and control section of Chinese preventive medicine association and the referees for their valuable comments and support. This work supported by the National Natural Science Foundation of China (NSFC)(No. 30571613).

References

1. T L Saaty (1980) The analytic hierarchy process. New York, McGraw-Hill
2. Lianfen Wang, Shubo Xu (1990) Analytic Hierarchy Process Foreword. Beijing, China Renmin University Press (in chinese)
3. Hao Wang, Da Ma (1993) Scale Evaluation and New Scale Methods. Systems Engineer-theory & Practice 13(9): 24-26(in chinese)
4. Zeshui Xu (2004) Uncertain multiple attribute decision making methods and its application. Beijing, Tsinghua University Press(in chinese)
5. Krug E G (2004) Injury surveillance is key to preventing injuries. The Lancet 364, Issue 9445, 30 October: 1563-1566
6. Shengyong Wang (2003) The proposal of bring injury prevention into national disease control work. Chinese Journal of Disease Control and Prevention 9(1): 93-95 (in chinese)

Decision-Making Rules Based on Belief Interval with D-S Evidence Theory

Lv Wenhong

Traffic Information Institute, Shandong University of Science and Technology
579 Qianwangang Road , Qingdao Economic & Technical Development Zone Qingdao ,
Shandong Province, China 266510
lwhgxh@sdust.edu.cn

Abstract: The uncertainty decision-making method is discussed in the paper. Pointing out the shortage of existing rule when decision information is incomplete, putting forward the colligation rule and the conception of unitary operator, the proof is given that the operator is in existence. It is an important tool when we transfer the frame of belief to the probability structure. Finally, a numerical example is given to illustrate the use of the calculation methods.

Keywords: Decision-making; evidence theory; incomplete information; unitary operator.

1 Introduction

Two important types of uncertain decision-making problem are always discussed: decision-making based on completely uncertainty information and decision-making on risking rules[1]. When we make decision on completely uncertain information, the probability of uncertainty results from the fact are unaware, when we make decision on risking rules, the probability of uncertainty results can be quantified by subjective probability. But in more circumstances, decision-makers maybe learn some of the uncertainty results distinctly, and ignorant of some others. It is unreasonable to express the ignorance (nope in inexistence) information with subjective probability. The Dempster-Shafer theory (evidence theory) originated from the concept of lower and upper probability induced by a multivalued mapping, the theory is able to express the ignorance of the information; Its combination tool is also very effectively. So it becomes available supplement to subjective probability method in some degree[1].

The transferable belief model of Dempster-Shafer is based on the appropriate basic probability assignment (BPA) to each proposition[2]. Because of the incompleteness of information, BPA can only be assigned on some certain range, to each proposition, these BPA combined to the interval whose two boundary of the interval are upper probability and lower probability using the Dempster rule. So we can say that the evidence model is the generalization of the probability model[1,3]. In this paper, we discuss the decision making rules based on the incompletely information.

In decision-making problem, the utility of the action is determined not only by the action itself but also determined by many external uncertain factors. These external

B.-Y. Cao (Ed.): Fuzzy Information and Engineering (ICFIE), ASC 40, pp. 619–627, 2007.
springerlink.com

factors are uncontrollable by decision-makers, so we use natural state (or abbreviated as the state) to describe all of external factors we can expected. We suppose, if the decision-makers know which natural state appears, namely, knowing the actual value of the external factor, they can confirm the consequence of any action. Suppose at the same time: decision-makers do not know the true state of the nature, they can only speculate on which state is likely to appear. To be simple, we suppose that there are limited kinds of incompatible possible states, they are $\Theta = \{\theta_1, \theta_2, \cdots, \theta_n\}$; suppose at the same time it has limited feasible actions, namely, action collection $A = \{a_1, a_2, \cdots, a_m\}$, decision makers have to select only one action; when we adopt action a_i and the true state is θ_j, the utility is u_{ij}. They are shown in table 1[4].

The purpose of decision-making is to select the optimum scheme from set A. Then, how can we judge what is the optimum scheme? In uncertain decision object, if we know the natural states of the decision object that may appear but do not know the possibility that each natural state may take place. We ask several experts to judge the possibility of different results. Therefore especially in some important decision problems, it is difficult to judge which scheme will get better utility, different decision makers' understanding about "good" is not unified, it must use some certain rule that the decision group can acceptable to make decision after synthesizing several experts' judgments.

Table 1. Uncertain decision-making with utility

state / action	θ_1	θ_2	…	θ_j	…	θ_n
a_1	u_{11}	u_{12}	…	u_{1j}	…	u_{1n}
a_2	u_{21}	u_{22}	…	u_{2j}	…	u_{2n}
⋮	⋮	⋮	…	⋮	…	⋮
a_i	u_{i1}	u_{i2}	…	u_{ij}	…	u_{in}
⋮	⋮	⋮	…	⋮	…	⋮
a_m	u_{m1}	u_{m2}	…	u_{mj}	…	u_{mn}

2 Decision Rule Based on the Belief Interval

In the value judgment in belief structure, the principles of value judgment based on evidence frame is: if belief function is described in the way of " income", we choose the scheme of higher trust value; if belief function is described in the way of "loss", we choose the scheme of lower trust value (that is we choose the scheme of highest income likelihood degree). When we choose from a lot of schemes, only judging intuitively is not enough, so we have to find a rational decision rule based on the belief function.

2.1 The Decision Rule on *Bel*

In evidence theory, the most often used decision rule is to choose the action of the greatest belief value. The rule is too simple because when we judge based on the boundaries of the belief interval [4,7], the existence of multi-elements is ignorance, so it can't always get a reasonable result. For example: in the same discriminating frame, the trust interval of scheme { θ_1 } is [0.3, 0.9], of scheme { θ_2 } is [0.35, 0.5]. According to the maximum Bel rule, the optimum scheme will be θ_2. The decision-maker only admits the certainty information but disregard the uncertainty caused by incomplete information.

2.2 The Decision Rule on *Pl*

The scheme can also be chosen by *Pl* rule[5]. In this way, the contribution of multi-elements propositions are considered, it can discern different propositions more effectively than the *Bel* rule. The maximum *Pl* rule regards the part that can't be confirmed as belief.

2.3 The Colligation Rule

The value of *Bel* describes the lowest trust degree of the given proposition, *Pl* value describes the highest degree of the given proposition. They show the trust degree of the given proposition from two points of view, so it is ex parte to make decision with only one of them. We should consider the contribution of them at the same time: if the value of *Bel* is the same, the proposition with greater value of *Pl* will obtain more allocation proportions; vice versa. So we propose the colligation rule based on two kinds of belief functions to compromise two decision rules aforesaid. Then we have:

$$FB(A) = Bel(A) + \alpha[Pl(A) - Bel(A)], 0 \le \alpha \le 1 \tag{1}$$

if $\alpha = 0$, $FB(A) = Bel(A)$

if $\alpha = 1$, $FB(A) = Pl(A)$

if $\alpha = \dfrac{Bel(A)}{Bel(A) + Bel(\overline{A})}$,

$$FB(A) = Bel(A) + \frac{Bel(A)}{Bel(A) + Bel(\bar{A})}[Pl(A) - Bel(A)] \quad (2)$$

or its equivalence:

$$FB(A) = Pl(A) - \frac{Bel(\bar{A})}{Bel(A) + Bel(\bar{A})}[Pl(A) - Bel(A)] \quad (3)$$

FB means final belief in this paper. It is used to describe the belief degree in final decision making.

According to this discriminating rule, the scheme with the higher trust value gets more plausibility in its FB. It agrees with people's cognitive experience. But the method needs that $Bel(A) \neq 0$, once the BPA of some scheme is 0, namely, $Bel(A) = 0$, then $FB(A) = 0$. In this situation, we usually use an appointed coefficient or the *Pl* rule.

The colligation rule is a compromise between the maximum *Bel* rule and the *Pl* rule.

2.4 Unitary Operator

When the three methods are regarded as consequences of possibility reasoning, there is always a situation that the sum is not 1. While judging in next step, we need normalization to satisfy: $\sum_{i=1}^{n} P(\theta_i) = 1$; then, can we try to get a feasible α, satisfy with $\sum_{i=1}^{n} FB(\theta_i) = 1$?

Theorem. On the frame of discernment: $\Theta = \{\theta_1, \theta_2, \cdots, \theta_n\}$, $m: 2^{\Theta} \to [0,1]$; $\sum m(B) = 1$. B is the focal element of the frame of discernment. Basic probability distributing of the focal element is: $m(B_i) = \beta_i$, $i = 1, 2, \cdots, q$; then there is a β, which satisfy with:

$$\sum_{i=1}^{n} FB(\theta_j) = 1,$$

$$FB(\theta_j) = Bel(\theta_j) + \beta[Pl(\theta_j) - Bel(\theta_j)], 0 \leq \beta \leq 1.$$

Proof: $\sum_{j=1}^{n} FB(\theta_i) = \sum_{j=1}^{n} \left\{ Bel(\theta_j) + \beta[Pl(\theta_j) - Bel(\theta_j)] \right\}$

$$= \sum_{j=1}^{n} \left\{ m(\theta_j) + \beta[Pl(\theta_j) - m(\theta_j)] \right\}$$

$$= \sum_{j=1}^{n} \left\{ m(\theta_j) + \beta[\sum_{\substack{\theta_j \in B_k \\ k=1,\cdots,q}} m(B_k) - m(\theta_j)] \right\}$$

$$= \sum_{j=1}^{n} \left\{ (1-\beta) m(\theta_j) + \beta[\sum_{\substack{\theta_j \in B_k \\ \theta_j \neq B_k \\ k=1,\cdots,q}} m(B_k)] \right\}$$

$$\because \sum_{j=1}^{n} \mathrm{FB}(\theta_{\mathrm{j}}) = 1 \quad \therefore \sum_{j=1}^{n} \left\{ (1-\beta) m(\theta_j) + \beta[\sum_{\substack{\theta_j \in B_k \\ \theta_j \neq B_k \\ k=1,\cdots,q}} m(B_k)] \right\} = 1$$

$$\beta \sum_{j=1}^{n} \sum_{\substack{\theta_j \in B_k \\ \theta_j \neq B_k \\ k=1,\cdots,q}} m(B_k) - \beta \sum_{j=1}^{n} m(\theta_j) = 1 - \sum_{j=1}^{n} m(\theta_j)$$

$$\beta = \frac{1 - \sum_{j=1}^{n} m(\theta_j)}{\sum_{j=1}^{n} \sum_{\substack{\theta_j \in B_k \\ \theta_j \neq B_k \\ k=1,\cdots,q}} m(B_k) - \sum_{j=1}^{n} m(\theta_j)} \tag{4}$$

β is called as unitary operator.

To all single-element collections, the evidence theory does not require the total of the distribution of the belief function is equal to 1, it can be smaller than 1, in discernment space, it is so-called the subaddition. The sum of plausible function $Pl(\theta_j)$ is often greater than 1 in discernment space; they form the effect of the upper and lower limits of probability in evidence theory. However, on the interval $\left[Bel(\theta_j) \quad Pl(\theta_j) \right]$, there must be an estimation value of every element between the two limits to describe the really state. The estimation need to satisfy: $\begin{cases} Bel(\theta_j) \leq FB(\theta_j) \leq Pl(\theta_j) \\ \sum_{i=1}^{n} FB(\theta_j) = 1 \end{cases}$. In this condition, the measurement we got is equivalent with the measurement of possibility.

3 Example[5]

Three factories affiliated to one company produce engines of the some type , in order to arrange the production plan of the next year, we have analyzed the market, all the possible states are as follows: $\Theta = \{$bad(θ_1) , common(θ_2) , preferable(θ_3) , excellent(θ_4) }.The scheme adopted by the company is respectively arrange A={one factory put into production(a_1) , two factories put into production(a_2) , three factories put into production(a_3) }, and its utility(unit: 10000$) is shown in table 2, try to choose the optimum scheme.

Table 2. The utility table of uncertain problems in example

	θ_1	θ_2	θ_3	θ_4
a_1	10	120	240	360
a_2	-80	200	360	480
a_3	-150	180	380	540

Table 3. Synthesized opinion of the market analysis personnel

$A_i \cap B_j$	$\{\theta_2\}$	$\{\theta_3\}$	$\{\theta_1,\theta_2\}$	$\{\theta_2,\theta_3\}$	$\{\theta_3,\theta_4\}$	$\{\theta_1,\theta_2,\theta_3\}$	$\{\theta_2,\theta_3,\theta_4\}$
	A_1	A_2	A_3	A_4	A_5	A_6	A_7
$m(A_i \cap B_j)$	1/6	1/12	1/12	1/9	1/6	1/18	1/3

Suppose the incomplete predicting information of two main markets (northern market and southern market) is as follows:

$$m_1(A_i)=\begin{cases}1/4 & A_1=\{\theta_1,\theta_2\}\\ 1/3 & A_2=\{\theta_2,\theta_3,\theta_4,\}\\ 1/4 & A_3=\{\theta_3,\theta_4,\}\\ 1/6 & A_4=\Theta\\ 0 & \text{others}\end{cases}$$

$$m_2(B_j)=\begin{cases}1/3 & B_1=\{\theta_1,\theta_2,\theta_3\}\\ 2/3 & B_2=\{\theta_2,\theta_3,\theta_4\}\\ 0 & \text{others}\end{cases}$$

a: Decision-making process with the maximum *Bel* rule

When we adopt the maximum *Bel* rule, we can get the optimum scheme as follows:

$$Bel(\{\theta_1\})=0\ ;\ Pl(\{\theta_1\})=1/12+1/18=5/36\ ;\ \overline{Bel}(\{\theta_1\})=31/36$$

$$Bel(\{\theta_2\})=1/6;\ Pl(\{\theta_2\})=3/4;\ \overline{Bel}(\{\theta_2\})=1/4$$

$$Bel(\{\theta_3\})=1/12;Pl(\{\theta_3\})=3/4;\overline{Bel}(\{\theta_3\})=1/4$$

$$Bel(\{\theta_4\})=0;\ Pl(\{\theta_4\})=1/2;\ \overline{Bel}(\{\theta_4\})=1/2$$

We make the belief value unitary:

$$p(\theta_1)=0;\quad p(\theta_2)=2/3;\quad p(\theta_3)=1/3;\quad p(\theta_4)=0$$

We calculate the expected utility[6] to get the final effect of each action according to table 2 :

$$E(a_1)=\frac{2}{3}\times120+\frac{1}{3}\times240=80;\quad E(a_2)=\frac{2}{3}\times200+\frac{1}{3}\times360=253.3;$$

$$E(a_3)=\frac{2}{3}\times180+\frac{1}{3}\times380=246.6;\quad a*=a_2$$

b: Decision-making process with the maximum *Pl* rule

When we adopt the maximum *Pl* rule, we can get the optimum scheme is θ_2 as follows:

we make the plausible value unitary :

$$p(\theta_1)=0.0649; \quad p(\theta_2)=0.3507; \quad p(\theta_3)=0.3507; \quad p(\theta_4)=0.2337$$

We calculate the expected utility to get the final effect of each action according to table 2 :

$$E(a_1)=211.033; \quad E(a_2)=303.376; \quad E(a_3)=312.855; \quad a*=a_3.$$

c: Decision-making process with the colligation rule

When we adopt the colligation rule, we can get the optimum scheme as follows:

$$FB(A)=Bel(A)+\alpha[Pl(A)-Bel(A)], 0\le\alpha\le 1$$

Because $Bel(\{\theta_1\})=0$; $Bel(\{\theta_4\})=0$, we adopt appointed coefficient $\alpha=0.5$ to calculate the final belief (FB).

$$FB(\{\theta_1\})=5/72; \quad FB(\{\theta_2\})=11/24; \quad FB(\{\theta_3\})=5/12;$$

$$FB(\{\theta_4\})=1/4$$

To get the probability estimate of the natural state, we make the FB unitary:

$$\mathrm{p}(\{\theta_1\})=0.058; \quad \mathrm{p}(\{\theta_2\})=0.384; \quad \mathrm{p}(\{\theta_3\})=0.349;$$

$$\mathrm{p}(\{\theta_4\})=0.209$$

To each scheme, we calculate the expected utility:

$$E(a_1)=214.18; \quad E(a_2)=298.12; \quad \underline{E_3=305.72}; \quad a^*=a_3$$

d: Decision-making process with unitary operator

$$\beta=\frac{1-\sum_{j=1}^{n} m(\theta_j)}{\sum_{\substack{\theta_j\in B_k\\ \theta_j\neq B_k\\ k=1,\cdots q}} m(B_k)}=\frac{27}{68} \qquad FB(\theta_1)=0.0551 \qquad FB(\theta_2)=0.3983$$

$$FB(\theta_3)=0.3481 \qquad FB(\theta_4)=0.1985$$

To each action, we calculate the expected utility:

$$E_1=203.35; \quad E_2=295.85; \quad \underline{E_3=302.90}; \quad a^*=a_3$$

4 Conclusion

The research about decision-making on interval has been attracted more and more attention in recent years. In this paper, we discuss the series decision rule on interval decision-making based on the d-s theory, put forward the conception of unitary operator, and proven is given that the operator is in existence. It is an important tool when we transfer the frame of belief to the probability structure. Finally, we illustrate the use of the calculation methods with a numerical example.

References

1. Yen John (1986) Evidential reasoning in expert systems.University of California, Berkeley:21-34.
2. Smets. P (1992) The transferable belief model and random sets. Int. J. Intell. Systems7: 37-46 .
3. Pan Wei, Wang Yangsheng.Yang Hongji (2004) Decision Rule Analysis of Dempster-Shafer Theory of Evidence .Computer Engineering and Application14: 14-17.
4. Yue Chaoyuan (2003) Theories and Methods of Decision Making . Beijing: Science Press: 194-221.
5. Duan Xinsheng (1993) Evidence Theory and Decision Making, Intelligence. Beijing: China Renmin University Press:32-48.
6. R.R.Yager (2004) Minimization of Regret Decision Making with Dempster-Shafer Uncertainty.2004 IEEE Fuzzy Systems Conference Proceedings. Budapest.Hungary:511-515.
7. R.R.Yager (1992) Decision making under Dempster-Shafer uncertainties. Int. J. general systems 20: 233-245.

An Optimization Model for Stimulation of Oilfield at the Stage of High Water Content

Song Kaoping[1], Yang Erlong[1], Jing nuan[2], and Liu Meijia[1]

[1] Key Laboratory of Enhanced Oil Recovery in Ministry of Education, Daqing Petroleum Institute, Daqing, Heilongjiang Province, 163318, P.R.C.
skp2001@sina.com

[2] Editorial Department of Oil Drilling & Production Technology, Research Institute of Oil Production Technology, Huabei Oil Field Corporation, Renqiu City, Hebei Province, 062552, P.R.C.
cyy_jingn@petrochina.com.cn

Abstract. This paper presents a method to predict pressure distribution and residual oil distribution for water flooding oilfield. To meet the demands of the integral design of oilfield development plan and the high-efficiency exploitation of oilfield, this paper proposes an optimization method for single stimulation measure and integral adjustment of a tract based on analysis of oilfield performance, economic evaluation and professional experience. This integral optimization method is helpful to decision making and optimize the adjustment project. Enforcement of the plan will help not only complete the mission of production, but can bring the maximum incomes under present conditions. So, this study is significant to improve the efficiency and effectiveness of field work.

Keywords: water flooding, stimulation, professional knowledge, optimization, fuzzy mathematics.

1 Introduction

There are plenty of research achievements both in China and in the other countries concerning the prediction of oilfield development performance and optimization of project of stimulation measures. Mature theories and methods have been set up, such as numerical simulation, material balance method, series of characteristic curves of displacement, production rate decline curve, fuzzy mathematics method,and neural network simulation. Lots of relative computer softwares have been worked out and in good application. However, most of these achievements and softwares have no standard database to support and have not yet been used to predict the performance indexes and optimize the simulation project integrally for a whole tract. A expert system with professional database in this aspect has not been worked out. Therefore, it is necessary to build up such a system and

B.-Y. Cao (Ed.): Fuzzy Information and Engineering (ICFIE), ASC 40, pp. 628–636, 2007.
springerlink.com

then use it to adjust water flooding in the later time of high water cut so as to integrate related databases and integrate mature performance analysis theory with experiential methods, periodically or nonperiodically analyze the whole tract and a well group by using computers and then make an optimization plan for the adjustment project in a long time consideration. This kind of optimization is of significant importance for improving the work efficiency and enhance the economic effect.

2 Source of Data for Professional System

10 types of basic databases are built up, such as information database of drilling, geology, and database of oil layers. At the same time, by consulting authority experts and referring to the abundant measured data of oilfield development in recent ten years, professional knowledge database of limit indexes for seven adjustment measures are created. The seven adjustment measures include fracturing, water shut off, additional perforation, changing high water cut oil wells into water injectors, shutting off the high water cut oil wells, enhancing the liquid production rate and profile control [1][2].

3 Method of Optimization for the Adjustment of Oilfield Development

Before optimizing the oilfield adjustment project, analysis and evaluation of the oilfield development performance should be done first for the purpose of knowing the fact objectively and thus making the most suitable and optimal project of adjustment[3].

4 Evaluation of Development Status of Oilfield

Among all of the indexes of oilfield development performance, residual oil and pressure are the most important two items. Residual oil is the material base for oilfield development and pressure is the crucial driving power for oil displacement.

4.1 Method for Predicting the Pressure of Single-Phase Flow

Studies conducted with VIP black oil simulator show that, during water flooding, if pressure of well point is specified, then at different water cut stage after water breakthrough, pressure distribution does not change obviously. So, for this reason, two-phase flow can be simplified as single-phase flow firstly and then predict the pressure distribution by using single-phase flow model.

Mathematical model for pressure distribution is as equation (1).

$$\frac{\partial}{\partial x}\left(\frac{K_x}{\mu}\frac{\partial P}{\partial x}\right)+\frac{\partial}{\partial y}\left(\frac{K_y}{\mu}\frac{\partial P}{\partial y}\right)+q_w=\varphi C\frac{\partial P}{\partial t} \tag{1}$$

The initial condition is as equation (2).

$$P(x, y, 0) = P_i \quad (0 \leq x \leq L_x) \quad (0 \leq y \leq L_y) \tag{2}$$

The external boundary constraint is that no flow occurs there and the internal boundary constraint is that pressure keeps constant.

By difference the above-mentioned partial differential equation and boundary constraints, we can get pressure equation group. Using the method of point relaxation, we can attain the pressure distribution.

4.2 Prediction of Residual Oil Saturation

The amount of injected water from one injector can be split into the oil wells connected to this injector according to the value of the influential factors such as resistance force of seepage, pressure gradient, etc.. The water production rate of one oil well is the same with the total water amount split to it from all the water injection wells connected to it.

On the basis of the theory of material balance, the method for predicting water saturation distribution can be set up.

The formula for calculating the water saturation of grid (i,j) is shown as equation (3).

$$S_w^{n+1} = S_w^n + \frac{W_l(1 - S_{wc} - S_{or}) - W_l(S_w^n - S_{wc})}{V_{Py}(1 - S_{wc} - S_{or}) + W_l} \tag{3}$$

This approach can avoid the shortcoming of conventional numerical simulation which often results in tremendous error of material balance during calculating water saturation.

5 Optimization of Adjustment Project by Fuzzy Theory

Fuzzy theory has been widely used in different areas, especially in oil fields. As a basis for possibility theory[4], fuzzy set concepts have been developed by many researchers, such as Dubois and Prade[5]. Based on probability and possibility[4][5][6], fuzzy random programming and random fuzzy programming, which combine randomness with fuzziness in parameters or coefficients, received much attention in recent years. Fuzzy dynamic programming proposed by Bellman and Zadeh[15],has been presented with various extensions and applications[7][8][9][10]. For the development of fuzzy programming, we may refer to the works such as Ichihashi and Kume[11], Lai and Hwang[12], Liu and Liu[13], and Luhandjula[14]. The concept of fuzzy random variable, which was first introduced by Kwakernaak[15], is a basic research tool. However, another important concept, random fuzzy variable, which was introduced by Liu[16], is necessary when various optimization models are formulated in random fuzzy decision systems, such as random fuzzy dependent-chance programming[17], and random fuzzy expected value models[18].

In this paper,according to the demand of designing oilfield development project and high-efficiently developing oilfield, under the guidance of oilfield performance analysis, evaluation of oilfield development indexes and professional knowledge, using fuzzy mathematical method[19] [20], the optimization models for the selection of wells and layers for the above-mentioned 7 stimulation measures are successfully constructed. With these models, the target of integral adjustment of a tract and the optimal adjustment project for a single well can be easily, rapidly and precisely determined. Reservoir is made up of many formations, as for any individual measurement, if the selected formation is different, the oil production and the investment for measurement must be different,too. Moreover, because of the difficulty to accurately get parameters (such as surface area of formation, density, porosity, permeability) and the complexity of oil, gas and water flowing in formations, when selecting formation for measurement, it is impossible to get reasonable formation only by individual factor. To contrast different indexes, the following fuzzy model is available. For the water shutoff formation, the optimization models are given as well. The method involves the following three steps.

5.1 The First Step

Firstly, wells and zones obviously unfit for water shutoff should be artificially picked out and eliminated. These wells and zones are classified as follows.

1. those very different from the above-mentioned indexes
2. those fit for other measures
3. those fit for neither mechanical water plugging nor chemical water plugging

Except for the three cases, water shutoff is available.

5.2 The Second Step

In the second step, first level judgement is conducted using fuzzy method.

Give points to every index for all wells according to the value. If value of the ith index of the jth well is d_j and the upper and the lower limit of it are d_{up} and d_{down} respectively, then the following two parameters can be calculated.

$$a = 0.5(d_{up} + d_{down})$$

$$b = \left[-(d_{up} - d_{down})/(4l \times n \times 0.5)\right]^{1/2}$$

Then tij and rij can be calculated respectively, as shown in (4) and (5).

$$t_{ij} = (d_i - a)^2 - b^2 \ln x \tag{4}$$

$$r_{ij} = \exp(-t_{ij}/b^2) \tag{5}$$

Then we obtain the fuzzy analogous matrix.

$$\mathrm{R}_1 = \begin{bmatrix} \mathrm{r}_{11},\mathrm{r}_{12},\dots\dots\dots,\mathrm{r}_{1n} \\ \mathrm{r}_{21},\mathrm{r}_{22},\dots\dots\dots,\mathrm{r}_{2n} \\ \mathrm{r}_{31},\mathrm{r}_{32},\dots\dots\dots,\mathrm{r}_{3n} \\ \mathrm{r}_{41},\mathrm{r}_{42},\dots\dots\dots,\mathrm{r}_{4n} \end{bmatrix} \begin{matrix} \mathrm{K} \\ \phi \\ \mathrm{S} \\ \mathrm{K_f} \end{matrix} \tag{6}$$

$$R_2 = \begin{bmatrix} r_{11}, r_{12}, \ldots\ldots\ldots, r_{1n} \\ r_{21}, r_{22}, \ldots\ldots\ldots, r_{2n} \\ r_{31}, r_{32}, \ldots\ldots\ldots, r_{3n} \\ r_{41}, r_{42}, \ldots\ldots\ldots, r_{4n} \end{bmatrix} \begin{matrix} H \\ N_p \\ S_w \\ Kh/u \end{matrix} \tag{7}$$

$$R_3 = \begin{bmatrix} r_{11}, r_{12}, \ldots\ldots\ldots, r_{1n} \\ r_{21}, r_{22}, \ldots\ldots\ldots, r_{2n} \\ r_{31}, r_{32}, \ldots\ldots\ldots, r_{3n} \\ r_{41}, r_{42}, \ldots\ldots\ldots, r_{4n} \end{bmatrix} \begin{matrix} P_e \\ Q \\ \eta \\ \Delta P \end{matrix} \tag{8}$$

Apply $B_i = A_i \times R_i$ to calculate B_i.

Reservoir property is $B_1 = (b_{11}, b_{12}, b_{13}, \ldots\ldots\ldots\ldots, b_{1n})$.

Oil related property is $B_2 = (b_{21}, b_{22}, b_{23}, \ldots\ldots, b_{2n})$.

Productivity related property is $B_3 = (b_{31}, b_{32}, b_{33}, \ldots\ldots.b_{3n})$.

Where $b_{ij} = \sum\limits_{k=1}^{4} A_{ik} r_{kj}$.

5.3 The Third Step

The third step is to execute the second level fuzzy judgement.

The fuzzy analogous matrix here can be formulated through the first level judgement.

$$R = \begin{bmatrix} B_1 \\ B_2 \\ B_3 \end{bmatrix} \tag{9}$$

Where, B_1= reservoir property; B_2= oil related property; B_3= productivity property

At last, based on the formula$B = A \times R$(Matrix multiplication), we can obtain (10).

$$B = \{b_1, b_2 \ldots\ldots, b_n\} \tag{10}$$

Where, $b_i = \sum\limits_{k=1}^{B} A_k b_{ki}$.

Then $\bar{b}_i$can be obtained through .

$$\bar{b}_i = b_i / \sum_{i=1}^{N} b_i \tag{11}$$

Final choice can be made according to the calculated$\bar{b}_i$ and its standard distribution. As for fracturing, additional perforation, turning high water cut oil wells into water injectors, shutting off the high water cut oil wells, enhancing the liquid production rate and profile control, the optimizations are similar to the above method.

6 Method of Integral Optimization

Firstly, we should sequence the well and layers for every single measure adjustment project according to the result of the optimization of single measure project and then, for every measure, calculate the amount of increased oil, the amount of decreased water production and the amount of increased water injection or decreased water injection. Secondly, we should calculate economic indexes of every single measure project and execute the integral optimization on the basis of the constructed integral optimization method. Finally, we can get several integrally optimal adjustment projects with a sequence.

No matter which measures are adopted, the managers must seek for the best incomes under the existing conditions. Thus the single object model is proposed as follows:

$$\max f = \sum_{j=1}^{7} \left(\sum_{i=1}^{X(j)} E\left(i, j\right) - X\left(j\right) P\left(j\right) \right) \tag{12}$$

s.t.

$$\sum_{j=1}^{7} X\left(j\right) P\left(j\right) \leq M \tag{13}$$

$$\sum_{i=1}^{7} X\left(j\right) \leq L \tag{14}$$

$$0 \leq X\left(j\right) \leq N\left(j\right) \tag{15}$$

where, M is the permitted total investment; L is the permitted total number of wells; $N\left(j\right)$ is the permitted number of wells for each measures; $P\left(j\right)$ is the unit cost of each measures; $E\left(i, j\right)$ is the increased earning of well i after taking measure j; $X\left(j\right)$ is the amount of wells taking measure j.

7 Diagram for the Idea of the Integral Optimization of Adjustment of Water Flooding Oilfield

The whole process of the integral optimization of the adjustment for water flooding oilfield can be illustrated by figure 1.

8 Applications of the Optimization Model

8.1 Working out of Development Plan

Based on the method established in this paper, the integral optimization software are compiled for setting down development plan of oil field. Users can obtain the optimal measures combinations with aimed well number and its measure

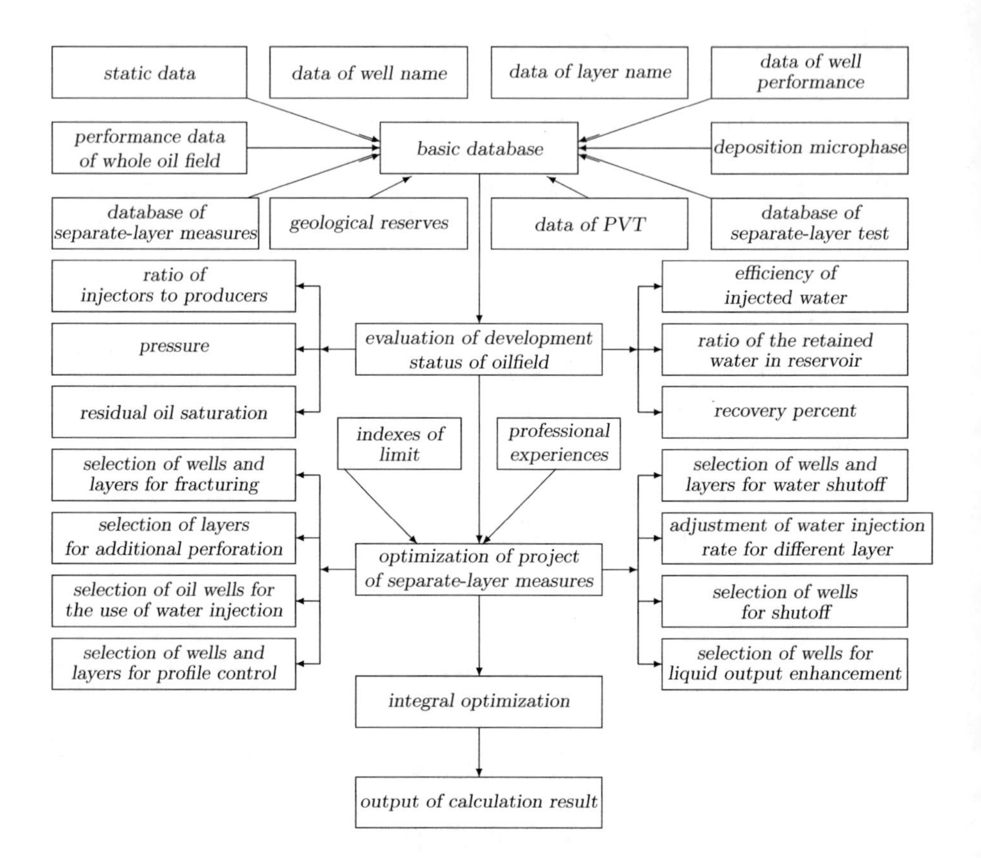

Fig. 1. The conception for globally optimizing the adjustment of water flooding oilfield

type under the permitted conditions. Enforcement of the plan will help not only complete the mission of production but bring the maximum incomes under present conditions. What'more, it can enhance the work efficiency of oil field workers greatly.

8.2 Application Effect of the Model

The methods established here are applied to set down development plan of Beisanxi tract of Daqing oil field in 2004.It only takes 10 days to present the integral adjustment plan of the Beisanxi tract, if the wok done artificially, at least 4 men working for 30 days is needed. Namely, the optimization method improves the compiling rate by 12 times.Compared to artificial plan, the coincidence rate of selecting wells for integral stimulation can reach 93%,that of selecting layers can reach over 85%.According to the development plan obtained by the method, 7 wells are fractured and 5,10,3,2,1,6 wells for water shut off, additional perforation, turning high water cut oil wells into water injectors, shutting off the high water cut oil wells, enhancing the liquid production rate and profile control measures respectively at the end of 2003. Up to January of 2005, Result shows that 7 fracturing wells increased oil production 6.7 tons per day averagely and after enhancing the liquid production rate the well selected increased liquid production 34 tons per day and oil production 7 tons per day. Other measures also obtained obvious effectiveness, which will not be mentioned in detail.

9 Conclusions

1) A mathematical model for the prediction of pressure and residual oil distribution for water flooding oilfield is constructed .
2) Professional knowledge databases for seven types of adjustment measures are build up.
3) Based on professional experiences and fuzzy theory, optimization models are set up for the selection of well and layers for seven types of adjustment measures.
4) According to the optimization result of single measure project, an global optimization method is constructed for integrally optimizing the adjustment measures for water flooding oil field.

References

1. Zhenqi Jia, et al.(1994), Design and Analysis of Oilfield Development, *Press of Harbin Industrial University*, 18-27
2. Xuefei Wang, et al.(1997), Method of Integral Adjustment for High Water Cut Oilfield Development, *Petroleum Industry Press*, 59-66
3. Qingnian Liu, et al.(1992), Optimization of Stimulation Measures, *Technology of Petroleum Drilling and Production*
4. Zadeh L.A.(1978), Fuzzy Sets as a Basis for a Theory of Possibility, Fuzzy Sets and Systems, Vol. 1, 3-28
5. Dubois D. and H. Prade(1988), Possibility Theory, Plenum, New York
6. G.H. Klir(1999), On Fuzzy-Fet Interpretation of Possibility Theory, Fuzzy Sets and Systems, Vol. 108, 263-373
7. Esogbue(1999), A.O, On the Computational Complexity of Some Fuzzy Dynamic Programs, Journal of Computers and Mathematics with Applications, Volume 37, Nos. 11-12,pp47-51
8. Bellman R.E. and L.A. Zadeh(1970), Decision Making in a Fuzzy Environment, Management Science, Vol. 17:4, pp. 141-164
9. Kacprzyk J(1997). Multistage Fuzzy Control: A Model-Based Approach to Control and Decision-Making, Willey, Chichester
10. Esogbue, A.O(2002), New Frontiers of Research in High Impact Dynamic Programming, Proceedings of the 9th Bellman Continuum International Workshop on Uncertain Systems and Soft Computing, July 24-27, Beijing, China
11. M. Inuiguchi, H. Ichihashi and Y. Kume(1993), Modality Constrained Programming Problems: A Unified Approach to Fuzzy Mathematical Programming Problems in the Setting of Prossibility Theory, Inform. Sciences, Vol. 67, 93-126
12. Y.-J. Lai and C.-L. Hwang(1992), Fuzzy Mathematical Programming: Methods and Applications, Springer-Verlag

13. B. Liu and Y.-K. Liu(2002), Expected Value of Fuzzy Variable and Fuzzy Expected Value Models, IEEE Trans. Fuzzy Syst., Vol. 10, No.4
14. M. K. Luhandjula(1996), Fuzziness and Randomness in an Optimization Framework, Fuzzy Sets and Systems, Vol. 77, 291-297
15. H. Lwakernaak(1978), Fuzzy Random Variables I. Definition and Theorems, Inform. Sciences, Vol. 15, 1-29
16. B. Liu(2002), Theory and Practice of Uncertain Programming , Physica-Verlag, Heidelberg
17. B. Liu(2002), Random Fuzzy Dependent-Chance Programming and Its Hybrid Intelligent Algorithm, Inform. Sciences, Vol.141, No. 3-4, 259-271
18. Y.-K. Liu and Liu B.(2001), Expected Value Operator of Random Fuzzy Variable and Random Fuzzy Expected Value Models, Technical Report
19. Kaufman, A.and Gupta, M.M.(1991), Fuzzy Mathematical Models in Engineering and Management Science, 2nd., North-Holland, Amsterdam
20. Liu, B., and Esogbue, A.O.(1996), Fuzzy Criterion Set and Fuzzy Criterion Dynamic Programming, *Journal of Mathmatical Analysis and Applications*, Vol.199, No.1, 293-311

New Research of Coefficient-Fuzzy Linear Programming

Pan Hua[1], Qiu Haiquan[2], Cao Bingyuan[3], and Hou Yongchao[4]

[1] Mathematics Department of Anhui Science and Technology University
Anhui Fengyang 233100
panhua1028@163.com

[2] Mathematics Department of Shantou University
Guangdong Shantou 515063
qiuhaiquan0902@163.com

[3] School of Math. and Inf. Sci. of Guangzhou University
Guangdong Guangzhou 510006
caobingy@163.com

[4] Mathematics Department of Shantou University
Guangdong Shantou 515063
yongchaohou1982@163.com

Abstract. In this paper, we make a further research on the coefficient-fuzzy linear programming. The constraint satisfactory function method and its deficiency are discussed. We introduced a new concept of the membership of constraint. Based on the new concept, the optimal solution varies with the membership of constraint. A new method is presented and an illustrative numerical example is provided to demonstrate the feasibility and efficiency of the proposed approach.

Keywords: Fuzzy Linear Programming, Membership Function, Satisfaction of Constraints, Optimal Solution.

1 Introduction

In this paper, we mainly discuss the following type of Fuzzy Linear Programming:

$$\begin{array}{lrl} (FLP) & \max & c^T x \\ & s.t. & \tilde{A}(x) \leq \tilde{b} \\ & & x \geq 0 \end{array}$$

Where c, x are $n-dimension$ crisp column vectors; $\tilde{A} = (\tilde{a}_{ij})_{m\times n}$, $\tilde{b} = (\tilde{b_i})_{m\times 1}$; $\tilde{a_{ij}}$, $\tilde{b_i}$ are triangular fuzzy numbers. A triangular fuzzy number is denoted as $\tilde{a} = (a_L, a, a_R)$, whose membership function can be denoted by:

$$\mu(x) = \begin{cases} \dfrac{x - a_L}{a - a_L}, & a_L \leq x \leq a; \\ 1, & x = a; \\ \dfrac{a_R - x}{a_R - a}, & a \leq x \leq a_R; \\ 0, & \text{otherwise.} \end{cases}$$

B.-Y. Cao (Ed.): Fuzzy Information and Engineering (ICFIE), ASC 40, pp. 637–643, 2007.
springerlink.com

The research on Fuzzy Linear Programming is becoming an active area in recent years. For solving coefficient-fuzzy linear programming, in most cases, we want to transform it into a classical programming. The most commonly used transform methods are as follows: Ranking fuzzy numbers method[8-9], Constraint satisfactory function method[1-3], Duality Theory[4-7], etc.

Ranking fuzzy numbers method is to establish a one-to-one correspondence between fuzzy numbers and real numbers according to the definite rule, then every fuzzy number is mapped to a point on the real line. Constraint satisfactory function method is a special method of comparing two fuzzy numbers, the relation between big and small is not considered absolutely, but regarded as a fuzzy membership function. The duality theory tried to import crisp (classical) linear programming theories into fuzzy linear programming, and received some similar theories of fuzzy linear programming.

Most researches consider the fuzzy linear programming optimal solution as a crisp number, but ignore the fuzziness of FLP. In this paper we discuss the relation between constraint coefficients and optimal solution. A new concept of FLP constraint membership function is proposed. Based on the new concept, an algorithm of confirming the FLP's constraint coefficients by solving FLP is presented.

2 Membership Function of FLP

A constraint satisfactory function method is to confirm the fulfillment of the relation between two fuzzy numbers in holistic. The definition of constraint satisfactory function was given by Cao[1], and an example was presented. In paper[2], Liu proposed another sort of constraint satisfactory function. If we only compare two fuzzy numbers, the satisfactory function method is effective enough and less strict as ranking fuzzy numbers method, so it could compare two fuzzy numbers very well. FLP does not only discuss two fuzzy numbers, and all fuzzy constraint coefficients will be crisp numbers of their fuzzy sets, but the constraint satisfactory function method does not take it into account. When the constraint satisfaction degree is 0, all constraint coefficients' membership degree are 0; and when all coefficients' membership degree are 1, the satisfactory degree is only 0.5, which is not the result we want. The constraint satisfactory function method focuses on the fulfillment of the relation between fuzzy numbers, ignoring the coefficients' membership degree. Figure 1 shows how the optimal value Z changes with the satisfaction degree p.

It is clear that the optimal value decreases with the satisfactory degree's increase. According to Cao's[1] symmetry method, the optimal value is the object function's intersection point with the object set $\{z \mid z = z1 + \lambda(z0 - z1), \lambda \in [0, 1]\}$, here $z1$ is an optimal value when the satisfaction degree is 1, and $z0$ is an optimal value when the satisfaction degree is 0. We can see it clearly in figure 2.

The solution obtained by constraint satisfactory function method has a low satisfaction degree as we see in figure 2. In fact, from Cao's[1] example we could

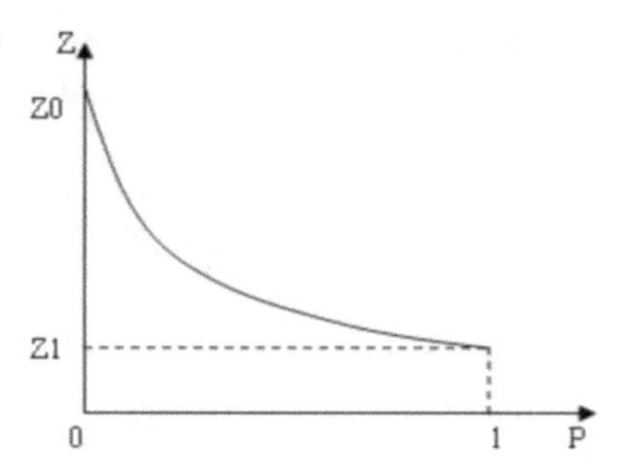

Fig. 1. Optimal Value Function

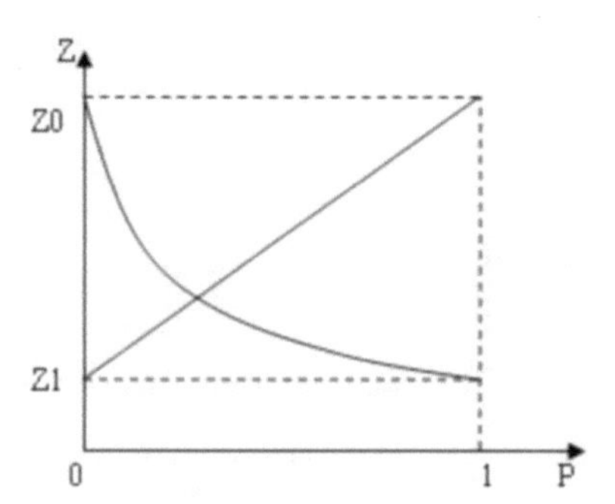

Fig. 2. Optimal Solution

see the final satisfaction degree is $\frac{1}{4}$. In order to solve this problem, we propose another concept of FLP's constraint membership function:

Definition 1. *Suppose $A = (a_{ij})_{m\times n}$, $b = (b_i)_{m\times 1}$ be a group constraint coefficients of FLP. Let μ_{ij} be the membership degree of a_{ij}, λ_i be the membership degree of b_i. Then $\mu = min\{\mu_{ij}, \lambda_i, i = 1, 2, ..., m.\ j = 1, 2, ..., n\}$ is called the membership degree of FLP's constraint under this group constraint coefficients.*

It is obviously that many cases have the same membership degree, here we only discuss the upper and lower bounds of FLP's optimal value that the constraints have the same membership degree. For a fuzzy number $\tilde{A}$, its α-cut $A_\alpha = \{x|x \in X, \mu_A(x) \geq \alpha\}$ is an interval. When A is on the left side of the inequality, the optimal value decreases with the value of A's increase; otherwise, if A is on the right side of the inequality, the optimal value increases with the value of A's increase. So, when constraint coefficient is on the left side of the inequality, we choose the lower bound value of its α-cut, and if it is on the right side, choose the upper bound value of its α-cut, then optimal value could receive the upper bound value on the membership degree of α, we denote this linear programming as $LP(\alpha)^R$ and denote the optimal value as Z_α^R. Contrarily, if constraint coefficient is on the left side of the inequality, we choose the upper bound value of its α-cut, and on the right ones choose the lower bound value of α-cut. The optimal value could receive the lower bound on the membership degree of α, so we denote this linear programming as $LP(\alpha)^L$and denote the optimal value as Z_α^L.

$$(LP(\alpha)^R) \quad \max \ c^T x \quad s.t. \ \tilde{A}_\alpha^L(x) \leq \tilde{b}_\alpha^R \quad x \geq 0$$

$$(LP(\alpha)^L) \quad \max \ c^T x \quad s.t. \ \tilde{A}_\alpha^R(x) \leq \tilde{b}_\alpha^L \quad x \geq 0$$

So all optimal values with the same membership degree α are in the interval $[Z_\alpha^L, Z_\alpha^R]$. Therefore, we could draw the optimal value's change with the constraint membership degree as following.

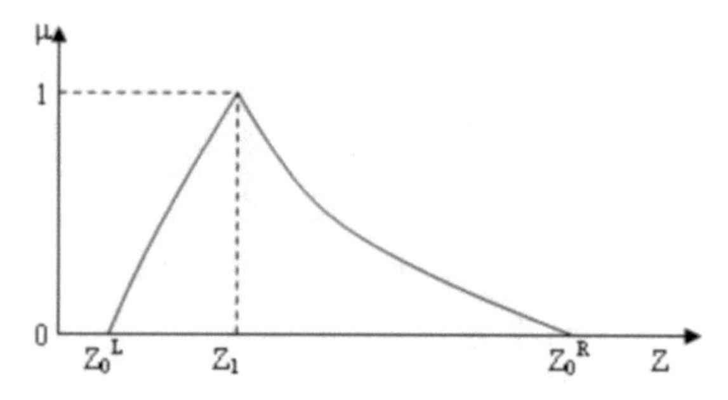

Fig. 3. FLP's constraint membership function

Compared with figure 1, the new definition of FLP's constraint membership function is more suitable for describing the optimal value's change with constraint coefficients.

3 Solution Method

3.1 Optimal Solution of FLP

Most of the cases, we only want to know the possible optimal value range of FLP, but researchers are apt to give a crisp optimal solution. A crisp number could give little information for making decisions. In this paper, we don't consider the optimal value of FLP as a crisp number. Instead, we use a fuzzy number to describe the optimal value. As we see in section 2, the membership of FLP's constraint we defined could do very well in describing the change of the optimal value. Because the object coefficients that discussed in this paper are crisp numbers, we regard constraint membership function of the FLP as the optimal value's membership function, and give decision makers a macroscopical knowledge of the solution. This type of optimization problem have been discussed in many papers, so we don't discuss it any more.

Here, we discuss another type of FLP's optimization problem. Suppose we could only know the range of some constraint coefficients, which could be denoted as fuzzy numbers. In order to obtain a better optimal solution, we should decide the constraint coefficients value by solving the FLP. For this problem, we choose the intersection point of the object function with the object set as an optimal solution. The object set is $\{z \mid z = z_0^L + \lambda(z_0^R - z_0^L), \lambda \in [0,1]\}$, where z_0^R is the upper and z_0^L is the lower bound of optimal value under the membership degree $\alpha = 0$. Finally, we get a membership degree and an optimal solution value by solving FLP, and then we can confirm the crisp numbers of the constraint coefficients.

3.2 Algorithm

Step 1: Select the last interval length l, solve linear programming $LP(0)^L$ and $LP(0)^R$, get optimal values Z_0^L and Z_0^R. Let $\lambda = 0.5$, $p = 0$, $q = 1$;
Step 2: . Solve linear programming $LP(\lambda)^R$, obtain the optimal value Z_λ^R;
If $|Z_\lambda^R - [z_0^L + \lambda(z_0^R - z_0^L)]| < l$, go to step 5;
otherwise, if $Z_\lambda^R > z_0^L + \lambda(z_0^R - z_0^L)$, go to step 3;
otherwise, if $Z_\lambda^R < z_0^L + \lambda(z_0^R - z_0^L)$, go to step 4;
Step 3: Let $q = \lambda$, $\lambda = 0.5(\lambda + p)$. Return to step 2;
Step 4: Let $p = \lambda$, $\lambda = 0.5(\lambda + q)$. Return to step 2;
Step 5: Optimal solution $Z = Z_\lambda^R$, $\lambda^* = \lambda$;
Optimal constraint coefficients $A^* = A_{\lambda^*}^L$, $b^* = b_{\lambda^*}^R$.

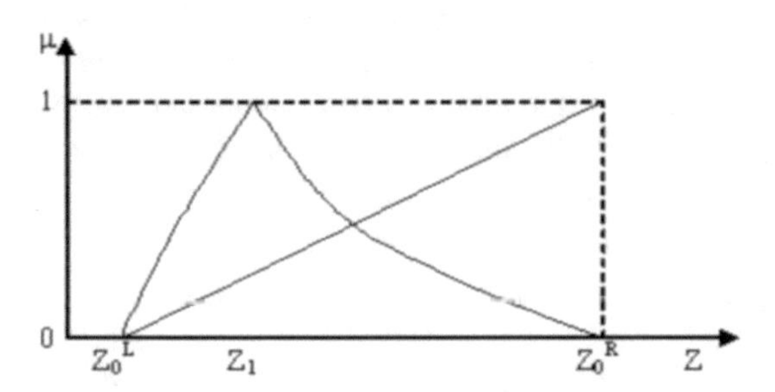

Fig. 4. new optimization solution

Because the object function is a normal fuzzy number, there must exist a pointZ_1 that $Z_0^L < Z_1 < Z_0^R$, and $\lambda_{Z_1} = 1$. The left branch of object function reach $\lambda = 1$ earlier than object set, so there is no intersection point between object set and the left branch of object function except Z_0^L. The intersection point of the object function with the object set is on the right branch of object function, we can see it clearly in figure 4.

4 Numerical Example

The following constraint coefficient-fuzzy linear programming is put forward from Cao[1]:

$$\begin{array}{ll} \max & f(x) = x_1 + x_2 \\ s.t. & \tilde{a}_{11}x_1 + \tilde{a}_{12}x_2 \leq b_1 \\ & \tilde{a}_{21}x_1 + \tilde{a}_{22}x_2 \leq b_2 \\ & x_1, x_2 \geq 0 \end{array}$$

Where $\tilde{a}_{11} = (1,2,3)$, $\tilde{a}_{12} = (0,1,2)$, $\tilde{a}_{21} = (0,1,2)$, $\tilde{a}_{22} = (1,2,3)$ are all triangular fuzzy numbers. $b = [3;3]$ are crisp numbers.

Select the last interval length $l = 0.01$; solve linear programming $LP(0)^L$ and $LP(0)^R$, obtain the optimal solution $Z_0^L = 1.2$ and $Z_0^R = 6$; Let $\lambda = 0.5$, $p = 0$, $q = 1$; Use the algorithm in part 3.2, we get the optimal solution value $Z = 3.2405$, $\lambda = 0.4258$; The optimal constraint coefficients $A^* = [1.4258, 0.4258; 0.4258, 1.4258]$.

In addition, suppose that constraint coefficients b_i are fuzzy numbers too. For example, $\tilde{b}_1 = (1,2,3)$, $\tilde{b}_2 = (2,3,4)$. Here $Z_0^L = 0.5$ and $Z_0^R = 7$; The optimal solution value $Z = 3.3025$, $\lambda = 0.4297$; The optimal constraint coefficients are $b^* = [2.5703; 3.5703]$, $A^* = [1.4297, 0.4297; 0.4297, 1.4297]$.

5 Conclusion

This paper discussed the optimal value change of the FLP with the constraint coefficients, proposed a new concept on constraint membership function of the FLP. Compared with the constraint satisfactory function, the new membership function could describe the optimal value's change better. Based on the new concept on the constraint membership function of FLP, we discussed the optimal solution, and suggested denoting it by using a fuzzy number . Besides, we also discussed how to confirm the constraint coefficients through solving FLP, and presented an algorithm. Finally, numerical examples illustrated the merits of the approach.

Acknowledgement

This work is supported by AHSTUSF grant number ZRC200696.

References

1. Cao Yongqiang, Qu Xiaofei(1995) A New Approach to Linear Programming with Fuzzy Coefficients. Journal of Systems and Engineering 10(1):21–31
2. Xinwang Liu(2001) Measuring the Satisfaction of Constraints in Fuzzy Linear Programming. Fuzzy Sets and Systems 122:263–275
3. Yang Jihui, Han Ying, Zhao Haiqing(2002) Improment on the Method for Solving Fuzzy Linear Programming Problems through the Satisfaction of Constraints. Journal of Liaoning Normal University(Natural Science Edition) 25(3):240–243
4. Jaroslav Ramik(2005) Duality in Fuzzy Linear Programming: Some New Concepts and Results. Fuzzy Optimization and Decision Making 4:25–39
5. Hsien-Chung Wu(2003) Duality in Fuzzy Linear Programming with Fuzzy Coefficients. Fuzzy Optimization and Decision Making 2:61–73

6. Cheng Zhang, Xuehai Yuan, E.Stanley Lee(2005) Duality in Fuzzy Mathematics Programming Problems with Coefficients. Computers and Mathematics with Application 49:1709–1730
7. Hsien-Chung Wu(2004) Duality in Fuzzy Optimization Problems. Fuzzy Optimization and Decision Making 3:345–365
8. Huang Zhengshu(1995) Fuzzy Linear Programming Problems with Fuzzy Coefficients. Mathematica Applicata 8(1):96–101
9. Guan Shijuan(2005) New Method for Ranking Fuzzy Numbers and It's Application in Fuzzy Linear Programming. Journal of Hainan Normal University(Natural Science) 18(2):104–110

Part VI

Artificial Intelligence

Robust Fuzzy Control for Uncertain Nonlinear Systems with Regional Pole and Variance Constraints

Miao Zhihong[1] and Li Hongxing[2]

[1] Department of Basic Science, The Chinese People's Armed Police Force Academy, Langfang, Hebei 065000, China
miaozhh@21cn.com
[2] Department of Mathematics, Beijing Normal University, Beijing, 100875, China
hxlqx@bnu.edu.cn

Abstract. For a class of uncertain nonlinear systems, a design method of robust fuzzy control systems with the state estimators is developed in this paper. This design method guarantees each local closed-loop systems poles within a specified disc and steady-state variances to be less than a set of given upper bounds. T-S fuzzy models with uncertainties are used as the model for the uncertain nonlinear systems. A sufficient condition for the existence of such robust fuzzy controllers is derived using the linear matrix inequality approach. Based on this, solving procedures for the feedback gains and the estimator gains are given.

Keywords: T-S Fuzzy Model, Covariance Control, Linear Matrix Inequalities.

1 Introduction

The design problem of robust fuzzy controller for uncertain nonlinear systems has drawn considerable attention in the past decade, since uncertainties are frequently a source of instability and encountered in various engineering systems [1, 2]. On the other hand, a good transient property and steady state performance are required in many applications [3, 4]. Therefore, how to design a controller for a uncertain nonlinear system so that the desired transient properties and steady-state performances are satisfied has been an important issue.

In this paper, the work of robust controller design with variance and disc pole constraints in [3, 5] is expanded to include nonlinear systems with parameters uncertainties and external disturbance using fuzzy control. Both transient response and steady-state performances are considered. A sufficient condition is derived so that the closed-loop system is asymptotically stable , and meets the pre-specified variance constraints and disc pole constraints, simultaneously. Moreover, the sufficient condition is expressed in LMI terms. Although the well-known separation property does not hold for this case, we can firstly design the feedback gains with the supposition that all the state variables could be measured, and then obtain the estimator gains using this solving procedures.

B.-Y. Cao (Ed.): Fuzzy Information and Engineering (ICFIE), ASC 40, pp. 647–656, 2007.
springerlink.com

2 Problem Description and Preliminaries

Consider a uncertain nonlinear system described by the following fuzzy IF-THEN rules [6]

$$\begin{aligned}&\textbf{Plant Rule i}: \text{ IF } z_1(t) \text{ is } M_{i1} \text{ and } \cdots \text{ and } z_s(t) \text{ is } M_{is}\\&\text{THEN } \dot{x}(t) = (A_i + \triangle A_i)x(t) + (B_i + \triangle B_i)u(t) + D_i w(t),\\&y(t) = C_i x(t), (i = 1, 2, \cdots, r)\end{aligned} \tag{1}$$

where M_{ij} is the fuzzy set, $x(t) = [x_1(t), x_2(t), \cdots, x_n(t)]^T \in R^n$ is the state vector, $z(t) = [z_1(t), z_2(t), \cdots, z_s(t)]^T$ is measurable system vector, $u(t) \in R^m$ is the control input vector, $y(t) \in R^p$ is the controled output vector, $w(t) \in R^l$ is a zero mean white noise process with identity covariance, $A_i, B_i$and C_i are constant matrices of appropriate dimension, and $\triangle A_i, \triangle B_i$ represent the time-varying parameter uncertainties. These parameter uncertainties are defined as follows:

$$[\triangle A_i \ \triangle B_i] = HF[E_{1_i} \ E_{2_i}] \tag{2}$$

where F is an uncertain matrix satisfying $FF^T \leq I$, and H, E_{1_i} and E_{2_i} are known constant matrices with appropriate dimensions.

For any current state vector $x(t)$ and input vector $u(t)$), the final output of the fuzzy system is inferred as follows:

$$\dot{x}(t) = \sum_{i=1}^{r} h_i(z(t))[(A_i + \triangle A_i)x(t) + (B_i + \triangle B_i)u(t) + D_i w(t)], \tag{3}$$

where $h_i(z(t)) = \frac{w_i(z(t))}{\sum_{i=1}^{r} w_i(z(t))}$, $w_i(z(t)) = \prod_{j=1}^{s} M_{ij}(z_j(t))$.

In practice, all or some of the state variables can not be measured. If each local linear plant models are observable, this problem can be overcome by introducing the fuzzy state estimators.

A full-dimensional fuzzy state estimator:

$$\begin{aligned}&\textbf{Estimator Rule i}: \text{ IF } z_1(t) \text{ is } M_{i1} \text{ and } \cdots \text{ and } z_s(t) \text{ is } M_{is}\\&\text{THEN } \ \dot{\hat{x}}(t) = A_i\hat{x}(t) + B_i u(t) - L_i(y(t) - \hat{y}(t)),\\&\hat{y}(t) = C_i\hat{x}(t), \ (i = 1, 2, \cdots, r)\end{aligned} \tag{4}$$

where L_i is $n \times p$ is the constant observer gain maxtrix, $\hat{x}$ is the estimated state vector, and $\hat{y}$ is the output of the fuzzy state estimator.

Therefore, the output of the full-dimensional fuzzy state estimator is inferred as follows:

$$\dot{\hat{x}}(t) = \sum_{i=1}^{r} h_i(z(t))[A_i\hat{x}(t) + B_i u(t) - L_i(y(t) - \hat{y}(t))], \tag{5}$$

A PDC-type T-S fuzzy controller which uses feedback of estimated state is composed of r control rules. And for any current estimated state vector $\hat{x}(t)$, the T-S fuzzy controller infers $u(t)$ as the output of the fuzzy controller as follows:

$$u(t) = \sum_{i=1}^{r} h_i(z(t))K_i\hat{x}(t), \tag{6}$$

Apply (6) to the system (3) and (5), the corresponding closed loop system is given by

$$\begin{aligned}\begin{bmatrix}\dot{x}(t)\\ \dot{\hat{x}}(t)\end{bmatrix} &= \sum_{i=1}^{r}\sum_{j=1}^{r} h_i(z(t))h_j(z(t))\{\begin{bmatrix}A_i+\triangle A_i & (B_i+\triangle B_i)K_j\\ -L_iC_j & A_i+L_iC_j+B_iK_j\end{bmatrix}\begin{bmatrix}x(t)\\ \hat{x}(t)\end{bmatrix}\\ &+\begin{bmatrix}D_i\\ 0\end{bmatrix}w(t)\}\end{aligned} \tag{7}$$

Denote $\tilde{x}(t) = \hat{x}(t) - x(t)$ and $X(t) = [x(t), \tilde{x}(t)]^T$, then the above equation can be transformed as

$$\dot{X}(t) = \sum_{i=1}^{r}\sum_{j=1}^{r} h_i(z(t))h_j(z(t))\{\bar{A}_{ij}X(t) + \bar{D}_iw(t)\} \tag{8}$$

where

$$\bar{A}_{ij} = \begin{bmatrix}A_i + B_iK_j + \Delta A_i + \Delta B_iK_j & B_iK_j + \Delta B_iK_j\\ -(\Delta A_i + \Delta B_iK_j) & A_i + L_iC_j - \Delta B_iK_j\end{bmatrix}, \bar{D}_i = \begin{bmatrix}D_i\\ -D_i\end{bmatrix}.$$

Therefore, the ith local closed-loop system which correspond to the rule i can be written as follows:

$$\dot{X}^i(t) = \bar{A}_{ii}X^i(t) + \bar{D}_iw(t), \tag{9}$$

If the ith local closed-loop system (9) is asymptotical stable for all admissible uncertainties, then the steady-state covariance of this system, $\mathcal{X}^i = \lim_{t\to\infty}\{E[X^i(t)X^{i^T}(t)]\}$ exists and statisfies [7]

$$\bar{A}_{ii}\mathcal{X}^i + \mathcal{X}^i\bar{A}_{ii}^T + \bar{D}_i\bar{D}_i^T = 0, \tag{10}$$

In this paper, $D(b, r)$ represents as a disc with the centre $-b + j0$ and the radius r, $D(\alpha, \beta)$ is also a disc and in right side of the disc $D(b, r)$ as shown in figure 1. $D(b, r)$ and $D(\alpha, \beta)$ can be combined as a pair of discs $(D(b, r), D(\alpha, \beta))$, where $\alpha = \beta + (b - r)$. Usually, we have to design the estimator to converge much faster than the state of a system. Therefore, for each local closed-loop systems, we can assign the disc $D(b, r)$ for the estimator poles and $D(\alpha, \beta)$ for the state closed-loop poles.

Why do we select this kinds of regions for poles placement ? One reason is that a disc region can be succinctly represented as a LMI condition. Another reason is that our algorith given in section 4 for solving state gain matrices and estimator gain matrices can be conveniently implemented. For the convenience of design, we need the following notations for the fuzzy system (3).

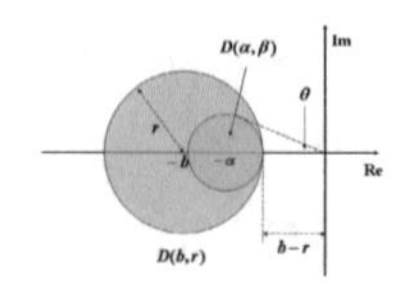

Fig. 1. A Pair of Discs $(D(b,r), D(\alpha,\beta))$

Definition 1. *The fuzzy system (3) is said to be a normal fuzzy system if there exist two constants* $c_1 > 0$,$c_2 > 0$ *such that*

$$c_1\|x(t)\| \leq \|z(t)\| \leq c_2\|x(t)\| \tag{11}$$

for all $t > 0$, *where* $\|\cdot\|$ *is some vector norm.*

It is obvious that if $z(t) = y(t), z(t) = x(t)$, the fuzzy system (3) is normal.

In this paper, we assume that the fuzzy system is normal and an equilibrium point of the closed-loop system (8) is $x(t) = (0, 0, \cdots, 0)^T$.

In order to place all poles of a local closed-loop system in a pre-specified region, we first given the following definition.

Definition 2. *For the ith local closed-loop system (9),*

$$d_i = m_{i1}^2 + m_{i2}^2 + \cdots + m_{is}^2 \tag{12}$$

is called as an off-centre degree, where m_{ij} *is the center of gravity of the membership function* M_{ij}.

From the above definition, we know that the off-centre degree of the local linear system which describes a plant's behavior at the equilibrium point x_0 is usually very small, and if the operating point([8]) of a local linear system has a long distance from the equilibrium point, then its off-centre degree d_i is large. In other word, the off-centre degree of a local linear system indicates a distance from the equilibrium point.

Usually, if the state of a system has a long distance from the equilibrium point, then we expect that the system has a fast response at the moment. On the opposite side, if the state of a system stands near to the equilibrium point, then overshoot should be avoided. Based on this fact, for each local closed-loop system, a pole region can be assigned according to its off-centre degree.

If we want to obtain a better transient, for instance, we can set the pole region far from the imaginary axis for the local linear system which has large off-centre degree, on the contrary, place the pole region near the imaginary axis. Therefore, if the off-centre degrees of all local linear systems has the following order

$$d_{j_1} \leq d_{j_2} \leq \cdots \leq d_{j_r}, \tag{13}$$

where $j_1, j_2, \cdots, j_r$ is a permutation of the sequence $1, 2, \cdots, r$, then we given a series of pairs of discs

$$
\begin{aligned}
&(D(b_{j_1}, r_{j_1}), D(\alpha_{j_1}, \beta_{j_1})), (D(b_{j_2}, r_{j_2}), D(\alpha_{j_2}, \beta_{j_2})), \cdots, \\
&(D(b_{j_r}, r_{j_r}), D(\alpha_{j_r}, \beta_{j_r}))
\end{aligned} \tag{14}
$$

as shown some in Figure 2, where we assume that $b_{j_1} - r_{j_1} \leq b_{j_2} - r_{j_2} \leq \cdots \leq b_{j_r} - r_{j_r}$ and $r_{j_1} \leq r_{j_2} \leq \cdots \leq r_{j_r}$. Conclusively, a disc pair $(D(b_i, r_i), D(\alpha_i, \beta_i))$ can be assigned to the i th local linear system.

Note: The above pole regions arrangement is one of the simple schemes. There may be more meticulous method to be continued study. Then the problem

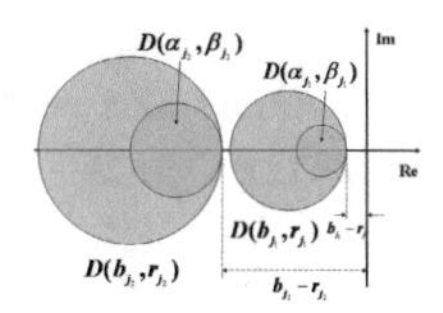

Fig. 2. Desired Pole Placement Regions

to be studied in this paper is the following: given system (3), constants σ_j's $(j = 1, 2, \cdots, 2n)$ and pairs of discs $(D(b_j, r_j), D(\alpha_j, \beta_j))$'s $(j = 1, 2, \cdots, r)$, design a estimator and a control law of the form (5) and (6), respectively, such that the following requirements are statisfied for all admissibles uncertainties:

(a) The closed-loop system (8) is asymptotically stable;
(b) For the ith local closed-loop system (9), all estimator poles lie within the pre-specified disc $D(b_i, r_i)$, and all state closed-loop poles lie within the pre-specified region $D(\alpha_i, \beta_i)$, $(i = 1, 2, \cdots, r)$.
(c) Each local closed-loop steady-state covariance $\mathcal{X}^i$ meets $[\mathcal{X}^i]_{jj} < \sigma_j^2, (j = 1, 2, \cdots, 2n, i = 1, 2, \cdots, r)$.

where $[\mathcal{X}^i]_{jj}$ means the jth diagonal element of $\mathcal{X}^i$, this is the steady-state variance of the jth state.

To obtain our main results, we first introduce two useful lemmas.

Lemma 1. *Given a disc $D(q_i, r_i)$ for the ith local system (9), $\sigma(\bar{A}_{ii}) \subset D(q_i, r_i)$ if there exists a symmetric positive definite matrix $\bar{P}$ such that*

$$
(\bar{A}_{ii} + q_i I)\bar{P}^{-1}(\bar{A}_{ii} + q_i I)^T - r_i^2 \bar{P}^{-1} + q_i \bar{D}_i \bar{D}_i^T < 0 \tag{15}
$$

In this case, the steady-state covariance matrix $\mathcal{X}^i$ exists and satisfies $\mathcal{X}^i < \bar{P}^{-1}$.

Proof: See [3]

A sufficient condition for stability of the closed-loop system (8) can be obtained in terms of Lyapunov's direct method and the following result can be reduced by the application of [6].

Lemma 2. *The closed-loop system (8) is globally asymptotically stable if there existsn a common symmetric positive definite matrix $\bar{P}$ such that*

$$
\bar{A}_{ii}^T \bar{P} + \bar{P} \bar{A}_{ii} < 0, \ i = 1, 2, \cdots, r \tag{16}
$$

and

$$(\bar{A}_{ij} + \bar{A}_{ji})^T \bar{P} + \bar{P}(\bar{A}_{ij} + \bar{A}_{ji}) < 0, \; 1 \le i < j \le r \tag{17}$$

Proof: See [6]

3 Main Results

According to the above section, the main task of control design problem is to select the feedback gains $K_i (i = 1, 2, \cdots, r)$ and the estimator gains $L_i (i = 1, 2, \cdots, r)$ such that the system (8) meets the requirements (a), (b) and (c).

To proceed, we need the following notations.

$$\begin{aligned}
&V_i = K_i P, \quad W_i = Q L_i, \\
&\Phi_i = A_i P + P A_i^T + B_i V_i + V_i^T B_i^T, \Psi_i = A_i^T Q + Q A_i + C_i^T W_i^T + W_i C_i, \\
&N_i = E_{i_1} P + E_{i_2} V_i, M_i = E_{i_2} K_i, \\
&\Phi_{ij} = (A_i + A_j) P + P (A_i + A_j)^T + (B_i V_j + B_j V_i) + (B_i V_j + B_j V_i)^T, \\
&\Psi_{ij} = (A_i + A_j)^T Q + Q (A_i + A_j) + (W_i C_j + W_j C_i) + (W_i C_j + W_j C_i)^T, \\
&N_{ij} = (E_{i_1} + E_{j_1}) P + E_{i_2} V_j + E_{j_2} V_i, M_{ij} = E_{i_2} K_j + E_{j_2} K_i. \\
&e_j = [0, \cdots, 1, \cdots, 0] \in R^n (i.e. \text{the } j \text{ th element of } e_j \text{ is 1, the others 0}).
\end{aligned} \tag{18}$$

where P and Q are $n \times n$ symmetric positive definite matrices.

Firstly, according to Lemma 2, the sufficient stability condition for system (8) can be obtained as follows.

Theorem 1. *If there exist common symmetric positive definite matrices P and Q, matrices K_i and $L_i (i = 1, 2, \cdots, r)$, constants $\lambda_i > 0 (i = 1, 2, \cdots, r)$ and $\lambda_{ij} > 0$ $(i, j = 1, 2, \cdots, r)$ such that*

$$\begin{bmatrix} \Phi_i + \lambda_i N_i^T N_i & B_i K_i & H & H \\ (B_i K_i)^T & \Psi_i + \lambda_i M_i^T M_i & -H & QH \\ H^T & -H^T & -\lambda_i I & 0 \\ H^T & -H^T Q & 0 & -\lambda_i I \end{bmatrix} < 0, \; (i = 1, 2, \cdots, r) \tag{19}$$

and

$$\begin{bmatrix} \Phi_{ij} + \lambda_{ij} N_{ij}^T N_{ij} & B_i K_j + B_j K_i & H & H \\ (B_i K_j + B_j K_i)^T & \Psi_{ij} + \lambda_{ij} M_{ij}^T M_{ij} & -H & QH \\ H^T & -H^T & -\lambda_{ij} I & 0 \\ H^T & -H^T Q & 0 & -\lambda_{ij} I \end{bmatrix} < 0, \quad (1 \le i < j \le r), \tag{20}$$

then the closed-loop system (8) is globally asymtotically stable for all admissiable uncertained F.

Secondly, in order that the closed-loop system (8) meets the desired transient and steady-state response, we can assign each local system pole in a disc and steady-state variance not greater than a upper bound. The following theorem can be obtained from (9) in Lemma 1.

Theorem 2. *For the ith local closed-loop system (8), given a disc $D(b_i, r_i)$, and constants σ_j $(j = 1, 2, \cdots, 2n)$, if there exist symmetric positive definite matrices P and Q, matrices K_i and L_i, and constants $\zeta_i > 0$ such that*

$$\begin{bmatrix} -P & * & * & * & * & * & * \\ 0 & -Q & * & * & * & * & * \\ (A_i + b_i I)P + B_i V_i & B_i K_i & -r_i^2 P & * & * & * & * \\ 0 & Q(A_i + b_i I) + W_i C_i & 0 & -r_i^2 Q & * & * & * \\ 0 & 0 & H^T & -H^T Q & -\zeta_i I & * & * \\ \zeta_i N_i & \zeta_i M_i & 0 & 0 & 0 & -\zeta_i I & * \\ 0 & 0 & D_i^T & D_i^T Q & 0 & 0 & -b_i^{-1} I \end{bmatrix} < 0. \tag{21}$$

where the asterisks replace the matrix blocks that are inferred readily by symmetry, and

$$[P]_{jj} < \sigma_j^2, \ (j = 1, 2, \cdots, n), \tag{22}$$

and

$$\begin{bmatrix} -\sigma_{n+j}^2 & e_j \\ e_j^T & -Q \end{bmatrix} < 0, \ (j = 1, 2, \cdots, n) \tag{23}$$

then its all poles lie with in $D(b_i, r_i)$, this is $\sigma(\bar{A}_{ii}) \subset D(b_i, r_i)$,and it satisfies the requirements (c), where $K_i = V_i P^{-1}$, $L_i = Q^{-1} W_i$.

Note that stable conditions (19), (20) in Theorem 1 imply that

$$\begin{bmatrix} \Phi_i + \lambda_i N_i^T N_i & H \\ H^T & -\lambda_i I \end{bmatrix} < 0, \ (i = 1, 2, \cdots, r) \tag{24}$$

and

$$\begin{bmatrix} \Phi_{ij} + \lambda_{ij} N_{ij}^T N_{ij} & H \\ H^T & -\lambda_{ij} I \end{bmatrix} < 0, \ (1 \le i < j \le r). \tag{25}$$

After some manipulation and by the Schur complement, (24) and (25) are equivalent to

$$\begin{bmatrix} \Phi_i & \lambda_i^{-1} H & N_i^T \\ H^t & -\lambda_i^{-1} I & 0 \\ N_i & 0 & -\lambda_i^{-1} I \end{bmatrix} < 0, \ (i = 1, 2, \cdots, r) \tag{26}$$

and

$$\begin{bmatrix} \Phi_{ij} & \lambda_{ij}^{-1}H & N_{ij}^T \\ \lambda_{ij}^{-1}H^T & -\lambda_{ij}^{-1}I & 0 \\ N_{ij} & 0 & -\lambda_{ij}^{-1}I \end{bmatrix} < 0, \ (1 \leq i < j \leq r), \tag{27}$$

respectively.

The pole constrant condition (21) in Theorem 2 implies that

$$\begin{bmatrix} -P & P(A_i+b_iI)^T+V_i^TB_i^T & PE_{i_1}^T+V_i^TE_{i_2}^T & 0 & 0 \\ (A_i+b_i)P+B_iV_i & -r_i^2P & 0 & \zeta_i^{-1}H & D_i \\ E_{i_1}P+E_{i_2}V_i & 0 & -\zeta_i^{-1}I & 0 & 0 \\ 0 & \zeta_i^{-1}H^T & 0 & -\zeta_i^{-1}I & 0 \\ 0 & D_i^T & 0 & 0 & -b_i^{-1}I \end{bmatrix} < 0. \tag{28}$$

On the other hand, if the control law of the system (8) is used as $u(t) = \sum_{i=1}^{r} h_i(z(t))K_ix(t)$ (without fuzzy estimator), then we can also obtain the following corollary through the same procedures as the proofs of Theorem 1 and Theorem 2.

Corollary 1. *For the system (8) in which the control law is used as* $u(t) = \sum_{i=1}^{r} h_i(z(t))K_ix(t)$ *, if there exist common symmetric positive definite matrices* P *and* Q*, matrices* $K_i(i = 1, 2, \cdots, r)$ *, and constants* $\lambda_i > 0(i = 1, 2, \cdots, r)$*,* $\lambda_{ij} > 0$ $(i, j = 1, 2, \cdots, r)$ *and* $\zeta_i > 0(i = 1, 2, \cdots, r)$ *such that (26),(27) and*

$$\begin{bmatrix} -P & P(A_i+\alpha_iI)^T+V_i^TB_i^T & PE_{i_1}^T+V_i^TE_{i_2}^T & 0 & 0 \\ (A_i+\alpha_i)P+B_iV_i & -\beta_i^2P & 0 & \zeta_i^{-1}H & D_i \\ E_{i_1}P+E_{i_2}V_i & 0 & -\zeta_i^{-1}I & 0 & 0 \\ 0 & \zeta_i^{-1}H^T & 0 & -\zeta_i^{-1}I & 0 \\ 0 & D_i^T & 0 & 0 & -\alpha_i^{-1}I \end{bmatrix} < 0, \ (i = 1, 2, \cdots, r) \tag{29}$$

and

$$[P]_{jj} < \sigma_j^2, \ (j = 1, 2, \cdots, n). \tag{30}$$

hold, then the system is asymtotically stable, all poles of the ith local closed-loop system (9), lie within the pre-specified disc $D(\alpha_i, \beta_i)$*, and the steady-state covariance* $\mathcal{X}^i$ *meets* $[\mathcal{X}^i]_{jj} < \sigma_j^2, (j = 1, 2, \cdots, n, i = 1, 2, \cdots, r)$*.*

From Theorem 1 and Theorem 2, we obtain the following main result of this section.

Theorem 3. *If there exist common symmetric positive definite matrices* P *and* Q*, matrices* K_i *and* $L_i(i = 1, 2, \cdots, r)$ *, and constants* $\lambda_i > 0(i = 1, 2, \cdots, r)$*,* $\lambda_{ij} > 0$ $(i, j = 1, 2, \cdots, r)$ *and* $\zeta_i > 0(i = 1, 2, \cdots, r)$ *such that inequalities (19), (20), (21), (22) (23) and (29) hold, then the closed-loop system (8) meets the requirements (a), (b) and (c) for all admissiable uncertained* F*.*

4 Solving Procedures

In this section, we present solving procedures for the feedback gains K_i's and estimator gains L_i's from the inequalities (19), (20), (21),(22),(23) and (29).

Although inequalities (19) and (20) can be easily converted into linear matrix inequalities (LMI's) in P, Q, V_i's , W_i's and K_i's by the Schur complement, it should be noted that (19), (20) and (21) can not be represented as LMI's without parameters K_i's. Hence, two differences K_i might be solved from the parameter $V_i(V_i = K_iP)$ and K_i since both V_i and K_i are appearred as parameters in (19) and (20). For this reasion, P, Q, V_i's , W_i's and K_i's can not be simultaneously solved from (19), (20), (21), (22), (23) and (29). However, we can obtain them by the following two-step procedures.

In the first step, note that the stable conditions (19), (20) in Theorem 1 imply that the conditions (26), (27). On the other hand, since $D(\alpha_i, \beta_i) \subset D(b_i, r_i)$, if P and K_i's satisfy the pole constraint conditions (29), then they also satisfy the pole constraint conditions (28).

It is obvious that the conditions (26), (27),(29) and (30) are linear matrix inequalities(LMI's) in P, V_i's, λ_i's , λ_{ij}'s and ζ_i's. Therefore, the unknown parameters P and V_i(thus $K_i = V_iP^{-1}$) can be solved from the conditions (26), (27),(29) and (30) by using a convex optimization technique. It means that in the first step we solve the parameters K_i's and P, equivalently, with the supposition that all the state variables could be measured.

In the second step, by substituting P and V_i's into (19), (20), (21) and (23), these inequalities become standard LMI's with unknown parameters Q, W_i's and λ_i's λ_{ij}'s and ζ_i's. Here, although the parameters λ_i's λ_{ij}'s and ζ_i's have be solved from the above step, these parameters can be treated as new variables again in this step. Similarly, we can easily solve Q, W_i (thus $L_i = Q^{-1}W_i$) from the LMI's (19), (20), (21) and (23).

It is obvious that the K_i's, L_i's, P and Q solved from the above two steps satisfly the inequalities (19), (20), (21), (22),(23) and (29), simultaneously. Hence, the close-loop system (8) designed by this method meets the requirements (a), (b) and (c).

From above discussion, we know that the well-known separation property does not hold for this case, but we can firstly design the feedback gains with the supposition that all the state variables could be measured, and then obtain the estimator gains using this solving procedures.

5 Conclusions

In this paper, the design problem of a robust fuzzy control system for a class of uncertain nonlinear systems with regional pole and variance constraints has been studied. In order to meet a satisfy dynamic performance, an arrangement plan of pole disc regions is given for each local linear system according to its off-centre degree. Based on the concept of PDC and the linear matrix inequality approach, a sufficient condition for the existance of such controller have been obtained. The solving procedures for the feedback gains and estimator gains have been presented.

References

1. Tanaka K, Sano M(1994) A robust stabilization problem of fuzzy control sysytems and its application to backing up control of a trucktrailer, IEEE Trans. Fuzzy Systems 2:119–134.
2. Tanaka K, Ikeda T, Wang HO(1996) Robust stabilization of a class of uncertain nonlinear systems via fuzzy control:quadratic stabilizability, H^{∞} control theory, and linear matrix inequalities, IEEE Trans. Fuzzy Systems 4: 1–13.
3. Yu L(2000) Robust control of linear uncertain systems with regional pole and variance constraints. International Journal of System Science 31(3):367–371.
4. Yu L, Chen GD and Yang MY(2002) Robust regional pole assignment of uncertain systems via output feedback controllers. Control Theory and Applications (In chinese) 19(2):244–246.
5. Kim K, Joh J, Langari R, Kwon W(1999) LMI-based design of Takagi-Sugeno fuzzy controllers for nonlinear dynamic systems using fuzzy estimators. International Journal of Fuzzy System 1(2):133–144.
6. Wang HO, Tanaka K, Griffin M(1995) An analytical framework of fuzzy modeling and design issues, Proc. American Control Conference, Seattle, Washington 2272–2276.
7. Kwakernaak H, and Sivan R(1972) Linear Optimal Control Systems. New York:Wiley
8. Marcelo CM, Teixeira and Stanislaw HZak(1999) Stabilizing controller design for uncertain nonlinear system using fuzzy models. IEEE Trans. on Fuzzy System 7(2): 133–142.

The Existence of Fuzzy Optimal Control for the Semilinear Fuzzy Integrodifferential Equations with Nonlocal Conditions

Young Chel Kwun[1,⋆], Jong Seo Park[2], and Jin Han Park[3,⋆⋆]

[1] Department of Mathematics, Dong-A University, Pusan 604-714, South Korea
yckwun@dau.ac.kr

[2] Department of Mathematics Education, Chinju National University of Education, Chinju 660-756, South Korea
parkjs@cue.ac.kr

[3] Division of Mathematical Sciences, Pukyong National University, Pusan 608-737, South Korea
jihpark@pknu.ac.kr

Abstract. This paper is concerned with fuzzy number whose values are normal, convex, upper semicontinuous and compactly supported interval in E_N. We study the existence of fuzzy optimal control for the Semilinear fuzzy integrodifferential equations with nonlocal conditions.

Keywords: Fuzzy number, Fuzzy solution, Nonlinear fuzzy control system, Cost function, Optimal solution, Optimal control.

1 Introduction

Generally, several systems are mostly related to uncertainty and unexactness. The problem of unexactness is considered in general exact science and that of uncertainty is considered as vagueness or fuzzy and accident. The problem of accident has been studied in probability theory and has much progress, but that of fuzzy is relatively new and has many possibilities on development.

The purpose of this note is to investigates the existence of fuzzy optimal control for the semilinear fuzzy integrodifferential equations with nonlocal conditions.

Let E_N be the set of all upper semicontinuous convex normal fuzzy numbers with bounded α-level intervals.

Diamond and Kloeden [2] proved the fuzzy optimal control for the following system:

$$\dot{x}(t) = a(t)x(t) + u(t), \quad x(0) = x_0$$

where $x(\cdot),\ u(\cdot)$ are nonempty compact interval-valued functions on E^1.

⋆ This paper was supported by Dong-A University Research Fund in 2006.

⋆⋆ Corresponding author.

B.-Y. Cao (Ed.): Fuzzy Information and Engineering (ICFIE), ASC 40, pp. 657–665, 2007.
springerlink.com

Kwun and Park [5] proved the fuzzy optimal control for the following system:

$$\dot{x}(t) = a(t)x(t) + f(t) + u(t), \quad x(0) = x_0$$

on assumption that for any given $T > 0$ there exists some compact interval valued function $M(T) = [M_l(T), M_r(T)]$ such that

$$M(T) = \int_0^T S(T-s)f(s)ds, \ [M(t)]^\alpha = [M_l^\alpha(T), M_r^\alpha(T)].$$

We consider the existence of fuzzy optimal control of the following semilinear fuzzy integrodifferential equation:

$$\begin{cases} \dot{x}(t) = A\big[x(t) + \int_0^t G(t-s)x(s)ds\big] + f(t, x(t)) + u(t), \\ x_0 = x(0) + g(t_1, t_2, \cdots, x(\cdot)), \quad \cdot \in \{t_1, t_2, \cdots, t_p\}, \end{cases} \tag{1}$$

where $A : I \to E_N$ is a fuzzy coefficient, E_N is the set of all upper semicontinuous convex normal fuzzy numbers with bounded α-level intervals, $f : I \times E_N \to E_N$ is a nonlinear continuous function, $G(t)$ is $n \times n$ continuous matrix such that $\frac{dG(t)x}{dt}$ is continuous for $x \in E_N$ and $t \in I$ with $\|G(t)\| \le k$, $k > 0$, $u : I \to E_N$ is control function and $g : I^p \times E_N \to E_N$ is a nonlinear continuous function. In the place of $\cdot$, we can replace the elements of the set $\{t_1, t_2, \cdots, t_p\}$, $0 < t_1 < t_2 \cdots < t_p \le T$, $p \in N$, the set of all natural numbers.

2 Preliminaries

A fuzzy subset of R^n is defined in terms of membership function which assigns to each point $x \in R^n$ a grade of membership in the fuzzy set. Such a membership function $m : R^n \to [0,1]$ is used synonymously to denote the corresponding fuzzy set.

Assumption 1. m maps R^n onto [0,1].

Assumption 2. $[m]^0$ is a bounded subset of R^n.

Assumption 3. m is upper semicontinuous.

Assumption 4. m is fuzzy convex.

We denote by E^n the space of all fuzzy subsets m of R^n which satisfy assumptions 1-4; that is, normal, fuzzy convex and upper semicontinuous fuzzy sets with bounded supports. In particular, we denoted by E^1 the space of all fuzzy subsets m of R which satisfy assumptions 1-4 [7].

A fuzzy number a in real line R is a fuzzy set characterized by a membership function m_a as $m_a : R \to [0,1]$. A fuzzy number a is expressed as $a = \int_{x \in R} m_a(x)/x$, with the understanding that $m_a(x) \in [0,1]$ represent the grade of membership of x in a and $\int$ denotes the union of $m_a(x)/x$'s [7].

Let E_N be the set of all upper semicontinuous convex normal fuzzy number with bounded α-level intervals. This means that if $a \in E_N$ then the α-level set

$$[a]^\alpha = \{x \in R : m_a(x) \ge \alpha, \ 0 < \alpha \le 1\}$$

is a closed bounded interval which we denote by

$$[a]^\alpha = [a_l^\alpha, a_r^\alpha]$$

and there exists a $t_0 \in R$ such that $a(t_0) = 1$ [7].

The support Γ_a of a fuzzy number a is defined, as a special case of level set, by the following

$$\Gamma_a = \{x \in R : m_a(x) > 0\}.$$

Two fuzzy numbers a and b are called equal $a = b$, if $m_a(x) = m_b(x)$ for all $x \in R$. It follows that

$$a = b \Leftrightarrow [a]^\alpha = [b]^\alpha \quad \text{for all } \alpha \in (0, 1].$$

A fuzzy number a may be decomposed into its level sets through the resolution identity

$$a = \int_0^1 \alpha[a]^\alpha,$$

where $\alpha[a]^\alpha$ is the product of a scalar α with the set $[a]^\alpha$ and $\int$ is the union of $[a]^\alpha$'s with α ranging from 0 to 1.

We define the supremum metric d_∞ on E^n and the supremum metric H_1 on $C(I : E^n)$ as follows.

Definition 1. [3] Let $a, b \in E^n$.

$$d_\infty(a, b) = \sup\{d_H([a]^\alpha, [b]^\alpha) : \alpha \in (0, 1]\},$$

where d_H is the Hausdorff distance.

Definition 2. [3] Let $x, y \in C(I : E^n)$.

$$H_1(x, y) = \sup\{d_\infty(x(t), y(t)) : t \in I\}.$$

Let I be a real interval. A mapping $x : I \to E_N$ is called a fuzzy process. We denote

$$[x(t)]^\alpha = [x_l^\alpha(t), x_r^\alpha(t)], \ t \in I, \ 0 < \alpha \leq 1.$$

The derivative $x'(t)$ of a fuzzy process x is defined by

$$[x'(t)]^\alpha = [(x_l^\alpha)'(t), (x_r^\alpha)'(t)], \quad 0 < \alpha \leq 1$$

provided that is equation defines a fuzzy $x'(t) \in E_N$.

The fuzzy integral

$$\int_a^b x(t)dt, \quad a, b \in I$$

is defined by

$$[\int_a^b x(t)dt]^\alpha = [\int_a^b x_l^\alpha(t)dt, \int_a^b x_r^\alpha(t)dt]$$

provided that the Lebesgue integrals on the right exist.

Definition 3. [1] The fuzzy process $x : I \to E_N$ is a solution of equations (1) without the inhomogeneous term if and only if

$$(\dot{x}_l^\alpha)(t) = \min\Big\{A_l^\alpha(t)\big[x_j^\alpha(t) + \int_0^t G(t-s)x_j^\alpha(s)ds\big],\ i,j = l,r\Big\},$$

$$(\dot{x}_r^\alpha)(t) = \max\Big\{A_r^\alpha(t)\big[x_j^\alpha(t) + \int_0^t G(t-s)x_j^\alpha(s)ds\big],\ i,j = l,r\Big\},$$

and

$$(x_l^\alpha)(0) = x_{0l}^\alpha - g_l^\alpha(t_1, t_2, \cdots, t_p, x(\cdot)),$$
$$(x_r^\alpha)(0) = x_{0r}^\alpha - g_r^\alpha(t_1, t_2, \cdots, t_p, x(\cdot)).$$

The following hypotheses and existence result are Balasubramaniam and Muralisakar's results [1].

(H1) The nonlinear function $g : I^p \times E_N \to E_N$ is a continuous function and satisfies the inequality

$$d_H([g(t_1, t_2, \cdots, t_p, x(\cdot))]^\alpha, [g(t_1, t_2, \cdots, t_p, y(\cdot))]^\alpha) \le c_1 d_H([x(\cdot)]^\alpha, [y(\cdot)]^\alpha)$$

for all $x(\cdot), y(\cdot) \in E_N$, c_1 is a finite positive constant.

(H2) The inhomogeneous term $f : I \times E_N \to E_N$ is a continuous function and satisfies a global Lipschitz condition

$$d_H([f(s, x(s))]^\alpha, [f(s, y(s))]^\alpha) \le c_2 d_H([x(s)]^\alpha, [y(s)]^\alpha)$$

for all $x(\cdot), y(\cdot) \in E_N$, and a finite positive constant $c_2 > 0$.

(H3) $S(t)$ is a fuzzy number satisfying for $y \in E_N$, $S'(t)y \in C^1(I : E_N) \bigcap C(I : E_N)$ the equation

$$\frac{d}{dt}S(t)y = A\big[S(t)y + \int_0^t G(t-s)S(s)yds\big]$$
$$= S(t)Ay + \int_0^t S(t-s)AG(s)yds, \quad t \in I,$$

such that

$$[S(t)]^\alpha = [S_l^\alpha(t), S_r^\alpha(t)]$$

and $S_i^\alpha(t)(i = l, r)$ is continuous. That is, there exists a constant $c > 0$ such that $|S_i^\alpha(t)| \le c$ for all $t \in I$.

Theorem 1. [1] *Let* $T > 0$, *and hypotheses* (H1)-(H3) *hold. Then for every* $x_0,\ g \in E_N$, *the fuzzy initial value problem* (1) *without control function has a unique solution* $x \in C(I : E_N)$.

3 Fuzzy Optimal Control

In this section, we consider the existence of fuzzy optimal control of the nonlinear fuzzy control system (1).

For $u, v \in E_N$,

$$\rho_2(u,v)^2 = \int_0^1 \int_{S^{n-1}} |s_u(\beta,x) - s_v(\beta,x)|^2 d\mu(x) d\beta,$$

where $\mu(\cdot)$ is unit Lebesgue measure on S^{n-1}.

In particular, define $\|u\| = \rho_2(u, \{0\})$. Observe that if $n = 1$ and $[u]^\beta = [u_l^\beta, u_r^\beta]$, then

$$\|u\|^2 = \int_0^1 ((u_l^\beta)^2 + (u_r^\beta)^2) d\beta.$$

Our problem is to minimize

$$J(u) = \frac{1}{2} \int_0^T \|u(t)\|^2 dt \tag{2}$$

subject to

$$x(T) \succeq_\alpha x^1, \ x^1 \in E_N. \tag{3}$$

For given u, the trajectory $x(t)$ is represented by

$$x(t) = S(t)(x_0 - g(t_1, t_2, \cdots, t_p, x(\cdot))) + \int_0^t S(t-s) f(s, x(s)) ds \tag{4}$$
$$+ \int_0^t S(t-s) u(s) ds,$$

for all $t \in [0, T]$, where $S(t)$ is satisfy (H3).

Theorem 2. (**Generalized Kuhn-Tucker Theorem** [2]) *Let f be a differentiable real function on R^m and $G_1, \cdots, G_k$ differentiable mappings from R^m into E^n_{Lip}. Suppose that ξ_0 minimizes $f(\xi)$ subject to $G_i(\xi) \succeq_\alpha 0$, $i = 1, \cdots, k$ and that ξ_0 is a regular point of the inequalities $G_i(\xi) \succeq_\alpha 0$, $i = 1, \cdots, k$. Then there exist dual functionals $\lambda_1^*, \cdots, \lambda_k^* \in \mathcal{P}_\alpha^\oplus$ such that the Lagrangian $f(\xi) + < G_1(\xi), \lambda_1^* > + \cdots + < G_k(\xi), \lambda_k^* >$ is stationary at ξ_0. Moreover, $\sum_{i=1}^k < G_i(\xi_0), \lambda_i^* >= 0$.*

Let P be the positive orthant in R^n. For a given $\alpha \in I$, defined $\mathcal{P}_\alpha \subset E^n_{Lip}$ by

$$\mathcal{P}_\alpha = \{u \in E^n_{Lip} : [u]^\alpha \subset P\}.$$

If $u \in \mathcal{P}_\alpha$, write $u \succeq_\alpha 0$ and if $u -_h v \succeq_\alpha 0$ write $u \succeq_\alpha v$, where $-_h$ is Hukuhara difference and $u \succeq_\alpha 0$ if and only if $u \geq 0$ with necessity α. The positive dual cone of $\mathcal{P}_\alpha$ is the closed convex cone $\mathcal{P}_\alpha^\oplus \subset E^{n^*}_{Lip}$, defined by

$$\mathcal{P}_\alpha^\oplus = \{p \in E_{Lip}^{n^*} :< u, p > \geq 0 \text{ for all } u \in \mathcal{P}_\alpha\},$$

where $< u, p >= p(u)$ is the value at $u \in E_{Lip}^n$ of the linear functional $p : E_{Lip}^n \to R$, the space of which is denoted by $E_{Lip}^{n^*}$.

The significance of $\mathcal{P}_\alpha^\oplus$ in our problem is that Lagrange multipliers for local optimization live in this dual cone. Local necessity conditions for (2) and (3) are more accessible when there is some notation of differentiation of the uncertain constraint functions $G = x(T) -_h x^1$. Let $\Pi : E_{Lip}^1 \to C(I \times S^0)$ be the canonical embedding where $S^0 = \{-1, +1\}$. The fuzzy function G is said to be (Fréchet) differentiable at ξ_0 if the map $\widehat{G} = \Pi \circ G$ is Fréchet differentiable at ξ_0. A point ξ_0 is said to be a regular point of the uncertain constraint $G(\xi) \succeq_\alpha 0$ if $G(\xi_0) \succeq_\alpha 0$ and there is $h \in R$ such that $\widehat{G}(\xi_0) + D\widehat{G}(\xi_0)h \succeq_\alpha 0$. Our constraint function be compact-interval valued, $G(\xi) = [G_l(\xi), G_r(\xi)]$, and $J : R^n \to R$. The support function of $G(\xi)$ is

$$\Pi\big(G(\xi)\big)(x) = S_{G(\xi)}(x) = \begin{cases} -G_l(\xi) & \text{if } \ x = -1, \\ +G_r(\xi) & \text{if } \ x = +1, \end{cases}$$

since $S^0 = \{-1, +1\}$. Then $\Pi \circ G = S_{G(\cdot)}$ is obviously differentiable if and only if G_l, G_r are differentiable, and $S'_{G(\xi)}(-1) = -\nabla G_l(\xi)$, $S'_{G(\xi)}(+1) = \nabla G_r(\xi)$. The element of $\mathcal{P}_\alpha^\oplus$ can be seen to be of the form $l_0\lambda_0 + l_{+1}\lambda_{+1} + l_{-1}\lambda_{-1}$, where l_i are nonnegative constants, the λ_i map S^0 to R. And $\lambda_{+1}(-1) = \lambda_{-1}(+1) = 0$, $\lambda_{+1}(+1) \geq 0$, $\lambda_{-1}(-1) \leq 0$ and $\lambda_0(-1) = \lambda_0(+1) \geq 0$. So each element of $\mathcal{P}_\alpha^\oplus$ acts like a triple of nonnegative constants $(\lambda_{-1}, \lambda_0, \lambda_{+1})$,

$$\lambda^*\big(S_{G(\xi)}(\cdot)\big) = (\lambda_{-1} - \lambda_0)G_l(\xi) + (\lambda_0 + \lambda_{+1})G_r(\xi),$$

which is always nonnegative since $\lambda_0\big(G_r(\xi) - G_l(\xi)\big) \geq 0$. If ξ_0 is a regular point which is a solution to the constrained minimization, the Kuhn-Tucker conditions, namely that there exists $\lambda^* \geq 0$ so that

$$\nabla J(u) + \lambda^*\big({S_{G(\xi_0)}}'(\cdot)\big) = 0,$$
$$\lambda^*\big(S_{G(\xi_0)}(\cdot)\big) = 0$$

can be written as

$$\nabla J(u) + (\lambda_{-1} - \lambda_0)\nabla G_l(\xi_0) + (\lambda_0 + \lambda_{+1})\nabla G_r(\xi_0) = 0,$$
$$(\lambda_{-1} - \lambda_0)G_l(\xi_0) + (\lambda_0 + \lambda_{+1})G_r(\xi_0) = 0$$

for some nonnegative reals $\lambda_{-1}, \lambda_0, \lambda_{+1}$.

This extends quite naturally to a fuzzy real number constraint with necessity α, as follows:

Define the function $G : R^n \to E^1$ by

$$[G(\xi)]^\alpha = [G_l^\alpha(\xi), G_r^\alpha(\xi)],$$

where for each $\xi \in R^n, G_l^\alpha(\xi)$ is monotone nondecreasing in α and $G_r^\alpha(\xi)$ is monotone nonincreasing in α (since $\alpha \leq \beta$ implies that $[G(\xi)]^\beta \subset [G(\xi)]^\alpha$).

Suppose further that $G_l^\alpha(\cdot)$ and $G_r^\alpha(\cdot)$ are differentiable in ξ for each $\alpha \in I$. Write, for each fixed α,

$$G(\xi) \succeq_\alpha 0 \text{ if and only if } [G(\xi)]^\alpha \geq 0.$$

Then, if ξ_0 is a regular point of the constraint $G(\xi) \succeq_\alpha 0$ minimizing $J(u)$, there exist nonnegative real numbers $\lambda_{-1}, \lambda_0, \lambda_{+1}$ satisfying

$$\nabla J(u) + (\lambda_{-1} - \lambda_0)\nabla_\xi G_l^\alpha(\xi_0) + (\lambda_0 + \lambda_{+1})\nabla_\xi G_r^\alpha(\xi_0) = 0,$$
$$(\lambda_{-1} - \lambda_0)G_l^\alpha(\xi_0) + (\lambda_0 + \lambda_{+1})G_r^\alpha(\xi_0) = 0.$$

Theorem 3. *There exists fuzzy control $u_0(t)$ for the fuzzy optimal control problem (2) and (3) such that*

$$\begin{aligned}
J(u_0) &= \min J(u) \\
&= \frac{1}{2T^2}\int_0^T \int_0^1 \Big[S_l^\beta(T-s)^{-2}\Big((x^1)_l^\beta - S_l^\beta(T)\big(x_{0l}^\beta - g_l^\beta(t_1,t_2,\cdots,t_p,x(\cdot))\big) \\
&\qquad - \int_0^T S_l^\beta(T-s) f_l^\beta(s,x(s))ds\Big)^2 \\
&\quad + S_r^\beta(T-s)^{-2}\Big((x^1)_r^\beta - S_r^\beta(T)\big(x_{0r}^\beta - g_r^\beta(t_1,t_2,\cdots,t_p,x(\cdot))\big) \\
&\qquad - \int_0^T S_r^\beta(T-s) f_r^\beta(s,x(s))ds\Big)^2\Big] d\beta dt
\end{aligned}$$

which is attained when

$$\begin{aligned}
L_{-1}(\beta) = \Big\{&(x^1)_l^\beta - S_l^\beta(T)\big(x_{0l}^\beta - g_l^\beta(t_1,t_2,\cdots,t_p,x(\cdot))\big) \\
&- \int_0^T S_l^\beta(T-s) f_l^\beta(s,x(s))ds\Big\} \times \frac{1}{TS_l^\beta(T-s)^2},
\end{aligned}$$

$$\begin{aligned}
L_{+1}(\beta) = \Big\{&(x^1)_r^\beta - S_r^\beta(T)\big(x_{0r}^\beta - g_r^\beta(t_1,t_2,\cdots,t_p,x(\cdot))\big) \\
&- \int_0^T S_r^\beta(T-s) f_r^\beta(s,x(s))ds\Big\} \times \frac{1}{TS_r^\beta(T-s)^2}.
\end{aligned}$$

Proof. For $\lambda^* \in \mathcal{P}_\alpha^\oplus$, λ^* is represented by three nonnegative, nondecreasing functions $\lambda_{-1}(\beta)$, $\lambda_0(\beta)$, $\lambda_{+1}(\beta)$ on $[\alpha,1]$, which are zero on $[0,\alpha]$, and

$$< u, \lambda^* >= \lambda^*(u) = \int_0^1 u_l^\beta d\big(\lambda_{-1}(\beta) - \lambda_0(\beta)\big) + u_r^\beta d\big(\lambda_0(\beta) + \lambda_{+1}(\beta)\big).$$

Regularity of the problem can be checked by considering it in $E_{Lip}^1([0,T]) \times E^1([0,T])$ and showing that the Fréchet derivative of the function determined by the functional is just the identity on E^1.

Further, for $h \in E^1_{Lip}([0,T])$,

$$DJ(u)\cdot h = \int_0^T \left[\int_0^1 (u_l^\beta(s) h_l^\beta + u_r^\beta(s) h_r^\beta) d\beta \right] ds,$$

clearly maps onto R. Then, since both the functional J and the constraint function are convex, the minimum J_0 of $J(u)$ can be represented in the dual form of the Lagrangian

$$J_0 = \max_{\lambda^* \succeq_\alpha 0} \min_u \Big(J(u) - \lambda^* \big(x(T) -_h x^1 \big) \Big), \tag{5}$$

where $x(T)$ is given by (4). The Hukuhara difference $x(T) -_h x^1$ exists, since it is the constraint under which minimization is taken, and has $\beta-$ level sets

$$\big[x(T) -_h x^1\big]^\beta = \big[x(T)\big]^\beta -_h \big[x^1\big]^\beta = \big[x_l^\beta(T) - (x^1)_l^\beta, x_r^\beta(T) - (x^1)_r^\beta\big].$$

The minimization over u in (5) is of the functional

$$\frac{1}{2}\int_0^T \int_0^1 (u_l^\beta(s)^2 + u_r^\beta(s)^2) d\beta ds - \int_0^T \int_0^1 \Big\{ S_l^\beta(T-s) u_l^\beta(s) d\big(\lambda_{-1}(\beta) - \lambda_0(\beta)\big) + S_r^\beta(T-s) u_r^\beta(s) d\big(\lambda_0(\beta) + \lambda_{+1}(\beta)\big) \Big\} ds \tag{6}$$

along with a term similar to the second in (6), but having

$$\Big(S_l^\beta(T)\big(x_{0l}^\beta - g_l^\beta(t_1,t_2,\cdots,t_p,x(\cdot))\big) + \textstyle\int_0^T S_l^\beta(T-s) f_l^\beta(s,x(s)) ds - (x^1)_l^\beta \Big) \times d\big(\lambda_{-1}(\beta) - \lambda_0(\beta)\big)$$
$$+ \Big(S_r^\beta(T)\big(x_{0r}^\beta - g_r^\beta(t_1,t_2,\cdots,t_p,x(\cdot))\big) + \textstyle\int_0^T S_l^\beta(T-s) f_r^\beta(s,x(s)) ds - (x^1)_r^\beta \Big) \times d\big(\lambda_0(\beta) + \lambda_{+1}(\beta)\big)$$

as integrand.

The usual type of calculus of variations argument shows that $\lambda_{-1}-\lambda_0$ and $\lambda_0 + \lambda_{+1}$ may be taken as differentiable functions of β, and we write the derivatives as $L_{-1}(\beta)$, $L_{+1}(\beta)$ respectively.

The minimum is then

$$-\frac{1}{2}\int_0^T \int_0^1 \{ S_l^\beta(T-s)^2 L_{-1}(\beta)^2 + S_r^\beta(T-s)^2 L_{+1}(\beta)^2 \} d\beta dt \tag{7}$$
$$+ \int_0^T T^{-1} \int_0^1 \Big[\Big((x^1)_l^\beta - S_l^\beta(T)\big(x_{0l}^\beta - g_l^\beta(t_1,t_2,\cdots,t_p,x(\cdot))\big)$$
$$- \int_0^T S_l^\beta(T-s) f_l^\beta(s,x(s)) ds \Big) L_{-1}(\beta) + \Big((x^1)_r^\beta - S_r^\beta(T)\big(x_{0r}^\beta$$
$$- g_r^\beta(t_1,t_2,\cdots,t_p,x(\cdot))\big) - \int_0^T S_r^\beta(T-s) f_r^\beta(s,x(s)) ds \Big) L_{+1}(\beta) \Big] d\beta dt$$

which is attained at the fuzzy control $u_0(t)$ defined by

$$\left[u_0(t)\right]^\beta = \left[S_l^\beta(T-s)L_{-1}(\beta),\ S_r^\beta(T-s)L_{+1}(\beta)\right],\ \ 0 \le \beta \le 1.$$

Maximizing (7) over L_{-1} and L_{+1} gives

$$\begin{aligned}
J_0 = \frac{1}{2T^2}\int_0^T\int_0^1 & \Big[S_l^\beta(T-s)^{-2}\big((x^1)_l^\beta - S_l^\beta(T)(x_{0l}^\beta - g_l^\beta(t_1,t_2,\cdots,t_p,x(\cdot)))\\
& - \int_0^T S_l^\beta(T-s)f_l^\beta(s,x(s))ds\big)^2\\
& + S_r^\beta(T-s)^{-2}\big((x^1)_r^\beta - S_r^\beta(T)(x_{0r}^\beta - g_r^\beta(t_1,t_2,\cdots,t_p,x(\cdot)))\\
& - \int_0^T S_r^\beta(T-s)f_r^\beta(s,x(s))ds\big)^2\Big]d\beta dt
\end{aligned}$$

which is attained when

$$\begin{aligned}
L_{-1}(\beta) = \Big\{ & (x^1)_l^\beta - S_l^\beta(T)(x_{0l}^\beta - g_l^\beta(t_1,t_2,\cdots,t_p,x(\cdot)))\\
& - \int_0^T S_l^\beta(T-s)f_l^\beta(s,x(s))ds\Big\} \times \frac{1}{TS_l^\beta(T-s)^2},
\end{aligned}$$

$$\begin{aligned}
L_{+1}(\beta) = \Big\{ & (x^1)_r^\beta - S_r^\beta(T)(x_{0r}^\beta - g_r^\beta(t_1,t_2,\cdots,t_p,x(\cdot)))\\
& - \int_0^T S_r^\beta(T-s)f_r^\beta(s,x(s))ds\Big\} \times \frac{1}{TS_r^\beta(T-s)^2}.
\end{aligned}$$

□

References

1. Balasubramaniam P, Murarisankar S (2004) Existence and uniqueness of fuzzy solution for semilinear fuzzy integrodifferential equations with nonlocal conditions. J. Computer & Mathematics with applications 47: 1115-1122.
2. Diamond P, Kloeden PE (1990), Optimization under uncertaintly. in: Bouchon-Meunier B, Yager BR (Eds), Proceedings 3rd. IPMU Congress, Paris, 247–249.
3. Diamond P, Kloeden PE (1994) Metric space of Fuzzy sets. World scientific, Singapore New Jersey London HongKong.
4. Hocking LM (1991) Optimal Control an introduction to the theory with applications. Oxford applied Mathematics and Computing Science Series Clarendon Press, Oxford.
5. Kwun YC, Park DG (1998) Optimal control problem for fuzzy differential equations. Proceedings of the Korea-Vietnam Joint Seminar, 103–114.
6. Luenberger DG (1969) Optimization by vector space methods. John Wiley and Sons, Inc..
7. Mizmoto M, Tanaka K (1979), Some properties of fuzzy numbers. Advaces in Fuzzy sets theory and applications, North-Holland Publishing Company, 153-164.

A PSO-Based Approach to Rule Learning in Network Intrusion Detection*

Chen Guolong[1,2,**], Chen Qingliang[1], and Guo Wenzhong[1]

[1] College of Mathematics and Computer Science, Fuzhou University, Fuzhou 350002, China
cgl@fzu.edu.cn, yiyeshucql@163.com, guowenzhong@fzu.edu.cn
[2] School of Computer Science, National University of Defense Technology, Changsha 410073, China

Abstract. The update of rules is the key to success for rule-based network intrusion detection system because of the endless appearance of new attacks. To efficiently extract classification rules from the vast network traffic data, this paper gives a new approach based on Particle Swarm Optimization (PSO) and introduces a new coding scheme called "indexical coding" in accord with the feature of the network traffic data. PSO is a novel optimization technique and has been shown high performance in numeric problems, but few researches have been reported in rule learning for IDS that requires a high level representation of the individual, this paper makes a study and demonstrates the performance on the 1999 KDD cup data. The results show the feasibility and effectiveness of it.

Keywords: Intrusion Detection; Particle Swarm Optimization (PSO); Rule Learning.

1 Introduction

With the rapid development of the Internet, its security has drawn more and more attentions of us. Various technologies have been developed and applied to protect it from attacks, such as firewall, message encryption, anti-virus software and intrusion detection. Intrusion Detection System (IDS) is a software system designed to identify and prevent the misuse of the computer networks and systems. It can be classified into misuse detection system and anomaly detection system [1,2]. Most of the current IDSs are rule-based, such as Bro, Snort and NFR. Rule-based analysis relies on sets of pre-defined rules that are provided by experts, automatically created by the system, or both. These rules are used to determine whether the attacks happened or not. Hence, the update of rules is critical to Rule-based IDS. However, it is a time-consuming and error-prone work for you to discover new rules from the vast network traffic data manually even though you are an expert. Therefore, numbers of approaches based on soft

* Supported by the National Natural Science Foundation of China under Grant No. 60673161, the Key Project of Chinese Ministry of Education under Grant No.206073, the Project supported by the Natural Science Foundation of Fujian Province of China(No.2006J0027).

** Corresponding author: Tel: 86-591-83712045, (0)13799998111.

B.-Y. Cao (Ed.): Fuzzy Information and Engineering (ICFIE), ASC 40, pp. 666–673, 2007.
springerlink.com

computing have been proposed for it, e.g. Decision Support System [3], Genetic Algorithm [4,5,6], Fuzzy Logic [7,8] and so on. Many of them are aimed to extract the rules from the vast data.

Particle Swarm Optimization (PSO), originally introduced by Eberhart and Kennedy [9] in 1995, is an optimization technique inspired by swarm intelligence and theory in general such as bird flocking, fish schooling and even human social behavior. It has been proved to be efficient at solving Global Optimization and Engineering Problems [10]. The advantages of PSO over many other optimization algorithms are its implementation simplicity and ability to converge to a reasonably good solution quickly. In the past several years, PSO has been successfully applied in many researches and application areas. It is demonstrated that PSO gets better results in a faster, cheaper way compared with other methods. However, PSO has rarely been applied in rule learning for network intrusion detection. This paper employs PSO algorithm and introduces a new coding scheme according to the feature of the network traffic data. The experimental results show that the proposed approach is effective and feasible.

2 Standard Particle Swarm Optimization

PSO is a swarm intelligence method, in that the population dynamics simulate the behavior of a "bird's flock", where social sharing of information takes place and individuals profit from the discoveries and previous experience of the other companions during the search for food. Thus, each companion in the population is assumed to "fly" over the search space looking for promising regions on the landscape.

Now some definitions will be explained for Standard Particle Swarm Optimization (SPSO): assuming that the search space is D-dimensional, the i-th particle of the swarm is represented by a D-dimensional vector $X_i = (x_{i1}, x_{i2}, \ldots, x_{iD})$ and the best particle in the swarm is denoted by the index g. The best previous position of the i-th particle is recorded and represented as $P_i = (p_{i1}, p_{i2}, \ldots, p_{iD})$, while the velocity for the i-th particle is represented as $V_i = (v_{i1}, v_{i2}, \ldots, v_{iD})$. Following these definitions, the particles are manipulated according to the following equations:

$$v_{id} = w \times v_{id} + c_1 \times r_1(p_{id} - x_{id}) + c_2 \times r_2(p_{gd} - x_{id}) \tag{1}$$

$$v_{id} = \begin{cases} v_{max} & v_{id} > v_{max} \\ -v_{max} & v_{id} < -v_{max} \end{cases} \tag{2}$$

$$x_{id} = x_{id} + v_{id} \tag{3}$$

Where $d = 1, 2, \ldots, D$; $i = 1, 2, \ldots, N$, N is the size of the population; w is the inertia weight; c_1 and c_2 are two positive constants; r_1 and r_2 are two random numbers in the range [0, 1]; v_{max} is used to control the velocity's magnitude [9].

The pseudo-code for SPSO is as follows:

```
Initiate_Swarm()
Loop
    For p = 1 to number of particles
    Evaluate(p)
    Update_past_experience(p)
    Update_neighborhood_best(p, k)
    For d = 1 to number of Dimentions
        Move(p, d)
    Until Criterion.
```

Fig. 1. SPSO algorithm

3 PSO for Rule Learning

3.1 Coding Scheme

In the approach, the population is a set of particles and the initial population is randomly selected from the training database. Each particle is a network connection re-cord derived from the database and stands for a rule. In the database, forty-one unique attributes are compiled from each connection record. These include symbolic attributes, such as "*protocol_type*", with values "TCP","$ICMP$", and "UDP", as well as continuous attributes, such as "*srv_count*", with integral values not less than 0. Each resulting connection record is then labeled as either normal or an attack type. The chosen attributes and the label make up of the D-dimensional space search.

In accord with the feature of the records, a new coding scheme is given in which uses a mapping function:

$$C_{new} = Index(A_i, C_{old}) \tag{4}$$

Where C_{old} stands for the original value of the i-th attribute and C_{new} stands for the coding of C_{old}, which actually equals to the index of C_{old} at all values of A_i. Then all the values of the attributes can be mapped to nonnegative integer.

For example, when A_i is "*protocol_type*", its values are "*icmp*","*tcp*" and "*udp*". Hence, the followings can be gained:

$Index(\text{"protocol\_type"}, \text{"icmp"}) = 0$,
$Index(\text{"protocol\_type"}, \text{"tcp"}) = 1$,
$Index(\text{"protocol\_type"}, \text{"udp"}) = 2$.

Following the mapping function, the i-th particle can be represented with an integer vector $X_i = [C_1, C_2, C_3, \ldots, C_m]$, where m is the number of the attributes. But the velocities of the particles are floating. According to equation (3), the new positions of the particles may become floating numbers, conflicting with our coding. Thus the Standard-PSO algorithm must be modified, and equation (3) must be changed as follows:

$$x_{id} = x_{id} + Round(v_{id}), \tag{5}$$

$$x_{id} = \begin{cases} MaxX[d], & if \quad x_{id} > MaxX[d], \\ 0, & if \quad x_{id} < 0 \end{cases} \tag{6}$$

Where $MaxX[d]$ stands for the max coded value of the d-th attribute. Equation (6) is used to control the position of each particle. If the particle flies out of the space, it will be pulled back to the nearest position.

Xiaodong Duan et al [15] have employed PSO for rule learning on some datasets. They proposed a coding scheme, which can be calculated as follows: Suppose that, a dataset D contains 2 condition-attributes $x\{x\_value1, x\_value2, x\_value3, x\_value4\}$ and $y\{y\_value1, y\_value2, y\_value3, y\_value4\}$, and a class-attribute $c\{class\_a, class\_b, class\_c, class\_d\}$. Then a rule:

If ($x = x\_value1$ or x_value4) and ($y = y\_value1$ or y_value2) then $c = class\_b$ can first be translated as a binary string (1001,011,01). Then it maps the binary string to an integral string (9, 3, 1).

This coding scheme performs well on some datasets [15], but it is not applicable on network intrusion detection data. For the data contains some discrete attributes which have a lot of distinct values. For example, the protocol attribute has 256 distinct values [16] and the service attributes has 65536 distinct values [17]. Hence, according to their coding scheme, the maximum value of protocol and service are 2256 and 265536, which can't be represented by a computer. So their method can't be applied to network intrusion detection. Fortunately, the problem doesn't exist in "indexical coding" scheme proposed in this paper.

3.2 Fitness Function

The fitness function is used to evaluate the quality of the particles. In this paper, the fitness is dependent upon how many attacks are correctly detected and how many normal connections are classified as attacks, which can be defined as:

$$Fitness(p) = \frac{a}{A} * 100\% - \frac{b}{B} * 100\% \tag{7}$$

Where, a is the number of correctly detected attacks, A the number of total attacks, b the number covered by this individual in the normal connections, and B the total number of normal connections.

Apparently, the range of fitness values for this function is over the closed interval [-1, 1] with -1 being the poorest fitness and 1 being the ideal.

3.3 Inertia Weight

The inertia weight w is considered to be crucial for the convergence of the algorithm. It is used to control the impact of the previous history of velocities on the current velocity of each particle. A large value facilitates global search, while a small one tends to facilitate local search. A good value can reduce the number of iteration required to get the optimum solution. People suggest that it is better to set a large value firstly, and then gradually decrease it. The two commonly used inertia weights are the random inertia weight: $w = 0.5 + \frac{Random}{2}$ [11] and the descending inertia weight: $w = (w_1 - w_2) \times \frac{MaxIte - Ite}{MaxIte} + w_2$[12].

3.4 Algorithm Description

The algorithm is showed in Fig.2

Algorithm: Rule set generation using PSO algorithm.
Input: Training data, Number of Iteration (Ite), parameters of PSO algorithm.
Output: rule sets.
Begin
  Normalize the training data using equation (4).
  Repeat
    Generate
    Repeat
      Compute the fitness $Q(p)$ of each particle p using equation (7).
      Update the pbest and gbest of each particle p.
      Update the velocity of each particle p using equations (1) and (2).
      Update the position of each particle p using equations (5) and (6).
    Until termination criteria are met.
    Add the gbest particle into the rule sets.
    Delete the training data covered by this rule.
  Untiltraining data = Φ.
End.

Fig. 2. PSO-based rule learning algorithm

4 Experiment Results

The approach is tested over the 1999 Knowledge Discovery in Database (KDD) Cup Data supplied by the DARPA [14] which is broadly used to evaluate IDSs.

The complete training database comprises approximately of five million connection records labeled by experts. The training step runs over a ten percent subset of the data, whose total number is 494021. Each connection record has forty-one attributes, but four attributes (*"protocol_type"*, *"root_shell"*, *"host_count"*, *"srv_count"*) and the label are chosen in these experiments. Then the rules are tested on the testing database, which has 311029 records.

In the first group of experiment, the system is trained with the training data and the parameters are set as: 40 particles, 50 generations, $c_1 = c_2 = 2.0$, $v_{max} = 4.0$. The random inertia weight is chosen. When the iteration reaches

10 and 50, the best rule is produced and tested on the testing database. Table1 shows the result, which can be concluded that as the evolution goes on, the fitness of the rule gets nearer to the ideal value of 1 in the training database, and so does it in the testing database.

Table 1. The fitness of the best rule in 10 and 50iteration

Iteration	Training Database	Testing Database
10	0.944413	0.788089
50	0.999888	0.998772

In the second group of experiment, the parameters are mostly maintained except the changing of w to the form of linear-descending and the number of iteration to 10. Table 2 shows the relevant result, which can be inferred that the linear-descending inertia weight gets better performance.

Table 2. The fitness of the best rule using different weights (10 iteration)

Type of w	Training Database	Testing Database
Random	0.944413	0.788089
Descending	0.999704	0.999000

Guan Jian and Liu Da-xin have presented an approach based on Genetic Algorithm to extract rules in IDS [5]. But they use binary coding scheme with an extended alpha-bet, which may consume more space if they choose more attributes to test. PSO algorithm has been employed into data mining by T. Sousa and A. Silva [13]. However, the methods used here are different from theirs in coding scheme. In Sousa's paper, they normalize all attribute values to the range $[0.0, t]$, with $0.0 < t < 1.0$, as the attributes might have three types: nominal, integer and real. But this paper normalizes the values to nonnegative integer, because the chosen attributes have limited values. Thus, if they are normalized to the floating type, they might be changed to unseen values during the evolution.

In the third experiment, the performance of this approach is compared with other algorithms. In Guan Jian's experiment [5], the size of a population is equal to 100 with the probability of crossover and mutation setting to 0.2 and 0.07. When the iteration reaches 100, the fitness of the generated best individual is shown in Table 3. This individual can be translated into this form:

If ($service = ecr\_i$) and ($host\_count >= 5$) and ($srv\_count >= 5$) then $attack = smurf$.

In our experiment, we maintain the parameters of the first experiment. And when the number of iteration reaches 50, the fitness of the best individual is shown in Table 4. (The chosen database is the same as Guan Jian's). This individual can be translated into this form:

If ($protocol\_type = icmp$) and($root\_shell = 0$) and ($host\_count >= 21$) and ($srv\_count >= 21$) then $attack = smurf$.

It is apparently that although the positive detection rate is lower than Guan Jian's, the false positive rate is also lower which leads to the higher value of the fitness. Hence it can be inferred that the approach proposed in this paper bears good performance as well as GA does in rule learning.

Table 3. The best individual in Guan Jian's paper[5] using GA(100 iteration)

Number of total PE	280790
Number of PE covered by the rule	280785
Positive detection rate	0.999982
Number of total NE	213231
Number of NE covered by the rule	65
False positive rate	0.000305
Value of the fitness function	0.999677

Table 4. The best individual using PSO and "CDR-FDR" fitness function (50 iteration)

Number of total PE	280790
Number of PE covered by the rule	280760
Positive detection rate	0.999893
Number of total NE	213231
Number of NE covered by the rule	1
False positive rate	0.000005
Value of the fitness function	0.999888

5 Conclusions

The performance of rule-based intrusion detection systems highly rely on the rules identified by security experts or automatically generated by the system. This paper first reviews the work relevant to intrusion detection system, and then presents an approach based on PSO with a new coding scheme to extract rules from the vast net-work traffic data. PSO has been proved to get a good performance in numeric problems, but few researches have been done in the rule learning of intrusion detection. This paper makes a research and reports the experimental results over the KDD'99 dataset. The results show that the proposed approach can achieve high detection rate with low false positive rate and can classify the type of attacks.

Future research will include improvement of the performance of the PSO algorithm especially in the adjustment of the parameters of it. In addition, the study of applying PSO to the anomaly detection system will be continued.

References

1. Bace R, Mell P (2001) Intrusion detection systems. NIST Computer Science Special Reports November SP800-31
2. Northcutt S, Novak J (2003) Network Intrusion Detection. New Riders, 3rd edition
3. Dasgupta D, Gonzalez FA (2001) An Intelligent Decision Support System for Intrusion Detection and Response. MMM-ACNS, Lecture Notes in Computer Science 2052: 1–14
4. Chittur A. Model Generation for an Intrusion Detection System Using Genetic Algorithms. http://www1.cs.columbia.edu/ids/publications/gaids-thesis01.pdf
5. Guan J, Liu D, Cui B (2004) An Induction Learning Approach for Building Intrusion Detection Models Using Genetic Algorithms. In: Proceedings of the 5th World Congress on Intelligent Control and Automation. Hangzhou, P.R. China: 15–19
6. Xiao T, Qu G, Hariri S, Yousif M (2005) An Efficient Network Intrusion Detection Method Based on Information Theory and Genetic Algorithm. In: Proceedings of the 24th IEEE International Performance Computing and Communications Conference (IPCCC '05). Phoenix, AZ, USA
7. Bridges SM, Vaughn RB (2000) Fuzzy Data Mining And Genetic Algorithms Applied to Intrusion Detection. In: Proceedings of 12th Annual Canadian Information Technology Se-curity Symposium: 109–122
8. Abadeh MS, Habibi J, Aliari S (2006) Using a Particle Swarm Optimization Approach for Evolutionary Fuzzy Rule Learning: A Case Study of Intrusion Detection, IPMU
9. Kennedy J, Eberhart RC (1995) Particle Swarm Optimization. In: Proceedings of the IEEE Int. Conf. Neural Networks: 1942–1948
10. Parsopoulos KE, Plagianakos VP, Magoulas GD, Vrahatis MN (2001) Streching Technique for Obtaining Gloabal Minimizers Through Particle Swarm Optimization. In: Proceedings Particle Swarm Optimization Workshop: 22–29.
11. Eberhart RC, Shi YH (2001) Tracking and optimizing dynamic systems with particle swarms. In: Proceedings of the IEEE congress on Evolutionary Computation. IEEE, Seoul, Korea: 94–97.
12. Shi YH, Eberhart RC (1998) A Modified Particle Swarm Optimizer. In: IEEE International Conference of Evolutionary Computation. IEEE, Piscataway, NJ: 69–73.
13. Sousa T, Silva A, Neves A. A Particle Swarm Data Miner: http://cisuc.dei.uc.pt/_binaries/615_pub_TiagoEPIA03.pdf.
14. 1999 KDD Cup competition: http://kdd.ics.uci.edu/databases/kddcup99/kddcup99.html.
15. Duan X, Wang C, Wang N, et al (2005) Design of Classifier Based on Particle Swarm Algorithm. Computer Engineering, China 31(20): 107–109
16. IP, Internet Protocol: http://www.networksorcery.com/enp/protocol/ip.htm
17. Port Numbers: http://www.iana.org/assignments/port-numbers.

Similarity Mass and Approximate Reasoning

Yalin Zheng[1,2], Huaqiang Yuan[1], Jing Zheng[3], Guang Yang[1], and Yongcheng Bai[4]

[1] Research Center of Intelligent Information Technology,
Department of Computer Science and Technology, Faculty of Software,
Dongguan University of Science and Technology, Dongguan 523808 , China
zheng-yl@mail.tsinghua.edu.cn

[2] State Key Laboratory for Intelligent Technology and Systems,
Department of Automation, Faculty of Information Science and Technology,
Tsinghua University, Beijing 100084, China
zheng-yl@mail.tsinghua.edu.cn

[3] Department of economy and commerce,
Dongguan University of Science and Technology, Dongguan 523808 , China
zhengj@dgut.edu.cn

[4] Institute of Information Science, Faculty of Mathematics and Information Science,
Shaanxi University of Science and Technology, Hanzhong 723001, China
qiaozhuanx@eyou.com

Abstract. Designing the approximate reasoning matching schemes and the corresponding algorithms with similarity relation $\mathbb{Q}$ instead of equivalence relation $\mathbb{Q}$ can reflect the nature of approximate reasoning and meet more application expectations. This paper bases on similarity relation $\mathbb{Q}$ and introduces the type V matching scheme and the corresponding approximate reasoning type V $\mathbb{Q}-$algorithm with the given input A^* and knowledge $A \to B$. We also present type V completeness and type V perfection of knowledge base K in $\mathbb{Q}-$logic $\mathbb{C}_{\mathbb{Q}}$ in this paper.

Keywords: Approximate Reasoning, Formula Mass, Knowledge Mass, $\mathbb{Q}-$logic, $\mathbb{Q}-$algorithm, $\mathbb{Q}-$completeness.

1 Introduction

For some applications of approximate reasoning, it is too restrict of using the equivalence $\mathbb{R}$ in formula set $F(S)$. If the approximate reasoning matching scheme is designed with the similarity $\mathbb{Q}$ in $F(S)$, it may be more reasonable to reflect the nature of approximate reasoning and meet the application expectations in many aspects. The difference is that the intersections of any components of family

$$F(S)/\mathbb{Q} = \{\langle A\rangle_{\mathbb{Q}} | A \in F(S)\}$$

which is made of corresponding formula mass

$$\langle A\rangle_{\mathbb{Q}} = \{B | B \in F(S), A\mathbb{Q}B\}$$

are unnecessarily to be empty in the situation of more general similarity $\mathbb{Q}$.

B.-Y. Cao (Ed.): Fuzzy Information and Engineering (ICFIE), ASC 40, pp. 674–684, 2007.
springerlink.com

2 The Construction of $\mathbb{Q}-$Formula Mass and $\mathbb{Q}-$Logic

In classic propositional logic $\mathbb{C} = (\hat{\mathbb{C}}, \tilde{\mathbb{C}})$, for any similarity relation $\mathbb{Q}$ in formula set $F(S)$, which satisfies

$$\langle A\rangle_{\mathbb{Q}} \to \langle B\rangle_{\mathbb{Q}} \subseteq \langle A \to B\rangle_{\mathbb{Q}}$$

for each implication $A \to B \in \mathscr{H}$
where

$$\langle A\rangle_{\mathbb{Q}} \to \langle B\rangle_{\mathbb{Q}} = \{A^* \to B^* | A^* \in \langle A\rangle_{\mathbb{Q}}, B^* \in \langle B\rangle_{\mathbb{Q}}\}$$

$(F(S), \mathbb{Q})$ is called $\mathbb{Q}-$*space*;

$$\mathbb{C}_{\mathbb{Q}} = (\hat{\mathbb{C}}, \tilde{\mathbb{C}}, \mathbb{Q})$$

is called $\mathbb{Q}-$*logic*.

In $\mathbb{Q}-$logic $\mathbb{C}_{\mathbb{Q}}$, for each formula $A \in F(S)$, $\langle A\rangle_{\mathbb{Q}}$ is called $\mathbb{Q}-$*formula mass* with the kernel of A, or just $\mathbb{Q}-$*mass* in short. Specially, for every theorem $A \to B \in \emptyset^{\vdash}$, $\langle A \to B\rangle_{\mathbb{Q}}$ is called $\mathbb{Q}-$*knowledge mass* with the kernel of knowledge $A \to B$.

3 Type V Simple Approximate Reasoning Based on $\mathbb{Q}-$Logic $\mathbb{C}_{\mathbb{Q}}$

Consider the simple approximate reasoning model in $\mathbb{Q}-$logic $\mathbb{C}_{\mathbb{Q}}$

$$\begin{array}{l} A \quad \to B \\ A^* \\ \hline \qquad\quad B^* \end{array} \tag{1}$$

where $A \to B \in \emptyset^{\vdash}$ is called *knowledge*; $A^* \in F(S)$ is called *input*; $B^* \in F(S)$ is called *output* or *approximate reasoning conclusion*. If input A^* is contained in $\mathbb{Q}-$formula mass $\langle A\rangle_{\mathbb{Q}}$ with the kernel of A, the antecedent of knowledge $A \to B$, which is

$$A^* \in \langle A\rangle_{\mathbb{Q}}$$

It means that input A^* can *activate* the knowledge $A \to B$. Regard

$$B^* \in \langle B\rangle_{\mathbb{Q}}$$

as the approximate reasoning conclusion, which is the *type V* $\mathbb{Q}-$*solution* of the simple approximate reasoning with regard to input A^* under knowledge $A \to B$.

If input A^* is not contained in $\mathbb{Q}-$formula mass $\langle A\rangle_{\mathbb{Q}}$ with the kernel of A, the antecedent of knowledge $A \to B$, that is

$$A^* \notin \langle A\rangle_{\mathbb{Q}}$$

It means that input A^* can not *activate* knowledge $A \to B$. And A^* is not considered in this situation, which means there is no type V $\mathbb{Q}-$solution with regard to input A^* under knowledge $A \to B$.

This algorithm is named *type V* $\mathbb{Q}-$ *algorithm* of simple approximate reasoning.

Theorem 3.1. In $\mathbb{Q}-$Logic $\mathbb{C}_\mathbb{Q}$, considering the simple approximate reasoning with regard to input A^* under knowledge $A \to B$, if B^* is one of the type V $\mathbb{Q}-$solutions, we have

$$A^* \to B^* \in \langle A \to B\rangle_\mathbb{Q}$$

It means that $A^* \to B^*$ is always contained in $\mathbb{Q}-$knowledge mass $\langle A \to B\rangle_\mathbb{Q}$ with the kernel of $A \to B$, that is to say, the approximate knowledge we obtained is $\mathbb{Q}-$similar to the standard knowledge $A \to B$.

Proof. Suppose the simple approximate reasoning with regard to input A^* under knowledge $A \to B$ has a type V $\mathbb{Q}-$solution B^*, then

$$A^* \in \langle A\rangle_\mathbb{Q}, B^* \in \langle B\rangle_\mathbb{Q}$$

therefore

$$A^* \to B^* \in \langle A\rangle_\mathbb{Q} \to \langle B\rangle_\mathbb{Q}$$

However, according to the construction of $\mathbb{Q}-$logic $\mathbb{C}_\mathbb{Q}$, we know

$$\langle A\rangle_\mathbb{Q} \to \langle B\rangle_\mathbb{Q} \subseteq \langle A \to B\rangle_\mathbb{Q}$$

therefore

$$A^* \to B^* \in \langle A \to B\rangle_\mathbb{Q}$$

□

Obviously, the solution domain, response domain and examination domain of the simple approximate reasoning type V $\mathbb{Q}-$algorithm with regard to input A^* under knowledge $A \to B$ are

$$\begin{aligned} Dom^{(V)}(A \to B) &= \langle A\rangle_\mathbb{Q} \\ Codom^{(V)}(A \to B) &= \langle B\rangle_\mathbb{Q} \\ Exa^{(V)}(A \to B) &= \langle A \to B\rangle_\mathbb{Q} \end{aligned}$$

Theorem 3.2. In $\mathbb{Q}-$logic $\mathbb{C}_\mathbb{Q}$, the type V $\mathbb{Q}-$algorithm of simple approximate reasoning(3.1) is a MP reappearance algorithm.

Proof. Because similarity $\mathbb{Q}$ is reflexive, which means

$$A\mathbb{Q}A,\ B\mathbb{Q}B$$

therefore

$$A \in \langle A\rangle_\mathbb{Q},\ B \in \langle B\rangle_\mathbb{Q}$$

□

Theorem 3.3. In $\mathbb{Q}-$logic $\mathbb{C}_\mathbb{Q}$, simple approximate reasoning type V $\mathbb{Q}-$algorithm is classic reasoning algorithm, if and only if similarity $\mathbb{Q}$ in $\mathbb{C}_\mathbb{Q}$ is the identity relation on formula set $F(S)$.

Proof. In $\mathbb{Q}-$logic $\mathbb{C}_\mathbb{Q}$, simple approximate reasoning type V $\mathbb{Q}-$algorithm is classic reasoning algorithm, if and only if "for each knowledge $A \to B$, it can be activate by input A^*, if and only if $A^* = A$", if and only if we have

$$\langle A\rangle_\mathbb{Q} = \{A\}$$

for every formula $A \in F(S)$, if and only if similarity relation $\mathbb{Q}$ of $F(S)$ is identity relation.

□

4 Type V Multiple Approximate Reasoning Based on $\mathbb{Q}-$Logic $\mathbb{C}_{\mathbb{Q}}$

Consider multiple approximate reasoning model in $\mathbb{Q}-$logic $\mathbb{C}_{\mathbb{Q}}$

$$\begin{array}{l} A_1 \rightarrow B_1 \\ A_2 \rightarrow B_2 \\ \cdots\;\cdots\cdots \\ A_n \rightarrow B_n \\ A^* \\ \hline \qquad\quad B^* \end{array} \tag{1}$$

where $K = \{A_i \rightarrow B_i | i = 1, 2, \cdots, n\} \subseteq \emptyset^{\vdash}$is called *knowledge base*; $A^* \in F(S)$ is called *input*; $B^* \in F(S)$ is called *output* or *approximate reasoning conclusion.*

If there exists an $i \in \{1, 2, \cdots, n\}$ which satisfies

$$A^* \in \langle A_i \rangle_{\mathbb{Q}}$$

we say input A^* *can activate* knowledge $A_i \rightarrow B_i$ in knowledge base K.

If there exists an $i \in \{1, 2, \cdots, n\}$ which satisfies

$$A^* \notin \langle A_i \rangle_{\mathbb{Q}}$$

we say input A^* *cannot activate* knowledge $A_i \rightarrow B_i$in knowledge base K.

Suppose there exists a non-empty subset E of $\{1, 2, \cdots, n\}$, for each $i \in E$, input A^* can activate knowledge $A_i \rightarrow B_i$ in knowledge base K; while for any $j \in \{1, 2, \cdots, n\} - E$, input A^* can not activate knowledge $A_j \rightarrow B_j$ in knowledge base K. Then let

$$B_i^* \in \langle B_i \rangle_{\mathbb{Q}}$$

and

$$B^* = \bigoplus_{i \subset E} B_i^*$$

which satisfy

$$B^* \in \bigcup_{i \in E} \langle B_i \rangle_{\mathbb{Q}}$$

where $\bigoplus_{i \in E} B_i^*$ is a certain aggregation of $\{B_i^* | i \in E\}$. B^*, the conclusion of approximate reasoning, is called multiple approximate reasoning *type V* $\mathbb{Q}-$*solution* with regard to input A^* under knowledge base $K = \{A_i \rightarrow B_i | i = 1, 2, \cdots, n\}$.

Suppose input A^* can not activate any knowledge $A_i \rightarrow B_i$ in knowledge base $K = \{A_i \rightarrow B_i | i = 1, 2, \cdots, n\}$, which means for each $i \in \{1, 2, \cdots, n\}$ we have

$$A^* \notin \langle A_i \rangle_{\mathbb{Q}}$$

Then, we say input A^* can not activate knowledge base $K = \{A_i \rightarrow B_i | i = 1, 2, \cdots, n\}$. In this case, we do not consider input A^*, that is, there dose not exist multiple approximate reasoning *type V* $\mathbb{Q}-$*solution* with regard to input A^* under knowledge base $K = \{A_i \rightarrow B_i | i = 1, 2, \cdots, n\}$.

This algorithm is called *type V $\mathbb{Q}-$algorithm*of multiple approximate reasoning.

Theorem 4.1. In $\mathbb{Q}-$logic $\mathbb{C}_\mathbb{Q}$, there exists type V $\mathbb{Q}-$solution of the multiple approximate reasoning with regard to input A^* under knowledge base $K = \{A_i \to B_i | i = 1, 2, \cdots, n\}$, if and only if there exists at least an $i \in \{1, 2, \cdots, n\}$ satisfied

$$A^* \in \langle A_i \rangle_\mathbb{Q}$$

that is, input A^* can activate at least one knowledge $A_i \to B_i$ in knowledge base K.

Theorem 4.2. In $\mathbb{Q}-$logic $\mathbb{C}_\mathbb{Q}$, there exists type V $\mathbb{Q}-$solution B^* of the multiple approximate reasoning with regard to input A^* under knowledge base $K = \{A_i \to B_i | i = 1, 2, \cdots, n\}$, if and only if

$$A^* \in \bigcup_{i=1}^{n} \langle A_i \rangle_\mathbb{Q}$$

.

Theorem 4.3. In $\mathbb{Q}-$logic $\mathbb{C}_\mathbb{Q}$, if there exists the multiple approximate reasoning type V $\mathbb{Q}-$solution B^* with regard to input A^* under knowledge base $K = \{A_i \to B_i | i = 1, 2, \cdots, n\}$, there exists at least an $i \in \{1, 2, \cdots, n\}$ satisfied

$$B^* \in \langle B_i \rangle_\mathbb{Q}$$

which means B^* is $\mathbb{Q}-$similar to B_i.

Theorem 4.4. In $\mathbb{Q}-$logic $\mathbb{C}_\mathbb{Q}$, if there exists the multiple approximate reasoning type V $\mathbb{Q}-$solution B^* with regard to input A^* under knowledge base $K = \{A_i \to B_i | i = 1, 2, \cdots, n\}$, we have

$$B^* \in \bigcup_{i=1}^{n} \langle B_i \rangle_\mathbb{Q}$$

Theorem 4.5. In $\mathbb{Q}-$logic $\mathbb{C}_\mathbb{Q}$, if there exists the multiple approximate reasoning V $\mathbb{Q}-$solution B^* with regard to input A^* under knowledge base $K = \{A_i \to B_i | i = 1, 2, \cdots, n\}$, there exists at least an $i \in \{1, 2, \cdots, n\}$ satisfied

$$A^* \to B^* \in \langle A_i \to B_i \rangle_\mathbb{Q}$$

that is, $A^* \to B^*$ is $\mathbb{Q}-$similar to knowledge $A_i \to B_i$.

Proof. Because there exists type V $\mathbb{Q}-$solution B^* of multiple approximate reasoning with regard to input A^* under knowledge base $K = \{A_i \to B_i | i = 1, 2, \cdots, n\}$, there is at least an $i \in \{1, 2, \cdots, n\}$ satisfied

$$A^* \in \langle A_i \rangle_\mathbb{Q}, \ B^* \in \langle B_i \rangle_\mathbb{Q}$$

therefore

$$A^* \to B^* \in \langle A_i \rangle_\mathbb{Q} \to \langle B_i \rangle_\mathbb{Q}$$

However, because of the construction of $\mathbb{Q}-$logic $\mathbb{C}_\mathbb{Q}$,

$$\langle A_i \rangle_{\mathbb{Q}} \to \langle B_i \rangle_{\mathbb{Q}} \subseteq \langle A_i \to B_i \rangle_{\mathbb{Q}}$$

therefore

$$A^* \to B^* \in \langle A_i \to B_i \rangle_{\mathbb{Q}}$$

□

Theorem 4.6. In $\mathbb{Q}-$Logic $\mathbb{C}_{\mathbb{Q}}$, if there exists multiple approximate reasoning type V $\mathbb{Q}-$solution B^* with regard to input A^* under knowledge base $K = \{A_i \to B_i | i = 1, 2, \cdots, n\}$, we have

$$A^* \to B^* \in \bigcup_{i=1}^{n} \langle A_i \to B_i \rangle_{\mathbb{Q}}$$

Theorem 4.7. In $\mathbb{Q}-$logic $\mathbb{C}_{\mathbb{Q}}$, the solution domain, response domain and examination domain of the multiple approximate reasoning type V $\mathbb{Q}-$algorithm with regard to input A^* under knowledge base $K = \{A_i \to B_i | i = 1, 2, \cdots, n\}$ are given by

$$\begin{aligned} Dom^{(V)}(K) &= \bigcup_{i=1}^{n} \langle A_i \rangle_{\mathbb{Q}} \\ Codom^{(V)}(K) &= \bigcup_{i=1}^{n} \langle B_i \rangle_{\mathbb{Q}} \\ Exa^{(V)}(K) &= \bigcup_{i=1}^{n} \langle A_i \to B_i \rangle_{\mathbb{Q}} \end{aligned}$$

5 Type V Completeness and Type V Perfection of Knowledge Base K in $\mathbb{Q}-$Logic $\mathbb{C}_{\mathbb{Q}}$

Endow the formula set $F(S)$ in classic propositional logic $\mathbb{C} = (\hat{\mathbb{C}}, \tilde{\mathbb{C}})$ with a similarity relation $\mathbb{Q}$, where for each implication $A \to B \in \mathscr{H}$ we have

$$\langle A \rangle_{\mathbb{Q}} \to \langle B \rangle_{\mathbb{Q}} = \langle A \to B \rangle_{\mathbb{Q}}$$

Then, $(F(S), \mathbb{Q})$ is called *regular* $\mathbb{Q}-$*space*, and $\mathbb{C}_{\mathbb{Q}} = (\hat{\mathbb{C}}, \tilde{\mathbb{C}}, \mathbb{Q})$ is called *regular* $\mathbb{Q}-$*logic*.

In $\mathbb{Q}-$logic $\mathbb{C}_{\mathbb{Q}}$, the type V approximate knowledge closure of knowledge base

$$K = \{A_i \to B_i | i = 1, 2, \cdots, n\}$$

is

$$K^{ⓑ} = \bigcup_{A \to B \in K} \langle A \to B \rangle_{\mathbb{Q}}$$

For each knowledge $A_i \to B_i$ in knowledge base K, $\mathbb{Q}-$knowledge mass $\langle A_i \to B_i \rangle_{\mathbb{Q}}$ with the kernel of $A_i \to B_i$ in $\mathbb{Q}-$space $(F(S), \mathbb{Q})$ is also called *type V approximate knowledge mass* with the kernel of $A_i \to B_i$.

So we say in $\mathbb{Q}-$logic $\mathbb{C}_{\mathbb{Q}}$, the type V approximate knowledge closure $K^{ⓑ}$ of knowledge base

$$K = \{A_i \to B_i | i = 1, 2, \cdots, n\}$$

is the union of finite number of type V knowledge mass $\langle A_i \to B_i \rangle_{\mathbb{Q}} (i = 1, 2, \cdots, n)$, that is

$$K^{ⓑ} = \bigcup_{i=1}^{n} \langle A_i \to B_i \rangle_{\mathbb{Q}}$$

In $\mathbb{Q}-$logic $\mathbb{C}_{\mathbb{Q}}$, knowledge base

$$K = \{A_i \to B_i | i = 1, 2, \cdots, n\}$$

is called *type V complete*, if type V approximate knowledge closure $K^{ⓑ}$ of K covers the theorem set $\emptyset^{\vdash}$, which means

$$\emptyset^{\vdash} \subseteq K^{ⓑ}$$

Knowledge base K is called *type V perfect*, if type V approximate knowledge closure $K^{ⓑ}$ of K is coincident with theorem set $\emptyset^{\vdash}$, that is

$$K^{ⓑ} = \emptyset^{\vdash}$$

Theorem 5.1. In $\mathbb{Q}-$logic $\mathbb{C}_{\mathbb{Q}}$, knowledge base

$$K = \{A_i \to B_i | i = 1, 2, \cdots, n\}$$

is type V complete, if and only if each theorem in $\emptyset^{\vdash}$ is contained in type V knowledge mass with the kernel of a knowledge in K.

Proof. K is type V complete, if and only if

$$\emptyset^{\vdash} \subseteq K^{ⓑ}$$

if and only if

$$\emptyset^{\vdash} \subseteq \bigcup_{i=1}^{n} \langle A_i \to B_i \rangle_{\mathbb{Q}}$$

if and only if for each theorem $A \to B \in \emptyset^{\vdash}$, there exists an $i \in \{1, 2, \cdots, n\}$ satisfied

$$A \to B \in \langle A_i \to B_i \rangle_{\mathbb{Q}}$$

□

Theorem 5.2. In regular $\mathbb{Q}-$logic $\mathbb{C}_{\mathbb{Q}}$, knowledge base

$$K = \{A_i \to B_i | i = 1, 2, \cdots, n\}$$

is type V complete, if and only if the antecedent of each theorem $A \to B$ in $\emptyset^{\vdash}$ can activate a certain knowledge in knowledge base K and the consequence is the multiple approximate reasoning type V $\mathbb{Q}-$solution with regard to input A under knowledge base K.

Proof. According to theorem 5.1, knowledge base K is type V complete, if and only if each of the theorems $A \to B$ in $\emptyset^{\vdash}$ must be contained in type V knowledge mass with the kernel of a certain knowledge in K, that is to say, there is an $i \in \{1, 2, \cdots, n\}$ satisfied

$$A \to B \in \langle A_i \to B_i \rangle_{\mathbb{Q}}$$

However, according to the construction of regular $\mathbb{Q}-$logic $\mathbb{C}_{\mathbb{Q}}$

$$\langle A_i \to B_i \rangle_{\mathbb{Q}} = \langle A_i \rangle_{\mathbb{Q}} \to \langle B_i \rangle_{\mathbb{Q}}$$

Therefore, knowledge base K is type V complete, if and only if for each theorem $A \to B$ in $\emptyset^{\vdash}$, there is an $i \in \{1, 2, \cdots, n\}$ satisfied

$$A \to B \in \langle A_i \rangle_{\mathbb{Q}} \to \langle B_i \rangle_{\mathbb{Q}}$$

that is,

$$A \in \langle A_i \rangle_{\mathbb{Q}}, \ B \in \langle B_i \rangle_{\mathbb{Q}}$$

It is equal to say A activates a certain knowledge $A_i \to B_i$ in knowledge base K according to type V $\mathbb{Q}-$algorithm and B can be taken as the multiple approximate reasoning type V $\mathbb{Q}-$solution with regard to A under knowledge base K.

□

Theorem 5.3. In regular $\mathbb{Q}-$logic $\mathbb{C}_{\mathbb{Q}}$, knowledge base

$$K = \{A_i \to B_i | i = 1, 2, \cdots, n\}$$

is type V perfect, if and only if K is type V complete, and $A^* \to B^*$ is always a theorem in logic $\mathbb{C}$ for each input A^* which can active the knowledge in knowledge base K according to type V $\mathbb{Q}-$algorithm, that is

$$A^* \to B^* \in \emptyset^{\vdash}$$

where B^* is multiple approximate reasoning type V $\mathbb{Q}-$solution with regard to input A^* under knowledge base K.

Proof $\Rightarrow$: Assume knowledge base K is type V perfect given by

$$K^{ⓑ} = \emptyset^{\vdash}$$

Because of $\emptyset^{\vdash} \subseteq K^{ⓑ}$, K is type V complete. Suppose A^* is an input which can activate the knowledge in K according to type V $\mathbb{Q}-$algorithm, there must exist an $i \in \{1, 2, \cdots, n\}$ satisfied

$$A^* \in \langle A_i \rangle_{\mathbb{Q}}, \ B^* \in \langle B_i \rangle_{\mathbb{Q}}$$

that is,

$$A^* \to B^* \in \langle A_i \rangle_{\mathbb{Q}} \to \langle B_i \rangle_{\mathbb{Q}}$$

where B^* is the multiple approximate reasoning type V $\mathbb{Q}-$solution with regard to input A^* under knowledge base K. But because of the construction of $\mathbb{Q}-$logic $\mathbb{C}_{\mathbb{Q}}$, we know

$$\langle A_i \rangle_{\mathbb{Q}} \to \langle B_i \rangle_{\mathbb{Q}} = \langle A_i \to B_i \rangle_{\mathbb{Q}}$$

Therefore

$$A^* \to B^* \in \langle A_i \to B_i \rangle_{\mathbb{Q}}$$

which means $A^* \to B^*$ is contained in knowledge mass $\langle A_i \to B_i \rangle_{\mathbb{Q}}$ with the kernel of the knowledge $A_i \to B_i$ in K. According to the construction of type V knowledge closure of K

$$K^{ⓑ} = \bigcup_{i=1}^{n} \langle A_i \to B_i \rangle_{\mathbb{Q}}$$

we know

$$A^* \to B^* \in \bigcup_{i=1}^{n} \langle A_i \to B_i \rangle_{\mathbb{Q}} = K^{ⓑ}$$

However, since that knowledge base K is type V perfect we know

$$K^{ⓑ} = \emptyset^{\vdash}$$

therefore

$$A^* \to B^* \in \emptyset^{\vdash}$$

Thus we have proved that $A^* \to B^*$ is always a theorem in logic $\mathbb{C}$.

$\Leftarrow$: Assume that knowledge K is type V complete, and for any input A^* which can activate the knowledge in K according to type V $\mathbb{Q}$−algorithm, $A^* \to B^*$ is always a theorem in logic $\mathbb{C}$, that is $A^* \to B^* \in \emptyset^{\vdash}$, where B^* is multiple approximate reasoning type V $\mathbb{Q}$−solution with regard to input A^* under knowledge base K.

Because K is type V complete, we know

$$\emptyset^{\vdash} \subseteq K^{ⓑ}$$

Now, we prove

$$K^{ⓑ} \subseteq \emptyset^{\vdash}$$

in fact, for any $A^* \to B^* \in K^{ⓑ}$, according to the construction of type V approximate knowledge closure of K

$$K^{ⓑ} = \bigcup_{i=1}^{n} \langle A_i \to B_i \rangle_{\mathbb{Q}}$$

we know there exists an $i \in \{1, 2, \cdots, n\}$ satisfies

$$A^* \to B^* \in \langle A_i \to B_i \rangle_{\mathbb{Q}}$$

that is, $A^* \to B^*$ is always contained in type V knowledge mass $\langle A_i \to B_i \rangle_{\mathbb{Q}}$ with the kernel of the knowledge $A_i \to B_i$ in K. But according to the construction of regular $\mathbb{Q}$−logic $\mathbb{C}_{\mathbb{Q}}$, we know

$$\langle A_i \to B_i \rangle_{\mathbb{Q}} = \langle A_i \rangle_{\mathbb{Q}} \to \langle B_i \rangle_{\mathbb{Q}}$$

therefore

$$A^* \to B^* \in \langle A_i \rangle_{\mathbb{Q}} \to \langle B_i \rangle_{\mathbb{Q}}$$

that is

$$A^* \in \langle A_i \rangle_{\mathbb{Q}},\ B^* \in \langle B_i \rangle_{\mathbb{Q}}$$

It means A^* is an input which can activate knowledge $A_i \to B_i$ in knowledge base K according to type V $\mathbb{Q}-$algorithm. B^* can be taken as the multiple approximate reasoning type V $\mathbb{Q}-$solution with regard to input A^* under knowledge K. From the assumption we know, $A^* \to B^*$ is always a theorem in logic $\mathbb{C}$, that is,

$$A^* \to B^* \in \emptyset^{\vdash}$$

Because of the arbitrariness of $A^* \to B^*$ in $K^{ⓑ}$, we know

$$K^{ⓑ} \subseteq \emptyset^{\vdash}$$

thereby we have proved knowledge base K is type V perfect, that is

$$K^{ⓑ} = \emptyset^{\vdash}$$

□

6 Conclusion

We do research on the approximate reasoning in logic frame. Based on the similarity relation $\mathbb{Q}$ in formula set $F(S)$, we illustrate the matching scheme and corresponding algorithms for approximate reasoning. We also discuss the type V completeness and type V perfection of knowledge base in $\mathbb{Q}-$logic $\mathbb{C}_{\mathbb{Q}}$. The construction of $\mathbb{Q}-$logic $\mathbb{C}_{\mathbb{Q}}$ ensures the approximate reasoning conclusion B^* is $\mathbb{Q}-$similar to B and the obtained approximate knowledge $A^* \to B^*$ is $\mathbb{Q}-$similar to $A \to B$. Using Similarity relation $\mathbb{Q}$ instead of equivalence relation $\Re$ will better reflect the nature of approximate reasoning and meet more application needs. We hope that the ideas, expressions and methods of this paper could become one of the logical standards of approximate reasoning.

References

1. Costas PP, Nikos IK (1993) A comparative assessment of measures of similiarity of fuzzy values. Fuzzy Sets and Systems, 56:171-174
2. Costas PP (1991) Value approximation of fuzzy Systems. Fuzzy Sets and Systems, 39: 111-115
3. Wang GJ (1998) Fuzzy continuous input-output controllers are universal approximators. Fuzzy Sets and Systems, 97:95-99
4. Wang GJ (1999) On the logic foundation of fuzzy reasoning. Information Science, 117: 47-88
5. Wang GJ, Wang H (2001) Non-fuzzy versions of fuzzy reasoning in classical logics. Information Sciences, 138:211-236
6. Wang GJ, Leung Y (2003) Intergrated semantics and logic metric spaces. Fuzzy Sets and Systems, 136:71-91
7. Meng GW (1993) Lowen's compactness in L-fuzzy topological spaces. Fuzzy Sets and Systems, 53:329-333

8. Meng GW (1995) On countably strong fuzzy compact sets in L-fuzzy topological spaces. Fuzzy Sets and Systems, 72:119-123
9. Bai SZ (1997) Q-convergence of ideals in fuzzy lattices and its applications. Fuzzy Sets and Systems, 92:357-363
10. Bai SZ (1997) Q-convergence of nets and week separation axiom in fuzzy lattices. Fuzzy Sets and Systems, 88:379-386
11. Dug HH, Seok YH (1994) A note on the value similarity of fuzzy systems variables. Fuzzy Sets and Systems, 66:383-386
12. Hyung LK, Song YS, Lee KM (1994) Similarity measure between fuzzy sets and between elements. Fuzzy Systems and Sets, 62:291-293
13. Chen SM, Yeh MS, Hsiao PY (1995) A comparison of similarity measures of fuzzy values. Fuzzy Sets and Systems, 72:79-89
14. Chen SM (1995) Measures of smilarity between vague sets. Fuzzy Sets and Systems, 74:217-223
15. Sudkamp T (1993) Similarity, interpolation, and fuzzy rule construction. Fuzzy Sets and Systems, 58:73-86
16. Kickert WJM, Mamdami EH (1978) Analysis of a fuzzy logic controller. Fuzzy Sets and Systems, 1:29-44
17. Czogala E, Leski J (2001) On eqivalence of approximate reasoning results using different interpretations of fuzzy if-then rules. Fuzzy Sets and Systems, 117:279-296
18. Zadeh LA (2003) Computing with words and perceptions – a paradigm shift in computing and decision ananlysis. Proceedings of International Conference on Fuzzy Information Processing, Tsinghua University Press, Springer Verlag
19. Ying MS (1994) A logic for approximate reasoning. The Journal of Symbolic Logic 59: 830-837
20. Ying MS (1992) Fuzzy reasoning under approximate match. Science Bulletin 37:1244-1245
21. Ying MS (2003) Reasoning about probabilistic sequential programs in a probabilistic logic. Acta Informatica 39:315-389
22. Zheng YL (2003) Stratified construction of fuzzy propositional logic. Proceedings of International Conference on Fuzzy Information Processing, Tsinghua University Press, Springer Verlag, 1-2:169-174
23. Zheng YL, Zhang CS, Yi X (2004) Mamdaniean logic. Proceedings of IEEE International Conference on Fuzzy Systems, Budapest, Hungary, 1-3:629-634

Intelligent Knowledge Query Answering System Based on Short Message

Ping Wang and Yanhui Zhu

Department of Computer Science and Technology, Hunan University of Technology,
Zhuzhou Hunan 412008, China
{hazellau,swayhzhu}@163.com

Abstract. Short message service (SMS) is a mechanism of delivery of short messages over the mobile networks. Because of its popularity, SMS is now a ubiquitous way of communicating. In order to improve its efficiency and intelligence, this paper examines methods for analyzing message entry techniques by using a query against a knowledge base which has been acquired background information. People can also use SMS to perform reasoning or discover implicit knowledge. This study proposes application and architecture of intelligent knowledge query answering system based on SMS. Short message processing module and knowledge query module are two main parts of this system. The first module includes correlative AT command & PDU data format analysis and processing flow of short message. The second module analyses query sentence by using natural language understanding technology which tries to understand query sentence by using a conceptual network model to organize knowledge. Conceptual network is adopted to systematically organize relationships among concepts and represent different features of each concept so as to substantially improve people's ability to get specific information in response to specific needs. Furthermore, the paper discusses PDU coding format and question analysis in detail.

Keywords: Query Answering System, Short Message Service, Conceptual Network.

1 Introduction

Internet provides abundant information, and mobile communication allows people to be free from fixed telephone line. A combination of mobile communication and Internet predicates that people can get information at any places any time. As a technique of mobile interconnection technology, GSM Short Message Service has seen unprecedented growth in the last few years and it is now applied broadly in many fields with its low cost and convenience [1-2].

People can acquire knowledge by using Short Message knowledge query system even being offline. If combined with the technology of natural language understanding at the same time, the system could understand the query intent of the user more easily and give the answers more exactly, which increases the system's practicability. In order to help people use SMS perform reasoning and derive added values, or discover implicit knowledge, we adopt a conceptual network combining techniques from knowledge representation and natural language processing to enable a computer to systematically organize relationships among concepts. In the conceptual network, each concept is represented by its different features such as attributes, behavior,

B.-Y. Cao (Ed.): Fuzzy Information and Engineering (ICFIE), ASC 40, pp. 685–692, 2007.
springerlink.com

description and all important relations with other concepts. Those features can be combined with search and retrieval techniques to improve retrieval effectiveness and intelligence.

2 System Framework

The system consists of two main modules: Short message processing module and knowledge query module. Fig. 1 shows the framework of the system.

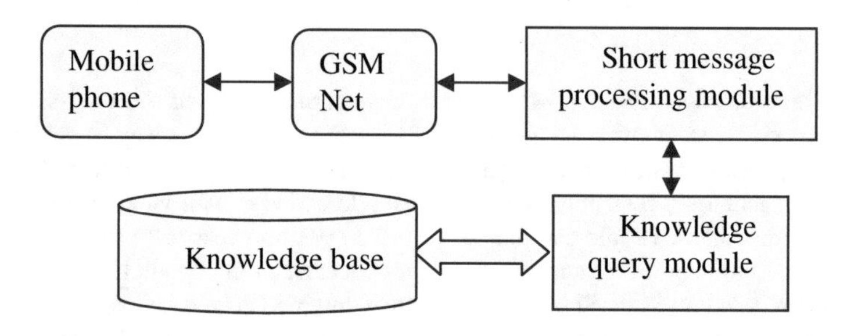

Fig. 1. System Framework

Short message processing module manages the receiving and sending of short message. In other words, it receives users' short message and draws the query content which is put into query module next. On the other hand, it can send the result from knowledge query module to user in form of short message. To the knowledge query module, it presides over getting the query intent of users, analyzing it by classifying and identifying the query type, then searching the appropriate answer from knowledge base and returning the results to short message processing module at last.

3 Short Message Processing Module

3.1 Hardware Condition

This system sends and receives Short Message by connecting GSM modem with computer serial port. GSM Modem' instruction is compatible of AT control command within SIM card inside. Its functions include facsimile, voice communication, SMS and so on. This method is suitable for corporation and individual as well.

3.2 Correlative AT Command and PDU Data Format Analyses

GSM Modem communicates with computer through AT command. AT commands begin with AT and end with Enter in addition to a sending symbol named "Ctrl+Z". Table 1 lists part of AT commands related with Short Message.

There are three modes of sending or receiving short message according to the time of appearing: block mode, text mode based on AT command and PDU mode based on

Table 1. Commonly used AT Commands

AT command	Function
AT+CMGF	Set the mode of Short Message: 0-PDU;1-Text
AT+CMGR	Read a Short Message
AT+CMGS	Send a Short Message
AT+CMGD	Delete a Short Message
AT+CSCA	Set or get the address of Short Message Center

AT command. Since block mode has been replaced by PDU mode and text mode doesn't support Chinese, this paper only introduces PDU mode. The following part will analysis PDU data format in detail.

First, an example of sending message is given below:

AT command: AT+CMGS=19<Enter>

PDU data:
>0891683108701305F011000D91683175149445F9000800044F60597D<Ctrl+Z>

According to the previous example, AT+CMGS is the command of sending a short message, 19 is the length of PDU data. PDU data is analyzed as follows:

08: the length of short message center number which unit is byte;

91683108701305F0: the first two numbers 91 represents char "+" and the latter part 8613800731500 is the number of short message center;

1100: flag which needn't change in general;

0D: the length of the receiving mobile phone number;

91683175149445F9: the number of receiving mobile phone; it has the same encoding mode as that of message center code;

00: flag bits;

08: encoding mode, 08 represents Unicode mode;

00: period of validity, 00 stands for 5 minutes;

04: the length of user's data;

4F60597D: the content of sending user's data Unicode which stands for "nihao";

Then, the module will analyze PDU data received. Assumed that the PDU data received is:

0891683108701305F0240BA13175149445F900085030321150922004 4F60597D,
then the analyses would be as follows:

08: the length of short message center number;

91683108701305F0: the number or short message center which is +8613800731500 after it is decoded;

24: Control bits;

0B: the length of sending mobile phone number;

A13175149445F9: sending mobile phone number which is 13574149549 after it is decoded;

00: flag bits;

08: encoding mode, 08 stands for Unicode code;

0303211509220: time of sending, decoding result is: 11:05:29 3-23, 2005, the last two bits stand for time zone;

04: the length of user's data;

4F60597D: the content of user's data which stands for "nihao"

3.3 Processing Flow of Short Message

The short message processing module processes short message using event-driven mode. When GSM modem receives a short message, it will process it in accordance with PDU mode, getting information of the sending content, time and the sender's phone number, decoding it and putting the processing result into knowledge query module. After getting the result from query module, it will send encoding PDU data to user and write the processing result into databases at last. Fig. 2 shows short message processing flowchart.

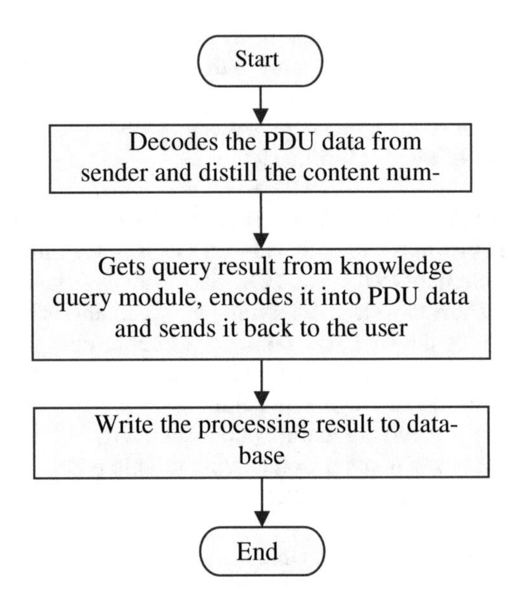

Fig. 2. Short Message Processing Flowchart

4 Knowledge Query Module

4.1 Query Mode of Users

When users query knowledge, they usually have some intention. The intention can be shown through the querying sentence pattern. The system provides users a way of studying by querying, allowing users to query answers in natural language for it is much simpler and natural. Answers are returned to users according to their query requirements so as to realize querying study in this way.

4.2 Knowledge Organization

Since traditional knowledge model has some disadvantages, this study adopts conceptual network model to organize knowledge. This model uses a five-attribute set {A, B, C, D, R} to express one concept. In this set, *A* represents attributes of concept, *B* represents concept's behavior, *C* represents concept name, *D* represents the description of concept and *R* represents all relations of concept. This model adopts the idea of OO(Object Oriented) and has practical meanings in applied fields because it can express an entity completely. A concept is expressed by its name and its meaning is presented through its attributes, behaviors, relations and description. Reference [3] gives the model's introduction in detail.

After knowledge is organized by a conceptual network, a question from user can be one among the following three types:

- Query concept description
- Query attributes or behavior of one or more concepts
- Query the relations among concepts

By classifying and identifying the query type, it is easier to guess user's query intention so as to descend the difficulty of natural language understanding.

4.3 Query Sentences Classification

The purpose of classifying sentences is to get the query intention of users more exactly, so that it can instruct parsing sentence. It is one crucial step of processing sentence [4]. This system adopts a conceptual network model to organize knowledge in that it has OO idea which can satisfy user's query demand basically. User's query could be one object which may be specified to its attribute, relations or description. According to the likely form of the user's question, the system collects and parses a certain amount of asking sentences, and then classify the query sentences into different templates(among those types *A* is attribute, *C* is concept, *A.V* is the value of attribute):

- Type 1: What is the A1, A2… of C1…?
- Type 2: Is A.V C'A ?
- Type 3: Which A of C is A.V ?
- Type 4: What is description of C?

The templates above only give the rough structure of the sentence. Certainly, one type has different kinds of modes to express by using the different words. It is similar to querying the relations of concept.

4.4 Examples of Parsing Query Sentence

Split sentence. *We use a professional split system and a domain dictionary which have remarks of concept words and attribute words, making the sentence parse easier. For example: "what is the ideal goal of image restoring?" the result after splitting is "what/ry is/v the ideal goal/A of/uj image restoring/C ?/w". from the result, we can get one concept word "image restoring" and one attribute word "ideal goal".*

Match sentence template. *The splitting sentence should be matched to a sentence template according to the types listed above. The method is illustrated as follows: firstly, a rules set is made according to syntax structure of each sentence template, then according to the rules set, the participial sentence is analyzed by a linear chart algorithm. The analysis success means that the matching is successful.*

Rules' establishment determines the degree of the matching success. For example, even some strict rules have definite meaning, it will make sentence matching fail because the sentence is not strictly in accordance with syntax. For example:

If the rules set is: {Cd->C uj; Np->Cd A; VP->v ry; S->NP VP}, then the sentence "the ideal goal of image restoring?" will fail matching because it hasn't "what" as its query word. If rule: VP->v is added to the set, then matching will be successful. So when the rules are made, they should be based on specific requirements. In reality, loosing rules could accept some irregular mistakes.

The linear chart algorithm described below is used to parse the sentence:

Basic data structure:

- Chart: {edge[i], i=1,2,…edge=<P1,P2,Label>}
- Agenda: Stack structure which stores the edges waiting for being added into chart
- Active arc: Store current parsing state, the state structure consists of three parts <P1, P2, rule>

Main algorithm steps:

- Clear agenda; store the sentence string to be parsed into the buffer.
- Repeat the following steps until buffer and agenda are both empty:

① If agenda is empty, then take a character from buffer and put the character, its starting position and ending position (P1,P2) into agenda;

② Pop the edge from agenda which starting position and ending position are (P1,P2) with tag "L" on the edge;

③ Check the rules set, find all rules like A->L β, each one adding a rule "A->L · β";

④ Pop the edge which tagged "L", add it to chart between P1 and P2;

⑤ Check all active arcs, if find a rule like <P0,P1, A->α · L β>, then add <P0,P2, A->α L · β> to active arc;

⑥ If there is a rule like A-> α L ·, then set starting position as P0, ending position as P2, and push <P0,P2,A> into agenda stack.

After parsing is finished, and if there is an edge which tag is "S" and cover all characters, then the parsing is successful. Fig.3 is a successful parsing example which result is displayed as a tree structure.

Query sentence's semantic parsing. *Semantic parsing aims at drawing the information from query sentence and the information is useful for getting the answer [5-6]. Useful information include sentence type, concept, center, focus point and predication according to the result of matching template. Sentence Type indicates the type the sentence belongs to; concept indicates range of the query; center of sentence indicates the center discussed in the sentence; focus indicates the place of answer; predication refers to the verb in the sentence. The following examples show some questions and their results after question analyzer processing:*

Question: What is research object of image restoring?

- Type: query attribute of concept
- Concept: image restoring

- Center: research object
- Focus: what
- Predication: is

Question: Is color image research object of color image processing?

- Type: affirm
- Concept: color image processing
- Center: research object
- Focus: Yes/No
- Predication: is

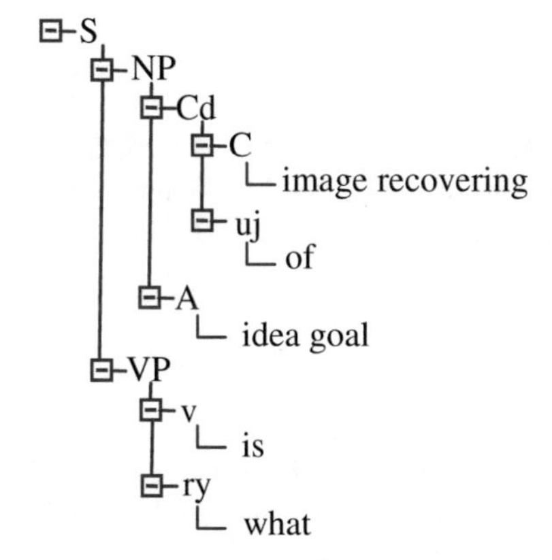

Fig. 3. An Example of Parsing Query Sentence *"what is the ideal goal of image recovering?"*

4.5 Drawing Answer

According to the result of template matching and semantic parsing, the appropriate query sentence is built to get the answer from knowledge database. Then the answer is organized, handed and given to short message module which will send the answer to users in the end. The following examples show some query sentences and their answers:

Query: What are the research object and ideal goal of image compressing and image restoring?

Answer: The research object of image restoring is image, the ideal goal of image restoring is to eliminate retrogress. The research object of image compressing is image data, the ideal goal of image compressing is to reduce the quantity of image data.

Query: Which attribute of image compress is compress with expense?

Answer: Compress with expense is one realizing means of image compress.

Query: What is compress without expense?

Answer: Compress without expense is a generic method for data compress.

5 Conclusion

Realizing knowledge query through the form of short message for users overcomes the limitation of time and space of studying. Knowledge query gets user's query intention by parsing and understanding question. At the same time, it descends the difficulty of natural language understanding by using conceptual network model to organize the knowledge. Moreover, it has constructed query templates and templates rules based on natural language understanding, which make the system gain a good effect. Even though, the maturity of the templates and rules of template should be based on parsing a great deal of language data, and this will be our future research direction.

Acknowledgements

This paper is sponsored by Natural Science Foundation of Hunan province (number: 05JJ30122). It is also a project supported by Scientific Research Fund of Hunan Provincial Education Department (number: 05C519).

References

1. YI Qing, SHI Zhiguo, WANG zhiliang, LI Qing (2003) Information Query System Based on GSM Short Message. Computer application research 2: 63-65
2. ZHU Huidong, HUANG Yan, LIU Xiangnan (2004) Application Design of Short Message in books query. Microcomputer and Application 4:47-49
3. LU Wenyan (2002) Modeling, realizing and application of Concept Net:[master's degree paper]. Changsha: Central South University
4. TANG Suqing (2003) Design of Knowledge Query for NKI-Tutor. Computer engineering 14:183-185
5. LU Zhijian, ZHANG Dongrong (2004) Question Interpretation for Chinese Question Answering. Computer engineering 18:64-65
6. YU Zhengtao, FAN Xiaozhong, KANG Haiyan (2004) Restricted Domain Question Answer System Based on Natural Language Understanding. Computer engineering 18:35-37

Comparison Study on Different Core Attributes

Bingru Yang[1], Zhangyan Xu[1,2], Wei Song[1], and Jing Gao[1]

[1] School of Information Engineering, University of Science and Technology Beijing, Beijing, 100083, China
bryang_kd@yahoo.com.cn

[2] Department of Computer, Guangxi Normal University, Guilin, 541004, China
xyzwlx72@yahoo.com.cn

Abstract. How to find the core attributes is important for attribute reduction based on rough set. However, there are few literatures discussing this topic, and most existing works are mainly based on Hu's discernibility matrix. Till now, there are three kinds of core attributes: Hu's core based on discernibility matrix (denoted by $Core1(C)$), core based on positive region (denoted by $Core2(C)$), and core based on information entropy (denoted by $Core3(C)$). Some researchers have been pointed out that these three kinds of cores are not equivalent to each other. Based on the above three kinds of core attributes, we at first propose three kinds of simplified discernibility matrices and their corresponding cores, which are denoted by $SDCore1(C)$, $SDCore2(C)$, and $SDCore3(C)$ respectively. And then it is proved that $Core1(C)$=$SDCore1(C)$, $Core2(C)$=$SDCore2(C)$, and $Core3(C)$=$SDCore3(C)$. Finally, based on three proposed simplified discernibility matrices and their corresponding cores, it is proved that $Core2(C) \subseteq Core3(C) \subseteq Core1(C)$.

Keywords: Rough set; HU's discernibility matrix; Positive region; Information entropy; Core; Simplified discernibility matrix.

1 Introduction

Attribute reduction is one of the most important issues of Rough Set theory [1, 2]. For many algorithms on attribute reduction, the core attributes are usually calculated at first, and then heuristic knowledge is combined with core attributes to find the minimal reduction. Based on discernibility matrix, Hu et al. [3] proposed an efficient algorithm for calculating core attributes, which reduced the computational complexity greatly. [4,5,6,7] have pointed out core attributes calculated by Hu's method are different to that found by algorithm based on positive region. New discernibility matrices and the corresponding core attributes are also constructed in these works. Furthermore, it is also proved that core attributes they proposed are equivalent to those based on positive region. Wang et al. [8,9,10] proposed the definition on attribute reduction based on information view, and the corresponding core attributes. Furthermore, they also proved that cores based on information entropy are different

B.-Y. Cao (Ed.): Fuzzy Information and Engineering (ICFIE), ASC 40, pp. 693–703, 2007.
springerlink.com

to that based on positive region, and also different to that based on Hu's discernibility matrix. In this paper, we at first construct three simplified discernibility matrices and their corresponding core attributes based on Hu's discernibility matrix, positive region, and information entropy. And these three kinds of core attributes are denoted by $SDCore1(C)$, $SDCore2(C)$, and $SDCore3(C)$ respectively. And then it is proved that $Core1(C)$=$SDCore1(C)$, $Core2(C)$ =$SDCore2(C)$, and $Core3(C)$=$SDCore3(C)$. Finally, based on three proposed simplified discernibility matrices and their corresponding cores, it is proved that $Core2(C) \subseteq Core3(C) \subseteq Core1(C)$ for inconsistent decision table. While for consistent decision tables, we have $Core2(C)$=$Core3(C)$= $Core1(C)$.

2 Preliminary

In this section, we introduce the basic concepts and correspondence definitions.

Definition 1[1,2]. A decision table is defined as $S = (U, C, D, V, f)$, where $U = \{x_1, x_2, ..., x_n\}$ is the set of objects, $C = \{c_1, c_2, ..., c_r\}$ is the set of condition attributes, D is the set of decision attributes, and $C \cap D = \emptyset$, $V = \cup V_a, a \in C \cup D$, where V_a is the value range of attribute a. $f : U \times C \cup D \rightarrow V$ is an information function, which assigns an information value for each attribute of each object, that is $\forall a \in C \cup D, f(x, a) \in V_a$ holds. Every attribute subset determines a binary discernibility relation $IND(P)$:

$$INP(P) = \{(x, y) \in U \times U | \forall a \in P, f(x, a) = f(y, a)\}$$

$IND(P)$ determines a partition of U, which is denoted by $U/IND(P)$ (in short U/P). Any element $[x]_P = \{y | \forall a \in P, f(y, a) = f(x, a)\}$ in U/P is called equivalent class.

Definition 2[1,2]. For a decision table $S = (U, C, D, V, f)$, $\forall R \in C \cup D, X \in U$, denote $U/R = \{R_1, R_2, ..., R_l\}$, then $R_{-}(X) = \bigcup\{R_i | R_i \in U/R, R_i \subseteq X\}$ is called lower approximation of X for R.

Definition 3[1,2]. For a decision table $S = (U, C, D, V, f)$, let $U/D = \{D_1, D_2, ..., D_k\}$ be the partition of D to U, and $U/C = \{C_1, C_2, ..., C_m\}$ be the partition of C to U, then $POS_C(D) = \bigcup_{D_i \in U/D} C_{-}(D_i)$ is called positive region of C on D. If $POS_C(D) = U$, then the decision table is called consistent, else it is called inconsistent.

Theorem 1. *For decision table $S = (U, C, D, V, f)$, there is*

$$POS_C(D) = \bigcup_{X \in U/C \wedge |X/D|=1} X.$$

Proof: *Obviously, theorem 1 can be proved directly according to definition 3.*

Definition 4. For decision table $S = (U, C, D, V, f)$, denote $U/C = \{[x'_1]_C, [x'_2]_C, ..., [x'_m]_C\}$, $U' = \{x'_1, x'_2, ..., x'_m\}$, and according to theorem 1, assume $POS_C(D) = [x'_{i_1}]_C \cup [x'_{i_2}]_C \cup ... \cup [x'_{i_t}]_C$ where $\{x'_{i_1}, x'_{i_2}, ..., x'_{i_t}\} \subseteq U'$ and $|[x'_{i_s}]_C/D| = 1 (s = 1, 2, ..., t)$, denote ,$U'_{pos} = \{x'_{i_1}, x'_{i_2}, ..., x'_{i_t}\}$, $U'_{neg} = U' - U'_{pos}$, then $S' = (U', C, D, V, f)$ is called simplified decision table.

3 Core Attributes of Simplified Discernibility Matrix Based on Hu's Discernibility Matrix

In this section, we propose some definitions and theorems about core attributes based on Hu's discernibility matrix.

Definition 5 [3]. For decision table $S = (U, C, D, V, f)$, $M = (m_{ij})$ is Hu's discernibility matrix, whose elements are:

$$m_{ij} = \begin{cases} \{a|a \in C, f(x_i, a) \neq f(x_j, a), f(x_i, D) \neq f(x_j, D)\} \\ \emptyset; else \end{cases} \tag{1}$$

Definition 6 [3]. For decision table $S = (U, C, D, V, f)$, let $M = (m_{ij})$ be Hu's discernibility matrix, then $Core1(C)$ is called the core of M, where $Core1(C) = \{a|a \in C, \exists m_{ij} \in M, m_{ij} = \{a\}\}$.

Definition 7. For decision table $S = (U, C, D, V, f)$, let $S' = (U', C, D, V, f)$ be simplified decision table, then we define simplified discernibility matrix based on Hu's discernibility matrix as $M'_1 = (m_{i'j'})$, and elements of M'_1 are:

$$m_{i'j'} = \begin{cases} \{a|a \in C, f(x'_i, a) \neq f(x'_j, a), if\ at\ most\ one\ of x'_i \\ and\ x'_j is\ in\ U'_{pos};\ or\ f(x'_i, a) \neq f(x'_j, a), \\ f(x'_i, D) \neq f(x'_j, D), if\ x'_i\ and\ x'_j\ are\ in\ U'_{pos};\} \\ \emptyset; else \end{cases} \tag{2}$$

Definition 8. For decision table $S = (U, C, D, V, f)$, let M'_1 be simplified discernibility matrix based on Hu's discernibility matrix, then $SDCore1(C)$ is called the core of M'_1, where $SDCore1(C) = \{a|a \in C, \exists m_{i'j'} \in M'_1, m_{i'j'} = \{a\}\}$.

Theorem 2. *For decision table $S = (U, C, D, V, f)$, we have*

$$Core1(C) = SDCore1(C).$$

Proof: *For any $c_k \in SDCore1(C)(1 \leq k \leq r)$,there exit x'_i, x'_j such that $m_{i'j'} = \{c_k\}$. If $x'_i, x'_j \in U'_{pos}$, according to definition 8, we have $f(x'_i, c_k) \neq f(x'_j, c_k), f(x'_i, c_s) = f(x'_j, c_s)(s \in \{1, 2, ..., r\} - \{k\})$, and $f(x'_i, D) \neq f(x'_j, D)$. Since $x'_i, x'_j \in U$, according to definition 6, we have $c_k \in Core1(C)$. If at most one of x'_i and x'_j is in U'_{pos}, according to definition 8, we have $f(x'_i, c_k) \neq f(x'_j, c_k)$ and $f(x'_i, c_s) = f(x'_j, c_s)(s \in \{1, 2, ..., r\} - \{k\})$. Since at most one of x'_i and x'_j is in U'_{pos}, so at least one of x'_i and x'_j is in U'_{neg}. Suppose $x'_i \in U'_{neg}$, if $f(x'_i, D) \neq f(x'_j, D)$, according to definition 6, we have $c_k \in Core1(C)$; if $f(x'_i, D) = f(x'_j, D)$, because $x'_i \in U'_{neg}$, there $\exists x''_i \in [x'_i]_C$ such that $f(x'_i, D) \neq f(x''_i, D)$, thus we have $f(x'_j, D) \neq f(x''_i, D)$ and $f(x''_i, c_k) \neq f(x'_j, c_k), f(x''_i, c_s) = f(x'_j, c_s)(s \in \{1, 2, ..., r\} - \{k\})$. Since $x''_i, x'_j \in U$, according to definition 6, we have $c_k \in Core1(C)$. Since c_k is selected arbitrarily, we have $SDCore1(C) \subseteq Core1(C)$.*

For any $c_k \in Core1(C)$ $(1 \leq k \leq r)$,there $\exists x_i, x_j$ such that $m_{ij} = \{c_k\}$. That is to say, there are $f(x_i, c_k) \neq f(x_j, c_k), f(x_i, c_s) = f(x_j, c_s)(s \in \{1, 2, ..., r\} - \{k\})$, and

$f(x_i, D) \neq f(x_j, D)$. So there $\exists x'_i, x'_j \in U$ such that $[x'_i]_C = [x_i]_C$ and $[x'_j]_C = [x_j]_C$. Hence we have $f(x'_i, c_k) \neq f(x'_j, c_k), f(x'_i, c_s) = f(x'_j, c_s)(s \in \{1, 2, ..., r\} - \{k\})$. If $[x'_i]_C \subseteq POS_C(D)$ and $[x'_j]_C \subseteq POS_C(D)$, then x'_i and x'_j is in U'_{pos}, and $f(x'_i, D) = f(x_i, D) \neq f(x_j, D) = f(x'_j, D)$, according to definition 8, we have $c_k \in SDCore1(C)$. If at most one of x'_i and x'_j is in U'_{pos}, obviously we have $f(x'_i, c_k) \neq f(x'_j, c_k), f(x'_i, c_s) = f(x'_j, c_s)(s \in \{1, 2, ..., r\} - \{k\})$, according to definition 8, we have $c_k \in SDCore1(C)$. Since c_k is selected arbitrarily, we have $Core1(C) \subseteq SDCore1(C)$.

According to the above discussion, we prove theorem 2.

4 Core Attributes of Simplified Discernibility Matrix Based on Positive Region

In this section, we propose some definitions and theorems about Core attributes based on positive region

Definition 9 [4]. For decision table $S = (U, C, D, V, f)$, $\forall b \in B \subseteq C$, if $POS_B(D) = POS_{B-\{b\}}(D)$, then b is unnecessary for D in B; else b is necessary for D in B. For $\forall B \subseteq C$, if every element in B is necessary for D, then B is independent to D. If for $\forall B \subseteq C, POS_B(D) = POS_C(D)$, and B is independent to D, then B is called the attribute reduction based on positive region of C to D .

Definition 10 [4]. For decision table $S = (U, C, D, V, f)$, let $Red2(C)$ be the set of all the attribute reduction based on positive region of C to D, then the core is defined as $Core2(C) = \bigcap_{B \in Red2(C)} B$.

Lemma 1. *For decision table S=(U, C, D, V, f), there is $\forall a \in Core2(C) \Leftrightarrow POS_{C-\{a\}}(D) \neq POS_C(D)$* [4,5].

Definition 11. For decision table $S = (U, C, D, V, f)$, let $S' = (U', C, D, V, f)$ be simplified decision table, then we define simplified discernibility matrix based on positive region as $M'_2 = (m_{i'j'})$, and elements of M'_2 are:

$$m_{i'j'} = \begin{cases} \{a|a \in C, f(x'_i, a) \neq f(x'_j, a), if\ only\ one\ of x'_i\ and\ x'_j is\ in\ U'_{pos}; \\ f(x'_i, a) \neq f(x'_j, a), f(x'_i, D) \neq f(x'_j, D), if\ x'_i\ and\ x'_j\ are\ in\ U'_{pos};\} \\ \emptyset; else \end{cases} \quad (3)$$

Definition 12. For decision table $S = (U, C, D, V, f)$, let M'_2 be simplified discernibility matrix based on positive region, then $SDCore2(C)$ is called the core of M'_2, where $SDCore2(C) = \{a|a \in C, \exists m_{i'j'} \in M'_2, m_{i'j'} = \{a\}\}$.

Theorem 3. *For decision table $S = (U, C, D, V, f)$, we have*

$$Core2(C) = SDCore2(C).$$

Proof: *For any $c_k \in SDCore2(C)(1 \leq k \leq r)$, there $\exists x'_i, x'_j$ such that $m_{i'j'} = \{c_k\}$. According to definition 12,if $x'_i, x'_j \in U'_{pos}$, we have $f(x'_i, c_k) \neq f(x'_j, c_k), f(x'_i, c_s) =$*

$f(x'_j, c_s)(s \in \{1,2,...,r\} - \{k\})$, and $f(x'_i, D) \neq f(x'_j, D)$. So there is $([x'_i]_C \cup [x'_j]_C) \subseteq [x'_i]_{C-\{c_k\}}$, while $f(x'_i, D) \neq f(x'_j, D)$. Thus $([x'_i]_C \cup [x'_j]_C) \cap POS_{C-\{c_k\}}(D) = \phi$. On the other hand,$([x'_i]_C \cup [x'_j]_C) \subseteq POS_C(D)$, thus $POS_{C-\{c_k\}}(D) \neq POS_C(D)$. According to lemma 1, we have $c_k \in Core2(C)$. If one of x'_i and x'_j is in U'_{pos}, the other is in U'_{neg}. Suppose $x'_i \in U'_{pos}$ and $x'_j \in U'_{neg}$, similarly there is $([x'_i]_C \cup [x'_j]_C) \subseteq [x'_i]_{C-\{c_k\}}$. Since $x'_j \in U'_{neg}$, so $([x'_i]_C \cup [x'_j]_C) \cap POS_{C-\{c_k\}}(D) = \phi$. On the other hand,$[x'_i]_C \subseteq POS_C(D)$, thus $POS_{C-\{c_k\}}(D) \neq POS_C(D)$. According to lemma 1, we have $c_k \in Core2(C)$. Since a is selected randomly, we have $SDCore2(C) \subseteq Core2(C)$.

For any $c_k \in Core2(C)(1 \leq k \leq r)$, according to lemma 1, $POS_{C-\{c_k\}}(D) \neq POS_C(D)$ holds. Thus there exits $x'_i \in U'_{pos}$ such that $[x'_i]_C \subseteq POS_C(D)$,but $[x'_i]_{C-\{c_k\}} \cap POS_{C-\{c_k\}}(D) = \phi$. So there at least exists $x'_j \in [x'_i]_{C-\{c_k\}} \wedge x'_j \notin [x'_i]_C (x'_j \in U')$ such that $f(x'_i, D) \neq f(x'_j, D)$. Since $x'_j \in [x'_i]_{C-\{c_k\}}$, thus $f(x'_i, b) = f(x'_j, b)(b \in C - \{c_k\})$. But $x'_j \notin [x'_i]_C$, we have $f(x'_i, c_k) \neq f(x'_j, c_k)$. If $x'_j \in U'_{neg}$, according to definition 11, we have $m_{i'j'} = \{c_k\}$. If $x'_j \in U'_{pos}$, since $f(x'_i, D) \neq f(x'_j, D)$, according to definition 11, we have $m_{i'j'} = \{c_k\}$. Hence $c_k \in SDCore2(C)$ holds. Since a is selected randomly, we have $Core2(C) \subseteq SDCore2(C)$.

According to the above discussion, we prove theorem 3.

5 Core Attributes of Simplified Discernibility Matrix Based on Information Entropy

In this section, we propose some definitions and theorems about Core attributes based on positive region.

Definition 13[8,9,10]. For decision table $S = (U, C, D, V, f)$, any attribute set $S \subseteq C \cup D$ (knowledge, equivalent relation cluster, $U/S = \{S_1, S_2, ..., S_t\}$) in U is a random variable in algebra formed by subsets of U, the probability distribution of the random variable can be determined by:

$$\left[S : P\right] = \begin{bmatrix} S_1 & S_1 & ... & S_t \\ p(S_1) & p(S_2) & ... & p(S_t) \end{bmatrix}$$

where $p(S_i) = |S_i|/|U|, i = 1, 2, ..., t$.

Definition 14[8,9,10]. For decision table $S = (U, C, D, V, f)$, the conditional entropy of set of decision attributes D ($U/D = \{D_1, D_2, ..., D_k\}$) to that of conditional attributes is defined as:

$$H(D|C) = -\sum_{i=1}^{m} p(C_i) \sum_{j=1}^{k} p(D_j|C_i) log p(D_j|C_i) \tag{4}$$

where $p(D_j|C_i) = \frac{|D_j \cap C_i|}{|C_i|}, i = 1, 2, ..., m; j = 1, 2, ..., k$.

Definition 15[8,9,10]. For decision table $S = (U, C, D, V, f)$ and $\forall b \in B \subseteq C$, if $H(D|B-\{b\}) = H(D|B)$, then b is unnecessary for D in B; else b is necessary for D in B. For $\forall B \subseteq C$, if every element in B is necessary for D, then B is independent to D.

Definition 16[8,9,10]**.** For decision table $S = (U, C, D, V, f)$, if $H(D|B) = H(D|C)$, and B is independent to D, then B is called the attribute reduction based on information entropy of C to D.

Definition 17[8,9,10]. For decision table $S = (U, C, D, V, f)$, let $Red3(C)$ be the set of all the attribute reduction based on information entropy of C to D, then the core is defined as $Core3(C) = \bigcap_{B \in Red3(C)} B$.

Definition 18. For decision table $S = (U, C, D, V, f)$ and $\forall B \subseteq C, \forall x \in U$, we define

$$\mu_B(x) = (\frac{|D_1 \cap [x]_B|}{|[x]_B|}, \frac{|D_2 \cap [x]_B|}{|[x]_B|}, ..., \frac{|D_k \cap [x]_B|}{|[x]_B|}) \tag{5}$$

$\mu_B(x)$ is called the probability distribution function of B on D.

Definition 19. For a decision table $S = (U, C, D, V, f), P, Q \subseteq (C \cup D)$, let $U/P = \{P_1, P_2, ..., P_t\}$, $U/Q = \{Q_1, Q_2, ..., Q_s\}$, we define the partition Q is coarser than the partition P , $U/P \preceq U/Q$, between partitions by

$$U/P \preceq U/Q \Leftrightarrow \forall P_i \in U/P \Rightarrow \exists Q_j \in U/Q \rightarrow P_i \subseteq Q_j \tag{6}$$

If $U/P \preceq U/Q$ and $U/Q \preceq U/P$, then we say $U/P = U/Q$. If $U/P \preceq U/Q$ and $U/P \neq U/Q$, then we say that U/Q is strictly coarser than U/Q , and write $U/P \prec U/Q$.

Theorem 4. *For decision table $S = (U, C, D, V, f), \forall Q \subseteq P \subseteq (C \cup D)$, we have $U/P \preceq U/Q$.*

Proof: *Suppose $U/P = \{P_1, P_2, ..., P_t\}$, $U/Q = \{Q_1, Q_2, ..., Q_s\}$, for any $P_i = [x]_P \in U/P$, since $Q \subseteq P$, then $P_i = [x]_P = \{y | y \in U, \forall a \in P, f(y, a) = f(x, a)\} \subseteq Q_j = [x]_Q = \{y | y \in U, \forall a \in Q, f(y, a) = f(x, a)\}$. Because P_i is selected randomly, we have $U/P \preceq U/Q$.*

Definition 20. For decision table $S = (U, C, D, V, f)$, let $S' = (U', C, D, V, f)$ be simplified decision table, then we define simplified discernibility matrix based on information entropy as $M'_3 = (m_{i'j'})$, and elements of M'_3 are:

$$m_{i'j'} = \{a | a \in C, f(x'_i, a) \neq f(x'_j, a), \mu(x'_i) \neq \mu(x'_j)\} \tag{7}$$

Definition 21. For decision table $S = (U, C, D, V, f)$, let M'_3 be simplified discernibility matrix based on information entropy , then $SDCore3(C)$ is called the core of M'_3, where $SDCore3(C) = \{a | a \in C, \exists m_{i'j'} \in M'_3, m_{i'j'} = \{a\}\}$.

Lemma 2. *Let U be universe of discourse, $A_1 = \{X_1, X_2, ..., X_r\}$ is a partition of equivalent relation A_1 on U, assume X_i and X_j are two equivalent blocks in A_1, $A_2 = \{X_1, X_2, ..., X_{i-1}, X_{i+1}, ...,$
$X_{j-1}, X_{j+1}, ..., X_r, X_i \cup X_j\}$ is a new partition formed by unifying X_i and X_j to $X_i \cup X_j$, $B = \{Y_1, Y_2, ..., Y_s\}$ is another partition of U, then*

$$H(B|A_2) \geq H(B|A_1) \tag{8}$$

The sufficient and necessary condition of $H(B|A_1) = H(B|A_2)$ *is that for any* $k(k = 1, 2, ..., s)$, $\frac{|X_i \cap Y_k|}{|X_i|} = \frac{|X_j \cap Y_k|}{|X_j|}$ *holds.* [8]

Theorem 5. *For decision table* $S = (U, C, D, V, f)$*, if* $\forall A_1 \subseteq A_2 \subseteq C$ *, then we have*

$$H(D|A_1) \geq H(D|A_2).$$

The sufficient and necessary condition of $H(D|A_1) = H(D|A_2)$ *is as follows: for any* $X_i, X_j \in U/A_2, X_i \neq X_j$ *, if* $(X_i \cup X_j) \subseteq Y \in U/A_1$*,for* $\forall r(r = 1, 2, ..., k)$, $\frac{|X_i \cap D_r|}{|X_i|} = \frac{|X_j \cap D_r|}{|X_j|}$ *always holds, where* $D_r \in U/D = \{D_1, D_2, ..., D_k\}$.

Proof: *We can prove theorem 5 easily based on theorem 4 and lemma 2.*

Theorem 6. *For decision table* $S = (U, C, D, V, f)$ *and* $\forall B \subseteq C$*, denote* $U/B = \{B_1, B_2, ..., B_s\}, U/C = \{C_1, C_2, ..., C_m\}$ *, then*

$$H(D|B) = H(D|C) \Leftrightarrow \forall x \in U \Rightarrow \mu_B(x) = \mu_C(x).$$

Proof: *When* $H(D|B) = H(D|C)$*, assume* $\exists x_0 \in U$ *such that* $\mu_B(x_0) \neq \mu_C(x_0)$*. That is to say, there is*

$$\left(\frac{|D_1 \cap [x_0]_B|}{|[x_0]_B|}, \frac{|D_2 \cap [x_0]_B|}{|[x_0]_B|}, ..., \frac{|D_k \cap [x_0]_B|}{|[x_0]_B|}\right)$$
$$\neq \left(\frac{|D_1 \cap [x_0]_C|}{|[x_0]_C|}, \frac{|D_2 \cap [x_0]_C|}{|[x_0]_C|}, ..., \frac{|D_k \cap [x_0]_C|}{|[x_0]_C|}\right)$$

Thus we have $[x_0]_B \neq [x_0]_C$*. According to theorem 4 and definition 19,* $[x_0]_B \supset [x_0]_C$ *holds. So let* $[x_0]_B = [x_0]_C \cup [x_1]_C \cup ... \cup [x_l]_C$ *(Where* $x_0, x_1, ..., x_l \in U, l \geq 1, \forall i \neq j(\imath, \jmath \in \{0, 1, ..., l\})$ *and* $[x_i]_C \cap [x_j]_C = \phi$ *), thus there at least exist* $(i_0, j_0 \in \{0, 1, ..., l\}), i_0 \neq j_0$ *, such that* $\mu_C(x_{i_0}) \neq \mu_C(x_{j_0})$*(else, obviously* $\mu_B(x_0) = \mu_C(x_0)$ *), i.e.*

$$\left(\frac{|D_1 \cap [x_{i_0}]_C|}{|[[x_{i_0}]_C|}, \frac{|D_2 \cap [x_{i_0}]_C|}{|[x_{i_0}]_C|}, ..., \frac{|D_k \cap [[x_{i_0}]_C|}{|[x_{i_0}]_C|}\right)$$
$$\neq \left(\frac{|D_1 \cap [x_{j_0}]_C|}{|[x_{j_0}]_C|}, \frac{|D_2 \cap [x_{j_0}]_C|}{|[x_{j_0}]_C|}, ..., \frac{|D_k \cap [x_{j_0}]_C|}{|[x_{j_0}]_C|}\right)$$

Based on theorem 5, it is obviously true that $H(D|B) > H(D|C)$ *. It conflicts with the hypothesis. Thus we have* $\forall x \in U \Rightarrow \mu_B(x) = \mu_C(x)$ *.*

When $\forall x \in U \Rightarrow \mu_B(x) = \mu_C(x)$ *, since* $B \subseteq C$*, according to the theorem 4, we have* $[x]_B = [x_1]_C \cup [x_2]_C \cup ... \cup [x_l]_C$ *(Where* $x_1, x_2, ..., x_l \in U, l \geq 1, \forall i \neq j(i, j \in \{1, 2, ..., l\})$ *and* $[x_i]_C \cap [x_j]_C = \phi$ *). According to definition 18, there is* $\mu_B(x) = \mu_C(x_1) = \mu_C(x_2) = ... = \mu_C(x_l)$ *. So*

$$\sum_{i=1}^{l} p([x_i]_C) \sum_{j=1}^{k} p(D_j|[x_i]_C) log p(D_j|[x_i]_C)$$

$$= \sum_{i=1}^{l} \frac{|[x_i]_C|}{|U|} \sum_{j=1}^{k} \frac{|D_j \cap [x_i]_C|}{|[x_i]_C|} log \frac{|D_j \cap [x_i]_C|}{|[x_i]_C|}$$

$$= \sum_{i=1}^{l} \frac{|[x_i]_C|}{|U|} \sum_{j=1}^{k} \frac{|D_j \cap [x]_B|}{|[x]_B|} log \frac{|D_j \cap [x]_B|}{|[x]_B|}$$

$$= \frac{|[x_1]_C \cup [x_2]_C \cup ... \cup [x_l]_C|}{|U|} \sum_{j=1}^{k} \frac{|D_j \cap [x]_B|}{|[x]_B|} log \frac{|D_j \cap [x]_B|}{|[x]_B|}$$

$$= \frac{|[x]_B|}{|U|} \sum_{j=1}^{k} \frac{|D_j \cap [x]_B|}{|[x]_B|} log \frac{|D_j \cap [x]_B|}{|[x]_B|}$$

So

$$H(D|C) = -\sum_{i=1}^{m} p(C_i) \sum_{j=1}^{k} p(D_j|C_i) log p(D_j|C_i)$$

$$= -\sum_{i=1}^{t} p(B_i) \sum_{j=1}^{k} p(D_j|B_i) log p(D_j|B_i) = H(D|B)$$

Thus we have $H(D|C) = H(D|B)$.

Lemma 3. *For decision table $S = (U, C, D, V, f)$, the sufficient and necessary condition of $\forall a \in Core3(C)$ is $H(D|(C - \{a\})) > H(D|C)$.* [8]

Theorem 7. *For decision table $S = (U, C, D, V, f)$, we have*

$$Core3(C) = SDCore3(C).$$

Proof: *For any $a \in Core3(C)$, we have $H(D|(C - \{a\})) > H(D|C)$ based on Lemma 3. According to theorem 6,there at least $\exists x_i \in U$ such that $\mu_{C-\{a\}}(x_i) \neq \mu_C(x_i)$. So $[x_i]_{C-\{a\}} \neq [x_i]_C$. Since $C - \{a\} \subseteq C$, we have $[x_i]_{C-\{a\}} \supseteq [x_i]_C$ according to the theorem 4. Then $\exists x_j \in [x_i]_{C-\{a\}} \wedge \mu_C(x_j) \neq \mu_C(x_i)$ (else, if $\forall x_j \in [x_i]_{C-\{a\}} \wedge \mu_C(x_j) = \mu_C(x_i)$, we have $\mu_{C-\{a\}}(x_i) = \mu_C(x_i)$ according to the proof of theorem 6). On the other hand, there exist $x'_j \in U' \wedge x'_j \in [x_j]_C, x'_i \in U' \wedge x'_i \in [x_i]_C$, thus $\mu_C(x'_j) \neq \mu_C(x'_i)$ holds; Meanwhile, since $x'_i, x'_j \in [x_i]_{C-\{a\}} \wedge [x'_i]_C \neq [x'_j]_C$, so $m_{i'j'} = \{a\}$ holds. Hence $a \in SDCore3(C)$. Since a is selected randomly, we have $Core3(C) \subseteq SDCore3(C)$.*

For any $a \in SDCore3(C)$, there exists $m_{i'j'} = \{a\}$,namely there exist $x'_i, x'_j \in U', x'_i, x'_j \in [x_i]_{C-\{a\}} \wedge [x'_i]_C \neq [x'_j]_C$ and $\mu_C(x'_j) \neq \mu_C(x'_i)$. So at least one of $\mu_{C-\{a\}}(x'_j) \neq \mu_C(x'_j)$ and $\mu_{C-\{a\}}(x'_i) \neq \mu_C(x'_i)$ holds (else, if $\mu_{C-\{a\}}(x'_j) = \mu_C(x'_j)$ and $\mu_{C-\{a\}}(x'_i) = \mu_C(x'_i)$, since $\mu_{C-\{a\}}(x'_j) = \mu_{C-\{a\}}(x'_i)$, we have $\mu_C(x'_j) = \mu_C(x'_i)$). According to theorem 6, we have $H(D|(C - \{a\}) \neq H(D|C)$. According to theorem 5, we have $H(D|(C - \{a\}) \geq H(D|C)$, thus $H(D|(C - \{a\}) > H(D|C)$. So we have $a \in Core3(C)$ according to Lemma 3. Since a is selected randomly, we have $SDCore3(C) \subseteq Core3(C)$.

Based on the above discussion, we have $SDCore3(C) = Core3(C)$.

6 Comparison Study on Three Kinds of Core Attributes

In this section, we discuss the relationship of three kinds of core attributes.

Theorem 8. *For decision table $S = (U, C, D, V, f)$, we have*

$$Core2(C) \subseteq Core3(C) \subseteq Core1(C).$$

Proof: *$\forall \phi \neq m_{i'j'} \in M'_2$, let $m_{i'j'} = \{a_1, a_2, ..., a_s\}(a_i \in C)$ and $U/D = \{D_1, D_2, ..., D_k\}$, then according to definition 11, we have:*

(1) $\exists x'_i, x'_j \in U'$, if $x'_i, x'_j \in U'_{pos}$, we have $f(x'_i, D) \neq f(x'_j, D), f(x'_i, a_l) \neq f(x'_j, a_l)(l = 1, 2, ..., s)$, and $f(x'_i, b) = f(x'_j, b)(b \in C - m_{i'j'})$. As $x'_i, x'_j \in U'_{pos}$, assume $[x'_i]_C \subseteq D_t = \{y | y \in U, f(y, D) = f(x'_i, D)\}$. For $\forall x_{i1}, x_{i1} \in [x'_i]_C$, we have

$$\mu_C(x_{i1}) = \mu_C(x_{i1}) = \mu_C(x'_i) = \frac{(0\ 0\ ...\ \ 0\ \ \ 1\ \ 0\ \ ...\ 0)}{1\ 2\ ...\ t-1\ t\ t+1\ ...\ k}$$

Assume $[x'_j]_C \subseteq D_r = \{y | y \in U, f(y, D) = f(x'_j, D)\}$. For $\forall x_{j1}, x_{j1} \in [x'_j]_C$, we have

$$\mu_C(x_{j1}) = \mu_C(x_{j1}) = \mu_C(x'_j) = \frac{(0\ 0\ ...\ \ 0\ \ \ 1\ \ 0\ \ ...\ 0)}{1\ 2\ ...\ r-1\ r\ r+1\ ...\ k}$$

Since $f(x'_i, D) \neq f(x'_j, D)$, thus $D_t \neq D_r$, then we have $\mu_C(x'_i) \neq \mu_C(x'_j)$. So we have $m_{i'j'} \in M'_3$ based on definition 20.

(2) if only one of x'_i and x'_j is in U'_{pos}, then we have $f(x'_i, a_l) \neq f(x'_j, a_l)(l = 1, 2, ..., s)$, and $f(x'_i, b) = f(x'_j, b)(b \in C - m_{i'j'})$. Suppose $x'_i \in U'_{pos}$ and $x'_j \in U'_{neg}$, then we can assume $[x'_i]_C \subseteq D_t = \{y | y \in U, f(y, D) = f(x'_i, D)\}$. Thus for $\forall x_{i1}, x_{i1} \in [x'_i]_C$, we have

$$\mu_C(x_{i1}) = \mu_C(x_{i1}) = \mu_C(x'_i) = \frac{(0\ 0\ ...\ \ 0\ \ \ 1\ \ 0\ \ ...\ 0)}{1\ 2\ ...\ t-1\ t\ t+1\ ...\ k}$$

On the other hand, $x'_j \in U'_{neg}$, so $[x'_j]_C \cap D_t = \phi$, thus we have $\mu_C(x'_i) \neq \mu_C(x'_j)$. So we have $m_{i'j'} \in M'_3$ based on definition 20.

According to the above discussion, for decision table $S = (U, C, D, V, f)$ and $\forall \phi \neq m_{i'j'} \in M'_2$, then $m_{i'j'} \in M'_3$. Thus to $\forall a \in SDCore2(C)$, there exists $\{a\} = m_{i'j'} \in M'_2$, so we have $\{a\} = m_{i'j'} \in M'_3$, namely $a \in SDCore3(C)$. Since a is selected randomly, we have $SDCore2(C) \subseteq SDCore3(C)$. According to theorem 3 and theorem 7, we have $Core2(C) \subseteq Core3(C)$.

For $\forall \phi \neq m_{i'j'} \in M'_3$, let $m_{i'j'} = \{a_1, a_2, ..., a_s\}(a_i \in C)$ and $U/D = \{D_1, D_2, ..., D_k\}$, then according to definition 20, $\exists x'_i, x'_j \in U'$ such that $\mu_C(x'_i) \neq \mu_C(x'_j)$, $f(x'_i, a_l) \neq f(x'_j, a_l)(l = 1, 2, ..., s)$, and $f(x'_i, b) = f(x'_j, b)(b \in C - m_{i'j'})$. If $x'_i, x'_j \in U'_{pos}$, assume $[x'_i]_C \subseteq D_t = \{y | y \in U, f(y, D) = f(x'_i, D)\}$ and $[x'_j]_C \subseteq D_r = \{y | y \in U, f(y, D) = f(x'_j, D)\}$. we have

$$\mu_C(x'_i) = \frac{(0\ 0\ \ldots\ \ 0\ \ 1\ \ 0\ \ \ldots\ 0)}{1\ 2\ \ldots\ t-1\ t\ t+1\ \ldots\ k} \text{ and } \mu_C(x'_j) = \frac{(0\ 0\ \ldots\ \ 0\ \ 1\ \ 0\ \ \ldots\ 0)}{1\ 2\ \ldots\ r-1\ r\ r+1\ \ldots\ k}$$

Since $\mu_C(x'_i) \neq \mu_C(x'_j)$ *, we have* $D_t \neq D_r$*, and accordingly* $f(x'_i, D) \neq f(x'_j, D)$*. Based on definition 7,* $m_{i'j'} \in M'_1$ *holds. If at most one of* x'_i *and* x'_j *is in* U'_{pos} *,since* $f(x'_i, a_l) \neq f(x'_j, a_l)(l = 1, 2, ..., s)$ *, and* $f(x'_i, b) = f(x'_j, b)(b \in C - m_{i'j'})$ *. we have* $m_{i'j'} \in M'_1$ *according to definition 7.*

Based on the above discussion, for decision table $S = (U, C, D, V, f)$ *and* $\forall \phi \neq m_{i'j'} \in M'_3$ *, then* $m_{i'j'} \in M'_1$*. Thus to* $\forall a \in SDCore3(C)$*, there exists* $\{a\} = m_{i'j'} \in M'_3$ *, so we have* $\{a\} = m_{i'j'} \in M'_1$ *, namely* $a \in SDCore1(C)$*. Since a is selected randomly, we have* $SDCore3(C) \subseteq SDCore1(C)$ *. According to theorem 2 and theorem 7, we have* $Core3(C) \subseteq Core1(C)$ *.*

According to the above discussion, we can prove theorem 8.

Theorem 9. *For decision table* $S = (U, C, D, V, f)$*, if* S *is consistent, we have* $Core2(C) = Core3(C) = Core1(C)$ *.*

Proof: $\forall \phi \neq m_{i'j'} \in M'_1$ *, let* $m_{i'j'} = \{a_1, a_2, ..., a_s\}(a_i \in C)$*. since* S *is consistent, thus* $POS_C(D) = U$ *, so we have* $U'_{pos} = U'$*. According to definition 7, there exist* $x'_i, x'_j \in U'_{pos} = U'$*, such that* $f(x'_i, D) \neq f(x'_j, D), f(x'_i, a_l) \neq f(x'_j, a_l)(l = 1, 2, ..., s)$*, and* $f(x'_i, b) = f(x'_j, b)(b \in C - m_{i'j'})$ *, so we have* $m_{i'j'} \in M'_2$*.That is to say,*$\forall \phi \neq m_{i'j'} \in M'_1$*, we have* $m_{i'j'} \in M'_1$*.Thus to* $\forall a \in SDCore1(C)$*, there exists* $\{a\} = m_{i'j'} \in M'_1$ *, so we have* $\{a\} = m_{i'j'} \in M'_2$*, namely* $a \in SDCore2(C)$*. Since a is selected randomly, we have* $SDCore1(C) \subseteq SDCore2(C)$*. According to theorem 2 and theorem 3, we have* $Core1(C) \subseteq Core2(C)$*.*

On the other hand, according to theorem 8, we have $Core2(C) \subseteq Core3(C) \subseteq Core1(C)$ *. So for consistent decision table,* $Core2(C) \subseteq Core3(C) \subseteq Core1(C) \subseteq Core2(C)$ *holds, namely* $Core2(C) = Core3(C) = Core1(C)$*.*

7 Conclusion

There are three kinds of core attributes: Hu's core based on discernibility matrix (denoted by $Core1(C)$), core based on positive region (denoted by $Core2(C)$), and core based on information entropy (denoted by $Core3(C)$). In this paper, we at first constructed three kinds of simplified matrices and their core attributes based on Hu's discernibility matrix, positive region and information entropy respectively. Secondly, it is proved that the three new cores are equivalent to those original ones accordingly. Thirdly, based on the proposed simplified matrices and cores, we proved that $Core2(C) \subseteq Core3(C) \subseteq Core1(C)$ for general decision table, while $Core2(C) = Core3(C) = Core1(C)$ for consistent decision table. Our construction illustrates that three kinds of core do have simplified discernibility matrices. Furthermore, according to the proof process, these three kinds of simplified matrices show the inclusion relation of three kinds of core attributes. These three simplified discernibility matrices introduce new idea to calculating core attributes. So developing new efficient algorithm for computing core attributes is our future work.

References

1. Pawlak Z(1995). Rough Set. Communication of the ACM,38(11): 89-95.
2. Pawlak Z(1998). Rough Set theory and Its Applications to Data Analysis. Cybernetics and Systems,29 (7): 661-688.
3. Hu XH, Cercne N(1995). Learning in Relational Databases: A Rough Set Approach. International Journal of Computational Intelligence, 11 (2): 323-338
4. Ye DY, Chen ZJ(2002). A New Discernibility Matrix and the Computation of a Core. Acta Electronica Sinica,30(7):1086-1088
5. Ye DY, Chen ZJ(2004). A New Binary Discernibility Matrix and the Computation of a Core. MINI-MICRO SYSTEMS, 25(6): 965-967
6. Yang M, Sun ZH(2004). Improvement of Discernibility Matrix and the Computation of a Core. Journal of Fudan University (Natural Science), 43 (5): 865-868
7. Zhao J, Wang GY(2003), Wu ZF. An Efficient Approach to Compute the Feature Core. MINI-MICRO SYSTEMS, 24(11): 1950-1953
8. Wang GY(2003). Calculation Method for Core Attributions of Decision Table. Chinese Journal of Computer,26(5): 611-615
9. Wang GY(2003). Rough Reduction in Algebra View and Information View. International Journal of Intelligent System, 18 (6): 679-688.
10. Wang GY, Yu H, Yang DC(2002). Decision Table Reduction Based on Conditional Information Entropy. Chinese Journal of Computers, 25(7):759-766
11. Hu KY, Lu YC and Shi CY(2001). Advances in rough set theory and its applications.Journal of tsinghua Univ(Sci. and Tech.), 41 (1) : 64-68
12. Liu SH, Sheng QJ(2003). Research on efficient algorithm for rough set method.Chinese Journal of Computer, 26(5):524-529
13. Du JL, Chi ZX, Zhai W(2003). An improved algorithm for reduction of knowledge based on significance of attribution. Mini-micro System, 24(6):976-978

Controllability for the Impulsive Semilinear Fuzzy Integrodifferential Equations

Jin Han Park[1], Jong Seo Park[2], Young Chel Ahn[3], and Young Chel Kwun[3,*]

[1] Division of Mathematical Sciences, Pukyong National University, Pusan 608-737, South Korea
jihpark@pknu.ac.kr
[2] Department of Mathematics Education, Chinju National University of Education, Chinju 660-756, South Korea
parkjs@cue.ac.kr
[3] Department of Mathematics, Dong-A University, Pusan 604-714, South Korea
yckwun@dau.ac.kr

Abstract. In this paper. we study the controllability for the impulsive semilinear fuzzy integrodifferential control system in E_N by using the concept of fuzzy number whose values are normal, convex, upper semicontinuous and compactly supported interval in E_N.

Keywords: Controllability, Impulsive semilinear fuzzy integrodifferential equations.

1 Introduction

Many authors have studied several concepts of fuzzy systems. Kaleva [3] studied the existence and uniqueness of solution for the fuzzy differential equation on E^n where E^n is normal, convex, upper semicontinuous and compactly supported fuzzy sets in R^n. Seikkala [7] proved the existence and uniqueness of fuzzy solution for the following equation:

$$\dot{x}(t) = f(t, x(t)) \ , \quad x(0) = x_0,$$

where f is a continuous mapping from $R^+ \times R$ into R and x_0 is a fuzzy number in E^1. Diamond and Kloeden [2] proved the fuzzy optimal control for the following system:

$$\dot{x}(t) = a(t)x(t) + u(t), \quad x(0) = x_0$$

where $x(\cdot), u(\cdot)$ are nonempty compact interval-valued functions on E^1. Kwun and Park [4] proved the existence of fuzzy optimal control for the nonlinear fuzzy differential system with nonlocal initial condition in E_N^1 using by Kuhn-Tucker theorems. Balasubramaniam and Muralisankar [1] proved the existence and uniqueness of fuzzy solutions for the semilinear fuzzy integrodifferential equation with nonlocal initial condition. Recently Park, Park and Kwun [6]

* Corresponding author. This paper is supported by Dong-A University Research Fund in 2006.

find the sufficient conditions of nonlocal controllability for the semilinear fuzzy integrodifferential equations with nonlocal initial conditions.

In this paper we prove the existence and uniqueness of fuzzy solutions and find the sufficient conditions of controllability for the following impulsive semilinear fuzzy integrodifferential equations:

$$\frac{dx(t)}{dt} = A\left[x(t) + \int_0^t G(t-s)x(s)ds\right] + f(t,x) + u(t), \quad t \in I = [0,T], \tag{1}$$

$$x(0) = x_0 \in E_N, \tag{2}$$

$$\triangle x(t_k) = I_k(x(t_k)), \quad k = 1, 2, \cdots, m, \tag{3}$$

where $A : I \to E_N$ is a fuzzy coefficient, E_N is the set of all upper semicontinuous convex normal fuzzy numbers with bounded α-level intervals, $f : I \times E_N \to E_N$ is a nonlinear continuous function, $G(t)$ is $n \times n$ continuous matrix such that $\frac{dG(t)x}{dt}$ is continuous for $x \in E_N$ and $t \in I$ with $\|G(t)\| \le k$, $k > 0$, $u : I \to E_N$ is control function and $I_k \in C(E_N, E_N)(k = 1, 2, \cdots, m)$ are bounded functions, $\triangle x(t_k) = x(t_k^+) - x(t_k^-)$, where $x(t_k^+)$ and $x(t_k^-)$ represent the left and right limits of $x(t)$ at $t = t_k$, respectively.

2 Existence and Uniqueness of Fuzzy Solution

In this section we consider the existence and uniqueness of fuzzy solutions for the impulsive semilinear fuzzy integrodifferential equation (1)-(3)($u \equiv 0$).

We denote the suprimum metric d_∞ on E^n and the suprimum metric H_1 on $C(I : E^n)$.

Definition 1. Let $a, b \in E^n$.

$$d_\infty(a,b) = \sup\{d_H([a]^\alpha, [b]^\alpha) : \alpha \in (0,1]\}$$

where d_H is the Hausdorff distance.

Definition 2. Let $x, y \in C(I : E^n)$.

$$H_1(x,y) = \sup\{d_\infty(x(t), y(t)) : t \in I\}.$$

Definition 3. The fuzzy process $x : I \to E_N$ is a solution of equations (1)-(2) without the inhomogeneous term if and only if

$$(\dot{x}_l^\alpha)(t) = \min\left\{A_l^\alpha(t)\left[x_j^\alpha(t) + \int_0^t G(t-s)x_j^\alpha(s)ds\right],\ i,j = l,r\right\},$$

$$(\dot{x}_r^\alpha)(t) = \max\left\{A_r^\alpha(t)\left[x_j^\alpha(t) + \int_0^t G(t-s)x_j^\alpha(s)ds\right],\ i,j = l,r\right\},$$

and

$$(x_l^\alpha)(0) = x_{0l}^\alpha, \quad (x_r^\alpha)(0) = x_{0r}^\alpha.$$

Now we assume the following:

(H1) The inhomogeneous term $f : I \times E_N \to E_N$ is a continuous function and satisfies a global Lipschitz condition

$$d_H([f(s,x(s))]^\alpha, [f(s,y(s))]^\alpha) \le c_1 d_H([x(s)]^\alpha, [y(s)]^\alpha),$$

for all $x(\cdot), y(\cdot) \in E_N$, and a finite positive constant $c_1 > 0$.

(H2) $S(t)$ is a fuzzy number satisfying for $y \in E_N$, $S'(t)y \in C^1(I : E_N) \bigcap C(I : E_N)$ the equation

$$\begin{aligned}\frac{d}{dt}S(t)y &= A\left[S(t)y + \int_0^t G(t-s)S(s)yds\right] \\ &= S(t)Ay + \int_0^t S(t-s)AG(s)yds, \quad t \in I,\end{aligned}$$

such that

$$[S(t)]^\alpha = [S_l^\alpha(t), S_r^\alpha(t)], \quad S(0) = I$$

and $S_i^\alpha(t)$ $(i = l, r)$ is continuous. That is, there exists a constant $c > 0$ such that $|S_i^\alpha(t)| \le c$ for all $t \in I$.

In order to define the solution of (1)-(3), we shall consider the space $\Omega = \{x : J \to E_N : x_k \in C(J_k, E_N), J_k = (t_k, t_{k+1}], k = 0, 1, \cdots, m$, and there exist $x(t_k^-)$ and $x(t_k^+)(k = 1, \cdots, m)$, with $x(t_k^-) = x(t_k)\}$.

Lemma 1. *If x is an integral solution of* (1)-(3) $(u \equiv 0)$, *then x is given by*

$$\begin{aligned}x(t) = S(t)x_0 &+ \int_0^t S(t-s)f(s,x(s))ds \\ &+ \sum_{0<t_k<t} S(t-t_k)I_k(x(t_k^-)), \text{ for t} \in \text{J}.\end{aligned} \tag{4}$$

Proof. Let x be a solution of (1)-(3). Define $\omega(s) = S(t-s)x(s)$. Then we have that

$$\begin{aligned}\frac{d\omega(s)}{ds} &= -\frac{S(t-s)}{ds}x(s) + S(t-s)\frac{x(s)}{ds} \\ &= -A[S(t)x + \int_0^t G(t-s)S(s)x(s)ds] + S(t-s)\frac{x(s)}{ds} \\ &= S(t-s)f(s,x(s)).\end{aligned}$$

Consider $t_k < t, k = 1, \cdots, m$. Then integrating the previous equation, we have

$$\int_0^t \frac{\omega(s)}{ds}ds = \int_0^t S(t-s)f(s,x(s))ds.$$

For $k = 1$,

$$\omega(t) - \omega(0) = \int_0^t S(t-s) f(s, x(s)) ds$$

or

$$x(t) = S(t)x_0 + \int_0^t S(t-s) f(s, x(s)) ds.$$

Now for $k = 2, \cdots, m$, we have

$$\int_0^{t_1} \frac{\omega(s)}{ds} ds + \int_{t_1}^{t_2} \frac{\omega(s)}{ds} ds + \cdots + \int_{t_k}^{t} \frac{\omega(s)}{ds} ds = \int_0^t S(t-s) f(s, x(s)) ds.$$

Then

$$\begin{aligned} &\omega(t_1^-) - \omega(0) + \omega(t_2^-) - \omega(t_1^+) + \cdots - \omega(t_k^+) + \omega(t) \\ &\quad = \int_0^t S(t-s) f(s, x(s)) ds \end{aligned}$$

if and only if

$$\omega(t) = \omega(0) + \int_0^t S(t-s) f(s, x(s)) ds + \sum_{0<t_k<t} [\omega(t_k^+) - \omega(t_k^-)].$$

Hence

$$x(t) = S(t)x_0 + \int_0^t S(t-s) f(s, x(s)) ds + \sum_{0<t_k<t} S(t-t_k) I_k(x(t_k^-)),$$

which proves the lemma. □

Assume the following:

(H3) There exists d_k, $k = 1, \cdots, m$, such that

$$d_H([I_k(x(t_k^-))]^\alpha, [I_k(y(t_k^-))]^\alpha) \leq d_k d_H([x(t)]^\alpha, [y(t)]^\alpha),$$

where $\sum_{k=1}^n d_k = \bar{d}$.

(H4) $c(c_1 T + \bar{d}) < 1$.

Theorem 1. *Let $T > 0$, and hypotheses* (H1)-(H4) *hold. Then, for every $x_0 (\in E_N)$, problem* (1)-(3) *has a unique solution $x \in \Omega$.*

Proof. For each $\xi(t) \in \Omega'$, $t \in J$, define

$$(\Phi\xi)(t) = S(t)x_0 + \int_0^t S(t-s) f(s, \xi(s)) ds + \sum_{0<t_k<t} S(t-t_k) I_k(\xi(t_k^-)).$$

Thus, $(\Phi\xi)(t) : J \to \Omega$ is continuous and $\Phi : \Omega \to \Omega$. It is obvious that fixed points of Φ are solution for the problem (1)-(3). For $\xi(t), \eta(t) \in \Omega$, we have

$$\begin{aligned}
&d_H([(\Phi\xi)(t)]^\alpha, [(\Phi\eta)(t)]^\alpha) \\
&\le cc_1 \int_0^t d_H([\xi(s)]^\alpha, [\eta(s)]^\alpha)ds + c \sum_{0<t_k<t} d_H([I_k(\xi(t_k^-))]^\alpha, [I_k(\eta(t_k^-))]^\alpha) \\
&\le cc_1 \int_0^t d_H([\xi(s)]^\alpha, [\eta(s)]^\alpha)ds + c\bar{d}\, d_H([\xi(t)]^\alpha, [\eta(t)]^\alpha).
\end{aligned}$$

Therefore,

$$\begin{aligned}
&d_\infty((\Phi\xi)(t), (\Phi\eta)(t)) = \sup_{\alpha\in(0,1]} d_H([(\Phi\xi)(t)]^\alpha, [(\Phi\eta)(t)]^\alpha) \\
&\le cc_1 \int_0^t \sup_{\alpha\in(0,1]} d_H([\xi(s)]^\alpha, [\eta(s)]^\alpha)ds + c\bar{d} \sup_{\alpha\in(0,1]} d_H([\xi(t)]^\alpha, [\eta(t)]^\alpha) \\
&= cc_1 \int_0^t d_\infty(\xi(s), \eta(s))ds + c\bar{d} d_\infty([\xi(t)]^\alpha, [\eta(t)]^\alpha).
\end{aligned}$$

Hence

$$\begin{aligned}
H_1(\Phi\xi, \Phi\eta) &= \sup_{t\in J} d_\infty((\Phi\xi)(t), (\Phi\eta)(t)) \\
&\le cc_1 \sup_{t\in J} \int_0^t d_\infty(\xi(s), \eta(s))ds + c\bar{d} \sup_{t\in J} d_\infty(\xi(t), \eta(t)) \\
&= c(c_1 T + \bar{d}) H_1(\xi, \eta).
\end{aligned}$$

By hypotheses (H4), Φ is a contraction mapping. By the Banach fixed point theorem, (4) has a unique fixed point $x \in \Omega$. □

3 Controllability

In this section, we show the controllability for the control system (1)-(3). The control system (1)-(3) is related to the following fuzzy integral system:

$$\begin{aligned}
x(t) = S(t)x_0 &+ \int_0^t S(t-s)f(s, x(s))ds \\
&+ \int_0^t S(t-s)u(s)ds + \sum_{0<t_k<t} S(t-t_k)I_k(x(t_k^-))
\end{aligned} \tag{5}$$

for $t \in J, t \ne t_k (k = 1, 2, \cdots, m)$, where $S(t)$ satisfies (H2).

Definition 4. The equation (5) is controllable if there exists $u(t)$ such that the fuzzy solution $x(t)$ of (5) satisfies $x(T) = x^1$, i.e., $[x(T)]^\alpha = [x^1]^\alpha$, where x^1 is target set.

We assume that the linear control system with respect to semilinear control system (5) is controllable. Then

$$x(T) = S(T)x_0 + \int_0^T S(T-s)u(s)ds + \sum_{0<t_k<T} S(T-t_k)I_k(x(t_k^-)) = x^1$$

and

$$[x(T)]^{\alpha} = [S(T)x_0 + \int_0^T S(T-s)u(s)ds + \sum_{0<t_k<t} S(T-t_k)I_k(x(t_k^-))]^{\alpha}$$

$$= \big[S_l^{\alpha}(T){x_0}_l^{\alpha} + \int_0^T S_l^{\alpha}(T-s)u_l^{\alpha}(s)ds + \sum_{0<t_k<t} S_l^{\alpha}(T-t_k){I_k}_l^{\alpha}(x(t_k^-)),$$

$$S_r^{\alpha}(T){x_0}_r^{\alpha} + \int_0^T S_r^{\alpha}(T-s)u_r^{\alpha}(s)ds + \sum_{0<t_k<t} S_r^{\alpha}(T-t_k){I_k}_r^{\alpha}(x(t_k^-))\big]$$

$$= [(x^1)_l^{\alpha}, (x^1)_r^{\alpha}].$$

Define the α-level set of fuzzy mapping $G : \tilde{P}(R) \to E_N$ by

$$G^{\alpha}(v) = \begin{cases} \int_0^T S^{\alpha}(T-s)v(s)ds, & v \subset \overline{\Gamma_u}\ , \\ 0, & \text{otherwise,} \end{cases} \tag{6}$$

where$\overline{\Gamma_u}$ is closure of support u. Then there exists $G_i^{\alpha}(i = l, r)$ such that

$$G_l^{\alpha}(v_l) = \int_0^T S_l^{\alpha}(T-s)v_l(s)ds, v_l(s) \in [u_l^{\alpha}(s), u^1(s)],$$

$$G_r^{\alpha}(v_r) = \int_0^T S_r^{\alpha}(T-s)v_r(s)ds, v_r(s) \in [u^1(s), u_r^{\alpha}(s)].$$

We assume that $G_l^{\alpha}, G_r^{\alpha}$ are bijective mapping. Hence α-level set of $u(s)$ is

$$[u(s)]^{\alpha} = [u_l^{\alpha}(s), u_r^{\alpha}(s)]$$

$$= [(\tilde{G}_l^{\alpha})^{-1}((x^1)_l^{\alpha} - S_l^{\alpha}(T)({x_0}_l^{\alpha}) - \sum_{0<t_k<T} S_l^{\alpha}(T-t_k){I_k}_l^{\alpha}(x(t_k^-))),$$

$$(\tilde{G}_r^{\alpha})^{-1}((x^1)_r^{\alpha} - S_r^{\alpha}(T)({x_0}_r^{\alpha}) - \sum_{0<t_k<T} S_r^{\alpha}(T-t_k){I_k}_l^{\alpha}(x(t_k^-)))].$$

Thus we can introduce $u(s)$ of semilinear system

$$[u(s)]^{\alpha} = [u_l^{\alpha}(s), u_r^{\alpha}(s)]$$

$$= [(\tilde{G}_l^{\alpha})^{-1}((x^1)_l^{\alpha} - S_l^{\alpha}(T)({x_0}_l^{\alpha}) - \int_0^T S_l^{\alpha}(T-s)f_l^{\alpha}(s, x(s))ds$$

$$- \sum_{0<t_k<T} S_l^{\alpha}(T-t_k){I_k}_l^{\alpha}(x(t_k^-))), (\tilde{G}_r^{\alpha})^{-1}((x^1)_r^{\alpha} - S_r^{\alpha}(T)({x_0}_r^{\alpha})$$

$$- \int_0^T S_r^{\alpha}(T-s)f_r^{\alpha}(s, x(s))ds - \sum_{0<t_k<T} S_r^{\alpha}(T-t_k){I_k}_l^{\alpha}(x(t_k^-)))].$$

Then substituting this expression into the equation (5) yields α-level of $x(T)$.

$$
\begin{aligned}
[x(T)]^\alpha = {} & [S_l^\alpha(T)(x_{0l}^\alpha) + \int_0^T S_l^\alpha(T-s) f_l^\alpha(s,x(s))ds \\
& + \sum_{0<t_k<T} S_l^\alpha(T-t_k) I_{kl}^\alpha(x(t_k^-)) + G_l^\alpha(\widetilde{G}_l^\alpha)^{-1}((x^1)_l^\alpha - S_l^\alpha(T)(x_{0l}^\alpha) \\
& \int_0^T S_l^\alpha(T-s) f_l^\alpha(s,x(s))ds - \sum_{0<t_k<T} S_l^\alpha(T-t_k) I_{kl}^\alpha(x(t_k^-)))ds, \\
& S_r^\alpha(T)(x_{0r}^\alpha) + \int_0^T S_r^\alpha(T-s) f_l^\alpha(s,x(s))ds \\
& + \sum_{0<t_k<T} S_r^\alpha(T-t_k) I_{kr}^\alpha(x(t_k^-)) + G_r^\alpha(\widetilde{G}_r^\alpha)^{-1}((x^1)_r^\alpha - S_r^\alpha(T)(x_{0r}^\alpha) \\
& \int_0^T S_r^\alpha(T-s) f_r^\alpha(s,x(s))ds - \sum_{0<t_k<T} S_r^\alpha(T-t_k) I_{kr}^\alpha(x(t_k^-)))ds] \\
= {} & [(x^1)_l^\alpha, (x^1)_r^\alpha] = [x^1]^\alpha.
\end{aligned}
$$

We now set

$$
\begin{aligned}
\Phi x(t) = {} & S(t)x_0 + \int_0^t S(t-s) f(s,x(s))ds + \sum_{0<t_k<t} S(t-t_k) I_k(x(t_k^-)) \\
& + \int_0^t S(t-s)\tilde{G}^{-1}(x^1 - S(t)x_0 - \int_0^t S(t-s) f(s,x(s))ds \\
& - \sum_{0<t_k<t} S(t-t_k) I_k(x(t_k^-)))ds,
\end{aligned}
$$

where the fuzzy mappings $\tilde{G}^{-1}$ satisfied above statements. Notice that $\Phi x(T) = x^1$, which means that the control $u(t)$ steers the equation (5) from the origin to x^1 in time T provided we can obtain a fixed point of nonlinear operator Φ.

Assume that the following hypotheses:

(H5) Linear system of equation (5) ($f \equiv 0$) is nonlocal controllable.

(H6) $\{c(\bar{d}(1+T) + c_1 T(1+cT))\} < 1$.

Theorem 2. *Suppose that* (H1)-(H6) *are satisfied. Then the equation* (5) *is controllable.*

Proof. We can easily check that Φ is continuous function from Ω' to itself. For $x, y \in \Omega'$,

$$
\begin{aligned}
& d_H([\Phi x(t)]^\alpha, [\Phi y(t)]^\alpha) \\
& = d_H([\int_0^t S(t-s) f(s,x(s))ds]^\alpha, [\int_0^t S(t-s) f(s,y(s))ds]^\alpha) \\
& \quad + d_H([\sum_{0<t_k<t} S(t-t_k) I_k(x(t_k^-))]^\alpha, [\sum_{0<t_k<t} S(t-t_k) I_k(y(t_k^-))]^\alpha) \\
& \quad + d_H([\int_0^t S(t-s)\widetilde{G}^{-1}(x^1 - S(T)x_0
\end{aligned}
$$

$$
\begin{aligned}
&-\int_0^T S(T-s)f(s,x(s))ds - \sum_{0<t_k<T} S(T-t_k)I_k(x(t_k^-)))ds]^\alpha, \\
&[\int_0^t S(t-s)\widetilde{G}^{-1}(x^1 - S(T)x_0 \\
&\quad -\int_0^T S(T-s)f(s,y(s))ds - \sum_{0<t_k<T} S(T-t_k)I_k(y(t_k^-)))ds]^\alpha) \\
&\leq c\bar{d}d_H([x(t)]^\alpha,[y(t)]^\alpha) + cc_1\int_0^t d_H([x(s)]^\alpha,[y(s)]^\alpha)ds \\
&+c(cc_1+\bar{d})\int_0^t\int_0^T d_H([x(r)]^\alpha,[y(r)]^\alpha)drds.
\end{aligned}
$$

Therefore,

$$
\begin{aligned}
d_\infty(\Phi x(t),\Phi y(t)) &= \sup_{\alpha\in(0,1]} d_H([\Phi x(t)]^\alpha,[\Phi y(t)]^\alpha) \\
&\leq c\bar{d}d_\infty(x(t),y(t)) + cc_1\int_0^t d_\infty(x(s),y(s))ds \\
&\quad +c(cc_1+\bar{d})\int_0^t\int_0^T d_\infty(x(r),y(r))drds
\end{aligned}
$$

Hence

$$
\begin{aligned}
H_1(\Phi x,\Phi y) &= \sup_{t\in[0,T]} d_\infty(\Phi x(t),\Phi y(t)) \\
&\leq C\{c(\bar{d}(1+T)+c_1T(1+cT))\}H_1(x,y).
\end{aligned}
$$

By hypotheses (H6), Φ is a contraction mapping. By the Banach fixed point theorem, (5) has a unique fixed point $x\in\Omega$. □

4 Example

Consider the semilinear one dimensional heat equation on a connected domain (0,1) for a material with memory, boundary condition $x(t,0)=x(t,1)=0$ and with initial condition $x(0,z)=x_0(z)$, where $x_0(z)\in E_N$. Let $x(t,z)$ be the internal energy and $f(t,x(t,z))=\tilde{2}tx(t,z)^2$ be the external heat. $\Delta x(t_k,z)=x(t_k^+,z)-x(t_k^-,z)$ is impulsive effect at $t=t_k(k=1,2,\cdots,m)$.

Let $A=\tilde{2}\frac{\partial^2}{\partial z^2}$, $\Delta x(t_k,z)=\Delta x(t_k)$, $x(t_k^+,z)-x(t_k^-,z)=I_k(x(t_k))$ and $G(t-s)=e^{-(t-s)}$, then the balance equation becomes

$$
\frac{dx(t)}{dt} = \tilde{2}[x(t) - \int_0^t e^{-(t-s)}x(s)ds] + \tilde{2}tx(t)^2 + u(t), t\in J, t\neq t_k, \tag{7}
$$

$$
x(0)=x_0\in E_N, \tag{8}
$$

$$
\Delta x(t_k) = I_k(x(t_k)), \qquad k=1,2,\cdots,m. \tag{9}
$$

The α-level set of fuzzy number $\tilde{2}$ is $[2]^\alpha = [\alpha+1, 3-\alpha]$ for all $\alpha \in [0,1]$. Then α-level sets of $f(t,x(t))$ is $[f(t,x(t))]^\alpha = t[(\alpha+1)(x_l^\alpha(t))^2, (3-\alpha)(x_r^\alpha(t))^2]$. Further, we have

$$\begin{aligned}
&d_H([f(t,x(t))]^\alpha, [f(t,y(t))]^\alpha)\\
&= d_H(t[(\alpha+1)(x_l^\alpha(t))^2, (3-\alpha)(x_r^\alpha(t))^2], t[(\alpha+1)(y_l^\alpha(t))^2, (3-\alpha)(y_r^\alpha(t))^2])\\
&= t\max\{(\alpha+1)|(x_l^\alpha(t))^2 - (y_l^\alpha(t))^2|, (3-\alpha)|(x_r^\alpha(t))^2 - (y_r^\alpha(t))^2|\}\\
&\leq 3T|x_r^\alpha(t) + y_r^\alpha(t)| \max\{|x_l^\alpha(t) - y_l^\alpha(t)|, |x_r^\alpha(t) - y_r^\alpha(t)|\}\\
&= c_1 d_H([x(t)]^\alpha, [y(t)]^\alpha),
\end{aligned}$$

where c_1 satisfies the inequality in hypothesis (H4). Then all the conditions stated in Theorem 1 are satisfied, so the problem (7)-(9) has a unique fuzzy solution.

Let initial value x_0 is $\tilde{0}$. Target set is $x^1 = \tilde{2}$. The α-level set of fuzzy number $\tilde{0}$ is $[\tilde{0}] = [\alpha-1, 1-\alpha], \alpha \in (0,1]$. We introduce the α-level set of $u(s)$ of equation (7)-(9).

$$\begin{aligned}
[u(s)]^\alpha &= [u_l^\alpha(s), u_r^\alpha(s)]\\
&= \Big[\tilde{G}_l^{-1}\big((\alpha+1) - S_l^\alpha(T)(\alpha-1) - \int_0^T S_l^\alpha(T-s)s(\alpha+1)(x_l^\alpha(s))^2 ds\\
&\qquad - \sum_{0<t_k<T} S_l^\alpha(T-t_k) I_{kl}^{\alpha}(x(t_k^-))\big),\\
&\quad \tilde{G}_r^{-1}\big((3-\alpha) - S_r^\alpha(T)(1-\alpha) - \int_0^T S_r^\alpha(T-s)s(3-\alpha)(x_r^\alpha(s))^2 ds\\
&\qquad - \sum_{0<t_k<T} S_r^\alpha(T-t_k) I_{kr}^{\alpha}(x(t_k^-))\big)\Big].
\end{aligned}$$

Then substituting this expression into the integral system with respect to (7)-(9) yields α-level set of $x(T)$.

$$\begin{aligned}
[x(T)]^\alpha &= \Big[S_l^\alpha(T)(\alpha-1) + \int_0^T S_l^\alpha(T-s)s(\alpha+1)(x_l^\alpha(s))^2 ds\\
&+ \sum_{0<t_k<T} S_l^\alpha(T-t_k) I_{kl}^{\alpha}(x(t_k^-))\\
&+ \int_0^T S_l^\alpha(T-s)(\tilde{G}_l^\alpha)^{-1}\Big((\alpha+1) - S_l^\alpha(T)(\alpha-1)\\
&- \sum_{0<t_k<T} S_l^\alpha(T-t_k) I_{kl}^{\alpha}(x(t_k^-)) - \int_0^T S_l^\alpha(T-s)s(\alpha+1)(x_l^\alpha(s))^2 ds\Big) ds,\\
&S_r^\alpha(T)((1-\alpha) + \int_0^T S_r^\alpha(T-s)s(3-\alpha)(x_r^\alpha(s))^2 ds\\
&+ \sum_{0<t_k<T} S_r^\alpha(T-t_k) I_{kr}^{\alpha}(x(t_k^-))
\end{aligned}$$

$$+\int_0^T S_r^\alpha(T-s)(\tilde{G}_l^\alpha)^{-1}\Big((3-\alpha) - S_r^\alpha(T)(1-\alpha)$$
$$-\sum_{0<t_k<T} S_r l^\alpha(T-t_k)I_{k_r}^\alpha(x(t_k^-)) - \int_0^T S_r^\alpha(T-s)S(3-\alpha)(x_r^\alpha(s))^2 ds\Big)ds\Big]$$
$$= [(\alpha+1),(3-\alpha)] = [\tilde{2}]^\alpha.$$

Then all the conditions stated in Theorem 2 are satisfied, so the system (7)-(9) is controllable on $[0,T]$.

References

1. Balasubramaniam P, Muralisankar S (2004) Existence and uniqueness of fuzzy solution for semilinear fuzzy integrodifferential equations with nonlocal conditions. Comput Math Appl 47: 1115–1122.
2. Diamand P, Kloeden PE (1994) Metric space of Fuzzy sets. World Scientific, Singapore New Jersey London HongKong.
3. Kaleva O (1987) Fuzzy differential equations. Fuzzy set and Systems 24: 301–317.
4. Kwun YC, Park DG (1989) Optimal control problem for fuzzy differential equations. Proceedings of the Korea-Vietnam Joint Seminar, 103–114.
5. Mizmoto M, Tanaka K (1979) Some properties of fuzzy numbers. Advances in Fuzzy Sets Theory and applications. North-Holland Publishing Company, 153–164.
6. Park JH, Park JS, Kwun YC (2006) Controllability for the semilinear fuzzy integrodifferential equations with nonlocal conditions. Lecture Notes in Artificial Intelligence 4223: 221–230.
7. Seikkala S (1987) On the fuzzy initial value problem. Fuzzy Sets and Systems 24: 319–330.

Remodeling for Fuzzy PID Controller Based on Neural Networks

Taifu Li[1], Yingying Su[2], and Bingxiang Zhong[1]

[1] Dept. of Electronic Information Engineering, Chongqing University of Science and Technology, Chongqing, 400050, China
litaifu@tom.com

[2] Dept. of Electronic Information and Automation, Chongqing Institute of Technology, Chongqing, 400050, China
yy_su2000@yahoo.com.cn

Abstract. Aimed at implementation puzzles of fuzzy PID controller because of computational complexity, in this paper, in order to reduce computational complexity and realize real-time control, authors utilized universal approximation of the NN to reconstruct an equivalent NN model to accurately approximate a known fuzzy PID controller. Consequently, authors simulated to control the same process model by the fuzzy PID controller and the remodeling NN with different reference inputs, respectively. Results show that control qualities from two different controllers were extremely similar. Therefore, the fuzzy PID controller can be replaced by a remodeling NN in purpose of reducing the computational complexity, dimensional disaster and improving the real-time performance.

Keywords: fuzzy PID; neural network; function approximation; remodeling; MATLAB.

1 Introduction

Thus far, PID controller that is feedback control method is the most widely applied, it is estimated that PID controller and its transmutations are still applied in more than 90 percents control circuits. There are many advantages in traditional PID control design, such as simpleness and low cost, so technology of PID pattern control is still in the highest flight in industrial process control, specially in chemistry engineering and metallurgy process control. It is no doubt that PID control is the most effective way in solving simple linear systems, but the control qualities are unsatisfactory when it comes to complex process, such as time delay, obvious oscillatory characteristics (with small damping complex poles), time-variant, nonlinear and MIMO systems. Therefore many researches are presented to improve it, such as self-regulating and adaptive PID controllers, and unconventional fuzzy PID controllers. Studies show that these controllers can't only solve problems of simple linear systems, but also effectively handle lots of controlled process which are complex, nonlinear, high at order, time-delay and so on[1]. Actually, fuzzy PID controller is adjustable gain or self-regulating

B.-Y. Cao (Ed.): Fuzzy Information and Engineering (ICFIE), ASC 40, pp. 714–725, 2007.
springerlink.com

PID. Three gain coefficients in fuzzy PID controller are all two-dimension fuzzy systems with respect to error e and change in error ec . It is difficult to directly implement with the algorithm because of its large computations in DSP, VHDL language, and MCU, etc, and it is also very poor in real-time performance by industrial computer[2]. So, now its better solution is to transform three fuzzy systems K_p, K_i, K_d into look up table in advance, but it is still hard to implement because of large computation for three fuzzy rule bases. Therefore,in order to utilize the advantages of fuzzy PID controller and effectively overcome its disadvantages, an equivalent NN model is trained to approach a fuzzy PID controller through lots of input and output data pairs from fuzzy PID controller. Authors tried to apply this method to reduce the computation complexity in fuzzy PID controller, and improve applications of fuzzy PID with hardware implementation in DSP, VHDL language, etc.

2 Fuzzy PID Controller

In the process of industrial production, the characteristic parameters or structures of many controlled plants would be varying resulted from changing load and disturbances. Nowadays, PID algorithm is still widely applied in industrial production and its parameter tuning is mainly implemented in manual. Combining the classical PID control with the advanced expert system, if all these tuning methods are described with fuzzy rules, PID optimal parameters could be automatically regulated with help of fuzzy inference engine. A fuzzy PID controller

takes error e and change in error ec as inputs, and it could meet requirements of PID parameters self-regulatings for different e and ec, whose structure is shown in Figure 1[3]. The PID controller is given by

$$u(k) = K_p e(k) + K_i \sum e(k) + K_d ec(k) \tag{1}$$

In a fuzzy PID, K_p,K_i and K_d are all the fuzzy systems with respect to error e and change in error ec. The fuzzy sets and domains of error e and change in error ec are defined in Figure 2, Figure 3[4]:
Where, K_p, K_i and K_d are all consisted of initial regulating parameters and modified parts which is fuzzy system with respect to error e and change in error ec.

$$\begin{aligned} K_p &= K_{p0} + \Delta K_p \\ &= K_{p0} + f_p(e, ec) \end{aligned} \tag{2}$$

$$\begin{aligned} K_i &= K_{i0} + \Delta K_i \\ &= K_{i0} + f_i(e, ec) \end{aligned} \tag{3}$$

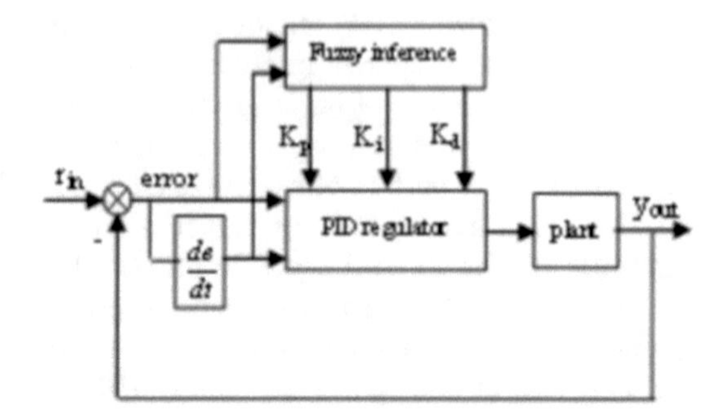

Fig. 1. Structure of fuzzy PID controller

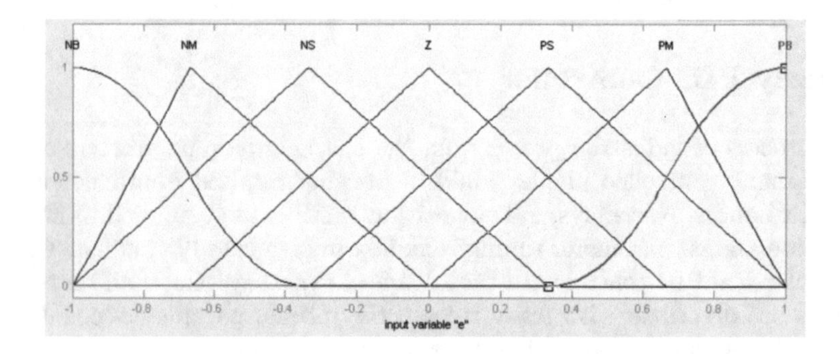

Fig. 2. Membership function of error$e$

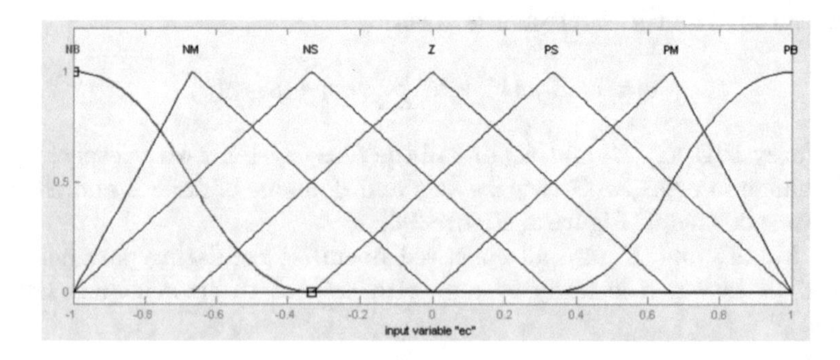

Fig. 3. Membership function of change in error ec

$$\begin{aligned} K_d &= K_{d0} + \Delta K_d \\ &= K_{d0} + f_d(e, ec) \end{aligned} \tag{4}$$

In which, domains and fuzzy sets of ΔK_p, ΔK_i and ΔK_d are shown in Figure 4, Figure 5 and Figure 6.

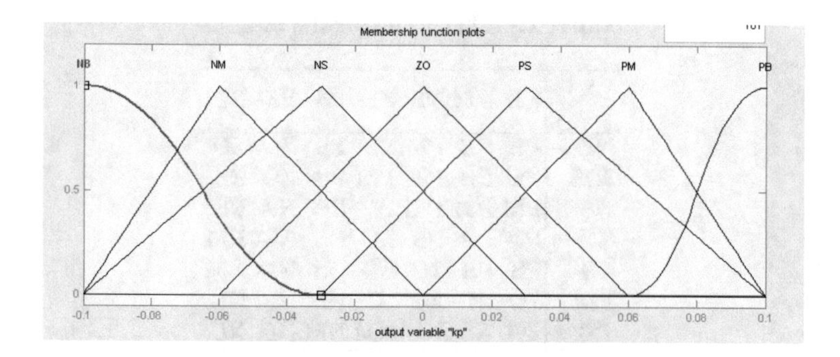

Fig. 4. Membership function of ΔK_p

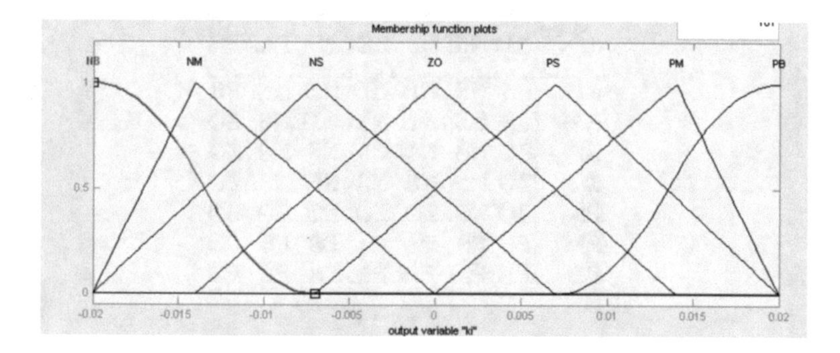

Fig. 5. Membership function of ΔK_i

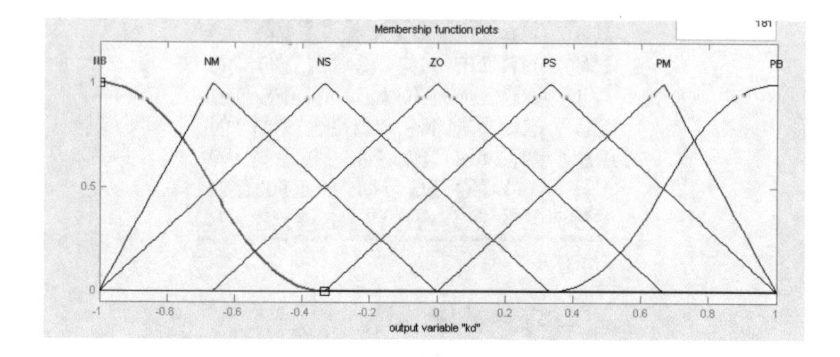

Fig. 6. Membership function of ΔK_d

The fuzzy rules of ΔK_p, ΔK_i and ΔK_d are shown in Table 1, Table 2 and Table 3:

Three fuzzy systems of ΔK_p, ΔK_i and ΔK_d are all chosen as singleton fuzzification, center average defuzzification, Mamdani product inference engine, correspondingly mathematic expressions are shown in equation (5), (6) and (7).

Table 1. Table 1 Fuzzy rules of ΔK_p

e \ ce	NB	NM	NS	ZE	PS	PM	PB
NB	PB	PB	PM	PM	PS	ZO	ZO
NM	PB	PB	PM	PM	PS	ZO	NS
NS	PM	PM	PM	PS	ZO	NS	NS
ZO	PM	PS	PS	ZO	NS	NM	NM
PS	PS	PS	ZO	NS	NS	NM	NM
PM	ZO	ZO	NS	NM	NM	NM	NB
PB	ZO	NS	NS	NM	NM	NB	NB

Table 2. Table 2 Fuzzy rules of ΔK_i

e \ ce	NB	NM	NS	ZE	PS	PM	PB
NB	PS	NS	NB	NB	NB	NM	PS
NM	PS	NS	NB	NM	NM	NS	ZO
NS	ZO	NS	NM	NM	NS	NS	ZO
ZO	ZO	NS	NS	NS	NS	NS	ZO
PS	ZO	ZO	ZO	ZO	ZO	ZO	ZO
PM	PB	PS	PS	PS	PS	PS	PB
PB	PB	PM	PM	PM	PS	PS	PB

Table 3. Table 3 Fuzzy rules of ΔK_d

e \ ce	NB	NM	NS	ZE	PS	PM	PB
NB	NB	NB	NM	NM	NS	ZO	ZO
NM	NB	NB	NM	NS	NS	ZO	ZO
NS	NB	NM	NS	NS	ZO	PS	PS
ZO	NM	NM	NS	ZO	PS	PM	PM
PS	NM	NS	ZO	PS	PS	PM	PB
PM	ZO	ZO	PS	PS	PM	PB	PB
PB	ZO	ZO	PS	PM	PM	PB	PB

$$\Delta K_p = f_p(e, ec) = \frac{\sum\limits_{l=1}^{49} \Delta \bar{K}_p^l \cdot \mu_{E^l}(e) \cdot \mu_{EC^l}(ec)}{\sum\limits_{l=1}^{49} \mu_{E^l}(e) \cdot \mu_{EC^l}(ec)} \tag{5}$$

$$\Delta K_i = f_i(e, ec) = \frac{\sum\limits_{l=1}^{49} \Delta \bar{K}_i^l \cdot \mu_{E^l}(e) \cdot \mu_{EC^l}(ec)}{\sum\limits_{l=1}^{49} \mu_{E^l}(e) \cdot \mu_{EC^l}(ec)} \tag{6}$$

$$\Delta K_d = f_d(e, ec) = \frac{\sum_{l=1}^{49} \Delta \bar{K}_d^l \cdot \mu_{E^l}(e) \cdot \mu_{EC^l}(ec)}{\sum_{l=1}^{49} \mu_{E^l}(e) \cdot \mu_{EC^l}(ec)} \tag{7}$$

3 Remodeling for an Equivalent NN of Fuzzy PID Controller

In order to model an equivalent NN of fuzzy PID controller, it is necessary to train NN with lots of data pairs, which are from the input/output of fuzzy PID controller. Usually, control output of general PID controller or fuzzy PID controller is all function with respect to error e, change in error ec and error integration $\sum e$. Here, the domain of $\sum e$ is chosen as $\{-100, -90, -80, \ldots\ldots, 80, 90, 100\}$ and domains of e, ec are both chosen as $\{-1, -0.9, -0.8, \ldots\ldots, 0.80, 0.9, 1\}$. In this way, every input has 21 data, considering all kinds of combination of input data, so there are 21× 21× 21=9261 different input data groups for the equivalent NN, correspondingly it may get 9261 output values. Meanwhile, to input the 9261 groups of data also may get 9261 output control values.

As following, we have to choose a suitable NN structure, in which input layer and output layer can be easily determined as three inputs and single output according to the mathematic expressions of fuzzy PID. Otherwise, according to Kolmogorov theorem, a BP NN with one hide-layer could approach every nonlinear continuous $R^n \to R^m$ mapping[5]. Thus, we chose a multi-layer feedforward NN with one hide-layer. How to select the node number of hide-layer is a very complex problem which can't be easily determined by a mathematic expression. It directly depends on the requirement of problems being solved and the number of input and output nodes. If the number of nodes is so many, it may be impossible to successfully train a NN, or can't identify samples which exceed to former input and output data pairs, and it is bad in fault tolerance. On the contrary, when the number of nodes is less, the learning time would be long and the error may not be controlled optimally. Therefore, authors chose a proper number of nodes as 22 on hide-layer with preferable training effects. So the structure of equivalent NN with 22 nodes in hide layer is determined, shown in Figure 7.

The transformation function of output layer neuron was "purelin" linear function. The hide layer transformation functions were chosen as "logsig", whose mathematic expression is given by equation (8):

$$f(x) = \frac{1}{1 + e^{-x}} \tag{8}$$

Training function is "trainlm", which applies negative gradient backpropagation algorithm of Levenberg-Marquard method[6]. All input data were normalized. The training process of the equivalent NN is shown in Figure 8.

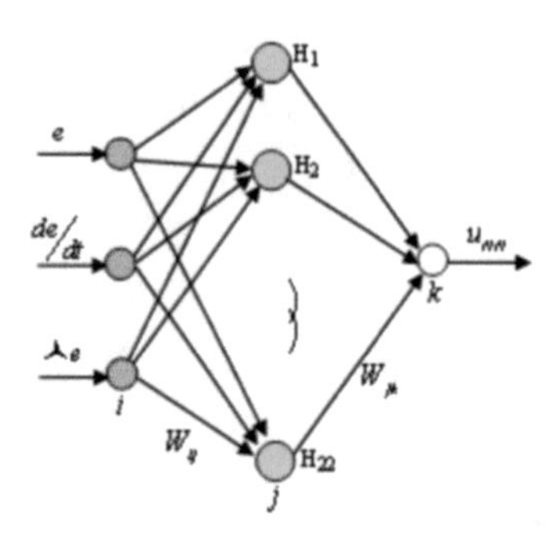

Fig. 7. The structure of equivalent NN model

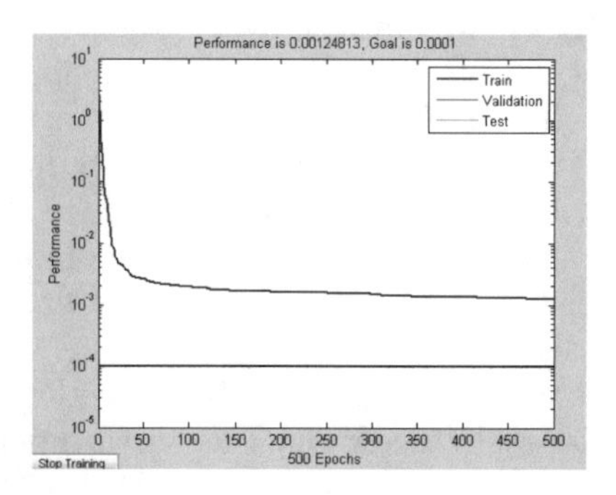

Fig. 8. The record of training process

Weight and threshold values of the equivalent NN were gained as following: w_{ij} (j = 1,2,3,.....21,22;i = 1)= {-1.1937, 14.8186, 0.8418, -4.8473, -0.7579, -3.9894, -6.4649, -2.6197, -2.8097, 2.7517, 6.4802, 3.8316, 1.9940, 0.1956, -0.8306, -0.1432, -1.7270, 3.3091, 3.9216, 2.3471, -1.8042, -5.0603};
w_{ij} (j= 1,2,3,.....21,22;i = 2) = {-1.6925, -5.5399, 0.7596, 1.0879, -0.7864, -15.2952, 10.4906, -2.1654, 1.9848, 5.9378, -10.6258, -1.5238, 3.3365, -3.4364, 3.6360, -0.0882, 3.6510, -4.0442, -5.1119, 4.8039, -25.0533, -19.7355};

w_{ij} (j= 1,2,3,.....21,22;i = 3) ={-0.0025, -0.0079, -0.0107, 0.0096, 0.0042, 0.0006, 0.0190, 0.0040, 0.0155, 0.0034, -0.0192, -0.0091, 0.0019, -0.0030, -0.0034, 0.0031, -0.0027, 0.0044, -0.7550, 0.0025, 0.0206, 0.0041};
w_{jk} (j= 1,2,3,.....21,22;k = 1) ={-13.871, -0.9063, 4.2437, 1.4971, 38.8504, -18.3030, -4.5422, 3.0385, 0.4725, 5.7586, -4.4200, 2.5598, 13.5060, -1.53453, 6.0997, -28.3032, -7.8694, -2.2617, 0.0023, -14.4635, -0.7464, 7.1322};
b_j (i= 1,2,3,.....21,22;) = {2.2530, -20.6660, -2.5367, 1.73623, 2.0121, 20.4910, -2.4047, -1.0459, 0.9300, 2.04569, 2.4201, -1.4426, 1.26189, -2.3060, -3.6105, 0.5319, -3.7039, 4.7482, 5.4961, 1.7211, -27.9237, 24.7829};

When it came to the end of training the equivalent NN, the agreement extent and precision were verified between fuzzy PID controller and the equivalent NN with same inputs, shown as u_{F-PID} and u_{NN} with different colors in Figure 9.

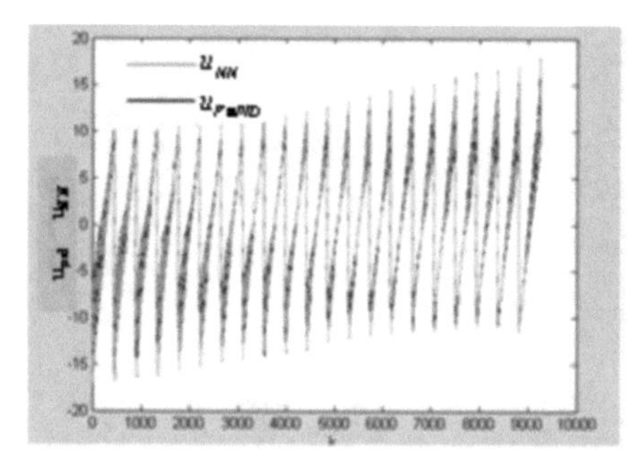

Fig. 9. Verification of equivalent NN

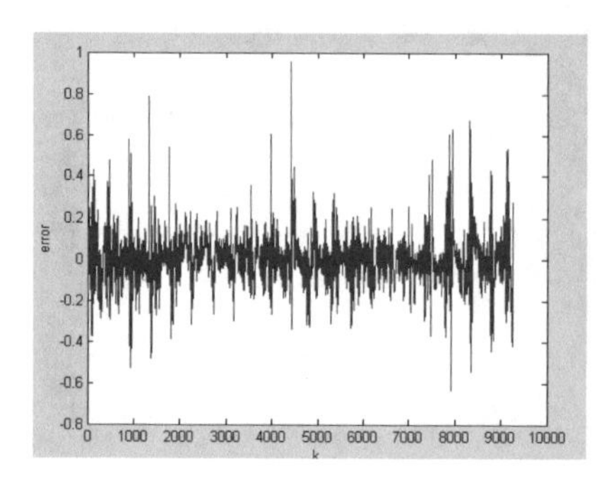

Fig. 10. Error curves of between u_{F-PID} and u_{NN}

As shown in Figure 9, the output between fuzzy PID controller and the equivalent NN model are mainly similar, which indicates that the equivalent NN can satisfy the precision requirement. But because of so many dense data in the figure, it was not easily to observe directly, so we plotted an approximation error curve of the equivalent NN, shown in Figure 10.

As shown in Figure 10, the error between fuzzy PID and equivalent NN was mastered within 0.2, which was less in a controller with output range of -20 to 20, which $(0.2/40) \times 100\% = 0.5\%$ error, so the fuzzy PID controller and the equivalent NN controller are almost same, and its precision can satisfy normal controllers.

4 Control Simulation

In order to farther verify the purpose of modeling an equivalent NN, we chose one controlled plant model, and controlled it with fuzzy PID and equivalent NNrespectively. The controlled plant is given by

$$G(s) = \frac{523500}{s^3 + 87.35s^2 + 10470s} \tag{9}$$

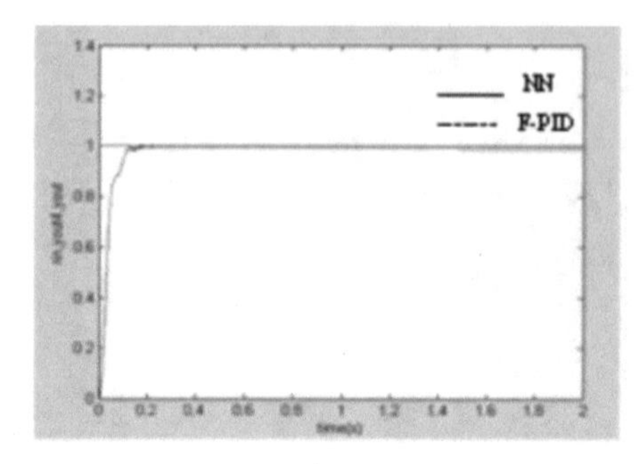

Fig. 11. The response curves from fuzzy PID and the equivalent NN in step input

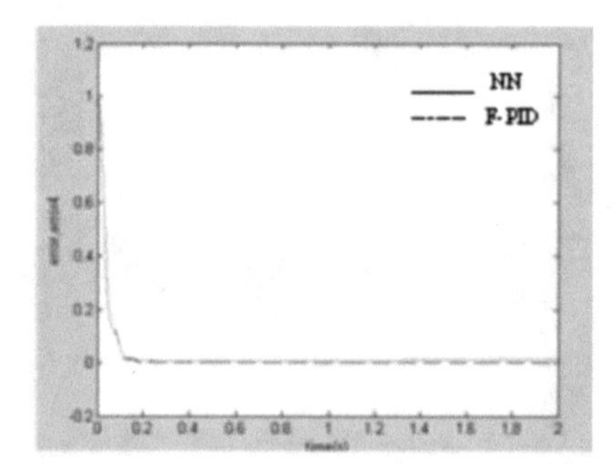

Fig. 12. The error curves from fuzzy PID and equivalent NN in step input

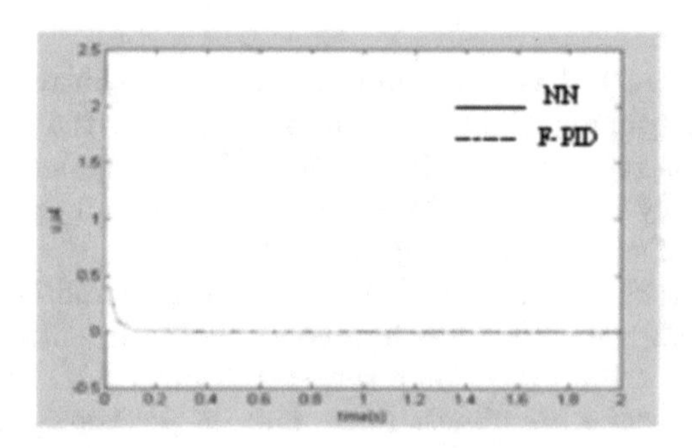

Fig. 13. The control output from fuzzy PID and equivalent NN in step input

The fuzzy PID controller and the equivalent NN model were given input signals as phase step, square wave and sinus with sample time 1ms, respectively. The simulation figure were shown in Figure 11 to Figure 19.

As shown in Figure 11, imaginal line and actual line respectively indicated the phase step control response curve of fuzzy PID and the equivalent NN. As shown in this figure, the two lines were almost coincident and difficult to

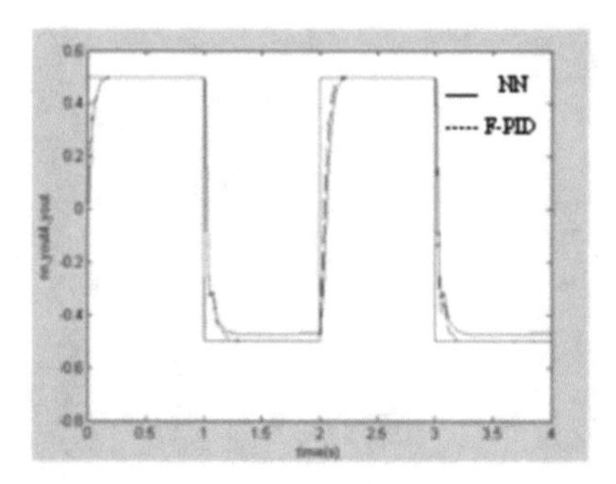

Fig. 14. The response curves from fuzzy PID and equivalent NN in square input

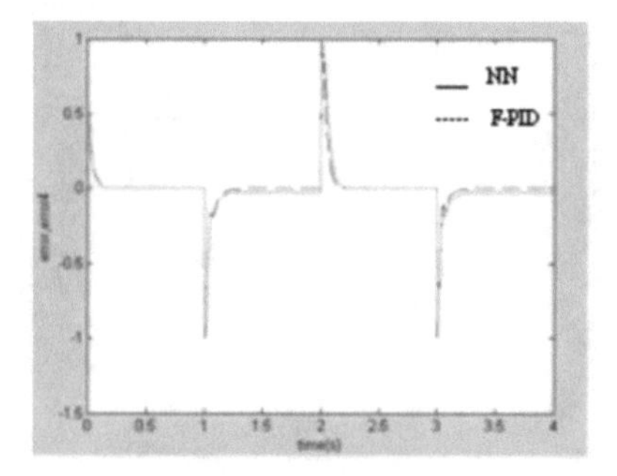

Fig. 15. The error curves from fuzzy PID and equivalent NN in square input

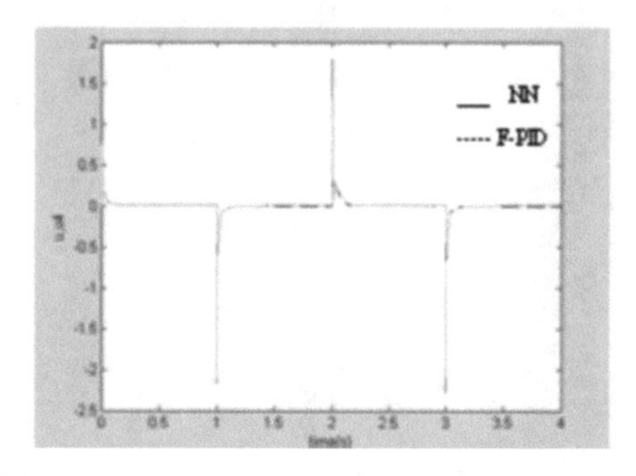

Fig. 16. The control output from fuzzy PID and equivalent NN in square input

distinguish, which indicated that the control effects between two type controllers were extremely equivalent.

As shown in Figure 12, imaginal line and actual line indicated the phase step response error curve of fuzzy PID and the equivalent NN, respectively. In the figure,the error curve between two different controllers were also extremely similar, and this farther verified the equivalent control effects between them.

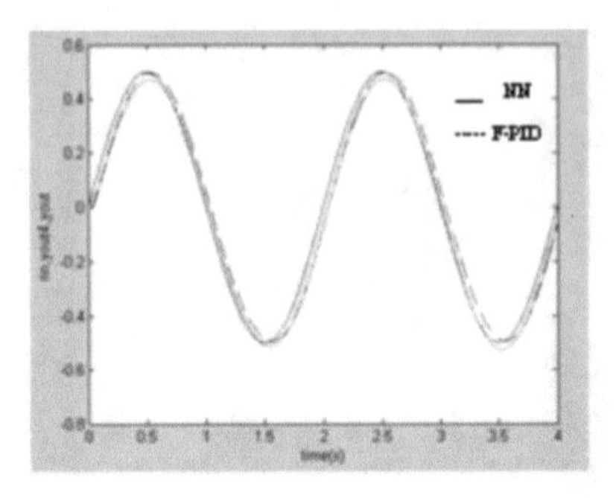

Fig. 17. The response curves from fuzzy PID and equivalent NN in sine input

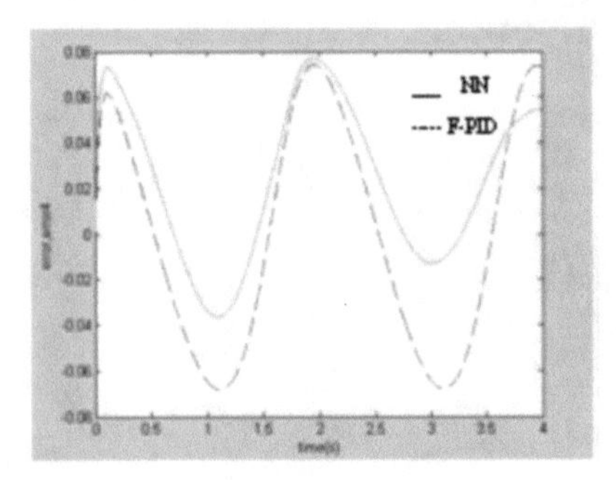

Fig. 18. The error curves from fuzzy PID and equivalent NN in sine input

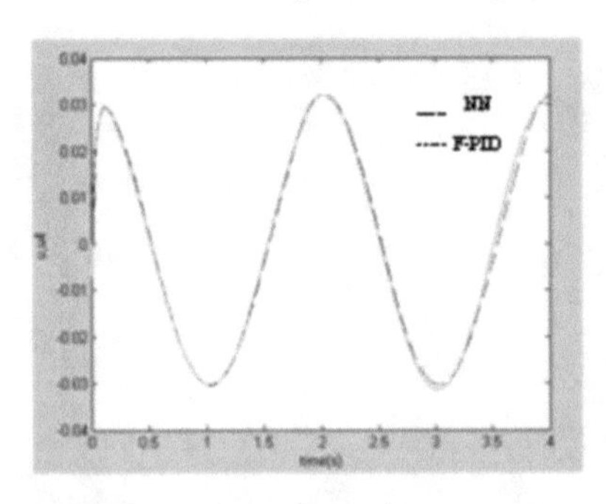

Fig. 19. The control output from fuzzy PID and equivalent NN in sine input

As shown in Figure 13, in like manner imaginal line and actual line indicated the phase step response output curve of fuzzy PID and the equivalent NN, respectively. In it, the imaginal line and actual line were nearly imminent, farther verified the equivalent control effects between them.

Following, the response curve of square wave and sinus were shown in Figure 14, Figure 15, Figure 16, Figure 17, Figure 18 and Figure 19. The actual line represented equivalent NN, and the imaginal line represented fuzzy PID controller, shown in these figures that the curve of response, error and control output were all extremely similar. These all showed that control effects between fuzzy PID controller and the equivalent NN were equivalent.

5 Conclusion

Throughout utilizing universal approximation of BP NN, an equivalent NN of fuzzy PID controller is modeled. The precision and simulation effects between these two controllers are verified similar. It is said that fuzzy PID controller can be equivalently replaced by a BP NN controller. In this way, an equivalent NN controller can solve hundreds of actual control problems instead of fuzzy PID controller. Thus, this study shows that a fuzzy PID controller can be equivalently modeled with universal approximation of NN. This method is used to take advantages of fuzzy PID control and overcome its disadvantages, reduce computation complexity of fuzzy PID mathematic model, and even improve the application on hardware with DSP, VHDL language, etc.

References

1. Zhao, Z.Y., M.Tomizuta, and S.Isaka(1993) Fuzzy Gain Scheduling of PID Controllers. IEEE Trans. on Systems, Man, and Cybernetics, 23(5):1392-1398.
2. Raju, G.V.S., and J. Zhou(1993) Adaptive Hierarchical Fuzzy Controller. IEEE Trans. on Systems, Man, and Cybernetics, 23(4):973-980.
3. Jinkun Liu(2004) Advanced PID Control with Matlab. Beijing: Publishing House of Electronics Industry.
4. Chao C T, Chen Y J, Teng C C.(1996) Simplification of fuzzy-neural systems using similarity analysis. IEEE Trans. On Systems, Man, and Cybernetics,26: 344-354
5. Wang L X.(1999) Analysis and design of hierarchical fuzzy systems. IEEE Trans. On Fuzzy Systems, 7: 617-624
6. Martin T. Hagan, Howard B.Demuth, Mark Beale(2002) Neural Network Design. Beijing: China Machine Press

Monitoring for Healthy Sleep Based on Computational Intelligence Information Fusion

Taifu Li[1], Dong Xie[1], and Renming Deng[2]

[1] College of Electronic Information Eng., Chongqing University of Science & Technology, Chongqing 400050, China
litaifu@tom.com

[2] College of Automation, Chongqing University, Chongqing 400044, China
dengrenming@sina.com

Abstract. Aimed at the problem of cold resulted from sleeping fidget about infants, this paper explored the mechanism of the sleeping fidget, and presented the relational model between the sleeping fidget and temperature, humidity, and their changes. Authors developed the data acquisition system of the temperature and humidity based on the digital sensor, acquired lots of data about the temperature/humidity and the corresponding fidget degree, designed the neural networks model construct through analyzing the relation among three kinds of data. On this basis, the NN model was trained with the experimental data based on MATLAB NN Toolbox. The model verification shows that the sleeping fidget model can effectively implement sleeping fidget monitoring.

Keywords: Multi-source Information Fusion, Fidget Model, Healthy Sleep, Neural Network.

1 Introduction

Healthy sleep is very important for an infant to ensure his good growth, in fact, an infant or baby is poor in self-nurse ability and necessary health knowledge, so his sleep needs nursing. In evening, while parents urge the baby to go to sleep as soon as possible, actually the baby wish to continue their amusements, he also has to unwillingly go to bed by force of command of his parents. At this time, the baby is still excitement, blood circulation is very acuteness, quantity of heat from body is also more, and the temperature at surroundings is higher than that of wee hours 4 to 5. However, parents worry about that the temperature of wee hours 4 to 5 is lower, the baby easily catch a cold, so the placement of the quilt is thicker relatively. At these conditions, the temperature balance in the cave of quilts is broken because producing quantity of heat is more than that of emitting, it will result that the temperature in the cave of quilts is rising. The biological control system in body will sweat to decrease the temperature of body in order to keep the temperature balance about $37^{O}C$. Furthermore, it will lead that the temperature and humidity in the cave of quilts increase together, so the baby is very fidget, then kicks off the quilts from the body in

B.-Y. Cao (Ed.): Fuzzy Information and Engineering (ICFIE), ASC 40, pp. 726–738, 2007.
springerlink.com

mistily, in this conditions, finally it is certain to catch a cold. Therefore, in order to avoid above-mentioned phenomena, it is very significant to monitoring the healthy sleep.

Sensor technology is rapid developing by fusing intelligent technology, nowadays, sensors trend multi-function and multi-variable measurement from single-function and single-object, unmeasurable fuzzy variable from measurable physical signal, systematization with network from isolated parts of an apparatus[1][2][3]. The intelligent sensor is closely depending on data fusion, so it can't only measure all kinds of physical parameters, but also get some unknown parameters based on known parameters. The development of data fusion also promotes that of the intelligent sensor at a certain extent. Multi-source information fusion with computational intelligence is an important method of data fusion, it utilizes multi-sensor data information to get objective state, is a kind of hierarchical automatic information processing[4], further produces new information which can't be get by any single sensor[5][6].

Therefore, the paper tried to apply multi-sensor information fusion and computational intelligence to research mathematic model of sleep fidget process. Its aim is to provide theoretical foundation to develop corresponding sensor hardware system, further monitor the sleep fidget and effectively prevent to catch a cold resulted from kicking off quilts.

2 Scheme of Monitoring Sleep Fidget

Temperature and humidity influence the degree of sleep fidget heavily, but there also is some individual difference for different human. If considering influence from temperature, humidity and their derivative at different order, the model can adapt individual difference for different one, therefore we may consider information fusion structure of input/output signal to monitor sleep fidget degree, shown in Fig.1.

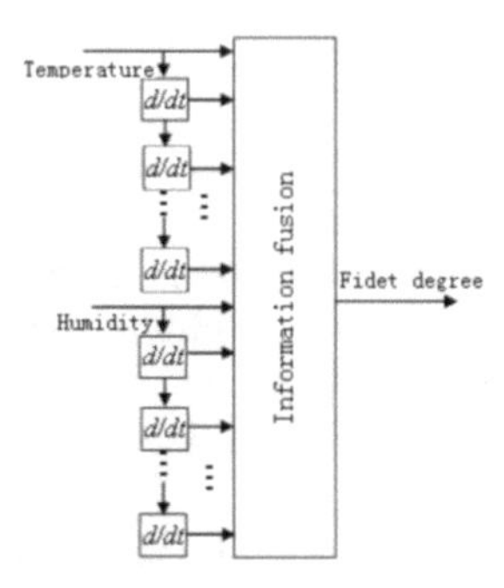

Fig. 1. Information fusion scheme I of fidget degree

In addition, in order to convenient implementation with computer, it also may adopt information fusion scheme, shown in Fig.2. If temperature, humidity represent different space information, then their time sequence signals represent different time information. In Fig.2, multi-sensor and multi-source information fusion may realize time-space information fusion.

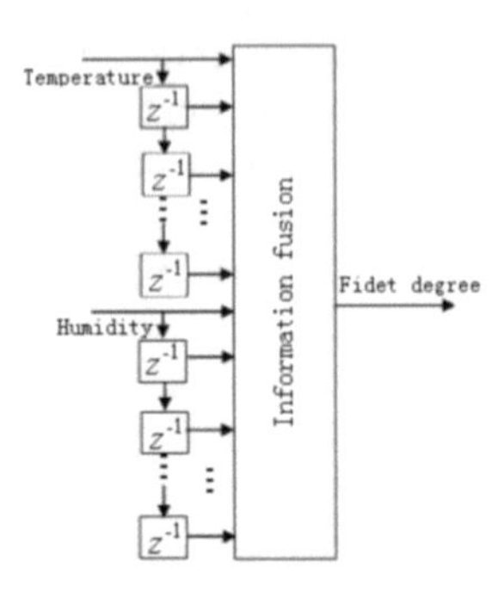

Fig. 2. Information fusion scheme II of fidget degree

Temperature and humidity signals both are measurable physical signals, but fidget degree hasn't accurate physical definition, is a fuzzy variable, and unmeasurable. In fact, the fidget is objective resulted from inadequate temperature and humidity. Here, we use membership function in fuzzy system theory, let the tested human record fidget degree at different time, the fidget degree belongs to interval [0, 1].

3 Information Fusion Strategy Based on NN

Observed the structure of sleep fidget monitoring, the key to implement the function is information fusion model. It is very difficult to get precision mathematic model of sleep fidget degree with respect to temperature, humidity and their time sequence signals. In intelligent sensor technology, the information fusion with computational intelligence is to fully simulate how the brain treat complex problem, and has widely researched by many experts. There are many intelligent methods of information fusion, in which, typical methods have Bayes Inference, Fuzzy System Theories, Expert System, Symbol Inference and Neural Networks.

NN(Neural Networks) had presented based on how to modern neural-biology and cognitive science treat information, and has powerful adaptive learning ability. NN is universal functional approximation which can approximate any nonlinear function in L_2 normalized function through training with lots of input/output data pairs.

In succession, we apply Peaks function in MATLAB inside to verify nonlinear function approximation ability of NN. The surface of the Peaks function is shown in Fig.3, its expression is given by

$$\begin{aligned} z = {} & 3 \cdot (1-x)^2 \cdot e^{-[x^2+(y+1)^2]} \\ & - 10 \cdot (\tfrac{1}{5}x - x^3 - y^5) \cdot e^{-(x^2+y^2)} - \tfrac{1}{3} \cdot e^{-[(x+1)^2+y^2]} \end{aligned} \tag{1}$$

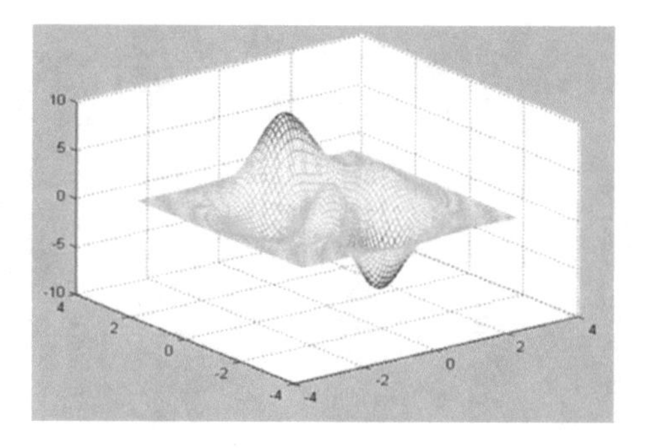

Fig. 3. The surface of peaks function

Selecting data [- 3: 0.1: 3] in X, Y axis to produce $61 \times 61 = 3721$ pairs training data $[x_p,\ y_p,\ z_p]$, $p = 1,\ 2,\ \cdots, 3721$. We selected the 2-5-1 structure of NN, and shown in Fig.4. Neurons in hide layer are non-symmetric Sigmoid function, and neurons in output layer are Purelin linear function. Traindx in MATLAB NN toolbox was selected as training algorithm. Stopping steps was chosen as 10000 by parameter epochs, and stopping condition of training error is given by

$$E = \frac{1}{P} \sum_{i=1}^{P} e_i^2 \leq \varepsilon = 0.001 \tag{2}$$

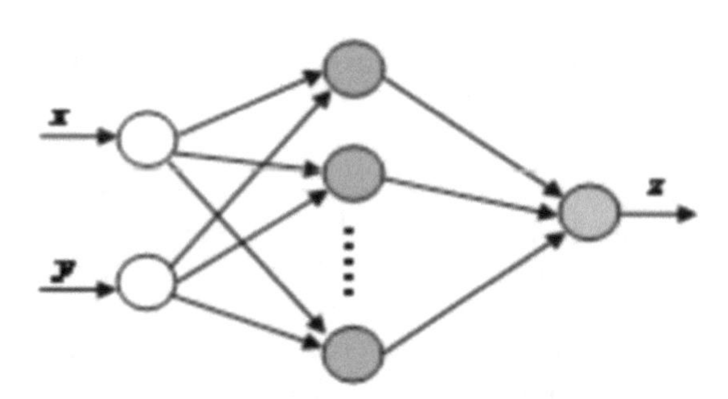

Fig. 4. The structure of NN

Fitted surface of NN for Peaks function is shown in Fig.5, the fitted error surface of NN is shown in Fig.6.

In many applications of multi-sensor information fusion, it is very strict in real-time performance, and sensor information processing system implemented by VHDL language may utilize high-speed and parallel processing of NN, so, NN was chosen as information fusion strategy.

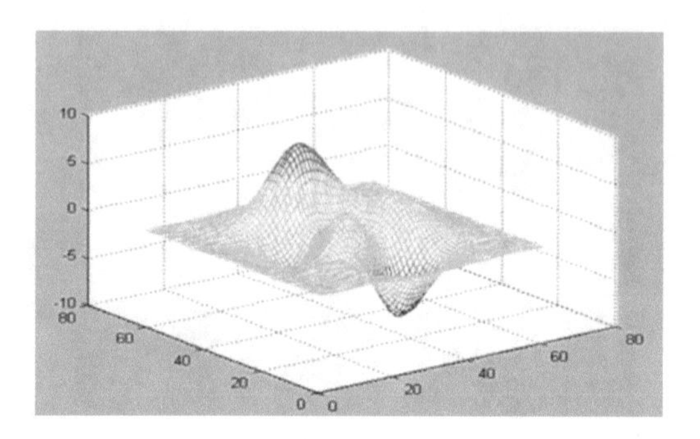

Fig. 5. Fitting Peaks function surface with NN

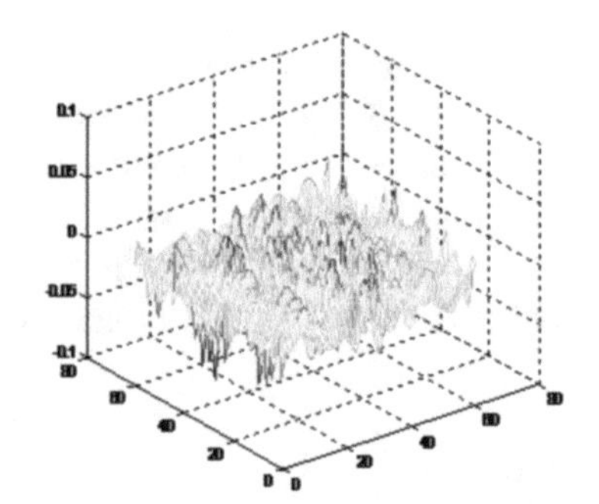

Fig. 6. Fitting error surface with NN

4 Experimental System for Input/Output Data Acquisition

In the experimental system, we chose SHT10 temperature and humidity sensor from Sensirion company. The sensor is integrated into one chip with temperature sensor, humidity sensor, signal amplifier, A/D convertor and I^2C bus together. The volume is very small (7.65mm × 5.08mm × 23.5mm), the I^2C bus is digital signal output interface, the resolution of humidity signal is 14 bit, the resolution of temperature signal is 12 bit. Aimed at characteristics of the temperature/humidity sensor, designed the experimental system shown in Fig.7.

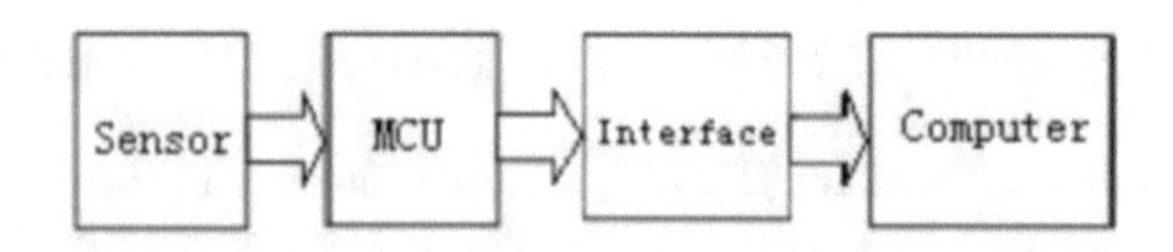

Fig. 7. Diagram of the experimental system

Observed Fig.7, the experimental system mainly consists of temperature/humidity sensor, MCU chip, RS-232 interface and computer. Temperature/humidity sensor contacts the measured object directly, the data of sensors is read by MCU, and transmitted into computer with RS-232 interface. Finally, the data will be read by special software in computer, and display it.

To construct information fusion model of sleep fidget degree with NN, it is necessary to use lots of input/output data pairs to train NN, the data pairs are temperature, humidity and fidget degree at corresponding time, respectively. Therefore, simulating sleep fidget process is very important. While a human is health and his emotion is normal, the fidget normally results from that the clothing is so thick, the temperature/humidity of surroundings is so high, physical exercise is so hard. Limited to experimental resource, considering the experimental feasibility and simulated comparability, authors studied out two kinds of experimental scheme.

(1) Warming feet experiment with hot water

Placing feet into the basin filled with hot water, the surrounding temperature of feet is increasing rapidly. In order to keep normal temperature of the feet about $37^O C$, the body control system tries to quicken blood circulation to take away superabundance quantity of heat with heat exchange, so the temperature of whole body is rising. In addition, quickening blood circulation results from more violent heart activities that will consume more biological energy, all these energy will translate into heat energy. In this way, whole body will feel fever, if the clothing is very thick, temperature will rise ($>37^O C$). So, the control system tries to sweat to decrease temperature, in a minute, the body inside clothing would feel fever and wet, fidget degree will rise rapidly. Therefore, the method can simulate fidget degree varying process and corresponding temperature/humidity. Warming feet experiment with hot water is shown in Fig.8.

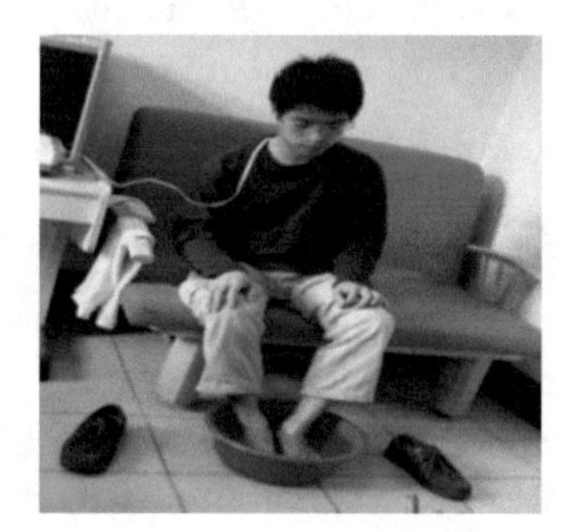

Fig. 8. Warming feet experiment with hot water

(2) Simulating sleep experiment

While a human sleeps on bed and placed with thicker quilts, whose life activity consumes energy, and produces heat energy continuously. Because the quilt and clothing is insulated material, it would result that the temperature inside quilts increases if the quilt is thicker, the emitting quantity of heat is less than that of producing. Consequently, the body control system has to sweat to decrease the body heat in order to ensure the balance between the emitting quantity and the producing quantity of heat. It would further lead that the humidity inside the quilts increases. In this way, the fidget would bring on. Therefore, the method can also simulate fidget degree varying process and corresponding temperature/humidity, simulating sleep experiment is shown in Fig.9.

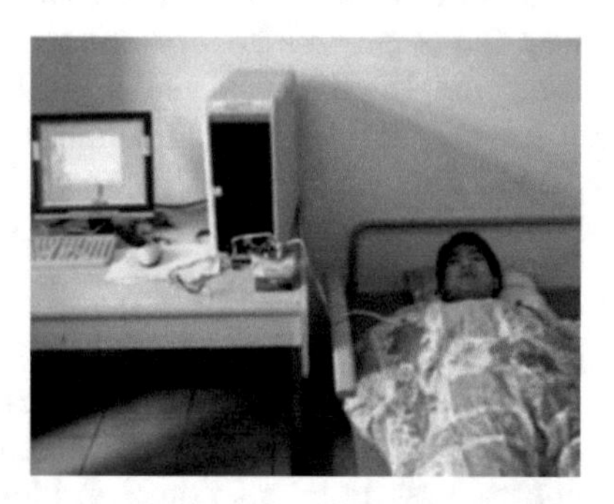

Fig. 9. Simulating sleep experiment

Temperature and humidity signals can get by sensors acquisition, but fidget degree is human's sense that can't get with instrument directly, only can record the sense of the experimenter, and may represent it with fuzzy language. In the process of experiment, referring to membership function and center average defuzzification in fuzzy system theory, the fidget degree is defined in interval[0,1], and described with five fuzzy sets such as comfortable, less comfortable, uncomfortable, more uncomfortable and fidget, respectively. The five sets have different center values which represent different fidget degree, shown in Table 1.

Table 1. Fuzzy measurement of fidget degree

Fidget degree	Comfortable	Less comfortable	uncomfortable	More uncomfortable	fidget
Center value	0.2	0.4	0.6	0.8	1

For every simulating experiment, we may get a batch of experimental data. Certainly, we may display these experimental curves through some software processing, shown in Fig.10.

5 NN Model of Monitoring Sleep Fidget

Fidget degree isn't only influenced with temperature and humidity, but also it is different because of different individuals, in addition, current fidget also is relations to a series of temperature and humidity at former time. Therefore, fidget

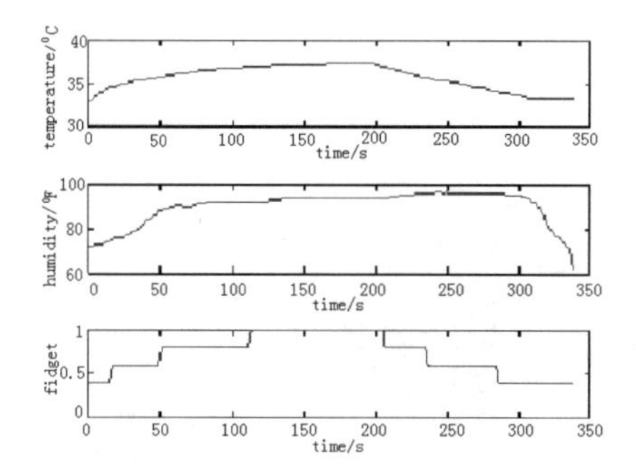

Fig. 10. Experimental curves of temperature, humidity and fidget

degree may be represented by nonlinear equation with respect to temperature and humidity at time sequence. Where, fidget degree is represented by F, temperature by T, humidity by H. So, nonlinear equation of fidget degree may be represented by

$$\begin{aligned} F(k) = f(T(k), T(k-1), \cdots, T(k-n), \\ H(k), H(k-1), \cdots, H(k-m)) \end{aligned} \tag{3}$$

Because Neural Networks is universal function approximation, which is powerful in nonlinear fitting ability, so we selected NN for information fusion strategy. If the NN is trained by lots of input/output experimental data, it could implement fitting task of $f(\cdot)$. To select neural networks structure is relative to actual application objects, but there isn't mature theory and method. Nowadays, multi-layer forward network is widely applied, its most training algorithm is BP learning, whose learning needs tutor signal and also belongs to gradient descent method. Two layers feedforward NN includes hide-layer and output layer, its neuron functions in hide-layer and output layer are normal nonsymmetrical sigmoid function. Studies show that multi-layer feedforward NN may approximate any objective function at any precision if the hide-layer includes enough neurons[7]. Therefore, neurons on hide-layer and output layer adopt nonsymmetrical sigmoid function. The time serial signals of temperature and humidity T(k), T(k+1), T(k+2), T(k+3), H(k), H(k+1), H(k+2) and H(k+3), are taken as inputs of the NN model, in which, k represents currently sampling time. Temperature and humidity is sequential 4 pieces of signals, its time interval is 6 second, designed NN is shown in Fig.11.

Input matrix P and target matrix T is

$$P = \begin{bmatrix} T(k) & T(k+1) & \cdots & T(k+n-3) \\ T(k+1) & T(k+2) & \cdots & T(k+n-2) \\ T(k+2) & T(k+3) & \cdots & T(k+n-1) \\ T(k+3) & T(k+4) & \cdots & T(k+n) \\ H(k) & H(k+1) & \cdots & H(k+n-3) \\ H(k+1) & H(k+2) & \cdots & H(k+n-2) \\ H(k+2) & H(k+3) & \cdots & H(k+n-1) \\ H(k+3) & H(k+4) & \cdots & H(k+n) \end{bmatrix} \tag{4}$$

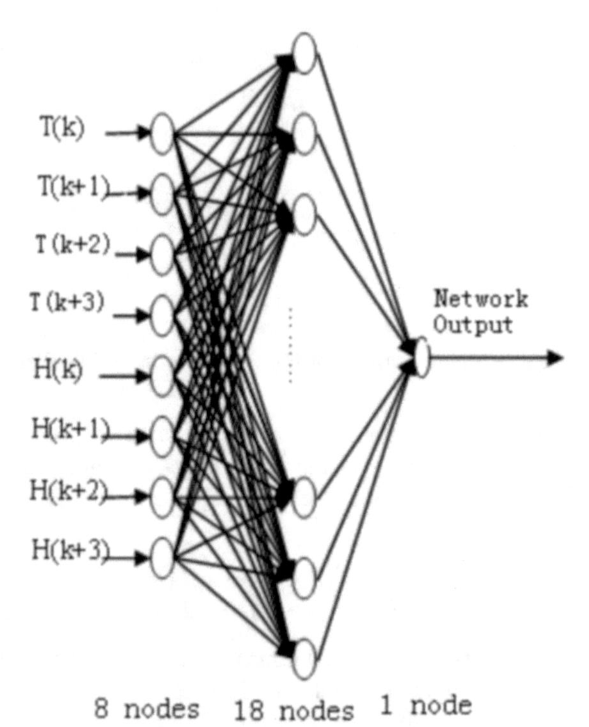

Fig. 11. Structure of NN model

$$T = [F(k+3), \cdots, F(k+n)] \tag{5}$$

Training samples of NN should cover the all changing range of sleeping fidget, and the distribution of training samples could reflect the total distribution of fidget degree. So, sample data should be from the whole process which is to begin from low temperature/humidity, pass fidget state, then return to comfortable state, and include all kinds of data in different experimental environment and experimental scheme. In the process of training, we selected 18 groups of experimental data including 1060 sample data pairs.

Because the quantity of input data in NN is very large, so it is necessary to normalize the input data before training networks. Authors selected Levenberg-Marquardt algorithm to train networks, updating parameters rule in L-M algorithm is given by

$$\Delta w = \left(J^T J + \mu I\right)^{-1} \cdot J^T e \tag{6}$$

With the increase of μ , the factor $J^T J$ in L-M algorithm might be ignored. Therefore, the learning process in L-M algorithm is mainly based on gradient descend method, i.e., the factor $\mu^T J^T e$. If recursion expands the error, then μ also will increase until the error trends constant. However, if μ is so big, the learning would stop because $\mu^T J^T e$ closes to zero. The phenomenon will occur when the minimum error is found. Advantages of the algorithm are rapid in convergence and small in error, so it is suitable to solve function approximation problem.

For the adaptive learning algorithm, we selected gradient descend method with momentum (LEARNGDM), which is faster in convergence than that of LEARNGD. Criterion function for training error performance is MSE function. Mathematic expression are given by

$$a_i(k) = f^2 \left\{ \sum_{j=1}^{s^2} \left[w_{i,j}^2(k) \bullet f^1 \left(\sum_{i=1}^{s^1} \left(iw_{i,j}^1(k) P_i + ib_i^1(k) \right) \right) + b_i^2(k) \right] \right\} \tag{7}$$

$$E(k) = E\left[e^2(k)\right] \approx \frac{1}{n} \sum_{i=1}^{n} \left(t_i - a_i(k)\right)^2 \tag{8}$$

In the experiment, selected learning function, training method and performance function in MATLAB NN toolbox is show in Fig.12.

Fig. 12. Application of NN toolbox

Shown as Fig.13, the performance function in final is 0.012912 through training of 5000 steps.

Finally, the model of sleep fidget monitoring is memorized with weight values and thresholds distributed in NN, in which, weight values matrix from input layer to hide layer are

lw{i, j}=[11.1003 -446.9839 399.0721 -307.7488 -85.9581 54.5748 22.3231 103.2223; -6.3943 18.2332 -3.4919 13.8771 -8.4868 -3.2378 -4.1011 -20.6852; 150.1909 146.209 -62.9091 127.1116 20.0697 -172.64 -28.6423 -45.7892; 138.1326 46.8546 -67.9057 -27.3218 135.0909 -181.3439 217.5256 -208.7609; 0.043606 -18.7625 18.8266 -13.3849 10.4996 3.9787 -3.3556 -7.2216; 79.5335 347.5518 -561.753 406.6844 -295.2914 24.2618 151.236 -10.0082; 5.6991 6.8569 7.5738 1.0031 28.2282 -34.0096 35.1728 -39.3461; 212.8717 354.1269 554.6582 -803.8437 653.1073 -494.318 -872.1543 670.229; -22.4919 9.7204 7.0935 -5.8631 54.0823 8.0735 -33.6536 9.7739; 6.9722 -18.2634 4.1325 -15.4476 6.424 4.8871 3.942 20.5277; -22.7846 9.8671 7.7873 -6.7399 54.704 7.3589 -34.2531 10.916; 1221.614

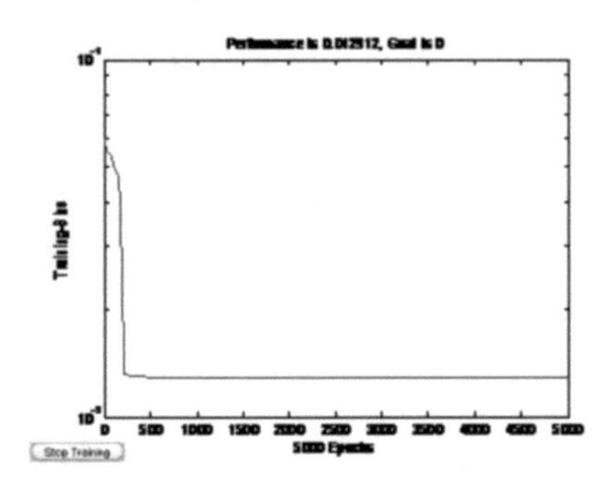

Fig. 13. The error curve of training process

-254.758 -898.856 41.6152 -516.1787 315.2132 153.6011 -11.2015; -42.8837 -7.9346 46.1546 -30.5327 38.6122 -37.1385 3.3858 14.1508; 1.1857 -4.4028 -2.3671 -2.3092 3.7511 -3.3037 3.5236 -7.0886; 2.2832 -3.8432 -0.63857 4.2195 -1.812 7.289 -3.9106 -0.38826; -7.7815 18.883 -4.9606 17.474 -4.2529 -7.0792 -3.7577 -20.8419; 326.241 -208.112 -29.9356 -170.1035 204.9379 -875.828 294.312 -131.4534; -335.489 449.218 316.4001 -623.249 -642.5914 610.4777 -490.0983 472.0787]

Thresholds in hide are

b{1}=[133.8259 16.9522 -256.0868 -57.3418 4.9686 48.1846 -3.5057 -240.9773 -19.2705 -16.2989 -19.3856 -33.0057 -3.3584 15.2148 3.6594 16.03 110.7886 207.3975]^{T}

Weight values in output layer are

lw{2, 1}=[0.57056 158.9478 0.5678 -0.67556 -1.3492 0.84072 0.79115 0.47579 41.8811 308.172 -41.798 0.48694 0.83004 -162.134 65.906 148.4792 1.0276 -0.65167]^{T}

The threshold in output layer is

b{2}=[-211.4469]

6 Precision Verification of Sleep Fidget Model

In order to verify the precision that NN model monitors sleep fidget, we selected additional several groups of experimental data which didn't be applied to train NN model. Here, introduced two groups of experimental verifications, we selected valid 54 pairs data in the first experimental process, whose the fidget degree from model computing is display with$''$ symbol, target output with $'_'$symbol. Model verification in the first experiment is shown in Fig.14, error-squared is $\sum_{i=1}^{54} e_i^2 = 0.293136164829912$.

We selected valid 75 pairs data in the second experimental process. Model verification is shown in Fig.15, error-squared is $\sum_{i=1}^{75} e_i^2 = 0.167694792107359$.

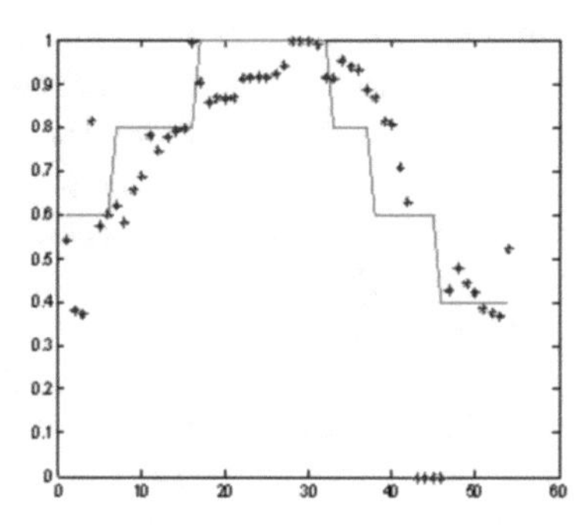

Fig. 14. Verification in the 1st experiment

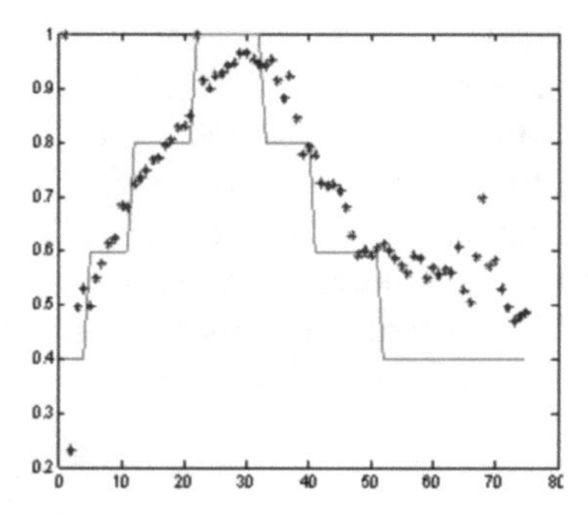

Fig. 15. Verification in the 2st experiment

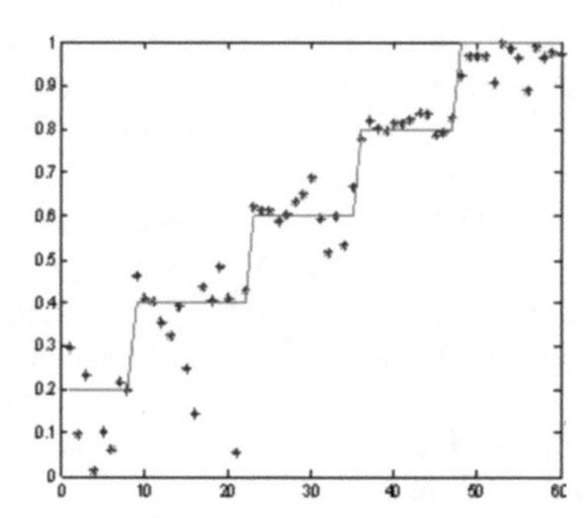

Fig. 16. Verification in sort ascending data

In addition, we also randomly selected 60 groups of input/output data pairs from other experimental data, and placed them in sort ascending of target values for verify precision of sleep fidget monitoring model, the result is shown in Fig.16.

Observed results of model precision verification, though there is some small difference between outputs from NN model and actual targets, and even obvious error in some individuals, but the idea that monitors sleep fidget has been satisfied. Further observing verification results, there are some little difference between outputs from NN model and actual targets while the sleep fidget is high. For example, it can completely monitor sleep fidget while the fidget degree is greater than 0.6.

7 Conclusions

In this paper, aimed at health problem from sleep fidget, authors explored the mechanism of sleep fidget, quantified the sleep fidget degree with fuzzy method, and presented theoretical and experimental scheme of the sleep fidget monitoring model based on NN information fusion. Then authors designed and developed the corresponding experimental system based on SHT10 temperature/humidity sensor from Sensirion Company, carried out about 40 groups of experiments, and trained NN model with experimental data. The verified results show that the model can effective monitor sleep fidget.

References

1. Robin R. Murphy(1996) Biological and Cognitive Foundations of Intelligent Sensor Fusion, IEEE Transactions on Systems, Man, and Cybernetics, Part A: Systems and Humans, 26(1): 42-51
2. Pablo H. Ibarguengoytia, Luis Enrique Sucar, Sunil Vadera(2001) Real Time Intelligent Sensor Validation, IEEE Transactions on Power Systems, 16(4):770-775
3. D.Rumelhart, J.McCelland(1986) Parallel Distributed Processing, MIT Press
4. R.C.Luo, Michael G.Kay(1989) Multisensor Integration and Fusion in Intelligent Systems. IEEE Transactions on Systems, Man, and Cybernetics, 19(5): 901-930
5. W. Elmenreich, S. Pitzek.(2001) The Time-Triggered Sensor Fusion Model. In Proceedings of the 5th IEEE International Conference on Intelligent Engineering Systems, Finland: 297-300
6. Wang Qi, Nie wei, Zhang zhaoli(1998) Data fusion and intelligent sensor system, Chinese Journal of Sensor technology, 17(6): 51-53
7. Martin T. Hagan, Howard B.Demuth, Mark Beale.(2002) Neural Network Design. Beijing: China Machine Press

Minimization of Mizumoto Automata

Mo Zhiwen and Hong Xiaolei

College of Mathematics and Software Science, Sichuan Normal University, Chengdu, China, 610066
Foundation item: Supported by the National Science Foundation of China (NO.10671030)
hxlei9@163.com

Abstract. In this paper, we introduce Mizumota automata whose initial and final states are fuzzy. Furthermore, the equivalence of the type of Mizumoto automaton(MA) and canonical fuzzy finite automaton(CA) with single initial state and deterministic transition function is established. Then we present a method to minimize the states of automata in canonical form.

Keywords: Mizumoto automata, canonical fuzzy finite automata, equivalence, minimization.

1 Introduction

Inspired by the theory of fuzzy sets introduced by Zadeh[1],[2], the concept of fuzzy automaton was originally introduced by Wee[3] in 1967. For a comprehensive overview of the area, the reader is refered to [4,5,6] which highlights the recent development and ongoing research. With the development of fuzzy automata theory, more recent contributions to this theory, significant for this work, are due to [7,8,9].

Fuzzy finite automata are used to design complex systems. For example, they are useful for a knowledge-based system designer since a knowledge-based system should solve a problem from fuzzy knowledge and should also provide the user with reasons for arriving at certain conclusion. A design tool is more valuable if there exit guidelines to assist the designer to come up with the best possible design. One of the major criteria for a best design is that it be minimal. It is not surprising to see a lot of research carried out working in this frame work. Minimization of fuzzy Mealy machine with output is discussed in [8]. Minimizing fuzzy finite automaton in canonical form is studied in [10], The state minimization problem for Mizumoto automata by using the method of quotient machines is solved in [11].

In this paper, we introduce Mizumoto automaton which is different from primary form. The significant difference attributes to fuzzy final states with degree. Furthermore, the equivalence of such Mizumoto automaton(MA) and canonical fuzzy finite automaton(CA) with single initial state and deterministic transition

B.-Y. Cao (Ed.): Fuzzy Information and Engineering (ICFIE), ASC 40, pp. 739–743, 2007.
springerlink.com

function is established. Then we present a method to minimize the states of automata in canonical form.

2 Mizumoto Automata and Their Equivalent Canonical Form

Definition 2.1[12]**.** A finite fuzzy automaton over the alphabet Σ is a system $A = \{S, \pi, \{F(\sigma)|\sigma \in \Sigma\}, \eta^G\}$,where

(1) $S = \{s_1, s_2, ..., s_n\}$ is a non-empty finite set of states.

(2) π is an $n-$dimensional fuzzy row vector, that is, $\pi = (\pi(s_1), \pi(s_2), ..., \pi(s_n))$, where $0 \leq \pi(s_i) \leq 1$, $1 \leq i \leq n$, and is called the initial state designator.

(3) G is a subset of S (the set of final states).

(4) $\eta^G = (\eta(s_1), \eta(s_2), ..., \eta(s_n))'$ is an $n-$dimensional column vector whose $i-$th component equal 1 if $s_i \in G$ and 0 otherwise, and is called the final state designator.

(5) For each $\sigma \in \Sigma$, $F(\sigma)$ is a fuzzy matrix of order n (the fuzzy transition matrix of A) such that $F(\sigma) = \| f_{s_i,s_j}(\sigma) \|$, $1 \leq i \leq n$, $i \leq j \leq n$.

Let element $f_{s_i,s_j}(\sigma)$ of $F(\sigma)$ be $f_A(s_i, \sigma, s_j)$, where $s_i, s_j \in S$ and $\sigma \in \Sigma$. The function f_A is a membership function of a fuzzy set in $S \times \Sigma \times S$; i.e.,$f_A : S \times \Sigma \times S \longrightarrow [0,1]$. f_A may be called the fuzzy transition function from state s to state t when the input is σ.

Now, we are able to introduce the concept of Mizumoto automaton(MA) with fuzzy final states.

Definition 2.2. Mizumoto automaton(MA) is a system $M = \{S, \pi, \{F(\sigma)|\sigma \in \Sigma\}, \eta^G\}$, where $S, \pi, F(\sigma)$ are as in definition 2.1 and the following hold.

(4)' $\eta^G = (\eta(s_1), \eta(s_2), ..., \eta(s_n))'$ is an $n-$dimensional column vector, where $0 \leq \eta(s_i) \leq 1$, $1 \leq i \leq n$.

Σ^* denotes the sets of all words of finite letters over Σ. Λ demotes the empty word. For any $x \in \Sigma^*$, $|x|$ stands for the length of x.

The function f_M is extend to $f_M : S \times \Sigma^* \times S \rightarrow [0,1]$.

if $x = \Lambda$, then $f_M(s, x, t) = \begin{cases} 1 \ if \ s = t \\ 0 \ if \ s \neq t \end{cases}$

if $x = x_1 x_2 \cdots x_n \in \Sigma^*$, then $f_M(s, x, t) = \bigvee_{s_1,\cdots,s_{n-1} \in S}[f(s, x_1, s_1) \wedge f(s, x_2, s_2) \wedge \cdots \wedge f(s_{n-1}, x_n, t)]$. Obviously, $F(x) = F(x_1) \circ F(x_2) \circ \cdots F(x_n)$.

Definition 2.3. Let $M = \{S, \pi, \{F(\sigma)|\sigma \in \Sigma\}, \eta^G\}$ be a MA. For any $x \in \Sigma^*$, the degree to which x is accept by M is $L_M(x) = \pi \circ F(x) \circ \eta^G$.

If $x = \sigma_1 \sigma_2 ... \sigma_m$, then $F(x) = F(\sigma_1) \circ F(\sigma_2) \circ ... \circ F(\sigma_m)$, where $\sigma_i \in \Sigma$, $1 \leq i \leq m$.

Consider a MA $M = \{S, \pi, \{F(\sigma)|\sigma \in \Sigma\}, \eta^G\}$, where $S = \{s_1, s_2, ..., s_n\}$. Here we give a method to build a canonical fuzzy finite automaton(CA) $N = \{P, \Sigma, \eta, p_0, \rho^G\}$ which is equivalent to M.

step 1. Let $A = \{elements\ of F(\sigma)|\sigma \in \Sigma\} \cup \{\pi(s_i)|s_i \in S\}$, where $i = 1, 2, ..., n$.

step 2. Let $P = \{(a_1, a_2, ..., a_n) | a_j \in A\}$, where $j = 1, 2, ..., n$.

step 3. Define $\eta : P \times \Sigma \longrightarrow P$ as $\eta((a_1, a_2, ..., a_n), \sigma) = (b_1, b_2, ..., b_n)$, where $b_i = \vee_{j=1}^{n} [a_j \wedge f_{s_j, s_i}(\sigma)]$.

step 4. Define $p_0 = (\pi(s_1), \pi(s_2), ..., \pi(s_n))$ and $\rho^G(a_1, a_2, ..., a_n) = \vee_{i=1}^{n} [a_i \wedge \eta^G(s_i)]$.

Definition 2.4. Let $N = \{P, \Sigma, \eta, p_0, \rho^G\}$ be a CA. For any $x \in \Sigma^*$, the degree to which x is accept by N is $L_N(x) = \rho^G(\eta(p_0, x))$.

Definition 2.5. Let $M = \{S, \pi, \{F(\sigma) | \sigma \in \Sigma\}, \eta^G\}$ be a MA. Let $N = \{P, \Sigma, \eta, p_0, \rho^G\}$ be a CA. M is equivalent to N ($M \equiv N$) if and only if the degree to which x is accept by M is the degree to which x is accept by N for any $x \in \Sigma^*$; i.e.,$L_M(x) = \pi \circ F(x) \circ \eta^G = \rho^G(\eta(p_0, x)) = L_N(x)$.

Theorem 2.6. $N = \{P, \Sigma, \eta, p_0, \rho^G\}$ defined above is well defined, and it is equivalent to MA M.

Proof. It is obvious that N is well defined. We now prove that $M \equiv N$. For any $x \in \Sigma^*$ and $p = (a_1, a_2, ..., a_n) \in P$, $\eta(p, x) = p \circ F(x)$, i.e.,the $i-$th coordinate $\eta(p, x)_i = \vee_{j=1}^{n} [a_j \wedge F_{s_j, s_i}(x)]$. we prove this relationship inductively by the length of x. Let $n = |x|$. If $n = 0$, then $\eta(p, \lambda) = p = p \circ F(\lambda)$ (obvious). Assume that for any x_1 of length $n - 1$, the relationship $\eta(p, x_1) = p \circ F(x_1)$ holds. Let $x = x_1\sigma, \sigma \in \Sigma$, then $\eta(p, x) = \eta(p, x_1\sigma) = \eta(\eta(p, x_1), \sigma) = \eta(p, x_1) \circ F(\sigma) = p \circ F(x_1) \circ F(\sigma) = p \circ (F(x_1) \circ F(\sigma)) = p \circ F(x_1\sigma) = p \circ F(x)$ by induction. Thus, for any $x \in \Sigma^*$, $L_N(x) = \rho^G(\eta(p_0, x)) = \rho^G(p_0 \circ F(x)) = p_0 \circ F(x) \circ \eta^G = \pi \circ F(x) \circ^G = L_M(x)$.

3 Minimization of NA

Definition 3.1. Let $N = \{P, \Sigma, \eta, p_0, \rho^G\}$ be a CA. For any $p_i, p_j \in P$, $i \neq j$, p_i is equivalent to $p_j (p_i \equiv p_j)$ if and only if for any $x \in \Sigma^*, \eta(p_i, x) = \eta(p_j, x)$ is satisfied.

For convenience, if $p_i \equiv p_j$, $i < j$, then $p_j \in [p_i]$.

Definition 3.2. Let $N' = \{P', \Sigma, \eta', p'_0, \rho'^G\}$ be a CA, N' is minimal if and only if states P'' of any CA $N'' = \{P'', \Sigma, \eta'', p''_0, \rho''^G\}$ which is equivalent to N' hold that $|P'| \leq |P''|$.

Given a CA $N = \{P, \Sigma, \eta, p_0, \rho^G\}$, let $P^a = \{\eta(p_0, x) : x \in \Sigma^*\}$. If $P = P^a$, then N is called accessible. The elements of P^a are called accessible states, and the elements of $P - P^a$ are called inaccessible elements.

The minimum NA N' such that $L_N = L_{N'}$ can be constructed as follows:

For convenience, let states in P be numbered from p_0 to p_{n-1}.

step 1. According to the transition of N, omit the inaccessible states.

step 2. Find equivalent states and construct minimum NA $N' = \{P', \Sigma, \eta', p'_0, \rho'^G\}$

Let $P' = \{[p_i] | p_i \in P\}$, $\eta'([p_i], \sigma) = [\eta(p_i, \sigma)]$, $p'_0 = [p_0]$, $\rho'^G([p_i]) = \rho(p_i)$.

Theorem 3.3. CA N' defined above is minimal and equivalent to N.

Proof. At first, we prove that $N' \equiv N$. For any $x \in \Sigma^*$, $L_{N'}(x) = \rho'^G(\eta'(p'_0, x)) = \rho'^G(\eta'([p'_0], x)) = \rho'^G([\eta'(p'_0, x)]) = \rho'^G(\eta(p_0, x)) = L_N(x)$. Thus, $N' \equiv N$.

Then we prove that N' is minimal. Assume that there is a CA $N'' = \{P'', \Sigma, \eta'', p''_0, \rho''^G\}$ such that $N'' \equiv N$ and $|P''| < |P|$. According to translativity of equivalence, we have $N'' \equiv N'$. Because $|P''| < |P|$, there exists $p'_i, p'_j \in P'$, $i < j$, such that $p'_i = p'_j$, i.e.,$\eta'(p'_i, x) = \eta'(p'_j, x)$ for any $x \in \Sigma^*$. Let $p'_i = [p_i]$, $p'_j = [p_j]$ by $N \equiv N'$, i.e.,$\eta'([p_i], x) = \eta'([p_j], x)$. According to the definition of N', we have $[\eta(p_i, x)] = [\eta(p_j, x)]$. It implies that $\eta(p_i, x) = \eta(p_j, x)$ for any $x \in \Sigma^*$, i.e.,$p_i \equiv p_j$. Thus, it is a contradiction.

Example 3.4. Let $M = \{S, \pi, \{F(\sigma)|\sigma \in \Sigma\}, \eta^G\}$ be a MA such that $S = \{s_1, s_2\}$, $\Sigma = \{a, b\}$, $\pi = (0.6, 1)$, $\eta^G = (1, 0.6)$, $F(a) = \begin{matrix} & s_1 & s_2 \\ s_1 & 0 & 0.6 \\ s_2 & 1 & 0.6 \end{matrix}$,

$F(b) = \begin{matrix} & s_1 & s_2 \\ s_1 & 0.6 & 1 \\ s_2 & 1 & 0 \end{matrix}$.

We construct the equivalent CA $N = \{P, \Sigma, \eta, p_0, \rho^G\}$. $P = \{p_1, p_2, p_3, p_4, p_5, p_6, p_7, p_8, p_9\}$, where $p_1 = (0, 0)$, $p_2 = (0, 0.6)$, $p_3 = (0, 1)$, $p_4 = (0.6, 0)$, $p_5 = (0.6, 0.6)$, $p_6 = (0.6, 1)$, $p_7 = (1, 0)$, $p_8 = (1, 0.6)$, $p_9 = (1, 1)$, $p_0 = p_6$, $\eta : P \times \Sigma \to P$ is given as follows.

η	a	b
p_1	p_1	p_1
p_2	p_5	p_4
p_3	p_8	p_7
p_4	p_2	p_5
p_5	p_8	p_5
p_6	p_8	p_8
p_7	p_2	p_6
p_8	p_5	p_6
p_9	p_8	p_9

, $\rho^G = \frac{0}{p_1} + \frac{0.6}{p_2} + \frac{0.6}{p_3} + \frac{0.6}{p_4} + \frac{0.6}{p_5} + \frac{0.6}{p_6} + \frac{1}{p_7} + \frac{1}{p_8} + \frac{1}{p_9}$.

According to η, we obtain accessible states $\{q_0, q_5, q_8\}$. By the method of minimization, the minimal CA $N' = \{P', \Sigma, \eta', p'_0, \rho'^G\}$ is defined as follows:

$P' = \{q_0, q_5, q_8\}$, $\Sigma = \{a, b\}$, $p'_0 = \{p_0\}$, $\rho'^G = \frac{0.6}{p_0} + \frac{0.6}{p_5} + \frac{1}{p_8}$,

η'	a	b
p_0	p_8	p_8
p_5	p_8	p_5
p_8	p_5	p_0

References

1. L.A.Zadeh (1965) Fuzzy sets. Information Control, 8:338–353.
2. L.A.Zadeh (1965) Fuzzy sets and systems. In Proc. symp. System Theory, Polytechnic Institute of Brooklyn, 29–37.

3. W.G.Wee (1967) On generalizations of adaptive algorithm and application of the fuzzy sets concept to pattern classification. Ph.D.Thesis, Purdue University.
4. A.Kandel, S.C.Lee (1980) Fuzzy Switching and Automata: Theory and Applications. New York, Crane Russak.
5. D.S.Malik, J.N.Mordeson (2000) Fuzzy Discrete Structures. Physica Verlag, New York.
6. J.N.Mordeson, D.S.Malik (2002) Fuzzy Automata and Languages: Theory and Applications. Chapman Hall/CRC, Boca Raton, London, New York, Wanskington.D.C.
7. Mingsheng Ying (2002) A formal model of computing with words. IEEE Transactions on Fuzzy Systems, 10:640–651.
8. Wei Cheng, Zhi-wen Mo (2004) Minimization algorithm of fuzzy automata. Fuzzy Sets and Systems, 4:439–448.
9. Tatjana Petkovic (2006) Congruences and homonorphisms of fuzzy automata. Fuzzy Sets and Systems,157:444–458.
10. Hsusan-Shih Lee (2000) Minimizing fuzzy finite automata. IEEE.
11. N.C.Basak, A.Gupta (2002) On quotient machines of fuzzy automaton and the minimal machine. Fuzzy Sets and Systems, 125:223–229.
12. Masaharu Mizumoto, Junichi Toyoda, Kohkichi Tanaka (1969) Some considerations on fuzzy automata. Journal of Computer and System Science, 3:409–422.

Transformation of Linguistic Truth Values During the Sensor Evaluation

Zhengjiang Wu[1], Xin Liu[2], Li Zou[1,3], and Yang Xu[1]

[1] Intelligent Control and Development Center, Southwest Jiaotong University, Chengdu, 610031, Sichuan, P.R. China
jiang2021987@163.com
[2] Mathematics College, Liaoning Normal University, Dalian, 116029, P.R. China
[3] School of Computer and Information Technology, Liaoning Normal University, Dalian, 116029, P.R. China

Abstract. During the sensor evaluation procedure, each valuator uses his/her own ordinary linguistic truth values for the same factor because of different preference. That will brings some disadvantages to aggregate the information. For a uniform criterion, the standard linguistic truth value set is proposed. Based on the former hypothesis of transformation models of linguistic truth values, four transformation models are discussed: the model of point to point, the model of fuzzy set to point, the model of point to fuzzy set and the model of fuzzy set to fuzzy set. An example is to analyze it.

Keywords: Linguistic truth value, Standard linguistic truth value, Sensor evaluation.

1 Introduction

In many industrial sectors such as food, cosmetic, medical, chemical, and textile, sensory evaluation is widely used for determining the quality of products, solving conflicts between customers and producers, developing new products, and exploiting new markets adapted to the consumer's preference [1, 2, 3].

It is inevitable to represent the evaluation variables with linguistic truth value during the sensor evaluation procedure [4]. Much of the evaluation variables can't be represented precisely because they are come from the reaction of sensor. The result can't be described with numerical value in order to keep all of the information. Linguistic truth values are more suitable to state the evaluation term than numerical value [5].

There are many groups of linguistic truth values in the procedure of sensor evaluation. Because the objects are evaluated by many valuators which level and aspect of knowledge are quite different, the factors of evaluation and the meaning of evaluation term are quite different in their evaluation system [6, 7]. Hence, before aggregating information, the linguistic truth values will be restated by some uniform criterion. The uniform criterion must be related to evaluation term. The criterion is called standard linguistic truth value (abbreviate it to SLTV) set, which is the evaluation terms used by some valuator during the

B.-Y. Cao (Ed.): Fuzzy Information and Engineering (ICFIE), ASC 40, pp. 744–750, 2007.
springerlink.com

evaluation procedure. But the SLTV set is usually set as a group of linguistic truth values by the evaluation organizer [8, 9].

As a Computer Aid Evaluation System, the information must be symbolized. In high-dimensional space, these linguistic truth values will be considered as a single point. Then we can coordinate their positions according to their relative position. For the convenience, in this paper, we suppose that the SLTV distributes averagely on its domain and has linear order relation [10].

In practice, the valuator has some perturbation when he gives his opinion. Because this perturbation abides by the statistical rules, we can constitute a membership function nearby the point to describe the position of this linguistic truth value. The degree of the membership is obtained through the statistic procedure.

In this paper, Section 1 is introduction. Section 2 is some hypotheses to construct the transformation models of linguistic truth value. In Section 3, four transformation models of linguistic truth value are given. (1) Point to point model: ordinary linguistic truth value (abbreviate it by OLTV) and SLTV are represented by a single point in this model; (2) Fuzzy set to point model: OLTV is represented by a fuzzy set and SLTV is representation by a point; (3) Point to fuzzy set model: OLTV is represented to a point and SLTV is represented to a fuzzy set; (4) Fuzzy set to fuzzy set model: SLTV and OLTV are both represented by fuzzy sets. An example based on above discussion is given in section 4.

2 Basic Concepts and Hypotheses

Hypothesis 1. *A linguistic truth value can be represented as a point in the high dimensional space.*

The distance of two linguistic truth values represents the dissimilarity measurement of them.

Hypothesis 2. *The SLTV set has linear order relation.*

If hypothesis 2 holds, linguistic truth value can be denoted on an axis. If there are incomparable elements then linguistic truth value need to decompose into chain structure.

Hypothesis 3. *The sets of SLTV and OLTV can be defined in a same measure space.*

If the sets of SLTV and OLTV are defined in different measure space, there exists a function such that the two measure space are isomorphic. If we use a fuzzy number to represent linguistic truth value, then this function can be extended by Extension Principle in order to deal with the case of fuzzy set. In fact, this transformation procedure between SLTV and OLTV must satisfy a series of conditions as follows:

1. The procedure needs to be monotone because the order must be consistent in two orientation spaces.

2. After the transformation the same dissimilarity need to be maintained. That is to say the rate of the distance between these linguistic truth values should be hold.
3. It is necessary that the function maps the domain (the coordinate interval which OLTV lies) onto the coordinate interval which SLTV lies.

Based on above three conditions, especially the dissimilarity between linguistic truth values is represented by numerical value. Hence we choose the same proportion zoom as transformation function.

Definition 1. *Let $f : X \to Y$, f induce a new mapping, denoted by F.*

$$\begin{aligned} &F : \mathscr{F}(X) \to \mathscr{F}(Y), \\ &A \to F(A), \\ &F(A)(y) = \bigvee_{x \in f^{-1}(y)} A(x), \end{aligned}$$

here abbreviate $f^{-1}(\{y\})$ to $f^{-1}(y)$.

If for coordinate X the membership function of linguistic truth value A is $\mu_A(x) = F(x)$ on the interval $[a, b]$ while the SLTV on $[c, d]$, then the membership function of the linguistic truth value A' after transformation is:

$$f(x) = \frac{d-c}{b-a} \times (x-a) + c,$$

$$\mu_A(y) = \bigvee_{y = \frac{d-c}{b-a} \times (x-a) + c} F(x) = F\left(\frac{b-a}{d-c} \times (x-c) + a\right). \quad (1)$$

Hypothesis 4. *Assume all the membership function of linguistic truth value is a triangular fuzzy function.*

3 Transformation Models

Let $S = \{S_1, S_2, \cdots, S_m\}$ be the set of standard linguistic values, $A = \{A_1, A_2, \cdots, A_n\}$ be a set of OLTV and assume they are distributed averagely on the coordinate axis. Here the membership function of linguistic truth value in S is defined on interval $[c, d]$ and the membership function of linguistic truth value in A is defined on the interval $[a, b]$. If these linguistic values are extended to triangular fuzzy set then the discourse domain is extended to $[a - e, b + e]$ or $[c - e, d + e]$, where e is a half span of linguistic truth value fuzzy set.

3.1 Point to Point Model

If the OLTV and the SLTV are represented to points then the dissimilarity can hold with some measure.

3.2 Fuzzy to Point Model

If the linguistic truth value is a triangular fuzzy number, then linguistic truth value in OLTV become a triangular fuzzy number after $F(x)$ maps linguistic truth value into $[c, d]$. Each SLTV has a membership degree to this fuzzy set.

Let the half span of OLTV T_i be e_i, $i = 1, 2, \cdots, m$ and the center of T_i be x_i, $x_i \in [a, b]$. The center of SLTV S_j is y_j, $y_j \in [c, d]$, $j = 1, 2, \cdots, n$. The similarity of T_i and S_j is denoted as μ_{ij}.

$$\mu_{ij} = \begin{cases} \frac{y_j}{e_i} - \frac{1}{e_i}\left[\frac{d-c}{b-a}(x_i - a) + c\right] + 1 & y_j \in \left[\frac{d-c}{b-a}(x_i - a) + c - e_i, \frac{d-c}{b-a}(x_i - a) + c\right] \\ -\frac{y_j}{e_i} + \frac{1}{e_i}\left[\frac{d-c}{b-a}(x_i - a) + c\right] + 1 & y_j \in \left[\frac{d-c}{b-a}(x_i - a) + c, \frac{d-c}{b-a}(x_i - a) + c + e_i\right] \\ 0 & otherwise \end{cases}$$

3.3 Point to Fuzzy Set Model

Similarity, let the center of T_i be x_i, $x_i \in [a, b]$, $i = 1, 2, \cdot, m$ and the half span of SLTV S_j be e_j and the center of S_j be y_j, $y_j \in [c, d]$. The similarity of T_i and S_j is denoted as μ_{ij}.

$$\mu_{ij} = \begin{cases} \frac{1}{e_i}\left[\frac{d-c}{b-a}(x_i - a) + c\right] - \frac{y_j}{e_j} + 1 & x_i \in \left[\frac{(b-a)(y_j - e_j - c)}{d-c} + a, \frac{(b-a)(y_j - c)}{d-c} + a\right] \\ -\frac{1}{e_i}\left[\frac{d-c}{b-a}(x_i - a) + c\right] + \frac{y_j}{e_j} + 1 & x_i \in \left[\frac{(b-a)(y_j - c)}{d-c} + a, \frac{(b-a)(y_j + e_j - c)}{d-c} + a\right] \\ 0 & otherwise \end{cases}$$

3.4 Fuzzy Set to Fuzzy Set Model

Let the set of OLTV be $T = \{T_1, T_2, \cdots, T_m\}$, where the half span of the linguistic truth value T_i is e_i, $i = 1, 2, \cdots, m$. Using formula (1) the set of linguistic truth value is redefined as $T' = \{T'_1, T'_2, \cdots, T'_m\}$ after transformation and the half span will change as $e'_i = \frac{d-c}{b-a} \times e_i$, and the center of linguistic value after transformation is $x_i \in [c, d]$, $i = 1, 2, \cdots, m$. The half span of SLTV S_j is e_j and the center of is y_j, $j = 1, 2, \cdots, n$. The similarity degree of T_i and S_j is denoted as μ_{ij}.

$$A_1 = \frac{[(e'_i + e_j) + (x_i - y_j)]^2}{2(e'_i + e_j)},$$

$$A_2 = \frac{[(e'_i - e_j) - (x_i - y_j)]^2}{2(e_j - e'_i)},$$

$$A_3 = \frac{[(e'_i - e_j) + (x_i - y_j)]^2}{2(e'_i - e_j)},$$

$$A_4 = \frac{[(e'_i + e_j) - (x_i - y_j)]^2}{2(e'_i + e_j)},$$

$$A_5 = \frac{[(e'_i - e_j) - (x_i - y_j)]^2}{2(e'_i - e_j)},$$

$$A_6 = \frac{[(e'_i - e_j) + (x_i - y_j)]^2}{2(e_j - e'_i)}.$$

Case I ($x_i = y_j$)

$$\mu_{ij} = 1.$$

Case II ($x_i < y_j$)
If $e'_i \leq e_j$ and $x_i - e'_i \leq y_j - e_j \leq x_i + e'_i \leq y_j + e_j$, then

$$\mu_{ij} = \frac{A_1}{e'_i}.$$

If $e'_i \leq e_j$ and $y_j - e_j \leq x_i - e'_i \leq x_i + e'_i \leq y_j + e_j$, then

$$\mu_{ij} = \frac{A_1 - A_2}{e'_i}.$$

If $e'_i > e_j$ and $x_i - e'_i \leq y_j - e_j \leq x_i + e'_i \leq y_j + e_j$, then

$$\mu_{ij} = \frac{A_1}{e_j}.$$

If $e'_i > e_j$ and $x_i - e'_i \leq y_j - e_j \leq y_j + e_j \leq x_i + e'_i$, then

$$\mu_{ij} = \frac{A_1 - A_3}{e_j}.$$

Case III ($x_i > y_j$)
If $e'_i \leq e_j$ and $y_j - e_j \leq x_i - e'_i \leq y_j + e_j \leq x_i + e'_i$, then

$$\mu_{ij} = \frac{A_4}{e'_i}.$$

If $e'_i \leq e_j$ and $y_j - e_j \leq x_i - e'_i \leq x_i + e'_i \leq y_j + e_j$, then

$$\mu_{ij} = \frac{A_4 - A_6}{e'_i}.$$

If $e'_i > e_j$ and $y_j - e_j \leq x_i - e'_i \leq y_j + e_j \leq x_i + e'_i$, then

$$\mu_{ij} = \frac{A_4}{e_j}.$$

If $e'_i > e_j$ and $x_i - e'_i \leq y_j - e_j \leq y_j + e_j \leq x_i + e'_i$, then

$$\mu_{ij} = \frac{A_4 - A_5}{e_j}.$$

4 Example

Assume a set of linguistic truth values have the same half span. Let the OLTV is $T = \{T_1, T_2, \cdots, T_5\}$, $e_T = 2$ and the center of T_i is $x_i = ii = 1, 2, \cdots, 5$. Let a $S = \{S_1, S_2, \cdots, S_7\}$, $e_S = 2$ and the center of the fuzzy set $y_j = j$, $j = 1, 2, \cdots, 7$. The similarity degree of T_i and S_j, $i = 1, 2, \cdots, 5$, $j = 1, 2, \cdots, 7$ are given as follows:

Table 1. Fuzzy Set to Point Model

	S_7	S_6	S_5	S_4	S_3	S_2	S_1
T_5	1	0.67	0.33	0	0	0	0
T_4	0.5	0.83	0.5	0	0	0	0
T_3	0	0.33	0.67	1	0.67	0.33	0
T_2	0	0	0	0.5	0.83	0.83	0.5
T_1	0	0	0	0	0.33	0.67	1

Table 2. Point to Fuzzy Set Model

	S_7	S_6	S_5	S_4	S_3	S_2	S_1
T_5	1	0.5	0	0	0	0	0
T_4	0.25	0.75	0.75	0.25	0	0	0
T_3	0	0	0.5	1	0.5	0	0
T_2	0	0	0	0.25	0.75	0.75	0.25
T_1	0	0	0	0	0	0.5	1

Table 3. Fuzzy Set to Fuzzy Set Model

	S_7	S_6	S_5	S_4	S_3	S_2	S_1
T_5	1	0.6	0.45	0.2	0.05	0	0
T_4	0.61	0.95	0.95	0.61	0.31	0.11	0.01
T_3	0.2	0.45	0.8	1	0.8	0.45	0.2
T_2	0.01	0.11	0.31	0.61	0.95	0.95	0.61
T_1	0	0	0.05	0.2	0.45	0.6	1

5 Conclusions

In this paper the standardization of linguistic truth value is obtained through SLTV. The comparability of the linguistic truth values can be described in four kinds of cases. The elements of SLTV group can be arbitrary number. The more SLTVs are, the more sensitive OLTV's position is. The standardization of the linguistic truth value can be used in the evaluation with linguistic truth value. We solve the problem that the information can't be aggregated because there are some different meanings of linguistic truth value in the process of evaluation. The further work is to describe the non-comparability of linguistic truth value since it is related to the distance of linguistic truth values. The method of transformation of linguistic truth values can also be used into pattern recognition [11, 12] and decision-making.

Acknowledgements

The authors are grateful to the referees for their valuable comments and suggestions. This work has been supported by the National Natural Science Foundation of China (Grant No. 60474022).

References

1. Zeng X., L. Koehl (2003) Representation of the subjective evaluation of fabric hand using fuzzy techniques, Int J Intelligent Systems, Vol.18, No.3, pp355-366.
2. Kawabata S, Niwa M (1996), Objective measurement of fabric hand, In: Raheel M, Dekker M (eds) Modern Textile Characterization Methods, pp 329-354.
3. Zeng X., Y. Ding, L. Koehl (2004) A 2-tuple fuzzy linguistic model for sensory fabric hand evaluation, in Intelligent sensory evaluation, Eds. Ruan D. and Zeng X., Springer (Berlin), February, pp.217-234.
4. F. Herrera, E. Herrera-Viedma (2000) Linguistic Decision Analysis: Steps For Solving Decision Problems Under Linguistic Information. Fuzzy Sets and Systems 115 67-82.
5. F. Herrera, E. Herrera-Viedma (2000) Luis Martinez. A Fusion Approach For Managing Multi-Granularity Linguistic Term Sets in Decision Making. Fuzzy Sets and Systems 114 43-58.
6. Zeng X., L. Koehl, M. Sahnoun, W.A. Bueno and M. Renner(2004) Integration of human knowledge and measured data for optimization of fabric hand, International Journal of General Systems, Vol.33, No.2-3, pp243-258.
7. S. J.Chen, C.L. Hwang (1992) Fuzzy Multiple Attribute Decision Making Methods and Application, Springer.Berlin.
8. Xianyi Zeng, Ludovic Koehl ,Zhengjiang Wu (2004) A Method for Sensory Evaluation Based on Fuzzy Satisfaction Degrees. International Conference on Service Systems and Service Management.
9. Zou K., Xu Y. (1989) Fuzzy Systems and Expert Systems, Publish House of Southwest Jiaotong University (in Chinese).
10. Levrat E., A. Voisin, S. Bombardier (1997) Subjective evaluation of car seat comfort with fuzzy set techniques, Int J Intelligent Systems, Vol.12, No.12, pp891-913.
11. Shoujue Wang, Biomimetic(Topolological) (2002) Pattern Recognition- A New Model of Pattern Recognition Theory and Its Applications, Chinese Journal of Electronics Vol 30(10), p1417-1420.
12. Shoujue Wang, Jiangliang Lai (2005) A More Complex Neuron in Biomimetic Pattern Recognition, the Second International Conference on Neural Networks and Brain. P1487-1489.

Guaranteed Cost Control for a Class of Fuzzy Descriptor Systems with Time-Varying Delay

Weihua Tian[1,2], Huaguang Zhang[1], and Ling Jiang[2]

[1] College of Information Science and Engineering, Northeastern University, Shenyang 110004, China
wendy_t@sohu.com; hg_zhang@21cn.com
[2] Department of Automatic Control, Shenyang Institute of Engineering, Shenyang 110136, China

Abstract. This paper presents a new guaranteed cost controller design approach for a class of Takagi-Sugeno (T-S) fuzzy descriptor systems with time-varying delay. Based on the relaxed quadratic stability condition and a linear quadratic cost function, the sufficient conditions for the existence of guaranteed cost controllers via state feedback are given in terms of linear matrix inequalities (LMIs). And the design of optimal guaranteed cost controller can be reduced to a convex optimization problem. At last, a numerical example is given to illustrate the effectiveness of the proposed method and the perfect performance of the optimal guaranteed cost controller.

Keywords: Fuzzy Descriptor Systems, Guaranteed Cost Control, Time-Varying Delay, Linear Matrix Inequalities (LMI).

1 Introduction

During the last two decades, fuzzy technique has been widely used in nonlinear system modeling, especially for systems with incomplete plant information. The well-known Takagi-Sugeno (T S) fuzzy model is a popular and convenient tool to approximate nonlinear systems. It is a nonlinear system described by a set of if-then rules, which give local linear representation of the underlining systems [1,2]. Recently, the descriptor system, which can describe a wider class of systems, including physical models and non-dynamic constraints, is paid a lot of attention. Therefore, it is meaningful to employ fuzzy descriptor model in control systems design. In [4,5], the fuzzy descriptor model is stated and the stability and stabilization problems of the systems are addressed. It is shown that the main feature of the fuzzy descriptor systems is it can reduce the number of LMI conditions for controller design. This rule reduction is an important issue for LMI-based control synthesis. Thereafter, Many further contributions have been made to the study of fuzzy descriptor systems[6, 7]. Very recently, some authors have paid their attention to control of nonlinear systems with time-delays by using T-S fuzzy descriptor models[8,12]. In addition to the simple stabilization, there have been various efforts to assign certain performance criteria when designing a controller, such as quadratic cost minimization, $H\infty$ norm minimization, pole placement, etc. Among them, the

B.-Y. Cao (Ed.): Fuzzy Information and Engineering (ICFIE), ASC 40, pp. 751–759, 2007.
springerlink.com

guaranteed cost control aims at stabilizing the systems while maintaining an adequate level of performance represented by a quadratic cost function [3,9]. Although it is an important problem to design a guaranteed cost controller for nonlinear systems, it seems that this field is still open to the control of fuzzy descriptor time-delay systems. In this paper, we mainly focus on the problem of fuzzy guaranteed cost control for a class of T-S fuzzy descriptor systems with time-varying delay. A linear quadratic cost function is used as a guaranteed performance index. The sufficient conditions for the existence of guaranteed cost controller are given in terms of LMIs. And a convex optimization problem with LMI constraints is formulated to design the optimal guaranteed cost controller which minimizes the upper bound of the quadratic performance index. The resulting fuzzy controller can not only guarantee that the closed-loop fuzzy system is quadratically stable, but also satisfy the quadratic performance index.

2 Preliminaries and Problem Formulation

In this section, we consider a class of fuzzy descriptor systems with time-varying delay described by the following fuzzy If-Then rules:

If ξ_1 is M_{1i} and … and ξ_p is M_{pi}, Then

$$E\dot{x}(t) = A_i x(t) + A_{1i} x(t-\tau(t)) + B_i u(t)$$
$$x(t) = \varphi(t) \qquad t \in [-\tau_0, 0],\ i = 1,2,\cdots,k, \tag{1}$$

where $x(t) \in R^n$ is the state, $u(t) \in R^m$ is the control input. The matrix $E \in R^{n\times n}$ is singular. $\tau(t)$ is the time-varying delay, $0 \le \tau(t) \le \tau_0 < \infty$ and $\dot{\tau}(t) \le d < 1$. A_i, A_{1i} and B_i are known real constant matrices with appropriate dimensions. Taking the weighted average of $E\dot{x}(t)$, $i = 1,2,\cdots,k$ as a defuzzification strategy, the final defuzzified output of the fuzzy model is derived as follows.

$$E\dot{x}(t) = \sum_{i=1}^{k} \lambda_i(\xi(t))[A_i x(t) + A_{1i} x(t-\tau(t)) + B_i u(t)]$$
$$x(t) = \varphi(t) \qquad t \in [-\tau_0, 0],\ i = 1,2,\cdots,k \tag{2}$$

where

$$\lambda_i(\xi(t)) = \frac{\Pi_{j=1}^{p} M_{ij}(\xi_j(t))}{\Sigma_{i=1}^{k} \Pi_{j=1}^{p} M_{ij}(\xi_j(t))},$$

and $M_{ij}(\xi_j(t))$ is the grade of membership of $\xi_j(t)$ in M_{ij}. $\lambda_i(\xi(t)) \ge 0$, $i = 1,\cdots,k$, and $\sum_{i=1}^{k} \lambda_i(\xi(t)) = 1\ \forall t$.

The unforced system of (2) is

$$E\dot{x}(t) = \sum_{i=1}^{k} \lambda_i(\xi(t))[A_i x(t) + A_{1i} x(t-\tau(t))]. \tag{3}$$

Definition 2.1. The fuzzy descriptor system (3) is regular if there exists $s \in C$ satisfying

$$\det\left[sE - \sum_{i=1}^{k} \lambda_i(\xi(t))(A_i + A_{1i}e^{-s\tau(s)})\right] \neq 0 \quad \forall t \geq 0.$$

Definition 2.2. The regular fuzzy descriptor system (3) is impulse free if

$$\deg\ \det\left[sE - \sum_{i=1}^{k} \lambda_i(\xi(t))(A_i + A_{1i}e^{-s\tau(s)})\right] = rank\ E\cdot$$

Definition 2.3. The regular and impulse free system (3) is quadratically stable if

$$\frac{dV(x(t))}{dt} \leq -\alpha\|x(t)\|^2, \ (\alpha > 0)$$

where,

$$V(x(t)) = x^T(t)E^T Px(t) + \frac{1}{1-d}\int_{-\tau(t)}^{t} x^T(\sigma)Sx(\sigma)d\sigma\cdot \tag{4}$$

The common nonsingular matrix P satisfies, $P \in R^{n\times n}$, $E^T P = P^T E \geq 0$. S is a symmetric positive definite matrix.

Given symmetric positive-definite matrices Q and R, we consider the cost function

$$J = \int_0^\infty \{x^T(t)Qx(t) + u^T(t)Ru(t)\}dt\cdot \tag{5}$$

Associated with the cost (5), the fuzzy guaranteed cost control is defined as follows.

Definition 2.4. For system (2) and cost function (5), if there exists a fuzzy control law $u(t)$ and a scalar J_0 such that the closed-loop value of the cost function (5) satisfies $J \leq J_0$, then J_0 is said to be a guaranteed cost and the control law $u(t)$ is said to be a guaranteed cost control law for (2).

Lemma 1[10]**.** For any matrices K_1, K_2 and K_3 of appropriate dimensions with $K_2 > 0$, we have

$$K_1^T K_3 + K_3^T K_1 \leq K_1^T K_2 K_1 + K_3^T K_2^{-1} K_3\cdot \tag{6}$$

The purpose of this paper is to develop a procedure to design a state feedback control law to make the system (2) to be quadratically stable, and the cost function J has a definite upper bound.

3 Main Results

Considering the fuzzy descriptor system (2), we design a parallel distributed compensation (PDC)controller. The fuzzy controller shares the same fuzzy sets with the fuzzy model in the premise parts and has local linear controllers in the consequent parts. The ith fuzzy rule of the fuzzy controller is of the following form.

If ξ_1 is M_{1i} and … and ξ_p is M_{pi}, Then
$u(t) = K_i x(t)$, $i = 1,\cdots,k$.
Hence, the overall fuzzy control law is represented by

$$u(t) = \sum_{i=1}^{k} \lambda_i(\xi(t)) K_i x(t), \tag{7}$$

where $K_i (i = 1,2.\cdots,k)$ are the local feedback gains. Substituting the control law (7) into system (2), the overall closed-loop system can be written as

$$E\dot{x}(t) = \sum_{i=1}^{k}\sum_{j=1}^{k} \lambda_i(\xi(t))\lambda_j(\xi(t)) \left[A_i x(t) + A_{1i} x(t-\tau(t)) + B_i K_j x(t)\right]$$
$$x(t) = \varphi(t),\ t \in \left[-\tau_0, 0\right],\ i = 1,2,\cdots,k. \tag{8}$$

Theorem 1. Consider the closed-loop system (8) associated with the cost function (5). Suppose there exist positive definite matrices U, X, W_{ij} and Y_i satisfying

$$X^T E^T = EX \geq 0 \tag{9}$$

$$\begin{bmatrix} \Phi_{ii} & A_{1i}U & X^T & X^T & Y_i^T \\ * & -U & 0 & 0 & 0 \\ * & * & -(1-d)U & 0 & 0 \\ * & * & * & -Q^{-1} & 0 \\ * & * & * & * & -R^{-1} \end{bmatrix} < 0, \quad 1 \leq i \leq k \tag{10}$$

$$\begin{bmatrix} \tilde{\Phi}_{ij} & (A_{1i} + A_{1j})U & X^T & X^T & Y_i^T & Y_j^T \\ * & -2U & 0 & 0 & 0 & 0 \\ * & * & -\frac{1-d}{2}U & 0 & 0 & 0 \\ * & * & * & -\frac{1}{2}Q^{-1} & 0 & 0 \\ * & * & * & * & -R^{-1} & 0 \\ * & * & * & * & * & -R^{-1} \end{bmatrix} \leq 0, \quad 1 \leq i < j \leq k \tag{11}$$

$$W = \begin{bmatrix} W_{11} & W_{12} & \cdots & W_{1k} \\ W_{12} & W_{22} & \cdots & W_{2k} \\ \vdots & \vdots & \ddots & \vdots \\ W_{1k} & W_{2k} & \cdots & W_{kk} \end{bmatrix} > 0, \tag{12}$$

where $\Phi_{ii} = A_i X + X^T A_i^T + B_i Y_i + Y_i^T B_i^T + W_{ii}$,
$\tilde{\Phi}_{ij} = (A_i + A_j)X + X^T(A_i + A_j) + B_i Y_j + B_j Y_i + Y_j^T B_i^T + Y_i^T B_j^T + 2W_{ij}$.
Then the control law (7) is a fuzzy guaranteed cost control law, and the cost function (5) has the upper bound

$$J_0 = x^T(0)E^T P x(0) + \frac{1}{1-d}\int_{-\tau_0}^{0} \varphi^T(s) U^{-1} \varphi(s) ds. \tag{13}$$

Proof: Choose a descriptor type of Lyapunov-Krasovskii functional in form of (4)

$$V(x(t)) = x^T(t)E^T Px(t) + \frac{1}{1-d}\int_{-\tau(t)}^{t} x^T(\sigma)Sx(\sigma)d\sigma .$$

For the simplicity, denote $\lambda_i(\xi(t))$ by λ_i and $x(t-\tau(t))$ by $x_\tau(t)$. By differentiating $V(x(t))$ along the trajectory of (8), we obtain

$$\dot{V}(x(t)) = (E\dot{x}(t))^T Px(t) + x^T(t)P^T E\dot{x}(t) + \frac{1}{1-d}x^T(t)Sx(t) - \frac{1-\dot{\tau}(t)}{1-d}x_\tau^T(t)Sx_\tau(t)$$

By lemma 1, we get

$$\begin{aligned}
\dot{V}(x(t)) \le & \sum_{i=1}^{k}\sum_{j=1}^{k}\lambda_i\lambda_j\Big\{x^T(t)\Big[(A_i+B_iK_j)^T P + P^T(A_i+B_iK_j) + \frac{S}{1-d}\Big]x(t) \\
& + x_\tau^T(t)A_{1i}^T Px(t) + x^T(t)P^T A_{1i}x_\tau(t) - x_\tau^T(t)Sx_\tau(t)\Big\} \\
\le & \sum_{i=1}^{k}\sum_{j=1}^{k}\lambda_i\lambda_j x^T(t)(\Theta_{ij} + \frac{S}{1-d} + P^T A_{1i}^T S^{-1}A_{1i}P + Q + K_i^T RK_j)x(t) \\
& - \sum_{i=1}^{k}\sum_{j=1}^{k}\lambda_i\lambda_j x^T(t)(Q + K_i^T RK_j)x(t) \\
\le & \sum_{i=1}^{k}\lambda_i^2 x^T(t)(\Theta_{ii} + \frac{S}{1-d} + P^T A_{1i}^T S^{-1}A_{1i}P + Q + K_i^T RK_i)x(t) \\
& + \sum_{i=1}^{k}\sum_{j>1}^{k}\lambda_i\lambda_j x^T(t)\Big[\tilde{\Theta}_{ij} + P^T(A_{1i}+A_{1j})^T(2S)^{-1}(A_{1i}+A_{1j})P\Big]x(t) \\
& - \sum_{i=1}^{k}\sum_{j=1}^{k}\lambda_i\lambda_j x^T(t)(Q + K_i^T RK_j)x(t),
\end{aligned}$$

where $\Theta_{ij} = P^T A_i + A_i^T P + P^T B_i K_j + K_j^T B_i^T P$,

$$\tilde{\Theta}_{ij} = \Theta_{ij} + \Theta_{ji} + \frac{2S}{1-d} + 2Q + K_i^T RK_i + K_j^T RK_j \cdot$$

Introducing new matrix variables T_{ij}, if the following conditions hold

$$\Theta_{ii} + \frac{S}{1-d} + P^T A_{1i}^T S^{-1}A_{1i}P + Q + K_i^T RK_i < -T_{ii}, \quad 1 \le i \le k, \tag{14}$$

$$\tilde{\Theta}_{ij} + P^T(A_{1i}+A_{1j})^T(2S)^{-1}(A_{1i}+A_{1j})P \le -2T_{ij}, \quad 1 \le i < j \le k. \tag{15}$$

Here,

$$T = \begin{bmatrix} T_{11} & T_{12} & \cdots & T_{1k} \\ T_{12} & T_{22} & \cdots & T_{2k} \\ \vdots & \vdots & \ddots & \vdots \\ T_{1k} & T_{2k} & \cdots & T_{kk} \end{bmatrix} > 0.$$

Then

$$\begin{aligned}
\dot{V}(x(t)) \le & -\sum_{i=1}^{k}\lambda_i^2 x^T(t)T_{ii}x(t) - 2\sum_{i=1}^{k}\sum_{j>1}^{k}\lambda_i\lambda_j x^T(t)T_{ij}x(t) \\
& - \sum_{i=1}^{k}\sum_{j=1}^{k}\lambda_i\lambda_j x^T(t)(Q + K_i^T RK_j)x(t)
\end{aligned}$$

$$=-\left[\lambda_1 x^T(t)\ \lambda_2 x^T(t)\ \cdots\ \lambda_k x^T(t)\right]T\left[\lambda_1 x^T(t)\ \lambda_2 x^T(t)\ \cdots\ \lambda_k x^T(t)\right]^T$$
$$-\sum_{i=1}^{k}\sum_{j=1}^{k}\lambda_i\lambda_j x^T(t)(Q+K_i^T RK_j)x(t). \tag{16}$$

Note that $X=P^{-1}$, $U=S^{-1}$, $W_{ij}=P^{-T}T_{ij}P^{-1}$ and $Y_i=K_iX$. Pre-and-post multiplying (10) by diag$\left[P^T\quad S\quad I\quad I\quad I\right]$ and $\left[P\quad S\quad I\quad I\quad I\right]$, and pre-and-post multiplying (11) by diag $\left[P^T\quad S\quad I\quad I\quad I\quad I\right]$ and $\left[P\quad S\quad I\quad I\quad I\quad I\right]$ respectively, then applying the Schur complement, we can easily get (14) and (15). Thereafter, (16) is obtained.

According to (16), there must have $\alpha>0$, such that

$$\dot{V}(x(t))\le-\sum_{i=1}^{k}\sum_{j=1}^{k}\lambda_i\lambda_j x^T(t)(Q+K_i^T RK_j)x(t)$$
$$=-(x^T(t)Qx(t)+u^T(t)Ru(t))\le-\alpha\|x(t)\|^2<0, \tag{17}$$

which implies that (8) is quadratically stable. Integrating (17) from 0 to T produces

$$\int_0^T(x^T(t)Qx(t)+u^T(t)Ru(t))dt\le-V(x(T))+x^T(0)E^TPx(0)$$
$$+\frac{1}{1-d}\int_{-\tau_0}^{0}\varphi^T(s)U^{-1}\varphi(s)ds.$$

Because $V(x(t))\ge0$ and $\dot{V}(x(t))<0$, there has $\lim_{T\to\infty}V(x(T))=C$, a nonnegative constant. Therefore, we can get the following inequality,

$$\int_0^\infty(x^T(t)Qx(t)+u^T(t)Ru(t))dt\le x^T(0)E^TPx(0)$$
$$+\frac{1}{1-d}\int_{-\tau_0}^{0}\varphi^T(s)U^{-1}\varphi(s)ds\cdot \tag{18}$$

Then $J\le J_0$, where

$$J_0=x^T(0)E^TPx(0)+\frac{1}{1-d}\int_{-\tau_0}^{0}\varphi^T(s)U^{-1}\varphi(s)ds.$$

This completes the proof.

In the following, the design of guaranteed cost control law which minimizes the upper bound of J, is reduced to a convex optimization problem with LMI constrains.

To solve the problem by using the Malab LMIs toolbox directly, we need to deal with the non-strict inequality constrain (9). Without the loss of generality, we assume that $E=\begin{bmatrix}I_r & 0\\ 0 & 0\end{bmatrix}$, where I_r is a $r\times r$ identity matrix and $rankE=r<n$. Do decomposition to the nonsingular matrix P, and get $P=\begin{bmatrix}P_1 & P_2\\ P_3 & P_4\end{bmatrix}$. Then the condition (9) is equivalent to $P=\begin{bmatrix}P_1 & 0\\ P_3 & P_4\end{bmatrix}$, where P_1 is a $r\times r$ symmetric positive definite matrix and P_4 is nonsingular. Therefore

$$X = P^{-1} = \begin{bmatrix} P_1^{-1} & 0 \\ -P_4^{-1}P_3P_1^{-1} & P_4^{-1} \end{bmatrix} := \begin{bmatrix} X_1 & 0 \\ X_3 & X_4 \end{bmatrix},\ X_1 = P_1^{-1} > 0.$$

At last, assume that the initial state $x(0) = \left[\varphi_1^T(0) \quad \varphi_2^T(0)\right]^T$, where the dimensions of $\varphi_1(0)$ and $\varphi_2(0)$ are compatible with the decomposition of the matrix *E*.

Theorem 2: Consider the fuzzy descriptor delay system (2) and the cost function (5), if the following optimization problem

$$\min_{X, Y_i, U, M} \beta + Trace(M) \tag{19}$$

$$s.t. \quad \text{i) LMIs (10), (11) and (12)}$$

$$\text{ii)} \begin{bmatrix} -\beta & \varphi_1^T(0) \\ \varphi_1(0) & -X_1 \end{bmatrix} < 0$$

$$\text{iii)} \begin{bmatrix} -M & N^T \\ N & -U \end{bmatrix} < 0$$

has a solution β, X, U, M and Y_i, $i, j = 1, 2, \cdots, k$, where N satisfies $NN^T = \int_{-\tau_0}^{0} \varphi(s)\varphi^T(s)ds$, then the control law in form of (7) is an optimal guaranteed cost control law.

Proof: By theorem 1, the control law in form of (7) constructed in terms of any feasible solution β, X, U, M and Y_i is a guaranteed cost control law for the system (2). Recall the assumption that the initial state $x(0) = \left[\varphi_1^T(0) \quad \varphi_2^T(0)\right]^T$, where the dimensions of $\varphi_1(0)$ and $\varphi_2(0)$ are compatible with the decomposition of the matrix *E*. Then from ii) in theorem 2, we have

$$x^T(0)E^T Px(0) = \varphi_1^T(0)X_1^{-1}\varphi_1(0) < \beta.$$

Next, by recalling $Trace(AB) = Trace(BA)$, and according to iii) in theorem 2, there has

$$\begin{aligned} \int_{-\tau_0}^{0} \varphi^T(s)U^{-1}\varphi(s)ds &= \int_{-\tau_0}^{0} Trace\left[\varphi^T(s)U^{-1}\varphi(s)\right]ds \\ &= Trace\left[NN^T U^{-1}\right] \\ &= Trace\left[N^T U^{-1} N\right] \\ &< Trace(M). \end{aligned}$$

Thus, it follows that

$$x^T(0)E^T Px(0) + \frac{1}{1-d}\int_{-\tau_0}^{0} \varphi^T(s)U^{-1}\varphi(s)ds < \beta + \frac{Trace(M)}{1-d}.$$

Therefore, the guaranteed cost controller with respect to (19) is an optimal guaranteed cost controller. This completes the proof.

4 Example

To demonstrate the effectiveness of the design procedure of the guaranteed cost controller, consider the following fuzzy descriptor delay system. It is supposed that x_1 is measurable online.

If x_1 is P, then

$E\dot{x}(t) = A_1 x(t) + A_{11} x(t - \tau(t)) + B_1 u(t)$

If x_1 is N, then

$E\dot{x}(t) = A_2 x(t) + A_{12} x(t - \tau(t)) + B_2 u(t)$

Here the membership functions of P and N are given as follows:

$$\lambda_1(x_1) = 1 - \frac{1}{1+e^{-2x_1}},\ \lambda_2(x_1) = \frac{1}{1+e^{-2x_1}}.$$

And $E = \begin{bmatrix} 1 & 0 \\ 0 & 0 \end{bmatrix}$, $A_1 = \begin{bmatrix} 0 & 1 \\ 1 & 2 \end{bmatrix}$, $A_2 = \begin{bmatrix} 0 & 1 \\ 1 & -1 \end{bmatrix}$, $B_1 = \begin{bmatrix} 1 \\ 2 \end{bmatrix}$, $B_2 = \begin{bmatrix} 1 \\ 3 \end{bmatrix}$,

$A_{11} = \begin{bmatrix} 0 & 0 \\ 0.2 & 0.1 \end{bmatrix}$, $A_{12} = \begin{bmatrix} 0 & 0 \\ 0.1 & 0.4 \end{bmatrix}$, $Q = I$, $R = I$, $\varphi(t) = \begin{bmatrix} -2.5 \\ 1 \end{bmatrix}$, $t \in [-\tau_0, 0]$,

$\tau(t) = 2 + 0.5\sin t$, $d = 0.5$.

First, by theorem 1 and solving the LMIs in (9), (10), (11) and (12), we obtain the two state feedback gains as follows:

$$K_1 = [-9.0499\ \ -16.1076],\ K_2 = [-11.0921\ \ -20.0221],$$

and the upper bound of the guaranteed cost is $J_0 = 27.4826$. Then according to theorem 2 and by solving the convex optimization problem with respect to (19), the two state feedback gains of the optimal guaranteed cost controller are obtained as

$$K_1 = [-4.2285\ -7.4418],\ K_2 = [-4.8279\ -9.9780],$$

and the minimized upper bound of the guaranteed cost is $J_0^* = 3.1625$.

5 Conclusions

The guaranteed cost control problem for a class of T-S fuzzy descriptor systems with time-varying delay is considered. Based on LMIs, the sufficient conditions of the existing of the guaranteed cost controller are given. And an optimal guaranteed cost controller which can make the system quadratically stable, as well as guarantee the cost has a minimized upper bound, is obtained by solving a convex optimization problem. An illustrative example is given to show the effectiveness of the proposed method.

Acknowledgements

This work was supported by the National Science Foundation of China (60534010, 60572070, 60521003, 60325311) and the Program for Cheung Kong Scholars and Innovative Research Team in Universities.

References

1. T. Takagi, M. Sugeno (1985) Fuzzy identification of systems and its application to modeling and control. IEEE Transaction on Systems, Man and Cybernetics 15: 116-132
2. K. Tanaka, M. Sugeno (1992) Stability analysis and design of fuzzy control systems. Fuzzy Sets and Systems 45: 135-156
3. Bing Chen, Xiaoping Liu (2005) Fuzzy Guaranteed Cost Control for Nonlinear Systems with Time-varying Delay. IEEE Transactions on Fuzzy Systems 13: 238-249
4. T. Taniguchi, K. Tanaka, K. Yamafuji, H.O. Wang (1999) Fuzzy descriptor systems: Stability analysis and design via LMIs. Proceedings of the American Control Conference, San Diego, California, 1827-1831
5. T. Taniguchi, K. Tanaka, H .O. Wang (2000) Fuzzy descriptor systems and nonlinear model following control. IEEE Transactions on Fuzzy Systems 8: 442-452
6. J. Yoneyama, I. Akira (1999) H∞control for Takagi-Sugeno fuzzy descriptor systems. Proceedings of IEEE Conference Systems, Man and Cybernetics, Tokyo, Japan 3: 28-33
7. Y. Wang, Q. L. Zhang, X. D. Liu (2002) Robustness design of uncertain discrete-time fuzzy descriptor systems with guaranteed admissibility. Proceedings of the American Control Conference, Anchorage, AK 1699-1704
8. Yan Wang, Zeng-Qi Sun, Fu-Chun Sun (2004) Robust Fuzzy Control of a class of Nonlinear Descriptor Systems with Time-Varying Delay. International Journal of Control, Automation, and Systems 2: 76-82
9. Y.S. Lee, Y.S. Moon, W.H. Kwon (2001) Delay-dependent guaranteed cost control for uncertain state-delay systems. Proceedings of American Control Conference, Arlington, VA 3376-3381
10. L. Xie, C. E. de Souza (1992) Robust H∞ control for linear systems with norm-bounded time-varying uncertainty. IEEE Transactions on Automatic Control 37: 1188–1191
11. L. Xie (1996) Output H∞ control of systems with parameter uncertainty. Int. J. Control 63: 741-750
12. Chong Lin, Qing-Guo Wang, Tong Heng Lee (2006) Stability and Stabilization of a class of Fuzzy Time-Delay Descriptor Systems. IEEE Transactions on Fuzzy Systems 14: 542-551

The Research and Simulation on the Walking Trajectory of the Hexapod Walking Bio-robot

Shangbin Run[1], Baoling Han[1], Qingsheng Luo[2], and Xiaochuan Zhao[2]

[1] School of Mechanical and Vehicular Engineering, Beijing Institute of Technology
[2] School of Mechatronic Engineering, Beijing Institute of Technology, Beijing 100081

Abstract. As the matter fact that there are many joints in the hexapod walking Bio-robot and the calculation of its walking trajectory, as well as the problem of the acceleration leap caused by adopting elementary function as the robot's feet trajectory. This paper puts forward using cubic spline curves as the robots feet trajectory and designing the robot's walking gait. Then the authors conducted kinematical simulation and analyses on the model of the bionic hexapod walking robot which have done with the use of Solidworks and MSC.ADAMS. Through the simulation experiments, we validated the applicability of designed triangle gait and the feasibility of using cubic spline curve as the feet trajectory. This paper shows principle, method and course on the trajectory simulation of hexapod walking Bio-robot in detail. The way to find a converse solution under the condition of ADAMS is found, which simplifies the theory calculation and improves design efficiency.

Keywords: hexapod walking Bio-robot, gait planning, kinematics simulation, virtual prototype, ADAMS.

1 Foreword

The bionic hexapod walking robot is a kind of special type robot that has many connecting bars and many DOF, it's kinematics and kinetic characteristics is very complicated. In order to improving the research and development level of these kind of robot, to construct a set of emulation system which can satisfy it's required kinematics and kinetic analysis is very necessary. As is now well known, ADAMS is a kind of real excellent software in the area of virtual prototype, it can construct emulation virtual prototype according to the actual motion of research subjects, and can analyse its operating characteristic and can improvement and optimize conveniently before the physical prototype is constructed. Because the 3d modeling function of ADAMS is relatively weakness, this article carries a movement emulation on the bionic hexapod walking robot in the way of adopting the combination of CAD software Solidworks and ADAMS, improved the research efficiency, saved research time and the development cost, realized a high quality , High Velocity , high efficiency and LOW-COST integrated design.

2 The Brief Introduction About the Integral Structure of the Bionic Hexapod Walking Robot

Figure 1 and figure 2 shows the integral structure molding and leg movement diagram of the bionic hexapod walking robot. Each legs of this kind of robot adopts special type

B.-Y. Cao (Ed.): Fuzzy Information and Engineering (ICFIE), ASC 40, pp. 760–767, 2007.
springerlink.com

structure which has three revolution joints. The root joint design to side-sway, hunting range is[-45°,45°], root length is 56mm; hip joint design to pitching, elevation limits is [-45°, 135°], femora length is 217.04mm; Knee joint design to bent crane, bent crane scope is [0°,135°], tibiae length is 244.04mm.Each joint is drive by the independent motor, the advantage lies in the structure tightly packed, the leg end can reach the movement space very flexible, and the movement is vivid. Owing to the leg joint of this robot is floating, it has a very strong pose recovery ability while it walks even if appears destabilization phenomena.

3 The Gait Theory Analyses on the Bionic Hexapod Walking Robot

The triangle gait a representative gait to realize walking. Its core thought is to divide the six legs of the robot into two groups(the front leg and , the hind leg on one side of the body and the middle leg on another side of the body) ,one group supports the body and pushes the robot goes forward(names support phase) ,the other group wiggles for the preparations of next step support (names wiggle phase) ,the whole movement course of the robot is the support phase and the wiggle phase mutually alternately, circulating process.

Fig. 1. The overall structure molding on the bionic hexapod walking robot

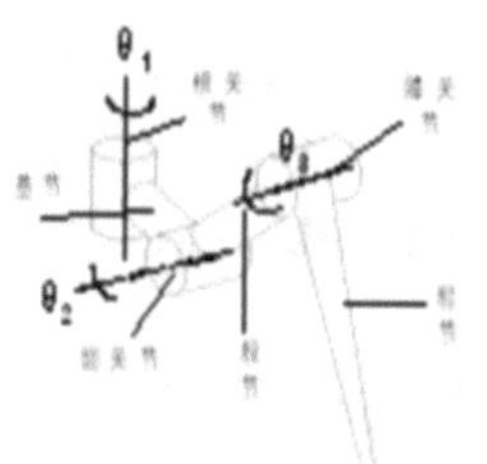

Fig. 2. The movement diagram of the bionic hexapod walking robot's leg

On the robot research area, the sequence set about the change of the support phase and the wiggle phase with time is named as gait. To the walking with uniform speed, its leg phase acts like periodic change rule. As a result of the gait is periodic change, we call its cycle gait. In a period T, the time of support phase is t ,the factor β of this leg is described as the follow formula[1]:

$$\beta = t/\mathrm{T} \tag{1}$$

In a periods of gait, the distance which walking robot's barycenter moves forward is called pace length s and the migration distance which the foot that situates support phase duration moves relative to body is called foot distance R. The relation between them described as follows:

$$R = \mathrm{s} \cdot \beta \tag{2}$$

For the convenience of description, we can give a serial number for each of the bionic hexapod walking robot's leg and give a grouping for them (as figure 3 shows) , among them: Number 1 , 3 , 5 legs for A group, number 2 ,4, 6 legs for B group . For the bionic hexapod walking robot, to design a simple beeline walking gait includes three steps as follows[2][3]:

(1) As is showed in figure 3a, robot hexa leg at the same time touchdown, gesture adjust;

(2) As is showed in figure 3b, each legs of B group supports and pushes the robot's body moving forward for half a step when each legs of A group lifts up.

(3) As is showed in figure 3c, each legs of B group lifts up when each legs of A group reaches to the ground, the transition between support leg and swinging leg finished, namely a movement process of period Completed, then repeat an above steps.

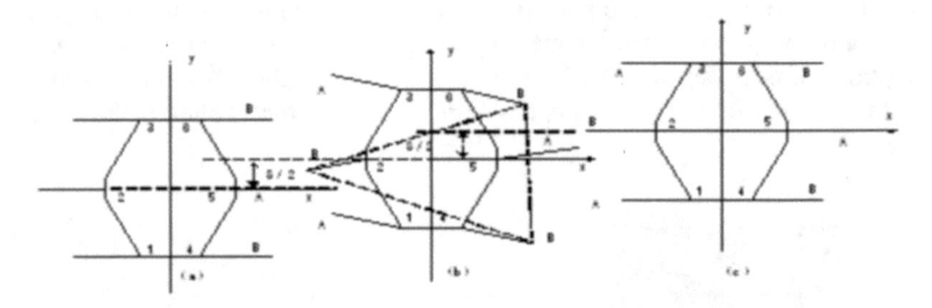

Fig. 3. Sketch map of the bionic hexapod walking robot trigonometry gait

4 The Trajectory Choices of Bionic Hexapod Walking Robot

After a lot of researches, the writer has found that the trajectory choices of robot's feet-tips are essential to their kinematics when the walking planning of robots is made. The consistency, stability, beauty together with the necessary power of driving torque in robot's walking process is contained by it. As for the Bionic Hexapod Robots, preferable feet-tip trajectory should be provided with better rising and landing, speeding and accelerating characteristics. Generally speaking, people usually use Elementary Functions to describe robot's termination trajectory. For example, monadic function, sine curve function, cycloid , parabola and so on. Under this condition, the description of robot's terminational trajectory with Elementary Functions will be inevitable resulting in the unexpected changes of acceleration which will further affect robot's stability in the walking process and result in overload of the driving machine drive motor. Some researchers have ever taken the polynomial interpolation method to solve the problem which can satisfy one order、two order derivative and continuity of the trajectory curve[4][5]. But because of too many sampling, this method can arouse an over highly order of polynomial interpolation and finally result in vibration[6].

Basing on the former factors, the present writer tries to take cubic spine curve as the Bionic Hexapod Walking Robot's feet-tip trajectory. In this way, it is not only content with the one order, two order derivative, continuity, but the trajectory is smooth with a lower order.

5 The Moving Trajectory Simulation of the Bionic Hexapod Walking Robot

On the base of analyzing the moving characteristic and simulation characteristic of the Bionic Hexapod Walking Robot seriously, the author draws up a flow chart of kinematics simulation which is showed as figure 4,the concrete job was done as the flow chart .

5.1 The Creation of Bionic Hexapod Walking Robot's Feet-Tip Trajectory Curve

To make the Bionic Hexapod Walking Robot walk as the gaits in chart 5, we choose the factor β =0.5, setting the walking period T =2s and the length of step S =200mm and design robot's feet-tip trajectory as the curve shown in chart 5. From chart 5, we can see that at this time the trajectory can be divided into five parts, in which parts AB to BC to CD to DE are a sway trajectory, whilst part AE is supporting phase trajectory. Every part is the cubic spline curve when it waves and continuous derivative at the joining points. In this experiment, we establish the robot's speed at the starting point A and the landing point E as zero. The data of starting point, landing point and middle point are as shown in chart 1. Basing on the data from chart 1, the writer uses spline function csape provided by MATLAB toolbox to insert the values of sampling pieces three times and receive the changing curves x(t),y(t) and y(x) as shown as in chart 6. From the received curves, we can see that they are continuously derivative, and the speed of starting point and landing point is zero. Next step, we can make curves x(t) and y(t) discretization and preserve as *.txt file for defining the feet-tip driver when ADAMS stimulates it. With the same method, we can insert values to the supporting phase, which will be not repeated here.

5.2 The Virtual Prototype Model Construction of Bionic Hexapod Walking Robot[7]

With the Solidworks software we set up the hexapod walking bio-robot's three dimension model. In order to avoid the complex simulation model, the authors simplified the model, leaving some key components, which can not influence the effect of simulation. After modeling, we saved the three dimension model as *.x_t document. Then we added restriction to three dimension model in ADMS. What we needed point out is as follows: the revolute should be added to the root joint, coax joint and knee joint respectively. Because of the robots symmetry structure, triangle gait and sameness between supporting legs and swinging legs, we only need to define swing phrase, support phrase and body these three driving parts so that the robot can move according to the expected trace.

Before we define the feet tip trace driving of the hexapod walking bio-robot, we must input the *.x_t document realized in the modeling process, so that cubic spline curve can be created. The specific method is as follows: In ADAMS

File→Import…→File Type→Test Data(*.*)→Creat Spline→File To Read→Choose the created document（ *.txt）→OK, then the cubic spline curves names spline_1can be realized automatically by the system。We can get other cubic spline curves in the same way.

The driving equations based what we mentioned above is as follows:

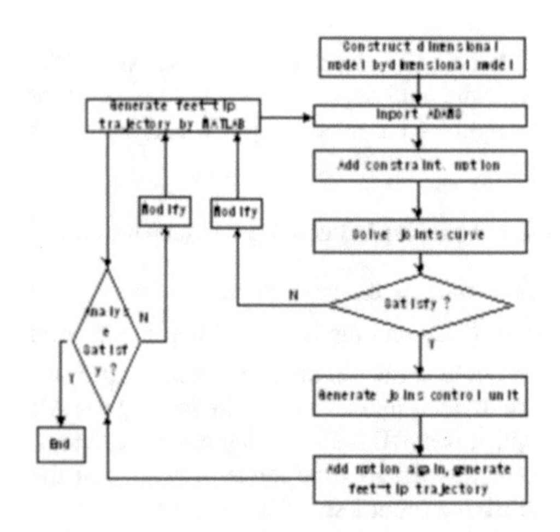

Fig. 4. The flow chart of kinematics simulation

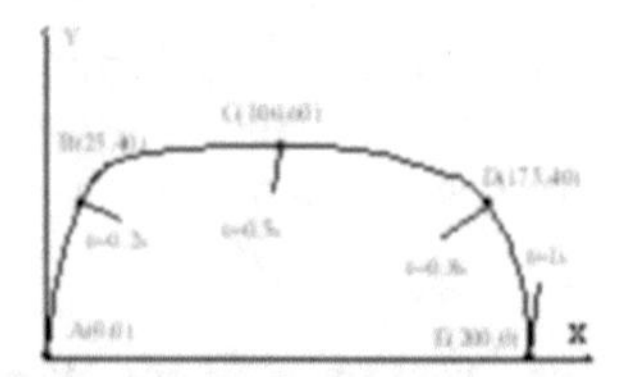

Fig. 5. The feet-tip trajectory of the bionic hexapod walking robot

Table 1. The data table satisfying predetermined trajectory

i	0	1	2	3	4
t_i	0	0.2	0.5	0.8	1
$X_i(t)$	0	25	100	175	200
$Y_i(t)$	0	40	60	40	0
V(x)	0				0
V(y)	0				0

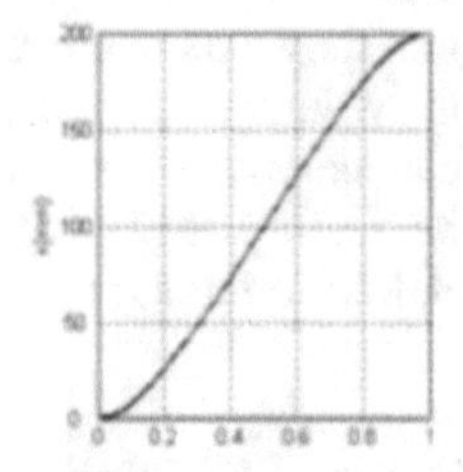

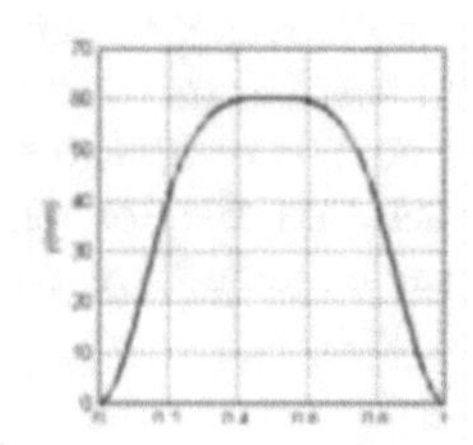

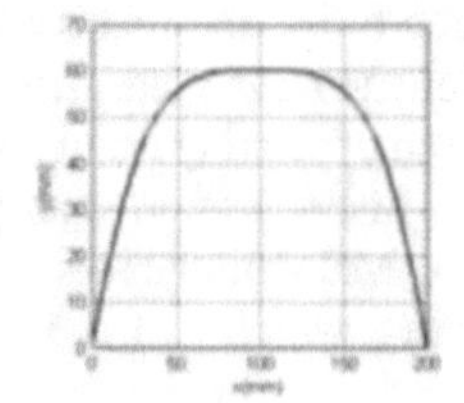

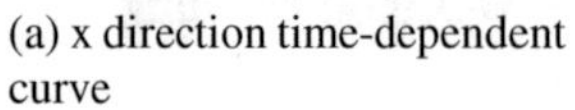

(a) x direction time-dependent curve (b) y direction time-dependent curve (c) x direction curve relates y direction

Fig. 6. The feet-tip curve after interpolation

x direction:
if(time:0,0,if(time-1:AKISPL(time,0,.jqr2.spline_2,0),200,if(time-2:200,200,if(time-3:AKISPL(time-2,0,.jqr2.spline_2,0)+200,400,if(time-4:400,400,400+AKISPL(time-4,0,.jqr2.spline_2,0))))))

y direction:
if(time:0,0,if(time-1:AKISPL(time,0,.jqr2.spline_1,0),0,if(time-2:0,0,if(time-3:AKISPL(time-2,0,.jqr2.spline_1,0),0,if(time-4:0,0,AKISPL(time-4,0,.jqr2.spline_1,0))))))

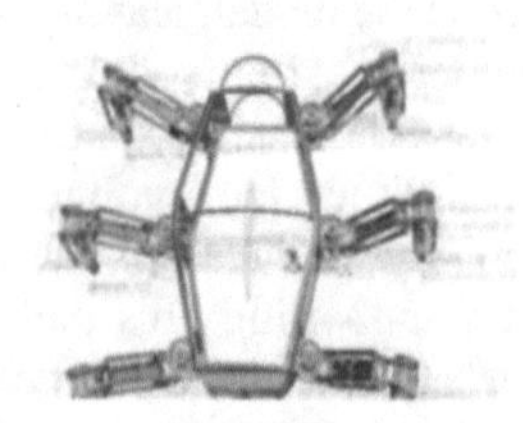

Fig. 7. The virtual prototype after adding constraint

z direction: 0

The driving program based what we mentioned above is as follows:

X direction:

if(time-1:0,0,if(time-2:AKISPL(time-1,0,.jqr2.spline_2,0),200,if(time-3:200,200,if(time-4:AKISPL(time-3,0,.jqr2.spline_2,0)+200,400,if(time-5:400,400,400+AKISPL(time-5,0,.jqr2.spline_2,0))))))

Y direction:

if(time-1:0,0,if(time-2:AKISPL(time-1,0,.jqr2.spline_1,0),0,if(time-3:0,0,if(time-4:AKISPL(time-3,0,.jqr2.spline_1,0),0,if(time-5:0,0,AKISPL(time-5,0,.jqr2.spline_1,0))))))

Z direction: 0

The corresponding robot body driving program is :

X direction: -100 * if(time:0,0,time)

y、Z direction and x、y、z direction's circumgyrate are all 0.

We can set the simulation time as 4s, the walking length is 1000 steps in each simulation.

After each simulation, open the disposal module, draw the joint 1's angle, angle speed, and angle acceleration, which can be seen from Fig.8,Fig9 and Fig.10. Because of the robots symmetry, the curves are same as joint 1's except the phase order, so we don't picture them here.

From the curves, we can see that, the changes of robot's each angle are smooth as well as the corresponding angle velocity. There is no abrupt change in the angle accelerations. In which, the angle acceleration of root joint and knee joint's peak is less than 500 rad/s2., and the peak of coax joint is less than 1500 rad/s2. Through simulation, it validates the applicability of designed triangle gait and the feasibility of using cubic spline curve as the feet trajectory.

5.3 Bionic Hexapod Walking Robot's Kinematic Inverse Solution

The so-called Kinematic inverse Solution is from Kinematic situation of the ending follower part to fix on the kinetic law of the driving part. The traditional method of Kinematic inverse Solution is setting up a D-H reference frame, and then using the changing of reference frame to get the result. The method is not only with cockamamie processes, complicated calculation, but also with a possibility to get more than one outcome. While with the MSC. ADAMS's Kinetic stimulation function, it is very convenient to get the answers of Kinematic inverse Solution. Every joint's moving curve can be obtained by defining the corner measurement of Robot's every joint. The writer has gotten the systematic kinematics input by defining on the turned angles in themselves reference frames of robot's root joint、hip joint、knee joint. The results of the turned angles by robot's root joint、hip joint、knee joint are betrayed as the corresponding curves in the following picture 8, 9 and 10. when we get the related results, we can preserve them as the sampling curves: genguanjie spline, kuanguanjie spline, xiguanjie spline. Then at the robot's root joint、hip joint、knee joint set up three moving constraint about the sampling curves: genguanjie _motion、 kuanguanjie

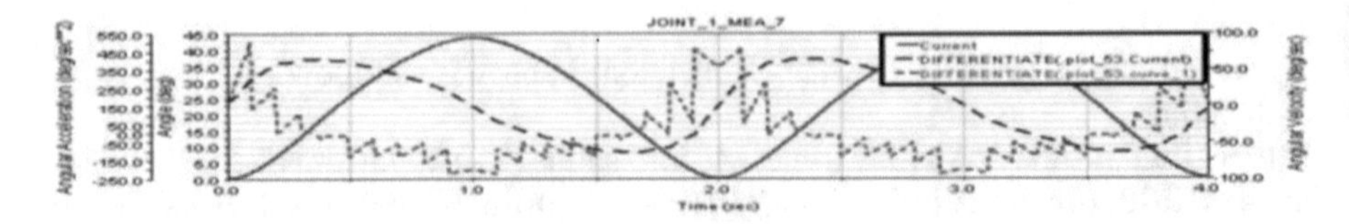

Fig. 8. The angle、angular velocity、angular acceleration time-dependent curve on root joint

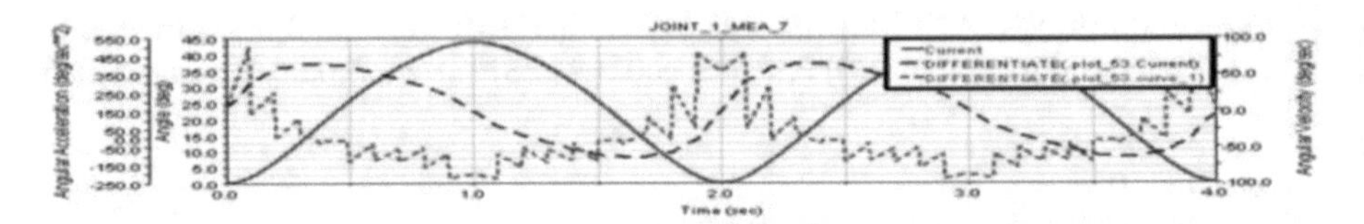

Fig. 9. The angle、angular velocity、angular acceleration time-dependent curve on hip joint

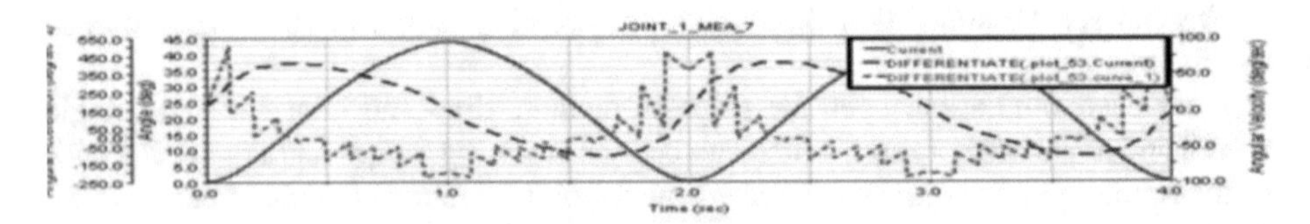

Fig. 10. The angle、angular velocity、angular acceleration time-dependent curve on knee joint

_motion and xiguanjie_motion. The movement function will be written out by using cubic spline curve. Under this condition, the respective movement functions of robot's root joint, hip joint, knee joint are:

AKISPL(time,0,genguanjie_spline, 0)*pi/180
AKISPL(time,0,kuanguanjie_spline, 0)*pi/180
AKISPL(time,0,xiguanjie_spline, 0)*pi/180

It should be attention that the unit of turning angle in movement functions is radian, so we should change the measured results into the radian units. Then, we can make the driving restrictions on robot's feet-tip points noneffective to realize Kinematic substitution. That is, taking the driving restriction on every joint substitutes for those which throwing on the ending parts[8]. At the same defining the contact between robot's feet-tip point and ground (the ground is substituted by a thin board in the condition of simulation). To prevent the phenomenon of skidding, it can be assumed that the frictional modulus is big enough. The simulated parameter can be further defined: setting the simulated time as 4 seconds, exporting gaits as 1000 gaits to simulate robot's movement and get the corresponding simulating movement flash as shown in figure 11. It can be seen from the simulating flash that Bionic Hexapod Walking Robot moves smoothly without the phenomena of tumble, slippage and interference. It can be further found that robot's feet-tip trajectory are completely tally with what the writer has designed (the curve in picture 5c) which testified the feasibility and validity of the method using simulation to solve the Robot's inverse Solution.

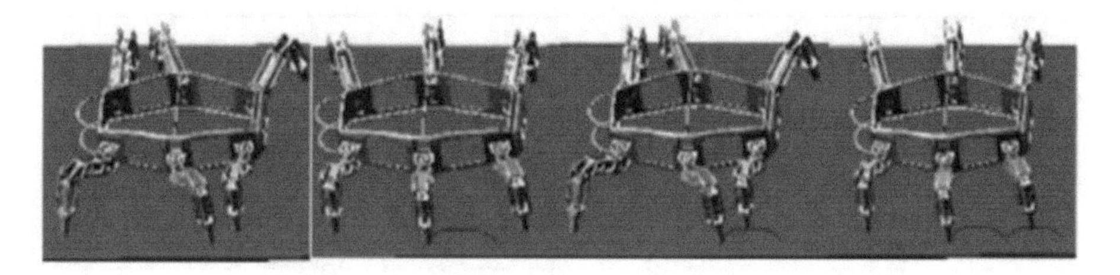

Fig. 11. Simulation animation access to object

6 End

This article conducted kinematical simulation and analyses on the model of the bionic hexapod walking robot which have done with the use of Solidworks and MSC.ADAMS. Through the simulation experiments, we validated the applicability of designed triangle gait and the feasibility of using cubic spline curve as the feet trajectory. In the course of simulation ,the author gets the angular curve on each joins through gait planning and kinematics simulation, avoided complicated calculation using conventional method to the bionic hexapod walking robot kinematics inverse solution , improved design efficiency and research and development level, which has many theoretical significance and utility value in the research of robotics area.

References

1. ZhengBang Gong, QingTe Wang,ZhenHua Chen,JinWu Qian(1995)Mechanial Design of Robot.Electronic Industry Press.
2. Lee Bo Hee, Lee In Ku(2001) The implementation of the gaitand body structure for hexapod robot .Industrial Eltronic.ISIS 2001,IEEE International Symposium on, 1959-1964.
3. Chen X D,Watanabe K and Izumi K(1999)Study on the con2trol algorithm of the transnational crawl for a quadrupedbot,systems,man and cybernetics.IEEE SMC,6:953-958
4. MuoNanWang,LiQuanWang,MengQinXin(2003)Kinematicsofbionicscrabwalkingprocess. HarbinEngineeringUniversity journal, 24(2):180-183.
5. JiaPin Chen(2001)Dynamic Quadruped Walking Control and Assisted by Learning.ShangHai JiaoTong UninVersityDoctor paper.
6. WeiPing Jia(2005)The Trajectory Project Research Of Stanford Manipulator's Moving Route.Mechanical Engineer,6:38-40.
7. JianRong Zheng(2001)MSC.ADAMS--The accidence and advance on virtual prototype.Mechanical Industry Press.
8. YanHua Xu,YanBin Ni,JiangPing Mei,Ke Li(2003)The kinematics simulation of a 2-DOF Parallel Robot .2003 MSC.Software chinese user collected papers.

Research of Group Decision Consensus Degree Based on Extended Intuitionistic Fuzzy Set

Tao li[1] and Xiao ciyuan[2]

[1] School of Postgraduate, Southwest Petroleum University, 610500
[2] Science Department, Southwest Petroleum University, 610500
taolilming@126.com, xcytlm2003@163.com

Abstract. The paper first put forward a definition of extended Intuitionistic Fuzzy Set,and then give the conception of the distance between two extended intuitionistic Fuzzy Set .By intuitionistic Fuzzy Set distance,the degree of agreement of each expert to the others is aggregated .Finally ,a numerical example is given to apply this model.

Keywords: extended intuitionistic fuzzy sets; intuitionistic fuzzy distance;consensus degree;average consensus degree;relative consensus degree.

1 Introduction

Group decision implies that experts play an evaluation on a series of projects according to some certain criterions ,and then choose a best one .however,there are always difference and agreement among experts.So how to solve the agreement and difference among experts and how to turn personal views into group's ones have become an importmant problem.In 1965,Zadeh put forward the theory of fuzzy which is widely applied to the research of group decision[1][2].In 1986,Atanassor extended the conception fuzzy set by putting forward the definition of intuitionistic fuzzy set(ie IFS)[3][4]. Based on the former ,paper [5]set up a method which is founded on the distance of intuitionistic fuzzy sets.This paper aims at futher extending the definition of IFS,and then renewing the distance between them,finally doing some research on group decision.

2 The Definition of Extended IFS

Provided X is a not null classic set, $A=\{<x,u_1(x),u_2(x),\ldots,u_n(x)>|x\in X\}$ is defined as extended IFS, n represent the dimsension of extended IFS。$U_i:X \rightarrow [0,1]$ and

Author introduction: Tao Li, applied mathmatics postgraduate in Southwest Petroleum University,china; research direction: Fuzzy Mathmatics. Xiao Ci Yuan, professor in Science Department of Southwest Petroleum University,china; research direction: Fuzzy Mathmatics.

B.-Y. Cao (Ed.): Fuzzy Information and Engineering (ICFIE), ASC 40, pp. 768–772, 2007.
springerlink.com

$\sum_{i=1}^{n} u_i(x)$ =1,$u_i(x)$indicates each intuitioistic index which belongs to set A。IFS[x] refers to the collective of intuitionistic sets.

It is obvious that Fuzzy set and traditional IFS can be considered as special extended IFS.

Extended IFS is more appropriate than the other two in describing each different state of people's decision 。For example, an evaluation on a school's study circumstance can be divided into five degrees:best,better,middle,common,bad.we can use $A=\{<x,u_1(x),u_2(x),\ldots,u_n(x)>|x\in X\}$ to describe it, the description is more fit than others.

3 The Definition of Distance of Extended IFS

Provided X is a not null classic set: $X=\{x_1,x_2,\ldots,x_m\}$, $B=\{<x,u_{B1}(x),u_{B2}(x),\ldots,u_{Bn}(x)>|x\in X\}$,the distance between A and B is defined as follow:

$$d_{IFS}(A,B)=\sqrt{\sum_{i=1}^{m}\sum_{j=1}^{n}(u_{Aj}(x_i)-u_{Bj}(x_i))^2}$$

The distance of extended IFS has the following character

(1) $d_{IFS}(A,B)\geq 0$

(2) $d_{IFS}(A,B)=d_{IFS}(B,A)$

(3) $d_{IFS}(A,A)=0$

(4) $d_{IFS}(A,C)\leq d_{IFS}(A,B)+d_{IFS}(B,C)$, $A,B,C\in IFS[X]$

(5)If $d_{IFS}(A,B)\leq d_{IFS}(A,C)$,it indicates that B is more closer to A than B

Method of aggregation of Opinion in Group Decision Making Based on extended Intuitionistic Fuzzy Set

Provided that expert set $E=\{e_1,e_2,\ldots,e_m\}$,project set $A=\{a_1,a_2,\ldots,a_r\}$,evaluation factor set $C=\{c_1,c_2,\ldots,c_s\}$.research on the agreement of group decision has the following steps:

(1) Expert k make an extended intuitionistic evaluation $P_{ij}^{k}=(U_{ij1}^{k}, U_{ij2}^{k},\ldots, U_{ijn}^{k})$ on project i according to the fact ,the expert's experience and preference,therefore we get an intuitionistic fuzzy evaluation matrix which is made by expert k on project set A

$$p^k = \begin{pmatrix} (u_{111}^k, u_{112}^k, \ldots, u_{11n}^k), (u_{121}^k, u_{122}^k, \ldots, u_{12n}^k), \ldots, (u_{1s1}^k, u_{1s2}^k, \ldots, u_{1sn}^k) \\ (u_{211}^k, u_{212}^k, \ldots, u_{21n}^k), (u_{221}^k, u_{222}^k, \ldots, u_{22n}^k), \ldots, (u_{2s1}^k, u_{2s2}^k, \ldots, u_{2sn}^k) \\ \cdots \\ (u_{r11}^k, u_{r12}^k, \ldots, u_{r1n}^k), (u_{r21}^k, u_{r22}^k, \ldots, u_{r2n}^k), \ldots, (u_{rs1}^k, u_{rs2}^k, \ldots, u_{rsn}^k) \end{pmatrix}$$

i=1, 2, ..., r;j=1,2, ...s;k=1,2, ...m

(2) Calculate the consensus degree of group decision between every two experts on project i according to the distance of extended IFS.

If evaluation factor set(C_1,C_2,C_3 ......, C_s) has weight w=(w_1,w_2......, w_s). $\sum_{j=1}^{n} w_s = 1$ .the distance between expert k and t on project I is defined as their consensus degree on project i

$$e_i^{kt} = \sqrt{\sum_{j=1}^{s} \omega_j ((u_{ij1}^k - u_{ij1}^t)^2 + (u_{ij2}^k - u_{ij2}^t)^2 + \cdots + (u_{ijn}^k - u_{ijn}^t)^2)}$$

(3) Construct a Consenus Degree Matrix AM^i of Project i

$$AM^i = \begin{pmatrix} o, e_i^{12}, \cdots e_i^{1j}, \cdots, e_i^{1m} \\ \cdots, \cdots, \cdots, \cdots, \cdots, \cdots \\ e_i^{k1}, e_i^{k2}, \cdots, e_i^{kj}, \cdots e_i^{km} \\ \cdots, \cdots, \cdots, \cdots, \cdots, \cdots \\ e_i^{m1}, e_i^{m2}, \cdots, e_i^{mj}, \cdots o \end{pmatrix}$$

(4) Calculate expert k's average and relative consenus degree of project i。Expert k's(k=1,2,3,…,m)average consenus degree of project i: $A(e_i^k) = \sum_{\substack{j=1 \\ j \neq k}}^{m} e_i^{kj}$ Expert k's(k=1,2,3,…,m)relative consenus degree of project I :$RAD(e_i^k) = \frac{A(e_i^k)}{\sum_{k=1}^{m} A(e_i^k)}$

(5) Acquire every expert's weight through AHP, and then obtain all expert's synthetic consenus degree of project i.

Provided that $\{e_1, e_2, \ldots, e_m\}$'s weight set is $(r_1, r_2, \ldots, r_m)$ and $\sum_{i=1}^{m} r_i = 1$, We can obtain all expert's synthetic consenus degree $e_i = \sum_{k=1}^{m} r_m RAD(e_i^k)$, i=1,2,…r

Make a sequence according to e_i.there is more consenus on the project with smallest synthetic consenus degree which indicates less difference among experts.

4 Example Analysis

Provided that expert set $E=\{e_1,e_2,e_3\}$,project $A=\{a_1,a_2\}$,evaluationary factor set $c=\{c_1,c_2,c_3\}$, the weight of $\{c_1, c_2, c_3\}$ is {0.3,0.5,0.2},the weight of $\{e_1,e_2,e_3\}$ is {0.42,0.25,0.33},three experts'extended intuitionistic fuzzy set matrixs on project a_1,a_2 are as follows:

$$p^1 = \begin{pmatrix} (0.1,0.7,0.2,0.1),(0.5,0.2,0.2,0.1),(0.3,0.3,0.2,0.2) \\ (0.3,0.4,0.1,0.2),(0.4,0.2,0.3,0.1),(0.5,0.3,0.1,0.1) \end{pmatrix}$$

$$p^2 = \begin{pmatrix} (0.3,0.4,0.2,0.1),(0.6,0.0,0.2,0.2),(0.3,0.4,0.2,0.1) \\ (0.4,0.2,0.3,0.1),(0.3,0.5,0.1,0.1),(0.7,0.1,0.2,0.0) \end{pmatrix}$$

$$p^3 = \begin{pmatrix} (0.2,0.4,0.1,0.3),(0.4,0.3,0.1,0.2),(0.3,0.3,0.3,0.1) \\ (0.3,0.5,0.2,0.0),(0.4,0.4,0.2,0.0),(0.4,0.3,0.2,0.1) \end{pmatrix}$$

we can calculate expert k、t's distance of extended intuitionistic fuzzy set on project a_1,a_2:e_1^{kt},e_2^{kt}.

$e_1^{12}=0.270, e_1^{13}=0.263, e_1^{23}=0.504; e_2^{12}=0.346, e_2^{13}=0.228, e_2^{23}=0.290$

So we get three experts's consensus degree matrix on project a_1,a_2

$$AM^1 = \begin{pmatrix} 0.000,0.270,0.263 \\ 0.270,0.000,0.504 \\ 0.263,0.504,0.000 \end{pmatrix} \qquad AM^2 = \begin{pmatrix} 0.000,0.346,0.228 \\ 0.346,0.000,0.290 \\ 0.228,0.290,0.000 \end{pmatrix}$$

Each expert's average consensus degree on project a_1:$A(e_1^1)=0.533$,$A(e_1^2)=0.774$,$A(e_1^3)=0.767$

Each expert's relative consensus degree on project a_1:

$RAD(e_1^1)=0.257$, $RAD(e_1^2)=0.373$,$RAD(e_1^3)=0.370$

All experts'synthetic consensus degree on project a_1:$e_1=0.42\times0.257+0.25\times0.373+0.33\times0.370=0.323$.

In the same way ,we can conclude $e_2=0.42\times0.332+0.25\times0.368+0.33\times0.300=0.33$.

$e_1 < e_2$,so we can conclude the experts'consensus degree on project a_1 is stronger than a_2.

5 Conclusions

How to describe fuzzy things appropriately is always an issue among us. The definition of extended IFS in this paper not only comes from fact but also has close relationship with fuzzy set and intuitionistic fuzzy set, therefore doing further reseach on it will have big significance.

References

[1] Zadeh lA.Fuzzy Sets[M].Inform and Control.1965,8:338-356

[2] Tanino T. Fuzzy preference orderings in group decision-making[J].Fuzzy Sets and Systems.1984.12 117—131

[3] Atanassov K. Intuitionistic Fuzzy Sets: Theory and Applications[M].Physica-Verlag. Heidelberg.1999

[4] Atanassov K. Intuitionistic fuzzy sets[J].Fuzzy Sets and Systems 20 (1986) 87—96

[5] Tan chunqiao,Zhang qiang.Aggegation of Opinion in group Decision Making on Intuitioistic Fuzzy Set[J]. Mathemmatics In Practice and Theory.2006.36(2):119-124

Part VII

Rough Sets and Its Application

A Novel Approach to Roughness Measure in Fuzzy Rough Sets

Xiao-yan Zhang[1] and Wei-hua Xu[2]

[1] Department of Mathematics and Information sciences, Guangdong Ocean University, Zhanjiang 524088, P.R. China
datongzhangxiaoyan@126.com

[2] Institute of Information and System Sciences, Xi'an Jiaotong University, Xi'an 710049, P.R. China
datongxuweihua@126.com

Abstract. A hot topic of rough set theory is the combination of fuzzy set theory and this one. Many researchers do the work by introducing the concept of level fuzzy sets. In this paper, we will study roughness measure in fuzzy rough set in this way, too. The lower and the upper approximation based on λ-level fuzzy set will be proposed, and their properties be discussed. Moreover we introduced the new method to the roughness measure in fuzzy rough sets according to these concepts.

Keywords: Fuzzy set, fuzzy rough set, λ-level fuzzy set, the lower and the upper approximation, roughness measure.

1 Introduction

Rough set theory [1,2], a new mathematical approach to deal with inexact, uncertain or vague knowledge, has recently received wide attention on the research areas in both of the real-life applications and the theory itself. It offers a mathematical model and tool for discovering hidden patterns in data, recognizing partial or total dependencies in data, removing redundant data, and many others [3-7]. Theory of fuzzy sets initiated by Zadeh [8] also provides useful methods of describing and modeling of vagueness in ill-defined environment.

It is a natural that Doubois and Prade [9] combined fuzzy sets and rough sets. Attempts to combine these two theories lead to some new notions [3,10,11], and some progresses were achieved [11-15]. The combination involves many types of approximations. The construction of fuzzy rough sets can be classified into two approaches, namely, functional approach and set approach [11]. At the same time, some researchers considered the combination of fuzzy sets and rough sets by introducing the concept of level fuzzy sets. It has been argued that benefits do exist in the use of level fuzzy sets over level sets [16-20]. The present study examines fundamental issues of rough sets by making use of level fuzzy sets. The lower and the upper approximations based on λ-level fuzzy set proposed, and some their properties are considered in fuzzy rough sets. Furthermore, we give the new method for the roughness measure in fuzzy rough set, and confirm our method by giving an example.

B.-Y. Cao (Ed.): Fuzzy Information and Engineering (ICFIE), ASC 40, pp. 775–780, 2007.
springerlink.com

2 Preliminaries

The following definitions and preliminaries are required in the sequel of our work. Next, we review them in brief.

The notion of fuzzy sets provides a convenient tool for representing vague concepts by allowing partial membership. In fuzzy systems, a fuzzy set can be defined using standard set operators.

Let U be a finite and non-empty set called universe. A fuzzy set of is defined by a membership function

$$A : U \to [0, 1].$$

The membership value may be interpreted in terms of the membership degree. And let $\mathcal{F}(U)$ denote the set of all fuzzy sets, i.e., the set of all functions from U to $[0, 1]$.

Definition 2.1. The λ-level set or λ-cut of a fuzzy set A in U, denoted by A_λ, comprise all elements of U whose degree of membership in A are all greater than or equal to λ, where $0 \leq \lambda \leq 1$. In other words,

$$A_\lambda = \{x \in U : A(x) \geq \lambda\}$$

is a non-fuzzy set, and be called the λ-level set or λ-cut. Moreover the set $\{x \in U : A(x) > 0\}$ is defined the supports of fuzzy set A, and denoted by $suppA$.

From above description, a crisp set can be regarded as a degenerated fuzzy set in which the membership is restricted to the extreme points {0,1} of [0,1]. Similarly, there are many definitions, i.e., complement, intersection, and union for fuzzy set. These detailed descriptions can be found in [8]. Doubois and Prade [9] combined fuzzy sets and rough sets in a fruitful way by defining rough fuzzy sets and fuzzy rough sets.

Definition 2.2. Let the pair (U, R) be a Pawlak approximation space, where U be a universe set and R be an equivalence relation or indiscernibility relation on U. Then for any non-empty fuzzy set $A \in \mathcal{F}(U)$, the fuzzy sets

$$\underline{R}(A)(x) = \min\{A(y) : y \in [x]_R\}, \quad x \in U$$

and

$$\overline{R}(A)(x) = \max\{A(y) : y \in [x]_R\}, \quad x \in U$$

are, respectively, called the lower and the upper approximation of A in (U, R), where $[x]_R$ denotes the equivalence class of the relation R containing the element x. Furthermore, A is called a definable fuzzy set if A satisfies $\underline{R}(A) = \overline{R}(A)$, otherwise A is called a fuzzy rough set.

Definition 2.3. Let the pair (U, R) be a Pawlak approximation space. For any non-empty fuzzy set $A \in \mathcal{F}(U)$, the lower and upper approximation with respect to parameters α, β, where $0 < \beta \leq \alpha < 1$, are defined respectively by

$$\underline{R}(A)_\alpha = \{x \in U : \underline{R}(A) \geq \alpha\},$$
$$\overline{R}(A)_\beta = \{x \in U : \overline{R}(A) \geq \beta\}.$$

Based on above definition, we can know that $\underline{R}(A)_\alpha$ is the set of all elements of U whose degree of membership in A certainly are not less than α, and $\overline{R}(A)_\beta$ is the set of all elements of U whose degree of membership in A possibly are not less than β.

Definition 2.4. Let the pair (U, R) be a Pawlak approximation space, and $\underline{R}(A)_\alpha$, $\overline{R}(A)_\beta$ the lower and the upper approximation with respect to parameters α, β $(0 < \beta \leq \alpha < 1)$. Then the roughness measure with respect to parameters α, β of $A \in \mathcal{F}(U)$ in (U, R) is defined by

$$\rho_A(\alpha, \beta) = 1 - \frac{|\underline{R}(A)_\alpha|}{|\overline{R}(A)_\beta|}.$$

When $\overline{R}(A)_\beta = \phi$, $\rho_A(\alpha, \beta) = 0$ is ordered.

There are many properties for the roughness measure with respect to α, β of fuzzy set A, which can be found in [7].

3 The Novel Approach to Roughness Measure in Fuzzy Rough Sets

In this section, concepts of the lower and the upper approximation based on λ-level fuzzy set are proposed in term of λ-level fuzzy set, and some properties and results are also discussed.

Definition 3.1. Let $A \in \mathcal{F}(U)$ be a fuzzy set, and $\lambda \in [0, 1]$. A λ-level fuzzy set, denoted by $\widetilde{A_\lambda}$, is a fuzzy set in universe U whose membership function is

$$\widetilde{A_\lambda} = \begin{cases} A(x), & \text{if } A(x) > \lambda. \\ 0, & \text{otherwise.} \end{cases}$$

From the above definition and some concepts in section 2, we can conclude that λ-level fuzzy sets are obtained by reducing parts of fuzziness or information holding in the original fuzzy sets. Obviously, we can obtain directly the following Properties.

Property 3.1. *$\widetilde{A_\lambda} \subseteq A$ and $supp\widetilde{A_\lambda} \subseteq suppA$ always hold, for any $\lambda \in [0, 1]$.*

Property 3.2. *If $\lambda_1 \leq \lambda_2$, then $\widetilde{A_{\lambda_2}} \subseteq \widetilde{A_{\lambda_1}}$ and $supp\widetilde{A_{\lambda_2}} \subseteq supp\widetilde{A_{\lambda_1}}$ always hold, for $\lambda_1, \lambda_2 \in [0, 1]$.*

These properties indicate that λ-level fuzzy sets are monotonic with respect to inclusion of fuzzy set. The supports of λ-level fuzzy sets are monotonic with respect to inclusion of set.

Let the pair (U, R)be a Pawlak approximation space, R be an equivalence relation in an universe U, and $A \in \mathcal{F}(U)$ be a fuzzy set. Consequently, we will give new definitions.

Definition 3.2. For any non-empty fuzzy set A the fuzzy sets

$$\underline{R_\lambda}(A)(x) = \min\{\widetilde{A_\lambda}(y) : y \in [x]_R\}, \quad x \in U$$

and

$$\overline{R_\lambda}(A)(x) = \max\{\widetilde{A_\lambda}(y) : y \in [x]_R\}, \quad x \in U$$

are, respectively, said the lower and the upper approximation based on λ-level fuzzy set of A in (U, R), where $[x]_R$ denotes the equivalence class of the relation R containing the element x.

According to these definitions, we obviously obtain the following.

Theorem 3.1. (1) *Let $\underline{R}(A)$ and $\overline{R}(A)$ be the lower and the upper approximation of fuzzy set A, $\underline{R_\lambda}(A)$and $\overline{R_\lambda}(A)$ be the lower and the upper approximation based on λ-level fuzzy set of A. Facts of $\underline{R_\lambda}(A) \subseteq \underline{R}(A)$ and $\overline{R_\lambda}(A) \subseteq \overline{R}(A)$ are true.*

(2) *If $\lambda_1 \leq \lambda_2$, then $\underline{R_{\lambda_2}}(A) \subseteq \underline{R_{\lambda_1}}(A)$ and $\overline{R_{\lambda_2}}(A) \subseteq \overline{R_{\lambda_1}}(A)$, for $\lambda_1, \lambda_2 \in [0, 1]$.*

From above Theorems, we can find that the lower and upper approximations based on λ-level fuzzy set of A are respectively parts of the lower and the upper approximation of A. It is a very useful property according to which we can get the modification of roughness measure of a fuzzy set A.

For the sake of convenience, we will use the following labels.

$$C\underline{R_\lambda}(A) = \{x \in U : \underline{R_\lambda}(A)(x) \neq 0\};$$
$$C\overline{R_\lambda}(A) = \{x \in U : \overline{R_\lambda}(A)(x) \neq 0\}$$

Theorem 3.2. *For $\lambda \in [0, 1]$, $C\underline{R_\lambda}(A) = \underline{R}(A)_\lambda$ and $C\overline{R_\lambda}(A) = \overline{R}(A)_\lambda$ always hold.*

Proof. We need only to prove $C\underline{R_\lambda}(A) = \underline{R}(A)_\lambda$ and another result can be proved similarly.

For any $x \in U$, if $x \in C\underline{R_\lambda}(A)$, then we have $\underline{R_\lambda}(A)(x) \neq 0$. Hence $\min\{\widetilde{A_\lambda}(y) : y \in [x]_R\} \neq 0$. On the other hand, from the definition of $\tilde{A}(\cdot)$, we can know $A(y) > \lambda, y \in [x]_R$.

So $\underline{R}(A)(x) = \min\{A(y) : y \in [x]_R\} > \lambda$, which means that $x \in \underline{R}(A)_\lambda$.

Conversely, for any $x \in U$ if $x \in \underline{R}(A)_\lambda$, then we have $\underline{R}(A)(x) = \min\{A(y) : y \in [x]_R\} > \lambda$. Hence $A(y) > \lambda, y \in [x]_R$, which implies that $\widetilde{A_\lambda}(y) = A(y) \neq 0, y \in [x]_R$.

So $\underline{R_\lambda}(A)(x) = \min\{\widetilde{A_\lambda}(y) : y \in [x]_R\} \neq 0$, that is to say $x \in C\underline{R_\lambda}(A)$.

Hence $C\underline{R_\lambda}(A) = \underline{R}(A)_\lambda$ holds.

Theorem 3.3. Let the pair (U, R) be a Pawlak approximation space, R be an equivalence relation, and $A \in \mathcal{F}(U)$. For parameters $0 < \beta \leq \alpha < 1$, we have

$$\rho_A(\alpha, \beta) = 1 - \frac{|C\underline{R_\alpha}(A)|}{|C\overline{R_\beta}(A)|}.$$

When $C\overline{R_\beta}(A) = \phi$, $\rho_A(\alpha, \beta) = 0$ is ordered.

Proof. It can be obtained directly from the Theorem 3.2 and Definition 2.4.

By comparing the Definition 2.4 and Theorem 3.3, it can be found easily that we replaced $\underline{R}(A)_\alpha$ and $\overline{R}(A)_\beta$ with $C\underline{R_\alpha}(A)$ and $C\overline{R_\beta}(A)$ respectively. We will compute conveniently the roughness measure of a fuzzy set A by making use of Theorem 3.3, because elements of $C\underline{R_\alpha}(A)$ and $C\overline{R_\beta}(A)$ are obtained easily. In fact, we only consider weather $\underline{R}(A)_\alpha$ and $\overline{R}(A)_\beta$ are zero when we using this theorem. Hence, this theorem provides a convenient method of computing roughness measure with respect to parameters, especially for a complicated system.

4 An Example

In this section, we indicate validity of this method by giving an example.

Example 4.1. Let $U = \{x_i : i = 1, 2, \cdots, 8\}$ be a set of students. They are parted four sections, which are

$$U/R = \{\{x_1, x_5\}, \{x_2\}, \{x_3, x_4, x_6\}, \{x_7, x_8\}\}$$

Suppose students' stature is a fuzzy set A, and let membership function of A be

$$A = \{\frac{x_1}{0.5}, \frac{x_2}{0.3}, \frac{x_3}{0.3}, \frac{x_4}{0.6}, \frac{x_5}{0.5}, \frac{x_6}{0.8}, \frac{x_7}{1}, \frac{x_8}{0.9}\}$$

If $\alpha = 0.6$ and $\beta = 0.4$, then we have

$$\begin{aligned}
\underline{R_{0.4}}(A) &= \{\frac{x_1}{0.5}, \frac{x_2}{0}, \frac{x_3}{0}, \frac{x_4}{0}, \frac{x_5}{0.5}, \frac{x_6}{0}, \frac{x_7}{0.9}, \frac{x_8}{0.9}\}; \\
\overline{R_{0.4}}(A) &= \{\frac{x_1}{0.5}, \frac{x_2}{0}, \frac{x_3}{0.8}, \frac{x_4}{0.8}, \frac{x_5}{0.5}, \frac{x_6}{0.8}, \frac{x_7}{1}, \frac{x_8}{1}\}; \\
\underline{R_{0.6}}(A) &= \{\frac{x_1}{0}, \frac{x_2}{0}, \frac{x_3}{0}, \frac{x_4}{0}, \frac{x_5}{0}, \frac{x_6}{0}, \frac{x_7}{0.9}, \frac{x_8}{0.9}\}; \\
\overline{R_{0.6}}(A) &= \{\frac{x_1}{0}, \frac{x_2}{0}, \frac{x_3}{0.8}, \frac{x_4}{0.8}, \frac{x_5}{0}, \frac{x_6}{0.8}, \frac{x_7}{1}, \frac{x_8}{1}\}.
\end{aligned}$$

So we can obtain

$$C\underline{R_{0.6}}(A) = \{x_7, x_8\}$$

and

$$C\overline{R_{0.4}}(A) = \{x_1, x_3, x_4, x_5, x_6, x_7, x_8\}.$$

Hence, we have

$$\begin{aligned}
\rho_A(0.4, 0.6) &= 1 - \frac{|C\underline{R_{0.6}}(A)|}{|C\overline{R_{0.4}}(A)|} \\
&= 1 - \frac{2}{7} \\
&= 0.714
\end{aligned}$$

5 Conclusions

Rough set theory has been regarded as a generalization of the classical set theory in one way. This is an important mathematical tool to deal with vagueness. Exact sets can be viewed as rough sets, conversely not. Clearly roughness induces fuzziness. In the present paper, we have introduced the lower and the upper approximation based on λ-level fuzzy set and studied their properties, and modified the roughness measure in fuzzy rough sets by making use of these concepts.

References

1. Pawlak. Z. (1982) Rough Sets, International Journal of Computer Information Sciences 11:145-172
2. Pawlak. Z. (1991) Rough Sets-Theoretical Aspects to Reasoning about Data, Kluwer Academic Publisher, Boston
3. Yao. Y.Y. (1997) Combination of Rough and Fuzzy Sets Based on -Level Sets, Rough Sets and Data Mining: Analysis for Imprecise Data, Lin. T.Y. and Cercone, N. (Ed.), Kluwer Academic, Boston
4. Yao. Y.Y. (1996) Two Views of The Theory of Rough Sets in Finite Universes, International Journal of Approximation Reasoning 15:291-317
5. Yao. Y.Y., Lin. T.Y. (1996) Generalization of Rough Sets Using Modal Logic, Intelligent Automation and Soft Computing, An International Journal 2:103-120
6. Zhang.W.X.,Wu.W.Z.,Liang.J.Y.,Li.D.Y. (2001) Theory and Method of rough sets. Science Press, Beijing
7. Zhang.W.X.,Liang.Y.,Wu.W.Z. (2003) Information Systems and Knowledge Discovery. Science Press, Beijing
8. Zadeh. L.A. (1965) Fuzzy Sets, Information and Control 8:338-353
9. Dubois. D., Prade. H. (1990) Rough Fuzzy Sets and fuzzy Rough Sets, International Journal of General Systems17:191-208
10. Cornelis. C., Cock, M.D. and Kerre. E.E. (2003) Intuitionistic Fuzzy Rough Sets: At the Crossroads of Imperfect Knowledge, Expert Systems 20:260-270
11. Radzikowska. A.M., Keere. E.E. (2002) A Comparative Study Rough Sets, Fuzzy Sets and Systems 126:137-156
12. Banerjee. M., Miltra. S., Pal. S.K. (1998) Rough fuzzy MLP: knowledge encoding and classification, IEEE Trans. Neural Network 9:1203-1216
13. Chakrabarty. K., Biswas. R., Nanda. S. (2000) Fuzziness in Rough Sets, Fuzzy Set and Systems 110: 247-251
14. Mordeson. J.N. (2001) Rough Sets Theory Applied to (Fuzzy) Ideal Theory, Fuzzy Sets and Systems 121:315-324
15. Sarkar. M. (2002) Rough-fuzzy Functions in Classification, Fuzzy Sets and Systems 132:353-369
16. Baets. B.D., Kerre, E. (1994) The Cutting of Compositions, Fuzzy Sets and Systems 62:295-309
17. Radecki. Z. (1997) Level Fuzzy Sets, Journal of Cybernet 7:189-198
18. Rembrand. B.R.B., Zenner and Rita. M.M. DE Caluwe (1984) A New Approach to Information Retrieval Systems Using Fuzzy Expressions, Fuzzy Sets and Systems17:9-22
19. Liu.W.N.,Yao.J.T., Yao.Y.Y. (2004) Rough Approximations under Level Fuzzy Sets, Lecture Notes in Artifical Intelligent 3066:78-83
20. Banerjee. M., Pal. S.K. (1996) Roughness of A Fuzzy Set, Information Sciences 1:235-264

Rough Communication of Dynamic Concept

Hong-Kai Wang[1], Jin-Jiang Yao[2], Pei-Jun Xue[3], and Kai-Quan Shi[4]

[1] School of Sciences, University of Jinan, 106 Jiwei Road, Jinan 250022, Shandong, China
wanghongkai@mail.sdu.edu.cn
[2] Department of Mathematics, Linyi Normal University, Linyi 276005, Shandong, China
pingerslim@sina.com
[3] School of Mathematics and System Sciences, Shandong University, 27 Shanda Road, Jinan 250100, Shandong, China
pjxue@sdu.edu.cn
[4] School of Mathematics and System Sciences, Shandong University, 27 Shanda Road, Jinan 250100, Shandong, China
shikq@sdu.edu.cn

Abstract. In rough communication, each agent taking part in rough communication may give new judge about the dynamic concept *X*. And this new information may be important. How to study the rough communication which concerns the useful subjective information is very important. The definition of rough communication of dynamic concept based on $\alpha -$ generation of two direction assistant sets is proposed in the paper. An example is presented to illustrate the reasonableness of the new definition.

Keywords: rough sets, rough communication, dynamic concept, two direction assistant sets, a generation.

1 Introduction

Rough sets theory, proposed by Z. Pawlak in 1982 [1], is a new mathematical tool to deal with vagueness and uncertainty. Rough sets theory has extensive applications in knowledge discovery, decision analysis, machine learning, pattern recognition and so on [2-6].

Rough communication, a proper tool of dealing with several information sources, is presented as a new extension of rough sets [7]. When dealing with several sources of information where only one concept is present, each agent has a different language and cannot provide precise communication to others, so it is called rough communication. It is done by defining the approximation of a rough set from one approximation space to another. By doing this we lose some information or precision of the concept and this results in a less precise or rougher concept.

For the problem of information loss in rough communication, [8] makes some quantization analysis for the problem of information loss in rough communication by defining the concepts of the degree of fidelity, the degree of distortion, and rough information flow. By means of the concepts of common knowledge and possible knowledge, [9] analyzes the concept translation result under different sequences of rough communication. It is of significance for our decision when we don't know the

B.-Y. Cao (Ed.): Fuzzy Information and Engineering (ICFIE), ASC 40, pp. 781–787, 2007.
springerlink.com

exact translation sequence. A new definition form of rough communication is proposed in [10], and the relation theorem between old-rough communication [7] and new-rough communication is obtained. There are $n!$ translation sequences in rough communication, which n agents take part in. In fact, with different paths the amount of the missed knowledge is varied. In [11-12] the global optimal translation sequence, in which every agent taking part in rough communication gets maximum information, and the result optimal translation sequence, in which the last agent taking part in rough communication gets maximum information, are defined and their corresponding mathematical models are given. In order to get the answer, simulated annealing algorithm is used for the two mathematical models. [13-14] propose the concept of rough communication of fuzzy concept. Global optimal translation sequence of fuzzy rough communication is also defined and its corresponding mathematical model is also given. By using $\alpha-$rough communication cut, the relation theorem between rough communication of fuzzy concept and that of classical concept is obtained.

Rough communication discussed in [7-12] is a rough communication of classical concept; rough communication discussed in [13-14] is a rough communication of fuzzy concept. Rough communication discussed in [7-14] is a rough communication of static concept. Now, we have some questions:

1. How a two direction dynamic concept [15–16] translate among multi-agents?
2. In rough communication [7–14], each agent taking part in rough communication is "static", that is, he (or she) deals with the concept completely depending on knowledge (equivalence relation), and no subjective information is involved in. How to study the rough communication which concerns the useful subjective information is very important.

The paper is organized as follows. In Section 2, we review the concepts of two direction S-rough sets and $\alpha-$generation of two direction assistant sets. In Section 3, the definition of rough communication of dynamic concept based on $\alpha-$generation of two direction assistant sets is proposed. An example is presented in Section 4 to illustrate the reasonableness of the definition. In Section 5, we make a summary for the paper.

Assumption: U is a nonempty finite universe, $[x]_R$ is the element equivalence class, and R is the element equivalence relation on U; $F, \overline{F}$ denote the element transfer families on U, where $F=\{f_1, f_2,\cdots, f_m\}$, $\overline{F}=\{\bar{f}_1, \bar{f}_2,\cdots, \bar{f}_n\}$, $\mathrm{F} = F\bigcup \overline{F}$ [15–16].

2 Two Direction S-Rough Sets and $\alpha-$Generation of Two Direction Assistant Sets

Definition 2.1. [15–16] Call $X^* \subset U$ two direction Singular Set on U, briefly called two direction S-set, if

$$X^* = X'\bigcup\{u \mid u\in U, u\,\overline{\in}\, X, f(u)=x\in X\}. \tag{1}$$

Call X' the loss set of $X \subset U$, if

$$X' = X - \{x \mid x \in X, \bar{f}(x) = u \,\overline{\in}\, X\}. \tag{2}$$

Call $X^{\bar{f}}$ $\bar{f}$-atrophy of $X \subset U$, if

$$X^{\bar{f}} = \{x \mid x \in X, \bar{f}(x) = u \,\overline{\in}\, X\} \tag{3}$$

Definition 2.2. [15–16] Suppose X^* be two direction S-sets on *U*, $X^* \subset U$, then call $(R,\mathrm{F})_\circ(X^*)$ lower approximation of two direction S-sets X^*, if

$$\begin{aligned}(R,\mathrm{F})_\circ(X^*) &= \bigcup[x] \\ &= \{x \mid x \in U, [x] \subseteq X^*\}.\end{aligned} \tag{4}$$

Call $(R,\mathrm{F})^\circ(X^*)$ upper approximation of two direction S-sets X^*, if

$$\begin{aligned}(R,\mathrm{F})^\circ(X^*) &= \bigcup[x] \\ &= \{x \mid x \in U, [x] \cap X^* \neq \phi\}.\end{aligned} \tag{5}$$

Where $\mathrm{F} = F \cup \overline{F}$, $F \neq \phi$, $\overline{F} \neq \phi$.

Definition 2.3. [15–16] Suppose X^* be two direction S-sets on U, $X^* \subset U$; $(R,\mathrm{F})_\circ(X^*)$ and $(R,\mathrm{F})^\circ(X^*)$ are lower approximation and upper approximation of X^*, respectively, and call

$$((R,\mathrm{F})_\circ(X^*), (R,\mathrm{F})^\circ(X^*)) \tag{6}$$

two direction S-rough sets of $X^* \subset U$.

Definition 2.4. [17–18] Suppose $A_s(X^*)$ be assistant sets of $((R,\mathrm{F})_\circ(X^*), (R,\mathrm{F})^\circ(X^*))$, $\forall \alpha_1 \in (0,1)$. Call $A_{s^\circ}^{\alpha_1}(X^*)$ α – upper generation of $A_s(X^*)$, if

$$A_{s^\circ}^{\alpha_1}(X^*) = \{x \mid u \in U, u \,\overline{\in}\, X, f(u) = x,\ \alpha_1 \leq \chi_X^{f(u)} < 1\}. \tag{7}$$

Definition 2.5. [17–18] Suppose $A_s(X^*)$ be assistant sets of $((R,\mathrm{F})_\circ(X^*), (R,\mathrm{F})^\circ(X^*))$, $\forall \alpha_2 \in (-1,0)$. Call $A_{s_\circ}^{\alpha_2}(X^*)$ α – lower generation of $A_s(X^*)$, if

$$A_{s_\circ}^{\alpha_2}(X^*) = \{y \mid x \in X, \bar{f}(x) = y, -1 \leq \chi_X^{\bar{f}(x)} < \alpha_2\}. \tag{8}$$

3 Rough Communication of Dynamic Concept

Suppose (U,R_1), (U,R_2), $\cdots$, (U,R_n) be *n* approximation spaces, which represent the knowledge of agent 1, agent 2, $\cdots$, agent *n* respectively. Therefore, according to Fig.1,

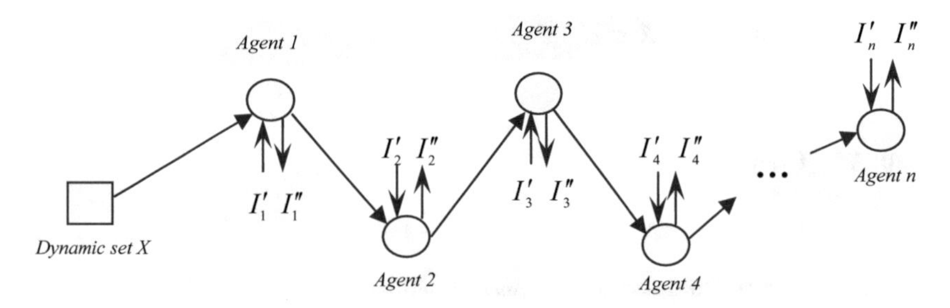

Fig. 1. A sequence on a population

agent1 has a direct translation of dynamic concept X and the knowledge of agent 2 about concept X is received through agent 1 and so on.

A family of n sets in which all of them are related to a same dynamic concept X, but by different agents, is called a sequence on population and is denoted by $DSEQ_X$:

$$DSEQ_X = \{(X_1^+, X_1^-), (X_2^+, X_2^-), \cdots, (X_n^+, X_n^-)\}. \tag{9}$$

And

$$\begin{aligned} X_1^+ &= \underline{R_1}(X \cup A_{s^\circ}^{\alpha_1}(X^*) - A_{s_\circ}^{\alpha_2}(X^*)) \cup I_1' - I_1'', \\ X_1^- &= U - \overline{R_1}(X \cup A_{s^\circ}^{\alpha_1}(X^*) - A_{s_\circ}^{\alpha_2}(X^*)) \cup I_1'' - I_1'; \\ X_k^+ &= \underline{R_k} X_{k-1}^+ \cup I_k' - I_k'', \ X_k^- = \underline{R_k} X_{k-1}^- \cup I_k'' - I_k' \qquad 2 \le k \le n . \end{aligned} \tag{10}$$

Where I_k' $(1 \le k \le n)$ is the set of subjective information of agent k in rough communication, and he (or she) thinks the information should belong to X ; I_k'' $(1 \le k \le n)$ is the set of subjective information of agent k in rough communication, and he (or she) thinks the information should not belong to X.

It is easy to get the following proposition:

Proposition 1. If $\alpha_1 = 1, \alpha_2 = -1$, and $I_k' = I_k'' = \phi$, then rough communication of dynamic concept based on $\alpha-$ generation of two direction assistant sets degenerates into rough communication [7–12] .

From Proposition 1, rough communication of dynamic concept based on $\alpha-$ generation of two direction assistant sets is the general form of rough communication [7–12] , and rough communication [7–12] is the special case of rough communication of dynamic concept based on $\alpha-$ generation of two direction assistant sets.

4 Example

Let $U = \{x_1, x_2, \cdots, x_{10}\}$, $X = \{x_1, x_2, x_3, x_7, x_8\}$, $\chi_X^{f(x_6)} = 0.9$, $\chi_X^{f(x_9)} = 0.3$, $\chi_X^{\bar{f}(x_3)} = -0.8$, $\chi_X^{\bar{f}(x_8)} = -0.4$. We have a population of three agents and we can form different sequences on it. And

agent1: $\{x_1,x_2\}, \{x_3\}, \{x_4\}, \{x_5,x_6\}, \{x_7,x_8\}, \{x_9,x_{10}\}$;
agent2: $\{x_1\}, \{x_2\}, \{x_3,x_4,x_5\}, \{x_6\}, \{x_7\}, \{x_8,x_9\}, \{x_{10}\}$;
agent3: $\{x_1,x_2,x_3\}, \{x_4,x_5\}, \{x_6\}, \{x_7\}, \{x_8,x_9\}, \{x_{10}\}$.

The translation sequence of rough communication of dynamic concept based on $\alpha-$ generation of two direction assistant sets is $DSEQ_X = \{(X_1^+, X_1^-), (X_2^+, X_2^-), (X_3^+, X_3^-)\}$. In rough communication, *agent*1 thinks x_3 should belong to X and x_4 should not belong to X; *agent*2 thinks x_2 should belong to X and x_6 should not belong to X; *agent*3 thinks x_6 should belong to X and x_8 should not belong to X. Choose $\alpha_1 = 0.8$, $\alpha_2 = -0.7$, and then

$$A_{s^\circ}^{\alpha 1}(X^*) = \{x_6\},\ A_{s_\circ}^{\alpha 2}(X^*) = \{x_3\};$$

$$X_1^+ = \underline{R_1}(X \cup A_{s^\circ}^{\alpha 1}(X^*) - A_{s_\circ}^{\alpha 2}(X^*)) \cup I_1' - I_1'' = \{x_1,x_2,x_7,x_8\} \cup \{x_3\} - \{x_4\}$$

$$= \{x_1,x_2,x_3,x_7,x_8\};$$

$$X_1^- = U - \overline{R_1}(X \cup A_{s^\circ}^{\alpha 1}(X^*) - A_{s_\circ}^{\alpha 2}(X^*)) \cup I_1'' - I_1' = \{x_3,x_4,x_9,x_{10}\} \cup \{x_4\} - \{x_3\}$$

$$= \{x_4,x_9,x_{10}\};$$

$$X_2^+ = \underline{R_2}X_1^+ \cup I_2' - I_2'' = \{x_1,x_2,x_7\} \cup \{x_2\} - \{x_6\} = \{x_1,x_2,x_7\};$$

$$X_2^- = \underline{R_2}X_1^- \cup I_2'' - I_2' = \{x_{10}\} \cup \{x_6\} - \{x_2\} = \{x_6,x_{10}\};$$

$$X_3^+ = \underline{R_3}X_2^+ \cup I_3' - I_3'' = \{x_7\} \cup \{x_6\} - \{x_8\} = \{x_6,x_7\};$$

$$X_3^- = \underline{R_3}X_2^- \cup I_3'' - I_3' = \{x_6,x_{10}\} \cup \{x_8\} - \{x_6\} = \{x_8,x_{10}\}.$$

So the translation sequence of one direction expansion rough communication of dynamic concept based on $\alpha-$ generation of assistant sets is:

$$(\{x_1,x_2,x_3,x_7,x_8\},\{x_4,x_9,x_{10}\}) \to (\{x_1,x_2,x_7\},\{x_6,x_{10}\}) \to (\{x_6,x_7\},\{x_8,x_{10}\}).$$

5 Conclusions

In rough communication [7–14], each agent taking part in rough communication is "static". That is to say, he (or she) deals with the concept completely depending on knowledge (equivalence relation), and no subjective information is involved in. In fact, each agent taking part in rough communication may give new judge about the dynamic concept X. And this new information may be important. How to make use of the new information is this paper's main intention. The definition of one direction expansion rough communication of dynamic concept based on $\alpha-$ generation of assistant sets is proposed in the paper. An example is presented to illustrate the reasonableness of the new definition.

Acknowledgement

This paper is supported by the Natural Science Foundation of Shandong Province (No.Y2006A12; Y2004A04), the Scientific Research Development Project of Shandong Provincial Education Department (No. J06P01) and the Doctoral Foundation of University of Jinan (No. B0633).

References

[1] Z. Pawlak (1982) Rough sets. International Journal of Computer and Information Sciences, 11: 341-356.

[2] Z. Pawlak, R. Slowinski (1994) Rough set approach to multi-attribute decision analysis. European Journal of Operational Research, 72: 443-459.

[3] R. R. Hashemi, L. A. Le. Blanc, C. T. Rucks, and A. Rajaratnam (1998) A hybrid intelligent system for predicting bank holding structures. European Journal of Operational Research, 109: 390-402,.

[4] A. I. Dimitras, R. Slowinski, R. Susmaga, and C. Zopounidis (1999) Business failure prediction using rough sets. European Journal of Operational Research, 114: 263-280.

[5] Z. Pawlak (2002) Rough sets, Decision algorithm and Bayes' theorem. European Journal of Operational Research, 136: 181-189.

[6] S. Asharaf, M. N. Murty (2003) An adaptive rough fuzzy single pass algorithm for clustering large data sets. Pattern Recognition, 36: 3015-3018.

[7] A.Mousavi, P.Jabedar-Maralani (2002) Double-faced rough sets and rough communication. Information Sciences, 148: 41-53.

[8] Hongkai Wang, Xiuhong Li, Kaiquan Shi (2005) Information Measure in Rough Communication. An International Journal: Advances in Systems Science and Applications, 5(4): 638-643.

[9] Yanyong GUAN, Hongkai Wang, Bingxue YAO (2006)Analysis of the Concept Translation Result in Rough Communication. Proceedings of 2006 International Conference on Artificial Intelligence, PP. 381-384.

[10] Hongkai Wang, Shuli Zhao (2006) A new definition form of rough communication. Computer science,, 33(9): 189-190. (in Chinese).

[11] Hongkai Wang, Yanyong Guan, Kaiquan Shi (2006) Simulated annealing algorithm for globaloptimal translation sequence in rough communication. Systems engineering-theory and practice, 26(9): 118-122. (in Chinese).

[12] Hongkai Wang, Xiuqiong Chen, Kaiquan Shi (2006) Simulated annealing algorithm for result Optimal translation sequence in rough communication. Fuzzy systems and mathematics, 20(5): 125-130. (in Chinese).

[13] Hongkai Wang, Peijun Xue, Kaiquan Shi (2005) The problem of rough communication of fuzzy concept. The 11 th Joint International Computer Conference, PP. 674-677.

[14] Hongkai Wang, Yanyong Guan, Bingxue Yao, Kaiquan Shi (2006) Fuzzy rough communication and optimization problem of its translation sequence. IEEE Proceedings of the fifth International Conference on Machine Learning and Cybernetics, PP. 2268-2273.

[15] Kaiquan Shi(2002)S-rough sets and its applications in diagnosis-recognition for disease.IEEE Proceedings of the first International Conference on Machine Learning and Cybernetics, PP. 50-54.

[16] Kaiquan Shi, T. C. Chang (2005) Two direction S-rough sets. International Journal of Fuzzy mathematics, 2: 335-349.
[17] Hongkai Wang, Haiqing Hu (2004) $\alpha -$ generation and $\alpha -$ generation theorem of assistant set of S-rough sets. Journal of Shandong University (natural science), 39(1): 9-14. (in Chinese)
[18] Hongkai Wang, Haiqing Hu (2004) $\alpha -$ generation granulation degree of assistant set of S-rough sets. Journal of Shandong University (natural science), 39(3): 32-36. (in Chinese)

Some Entropy for Rough Fuzzy Sets

Zhang Chengyi[1,2], Qu Hemei[2], Chen Guohui[1], and Peng Dejun[1]

[1] *Dept. of Math., Hainan Normal University, 571158 , Hainan, Haikou, P.R. of China
chengyizhang@hainnu.edu.cn
[2] Dept. of Math., Huanghuai College, Henan, Zhumadian, 463000, P.R. of China

Abstract. The entropy, fuzzy positive entropy and fuzzy negative entropy measure for rough fuzzy sets are proposed. Some properties of entropy measure for rough fuzzy sets are discussed. It is also shown that the proposed measure can be defined in terms of the ratio of rough fuzzy cardinalities.

Keywords: rough fuzzy sets, Cardinality of rough fuzzy sets, Fuzzy positive entropy, fuzzy negative entropy.

1 Introduction

Fuzzy entropy, a measure of fuzziness is first mentioned in 1965 by Zadeh[1]. The name entropy was chosen due to an intrinsic similarity of equations to the ones in the Shannon entropy [2]. However, the two functions measure fundamentally different types of uncertainty. Basically, the Shannon entropy measures the average uncertainty in bits associated with the prediction of outcomes in a random experiment.

In 1972, De Luca and Termini [3] introduced some requirements which capture our intuitive comprehension of the degree of fuzziness. Kaufmann [4] proposed to measure the degree of fuzziness of any fuzzy set A by a metric distance between its membership function and the membership function (characteristic function) of its nearest crisp set. Another way given by Yager [5] was to view the degree of fuzziness in terms of a lack of distinction between the fuzzy set and its complement. B.Kosko [6]-[8] investigated the fuzzy entropy in relation to a measure of subsethood:let a denote the fuzzy distance between a fuzzy set and the nearest crisp set, let b denote the fuzzy distance between a fuzzy set and the farthest crisp set; then the fuzzy entropy equals the ratio of a to b. Recently,using Kosko's subsethood measure, Szmidt and Kacprzyk [9] give a non-probabilistic-type entropy measure for intuitionistic fuzzy sets introduced by Atanassov [10]-[12], based on a geometric interpretation of intuitionistic fuzzy sets.

Pawlak proposed rough set [13] in 1982. Although fuzzy sets and rough sets are methods to handle vague and inexact information, their starts and emphases

* Project supported by NSF of China(60364001,70461001) and Hainan(80401).

are different. The fuzzy sets emphasize on the morbid definition of the boundary of sets, in which the relations of "belong to" and "not belong to" between elements and sets in the classical set theory are characterized by the membership degree; while rough sets use the approximation given the equivalent relation of the classical sets to research the indistinguishableness of elements. Two theories from different points of view may complementary each other. Combining fuzzy sets with the rough sets, Dubios D and Prade H [14] proposed fuzzy rough sets(FRSs) and rough fuzzy sets(RFSs) in 1990. In the article [15], based on the background of vague information handling, the fuzzy positive entropy and fuzzy negative entropy of vague set are given, and its properties are discussed. In [17], we discussed the relations of FRSs , RFSs and Intuitionistic fuzzy sets(IFSs) and their similarity measures.

In this paper, we have introduced the entropy, fuzzy positive entropy and fuzzy negative entropy measure for rough fuzzy sets. It can make the expression of fuzzy information more clear. Some properties of entropy measure for rough fuzzy sets are discussed. It is also shown that the proposed entropy measure can be defined in terms of the ratio of rough fuzzy cardinalities.

2 Rough Fuzzy Sets

2.1 The Expression of Rough Fuzzy Sets

In this section, we will present those aspects of rough fuzzy sets which will be needed in our next discussion.

Definition 2.1 [13]. Let U be a non-empty universe of discourse and R an equivalent relation on U, which is called an indistinguishable relation, $U/R = \{X_1,\ X_2 \cdots, X_n\}$ is the all the equivalent class derived from R. $W = (U, R)$ are called an approximation space. $\forall X \subseteq U$, suppose $X_L = \{x \in U \mid [x] \subseteq X$ and $X_U = \{x \in U \mid [x] \cap X \neq \emptyset\}$, a set pairs $(X_L, X_U)$are called rough set, and denoted as $X = (X_L, X_U)$; X_L and X_U are the lower approximation and the upper approximation of X on W respectively.

Definition 2.2 [14]. Let $W = (U, R)$ be an approximation space. Theupper approximation and the lower approximation of the fuzzy set B in U are denoted as $\overline{B}$ and $\underline{B}$ respectively, and defined to be the fuzzy set $\overline{B}, \underline{B} : U/R \mapsto [0,1]$ in $U/R = \{X_1, X_2, \cdots, X_n\}$, such that

$$\overline{B}(X_i) = \sup_{x \in X_i} B(x),\ \underline{B}(X_i) = \inf_{x \in X_i} B(x),\ i = 1, 2, \cdots, n \tag{1}$$

then $A = (\overline{B}, \underline{B})$ is called a RFS. It is clear that

$$\underline{B}(X_i) \leq \overline{B}(X_i),\ \forall X_i \in U/R \tag{2}$$

We can define two fuzzy sets as following: $\underline{A}(x) = \underline{B}(X_i)(x \in X_i)$, $\overline{A}(x) = \overline{B}(X_i)(x \in X_i)$, then $A = (\underline{A}, \overline{A})$ is a RFS on U and also can be written as:

$A = \{x =< \underline{A}(x), \overline{A}(x) > | x \in U\}$. $A^c = (\underline{A}^c, \overline{A}^c)$ is denoted the complement set of $A = (\underline{A}, \overline{A})$, where

$$\underline{A}^c(x) = 1 - \overline{A}(x) = 1 - \sup_{x \in X_i} B(x) = \inf_{x \in X_i} (1 - B(x)) \quad x \in U \tag{3}$$

$$\overline{A}^c(x) = 1 - \underline{A}(x) = 1 - \inf_{x \in X_i} B(x) = \sup_{x \in X_i} (1 - B(x)) \quad x \in U \tag{4}$$

Then $(A^c)^c = A$

For each RFS in U, we will call

$$\pi_A(x) = 1 - \overline{A}(x) \tag{5}$$

the non-membership degree of x to A.

The family of all RFSs in U is denoted by $RFS(U)$. The operations of RFSs may be seen in [14].

When $A = \{\langle \underline{A}(x), \overline{A}(x) \rangle | x \in U\} = \{\langle 1, 1 \rangle | x \in U\}$, A is a crisp set in U, it is meant as $x \in A$; Similarly, when $A = \{\langle \underline{A}(x), \overline{A}(x) \rangle | x \in U\} = \{\langle 0, 0 \rangle | x \in U\}$, A is a crist set in U, it is meant as $x \notin A$; When $\underline{A} = \overline{A}(x)$, $\forall x \in U$, it is meant as A is a fuzzy set.

2.2 The Cardinalities of RFSs

In our further considerations on entropy for RFSs, besides the distances the concept of cardinalities of RFS will also be useful.

Definition 2.3. Let $A \in RFS(U)$. First, we define the following two cardinalities of a RFS:
(1)the minimum cardinality of A is defined as:

$$Card_{min}(A) = \sum_{x \in U} \underline{A}(x) \tag{6}$$

(2) the maximum cardinality of A is defined as:

$$Card_{max}(A) = \sum_{x \in U} \overline{A}(x) \tag{7}$$

Then the cardinality of a RFS is defined as the interval

$$CardA = [\sum_{x \in U} \underline{A}(x), \sum_{x \in U} \overline{A}(x)] \tag{8}$$

And clearly, for

$$A^c = \{(x, \underline{A}^c(x), \overline{A}^c(x)) | \forall x \in U\} = \{(x, 1 - \underline{A}^c(x), 1 - \overline{A}^c(x)) | \forall x \in U\}$$

we have

$$CardA^c = [\sum_{x \in U} \underline{A}^c(x), \sum_{x \in U} \overline{A}^c(x)] = [|U| - \sum_{x \in U} \underline{A}(x), |U| - \sum_{x \in U} \overline{A}(x)] \tag{9}$$

where $|U|$ is the cardinality of U.

3 Entropy for RFSs

De Luca and Termini [3] first axiomatized non-probabilistic entropy. The De Luca Termini axioms formulated for RFSs are intuitive and have been widely employed in the fuzzy literature. They were formulated in the following way. Let E be a set-to-point mapping $E : F(U) \mapsto [0, 1]$. Hence E is a fuzzy set defined on fuzzy sets and $F(U)$ is the family of all fuzzy sets in U. E is an entropy measure if it satisfies the four De Luca and Termini axioms:
1:

$$E(A) = 0 \text{ iff } A \in 2^U \ (A \text{ non} - \text{fuzzy}) \tag{10}$$

2:

$$E(A) = 1 \text{ iff } \mu_A(x) = 0.5 \ for \ all \ x \in U \tag{11}$$

3: $E(A) \leq E(B)$ if A is less fuzzy than B, i.e. if

$$\mu_A(x) \leq \mu_B(x) \text{ when } \mu_B(x) \leq 0.5; \ \mu_B(x) \leq \mu_A(x) \text{ when } \mu_B(x) \geq 0.5 \tag{12}$$

4:

$$E(A) = E(A^c) \tag{13}$$

Since the De Luca and Termini axioms (10)-(13) were formulated for fuzzy sets, they are expressed for the RFSs as follows:

Definition 3.1. Let E be a set-to-point mapping $E : RFS(U) \mapsto [0, 1]$. E is an entropy measure if it satisfies the four axioms:
EP1

$$E(A) = 0 \ if \ A \in 2^x \ (A \text{ non} - \text{fuzzy}) \tag{14}$$

EP2

$$E(A) = 1 \text{ iff } \underline{A}^c(x) = 1 - \overline{A}^c(x) \ for \ all \ x \in U \tag{15}$$

EP3

$$E(A) \leq E(B) \ if \ \underline{A}(x) \leq \underline{B}(x) \text{ and } \overline{A}(x) \leq \overline{B}(x), \text{ when } \underline{B}(x) + \overline{B}(x) \leq 1$$

or

$$E(A) \leq E(B) \text{ if } \underline{A}(x) \geq \underline{B}(x) \text{ and } \overline{A}(x) \geq \overline{B}(x), \text{ when } \underline{B}(x) + \overline{B}(x) \geq 1 \tag{16}$$

EP4

$$E(A) = E(A^c) \tag{17}$$

Applying the Hamming distance (or other most widely used distances for fuzzy sets A, for example, the Euclidean distance), we can define the entropy for RFSs:

$$E(A) = \frac{1}{Card(U)} \sum_{x \in A_U} E(x) \tag{18}$$

If there are n points belonging to X, we have

$$E(A) = \frac{1}{n}\sum_{i=1}^{n} E(x_i) \tag{19}$$

Where

$$E(x) = \begin{cases} \dfrac{\overline{A}(x)}{1-\underline{A}(x)} & if\ \underline{A}(x)+\overline{A}(x) \le 1 \\ \\ \dfrac{1-\underline{A}(x)}{\overline{A}(x)} & if\ \underline{A}(x)+\overline{A}(x) \ge 1 \end{cases} \tag{20}$$

Theorem 3.1. Let $A \in RFS(U)$. Then $E(A) = \frac{1}{Card(U)} \sum_{x \in A_U} E(x)$ is an entropy for a RFS.

Proof. (1) and (2) are clear.
(3) For $\underline{B}(x) + \overline{B}(x) \le 1$, if $\underline{A}(x) \le \underline{B}(x)$ and $\overline{A}(x) \le \overline{B}(x)$ then

$$\underline{A}(x) + \overline{A}(x) \le 1$$

Since $\dfrac{\overline{A}(x)}{1-\underline{A}(x)} \le \dfrac{\overline{B}(x)}{1-\underline{A}(x)} \le \dfrac{\overline{B}(x)}{1-\overline{B}(x)}$ thus

$$E(A) = \frac{1}{Card(U)}\sum_{x\in U}\frac{\overline{A}(x)}{1-\underline{A}(x)} \le \frac{1}{Card(U)}\sum_{x\in U}\frac{\overline{B}(x)}{1-\overline{B}(x)} = E(B) \tag{21}$$

For $\underline{B}(x) + \overline{B}(x) \ge 1$, if $\underline{A}(x) \ge \underline{B}(x)$ and $\overline{A}(x) \ge \overline{B}(x)$ then

$$\underline{A}(x) + \overline{A}(x) \ge 1$$

then

$$E(A) = \frac{1}{Card(U)}\sum_{x\in U}\frac{1-\underline{A}(x)}{\overline{A}(x)} \le \frac{1}{Card(U)}\sum_{x\in U}\frac{1-\overline{B}(x)}{\overline{B}(x)} = E(B) \tag{22}$$

(4) If $\underline{A}(x) + \overline{A}(x) \ge 1$ then $\underline{A}^c(x) + \overline{A}^c(x) \le 1$,

$$E(A^c) = \frac{1}{Card(U)}\sum_{x\in U}\frac{\overline{A}^c(x)}{1-\underline{A}^c(x)} = \frac{1}{Card(U)}\sum_{x\in U}\frac{1-\underline{A}(x)}{\overline{A}(x)} = E(A) \tag{23}$$

If $\underline{A}(x) + \overline{A}(x) \le 1$, then $\underline{A}^c(x) + \overline{A}^c(x) \ge 1$

$$E(A^c) = \frac{1}{Card(U)}\sum_{x\in U}\frac{1-\underline{A}^c(x)}{\overline{A}^c(x)} = \frac{1}{Card(U)}\sum_{x\in U}\frac{\overline{A}(x)}{1-\underline{A}(x)} = E(A) \tag{24}$$

So $E(A)$ is an entropy for RFSs.

Following, we discuss the relation between the entropy and cardinality of a RFS.

Let $x = \langle \underline{A}(x), \overline{A}(x) \rangle$ we call $x^c = \langle 1 - \overline{A}(x), 1 - \underline{A}(x) \rangle = \langle \underline{A}^c(x), \overline{A}^c(x) \rangle$ the complement of x. We define

$$x \cap x^c = \langle \min(\underline{A}(x), \underline{A}(x^c)), 1 - \max(\underline{A}(x), \underline{A}(x^c)) \rangle \tag{25}$$

$$x \cup x^c = \langle \max(\underline{A}(x), \underline{A}(x^c)), 1 - \min(\underline{A}(x), \underline{A}(x^c)) \rangle \tag{26}$$

Then since $1 - \max(\underline{A}(x), \underline{A}(x^c)) = 1 - \max(1 - \overline{A}(x), \underline{A}(x)) = \min(\overline{A}(x), 1 - \underline{A}(x)) \geq \min(\underline{A}(x), 1 - \overline{A}(x))$ thus $(x \cap x^c) \in RFS(U)$. Similarly, $(x \cup x^c) \in RFS(U)$

Theorem 3.2. Let $A \in RFS(U)$ and $E(A)$ be the entropy for RFS in Theorem 3.1. Then

$$E(A) = \frac{1}{Card(U)} \sum_{x \in U} \frac{Card_{max}(x \cap x^c)}{Card_{max}(x \cup x^c)} \tag{27}$$

Proof
(1) If $\underline{A}(x) + \overline{A}(x) \geq 1$, then

$$\underline{A}(x) \geq 1 - \overline{A}(x) = \underline{A}(x^c)$$

$$x \cap x^c = \langle \underline{A}(x^c), 1 - \underline{A}(x) \rangle = \langle \underline{A}(x^c), \overline{A}(x^c) \rangle = x^c$$

$$x \cup x^c = \langle \underline{A}(x), 1 - (1 - \overline{A}(x)) \rangle = \langle \underline{A}(x), \overline{A}(x) \rangle = x$$

So

$$Card_{\max}(x \cap x^c) = Card_{\max}(x^c) = 1 - \underline{A}(x)$$

$$Card_{\max}(x \cup x^c) = Card_{\max}(x) = \overline{A}(x)$$

Thus for $\underline{A}(x) + \overline{A}(x) \geq 1$

$$E(A) = \frac{1}{Card(U)} \sum_{x \in U} \frac{1 - \underline{A}(x)}{\overline{A}(x)} = \frac{1}{Card(U)} \sum_{x \in U} \frac{Card_{max}(x \cap x^c)}{Card_{min}(x \cup x^c)}$$

(2) If $\underline{A}(x) + \overline{A}(x) < 1$ then

$$\underline{A}(x) < 1 - \overline{A}(x) = \underline{A}(x^c)$$

And

$$x \cap x^c = \langle \underline{A}(x), 1 - \underline{A}(x^c) \rangle = \langle \underline{A}(x), \overline{A}(x) \rangle = x$$

$$x \cup x^c = \langle \underline{A}(x^c), 1 - (1 - \underline{A}(x)) \rangle = x^c$$

Thus for $\underline{A}(x) + \overline{A}(x) \leq 1$

$$E(A) = \frac{1}{Card(U)} \sum_{x \in U} \frac{\overline{A}(x)}{1 - \underline{A}(x)} = \frac{1}{Card(U)} \sum_{x \in U} \frac{Card_{max}(x \cap x^c)}{Card_{min}(x \cup x^c)}$$

Example 3.1. Let

$$U = \{x_1, x_2, x_3, x_4, x_5\} = \{\langle \frac{3}{4}, \frac{5}{6}\rangle, \langle \frac{1}{4}, \frac{1}{2}\rangle, \langle \frac{1}{3}, \frac{1}{2}\rangle, \langle \frac{0}{,} \frac{1}{3}\rangle, \langle \frac{1}{2}, \frac{3}{4}\rangle\} \in RFS(U)$$

We can calculate the entropy $E(A) = 0.51$ for A using Eq (20) and Eq (27) respectively.

Example 3.2. Suppose that $x = \langle \frac{3}{4}, \frac{5}{6}\rangle$ and $y = \langle \frac{1}{6}, \frac{1}{4}\rangle \in A$, then due to Eq(20), we obtain $E_A(x) = 0.3 = E_A(y)$. This results indicate: although x and y are two different rough fuzzy values, (we have $\underline{A}(x) + \overline{A}(x) \geq 1$ for x and we have $\underline{A}(x) + \overline{A}(x) \leq 1$ for y) maybe exist $E_A(x) = E_A(y)$. It shows Eq(21) couldn't distinguish the above two different rough fuzzy valuesTherefore, we need define more elaborate fuzzy entropy.

Definition 3.2. Let $A \in RFS(U)$ and $x = \langle \underline{A}(x), \overline{A}(x)\rangle$ then
(1) The fuzzy positive entropy of x for A is defined as

$$E_A^P(x) = \frac{\overline{A}(x) - \underline{A}(x)}{\overline{A}(x)} \tag{28}$$

(2) The fuzzy negative entropy of x for A is defined as

$$E_A^N(x) = \frac{\overline{A}(x) - \underline{A}(x)}{1 - \overline{A}(x)} \tag{29}$$

It is clear that $0 \leq E_A^P \leq 1$ and $0 \leq E_A^N(x) \leq 1$. And

$$E^P(A) = \frac{1}{Card(U)} \sum_{x \in A} E_A^P(x) \tag{30}$$

$$E^N(A) = \frac{1}{Card(U)} \sum_{x \in A} E_A^N(x) \tag{31}$$

Example 3.3. Suppose that x and y are above Ex.3.2. Then due to Eq.(28) and Eq.(29),we obtain

$$E_A^P(x) = \frac{1}{10},\ E_A^N(x) = \frac{1}{3},\ E_A^P(y) = \frac{1}{3},\ E_A^N(y) = \frac{3}{10}$$

Although $E_A(x) = E_A(y)$, but $\underline{A}(x) + \overline{A}(x) \geq 1$ implies $E_A^P(x) \leq E_A^P(y)$ and $E_A^N(x) \geq E_A^N(y)$. Which show: the degree of fuzziness of x is mainly caused by the absence of negative degree, whereas the degree of fuzziness of y is mainly caused by the absence of positive degree.

The relation among $E_A(x)$, $E^P(A)$, and $E^N(A)$ can be described by following theorem.

Theorem 3.2. Let $A \in RFS(U)$ and $x = \langle \underline{A}(x), \overline{A}(x) \rangle$. Then

$$E(x) = \begin{cases} \dfrac{E_A^P(x)}{E_A^N(x)} & if\ \underline{A}(x) + \overline{A}(x) \geq 1 \\ \\ \dfrac{E_A^N(x)}{E_A^P(x)} & if\ \underline{A}(x) + \overline{A}(x) \leq 1 \end{cases} \tag{32}$$

Definition 3.3. Let $A \in RFS(U)$. Then
(1) The fuzzy positive entropy for A is defined as

$$E^P(A) = \frac{1}{Card(U)} \sum_{x \in A} E_A^P(x) = \frac{1}{Card(U)} \sum_{x \in A} \frac{\overline{A}(x) - \underline{A}(x)}{\overline{A}(x)} \tag{33}$$

(2) The fuzzy negative entropy for A is defined as

$$E^N(A) = \frac{1}{Card(U)} \sum_{x \in A} E_A^N(x) = \frac{1}{Card(U)} \sum_{x \in A} \frac{\overline{A}(x) - \underline{A}(x)}{1 - \underline{A}(x)} \tag{34}$$

(3) The entropy for A is defined as

$$E(A) = \frac{1}{Card(U)} \sum_{x \in A} E_A(x) \tag{35}$$

Theorem 3.3. Let $A \in RFS(U)$, $E^P(A)$, $E^N(A)$ and $E(A)$ be above. Then
(1)

$$E^P(A) = E^N(A^c) \tag{36}$$

(2)

$$E^N(A) = E^P(A^c) \tag{37}$$

(3)

$$E(A) = E(A^c) \tag{38}$$

Proof
(1)

$$E^N(A^c) = \frac{1}{Card(U)} \sum_{x \in A} \frac{\overline{A}^c(x) - \underline{A}^c(x)}{1 - \underline{A}^c(x)} = \frac{1}{Card(U)} \sum_{x \in A} \frac{\overline{A}(x) - \underline{A}(x)}{\overline{A}(x)} = E^P(A)$$

(2) Similarly to (1).
(3) Suppose that $x = \langle \underline{A}(x), \overline{A}(x) \rangle$. If $\underline{A}(x) + \overline{A}(x) \leq 1$, then $E_A^P(x) > E_A^N(x)$.

It implies that

$$E_A(x) = \frac{\overline{A}(x)}{1 - \underline{A}(x)} = \frac{1 - \underline{A}^c(x)}{\overline{A}^c(x)} = E_{A^c}(x)$$

Similarly if $\underline{A}(x) + \overline{A}(x) \geq 1$, we have

$$E_A(x) = \frac{1 - \underline{A}(x)}{\overline{A}(x)} = \frac{\overline{A}^c(x)}{1 - \underline{A}^c(x)} = E_{A^c}(x)$$

Thus $E(A) = E(A^c)$.

Theorem 3.4. Let $A \in RFS(U)$, $E^P(A), E^N(A)$ and $E(A)$ be above.
(1) If $\underline{A}(x) \leq \underline{A}(y)$ and $\overline{A}(x) = \overline{A}(y)$, then we have

$$E_A^P(x) \geq E_A^P(y),\ E_A^N(x) \geq E_A^N(y) \tag{39}$$

(2) If $\underline{A}(x) = \underline{A}(y)$ and $\overline{A}(x) \geq \overline{A}(y)$,then we have

$$E_A^P(x) \geq E_A^P(y),\ E_A^N(x) \geq E_A^N(y) \tag{40}$$

(3) If $x \leq y$ i.e $\underline{A}(x) \leq \underline{A}(y)$ and $\overline{A}(x) \geq \overline{A}(y)$, then we have

$$E_A^P(x) \geq E_A^P(y),\ E_A^N(x) \geq E_A^N(y) \tag{41}$$

Proof. (1) Suppose that $\underline{A}(x) \leq \underline{A}(y)$ and $\overline{A}(x) = \overline{A}(y)$, then

$$E_A^P(x) = \frac{\overline{A}(x) - \underline{A}(x)}{\overline{A}(x)} \geq \frac{\overline{A}(y) - \underline{A}(y)}{\overline{A}(y)} = E_A^P(y)$$

Since the function $y = \frac{a-x}{1-x}$ decreases when $x \neq 1$ and $a \leq 1$, then

$$E_A^N(x) = \frac{\overline{A}(x) - \underline{A}(x)}{1 - \underline{A}(x)} \geq \frac{\overline{A}(y) - \underline{A}(y)}{1 - \underline{A}(y)} = E_A^N(y)$$

(2) Suppose that $\underline{A}(x) = \underline{A}(y)$ and $\overline{A}(x) \geq \overline{A}(y)$, then since the function $y = \frac{x-a}{x}$ increases when $a \geq 0$, then

$$E_A^P(x) = \frac{\overline{A}(x) - \underline{A}(x)}{\overline{A}(x)} \geq \frac{\overline{A}(y) - \underline{A}(y)}{\overline{A}(y)} = E_A^P(y)$$

And

$$E_A^N(x) = \frac{\overline{A}(x) - \underline{A}(x)}{1 - \underline{A}(x)} \geq \frac{\overline{A}(y) - \underline{A}(y)}{1 - \underline{A}(y)} = E_A^N(y)$$

(3) If $x \leq y$, suppose that $Z = \langle \underline{A}(z), \overline{A}(z) \rangle$, where $\underline{A}(z) = \underline{A}(y)$ and $\overline{A}(z) = \overline{A}(y)$. Thus

$$E_A^P(x) \geq E_A^P(z) \geq E_A^P(y),\ E_A^N(x) \geq E_A^N(z) \geq E_A^N(y)$$

4 Conclusions

We have introduced the entropy ,fuzzy positive entropy and fuzzy negative entropy measure for rough fuzzy sets. It can make the expression of fuzzy information more clear. Some properties of entropy measure for rough fuzzy sets are discussed. It is also shown that the proposed entropy measure can be defined in terms of the ratio of rough fuzzy cardinalities.

References

1. L.A. Zadeh (1965) Fuzzy Sets and Systems, in: Proc. Symp. on Systems Theory, Polytechnic Institute of Brooklyn, New York, pp. 29 -37.
2. E.T. Jaynes, Where do we stand on maximum entropy? in: Levine, Tribus (Eds.), The Maximum Entropy Formalism, MIT Press,Cambridge, MA.
3. A. De Luca, S. Termini (1972) A definition of a non-probabilistic entropy in the setting of fuzzy sets theory, Inform. and Control 20: 301-312.
4. A. Kaufmann (1975) Introduction to the Theory of Fuzzy Subsets vol. 1: Fundamental Theoretical Elements, Academic Press, New York.
5. R.R. Yager (1979) On the measure of fuzziness and negation. Part I: Membership in the unit interval, Internat. J. General Systems 5: 189 -200.
6. B. Kosko (1986) Fuzzy entropy and conditioning, Inform. Sci. 40 (2): 165 -174.
7. B. Kosko (1990) Fuzziness vs. probability, Internat. J. General Systems 17 (23): 211-240.
8. B. Kosko (1997) Fuzzy Engineering, Prentice-Hall, Englewood Cliffs, NJ.
9. E. Szmidt and J. Kacprzyk (2001) Entropy for intuitionistic fuzzy sets, Fuzzy Sets and Systems 118 (3):467-477.
10. K. Atanassov (1986) Intuitionistic fuzzy sets, Fuzzy Sets and Systems 20 (1): 87 - 96.
11. K. Atanassov (1994) New operations defined over the intuitionistic fuzzy sets, Fuzzy Sets and Systems 61 (2): 137-142.
12. K. Atanassov (1999) Intuitionistic Fuzzy Sets. Theory and Applications, Physica-Verlag,Heidelberg /New York.
13. Pawlak Z (1982) Rough Sets[J], Inter.J Inform.Comput.Sci. ,11(5): 341-356.
14. Dubios D and Prade H (1990) Rough fuzzy sets and fuzzy rough sets, International Journal of General Systems17:190-209.
15. Zhang Chengyi, Dang Pingan,Li dongya (2004) "Note on "Fuzzy entropy of vague sets and its construction method" Computer Applications and Software, 21, No. 5 :27 104. (in chinese)
16. Zhang Chengyi,et.al (2005) On measures of similarity between fuzzy rough sets, International Journal of Pure and Applied Mathematics, 10(4),451-460.
17. Zhang Chengyi and Fu Haiyan (2006) Similarity Measures on Three Kinds of Fuzzy Sets, Pattern Recognition Letters 27: 1307-1317.

A Fuzzy Measure Based on Variable Precision Rough Sets

Shen-Ming Gu[1,2], Ji Gao[2], and Xiao-Qiu Tan[1]

[1] Information School, Zhejiang Ocean University, Zhoushan, Zhejiang 316004, PR China
gsm@zjou.edu.cn, tanxq@zjou.edu.cn
[2] Artificial Intelligence Institute, Zhejiang University, Hangzhou, Zhejiang 310027, PR China
gaoji@mail.hz.zj.cn

Abstract. A variable precision rough set(VPRS) is an extension of a Pawlak rough set. By setting a threshold β, VPRS loosens the strict definition of approximate boundary in Pawlak rough sets. This paper deals with uncertainty of rough sets based on the VPRS model. A measure is first defined to characterize fuzziness of a set in an information system. A pair of lower and upper approximations based on the fuzzy measure are then defined. Properties of the fuzzy measure and approximations are also examined.

Keywords: Fuzzy measure; rough sets; variable precision rough sets.

1 Introduction

Rough set theory, proposed by Pawlak [11], is an extension of set theory for the study of intelligent systems characterized by insufficient and incomplete information. The concepts of lower and upper approximations being used in rough set theory, knowledge hidden in information systems may be unravelled and expressed in the form of decision rules [2, 8]. With more then twenty years development, rough set theory has been applied successfully in machine learning, pattern recognition, knowledge acquisition and decision analysis, etc [12, 16, 17, 21].

An object classified by using Pawlak rough set is assumed that there is a completely certainty that it is a correct classification by an equivalence relation [11]. Namely, an object belongs to or not to a classification. An object cannot be classified in a level of confidence in its correct classification.

In practice, it seems that admitting some level of uncertainty in the classification process may lead to a deeper understanding and a better utilization of properties of the data being analyzed. To deal with uncertainty in such a case, an extended variable precision rough set(VPRS) model was proposed by Ziarko [22], by setting a threshold β, VPRS loosens the strict definition of approximate boundary in Pawlak rough sets for probabilistic classification. Quite different in Pawlak rough sets, when an object is classified in VPRS, there is a level of confidence in its correct classification, which helps to discover related knowledge from non-related data.

B.-Y. Cao (Ed.): Fuzzy Information and Engineering (ICFIE), ASC 40, pp. 798–807, 2007.
springerlink.com

The theory of fuzzy set initiated by Zadeh [19] is another tool to deal with uncertainty and vagueness. Dubois and Prade [5] made an investigation on relationship between rough set theory and fuzzy set theory, and reported that they are not rival theories and they aim at two different purposes. Many researchers studied rough sets in fuzzy environment and obtained the concepts of rough fuzzy sets and fuzzy rough sets [1, 4, 5, 6, 7, 10, 12, 13, 14, 15, 16].

In this paper, a fuzzy measure of knowledge is introduced based on VPRS, a pair of lower and upper approximation operators are described in fuzzy sets and some properties of the fuzzy measure and approximations are examined.

2 Basic Notions Related to Rough Sets

2.1 Pawlak Rough Sets

Let U be a nonempty finite set of objects called universe, and let R be an equivalence binary relation on U, i.e., R is reflexive, symmetric, and transitive. The pair (U, R) is called a Pawlak approximation space. The equivalence relation R partitions U into disjoint subsets called equivalence classes. If two elements $x, y \in U$ belong to the same equivalence class, we say that x and y are indiscernible. Given an arbitrary set $X \subseteq U$, it may be impossible to describe X precisely by using the equivalence classes of R. In this case, one may characterize X by a pair of lower and upper approximations [11]:

$$\underline{R}(X) = \{x \in U : [x]_R \subseteq X\} \tag{1}$$

$$\overline{R}(X) = \{x \in U : [x]_R \cap X \neq \emptyset\} \tag{2}$$

where $[x]_R = \{y \in U : (x, y) \in R\}$ is the R equivalence class containing x. The pair $(\underline{R}(X), \overline{R}(X))$ is called the Pawlak rough sets of X with respect to (w.r.t.) (U, R).

2.2 Information Systems

The notion of information systems provides a convenient tool for the description of objects in terms of their attribute values. An information system is a pair (U, A), where U is a nonempty finite set of objects called universe and A is a nonempty finite set of attributes, such that $a : U \to V_a$ for any $a \in A$, i.e., $a(x) \in V_a$, where V_a is the domain of attribute a.

Each subset of attributes $B \subseteq A$ determines an indiscernibility relation as follows:

$$R_B = \{(x, y) \in U \times U : a(x) = a(y), \forall a \in B\} \tag{3}$$

R_B partitions U into a family of disjoint subsets U/R_B called a quotient set of U:

$$U/R_B = \{[x]_B : x \in U\} \tag{4}$$

Let $X \subseteq U$, $B \subseteq A$, one can characterize X by a pair of lower and upper approximations w.r.t the knowledge derived from attribute set B [8, 18, 21]:

$$\underline{R_B}(X) = \{x \in U : [x]_B \subseteq X\} = \bigcup\{[x]_B : [x]_B \subseteq X\} \quad (5)$$

$$\overline{R_B}(X) = \{x \in U : [x]_B \cap X \neq \emptyset\} = \bigcup\{[x]_B : [x]_B \cap X \neq \emptyset\} \quad (6)$$

The lower approximation $\underline{R_B}(X)$ is the set of objects that belong to X with certainty, while the upper approximation $\overline{R_B}(X)$ is the set of objects that possibly belong to X. The pair $(\underline{R_B}(X), \overline{R_B}(X))$ is referred to as the rough sets of X w.r.t. B.

An information system $S = (U, A)$ is referred to as a decision table if $A = C \cup D$ and $C \cap D = \emptyset$, in this case, C is called the conditional attribute set and D is called the decision attribute set. If $D = \{d\}$, then d is called the decision. If $R_C \subseteq R_D$, then $S = (U, A)$ is consistent, otherwise it is inconsistent [2, 20].

Example 2.1. Table 1 illustrates a decision table $S = (U, A)$, where $U = \{x_1, x_2, \ldots, x_7\}$, the conditional attribute set $C = \{a_1, a_2, \ldots, a_5\}$, the decision is d.

Table 1. A decision table

U	a_1	a_2	a_3	a_4	a_5	d
x_1	0	0	0	1	0	T
x_2	0	0	0	1	0	F
x_3	1	0	1	1	1	T
x_4	0	0	0	1	0	F
x_5	1	1	1	0	0	F
x_6	0	1	1	1	1	T
x_7	1	1	1	0	1	F

In the terms of binary relation R_C derived from the conditional attribute set, the universe U is partitioned into the following equivalence classes:

$$U/C = \{X_1, X_2, X_3, X_4, X_5\} \quad (7)$$

where $X_1 = \{x_1, x_2, x_4\}$, $X_2 = \{x_3\}$, $X_3 = \{x_5\}$, $X_4 = \{x_6\}$, $X_5 = \{x_7\}$.

$$U/\{d\} = \{D_T, D_F\} \quad (8)$$

where $D_T = \{x_1, x_3, x_6\}$, $D_F = \{x_2, x_4, x_5, x_7\}$.

2.3 Variable Precision Rough Sets

Let $S = (U, A)$ be a decision table, $A = C \cup D$, $X \subseteq U$, $B \subseteq C$, $0.5 < \beta \leq 1$, the β-lower approximation and β-upper approximation of X in S are defined [22], respectively, by

$$\underline{R_B^{\beta}}(X) = \bigcup\{[x]_B : \frac{|[x]_B \cap X|}{|[x]_B|} \geq \beta\} \tag{9}$$

$$\overline{R_B^{\beta}}(X) = \bigcup\{[x]_B : \frac{|[x]_B \cap X|}{|[x]_B|} > 1 - \beta\} \tag{10}$$

where $|X|$ denotes the cardinality of the set X.

The measure of quality of classification w.r.t. B is defined [21] by

$$\gamma^{\beta}(B, D) = \frac{|\cup\{[x]_B : \frac{|[x]_B \cap X|}{|[x]_B|} \geq \beta\}|}{|U|} \tag{11}$$

The value $\gamma^{\beta}(B, D)$ measures the proportion of objects in the universe for which classification is possible at the specified value of β.

A β-reduct B is a minimal subset of C which keeps the quality of classification unchanged at the specified value β. Thus a β-reduct B has two properties [3, 9]:

(1) $\gamma^{\beta}(C, D) = \gamma^{\beta}(B, D)$;

(2) No proper subset of B can give the same quality of classification at the same β value.

Example 2.2. β-reducts and quality of classification computed by using the same method as depicted in Table 1 are shown in Table 2.

Table 2. β-reducts and quality of classification

β-reduct	Quality of classification	β
$\{a_1, a_2\}$	1.00	0.67
$\{a_1, a_3\}$	1.00	0.67
$\{a_3, a_4\}$	0.57	1.00

3 A Fuzzy Measure Based on VPRS

3.1 Basic Notions of Fuzzy Sets

Let U be a finite and non-empty set called universe. A fuzzy subset A of U is defined by a membership function [19]:

$$\mu_A : U \to [0, 1] \tag{12}$$

Fuzzy set inclusion and equality can be defined component-wise. A fuzzy set A is a subset of fuzzy set B, written $A \subseteq B$, if and only if $\mu_A(x) \leq \mu_B(x)$ for all $x \in U$. Fuzzy set A is equal to fuzzy set B, written $A = B$, if and only if $\mu_A(x) = \mu_B(x)$ for all $x \in U$.

The definitions for fuzzy set complement, intersection, and union in the sense of Zadeh [19], are respectively defined as follows:

$$\mu_{\neg A}(x) = 1 - \mu_A(x) \tag{13}$$

$$\mu_{A \cap B}(x) = \min(\mu_A(x), \mu_B(x)) \tag{14}$$

$$\mu_{A \cup B}(x) = \max(\mu_A(x), \mu_B(x)) \tag{15}$$

A crisp subset of U may be viewed as a degenerated fuzzy set. In this case, the membership function is the characteristic function taking only two values, 0 and 1. The min-max fuzzy set theoretic operators are reduced to the classical set-theoretic operators when characteristic functions are used.

3.2 A Fuzzy Measure Based on VPRS

Suppose $S = (U, A)$ is an information system, and R is the equivalence relation on U induced by A, i.e., $R = R_A$. The equivalence relation R generates a partition U/R of the universe U, i.e., a family of disjoint subsets of the universe known as equivalence classes. We see that each element $x \in U$ belong to one and only one equivalence class $[x]$, where $[x] = \{y \in U : (x, y) \in R\}$. For a subset $X \subseteq U$, a membership function can be defined [14, 18] as follows:

$$\mu_X(x) = \frac{|[x] \cap X|}{|[x]|}, \quad x \in U \tag{16}$$

Obviously, $0 \leq \mu_X(x) \leq 1, \forall x \in U$. By the definition, the membership values are all rational numbers. If two elements are in the same equivalence class they must have the same degree of membership. The value $\mu_X(x)$ may be interpreted as the conditional probability, that is, an arbitrary element belongs to X given that the element belongs to $[x]$. The set X is called a generating set of the membership μ_X.

With the membership function $\mu_X(x)$, we can define a fuzzy set F_X on U as follows:

$$F_X = \{\mu_X(x)/x : \mu_X(x) = \frac{|[x] \cap X|}{|[x]|}, x \in U\} \tag{17}$$

For each element $x \in U$, if $\mu_X(x) = 0$, then x does not belong to F_X, and if $0 < \mu_X(x) < 1$, then x possibly belongs to F_X, in this case, the possibility of x belonging to F_X is $\mu_X(x)$.

Example 3.1. The sets D_T and D_F are from Example 2.1. By the definition of the membership functions, fuzzy sets F_{D_T} and F_{D_F} are given, respectively, by

$$F_{D_T} = \{0.33/x_1, 0.33/x_2, 1/x_3, 0.33/x_4, 0/x_5, 1/x_6, 0/x_7\},$$

$$F_{D_F} = \{0.67/x_1, 0.67/x_2, 0/x_3, 0.67/x_4, 1/x_5, 0/x_6, 1/x_7\}.$$

Proposition 1. Let $S = (U, A)$ be an information system, and $X, Y \subseteq U$, if $X \subseteq Y$, then $F_X \subseteq F_Y$.

Proof. $\forall x \in U$, since $X \subseteq Y$, we have $|[x] \cap X| \leq |[x] \cap Y|$, then

$$\frac{|[x] \cap X|}{|[x]|} \leq \frac{|[x] \cap Y|}{|[x]|}. \tag{18}$$

That is $\mu_X(x) \leq \mu_Y(x)$, for all $x \in U$, therefore $F_X \subseteq F_Y$.

Proposition 2. Let $S = (U, A)$ be an information system, $\forall X, Y \subseteq U$, then

$$F_{X \cap Y} \subseteq F_X \cap F_Y; \tag{19}$$

$$F_{X \cup Y} \supseteq F_X \cup F_Y. \tag{20}$$

Proof. (19) $\forall x \in U$,

$$\begin{aligned}
\mu_{F_{X \cap Y}}(x) &= \frac{|[x] \cap (X \cap Y)|}{|[x]|} \\
&= \frac{|([x] \cap X) \cap ([x] \cap Y)|}{|[x]|} \\
&\leq \frac{\min\{|[x] \cap X|, |[x] \cap Y|\}}{|[x]|} \\
&= \min\{\frac{|[x] \cap X|}{|[x]|}, \frac{|[x] \cap Y|}{|[x]|}\} \\
&= \min\{\mu_{F_X}(x), \mu_{F_Y}(x)\} \\
&= \mu_{F_X \cap F_Y}(x).
\end{aligned}$$

Therefore $F_{X \cap Y} \subseteq F_X \cap F_Y$.

(20) It is similar to the Proof of (19).

Suppose $S = (U, A)$ is an information system, $X \subseteq U$, and F_X is the fuzzy set of U induced from X. Then lower and upper approximations of F_X w.r.t. (U, R) are respectively defined [5, 10, 15] by

$$\mu_{\underline{F_X}}(x) = \inf\{\mu_X(y) : y \in [x]\}, \forall x \in U \tag{21}$$

$$\mu_{\overline{F_X}}(x) = \sup\{\mu_X(y) : y \in [x]\}, \forall x \in U \tag{22}$$

Moreover, if $0.5 < \beta \leq 1$, then the β-lower approximation and β-upper approximation of F_X w.r.t. (U, R) are defined as follows:

$$\underline{F_X^\beta} = \bigcup\{[x] : \mu_{\underline{F_X}}(x) \geq \beta,\ x \in U\} \tag{23}$$

$$\overline{F_X^\beta} = \bigcup\{[x] : \mu_{\overline{F_X}}(x) > 1 - \beta,\ x \in U\} \tag{24}$$

The β-lower approximation $\underline{F_X^\beta}$ of fuzzy F_X can be interpreted as the union of the equivalence classes of the elements whose degree of confidence belonging to F_X is not less than β, where the β-upper approximation $\overline{F_X^\beta}$ can be interpreted as the union of the equivalence classes of the elements whose membership degree belonging to F_X is grater than $1-\beta$.

Proposition 3. Let $S=(U,A)$ be an information system, $X \subseteq U$, F_X is the fuzzy set induced from X, then

$$\underline{F_X^\beta} \subseteq \overline{F_X^\beta}. \tag{25}$$

Proof. $\because 0.5 < \beta \leq 1, \therefore 0 \leq 1-\beta < 0.5$, then

$$\begin{aligned}
\underline{F_X^\beta} &= \bigcup\{[x] : \mu_{\underline{F_X}}(x) \geq \beta,\ x \in U\} \\
&= \bigcup\{[x] : \inf\{\mu_{(F_X)}(y) : y \in [x]\} \geq \beta,\ x \in U\} \\
&\subseteq \bigcup\{[x] : \sup\{\mu_{(F_X)}(y) : y \in [x]\} \geq \beta,\ x \in U\} \\
&\subseteq \bigcup\{[x] : \sup\{\mu_{(F_X)}(y) : y \in [x]\} > 1-\beta,\ x \in U\} \\
&= \bigcup\{[x] : \mu_{\overline{F_X}}(x) > 1-\beta,\ x \in U\} \\
&= \overline{F_X^\beta}.
\end{aligned}$$

Proposition 4. Let $S=(U,A)$ be an information system, and $X, Y \subseteq U$, F_X and F_Y are the fuzzy sets induced from X and Y, if $X \subseteq Y$, then

$$\underline{F_X^\beta} \subseteq \underline{F_Y^\beta} \tag{26}$$

$$\overline{F_X^\beta} \subseteq \overline{F_Y^\beta}. \tag{27}$$

Proof. (26) $X \subseteq Y$, according Proposition 1, we have $F_X \subseteq F_Y$, i.e. $\forall x \in U$, $\mu_X(x) \leq \mu_Y(x)$, then,

$$\begin{aligned}
\underline{F_X^\beta} &= \bigcup\{[x] : \mu_{\underline{F_X}}(x) \geq \beta,\ x \in U\} \\
&= \bigcup\{[x] : \inf\{\mu_X(y) : y \in [x]\} \geq \beta,\ x \in U\} \\
&\subseteq \bigcup\{[x] : \inf\{\mu_Y(y) : y \in [x]\} \geq \beta,\ x \in U\} \\
&= \bigcup\{[x] : \mu_{\underline{F_Y}}(x) \geq \beta,\ x \in U\} \\
&= \underline{F_Y^\beta}.
\end{aligned}$$

(27) It is similar to the Proof of (26).

Proposition 5. Let $S=(U,A)$ be an information system, $X, Y \subseteq U$, F_X and F_Y are the fuzzy sets induced from X and Y, then

$$\underline{(F_X \cap F_Y)^\beta} = \underline{F_X^\beta} \cap \underline{F_Y^\beta} \tag{28}$$

$$\overline{(F_X \cup F_Y)^\beta} = \overline{F_X^\beta} \cup \overline{F_Y^\beta} \tag{29}$$

Proof. (28) $\underline{(F_X \cap F_Y)^\beta} = \bigcup\{[x] : \mu_{\underline{(F_X \cap F_Y)}}(x) \geq \beta,\ x \in U\}$

$$
\begin{aligned}
&= \bigcup\{[x] : \inf\{\mu_{(F_X \cap F_Y)}(y) : y \in [x]\} \geq \beta,\ x \in U\} \\
&= \bigcup\{[x] : \inf\{\min\{\mu_{(F_X)}(y), \mu_{(F_Y)}(y)\} : y \in [x]\} \geq \beta,\ x \in U\} \\
&= \bigcup\{[x] : \min\{\inf\{\mu_X(y) : y \in [x]\}, \inf\{\mu_Y(y) : y \in [x]\}\} \geq \beta,\ x \in U\} \\
&= \bigcup\{[x] : \min\{\mu_{\underline{F_X}}(x), \mu_{\underline{F_Y}}(x)\} \geq \beta,\ x \in U\} \\
&= \bigcup\{[x] : \mu_{\underline{F_X}}(x) \geq \beta, \mu_{\underline{F_Y}}(x) \geq \beta,\ x \in U\} \\
&= \bigcup\{\{[x] : \mu_{\underline{F_X}}(x) \geq \beta\} \cap \{[x] : \mu_{\underline{F_Y}}(x) \geq \beta\},\ x \in U\} \\
&= \{\bigcup\{[x] : \mu_{\underline{F_X}}(x) \geq \beta,\ x \in U\}\} \cap \{\bigcup\{[x] : \mu_{\underline{F_Y}}(x) \geq \beta,\ x \in U\}\} \\
&= \underline{F_X^\beta} \cap \underline{F_Y^\beta}.
\end{aligned}
$$

(29) It is similar to the Proof of (28).

Proposition 6. Let $S = (U, A)$ be an information system, and $X, Y \subseteq U$, F_X and F_Y are the fuzzy sets induced from X and Y, then

$$\overline{(F_X \cap F_Y)^\beta} \subseteq \overline{F_X^\beta} \cap \overline{F_Y^\beta} \tag{30}$$

$$\underline{F_X^\beta} \cup \underline{F_Y^\beta} \subseteq \underline{(F_X \cup F_Y)^\beta} \tag{31}$$

Proof. (30) $\overline{(F_X \cap F_Y)^\beta} = \bigcup\{[x] : \mu_{\overline{(F_X \cap F_Y)}}(x) > 1 - \beta,\ x \in U\}$

$$
\begin{aligned}
&= \bigcup\{[x] : \sup\{\mu_{(F_X \cap F_Y)}(y) : y \in [x]\} > 1 - \beta,\ x \in U\} \\
&= \bigcup\{[x] : \sup\{\min\{\mu_{(F_X)}(y), \mu_{(F_Y)}(y)\} : y \in [x]\} > 1 - \beta,\ x \in U\} \\
&\subset \bigcup\{[x] : \min\{\sup\{\mu_X(y) : y \in [x]\}, \sup\{\mu_Y(y) : y \in [x]\}\} > 1 - \beta,\ x \in U\} \\
&= \bigcup\{[x] : \min\{\mu_{\overline{F_X}}(x), \mu_{\overline{F_Y}}(x)\} > 1 - \beta,\ x \in U\} \\
&= \bigcup\{[x] : \mu_{\overline{F_X}}(x) > 1 - \beta, \mu_{\overline{F_Y}}(x) > 1 - \beta,\ x \in U\} \\
&= \bigcup\{\{[x] : \mu_{\overline{F_X}}(x) > 1 - \beta\} \cap \{[x] : \mu_{\overline{F_Y}}(x) > 1 - \beta\},\ x \in U\} \\
&= \{\bigcup\{[x] : \mu_{\overline{F_X}}(x) > 1 - \beta,\ x \in U\}\} \cap \{\bigcup\{[x] : \mu_{\overline{F_Y}}(x) > 1 - \beta,\ x \in U\}\} \\
&= \overline{F_X^\beta} \cap \overline{F_Y^\beta}.
\end{aligned}
$$

(31) It is similar to the Proof of (30).

4 Conclusion

We have introduced in this paper a fuzzy measure based on VPRS model, and we've also defined the concepts of β-lower approximation and the β-upper approximation by using the fuzzy measure. The β-lower approximation and the

β-upper approximation not only can be interpreted clearly, but also have good properties in intelligence information systems. The results may be useful for the applications of the fuzzy set theory and rough set theory.

Acknowledgments

This work is supported by a grant from the National Natural Science Foundation of China (No.60373078, No.60673096), and also supported by Scientific Research Fund of Zhejiang Provincial Education Department in China (No.20050125).

References

1. Banerjee, M., Pal, S.K.(1996) Roughness of a fuzzy set, Information Science, 93: 235–246
2. Bazan, J.A.(1998) Comparison of dynamic and non-dynamic rough set methods for extracting laws from decision tables, in: Polkowski, L., Skowron A. (Eds.), Rough Sets in Knowledge Discovery 1, Physica-Verlag, Heidelberg
3. Beynon, M.(2001) Reducts within the variable precision rough sets model: A further investigation, European Journal of Operational Research, 134: 592–605
4. Chakrabarty, K., Biswas, R., and Nanda, S.(2000) Fuzziness in rough sets, Fuzzy Sets and Systems, 110: 247–251
5. Dubois, D., Prade, H.(1990) Rough fuzzy sets and fuzzy rough sets, International Journal of General System, 17: 191–208
6. Dubois, D., Prade, H.(1987) Twofold fuzzy sets and rough sets–some issue in knowledge representation, Fuzzy Sets and Systems, 23: 3–18
7. Hong, T.P., Wang, T.T., and Wang, S.L. et al.(2000) Learning a coverage set of maximally general fuzzy rules by rough sets, Expert System with Applications, 19: 97–103
8. Kryszkiewicz, M.(1999) Rules in incomplete information systems. Information Sciences 113: 271–292
9. Mi, J.S., Wu, W.Z., Zhang, W.X.(2004) Approaches to knowledge reductions based on variable precision rough sets model. Information Sciences 159: 255–272
10. Nakamura, A.(1988) Fuzzy rough sets, Note on Multiple-valued Logic in Japan, 9: 1–8
11. Pawlak, Z.(1982) Rough sets, International Journal of Computer and Information Science, 11: 341–356
12. Pawlak, Z.(1991) Rough sets: Theoretical aspects of reasioning about data, Kuwer Academic Publishers, Boston
13. Pawlak, Z.(1985) Rough sets and fuzzy sets, Fuzzy Sets and Systems, 17: 99–102
14. Pawlak, Z., Skowron, A.(1994) Rough membership functions, in: Yager, R.R., Fedrizzi, M., Kacprzyk, J. (Eds.), Advances in the Depster-Shasets Theory of Evidence, John Wiley and Sons, New York
15. Wu, W.Z., Mi, J.S., and Zhang, W.X.(2003) Generalized fuzzy rough sets, Information Sciences, 151: 263–282
16. Wu, W.Z., Zhang, W.X., Li, H.Z.(2003) Knowledge acquisition in incomplete fuzzy information systems via rough set approach. Expert Systems 20: 280–286
17. Yao, Y. Y.(1998) Generalized rough set models, in: Polkowski, L., Skowron, A. (Eds.), Rough Sets in Knowledge Discovery: 1. Methodology and Applications, Physica- Verlag, Heidelberg

18. Yao, Y.Y., Wong, S.K.M.(1992) A decision theoretic framework for approximating concepts, International Jourmal of Man-machine Studies, 37: 793–809
19. Zadeh, L.A.(1965) Fuzzy sets, Information and Control, 8: 338–353
20. Zhang, W.X., Mi, J.S., Wu, W.Z.(2003) Approaches to knowledge reductions in inconsistent systems. International Journal of Intelligent Systems, 21: 989–1000
21. Zhang, W.X., Wu, W.Z. Liang, J.Y., Li, D.Y.(2001) Rough Set Theory and Approaches, Science Press, Beijing
22. Ziarko, W.(1993) Variable precision rough set model, Journal of Computer and System Science, 46: 39–59

Rough Sets of System

Dong Qiu and Lan Shu

School of Applied Mathematics,
University of Electronic Science and Technology of China,
Chengdu, Sichuan, 610054, P.R. China
qiudong77@sohu.com, shul@uestc.edu.cn

Abstract. The classical rough set theory mostly emphasizes the differentiation among the elements of the universe, but does not investigate the structure of the universe. This paper extends rough set theory from the set to the system, making it not only can well classify elements of the universe but also can reveal the interrelation among elements of the universe. Then some properties about the approximation operators are discussed. Finally, some examples of rough sets of system are given.

Keywords: Rough sets, system, lower system approximation, upper system approximation, lower approximation, upper approximation.

1 Introduction

In Pawlak's rough set model [1, 2] and its many extensions such as [3, 4, 5, 6, 7], the classical rough set theory is based on the ordinary set called the universe. This theory assumes that we have initially some information about elements of the universe. Because some elements are indiscernible in view of the available information, the information associated with elements of the universe generates an indiscernibility relation on its elements. When our knowledge increases to such an extent that all elements can be discerned, there would not be indiscernible phenomenon on the universe.

However, we usually need to research the universe that has some structure (or interrelation) existing among its elements, such as a group or a ring in the algebra, a group of physiologically or anatomically complementary organs or parts, a group of interacting mechanical or electrical components. Therefore we think that our comprehension of objects actually contains two aspects, the differentiation and the structure (or interrelation) among the objects. The information about the universe not only reflects the differentiation but also the structure (or interrelation). The vagueness of the set causing by imperfect knowledge about the universe also contains the two aspects. In other words, because of the lack of information, the differentiation can not be exactly discerned, and for the same reason, the structure can not be exactly discerned either. As usual, the rough set theory mostly emphasizes the differentiation, but does not investigate the structure. Therefore it is necessary to extend the rough set theory so that it can take both aspects into consideration.

B.-Y. Cao (Ed.): Fuzzy Information and Engineering (ICFIE), ASC 40, pp. 808–815, 2007.
springerlink.com

In 1996, Pawlak.Z extended rough set theory to real line [8], this is a meaningful try. In this paper, we will establish rough set theory based on the general case, and discuss some of its properties.

2 Rough Sets of System

A *system* is a set of interacting, interrelated, or interdependent elements forming a complex whole. The *structure* of the system is a way in which elements are arranged or put together to form the system. A system is written as (X, f),where X is the set of elements of the system (note that generally X is not empty) and f is the structure of the system. Those above mentioned examples are just systems.

Let (X, f) be a system and $x \in X$. $x \in (X, f)$ means x is an *element* of the system (X, f) .For $A \subseteq X$, $A \subseteq (X, f)$ means A is a *part* of the system (X, f). If A is a nonempty subset of X and (A, f) is also a system, then (A, f) is called the *subsystem* of (X, f) denoted by $(A, f) \subseteq (X, f)$. For the empty set $\emptyset$, although it is unmeaning, we still define that $(\emptyset, f)$ is a subsystem of (X, f).

For $A \subseteq (X, f)$, if there exists B, which satisfies $A \subseteq B, (B, f) \subseteq (X, f)$,moreover, any C, which satisfies $A \subseteq C \subset B$, is not the set of elements of a subsystem of (X, f), then the system (B, f) is called a *spanning system* of the set A and the set of all spanning systems of A is denoted by A^s . If A^s contains only one element then we write the element as A^s without confusion.

If for any subset of X, it has only one spanning system, (X, f) is called the *accurate* system. If for any subset of X, it has at least one spanning system and there exists a subset of X, which has more than one spanning system, (X, f) is called the *inaccuracy* system. From now on, Suppose that the system is always one of above two kinds of system.

Let $(A, f) \subseteq (X, f), (B, f) \subseteq (X, f)$. $A - B = A - A \cap B$ is called the *characteristic* set of A contrasted to B and $A \cap B$ is called the *similarity* set of (A, f) and (B, f). Furthermore, let $\{(A_i, f)\}_{i \in I}$ is a collection of subsystems. $A_i - \bigcap_{j \neq i} A_j$ is called the *characteristic* set of (A_i, f) contrasted to the others and $\bigcap_{i \in I} A_i$ is called the *similarity* set of the systems $\{(A_i, f)\}_{i \in I}$.

Let (U, f) be an accurate system called the *universe* and R be a binary relation on U. By $R(x)$ called indiscernibility class of x we mean the set of all y such that xRy. For simplicity, we assume that R is an equivalence relation. The relation will be referred to as an indiscernibility relation. The *quotient set* of the set U by the relation R $\{R(x) : x \in U\}$ will be denoted by U/R. If there is a structure on U/R that is induced by f, which is denoted by f/R. In other words, if $(U/R, f/R)$ is a system, we will call $(U/R, f/R)$ a *quotient system* of the system (U, f) with respect to R.

Remark 1. *Not all equivalence relations can induce the quotient system.*

For example, in general group G, any subgroup A can induce a partition of G which is probably not a quotient group, and the partition is a quotient group iff A is normal subgroup of G.

Let $X \subseteq U$. If $(A, f/R) \subseteq (U/R, f/R)$ and $X \subseteq \bigcup_{R(x)\in A} R(x)$ and moreover, any C, which satisfies $X \subseteq \bigcup_{R(x)\in C} R(x)$ and $C \subset A$, is not the set of elements of a subsystem of $(U/R, f/R)$, then the system $(A, f/R)$ is called a *spanning system of the set* X *with respect to* R. The set of all spanning systems of X with respect to R denoted by $R^s(X)$ is called the $R-$ *upper system approximation of the set* X.

Define the $R-$ *image of the set* X by $R(X) = \{R(x) : x \in X\}$.
Let $A, B \subseteq U$,

$$\bigcup_{R(x)\in R(A)\cap R(B)} R(x)$$

is called the *similarity set of* A *and* B *with respect to*R and

$$\bigcup_{R(x)\in R(A)-R(B)} R(x)$$

is called the *characteristic set of* A *contrasted to* B *with respect to* R.

Whether the elements in exactly belong to$A \cap B$,$A - B$, $B - A$ or none of them is depended on more knowledge about the universe. This means that the similarity set of A and B with respect to R is vaguer than the characteristic set of A contrasted to B with respect to R.

Define the $R-$ *lower system approximation of the set* X by

$$R_s(X) = \left\{ (A, f/R) : (A, f/R) \subseteq (U/R, f/R), \bigcup_{R(x)\in A} R(x) \subseteq X \right\}.$$

Now we will define two basic operations on the system in the rough set theory, called the $R-$ *lower* and the $R-$*upper approximation.* The first operation is

$$\underline{R}(X) = \bigcup_{R(x)\in \cup A_i} R(x),$$

where A_i is the subset of U/R and satisfies $(A_i, f/R) \in R_s(X)$.
The second operation is

$$\overline{R}(X) = \bigcup_{R(x)\in \cap B_i} R(x),$$

where B_i is the subset of U/R and satisfies $(B_i, f/R) \in R^s(X)$.

The difference between the upper and the lower approximation will be called the $R-$ *boundary of* X and will be denoted by $BN_R(X)$, i.e. $BN_R(X) = \overline{R}(X) - \underline{R}(X)$.

If the boundary region is the empty set, X can be "observed" *exactly* through the indiscernibility R, and in the opposite case the set X can be "observed" *roughly* only due to the indiscernibility R. The former sets are *crisp (exact)*, whereas the later are *rough (inexact)*,with respect to indiscernibility R.

3 Discussion About the Rough Sets of System

In the universe (U, f), the indiscernibility relation in certain sense describes our lack of knowledge about the universe. The indiscernibility classes of the indiscernibility relation, called granules generated by R, represent elementary portion of knowledge we are able to perceive due to R. Thus in view of the indiscernibility relation, in general, we are not an able to exactly observe the system (U, f) and are forced to reason only about the accessible granules of knowledge, the quotient system $(U/R, f/R)$. In other words, in view of the available information, we can see the differentiation among elements of U as exactly as the differentiation among the elements of U/R and the structure f as exactly as f/R. Furthermore $(U/R, f/R)$ may not be an accurate system.

For $X \subseteq U$, we not only want to discern its elements, but also want to discern the structure among them. In other words, in the universe (U, f), we want to exactly understand the system X^s generated by X.

But in view of the available information, we only can get its $R-$ lower and the $R-$ upper approximation sets. This means that if we "see" the set X through the information, which generates the indiscernibility relation R, only the above approximations of X can be "observed", but not the set X.

As we can see from their definitions, the $R-$ lower and the $R-$ upper approximations are expressed in terms of granules of knowledge. The $R-$ lower approximation of X is union of all elements of all systems $(A_i, f/R)$ that belong to $R_s(X)$. This means that in view of the available information we can certainly discern that some of elements of X can form the system which belongs to $R_s(X)$. The upper approximation is union of all elements of similarity set of spanning systems of X. This means that the system generated by X is probably one of the systems which belongs to $R^s(X)$ in view of the available information.

Now let us analyze the relation between the element of U and the set X.

Case 1.$R(x) \in \underline{R}(X)$. It means that in view of the available information, $x \in X$ and it together with some other elements of X can form the system which belongs to $R_s(X)$.

Case 2: $R(x) \subseteq X, R(x) \cap \underline{R}(X) = \emptyset$. It means that in view of the available information, $x \in X$ and x belongs to the similarity set of the systems in $R^s(X)$.

Case 3: $R(x) \cap X \neq \emptyset, R(x) \not\subset X, R(x) \subseteq \overline{R}(X)$. It means that in view of the available information x belongs to similarity set of the system in $R^s(X)$ and is possibly classified as X.

Case 4: $R(x) \cap X = \emptyset, R(x) \subseteq \overline{R}(X)$. It means that in view of the available information x is in the similarity set of the systems in $R^s(X)$and x is certainly not in X.

Case 5: $R(x)\cap\overline{R}(X) = \emptyset, R(x) \subseteq (A_i, f/R), (A_i, f/R) \in A \subset R^s(X)$. it means that in view of the available information x is not in X, and x is in the similarity set of the system in A and is certainly not in the system in $R^s(X) - A$.

Case 6: $R(x) \cap \overline{R}(X) = \emptyset, R(x) \cap \bigcup_{R(x) \in \cup A_i} R(x) = \emptyset, (A_i, f/R) \in R^s(X)$. It means that in view of the available information x is not in X and is not in any system in $R^s(X)$.

Remark 2. *In those case 2-4, the relation between X and x and the relation between x and the spanning system of X are vaguer than the other cases. In cases 2-4 x is just in $BN_R(X)$. When $(U/R, f/U)$ is an accurate system, $R^s(X)$ will be simplified into a set containing one element and case 5 will disappear.*

4 Some Properties of Rough Sets of System

With the operations of the $R-$ lower and the $R-$ upper approximation defined on system universe as in Section 2, we extend some of the basic properties which hold for rough sets of set to rough sets of system.

Theorem 1. *The $R-$ lower and the $R-$ upper approximation operators have the following properties:*
1)$\underline{R}(X) \subseteq X \subseteq \overline{R}(X)$, 2)$\underline{R}(\emptyset) = \overline{R}(\emptyset) = \emptyset, \underline{R}(U) = \overline{R}(U) = U$,
3)$\underline{R}(\underline{R}(X)) = \underline{R}(X) \subseteq \overline{R}(\underline{R}(X))$, 4)$\underline{R}(\overline{R}(X)) \subseteq \overline{R}(X) = \overline{R}(\overline{R}(X))$.

Proof. According to the definitions of the $R-$ lower and the $R-$ upper approximation, we get 1). Because $(\emptyset, f/R) \in R^s(\emptyset), (\emptyset, f/R) \in R_s(\emptyset)$ and $(U/R, f/R) \in R^s(U)$, $(U/R, f/R) \in R_s(U)$, 2) follows quickly from the definitions of the $R-$ lower and the $R-$ upper approximation. Owning to the property 1) and the fact $R_s(X) = R_s(R_s(X))$, we have 3). $\underline{R}(\overline{R}(X)) \subseteq \overline{R} \subseteq \overline{R}(\overline{R}(X))$ follows from 1). By the definition of $R^s(X)$, $R^s(X) = R^s(R^s(X))$ holds. Then $\overline{R}(X) = \overline{R}(\overline{R}(X))$,that is, 4) holds. □

Remark 3. *When $(U/R, f/R)$ is an accurate system, there is only one spanning system of any set. Then $R_s(R^s(X)) = R^s(X)$ holds. Therefore $\underline{R}(\overline{R}(X)) = \overline{R}(X) = \overline{R}(\overline{R}(X))$ holds.*

In a system (U, f), for $X \subseteq (U, f)$, we define

$$P(U, f) = \{(A, f) : (A, f) \subseteq (U, f)\},$$
$$P(X) = \{(A, f) : X \subseteq (A, f)\}.$$

It is obvious$P(U, f)$ is a partially ordered set in view of $\subseteq$.

Lemma 1. *In a system (U, f), if any complete ordered subset of $P(U, f)$ has the minimum element. Let$X \subseteq (U, f), (Y, f) \subseteq (U, f)$. Then $(Y, f) \in X^s$ iff for any complete ordered subset of $P(X)$, which contains (Y, f), (Y, f) is the minimum element of the set.*

Proof. Let $(Y, f) \in X^s$, W is a complete ordered subset of $P(X)$, which contains (Y, f). By the definition of spanning system, we get that (Y, f) is the minimum element of W.

We now turn to the converse. Let (Y, f) satisfy the hypothesis in the theorem. Suppose $(Y, f) \notin X^s$. Thanks to $X \subseteq (Y, f)$, we drive that there is a system (Z, f), which satisfies $X \subseteq (Z, f) \subset (Y, f)$. Consider $\{(Z, f), (Y, f)\}$. By the hypothesis, we have $(Y, f) \subseteq (Z, f)$. Contradiction. Thus we conclude that $(Y, f) \in X^s$. □

Corollary 1. *In a system (U, f), for any subset W of $P(U, f)$, W has the minimum element, then (U, f) is a accurate system and for any $X \subseteq U$, $X^s = \bigcap A_i$, where $X \subseteq (A_i, f)$.*

Proof. Let $X \subseteq U$. For the set $P(X)$, according to the given condition, the spanning set of X must be the unique minimum element of $P(X)$, and $X^s = \bigcap A_i$, where $X \subseteq (A_i, f)$. □

Theorem 2. *Let $(U/R, f/R)$ be a quotient system of the universe (U, f) with respect to R. If $(U/R, f/R)$ satisfies the condition in lemma 1, then the $R-$ lower and the upper $R-$ approximation operators have the following properties:*
1) $X \subseteq Y$ *implies* $\underline{R}(X) \subseteq \underline{R}(Y)$ *and* $\overline{R}(X) \subseteq \overline{R}(Y)$,
2) $\overline{R}(X \cup Y) \supseteq \overline{R}(X) \cup \overline{R}(Y)$, 3) $\underline{R}(X \cap Y) \subseteq \underline{R}(X) \cap \underline{R}(Y)$,
4) $\underline{R}(X \cup Y) \supseteq \underline{R}(X) \cup \underline{R}(Y)$, 5) $\overline{R}(X \cap Y) \subseteq \overline{R}(X) \cap \overline{R}(Y)$.

Proof. When $X \subseteq Y$, $R_s(X) \subseteq R_s(Y)$ and $\underline{R}(X) \subseteq \underline{R}(Y)$ follow from the definition of the $R-$ lower system approximation. For any $(A, f/R) \in R^s(Y)$, by the lemma 1 and $X \subseteq Y \subseteq (A, f/R)$, we get a system $(B_A, f/R) \in R^s(X)$, which satisfies $(B_A, f/R) \subseteq (A, f/R)$. Then

$$\overline{R}(X) \subseteq \bigcup_{R(x) \in \bigcap B_A} R(x) \subseteq \overline{R}(Y),$$

where A runs through the index set $R^s(Y)$. Therefore 1) holds.

Because $X \subseteq X \cup Y, Y \subseteq X \cup Y$ and $X \cap Y \subseteq X, X \cap Y \subseteq Y$, the 2)-5)follow form the property 1). □

5 Some Examples of Rough Sets of System

5.1 Rough Sets on the Real Line

In [8], Pawlak.Z gave the definitions of rough sets on $\mathbb{R}^+$ the set of nonnegative reals.

Let $S \subseteq \mathbb{R}^+$ be the following sequence of reals $x_1, x_2, ..., x_i, ...$ such that $0 \leq x_1 < x_2 < ... < x_i$. S will be called a categorization of $\mathbb{R}^+$ and the ordered pair$A = (\mathbb{R}^+, S)$ will be referred as an approximation space. The partition $\pi(S)$ on $\mathbb{R}^+$ induced by the sequence S defined as $\pi(S) = \{0, (0, x_1), x_1, (x_1, x_2), x_2, ..., x_i, (x_i, x_{i+1}), x_{i+1}...\}$, where (x_i, x_{i+1}) denotes an open interval. Denotes the block of the partition $\pi(S)$ containing x by $S(x)$. In particular, if $x \in S$ then $S(x) = \{x\}$. He has focus on approximating closed intervals of the form $< 0, x >= Q(x)$ for any $x \in \mathbb{R}^+$.

Suppose there is an approximation space $A = (\mathbb{R}^+, S)$. The $S-$ lower and the $S-$ upper approximations of defined respectively by

$$S_*(Q(x)) = \{y \in R^+ : S(y) \subseteq Q(x)\},$$
$$S^*(Q(x)) = \{y \in R^+ : S(y) \cap Q(x) \neq \emptyset\}.$$

The above definitions of approximations of interval $< 0, x >$ can be understood as approximations of the real number x which are simple the ends of the interval $S(x)$. In other words given any real number x and a set of reals S, by the $S-$ lower and the $S-$ upper approximations of x be defined below:

$$S^*(x) = Sup\,\{y \in S : y \leq x\},$$
$$S_*(x) = Inf\,\{y \in S : x \leq y\}.$$

One has $S(x) = (S_*(x), S^*(x))$.
If defined a partial order on $\pi(S)$ and denoted by $\leq /s$, by

(1) $(x_i, x_{i+1}), (x_j, x_{j+1}) \in \pi(S)$,
$(x_i, x_{i+1}) \leq /s(x_j, x_{j+1}) \Leftrightarrow x_{i+1} \leq x_j$;
(2) $x_i, (x_j, x_{j+1}) \in \pi(S)$,
$x_i \leq /s(x_j, x_{j+1}) \Leftrightarrow x_i \leq x_j; (x_j, x_{j+1}) \leq /sx_i \Leftrightarrow x_{j+1} \leq x_i$;
(3) $x_i, x_j \in \pi(S), x_i \leq /sx_j \Leftrightarrow x_i \leq x_j$.

It is easily proved that $\pi(S)$ is a complete ordered set with the relation $\leq /s$.
Because $\mathbb{R}^+$ is a complete ordered set with the relation $\leq$, S can be viewed as an indiscernibility relation defined on $\mathbb{R}^+$, and $\leq /s$ induced by $\leq$ is a structure on $\mathbb{R}^+/S = \pi(S)$, the rough sets on real line actually is rough sets on the system $(\mathbb{R}^+, \leq)$,the approximation space $A = (\mathbb{R}^+, S)$ is just the quotient system$(\mathbb{R}^+/S, \leq /S)$.
Let $y_1, y_2 \in (\mathbb{R}^+, \leq)$, we want to contrast y_1 with y_2, in view of the available information of $(\mathbb{R}^+, \leq)$, we only can get $S(y_1), S(y_2)$. If $S(y_1) \leq /sS(y_2)$ or $S(y_1) = S(y_2)$ then we can understand $y_1 \leq y_2$ or $y_1 = y_2$ with respect to the relation S.

5.2 Rough Sets of the Additive Group of Integers

Let $(\mathbb{Z}, +)$ be the additive group of integers and let $M = (A, +)$ be a subgroup of $(\mathbb{Z}, +)$. Because $(\mathbb{Z}, +)$ is an Abelian group, then M is a normal subset of and there is a quotient group $(\mathbb{Z}/M, +/M)$ with respect to M. Now we can get the rough set theory of $(\mathbb{Z}, +)$ as the discussion in section 2.
For example, let $A = 10\mathbb{Z} = \{10n : n \in \mathbb{Z}\}$, then the quotient group $\mathbb{Z}_{10} = \{10\mathbb{Z}, 10\mathbb{Z} + 1, 10\mathbb{Z} + 2, ...10\mathbb{Z} + 9\}$ is induced by $M = (10\mathbb{Z}, +)$.
Is $1 + 23 = 14$ true? In view of the available information M, we only can get that $1 \in 10\mathbb{Z}+1, 23 \in 10\mathbb{Z}+3, 14 \in 10\mathbb{Z}+4$ and $(10\mathbb{Z}+1)+/M(10\mathbb{Z}+3) = 10\mathbb{Z}+4$, then we can understand $1 + 23 = 14$ is true with respect to the relation M.
Assuming that we have more detail about the universe $(\mathbb{Z}, +)$, for instance, we have $M = (100\mathbb{Z}, +)$. Then we get the quotient group $\mathbb{Z}_{100}$. Now $1 \in 100\mathbb{Z} + 1, 23 \in 100\mathbb{Z} + 3, 14 \in 100\mathbb{Z} + 14$ and $(100\mathbb{Z} + 1) + /M(100\mathbb{Z} + 3) = 100\mathbb{Z} + 4 \neq 100\mathbb{Z}+14$, we can understand $1+23 = 14$ is not true with respect to the available information M.

6 Conclusions

As a suitable mathematical model to handle partial knowledge in data bases, rough set theory is emerging as a powerful theory and has found its successive applications in the fields of artificial intelligence such as pattern recognition, machine learning, and automated knowledge acquisition.

The classical rough set theory focuses on the classification. In this paper we have presented a rough sets theory of the system which focuses on the classification as well as the interrelation among the classes. It well reveals the differentiation and the interrelation among the elements of the universe and it extends the scope of the objects of the research on rough sets and extends the applied field of rough sets. We need to point out that the assumptions made in this paper, which would be obviously true when the universe consist of finite elements, are just for general discussion. To understand and apply better the rough sets on system further study is necessary.

Acknowledgements

This work was supported by the National Natural Science Foundation of China (Grant no.10671030) and the Fostering Plan for Young and Middle Age Leading Researchers in University of Electronic Science and Technology of China(Grant no.Y02018023601033).

References

[1] Pawlak Z (1982) Rough sets. Internal Joural of Computer and Information Science 11: 341-356
[2] Pawlak Z (1991) Rough sets: Theoretical Aspects of Reasoning about Data. Kluwer Academic Publishers, Boston
[3] Yao YY (1996) Two views of the theory of rough sets in finite universes. International Joural of Approximate Reasoning 15: 291-317
[4] Yao YY, Lin TY (1996) Generalization of rough sets model using logic. Intelligent Automation and Soft Computing, an International Joural 2: 103–120
[5] Yao YY (1998) Relational interpretations of neighborhood operators and rough set approximation operators. Information Sciences 111: 239-259
[6] Slowinski R, Vanderpooten D (2000) A Generalized definition of rough approximations based on similarity. IEEE Transactions on Knowledge and Data Engineering 12: 331-336
[7] Salvatore G, Benedetto M, Slowinski R (2002) Rough approximation by dominance relations, in: A Rough sets Approach to Knowledge Discovery. International Journal of Intelligent Systems 17: 153-171
[8] Pawlak Z (1996) Rough sets, rough relations and rough functions. Fundamenta Informaticae 27: 103-108

Synthesis of Concepts Based on Rough Set Theory

Dayong Deng[1,2], Houkuan Huang[1], and Hongbin Dong[1,3]

[1] School of Computer and Information Technology, Beijing Jiaotong University, Beijing, PR China, 100044
dydeng_bjtu@yahoo.com.cn, hkhuang@center.njtu.edu.cn
[2] Zhejiang Normal University, Jinhua, PR China, 321004
[3] Harbin Normal University, Harbin, PR China, 150025
donghongbin@263.net

Abstract. Rough set theory usually deals with how a concept is represented with some granules. In this paper we extend rough set theory, and use it to deal with how a series of concepts are represented with a granule. We use it to seek the most possible determined reasons and the most possible reasons from decision systems, and also to reduce decision tables. This may be a new aspect of rough set research, and may extend application fields of rough set theory.

1 Introduction

The term "granular computing" was first suggested by T.Y.Lin, and it is difficult to have a precise and uncontroversial definition[3, 10]. Zadeh identified three basic concepts that underlie human cognition, namely, granulation, organization, and causation[3, 4]. Basic ideas of granular computing have been explored in many fields, such as interval analysis, quantization, cluster analysis, and rough set theory etc [1, 2, 3].

Rough set theory is a valid mathematical tool, which deals with imprecise, incomplete or inconsistent knowledge[5, 6, 7]. There are several rough set models, including Pawlak's rough set model and variable precise rough set model[8, 9] etc. Researchers have got many results from various rough set models, some of them have a widely application[7]. But almost all of the results achieved deal with how a concept is divided into smaller granules. In this paper, we use ideas of rough set theory to deal with how a series of concepts in information systems are organized to be one. We mainly dispose how to combine a series of concepts into a granule with the ideas of rough set theory, and how to reduce decision tables with these ideas. This new method extends rough set theory. Hence, we could apply rough set theory to both granulation and organization.

The rest of the paper is arranged as follows: First, we introduce basic ideas of rough set theory in section 2. Second, we use ideas of rough set theory to deal with how a series of concepts in information systems are organized to be one in section 3. Third, we use the ideas of synthesis of concepts based on rough

B.-Y. Cao (Ed.): Fuzzy Information and Engineering (ICFIE), ASC 40, pp. 816–824, 2007.
springerlink.com

set theory to seek the most determined possible reasons and the most possible reasons from decision systems, and to reduce decision tables in section 4. At last, we draw some conclusions in section 5.

2 Rough Sets

Rough set theory was proposed by Z.Pawlak[5, 6, 7, 11] in the context of knowledge represented as an equivalence relation R on a set U of entities. Equivalence classes of R contain entities that are indiscernible with respect to R, and a concept (a subset of the set U) X is said to be exact in case that it is a union of a family of equivalence classes of R; otherwise, X is said to be rough.

An information system is symbolically represented as a pair $IS = (U, A)$. The symbol U denotes a set of objects, and the symbol A denotes a set of attributes. Each pair (object, attribute) is uniquely assigned a value: given $a \in A$, $u \in U$, the value $a(u)$ is an element of the value set V.

Each entity (object) $u \in U$ is represented in the information system IS by its information set $Ins_A(u) = \{(a, a(u)) : a \in A\}$, that corresponds to the u-th row of the data table IS. Two objects u, v may have the same information set: $Ins_A(u) = Ins_A(v)$, in this case they are said to be A-indiscernible ; the relation $IND(A) = \{(u, v) : Ins_A(u) = Ins_A(v)\}$ is said to be A-indiscernibility relation, which is an equivalence relation. The symbol $[u]_A$ de-notes the equivalence class of the relation $IND(A)$ containing u.

It follows that a concept X is A-exact if and only if X is a union of equivalence classes: $X = \bigcup\{[u]_A : u \in U\}$.

Given a set B of attributes, and a concept $X \subseteq U$, that is not B-definable, there exists $u \in U$ with neither $[u]_B \subseteq X$ nor $[u]_B \subseteq U \setminus X$. Thus, B-definable sets $B_*(X) = \{u \in U : [u]_B \subseteq X\}$and $B^*(X) = \{u \in U : [u]_B \bigcap X \neq \phi\}$ are distinct, and $B_*(X) \subseteq X \subseteq B^*(X)$.The set $B_*(X)$ is the lower B-approximation to X whereas $B^*(X)$ is the upper B-approximation to X. The concept X is said to be B-rough.

3 Synthesis of Concepts Based on Rough Set Theory

A information system $IS = (U, A)$ determines a granule system $GS = \{[x]_B : B \subseteq A, x \subseteq U\}$, in which there are two special elements, the empty set ϕ and the universe set U. In the following paragraphs, we suppose that the granule system is GS or its subset, but we also represent it with GS. This is to say, we could change granule systems in terms of our research.

Given a series of concepts $X_i \subseteq U$ (i=1,2,...,n), we could abstract them and use a granule in GS to represent them approximately. Suppose $X = Syn(X_1, X_2, \ldots, X_n)$, where the symbol Syn denotes the operators among these concepts, and it may be union, intersection, complement or composite of them. We could choose operators according to our application. In the sequel, the symbol X usually denotes $Syn(X_1, X_2, \ldots, X_n)$, or else we will make a statement.

Definition 3.1. The *upper synthesis approximation*(*USA* in short) to the series of concepts X_i(i=1,2,...,n) is the minimum granule in *GS* which contains *X*.

$$USA(X) = Inf\{Y : X \subseteq Y, Y \in GS\} \tag{1}$$

Definition 3.2. The *lower synthesis approximation*(*LSA* in short) to the series of concepts $X_i(i = 1, 2, \ldots, n)$ is the maximum granule in *GS* which is contained in *X*.

$$LSA(X) = Sup\{Y : Y \subseteq X, Y \in GS\} \tag{2}$$

In definition 3.1 and 3.2, we usually compute $USA(X)$ or $LSA(X)$ according to the cardinality of set in the sequel. If the conditions of problems are changed, we may change the methods of Sup and Inf.

Example 1. Table 1 is an information system *IS*. The granule system which the in-formation system *IS* determines is $GS = \{[x]_B : B \subseteq A, x \in U\} = \{\phi, \{x_1, x_5\}, \{x_2\}, \{x_3\}, \{x_4\}, \{x_6\}, \{x_7\}, \{x_8\}, \ldots, U\}$. Suppose there are two concepts $X_1, X_2 \subseteq U$, $X_1 = \{x_1, x_5\}$, $X_2 = \{x_2, x_3\}$, then $LSA(X_1 \bigcup X_2) = LSA(\{x_1, x_2, x_3, x_5\}) = [x_1]_{\{a_2\}} = \{x_1, x_3, x_5\} \in GS$, $USA(X_1 \bigcup X_2) = USA(\{x_1, x_2, x_3, x_5\}) = U \in GS$.

Table 1. Information system *IS*

U	a1	a2	a3	a4
x1	0	0	3	1
x2	1	1	2	1
x3	2	0	1	0
x4	0	2	1	0
x5	0	0	3	1
x6	3	1	2	1
x7	3	2	0	0
x8	1	2	0	0

For a series of concepts $X_i(i = 1, 2, \ldots, n)$, the USA and the *LSA* of them are approximately abstract representation of the concepts. In an information system the former is only one, but the later may be more than one(Example 2). The reason for it is that *GS* is algebraically closed for the operator $\bigcap$ but not for the operator $\bigcup$.

The synthesis of $X_i(i = 1, 2, \ldots, n)$ is called to be crisp(or exact) if

$$USA(X) = LSA(X) \tag{3}$$

Definition 3.3. The *boundary synthesis region*(*BSR* in short) of X_i(i=1,2,,n) is the result that their *USA* minuses their *LSA*.

$$BSR(X) = USA(X) - LSA(X) \tag{4}$$

Definition 3.4. The *accuracy degree* of the synthesis of $X_i (i = 1, 2, \ldots, n)$ is defined by

$$\alpha = \frac{card(LSA(X))}{card(USA(X))} \quad (5)$$

where $card(X)$ is the cardinality of the set X.

Example 2. Suppose $U = \{1, 2, 3, 4, 5, 6, 7, 8, 9\}$ is a universe of discourse. R is an equivalence relation on U, $U/R = \{\{1, 3, 5\}, \{2\}, \{4, 6, 8, 9\}, \{7\}\}$. Assume $X_1 = \{1, 2\}$, $X_2 = \{7\}$, then $USA(X_1 \bigcup X_2) = U$, $LSA(X_1 \bigcup X_2) = \{2\}\text{or}\{7\}$, and $\alpha = 1/9$.

In a decision system $DS = (U, C, D)$, suppose $GS = U/D$, For $\forall [x]_C \subseteq U$, if $USA([x]_C) = [x]_D$, then the decision system is consistent, otherwise it is inconsistent.

There are differences between LSA, USA and lower approximation, upper approximation. The purpose of LSA and USA is to represent a synthesis of concepts with a granule, while the purpose of lower approximation and upper approximation is to represent a concept with multi-granules. The result of LSA or USA is only one granule(maybe the result of LSA has alternatives), whereas the result of lower approximation or upper approximation is the union of granules.

In the following paragraph, we give an algorithm to count LSA and USA.

Input: a granule system $GS = \{[x]_B : B \subseteq A, x \in U\}$(the information system is $IS = (U, A)$) and a synthesis of concepts X.

Output: $LSA(X)$ and $USA(X)$.

The algorithm is as follows:

```
    LSA(X) = φ;
    USA(X) = U;
    GS1 = {[x]_B : B ⊆ A, x ∈ X}; // GS1 is a subset of GS
    While(g ∈ GS1)                //GS1 ≠ φ
    { if ((g ⊆ X) &&(card(g) > card(LSA(X)))
        LSA(X) = g;
    if((g ⊇ X) &&(card(g) < card(USA(X)))
        USA(X) = g;
    GS1 = GS1 − {g};
    }
```

The strategy of the algorithm is blind search. If there is some heuristic information, its efficiency could be improved. In the sequel, we discuss some properties of LSA and USA.

Proposition 1. There are some properties for USA and LSA:

1. For$\forall X \in GS$, $USA(X) = LSA(X) = X$
2. $LSA(X) \subseteq X \subseteq USA(X)$, for $\forall X = Syn(X_1, X_2, \ldots, X_n) \subseteq U(i = 1, 2, \ldots, n)$.
3. Assume $X = Syn_1(X_1, X_2, \ldots, X_n)$, $Y = Syn_2(Y_1, Y_2, \ldots, Y_m)$, $X \subseteq Y$, whereX_i, $Y_j \subseteq U$, then $card(LSA(X)) \leq card(LSA(Y))$ $USA(X) \subseteq USA(Y)$.
4. $B \subseteq A \Rightarrow card(LSA_B(X)) \leq card(LSA_A(X))$ $B \subseteq A \Rightarrow USA_A(X)) \subseteq USA_B(X)$, where A, B are sets of attributes.

5. The synthesis of concepts X_i(i=1,2,...,n) is crisp iff its accuracy degree $\alpha = 1$.
6. $LSA(USA(X)) = USA(X)$.
7. $USA(USA(X)) = USA(X)$.
8. $USA(LSA(X)) = LSA(X)$.
9. $LSA(LSA(X)) = LSA(X)$.

Above properties are easy to understand, here we only prove 3.

Proof. $LSA(X) \subseteq X \subseteq Y$, $LSA(Y) = Sup\{Z : Z \subseteq Y, Z \in GS\}$,Therefore, $card(LSA(X)) \leq card(LSA(Y))$.

$X \subseteq Y \subseteq USA(Y)$, $USA(X) = Inf\{Z : X \subseteq Z, Z \in GS\}$. Suppose $USA(X) \subseteq USA(Y)$does not hold, we have $X \subseteq USA(X) \bigcap USA(Y) \subseteq USA(X)$, but $USA(X) \bigcap USA(Y) \in GS$, hence, $USA(X) = USA(X) \bigcap USA(Y) \subseteq USA(Y)$.

Proposition 2. Given $X = Syn(X_1, X_2, \ldots, X_n) \subseteq U$, we have $P(USA(X)|X) = 1$ and $P(X|LSA(X)) = 1$, where $P(X|Y) = \frac{card(X \bigcap Y)}{card(Y)}$.

Proof. It could be got directly from the definitions of $USA(X)$ and $LSA(X)$.

We could redefine the $USA(X)$ and $LSA(X)$ as follows:

$$USA(X) = Inf\{Y : P(Y|X) = 1, Y \in GS\} \tag{6}$$

$$LSA(X) = Sup\{Y : P(X|Y) = 1, Y \in GS\} \tag{7}$$

The above definitions show that $LSA(X)$ is the maximum granule which supports the concept X completely in the GS, and that $USA(X)$ is the minimum granule in the GS which is completely supported by the concept X.

Proposition 3

$$X \subseteq Y \Rightarrow LSA(X) \subseteq LSA(Y) \tag{8}$$

$$B \subseteq A \Rightarrow LSA_B(X)) \subseteq LSA_A(X) \tag{9}$$

do not hold, where A, B are sets of attributes. The reason for those is the relations among granules in GS are only partial order for the operator $\subseteq$.

Proposition 4

$$USA(X) \cup USA(Y) \subseteq USA(X \cup Y) \tag{10}$$

$$USA(X) \cap USA(Y) \supseteq USA(X \cap Y) \tag{11}$$

Proof. $X \subseteq X \cup Y \Rightarrow USA(X) \subseteq USA(X \cup Y)$,$Y \subseteq X \cup Y \Rightarrow USA(Y) \subseteq USA(X \cup Y)$.Hence,$USA(X) \cup USA(Y) \subseteq USA(X \cup Y)$. Similarly,we can prove the equation(11).

4 Reduction Based on the Ideas of Synthesis of Concepts

The application of rough set theory is mainly to get reducts from decision tables. It is an efficient method of reduction in decision systems, but at least there is a fault when we use it to get reducts: the more coarser the partition is, the more possible the attrib-ute is eliminated. This is to say, the reducts of classical rough set theory may be lose the most possible reasons for a conclusion.

Example 3. Table 2 is a decision system $DS = (U, C, D)$, the set of condition attributes $C = \{a_1, a_2, a_3\}$, the set of decision attributes $D = \{a_4\}$. We may get the minimum reducts $\{a_1\}$, and eliminate the attribute a_3.

Table 2. A decision system

U	a1	a2	a3	a4
x1	0	0	1	1
x2	1	1	1	1
x3	2	0	0	0
x4	3	2	0	0
x5	4	0	1	1
x6	5	1	1	1
x7	6	2	0	0
x8	7	2	0	0

In the sequel, we would like to use the above ideas to get the most possible deter-mined reasons for a decision from a decision system(or a decision table), and to dis-cuss reduction based on synthesis of concepts.

Definition 4.1. Suppose a decision system $DS = (U, C, D)$, where U is a universe of discourse, C is the set of condition attributes, D is the set of decision attributes. $GS = \{[x]_B : B \subseteq C, x \in U\}$, $U/D = \{Y_1, Y_2, \ldots, Y_n\}$. The *positive region* based on synthesis of concepts(*SPOS* in short) is defined as

$$SPOS_C(D) = \bigcup_{i=1}^{n} LSA(Y_i) \tag{12}$$

Definition 4.2. In a decision system $DS = (U, C, D)$, $S \subseteq C$ is called a *reduct of positive region* based on synthesis of concepts, if it satisfies the following two conditions:

(1) $SPOS_S(D) = SPOS_C(D)$
(2) $\forall B \subset S, SPOS_B(D) \neq SPOS_S(D)$

The set of reducts is denoted by $SRED_C(D)$.

Definition 4.3. $\bigcap SRED_C(D)$ is called the core of reducts of positive region based on synthesis of concepts.

Example 4. Suppose Table 1 is a decision system $DS = (U, C, D)$, the set of condition attributes $C = \{a_1, a_2, a_3\}$, the set of decision attributes $D = \{a_4\}$, the granule system $GS = \{[x]_B : B \subseteq C, x \in U\}$. $U/D = \{\{x_1, x_2, x_5, x_6\}, \{x_3, x_4, x_7, x_8\}\}$, $LSA(\{x_1, x_2, x_5, x_6\}) = \{x_2, x_6\}$ or $\{x_1, x_5\}$, $LSA(\{x_3, x_4, x_7, x_8\}) = \{x_4, x_7, x_8\}$. From the two LSAs we could get the main possible determined reasons for results: the main reasons for $a_4 = 0$ are $a_2 = 1$, or $a_3 = 3$, or $a_1 = 0$ and $a_2 = 0$, the main reason for $a_4 = 1$ is $a_2 = 2$.

Example 5. In Table 2, we could easily get the reduct $\{a_3\}$ and the core $\{a_3\}$.

Example 6. Table 3 shows a data set containing seven objects $U = \{1, 2, 3, 4, 5, 6, 7\}$, two condition attributes $C = \{Systolic\ pressure(SP), Diastolic\ pressure\ (DP)\}$, and a decision attribute $D = \{Blood\ pressure(BP)\}$. All of the three attributes have three possible values:{Low(L), Normal(N), High(H)}.

Table 3. The data set for example 6

U	SP	DP	BP
1	L	N	L
2	H	N	H
3	N	N	N
4	L	L	L
5	H	H	H
6	N	H	H
7	N	L	N

The equivalence partitions for condition attributes can be derived as follows:
$U/\{SP\} = \{\{1, 4\}, \{2, 5\}, \{3, 6, 7\}\}$,
$U/\{DP\} = \{\{1, 2, 3\}, \{4, 7\}, \{5, 6\}\}$, and
$U/\{SP, DP\} = \{\{1\}, \{2\}, \{3\}, \{4\}, \{5\}, \{6\}, \{7\}\}$
Hence the granule system
$GS = \bigcup\{U/\{SP\}, U/\{DP\}, U/\{SP, DP\}\}$
$= \{\{1, 4\}, \{2, 5\}, \{3, 6, 7\}, \{1, 2, 3\}, \{4, 7\}, \{5, 6\}, \{1\}, \{2\}, \{3\}, \{4\}, ,\{5\}, \{6\}, \{7\}\}$.
The equivalence partition for the decision attribute can be derived as follows:
$U/\{BP\} = \{\{1, 4\}, \{2, 5, 6\}, \{3, 7\}\}$
For $\forall X \in U/\{BP\}$, we could count their LSA respectively,
$LSA(\{1, 4\}) = \{1, 4\}$,
$LSA(\{2, 5, 6\}) = \{2, 5\}$or {5,6},and
$LSA(\{3, 7\}) = \{3\}$ or $\{7\}$
In terms of definition 4.2 and definition 4.3 we could get both the reduct and the core are $\{SP,\ DP\}$. We could explain the above result: the main causes of BP are SP and DP. We could explain it in detail further,

$LSA(\{1, 4\}) = \{1, 4\}$ means the low of blood pressure is mainly due to the low of systolic pressure.

$LSA(\{2, 5, 6\}) = \{2, 5\}$,or {5,6} means the high of blood pressure is mainly due to the high of systolic pressure or the high of dias -tolic pressure.

$LSA(\{3, 7\}) = \{3\}$ or $\{7\}$ means the normal of blood pressure is mainly due to both systolic pressure and diastolic pressure are normal, or systolic pressure is normal and diastolic pres -sure is low.

There are differences between the reduction of synthesis of concepts and that of classical rough set theory. At first, their purposes are different. The former is to find the most possible determined reasons, while the later is to seek the optimal rules. Sec-ond, the results of reduction may be different.

5 Conclusion and Further Research

Rough set theory is an important method of granulating information systems. But it is usually applied to divide a concept into parts. In this paper, we use rough set theory to synthesize a series of concepts into a granule, and present an algorithm to count the LSA and USA of a concept X(it may be a synthesized concept). The new method extends rough set theory, and may extend its application fields, such as words computing, machine learning, reasoning, expert systems and pattern recognition etc. We will investigate the new method and its application in our future work.

In real-world application, the granule system may not be what we have supposed above, and it may be generated by similar relation or tolerance relation, we would like to research how to synthesize a series of concepts into a granule based on these relations further. Because the data of real world maybe contain some noises, we should use variable precise rough set theory or other theory to solve them, this question is also in our further research.

Acknowlegement

This research was partially supported by the Natural Science Foundation of Heilongjiang Province, China(Grant No.F200605).

References

1. Yao, Y. Y.(2004) Granular Computing, Computer Science (China) 31(10.A), 1-5.
2. Yao, Y. Y., Yao, J. T.(2002) Granular Computing as a Basis for Consistent Classification Problems. In: PAKDD 2002 Workshop entitled" Towards Foundation of Data Mining", Communications of Institute of Information and Computing Machinery 5(2):101-106.
3. Yao, Y. Y.(2004) A Partition Model of Granular Computing. LNCS Transactions on Rough Sets 3100(1):232-253.
4. Zadeh, L. A.(1998) Some Reflections on Soft Computing, Granular Computing and Their Roles in the Conception, Design and Utilization of Information/Intelligent Systems. Soft Computing 2(1):23-25.
5. Pawlak, Z.(1982) Rough Sets.International Journal of Computer and Information Science 11(5)341-356.
6. Pawlak, Z.(1991) Rough sets-Theoretical Aspect of Reasoning about Data. Kluwer Academic Publishers.
7. Pawlak, Z.: Rough Sets. http://pizarro.fll.urv.es/continguts/linguistica/proyecto/reports/rep29.doc.

8. Ziarko, W.(1993) Variable Rough Set Model. Journal of Computer and System Sci-ences 46(1):39-59.
9. Katzberg, J. D., Ziarko, W.(1996) Variable Precision Extension of Rough Set. Foundamenta Informaticae 27(2-3):155-168.
10. Zadeh, L. A.(1997) Towards a theory of fuzzy information granulation and its centrality in human reasoning and fuzzy logic. Fuzzy Sets and Syatems 90(2): 111-127.
11. Polkowski, L.(2004) Toward Rough Set Foundations.Mereological Approach. In Proceedings of the 4th International Conference on Rough Sets and Current Trends in Computing(RSCTC2004),Sweden,8-25.

γ-Tolerance Relation-Based RS Model in IFOIS

Lujin Tang[1] and Dakuan Wei[2]

[1] Department of Mathematics and computing Science, Hunan University of Science and Engineering, Yongzhou, Hunan, 425100, P.R. China
tanglujin@126.com

[2] College of Information Engineering, Hunan University of Science and Engineering, Yongzhou, Hunan, 425100, P.R. China
weidakuan126.com

Abstract. The traditional rough set (RS) theory is a powerful tool to deal with complete information system, and its performance to process incomplete information system is weak, especially, its effect of combining the incomplete information system with fuzzy objective information system is weaker. The paper improves the tolerance relation proposed by M.Kryszkiewcz to obtain the $\gamma-tolerance$ relation and $\gamma-tolerance$ classes, presents the concept of the incomplete and fuzzy objective information system (IFOIS in short), and gives its rough set model based on the $\gamma-tolerance$ relation, i.e., the rough set model in incomplete and fuzzy objective information system. Finally, the concept of precision reduction is defined, and the corresponding algorithm is provided.

Keywords: Rough Set, $\gamma-tolerance$ Relation, Incomplete Information System, Fuzzy Objective Information System, Precision Reduction.

1 Introduction

Rough set theory, introduced by Pawlak, is a powerful mathematical tool to handle imprecision, uncertainty, and vagueness. It has attracted the attention of researchers and practitioners all over the world who are contributed to its development and applications during the last decade. It has also been applied successfully in many fields, such as pattern recognition, data mining, decision analyzing, machine learning, knowledge acquiring, approximation reasoning, etc..

The investigative object for the topical Pawlak rough set theory is complete information table, and domain of conditional attribute values and domain of decision attribute values are all complete and traditional Contor sets. And in real world applications, there exist many such information tables in which partial conditional attribute values are unknown for a certain piece of information vacating or missing; the classical rough set theory can't cope with the kinds of incomplete information system, however many scholars have been studying them[1-11]; While the whole conditional attribute values are known and the decision attribute values are fuzzy, many researchers do likewise [12-18]. So far, few scholars have been studying the information system that the domain of conditional attribute values is incomplete and the domain of decision attribute values is fuzzy.

B.-Y. Cao (Ed.): Fuzzy Information and Engineering (ICFIE), ASC 40, pp. 825–836, 2007.
springerlink.com

In our former papers [19,20], we have discussed the IFOIS based on incomplete information system and fuzzy objective information system, although the way employed the tolerance relation and tolerance classes in those papers are good kind of method for establishing the rough set model, classification is quite coarse for many intersectional elements between tolerance classes. In order to overcome the problem, $\gamma - tolerance$ relation is proposed, furthermore, the concept of the IFOIS under $\gamma - tolerance$ relation is defined, and its rough set model is also suggested, which is the generalization of the rough set model in complete and fuzzy objective information system and the Pawlak rough set model. Finally, we give the definition of precision reduction and its algorithm in IFOIS.

2 Basic Theories

In this section, we briefly describe the basic concepts and correlative contents about incomplete information system and fuzzy objective information system.

2.1 Incomplete Information System

Let $S = (U, A, V, F)$ be an information system,Where U is a nonempty finite set of objects called universe discourse, A is a nonempty finite set of conditional attributes; and for every $a \in A$, such that $f : U \rightarrow V_a$, $V = \cup_{a \in A} V_a$, where V_a is called the value set of attribute a.

Definition 1. If some of the precise attribute values in an information system are unknown, i.e., missing or known partially, then such a system is called an incomplete information system and is still denoted with convenience by the original notation $S = (U, A, V, F)$. That is, if there exist at least a attribute $a \in B \subseteq A$, such that V_a includes null values, then the system is called an incomplete information system, the assign $*$ usually denotes null value. Otherwise the system is called a complete information system.

Many scholars have been deeply researching on rough set methods of such the system and have got a lot of excellent conclusions in recent years.

Definition 2. Let $S = (U, A, V, F)$ be an incomplete information system, and $T(B) = \{(x, y) \in U \times U : \forall b \in B, b(x) = b(y) \vee b(x) = * \vee b(y) = *\}$, where $B \subseteq A$, then $T(B)$ is called a M.Kryszkiewcz's tolerance relation or called a tolerance relation in short [5-6].

Obviously, $T(B) = \cap_{b \in B} T(\{b\})$ holds, T is reflexive and symmetric , but not transitive.

Let $T_B(x) = \{y \in U : (x, y) \in T(B)\}$, and then $T_B(x)$ is called the tolerance class of the object $x \in U$ with respect to the set $B \subseteq A$ of conditional attributes. $T_B(x)$ is constituted by all objects in universe U which is possibly indiscernible with x.

Suppose that $U/T(B) = \{T_B(x) : x \in U\}$ represents classifications, then the elements in $U/T(B)$ are tolerance classes. Generally, the tolerance class in

$U/T(B)$ don't consist the partition of U, they may be subsets of each other or may overlap, but form a cover, i.e., $\cup U/T(B) = U$.

Definition 3. Let $S = (U, A, V, F)$ be an incomplete information system, and $X \subseteq U, B \subseteq A$, the upper-approximation and lower-approximation of X with regard to set B of conditional attributes under the tolerance relation T can be defined as:

$$\overline{T_B}(X) = \{x \in U : T_B(x) \cap X \neq \Phi\} = \cup\{T_B(x) : x \in X\},$$
$$\underline{T_B}(X) = \{x \in U : T_B(x) \subseteq X\} = \{x \in X : T_B(x) \subseteq X\}$$

respectively.

Lemma 1. Let $S = (U, A, V, F)$ be an incomplete information system, and $X \subseteq U$, $C \subseteq B \subseteq A$. Then: (1) $\underline{T_B}(X) \subseteq X \subseteq \overline{T_B}(X)$; (2) $\overline{T_B}(X) \subseteq \overline{T_C}(X)$ and $\underline{T_B}(X) \supseteq \underline{T_C}(X)$.

2.2 Fuzzy Objective Information System

Definition 4. Suppose that $S = (U, A, D, F)$ is an information system, where U, A *and* D are a nonempty finite set of objects, conditional attributes and decision attributes respectively. For every $d \in D$, such that there exists $f_d : U \rightarrow [0, 1]$, where $f_d \in F$, then $S = (U, A, D, F)$ is called fuzzy objective information system [12].

Definition 5. Let W be a fuzzy set defined on universal set U, for every $\alpha \in [0, 1]$, then

$$W_\alpha = \{x \in U : W(x) \geq \alpha\} \textit{ and } W_{\alpha+} = \{x \in U : W(x) > \alpha\}$$

are called $\alpha - cut$ set and strong $\alpha - cut$ set of W separately [12].

Hereafter, $P(U)$ and $F(U)$ always represent all the traditional subsets and fuzzy sets of U respectively.

Definition 6. Let (U, R) be a Pawlak approximation space, $[x]_R$ be equivalence class involving x, where R indicates equivalence relation on U. For each $W \in F(U)$, the upper-approximation and lower-approximation of fuzzy set W with respect to approximation space (U, R), denoted by $\overline{R}(W)$ and $\underline{R}(W)$ respectively, are defined as follows:

$$\overline{R}(W)(x) = \max\{W(y) : y \in [x]_R\}, \underline{R}(W)(x) = \min\{W(y) : y \in [x]_R\}.$$

Similarly, the upper-approximation and lower-approximation of $\alpha - cut$ set W_α with respect to approximation space (U, R), denoted by $\overline{R}(W_\alpha)$ and $\underline{R}(W_\alpha)$ respectively, are defined as follows [13]:

$$\overline{R}(W_\alpha) = \{x : [x]_R \cap W_\alpha \neq \Phi\}, \underline{R}(W_\alpha) = \{x : [x]_R \subseteq W_\alpha\}$$

Evidently, for $\alpha > \beta$, such that $\overline{R}(W_\alpha) \subseteq \overline{R}(W_\beta)$ and $\underline{R}(W_\alpha) \subseteq \underline{R}(W_\beta)$.

3 $\gamma - Tolerance$ Relation

From the definition of tolerance relation, we can see that if the attribute values of two objects with respect to every attribute in conditional attribute set are equivalent or the attribute values of at least one object include null values, then the two objects must belong to the same tolerance class; in other words, the null may take arbitrary values. Therefore such a tolerance class is not totally suit for the reality. In this section, we are going to improve the tolerance relation.

Let $S = (U, A, V, F)$ be an incomplete information system, $b \in B \subseteq A, V_b = \{b(x) : b(x) \neq *, x \in U\}$, and it be possible that V_b contains the same elements, $|V_b|$ represents the number of all the elements in V_b, $|b(x)|$ indicates the number of the non-null attribute value $b(x)$.

Definition 7. Let $S = (U, A, V, F)$ be an incomplete information system, $x, y \in U, b \in B \subseteq A$, and there exist at least a $b(z) \neq *$ in V_b, then the probability that $b(x)$ equals $b(y)$, denoted by $p_b(x, y)$, is defined as follows:

$$p_b(x,y) = \begin{cases} 1, & b(x) \neq * \wedge b(y) \neq * \wedge b(x) = b(y) \\ \frac{|b(y)|}{|V_b|}, & b(x) = * \wedge b(y) \neq * \\ \frac{|b(x)|}{|V_b|}, & b(y) = * \wedge b(x) \neq * \\ 1, & b(x) = * \wedge b(y) = * \\ 0, & b(x) \neq * \wedge b(y) \neq * \wedge b(x) \neq b(y) \end{cases}$$

Definition 8. Let $S = (U, A, V, F)$ be an incomplete information system, $x, y \in U, b \in B \subseteq A$, the probability that x and y have the same attribute values on B, denoted by $p_B(x, y)$, is defined as follows: $p_B(x, y) = \prod_{b \in B} p_b(x, y)$.

$p_B(x, y)$ is also called the probability that x equals y on B.

Evidently, if $C \subseteq B$, then $p_B(x, y) \leq p_C(x, y)$.

Similarity to the paper [8], a section-value γ $(0 \leq \gamma \leq 1)$ can be used to ascertain a tolerance relation in the incomplete information system. As the probability that x is equal to y on B is not less than γ, we believe that x and y belong to the same tolerance class.

Definition 9. Let $S = (U, A, V, F)$ be an incomplete information system, $x, y \in U, B \subseteq A$. Then: $T^\gamma(B) = \{(x, y) : x \in U \wedge y \in U \wedge p_B(x, y) \geq \gamma\}$ is called $\gamma - tolerance$ relation, and $T_B^\gamma(x) = \{y : y \in U \wedge (x, y) \in T^\gamma(B)\}$ is named $\gamma - tolerance$ class of x.

From definition 8 and 9, we can know that $\gamma - tolerance$ relation and $\gamma - tolerance$ class of x are equivalence relation and equivalence class when the system S is a complete information system and $\gamma = 1$ respectively.

Suppose that $U/T^\gamma(B) = \{T_B^\gamma(x) : x \in U\}$ stands for classifications, then the elements in $U/T^\gamma(B)$ are $\gamma - tolerance$ classes. In general, $\gamma - tolerance$ classes in $U/T^\gamma(B)$ don't consist the partition of U, but form a cover, i.e., $\cup U/T^\gamma(B) = U$.

Under $\gamma - tolerance$ relation, $T_B^\gamma(x) = \{y : y \in U \wedge (x, y) \in T^\gamma(B)\}$ implies the maximal set of objects that are possibly indiscernible by B with x.

From definition 9, we also easily obtain the following two simple conclusions.

Theorem 1. $\gamma - tolerance$ relation has reflexivity and symmetry property, but may not have transitivity property.

Theorem 2. Let $S = (U, A, V, F)$ be an incomplete information system, $C \subseteq B \subseteq A$, $\gamma \in [0, 1]$. The following properties hold: (1) $T^{\gamma}(B) \subseteq T^{\gamma}(C)$; (2) $T^{\gamma}(B) = \cap_{b \in B} T^{\gamma}(\{b\})$; (3) $T^{\gamma}(B) \subseteq T(B)$.

From above definition 9, theorem 1 and 2, there are the following theorems for $\gamma - tolerance$ relation.

Theorem 3. Let $S = (U, A, V, F)$ be an incomplete information system, $B \subseteq A$, $0 \leq \gamma_1 < \gamma_2 \leq 1$. Then $T_B^{\gamma_1}(x) \supseteq T_B^{\gamma_2}(x)$, for each $x \in U$.

Proof. Let $x_0 \in T_B^{\gamma_2}(x)$. Suppose that $x_0 \notin T_B^{\gamma_1}(x)$, then $p_B(x, x_0) \geq \gamma_2$ and $p_B(x, x_0) < \gamma_1$, Further, there would be $\gamma_2 < \gamma_1$ which contradicts the condition $\gamma_1 < \gamma_2$, therefore, $x_0 \in T_B^{\gamma_1}(x)$. For this reason, the conclusion in theorem 3 holds.

Theorem 4. Let $S = (U, A, V, F)$ be an incomplete information system, $C \subseteq B \subseteq A$, $\gamma \in [0, 1]$. Then $T_B^{\gamma}(x) \subseteq T_C^{\gamma}(x)$.

Proof. Assume that $x_0 \in T_B^{\gamma}(x)$, we then have $P_B(x_0, x) \geq \gamma$. Since $C \subseteq B \subseteq A$. hence $p_B(x_0, x) \leq p_C(x_0, x)$, or equivalently, $p_C(x_0, x) \geq \gamma$, namely $x_0 \in T_C^{\gamma}(x)$. Consequently, $T_B^{\gamma}(x) \subseteq T_C^{\gamma}(x)$, which completes the proof.

Theorem 5. Let $S = (U, A, V, F)$ be an incomplete information system, $B \subseteq A$. Such that $T_B^{\gamma}(x) \subseteq T_B(x)$ for every $\gamma \in [0, 1]$. Furthermore, $T_B^{\gamma}(x) \cap T_B^{\gamma}(y) \subseteq T_B(x) \cap T_B(y)$

Proof. Form definition 9 and theorem 2(3), we can directly gain the consequence.

Theorem 5 shows that the number of intersectional elements between $\gamma - tolerance$ classes is not greater than that of intersectional elements between tolerance classes.

4 Rough Set Model in Incomplete and Fuzzy Objective Information System

4.1 The Concept of Incomplete and Fuzzy Objective Information System [19]

What we shall research is fuzzy objective information system in an incomplete approximation space that is combined with the fuzzy objective information system.

Definition 10. Incomplete and fuzzy objective information system, denoted still by S , is defined as follows:

$$S = (U, A, V, F; D, W, G). \text{ where:}$$

(1) U is a nonempty finite set of objects called universe of discourse, i.e., $U = \{x_1, x_2, ..., x_n\}$;

(2) A is a nonempty finite set of conditional attributes, i.e., $A = \{a_1, a_2, ..., a_m\}$;

(3) $F = \{f_1, f_2, ..., f_m\}$ is a set of conditional attribute mappings, $f_j : U \to V_j$, $j \leq m$, and there exists at least a V_j including null values, $V = \cup_{1 \leq j \leq m} V_j$;

(4) D is a nonempty finite set of decision attributes, i.e.,$D = \{d_1, d_2, ..., d_p\}$;

(5) $G = \{g_1, g_2, ..., g_p\}$ is a set of fuzzy decision mappings, $g_k : U \to W_k$, $W_k \in F(U)$, $W = \cup_{1 \leq k \leq p} W_k$, and $W_k(x_i) \in [0, 1]$, $i \leq n$, $k \leq p$.

As $p = 1$ and let $D = \{d\}, G = \{g\}, W \in F(U)$, then $S = (U, A, V, F; \{d\}, W, \{g\})$, the situation is mainly discussed in this paper. The other case ($p > 1$) will be discussed in continuation papers.

Example 1. Table 1 gives an example of incomplete and fuzzy objective information system(IFOIS in short) $S = (U, A, V, F; \{d\}, W, \{g\})$, where $U = \{x_1, x_2, ..., x_9\}$ is the set of objects, $A = \{a_1, a_2, a_3\}$ is the set of conditional attributes, d is an decision attribute, $*$ represents null values.

Table 1. IFOIS

	x_1	x_2	x_3	x_4	x_5	x_6	x_7	x_8	x_9
a_1	1	1	3	1	*	3	3	3	2
a_2	2	*	2	2	2	1	2	1	3
a_3	1	3	3	*	1	*	*	2	*
d	0.8	0.6	0.9	0.7	0.8	0.5	0.7	0.6	0.5

If $\gamma = 0.2, 0.3, 0.4$ respectively, then all the $\gamma - tolerance$ classes of objects in U are given as follows (table 2).

Table 2. $\gamma - tolerance$ classes ($\gamma = 0.2, 0.3, 0.4$)

	$T_A^\gamma(x_1)$	$T_A^\gamma(x_2)$	$T_A^\gamma(x_3)$	$T_A^\gamma(x_4)$	$T_A^\gamma(x_5)$	$T_A^\gamma(x_6)$	$T_A^\gamma(x_7)$	$T_A^\gamma(x_8)$	$T_A^\gamma(x_9)$
0.2	$\{x_1, x_4, x_5\}$	$\{x_2, x_4\}$	$\{x_3, x_7\}$	$\{x_1, x_2, x_4\}$	$\{x_1, x_5, x_7\}$	$\{x_6, x_8\}$	$\{x_3, x_5, x_7\}$	$\{x_6, x_8\}$	$\{x_9\}$
0.3	$\{x_1, x_4, x_5\}$	$\{x_2\}$	$\{x_3, x_7\}$	$\{x_1, x_4\}$	$\{x_1, x_5\}$	$\{x_6\}$	$\{x_3, x_7\}$	$\{x_8\}$	$\{x_9\}$
0.4	$\{x_1, x_4\}$	$\{x_2\}$	$\{x_3, x_7\}$	$\{x_1, x_4\}$	$\{x_5\}$	$\{x_6\}$	$\{x_3, x_7\}$	$\{x_8\}$	$\{x_9\}$

From above table 2, we can see the number of elements in the same $\gamma - tolerance$ classes is decreasing as γ is gradually increasing, this is consistent with theorem 3.

4.2 Rough Set Model

The rough set model considered afterwards is in an incomplete and fuzzy objective information system based on $\gamma - tolerance$ relation.

Definition 11. Let $S = (U, A, V, F; \{d\}, W, \{g\})$ be an incomplete and fuzzy objective information system, $B \subseteq A$, $T^\gamma(B)$ is a $\gamma - tolerance$ relation with regard to B in incomplete space (U, A, V, F), $T_B^\gamma(x) = \{y \in U : (x, y) \in T^\gamma(B)\}$ is a $\gamma - tolerance$ class contained x. For every fuzzy set $W \in F(U)$, let:

$$\overline{T_B^\gamma}(W)(x) = max\{W(y) : y \in T_B^\gamma(x)\}$$
$$\underline{T_B^\gamma}(W)(x) = min\{W(y) : y \in T_B^\gamma(x)\}$$

where $W(y)$ represents the membership value of the object y. Then $\overline{T_B^\gamma}(W)$ and $\underline{T_B^\gamma}(W)$ are said to be fuzzy upper-approximation and fuzzy lower-approximation of W based on $\gamma - tolerance$ relation in incomplete space (U, A, V, F) (shortly, to be fuzzy upper-approximation and fuzzy lower-approximation of W) respectively.

$$\overline{T_B^\gamma} : F(U) \to F(U) \ \ and \ \ \underline{T_B^\gamma} : F(U) \to F(U)$$

are called incomplete fuzzy upper-approximation and lower-approximation operators separately. Evidently, $\overline{T_B^\gamma}(W)$ and $\underline{T_B^\gamma}(W)$ are a pair of fuzzy sets.

From definition 11 known, when (U, A, V, F) is complete approximation space, $B \subseteq A$, and $T(B)$ becomes equivalence relation R_B (just now $\gamma = 1$), $\overline{T_B^\gamma}(W)(x)$ and $\underline{T_B^\gamma}(x)$ are:

$$\overline{T_B^\gamma}(W)(x) = \overline{R_B}(W)(x) = max\{W(y) : y \in [x]_{R_B}\} \ and$$
$$\underline{T_B^\gamma}(W)(x) = \underline{R_B}(W)(x) = min\{W(y) : y \in [x]_{R_B}\} \ respectively.$$

Namely, $\overline{T_B^\gamma}(W)$ and $\underline{T_B^\gamma}(W)$ are fuzzy upper-approximation and fuzzy lower-approximation of W in complete space respectively [12].

Furthermore, if W is a classical Contor set, then:

$\overline{T_B^\gamma}(W)(x) = \overline{R_B}(W)(x) = 1 \Leftrightarrow [x]_{R_B} \cap W \neq \Phi \Leftrightarrow \overline{T_B^\gamma}(W) = \overline{R_B}(W) = \{x \in U : \overline{R_B}(W)(x) = 1\} = \{x \in U : [x]_{R_B} \cap W \neq \Phi\}$,

$\underline{T_B^\gamma}(W)(x) = \underline{R_B}(W)(x) = 1 \Leftrightarrow [x]_{R_B} \subset W \Leftrightarrow \underline{T_B^\gamma}(W) = \underline{R_B}(W) = \{x \in U : \underline{R_B}(W)(x) = 1\} = \{x \in U : [x]_{R_B} \subset W\}$

Namely, $\overline{T_B^\gamma}(W)$ and $\underline{T_B^\gamma}(W)$ are the classical Pawlak upper-approximation and lower-approximation respectively.

Example 2 (continued example 1). Table 1 gives a data-table of incomplete and fuzzy object information system, and table 2 provides the relevant $\gamma - tolerance$ classes ($\gamma = 0.2, 0.3, 0.4$ $separately$). Where $W = 0.8/x_1 + 0.6/x_2 + 0.9/x_3 + 0.7/x_4 + 0.8/x_5 + 0.5/x_6 + 0.7/x_7 + 0.6/x_8 + 0.5/x_9$.

Then we obtain according to definition 11:

$\overline{T_A^{0.2}}(W) = 0.8/x_1 + 0.7/x_2 + 0.9/x_3 + 0.8/x_4 + 0.8/x_5 + 0.6/x_6 + 0.9/x_7 + 0.6/x_8 + 0.5/x_9$,

$\underline{T_A^{0.2}}(W) = 0.7/x_1 + 0.6/x_2 + 0.7/x_3 + 0.6/x_4 + 0.7/x_5 + 0.5/x_6 + 0.7/x_7 + 0.5/x_8 + 0.5/x_9$,

$\overline{T_A^{0.3}}(W) = \overline{T_A^{0.4}}(W) = 0.8/x_1 + 0.6/x_2 + 0.9/x_3 + 0.8/x_4 + 0.8/x_5 + 0.5/x_6 + 0.9/x_7 + 0.6/x_8 + 0.5/x_9$,

$\underline{T_A^{0.3}}(W) = \underline{T_A^{0.4}}(W) = 0.7/x_1 + 0.6/x_2 + 0.7/x_3 + 0.7/x_4 + 0.8/x_5 + 0.5/x_6 + 0.7/x_7 + 0.6/x_8 + 0.5/x_9$.

Theorem 6. Let $S = (U, A, V, F; \{d\}, W, \{g\})$ be an incomplete and fuzzy objective information system, $W \in F(U)$, $B \subseteq A$, and for each $\gamma \in [0,1]$ then incomplete and fuzzy upper-approximation operator $\overline{T_B^\gamma}$ and lower-approximation operator $\underline{T_B^\gamma}$ have the following properties:

(1) $\underline{T_B}(W) \subseteq \underline{T_B^\gamma}(W) \subseteq W \subseteq \overline{T_B^\gamma}(W) \subseteq \overline{T_B}(W)$;

(2) $If\ W_1 \subset W_2,\ Then:\ \overline{T_B^\gamma}(W_1) \subseteq \overline{T_B^\gamma}(W_2)$, $\underline{T_B^\gamma}(W_1) \subseteq \underline{T_B^\gamma}(W_2)$;

(3) $\overline{T_B^\gamma}(\overline{T_B^\gamma}(W)) \supseteq \overline{T_B^\gamma}(W) \supseteq \underline{T_B^\gamma}(\overline{T_B^\gamma}(W))$, $\underline{T_B^\gamma}(\underline{T_B^\gamma}(W)) \subseteq \underline{T_B^\gamma}(W) \subseteq \overline{T_B^\gamma}(\underline{T_B^\gamma}(W))$;

(4) $\overline{T_B^\gamma}(W) = (\underline{T_B^\gamma}(W)^C)^C$, $\underline{T_B^\gamma}(W) = (\overline{T_B^\gamma}(W)^C)^C$.

where T_B represents Kryszkiewcz's tolerance relation with respect to B.

Proof. (1) From each $x \in U$, from definition 11:

$\overline{T_B^\gamma}(W)(x) = max\{W(y) : y \in T_B^\gamma(x)\}$, $\underline{T_B^\gamma}(W)(x) = min\{W(y) : y \in T_B^\gamma(x)\}$;

In the addition, we also have:

$\overline{T_B}(W)(x) = max\{W(y) : y \in T_B(x)\}$, $\underline{T_B}(W)(x) = min\{W(y) : y \in T_B(x)\}$. and for each $\gamma \in [0,1]$, such that $T_B^\gamma(x) \subseteq T_B(x)$. Therefore,

$\underline{T_B}(W)(x) \subseteq \underline{T_B^\gamma}(W)(x) \subseteq W(x) \subseteq \overline{T_B^\gamma}(W)(x) \subseteq \overline{T_B}(W)(x)$, Then the conclusion holds.

(2) Similar to the proof of (1).

(3) From the results of (1) and (2), we can immediately gain the proof.

(4) $\underline{T_B^\gamma}(W)^C(x) = min\{(W)^C(y) : y \in T_B^\gamma(x)\} = min\{1 - W(y) : y \in T_B^\gamma\}$ $=1 - max\{W(y) : y \in T_B^\gamma(x)\} = 1 - \overline{T_B^\gamma}(x)$. Consequently $\overline{T_B^\gamma}(W) = (\underline{T_B^\gamma}(W)^C)^C$.

Analogically, the second formula in (4) can be proved.

5 Precision Reduction in Incomplete and Fuzzy Objective Information System

5.1 Basic Theory of Precision Reduction

First of all, the two conclusions, which are related to precision reduction, are introduced.

Theorem 7. Let $S = (U, A, V, F; \{d\}, W, \{g\})$ be an incomplete and fuzzy objective information system, $W \in F(U)$, $B \subseteq A$, $\gamma \in [0,1]$, $0 \leq \alpha, \beta \leq 1$ and assume that

$\overline{T_B^\gamma}(W)_\beta = \{x \in U : \overline{T_B^\gamma}(W)(x) \geq \beta\}$, $\underline{T_B^\gamma}(W)_\alpha = \{x \in U : \underline{T_B^\gamma}(W)(x) \geq \alpha\}$.

Then:

(1) $\overline{T_B^\gamma}(W)_\beta$ *and* $\underline{T_B^\gamma}(W)_\alpha$ with respect to β and α monotonously decrease respectively;

(2) $\underline{T_B^\gamma}(W)_\alpha \subseteq W_\alpha \subseteq \overline{T_B^\gamma}(w)_\alpha$;

(3)$\alpha \geq \beta \Rightarrow \underline{T_B^\gamma}(W)_\alpha \subseteq \overline{T_B^\gamma}(w)_\beta$.

Proof. The proofs of (1), (2) and (3) are directly derived from the representations of $\overline{T_B^\gamma}(W)_\beta$ *and* $\underline{T_B^\gamma}(W)_\alpha$.

Theorem 8. Let $S = (U, A, V, F; \{d\}, W, \{g\})$ be an incomplete and fuzzy objective information system, $W \in F(U)$, $C \subseteq B \subseteq A$,$\gamma \in [0,1]$, $0 \leq \alpha, \beta \leq 1$.Then: (1) $\overline{T_B^\gamma}(W)_\beta \subseteq \overline{T_C^\gamma}(W)_\beta$; (2) $\underline{T_C^\gamma}(W)_\alpha \subseteq \underline{T_B^\gamma}(W)_\alpha$.

Proof. (1) From the given conditions $C \subseteq B \subseteq A$ and theorem 4, for every $x \in U$, such that $T_C^\gamma(x) \supseteq T_B^\gamma(x)$. Hence,

$max\{W(y) : y \in T_C^\gamma(x)\} \geq max\{W(y) : y \in T_B^\gamma\}(x)$. That is,

$\overline{T_C^\gamma}(w)(x) \geq \overline{T_B^\gamma}(W)(x) \Rightarrow \overline{T_C^\gamma}(W) \supseteq \overline{T_B^\gamma}(W) \Rightarrow \overline{T_C^\gamma}(w)_\beta \supseteq \overline{T_B^\gamma}(W)_\beta$.
(2) The proof of (2) is similar to that of above (1).

Definition 12. Let $S = (U, A, V, F; \{d\}, W, \{g\})$ be an incomplete and fuzzy objective information system, $W \in F(U)$, $C \subseteq B \subseteq A$,$\gamma \in [0,1]$, $0 \leq \beta \leq \alpha \leq 1$. Then the precision degree and coarse degree of W with regard to (α, β), denoted by $\alpha_B^\gamma(\alpha, \beta)$ and $\rho_B^\gamma(\alpha, \beta)$ respectively, are defined as follows:

$$\alpha_B^\gamma(\alpha, \beta) = \frac{|\underline{T_B^\gamma}(W)_\alpha|}{|\overline{T_B^\gamma}(W)_\beta|}, \quad \rho_B^\gamma(\alpha, \beta) = 1 - \alpha_B^\gamma(\alpha, \beta) = 1 - \frac{|\underline{T_B^\gamma}(W)_\alpha|}{|\overline{T_B^\gamma}(W)_\beta|}.$$

And ruled: $\alpha_B^\gamma(\alpha, \beta) = 1$ and $\rho_B^\gamma(\alpha, \beta) = 0$ for $\overline{T_B^\gamma}(W)_\beta = \Phi$.

Obviously, $0 \leq \alpha_B^\gamma(\alpha, \beta) \leq 1$, $0 \leq \rho_B^\gamma(\alpha, \beta) \leq 1$; $\alpha_B^\gamma(\alpha, \beta)$ is monotonously non-increasing w.r.t. α and is monotonously non-decreasing w.r.t. β; $\rho_B^\gamma(\alpha, \beta)$ is monotonously non-decreasing w.r.t. α and is monotonously non-increasing w.r.t. β.

According to theorem 8 and definition 12, the following theorem holds.

Theorem 9. Let $S = (U, A, V, F; \{d\}, W, \{g\})$ be an incomplete and fuzzy objective information system, $W \in F(U)$, $B \subseteq A$,$\gamma \in [0,1]$, $0 \leq \beta \leq \alpha \leq 1$. Then:

$$\alpha_A^\gamma(\alpha, \beta) \geq \alpha_B^\gamma(\alpha, \beta), \quad \rho_A^\gamma(\alpha, \beta) \leq \rho_B^\gamma(\alpha, \beta).$$

Definition 13. Let $S = (U, A, V, F; \{d\}, W, \{g\})$ be an incomplete and fuzzy objective information system, $W \in F(U)$, $B \subseteq A$,$\gamma \in [0,1]$, $0 \leq \beta \leq \alpha \leq 1$.If B satisfies the equation $\alpha_A^\gamma(\alpha, \beta) = \alpha_B^\gamma(\gamma, \beta)$, and the equation is invalid for the arbitrary a proper subset of B. Then B is said to be (α, β) precision reduction in incomplete and fuzzy objective information system S .

By definition 13, precision reduction ensures the ratio of the number of the certain objects that membership values are not less than α to that of the possible objects that membership values are not less than β is invariable.

Theorem 10. Let $S = (U, A, V, F; \{d\}, W, \{g\})$ be an incomplete and fuzzy objective information system, $W \in F(U)$, $B \subseteq A$,$\gamma \in [0,1]$, $0 \leq \beta \leq \alpha \leq 1$. Then B is (α, β) precision reduction in incomplete and fuzzy objective information system S if and only if B is the minimal proper subset of conditional attributes and simultaneously satisfies the two equations

$$\overline{T_B^\gamma}(W)_\beta = \overline{T_A^\gamma}(W)_\beta \ \ and \ \ \underline{T_B^\gamma}(W)_\alpha = \underline{T_A^\gamma}(W)_\alpha.$$

5.2 Precision Reduction Algorithm

To further illustrate the idea of precision reduction, we shall give an algorithm for finding the precision reduction and a continual example of the former example 1 and 2.

Reduction Algorithm

Input: $S = (U, A, V, F; \{d\}, W, \{g\})$, the valid values of γ, α and β $(0 \leq \gamma \leq 1,\ 0 \leq \beta \leq \alpha \leq 1)$.

Output: The system precision reduction B $(B \subseteq A)$.

Step 1. Calculate the subsets of conditional attributes A: $B_1, B_2, ..., B_r$, where $|B_k| = |A| - 1$ $(k = 1, 2, ..., r)$;

Step 2. Calculate $\alpha_A^\gamma(\alpha, \beta)$(i.e., find $\gamma - tolerance$ classes, every pair of fuzzy upper-approximation and lower-approximation of W and $\alpha_A^\gamma(\alpha, \beta)$ in turn);

Step 3. Calculate $\alpha_{B_k}^\gamma(\alpha, \beta)$ $(k = 1, 2, ..., r)$. If $\alpha_{B_k}^\gamma(\alpha, \beta) < \alpha_A^\gamma(\alpha, \beta)$, then stop; Otherwise, $\alpha_{B_k}^\gamma(\alpha, \beta) = \alpha_A^\gamma(\alpha, \beta)$, go to step 1 (consider continuously the corresponding precision degree of the subsets of B_k until finding a $B \subseteq B_k \subseteq A$ satisfies that$\alpha_B^\gamma(\alpha, \beta) = \alpha_{B_k}^\gamma(\alpha, \beta) = \alpha_A^\gamma(\alpha, \beta)$ and $\alpha_C^\gamma(\alpha, \beta) \neq \alpha_{B_k}^\gamma(\alpha, \beta)$ for every $C \subset B_k \subseteq A$).

Example 3. Example 1 presents various $\gamma - tolerance$ classes $(\gamma = 0.2, 0.3, 0.4)$, example 2 gives three pairs of fuzzy upper-approximation and lower-approximation of W as $\gamma = 0.2, 0.3, 0.4$ respectively. We shall only consider how to find the precision reduction as $\gamma = 0.2,\ \alpha = \beta = 0.6$.

Evidently, $\alpha_A^\gamma(\alpha, \beta) = 3/4$.

Let $B_1 = \{a_1, a_2\}$, $B_2 = \{a_1, a_3\}$, $B_3 = \{a_2 a_3\}$. Analogue to finding $\alpha_A^\gamma(\alpha, \beta)$, we can gain

$\alpha_{B_1}^\gamma(\alpha, \beta) = 3/4$, $\alpha_{B_2}^\gamma(\alpha, \beta) = 3/8$, $\alpha_{B_3}^\gamma(\alpha, \beta) = 3/4$.

Furthermore, we can still find

$\alpha_{\{a_1\}}^\gamma(\alpha, \beta) = 3/8$, $\alpha_{\{a_2\}}^\gamma(\alpha, \beta) = 3/4$, $\alpha_{\{a_3\}}^\gamma(\alpha, \beta) = 1/8$.

For this reason, as $\gamma = 0.2$, $\{a_2\}$ is $(0.6, 0.6)$ precision reduction. Of course, we can also consider finding other precision reductions while γ, α and β take other valid values.

6 Conclusion

The $\gamma - tolerance$ relation and $\gamma - tolerance$ class based on the probability that objects equal on the set of conditional attributes are first proposed in this paper, with variation of γ value, intersectional elements between $\gamma - tolerance$ classes are variable, for this reason, the $\gamma - tolerance$ relation is superior to tolerance relation in some way. And then, we also present the concept of the incomplete and fuzzy objective information system and its rough set model under $\gamma - tolerance$ relation, which is the generalization of the rough set model in complete and fuzzy objective information system and the Pawlak rough set model. Finally, we give the definition of precision reduction and its algorithm in incomplete and fuzzy objective information system. We will focus on knowledge reduction of the system with changing of γ , α *and* β in our next work.

Acknowledgements

The authors would like to thank the anonymous referees for their constructive comments. The paper is partially supported by the Nature Science Foundation of China (70571032) and the Scientific Research Foundation of Hunan Provincial Education Department (06C367).

References

1. Wang G.Y. *Rough Set Theory and Knowledge Discovery.* Xi'an: Xi'an JiaoTong University Press (2001).
2. Wang G.Y, Wu Y, Fisher. P. S. Rule Generation Based on Rough Set Theory. In Data Mining and Knowledge: Theory, Tools, and Technology II, Belur V. Dasarathy, Editor, Proceedings of SPIE **Vol.4057** (2000) 181-189.
3. Wu W.Z, Mi J.S, Zhang W.X. A New Rough Set Approach to Knowledge Discovery in Incomplete Information System. Proceedings of the Second International Conference on Machine Learning and Cybernetics, Xi'an, 2-5 November 2003 (1713-1718).
4. Liu Q. *Rough Set and Rough Reasoning.* Beijing: Science Press (2001).
5. Kryszkiewicz M. Rough set approach to incomplete information system. *Information Sciences*, **112** (1998) 39-49.
6. Marzena Kryszkiewicz. Rules in incomplete information systems. *Information Sciences*, **113** (1999) 271-292.
7. Zhang H.Y, Liang J.Y. Variable Precision Rough Set Model and a Knowledge Reduction Algorithm for Incomplete Information System. *Computer science*, **Vol.30** (4),2003, 153-155.
8. He W, Liu C-Y, Zhao J, Li H. An Algorithm of Attributes Reduction in Incomplete Information System. *Computer science*, **Vol.31** (2),2004, 153-155.
9. Huang B, Wei D.K, Zhou X.Z. An algorithm for Maximum Distribution Reductions and Maximum Distribution Rules in Information system. *Computer science*, **Vol.31** (10A),2004, 80-83.
10. Huang B, Zhou X.Z. Extension of Rough Set Model Based on Connection Degree under Incomplete Information System. *Systems Engineering –Theory and Practice.* **1** (2004) 88-92.
11. Roman Slowinski, Danel Vanderpooten. A Generalized Definition of Rough Approximation Based on Similarity. IEEE TRANSACTIONS ON KNOWLEDGE AND DATA ENGINEERING, **Vol.12** (2), 2000, 331-336.
12. Zhang W.X, Leung Y,Wu W.Z. *Information System and Knowledge Discovery.* Beijing: Science Press (2003).
13. D.Dubois, H.Prade. Rough fuzzy sets and fuzzy rough sets. *International Journal of Systems*, **17** (1990) 191-208.
14. Qiang Shen, Alexios chouchoulas. A rough-fuzzy approach for generating classification rules. *Pattern Recognition*, **35** (2002) 2425-2438.
15. Manish Sarkar. Rough-fuzzy functions in classification. *Fuzzy Sets and Systems*, **132** (2002) 353-369.
16. Wu W.Z, Zhang W.X, Xu Z.B. Characterizating Rough Fuzzy Set in Constructive and Axiomatic Approaches. *Chinese Journal of Computer*, **Vol.27** (2),2004, 197-203.
17. Liu G.L. The Axiomatic Systems Rough Fuzzy Set on Fuzzy Approximation Space. *Chinese Journal of Computer*, **Vol.27** (9),2004, 1187-1191.

18. Zhang M, Li H.Z, Zhang W.X. Rough Set Theory on Fuzzy Information System. *Fuzzy Systems and Mathematics*, **Vol.16** (3),2002, 44-49.
19. Wei D.K, Huang B, Zhou X.Z. Rough Set Model and Knowledge Reduction in Incomplete and Fuzzy Objective Information system. *Computer Engineering.* **Vol.32** (8), 2006, 48-51.
20. Wei D.K. Similarity Relation-Based Knowledge Reduction in Incomplete and Fuzzy Decision Information Systems. *Joural of Nature Science of Hunan Normal University.* **Vol.29**(2),2006,18-23

Granular Ranking Algorithm Based on Rough Sets

Zhen Cao and Xiaofeng Wang

Information Engineering College, Shanghai Maritime University, Shanghai, China 200135
caozhendsg@hotmail.com

Abstract. Compared with the rule form in traditional data mining techniques, expressing knowledge in the form of rank list can avoid many disadvantages, and may be applied for the investigation of targeted marketing widely, identifying potential market values of customers or products. Based on the rough sets theory in granular computing, this paper proposes a *Granular Ranking Algorithm* with the time complexity $O(nm)$, gives the framework of algorithm and the concrete algorithm steps. The core of new algorithm is the construction of *Granular Ranking Function* $r_G(x)$, which guides instances in the testing dataset finish ranking. The ranked result has a strong readability. The new algorithm improves the computation efficiency further relative to existing algorithms, e.g. the *Market Value Function*. The experiment result shows that the computation accuracy of granular ranking algorithm approaches to the market value function. Meanwhile, the time consumption of the former one is much less than the latter.

Keywords: Rough sets; decision table; granule; ranking.

1 Introduction

Traditional data mining techniques, e.g. association rules, may be applied for the investigation of targeted marketing [1-3] widely, identifying potential market values of customers or products. However, there may be some difficulties with these techniques. One may produce too many or too few rules; it is an uneasy job to define the threshold; expressing knowledge in the form of rules may have a bad readability. Rank list is a good alternative solution. Therefore, some specialists and scholars have made elementary investigations, proposing the *Market Value Function* (*MVF*) [5,6,8,9].

Suppose there is a health club. The market value function, which is constructed by members' information in the training instance set, computes the probability, whether new members in the testing instance set will join the health club, and ranks the testing set in the light of the probability of every member. A threshold can be defined for the ranked list according to various criteria. Members, whose probabilities are higher than the threshold, can be preserved. The rest can be excluded. The health club sends advertisement only to the preserved members, which can reduce the advertisement cost, and improve the advertisement efficiency.

B.-Y. Cao (Ed.): Fuzzy Information and Engineering (ICFIE), ASC 40, pp. 837–845, 2007.
springerlink.com

Literature [8] defines the Market Value Function as follows:

$$r(x) = \sum_{a \in At} w_a u_a (I_a(x)) \tag{1}$$

w_a is the weight function of attribute $a \in At$. The utility function $u_a : V_a \rightarrow R$ expresses a mapping to real number R when $a \in At$ takes value $v \in V_a$. Since the computation of weight functions w_a and utility functions u_a is based on information entropy and bayes classification respectively, the entire process is very complex and spends a lot. Furthermore, the result can not express the authentic market value sometimes. Some people, who don't want to join the club, may not be excluded, since their function value is higher than the threshold, vice versa. These problems are waiting for solutions.

This paper proposed a *Granular Ranking Algorithm* (*GRA*) based on the rough sets theory in granular computing. In the universe, suppose instances are in different equivalence classes, i.e. granules. The granular ranking algorithm computes a ranking function value for every equivalence class. Instances in the unknown instance set are ranked according to the ranking function value of the equivalence class, which the instance belongs to. Compared with traditional data mining techniques, the granular ranking algorithm preserves many advantages of rank list. Moreover, compared with the market value function, this algorithm can reduce the spending of computation and improve the accuracy of the ranked result.

The rest of the paper is organized as follows. Section 2 discusses the essential thought of granular ranking algorithm, gives the framework of this algorithm and the concrete algorithm steps. In section 3, we choose a breast cancer data set, and make the comparison test, adopting the market value function and the granular ranking algorithm respectively. Section 4 makes a conclusion of this paper.

2 The Granular Ranking Algorithm

2.1 The Thought of Algorithm Designing

Let U be a finite universe of objects. Formally, an information table is a quadruple [7]:

$$S = (U, At, \{V_a \mid a \in At\}, \{I_a \mid a \in At\}),$$

where

U is a finite nonempty set of objects,

At is a finite nonempty set of attributes,

V_a is a nonempty set of values for $a \in At$,

$I_a : U \rightarrow V_a$ is a mapping function from object in U to one value in V_a.

$At = C \cup D, C \cap D = \phi$. C is condition attribute set, D is decision attribute set. Therefore, S is also a decision table, $D = \{d\}, |V_d| = 2, V_d = \{T, F\}$. Instances, whose decision attribute value is T, are positive samples; instances, whose decision attribute value is F, are negative samples. The universe U is consequently divided into three pair-wise disjoint classes, $U = P \cup N \cup D$. The sets P, N and D are called positive, negative, and unknown sample set, respectively.

Let $R = IND(C) = \{(u, u') \in U \times U : \forall a \in C, a(u) = a(u')\}$. Then, R is an equivalence relation, $K = (U, R)$ is an approximate space.

Let $X = \{x \mid x \in U, I_d(x) = T\}$. Sequentially, R-lower approximation of X $R\_(X) = \{x \mid x \in X, [x]_R \subseteq X\}$, R-upper approximation of X $R^{-}(X) = \{x \mid x \in U, [x]_R \cap X \neq \phi\}$. $BN_R(X) = R^{-}(X) - R\_(X)$ is the R-boundary region of X. $POS_R(X) = R\_(X)$, $NEG_R(X) = U - R^{-}(X)$.

There are three kinds of equivalence class, i.e. granule, in the ranked list:

1. $[x]_R^1 = \{y \mid x \in POS_R(X), yRx\}$, the granule whose instances are all positive samples;
2. $[x]_R^2 = \{y \mid x \in BN_R(X), yRx\}$, the granule whose instances are partially positive samples and partially negative samples;
3. $[x]_R^3 = \{y \mid x \in NEG_R(X), yRx\}$, the granule, whose instances are all negative samples.

According to the definition of market value function, since instances in a granule (equivalence class) have the same condition attribute values, these instances also have the same market value $r(x)$. If granules of $[x]_R^2$, $[x]_R^3$ have high $r(x)$ values, they will exist at the top of the ranked list. Contrarily, if granules of $[x]_R^1$ have low $r(x)$ values, they will be in the middle or at the bottom of the ranked list. That is the reason why negative samples may exist at the top and positive samples exist at the bottom of the ranked list respectively.

Suppose we can unconditionally position granules $[x]_R^1$ and granules $[x]_R^3$ at the top and bottom of the ranked list respectively, this phenomenon can be avoided. The result will also be more satisfactory. Here, we refer to the *Rough Membership Function* [4]:

$$\mu_X^R(x) = \frac{card(X \cap [x]_R)}{card([x]_R)} \tag{2}$$

According to the definition, instances in the same granule have equal rough membership function values. For instance x in granule $[x]_R^1$, $\mu_X^R(x) = 1$; for instances in granule $[x]_R^2$, $0 < \mu_X^R(x) < 1$; for instances in granule $[x]_R^3$, $\mu_X^R(x) = 0$. If we take the rough membership function as the ranking criterion, the previous supposition can be realized.

Furthermore, when two granules have the equal rough membership function value, the ranked result may be optimized, if we position the granule with bigger number of positive samples in the front of the other.

Let $\sigma_X^R(x) = card(X \cap [x]_R)$ expresses the number of positive samples in the granule $[x]_R$. The *Granular Ranking Function* (*GRF*) can be defined as follows:

$$r_G(x) = \sigma_X^R(x) \times \mu_X^R(x) \tag{3}$$

$$= card(X \cap [x]_R) \times \frac{card(X \cap [x]_R)}{card([x]_R)}$$

$$= \frac{card(X \cap [x]_R)^2}{card([x]_R)}$$

In a random granule, since instances have the same rough membership function value $\mu_X^R(x)$ and the same number of positive samples $\sigma_X^R(x)$, the granular ranking function value of any instance is also equal.

Let $m = card(X)$, for instance x in granule $[x]_R^1$, $1 \le r_G(x) \le m$; for instances in granule $[x]_R^2$, $0 < r_G(x) < m$; for instances in $[x]_R^3$, $r_G(x) = 0$.

In a random granule, instances have the same value of $\mu_X^R(x)$, $\sigma_X^R(x)$ and $r_G(x)$. Therefore, let $g = [x]_R$. $\mu_X^R(g)$, $\sigma_X^R(g)$ and $r_G(g)$ express the rough membership function , the number of positive samples and the granular ranking function value of any instance in granule *g*, respectively.

2.2 The Framework of Algorithm

The entire granular ranking algorithm is composed of two parts, training and testing. In training part, the *Granular Training Algorithm* (*GTA*) scans the training dataset, whose decision attribute is known, acquiring granular information, constructs the granule set. In testing set, the *Ranking Algorithm* (*RA*) scans the testing dataset, and ranks it according to the information in the granule set.

2.3 The Description of Granular Ranking Algorithm

2.3.1 The Granular Training Algorithm

Let U_G be a finite nonempty set of granules. For a random granule $g \in U_G$, set:

1. $countT(g) = \sigma_X^R(g)$
2. $countF(g) = card(g) - \sigma_X^R(g)$
3. $proportion(g) = \mu_X^R(g) = countT(g)/(countT(g) + countF(g))$
4. $r_G(g) = countT(g) \times proportion(g)$

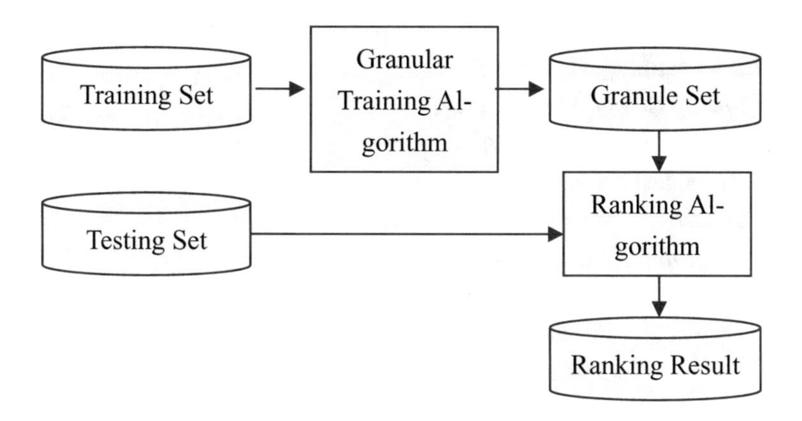

Fig. 1. The Framework of Algorithm

Algorithm 1: Granular Training Algorithm
Input: Training dataset $T = (U, A)$,
where $A = C \cup D, C \cap D = \phi, D = \{d\}$;
Output: Granule set $G = (U_G, A')$,
where $A' = C \cup \{countT, countF, proportion, r_G\}$;
Step 1: *For* each instance x in the training dataset U
For each $a \in C$, read $I_a(x)$;
If there is a corresponding granule g, which has the same attribute values with the instance x, in the granule set G
If $I_d(x) = T$
$g.countT$ ++;
Else
$g.countF$ ++
Else
Insert g', whose attributes have the same values with the attributes of instance x, into the granule set G;
If $I_d(x) = T$
$g'.countT = 1$, $g'.countF = 0$;
Else
$g'.countT = 0$, $g'.countF = 1$;
Step 2: *For* each granule g in the granule set G
$g.proportion = g.countT / (g.countT + g.countF)$;
$g.r_G = g.countT \times g.proportion$

2.3.2 The Ranking Algorithm

Algorithm 2: Ranking Algorithm

Input 1: Testing dataset $T'=(U',A)$,

Input 2: Granule set $G=(U_G,A')$,

Output: The ranked list of testing dataset

Step 1: *For* each instance *x* in the testing dataset U'

For each $a \in C$, read $I_a(x)$;

If there is a corresponding granule *g*, which has the same attribute values with the instance *x*, in the granule set *G*

$x.r = g.r_G$;

Else

$x.r = 0$;

Step 2: Rank the testing dataset by descending order of $x.r$

2.4 The Complexity of Algorithm

The number of condition attributes $|C|=m$. Let T_1, T_2 denote the time complexity of granular training algorithm and ranking algorithm respectively. For the granular training algorithm, let n be the number of instances in the training dataset, the time complexity of training algorithm $T_1 = O(nm)$. For the ranking algorithm, let n' be the number of instances in the testing dataset, suppose we use the *Quick Sort Algorithm*, the time complexity of ranking algorithm $T_2 = O(n'm) + O(n'\log_2 n')$.

3 Experiment

3.1 Dataset

The experiment chooses a breast cancer dataset with 9 condition attributes and 699 records, which was provided by Dr. William H. Wolberg from the University of Wisconsin Hospitals, Madison. The decision attribute divide the dataset into two classes: *benign* and *malignant*. Let benign instances be positive samples, and malignant instances be negative samples. Since we don't focus on data cleaning, 16 records missing attribute values are deleted. We pick up the preceding 600 records from the remaining 683 records as training dataset, and the rest as testing dataset.

Every attribute of original training dataset has the integral value range between 1 and 10. We first make the testing experiment based on the original training dataset. Then, the value range is generalized to the integer between 1 and 5, 1 and 3, 1 and 2. And make the test respectively again.

3.2 Hit Rate

In order to define evaluation criterion, we refer to the concept *Hit Rate*. Let L be a segment of the ranked result. Let $M = \{x \mid x \in L, I_d(x) = T\}$, the *Hit Rate*:

$$h = \frac{card(M)}{card(L)} \tag{4}$$

The ranked list is divided into 10-equal deciles. The hit rate of the preceding 10 percent, 20 percent, up to 100 percent of the ranked result are computed respectively.

3.3 The Result of Experiment

The experiment chooses the market value function and the granular ranking algorithm as the experiment object. The market value function picks up u_a^1、w_a^1 [8] as the utility function and the weight function respectively.

The experiment result shows that the granular ranking algorithm has an obvious advantage over the market value function in the computation efficiency. In the granular training algorithm, the average time consumption of market value function is about 4.5 times longer than that of granular ranking algorithm. And in the ranking algorithm, the average time consumption of the market value function is around 2.6 times longer than that of granular ranking algorithm (see Fig#2).

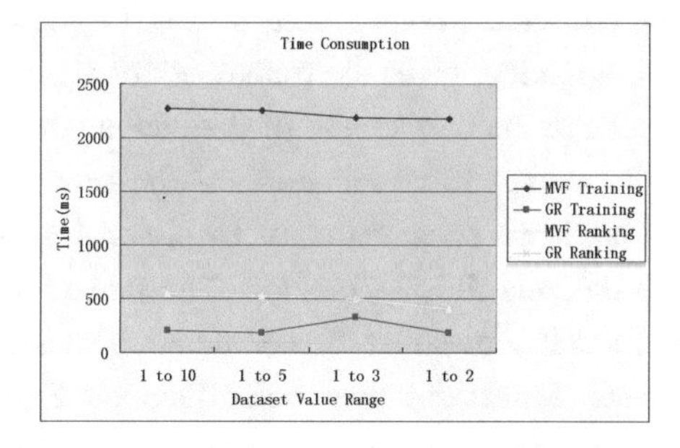

Fig. 2. Time Consumption

In the hit rate aspect, the granular ranking algorithm has the same hit rate with the market value function in the preceding 50 percent samples of ranked list. In the back 50 percent samples of ranked list, the hit rate of granular ranking algorithm is a little lower than that of market value function. But, when generalize the attribute value range to a higher level, the hit rate of granular ranking algorithm approaches to that of market value function gradually. When the attribute values are generalized to the integer between 1 and 2, the hit rate of granular ranking function and market value function are superposed (see Fig#3).

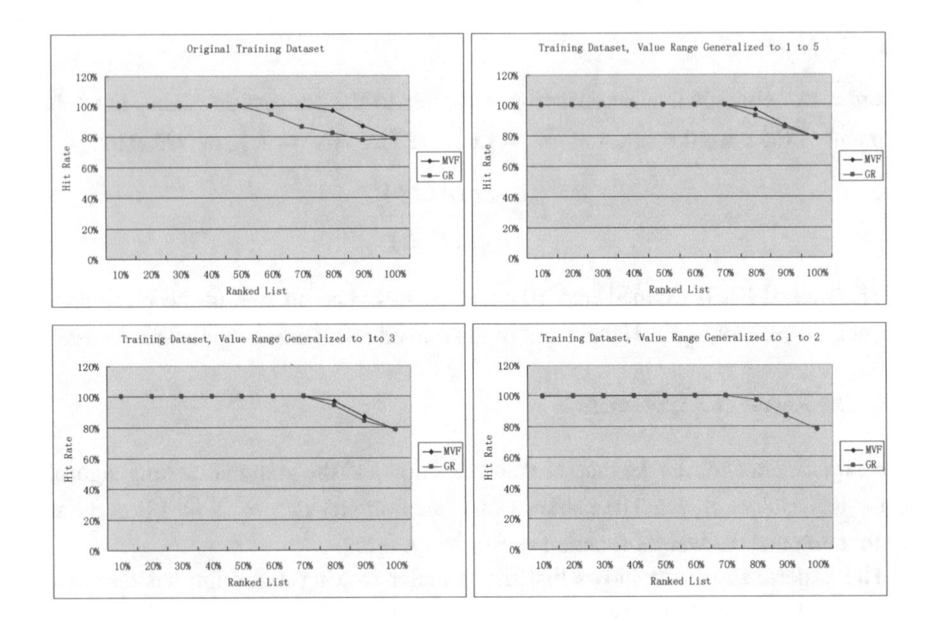

Fig. 3. Hit Rate

4 Conclusion

This paper proposes the *Granular Ranking Algorithm* (*GRA*) based on the basic rough sets theory in granular computing, gives the framework of algorithm and the concrete algorithm steps, and validate the correctness of the new algorithm through testing a standard dataset with the market value function and the new algorithm. The core of granular ranking algorithm is the construction of *Granular Ranking Function* $r_G(x)$, which guides unknown datasets finishes ranking. The ranked list has a strong readability. The experiment result shows that the computation accuracy of granular ranking algorithm approximates to that of market value function. Furthermore, the hit rate of ranked result can be improved further, increasing the quantity of samples in training datasets. In computation efficiency, however, the former one has an obvious advantage over the latter.

The new algorithm can also be applied for the investigation of targeted marketing, identifying potential market values of customers or products.

For the similar algorithms, if the accuracy of approximation [4] $\alpha_R(X) = R_\_(X) / R^-(X)$ is low, the boundary region $BN_R(X)$ is large, and the second kind of granule $[x]_R^2$ is excessive. Consequently, the hit rate in the middle of ranked result is low. Fortunately, it can be solved by attribute generalization, i.e. optimizing the division of equivalence classes. Attribute generalization includes attribute reduction and attribute value generalization. How to use these two methods to optimize the division of equivalence classes is the key work during the following investigation.

Acknowledgements

The work was supported by Shanghai Leading Academic Discipline Project, Project Number: T0602 and the Science and Technology Foundation of Shanghai Municipal Education Committee.

References

[1] Agrawal R, Imielinski T, Swami A (1993) Mining association rules between sets of items in large databases. In: Proceedings of the ACM SIGMOD International Conference on the Management of Data, pp207-216

[2] Han JW, Kamber M (2001) Data mining concepts and techniques. Morgan Kaufmann Publisher

[3] Ling CX, Li C (1998) Data mining for direct marketing: problems and solutions. In: Proceedings of KDD'98, pp73-79

[4] Liu Q (2001) Rough sets and rough reasoning. Science Press, Beijing

[5] Salton G, McGill MH (1983) Introduction to modern information retrieval. McGraw-Hill, New York

[6] Sparck Jones K, Willett P (1997) Readings in information retrieval. Morgan Kaufmann

[7] Yao YY (2001) On modeling data mining with granular computing. Proceedings of COMPSAC'01, pp638–643

[8] Yao YY, Zhong N (2001) Mining market value functions for targeted marketing. In: Proceedings of COMPSAC'01, pp517–522

[9] Yao YY, Zhong N, Huang JJ, Ou CX, Liu CN (2002) Using market value functions for targeted marketing data mining. In: International Journal of Pattern Recognition and Artificial Intelligence, 16(8), pp1117-1131

Remote Sensing Image Classification Algorithm Based on Rough Set Theory

Guang-Jun Dong[1,2], Yong-Sheng Zhang[2], and Yong-Hong Fan[2]

[1] Guangdong Public Laboratory of Environmental Science and Technology, Guangzhou, 510650 P.R. China
[2] Information Engineering University Institute of Surveying and Mapping, Postfach 66, 450052 Zhengzhou, China
topd@163.com

Abstract. Rough sets theory is a relatively new soft computing tool to deal with vagueness and uncertainty. Considering the feature of remote sensing images and the basic theory and applications of rough sets, we put forward a remote sensing image classification algorithm based on rough set theory. In this article we first introduce the basic theory and character of rough sets and its applications in recent years are also pointed out. Then the theory of rough sets is introduced into the processing of remote image classify. Experiment research and classification effects are showed in this article about the new technology and it seems innovational and useful.

Keywords: Rough sets, Image Classification.

1 Introduction

Remote sensing image classification is one of the important application worlds for remote sensing technique, but also is a form of showing the property of image objects directly and vividly. For several decades a good harvest has been gained, but relative to multitudinous images and given application, few image classifications can obtain perfect precision and speed, which is also emphases and hotspot in the remote sensing world. On the hand, with more and more spatiotemporal dimensions of remote sensing data, classical algorithms have gradually exposed some weaknesses, for instance, multi-source and multi-dimensional data may not have normal distribution feature; establishing initial condition is difficult and in most cases discrete data (ground truth data) has no statistical meaning. So in recent year artificial neural network technique has being applied to Remote sensing image classification.

Rough sets theory is a new mathematical tool to deal with problems on vagueness and uncertainty [1]. It is regarded as a soft computation method. The theory of rough sets can be viewed as an extension of the "classical" set theory by incorporating the model of knowledge into its formalism, thus allowing for representing sets approximately in terms of the available context knowledge. Such a representation, in general,

B.-Y. Cao (Ed.): Fuzzy Information and Engineering (ICFIE), ASC 40, pp. 846–851, 2007.
springerlink.com

leads to approximate decision logics in which uncertainty is its natural component, reflecting the imperfections of the context knowledge. The incorporation of the knowledge model into fundamental set theory opens up new possibilities in machine learning, pattern classification, control systems, data mining, medical diagnosis and in variety of other areas dealing with complex objects, systems or natural phenomena [2, 3, 4]. In remote sensing data processing, the superior characteristic is the redundancy and complementary. As rough sets theory is a effective tool to eliminate redundancy and improve efficiency, so it has a immensurable potential in remote sensing data processing.

We start this paper with a short overview of some rough set theories involved (Section 2). Then the relation of rough set theory and remote sensing image classification technique is detailed studied. Finally a new multi-spectral remote sensing image classification algorithm and application model based on rough sets is presented. The results of experiment show that this method has outstanding advantage of processing rapidity and classification and potential of rough set applied in image computing.

2 Principles of Rough Sets

The theory of rough set is an extension of set theory, in which a subject of a universe is described by a pair of ordinary sets called the low and upper approximations. A key notion in Pawlak rough set model is an equivalence relation. The equivalence classes are the building blocks for the construction of the low and upper approximations. It is a concept that has many applications in machine learning, pattern recognition, decision support system, expert systems, data analysis, and data mining.

Let S={U, C, D, V, f} be a knowledge representation system, where U is a nonempty set of objects (i.e., $U=\{u_1,u_2 \cdots u_n\}$), C is a nonempty set of conditional attributes, and D is a nonempty set of decision attributes. A is the set of all attributes and is defined as: $C \cap D=\phi$, , $C \cup D=A$.

Let $V=\cup\{V_a \mid a\in A\}$, where V_a is a finite attribute domain and the elements of V_a are called values of attribute. f is an information function such that $f(U_I,A) \in V_a$, for every $a\in A$ and $u_i \in U$. Every object that belongs to U is associated with a set of values corresponding to the condition attributes C and decision attributes D.

Suppose B is a nonempty subset of A, u_i, u_j are members of U, and R is an equivalence relation over $U(R=U\times U)$. Define a binary relation, called an indiscernibility relation as

$$IND(B)=\{(u_i,u_j)\in R \mid u_i,u_j \in U, a\in B, f(u_i,a)=f(u_j,a)\} \tag{2.1}$$

It is believed that u_i and u_j are indiscernible by a set of condition attributes B. The indiscernibility relation partitions U into equivalence classes.

Equivalence-classes of the relation R are called elementary sets in an approximation space. For any object $X \in U$, the equivalence classes of the relation R containing X are denoted $IND(R)$. Then the lower approximation and the upper approximation, are respectively defined as

$$R\_(X) = \bigcup\{U_i \in U \mid IND(R) : U_i \subseteq X\}$$
$$R^{-}(X) = \bigcup\{U_i \in U \mid IND(R) : U_i \bigcap X \neq \phi\} \quad (2.2)$$

X is rough with respect to $IND(R)$ if $R\_(X) \neq R^{-}(X)$ and a subset defined with the lower approximation and upper approximation is called Rough Set.

3 Remote Sensing Image Classification Model Based on Rough Set Theory

Image classification can use spectrum characteristic, spatial information and temporal information to realize object recognition and classification. Now classifications mostly partition character space into subspaces according to mathematical statistics of ground feature unit and class each unit to every subspace. In fact, as a result of localization of remote sensing information transmission and complexity of information, remote information has uncertainty and traditional classifications always obtain low accuracy. The image classification method based on rough set or fuzzy set has become an important development trend.

Based on obtaining decision rules and inference strategy of rough set, we bring out the new image classification. The remote sensing image classification algorithm based on rough set is described as follows:

Step 1) Obtain class knowledge: In this method, we first think images as knowledge system and make image preprocessing, then we select certain character as attribute and form a information decision table. The image characters include contrast, mean value, variance, entropy, and gradient.

Step 2) Make rule table discreteness: Use equidistance partition algorithm or equifrequency partition algorithm to disperse the rule table.

Step 3) Obtain the class rule method by using the computation of minimal decision rules based on the rough sets.

Step 4) Make image classification: The process of classification is divided into training model and execute model. In training model the rules are made from decision table based on rough set theory and its reduction algorithm. In execute model the unclassed units are dealed with according to rule sets and the classification accuracy is also evaluated.

The procedure flow for remote sensing Image classification model based on rough set is as follows:

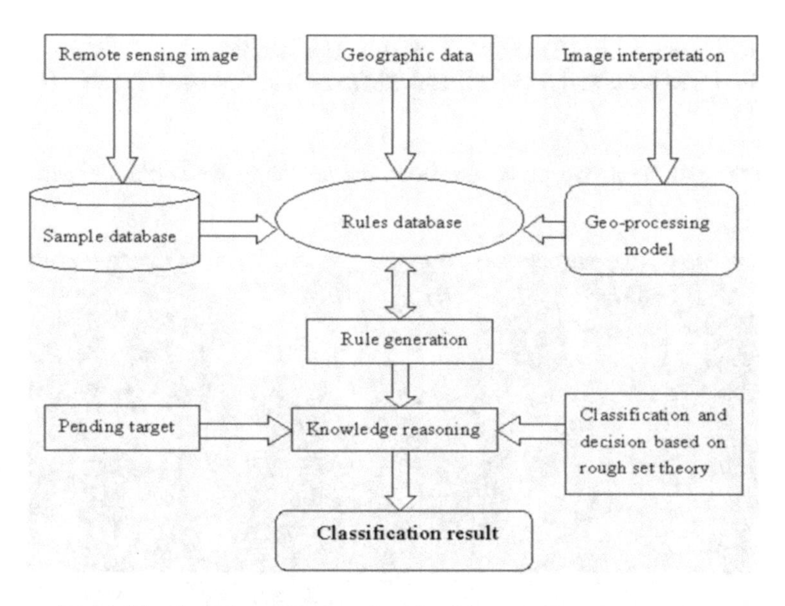

Fig. 1. The flow chart of Image classification model based on rough set

4 Results of Experiment and Analysis

We select one TM image from ERDAS IMAGE SYSTEM and the size is 400*500 pixel. From visual interpretation, we division it into six ground feature types:U1 represents habitation, U2 represents dry land, U3 represents swamp, U4 represents forest land, U5 represents irrigated field, U6 represents grass land. Then 1600 samples are selected and their values of all bands are made as condition attributes. The decision table is formed as follows:

Table 1. Samples spectrum value table

U	TM2	TM3	TM4	TM5	TM7
Habitation(U1)	38	44	53	81	47
dry land(U2)	40	45	82	102	49
swamp(U3)	39	37	17	8	5
forest land(U4)	33	33	63	71	30
irrigated field(U5)	31	28	86	63	22
grass land(U6)	40	44	39	42	19
⋮	⋮	⋮	⋮	⋮	⋮

Then we make attribute reduction to the decision table based on rough set theory and get the rules as follows:

(1) IF (TM5>80 and TM5<100) THEN U is habitation

(2) IF (TM4>60 and TM4<80 and TM7>20 and TM7<40 THEN U is forest land

……

According to the above procedure flow, the results of water class experiment are shown:

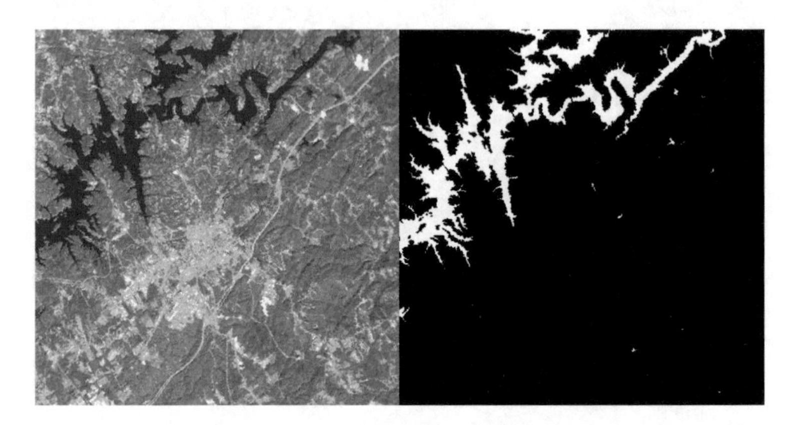

Fig. 2. Result of rough set classification

With comparison to the original image, we select 1000 test samples to evaluate classified result and the classification error matrix table is calculated:

Table 2. Result comparison among different methods to classification

class fact	U1	U2	U3	U4	U5	U6	总计
U1	120	6	1	12	0	0	139
U2	10	114	2	4	0	0	120
U3	2	5	127	5	3	17	159
U4	0	10	7	170	11	9	207
U5	0	0	7	4	171	10	192
U6	0	0	7	9	3	164	183
total	132	125	151	204	188	200	1000

From the result of TM image classification, the quality of image classifying is good enough and meets engineering requirements in this new method.

5 Conclusion

The image classification based on rough set is not only a classification but also can make feature extraction and knowledge mining. It integrates the traditional geo-processing and artificial intelligence technology. So it can improve the classification accuracy by reducing data redundancy based on attribute reduction of rough set and provide a new method for image intellectualized processing by mining knowledge from remote sensing image or database.

Ackowledgements

The research is supported by the Open Research Fund of State Key Laboratory of Information Engineering in Surveying, Mapping and Remote Sensing and Guangdong Public Laboratory of Environmental Science and Technology(060203).

References

[1]. Pawlak Z (1982) Rough sets. International Journal of Information and Computer Science, 11: 341-356
[2]. Hu Xiaohua (1996) Mining knowledge rules from databases-a rough set approach. Proc. of IEEE International Conference on Data Engineering, Los Alamitos, 10:96-105
[3]. Jelonek J (1995) Rough set reduction of attributes and their domains for neural networks. Computational Intelligence, 11:339-347
[4]. Slowinski (1996) Rough set reasoning about uncertain data. Fundamenta Informaticae, 27:229-243

Topological Relations Between Vague Objects in Discrete Space Based on Rough Model

Zhenji Gao[1,2], Lun Wu[1], and Yong Gao[1]

[1] School of Earth and Space Sciences, Peking University, Beijing, China, 100871
gaozhenji@gmail.com, lwu@urban.pku.edu.cn, gaoyong@pku.edu.cn
[2] Shenzhen Graduate School, Peking University, Shenzhen, Guangdong, China, 518055
gaozj@szpku.edu.cn

Abstract. Rough model is proposed to model vague geographic objects based on the definition of rough regions, lower approximation regions, upper approximation regions and boundary regions. With the combination of rough model and RCC-D-8, which is a discrete version of RCC extended into two-dimensional discrete space ($\mathbb{Z}^2$), 599 topological relations are got between simple vague geographic objects in raster regions described by intersection and containing matrix.

Keywords: Rough Model, Discrete Space, RCC-D-8, Vague Objects, Topological Relations.

1 Introduction

Research of topological relations between vague objects is one of the key parts in GIS modeling. Currently, the research of topological relations between vague objects is mainly focused on objects in continuous space ($\mathbb{R}^n$) and the main models are Egg-Yolk model [1], broad boundary model [2], fuzzy model [3-4] and probability model [5] whereas the research in discrete space is relatively few. As problems exist in continuous space model in the application of GIS [6], the research of discrete space model, data models and topological relations in discrete space becomes pressing.

There are three main discrete space model, i.e. graph-based model [7], cell complex [8] and oriented matroids[9]. The raster model of GIS is one kind of graph-based model. Egenhofer and Sharma [10] got 16 topological relations between two simple crisp regions in raster regions based on 9 intersection model after the definition of the interior, outer and boundary of a simple crisp region. The result of topological relations in raster space is 8 more than that in continuous space. This makes difficult the application of intersection model to vague regions in raster space.

The Region Connection Calculus (RCC) by Randell, Cui, and Cohn [11] is a fully axiomatized first-order theory based on regions for representing topological relationships between spatial entities and is a promising basis for a treatment of vague regions [12]. Galton defined the RCC system which is called RCC-D in graph-based space based on two primitive relations containment and adjacency [13].

B.-Y. Cao (Ed.): Fuzzy Information and Engineering (ICFIE), ASC 40, pp. 852–861, 2007.
springerlink.com

Roy and Stell proposed the RCC suitable to cell complex based on connection algebra [14].Li Sanjiang and Ying Mingsheng established GRCC, a formal theory accommodating both discrete and continuous spatial information, based on two primitives, the mereological notion of part and the topological notion of connection [15].

The mathematic model of the description and representation of topological relations between vague regions in continuous space mainly based on fuzzy sets and inspiriting results are achieved. Howerver, the fuzzy membership function is too difficult to decide and mostly experience-judged, the spreading of the research results to specific applications is hindered to some extent. Rough sets is a new theory to deal with uncertain and imperfect information [16].It needs no prior knowledge or information except the target data. This is its main advantage over fuzzy sets and Dempster-Shafer evidence theory. The application of rough sets to geosciences attracts the attention of many researchers. Shuliang WANG et al [17] proposed rough spatial entities, rough relationships and rough algorithms. Beauboue and Petry [18-21] studied the uncertainty management and data mining including vague regions in spatial database based on rough sets. They use rough sets to express topological relationships and concepts defined with the egg-yolk model. Shihong DU et al [22-23] deals with direction relations between fuzzy objects and direction relation representations based on rough sets, variable precision rough sets are proposed together with the rough reasoning of direction relations. Bittener and Stell[24-26] described vagueness with stratified rough sets and applied rough sets to approximate spatial reasoning and rough location.

Due to complexity and difficulties of the research, vague regions in this paper are confined to:

(1) The vague regions have vague or indeterminate boundaries but with a know location.
(2) The vague regions are finite sets of two-dimensional discrete space.
(3) The vague regions are simple vague regions without convex or hole.

2 Rough Model

2.1 The Definition of Region and Boundary of Raster Space

2.1.1 Simple Boundary and Simple Region

Definition 1. The connected sets with same attribute value in two-dimensional discrete space $\mathbb{Z}^2$ are called a region. Here connection is defined as there exits connected path and shortest path between any two grids.

Definition 2. The grids without entire adjacent grids (the entire adjacent grids of one grid) of a region form the boundary in two-dimensional discrete space $\mathbb{Z}^2$.

Definition 3. The boundary of a simple region in $\mathbb{Z}^2$ meets the following requirements is called a simple boundary

(a) It is a closed curve.
(b) Each grid of the connected path of the curve appears only once.

Figure1 shows different simple boundaries (the region of oblique line) defined by different adjacent relations.

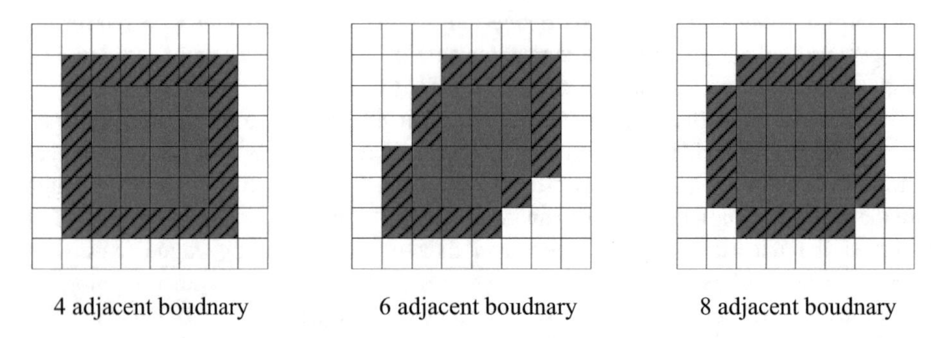

Fig. 1. Simple Region and its Boundaries in Raster Space

A region with simple boundary in $\mathbb{Z}^2$ is called a simple region. Figure1 shows simple regions (grey part) defined by different adjacent relations. To lower the descriptive complexity of regions, all the regions discussed below are defined based on 4 adjacent relations.

2.1.2 Complex Boundaries and Regions

A region without a simple boundary in $\mathbb{Z}^2$ is called a complex region. Figure2 shows three complex regions (grey part) and their boundaries (regions of oblique line). The boundaries don't meet the requirement that each grid of the connected path of the curve passes only once.

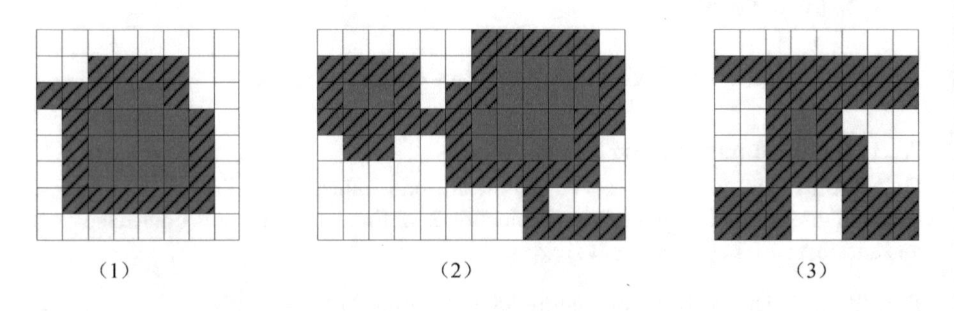

Fig. 2. Complex Region and its Boundaries in Raster Space

2.2 Definition of Rough Model

2.2.1 Basic Definition

Definition 4. Let $X \subseteq \mathbb{Z}^2$ and ϑ be an equivalence relation of the attribute function $A = f(x, y)$ on X . Then X can be divided into one partition based on ϑ and the partition is called basic region partition, notated by *P*

$$P = \{[a]_\vartheta \mid a \in X\} \tag{1}$$

where $[a]_\vartheta$ denotes a category in ϑ containing an element $a \in X$ and is called the basic region of X . In figure 1, basic region a,b,c,d,e,f,g,h,i forms the basic region partition on *X*, i.e. $P_X = \{a,b,c,d,e,f,g,h,i\}$.

Accordingly, let $X \subseteq \mathbb{Z}^2$ and ϑ_i be a family of equivalence relation of the attribute function $A = f(x, y)$ on X . Then X can be divided into a family of partitions based on ϑ_i and the partitions are called partitions cluster, notated by *U*, i.e.

$$U = \bigcup_{i=1}^{n} P_i \tag{2}$$

Let $R \subseteq X \subseteq \mathbb{Z}^2$ and ϑ be an equivalence relation of the attribute function $A = f(x, y)$ on X . If *R* cannot be represented by the union of some basic regions of *X*, then *R* is said to be a rough region on ϑ and is donated by *R*. A Rough set *R* can be described as:

$$\exists R\ , R \not\subset \psi([a]_\vartheta) \wedge R \subseteq X \subseteq \mathbb{Z}^2 \tag{3}$$

where $\psi([a]_\vartheta)$ is the power set of basic regions. The circular area of figure3 is a rough set defined on *X*.

Definition 5. Upper approximate region.

Let $R \subseteq X \subseteq \mathbb{Z}^2$ and ϑ be an equivalence relation of the attribute function $A = f(x, y)$ on *X* and *R* is a rough set on ϑ. The upper approximate region of *R* is described as:

$$\vartheta^{-}(R) = \bigcup\{Y \in \mathbb{Z}^2 / \vartheta : Y \cap R \neq \Phi\} \tag{4}$$

The upper approximate region is regarded as union of the basic regions intersected with the rough set *R*. Figure3(3) is the upper approximate region of rough set *R* expressed as circular area in Figure3(1).

Definition 6. Lower approximate region.

Let $R \subseteq X \subseteq \mathbb{Z}^2$ and ϑ be an equivalence relation of the attribute function $A = f(x, y)$ on *X* and *R* is a rough set on ϑ. The lower approximate region of *R* is described as:

$$\vartheta_{-}(R) = \bigcup\{Y \in \mathbb{Z}^2 / \vartheta : Y \subseteq R\} \tag{5}$$

The lower approximate region is regarded as union of the basic regions contained in the rough set *R*. Figure3(2) is the lower approximate region of rough set *R* expressed as circular area in Figure3(1).

Definition 7. Boundary region

Let $R \subseteq X \subseteq \mathbb{Z}^2$ and ϑ be an equivalence relation of the attribute function $A = f(x, y)$ on X and *R* is a rough set on ϑ. The boundary region of *R* is described as:

$$BN_R(R) = \vartheta^-(R) - \vartheta_-(R) \tag{6}$$

Figure3 (4) is the boundary region of rough set *R* expressed as circular area in Figure3 (1).

The vague region described and represented by rough regions is called rough model. As shown in figure3, the basic region $\{e\}$ is the lower approximate region of vague region *R* .Basic region $\{e, d, f, i, j\}$ is the upper approximate region of *R* and basic region $\{d, f, i, j\}$ is the boundary region of R .

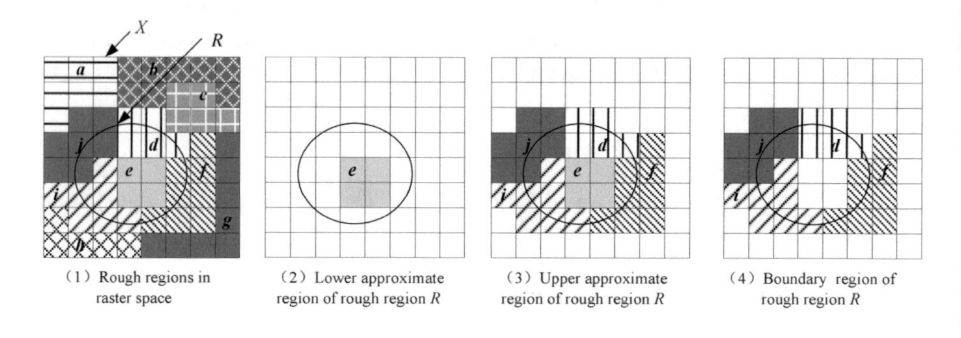

Fig. 3. Rough Region in Raster Space

2.2.2 Types of Rough Regions

Let ρ be the roughness of a rough set *R*. ρ can be defined as:

$$\rho_\vartheta(R) = 1 - \frac{card\ R_-}{card\ R^-} \tag{7}$$

According to (7), if ρ =0, $card\ R_- = card\ R^-$ then $R_- = R^-$, the rough set *R* becomes crisp set. If $0 < \rho < 1$=0, $card\ R_- < card\ R^-$ then $R_- \subset R^-$, the rough set *R* is rough.

(a) C-C pair (Crisp-Crisp regions)

$\forall R_1, R_2 \subset X$， $\exists \rho_1 = 0 \wedge \rho_2 = 0$， then R_1 and R_2 form a C-C pair.

(b) C-R pair (Crisp-Rough regions)

$\forall R_1, R_2 \subset X$, $\exists(\rho_1 = 0 \wedge < 0 < \rho_2 < 1) \vee (\rho_2 = 0 \wedge < 0 < \rho_1 < 1)$, then R_1 and R_2 form a Crisp-Rough pair.

(c) R-R pair (Rough-Rough regions)

$\forall R_1, R_2 \subset X$, $\exists(0 < \rho_1 < 1 \wedge 0 < \rho_2 < 1)$, then R_1 and R_2 form a Rough-Rough pair.

3 Study of Topological Relations Between Vague Objects Based on Rough Model and RCC-D-8

3.1 The Constraints of Lower and Upper Approximate Regions

The topological relations between two regions are described as 8 JEPD (jointly exhaustive and pairwise disjoint) sets based on RCC-D-8. The JEPD sets include *DC* (*Disconnected*), *EC* (*Externally Connected*), *PO* (*Partial Overlap*), *TPP* (*Tangential Proper Part*), *TPPI* (*Inverse of Tangential Proper Part*), *NTPP* (*Non- Tangential Proper Part*), *NTPPI* (*Inverse of Non-Tangential Proper Part*) and *EQ* (*Equal*).

The constraints of upper and lower approximate regions helps to reduce the research scope and exclude some meaningless topological relations. This makes possible that we focus main efforts on significant topological relations. For example, if $DC(R_1^-, R_2^-)$ is true then $DC(R_1^-, R_{2-})$ is definitely true because $R_{2-} \subset R_2^-$ and there are no other JEPD relations between R_1^- and R_{2-} any more. Another example is if $PO(R_{1-}, R_2)$ is true then $DC(R_1^-, R_2^-)$ is definitely false. The constraints of lower and upper approximate regions makes some topological relations are mutual exclusive (see Table1).

Table 1. Constraints of upper and lower approximate regions between vague regions based on RCC-D-8

Constraints under RCC-D-8	
$DC(R_1^-, R_2^-) \Rightarrow DC(R_{1-}, R_2^-)$	$EQ(R_1^-, R_2^-) \Rightarrow TPPI(R_{1-}, R_2^-) \vee NTPPI(R_{1-}, R_2^-)$
$NTPP(R_1^-, R_2^-) \Rightarrow NTPP(R_{1-}, R_2^-)$	$EQ(R_1^-, R_{2-}) \Rightarrow \{NTPP(R_1^-, R_2^-) \wedge NTPP(R_{1-}, R_2^-) \wedge NTPP(R_{1-}, R_{2-})\}$ $\vee\{TPP(R_1^-, R_2^-) \wedge TPP(R_{1-}, R_2^-) \wedge TPP(R_{1-}, R_{2-})\}$
$NTPP(R_1^-, R_{2-}) \Rightarrow NTPP(R_{1-}, R_{2-})$	$EQ(R_1^-, R_2^-) \Rightarrow TPP(R_{2-}, R_1^-) \vee NTPP(R_{2-}, R_1^-)$
$NPPI(R_1^-, R_2^-) \Rightarrow NPPI(R_{1-}, R_2^-)$	$EQ(R_{1-}, R_2^-) \Rightarrow \{TPPI(R_1^-, R_2^-) \wedge TPPI(R_1^-, R_{2-}) \wedge TPPI(R_{1-}, R_{2-})\} \vee$ $\{NTPPI(R_1^-, R_2^-) \wedge NTPPI(R_1^-, R_{2-}) \wedge NTPPI(R_{1-}, R_{2-})\}$
$NTPPI(R_1^-, R_{2-}) \Rightarrow NTPPI(R_{1-}, R_{2-})$	$EQ(R_{1-}, R_{2-}) \Rightarrow NTPP(R_{1-}, R_2^-) \wedge NTPPI(R_1^-, R_{2-})$
	$PO(R_{1-}, R_2^-) \vee PO(R_{2-}, R_1^-) \Rightarrow \neg DR(R_1^-, R_2^-)$

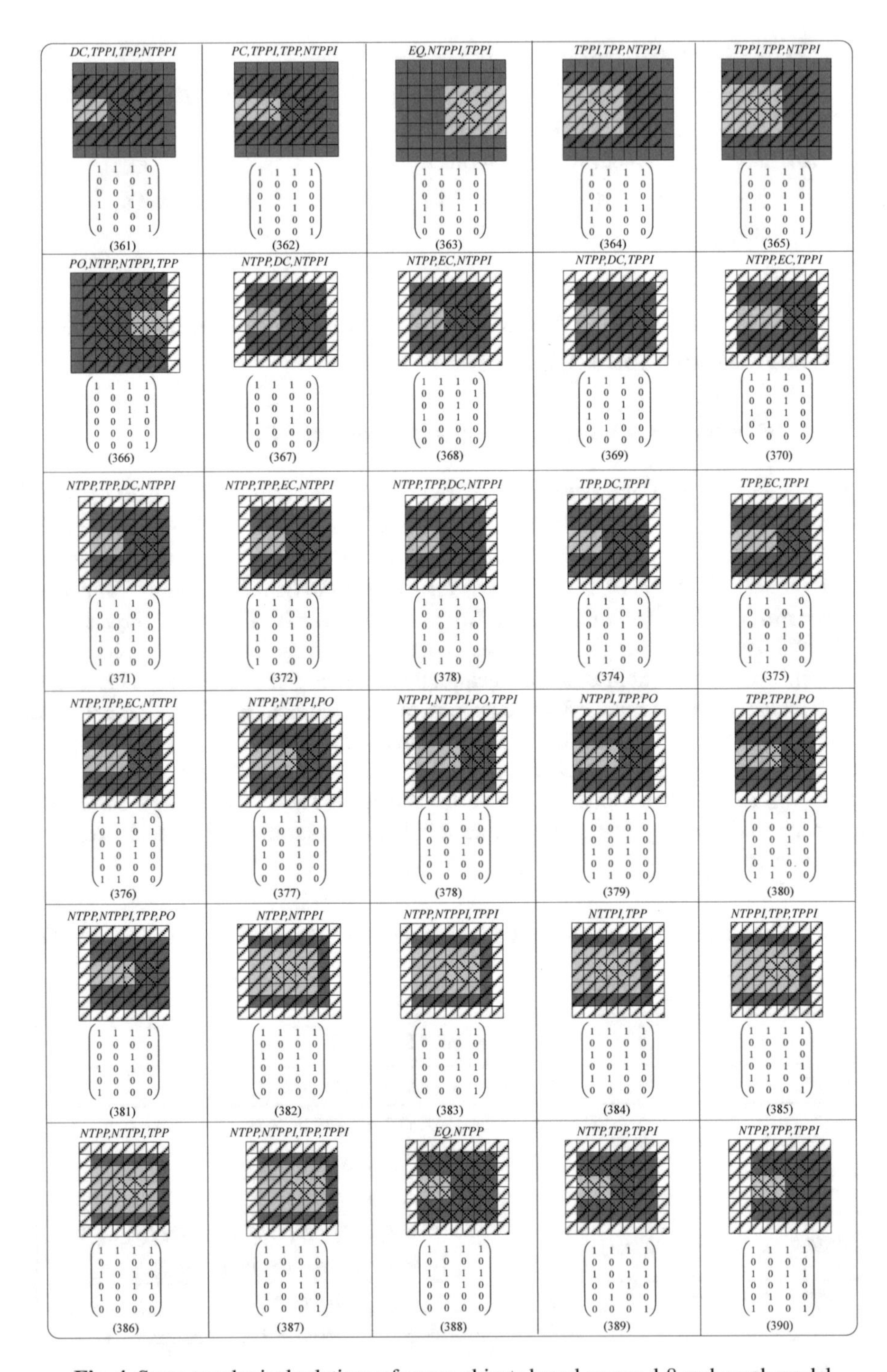

Fig. 4. Some topological relations of vague objects based on rcc-d-8 and rough model

3.2 Study of Topological Relations

The topological relations between two vague regions can be described as a 4 tuple matrix composed of two lower approximate regions and two upper approximate regions as shown in

$$\begin{pmatrix} f(R_1^-, R_2^-) & f(R_1^-, R_{2-}) \\ f(R_{1-}, R_2^-) & f(R_{1-}, R_{2-}) \end{pmatrix} \tag{8}$$

where $f : R \times R \rightarrow \{DC, EC, PO, TPP, NTPP, PPI, NTPPI, EQ\}$ is the JEPD mapping function between any approximate region pair. According to (8), the amount of entire topological relations between R-R pair, C-R pair and C-C pair is 8^4=4096, 8^2×2=128 and 8^1=8respectively and totals 4232.

In order to give a formal description, the intersection and inclusion matrix is introduced to describe the any possible topological relation. Each existent topological relation can be represented by the intersection, inclusion, tangential ($\oplus$) and adjacent ($\boxplus$) matrix composed of lower and upper approximate regions of the two vague regions as shown in (9).

$$\begin{pmatrix} R_1^- \cap R_2^- & R_1^- \cap R_{2-} & R_{1-} \cap R_2^- & R_{1-} \cap R_{2-} \\ R_2^- \boxplus R_1^- & R_2^- \boxplus R_{1-} & R_{2-} \boxplus R_1^- & R_{2-} \boxplus R_{1-} \\ R_1^- \subseteq R_2^- & R_1^- \subseteq R_{2-} & R_{1-} \subseteq R_2^- & R_{1-} \subseteq R_{2-} \\ R_2^- \subseteq R_1^- & R_2^- \subseteq R_{1-} & R_{2-} \subseteq R_1^- & R_{2-} \subseteq R_{1-} \\ R_1^- \oplus R_2^- & R_1^- \oplus R_{2-} & R_{1-} \oplus R_2^- & R_{1-} \oplus R_{2-} \\ R_2^- \oplus R_1^- & R_2^- \oplus R_{1-} & R_{2-} \oplus R_1^- & R_{2-} \oplus R_{1-} \end{pmatrix} \tag{9}$$

If 1 is used to represent the true status and 0 to the false status of intersection of any two approximate regions, then each topological relation can be described by a 4×6 matrix. Theoretically, the matrix can describe 2^{24}= 16777216 kinds of topological relations and this covers all the possible 4232 kinds of topological relations based RCC-D-8.

599 topological relations are got between two vague regions based on RCC-D-8 and the constraint of the lower and upper approximate regions. Of which there are 505 topological relations between R-R pair, 86 between C-R pair and 8 between C-C pair. Some topological relations and matrix description are exemplified in Figure4.

4 Conclusions

The rough model is a mathematical model based on rough sets theory to describe vague regions in discrete space. It moves the egg() and yolk() of Egg-Yolk model to the equivalence relation (a group of rules or constraints) of rough sets. The equivalence relation is decided by rules and requirements of specific application and can be calculated in computer without any prior knowledge. The combination of RCC with rough model is a promising model to deal with vague regions in discrete space. The topological relations between vague regions are fully studied based on RCC-D-8 and rough model. Finally, 599 groups of topological relations are described by 4×6 intersection, inclusion, tangential and adjacent matrix. The research scope of this paper is mainly focused in graph-based discrete space. Research efforts in cell complex and oriented matroid model of discrete space based on RCC-D and rough model are worth expecting.

References

1. A.G. Cohn and N.M.Gotts (1996) *Geographic Objects with Indeterminate Boundaries*, GISDATA Series, vol. 2, Taylor & Francis:171-187
2. Eliseo Clementini and Paolino Di felice (1996) Geographic Objects with Indeterminate Boundaries, GISDATA Series, vol. 2, Taylor & Francis:155-169
3. Zhan FB. (1997) Topological relations between fuzzy regions. In: Bryant B, Carroll J, Oppenheim D, Hightower J, George KM, eds. *Proceedings of the 1997 ACM Symposium on Applied Computing*. New York: ACM Press: 192~196
4. Zhan FB (2001) A fuzzy set model of approximate linguistic terms in descriptions of binary topological relations between simple regions. In: Matsakis P, Sztandera LM, eds. *Applying Soft Computing in Defining Spatial Relations.* Heidelberg: Physica-Verlag:179~202
5. Winter S (1994) Uncertainty of topological relations in GIS. In: Ebner H, Heipke C, Eder K, eds. *Proceedings of ISPRS Commission III Symposium: Spatial Information from Digital Photogrammetry and Computer Vision*. Bellingham: SPIE: 924~930
6. John Stell, Julian Webster (2004) Oriented Matroids as a Foundation for Space. *GIScience'04.*
7. Anthony J. Roy and John G. Stell, 2002, A Qualitative Account of Discrete Space. GIScience 2002, Springer-Verlag Berlin Heidelberg: 276–290
8. Kovalevsky V. A. (1989) Finite topology as applied to image analysis. Computer Vision, Graphics and Image Processing, 46: 141–161
9. John Stell, Julian Webster (2004) Oriented Matroids as a Foundation for Space. *GIScience'04.*
10. Max J. Egenhofer, Jayant Sharma (1993) Topological Relations Between Regions in $\mathbb{R}^2$ and $\mathbb{Z}^2$, Advances in Spatial Databases-Third International Symposium on Large Spatial Databases, Lecture Notes in Computer Science, Vol.692, Springer-Verlag:316-336
11. Randell, D.A., Cui, Z, and Cohn, A.G. (1992) A spatial logic based on regions and connections, Proceedings 3rd International Conference on Knowledge Representation and Reasoning, Morgan Kaufmann, San Mateo:162-176
12. A. G. Cohn and N. M. Gotts (1994) Spatial Regions with Undetermined Boundaries, Proceedings of Gaithesburg Workshop on GIS, ACM.
13. Antony Galton (1999) The Mereotopology of Discrete Space. COSIT'99,Springer-Verlag Berlin Heidelberg: 251-266
14. Anthony J. Roy and John G. Stell (2002) A Qualitative Account of Discrete Space. GIScience 2002, Springer-Verlag Berlin Heidelberg: 276–290
15. Sanjiang Li , Mingsheng Ying (2004) Generalized Region Connection Calculus, Artificial Intelligence 160: 1–34

16. Zdzistaw Pawlak(1982) Rough Sets, International Journal of Computer and Information Sciences, Vol.11, No.5, 1982
17. Shuliang WANG, Deren LI, Wenzhong SHI et al (2002) IAPRS, VOLUME XXXIV, PART2, COMMISSION II, Xi'an,Aug.20-23
18. Theresa Beaubouef and Federick E. Petry (2002) A rough foundation for spatial data mining Involving Vague regions. FUZZ-IEEE'02, as part of WCCI'02: 767-772
19. Theresa Beaubouef and Federick E. Petry (2001) Vague regions and spatial relationships: A rough set approach, Fourth International Conference on Computational Intelligence and Multimedia Applications (ICCIMA'01):313-318
20. Theresa Beaubouef and Federick E. Petry (2001) Vagueness in spatial data: rough sets and egg-yolk approaches. 14th International Conference on Industrial & Engineering Applications of Artificial Intelligence (IEA/AIE 2001)
21. T. Beaubouef and J. Breckenridge (2000) Rough Set Based Uncertainty Management for Spatial Databases and Geographical Information Systems, in Soft Computing in Industrial Applications (ed. Y. Suzuki), Springer-Verlag.
22. DU Shihong ,WANGQiao ,LI Shun et al (2004) The Research of Rough Expression of Fuzzy Objects and their Spatial Relations, *Journal of Remote Sensing*, China, Vol.8, No.1:1-7
23. DU Shi2hong, WANG Qiao, WEI Bin et al (2003) Spatial Orientational Relations Rough Reasoning, *ACTA GEODAETICA et CARTOGRAPHICA SINICA,*Vol.32, No.4:334-338
24. Bittner, T. and Stell, J. (2003) Stratified rough sets and vagueness, In Kuhn, W. and Worboys, M. and Timpf, S. (ed.): *Spatial Information Theory. Cognitive and Computational Foundations of Geographic Information Science. International Conference COSIT'03* :286-303
25. Bittner, T and Stell, J. (2000) Rough sets in Approximate spatial reasoning. *Proceedings of RSCTC'2000 , Lecture Notes in Artificial Intelligence,* Berlin-Heidelberg, Springer-Verlag:445-453
26. Bittner, T., & Stell, J.G. (2002) Vagueness and Rough Location. *Geoinformatica*, Vol. 6:99-121

Part VIII

Application in Fuzzy Mathematics and Systems

A Least Squares Fuzzy SVM Approach to Credit Risk Assessment

Lean Yu[1,2], Kin Keung Lai[2], Shouyang Wang[1], and Ligang Zhou[2]

[1] Institute of Systems Science, Academy of Mathematics and Systems Science, Chinese Academy of Sciences, Beijing 100080, China
{yulean,sywang}@amss.ac.cn

[2] Department of Management Sciences, City University of Hong Kong, 83 Tat Chee Avenue, Kowloon, Hong Kong
{msyulean,mskklai,mszhoulg}@cityu.edu.hk

Abstract. The support vector machine (SVM) is a class of powerful classification tools that have many successful applications. Their classification results usually belong to either one class or the other. But in many real-world applications, each data point no more exactly belongs to one of the two classes, it may 70% belong to one class and 30% to another. That is, there is a fuzzy membership associated with each data. In such an environment, fuzzy SVM (FSVM), which treats every sample as both positive and negative classes with the fuzzy membership, were introduced. In this way the FSVM will have more generalization ability, while preserving the merit of insensitive to outliers. Although the FSVM has good generalization capability, the computational complexity of the existing FSVM is rather large because the final solution is obtained from solving a quadratic programming (QP) problem. For reducing the complexity, this study proposes a least squares method to solve FSVM. In the proposed model, we consider equality constraints instead of inequalities for the classification problem with a formulation in a least squares sense. As a result the solutions follow directly from solving a set of linear equations instead of QP thus reducing the computational complexity greatly relative to the classical FSVM. For illustration purpose, a real-world credit risk assessment dataset is used to test the effectiveness of the LS-FSVM model.

Keywords: Least Squares Fuzzy Support Vector Machine, Credit Risk Assessment.

1 Introduction

Credit risk assessment has been the major focus of financial and banking industry due to recent financial crises and regulatory concern of Basel II [1]. Since the seminal work of Altman [2] was published, many different techniques, such as discriminant analysis [2], logit analysis [3], probit analysis [4], linear programming [5], integer programming [6], k-nearest neighbor (KNN) [7] and classification tree [8] have widely been applied to credit risk assessment tasks. Recently, tools taken from artificial intelligent (AI) area such as artificial neural networks (ANN) [9-10], genetic algorithm (GA) [11-12] and support vector machine (SVM) [1, 13-15] have also been employed. The empirical results revealed that the AI techniques are advantageous to traditional statistical models and optimization techniques for credit risk evaluation.

B.-Y. Cao (Ed.): Fuzzy Information and Engineering (ICFIE), ASC 40, pp. 865–874, 2007.
springerlink.com

Almost all classification methods can be used to assess credit risk, some hybrid and combined (or ensemble) classifiers, which integrate two or more single classification methods, have shown higher correctness of predictability than any individual methods. Combined or ensemble classifier research is currently flourishing in credit scoring. Recent examples are neural discriminant technique [16], neuro-fuzzy [17-18], fuzzy SVM [19], evolving neural network [20], and neural network ensemble [21]. Some comprehensive literature about credit risk assessment or credit scoring can be referred to two recent surveys [22-23] for more details.

Motivated by hybrid and ensemble models, a new credit risk classification technique called least squares fuzzy SVM (LS-FSVM) is proposed to discriminate good creditors from bad ones. The fuzzy SVM (FSVM) was first proposed by Lin and Wang [24] and it has more suitability in credit risk assessment. The main reason is that in credit risk assessment areas we usually cannot label one customer as absolutely good who is sure to repay in time, or absolutely bad who will default certainly, the FSVM treats every sample as both positive and negative classes with the fuzzy membership. By this way the FSVM will have more generalization ability, while preserving the merit of insensitive to outliers. Although the FSVM has good generalization capability, the computational complexity of the existing FSVM is rather difficult because the final solution is derived from a quadratic programming (QP) problem. For reducing the complexity, this study proposes a least squares solution to FSVM. In the proposed model, we consider equality constraints instead of inequalities for the classification problem with a formulation in least squares sense. As a result the solutions follow directly from solving a set of linear equations, instead of QP from the classical FSVM approach [24], thus reducing the computational complexity relative to the classical FSVM. The main motivation of this study is to formulate a least squares version of FSVM for binary classification problems and to apply it to the credit risk evaluation field and meantime, to compare its performance with several typical credit risk assessment techniques.

The rest of this study is organized as follows. Section 2 illustrates the methodology formulation of LS-FSVM. In Section 3, we use a real-world credit dataset to test the classification potential of the LS-FSVM. Section 4 concludes the study.

2 Methodology Formulation

In this section, we first present a brief introduction on SVM in classification problems [25]. Then a Fuzzy SVM model [24] is also briefly reviewed. Motivated by Lai et al. [1] and Suykens and Vandewalle [26], a least squares FSVM model is formulated.

2.1 SVM (By Vapnik [25])

Given a training dataset $\{x_i, y_i\}(i = 1, \ldots, N)$ where $x_i \in R^N$ is the ith input pattern and y_i is its corresponding observed result, and it is a binary variable. In credit risk evaluation models, x_i denotes the attributes of applicants or creditors; y_i is the observed result of timely repayment. If the customer defaults,

y_i =1, else y_i = −1. SVM first maps the input data into a high-dimensional feature space through a mapping function $\phi(\cdot)$ and finds the optimal separating hyperplane with the minimal classification errors. The separating hyperplane can be represented as follows:

$$z(x) = w^T \phi(x) + b = 0 \tag{1}$$

where $\boldsymbol{w}$ is the normal vector of the hyperplane and b is the bias that is a scalar.

Suppose that $\phi(\cdot)$ is a nonlinear function that maps the input space into a higher dimensional feature space. If the set is linearly separable in this feature space, the classifier should be constructed as follows:

$$\begin{cases} w^T \phi(x_i) + b \geq 1 & if \quad y_i = 1 \\ w^T \phi(x_i) + b \leq -1 & if \quad y_i = \text{ - } 1 \end{cases} \tag{2}$$

which is equivalent to

$$y_i(w^T \phi(x_i) + b) \geq 1 \quad for \quad i = 1, \ldots, N \tag{3}$$

In order to deal with data that are not linearly separable, the previous analysis can be generalized by introducing some nonnegative variables $\xi_i \geq 0$ such that (3) is modified to

$$y_i[w^T \phi(x_i) + b] \geq 1 - \xi_i \quad for \quad i = 1, \ldots, N \tag{4}$$

The nonzero ξ_i in (4) are those for which the data point x_i does not satisfy (3). Thus the term $\sum_{i=1}^{N} \xi_i$can be thought of as some measures of the amount of misclassifications.

According to the structural risk minimization principle, the risk bound is minimized by formulating the following optimization problem

$$\begin{array}{ll} \text{Minimize} & \Phi(w, b, \xi_i) = \frac{1}{2} w^T w + C \sum_{i=1}^{N} \xi_i \\ \text{Subject to:} & y_i(w^T \phi(x_i) + b) \geq 1 - \xi_i, \quad \text{for} \quad i = 1, \ldots, N \\ & \xi_i \geq 0, \text{ for } \quad i = 1, \ldots, N \end{array} \tag{5}$$

where C is a free regularization parameter controlling the trade-off between margin maximization and tolerable classification error.

Searching the optimal hyperplane in (5) is a QP problem. By introducing a set of Lagrangian multipliers α_i and β_i for constraints in (5), the primal problem in (5) becomes the task of finding the saddle point of the Lagrangian function, i.e.,

$$\begin{aligned} L(w, b, \xi_i; \alpha_i, \beta_i) = & \tfrac{1}{2} w^T w + C \textstyle\sum_{i=1}^{N} \xi_i \\ & - \textstyle\sum_{i=1}^{N} \alpha_i [y_i(w^T \phi(x_i) + b) - 1 + \xi_i] - \sum_{i=1}^{N} \beta_i \xi_i \end{aligned} \tag{6}$$

Differentiate (6) with w, b, and ξ_i, we can obtain

$$\begin{cases} \frac{d}{dw} L(w, b, \xi_i; \alpha_i, \beta_i) = w - \sum\limits_{i=1}^{N} \alpha_i y_i \phi(x_i) = 0 \\ \frac{d}{db} L(w, b, \xi_i; \alpha_i, \beta_i) = - \sum\limits_{i=1}^{N} \alpha_i y_i = 0 \\ \frac{d}{d\xi_i} L(w, b, \xi_i; \alpha_i, \beta_i) = C - \alpha_i - \beta_i = 0 \end{cases} \tag{7}$$

To obtain a solution of α_i, the dual problem of the primal problem (5) becomes

$$\begin{array}{l} \text{Maximize } Q(\alpha) = -\frac{1}{2}\sum_{i=1}^{N}\sum_{j=1}^{N}\alpha_i\alpha_j y_i y_j \phi(x_i)^T\phi(x_j) + \sum_{i=1}^{N}\alpha_i \\ \text{Subject to: } \sum_{i=1}^{N}\alpha_i y_i = 0,\ 0 \le \alpha_i \le C, \quad i = 1, \ldots, N \end{array} \tag{8}$$

The function $\phi(x)$ in (8) is related then to $K(x_i, x_j)$ by imposing

$$\phi(x_i)^T\phi(x_j) = K(x_i, x_j) \tag{9}$$

which is motivated by Mercer's Theorem [25], $K(x_i, x_j)$ is the kernel function in the input space that computes the inner product of two data points in the feature space. According to the Kuhn-Tucker (KT) theorem [27], the KT conditions are defined as

$$\alpha_i[y_i(w^T\phi(x) + b) - 1 + \xi_i] = 0, i = 1, 2, \cdots, N \tag{10}$$

$$(C - \alpha_i)\xi_i = 0, i = 1, 2, \cdots, N \tag{11}$$

From this equality it comes that the only nonzero values α_i in (10) are those for which the constraints (4) are satisfied with the equality sign. The data points x_i (i=1,2,..., s) corresponding with α_i¿0 are called support vectors (SVs). But there are two types of SVs in a nonseparable case. In the case 0¡α_i¡C, the corresponding support vector x_i satisfies the equalities $y_k[w^T\phi(x_k) + b] = 1$ and ξ_i=0. In the case $\alpha_i = C$, the corresponding $\xi_i$is not zero and the corresponding support vector x_i does not satisfy (2). We refer to such SVs as errors. The data points x_i (i=1, 2,..., s) corresponding with α_i=0 are classified correctly.

From (7), the optimal solution for the weight vector is given by

$$w = \sum_{i=1}^{N_s}\alpha_i y_i \phi(x_i) \tag{12}$$

where N_s is the number of SVs. Moreover, in the case of 0¡α_i¡C, we have ξ_i=0 in terms of KT condition in (11), thus one may determine the optimal bias b by taking any data point in the dataset. However, from the numerical perspective it is better to take the mean value of b resulting from such data points in the data set. Once the optimal pair (w, b) is determined, the decision function of SVM is obtained as

$$z(x) = sign(\sum_{i=1}^{N_s}\alpha_i y_i K(x_i, x_j) + b) \tag{13}$$

2.2 FSVM (By Lin and Wang [24])

SVM is a powerful tool for solving classification problems [25], but there are still some limitations of this theory. From the training dataset and formulation discussed above, each training point belongs to either one class or the other. But in many real-world applications, each training data points no more exactly belongs to one of the two classes, it may 80% belong to one class and 20% be meaningless. That is to say, there is a fuzzy membership $\{\mu_i\}_{i=1}^{N} \in [0, 1]$ associated with each training data point x_i. In this sense, FSVM is an extension of SVM that takes into consideration the different significance of the training samples. For FSVM, each training sample is associated with a fuzzy membership value $\{\mu_i\}_{i=1}^N \in [0,1]$. The membership value μ_i reflects the confidence degree of the data points. The higher its value, the more confident we are about its class label. Similar to SVM, the optimization problem of the FSVM is formulated as follows [24]:

$$\begin{array}{ll} \text{Minimize} & \Psi(w,b,\xi_i,\mu_i) = \frac{1}{2}w^T w + C\sum_{i=1}^N \mu_i\xi_i \\ \text{Subject to:} & y_i(w^T\phi(x_i)+b) \geq 1-\xi_i \quad \text{for} \quad i=1,\ldots,N \\ & \xi_i \geq 0 \text{ for } \quad i=1,\ldots,N \end{array} \tag{14}$$

Note that the error term ξ_i is scaled by the membership value μ_i. The fuzzy membership values are used to weigh the soft penalty term reflects the relative confidence degree of the training samples during training. Important samples with larger membership values will have more impact in the FSVM training than those with smaller values.

Similar to the Vapnik's SVM, the optimization problem of FSVM can be transformed into the following dual problem:

$$\begin{array}{ll} \text{Maximize} & W(\alpha) = -\frac{1}{2}\sum_{i=1}^N\sum_{j=1}^N \alpha_i\alpha_j y_i y_j K(x_i,x_j) + \sum_{i=1}^N \alpha_i \\ \text{Subject to:} & \sum_{i=1}^N \alpha_i y_i = 0, \ 0 \leq \alpha_i \leq \mu_i C, \quad i=1,\ldots,N \end{array} \tag{15}$$

In the same way, the KT conditions are defined as

$$\alpha_i[y_i(w^T\phi(x)+b)-1+\xi_i] = 0, i=1,2,\cdots,N \tag{16}$$

$$(\mu_i C - \alpha_i)\xi_i = 0, i=1,2,\cdots,N \tag{17}$$

The data point x_i corresponding with α_i¿0 is called a support vector. There are also two types of SVs. The one corresponding with $0 < \alpha_i < \mu_i C$ lies on the margin of the hyperplane. The one corresponding with $\alpha_i = \mu_i C$ is misclassified.

Solving (15) will lead to a decision function similar to (13), but with different support vectors and corresponding weights α_i. An important difference between SVM and FSVM is that the data points with the same value of α_i may indicated a different type of SVs in FSVM due to the membership factor μ_i [24].

2.3 Least Squares FSVM

In both SVM and FSVM, the solution can be obtained by solving some QP problem. The deadly problem of the method is that it is difficult to find the solution by QP when we face some large-scale real-world problems. Motivated by Lai et al. [1] and Suykens and Vandewalle [26], a least squares FSVM (LS-FSVM) model is introduced by formulating the classification problem as

$$\begin{array}{ll} \text{Minimize} & \varphi(w,b,\xi_i,\mu_i) = \frac{1}{2}w^T w + \frac{C}{2}\sum_{i=1}^N \mu_i\xi_i^2 \\ \text{Subject to:} & y_i(w^T\phi(x_i)+b) = 1-\xi_i \quad \text{for} \quad i=1,\ldots,N \end{array} \tag{18}$$

One can defines the Lagrangian function

$$L(w,b,\xi_i;\alpha_i) = \frac{1}{2}w^T w + \frac{C}{2}\sum\nolimits_{i=1}^{N} \mu_i \xi_i^2 - \sum\nolimits_{i=1}^{N} \alpha_i [y_i(w^T\phi(x_i)+b) - 1 + \xi_i] \tag{19}$$

where α_i are Lagrangian multipliers, which can be either positive or negative now due to the equality constraints as follows from the KT conditions [27].

The optimal conditions are obtained by differentiating (19)

$$\begin{cases} \frac{d}{dw}L(w,b,\xi_i;\alpha_i) = w - \sum\limits_{i=1}^{N} \alpha_i y_i \phi(x_i) = 0 \\ \frac{d}{db}L(w,b,\xi_i;\alpha_i) = -\sum\limits_{i=1}^{N} \alpha_i y_i = 0 \\ \frac{d}{d\xi_i}L(w,b,\xi_i;\alpha_i) = \mu_i C\xi_i - \alpha_i = 0 \\ \frac{d}{d\alpha_i}L(w,b,\xi_i;\alpha_i) = y_i[w^T\phi(x_i)+b] - 1 + \xi_i = 0 \end{cases} \tag{20}$$

From (20) we can obtain the following representation

$$\begin{cases} w = \sum_{i=1}^{N} \alpha_i y_i \phi(x_i) \\ \sum_{i=1}^{N} \alpha_i y_i = 0 \\ \alpha_i = \mu_i C \xi_i \\ y_i[w^T\phi(x_i)+b] - 1 + \xi_i = 0 \end{cases} \tag{21}$$

In a matrix form, these optimal conditions in (21) can be expressed by

$$\begin{bmatrix} \mathbf{\Omega} & \mathbf{Y} \\ \mathbf{Y}^T & 0 \end{bmatrix} \begin{bmatrix} \alpha \\ b \end{bmatrix} = \begin{bmatrix} \mathbf{1} \\ 0 \end{bmatrix} \tag{22}$$

where Ω, $\mathbf{Y}$, and $\mathbf{1}$ are, respectively,

$$\mathbf{\Omega}_{ij} = y_i y_j \phi(x_i)^T \phi(x_j) + (\mu_i C)^{-1} I \tag{23}$$

$$\mathbf{Y} = (y_1, y_2, \cdots, y_N)^T \tag{24}$$

$$\mathbf{1} = (1, 1, \cdots, 1)^T \tag{25}$$

From (23), Ω is positive definite, thus the α can be obtained from (22), i.e.,

$$\alpha = \mathbf{\Omega}^{-1}(\mathbf{1} - b\mathbf{Y}) \tag{26}$$

Substituting (26) into the second matrix equation in (22), we can obtain

$$b = \frac{\mathbf{Y}^T\mathbf{\Omega}^{-1}\mathbf{1}}{\mathbf{Y}^T\mathbf{\Omega}^{-1}\mathbf{Y}} \tag{27}$$

Here, since Ω is positive definite, Ω^{-1} is also positive definite. In addition, since Y is a non zero vector, $\mathbf{Y}^T\Omega^{-1}\mathbf{Y}$¿0. Thus, b is always obtained. Substituting (27) into (26), α can be obtained.

Hence, the separating hyperplane can be found by solving the linear set of Equations (22)-(25) instead of quadratic programming thus reducing the computational complexity especially for large-scale problems. This is a distinct advantage over standard SVM and FSVM model.

3 Experiment Analysis

In this section, a real-world credit dataset is used to test the performance of LS-FSVM. For comparison purposes, linear regression (LinR) [1], logistic regression (LogR) [1], artificial neural network (ANN) [1], Vapnik's SVM [25], Lin & Wang's FSVM [24] and LSSVM [1, 26] are also conducted the experiments.

The dataset in this study is from the financial service company of England, obtaining from accessory CDROM of Thomas, Edelman and Crook [28]. The dataset includes detailed information of 1225 applicants, in which including 323 observed bad creditors. Every applicant includes 14 variables: (1) Year of birth; (2) Number of children; (3) Number of othe r dependents; (4) Is there a home phone; (5) Applicant's income; (6) Applicant's employment status; (7) Spouse's income; (8) Residential status; (9) Value of home; (10) Mortgage balance outstanding; (11) Outgoings on mortgage or rent; (12) Outgoings on loans; (13) Outgoings on hire purchase; (14) Outgoings on credit cards.

In this experiment, LS-FSVM, FSVM, LSSVM and SVM models use RBF kernel to perform classification task. In the ANN model, a three-layer back-propagation neural network with 10 TANSIG neurons in the hidden layer and one PURELIN neuron in the output layer is used. The network training function is the TRAINLM. Besides, the learning rate and momentum rate is set to 0.1 and 0.15. The accepted average squared error is 0.05 and the training epochs are 1600. The above parameters are obtained by trial and error. The experiment runs by Matlab 7 with statistical toolbox, NNET toolbox and LS-SVM toolbox provided by [26]. In addition, three evaluation criteria, Type I accuracy, Type II accuracy and total accuracy [1] is used.

To show its ability of LS-FSVM in discriminating potentially insolvent creditors from good creditors, we perform the testing with LS-FSVM at the beginning. This testing process includes five steps. First of all, we triple every observed bad creditor to make the number of observed bad nearly equal the number of

Table 1. The credit risk evaluation result by LSSVM

Experiment No.	Type I (%)	Type II (%)	Total (%)
1	81.56	93.81	88.54
2	86.98	95.14	92.53
3	82.02	91.84	88.49
4	79.81	98.36	93.53
5	87.77	94.03	92.24
6	81.56	96.85	92.38
7	79.19	93.05	87.41
8	85.69	92.11	88.86
9	79.27	89.33	87.63
10	82.45	96.58	90.45
Mean	82.63	94.11	90.21
Stdev	3.14	2.71	2.29

observed good. Second we preprocess the dataset so that the mean is 0 and the standard deviation is 1. Third the dataset is randomly separated two parts, training samples and evaluation samples, 1500 and 371 samples respectively. Fourth, fuzzy membership is generated by linear transformation function proposed by [19] in terms of initial score by expert's experience. Finally we train the FSVM classifier and evaluate the results. The above five steps are repeated 10 times to evaluate its robustness and Table 1 reports the results.

From Table 1, we can find the LS-FSVM has a strong classification capability. In the 20 experiments, Type I accuracy, Type II accuracy and total accuracy are 82.63%, 94.11% and 90.21%, respectively, in the mean sense. Furthermore, the standard devation is rather small, revealing that the robustness of the LS-FSVM is good. These results imply that the LSSVM is a feasible credit risk evaluation technique.

For further illustration, LS-FSVM's power of classification is also compared with other six commonly used classifiers: liner regression (LinR), logistics regression (LogR), artificial neural network (ANN), Vapnik's SVM, FSVM and LSSVM. The results of comparison are reported in Table 2.

Table 2. The comparison with other classifiers

Methods	Type I (%)	Type II (%)	Total (%)
LinR	52.87	43.48	50.22
LogR	60.08	62.29	60.66
ANN	56.57	78.36	72.24
SVM	70.13	83.49	77.02
LSSVM	79.37	93.27	89.16
FSVM	80.08	92.86	88.38
LS-FSVM	82.63	94.11	90.21

As can be seen from Table 2, we can find the following several conclusions:

(1) For type I accuracy, the LS-FSVM is the best of all the listed approaches, followed by the FSVM, LSSVM, Vapnik's SVM, logistics regression, artificial neural network model, and linear regression model, implying that the LS-FSVM is a very promising technique in credit risk assessment. Particularly, the performance of two fuzzy SVM techniques (FSVM [24] and LS-FSVM [1, 26]) is better than that of other listed classifiers, implying that the fuzzy SVM classifier may be more suitable for credit risk assessment tasks than other deterministic classifiers, such as LogR.

(2) For Type II accuracy and total accuracy, the LS-FSVM and LSSVM outperforms the other five models, implying the strong capability of least squares version of SVM model in credit risk evaluation. Meantime, the proposed LS-FSVM model seems to be slightly better LSSVM, revealing that the LS-FSVM is a feasible solution to improve the accuracy of credit risk evaluation.

Interestedly, the performance of the FSVM is slightly worse than that of the LSSVM, the main reasons leading to this are worth exploring further.

(3) From the general view, the LS-FSVM dominates the other six classifiers, revealing the LS-FSVM is an effective tool for credit risk evaluation.

4 Conclusions

In this study, a powerful classification method, least squares fuzzy support vector machine (LSFSVM), is proposed to assess the credit risk problem. Through the least squares method, a quadratic programming problem of FSVM is transformed into a linear equation group problem successfully and thus reducing the computational complexity greatly. With the practical data experiment, we have obtained good classification results and meantime demonstrated that the LS-FSVM model can provide a feasible alternative solution to credit risk assessment. Besides credit risk evaluation problem, the proposed LS-FSVM model can also be extended to other applications, such as consumer credit rating and corporate failure prediction.

Acknowledgements

This work described here is partially supported by the grants from the National Natural Science Foundation of China (NSFC No. 70221001, 70601029), the Chinese Academy of Sciences (CAS No. 3547600), the Academy of Mathematics and Systems Sciences (AMSS No. 3543500) of CAS, and the Strategic Research Grant of City University of Hong Kong (SRG No. 7001806).

References

1. Lai KK, Yu L, Zhou LG, Wang SY (2006) Credit risk evaluation with least square support vector machine. Lecture Notes in Artificial Intelligence 4062: 490-495
2. Altman EI (1968) Financial ratios, discriminant analysis and the prediction of corporate bankruptcy. Journal of Finance 23: 89-609
3. Wiginton JC (1980) A note on the comparison of logit and discriminant models of consumer credit behaviour. Journal of Financial Quantitative Analysis 15: 757-770
4. Grablowsky BJ, Talley WK (1981) Probit and discriminant functions for classifying credit applicants: A comparison. Journal of Economic Business 33: 254-261
5. Glover F (1990) Improved linear programming models for discriminant analysis. Decision Science 21: 771-785
6. Mangasarian OL (1965) Linear and nonlinear separation of patterns by linear programming. Operations Research 13: 444-452
7. Henley WE, Hand DJ (1996) A k-NN classifier for assessing consumer credit risk. Statistician 45: 77-95
8. Makowski P (1985) Credit scoring branches out. Credit World 75: 30-37
9. Malhotra R, Malhotra DK (2003) Evaluating consumer loans using neural networks. Omega 31: 83-96

10. Lai KK, Yu L, Wang SY, Zhou LG (2006) Neural network meta-learning for credit scoring. Lecture Notes in Computer Science 4113: 403-408
11. Chen MC, Huang SH (2003) Credit scoring and rejected instances reassigning through evolutionary computation techniques. Expert Systems with Applications 24: 433-441
12. Varetto F (1998) Genetic algorithms applications in the analysis of insolvency risk. Journal of Banking and Finance 22: 1421-1439
13. Van Gestel T, Baesens B, Garcia J, Van Dijcke P (2003) A support vector machine approach to credit scoring. Bank en Financiewezen 2: 73-82
14. Huang Z, Chen HC, Hsu CJ, Chen WH, Wu SS (2004) Credit rating analysis with support vector machines and neural networks: A market comparative study. Decision Support Systems 37: 543-558
15. Lai KK, Yu L, Huang W, Wang SY (2006) A novel support vector machine metamodel for business risk identification. Lecture Notes in Artificial Intelligence 4099: 480-484
16. Lee TS, Chiu CC, Lu CJ, Chen IF (2002) Credit scoring using the hybrid neural discriminant technique. Expert Systems with Application 23(3): 245-254
17. Piramuthu S (1999) Financial credit-risk evaluation with neural and neurofuzzy systems. European Journal of Operational Research 112: 310-321
18. Malhotra R, Malhotra DK (2002) Differentiating between good credits and bad credits using neuro-fuzzy systems. European Journal of Operational Research 136: 190-211
19. Wang YQ, Wang SY, Lai KK (2005) A new fuzzy support vector machine to evaluate credit risk. IEEE Transactions on Fuzzy Systems 13(6): 820-831
20. Smalz R, Conrad M (1994) Combining evolution with credit apportionment: A new learning algorithm for neural nets. Neural Networks 7: 341-351
21. Lai KK, Yu L, Wang SY, Zhou LG (2006) Credit risk analysis using a reliability-based neural network ensemble model. Lecture Notes in Computer Science 4132: 682-690
22. Thomas LC (2002) A survey of credit and behavioral scoring: Forecasting financial risk of lending to consumers. International Journal of Forecasting 16: 149-172
23. Thomas LC, Oliver RW, Hand DJ (2005) A survey of the issues in consumer credit modelling research. Journal of the Operational Research Society 56: 1006-1015
24. Lin CF, Wang SD (2002) Fuzzy support vector machines. IEEE Transaction on Neural Networks 13: 464-471
25. Vapnik V (1995) The nature of statistical learning theory. Springer, New York
26. Suykens JAK, Vandewalle J (1999) Least squares support vector machine classifiers. Neural Processing Letters 9: 293-300
27. Fletcher R (1987) Practical methods of optimization. John Wiley & Sons, New York
28. Thomas LC, Edelman DB, Crook JN (2002) Credit scoring and its applications. Society of Industrial and Applied Mathematics, Philadelphia

Similarity Measures on Interval-Valued Fuzzy Sets and Application to Pattern Recognitions

Hongmei Ju[1] and Xuehai Yuan[2]

[1] Information Institute, Beijing Wuzi University, Beijing, China
jhm2000@sohu.com
[2] School of Mathematics, Liaoning Normal University, Dalian, Liaoning, China
yuanxuehai@yahoo.com.cn

Abstract. The concept of interval valued fuzzy sets ($IVFSs$), proposed by Dubois et al and Gorza lczany, has gained attention from researchers for its applications in various fields. Although many measures of similarity between FSs have been proposed in the literature, those measures cannot deal with the similarity measures between $IVFSs$. In this paper, the definition of the degree of similarity between $IVFSs$ is introduced. Then similarity measures between $IVFSs$ are proposed and corresponding proofs are given. Finally, the similarity measures are applied to pattern recognitions.

Keywords: Interval-valued fuzzy sets; similarity measures; pattern recognition.

1 Introduction

The theory of fuzzy sets, proposed by Zadeh (1965) has gained successful applications in various fields. Measures of similarity between fuzzy sets, as an important content in fuzzy mathematics, have gained attention from researchers for their wide applications in real world. Based on similarity measures that are very useful in some areas, such as pattern recognition, machine learning, decision making and market prediction, many measures of similarity between fuzzy sets have been proposed and researched in recent years. For example, Chen (1994) proposed a similarity function F to measure the degree of similarity between fuzzy sets. Then Chen (1995) also made a comparison of similarity measures of fuzzy values. Wang (1997) proposed new fuzzy similarity measures on fuzzy sets and elements.

Dubois et al and Gorza lczany proposed the concept of interval-valued fuzzy sets ($IVFSs$). In this paper, we introduce the concept of the degree of similarity between $IVFSs$, present several new similarity measures for measuring the degree of similarity between $IVFSs$, which may be finite or continuous, and give corresponding proofs of these similarity measure and discuss application of the similarity measures between $IVFSs$ to pattern recognition problems.

B.-Y. Cao (Ed.): Fuzzy Information and Engineering (ICFIE), ASC 40, pp. 875–883, 2007.
springerlink.com

2 Degree of Similarity Between IVFSs and Similarity Measures

Definition 1.(Dubois et al [5] and Gorza lczany [6]) Let X be an set, $L = \{[a^-,\ a^+]|0 \le a^- \le a^+ \le 1\}$, an interval-valued fuzzy set A in X is an expression A given by

$$A : X \to L$$
$$x \mapsto [A^-(x), A^+(x)]$$

Obviously, every fuzzy set $\tilde{A}$ corresponds to the following interval-valued fuzzy set

$$\tilde{A} : X \to L$$
$$x \mapsto [\tilde{A}(x), \tilde{A}(x)] = \tilde{A}(x)$$

We will represent as $IVFS(X)$ the set of all the interval-valued fuzzy sets in X.

Definition 2. If A and B are two $IVFSs$ in X, then $A \subseteq B \Leftrightarrow A^-(x) \le B^-(x)$ and $A^+(x) \le B^+(x)$.

Definition 3. Denote $S : IVFS(X) \times IVFS(X) \to [0,1]$. S(A, B) is said to be the degree of similarity between A and $B \in IVFS(X)$ if $S(A,B)$ satisfy the properties $(1-4)$
(P1) $0 \le S(A,B) \le 1$
(P2) $S(A,B) = 1$ if $A = B$
(P3) $S(A,B) = S(B,A)$
(P4) $S(A,C) \le S(A,B)$ and $S(A,C) \le S(B,C)$ if $A \subseteq B \subseteq C$, $C \in IVFS(X)$

Definition 4. For every IVFSA, let

$$\varphi_A(i) = \frac{A^-(x_i) + A^+(x_i)}{2}, \tag{1}$$

$$where\ x_i \in X = \{x_1, x_2, ..., x_n\},$$

$$then\ S_{d_1}^p(A,B) = 1 - \frac{1}{\sqrt[p]{n}} \sqrt[p]{\sum_{i=1}^{n} |\varphi_A(i) - \varphi_B(i)|^p}\ , \tag{2}$$

where the $IVFSB = \{[B^-(x_i),\ B^+(x_i)]|x_i \in X\}$ and $1 \le p < +\infty$

Theorem 1. $S_{d_1}^p(A,B)$ is the degree of similarity between two $IVFSs$ in $X = \{x_1, x_2, ..., x_n\}$.

Proof. Obviously, $S_{d_1}^p(A,B)$ satisfies (P1-P3) of Definition 2.

In the following, $S^p_{d_1}(A,B)$ will be proved to satisfy (P4).
For any $C = \{[C^-(x_i),\ C^+(x_i)]|x_i \in X\}$,
$A \subseteq B \subseteq C$, We have $\varphi_A(i) \leq \varphi_B(i) \leq \varphi_C(i)$,
Hence $\varphi_C(i) - \varphi_A(i) \geq \varphi_B(i) - \varphi_A(i) \geq 0$.
So we have

$$\sqrt[p]{\sum_{i=1}^{n} |\varphi_A(i) - \varphi_C(i)|^p} \geq \sqrt[p]{\sum_{i=1}^{n} |\varphi_A(i) - \varphi_B(i)|^p}$$

It follows that $S^p_{d_1}(A,C) \leq S^p_{d_1}(A,B)$.
In a similar way, we can prove that $S^p_{d_1}(A,C) \leq S^p_{d_1}(B,C)$.
Therefore, $S^p_{d_1}(A,B)$ satisfies (P4) of Definition 2.
So we have finished the proof of the theorem.

Obviously, when p=1, (2) can be written as

$$S^p_{d_1}(A,B) = 1 - \frac{\sum_{i=1}^{n} |\varphi_A(i) - \varphi_B(i)|}{n}$$

Definition 5. For $IVFSA = \{[A^-(x_i),\ A^+(x_i)]|x \in [a,b]\}$ and $B = \{[B^-(x_i), B^+(x_i)]|x \in [a,b]\}$, then

$$S^p_{c_1}(A,B) = 1 - \frac{1}{\sqrt[p]{b-a}} \sqrt[p]{\int_a^b |\varphi_A(x) - \varphi_B(x)|^p \, dx}, \tag{3}$$

$$where\ \varphi_A(x) = \frac{A^-(x) + A^+(x)}{2},$$

$$\varphi_B(x) = \frac{B^-(x) + B^+(x)}{2},\ x \in [a,b]$$

Theorem 2. $S^p_{c_1}(A,B)$ is the degree of similarity between the two $IVFSsA$ and B in $X = [a,b]$.

Proof. The proof is similar to that of Theorem 1.

When we consider the weight of the element$x \in X$, we present the following weighted similarity measures between $IVFSs$.

Definition 6. Assume that the weight of $x_i \in X = \{x_1, x_2, ..., x_n\}$ is μ_i (i=1,2,...,n), where $0 \leq \mu_i \leq 1$, $\Sigma_{i=1}^{n}\mu_i = 1$, then

$$S^p_{d_1\mu}(A,B) = 1 - \sqrt[p]{\sum_{i=1}^{n} \mu_i |\varphi_A(i) - \varphi_B(i)|^p}, \tag{4}$$

Theorem 3. $S^p_{d_1\mu}(A,B)$ is the degree of similarity between two $IVFSs$ in $X = \{x_1, x_2, ..., x_n\}$.

Proof. The proof is similar to that of Theorem 1.
Obviously, if $\mu_i = \frac{1}{n}$ (i=1, 2, ..., n), formula (4) becomes formula (2). So (2) is only a special case of (4).

Definition 7. Assume that the weight of $x \in X = [a,b]$ is $\mu(x)$, where $0 \leq \mu(x) \leq 1$, $\int_a^b \mu(x)\, dx = 1$, then

$$S^p_{c_1\mu}(A,B) = 1 - \sqrt[p]{\int_a^b \mu(x)|\varphi_A(i) - \varphi_B(i)|^p\, dx}, \tag{5}$$

Theorem 4. $S^p_{c_1\mu}(A,B)$ is the degree of similarity between two $IVFSs$ in $X = [a,b]$.

Proof. The proof is similar to that of Theorem 1.
Obviously, if $\mu(x) = \frac{1}{b-a}$ (i=1, 2, ..., n) for any $x \in X = [a,b]$, formula (5) becomes formula (3). So (3) is only a special case of (5).

Definition 8. Let

$$T_{AB^-}(i) = \frac{|A^-(x_i) - B^-(x_i)|}{2},$$

$$\textit{and } T_{AB^+}(i) = \frac{|A^+(x_i) - B^+(x_i)|}{2},$$

$$\textit{then } S^p_{d_2}(A,B) = 1 - \frac{1}{\sqrt[p]{n}} \sqrt[p]{\sum_{i=1}^{n} (T_{AB^-}(i) + T_{AB^+}(i))^p}, \tag{6}$$

Theorem 5. $S^p_{d_2}(A,B)$ is the degree of similarity between two $IVFSs$ in $X = \{x_1, x_2, ..., x_n\}$.

Proof. Obviously, $S^p_{d_2}(A,B)$ satisfies (P1-P3) of definition 2.
In the following, $S^p_{d_2}(A,B)$ will be proved to satisfy (P4).
$Since A \subseteq B \subseteq C$,
there exist $A^-(x_i) \leq B^-(x_i) \leq C^-(x_i)$,
and $A^+(x_i) \leq B^+(x_i) \leq C^+(x_i)$.

$$S^p_{d_2}(A,C) = 1 - \frac{1}{\sqrt[p]{n}} \sqrt[p]{\sum_{i=1}^{n} (T_{AC^-}(i) + T_{AC^+}(i))^p},$$

$$S_{d_2}^p(A,B) = 1 - \frac{1}{\sqrt[p]{n}} \sqrt[p]{\sum_{i=1}^{n} (T_{AB^-}(i) + T_{AB^+}(i))^p}$$

$$(T_{AB^-}(i) + T_{AB^+}(i)) - (T_{AC^-}(i) + T_{AC^+}(i)) \leq 0,$$

$$then\ S_{d_2}^p(A,C) \leq S_{d_2}^p(A,B).$$

In the similar way, it is easy to prove $S_{d_2}^p(B,C) \leq S_{d_2}^p(A,B)$.
In this definition, we make use of two end points of the subinterval in $IVFSs$ to define similarity measures which contains more information than definition 4. In other words, the definition 8 can avoid the problem of each interval having equal median values in definition 4 sometimes.

Definition 9. When the universe of discourse is continuous, we can obtain the following similar result,then

$$S_{c_2}^p(A,B) = 1 - \frac{1}{\sqrt[p]{b-a}} \sqrt[p]{\int_a^b (T_{AB^-}(x) + T_{AB^+}(x))^p \, dx}, \tag{7}$$

Theorem 6. $S_{c_2}^p(A,B)$ is the degree of similarity between two $IVFSs$ in $X = [a,b]$.

Proof. The proof is similar to that of Theorem 5.
However, the elements in the universe may have different importance in pattern recognition. We should consider the weight of the elements so that we can obtain more reasonable results in pattern recognition.

Definition 10. Assume that the weight of $x_i \in X = \{x_1, x_2, ..., x_n\}$ is μ_i, (i=1,2,...,n), where

$$0 \leq \mu_i \leq 1, \Sigma_{i=1}^n \mu_i = 1,$$

$$then S_{d_2\mu}^p(A,B) = 1 - \sqrt[p]{\sum_{i=1}^{n} \mu_i (T_{AB^-}(x) + T_{AB^+}(x))^p}, \tag{8}$$

Theorem 7. $S_{d_2\mu}^p(A,B)$ is the degree of similarity between two $IVFSs$ in $X = \{x_1, x_2, ..., x_n\}$

Proof. The proof is similar to that of Theorem 5.
Obviously, if $\mu_i = \frac{1}{n}$ (i=1, 2, ..., n), formula (8) becomes formula (6). So (6) is only a special case of (8).

Definition 11. Assume that the weight of $x \in X = [a,b]$ is $\mu(x)$, where $0 \leq \mu(x) \leq 1$, $\int_a^b \mu(x)\, dx = 1$. then

$$S^p_{c2\mu}(A,B) = 1 - \sqrt[p]{\int_a^b \mu(x)(T_{AB^-}(x) + T_{AB^+}(x))^p \, dx}, \quad (9)$$

Theorem 8. $S^p_{c2\mu}(A,B)$ is the degree of similarity between two $IVFSs$ in $X = [a,b]$.

Proof. The proof is similar to that of Theorem 5.
Obviously, if $\mu(x) = \frac{1}{b-a}$ (i=1, 2, ..., n) for any $x \in X = [a,b]$, formula (9) becomes formula (7). So (7) is only a special case of (9).

In order to further deal with the question, we should make full of known information on $IVFSs$ such as the length of the subinterval, the median value of the subinterval to define similarity measures. Only in this way can patterns be recognized more accurately.

Now the interval $[A^-(x_i),\ A^+(x_i)]$ is divided into two subintervals. Then the two subintervals are denoted as $[A^-(x_i),\ \varphi_A(x_i)]$, $[\varphi_A(x_i),\ A^+(x_i)]$.

And then let

$$m_A(x_i) = \frac{A^-(x_i) + \varphi_A(x_i)}{2} \quad (10)$$

$$n_A(x_i) = \frac{\varphi_A(x_i) + A^+(x_i)}{2} \quad (11)$$

In a similar way, we can obtain the following results for $IVFSB$.

$$m_B(x_i) = \frac{B^-(x_i) + \varphi_B(x_i)}{2} \quad (12)$$

$$n_B(x_i) = \frac{\varphi_B(x_i) + B^+(x_i)}{2} \quad (13)$$

Definition 12. Assume that $m_A(x_i)$, $n_A(x_i)$, $m_B(x_i)$, $n_B(x_i)$ are obtained by (10)-(13). Let

$$T_{s1}(i) = \frac{|m_A(x_i) - \ m_B(x_i)|}{2}$$

$$T_{s2}(i) = \frac{|n_A(x_i) - \ n_B(x_i)|}{2}$$

$$then\ S^p_{d_3}(A,B) = 1 - \frac{1}{\sqrt[p]{n}} \sqrt[p]{\sum_{i=1}^{n} (T_{s1}(i) + T_{s2}(i))^p}, \quad (14)$$

Theorem 9. $S^p_{d_3}(A,B)$ satisfies (P1-P3).

Proof. Obviously.

Definition 13. When the universe of discourse is continuous, we can obtain the following similar result.

$$S^p_{c_3}(A,B) = 1 - \frac{1}{\sqrt[p]{b-a}} \sqrt[p]{\int_a^b (T_{s1}(x) + T_{s2}(x))^p \, dx}, \tag{15}$$

Theorem 10. $S^p_{c_3}(A,B)$ satisfies (P1-P3).

Proof. Obviously.

Definition 14. Assume that the weight of $x_i \in X = \{x_1, x_2, ..., x_n\}$ is μ_i(i=1, 2, ..., n),where

$$0 \leq \mu_i \leq 1, \Sigma^n_{i=1}\mu_i = 1, then$$

$$S^p_{d_3\mu}(A,B) = 1 - \sqrt[p]{\sum_{i=1}^{n} \mu_i (T_{s1}(x) + T_{s2}(x))^p}, \tag{16}$$

Theorem 11. $S^p_{d_3\mu}(A,B)$ satisfy (P1-P3).

Proof. Obviously.
Obviously, if $\mu_i = \frac{1}{n}$ (i=1, 2, ..., n), formula (16) becomes formula (14). So (14) is only a special case of (16).

Definition 15. Assume that the weight of $x \in X = [a,b]$ is $\mu(x)$, where $0 \leq \mu(x) \leq 1$, $\int_a^b \mu(x)\, dx = 1$, then

$$S^p_{c_3\mu}(A,B) = 1 - \sqrt[p]{\int_a^b \mu(x)(T_{s1}(x) + T_{s2}(x))^p \, dx}, \tag{17}$$

Theorem 12. $S^p_{c_3\mu}(A,B)$ satisfy (P1-P3).

Proof. Obviously.
Obviously, if $\mu(x) = \frac{1}{b-a}$ (i=1,2,...,n) for any $x \in X = [a,b]$, formula (17) becomes formula (15). So (15) is only a special case of (17).

3 Applications of the Similarity Measures to Pattern Recognitions

Assume that there exist m patterns which are represented by $IVFSs\ A_j = \{[A^-_j(x_i),\ A^+_j(x_i)] | x_i \in X\} \in IVFS(X)$, (j=1, 2, ..., m). Suppose that there be a sample to be recognized which is represented by a $IVFSB_j = \{[B^-_j(x_i),\ B^+_j(x_i)] | x_i \in X\}$.

$$Set\ S^p_w(A_{i_0}, B) = \max_{1 \le i \le m} \{S^p_w(A_i, B)\},$$

$$Set\ S^p_{w'}(A_{i_0}, B) = \max_{1 \le i \le m} \{S^p_{w'}(A_i, B)\},$$

$$where\ X = \{x_1, x_2, ..., x_n\}\ ,\ w = \{d_1\mu,\ d_2\mu, d_3\mu\}, \tag{18}$$

$$where\ X = [a, b]\ ,\ w^{'} = \{c_1\mu,\ c_2\mu, c_3\mu\}, \tag{19}$$

According to the principle of the maximum degree of similarity between $IVFSs$, we can decide that sample B belongs to the pattern A_{i_0}.

In the following, two fictitious numerical examples are given to show applications of the similarity measures to pattern recognition problems.

Example 1. Let three patterns be represented by $IVFSs$ in $X = \{x_1, x_2, x_3\}$

$$A_1 = \{[0.9, 1.0], [0.8, 0.9], [1.0, 1.0]\},$$

$$A_2 = \{[1.0, 1.0], [0.6, 0.8], [0.8, 1.0]\},$$

$$A_3 = \{[0.7, 0.9], [1.0, 1.0], [0.8, 1.0]\}, respectively.$$

Consider a sample $B \in IVFSs$ which will be recognized, where

$$B = \{[0.8, 0.9], [0.5, 0.7], [0.6, 0.8]\}.$$

Assume that weights of x_i are $\frac{1}{3}$ (i=1,2,3). Choose p=1. By applying (2), (6), (14) ,we have

$$S^1_{d_1}(A_1, B) = 0.78\ ,\ S^1_{d_1}(A_2, B) = 0.85\ ,\ S^1_{d_1}(A_3, B) = 0.78$$

$$S^1_{d_2}(A_1, B) = 0.78\ ,\ S^1_{d_2}(A_2, B) = 0.85\ ,\ S^1_{d_2}(A_3, B) = 0.78$$

$$S^1_{d_3}(A_1, B) = 0.78\ ,\ S^1_{d_3}(A_2, B) = 0.85\ ,\ S^1_{d_3}(A_3, B) = 0.82$$

Similarly, choosing p=2 and using (2), (6),(14) we obtain

$$S^2_{d_1}(A_1, B) = 0.77\ ,\ S^2_{d_1}(A_2, B) = 0.84\ ,\ S^2_{d_1}(A_3, B) = 0.74$$

$$S^2_{d_2}(A_1, B) = 0.77\ ,\ S^2_{d_2}(A_2, B) = 0.84\ ,\ S^2_{d_2}(A_3, B) = 0.74$$

$$S^2_{d_3}(A_1, B) = 0.77\ ,\ S^2_{d_3}(A_2, B) = 0.84\ ,\ S^2_{d_3}(A_3, B) = 0.84$$

According to (18), we follow that B should belong to A_2 for both p=1 and p=2. Obviously, the sample B belongs to the pattern A_2 too.

Example 2. Let two patterns be represented by $IVFSs$

$$A_1 = \{[A_1^-(x),\ A_1^+(x)] | x \in [1, 3]\},$$

$$A_2 = \{[A_2^-(x),\ A_2^+(x)] | x \in [1, 3]\}, respectively, where$$

$$A_1^-(x) = 0.6(x-1),\quad 1 \le x\ < 3,\qquad A_1^+(x) = 0.8(x-1),\quad 1 \le x\ < 3,$$

$$A_2^-(x) = 0.4x + 0.1, \quad 1 \leq x < 3, \qquad A_2^+(x) = 0.8x + 0.3, \quad 1 \leq x < 3.$$

Consider a sample $B \in IVFSs$ which will be recognized, where $B = \{[B^-(x), B^+(x)]$, where

$$B^-(x) = 0.3x + 0.2\ , \quad 1 \leq x < 3, \qquad B^+(x) = 0.9x - 0.4\ , \quad 1 \leq x < 3.$$

Assume that weights of x are $\frac{1}{3}$, $1 \leq x < 3$ .Choose p=1. By applying (3), (7), (15) ,we have

$$S_{c_1}^1(A_1, B) = 0.40\ ,\ S_{c_1}^1(A_2, B) = 0.70$$

$$S_{c_2}^1(A_1, B) = 0.59\ ,\ S_{c_2}^1(A_2, B) = 0.70$$

$$S_{c_3}^1(A_1, B) = 0.60\ ,\ S_{c_3}^1(A_2, B) = 0.70$$

According to (19), we follow that B should belong to A_2 for p=1, other cases are discussed in a similar way.

References

1. Zadeh LA(1965) Fuzzy sets. Information and Control 8(3): 338-353
2. Zadeh LA(1975) The concept of a linguistic variable and its application to approximate reasoning. Information Science 8(2): 199-249, 8(3): 301-357, 9(1): 43-80
3. Zwich, R., Larlstein, E., Budescu, D.(1987) Measures of similarity among fuzzy sets: A comparative analysis. Internet J. Approx. Reason 1: 221-242
4. K.Atanassov(1986) Intuitionistic fuzzy sets. Fuzzy Sets and Systems 20: 87-96
5. K.Atanassov(1989) More on intuitionistic fuzzy sets. Fuzzy Sets and Systems 33: 37-46
6. K.Atanassov(1994a) Operators over interval valued intuitionistic fuzzy sets. Fuzzy Sets Systems 64(2): 159-174
7. Dubios D, Ostasiewicz W, Prade H(2000) Fuzzy sets: history and basic notions In : Dubois D, Prade H. Fundamentals of Fuzzy Sets pp. 80-93
8. Gorza Iczany(1987) A method of inference in approximate reasoning based on interval valued fuzzy set. Fuzzy Sets and Systems 21(1): 1-17
9. Gau, W.L.,Buehere, D.J.(1993) Vagye sets. IEEE Trans. Systems Man Cybernet 23(2): 610-614
10. Hyung, L.K., Song, Y.S., Lee, K.M.(1994) Similarity measures between fuzzy sets and between element. Fuzzy Sets and Systems 62(3): 291-293
11. Wang, W.-J.(1997) New similarity measures on fuzzy sets and on elements. Fuzzy Sets Sestems 85(3): 305-309
12. Chen, S.M.(1994) A weighted fuzzy reasoning algorithm for medical diagnosis. Decis. Support Systems 37-43

Updating of Attribute Reduction for the Case of Deleting

Ming Yang, Ping Yang, and Gen-Lin Ji

Department of Computer Science, Nanjing Normal University Nanjing, 210097, P.R. China
m.yang@njnu.edu.cn

Abstract. Knowledge reduction is an important issue when dealing with huge amounts of data. Rough set theory provides two fundamental concepts to deal with this special problem: attribute reduction and core, where attribute reduction is the key problem. There are a lot of methods available for attribute reduction, but very little work has been done for attribute reduction for the cases of deleting and modifying. Therefore, in this paper, we introduce a new updating method based on discernibility matrix for attribute reduction for the case of deleting. When a few old objects are deleted from a decision information system, a core and a new attribute reduction can be obtained quickly because it only deletes a few old rows and columns, or inserts a few rows and updates corresponding columns when updating the discernibility matrix.

Keywords: Rough Sets,Discernibility Matrix, core,Attribute Reduction, Updating.

1 Introduction

Rough Sets (RS) theory was proposed by Professor Z.Pawlak[1], which is a valid mathematical theory to deal with imprecise, uncertain, and vague information. It has been widely applied in many fields as machine learning[2], data mining[3], intelligent data analyzing and control algorithm acquiring[4],etc. The theory provides the attribute reduction and the core concepts for knowledge reduction.

Rough set theory provides two fundamental concepts to deal with this special problem: attribute reduction and core. There are a lot of methods available for core[5-6] and attribute reduction[7-10].Unfortunately, most of them are designed for the off-line data processing. However, many real applications need to process data dynamically due to inserting, deleting, and modifying of data. So, many researchers in knowledge discovery fields suggest that knowledge acquisition algorithms should better be dynamic[11-14]. In paper[15],an incremental attribute reduction algorithm is developed. It can only process information systems without decision attribute. However, most of real information systems are decision information system. Hence, dynamic attribute reduction in decision information

B.-Y. Cao (Ed.): Fuzzy Information and Engineering (ICFIE), ASC 40, pp. 884–893, 2007.
springerlink.com

system would be more important. Some methods were introduced for incremental attribute reduction in decision information system in paper[16-17]. Among them, paper[19] is more efficient as compared to the paper[16-18] according to the preliminary experiments, but it is only suitable to the case of inserting as the other methods.

As pointed out in [14], we need to maintain the attribute reductions discovered in a decision table in the case including insertion, deletion, and modification of objects in the decision table. However, many researchers mostly focus on the case of inserting, very litter work concerns with the cases of deleting and modifying, this is our motivation of this study. In this paper, we only introduce a new updating method based on discernibility matrix for attribute reduction for the case of deleting because modifying operation can be easily implemented by deleting and inserting operation as the existing database system. The method of this paper can get a core and a new attribute reduction for a decision information system quickly after some old objects are deleted from a decision information system because it only deletes a few old rows and columns, or inserts a few rows and updates corresponding columns when updating the discernibility matrix. Clearly, this algorithm is a complement of the existing incremental algorithms for attribute reduction, not instead of them.

The rest of this paper is organized as follows. In Section 2, some basic concepts on the Rough Sets theory are briefly introduced. In Section 3, an updating algorithm of a core is derived from the original computing algorithms of a core based on discernibility matrix. The updating principle and algorithm of attribute reduction are presented in Section 4 and Section 5 respectively. Finally, Section 6 gives our conclusions.

2 Preliminaries

For the convenience of description, some concepts of decision information systems in Rough Sets theory are introduced here at first.

Definition 1. (decision information systems[4,19-20]) A decision information system is defined as $S =< U, Q, V, f >$, where U is a non-empty finite set of objects, called universe, Q is a non-empty finite set of attributes, $Q = C \bigcup D$, where C is the set of condition attributes and D is the set of decision attributes, $D \neq \emptyset$.$V = \bigcup_{a \in Q} V_a$, and V_a is the domain of the attribute a.$f : U \times Q \to V$ is a total function such that $f(x_i, a) \in V_a$ for every $a \in Q, x_i \in U, i = 1, 2, ..., card(U)$. Throughout this paper, $\emptyset$ denotes the empty set, and $card(x)$ denotes the function that returns the cardinality of the argument set x.

With any $P \subseteq Q$ there is an associated equivalence relation $IND(P)$: $IND(P) = \{(x, y) \in U^2 | \forall a \in P, a(x) = a(y)\}$. The partition of U, generated by $IND(P)$ is denoted as $U/IND(P)$. If $(x, y) \in IND(P)$, then x and y are indiscernible by attributes from P. The equivalence classes of the P-indiscernibility relation are denoted $[x]_P$.

Definition 2. (consistency and inconsistency of $[x]_P$ [19-20]) Given a decision information system S. An equivalence class $[x]_P \in U/IND(P)(P \subseteq C, x \in U)$ is consistent iff all its objects have the same decision value. Otherwise, it is inconsistent.

Definition 3. (lower-approximation, positive region and approximation quality [4-6,19-20]) Given a decision information system S, for any subset $X \subseteq U$ and indiscernibility relation $IND(B)$, the B lower-approximation of X is defined as: $B\_(X) = \{x | [x]_B \subseteq X\}$.For $P \subseteq Q, B \subseteq Q$,the P positive region of B is defined as :$Pos_p(B) = \bigcup_{X \in U/IND(B)} P\_(X)$; the P-approximation quality of B is defined as :

$$\gamma_P(B) = |Pos_P(B)|/|U|, \ namely \ \gamma_P(B) = |\bigcup_{X \in U/IND(B)} P\_(X)|/|U|.$$

Definition 4.(candidate attribute reduction, attribute reduction and core[4-6,20]) Given a decision information system S. A candidate attribute reduction R is defined as a subset of the conditional attribute set C such that $\gamma_R(D) = \gamma_C(D)$; further, if R is a candidate attribute reduction and $\gamma_S(D) \neq \gamma_C(D)$ holds for any $S \subset R$, then R is an attribute reduction. The intersection of all the sets of attribute reduction is called the core, it is denoted by $core(C, U)$.

For simplicity of description, the set of all attribute reductions and that of all candidate attribute reductions for a decision information system S is denoted by $R(U)$ and $CR(U)$, respectively.

Proposition 1[9]**.** For a decision information system S, let $P \subseteq C$, if P is a candidate attribute reduction, and $P = Core(P, U)$, then P is an attribute reduction.

3 Improvement of Discernibility Matrix and Updating of a Core

There are a lot of methods available for attribute reduction. Among them, the previous methods that use a core as starting point are of considerable benefits, because they could improve the performance efficiently. Similarly, an efficiently updating method of a core can be also regarded as one of dynamically updating strategies of attribute reduction. In paper[5-6], an updating algorithm of a core based on an improved discernibility matrix(see, e.g., [5,6] or the following Def. 5) was proposed, which is suitable to both consistent and inconsistent decision information systems and has high efficiency. However, it is not appropriate for the case of deleting.

Definition 5[5]**.** Given a decision information system S, a discernibility matrix $M1 = \{m_{ij}\}$ of S is a $|U1| \times |U|$ matrix with entries defined as:

$$m_{ij} = \begin{cases} \{a \in C : f(x_i, a) \neq f(x_j, a)\}, \text{ if } f(x_i, D) \neq f(x_j, D) \text{ for } x_i \in U1, x_j \in U1; \\ \{a \in C : f(x_i, a) \neq f(x_j, a)\}, \text{ for } x_i \in U1, x_j \in U2; \\ \emptyset \qquad\qquad\qquad\qquad\qquad \text{otherwise} \end{cases}$$

where $U1 = Pos_C(D), U2 = U - U1$,$U1$ is also called the set of all consistent objects of S.

By Definition 5, a theorem was presented as below in paper[5-6], which can efficiently judge whether an element in discernibility matrix belongs to a core or not.

Theorem 1. [5−6] *Given a decision information system $S, M1$ is a discernibility matrix of S (that is defined as definition 5), let $IDM(C, M1) = \{m_{ij} : m_{ij} \in M1$ and $card(m_{ij}) = 1\}$, $IDM(C, M1) = Core(C, U)$ holds, that is, a condition attribute a belongs to $Core(C, U)$ iff $\{a\}$ is an element of M1.*

Given a decision information system S, let $U1 = Pos_C(D), U2 = U - U1$. Suppose an object x in U is deleted, the discernibility matrix of S' ($S' =< U - \{x\}, C \bigcup D, V, f >$) is denoted as $M1(x)$. Similar to [6], updating of a core can be shifted the problem of how to update the discernibility matrix $M1$ of S for getting $M1(x)$ after an object x is deleted, and an updating algorithm of a *core* is introduced, it is designed according to two cases, namely $x \in U1$ or $x \in U2$.

Algorithm 1. UAC(Updating Algorithm of a Core)
Input: (1) $U1 = Pos_C(D), U2 = U - U1$, $M1$ is the discernibility matrix of a decision information system $S(S =< U, C \bigcup D, V, f >)$;
(2) an object x is deleted from U.
Output:a core of $S'(S' =< U - \{x\}, C \bigcup D, V, f >)$,namely $Core(C, U - \{x\})$.

1. if $x \in U1$,then delete the row and column that relate to x in $M1$,and let $U1 = U1 - \{x\}, U = U - \{x\}$; goto 3.
2. if $x \subset U2$,
   2.1 if $card([x]_C) > 2$, then delete the column concerning x in $M1$, let $U2 = U2 - \{x\}, U = U - \{x\}$; goto 3.
   2.2 if $card([x]_C) = 2$,let object y be inconsistent with x, then delete the column involving x in $M1$, and add a new row relating to the object y in $M1$. Let $U2 = U2 - \{x\}, U = U - \{x\}$;goto 3.
3. $M1(x)$ is gotten by the case 1 and case 2 , hence $Core(C, U - \{x\})$ can be acquired directly by Theorem 1.
4. return $Core(C, U - \{x\})$.

Because a few rows or columns of the given discernibility matrix need to be adjusted, thus Algorithm 1 has better performance as compared to the non-updating algorithms intuitively. In this paper, the real objective of updating of a core is to develop an efficient updating algorithm of attribute reduction. So, in Section 4 and section 5, the updating principle and algorithm of attribute reduction are presented by utilizing the gotten core for the case of deleting and the known set of attribute reductions.

4 Updating Principle of Attribute Reduction

By Definition 5, all objects of a decision information system can be divided into two parts: the set of all consistent objects U1 and the set of all inconsistent objects U2. Therefore, updating of attribute reduction for the case of deleting will be the following two cases:

(1) the deleted objects originates from U1;
(2) some objects in U2 are deleted. Section 4 focuses on the updating strategies of the old attribute reductions when the consistent objects or the inconsistent objects are deleted.

Lemma 1. *Given a decision information system S, assume an object $x(x \in U)$is deleted, let P be a subset of the set of condition attributes C, $M1(x)$ be the discernibility matrix by algorithm 1, $U1 = Pos_C(D), U2 = U - U1$. For a given object y in U1, if there exists an element m_{ij} in the row that relates to y in $M1(x)$ such that $m_{ij} \bigcap P = \emptyset$ and $m_{ij} \neq \emptyset$,then both $[y]_p \bigcap \psi_s \neq \emptyset$ and $[y]_p \bigcap \psi_t \neq \emptyset$ hold.*

Proof. Suppose the object concerning the column of m_{ij} is z. If $z \in U1$ then $f(y, D) \neq f(z, D)$ holds by $m_{ij} \neq \emptyset$, namely $[y]_D \neq [z]_D$, and $f(y, b) = f(z, b)$ holds for every $b \in P$ by $m_{ij} \bigcap P = \emptyset$. Set $\psi_s = [y]_D$ and $\psi_t = [z]_D$,$[y]_P \bigcap \psi_s \neq \emptyset$ and $[y]_P \bigcap \psi_t \neq \emptyset$ hold. Similarly, the conclusion holds as well when $z \in U2$.

Lemma 2. *Given a decision information system S, assume an object $x(x \in U)$is deleted, let P and $M1(x)$ be a subset of the set of condition attributes C and the discernibility matrix by algorithm 1 respectively. Set $U1 = Pos_C(D)$and $U2 = U - U1$. For a given object y in U1, there are two equivalence classes $\psi_s \in U/IND(D), \psi_t \in U/IND(D)$ such that $[y]_P \bigcap \psi_s \neq \emptyset$ and $[y]_P \bigcap \psi_t \neq \emptyset$. If there exists an element m_{ij} in the row involving the object y in $M1(x)$ such that $P \bigcap m_{ij} = \{a\}$, the condition attribute a belongs to $Core(P, U)$.*

Proof. Similar to the proof of Theorem 1.

It is obvious that $Core(P, U)$ can be obtained for any subset P of condition attributes C by Lemma 1 and Lemma 2.

Theorem 2. *Given a decision information system S, set $U1 = Pos_C(D)$ and $U2 = U - U1$ .Assume an object $x(x \in U1)$is deleted. If P is an attribute reduction,namely $P \in R(U)$,we can get the following two conclusions.*

(1) if $card([x]_P) = 1$, then P is a candidate attribute reduction of $S'(S' =< U - \{x\}, C \bigcup D, V, f >)$, namely $P \in CR(U - \{x\})$.
(2) if $card([x]_P) > 1$, then P is an attribute reduction of S', namely $P \in R(U - \{x\})$.

Proof. Because x is an object of $U1$ and P is an attribute reduction, hence $Pos_P(D) = Pos_C(D)$ also holds in S'. Further, if $card([x]_P) = 1$, perhaps there exists a subset B of P and $B \neq P$ such that $Pos_B(D = Pos_P(D)$ holds(see the case 1 of example 1).Hence, the conclusion of (1) holds, that is, P is a candidate attribute reduction.

If $card([x]_P) > 1$, then there exists an object y that belongs to $U1$ and $y \neq x$ such that $[x]_P = [y]_P$ holds because of $x \in U1$. Assume E is a subset of P and $E \neq P$ such that $Pos_E(D) = Pos_P(D)$ holds in S', x must belong to $[y]_E$, that is, E is an attribute reduction of S, this contradicts with that P is an attribute reduction of S. So, the conclusion of (2) also holds.

By Theorem 2, it is seen that the original attribute reduction becomes a new attribute reduction or a candidate attribute reduction when one object of consistent objects is deleted.

Theorem 3. *Given a decision information system S, set $U1 = Pos_C(D)$ and $U2 = U - U1$.Assume an object $x(x \in U2)$is deleted. If P is an attribute reduction, namely $P \in R(U)$,we can get the following two conclusions.*

(1) if $card([x]_C) = 2$, then there will be the following two cases.
- *(1.1) if $card([x]_P) = 2$, then P is an attribute reduction of S', namely $P \in R(U - \{x\})$.*
- *(1.2) if $card([x]_P) > 2$, then P is not an attribute reduction of S', namely $P \notin R(U - \{x\})$.*

(2) if $card([x]_C) > 2$, then P is an attribute reduction of S', namely $P \in R(U - \{x\})$.

Proof. (1) If $card([x]_C) = 2$ and $card([x]_P) = 2$, then there exists an object y that belongs to $U2$ and $y \neq x$ such that $[z]_P \neq [x]_P$ for any $z \in U1$, and $[x]_P = [y]_P$ and $[x]_C = [y]_C$ hold. Clearly, it is easily verified that $Pos_P(D) = Pos_C(D)$ and $Pos_E(D) \neq Pos_P(D)$ for any subset E of P when the object x is deleted from $U2$, hence P is also a new attribute reduction, namely (1.1) holds. Similarly, the conclusion of (1.2) can be easily proved.
(2) It is obvious by definition 3 and definition 4 because $U1$ keeps invariant while the inconsistent object x is removed from $U2$.

In Section 5, we will introduce an updating algorithm of attribute reduction by Lemma 1,Lemma 2, and Theorem 2 and Theorem 3.

5 Updating Algorithm of Attribute Reduction Based on Discernibility Matrix

As analyzed above, an updating algorithm for attribute reduction is as below when the old objects are deleted.

Algorithm 2. $UAARD$(Updating Algorithm of Attribute Reduction based on Discernibility matrix)

Input:

(1) $U1 = Pos_C(D)$,$U2 = U - U1$, $M1$ is the discernibility matrix of a decision information system $S(S =< U, C \bigcup D, V, f >)$;
(2) an object x is deleted from U;
(3) P is an attribute reduction of S.

Output: the attribute reduction result P of the new decision information system $S^{'}(S^{'} =< U - \{x\}, C \bigcup D, V, f >)$.

1. By Algorithm 1, update the $M1$ and get a new discernibility matrix $M1(x)$ and a new core $Core(C, U - \{x\})$;
2. if $x \in U1$,
   2.1 if $card([x]_P) > 1$, then $P \in R(U - \{x\})$ by Theorem 2;goto 6.
   2.2 if $card([x]_P) = 1$, then $P \in CR(U - \{x\})$ by Theorem 2;goto 4.
3. if $x \in U2$,
   3.1 if $card([x]_C) > 2$, then $P \in R(U - \{x\})$ by Theorem 3; goto 6.
   3.2 if $card([x]_C) = 2$,
      3.2.1 if $card([x]_P) = 2$, then $P \in R(U - \{x\})$ by Theorem 3; goto 6.
      3.2.2 if $card([x]_P) > 2$, then $P \notin R(U - \{x\})$ by Theorem 3; goto 5.
4. for $j = 1$ to $card(P)$ do
   4.1 $T = P$;
   4.2 Let c_j be $j - th$ attribute in P;
   4.3 $T = T - \{c_j\}$;
   4.4 if $T = Core(T, U - \{x\})$, then $P = T$, goto 6.//by Lemmas 1 and 2
5. if $P = Core(P, U - \{x\})$, then goto 6; otherwise, the following steps are executed.
   5.1 $T = Core(C, U - \{x\})$;
   5.2 repeat the following 5.3 and 5.4 until $T = Core(T, U - \{x\})$;$P = T$.//Ref.[9]
   5.3 select the important condition attribute c_l from the remainder condition attribute $set(C - T)$).
   5.4 $T = T \bigcup \{c_l\}$;
6. return P.

Clearly, in most cases, a new attribute reduction can be directly acquired from the old attribute reduction by Algorithm 2. In the worse case, that is, the original attribute reduction is not a new attribute reduction, a new attribute reduction can be also derived from the Core quickly.To illustrate the operation of Algorithm 2, the following Example 1 is given.

Example 1. Let S be a decision information system (Table 1), it consists of four condition attributes $\{c_1, c_2, c_3, c_4\}$,one decision attribute d and eight objects,in which the set of all consistent objects $U1$ and the set of all inconsistent objects $U2$ are $\{x_3, x_4, x_5\}$ and $\{x_1, x_2, x_6, x_7, x_8\}$, respectively. According to Definition 5, the discernibility matrix M1 of S is as follows.

	$x3$	$x4$	$x5$	$x1$	$x2$	$x7$	$x8$	$x9$
$x3$	$\emptyset$	$\{c_1, c_3\}$	$\{c_1, c_4\}$	$\{c_2\}$	$\{c_2\}$	$\{c_2, c_3, c_4\}$	$\{c_2, c_3, c_4\}$	$\{c_2, c_3, c_4\}$
$x4$	$\{c_1, c_3\}$	$\emptyset$	$\emptyset$	$\{c_1, c_2, c_3\}$	$\{c_1, c_2, c_3\}$	$\{c_1, c_2, c_4\}$	$\{c_1, c_2, c_4\}$	$\{c_1, c_2, c_4\}$
$x5$	$\{c_1, c_4\}$	$\emptyset$	$\emptyset$	$\{c_1, c_2, c_4\}$	$\{c_1, c_2, c_4\}$	$\{c_1, c_2, c_3\}$	$\{c_1, c_2, c_3\}$	$\{c_1, c_2, c_3\}$

Table 1. A decision information system S

objects	*attributes*				
	c_1	c_2	c_3	c_4	d
$x1$	1	0	1	0	2
$x2$	1	0	1	0	1
$x3$	1	1	1	0	3
$x4$	0	1	0	0	2
$x5$	0	1	1	1	2
$x6$	1	0	0	1	1
$x7$	1	0	0	1	2
$x8$	1	0	0	1	3

By Theorem 1, the Core of S is $\{c_2\}$ and the set of all attribute reductions of S is $R(U) = \{\{c_1, c_2\}, \{c_2, c_3, c_4\}\}$.

The operation of Algorithm 2 can be explained by deleting different objects of S. For saving space, only the following three cases are presented.

Case 1.Assume the object $x3$ is deleted, by Algorithm 1,the new discernibility matrix $M1(x)$ deriving from M1 is as follows.

	$x4$	$x5$	$x1$	$x2$	$x6$	$x7$	$x8$
$x4$	$\emptyset$	$\emptyset$	$\{c_1, c_2, c_3\}$	$\{c_1, c_2, c_3\}$	$\{c_1, c_2, c_4\}$	$\{c_1, c_2, c_4\}$	$\{c_1, c_2, c_4\}$
$x5$	$\emptyset$	$\emptyset$	$\{c_1, c_2, c_4\}$	$\{c_1, c_2, c_4\}$	$\{c_1, c_2, c_3\}$	$\{c_1, c_2, c_3\}$	$\{c_1, c_2, c_3\}$

Suppose we choose the attribute reduction $P(P = \{c_1, c_2\})$, it is obvious that P is a candidate reduction by Algorithm 2 because of $x3 \in U1$ and $card([x3]_P) = 1$. Hence, a new attribute reduction $\{c_1\}$ can be computed by the step 4 of Algorithm 2.

Case 2. Suppose the object $x4$ is deleted, the new discernibility matrix $M1(x)$ deriving from $M1$ is as follows by Algorithm 1.

	$x3$	$x5$	$x1$	$x2$	$x6$	$x7$	$x8$
$x3$	$\emptyset$	$\{c_1, c_4\}$	$\{c_2\}$	$\{c_2\}$	$\{c_2, c_3, c_4\}$	$\{c_2, c_3, c_4\}$	$\{c_2, c_3, c_4\}$
$x5$	$\{c_1, c_4\}$	$\emptyset$	$\{c_1, c_2, c_4\}$	$\{c_1, c_2, c_4\}$	$\{c_1, c_2, c_3\}$	$\{c_1, c_2, c_3\}$	$\{c_1, c_2, c_3\}$

If the attribute reduction $P(P = \{c_1, c_2\})$ is selected, P is also a new attribute reduction by the step 2.1 of Algorithm 2 due to $x4 \in U1$ and $card([x4]_P) = 2$.

Case 3. Let the object $x7$ be deleted, the new discernibility matrix $M1(x)$ deriving from $M1$ is as follows by Algorithm 1.

	$x3$	$x4$	$x5$	$x1$	$x2$	$x8$	$x9$
$x3$	$\emptyset$	$\{c_1, c_3\}$	$\{c_1, c_4\}$	$\{c_2\}$	$\{c_2\}$	$\{c_2, c_3, c_4\}$	$\{c_2, c_3, c_4\}$
$x4$	$\{c_1, c_3\}$	$\emptyset$	$\emptyset$	$\{c_1, c_2, c_3\}$	$\{c_1, c_2, c_3\}$	$\{c_1, c_2, c_4\}$	$\{c_1, c_2, c_4\}$
$x5$	$\{c_1, c_4\}$	$\emptyset$	$\emptyset$	$\{c_1, c_2, c_4\}$	$\{c_1, c_2, c_4\}$	$\{c_1, c_2, c_3\}$	$\{c_1, c_2, c_3\}$

Clearly, the old attribute reduction $\{c_1, c_2\}$ is also a new attribute reduction by the step 3.1 of Algorithm 2.

Intuitively, in most cases, it is more likely that a new attribute reduction can be derived from the original attribute reduction quickly.

Some preliminary experimental results for comparing with the existing non-updating algorithms show that the algorithm of this paper is efficient and flexibility. For limitation of space, however, this paper only introduces the basic principle of updating the attribute reductions discovered, the experimental results and further experiments will be presented in future work.

6 Conclusions

Attribute reduction is a key problem for Rough Sets based knowledge acquisition. There are a lot of methods available for attribute reduction, but very little work has been done for attribute reduction for the case of deleting. Therefore, in this paper, we introduce a new updating method based on discernibility matrix for attribute reduction for the case of deleting, a new attribute reduction can be derived from the original attribute reduction quickly.

Now, we are comparing our algorithm with the existing non-updating algorithms using different data sets. At the same time, in order to further improve the performance of updating attribute reductions discovered ,we are searching for other storage methods for reducing the space complexity of discernibility matrix.

Acknowledgments. We would like to thank the reviewers for their valuable suggestions for improving the presentation of this paper.Meanwhile we also thank National Natural Science Foundation of P.R.China and of Jiangsu under Grant Nos.70371015 and BK2005135, and National Science Research Foundation of Jiangsu Province under grant No.05KJB520066 for partial supports, respectively.

References

1. Pawlak Z(1982) Rough sets. International Journal of Information and Computer Science 11:341-356
2. Swiniarski RW,Skowron A(2003) Rough set methods in feature selection and recognition. Pattern Recognition Letters 24:833-849
3. Li JY,Cercone N(2005) A Rough Set Based Model to Rank the Importance of Association Rules. Lecture Notes in Computer Science 3642:109-118
4. Wang GY(2001) Rough Set Theory and Knowledge Acquisition. Xi'an Jiaotong University Press,China
5. Yang M,Sun ZH(2004) Improvement of Discernibility Matrix and the Computation of a Core. Journal of Fudan University (Natural Science) 43:865-868
6. Yang M(2006) An Incremental Updating Algorithm of the Computation of a Core Based on the improved Discernibility Matrix. Chinese Journal of Computers 29:407-413

7. Skowron A,Rauszer C(1992) The Discernibility Matrices and Functions in Information Systems. Intelligent Decision SupportHandbook of Applications and Advances of Rough Sets Theory,Kluwer Academic Publishers 331-362
8. Hu XH, Cercone N(1995) Learning in relational databases: A rough set approach. Computational Intelligence: An International Journal 11:323-338
9. Ye DY(2002) An Improvement to Jelonek's Attribute Reduction Algorithm. Acta Electronica Sinica 18:81-82
10. Wang J,Wang Ju(2001) Reduction algorithm based on discernibility matrix the ordered attributes method. Journal of Computer Science and Technology 16:489-504
11. Fayyad UM,Shapiro LP,Smyth P,Uthurusamy R(1996) Advances in Knowledge Discovery and Data Mining. Menlo Park(ed.),AAAI/MIT Press,California
12. Yang M,Sun ZH, Song YQ(2004) Fast updating of globally frequent itemsets. Journal of Software 15:1189-1197
13. Yang M,Sun ZH(2003) An Incremental Updating Algorithm Based on Prefix General List for Association Rules. Chinese Journal of Computers26:1318-1325
14. Cheung D, LEE S, Kao B(1997) A general incremental technique for maintaining discovered association rules. Advanced Database Research and Development Series 6:185-194
15. Liu ZT(1999) An Inremental Arithmetic for the Smallest Reduction of Attributes. Acta Electronica Sinica 27:96-98
16. Ziarko W,Shan N(1995) Data-Based Acquisition and Incremental Modification of Classification Rules. Computational Intelligence 11:357-370
17. Susmaga R(1998) Experiments in Incremental Computation of Reducts. Rough Sets in Data Mining and Knowledge Discovery, Springer-Verlag,Berlin 530-553
18. Orlowska M, Orlowski M(1992) Maintenance of Knowledge in Dynamic Information Systems. Intelligent Decision Support:Handbook of Applications and Advances of the Rough Set Theory, Kluwer Academic Publishers 315-330
19. Hu F,Wang GY,Huang H,Wu Y(2005) Incremental attribute reduction based on elementary sets. Lecture Notes in computer science 3641:185-193
20. Jensen R,Shen Q(2004) Semantics-Preserving Dimensionality Reduction: Rough and Fuzzy-Rough-Based Approaches. IEEE Transactions on Knowledge and Data Engineering 16:1457-1471

The Research of the Exposition Economy Model

Yuhong Li[1] and Jiajun Lai[2]

[1] Business School of Jinggangshan University, Ji'an, Jiangxi, 343009, P.R. China
homer2005@163.com
[2] Intelligent Control Development Center,
Southwest Jiaotong University, Chengdu, 610031, P.R. China
eagles_laijiajun7309@126.com

Abstract. The development of Exposition Economy has greatly influenced over the global economy. Having done little research work of Exposition Economy, the writer tried to build an economical model through the theory of Exposition multiplier, the urban development potential and the pull on the national economy, as well as through the knowledge relevant to econometrics, hoping to explain the reaction over the national economy.

Keywords: exposition economy exposition multiplier development potential model Exposition.

1 Introduction

Exposition Economy, just as the name implies, is a kind of phenomenon or action, which can promote products and bring out direct or indirect economic effect via holding meeting and exhibitions. It's also called Exposition Industry or Exposition Market. Exposition Economy comes out of the Market Economy, it's been hundreds of years already overseas. From the end of last century, the high-tech revolution developed more swiftly, products updated faster, the communication of high-tech as well as the expansion of trade activities have provided the countries all over the world with chances and challenges. Meanwhile, economic globalization trend is getting more and more evident, many countries and enterprises are targeting at penetrating into others to expand bigger market. So, they are groping for markets, expanding exports, new forms and chances of exporting high-tech and management. Under this situation, the exposition, which converges sorts of product shows, trades and promotion of economic operation, is developing quickly.

Well, how does Exposition Economy impact the economy in a country or an area to some extent? How can we qualify and quantify it? Hereinafter, the following are to be researched.

2 Investment Multiplier of Exposition

Exposition Economy causes multiplier effect for its coherent related actions. However, how does the multiplier reaction reflect that exposition investment has the function to bring out added value and chaining promoting reaction on the economy of a country or

B.-Y. Cao (Ed.): Fuzzy Information and Engineering (ICFIE), ASC 40, pp. 894–901, 2007.
springerlink.com

an area through the primary allocation and the reallocation? To get the answer to it, *Investment Multiplier of Exposition* is cited to gauge.

As we know, Keynes thought that the income would get multiplied when investment is increased. If K is cited to express this multiplier, then we call it the investment multiplier. that is to say, K is the rate of ΔY the variation of the income and ΔI the variation of the investment expenditure that has brought out the former. That is to say,

$$K = {}^{\Delta Y}\!/_{\Delta I}$$

The course to deduce K follows the case: to increase $ 10 billion to purchase factors of production, it will flow into the factor owners by the forms of wages, incomes, profits and rents, of course, their income is added $ 10 billion. This amount of money is the first amount of added national income. The marginal propensity to consume is supposed to be 0.8, there would be $ 8 billion used to purchase consumer goods and this amount of money again flow into the owners who possess the factors used for new products, therefore, $ 8 billion is another amount of added national income, this is the second amount of added national income. Likewise, the consumers would spend $ 6.4 billion in purchasing consumer goods; as a result, the total social demand is obtaining another $ 6.4 billion. In this way, the national income would be added finally as the following:

$$100 + 100\times 0.8 + 100\times 0.8\times 0.8 + \ldots\ldots + 100^{\times}0.8^{n-1}$$

$$=100\times 1/(1-0.8)$$

$$=500$$

This case illustrates that the mentioned K is 1/(1-Marginal Propensity To Consume).

That is to say,

K= 1/ (1 - MPC) or K = 1/ MPS, MPS is marginal propensity to save, and MPS = 1 - MPC

Commonly, it's written in the form K = 1/ (1- β), β is *Marginal Propensity To Consume*.

We can see that K depends upon β, the bigger the β, the bigger the K will be. They both vary towards the same direction.

Now if △Y is cited to gauge the increment of national income, △I is used to gauge the increment of *exposition investment*, we can build a model for the multiplier of exposition investment

$$K = {}^{\Delta Y}\!/_{\Delta I} \text{ or } \Delta Y = K \times \Delta I$$

Obviously, the national income Y is the function of the exposition investment I, that is to say,

$$Y = K \times I$$

It's easy to see that as long as the exposition investment flows into the countries or areas where the expositions are hold, it will directly or indirectly impact the departments where means of livelihood and producer goods are produced, even finally the over-all social reaction gets increased through the chain reactions of the social economic activities. If part of the capital of some place is saved or used to import goods. That means the capital would leave for other areas or flow into foreign countries, as a result, the essential exposition investment in the place has to decline. And, the bigger the marginal propensity to save for other use or the marginal propensity to import, the smaller the multiplier reaction is.

Under this principle, *the model for exposition multiplier* would be modified to

$$K = 1/(1 - MPC) \text{ or } K = 1/MPS$$

(MPC is marginal propensity to consume exposition; MPS is marginal propensity to save)

3 The Value of Development Potential of a City's Exposition Economy

The development of a country or an area is related to its infrastructure besides the scale of investment and the scale of consumption. To be concrete, it's related to its development potential.

If experts can appraise all the items related to the development of Exposition Economy of all the countries or cities by Delphi Method, and get their weight coefficients respectively, [1] likewise, to calculate the primary value S_{i0} by Delphi Method, then give it a perfect score Q_i if it's ranked within the top five ones, thus, the primary value S_{i0} appraised by experts could be converted into a standard value S_i ($S_i = S_{i0}/Q_i$). Finally, we can compute *the value of development potential of a city's exposition economy* via the following formula,

$$D = \sum W_i \times S_i (i = 1, 2, ..., n)$$

W_i is the weight coefficient of related item; also appraised by experts using Delphi Method (shown as the following form)

1st stage indicator	Weight coefficient W_i	2nd stage indicator	Weight coefficient W_i
A1 economic level	1.8	B1 GNP	0.7
		B2 gross industrial products	0.7
		B3 fiscal revenue	0.4
A2 trade development level	1.2	B4 value of retail sales	0.7
		B5 trade market turn-over	0.5
A3 situation of the exposition	2.5	B6 total exposition square	1.0
		B7 concentration of phosphate	0.8
		B8 civilization	0.7
A4 tour reception	1.6	B9 tour income	0.9
		B10 tour reception men times	0.7
A5 traffic situation	1.2	--	--
A6 zone situation	0.7	--	--
A7 social undertakings	1.0	--	--

With this model, we can easily qualify and quantify the impact of the exposition economy on the economy of a country or an area; of course, the numerical value can be given exactly.

4 The Pull Model of Exposition Economy on National Economy

The development of Exposition Economy in a country, a city or an area, generally speaking, will pull on its national income to some extent. When a mount of variation of exposition investment is added, the national income will immediately gets multiplied.

Likewise, if the development potential of a country, a city or an area, is fine enough, the exposition economy will multiply the potential pull reaction on the national economy as the investment multiplier does, the result is the national income variation multiplies the development potential, that is to say, the pull value of the exposition economy on national economy consists of two, one is the vanue of the investment multiplier reaction, the other is the one of the development potential,.

Synthesizing the above researches, if **G** is cited to gauge the vanue of the final impact of exposition economy on national economy, that is to say**,** **G** is cited to gauge the pull value of exposition economy on national economy**,** **Y** is cited to gauge the value of the national income variation caused by investment multiplier reaction and **D** is cited to gauge the Value of Development Potential of a City's Exposition Economy, then the potential national income variation will be **D×Y**, therefore, we can construct a model for the impact of exposition economy on national economy correspondingly as the following:

$$G = Y + D \times Y$$

$$= K \times I + D \times K \times I$$

$$= (I + D) \times K \times I$$

$$= (1 + \sum W_i \times S_i) \times I / (1 - MPC)$$

From this model, the pull value **G** of exposition economy on national economy is the function of the factors, exposition investment multiplier k, or the marginal propensity to consume exposition MPC and the exposition development potential D, any of the factors can cause the value G to vary. **（D+1）/（1—MPC）** is called the pull coefficient. Obviously, when the factors all become perfect, the pull reaction will be ideal.

This model is the final target to be researched by the author. With this model now we can find out the answers to the questions listed in the front introduction. This model will not only present the qualitative relationship but also the quantitative relationship among the pull value of exposition economy on national economy **G**, the value of the national income variation **Y** caused by the exposition investment and the value of development potential of a city's exposition economy **D**. If a country or a region is defined, the marginal propensity to consume exposition MPC of it is correspondingly definite, so the pull value **G** of exposition economy on its national economy will only depend on the amount of its exposition investment **I**, but if the conditions of the country or the region get better, that is to say, the development potential get better, as a result, the pull value **G** will get higher correspondingly. The great function of this model will be showed clearly by the following analyses.

5 The Theoretical Analysis of the Model

That's great to construct the model, and now we can analysis how exposition economy impact our national economy. It will be presented clearly as the following.

One thing, from this model, we can see that the pull coefficient (**D+1**) × **K** can be computed easily as soon as the development potential D and the multiplier K are got, and simultaneously, the pull value G of exposition economy on national economy. Thus, D and K confine G.

From the formula **K = 1 / (1 - MPC)**, k is decided by MPC, the bigger the MPC, the less the economic loss will be, the smaller the MPC, the less of the income used for consumption will be, as a result, there will be more economic loss. Then, K gets smaller.

For instance, in our country, if MPC of exposition is 0.8, that means 80% of the national income will be used for exposition consumption, only 20% of that will be saved for other uses. From the formula **K = 1 / (1 - MPC)**, K is decided to be 5, that is to say, *only the exposition economy activity* will make the national income 5 times the original.

We can have the conclusion, when a mount of exposition investment increased in a country, a city or an area, correspondingly, as the multiplier reaction, that area will get its national economy multiplied as the K, the total national income growth will be multiplied finally. Meanwhile, the impact of exposition economy on national economy is reflected. In fact, similar impact from other multipliers like business income multiplier, resident income multiplier, government income multiplier, employment multiplier and consumption multiplier etc. exists.

The other thing, according to the model **G =** (**1 + D**) ×**K**×**I**, further conclusion can be got that if D is big enough, then the multiplier reaction will be greater. So, when an area has good potential to develop exposition economy, and we increase exposition investment, the exposition economy will boom, in return, it will greatly pull on the national economy of the area.

So, we must do best to create good situation for the development of exposition economy to promote the consumers to come for exposition consumption. Then improving the service level is necessary, including repast, hotels, traffic, physical distribution, communications, translation, reasonable rent, perfect exposition management, and etc. to attract consumers, to enforce the international competitive strength of the country, finally to increase the exposition consumption.

The pull reaction of exposition economy results in its chain effect, while meetings or expositions are held in the exhibition hall or center, the attendants will greatly benefit the local hotels, traffic, repast, recreation facilities, and etc. Besides, in order to hold each exposition, there must be huge investment and consumption as well. Thus, every department will bring their relevant units huge investment and consumption, that is to say, there must be multiplier reaction respectively. Therefore, it's said world widely that the rate of the chain effect is 1:9,that is to say, when one yuan is invested, the final social economy will benefit nine yuan for its multiplier reaction, then the income derives from exposition economy will flow into the departments like, tour, exhibition company, retails, finance and etc.

Well, the departments, which were impacted by exposition economy, don't only include the above ones mentioned, but also other relevant ones like environment protection, advertisements for the products in a city and etc. furthermore, the industry configuration of a city or a place will be optimized. For the reason, many cities develop their exposition economy as mainstay industry pulling on other relevant industries finally. It has been the new economy growth point in the cities like Hanover, Munich, Chicago, Paris, London, Milan and Chinese Hong Kong as examples.

As a kind of service industry, exposition can absorb lots of employees, likewise, due to directly or in directly relevant industries, the multiplier reaction of course effects exceedingly. In Europe, it's said that every 20-exposition attendants can make an employee opportunity. According to statistics, the exhibition industry created 1570 million employee opportunities for America in 1994 and 100 million for Hanover as well.

From the above analyses, we can easily realize the meaning of developing exposition. As expositions are mainly held in cities, subsequently, many cities regard developing exposition economy as one of their strategies to develop city economy. With the chain reaction, exposition economy promotes relevant industries booming as well as the city economy.

With the pull theory, the writer has learned the exposition, which will be held in Shanghai in 2010. In order to hold the exposition, the planned exposition investment will be $3.3 billion equivalent to RMB 25 billion, if not considering the exchange rate and the inflation, by then, each ticket sells 170 Yuan, and 43 million tickets are sold out, then the ticket income will be 7.31 billion. Suppose 60% of the attendants dine once spending 30 yuan, there will be 0.9 billion repast income, 90% of them spend average 10 yuan on beverage, the income will be 0.45 billion, and 30%of the audience spend average 30 yuan on tokens, the income will be 0.45 billion, in this matter, Shanghai exposition can directly gain 9.11 billion of proceeds of sale. It's predicted by officers that the future exposition will bring Shanghai 15 billion yuan for relevant industries. [2] Meanwhile, pulling the over-all social economy growth by 0.6% higher than that 0.3% of growth drives from the Olympic game, which will be held in 2008. Statistics show the exposition investment will bring 12.9 billion yuan for Shanghai by then. That is to say, the future exposition will bring shanghai total 37 billion. From the model **G =（1 + D）×K×I** we can get the pull coefficient **（1 + D）×K** is 370/33 = 11.2 under this situation, it's just 11.2 times the original investment. It's bigger than the coefficient 10:1 of the best-developed exposition economy in Germany. [3]In short, $1 of investment in shanghai will finally bring relevant industries $11.2 of virtual outcome correspondingly.

6 Conclusion

The author has originally tried to construct a model for Exposition Economy by researching its relevant factors. This model can easily explain how Exposition Economy impacts the national economy in a country, a city or an area to some extent, and the quantity, as well as the quality, of the impact of Exposition Economy on national economy. It tells us that the development of modern Exposition Economy depends on the consideration of the governments, including the investment and the management, especially the management is more important. The investment brings great multiplier reaction; the management makes the potential development of

Exposition Economy possible. When such conception exists and is penetrated into all exposition organizations, the Exposition Economy will have a long term, stable and healthy development, and it will finally bring positive impact on national economy of a country, a city or an area.

References

[1] MaYong. *Theory method, cases of exposition management. P.103.* Advanced Education press, 2003
[2] Zhou Bing. *Exposition Generality.P.97,* Li Xin Account Press, *2004*
[3] Huang Hai. *The lecture at "2003 China Exposition Economy Forum", December 21, 2003*
[4] Wang Shubin and Guo Min. *Researching and Thinking of the City Exposition Economy Issues.* Jiangsu *Business Discussion, No.12, 2002*
[5] Chen Xiangjun. Developing Exposition Economy Cultivating New Economy Growth Point, Jiangsu Business Discussion, No.7, 2000
[6] Zhen Jianyu. Current Situation and Development Trend Of Exposition Economy in Shanghai, Tourism Tribune Bimonthly, No.11, 2000
[7] Hu Xiao. Exposition Economy and the Urban Development, Researching of Economic Issues, No.6, 2002

A Region-Based Image Segmentation Method with Kernel FCM

Ai-min Yang[1,2], Yong-mei Zhou[1], Xing-guang Li[1], and Min Tang[2]

[1] School of Informatics, Guangdong University of Foreign Studies, Guangzhou, Guangdong, 510420, China
{amyang18,yongmeizhou}@163.com
[2] School of Computation and Communication, Hunan University of Technology, Wenhua Road, Zhuzhou, Hunan, 412008, China

Abstract. A region-based image segmentation method with kernel fuzzy c-means clustering (FCM) is proposed. This method firstly extracts color, texture, and location features for each pixel by selecting suitable color space. Then, according to the feature vectors, pixeles are clustering by kernel FCM.The number and the center of clusters are decided by the proposed selecting the optimal cluster number method.Each pixel is grouped and labeled according to its membership degree. Finally, according to the pixels' neighbor connection theory, the regions with the same label are segmented again.

Keywords: Image Segmentation, Kernel Fuzzy C-means Clustering, Feature Extraction, Region Description.

1 Introduction

The task of image segmentation is dividing an image into different regions which are non-intersect each others.image segmentation is an important research field for current image processing and pattern recognition[1]. Since fuzzy set theory can describe and process uncertain problems well,it is wildly applied for image segmentation[1-2].Kernel function is used to map patterns in the initial space to high dimension feature space by an implicit non-linear function[3].This paper introduces a region-based color image segmentation method with kernel fuzzy c-means clustering(FCM).This method firstly extracts color, texture, and location features for each pixel to form integrated feature vectors by selecting suitable color space.Then, the integrated feature vectors are clustering with kernel FCM.The number and the center of clusters are decided by the proposed method of selecting clustering number.Each pixel is grouped and labeled according to its membership degree.Finally,the regions with the same label are segmented again according to the pixels' neighbor connection theory and features which describe regions are presented.

In section 2,the process of color image segmentation is introduced. Section 3 is about extraction of integrated features based on pixels.Section 4 is region segmentation with kernel FCM. In section 5, the experiment result is introduced.Finally, conclusion and future work are discussed.

B.-Y. Cao (Ed.): Fuzzy Information and Engineering (ICFIE), ASC 40, pp. 902–910, 2007.
springerlink.com

2 Region Segmentation Process

The Region Segmentation process is as Fig.1.The first step is extracting the integrated features which include three color features (L, a, b), three texture features (contrast, anisotropy, polarity) and two location features for each pixel.Color and location features can be got easily and the main task is how to extract texture features. The second step is to cluster pixels with kernel FCM and determine the optimal number of clusters and center for each cluster by the proposed method of selecting optimal cluster number. For the third step, each pixel is grouped and labeled according to its membership degree. Then the regions with the same label are segmented again according to the neighbor connection theory for pixels.The last step is region feature description, which is extracted features for each region.

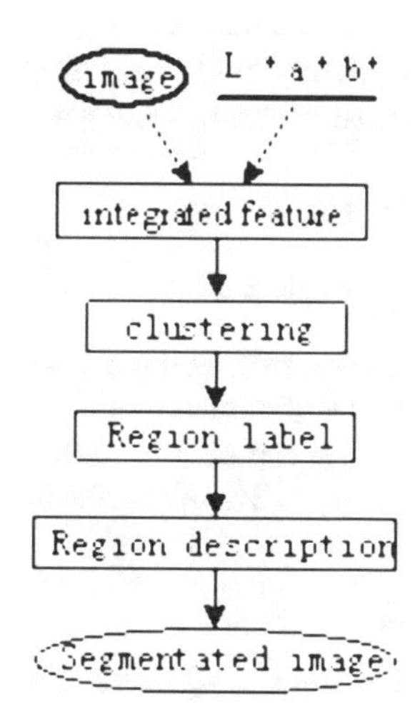

Fig. 1. Flow Chart of Image Segmentation

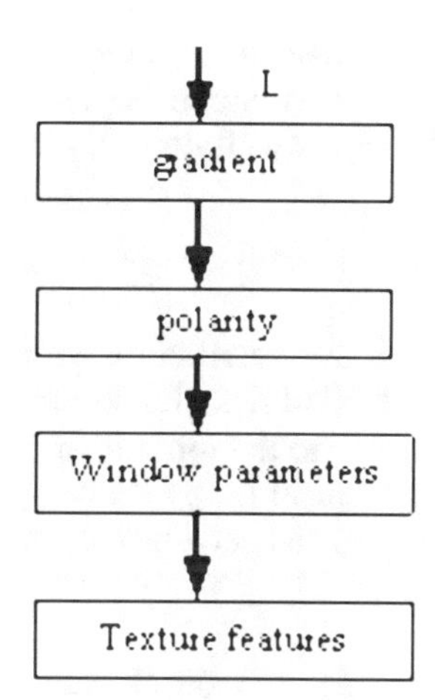

Fig. 2. Process of Texture extraction

3 Integrated Feature Extraction

Texture is a local neighborhood property.The selection of scale directly impacts the precision of texture representation. In this paper, texture features are contrast, polarity, anisotropy[4]. According to the component L in L*a*b color space and the analysis of polarity feature using different scale parameters of Gauss function, we can get the optimal scale parameters, then, calculate contrast and anisotropy. The process is as Fig.2.

3.1 Extracting Texture Features

In this paper, polarity is the proportion of gradient of all the pixels with same direction in neighborhood. Its formal definition is as Eq. (1).

$$p_\sigma = \frac{|C_+ - C_-|}{C_+ + C_-} \tag{1}$$

Where C_+, C_- follows as Eq.(2)

$$C_+ = \sum_{x,y\in\Omega} w_\sigma(x,y)\left[\nabla G\cdot\vec{n}\right]_+, C_- = \sum_{x,y\in\Omega} w_\sigma(x,y)\left[\nabla G\cdot\vec{n}\right]_-, \nabla G = \begin{bmatrix} G_x \\ G_y \end{bmatrix} \tag{2}$$

Where, G_x, G_y is gradient of L in x (row)and y (column) axis respectively in L*a*b color space. $\vec{n}$ is a orthogonal unit vector with ϕ which is the most coincident direction of gradient (Gx, Gy) ranging from 0 to 180°. ϕ is discrete value 22.5°*i (i=0, 1,…, 7). $[.]_+$, $[.]_-$ represent the product of vectors for non-negative and negative respectively. Ω is gradient vectors (gradient pool) of Gauss window.$w_\sigma(x, y)$ is Gauss value when width of gradient vectors isσin gradient pool. C_+, C- are the number of gradient vectors in direction of positive and negative respectively.$p_\sigma\in[0,1]$ changes with σ.According to definition 1, p_σ is related with Gauss parameter σ. Gauss window parameter is defined as follows Eq.(3).

$$\text{WindowSize} = [k^2+1]\times[k^2+1], k=0,1,\ldots,K$$
$$K = \left\lfloor \sqrt{\min(m,n)/2-1} \right\rfloor, \quad and \quad K \le 10 \tag{3}$$

Where, $\lfloor\cdot\rfloor$ is an operation for getting the minimal integer, m*n is the number of pixels. $\sigma=\sigma_k=k/2(k=0,1,\ldots,K)$.So we can get the window parameter of Gauss function.According to the formula of p_σ and $\sigma=2*\sigma_k$, $p'_{\sigma k}(x,y)$ can be calculated. $p'_{\sigma k}(x,y)$ is polarity of pixel (x, y) with σ_k parameter. For k=0, 1,…, K, select the first k that satisfies Eq.(4) and get window parameter $\sigma(\sigma=k/2)$.

$$p'_{\sigma(k+1)}\text{-}p'_{\sigma k}\le 2\% \tag{4}$$

If p_σ is close to 1 for all values of σ, the region is edge. If p_σ decreases with the increase of σ, the region is texture region. Because the increase of σ makes the window of Gauss function larger and the directions of pixel gradient vectors more various in the window. If p_σ has a constant value for all σ, the region is uniform region, and the brightness of his region is constant. From above calculation, for each pixel, window parameter σ_k is determined and can be used to calculate polarity. It can also determine contrast (con) and anisotropy (ani) using the second moment matrix(SMM)[5] with σ'. The SMM of a pixel is as Eq.(5)

$$SMM_{\sigma'}(x,y) = w_{\sigma'}(x,y) * (\nabla G)(\nabla G)^T \tag{5}$$

Where, $SMM_{\sigma'}(x, y)$ is the second moment matrix of pixel (x, y) and a symmetrical half-positive matrix. Other symbols have the same meaning in Eq.(2).If the characteristic values of $SMM_{\sigma'}(x, y)$ are λ_1 and $\lambda_2(\lambda_1\geqslant\lambda_2)$,Contrast and anisotropy are calculated according to Eq.(6).

$$con = 2\sqrt{\lambda_1+\lambda_2}, \quad ani = 1-\lambda_2/\lambda_1 \tag{6}$$

3.2 Extracting Other Features

If the number of pixels for an image is m*n, location features X, Y of pixel (x, y) is as follows.

$$X=w_{xy}*(x/\max(m,n)),Y= w_{xy}*(y/\max(m,n))$$

Where,max(.) is the max operation, w_{xy} can be thought as the weights of location features ($w_{xy} \in (0,1]$).Here, w_{xy}=1.Color features(labL, labA, labB) are as Eq.(7).

$$labL = L/100, labA = (a+50)/100, labB = (b+50)/100 \quad (7)$$

Where, L, a, b is primitive component of an image for L*a*b color space.So, the feature vector of pixel (I) is as Eq.(8).

$$I = (labL, labA, labB, p_\sigma, con, ani, X, Y) \quad (8)$$

4 Region-Based Color Image Segmentation

In order to segment an image based on region, we use kernel fuzzy c-means clustering method to cluster with image pixel features. Let an image data set be presented as Eq.(9).

$$\{I_k | I \in R^d\}_{k=1}^{N}, \ 2 \le c \le Cn = \lfloor \sqrt{N}/3 \rfloor \text{ and } c \leqslant 8 \quad (9)$$

Where, d is dimension of I(here,d = 8),N is the number of pixels.Let c be the number of clusters. symbol $\lfloor \cdot \rfloor$ means to get the minimal integer.

Firstly, we introduce the standard Fuzzy c-means (FCM) method.

(1) Target function of standard FCM algorithm is as Eq.(10).

$$J_m = \sum_{i=1}^{c}\sum_{k=1}^{N} u_{ik}^m \|I_k - CC_i\|^2 \quad (10)$$

Where, $\{CC_i\}_{i=1}^{c}$ is the center of each cluster, μ_{ik} is membership degree(it expresses the degree which pixel k belongs to cluster i. The constraint condition (Eq.(11)) is satisfied.

$$\left\{ u_{ik}^m \in [0,1] \middle| \sum_{i=1}^{c} u_{ik}^m = 1, \forall k \quad and \quad 0 < \sum_{k=1}^{N} u_{ik}^m < N, \forall i \right\} \quad (11)$$

Where, parameter m is the weight exponent.

(2) Kernel Fuzzy c-means clustering (KFCM) method.

The target function of KFCM algorithm is as Eq.(12).

$$J_m = \sum_{i=1}^{c}\sum_{k=1}^{N} u_{ik}^m \|\Phi(I_k) - \Phi(CC_i)\|^2 \quad (12)$$

$$\begin{aligned}\|\Phi(I_k)-\Phi(CC_i)\|^2 &= (\Phi(I_k)-\Phi(CC_i))^T(\Phi(I_k)-\Phi(CC_i)) \\ &= \Phi(I_k)^T\Phi(I_k)-\Phi(CC_i)^T\Phi(I_k)- \\ &\quad \Phi(I_k)^T\Phi(CC_i)+\Phi(I_i)^T\Phi(CC_i) \\ &= K(I_k,I_k)+K(CC_i,CC_i)-2K(I_k,CC_i)\end{aligned}$$

Where, Φ is an implicit non-linear mapping function, here,Gauss RBF function is selected as kernel function.For Gauss RBF kernel function, $K(x,x)=1$, Eq.(12) can be written as Eq.(13).

$$J_m = 2\sum_{i=1}^{c}\sum_{k=1}^{N} u_{ik}^m (1-K(I_k,CC_i)) \tag{13}$$

So, optimal problem above is transformed to seek the minimal value for Eq. (14).

$$\min_{\mu_{ik},\{CC_i\}_{i=1}^c} J_m \tag{14}$$

We can get the update representation of membership degree and center of cluster with Lagrange and the constraint condition of Eq. (15).

$$\begin{aligned} u_{ik}^m &= \frac{(1-K(I_k,CC_i))^{-1/(m-1)}}{\sum_{j=1}^{c}(1-K(I_k,CC_j))^{-1/(m-1)}} \\ CC_i &= \frac{\sum_{k=1}^{n} u_{ik}^m K(I_k,CC_i) I_k}{\sum_{k=1}^{n} u_{ik}^m K(I_k,CC_i)} \end{aligned} \tag{15}$$

The KFCM algorithm is as follows.
Input: the number of clusters c, max iteration times t_{max}, and m>1, $\varepsilon>0$, δ is the parameter of kernel function.
Output:CC_i and μ_{ik},i=1,2,…,c,k=1,2,…,N.
Step 1: initialization of clustering center CC^0_i, calculation initial u_{ik}^0.
Step 2: for t=1, 2,…,t_{max}, loop calculation as the following order.
(a)update all clustering centers and all membership degree by Eq.(16).
(b)calculate $E^t = \max_{i,k}\left|u_{ik}^t - u_{ik}^{t-1}\right|$. If $E^t \leqslant \varepsilon$, stop the loop, else go to (a).
Step 3: output CC_i and μ_{ik}.

4.1 The Number of the Optimal Clusters

KFCM can determine the center of each cluster under knowing the number of cluster. How to determine parameter c is a key problem for region-based image segmentation.Here,we adopt the following process to decide c.First, make sure all the possible values of c according to Eq.(9).Then, calculate all the center and membership degree respectively by KFCM algorithm. Finally, determine c by our proposed method which selects optimal c using average scatter degree(S) within clusters and average divided degree(D) among clusters.The definition of S and D is as Eq.(16).

$$S=\frac{1}{c}\sum_{i=1}^{c}\frac{1}{N_i}\sum_{k=1}^{N_i}\left\|\Phi(I_k)-\Phi(CC_i)\right\|, D=\frac{1}{c}\sum_{i=1}^{c}\frac{1}{c-1}\sum_{k',k'\neq k}^{c}\left\|\Phi(CC_k)-\Phi(CC_{k'})\right\| \quad (16)$$

Where, N_i is the number of pixels in region i. S represents different degree among pixels,the smaller S, the more similar features of pixels. D represents distance among cluster centers, the larger D, the larger different clusters[6]. So, clustering with small S and large D has optimal c.

$$c=\arg\min_{c\in\{2,3,\ldots,Cn\}} r_c=\arg\min_{c} {S_c}/{D_c} \quad (17)$$

4.2 Label and Segment Image

The labeling and segmenting image region contains three processes.

Process 1: after determination of c and membership degree of each cluster for pixel, initial region label follows Eq.(18). Fig.3(b) (Number is serial number of region) is a result of labeling an image.

$$I_k \in \arg\max_{i\in\{1,2,..,c\}}\{\mu_{ik}\} \quad (18)$$

Process 2: region label in process 1 is based on the feature space of pixels, and the pixels with the same label (the same clustering) may not be neighbor in 2-D space, such as label 2 in Fig.3(b). So, these regions are re-labeled in this process, as Fig.3(c). In addition, morphologic methods (dilation and erosion) are used to process the edge of regions.

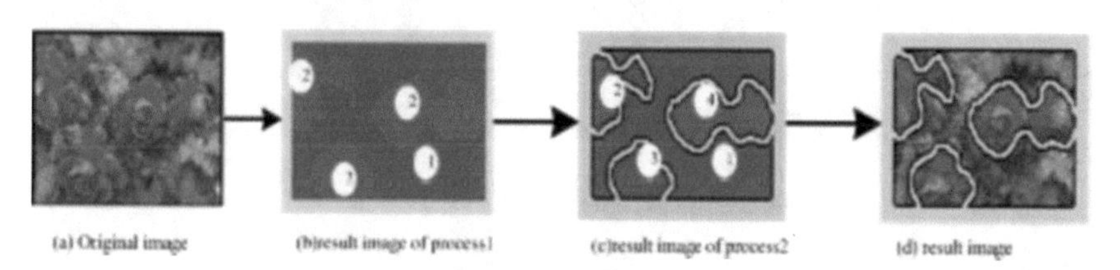

Fig. 3. An example for label and segmentation image

Process 3: the main task in this process is to describe these segmented regions to save and apply the results. The following features are extracted.

(1) Region texture features.They mainly include average contrast ($\overline{con}$) and average anisotropy ($\overline{ani}$) (as Eq.(19))for every region.

$$\overline{con}=\frac{1}{n_\Omega}\sum_{i=1,I_i\in\Omega}^{n_\Omega} I_i(con), \quad \overline{ani}=\frac{1}{n_\Omega}\sum_{i=1,I_i\in\Omega}^{n_\Omega} I_i(ani) \quad (19)$$

Where, n_Ω is the number of pixels in region Ω, $I_i(con)$, $I_i(ani)$ is contrast and anisotropy of pixel i.In some applications, texture features are regarded as a variable. According to Eq.(20),texbin feature which is called text bins[7] is getted.

$$texbin = (cbin - 1) * m_0 + abin\text{ , } cbin = \left\lceil \tfrac{con}{k_1} * m_1 \right\rceil (if\,0, cbin = 1)$$

$$abin = \left\lceil \tfrac{ani}{k_2} * m_2 \right\rceil (\text{ if } 0, \quad abin = 1) \tag{20}$$

Where, $\lceil \cdot \rceil$ is an operation for getting max integer, m_0=4, m_1=5, m_2=4, k_1=0.6, k_2=0.8.

(2) Region location features(as Eq.(21)). They is the position of average row ($\overline{X}$) and column ($\overline{Y}$) direction of region pixels.

$$\overline{X} = \frac{1}{n_\Omega} \sum_{i,I_i}^{n_\Omega} I_i(X), \overline{Y} = \frac{1}{n_\Omega} \sum_{i,I_i}^{n_\Omega} I_i(Y) \tag{21}$$

The following is a method that quantizes two features into one called location bins (locbin,as Eq.(22)). Where, m_0=m_1=m_2=3.

$$locbin = (ybin - 1) * m_0 + xbin$$
$$xbin = \left\lceil \overline{X} * m_1 \right\rceil \quad (if\,0, \quad xbin = 1) \tag{22}$$
$$ybin = \left\lceil \overline{Y} * m_2 \right\rceil \quad (if\,0, \quad ybin = 1)$$

(3) Region shape features.We adopt the method of ellipse Fourier to describe the features of segmented regions. First, we get the pixels on the region edge, and get the boundary contour and boundary function after smoothing. Then, the contour is described by several ellipses and processed with Fourier transformation.Fourier coefficient is regarded as region shape features. Fig.4 is an example for Fig.3(c).

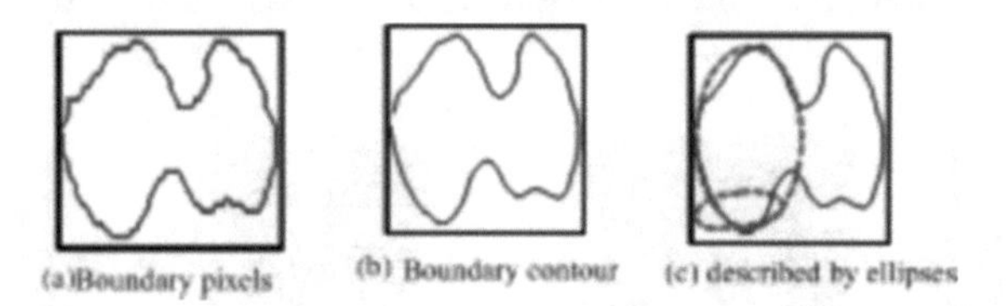

Fig. 4. Description of region shape features

Each ellipse is described by four Fourier coefficients. The region in Fig.4 is described by 8 ellipses and the coefficients are as Table 1.

Table 1. Region shape feature (Fourier coefficients)

	$Coeff_1$	$Coeff_2$	$Coeff_3$	$Coeff_4$
1	28.0776	0.4864	1.0636	-12.839
2	-0.5594	0.9960	0.0047	0.8294
3	2.8840	0.9845	0.5388	-5.1356
4	0.0295	0.4106	0.1147	-1.3195
5	0.5643	-0.0027	-1.1102	1.0730
6	-0.0247	-0.0611	0.0792	0.2543
7	-0.0640	-0.0146	0.7005	-0.4875
8	-0.3510	-0.0622	0.4951	0.1981

(4) Region color features(as Eq.(23)). Color features in each segmented region are represented by the average value of each component of pixels in L*a*b color space.

$$\overline{labL} = \frac{1}{n_\Omega} \sum_{i=1, I_i \in \Omega}^{n_\Omega} I_i(\text{labL}), \quad \overline{labA} = \frac{1}{n_\Omega} \sum_{i=1, I_i \in \Omega}^{n_\Omega} I_i(\text{labA})$$
$$\overline{labB} = \frac{1}{n_\Omega} \sum_{i=1, I_i \in \Omega}^{n_\Omega} I_i(\text{labB}) \quad (23)$$

In addition, color histogram (c_h) is provided for each region(quantization scheme is 6*6*6 bins).

(5) Global features. We have following methods to generate global features, i.e. global color histogram (g_c_hist), global contrast histogram (g_con_hist), global anisotropy histogram (g_ani_hist). The quantization scheme of g_c_hist is 6*6*6 bins, and that of g_con_hist and g_ani_hist is 20 bins.

5 Experiment Results

The proposed image segmentation method has been implemented with Matlab 7.0 for windows XP system. We adopt Gauss RBF function as the kernel function. Fig.5 is some samples of segmentation and parameters. Where, C is the number of clusters, S_n is the number of segmented regions, δ is the parameter of kernel function. Table 2 and Fig.6 describe the region and global features of a dog in Fig. 5(a).

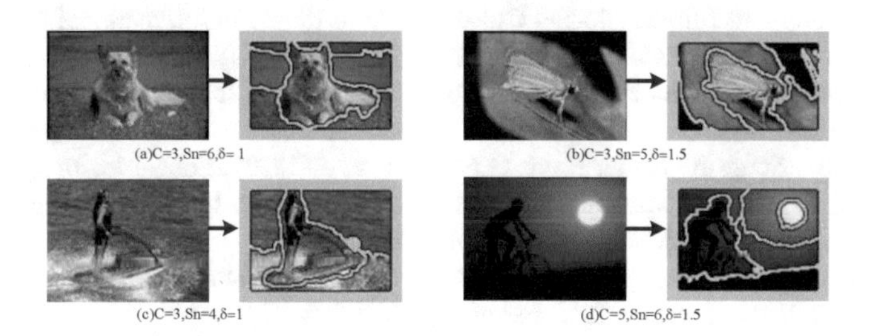

Fig. 5. Image segmentation and parameters

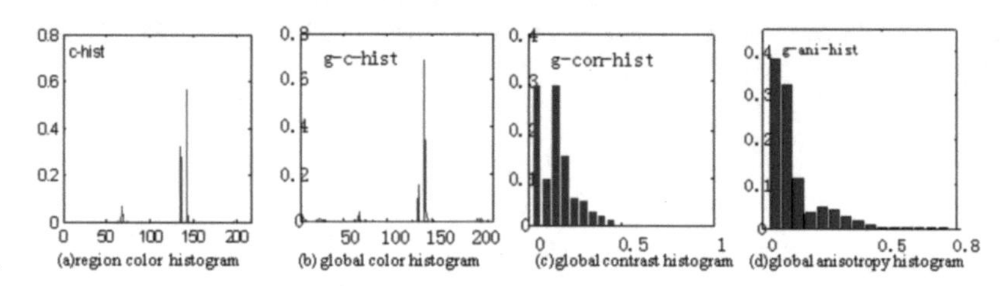

Fig. 6. Region and global histograms

Table 2. Region features of a dog in Fig 5(a)

$\overline{con}$	$\overline{ani}$	tex bin	$\overline{X}$	$\overline{Y}$	loc bin	$\overline{labL}$	$\overline{labA}$	$\overline{labB}$
0.088	**0.032**	**1**	**61.87**	**20.71**	**372**	**5.11**	**-16.16**	**21.82**

Note: region shape Fourier coefficients is similar to Table 1.

6 Conclusion and Future Work

Color image segmentation is the basic work in image understanding, contented-based image retrieval.In this paper,we propose a region-bassed color image segmentation method with kernel fuzzy c-means clustering algorithm. Plenty of features which describe regions are provided for other applications. Experiment shows this method can be applied widely.

In the future, we will do further research about how to select kernel function and its parameters, and apply this method to content-based image and video query systems.

References

1. LIN Kai-yan,WU Jun-hui,XU Li-hong(2005)A Survey on Color Image Segmentation Techniques.Journal of Image and Graphics,10:1-10
2. M.Ameer Ali, Laurence S. Dooley, Gour C. Karmakar(2005)Fuzzy Image Segmentation Combing Ring and Elliptic Shaped Clustering Algorithms. In:International Conference on Information Technology:Coding and Computing (ITCC'05),USA,pp,118-122
3. Girolami M(2002) Mercer kernel-based clustering in feature space. IEEE Trans Neural Networks,13:780-784
4. Chad Carson, Serge Belongie, Hayit Greenspan et al(2002) Blobworld: Image Segmentation Using Expectation-Maximization and Its Application to Image Querying.IEEE Trans. on Pattern Analysis and Machine Intelligence,24: 1026-1038
5. Wolfgang Forstner(1994) A Framework for Low-Level Feature Extraction. In: Computer Vision - ECCV '94,, Sweden: J. O. Eklundh,pp,383-394
6. Dae-Won Kim,Kwang H. Lee,Doheon Lee(2004) A novel initialization scheme for the fuzzy c-means algorithm for color clustering.Pattern Recognition Letters, 25:227-237
7. J.R. Smith,S.-F(1995) Chang. Single color extraction and image query. In:Proceedings of the 2nd IEEE International Conference on Image Processing, IEEE Signal Processing Society, B. Liu, editor,pp,528-531

An Efficient Threshold Multi-group-Secret Sharing Scheme*

Hui-Xian Li[1], Liao-Jun Pang[2], and Wan-Dong Cai[1]

[1] Dept of Comp Sci. and Eng., Northwestern Polytechnical Univ.
Xi'an, China 710072
{lihuixian,caiwd}@nwpu.edu.cn
[2] The Ministry of Edu. Key Lab. of Computer Networks and Information Security Xidian Univ., Xi'an, China 710071
ljpang@mail.xidian.edu.cn

Abstract. In this paper, a novel threshold multi-group-secret sharing scheme is proposed based on Chan *et al.*'s scheme. Multiple groups of secrets are packed into a group of large secrets by using the Chinese Remain Theorem, and then shared by constructing a secret polynomial such that its coefficients are those large secrets. In the proposed scheme, the secret distribution procedure is needed only once to share multiple groups of secrets, which reduces the amount of computation largely. Moreover, each group of secrets has a different threshold access structure and includes a distinct number of secrets. Analysis results show that the proposed scheme needs fewer public values and is higher in efficiency and easier in implementation than existing schemes, especially for sharing many groups of secrets, which makes it more practical in practice.

Keywords: Secret sharing, Threshold, Multi-group-secret sharing, Chinese Remain Theorem.

1 Introduction

A secret sharing scheme is a technique to share a secret among a group of participants. It plays an important role in protecting important information from being lost, destroyed, modified, or into wrong hands. Secret sharing has been an important branch of modern cryptography and secret sharing systems have been widely used in network security [1]. The first secret sharing scheme, i.e. the (t, n)-threshold scheme, was introduced by Shamir [2] and Blakley [3] independently in 1979, respectively. In 1983, Asmuth and Bloom [4] proposed a congruence class scheme based on Chinese Remain Theorem. Shamir's scheme is based on the theory of interpolation polynomials over the number field, and it is also a perfect scheme.

A multiple secret sharing scheme is a natural extension of a secret sharing scheme [5]. In a multiple secret sharing scheme, $p(p \geq 1)$ secrets are to be shared among participants. A threshold multiple secret sharing scheme is a special multiple secret sharing scheme, such as Harn's scheme [6]. In 2000, Chien *et al.* [7] proposed a new type of threshold multi-secret sharing scheme based on the systematic block codes, in

* This work is supported by the National Science Foundation for Post-doctoral Scientists of China (Grant No. 20060401008).

B.-Y. Cao (Ed.): Fuzzy Information and Engineering (ICFIE), ASC 40, pp. 911–918, 2007.
springerlink.com

which $p(p \geq 1)$ secrets instead of only one secret can be shared in each sharing session. Their scheme can find important applications especially in sharing a large secret [8]. Later in 2005, Li [9], Pang and Chan [10] proposed an alternative implementation of Chien *et al.*'s scheme, respectively. However, only one group of secrets is shared in a sharing session of these multi-secret sharing schemes. If m ($m>1$) groups of secrets are shared, m sharing sessions are needed, which needs a large amount of computation. In this paper, we proposed a threshold multi-group-secret sharing scheme based on Chan *et al.*'s scheme [10]. Analyses show that our scheme is higher in efficiency and more practical than existing multi-secret sharing schemes.

The rest of this paper is organized as follows. In Section 2, we shall describe the proposed scheme. In Section 3, we give a numerical example of our scheme. In Section 4, we shall make an analysis and discussion on our scheme. Finally, we shall present conclusions in Section 5.

2 The Proposed Scheme

The proposed scheme is mainly based on the Chinese Remain Theorem [12], and Lagrange interpolation operation [2], and it is an extension of Chan *et al.*'s scheme. We will describe the proposed scheme in three parts: system parameters, secret distribution, and secret reconstruction.

2.1 System Parameters

Suppose that there are n participants $P = \{P_1, P_2, \ldots, P_n\}$ in the system and $D(D \notin P)$ denotes a dealer. In the whole paper, we assume that the dealer is a trusted third party. The dealer will generate a shadow for each participant, and it is called the master shadow in our scheme. At the proper time, participants in an authorized subset, who want to compute some group of secrets, only poll their sub-shadows, which are computed according to their master shadows, but not their master shadows. Let K, an integer set, be the secret space, and $\{k_{1,1}, k_{1,2}, \ldots, k_{1,p_1}\}$, $\{k_{2,1}, k_{2,2}, \ldots, k_{2,p_2}\}$, …, $\{k_{m,1}, k_{m,2}, \ldots, k_{m,p_m}\} \subset K$ be m groups of secrets to be shared among P. Each group of secrets, $\{k_{i,1}, k_{i,2}, \ldots, k_{i,p_i}\}$, is shared according to a distinct (t_i, n)-threshold access structure for all $i = 1, 2, \ldots, m$ where $t_i \leq n$. In order to construct a single polynomial in secret distribution, we must sort the threshold values in ascending order, and then sort the groups of secrets in the same order. Here, we assume that $t_1 \leq t_2 \leq \ldots \leq t_m$, thus t_i $(1 \leq i \leq m)$ is corresponding to the ith group of secrets, $G_i = \{k_{i,1}, k_{i,2}, \ldots, k_{i,p_i}\}$. For reasons of briefness, we assume that $p_1 \leq p_2 \leq \ldots \leq p_m$. In our scheme, a notice board [8] is necessary, on which the dealer can publish the public information. Only the dealer can modify and update the content of the notice board, and the others can only read and download the public information.

2.2 Secret Distribution

The dealer executes the following steps to implement the secret distribution. In the following, we will describe the secret distribution in two cases.

(Case 1) If $p_m > t_1$, the dealer performs the following steps:

(i) Randomly choose m distinct prime numbers $q_1, q_2, \ldots, q_m$ such that $q_1<q_2<\ldots<q_m$, $k_{i,j} < q_i$ for all $i = 1, 2, \ldots, m$ and $j = 1, 2, \ldots, p_i$, and $p_m - t_1 + 1 + n < q_1$.

(ii) Randomly choose n distinct integers $x_1, x_2, \ldots, x_n$ from the interval of $[p_m - t_1 + 1, q_1]$ as the public identifiers of participants $P_1, P_2, \ldots, P_n$, respectively.

(iii) Compute the solutions $a_0, a_1, \ldots, a_{p_m-1}$ of the following p_m groups of simultaneous congruences by using Chinese Remain Theorem [10]:

$$\begin{cases} a_0 \equiv k_{1,1} \bmod q_1 \\ a_0 \equiv k_{2,1} \bmod q_2 \\ \quad\vdots \\ a_0 \equiv k_{m,1} \bmod q_m \end{cases}, \quad \begin{cases} a_1 \equiv k_{1,2} \bmod q_1 \\ a_1 \equiv k_{2,2} \bmod q_2 \\ \quad\vdots \\ a_1 \equiv k_{m,2} \bmod q_m \end{cases}, \quad \ldots, \quad \begin{cases} a_{p_1-1} \equiv k_{1,p_1} \bmod q_1 \\ a_{p_1-1} \equiv k_{2,p_1} \bmod q_2 \\ \quad\vdots \\ a_{p_1-1} \equiv k_{m,p_1} \bmod q_m \end{cases},$$

$$\begin{cases} a_{p_1} \equiv 0 \bmod q_1 \\ a_{p_1} \equiv k_{2,p_1+1} \bmod q_2 \\ \quad\vdots \\ a_{p_1} \equiv k_{m,p_1+1} \bmod q_m \end{cases}, \quad \begin{cases} a_{p_1+1} \equiv 0 \bmod q_1 \\ a_{p_1+1} \equiv k_{2,p_1+2} \bmod q_2 \\ \quad\vdots \\ a_{p_1+1} \equiv k_{m,p_1+2} \bmod q_m \end{cases}, \quad \ldots, \quad \begin{cases} a_{p_2-1} \equiv 0 \bmod q_1 \\ a_{p_2-1} \equiv k_{2,p_2} \bmod q_2 \\ \quad\vdots \\ a_{p_2-1} \equiv k_{m,p_2} \bmod q_m \end{cases}, \quad \vdots$$

$$\vdots \qquad\qquad \vdots$$

$$\begin{cases} a_{p_{m-1}} \equiv 0 \bmod q_1 \\ \quad\vdots \\ a_{p_{m-1}} \equiv 0 \bmod q_{m-1} \\ a_{p_{m-1}} \equiv k_{m,p_{m-1}+1} \bmod q_m \end{cases}, \quad \begin{cases} a_{p_{m-1}+1} \equiv 0 \bmod q_1 \\ \quad\vdots \\ a_{p_{m-1}+1} \equiv 0 \bmod q_{m-1} \\ a_{p_{m-1}+1} \equiv k_{m,p_{m-1}+2} \bmod q_m \end{cases}, \ldots, \begin{cases} a_{p_m-1} \equiv 0 \bmod q_1 \\ \quad\vdots \\ a_{p_m-1} \equiv 0 \bmod q_{m-1} \\ a_{p_m-1} \equiv k_{m,p_m} \bmod q_m \end{cases}.$$

Note that $a_j \equiv 0 \bmod q_k$ for all $k = 1, 2, \ldots, i$ and $j \geq p_i$.

(iv) For $i = 1, 2, \cdots, m-1$, compute $b_{t_i}, b_{t_i+1}, \cdots, b_{t_{i+1}-1}$, as follows:

for each $j = t_i, t_i+1, \ldots, t_{i+1}-1$, compute $b_j = c_j \times r_j \times \prod_{k=1}^{i} q_k \bmod M$, where $M = \prod_{i=1}^{m} q_i$, c_j is randomly selected from $\{0, 1, 2, \ldots, q_i-1\}$, and r_j is a random integer.

Construct a $(p_m + t_m - t_1 - 1)$th degree polynomial $H(x) = a_0 + a_1 x + \ldots + a_{p_m-1} x^{p_m-1} + b_{t_1} x^{p_m} + b_{t_1+1} x^{p_m+1} + \ldots + b_{t_m-1} x^{p_m+t_m-t_1-1}$. Note that $b_j \equiv 0 \bmod q_k$ for all $k = 1, 2, \ldots, i$ and $j \geq t_i$.

(v) For $i = 1, 2, \ldots, n$, compute the ith master shadow $y_i = H(x_i) \bmod M$.

(vi) For $i = 1, 2, \ldots, p_m - t_1$, compute the public value $d_i = H(i) \bmod M$.

(vii) Send each y_i to the corresponding participant P_i $(1 \leq i \leq n)$ over a secure channel, and publish $q_1, q_2, \ldots, q_m$ and the public values $d_1, d_2, \ldots, d_{p_m-t_1}$ in any authenticated manner such as that in [11].

(Case 2) If $p_m \leq t_1$, the dealer performs the following steps:

(i) Randomly choose m distinct prime integers $q_1, q_2, \ldots, q_m$ such that $q_1<q_2<\ldots<q_m$, $k_{i,j} < q_i$ for all $i = 1, 2, \ldots, m$ and $j = 1, 2, \ldots, p_i$, $n < q_1$ and $p_m < q_1$.

(ii) Randomly choose n distinct integers $x_1, x_2, \ldots, x_n$ from the interval of $[1, q_1]$ as the public identifiers of participants, respectively.

(iii) Compute the solutions $a_0, a_1, \ldots, a_{p_m-1}$ of the p_m groups of simultaneous congruences as described in step (iii) of case 1.

(iv) Randomly choose $(t_1 - p_m)$ integer $a_{p_m}, a_{p_m+1}, \ldots, a_{t_1-1} \in_R \mathbb{Z}_M$, then for $i = 1, 2, \cdots, m-1$, compute $a_{t_i}, a_{t_i+1}, \cdots, a_{t_{i+1}-1}$ as follows:

for each $j = t_i, t_i+1, \ldots, t_{i+1}-1$, compute $a_j = c_j \times r_j \times \prod_{k=1}^{i} q_k \bmod M$.

Construct a (t_m-1)th degree polynomial $H(x) = a_0 + a_1 x + \ldots + a_{t_m-1} x^{t_m-1}$.

Note that $a_j \equiv 0 \bmod q_k$ for all $k = 1, 2, \ldots, i$ and $j \geq t_i$.

(v) For $i = 1, 2, \ldots, n$, compute the ith master shadow $y_i = H(x_i) \bmod M$.

(vi) Send each y_i to the corresponding participant P_i $(1 \leq i \leq n)$ over a secure channel, and publish $q_1, q_2, \ldots, q_m$ in any authenticated manner such as that in [11].

2.3 Secret Reconstruction

Without loss of generality, we assume that an authorized subset of $A = \{P_1, P_2, \ldots, P_{t_i}\}$ want to recover the ith group of secrets with the (t_i, n)-threshold structure, G_i. In the following, we will introduce the secret reconstruction process in two cases.

(Case 1) $p_m > t_1$. It is observed that $H_i(x) = H(x) \bmod q_i \in \mathbb{Z}_{q_i}[x]$ is a polynomial of degree equal to or less than $p_m + t_i - t_1 - 1$ and $k_{i,j} = a_{j-1} \bmod q_i$ for all $j = 1, 2, \ldots, p_i$. According to the public values $d_1, d_2, \ldots, d_{p_m-t_1}$, one can compute $d_j' \equiv d_j \bmod q_i$ $(1 \leq j \leq p_m - t_1)$ and get $(p_m - t_1)$ co-ordinates, $(1, d_1')$, $(2, d_2')$, $\ldots$, $(p_m - t_1, d_{p_m-t_1}')$. Participants in A polling their sub-shadows $y_{i,j} \equiv y_j \bmod q_i$ $(1 \leq j \leq t_i)$ can get t_i co-ordinates, $(x_1, y_{i,1})$, $(x_2, y_{i,2})$, $\ldots$, (x_{t_i}, y_{i,t_i}). Thus, these participants can obtain altogether $(p_m + t_i - t_1)$ co-ordinates. So they can reconstruct the ith group of secrets via the following formula:

$$\begin{aligned} H_i(x) &= \sum_{j=1}^{t_i} y_{i,j} \prod_{h=1,h\neq j}^{t_i} \frac{x - x_h}{x_j - x_h} \prod_{u=1}^{p_m-t_1} \frac{x-u}{x_j - u} + \sum_{j=1}^{p_m-t_1} d_j' \prod_{u=1,u\neq j}^{p_m-t_1} \frac{x-u}{j-u} \prod_{h=1}^{t_i} \frac{x-x_h}{j-x_h} \bmod q_i \\ &= a_0 + a_1 x + \cdots + a_{p_i-1} x^{p_i-1} + b_{t_1} x^{p_m} + \cdots + b_{t_i-1} x^{p_i+t_i-t_1-1} \bmod q_i \end{aligned} \tag{2.1}$$

From the first p_i coefficients of $H_i(x)$, the ith group of secrets can be computed as $k_{i,1} = a_0 \bmod q_i$, $k_{i,2} = a_1 \bmod q_i$, $\ldots$, $k_{i,p_i} = a_{p_i-1} \bmod q_i$.

(Case 2) $p_m \leq t_1$. It is observed that $H_i(x) = H(x) \bmod q_i \in \mathbb{Z}_{q_i}[x]$ is a polynomial of degree no greater than $t_i - 1$ and $k_{i,j} = a_{j-1} \bmod q_i$ for all $j = 1, 2, \ldots, p_i$. Participants in A polling their sub-shadows $y_{i,j} \equiv y_j \bmod q_i$ $(1 \leq j \leq t_i)$ can get t_i co-ordinates, $(x_1, y_{i,1})$, $(x_2, y_{i,2})$, $\ldots$, (x_{t_i}, y_{i,t_i}). Thus, these participants can reconstruct the ith group of secrets via the following formula:

$$\begin{aligned} H_i(x) &= \sum_{j=1}^{t_i} \left(y_{i,j} \prod_{h=1,h\neq j}^{t_i} \frac{x - x_h}{x_j - x_h} \right) \bmod q_i \\ &= a_0 + a_1 x + \cdots + a_{p_i-1} x^{p_i-1} + a_{p_m} x^{p_m} + \cdots + a_{t_i-1} x^{t_i-1} \bmod q_i \end{aligned} \tag{2.2}$$

From the first p_i coefficients of $H_i(x)$, the ith group of secrets can be computed as $k_{i,1} = a_0 \bmod q_i$, $k_{i,2} = a_1 \bmod q_i$, …, $k_{i,p_i} = a_{p_i-1} \bmod q_i$.

3 Numerical Example

In this section, an example is showed to expound our scheme further.

Example 1. Let n=5 and $P = \{P_1, P_2, P_3, P_4, P_5\}$ denote a set of five participants. There are three groups of secrets, {10, 4}, {4, 15}, {27, 9, 28} among P such that each group is shared according to $(t_i, 5)$-threshold access structure for i=1, 2, 3 and $t_i <$ 5. The threshold values are $t_1 = 2$, $t_2 = 3$ and $t_3 = 4$, respectively. The numbers of each group are $p_1 = 2$, $p_2 = 2$ and $p_3 = 3$, respectively.

First, we sort the thresholds in ascending order, that is, $t_1 < t_2 < t_3$. We also sort the groups of secret in the same order as $G_1 = \{10, 4\}$, $G_2 = \{4, 15\}$, $G_3 = \{27, 9, 28\}$.

(1) Secret Distribution

(i) Randomly choose 3 distinct primes q_1=17, q_1=23, q_2= 29.

(ii) For $p_3 = 3 > t_1 = 2$, then randomly choose 5 distinct integers x_1 =12, x_2 =5, x_3 =3, x_4 =11, x_5 =6 from the interval of [2, 17] as the public identifiers of participants $P_1, P_2, \ldots, P_5$, respectively.

(iii) Let a_0, a_1, a_2 be the solutions of the following 3 groups of simultaneous congruences, respectively:

$$\begin{cases} a_0 \equiv 10 \bmod 17 \\ a_0 \equiv 4 \bmod 23 \\ a_0 \equiv 27 \bmod 29 \end{cases}, \quad \begin{cases} a_1 \equiv 4 \bmod 17 \\ a_1 \equiv 15 \bmod 23, \\ a_1 \equiv 9 \bmod 29 \end{cases} \quad \begin{cases} a_2 \equiv 0 \bmod 17 \\ a_2 \equiv 0 \bmod 23 \\ a_2 \equiv 28 \bmod 29 \end{cases}.$$

We can gets $a_0 = 27, a_1 = 38, a_2 = 782$.

(iv) We have $M = \prod_{i=1}^{3} q_i = 17 \times 23 \times 29 = 11339$. For i = 1, 2, the coefficients $b_{t_i}, b_{t_i+1}, \ldots, b_{t_{i+1}-1}$ are computed as follows:

when $j = t_1$ =2, randomly choose $c_2 = 16 \in \mathbb{Z}_{q_1}$ and $r_2 = 42$, and compute $b_2 = c_2 \times r_2 \times \prod_{k=1}^{1} q_k \bmod M = 16 \times 42 \times 17 \bmod 11339 = 85$;

when $j = t_2$ =3, randomly choose $c_3 = 18 \in \mathbb{Z}_{q_2}$ and $r_3 = 21$, and compute $b_3 = c_3 \times r_3 \times \prod_{k=1}^{2} q_k \bmod M = 18 \times 21 \times 17 \times 23 \bmod 11339 = 391$;

We have $p_m + t_m - t_1 - 1 = 3 + 4 - 2 - 1 = 4$. Construct a fourth degree polynomial $H(x) = a_0 + a_1 x + a_2 x^2 + b_2 x^3 + b_3 x^4 = 27 + 38x + 782x^2 + 85x^3 + 391x^4$.

(v) For i = 1, 2, …, 5, the master shadow y_i of each participant P_i is computed as $y_1 = H(12) \bmod 11339 = 10904$, $y_2 = H(5) \bmod 11339 = 2631$, $y_3 = H(3) \bmod 11339 = 7128$, $y_4 = H(11) \bmod 11339 = 2536$, $y_5 = H(6) \bmod 11339 = 9231$.

(vi) Compute the public value $d_1 = H(1) \bmod 11339 = 1323$;

(vii) Send y_1= 10904, y_2= 2631, y_3 = 7128, y_4= 2536, y_5=9231 to corresponding participants P_1, P_2, P_3, P_4, P_5 over a secure channel, and publish q_1=17, q_2= 23, q_3= 29 and d_1=1323.

(2) Secret Reconstruction

Assume that an authorized subset of $A = \{P_1, P_2\}$ want to recover the first group of secrets $G_1 = \{10, 4\}$ with the (2, 5)-threshold structure. In the following, we will introduce the secret reconstruction in two cases. Participants in A polling their sub-shadows $y_{1,1} = y_1 \bmod q_1 = 10904 \bmod 17 = 7$ and $y_{1,2} = y_2 \bmod q_1 = 2631 \bmod 17 = 13$ can get 2 co-ordinates, (12, 7), (5, 13). According to the public value $d_1 = 1323$, one can compute $d_1' = d_1 \bmod 17 = 1323 \bmod 17 = 14$ and get 1 co-ordinates, (1,14). Thus, these participants can obtain altogether 3 co-ordinates. So they can reconstruct the polynomial $H_1(x)$ as follows:

$$H_1(x) = \left(\sum_{j=1}^{2} y_{1,j} \left(\prod_{h=1,h\neq j}^{2} \frac{x - x_h}{x_j - x_h} \right) \frac{x-1}{x_j - 1} \right) + d_1' \prod_{h=1}^{2} \frac{x - x_h}{1 - x_h} \bmod 17$$

$$= 7 \times \frac{x-5}{12-5} \times \frac{x-1}{12-1} + 13 \times \frac{x-12}{5-12} \times \frac{x-1}{5-1} + 14 \times \frac{x-12}{1-12} \times \frac{x-5}{1-5} \bmod 17$$

$$\overset{(*)}{=} 7 \times 5 \times 14 \times (x-5) \times (x-1) + 13 \times 12 \times 13 \times (x-12) \times (x-1)$$

$$+ 14 \times 14 \times 13 \times (x-12) \times (x-5) \bmod 17$$

$$= 10 + 4x$$

Where the step (∗) comes from the fact that division over number field can be calculated by multiplying the dividend by the multiplicative inverse of the divisor [12], that is, $7^{-1} = 5 \bmod 17$, $11^{-1} = 14 \bmod 17$, $(-7)^{-1} = 12 \bmod 17$, $4^{-1} = 13 \bmod 17$. From the first two coefficients of $H_1(x)$, the first group of secrets can be computed as $k_{1,1} = 10$, $k_{1,2} = 4$.

4 Analyses and Discussions

4.1 Security Analysis

In this section, we will analyze the security of the proposed scheme. From the subsection 2.2, when $p_m > t_1$, a polynomial $H_i(x) = H(x) \bmod q_i$ $(1 \le i \le m)$ of degree no greater than $(p_m + t_i - t_1 - 1)$ is needed to be reconstructed in the recovery of each group of secrets, G_i. We know that $H_i(x) (1 \le i \le m)$ can be uniquely determined by exactly $(p_m + t_i - t_1)$ distinct co-ordinates which satisfy the polynomial. According to the public values, one can get $(p_m - t_1)$ co-ordinates. So any t_i out of n participants pooling their sub-shadows can reconstruct $H_i(x) (1 \le i \le m)$; any set of $(t_i - 1)$ or more less participants cannot recover the polynomial $H_i(x) (1 \le i \le m)$. So when $p_m > t_1$ each resultant threshold secret sharing scheme satisfies the security of Shamir's threshold scheme. When $p_m \le t_1$, a polynomial $H_i(x) = H(x) \bmod q_i$ $(1 \le i \le m)$ of degree no greater than $(t_i - 1)$ is needed to be reconstructed in the recovery of each group of secrets, G_i. We know that $H_i(x)$ $(1 \le i \le m)$ can be uniquely determined by exactly t_i distinct co-ordinates which satisfy the polynomial. So any t_i out of n participants pooling their sub-shadows can reconstruct $H_i(x)$ $(1 \le i \le m)$; any set of $(t_i - 1)$ or more less participants cannot recover the polynomial $H_i(x)$ $(1 \le i \le m)$. Thus, each resultant threshold secret sharing scheme satisfies the security of Shamir's threshold scheme when $p_m \le t_1$. Through the above analyses, we conclude that the security of our scheme is the same as that of Shamir's scheme, that is, our scheme is unconditionally secure.

In our scheme, an authorized subset A can recover the ith group of secrets G_i. They can also reconstruct the first i–1 groups of secrets, i.e. $G_1, G_2, \ldots, G_{i-1}$. But this is not a security problem because the subset A is also a member of the (t_j, n)-threshold access structure for all $j < i$. Suppose that A is an authorized subset of (t_i, n)-threshold access structure but not an authorized subset of (t_{i+1}, n)-threshold access structure. Then A cannot reconstruct the $(i+1)$th group of secrets because they need to guess the first p_{i+1} coefficients of the polynomial $H_{i+1}(x) = H(x) \bmod q_{i+1}$. This is no better than that they guess sub-shadows of $H_{i+1}(x)$. So the proposed scheme is a perfect one.

4.2 Performance Analysis

In this section, we will evaluate the performance of the proposed scheme. When one group of secrets is shared, the performance of the proposed scheme is the same as that of Chan *et al.*'s scheme, that is to say, each scheme needs one secret distribution procedure. But when m groups of secrets with different threshold access structures, G_1, $G_2, \ldots, G_m$, are shared, m secret distribution procedures are needed in Chan *et al.*'s scheme. It is easy to see that the amount of computation in Chan *et al.*'s scheme is very large for sharing multiple groups of secrets. However, in the proposed scheme, still only one secret distribution procedure is needed to share m groups of secrets. Therefore, our scheme is easier in implementation and needs less amount of computation than Chan *et al.*'s scheme. For the same reason, our scheme also has advantages over Chien *et al.*'s, Li *et al.*'s and Pang *et al.*'s schemes.

The number of public values is also an important parameter that determines the performance of a scheme, because it affects the storage and communication complexity of the scheme [9]. In Table 1, we list numbers of public values with the same bit length needed in each scheme for sharing m groups of secrets described in subsection 2.1.

Table 1. The number of public values in each scheme

Scheme	Public values	
The proposed scheme	$p_m \le t_1$	m
	$p_m > t_1$	$p_m - t_1 + m$
Chan *et al.*'s scheme	$\sum_{i=1}^{m} p_i$	
Chien *et al.*'s scheme	$m + \sum_{i=1}^{m}(n + p_i - t_i)$	
Li *et al.*'s scheme	$m + \sum_{i=1}^{m}(n + p_i - t_i)$	
Pang *et al.*'s scheme	$m + \sum_{i=1}^{m}(n + p_i - t_i)$	

From Table 1, when $p_m \le t_1$, only m public values are published in our scheme, $\sum_{i=1}^{m} p_i$ public values are published in Chan *et al.*'s scheme, $m + \sum_{i=1}^{m}(n + p_i - t_i)$ public values are published in Chien *et al.*'s, Li *et al.*'s and Pang *et al.*'s schemes. It is obvious that $m + \sum_{i=1}^{m}(n + p_i - t_i) > \sum_{i=1}^{m} p_i > m$. When $p_m > t_1$, only $p_m - t_1 + m$ public values are published in our scheme, $\sum_{i=1}^{m} p_i$ public values are published in Chan *et al.*'s scheme and $m+\sum_{i=1}^{m}(n+p_i-t_i)$ are published in Chien *et al.*'s, Li *et al.*'s and Pang *et al.*'s schemes. Since we have assumed that $1\leq t_i \leq n$ for all i = 1, 2, ..., m, we have $m+\sum_{i=1}^{m}(n+p_i-t_i) = m + p_m - t_1 + 2n + p_1 - t_m + \sum_{i=2}^{m-1}(n+p_i-t_i) > \sum_{i=1}^{m} p_i > p_m - t_1 + m$. So the number of public values needed in our scheme is much fewer than those of the other four schemes in any cases.

Through the above the discussions, we can draw a conclusion that our scheme is more efficient than the other four schemes.

5 Conclusions

In this paper, we proposed a new threshold multi-group-secret sharing scheme based on Chan *et al.*'s scheme. It can pack multiple groups of secrets with different threshold access structures into one group of big secrets, and share them by using a single secret distribution. It allows recovering a group of secrets in a secret reconstruction procedure. The dealer can dynamically determine the number of secrets in each group. Analyses show that our scheme is unconditionally secure. Compared with the existing schemes, the proposed scheme needs fewer public values and less amount of computation for sharing multiple groups of secrets, which makes our scheme capable of provide more practicability for many applications.

References

1. Desmedt Y (1997) Some recent research aspects of threshold cryptography. In: Okamoto R, Davida G, Mambo M (eds) Information Security. Springer-Verlag, Berlin, pp 158-173
2. Shamir A (1979) How to share a secret. Communications of the ACM 22: 612-613
3. Blakley G (1979) Safeguarding cryptographic key. In: Merwin RE, Zanca JT, Smith M (eds) Proceedings of AFIPS. AFIPS Press, New York, pp 313-317
4. Asmuth C, Bloom J (1983) A modular approach to key safeguarding. IEEE Transactions on Information Theory IT-29: 208-210
5. Crescenzo GD (2003) Sharing one secret vs. sharing many secrets. Theoret. Comput. Sci. 295: 123-140
6. Harn L (1995) Efficient sharing (broadcasting) of multiple secrets. IEE Proceedings—Computers and Digital Techniques 142: 237-240
7. Chien HY, Jan JK, Tseng YM (2000) A practical (*t*, *n*) multi-secret sharing scheme. IEICE Transactions on Fundamentals E83-A: 2762-2765
8. Li HX, Cheng CT, Pang LJ (2005) A new (*t*, *n*)-threshold multi-secret sharing scheme. In: Hao Y, Liu J, Wang Y, etc (eds) Computational Intelligence and Security. Springer-Verlag, Berlin, pp 421-426
9. Pang LJ, Wang YM (2005) A new (*t*, *n*) multi-secret sharing scheme based on Shamir's secret sharing. Appl. Math. & Compt 167: 840-848
10. Chan CW, Chang CC (2005) A scheme for threshold multi-secret sharing. Appl. Math. & Compt 166:1-14
11. ElGamal T (1985) A public-key cryptosystem and a signature scheme based on discrete logarithms. IEEE Trans. on Info. Theory IT-31: 469-472
12. Melvyn B Nathanson (2000) Elementary methods in number theory. Springer, New York

A Morphological Approach for Granulometry with Application to Image Denoising

Tingquan Deng[1] and Yanmei Chen[2]

[1] College of Science, Harbin engineering University
Harbin 150001 P.R. China
Deng.Tq@hrbeu.edu.cn , Tq_deng@163.com

[2] Department of Mathematics, Harbin Institute of Technology
Harbin, 150001 P.R. China
Chen.yanmei@163.com

Abstract. Granulometry formalizes the intuitive geometric notion of a sieving process. It was initially set oriented to extract size distribution from binary images, and has been extended to function operators to analyze and extract texture features in grey-scale images. In this paper, we study and establish granulometry with respect to grey-scale morphological operators based on fuzzy logic. We discuss applications of the granulometry in image analysis. A numerical experiment shows that granulometry is a powerful tool for image denoising for image analysis and processing.

Keywords: Fuzzy logic, morphological operators, granulometry, image denoising, image analysis.

1 Introduction

Granulometry was initially developed by G. Matheron [9] in the study of random sets. It formalizes the intuitive physical process of sieving and extracts size distribution from binary images [10]. It is one of the most practical tools in image denoising and depressing with increasing sizes of mesh of a sieve, and has wide application in texture analysis, image segmentation and object recognition [11, 1, 6, 2, 7, 12, 4].

Granulometry is closely connected with the theory of convex sets and can be formalized with mathematical morphology. A morphological granulometry is essentially a sequence of morphological openings by a series of structuring elements. By performing a series of morphological opening with increasing structuring element size we can obtain a granulometry function which maps each structuring element size to the number of image pixels removed during the opening operations with the corresponding structuring element.

Granulometry has been extended to function by using the morphological operators provided by umbra approach as a tool for texture analysis by Chen et al [3]. In the processing of grey-scale images, morphological operators based on fuzzy logic have more flexility [5]. In this paper, we establish a framework for

B.-Y. Cao (Ed.): Fuzzy Information and Engineering (ICFIE), ASC 40, pp. 919–929, 2007.
springerlink.com

granulometry with the use of the concept of convex fuzzy sets in fuzzy logic and present an application of granulometry in grey-scale image noising.

2 Fuzzy Logical Operators

Definition 1. *Let $C : [0,1]^2 \to [0,1]$ be an increasing mapping satisfying $C(1,0) = C(0,1) = 0$ and $C(1,1) = 1$, then it is called a conjunction. Let I be a mapping from $[0,1]^2$ to $[0,1]$ with non-increasing in its first argument and non-decreasing in its second, $I(0,0) = I(1,1) = 1$ and $I(1,0) = 0$, then it is called an implication.*

If C is a commutative and associative conjunction satisfying the boundary condition $C(1,s) = s$ for all $s \in [0,1]$, it is called a t-norm.

The following lists some well-known t-norms and conjunctions. The *Gödel-Brouwer* t-norm $C(s,t) = \min(s,t)$, *Lukasiewicz* t-norm $C(s,t) = \max(0, s+t-1)$, *Kleene-Dienes* conjunction $C(s,t) = \begin{cases} 0 & s+t \le 1 \\ t & s+t > 1 \end{cases}$, *Reichenbach* conjunction $C(s,t) = \begin{cases} 0 & s+t \le 1 \\ (s+t-1)/s & s+t > 1 \end{cases}$, and the *Hamacher family* of conjunctions $C_p(s,t) = \frac{(1-p)st+pt}{p+(1-p)(1-t+st)}$ $(0 \le p \le 1)$ or $\frac{t}{p+(1-p)(s+t-st)}$ $(p \ge 1)$ when $s > 0$ and $C_p(0,t) = 0$.

Definition 2. *A conjunction C and an implication I on $[0,1]$ are said to form an adjunction if*

$$C(r,s) \le t \iff s \le I(r,t) \tag{1}$$

for each $r \in [0,1]$ and for arbitrary $s,t \in [0,1]$.

With the adjunction relation between C and I the following propositions have been proved in [5].

Proposition 1. *Let (I,C) be an adjunction, then for any $\{s_i\} \subseteq [0,1]$ and $t \in [0,1]$, $C(t, \vee_i s_i) = \vee_i C(t, s_i)$ and $I(t, \wedge_i s_i) = \wedge_i I(t, s_i)$. Furthermore, if C is commutative, then $I(\vee_i s_i, t) = \wedge_i I(s_i, t)$, where $\vee$ and $\wedge$ denote the supremum and infimum, respectively.*

Proposition 2. *Let (I,C) be an adjunction, then for arbitrary $r,s,t \in [0,1]$, $C(C(r,s),t) = C(s, C(r,t)) \iff I(r, I(s,t)) = I(C(r,s),t)$. Furthermore, if C is commutative and associative, then $I(r, I(s,t)) = I(C(r,s),t)$.*

3 Operations of Fuzzy Sets

Let U be a linear space or a subspace of $\mathbf{R}^n$, called the universe of discourse. A fuzzy set F on U means a mapping $F : U \to [0,1]$, determined by its membership function $F(x)$. Let $\mathcal{F}(U)$ and $\mathcal{P}(U)$ denote the family of all fuzzy sets on U and the power set of U, respectively.

Definition 3. *Let $F \in \mathcal{F}(U)$ and $\{F_\lambda\}_{\lambda \in (0,1)} \subseteq \mathcal{P}(U)$, if*

(1) $0 < \lambda < \mu < 1 \Rightarrow F_\mu \subseteq F_\lambda$,
(2) $F(x) = \vee\{\lambda \in (0,1) : x \in F_\lambda\},\ x \in U$,

then $\{F_\lambda\}_{\lambda \in (0,1)}$ is called a set representation *of F.*

Definition 4. *For $F \in \mathcal{F}(U)$, $\lambda \in [0,1]$, the set $[F]_\lambda = \{x \in U : F(x) \geq \lambda\}$ and $[F]^\lambda = \{x \in U : F(x) > \lambda\}(\lambda \neq 1)$ are called the cut set and strict cut set, respectively, of F at level λ.*

Proposition 3. *Both $\{[F]_\lambda\}_{\lambda \in [0,1]}$ and $\{[F]^\lambda\}_{\lambda \in [0,1)}$ are the set representations of fuzzy set F.*

Proposition 4. *$\{F_\lambda\}_{\lambda \in (0,1)}$ is a set representation of fuzzy set F if and only if for all $\lambda \in (0,1)$,*

$$[F]^\lambda \subseteq F_\lambda \subseteq [F]_\lambda\,. \tag{2}$$

Proof. $\Rightarrow$ If $\{F_\lambda\}_{\lambda \in (0,1)}$ is a set representation of F, then $x \in [F]^\lambda \Rightarrow F(x) > \lambda \Rightarrow \vee\{r \in (0\,,1) : x \in F_r\} > \lambda \Rightarrow \exists \mu \in (\lambda\,,1)$ such that $x \in F_\mu \Rightarrow x \in F_\lambda \Rightarrow F(x) = \vee\{r \in (0\,,1) : x \in F_r\} \geq \lambda$, so $x \in [F]_\lambda$.

$\Leftarrow$ Suppose that there exists a family of crisp sets $\{F_\lambda\}_{\lambda \in (0\,,1)}$ such that $[F]^\lambda \subseteq F_\lambda \subseteq [F]_\lambda$. If $0 < \lambda < \mu < 1$, then $x \in F_\mu \Rightarrow x \in [F]_\mu \Rightarrow F(x) \geq \mu > \lambda \Rightarrow x \in [F]^\lambda \Rightarrow x \in F_\lambda$. Thus $F_\mu \subseteq F_\lambda$.

Defining fuzzy sets $\lambda[F]_\lambda$, $\lambda[F]^\lambda$ and λF_λ as $\lambda[F]_\lambda(x) = \begin{cases} \lambda & x \in [F]_\lambda \\ 0 & \text{otherwise} \end{cases}$, $\lambda[F]^\lambda(x) = \begin{cases} \lambda & x \in [F]^\lambda \\ 0 & \text{otherwise} \end{cases}$, and $\lambda F_\lambda(x) = \begin{cases} \lambda & x \in F_\lambda \\ 0 & \text{otherwise} \end{cases}$, respectively, then $\lambda[F]^\lambda \subseteq \lambda F_\lambda \subseteq \lambda[F]_\lambda$. Thus, by the Decomposition Theorem, we have that

$$F = \vee_{\lambda \in [0\,,1)}\lambda[F]^\lambda \subseteq \vee_{\lambda \in (0\,,1)}\lambda F_\lambda \subseteq \vee_{\lambda \in [0\,,1]}\lambda[F]_\lambda = F\,. \tag{3}$$

Therefore, for any $x \in U$,

$$F(x) = \vee_{\lambda \in (0\,,1)}(\lambda F_\lambda)(x) = \vee\{\lambda \in (0\,,1) : x \in F_\lambda\}\,. \tag{4}$$

Proposition 5. *If the family of crisp sets $\{F_\lambda\}_{\lambda \in (0\,,1)}$ is the set representation of fuzzy set F as well as fuzzy set G, then $F = G$.*

Proposition 6. (Extension Principle) *Let $f : U^2 \to U$ be a mapping, C be a conjunction on $[0\,,1]$, and $F_1\,,F_2 \in \mathcal{F}(U)$ be two fuzzy sets, then $f(F_1\,,F_2)$ is a fuzzy set on U. Moreover, for arbitrary $y \in U$,*

$$f(F_1\,,F_2)(y) = \vee_{f(x_1\,,x_2)=y\,,(x_1\,,x_2)\in U^2}C(F_1(x_1)\,,F_2(x_2))\,. \tag{5}$$

If there doesn't exist $(x_1\,,x_2) \in U^2$ such that $f(x_1\,,x_2) = y$, then $f(F_1\,,F_2)(y) = 0$.

Proposition 7. *Let $f : U^2 \to U$ be a non-decreasing mapping, C be a conjunction on $[0\,,1]$ satisfying $C(s\,,t) \leq \min(s\,,t)$ and $C(s\,,s) \geq s$ for arbitrary $s\,,t \in [0\,,1]$, and $F_1\,,F_2 \in \mathcal{F}(U)$ be two fuzzy sets, then*

(1) $[f(F_1, F_2)]^\lambda = f([F_1]^\lambda, [F_2]^\lambda) = \{f(x, y) : x \in [F_1]^\lambda, y \in [F_2]^\lambda\}, \lambda \in [0, 1)$.
(2) $[f(F_1, F_2)]_\lambda \supseteq f([F_1]_\lambda, [F_2]_\lambda) = \{f(x, y) : x \in [F_1]_\lambda, y \in [F_2]_\lambda\}, \lambda \in [0, 1]$.
(3) *If* $\{G_\lambda\}_{\lambda \in (0,1)}$ *and* $\{H_\lambda\}_{\lambda \in (0,1)}$ *are the set representations of* F_1 *and* F_2*, respectively, then* $\{f(G_\lambda, H_\lambda)\}_{\lambda \in (0,1)}$ *is a set representation of* $f(F_1, F_2)$.

Proof. (1) For $\lambda \in [0, 1)$ and $y \in U$,

$$\begin{aligned} y \in [f(F_1, F_2)]^\lambda &\iff f(F_1, F_2)(y) > \lambda \\ &\iff \vee_{f(x_1, x_2)=y, (x_1, x_2) \in U^2} C(F_1(x_1), F_2(x_2)) > \lambda \\ &\iff \exists (x_1, x_2) \in U^2, \ni f(x_1, x_2) = y \,\&\, C(F_1(x_1), F_2(x_2)) > \lambda \\ &\iff f(x_1, x_2) = y \,\&\, F_1(x_1) > \lambda \,\&\, F_2(x_2) > \lambda \\ &\iff f(x_1, x_2) = y \,\&\, x_1 \in [F_1]^\lambda \,\&\, x_2 \in [F_2]^\lambda \\ &\iff f(x_1, x_2) = y \in f([F_1]^\lambda, [F_2]^\lambda). \end{aligned}$$

(2) Let $y \in U$, then

$$\begin{aligned} y \in f([F_1]_\lambda, [F_2]_\lambda) &\iff \exists x_1 \in [F_1]_\lambda, x_2 \in [F_2]_\lambda, \ni f(x_1, x_2) = y \\ &\iff f(x_1, x_2) = y, F_1(x_1) \geq \lambda, F_2(x_2) \geq \lambda \\ &\iff f(x_1, x_2) = y, C(F_1(x_1), F_2(x_2)) \geq C(\lambda, \lambda) = \lambda \\ &\implies f(F_1, F_2)(y) \geq \lambda \\ &\iff y \in [f(F_1, F_2)]_\lambda. \end{aligned}$$

(3) It's sufficient to prove that

$$[f(F_1, F_2)]^\lambda \subseteq f(G_\lambda, H_\lambda) \subseteq [f(F_1, F_2)]_\lambda. \tag{6}$$

For $\lambda \in (0, 1), [F_1]^\lambda \subseteq G_\lambda \subseteq [F_1]_\lambda$ and $[F_2]^\lambda \subseteq H_\lambda \subseteq [F_2]_\lambda$. In term of the results in (1), (2), as well as the monotone of f, we have that

$$[f(F_1, F_2)]^\lambda = f([F_1]^\lambda, [F_2]^\lambda) \subseteq f(G_\lambda, H_\lambda) \subseteq f([F_1]_\lambda, [F_2]_\lambda) \subseteq [f(F_1, F_2)]_\lambda.$$

It has been proved that if the conjunction C satisfies the conditions in Proposition 7, then $C = \min$.

In Proposition 6 let $f(x_1, x_2) = x_1 + x_2$, the following definition is implied.

Definition 5. *Let* C *be a conjunction and* I *be an implication on* $[0, 1]$, $F, G \in \mathcal{F}(U)$*, the* C*-Minkowski addition of* F *and* G *is defined by*

$$(F \oplus_C G)(x) = \mathcal{D}_G(F)(x) = \vee_{y \in U} C(G(x - y), F(y)), x \in U. \tag{7}$$

Meanwhile, the I*-Minkowski substraction of* F *and* G *is defined by*

$$(F \ominus_I G)(x) = \mathcal{E}_G(F)(x) = \wedge_{y \in U} I(G(y - x), F(y)), x \in U. \tag{8}$$

According to Proposition 1 and Proposition 2, the following results can be proved [5].

Proposition 8. *The pair* (I, C) *forms an adjunction if and only for arbitrary* $G \in \mathcal{F}(U)$, $(\mathcal{E}_G, \mathcal{D}_G)$ *is an adjunction. In which case,* $\mathcal{D}_G(F)$ *is called a dilation of fuzzy set* F *with respect to the fuzzy set (structuring element)* G *and* $\mathcal{E}_G(F)$ *is called an erosion of* F *with respect to this structuring element.*

Proposition 9. *If a conjunction* C *is associative and continuous in its first argument, then for arbitrary* $G, H \in \mathcal{F}(U)$,

$$\mathcal{D}_G \mathcal{D}_H = \mathcal{D}_{\mathcal{D}_G(H)} . \tag{9}$$

Proposition 10. *Let* (I, C) *be an adjunction, if* C *is a commutative and associative conjunction, then for arbitrary* $G, H \in \mathcal{F}(U)$,

$$\mathcal{E}_G \mathcal{E}_H = \mathcal{E}_{\mathcal{D}_G(H)} . \tag{10}$$

4 Convex Fuzzy Sets

The concept of convex fuzzy sets is an important one in the theories of fuzzy sets and fuzzy logic. Convex sets and convex fuzzy sets can find wide applications in optimization and image processing. Firstly, we recall the concept of convex sets.

Definition 6. *A set* $A \in \mathcal{P}(U)$ *is called convex if for any* $x, y \in A$, $r \in [0, 1]$, $rx + (1 - r)y \in A$.

Proposition 11. *A compact set* $A \in \mathcal{P}(U)$ *is convex if and only if* $pA \oplus qA = (p + q)A$ *for arbitrary* $p, q > 0$, *where* $pA = \{px \mid x \in A\}$.

In the equality $pA \oplus qA = (p + q)A$, taking $r = p/(p + q)$, one obtains that A is a convex set if and only if $rA \oplus (1 - r)A = A$ for any $r \in [0, 1]$.

Similarly, we may discuss convexity of fuzzy sets.

Definition 7. *Let* $F \in \mathcal{F}(U)$, $p \geq 0$, *define fuzzy set* pF *as follows:* $pF(x) = F(x/p)$ *for any* $x \in U$. *If* $p = 0$, *we suppose that* $0F(x) = \begin{cases} 1 & x = 0 \\ 0 & \text{otherwise} \end{cases}$.

Definition 8. *Let* $F \in \mathcal{F}(U)$, *if for arbitrary* $x, y \in U$ *and* $r \in [0, 1]$, $F(rx + (1 - r)y) \geq \min(F(x), F(y))$, *then* F *is called a convex fuzzy set.*

Proposition 12. $F \in \mathcal{F}(U)$ *is convex if and only if for arbitrary* $p > 0$, $pF \in \mathcal{F}(U)$ *is convex.*

Proof. For $x, y \in U$ and $r \in [0, 1]$, according to the inequality $pF(rx+(1-r)y) = F(rx/p + (1 - r)y/p) \geq \min(F(x/p), F(y/p)) = \min(pF(x), pF(y))$, the result is clear.

Proposition 13. *The following three statements are equivalent.*

(1) $F \in \mathcal{F}(U)$ *is a convex fuzzy set.*
(2) *For any* $\lambda \in [0, 1]$, $[F]_\lambda$ *is a convex set.*
(3) *For any* $\lambda \in [0, 1)$, $[F]^\lambda$ *is a convex set.*

Proposition 14. *Let $F \in \mathcal{F}(U)$ be a convex fuzzy set, and let $p > 0$, then $[pF]_\lambda = p[F]_\lambda\,(\lambda \in [0\,,1])$ and $[pF]^\lambda = p[F]^\lambda\,(\lambda \in [0\,,1))$.*

Proof. $x \in [pF]_\lambda \iff pF(x) \geq \lambda \iff F(x/p) \geq \lambda \iff x/p \in [F]_\lambda \iff x \in p[F]_\lambda$. From which, the first equation holds.

The second equation can be proved in the same way.

Proposition 15. *Let $F \in \mathcal{F}(U)$ be a convex fuzzy set, $p\,,q > 0$, then for all $\lambda \in [0\,,1]$, $[pF]_\lambda \oplus [qF]_\lambda = [(p+q)F]_\lambda$ and $[pF]^\lambda \oplus [qF]^\lambda = [(p+q)F]^\lambda$ $(\lambda \neq 1)$, where $A \oplus B = \{x + y : x \in A\,, y \in B\}$.*

Proposition 16. *If $F \in \mathcal{F}(U)$ is a convex fuzzy set, then for arbitrary $p\,,q > 0$, $pF \oplus_{\min} qF = qF \oplus_{\min} pF = (p+q)F$.*

Proof. Clearly, $pF \oplus_{\min} qF = qF \oplus_{\min} pF$. We have to prove that $pF \oplus_{\min} qF = (p+q)F$.

Let $\{G_\lambda\}_{\lambda\in(0\,,1)}$ and $\{H_\lambda\}_{\lambda\in(0\,,1)}$ be the set representations of fuzzy sets pF and qF, respectively, then

$$[(p+q)F]^\lambda \subseteq G_\lambda \oplus H_\lambda \subseteq [(p+q)F]_\lambda\,, \tag{11}$$

which means that the family $\{G_\lambda \oplus H_\lambda\}_{\lambda\in(0\,,1)}$ is the set representation of fuzzy set $(p+q)F$.

On the other hand, according to Proposition 7 $\{G_\lambda \oplus H_\lambda\}_{\lambda\in(0\,,1)}$ is the set representation of fuzzy set $pF \oplus_{\min} qF$. Therefore $pF \oplus_{\min} qF = (p+q)F$.

Definition 9. *Let C be a conjunction and $F \in \mathcal{F}(U)$ be a fuzzy set. For arbitrary $x \in U$ and $p\,,q > 0$, if*

$$F(x/(p+q)) = \vee_{h\in U} C(F((x-h)/p)\,, F(h/q)) \tag{12}$$

or equivalently

$$F(x) = \vee_{(x_1\,,x_2)\in U^2\,,px_1+qx_2=(p+q)x} C(F(x_1)\,, F(x_2))\,, \tag{13}$$

then F is called a C-convex fuzzy set.

Proposition 17. *Let C be a conjunction, then $F \in \mathcal{F}(U)$ is a C-convex fuzzy set if and only if $pF \oplus_C qF = (p+q)F$ for arbitrary $p\,,q > 0$. Furthermore, F is a C-convex fuzzy set if and only if, for any $r \in [0\,,1]$, $F = rF \oplus_C (1-r)F$.*

Proposition 18. *Let $F \in \mathcal{P}(U)$, if F is a C-convex fuzzy set, then F is a convex set.*

Proposition 19. *The* min*-convexity of a fuzzy set is equivalent to its convexity.*

Proposition 20. *If a conjunction C is associative and continuous in its first argument satisfying $C(s\,,1) = s\,(s \in [0\,,1])$, and $F \in \mathcal{F}(U)$. Then, G is a C-convex fuzzy set if and only if for arbitrary $p\,,q > 0$,*

$$\mathcal{D}_{pG}\mathcal{D}_{qG} = \mathcal{D}_{(p+q)G}\,. \tag{14}$$

Proof. $\Rightarrow$ Let $F, G \in \mathcal{F}(U)$, $p, q > 0$, then $\mathcal{D}_{pG}\mathcal{D}_{qG}(F) = (F \oplus_C qG) \oplus_C pG = F \oplus_C (qG \oplus_C pG) = F \oplus_C (q+p)G = F \oplus_C (p+q)G = \mathcal{D}_{(p+q)G}(F)$.

$\Leftarrow$ In $F \oplus_C (pG \oplus_C qG) = F \oplus_C (p+q)G$ taking F as: $F(x) = \begin{cases} 1 & x = 0 \\ 0 & \text{otherwise} \end{cases}$, $x \in U$, then we have that

$$\begin{aligned} (F \oplus_C (pG \oplus_C qG))(x) &= \vee_{y \in U} C((pG \oplus_C qG)(x-y), F(y)) \\ &= C((pG \oplus_C qG)(x), 1) = (pG \oplus_C qG)(x), \\ (F \oplus_C (p+q)G)(x) &= \vee_{y \in U} C((p+q)G(x-y), F(y)) \\ &= C((p+q)G(x), 1) = (p+q)G(x). \end{aligned}$$

Hence, $pG \oplus_C qG = (p+q)G$.

Proposition 21. *Let C be a commutative and associative conjunction on $[0, 1]$ and (I, C) be an adjunction. If $G \in \mathcal{F}(U)$ is a C-convex fuzzy set, then for arbitrary $p, q > 0$,*

$$\mathcal{E}_{pG}\mathcal{E}_{qG} = \mathcal{E}_{(p+q)G}. \tag{15}$$

5 Granulometry

Granulometry is a mathematical characteristics for modeling the physical processing of sieving. It is a useful tool in image processing, especially for the design of noise suppressor. Firstly, we present the concept of openings [8].

Definition 10. *Let α be a mapping on $\mathcal{F}(U)$, if it is*

(1) *increasing: $\alpha(F) \subseteq \alpha(G)$ if $F \subseteq G$ for $F, G \in \mathcal{F}(U)$;*
(2) *anti-extensive: $\alpha(F) \subseteq F$ for arbitrary $F \in \mathcal{F}(U)$; and*
(3) *idempotent: $\alpha\alpha(F) = \alpha(F)$ for arbitrary $F \in \mathcal{F}(U)$,*

then it is called an opening on $\mathcal{F}(U)$.

In image analysis and processing, there are many opening operators used. $\alpha(X) = X \circ A = (X \ominus A) \oplus A$ is an opening, called structuring opening, for binary images, where A is a set, called a structuring element.

Let $T = (U, V)$ be a graph and k be a positive integer, then $\theta(A) = \begin{cases} A, & |A| \geq k \\ \emptyset, & \text{otherwise} \end{cases}$ is an opening, called an area opening. The operator $\alpha_p(X) = \{Y \in Conn(X) : Y_h \notin pA, h \in U\}$ is also an opening, called connected opening, where $Conn(X)$ denotes the connectivity class of $X \in \mathcal{P}(U)$ [6].

Let $\mathcal{C}$ be a connectivity class of $\mathcal{F}(U)$, $\alpha(F) = \cup\{G \in \mathcal{C} \mid G \subseteq F\}$ is also an opening on $\mathcal{F}(U)$.

It is easily proved that if $\mathcal{D}_G$ is a dilation on $\mathcal{F}(U)$ and $\mathcal{E}_G$ is an erosion on $\mathcal{F}(U)$, then their composition $\mathcal{D}_G\mathcal{E}_G$ is an opening for arbitrary $G \in \mathcal{F}(U)$.

Definition 11. *A family of opening $\{\alpha_p\}_{p>0}$ on $\mathcal{F}(U)$ is called a granulometry if*

$$\alpha_p\alpha_q = \alpha_q\alpha_p = \alpha_p, p \geq q > 0. \tag{16}$$

A granulometry $\{\alpha_r\}_{p>0}$ is called a Minkowski granulometry if for arbitrary $p>0$, the opening α_p is translation invariant, and $\alpha_{pG}(pF) = p\alpha_G(F)$ for arbitrary $F \in \mathcal{F}(U)$.

Proposition 22. *Suppose that $\{\alpha_p\}_{p\in P}$ is a family of openings, then the following assertions are equivalent, for all $p > q > 0$,*

(1) $\alpha_p \leq \alpha_q$;
(2) $\alpha_p\alpha_q = \alpha_q\alpha_p = \alpha_p$.

Proof. (1) $\Rightarrow$ (2) : $\alpha_p\alpha_q \leq \alpha_p\,, \alpha_q\alpha_p \leq \alpha_p$. And moreover, $\alpha_p\alpha_q \geq \alpha_p\alpha_p = \alpha_p\,, \alpha_q\alpha_p \geq \alpha_p\alpha_p = \alpha_p$.

(2) $\Rightarrow$ (1) : By the anti-extensive of openings, it is clear that $\alpha_p = \alpha_p\alpha_q \leq \alpha_q$.

If $\{\alpha_p\}_{p>0}$ is a collection of openings, let $g_p = \cup_{p\leq q}\alpha_q$, then $\{g_p\}_{p>0}$ is a granulometry. Furthermore, if for arbitrary nonempty index set J, $\{\alpha_p^{(j)}\}_{p>0}$ is a granulometry for every $j \in J$, then $\{g_p = \cup_{j\in J}\alpha_p^{(j)}\}_{p>0}$ is also a granulometry.

Proposition 23. *Let $(I\,, C)$ be an adjunction, if C is commutative and associative, and $G \in \mathcal{F}(U)$ is a C-convex fuzzy set, then $\{\alpha_p = \mathcal{D}_{pG}\mathcal{E}_{pG}\}_{p>0}$ is a granulometry. Moreover, $\{\alpha_p\}_{p>0}$ is a Minkowski granulometry if and only if G is a C-convex fuzzy set.*

Proof. From the adjunction of $(\mathcal{E}_{pG}\,, \mathcal{D}_{pG})$, we have that $\mathcal{D}_{pG}\mathcal{E}_{pG}\mathcal{D}_{pG} = \mathcal{D}_{pG}$ and $\mathcal{E}_{pG}\mathcal{D}_{pG}\mathcal{E}_{pG} = \mathcal{E}_{pG}$. Furthermore, for arbitrary $p > q > 0$, $\mathcal{D}_{pG} = \mathcal{D}_{qG}\mathcal{D}_{(p-q)G}$ and $\mathcal{E}_{pG} = \mathcal{E}_{(p-q)G}\mathcal{E}_{qG}$. Thus

$$\begin{aligned}
\alpha_p\alpha_q &= \mathcal{D}_{pG}\mathcal{E}_{pG}\mathcal{D}_{qG}\mathcal{E}_{qG} = \mathcal{D}_{pG}\mathcal{E}_{(p-q)G}\mathcal{E}_{qG}\mathcal{D}_{qG}\mathcal{E}_{qG} \\
&= \mathcal{D}_{pG}\mathcal{E}_{(p-q)G}\mathcal{E}_{qG} = \mathcal{D}_{pG}\mathcal{E}_{pG} = \alpha_p\,, \\
\alpha_q\alpha_p &= \mathcal{D}_{qG}\mathcal{E}_{qG}\mathcal{D}_{pG}\mathcal{E}_{pG} = \mathcal{D}_{qG}\mathcal{E}_{qG}\mathcal{D}_{qG}\mathcal{D}_{(p-q)G}\mathcal{E}_{pG} \\
&= \mathcal{D}_{qG}\mathcal{D}_{(p-q)G}\mathcal{E}_{pG} = \mathcal{D}_{pG}\mathcal{E}_{pG} = \alpha_p\,.
\end{aligned}$$

If $p = q > 0$, it's natural that $\alpha_p\alpha_q = \alpha_q\alpha_p = \alpha_p$.

For $F \in \mathcal{F}(U)$, $h \in U$, note that $F_h(x) = F(x-h)\,(x \in U)$, we have that

$$\begin{aligned}
\alpha_p(F_h) &= (\mathcal{D}_{pG}\mathcal{E}_{pG})(F_h) = \mathcal{D}_{pG}(\mathcal{E}_{pG}(F_h)) \\
&= \mathcal{D}_{pG}((\mathcal{E}_{pG}(F))_h) = (\mathcal{D}_{pG}(\mathcal{E}_{pG}(F)))_h = (\alpha_p(F))_h\,.
\end{aligned}$$

Furthermore, for arbitrary $x \in U$,

$$\begin{aligned}
\alpha_p(pF)(x) &= \mathcal{D}_{pG}\mathcal{E}_{pG}(pF)(x) = \mathcal{D}_{pG}(\mathcal{E}_{pG}(pF)(x)) \\
&= \vee_{y\in U}C(pG(x-y)\,, \wedge_{z\in U}I(pG(z-y)\,, pF(z))) \\
&= \vee_{y\in U}C(G(x/p-y/p)\,, \wedge_{z\in U}I(G(z/p-y/p)\,, F(z/p))) \\
&= \vee_{y'\in U}C(G(x/p-y')\,, \wedge_{z'\in U}I(G(z'-y')\,, F(z'))) \\
&= \mathcal{D}_G\mathcal{E}_G(F)(x/p) \\
&= \alpha_G(F)(x/p) \\
&= p\alpha_G(F)(x)\,.
\end{aligned}$$

Therefore, $\{\alpha_p\}_{p>0}$ is a Minkowski granulometry.

The reverse statement can be proved easily.

Proposition 24. *Assume that conjunction C is commutative and associative, and G is a C-convex fuzzy set. For every positive integer $n \in Z^+$, let $\alpha_n = \mathcal{D}^n_G \mathcal{E}^n_G$, then $\{\alpha_n\}_{n \in Z^+}$ is a Minkowski granulometry.*

In the crisp case, the granulometry reduces to that presented in [6].

Proposition 25. *Given a compact structuring element $A \in \mathcal{P}(U)$, the openings $\alpha_p(X) = X \circ pA$ define a Minkowski granulometry if and only if A is convex.*

Granulometry performs the process of sieving with increasing sizes of structuring elements. It is a powerful tool for processing problems concerning the geometric features and granular analysis of an image. For a given structuring element, an object in an image can be extracted or removed by choosing appropriate size of a structuring element. In the following section, we presented an example for granulometry in the application of image denoising.

6 Experiment Results

In this section, we present an numerical example to show the application of granulometry in image denoising. Here we only show the application of the result in Proposition 23. The other formalizations of granulometry have also perfect effects in this application.

Fig. 1. Original images and the corrupted image $((a), (b))$

Fig. 2. The output results $((a), (b))$

We choose the conjunction $C(s\,,t) = \begin{cases} min(2s\,,1)\cdot min(2t\,,1)/2\,, & s\cdot t < 1/2 \\ s\cdot t\,, & s\cdot t \geq 1/2 \end{cases}$ and a fuzzy set G, presented by a matrix as follow

$$G = \begin{pmatrix} 0.1 & 0.7 & 0.1 \\ 0.7 & 1 & 0.7 \\ 0.1 & 0.7 & 0.1 \end{pmatrix}.$$

We take the 256×256 image (Fig. 1(a)) as our experimental object and add 10% salt-pepper noise as well as 10% random noise (Fig. 1(b)) in it. By computation, we obtain that α_1 and α_2, shown as Fig. 2(a) and Fig. 2(b), respectively.

From the simulated result, we see that there are less noises in Fig. 2(a) as well as in Fig. 2(b). The 'salt and pepper' noises are removed well, either the random noises.

In the processing of denoising, the edges of objects in images have been fuzzified. In combination with other edge smoothing algorithms, it is believed that the image will be recovered very well.

7 Conclusion

Granulometry is one of powerful tools in mathematical morphology for detecting size distribution of objects in image analysis and processing. It is developed based on the theory of convex set and is used for image denoising and object recognition. In this paper, we analyzed the the convexity of fuzzy sets and introduced the concept of C-convex fuzzy set. With this notion, we discussed the fundamental morphological operators based on fuzzy logic. After that, a framework of granulometry was established regarding a family of morphological opening based on fuzzy logic.

Image denoising is a hard nut to crack in image analysis. There is not an efficient method to remove the random noise yet. We presented an example for the approach to granulometry in image denoising. The experiment image contains salt and pepper noises as well as random noises. By using granulometric approach, a satisfactory result is obtained. It is proved that granulometry is a practical tool for image denoising.

Acknowledgements

This work is partially supported by the postdoctoral science-research developmental foundation of Heilongjiang province (LBH-Q05047), the fundamental research foundation of Harbin Engineering University (HEUFT05087) and the the Multi-discipline Scientific Research Foundation of Harbin Institute of Technology (HIT.MD2001.24).

References

1. Balagurunathan Y, Dougherty ER (2003) Granulometric parametric estimation for the random boolean model using optimal linear filters and optimal structuring elements. Pattern Recognition Letters 24: 283–293
2. Braga-Neto UM, Goutsias J (2004) Grayscale level connectivity: Theory and applications. IEEE Transactions on Image Processing 13: 1567–1580
3. Chen Y, Dougherty ER (1994) Gray-scale morphological granulometric texture classification. Optical Engineering 33: 2713–2722
4. de Moraes RM, Banon GJ, Sandri SA (2002) Fuzzy expert systems architecture for image classification using mathematical morphology operators. Information Sciences 142: 7–21
5. Deng TQ, Heijmans H (2002) Grey-scale morphology based on fuzzy logic. Journal of Mathematical Imaging and Vision 16: 155–171
6. Goutsias J, Heijmans H (2000) Fundamenta morphologicae mathematicae. Fundamenta Informaticae 41: 1–31
7. Hanbury AG, Serra J (2001) Morphological operators on the unit circle. IEEE Transactions on Image Processing 10: 1842–1850
8. Heijmans H (1994) Morphological image operators. Academic Press, Boston
9. Matheron G (1975) Random sets and integral geometry. John Wiley & Sons, New York
10. Serra J (1982) Image analysis and mathematical morphology. Academic Press, London
11. Serra J (2006) A lattice approach to image segmentation. Journal of Mathematical Imaging and Vision 24: 83–130
12. Mukherjee DP, Acton ST (2000) Scale space classification using area morphology. IEEE Transactions on Image Processing 9: 623–635

A Hybrid Decision Tree Model Based on Credibility Theory

Chengming Qi

College of Automation, Beijing Union University, China, 100101
Qicm123@163.com

Abstract. Decision-Tree (DT) is a widely-used approach to retrieve new interesting knowledge. Fuzzy decision trees (FDT) can handle symbolic domains flexibly, but its preprocess and tree-constructing are much costly. In this paper, we propose a hybrid decision tree (HDT) model by introducing credibility entropy into FDT. Entropy of multi-valued and continuous-valued attributes are both calculated with credibility theory, while entropy of other attributes is dealt with general Shannon method. HDT can decrease the cost of preprocess and tree-constructing significantly. We apply the model into geology field to find out the factors which cause landslide. Experiment results show that the proposed model is more effective and efficient than fuzzy decision tree and C4.5.

Keywords: Classification, Credibility Theory, Fuzzy Decision Tree, Fuzzy Entropy.

1 Introduction

Classical crisp decision trees are widely applied in pattern recognition, machine learning and data mining. Decision tree is introduced by Quinlan for inducing classification models [1]. It classifies a sample by propagating the sample along a path, from the root down to a leaf, which contains the classification information.

From a training set, inductive learning enables us to extract knowledge. Such knowledge is used to associate any forthcoming description with a correct decision. For instance, an inductive learning process is the construction of a decision tree that is used to classify unknown descriptions. Decision tree is a natural and understandable representation of knowledge. Each node in such a tree is associated with a test on the values of an attribute. Each edge is labeled with a particular value of the attribute, and each leaf of the tree is associated with a value of the class[1]. However, the decision with the previous description might vary greatly, though value of attributes change slightly. It is why to introduce fuzziness into decision trees [2].

Fuzzy decision tree is a more general means to represent knowledge. It enables us to use both numerical and symbolic values in representing fuzzy modalities, during either the learning phase (construction of a tree) or the generalization phase. Moreover, Bouchon-Meunicr and Marsala argue that a FDT is equivalent to a set of fuzzy rules [3]. And such a kind of induced rules can be introduced to optimize the query process of the database or to deduce decisions from data [4].

B.-Y. Cao (Ed.): Fuzzy Information and Engineering (ICFIE), ASC 40, pp. 930–939, 2007.
springerlink.com

The reminder of this paper is arranged as follows. Section 2 introduces credibility theory. Section 3 proposes a hybrid model by introducing credibility entropy into FDT. Section 4 illustrates the experiment result and compares to other methods. Finally, conclusion is in section 5.

2 Credibility Theory

Credibility theory was founded by Liu [5] in 2004 as a branch of mathematics for studying the behavior of fuzzy phenomena.

The concept of fuzzy set was initiated by Zadeh [6] via membership function in 1965. In order to measure a fuzzy event, Zadeh [7] proposed the concept of possibility measure in 1978. Although possibility measure has been widely used, it has no self-duality property. However, a self-dual measure is absolutely needed in both theory and practice. To define a self-dual measure, the concept of credibility measure was presented in 2002[8]. After that, Li and Liu [9] gave a sufficient and necessary condition for credibility measure.

Let ξ be a fuzzy variable with membership function μ, then its α-level set is

$$\xi_\alpha = \{x | x \in \Re, \mu(x) \geq \alpha\} \tag{1}$$

Let ξ be a fuzzy variable with membership function μ. Then for any set B of real numbers, we have

$$Cr\{\xi \in B\} = \frac{1}{2}[\sup_{x \in B} \mu(x) + 1 - \sup_{x \in B^c} \mu(x)] \tag{2}$$

where B^c is complement of set B.

Let $\xi_1, \ldots, \xi_n$ be independent fuzzy variable with membership function $\mu_1, \ldots, \mu_n$, respectively, and $f : \Re^n \longrightarrow \Re^m$ a function. Then for any set $B \in \Re^m$, the credibility $Cr\{f(\xi_1, \ldots, \xi_n) \in B\}$ is

$$\frac{1}{2}[\sup_{f(x_1,x_2,\ldots,x_n) \in B} \min_{1 \leq i \leq n} \mu_i(x_i) + 1 - \sup_{f(x_1,x_2,\ldots,x_n) \in B^c} \min_{1 \leq i \leq n} \mu_i(x_i)] \tag{3}$$

Fuzzy entropy is a measure of uncertainty and has been studied by many researchers such as Lao [10], Yager [11, 12, 13, 14], Liu [15], Pal and Pal [16]. A survey on fuzzy entropy can be found in Pal and Eezdek [17].

Especially, when ξ is a fuzzy set taking values x_i with membership degrees, $i = 1, \ldots, n$, respectively, De Luca and Termini [18] defined its entropy as $S(\mu_1) + S(\mu_2) + \cdots + S(\mu_n)$, where $S(t) = -t \ln t - (1-t) \ln 1 - t$. It is easy to verify that the function $S(t)$ is symmetrical about $t = 0.5$, strictly increases on the interval [0, 0.5], strictly decreases on the interval [0.5, 1], and reaches its unique maximum $\ln t$ at $t = 0.5$.

The entropy by De Luca and Termini [18] characterizes the uncertainty resulting primarily from the linguistic vagueness rather than resulting from information deficiency, and vanishes when the fuzzy variable is an equipossible one. However, see hope that the entropy is 0 when the fuzzy variable degenerates to a crisp number, and is maximum when fuzzy variable is an equipossible one. In order to meet such a requirement, Li and Liu [19] provided a new definition of fuzzy entropy to characterize the uncertainty resulting from information deficiency which is caused by the impossibility to predict the specified value that a fuzzy variable takes.

Let ξ be a discrete fuzzy variable taking values in $\{x_1, x_2, \ldots\}$. Then its entropy is defined by

$$H[\xi] = \sum_{i=1}^{\infty} S(Cr\{\xi = x_i\}) \tag{4}$$

where $S(t) = -t \ln t - (1-t) \ln 1 - t$.

It is clear that the entropy depends only on the number of values and their credibility, but not on the actual values that the fuzzy.

Let ξ be a discrete fuzzy variable taking values in $\{x_1, x_2, \ldots\}$.Then $H[\xi] \geq 0$ and equality holds if and only if ξ is essentially a crisp number.

Thus theorem states that the entropy of a fuzzy variable reaches its minimum 0 when the fuzzy variable degenerates to a crisp number. In this case, there is no uncertainty.

For example, let ξ be a trapezoidal fuzzy variable *(a, b, c, d)*. Then its entropy is $H[\xi] = (d-a)/2 + (\ln 2 - 0.5)(c-b)$.

Given some constraints, there are usually multiple compatible membership functions. For this case, the maximum entropy principle attempts to select the membership function that maximizes the value of entropy and satisfies the prescribed constraints.

3 Hybrid Decision Trees Model

In FDT, The fuzzy sets of all attributes are user-defined. Each attribute is considered as a linguistic variable. In our opinion, some attributes are not necessary being so. In this section, a hybrid decision tree (HDT) model is proposed by introducing credibility entropy into FDT. Entropy of multi-valued and continuous-valued attributes are both calculated with credibility theory, while entropy of other attributes is dealt with general Shannon method.

3.1 Fuzzification of Numerical Numbers

A decision tree is equivalent to a set of IF...THEN rules. Each path in the tree is associated with a rule, where premises are composed by the tests of the encountered nodes, and the conclusion of the rule is composed by the class associated with the leaf of the path:

$$R_i : if\, A_1\, is\, v_{i1}\, and \ldots and\, A_n\, is\, v_{in}\, then\, c = c_k\, with\, the\, degree\, p \tag{5}$$

Edges can be labeled by so-called numeric-symbolic values. A numeric-symbolic attribute is an attribute whose values can be either numeric or symbolic linked with a numerical variable. Such kind of values leads to the generalization of decision trees into fuzzy decision trees [20].

FDT enable us to use numeric-symbolic values either during their construction or when classifying new cases. As with crisp decision trees, fuzzy decision tree induction involves the recursive partitioning of training data in a top-down manner.

Fuzzification is a process of fuzzifying numerical numbers into linguistic terms, which is often used to reduce information overload in human decision making process. The Earthquake Magnitude, for example, may be perceived in linguistic terms such as high, medium and low. One way of determining membership functions of these linguistic terms is by expert opinion or by people's perception. Yet another way is by statistical methods [21]. An algorithm for generating certain type of membership functions, which is based on self-organized learning [22], can be found in [23].

Based on Credibility theory, by a trapezoidal fuzzy variable we mean the fuzzy variable fully determined by the quadruplet *(a; b; c; d)* of crisp numbers with $a < b < c < d$, whose membership function is given by [24]

$$f(x) = \begin{cases} \frac{x-a}{b-a} & \text{if } a < x < b; \\ 1 & \text{if } b < x < c; \\ \frac{x-d}{c-d} & \text{if } c < x < d; \\ 0 & \text{otherwise.} \end{cases} \quad (6)$$

Consider Earthquake Magnitude attribute from Table 1 has three linguistic terms, we can find cut points as Fig.1.

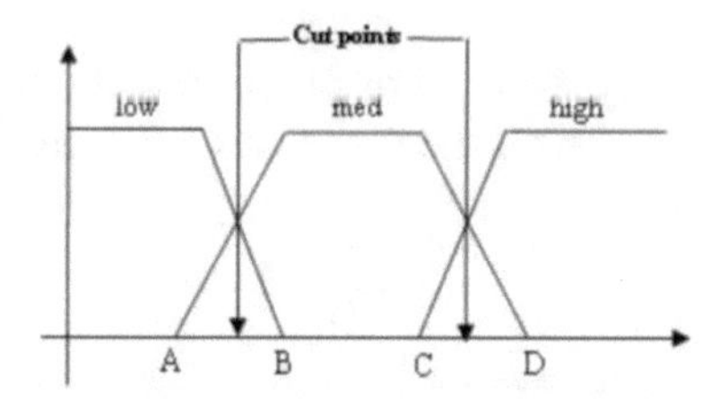

Fig. 1. Earthquake attribute membership function

Fig.1 depicts the overlapping membership functions. A point "near" to a cut point belongs to two fuzzy sets with a membership grade less than 1 and greater than 0 for both membership functions. Assuming that the pair (cut point; 0.5) is the point of intersection of two membership functions, only one additional point on each side of the trapezoid is necessary to determine the membership functions. The corner points of the trapezoid are calculated as follows. Value B is equal to the first attribute value greater than the cut point (all attribute values are sorted). Likewise, A is equal to the attribute value left of the cut point.

It is obvious that the three linguistic terms can be described as Low, Medium and High. Table 2 shows the membership degree of the attribute Earthquake Magnitude belonging to the three membership functions. Similarly we can find membership function for others attributes.

3.2 Hybrid Decision Trees Model

Let $I=(U,A)$ be an information system, where U is a non-empty set of N finite objects and A is a non-empty finite set of n attributes,. An attribute A^k takes m_k values of fuzzy subsets$\{A_1^k, A_2^k, \ldots, A_n^k\}$. Based on the attributes, an object is classified into C fuzzy subsets$\{\omega_1, \omega_2, \ldots, \omega_c\}$. The fuzzy entropy for a subset can be defined as (4).

It is well known that the size of the overall tree strongly influences the generalization performance obtained from the tree based classifier [25]. As with other empirical methods of model construction, several approaches have been used in the context of decision trees to produce trees of overall small size and depth. Pruning is the removal of sub-trees of the tree that have little statistical validity [26, 27] thereby obtaining a tree with smaller size. Pre-pruning and post-pruning are two standard methods for pruning a decision tree. Post-pruning are based on removing a node (sub-tree) after the tree has been constructed. Our algorithm adopts the latter.

Assume that these linguistic terms is considered as a set of data D, where each data has l numerical values for n attributes $A^{(1)}, A^{(2)}, \ldots\ldots, A^{(n)}$.We take the reasonable sample size from n attributes $A^{(1)}, A^{(2)}, \ldots\ldots, A^{(n)}$.We apply the algorithm proposed in [20],and the primary steps are listed as follows.

1. Generate the root node that has a set of all data, i.e., a fuzzy set of all data with the membership value 1.
2. If a node O with a fuzzy set of data D satisfies the following conditions:
   - the proportion of a data set of a class C_k is greater than or equal to a threshold θ_r , that is
   $$\frac{|D^{C_k}|}{|D|} \geq \theta_r \tag{7}$$
   - there are no attribute for more classifications, then it is a lead node and assigned by the class name.
3. Test for leaf node (see section 3 for three conditions)
4. (a)Find a test attribute
   (b) Divide the data according to this attribute
   (c) Generate new nodes for fuzzy subsets
5. Recursion of the process for the new nodes from point 2

In our method only point 4(a) is modified as follows:

For attributes A_i $(i = 1,2,\ldots,l)$, find the best cut point, according to (4), compute the fuzzy entropy of continuous-valued and multi-valued, compute Shannon entropy of others, ..., and select the test attribute A_{min} that minimizes them.

The number of cut points produced by our method may be very large if the attribute is not very informative. Catlett [28] proposed three criteria to stop the recursive splitting. The most important one is to limit the number of cut points (intervals) by the user.

4 Experimental Results and Discussion

In this section, we apply the above approach on the landslide database to find out the factors which cause landslide, and we obtain some fuzzy rules finally.

4.1 Experimental Results

The data consisted of 2300 cases. Of these cases, 237 were stable (class 1), 775 were hypo-stable (class 2), and 1288 of them were unstable (class 3). Nine attributes associated with each of the 2300 samples are listed in Table 1.

Table 1. List of Attributes

Attribute Number	Attribute Name	Value
X1	lithology	1— rocky layer
		0— incompact layer
X2	angle of dip	0-90
X3	earthquake magnitude	4-8
X4	original gradient	0-90
X5	slope shape	1— concave
		0— convex
X6	maximum amount of rain	50-238
X7	microtopography	1— steep incline
		0— gentle slope
X8	physical factor	1— weathering
		0— flood erosion
X9	human factor	1— vegetation deterioration
		0— excessive cutting slope

Three classes with labels A, B, and C denote stable, hypo-stable, and unstable, respectively. The numbers in Table 2 are the membership values. Membership values are not probabilities, thus, it is not necessary for membership values of all linguistic terms of an attribute to add to one.

The construction of the tree, based on our algorithm proceeds as follows. At the root node, the fuzzy entropy for the attributes Angle of dip, Earthquake magnitude (which is denoted by *Earthquake Mag.* in Table 2) and Original gradient based on (4) is

$$\begin{aligned} &H[angle\quad of\quad dip] = 0.4317 \\ &H[earthquake\quad magnitude] = 0.3863 \\ &H[original\quad gradient] = 1.0651 \end{aligned} \qquad (8)$$

Table 2. Membership Values of Some Attributes (part)

Angle of Dip			Earthquake Mag.			Original Gradient			Stable State		
large	mid	small	high	med	low	large	mid	small	A	B	C
0.0	0.0	1.0	0.0	1.0	0.0	0.0	0.2	0.8	0.9	0.2	0.0
0.0	0.1	0.8	0.0	1.0	0.0	0.0	0.3	0.7	0.9	0.2	0.0
0.0	0.3	0.0	0.0	0.4	0.7	0.0	0.2	0.8	0.8	0.2	0.0
0.2	0.6	0.1	0.0	0.4	0.7	0.0	0.3	0.8	0.2	0.9	0.0
0.0	0.1	0.8	0.0	1.0	0.0	0.0	0.2	0.7	0.2	0.9	0.0
0.5	0.3	0.0	0.0	0.4	0.7	0.0	0.4	0.6	0.1	0.8	0.0
0.0	0.1	0.8	0.0	0.4	0.7	0.0	0.2	0.8	0.2	0.9	0.0
e0.7	0.4	0.0	1.0	0.0	0.0	0.2	0.7	0.1	0.0	0.1	0.9
e0.1	0.6	0.4	0.4	0.7	0.0	0.1	0.6	0.4	0.0	0.2	0.8
0.0	0.4	0.6	0.4	0.7	0.0	0.2	0.8	0.1	0.0	0.1	0.9

And compute Shannon entropy of others attribute, for example, Slope shape and Lithology:

$$\begin{aligned} H[slope \quad shape] &= 0.4465 \\ H[lithology] &= 0.7725 \end{aligned} \tag{9}$$

Thus, using (8) and (9), we choose the attribute Earthquake Magnitude, which has the lowest fuzzy entropy at the root node. As mentioned in section 4, repeating this further for each of the child nodes, we arrive at the decision tree as shown in Fig.2.

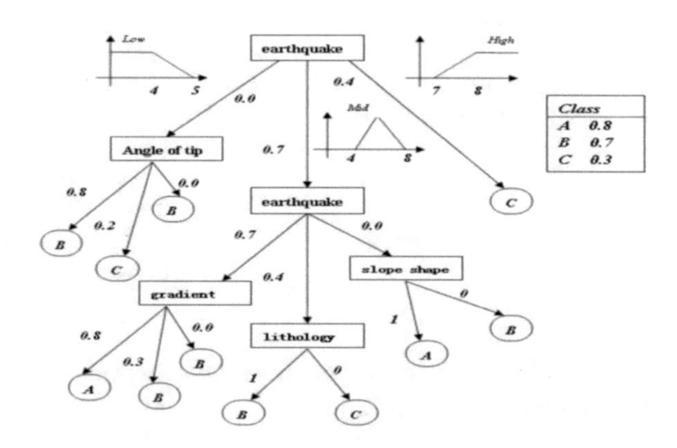

Fig. 2. Binary tree learned from 2300 cases

This implies that the value at the parent node was greater than 1; however, the value at each the child node is less than 1. This example illustrates that choosing an attribute based on jointly considering the node splitting criterion (information gain here) and classifiability of the instances along all the child branches can lead to smaller decision trees.

We can extract some rules from the above FDT like (5):

Rule 1: IF (Earthquake Magnitude =Low) AND (Angle of dip =Small) THEN (Stable state=hypo-stable with the degree 0.8) and (Stable state=unstable with the degree 0.2)

Rule 2: IF (Earthquake Magnitude =Medium) AND (Earthquake Magnitude =Low) AND (Original Gradient =Low) THEN (Stable state=hypo-stable with the degree 0.8) and (Stable state=unstable with the degree 0.3)

Rule 3: IF (Earthquake Magnitude =Medium) AND (Earthquake Magnitude = Medium) AND (Lithology = incompact layer THEN (Stable state=hypo-stable with the degree 0.8)

Rule 4: IF (Earthquake Magnitude =Medium) AND (Earthquake Magnitude = Low) AND (slope shape= concave) THEN (Stable state= stable with the degree 0.8)

Rule 5: IF (Earthquake Magnitude =High) THEN (Stable state=unstable with the degree 0.3)

These rules constitute a useful model to predict the degree of landslide effectively.

4.2 Discussion

The heuristic information of fuzzy ID3 is based on the minimum fuzzy entropy. The fuzzy entropy defined on a discrete fuzzy set refers to the fuzziness (vagueness measurement)[18], similar to Shannon's entropy measure of randomness. The proposed method calculates entropy of multi-valued and continuous-valued attributes based on credibility theory, and calculates Shannon entropy of other attributes.

For landslide database, we separate the data into two sets, one is the training set which contains 70% of all the cases and the other is the testing set which contains the remains 30%. We compare our algorithm with Fuzzy-ID3(FDC) and C4.5 in three facets: number of leaf nodes, average depth and testing accuracy And the result is shown in Table 3.

Table 3. Average Percentage Accuracy

Method	Size	Classification Rate
HDT	6.7	80.3%
FDC	8.4	76.5%
C4.5	10.2	72.9%

The size is the average number of paths of the built trees, and the classification rate is the number of test examples which are well classified by means of the built tree. It can be observed that our approach performed better than Fuzzy-ID3 and C4.5 in the above cases. It achieved an average accuracy of 80.3%, and it is better than Fuzzy-ID3 by 3.8% and C4.5 by 7.4%.

5 Conclusion

In this paper, we propose a hybrid decision trees model to classification trees. A new splitting criteria is developed based on credibility entropy with a fuzzy cumulative distribution function. The criteria adopt minimum classification information entropy to select expanded attributes. We apply HDT into geology field to find out the factors which cause the landslides from sample database. The rules can be expressed as fuzzy rules in which the number and size of the membership functions are automatically determined. The experiment result shows that the hybrid model is efficient and effective.

In future works, we will analyze the stability of the proposed hybrid model in constructing fuzzy decision trees. We will also apply the model to many other application fields.

References

1. Quinlan(1986) Induction on decision trees.Machine Learning,1(1):81-106
2. B. Bouchon-Meunier, C. Marsala, and M. Ramdani(1997) Learning from imperfect data. In D. Dubois, H. Prade, and R. R.Yager editors, Fuzzy Information Engineering: a Guided Tour of Applications, pages 139-148. John Wileys and Sons
3. B. Bouchon-Meunicr and C. Marsala(1999) Learning fuzzy decision rules. In D. Dubois J. Bczdek and H. Prade, editors, Fuzzy Sets in Approximate Reasoning and Information Systems, volume 3 of Handbook of Fuzzy Sets, chapter 4. Kluwer Academic Publisher
4. B. Lent, 11. Swami, and J. Widom(1997) Clustering association rule. In Proceedinds of the 13th International Conference on Data Engineering,pages 220-231, Birmingham, UK. IEEE Computer Society Press
5. Liu B(2004) Uncertainty Theory: An Introduction to its Axiomatic Foundations,Springer-Verlag, Berlin
6. Zadeh LA(1965) Fuzzy sets, Information and Control, Vol.8, 338-353
7. Zadeh LA(1978) Fuzzy sets as a basis for a theory of possibility, Fuzzy Sets and Systems, Vol.1, 3-28
8. Liu B, and Liu YK(2002) Expected value of fuzzy variable and fuzzy expected value models, IEEE Transactions on Fuzzy Systems, Vol.10, No.4,445-450
9. Li X, and Liu B(2006) New independence definition of fuzzy random variable and random fuzzy variable, World Journal of Modelling and Simulation
10. Loo SG(1977) Measures of fuzziness, Cybernetica,Vol.20, 201-210
11. Yager RR(1979) On measures of fuzziness and negation, Part I: Membership in the unit interval, International Journal of General Systems,Vol.5, 221-229
12. Yager RR(1980) On measures of fuzziness and negation, Part II: Lattices, Information and Control,Vol.44, 236-260
13. Yager RR(1983) Entropy and specificity in a mathematical theory of evidence, International Journal of General Systems,Vol.9, 249-260
14. Yager RR(1995) Measures of entropy and fuzziness related to aggregation operators, Information Sciences,Vol.82, 147-166
15. Liu XC(1992) Entropy, distance measure and similarity measure of fuzzy sets and their relations, Fuzzy Sets and Systems,Vol.52, 305-318

16. Pal NR(1992) and Pal SK, Higher order fuzzy entropy and hybrid entropy of a set, Information Sciences,Vol.61, 211-231
17. Pal NR(1994) and Bezdek JC, Measuring fuzzy uncertainty, IEEE Transactions on Fuzzy Systems,Vol.2, 107-118
18. De Luca A, and Termini S(1972) A definition of nonprobabilistic entropy in the setting of fuzzy sets theory, Information and Control,Vol.20, 301-312
19. Li P, and Liu B(2005) Entropy of credibility distributions for fuzzy variables, Technical Report
20. M. Umano, H. Okamoto, l. Hatono, H. Tamura, F. Kawachi, S. Umedzu, and J. Kinoshita(1994) Fuzzy decision trees by fuzzy ID3 algorithm and its application to diagnosis systems. In Proceedings of the 3rd IEEE Conference on Fuzzy Systems, volume 3, pages 2113-2118, Orlando
21. M. R. Civanlar and H. J. Trussell(1986) "Constructing membership functions using statistical data," Fuzzy Sets and Systems,vol. 18, pp. 1-14
22. T. Kohonen(1988) "Self-Organization and Association Memory "Springer, Berlin
23. Y. Yuan and M. J. Shaw(1995) "Induction of fuzzy decision trees," Fuzzy Sets and Systems,vol. 69, pp. 125-139
24. K. J. Cios and L. M. Sztandera(1992) "Continuous ID3 algorithm with fuzzy entropy measures," Proc. IEEE Int. Conf. Fuzzy Syst,pp. 469-476
25. L. Breiman, J. H. Friedman, J. A. Olshen, and C. J. Stone(1984)Classification and Regression Frees,Wadsworth International Group, Behnont, CA
26. J. Mingers(1989) "An empirical comparison of pruning methods for decision-tree induction," Machine Learning ,vol. 4, pp. 227-243
27. J. Frnkranz(1997) "Pruning algorithms for rule learning," Machine Learning,vol. 27, pp. 139-172
28. J. Catlett(1991) On changing continuous attributes into ordered discrete attributes. In Y. Kodratoff, editor proc. of 5th European Working Sessions on Learning,pages 164-178. Springer, Berlin, Heidelherg, Porto, Portugal

A Region-Based Image Retrieval Method with Fuzzy Feature

Min Tang[1], Ai-min Yang[1,2], Ling-min Jiang[2], and Xing-guang Li[2]

[1] School of Computation and Communication, Hunan University of Technology, Wenhua Road, Zhuzhou, Hunan, China, 412008
monicabey@ 126.com, amyang18@163.com
[2] School of Informatics, Guangdong University of Foreign Studies, Guangzhou, Guangdong, China, 510420

Abstract. A novel method is proposed in the paper, called A Region-Based Image Retrieval Method with Fuzzy Features(RBIRFF), which partitions images into regions firstly, then computes each region fuzzy features for color, texture and shape with fuzzy set theory, finally computes similarity between images and gets the retrieval results according to the similarity sequences. The design and realization of RBIRFF system is also introduced in the paper, and its retrieval effectiveness is evaluated by experiments.

Keywords: Content-Based Image Retrieval, Fuzzy Color Histogram, Fuzzy Texture Features, Fuzzy Shape Features, Similarity Computation.

1 Introduction

Content-Based Image Retrieval(CBIR) is on the basis of image itself vision features to describe image, to set up feature vector and retrieve by these similarities between images.Berkeley Blobworld[1], UCSB Netral[2], Columbia VisualSEEK[3], and Standford IRM[4], of which are the classic region based CBIR system which require significant user interaction in defining or selecting region features. Wang et al.[5] recently proposed an integrated region matching scheme called IRM for CBIR, which allows for matching a region in one image to several regions of another image. This scheme decreases the impact of inaccurate region segmentation. Nevertheless, the color representation of each region is simplistic such that much of the rich color information in a region is lost, as it fails to explicitly capture the "inaccuracy" of the color features exhibited by the fuzzy nature in the feature extraction.

To accommodate the imprecise image segmentation and uncertainty of human perception, propose to fuzzify each region generated in image segmentation using a parameterized membership function. The feature vector of a block belongs to multiple regions with different degrees of membership as opposed to the classic region representation, in which a feature vector belongs to exactly one region.

A region-based image retrieval method with fuzzy features is proposed in the paper. The method is, firstly partition images into blocks, and extract color and texture features of each block in the L*a*b* space, secondly apply K-mean cluster algorithm

B.-Y. Cao (Ed.): Fuzzy Information and Engineering (ICFIE), ASC 40, pp. 940–948, 2007.
springerlink.com

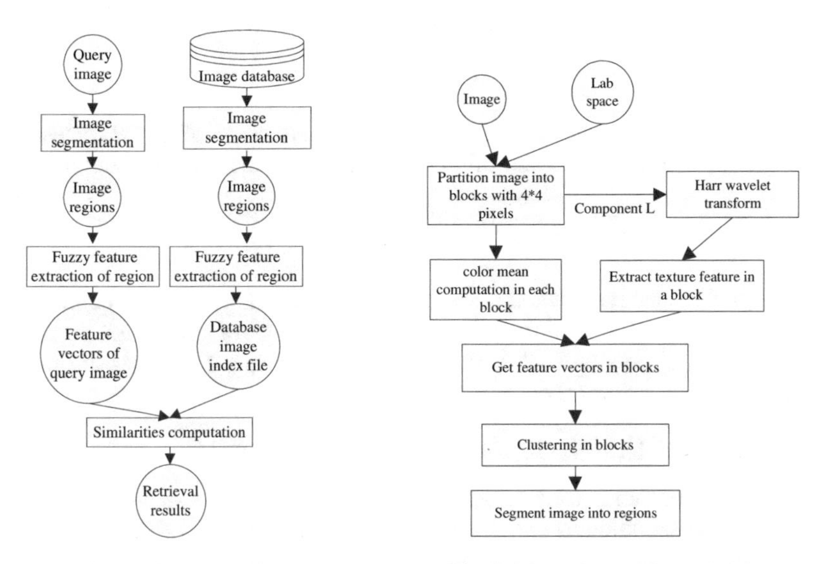

Fig. 1. Flow Chart of RBIRFF

Fig. 2. Flow chart of Segmentation

to segment images into regions, thirdly extract fuzzification of the color, texture and shape features with fuzzy set theory, finally, apply similarity computation to measure the similarities between a query image and images in image database, and show the results by similarities descend. Fig.1 shows the flow of the whole RBIRFF.

The remainder is organized as follows. Section 2 describes how to segment images into regions. Region fuzzy features extraction is introduced in Sect.3. The region matching and similarity computation are discussed in Sect.4. Sect.5 presents the experiment results.

2 Image Segmentation

This section will describe image segmentation process in detail. Fig.2 shows the flow of the segmentation.Many researches show that, the result of segmentation with color and texture is better than that of segmentation with either feature, so, this paper segments image by extracting color and texture features of image in L*a*b* color space.

To segment an image, first of all, partition the image into blocks with 4*4 to compromise between texture effectiveness and computation time. Then extract a feature vector consisting of six features from each block, block i is $\{b_i^{C_1}, b_i^{C_2}, b_i^{C_3}, b_i^{T_1}, b_i^{T_2}, b_i^{T_3}\}$,where, $\{b_i^{C_1}, b_i^{C_2}, b_i^{C_3}\}$ are average color components and $\{b_i^{T_1}, b_i^{T_2}, b_i^{T_3}\}$ are texture features, which represent energy in the high frequency bands of the Harr wavelet transform. After a one-level wavelet transform, a 4*4 block is decomposed into four frequency bands; each band contains 2*2 coefficients.

Suppose coefficients of block i in HL band are $\{T_{i,11}^{HL}, T_{i,12}^{HL}, T_{i,21}^{HL}, T_{i,22}^{HL}\}$, compute the feature as Eq.(1). The other two texture features are gotten from band LH and band HH by the similar computation.

$$b_i^{T_1} = \sqrt{\tfrac{1}{4}\left(\sum_{j=1}^{2}\sum_{k=1}^{2}\left(T_{i,jk}^{HL}\right)^2\right)} \tag{1}$$

After we get feature vectors for all blocks, a color-texture weighted L2 distance metric is as Eq.(2) in the k-means algorithm used to describe distance between blocks.

$$D_b^{pq} = \sqrt{\omega_c \sum_{j=1}^{3}\left(b_p^{C_j} - b_q^{C_j}\right)^2 + \omega_t \sum_{j=1}^{3}\left(b_p^{T_j} - b_q^{T_j}\right)^2} \tag{2}$$

where p, q are respectively block p and q. ω_c and ω_t are the weights specified in specific experiments. Color influences segmentation more than texture so that $\omega_c = 0.70$ and $\omega_t = 0.30$ though a lot of trials.After clustering, K regions are obtained, and each region feature vector is Eq.(3).

$$R_n = [R_n^{C_1}, R_n^{C_2}, R_n^{C_3}, R_n^{T_1}, R_n^{T_2}, R_n^{T_3}]^T \; (n = 1, \ldots, K) \tag{3}$$

The k-means algorithm is used to cluster the feature vectors into several classes with every class corresponding to one region in the segmented image. The k-means algorithm does not specify how many clusters to choose. We adaptively select the number of clusters K by gradually increasing K until a stop criterion is met.Suppose image block sets are $\{b_i\}(i=1,\ldots,L)$ (Eq.(4))with each block feature vector.

$$b_i = [b_i^{C_1}, b_i^{C_2}, b_i^{C_3}, b_i^{T_1}, b_i^{T_2}, b_i^{T_3}]^T \tag{4}$$

$$D(k) = \sum_{i=1}^{L} \min_{1 \le j \le k} \sqrt{\sum_{m=1}^{6} (b_{im} - \hat{b}_{jm})^2} \tag{5}$$

The goal of the k-means algorithm is to partition the sets into k groups with means $\left\{\hat{b}_1, \hat{b}_2, \ldots, \hat{b}_C\right\}$, such that Eq.(5) is minimized. We start with k=2 and stop increasing k if one of the following conditions is satisfied.

(1)The first derivative of distortion with respect to k, D(k)-D(k-1), is below a threshold with comparison to the average derivative at k=2,3.The threshold determines overall time to segment images and it can be adjusted according to the experimental.(2)The number k exceeds an upper bound. We allow an image to be segmented into a maximum of 16 segments.

3 Region Fuzzy Feature Extraction

Considering the typical uncertainty stemming from color quantization and human perception, we develop a modified color histogram using the fuzzy technique[6] to accommodate the uncertainty.

In our system, color space is quantized into 500 bins by using uniform quantization(L* by 5, a* by 10, b* by 10). Suppose color bins as the U^c, then $\mu_c : U^c \rightarrow [0,1]$, $\mu_c(c')$ is the membership degree of c' to c. We can get the conclusion that membership degree is affected by distance, the larger the distance, the lower the similarity.We adopt Cauchy function as membership function.The membership function for bins is as Eq.(6).

$$\mu_c(c') = 1\Big/\left(1+\left(d(c,c')/\sigma\right)^{\alpha}\right) \tag{6}$$

Where d is the Euclidean distance between color c and c' and σ is the average distance between colors, its computation is as Eq.(7)

$$\sigma = \frac{2}{B(B-1)}\sum_{i=1}^{B-1}\sum_{k=i+1}^{B} d(c,c') \tag{7}$$

Where B is the number of bins in the color partition. The average distance between colors is used to approximate the width of the fuzzy membership function. The average retrieval accuracy changes insignificantly when α is in the interval [0.7,1.5] but degrades rapidly outside the interval. We set $\alpha = 1$ to simplify the computation.

This fuzzy color model enables us to enlarge the influence of a given color to its neighboring colors according to the uncertainty principle and the perceptual similarity.This means that each time a color c is found in an image, it influences all the quantized colors according to their resemblance to the color c. Numerically, this could be expressed as Eq.(8)

$$h_{n2} = \sum_{c' \in U^c} h_{n1}(c')\mu_c(c') \tag{8}$$

Where U^c is the color universe of images, 218 colors in this paper, and $h_{n1}(c')$ is the normalized conventional color histogram. Finally, the normalized fuzzy color histogram is computed as Eq.(8)

$$h_n(c) = \frac{h_{n2}(c)}{\max_{c' \in U^c} h_{n2}(c')} \tag{8}$$

We take each region as a fuzzy set of blocks. To propose a unified approach consistent with the fuzzy color histogram representation, we again use the Cauchy function to be the fuzzy membership function for computing fuzzy texture features by Eq.(9)

$$\mu_n(f_j) = 1\Big/\left(1+\left(d\left(f_j - \hat{f}_n\right)/\sigma\right)^{\alpha}\right) \tag{9}$$

where $f_j = \left[b_j^{T_1}, b_j^{T_2}, b_j^{T_3}\right]^T$ is the texture feature vector of block j, $\hat{f}_n = \left[R_n^{T_1}, R_n^{T_2}, R_n^{T_3}\right]^T$ is the average texture feature vector of region n, d is the Euclidean distance between $\hat{f}_n$ and feature f_j, andσrepresents the average distance for texture features among the cluster centers obtained from the k-means algorithm.σis defined as Eq.(10)

$$\sigma = \frac{2}{M(M-1)}\sum_{n=1}^{M-1}\sum_{k=n+1}^{M}\left\|\hat{f}_n - \hat{f}_k\right\| \tag{10}$$

Where M is the number of regions in a segmented image and $\hat{f}_n$ is the average texture feature vector of region n. With this block membership function, the fuzzified texture property of region n is as Eq.(11)

$$\vec{f}_n^{\,T} = \sum_{f_j \in U^T} f_j \mu_n(f_j) \tag{11}$$

WhereU^T is the feature space composed of texture features of all blocks.

After the segmentation, three additional features are determined for each region to describe the shape property.For a region n in the 2D Euclidean integer space, its normalized inertia of order p(p=1,2,3) is as Eq.(12).

$$l(n,p) = \sum_{(x,y):(x,y)\in V} [(x-\hat{x})^2 + (y-\hat{y})^2]^{p/2} \Big/ [N(n)^{1+p/2}] \tag{12}$$

Where,N(n) is the number of pixels in region n, (x, y) is the centroid of each block in region n and $(\hat{x}, \hat{y})$ is the centroid of region n. The minimum normalized inertia is achieved by spheres.Denoting the pth-order normalized inertia of spheres as Lp, define the following three features to describe the shape of each region as Eq.(13).

$$S_{n1} = l(n,1)/L_{n1}, S_{n2} = l(n,2)/L_{n2}, S_{n3} = l(n,3)/L_{n3} \tag{13}$$

Since we extracted fuzzification of color and texture in each region, compute fuzzy shape feature for each region in a similar fashion. Based on the shape membership function $\mu_n(f_j)$, fuzzy shape features for a region are gotten by Eq. (14).

$$f\_l(n,p) = \sum_{f_j \in U^s} [(f_{jx} - \hat{x})^2 + (f_{jy} - \hat{y})]^{p/2} \mu_n(f_j) \Big/ [V]^{1+p/2} \tag{14}$$

Where $f_{jx} and f_{jy}$ are the x and y coordinates of the block with the shape feature f, respectively; $\hat{x}$ *and* $\hat{y}$ are the x and y central coordinates of region n.V is the number of blocks in an image, and U^S is the block shape feature space in an image.Based on the last two equations, determine the fuzzified shape feature of each region, denoted as Eq.(15).

$$f\_S_{n1} = f\_l(n,1)/L_{n1}, f\_S_{n2} = f\_l(n,2)/L_{n2}, f\_S_{n3} = f\_l(n,3)/L_{n3} \tag{15}$$

Then fuzzy shape feature of region n is Eq.(16).

$$\vec{f}_n^{\,S} = [f\_S_{n1}, f\_S_{n2}, f\_S_{n3}]^T \tag{16}$$

4 Similarities Between Images Computation

Once we have fuzzified representations for color, texture, and shape features, we record the following information as its indexed data for each region: (1) fuzzy color histogram $h_n(c)$.(2) fuzzy texture feature $\vec{f}_n^{\,T}$.(3) fuzzy shape feature $\vec{f}_n^{\,S}$.(4) relative size of region to whole image ω_n.and (5) central coordinate of region area $(\hat{x}_n, \hat{y}_n)$. Such indexed data of all regions in an image are recorded as the signature of the image. Based on theses features, define similarity computation for color, texture, shape for region p and q with L2 distance function.

For the similarity computation between region p and q, compute fuzzy texture, shape, color feature by L2 distance as in sequences (see Eq.(17)).

$$d_T^{pq} = \left\| \vec{f}_p^{\,T} - \vec{f}_q^{\,T} \right\|, d_S^{pq} = \left\| \vec{f}_p^{\,S} - \vec{f}_q^{\,S} \right\|, d_c^{pq} = \sqrt{\sum_{i=1}^{B} \left[h_p(n) - h_q(n) \right]^2 \Big/ B} \tag{17}$$

The intercluster distance on color and texture between region p and q is defined as Eq.(18).

$$d_{CT}^{\;pq} = \sqrt{\left(d_C^{\;pq}\right)^2 + \left(d_T^{\;pq}\right)^2} \tag{18}$$

The overall distance between two regions is defined as Eq.(19)

$$DIST(p,q) = \omega d_{CT}^{\;pq} + (1-\omega) d_S^{\;pq} \tag{19}$$

Where ω is the weight, in this paper, $\omega = 0.7$ to influence more on color and texture than shape. we construct an image distance measure through the following steps. Suppose that we have M regions in image 1 and N regions in image 2.

Step1: Determine the distance between one region in image 1 and all regions in image 2, $R_{i,\mathrm{Image}2} = \min\limits_{1 \le j \le N} DIST(i,j)$.

Step2: Determine the distance between one region in image 2 and all regions in image 1, $R_{j,\mathrm{Image}1} = \min\limits_{1 \le i \le M} DIST(j,i)$.

Step3: After obtaining the M + N distances, define the distance between the two images (1 and 2) as Eq.(20).

$$Dist(1,2) = \left(\sum_{i=1}^{M} \omega_{1i} R_{i,\mathrm{Image}2} + \sum_{j=1}^{N} \omega_{2j} R_{j,\mathrm{Image}1} \right) \Big/ 2 \tag{20}$$

Where ω_{1i} is the ratio of the number of blocks in region i and the total number of blocks in image 1. ω_{2j} is defined similarly. For each query q, Dist (q, d) is determined for each image d in the database and relevant images are retrieved through sorting the similarities Dist (q, d).Similarity measuring in our system is the similarity between query image and each image in database.We transfer distance into similarity as Eq.(21).

$$sim(1,2) = \beta / \left(DistIge(1,2) + \beta \right) \tag{21}$$

Where β is an adaptive parameter. This transform is one of the changes from distance to similarity. Through a lot of trials, β is 100 in our system.

5 Experiment Results and Analysis

Image database is gotten from http://www.fortune.binghamton.edu/ download.html, with corel including 10000 images. We use these images for testing RBIRFF which include people, nature, animal, vehicles and so on with each category including

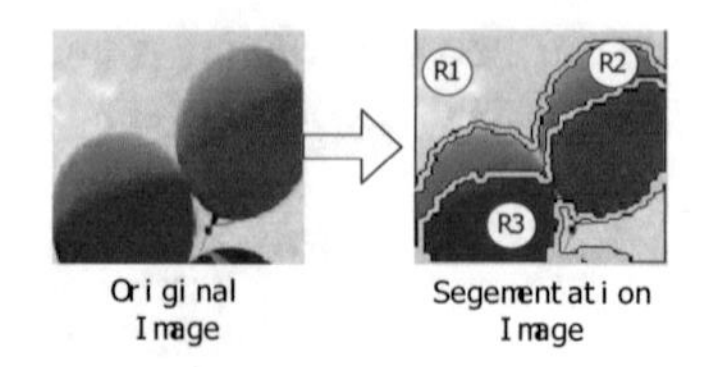

Fig. 3. An Example of Image Segmentation Result

Table 1. Color and Texture Features in Blocks for Segmentation

$b_i^{C_1}$	$b_i^{C_2}$	$b_i^{C_3}$	$b_i^{T_1}$	$b_i^{T_2}$	$b_i^{T_3}$
90.534	0.75894	-9.0163	0	0.06329	0
90.336	-0.38756	-8.9684	0	0.12337	0
90.559	-1.2531	-8.8285	0	0	0
...	...	...	...	...	...

Table 2. Color and Texture Features in Regions

	$R_n^{C_1}$	$R_n^{C_2}$	$R_n^{C_3}$	$R_n^{T_1}$	$R_n^{T_2}$	$R_n^{T_3}$
R_1	47.284	69.864	21.982	0.53764	0.55752	0.24989
R_2	68.5	43.178	3.7939	1.9395	2.2291	0.90046
R_3	89.311	1.8258	-8.4966	1.074	1.0408	0.42862

Table 3. Fuzzy Color and Texture Features in Regions

	ω	$\vec{f}_1^{\,T}$	$\vec{f}_1^{\,T}$	$\vec{f}_2^{\,T}$	$\vec{f}_3^{\,T}$	$\vec{f}_1^{\,S}$	$\vec{f}_2^{\,S}$	$\vec{f}_3^{\,S}$
R_1	0.4209	336.01	336.01	315.79	110.18	0.82338	3.4588	15.358
R_2	0.19629	274.77	274.77	275.29	102.18	0.26449	0.9771	3.689
R_3	0.38281	336.05	336.05	317.97	110.3	0.82109	2.7502	9.4637

Table 4. Fuzzy Color Histogram in Regions

	bin1	bin2	bin3	bin4	bin5	bin1	bin2	bin3	...
$h_1(c)$	0.6479	0.6479	0.6404	0.68808	0.70066	0.69128	0.63566	0.68731	...
$h_2(c)$	0.53241	0.53241	0.53626	0.55942	0.58204	0.58826	0.51316	0.55747	...
$h_3(c)$	0.56383	0.56383	0.56621	0.59088	0.61219	0.61641	0.54343	0.58795	...

different number of images, 25-300. We randomly select 50 query images. If a retrieval image belongs to the category of query image, consider them match.We implemented the system on Pentium IV Celeron 1.7G computer with 512 M memory, in Windows XP, with Matlab7.0.Fig.3 shows the segmentation result for example image. Table 1-table 4 show some useful data of experiment.

The retrieval interface of the prototype system is shown in Fig.4.

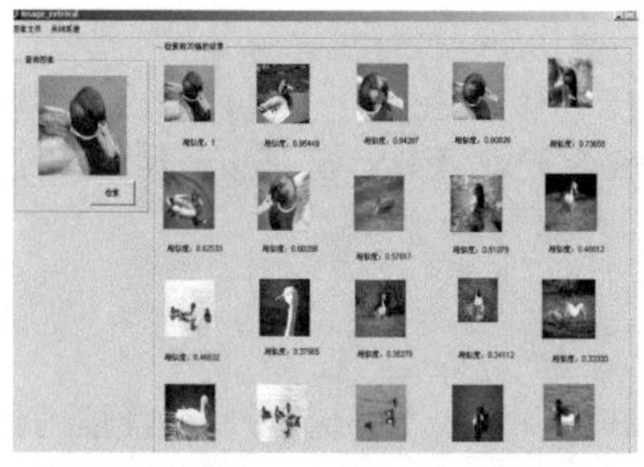

(a) Duck:20 matches out of 20.

(b) cherry:16 matches out of 20.

Fig. 4. Retrieval Results of Two Query Images

To evaluate our approach more quantitatively, we compare RBIRFF with CLEAR[7] and UFM[8].Retrieval effectiveness is measured by precision and recall metrics. The average precision results of 50 query images for different number of retrieval images are recorded in Fig.5, which demonstrates that the retrieval precision of RBIRFF is generally superior to that of UFM and CLEAR.The average recalls of RBIRFF, CLEAR and UFM are comparable and the advantages of RBIRFF to CLEAR and UFM are shown more clearly when the number of images retrieval on which we calculate the average recall statistics increase.

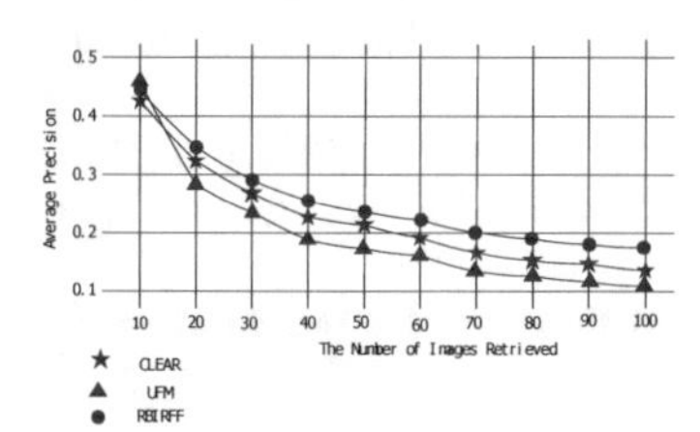

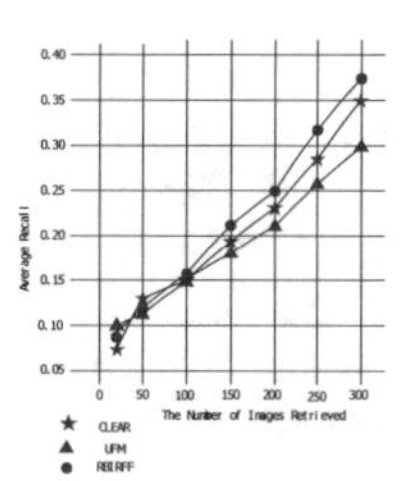

Fig. 5. The Comparison of Average Precision and Average Recall among RBIRFF, CLEAR and UFM

6 Conclusions and Further Work

A new image retrieval method is described, which is compared with exists systems, whose advantages are: (1) RBIRFF proposed applies fuzzy set model to regional color histograms as well as texture and shape representations to improve the retrieval effectiveness; (2) RBIRFF develops regional similarity based on all the feature components without efficiency loss.

In the further work, (1) high-dimension research,for the large image database, lots of images require setting up multi-level high-dimension index to improve retrieval efficiency.(2) relevance feedback research, It needs to develop an algorithm to infer user preference in order to tailor to the intended retrieval from a sparse distribution of these "qualitative" data.

Acknowledgments

This paper is supported by Provincial Natural Science Foundation of Hunan (05JJ40101).

References

1. Carson, C., Thomas, M., Belongie(1999)Blobworld: a system for region-based image indexing and retrieval. In: Proceedings of the 3rd International Conference on Visual Information Systems, Amsterdam, 509-516
2. Ma, W.Y., Manjunath, B(1997)Netra: a toolbox for navigating large image databases.In: Proceedings of the IEEE International conference on Image Processing, Santa Barbara, CA, pp,568-571.
3. Smith J,Chang S F(1996)VisualSEEk: A Fully Automated Content-Based Image Query System.In Proceedings of the Fourth ACM Multimedia Conference (MULTIMEDIA 96),New York: ACM Press,87-98
4. James Z. Wang, Jia Li, Gio Wiederhold2001)SIMPLIcity: Semantics-Sensitive Integrated Matching for Picture Libraries.IEEE Trans.On PAMI, 9: 947-963
5. James Z.Wang, Yanping Du(2001)Scalable integrated region-based image retrieval using IRM and statistical clustering.In: Proceedings of the ACM and IEEE Joint Conference on Digital Libraries, Roanoke, pp, 268-277
6. Vertan, C., Boujemaa, N.(2000)Embedding fuzzy logic in content based image retrieval.In: Proceedings of the 19th International Meeting of the North America Fuzzy Information Processing Society, Atlanta, pp, 85-89
7. R. Zhang and Z. Zhang(2003)Using Region Analysis Technique to Efficiently Retrieve Online Image Databases.In: proceedings of the SPIE's International Conference on Electronic Imaging 2003, Santa Clara, CA, Jan.,pp, 198-207
8. Chen, Y., Wang, J.Z.(2002)A region-based fuzzy feature matching approach to content-based image retrieval. IEEE Trans. Pattern Anal. Mach. Intell.,Vol.24,9:1252-1267

Association Rule Mining of Kansei Knowledge Using Rough Set

Shi Fuqian[1,2], Sun Shouqian[2], and Xu jiang[2]

[1] Wenzhou Medical College
sfq@wzmc.net
[2] College of Computer Science, Zhejiang University
xujzju@163.com

Abstract. In the field of Kansei Engineering, Semantic Differential (SD) is a typical method in evaluating conceptual products. We design a WEB-based kansei questionnaire system to generate a Decision Table(DT) which is composed of typical form features(condition attributes) and kansei words(decision attributes) through SD methods; by using statistical tools, frequent records are stored to DT as decision rules which are indexed by kansei word. First, some rules which have important contributions to the corresponding kansei evaluation will be extracted through the attribute reduction algorithm of Rough Set Theory(RST). Second, the size of decision table has been reduced through association rule mining based on Rough set and rules joining operating; finally, strong association rule set which describe the relation between the key form feature and the corresponding kansei word is generated. The proposed method has been successfully implemented in cell phone design case.

Keywords: Kansei Engineering Rough Set Theory Association Rule Mining.

1 Introduction

With the development of the product design technologies, many similar products will be functionally equivalent therefore customers find it is difficult to distinguish and choose many product offerings. Hence, purely design, such as functional design, ergonomic design have no longer empower a competitive edge because the product technologies turn to be mature which competitors can quickly catch up with. And now, satisfying customer needs has become a great concern of almost every company. While there are various customer needs, the functional and affective needs have been recognized to be as the primary importance for customer satisfaction. In particular, mass customization and personalization are increasingly being accepted as important instruments for firms to gain competitive advantages.

In this regard, customers' affective needs must be considered in designing products. Affect is said to be a customer's psychological response to the perceptual design details (e.g. styling) of the product. Many approaches have received a

B.-Y. Cao (Ed.): Fuzzy Information and Engineering (ICFIE), ASC 40, pp. 949–958, 2007.
springerlink.com

significant amount of interests and been applied in the discovery and acquisition of the customer's personal impression regarding to products' emotional impact, such as Semantic Differential Methods[1], emotional design[2] and Kansei Engineering which study on individuality of humans' subjectivity[3][4]. A lot of studies on the utilization of humans' Kansei for marketing or product developments have been reported. Semantic Differential (SD) method proposed in this paper is useful to quantify individual impressions and can measure individual impressions of objects by using multiple pairs of adjectives (impression words) with antithetical meanings. SD data obtained from many subjects are generally classified by similarity/dissimilarity; and then stratified analysis is applied to the data[5].

KDD(Knowledge Discovery in Database) based on Kansei knowledge and data mining such as association rule mining technologies are recently developed with the emerging and development of the Database and Artificial Intelligence technologies which have been applied in machine learning, compute visualization, pattern recognition and statistics. It extracts valuable knowledge from the database using machine learning methods[6][7].

2 Frameworks

Web-based kansei questionnaire is established to acquire user's subjective preference information. In data processing phase, the statistical methods are used to count the number of frequent records which appear on the same evaluation matrix; When the number achieve to a certain threshold, the corresponding record is stored to Decision Table (DT) whose condition attributes are product form features and decision attributes are kansei words. Rough-set-attributes-reduction algorithms are applied to extract the key form features which have much important contributions to the relevant kansei word; system deletes the redundant attributes. Then, we re-define the rules space and the corresponding indicators in association rule algorithm to mine the Strong Rule Set which denotes the extent of relation between local form features and kansei evaluation of integrated production. The proposed methods are shown in Fig.1.

3 Rough Set Theory

3.1 Review and Background

Rough set theory(RST) was introduced in 1982 by Z.Pawlak[8]and it has been successfully applied in such fields as artificial intelligence, knowledge discovery[9], data mining[10], pattern recognition,fault diagnose and insurance market[11] in the recent 20 years.It represents an objective approach to imperfections in data.All computations are performed directly on data sets, i.e., no feedback from additional experts is necessary. Thus, there is no need for any additional information about data such as, a probability distribution function in statistics, a grade of membership from fuzzy set theory, etc. (Grzymala-Busse,1988). In this

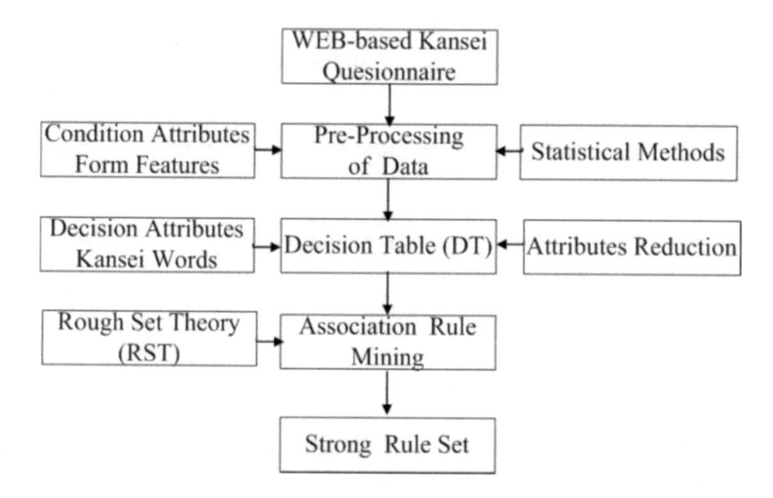

Fig. 1. The framework of Association Rule Mining of Kansei Knowledge Using Rough Set

theory,knowledge is regarded as partition of the Universe by defining the knowledge from a new angle of view. Knowledge is discussed by equivalence relation in algebra. RST may be applied to consistent data (without conflicting cases) to study relations between attributes.Inconsistent data sets are handled by rough set theory using lower and upper approximations for every concept.These approximations are definable using existing attributes. Furthermore, from concept lower and upper approximations certain and possible rule sets are induced.

3.2 Definition

Definition 3.2.1: Let U denote the non-empty set of all cases called Universe, if R is an Equivalence Relation on U, then U/R is a Set Partition of U;let $[X]_R$ denote the Equivalence Class of R including X, or subset X belongs to a "Category" of R.

Definition 3.2.2: If R is a partition on U, Equivalence Relation $R = \{X_1, X_2, ..., X_n\}$ which is marked by (U, R) is defined as an Approximation Space.

Definition 3.2.3: If $P \subset R$,then $\cap P$ (intersections of all Equivalence Relation in P) is also an Equivalence Relation and is an Indiscernibility Relation on P, marked by $ind(P)$.

Definition 3.2.4: Let $X \subseteq U$,and R is an Equivalence Relation. When X is combine of some basic Category on R, we call X is R definable, or else X is R indefinable; R definable set is the subset of U, and be exactly defined in repository, called R Exact Sets, contrarily called R Rough Sets.

Supposed repository $K = (U, R)$ has all subset $X \in U$ and an Equivalence Relation $R \in ind(K)$, so it can make set partition on X according to the basic sets on R.

Definition 3.2.5: Lower Approximation of X is defined to be the maximal definable set of X in R:

$$R_{-}(X) = \{x \in U : [x]_R \subset X\} = \cup\{Y \in U/R : Y \subset X\} \tag{1}$$

Definition 3.2.6: Upper Approximation of X is defined to be the minimal definable set of X in R:

$$R^{-}(X) = \{x \in U : [x]_R \subset X\} = \cup\{Y \in U/R : Y \cap X \neq \emptyset\} \tag{2}$$

Definition 3.2.7: R boundary set of X is defined to be $BN_R(X) = R^{-}(X) - R_{-}(X)$; $Pos_R(X) = R_{-}(X)$ denoting R positive region of X. Let $Neg_R(X) = U - Pos_R(X)$ be R negative region of X, $BN_R(X)$ be the boundary region of X. We know that if boundary region is a empty set,then X is a R definable set;

Definition 3.2.8: Precision is defined to be:

$$Dr(X) = Card(R_{-}(X))/Card(R^{-}(X)) \tag{3}$$

where Card(X) denotes the base of set X, and $X \neq \emptyset$.

Definition 3.2.9: If R is a Equivalence Relation, and $r \in R$, when $ind(R) = ind(R-r)$, we say r is R-dispensable, or else is R-indispensable; when $\forall r \in R$,R indispensable, then R is independent.

Definition 3.2.10: When Q is independent, and $ind(Q) = ind(P)$, then, $Q \subset P$ is a reduction of P, the core of P which is marked by Core(P) are composed of all indispensable sets in P; i.e. if Red(P) is a reduction of P, then $Core(P) = \cup Red(P)$.

Definition 3.2.11: Knowledge Representation System(KRS) is defined to be $< U, C, D, V, f >$ where U is cases set, $C \cup D = A$ is attribute set, C and D is respectively defined as condition set and decision set. $V = \cup_{a \in A} V_a$ is a set of attribute set; V_a denotes the region of attribute $a \in A$, $f : U_x \rightarrow V$ is a information function which denotes the attribute value of every X in U; its table is called as KRS.

3.3 Reduction and Core Computing

The reduction of decision table includes attribute reduction and value reduction; some condition attributes which have no impact on the final decision are deleted through attribute reduction algorithm. And the decision table after processing have fewer condition attributes. The following steps show the details:

Step 1: Delete reduplicate row.

Step 2: Let $c_1, c_2, ..., c_n$ be subset of attribute set C; calculate respectively $A_i = U/ind(c_i)$, $E = U/ind(D)$, where D is the decision subset.

Step 3: Let $R = \{A_i|i = 1, ..., n\}$, calculate $U/ind(R)$

Step 4: Classify by R to get $Pos_R(E)$.

Step 5: Calculate $U/ind(R - A_i)$, and compare its result with $Pos_{(R-A_i)}(E)$; if they are equal,then the corresponding attribute A_j is dispensable,and delete the attribute col from DT; if they are not equal, then the corresponding attribute is indispensable.

Step 6: Calculate the Core which is defined to be all indispensable condition attributes.

4 Association Rule Mining Based on Rough Set

4.1 Association Rule Definition

Association Rule: in Decision Table(DT), $c \in C$ is condition attribute; $d \in D$ is decision attribute; rule $c \Rightarrow d$ denotes the relation of c and decision; it shows the c's occurrence leads to d's occurrence.

Support: the frequency of some item-sets in database which is expressed by $Support(X)$. Where X is a item-set. The higher Support in this project indicates the higher popularity in the database. In this paper,Support of Rule is expressed by $Support(c \Rightarrow d)$; we define the support of association rule of condition attribute and decision attribute using statistic methods,the formula is:

$$Support(c \Rightarrow d) = \frac{Count(c \wedge d)}{Count(CD|D_i)} \tag{4}$$

where $Count(CD|D_i)$ denotes count the record number of decision table where the subset of decision attribute set-D is D_i.

Minimum Support: in association rule mining, a certain item-set is not representative if its Support is too low. So,before using the rules mining algorithm, system will set a threshold value of the support which is called Minimum Support.

Confidence: if association rule is $c \Rightarrow d$, the confidence is a conditional probability of d's occurrence on the condition of c's occurrence which defined as:

$$Confidence(c \Rightarrow d) = \frac{Count(c \wedge d)}{Count(c)} \tag{5}$$

Minimum Confidence: similar as Minimum Support.

Certain Rule Set: if Support $(c \Rightarrow d) \geq$ Minimum-Support and Confidence $(c \Rightarrow d) \geq$ Minimum-Confidence, then we say rule $c \Rightarrow d$ is a certain rule.

Strong Rule Set: joined by certain rules using Apriori algorithm which is described in section 4.2

4.2 Association Rule Algorithm Using Rough Set

Let decision table be DT, the thresholds of the ith-rank rule are Minimum-Support(i) and Minimum-Confidence(i), system output rule set R_k, algorithm is designed as the following based on Apriori algorithm:

Step 1: Attribute Reduction in DT.(Section 3.3)

Step 2: k=1

Step 3: Calculate the Support and Confidence of every rule in candidate set C_k.

Step 4: If Support of rule is less than the Minimum-Support(i) and Confidence is less than the Minimum-Confidence(i)then delete it from C_k; or else reserve and store it to R_k.

Step 5: Expand C_k to $C_k + 1$. Scan $C_k$join $k1$ attributes to C_k to become $C_k + 1$; Support and Confidence of rule after joining is shown as

$$Support(c \Rightarrow d, e \Rightarrow f) = min\{Support(c \Rightarrow d), Support(e \Rightarrow f)\} \quad (6)$$

$$Confidence(c \Rightarrow d, e \Rightarrow f) = min\{Confidence(c \Rightarrow d), Confidence(e \Rightarrow f)\} \quad (7)$$

then we can delete some rules using Minimum-Support(i+1) and Minimum Confidence(i+1).

Step 6: k=k+1. If C_{k+1} is a empty set, then goto step 7, else Goto step 3

Step 7: End.

5 Case Study

5.1 Product Knowledge Representation System Construction

We applied the proposed methods in cell phone design case. Let product form feature set be: $C = \{C1, C2, C3, C4, C5\}$ which are Body shape, Screen, Sound hole, Direct Key and Press Key; we give three representative form features of every product form(shown in Fig .2.);

Let kansei words set be $D = \{D1, D2, D3, D4, D5\}$ which are Deluxe, Cute, Light, Sporty and Scientific; then the condition attribute set is C, decision set is D, we acquire data through WEB-based kansei questionnaire system. 500 testees are invited to evaluate a integrated product which is combined of every attribute. The record which is generated by testees are collected automatically by the same both the choice of condition attributes(form features) and the kansei word evaluation; if the number of record is bigger than 30, we regard this evaluation as more representative and store this record as a decision rule to decision table. Finally, we get the decision table about Kansei questionnaire; to facilitate the description, we only list such rules where $D = D1$(shown in Table 1.)

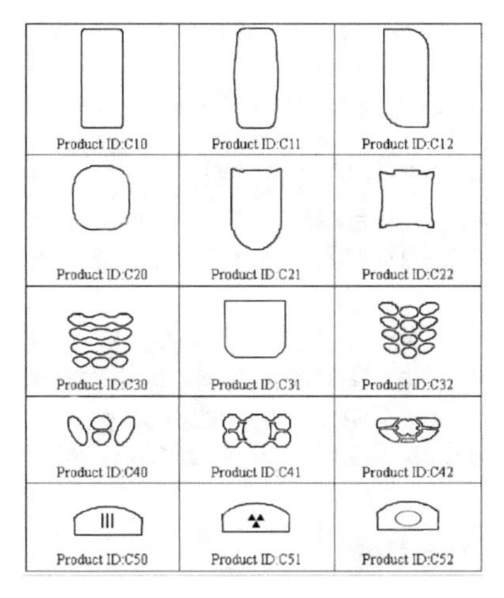

Fig. 2. Typical form feature list for Kansei evaluation

Table 1. Decision Table

U	$C1$	$C2$	$C3$	$C4$	$C5$	$D1$	U	$C1$	$C2$	$C3$	$C4$	$C5$	$D1$
1	1	2	0	1	0	0	8	2	1	0	2	2	0
2	0	1	1	0	1	1	9	1	0	2	0	0	1
3	0	1	1	0	0	0	10	2	0	1	1	2	0
4	0	1	0	0	0	1	11	0	0	1	1	2	1
5	1	2	2	1	1	0	12	2	2	0	2	0	0
6	1	2	2	2	1	1	13	0	0	1	0	2	1
7	1	0	2	0	1	1	14	1	1	2	1	0	0

5.2 Attributes Reduction

Firstly, we take the attribute reduction on decision table using rough set theory, extract some key form features which have much important contribution to the integrated product, calculated as:

$Z1 = U/ind(C1) = \{\{2,3,4,11,13\},\{1,5,6,7,9,14\},\{8,10,12\}\}$
$Z2 = U/ind(C2) = \{\{7,9,10,11,13\},\{2,3,4,8,14\},\{1,5,6,12\}\}$
$Z3 = U/ind(C3) = \{\{1,4,8,12\},\{2,3,10,11,13\},\{5,6,7,9,14\}\}$
$Z4 = U/ind(C4) = \{\{2,3,4,7,9,13\},\{1,5,10,11,14\},\{6,8,12\}\}$
$Z5 = U/ind(C5) = \{\{1,3,4,9,12,14\},\{2,5,6,7\},\{8,10,11,13\}\}$
$Z6 = U/ind(D1) = \{\{1,3,5,8,10,12,14\},\{1,2,4,6,7,9,11,13\}\}$

Let $R = \{Z1, Z2, Z3, Z4, Z5\}$, we calculate:

$U/ind(R) = \{\{1\},\{2\},\{3\},\{4\},\{5\},\{6\},\{7\},\{8\},\{9\},\{10\},\{11\},\{12\},$ $\{13\},\{14\}\}$, $Pos_R(Z6)$={1,2,3,4,5,6,7,8,9,10,11,12,13,14}

$U/ind(R-Z1) = \{\{1\},\{2\},\{3\},\{4\},\{5\},\{6\},\{7\},\{8\},\{9\},\{10,11\},\{12\},$ $\{13\},\{14\}\}$

$Pos_{(R-Z1)}(Z6) = \{\{1\},\{2\},\{3\},\{4\},\{5\},\{6\},\{7\},\{8\},\{9\},$ $\{12\},\{13\},\{14\}\} \neq Pos_R(Z6)$, so $Z1$ is indispensable.

$U/ind(R-Z2) = \{\{1\},\{2\},\{3\},\{4\},\{5\},\{6\},\{7\},\{8\},\{9\},\{10,11\},$ $\{12\},\{13\},\{14\}\}$

$Pos_{(R-Z2)}(Z6) = \{\{1\},\{2\},\{3\},\{4\},\{5\},\{6\},\{7\},\{8\},\{9\},\{10\},\{11\},$ $\{12\},\{13\},\{14\}\} = Pos_R(Z6)$. Therefore, Z2 is dispensable.

By calculating $Pos_{(R-Z3)}$,$Pos_{(R-Z4)}$,$Pos_{(R-Z5)}$,we know $Z3, Z4, Z5$ are indispensable. Hence,we delete the C2 col. and $Core = \{Z1, Z3, Z4, Z5\}$

5.3 Strong Rule Extracting

We assign a binary form (Ci, v) to denote attribute Ci's value as v and decision attribute $D1$'s value as dv; then the rule is transformed by $(Ci, v) \Rightarrow (D1, dv)$. Using definition and formula which is described in section 4.1, we get 24 rules called 1-rank association rule (Table .2.), if a rule's support(Formula (4)) is less than the Minimum-Support(1), we needn't continue to calculate its Confidence(Formula (5)); and we assign Minimum-Support(1) be $\frac{3}{14}$,Minimum-Confidence(1) be 0.55.

Table 2. 1-rank association rule

Rule	Sup.	Conf.	Op.	Rule	Sup.	Conf.	Op.
$(C1,0) \Rightarrow (D1,0)$	$\frac{1}{14}$	-	Delete	$(C4,0) \Rightarrow (D1,0)$	0	-	Delete
$(C1,1) \Rightarrow (D1,0)$	$\frac{3}{14}$	$\frac{3}{6}$	Delete	$(C4,1) \Rightarrow (D1,0)$	$\frac{4}{14}$	$\frac{4}{5}$	**Reserve**
$(C1,2) \Rightarrow (D1,0)$	$\frac{3}{14}$	1	**Reserve**	$(C4,2) \Rightarrow (D1,0)$	$\frac{2}{14}$	-	Delete
$(C1,0) \Rightarrow (D1,1)$	$\frac{4}{14}$	$\frac{4}{5}$	**Reserve**	$(C4,0) \Rightarrow (D1,1)$	$\frac{5}{14}$	$\frac{5}{6}$	**Reserve**
$(C1,1) \Rightarrow (D1,1)$	$\frac{3}{14}$	$\frac{3}{6}$	Delete	$(C4,1) \Rightarrow (D1,1)$	$\frac{1}{14}$	-	Delete
$(C1,2) \Rightarrow (D1,1)$	0	-	Delete	$(C4,2) \Rightarrow (D1,1)$	$\frac{1}{14}$	-	Delete
$(C3,0) \Rightarrow (D1,0)$	$\frac{3}{14}$	$\frac{3}{4}$	**Reserve**	$(C5,0) \Rightarrow (D1,0)$	$\frac{4}{14}$	$\frac{4}{6}$	**Reserve**
$(C3,1) \Rightarrow (D1,0)$	$\frac{2}{14}$	-	Delete	$(C5,1) \Rightarrow (D1,0)$	$\frac{1}{14}$	-	Delete
$(C3,2) \Rightarrow (D1,0)$	$\frac{2}{14}$	-	Delete	$(C5,2) \Rightarrow (D1,0)$	$\frac{2}{14}$	-	Delete
$(C3,0) \Rightarrow (D1,1)$	$\frac{2}{14}$	-	Delete	$(C5,0) \Rightarrow (D1,1)$	$\frac{2}{14}$	-	Delete
$(C3,1) \Rightarrow (D1,1)$	$\frac{3}{14}$	$\frac{3}{5}$	**Reserve**	$(C5,1) \Rightarrow (D1,1)$	$\frac{3}{14}$	$\frac{3}{4}$	**Reserve**
$(C3,2) \Rightarrow (D1,1)$	$\frac{2}{14}$	-	Delete	$(C5,2) \Rightarrow (D1,1)$	$\frac{2}{14}$	-	Delete

From Table.2., we reserve 8 rules and continue to generate 2-rank association rules which perhaps have $C_4^2 \times C_2^1 \times C_2^1 = 24$ rules. If there have two rules such as ,$(C1, 0) \Rightarrow (D1, 1)$ and $(C3, 1) \Rightarrow (D1, 1)$, we can join them using algorithm which is described in section 4.2 as $(C1, 0), (C3, 1) \Rightarrow (D1, 1)$; its Support and Confidence are calculated by Formula (6),(7).According to the Minimum-Support(2) and Minimum-Confidence(2), we can delete some rules.Similarly,if there have two 2-rank association rules such as,$(C1, 0), (C3, 1) \Rightarrow (D1, 0)$ and $(C3, 1), (C5, 0) \Rightarrow (D1, 0)$, we can join them as $(C1, 0), (C3, 1), (C5, 0) \Rightarrow (D1, 0)$, continue this processing till it can't generate new rules. Finally, we get Strong Rules Set shown in Table.3.

Table 3. Strong Rule Set

$(C1, 2), (C4, 1), (C5, 0) \Rightarrow (D1, 0)$	$(C2, 2), (C4, 1), (C5, 0) \Rightarrow (D4, 0)$
$(C1, 0), (C4, 0), (C5, 0) \Rightarrow (D1, 1)$	$(C1, 2), (C4, 2), (C5, 0) \Rightarrow (D4, 1)$
$(C2, 0), (C3, 0), (C4, 1) \Rightarrow (D2, 0)$	$(C1, 2), (C3, 1), (C4, 0) \Rightarrow (D5, 0)$
$(C2, 1), (C3, 1), (C5, 1) \Rightarrow (D2, 1)$	$(C1, 1), (C2, 1), (C3, 2) \Rightarrow (D5, 1)$
$(C3, 0), (C4, 1), (C5, 1) \Rightarrow (D3, 0)$	-
$(C1, 2), (C3, 1), (C5, 0) \Rightarrow (D3, 1)$	-

6 Conclusion and Future Works

Under normal circumstances, there are many ways about Attribute Reduction using Rough Set. The proposed algorithm in this paper applies association rule mining technology to extract users' kansei knowledge. The key to this method is data preprocessing which makes every record in Decision Table be agreed by a large number of users. On this basis, the application of rough set theory can be better applied. In addition, The re-definition of association rules, especially, the Support and Confidence is proved as much which is feasible in cell phone design case. Using Rough set in the association rule mining is effective to extract Strong rule set, which helps designers and enterprises to make a timely response to users' demands. The next work is to conduct a comprehensive survey system to acquire more information; and the association rule algorithm also need to be improved.

Foundation item: Project supported by the National Natural Science Foundation, China(No.60475025), the Specialized Research Fund for the Doctoral Program of Higher Education, China(No.20050335096).

References

1. Osgood, C., Suci, G.,& Tannenbaum, P.H. The measurement of Meaning[M]. Urbana:University of Illinois Press.1957.
2. Picard, R.W. Affective Computing[M], Cambridge:MIT Press, 1997.

3. Nagamachi, Kansei engineering and comfort ,International Journal of Industrial Ergonomics, Volume 19, Issue 2, 1997(2): 79-80
4. Mitsuo Nagamachi ,Kansei engineering as a powerful consumer-oriented technology for product development . Applied Ergonomics, Volume 33, Issue 3, 2002(5): 289-294
5. Shang H. Hsu, Ming C. Chuang. A semantic differential study of designers' and users' product form perception[J].International Journal of Industrial Ergonomics.2000(25):375-391.
6. Sun-mo Yang, Mitsuo Nagamachi and Soon-yo Lee,Rule-based inference model for the Kansei Engineering System .International Journal of Industrial Ergonomics, Volume 24, Issue 5, 14 1999(9): 459-471
7. Jianxin (Roger) Jiao, Yiyang Zhang and Martin Helander . A Kansei mining system for affective design Expert Systems with Applications, Volume 30, Issue 4, 2006(3): 658-673
8. Pawlak, Z. Rough Sets. International Journal of Computer and information sciences, 1982(11):341-356
9. Pawlak and Andrzej Skowron, Rough sets and Boolean reasoning . Information Sciences, Volume 177, Issue 1, 2007(1): 41-73
10. Pattaraintakorn, N. Cercone and K. Naruedomkul ,Rule learning: Ordinal prediction based on rough sets and soft-computing .Applied Mathematics Letters, Volume 19, Issue 12, 2006(12): 1300-1307
11. Jhieh-Yu Shyng, Fang-Kuo Wang, Gwo-Hshiung Tzeng and Kun-Shan Wu ,Rough Set Theory in analyzing the attributes of combination values for the insurance market ,Expert Systems with Applications, Volume 32, Issue 1, 2007(1): 56-64

A Survey of Fuzzy Decision Tree Classifier Methodology

Tao Wang[1], Zhoujun Li[2], Yuejin Yan[1], and Huowang Chen[1]

[1] Computer School, National University of Defense Technology, Changsha, 410073, China
[2] School of Computer Science & Engineering, Beihang University, Beijing, 100083, China
InsistStar@nudt.edu.cn

Abstract. Decision-tree algorithms provide one of the most popular methodologies for symbolic knowledge acquisition. The resulting knowledge, a symbolic decision tree along with a simple inference mechanism, has been praised for comprehensibility. The most comprehensible decision trees have been designed for perfect symbolic data. Over the years, additional methodologies have been investigated and proposed to deal with continuous or multi-valued data, and with missing or noisy features. Recently, with the growing popularity of fuzzy representation, some researchers have proposed to utilize fuzzy representation in decision trees to deal with similar situations. This paper presents a survey of current methods for FDT(Fuzzy Decision Tree)designs and the various existing issues. After considering potential advantages of FDT`s over traditional decision tree classifiers, the subjects of FDT attribute selection criteria, inference for decision assignment, and decision and stopping criteria are discussed. To be best of our knowledge, this is the first overview of fuzzy decision tree classifier.

1 Introduction

Decision trees are one of the most popular methods for learning and reasoning from feature-based examples. They have undergone a number of alternations to deal with language and measurement uncertainties. Fuzzy decision trees (FDT) is one of such extensions, it aims at combining symbolic decision trees with approximate reasoning offered by fuzzy representation. The intent is to exploit complementary advantages of both: popularity in applications to learning from examples and high knowledge comprehensibility of decision trees, ability to deal with inexact and uncertain information of fuzzy representation [2].

In the past, there are roughly a dozen publications on this field. Ichihashi *et al.* [6] extract fuzzy reasoning rules viewed as fuzzy partitions. An algebraic method to facilitate incremental learning is also employed. Xizhao and Hong [7] discretize continuous attributes using fuzzy numbers and possibility theory. Pedrycz and Sosnowski [8], on the other hand, employ context-based fuzzy clustering for this purpose. Yuan and Shaw [9] induce a fuzzy decision tree by reducing classification ambiguity with fuzzy evidence. The input data is fuzzified using triangular membership functions around cluster centers obtained using Kohonen's feature map [24]. Wang *et al.* [10] present optimization principles of fuzzy decision trees based on minimizing the total number and average depth of leaves, proving that the algorithmic complexity of constructing a minimum tree is NP-hard. Fuzzy entropy and classification ambiguity are minimized at node level, and fuzzy clustering is used to merge branches.

B.-Y. Cao (Ed.): Fuzzy Information and Engineering (ICFIE), ASC 40, pp. 959–968, 2007.
springerlink.com

The organization of the paper is as follows: Section 2 contains the preliminaries, definitions, and terminologies needed for later sections; section 3 explains the motivations behind FDT's and their potential use and drawbacks; section 4 addresses the problems of attribute selection criteria, inference for decision assignment and stopping criteria. Summary and conclusions are provided in section 5.

2 Preliminaries

We briefly describe some necessary terminology for describing fuzzy decision trees.

1) The set of fuzzy variables is denoted by $V = \{V_1, V_2, ... V_n\}$.
2) For each variable $V_i \in V$
   I. Crisp example data is $u^i \in U_i$
   II. D_i denotes the set of fuzzy terms
   III. v_p^i denotes the fuzzy tem p for the variable V_i (e.g., v_{Low}^{Income} ,as necessary to stress the variable or with anonymous values-otherwise p alone may be used)
3) The set of fuzzy terms for the decision variable is denoted by D_c.
4) The set of training examples is $E = \{e_j | e_j = (u_j^i, ... u_j^n, y_j)\}$,where y_j is the crisp classification. Confidence weights of the training examples are denoted by $W = \{w_j\}$,where w_j is the weight for $e_j \in E$.
5) For each node N of the fuzzy decision tree
   I. F^N denotes the set of fuzzy restrictions on the path leading to N.
   II. V^N is the set of attributes appearing on the path leading to $N : V^N = \{V_i | \exists p([V_i is v_p^i] \in F^N)\}$
   III. $x^N = \{x_j^N\}$ is the set of memberships in N for all the training examples
   IV. $N | v_p^i$ denotes the particular child of node N created by using V_i to split N and following the edge $v_p^i \in D_i$
   V. $S_{V_i}^N$ denotes the set of N `s children when $V_i \in (V - V^N)$ is used for the split. Note that $S_{V_i}^N = \{(N | v_p^i) | v_p^i \in D_i^N, D_i^N = \{v_p^i \in D_i | \exists (e_j \in E)(x_j^N > 0 \wedge \mu_p^i(u_j^i) >)\}$;in other words, there are no nodes containing no training examples, and thus some linguistic terms may not be used to create subtrees.
   VI. P_k^N denotes the example count for decision $v_k^c \in D_c$ in node N .It is important to note that unless the sets are such that the sum of all

memberships for any u is 1, $P_k^N \neq \sum_{v_p^i \in D_i} P_k^{N|v_p^i}$;that is ,the membership sum from all children of N can differ from that of N ;this is due to fuzzy sets; the total membership can either increase or decrease while building the tree

VII. P^N and I^N denote the total example count and information measure for node N

VIII. $G_i^N = I^N - I^{S_{Vi}^N}$ denotes the information gain when using V_i in N ($I^{S_{Vi}^N}$ is the weighted information content)

6) α denotes the area; ς denotes the centroid of a fuzzy set.

7) $I(x^N)$ denotes the entropy of the class distribution w.r.t. the fuzzy example set x^N in node N . $I(x^N | A_i)$ is the weighted sum of entropies from all child nodes, if A_i is used as the test attribute in node N .

8) $Gain(x^N, A_i) = I(x^N) - I(x^N | A_i)$ is the information gain w.r.t. attribute A_i ,which is the first of the two attribute selection measures we consider.

9) $SplitI(x^N, A_i)$ denotes the split information---the entropy w.r.t. the value distribution of attribute A_i (instead of the class distribution).

10) $GainR(x^N, A_i) = Gain(x^N, A_i) / SplitI(x^N, A_i)$ is the information gain ratio w.r.t. attribute A_i ,which is the second attribute selection measure we consider.

3 Potentials and Problems with Fuzzy Decision Tree Classifiers

Decision trees are one of the most popular choices for learning and reasoning from feature-based examples. Fuzzy decision tree (FDT) aims at combining symbolic decision trees with approximate reasoning offered by fuzzy representation. It is attractive for the following reasons [14]:

1) An apparent advantage of fuzzy decision trees is that they use the same routines as symbolic decision trees (but with fuzzy representation). This allows for utilization of the same comprehensible tree structure for knowledge understanding and verification. This also allows more robust processing with continuously gradual outputs. Moreover, one may easily incorporate rich methodologies for dealing with missing features and incomplete trees. For example, suppose that the inference descends to a node, which does not have a branch (maybe due to tree pruning, which often improves generalization properties) for the corresponding feature of the sample. This dandling feature can be fuzziffied and then its match to

fuzzy restrictions associated with the available branches provides better than uniform discernibility among those children.

2) Fuzzy decision trees can process data expressed with symbolic, numerical values (more information) and fuzzy terms. Because fuzzy restrictions are evaluated using fuzzy membership functions, this process provides a linkage between continuous domain values and abstract features
3) Fuzzy sets and approximate reasoning allow for processing of noisy, inconsistent and incomplete data. It is more accurate than standard decision trees.

The possible drawbacks of FDT, on the other hand, are:

1) From a computational point of view, the method is intrinsically slower than crisp tree induction. This is the price paid for having a more accurate but still interpretable classifier.
2) FDT does not seem to offer fundamentally new concepts or induction principles for the design of learning algorithms, e.g., to the ideas of resampling and ensemble learning (like bagging and boosting) or the idea of margin maximization underlying kernel-based learning methods. [30] doubts that FDT will be very conductive to generalization performance and model accuracy.

4 Special Issues of a Fuzzy Decision Tree Classifier

Like classical decision trees with the ID3 algorithm, fuzzy decision trees are constructed in a top-down manner by recursive partitioning of the training set into subnets. Here, we just list some special issues of fuzzy decision trees:

1) Attribute selection criteria in fuzzy decision trees

A standard method to select a test attribute in classical decision tree is to choose the attribute that yields the highest information gain. When it goes to fuzzy decision trees, it'll lead to some problem, some modifications and enhancements of the basic algorithm have been proposed and studied.

2) Inference for Decision Assignment

The inference procedure is an important part of FDT, and it is different from traditional DT.

3) Stopping criteria

Usually classical tree learning is terminated if all attributes are already used on the current path; or if all examples in the current node belong to the same class. In FDT an example may occur in any node with and membership degree. Thus general more examples are considered per node and fuzzy trees are usually larger than classical trees. To solve this problem, there are many methods mentioned.

4.1 Attribute Selection Criteria in Fuzzy Decision Trees

One important aspect of tree induction is the choice of feature at each stage of construction. If weak features are selected, the resulting decision tree will be meaningless and will exhibit poor performance. Several methods have been proposed for attribute selection criteria in fuzzy decision trees. They can be categorized to five kinds.

4.1.1 Attribute Selection Criteria Based on Information Gain

Tests performed on data (corresponding to selecting attributes to be tested in tree nodes) are decided based on some criteria. The most commonly used is information gain, which is computationally simple and shown effective: select an attribute for testing (or a new threshold on a continuous domain) such that the information difference between that contained in a given node and in its children nodes is maximized. The information content is measured according to [2]: $I^N = -\sum_{2=1}^{|Dc|}(\frac{P_k^N}{P^N}\cdot\log\frac{P_k^N}{P^N})$, where $P_k^N = \sum_{j=1}^{|E|} f_2(x_j^N, \mu_{v_k^c}(y_j))$, $p^N = \sum_{k=1}^{|Dc|} P_k^N$.

4.1.2 Attribute Selection Criteria Based on Gain Ratio

Gain ratio is used in classical decision trees to select the test attribute in order to reduce the natural bias of information gain, i.e., the fact that it favors attributes with many values (which may lead to a model of low predictive power). In FDT, fuzzy partitions are created for all attributes before the tree induction. To keep the tree simple, usually each partition possesses as few fuzzy sets as possible. The information gain ratio is measured according to [3]: $GainR(x^N, A_i) = Gain(x^N, A_i) / SplitI(x^N, A_i)$.

4.1.3 Attribute Selection Criteria Based on Extended Information Measure

As mentioned in [16], negative information gain (ratio) can occur in FDT, so they suggest a different way of computing information measure in FDT to make information gain ratio applicable as a selection measure.

To determine best test attribute, [16] create a fuzzy contingency table (see Table 1) for each candidate A in node N, from which the information measure for attribute A can be computed. They use 9 steps to calculate gain ratio of an attribute, and point out it will not be negative. The detail is in [16].

Table 1. A fuzzy contingency table

$N(A)$	C_1	C_2	$ASum$
a_1	$Z_{C_1}^{N\|a_1}$	$Z_{C_2}^{N\|a_1}$	$Z^{N\|a_1}$
a_2	$Z_{C_1}^{N\|a_2}$	$Z_{C_2}^{N\|a_2}$	$Z^{N\|a_2}$
$CSum$	$Z_{C_1}^N$	$Z_{C_2}^N$	Z^N

4.1.4 Attribute Selection Criteria Based on Yuan & Shaw's Measure

In [9], a method is introduced to construct FDT by means of a measure of classification ambiguity as a measure of discrimination. This measure is defined from both a measure of fuzzy subset hood and a measure of non-specificity. The measure of ambiguity $H_Y(C|A_j)$ was introduced by[9]to measure the discriminating power of attribute A_j with regard to $C: H_Y(C|A_j) = \sum_{t=1}^{mj} w(v_l).G_Y(v_l)$ (1)

where $w(v_l) = \frac{M(v_l)}{\sum_{t=1}^{mj} M(v_l)}$, $M(v_l) = \sum_{x \in X} \mu_{v_l}(x)$,and the details of $G_Y(v_l)$ `s calculation can be found in[9].

4.1.5 Attribute Selection Criteria Based on Fuzzy-Rough Sets

It has been shown that fuzzy-rough metric is a useful gauger of (discrete and real-valued) attribute information content in datasets. This has been employed primarily within the feature selection task, to benefit the rule induction that follows this process [22].

Fuzzy-rough measure is comparable with the leading measures of feature importance. Its behavior is quite similar to the information gain and gain ratio metrics. The results show that the fuzzy-rough measure performs comparably to fuzzy ID3 for fuzzy datasets, and better than it for crisp data [22].

4.2 Inference for Decision Assignment

The most profound differences between FDT and DT are in the process of classifying a new sample. These differences arise from the fact that

1) FDT have leaves that are more likely to contain samples of different classes (with different degrees of match).
2) The inference procedure is likely to match the new sample against multiple leaves, with varying degrees of match.

To account for these potential problems, a number of inference routines have been proposed. Some inferences follow the idea of classifying a new sample directly from the built FDT; others generate rules first, and then utilize these rules to classify new samples.

4.2.1 Classifying New Sample Depending on FDT

To decide the classification assigned to a sample using FDT, we have to find leaves whose restrictions are satisfied by the sample, and combine their decisions into a single crisp response. Such decisions are very likely to have conflicts, found both in a single leaf (non-unique classification of its examples) and across different leaves (different satisfied leaves have different examples, possibly with conflicting classifications).

To define such a decision procedure, [2] define four operators: g_0, g_1, g_2, g_3. Let L be the set of leaves of the FDT. $l \in L$ be a leaf. Each leaf is restricted by. First, a sample e_j is computed the satisfaction of each restriction $[V_i \ is \ v_p^i] \in F^l$ (using g_0), and all results are combined with g_1 to determine to what degree the combined restrictions are satisfied. g_2 then propagates this satisfaction to determine the level of satisfaction of that leaf. A given leaf by itself may contain inconsistent information if it contains examples of different classes. Different satisfied leaves must also have their decisions combined. Because of this two level disjunctive representation, the choice of g_3 is distributed into these two levels.

4.2.2 Classifying New Sample Depending on Rule Base

The most frequent application of FDT is the induction or the adaptation of rule-based models.

A plenty of methods has been developed for inducing a fuzzy rule base from the data given. Among them [14,16,21,32,35], the main difference concerns the way in which individual rules or their condition parts are learned. One possibility is to identify regions in the input space that seem to be qualified to form the condition part of a rule. This can be done by looking for clusters using clustering algorithms, or by identifying hyperboxes in the manner of so-called covering (separate and conquer) algorithms. By projecting the regions thus obtained onto the various dimensions of the input space, rule antecedents of the form $X \in A$ are obtained, where X is an individual attribute and A is a fuzzy set (the projection of the fuzzy region). The condition part of the rule is then given by the conjunction of these antecedents. This approach is relatively flexible, though it suffers from the disadvantage that each rule makes use of its own fuzzy sets. Thus, the complete rule base might be difficult to interpret [30].

An alternative is to proceed from a fixed fuzzy partition for each attribute, i.e., a regular "fuzzy grid" of the input space, and to consider each cell of this grid as a potential antecedent part of a rule. This approach is advantageous from an interpretability point of view. On the other hand, it is less flexible and may produce inaccurate models when the one-dimensional partitions define a multi dimensional grid that does not reflect the structure of the data [30].

Recently, an important improvement in the field of fuzzy rule learning is hybrid methods that combine FDT with other methodologies, notably evolutionary algorithms and neural networks [6,13,22,31,32]. For example, evolutionary algorithms are often used in order to optimize a fuzzy rule base or for searching the space of potential rule bases in a systematic way [36]. Quite interesting are also neuro-fuzzy methods [28]. For example, one idea is to encode a fuzzy system as a neural network and to apply standard methods (like back propagation) in order train such a network. This way, neuro-fuzzy systems combine the representational advantages of fuzzy systems with the flexibility and adaptivity of neural networks.

4.3 Stopping Criteria

One of the challenges in fuzzy decision tree induction is to develop algorithms that produce fuzzy decision trees of small size and depth. In part, smaller fuzzy decision trees lead to lesser computational expense in determining the class of a test instance.

More significantly, however, larger fuzzy decision trees lead to poorer generalization performance. Motivated by these considerations, a large number of algorithms have been proposed toward producing smaller decision trees. Broadly these may be classified into three categories [13].

1) The first category includes those efforts that are based on different criteria to split the instances at each node. Some examples of the different node splitting criteria include entropy or its variants, the chi-square statistic, the G statistic, and the GINI index of diversity. Despite these efforts, there appears to be no single node splitting that performs the best in all cases; nonetheless there is little doubt that random splitting performs the worst. And for FDT, usually adds a condition, namely whether the information measure is below a specified threshold. The threshold defined here enables a user to control the tree growth, so that unnecessary nodes are not added.
2) The second category is based on pruning a decision tree either during the construction of the tree or after the tree has been constructed. In either case, the idea is to remove branches with little statistical validity.
3) The third category of efforts toward producing smaller fuzzy decision trees is motivated by the fact that a locally optimum decision at a node may give rise to the possibility of instances at the node being split along branches, such that instances along some or all of the branches require a large number of additional nodes for classification. The so-called look-ahead methods attempt to establish a decision at a node by analyzing the classifiability of instances along each of the branches of a split. Surprisingly, mixed results (ranging from look-ahead makes no difference to look-ahead produces larger trees]) are reported in the literature when look-ahead is used.

5 Summary and Conclusions

FDT combines fuzzy theory with classical decision trees in order to learn a classification model, which is able to handle vagueness and also comprehensible.

In the past, different authors introduced several variants of fuzzy decision trees. Boyen and Wenkel [1] presented the automatic induction of binary fuzzy trees based on a new class of discrimination quality measures. Janikow [2] adapted the well-known ID3 algorithm so that it works with fuzzy sets.

To be best of our knowledge, in spite of these papers in this field, there isn't a survey on this field. The aim of this paper is to give a global overview of FDT. We categories the main research issues of FDT into three parts: attribute selection criteria, inference procedure and stopping criteria.

After several years of intensive research of FDT classification has reached a somewhat mature state, and a lot of quite sophisticated algorithms is now available. A significant improvement of the current quality level can hardly be expected. In the future, some other research issues of FDT should be paid more attention, such as incremental fuzzy decision trees and FDT on data streams mining.

Ackowledgements

The National Science Foundation of China under Grants No. 60573057, 60473057 and 90604007 supported this work.

References

[1] X.P. Boyen and L. Wenke (1995) "Fuzzy Decision Tree Induction for Power System Security Assessment," IFAC Symposium on control of power plants and power systems. Mexico.

[2] C.Z. Janikow (1998) "Fuzzy Decision Trees: Issues and Methods," IEEE Transactions on Systems, Man and Cybernetics, 28(1): 1–14. IEEE Press, Piscataway, NJ, USA..

[3] J.R. Quinlan (1993) "C4.5: Programs for Machine Learning," Morgan Kaufman, San Mateo, CA, USA.

[4] S.R Safavian and D.Landgrebe (1991) "A Survey of Decision Tree Classifier Methodology," IEEE Transactions on Systems, Man and Cybernetics, vol.21, No.3, pp. 660-674, May.

[5] C.Olaru and L.Wehenkel (2003) "A Complete Fuzzy Decision Tree Technique," Fuzzy Sets System.138: 221-254.

[6] H. Ichihashi, T. Shirai, K. Nagasaka, and T. Miyoshi (1996) "Neuro fuzzy ID3: A method of inducing fuzzy decision trees with linear programming for maximizing entropy and algebraic methods," Fuzzy Sets System, vol. 81,no. 1, pp. 157–167.

[7] W. Xizhao and J. Hong (1998) "On the handling of fuzziness for continuous valued attributes in decision tree generation," Fuzzy Sets System, vol. 99,pp. 283–290.

[8] W. Pedrycz and A. Sosnowski (2000) "Designing decision trees with the use of fuzzy granulation," EEE Transactions Systems, Man and Cybernetics, vol. 30, pp.151–159, Mar.

[9] Y. Yuan and M. J. Shaw (1995) "Induction of fuzzy decision trees," Fuzzy Sets System, vol. 69, pp. 125–139.

[10] X. Wang, B. Chen, G. Qian, and F. Ye (2000) "On the optimization of fuzzy decision trees," Fuzzy Sets System, vol. 112, pp. 117–125.

[11] I. J. Chiang and J. Y. J. Hsu (1996) "Integration of fuzzy classifiers with decision trees," in Proc. Asian Fuzzy System. Symp, pp. 266–271.

[12] I. Hayashi, T. Maeda, A. Bastian, and L. C. Jain (1998) "Generation of fuzzy decision trees by fuzzy ID3 with adjusting mechanism of and/or operators," in Proc. Int. Conf. Fuzzy System, pp. 681–685.

[13] M.Dong and R.Kothari (2001) "Look-Ahead Based Fuzzy Decision Tree Induction," EEE Transactions Systems, vol. 9,no.3.june.

[14] C.Z. Janikow (1996) "Exemplar Learning in Fuzzy Decision Trees," Proceedings of FUZZY-IEEE, pp. 1500-1505. Invited by Prof. Bouchon-Meunier.

[15] B. Bouchon-Meunier and C. Marsala (2003) "Measures of discrimination for the construction of fuzzy decision trees", In Proc. of the FIP'03 conference, Beijing, China, pp. 709-714.

[16] X.Wang and C. Borgelt (2004) "Information Measures in Fuzzy Decision Trees," Proc. 13th IEEE International Conference on Fuzzy Systems (FUZZ-IEEE'04, Budapest, Hungary), vol. 1,pp. 85-90.

[17] R. Jensen and Q. Shen (2004) "Semantics-Preserving Dimensionality Reduction: Rough and Fuzzy-Rough Based Approaches," IEEE Transactions on Knowledge and Data Engineering, vol. 16, no. 12, pp. 1457-1471.

[18] D. Dubois and H. Prade (1980) " Fuzzy Sets and Systems: Theory and Applications," Academic Press, New York.
[19] M. Guetova, S. Hölldobler, and H. Stör (KI 2002) "Incremental Fuzzy Decision Trees," 25th German Conference on Artificial Intelligence.
[20] J.Zeidler and M.Schlossor (1996) "Continuous-Valued Attributes in Fuzzy Decision Trees," IPMU 96, pp 395-400, 36.
[21] C.Z. Janikow and M. Faifer (1999) "Fuzzy Partitioning with FID3.1", Proc. of the 18th International Conference of the North American Fuzzy Information Processing Society. New York.
[22] R.Jensen and Q.Shen(UKCI 2005) "Fuzzy-Rough Feature Significance for Fuzzy Decision Trees," The 5th annual UK Workshop on Computational Intelligence, Birkbeck, university of London.
[23] Pedrycz, W. and Sosnowski, Z.A., "c-fuzzy Decision Trees," IEEE Transactions Systems, Man and Cybernetics, Part C.
[24] T. Kohonen(1989)"Self-Organization and Associative Memory," Berlin, Germany: Springer-Verlag.
[25] Y. Peng and P. Flach (2001) " Soft Discretization to Enhance the Continuous Decision Tree Induction," Aspects of Data Mining, Decision Support and Meta-Learning, Christophe Giraud-Carrier, Nada Lavrac and Steve Moyle, editors, pages 109--118. ECML/PKDD'01 workshop notes, September.
[26] Xavier Boyen, Louis Wehenkel (1999) "Automatic induction of fuzzy decision trees and its application to power system security assessment," Fuzzy Sets and Systems, Volume 102, page 3-19 .
[27] C. Marsala, B. Bouchon-Meunier (1997) "Choice of a method for the construction of fuzzy decision trees," Proc. 12th IEEE Int. Conf. on Fuzzy Systems, St. Louis, MI, USA, 2003.
[28] D. Nauck, F. Klawonn, and R. Kruse, "Foundations of Neuro-Fuzzy Systems," Wiley and Sons, Chichester, UK.
[29] Rudolf Kruse, Detlef Nauck, and Christian Borgelt, "Data mining with fuzzy methods: status and perspectives," EUFIT '99.
[30] Eyke Hullermeier (2005) "Fuzzy Methods in Machine Learning and Data mining: Status and Prospects," Sets and Systems 156(3), 387-407.
[31] C. Olaru, L. Wehenkel (2000) "On neurofuzzy and fuzzy decision tree approaches," Uncertainty and Fusion Eds.: B. Bouchon-Meunier, R. R. Yager, L. A. Zadeh, Kluwer Academic Publishers, pp. 131-145.
[32] Sushmita Mitra, Senior Member, IEEE, Kishori M. Konwar, and Sankar K. Pal. "Fuzzy Decision Tree, Linguistic Rules and Fuzzy Knowledge-Based Network: Generation and Evaluation," IEEE Transactions on Systems, Man and Cybernetics, vol. 32, no. 4, november.
[33] J.R.Quilan (1986) "Induction on Decision Trees," Machine learning, vol.1, pp.81-106.
[34] Tzung-Pei Hong, Chai-Ying Lee, "Learning Fuzzy Knowledge from Training Examples," Conference on Information and Knowledge Management Proceedings of the seventh international conference on Information and knowledge management, Bethesda, Maryland, United States Pages: 161 - 166.
[35] O. Cordon, F. Gomide, F. Herrera, and F. Hoffmann and L. Magdalena (2004) "Ten years of genetic fuzzy systems: current framework and new trends," Fuzzy Sets and Systems, 141(1): 5,31.

The Optimization for Location for Large Commodity's Regional Distribution Center

Yubin Zhong

School of Mathematics and Information Sciences, Guangzhou University,
Guangzhou Guangdong 510006, P.R. China
Zhong_yb@163.com

Abstract. This dissertation adopts programming theory to carry on the demonstration of the logistics distribution system of a large commodity manufacturer in Guangzhou. Aiming at solving the layout of location on the regional distributing center (RDC) of the actual logistics distribution system of the company, a programming model on the optimization of location is put forward. This model is applied to solve the location programming problem with the help of MATLAB6.5 tool, and a decisive supporting system that can examine the accuracy of the model result automatically is designed through the application of C++ BUILDER6.0. It is of great significance of optimization and making decision to the operating of logistics.

Keywords: the logistics distribution; the Regional Distributing Center (RDC); Fuzzy Hierarchy Synthetic Evaluating, the optimization model for location, decisive supporting system.

While building a new logistics node in logistics network, under the condition of considering the investment of fundamental construction, the investment of fundamental construction is taken into consideration and programming theory is applied in study of the location and the arrangement of logistic node.

Followings are the basic train of thought in its application. Firstly, analyzing and summing up the total expenditure of the logistic node and some restrict conditions in order to build up a reasonable layout model. Then, making use of the mathematics software MATLAB6.5 tool to solve the model, and finding out the best location for layout of logistics node, which possesses significant meaning for the decision of operation and the logistic optimization. Under the guide of programming theory, this dissertation makes research on the reasonable location and the layout of the logistic node, carries on some study that aims at the layout of location on the regional distributing center of the actual logistics distribution system of the company and gets the following optimization of the project.

1 Problem Statement

There has a large international Commodity manufacturer in Guangzhou, producing various commodities and taking up a huge proportion of the market sale. And it is the

B.-Y. Cao (Ed.): Fuzzy Information and Engineering (ICFIE), ASC 40, pp. 969–979, 2007.
springerlink.com

only regional Distribution Center in Guangzhou that serves the whole regional in South China with commodity. Because of the continuous extension of the company business, the company has already been overloading, based on its existing infrastructure and scales. Therefore, it needs taking some measures to improve the operation of the RDC, such as consummating and developing the present operation of the RDC in Guangzhou, or replacing it with a new RDC center of South China in Guangzhou to satisfy the company's need in South China in the following six years after 2005. Now this dissertation focus on the Logistics Distribution System of the product currently (see the Figure 1). The principle of the lowest logistic Distribution cost is to determine a new RDC layout within the region of South China in Guangzhou, finally to make a comprehensive decision and to select the best location for the RDC.

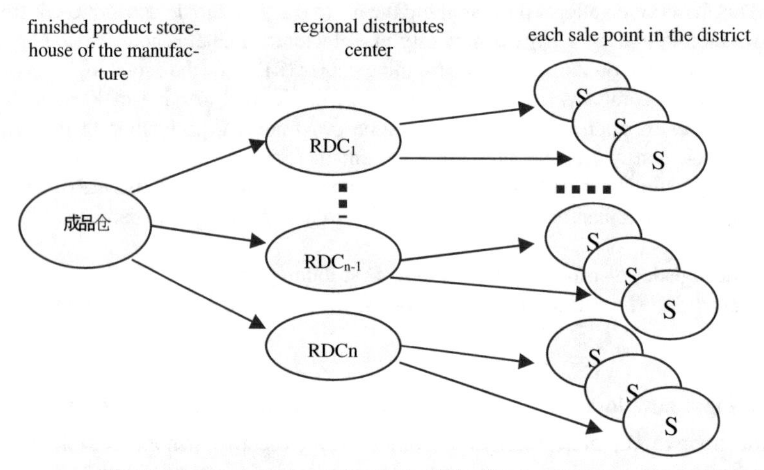

Fig. 1. The sketch map of Logistics Distribution System of the enterprise

2 The Model Hypothesis and Data Processing

In order to make the decision about the location for optimization of the regional distribute center more scientific, before building up the principle and calculating algorithm of the optimization location programming models, it is necessary to hypo the size the model reasonably, process the data appropriately, and elucidate the homologous sign.

(1)Suppose that the logistic demands of each sale point changes slightly during the period of the decision for location, that is to say, assume that the average amount of logistic of RDC per month is fixed.

(2)Suppose that research is made on distance, it is relatively reasonable that the error is within 2 kilometers, which will not affect the conclusion of the topic.

(3)Suppose that the land purchasing expenses is taken into consider in terms of the fee of candidate fundamental construction, other factors can be excluded; or suppose that other factors have similar influence on the fee of fundamental construction.

(4)Suppose that the expenses of each RDC candidate location are fixed during the period of making decision for location.

The elucidation of the sign is:

α_n : The amount of logistics per month from the finished product storehouse of the manufacture to the RDC candidate location, β_n : The transportation distance from the finished product storehouse of the manufacture to the RDC candidate location, γ_n : The transporting unit price from the finished product storehouse of the manufacture to the RDC candidate location, $x_n = \begin{cases} 1 & \text{mean at construct in } R_n \\ 0 & \text{mean at construct in } R_n \end{cases}$, TC : The Logistic Distribution cost of the company business operation through the RDC, R_n : The candidate ground n of RDC, S_n : The n sale-points, N_{S_n} : The average demand of logistics per month for RDC in Guangzhou, C_{MR_n} : The cost for transportation from the finished product storehouse of the manufacture to the RDC candidate location, $C_{R_nS_n}$: The cost for distribution from each sale-point S to the RDC candidate location, C_{R_n} : The fixed expenses of construction and operation of the RDC candidate location.

3 The Theory and the Algorithm of Optimization Model for Location

(1) The basic train of thought for selecting the location

According to Fuzzy Hierarchy Synthetic Evaluating Model in reference [1]: $B = A \circ R$. As to the location programming of Guangzhou RDC of the company in South China, after the analysis of some decisive factors of the location, such as the obtainment of the location, the expense of land, convenience for transportation, natural condition, factor of labor force, distance of the market, 4 candidate locations are found for suited to be the location of the RDC in Guangzhou. Plus the RDC operating location in Guangzhou, there have 5 candidate locations totally, marked respectively as R_0、R_1、R_2、R_3、R_4. According to principles of the amount of logistics from the finished product storehouse of the manufacture to the RDC location in Guangzhou, the transportation distance and the minimum coat for transportation in unit, the amount of logistics from 5 RDC candidate in Guangzhou to the major sale-points in the region of South China, the transportation distance and the cost for transportation in unit as well as the fixed expenses of construction and operation of the RDC candidate locations in Guangzhou should be figured out. Through the application of the optimized location programming model constructed below, the Logistic Distribution cost in Guangzhou's RDC can be solved, thus, the optimization of location can be resolved.

(2) The optimized target

Specifically speaking, the optimized target is to make the Logistic Distribution cost lowest, which flows through the RDC in Guangzhou. The Logistic Distribution cost can be expressed as: The Logistic Distribution cost of the RDC in Guangzhou= the cost for transportation from the finished product storehouse to the RDC in Guangzhou + the cost for distribution from the RDC in Guangzhou to every sale-spot + the fixed expenses for operation of the RDC in Guangzhou. That's to say: $TC = C_{MR} + C_{RS} + C_R$

(3) The model of optimized location program

$$\text{Min}:\ TC(x_n) = C_{MR} + C_{RS} + C_R = \sum_{n=0}^{4} (C_{MR_n} + C_{R_nS} + C_{R_n})x_n$$

$$\text{S. T.} \begin{cases} \sum_{n=0}^{4} x_n = 1 \\ x_n = \begin{cases} 1 & \text{1----}R_n \text{ is on building} \\ 0 & \text{0----}R_n \text{ is off building} \end{cases} \\ C_{MR_n} = \alpha_n \bullet \beta_n \bullet \gamma_n \end{cases}$$

(4) The solution of model

1) According to table 1, the cost for transportation from the finished product storhouse to the RDC candidate location in Guangzhou C_{MR_n} can be calculated.

Table 1. The statistics of the cost for transportation

The finished product storehouse to each candidate location	operating RDC	The candidate location 1	The candidate location 2	The candidate location 3	The candidate location 4
	R_0	R_1	R_2	R_3	R_4
The logistics quantity(ton) per month	3004.2	3004.2	3004.2	3004.2	3004.2
The transportation distance(kilometer)	3.0	4.0	44.0	46.0	43.0
The unit price for transportation($/ t/ k)	0.5	0.5	0.5	0.5	0.5

The amount of logistics per month is obtained from the data that have existed. The transportation distance is set after the 4 candidate locations are established and is calculated by the Chinese electronics map software. The error margin does not exceed the scope of error margin set by the research of these items. The unit price for transportation is set by the existing data of the cost for transportation.

The cost for transportation from the finished product storehouse to every RDC candidate location C_{MR_n} = The amount of logistics per month α_n from the finished product storehouse to every RDC candidate location n × The transportation distance β_n from the finished product storehouse to every RDC candidate location n × The unit price for transportation γ_n from the finished product storehouse to every RDC candidate location n, or $C_{MR_n} = \alpha_n \bullet \beta_n \bullet \gamma_n$, thus:

$$C_{MR_0} = \alpha_0\beta_0\gamma_0 = 3004.2*3*0.5=4506.3 \text{ yuan}$$
$$C_{MR_1} = \alpha_1\beta_1\gamma_1 = 3004.2*4*0.5=6008.4 \text{ yuan}$$
$$C_{MR_2} = \alpha_2\beta_2\gamma_2 = 3004.2*44*0.5=66092.4 \text{ yuan}$$
$$C_{MR_3} = \alpha_3\beta_3\gamma_3 = 3004.2*46*0.5=69096.6 \text{ yuan}$$
$$C_{MR_4} = \alpha_4\beta_4\gamma_4 = 3004.2*43*0.5= 64590.3 \text{ yuan}$$

and α_n : The amount of logistics per month from the finished product storehouse to every RDC candidate location n , β_n : The transportation distance from the finished product storehouse to every RDC candidate location n, γ_n: The unit price for transportation from the finished product storehouse to every RDC candidate location n

2) According to the table 2, the cost for distribution C_{R_nS} from every RDC candidate location to all the sale-spots in order can be obtained.

The sale-spots are the major sale-spots obtained by the analysis of the existing data. N_{S_n} represents the average amount of logistics of the RDC in Guangzhou demanded by the sale-spots S_n per month. It is obtained through the existing data. The transportation distance is set after the 4 candidate locations are established and is calculated by the Chinese electronics map software. The error margin does not exceed the scope of error margin set by the research of these items. The unit price for transportation P_{RS} is 0.8 yuan/ton per kilometer. .

The cost for distribution C_{R_nS} from every RDC candidate location to all the sale-spots in order = the sum of cost of distribution of the sale-spots in the same district. And the cost of distribution in the same district = the average amount of logistics of the RDC in Guangzhou N_{S_n} demanded by the sale-spots S_n per month × The cost for transportation from the finished product storehouse to every RDC candidate location C_{MR_n} × The unit price for transportation P_{RS}, therefore,

$$C_{R_0S} = \sum_{n=1}^{35} (N_{Sn} * D_{R_0Sn} * P_{RS}) = 613977.4 \text{ yuan}$$

$$C_{R_1S} = \sum_{n=1}^{35} (N_{Sn} * D_{R_1Sn} * P_{RS}) = 617084.2 \text{ yuan}$$

$$C_{R_2S} = \sum_{n=1}^{35} (N_{Sn} * D_{R_2Sn} * P_{RS}) = 616551.2 \text{ yuan}$$

$$C_{R_3S} = \sum_{n=1}^{35} (N_{Sn} * D_{R_3Sn} * P_{RS}) = 623044.7 \text{ yuan}$$

$$C_{R_4S} = \sum_{n=1}^{35} (N_{Sn} * D_{R_4Sn} * P_{RS}) = 625739.3 \text{ yuan}$$

N_{S_n}: The average amount of logistics of the RDC in Guangzhou demanded by the sale-spots S_n per month, C_{MR_n}: The cost for transportation from the finished product storehouse to every RDC candidate location n, $C_{R_nS_n}$: The cost of distribution from the RDC candidate location n to each sale-spot S, Here the unit price for transportation P_{RS} is 0.8 yuan/ton per kilometer

Table 2. The statistics of the distribution coast

Table 2: the statistics of the cost for distribution from every RDC candidate location to all the sale-spots (The unit price for transportation P_{RS} is 0.8 yuan/ton per kilometer)							
The sale-spots		N_{S_n} (Ton)	R_0 Distance (kilometer)	R_1 Distance (kilometer)	R_2 Distance (kilometer)	R_3 Distance (kilometer)	R_4 Distance (kilometer)
S1	SHENZHEN	495.8	112	120	150	153	150
S2	ZHUHAI	35.4	144	148	183	185	135
S3	SHUNDE	21.4	63	63	61	62	46
S4	ZHONGSHAN	58.8	110	112	107	106	90
S5	PANYU	5.8	41	40	48	46	30
S6	JIANMEN	16.7	144	140	119	112	92
S7	ZHAOQING	3.2	133	133	122	117	111
S8	HUIZHOU	54.3	139	138	147	156	159
S9	GUANGZHOU	769.5	31	30	15	16	12
S10	LUFENG	7.6	300	294	304	314	335
S11	DONGGUAN	121.6	31	33	74	76	74
S12	XINTAN	22.8	13	15	52	54	52
S13	FOSHAN	77.4	55	54	40	35	23
S14	NANHAI	14.0	51	51	38	32	20
S15	QINGYUAN	21.1	107	109	62	61	85
S16	KAIPING	15.9	151	151	129	123	130
S17	CHAOYANG	10.7	416	417	428	437	459
S18	PUNING	24.1	367	368	379	388	410
S19	ZHANJIANG	54.7	466	464	440	441	442
S20	JIEYANG	31.4	477	478	491	502	535
S21	SHAOGUANG	407.4	260	261	239	244	261
S22	MEIZHOU	23.6	405	407	423	433	434
S23	SHANTOU	45.2	440	441	459	468	485
S24	XIAMEN	214.6	688	690	707	717	734
S25	YULIN	25.8	460	458	423	428	424
S26	NANNING	60.0	721	719	702	695	685
S27	GUILIN	64.2	601	629	619	604	599
S28	LIUZHOU	57.2	659	657	632	630	627
S29	GANZHOU	35.3	460	459	431	442	455
S30	HAIKOU	42.9	705	702	691	683	601
S31	ZHANGZHOU	29.1	666	667	683	693	709
S32	QUANZHOU	68.0	750	751	766	774	793
S33	CHANGSHA	9.7	724	713	694	702	722
S34	FUZHOU	55.1	962	928	938	948	964
S35	NANCHANG	3.9	903	902	878	885	906

3) The expense of land chosen as the RDC candidate location.

The serial number of each RDC candidate location is the fixed number in the region of Guangzhou. The expense of land is based on the reference of the present price of the land in each district of Guangzhou and determined by the relationship between them.

Table 3. The expenses for construction of the RDC candidate location

RDC candidate location	The RDC now operating	Candidate location 1	Candidate location 2	Candidate location 3	Candidate location 4
the serial number of the location	The RDC now operating	AP090304W 2	AB190648 W1	AB2305016 W1	AF0607015 W1
Price of the land(10^3 $ / acre)	16	15.8	16.5	16.5	15.7
The expense of the land	1792	1769.6	1848.0	1848.0	1758.4

4) The fixed expense of each RDC candidate location C_{R_n} .

The operation expenses per month of each RDC candidate location is calculated according to the data obtained from the existing RDC.

Table 4. The budget fixed expenses (ten thousand dollars)

Table 4: The budget fixed for construction and operation of each RDC candidate location (ten thousand dollars)					
The chosen location of RDC	The RDC now operating	Candidate location 1	Candidate location d 2	Candidate location 3	Candidate location 4
The basic construction expenses	1792.0	1769.6	1848.0	1848.0	1758.4
The operating expenses per month	135.0	135.0	135.0	135.0	135.0
Summation	1927.0	1904.6	1983.0	1983.0	1893.4

The fixed expenses of each RDC candidate location C_{R_n} = the expenses for the basic construction + the operating expenses per month, thus:

$$C_{R_0} = 1927\times 10000 = 19270000 \text{ (yuan)},$$
$$C_{R_1} = 1904.6\times 10000 = 19046000 \text{ (yuan)}$$
$$C_{R_2} = 1983\times 10000 = 19830000 \text{ (yuan)}$$
$$C_{R_3} = 1983\times 10000 = 19830000 \text{ (yuan)}$$
$$C_{R_4} = 1893.4\times 10000 = 18934000 \text{ (yuan)}$$

5) Solving the target function

$$\text{Min: } TC(x_n) = C_{MR} + C_{RS} + C_R = \sum_{n=0}^{4} (C_{MR_n} + C_{R_nS} + C_{R_n})x_n$$

With the formula above, plus the parameter from step 1) to 4), the optimal location can be obtained with the help of mathematics software, MATLAB6.5 tools to solve the calculation. Thus, this kind of method of choosing the best location is called the optimized location programming method.

4 The Calculation Result and Processing

With the above preparation, mathematics software MATLAB6.5 tools can figure out the calculation. And the result is:

$$\text{Min } TC\ (X_n) = TC\ (X_1) = 19669092.5 \text{ (Yuan)}$$

Then, the RDC candidate location 1 is the location chosen as the lowest cost of the logistics distribution, that is to say, the candidate location 1 is the optimal location in the RDC candidate locations.

In order to test the accuracy of the result of the above-mentioned model, the model is used as the knowledge base of the software system. According to the systematic procedure flow chart (figure 2), the C++ the BUILDER6.0 is applied to write a decisive support system. And the characteristic of this system is to put the data into the procedure from EXCEL, and to operate on them so as to let the data process automatically. It can also reduce the time of the processing by man, as well as avoid the calculation mistake caused by man. It improves the management level and ability of the manufacturer. And it plays a dominant role in enhancing the manufacturer's competence of logistics operation and management. It provides practical methods for later logistics operation and management.

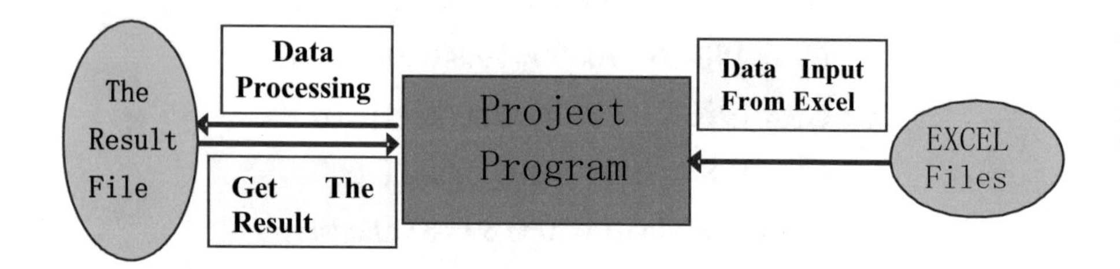

Fig. 2. System flow chart

5 Conclusion

(1)The solution of optimizing the location of the regional distribution center represents the combination of qualitative method and quantitative method in logistics operating decision. That's to say, first, according to several decisive factors of choosing the location, such as the obtainment of the candidate land, the cost for the land, the convenience for transportation, the natural condition, the labor force and the distance of the market and through Fuzzy Hierarchy Synthetic Evaluating Method, several candidate location can be got as RDD candidate location. Then the model of the optimal location programming can be established by the analysis of quantitative method. Finally, the optimal location is found after the calculation of the mode.

(2) Through the solution of choosing the optimal location based on the mathematics model, we can also discover that as for deciding the location of the logistics node of the Regional Distribution Center, it can't easily take the construction cost of logistics node as reference, but should be decided by the total cost of the logistics system, even the total cost of the supply chain of the whole business operation, as logistics node etc. is only a logistics system or part of the supply chain,. Only when the total cost of the whole logistics system is most appropriate can the competition ability of the supply chain of the logistics be promoted and can the positive social benefit of the logistics system be carried out. As the RDD location mention above, if the RDC construction expenses is only consideration for making decision, candidate location 4 (the construction expenses is minimum)will be chosen, however, candidate location 4 is in fact not the best as the total cost of the logistics distribution flowing through Guangzhou is not superior, compared with other candidate locations. Candidate location 1 can realize the lowest cost of the logistics distribution.

(3) The data which have put forward above and involved in tried to solving the problem of selecting the location of the religion distribution center are based on the practice observation of the modern logistics operation personally. According to the

data collected with the guide of the theories of logistics and OS, further research is made. The result and the project have practical meaning for the optimization of the logistics operation and making final decision.

(4)The software system for optimal location of regional distribution center we designed is a widely used decisive support system. It can be represented in construction as putting the information into different categories and flexible combination. In function, the system for selecting the optimal location can be extended as a system for logistics operation and management that can be widely applied. The customer can design his own index sign system according to his scheduled purpose, and input the relative data, bringing them into the index sign system database.

(5)The logistics and OS have close relationship. The theoretically method of the OS is a powerful tool for the logistics operation and managements, in other words, the problems of many optimization programming, the prediction of logistics, the logistics project and operation results in modern logistics operation practice can be applied by the OS to solve.

Acknowledgement

This subject is supported by High-education Science and Technique Project of Guangzhou, the People's Republic of China (No.2056).

References

1. Zhong Yubin(2005). The FHSE Model of Software System for Synthetic Evaluating Enterprising[J]. Journal of Guangzhou University.4(4):316-320.
2. Zhong Yubin(1994). Imitating and Forecast of Economic Systems of District Industry of Central District of City[J], Journal of Decision Making and Decision Support Systems, 4(4): 91-95.
3. Zhong Yubin, Luo Chengzhong(1991). Reforming Plan of Gray-system Modeling [J], Systems Engineering, 9 (6): 13-17
4. Hu Yunquan(2003). The OS Course[M], Beijing: QinHua University Press
5. Cheng Guoquan(2003). The logistics facilities programming and designing [M], China supplies Press
6. Wang Yan. Jiang Xiaomei(2004). The whole-distance programming of distribution center[M], China Machine Industry Press
7. Qin Mingshen(2003). The technique of the logistics decision analyzes[M], Chinese supplies Press
8. Wang Moran(2003). MATLAB and science calculation [M], the electronics industry Press
9. Qian Neng(2003). C++ program design course[M], Beijing: QinHua University Press

Fuzzy Evaluation of Different Irrigation and Fertilization on Growth of Greenhouse Tomato

Na Yu, Yu-long Zhang*, Hong-tao Zou, Yi Huang, Yu-ling Zhang, Xiu-li Dang, and Dan Yang

College of land and Environment, Liaoning Key Laboratory of Agricultural Resources and Environment, Shenyang Agricultural University
Liaoning, ZIP 110161, P.R. China
sausoilyn@163.com
*ylzsau@163.com

Abstract. Based on experiment of plastic mulching and drip irrigation in greenhouse, nine treatments of different irrigation and fertilization were evaluated with multi-factor fuzzy synthesis evaluation method. Linear function was selected as membership function in fuzzy synthesis evaluation model. The weighting of factors set which included the yield, fruit qualities, water use efficiency and fertilizer utilization efficiency was confirmed by principal component analysis method. The evaluation result showed that the optimum treatment of integral indicators that fertilizer amount was pure N 337.5 $kg{\cdot}ha^{-1}$ and pure K_2O 337.5 $kg{\cdot}ha^{-1}$, lower and high limit of irrigation were respective 40 kPa and 6 kPa.

Keywords: Different irrigation and fertilization treatment; Greenhouse tomato; Principal component analysis; Fuzzy synthetic evaluation.

1 Introduction

With the quick development of protected agriculture, some problems bringing by plentiful fertilization and excessive irrigation were serious increasingly such as soil degradation, poor crop quality etc [1-2]. Efficient use of water and fertilizer were becoming important particularly in protected production where irrigation was an unique water source in protected field. Improper irrigation and fertilization management was also major contributor to water contamination and water resource shortage [3-4]. Though water-saving irrigation techniques such as the drip irrigation has relieved above problems a certain degree in protected agriculture, it also caused low yield and quality of agricultural products because of lacking of the relevant irrigation and fertilization indexes. Rational irrigation and fertilization management has already become the key technology of high-quality

* Corresponding author.

B.-Y. Cao (Ed.): Fuzzy Information and Engineering (ICFIE), ASC 40, pp. 980–987, 2007.
springerlink.com

and environment-friendly vegetables production, prevention and control degradation of protected soil [5]. In protected production, irrigation and fertilization or their interactions were usually evaluated in terms of crop yield, few studies used multi-index simultaneity to identify the optimum combination scheme of irrigation and fertilization [6-9]. Only a few studies have documented the method of multi-factors synthesis evaluation to evaluate the irrigation and fertilization management on growth and development of crop in agricultural field [10-11].

The fuzzy set theory was introduced by Zadeh [12], however, it became popular after 1980s and were used mainly by researchers in electronics and computer engineering. Nowadays, an immense number of studies were carried out using this method [13-14]. Fuzzy theory was designed to interpret the uncertainties of the real situations. Fuzzy evaluation methods processed all the components according to predetermined weights and decrease the fuzziness by using membership functions [15-18], therefore sensitivity was quite high compared to other index evaluation techniques. So the objectives of the research were to evaluate the combination strategy of optimizing irrigation and fertilization management by fuzzy synthesis evaluation method in the plastic mulching and drip fertigation greenhouse production of high-yield and high-quality tomato [19-20]. The research also could support the theory for rational irrigation and fertilization management under the condition of agricultural water-saving irrigation, and inject new content in solving the agricultural problem by using the fuzzy theory.

2 Materials and Methods

2.1 Experimental Materials

The experiment was conducted in a greenhouse at scientific research base of Shenyang Agricultural University in China. The experimental soil was a tillage meadow soil, Table 1. shows the characteristics of the experimental soil, a routine method was used as the determination method of experimental soil [21].

The experimental plant was tomato (Lycopersicon esculentum), tomato species is Liao-yuan-duo-li. Tomato seeds were planted by soilless seedling on March 8, 2002. Tomato seedlings were transplanted on April 20, 2002 and harvested on July 19, 2002.

2.2 Experimental Design

Implemental scheme of three levels of two factors as shown in Table 2. were to be completely randomized in these designs.

2.3 Experimental Method

The experiment consisted of nine treatments replicated two times. The plot size of each replication was 6.075 m^2 consisting of two 0.563 m wide beds, 5.4 m long. The tomatoes were planted at a density of approximately 30000 seeds$\cdot ha^{-1}$. Each plot had isolation belt, plastic film was placed vertically to a depth of 60 cm to prevent water and fertilizer flow between the treatments before transplant.

Table 1. Physical and chemical properties of experimental soil

pH	Orgamic matter $(g{\cdot}kg^{-1})$	Alkali-hydrolyzable N $(mg{\cdot}kg^{-1})$	Available N $(mg{\cdot}kg^{-1})$	Available K $(mg{\cdot}kg^{-1})$
6.55	19.48	88.10	175.10	251.40

Total N $(g{\cdot}kg^{-1})$	Total P $(g{\cdot}kg^{-1})$	Total K $(g{\cdot}kg^{-1})$	Soil bulk density $(g{\cdot}cm^{-3})$	Field capacity $(cm^3{\cdot}cm^{-3})$
1.91	1.70	14.79	1.44	0.4154

Soil texture(%)			
>0.25mm	0.25-0.02mm	0.02-0.002mm	>0.002mm
16.68	43.76	21.12	18.44

Table 2. Experimental design scheme

Treatment	Implemental scheme	
	$Dosage of pure N/Dosage of pure K_2O$ $(kgN{\cdot}ha^{-1})/(kgK_2O{\cdot}ha^{-1})$	Soil water suction (kPa)
1	75.00/75.00(Low)	20
2	337.50/337.50(Middle)	20
3	600.00/600.00(High)	20
4	75.00/75.00(Low)	40
5	337.50/337.50(Middle)	40
6	600.00/600.00(High)	40
7	75.00/75.00(Low)	60
8	337.50/337.50(Middle)	60
9	600.00/600.00(High)	60

The operation were controlled by distinguishing surface soil from subsoil, and soil were backfilled strictly according to original soil layer. Soil preparation were conducted with rolling cultivator, the depth of work was 30cm, having ridge forming, tomato seedlings were transplanted with 30 cm spacing within rows and 55cm between rows.

The fertilizer (N, P, K) applied in form of urea, superphosphate, potassium sulphate. The amounts of nitrogen and potassium fertilizer were divided into four equal parts. The first part was applied as basal fertilizer, the others was applied at three fruit-swelling periods through the irrigation system, the allocation proportion of complementary fertilizer was 2:2:1. The amounts of phosphate fertilizer applied equally to all plots, the amounts added were P_2O_5 217.5 $kg{\cdot}ha^{-1}$. All the phosphate fertilizer were applied as basal fertilizer.

All tomato seedlings were irrigated after transplanting, and irrigated at reviving stage secondly, then the differential treatments began. Tensiometers (ICT, Australia) were installed in all plots shortly after transplant. Tensiometers were vertically inserted adjacent to the drip belt midway between two plants, with the porous cups positioned at a depth of 0.3 m. Tensiometer showed the change of soil water potentials, and irrigation time was confirmed. The soil water content of high irrigation limits is field capacity, soil water suction was corresponding 6 kPa, low irrigation limits was calculated according to predefining soil water suction (table 2), i.e. irrigation was applied when the reading of tensiometer was reached the designed soil suction (at 8:00 am, at soil depth of 30 cm), according as the reading, soil water characteristic curve equation ($y = 0.4883x^{-0.0876}$, $r = 0.993^{**}$, n=9, x: soil water suction (kPa); y: soil volumetric water content($cm^3{\cdot}cm^{-3}$)) was used to convert soil water suction into soil water content, and a single irrigation amount was calculated for each plot based on measurements of soil water suction according as the following formula:

$$\mathrm{Q}\ (m^3) = (\theta_F - \theta_S)\times H\times R\times S$$

Q: single irrigation amount (m^3/plot), θ_F and θ_S were the soil water content of high and lower irrigation limits, H was designed moisting layer of soil (m), H=0.3m; R was percentage of wetted area, R=0.7; S was area of plot (m^2).

A drip irrigation set was designed and installed at the experimental plot. After transplanting the tomatoes, the drip belts were put on the ridge, dripper and rootage of tomato plant were opposite of each other and insured 10cm distance between them. After installation it was tested for emission uniformity which was acceptable. Then mulching the plastic film, the experiments were carried out under the condition of plastic mulching and drip fertigation subsequently.

2.4 Sampling Collection and Analysis

The tomatoes were reserved for three fringes, the tomatoes were pinched off young shoots and were recorded the fruit yields in whole planting stage. Water table was used to record the irrigation amounts. Ten tomato fruits were collected from each plant with same position at full fruit period, and were blended by beating for standby. The measured indexes of fruit quality and the methods were as follows: vitamin C content was determined by 2, 6-dichlorphenal-indophenol method, organic acid was titrated by alkali-titration method, soluble sugar content was determined by anthrone-colorimetric method [22].

3 Result and Analysis

3.1 Confirm of Model Fuzzy Synthetic Evaluation

The evaluation was performed by fuzzy synthesis evaluation model $M = (\vee, \cdot)$, synthesis evaluation space was triple (U, V, R). Among it,

$S = \{x_1, x_2 \cdots x_m\}$ is scheme set,
$U = \{u_1, u_2 \cdots u_n\}$ is factors set of model,
$V = \{v_1, v_2 \cdots v_n\}$ is evaluation set,
$\tilde{R} \in F(U \times V)$ is fuzzy evaluation matrix.

$r_{ij}(i = 1, 2, \cdots m, j = 1, 2 \cdots, n)$ is value of evaluation, mean the membership degree of ith factor to jth grade, then fuzzy array matrix

$$r_j = (r_{1j}, r_{2j}, \cdots, r_{nj})^T (j = 1, 2, \cdots, m)$$

could constitute a fuzzy relation from U to $\{v_j\}$, all the fuzzy array matrix constituted a fuzzy evaluation matrix $\tilde{R}$.

$$\tilde{R} = \left(r_1, r_2 \cdots r_n\right)^T = \begin{pmatrix} r_{11} & r_{12} & \cdots & r_{1m} \\ r_{21} & r_{22} & \cdots & r_{2m} \\ \cdots & \cdots & \cdots & \cdots \\ r_{n1} & r_{n2} & \cdots & r_{nm} \end{pmatrix}$$

$$\tilde{A} = (a_1, a_2 \cdots a_n), \ \sum_{i=1}^{n} a_i = 1, \ a_i > 0$$

vector $\tilde{A}$ is the weighting of evaluation factors, result of evaluation $\tilde{B} = \tilde{A} \circ \tilde{R} \in F(V)$.

3.2 The Solve of Fuzzy Synthetic Evaluation Model

In order to evaluate the applicability of each treatment comprehensively, the experiment overall evaluated experimental factor (irrigation, fertilization) and yield, quality index of tomato growth, six factors were selected,

u_1: tomato yield $(kg{\cdot}ha^{-1})$,
u_2: vitamin C content $(mg{\cdot}100g^{-1})$,
u_3: soluble sugar content (%),
u_4: sugar/acid,
u_5: water use efficiency $(kg{\cdot}m^3)$,
u_6: fertilizer$(N + K)$production efficiency $(kg{\cdot}kg^{-1})$.

The six indexes constituted the factors set of fuzzy synthesis evaluation model, Suppose that factors set is

$$U = \{u_1, u_2, u_3, u_4, u_5, u_6\}.$$

In experimental analysis, evaluation set was $V = \{v_1\}$, under the experimental condition, according to max membership rule finally, the different irrigation and fertilization treatments were sorted and confirmed the optimal treatment scheme. In computational process, membership function of linear function can be mathematically formulated as:

$$b_i = \begin{cases} 0, & x \leq x_{\min} \\ \dfrac{|x - x_{\min}|}{|x_{\max} - x_{\min}|}, & x_{\min} < x < x_{\min} \\ 1, & x > x_{\max} \end{cases}$$

Where $x_{\min}$ and $x_{\max}$ were the lower and upper boundaries of the measured values of each experimental factor. The weighting a_i $(1 \leq i \leq 6)$ of each factor was confirmed in the process of synthesis evaluation, method of principal component analysis was used to gain the weighting $\tilde{A}$. Weight determination is a complex issue. Weight factors were common distributed according to the knowledge and experience of the experts. But the relevant research was seldom reported the weight, the method of principal component analysis could objectively confirm the principal component by reducing the subjective influence. The score of principal component could be transformed the weighting [23]. After linear transformation, the score of principal component(total variance $> 85\%$) could get the weighting of each factor. Fuzzy max- product operator was selected to progress synthesis operation in this model, that is, "$\circ$" was max-product operator $(\vee, \cdot)$. Finally, evaluation result was obtained by maximum membership rule. It was v_i corresponding treatment of irrigation and fertilization which is the optimum combination scheme in local condition.

3.3 The Result of Fuzzy Synthesis Evaluation Model

According to the measured values of factors set with different irrigation and fertilization treatments, after calculating by membership function, got the fuzzy evaluation matrix as follows:

$$\tilde{R} = \begin{pmatrix} 0.5450 & 0.5986 & 0.0000 & 0.2584 & 1.0000 & 0.0597 & 0.2579 & 0.2009 & 0.1125 \\ 0.6166 & 1.0000 & 0.5815 & 0.0239 & 0.6373 & 0.8225 & 0.0000 & 0.3582 & 0.3362 \\ 0.7244 & 0.4257 & 0.0486 & 1.0000 & 0.9559 & 0.0000 & 0.1995 & 0.7110 & 0.3635 \\ 1.0000 & 0.4439 & 0.4361 & 0.4456 & 0.5240 & 0.0000 & 0.3117 & 0.5128 & 0.1811 \\ 0.0869 & 0.1997 & 0.7231 & 0.8759 & 0.9942 & 0.6493 & 0.0000 & 0.9432 & 1.0000 \\ 1.0000 & 0.0367 & 0.0000 & 0.9356 & 0.0412 & 0.0002 & 0.9355 & 0.0323 & 0.0004 \end{pmatrix}$$

Principal component analysis was performed for each factor of different irrigation and fertilization treatments, the score of The first principal component(occupied total variance$> 99\%$) traversed the linear transformation, the weighting of each factor was obtained.

$$\tilde{A} = (0.7257, 0.0549, 0.0548, 0.0548, 0.0551, 0.0548).$$

Fuzzy max-product operator $(\vee, \cdot)$ was selected to progress synthesis operation, the following fuzzy synthesis operation result could be obtained based on the weighting value $\tilde{A} = \{a_1, a_2 \cdots a_6\}$ and fuzzy evaluation matrix $\tilde{R}$:

$$\begin{aligned} \tilde{B} &= \tilde{A} \circ \tilde{R} \\ &= (0.3955, 0.4344, 0.0398, 0.1875, 0.7257, 0.0451, 0.1872, 0.1458, 0.0816) \end{aligned}$$

Consequently, the membership degree of evaluation set was obtained. According to maximum membership principle, scheme set ranking of different irrigation and fertilization treatments was obtained as follows.

$$x_5 \succ x_2 \succ x_1 \succ x_4 \succ x_7 \succ x_8 \succ x_9 \succ x_6 \succ x_3$$

The result showed the fifth treatment had the maximum membership degree, which was the optimal combination scheme. That is, in the local condition of climate and soil and under the condition of plastic mulching and drip irrigation, optimum scheme of synthesized character was the treatment which fertilizer amount was pure N 337.5 $kg{\cdot}ha^{-1}$ and pure K_2O 337.5 $kg{\cdot}ha^{-1}$, lower and high limit of irrigation were respective 40 kPa and 6 kPa in greenhouse tomato cultivation. The research result was consistent with the existing research [24].

4 Conclusion

The effect of fertilizer amounts and low irrigation limit on tomato growth and development was studied with the method of plot cultivation experiment of plastic mulching and drip irrigation in greenhouse in Shenyang meadow soil region, the optimum scheme of nine different irrigation and fertilization treatments was obtained based on the fuzzy synthesis evaluation, which fertilizer amount was pure N 337.5 $kg{\cdot}ha^{-1}$ and pure K_2O 337.5 $kg{\cdot}ha^{-1}$, lower and high limit of irrigation were respective 40 kPa and 6 kPa.

Acknowledgments

I am grateful to the referees for their valuable comments and suggestions. This work supported by the key project of Agricultural water-saving irrigation by Liaoning Province of China (No.2001212001) and Program Funded by Liaoning Province Education Administration (No.05L386), and by Doctoral Startup Fund Project of Liaoning Province (No.20061043).

References

1. Dehghanisanij H, Agassi M, Anyoji H, et al (2006) Improvement of saline water use under drip irrigation system. Agricultural Water Managementa 85: 233-242
2. W P de Clercqa, M Van Meirvenne (2005) Effect of long-term irrigation application on the variation of soil electrical conductivity in vineyards. Geoderma 128: 221-233
3. Li J S, Li B, Rao M J (2005) Spatial and temporal distributions of nitrogen and crop yield as affected by nonuniformity of sprinkler fertigation. Agricultural Water Management 76: 160-180
4. Santos D V, Sousa P L, Smith R E (1997) Model simulation of water and nitrate movement in a level-basin under fertigation treatments.Agricultural Water Management 32: 293-306

5. Yu N, Zhang Y L, Huang Y, et al (2003) Interactive effect between water and fertilizer coupling on tomato cultivation under drip fertilization in greenhouse. Chinese Journal of Soil Science 34(3): 179-183 (in chinese)
6. Yao J, Zou Z R, Yang M (2004) Effect of water-fertilizer coupling on yield of cantaloupes. Acta Botanica Boreali-Occidentalia Sinica 24(5): 890-894
7. Liu F Z, Zhang Z B, He C X (2001) The Mathematical Model on Effects among 3 Factors of Water, K_2O, N and Yield of Tomato in Lean-to Solar Greenhouse. China Vegetables (1): 31-33 (in chinese)
8. Thomas L T, Scott A W, J W, et al (2003) Fertigation frequency for subsurface drip-irrigated broccoli. Soil Science Society of America Journal 67(3): 910-918
9. Thomas L T, Thomas A D, Ronald E G, et al (2002) Subsurface Drip Irrigation and Fertigation of Broccoli: I. Yield, Quality, and Nitrogen Uptake. Soil Science Society of America Journal 66: 186-192
10. Hao S R, Guo X P, Wang W M, et al (2004) The Compensation Effect of Rewatering in Subjecting to Water Stress on Leaf Area of Rice. Journal of Irrigation and Drainage 24(4): 29-32 (in chinese)
11. Yang C M, Yang L Z, Yan T M, et al (2004) Effects of nutrient and water regimes on paddy soil quality and its compre-hensive evaluation in the Taihu Lake Region. Acta Ecologica Sinica 24(1): 63-70 (in chinese)
12. Zadeh, L A (1965) Fuzzy sets. Information and Control 8: 338-353
13. Janikow C J (1998) Fuzzy decision treesIssues and methodsIEEE Transaction on SystemMan and Cybernetics part B: Cybernetics 28(1): l-l4
14. Fang S C, Wang D W (1997) Fuzzy Mathematics and Fuzzy Optimization Beijing, Science Press (in chinese)
15. Guleda O E, Ibrahim D, Halil H (2004) Assessment of urban air quality in Istanbul using fuzzy synthetic evaluation. Atmospheric Environment 38: 3809-3815
16. Rehan S, Manuel J R (2004) Fuzzy synthetic evaluation of disinfection by-productsa risk-based indexing system. Journal of Environmental Management 73: 1-13
17. Kuo Y F, Chen P C (2006) Selection of mobile value-added services for system operators using fuzzy synthetic evaluation. Expert Systems with Applications 30: 612-620
18. Rehan S, Tahir H, Brian V, et al (2004) Risk-based decision-making for drilling waste discharges using a fuzzy synthetic evaluation technique. Ocean Engineering 31: 1929-1953
19. Fu Q (2006) Data Processing Method and its Agricultural Application. Beijing, Science Press (in chinese)
20. Xiao S X, Wang P Y, Lv E L (2004) The Application of Fuzzy Mathematics in Building Water Resources Project. Beijing, The People's Traffic Press (in chinese)
21. Lv R K (2000) Analysis Methods of Soil and Agricultural Chemistry. Beijing, China Agriculture Sicientech Press (in chinese)
22. Zhang X Z, Chen F Y, Wang R F (1999) Plant physiology experimental technique. Shenyang, Liaoning Science Technique Press (in chinese)
23. Hotelling H (1933) Analysis of a complex of statistical variables into principal components. Journal of Educational Psychology 24: 417-441
24. Yu N, Zhang Y L, Zou H T, et al (2006) Effects of different water and fertilization treatments on yield and fruit quality of tomato with plastic mulching and drip irrigation in greenhouse. Agricultural Research in the Arid Areas 24(1): 60-64 (in chinese)

The Solution of Linear Programming with LR-Fuzzy Numbers in Objective Function

Qianfeng Cai[1,2], Zhifeng Hao[1], and Shaohua Pan[1]

[1] Department of Computer Science and Engineering, South China University of Technology, Guangzhou 516040, China
{mazfhao,shhpan}@scut.edu.cn

[2] Department of Applied Mathematics, Guangdong University of Technology, Guangzhou 510060, China
caiqianfeng@163.com

Abstract. This paper considers a class of fuzzy linear programming (FLP) problems where the coefficients of the objective function are characterized by LR-fuzzy intervals or LR-fuzzy numbers with the same shapes. The existing methods mainly focus on the problems whose fuzzy coefficients are linear shape. In this paper, we are interested in the solution of FLP problems whose fuzzy coefficients are nonlinear shapes. Specifically, we show that the FLP problem can be transformed into a multi-objective linear programming problem with four objectives if LR-fuzzy numbers have the same shape, and therefore the optimal solutions can be found by the relationships between FLP problems and the resulting parameter linear programming problems. An example is also presented to demonstrate that the method is invalid if fuzzy coefficients have different shapes. Then, we discuss the relationships among FLP problems, possibility and necessity maximization problems, and obtain a conclusion that all Pareto optimal solutions of the possibility-necessity approaches are the subset of the weak dominated solutions of the FLP problem. Finally, one numerical example is given to illustrate the solution procedure.

Keywords: Fuzzy number, Linear Programming, Multiobjective programming, Fuzzy max order.

1 Introduction

It often happens that the coefficients of a real linear programming problem (LP) are imprecise because the data of the problem is vague or ambiguously knows to the experts. In that case, fuzzy numbers based on fuzzy set theory are appropriate to represent the vague-magnitudes.

In past thirty years, the solution of FLP problems with fuzzy coefficients in objective function has been studied in theory and many algorithms have been found to solve it [3]. Tanaka, Ichihashi and Asai [8] formulated the FLP problem as a parametric linear programming problem and proposed a "compromise solution". Rommelfanger, Hanuscheck and Wolf [6] presented a compromise solution with the α-level related pair formation. Sakawa and Yano [7] provided α-pareto-optimal solution by restricting the coefficients $\tilde{c_j}$ to α-level-sets. While Luhand-

B.-Y. Cao (Ed.): Fuzzy Information and Engineering (ICFIE), ASC 40, pp. 988–999, 2007.
springerlink.com

jula [4] described $\beta-$possibility efficient solution and formulated FLP problem as semi-infinite linear programming problems with infinitely many objective functions. Recently, Maeda [5] formulated the FLP problem as a bi-criteria linear programming problem, however, Maeda's study is only appropriate to problems in which the coefficients of objective functions are triangular fuzzy numbers. Zhang, Wu and Lu [9] formulated fuzzy linear programming (FLP) with fuzzy coefficients in the objective function as a multi-objective linear programming (MOLP) problem with four-objective functions for any form fuzzy numbers. In practice, the conclusion is only valid under the condition that the fuzzy coefficients have the same reference functions. This paper presents an example to explain this result.

In this paper, we focus on the linear programming problem with fuzzy coefficients in the objective function. we work on the assumption that fuzzy coefficients are modeled by LR-fuzzy intervals or LR-fuzzy numbers. The paper is constructed as follows. Section 2 gives the basic definitions and concepts on fuzzy numbers. Section 3 proves FLP problems can be translated into MOLP problems with four objectives if fuzzy coefficients have the same reference functions, gives an examples to explain that this conclusion does not hold if fuzzy coefficients have different shape, and presents the methods to find the optimal solutions. Section 4 discusses the relationship among FLP problems, possibility and necessity maximization problems, and generalizes the result [5] into nolinear LR-fuzzy numbers and LR-fuzzy intervals. Section 5 presents one example to illustrate the solving procedure.

2 Preliminaries

Let $x = (x_1, x_2, \cdots, x_n)^T, y = (y_1, y_2, \cdots, y_n)^T \in R^n$. We write $x \geqq y$ iff $x_i \geqq y_i$, $\forall i = 1, 2, \cdots, n$; $x \geq y$ iff $x \geqq y$ and $x \neq y$; and $x > y$ iff $x_i > y_i$, $\forall i = 1, 2, \cdots, n$.

Definition 2.1. *Let L be a function from $(-\infty, +\infty)$ to $[0, 1]$ satisfying the following conditions;*

(i)$L(x) = L(-x)$, $\forall x \in (-\infty, +\infty)$,
(ii) $L(0) = 1$,
(iii)$L(\cdot)$ is nonincreasing and upper semi-continuous on $[0, +\infty)$,
(iv) Either $L(1) = 0$ or $\lim_{x\to+\infty} L(x) = 0$,
Then the function L is called a reference function.

Definition 2.2. *A fuzzy interval $\tilde{a}$ whose membership function $\mu_{\tilde{a}}$ can be described by means of two reference functions L and R with the following form:*

$$\mu_{\tilde{a}}(x) = \begin{cases} L(\frac{m-x}{\alpha}) & x \leq m, \alpha > 0, \\ 1 & m \leq x \leq n, \\ R(\frac{x-n}{\beta}) & x \geq n, \beta > 0, \end{cases}$$

is called an LR fuzzy interval, and α, β are the left and right spreads, respectively. The fuzzy interval can be written as $\tilde{a} = (m, n, \alpha, \beta)$. If $m = n$, it is often written *as* $\tilde{a} = (m, \alpha, \beta)$ *which is called an LR fuzzy number. If* $L(1) = 0$ *and* $R(1) = 0$, *it is called an finite fuzzy interval. If* $\lim_{x\to\infty} L(x) = 0$ *and* $\lim_{x\to\infty} R(x) = 0$, *it is called an infinite fuzzy interval.*

Let $\mathscr{F}_{LR}$ and $\mathscr{F}^*_{LR}$ be the set of all LR-fuzzy intervals or fuzzy numbers with the same reference functions and the set of all finite LR-fuzzy intervals or fuzzy numbers with the same reference functions on R^n respectively.

Definition 2.3. *The* $\lambda-$*level set (*λ*-cut) of a fuzzy interval* $\tilde{a}$ *is a crisp subset of* R *and is defined by* $a_\lambda = \{\, x \,|\, \mu_{\tilde{a}}(x) \geqq \lambda \text{ and } x \in R\,\}$. *Hence,* $\lambda-$*level set of* $\tilde{a}$ *is denoted by* $[a^L_\lambda, a^R_\lambda]$, *where* $a^L_\lambda \equiv inf\, a_\lambda$ *and* $a^R_\lambda = sup\, a_\lambda$.

Dubois [1] provided a partial relation in the following definition:

Definition 2.4. *For any LR fuzzy intervals* $\tilde{a}$, $\tilde{b} \in \mathscr{F}_{LR}$, *The following order relations*

$$\begin{aligned} \tilde{a} \gtrsim \tilde{b} \; iff (a^L_\lambda, a^R_\lambda)^T \geqq (b^L_\lambda, b^R_\lambda)^T, \forall\lambda \in [0,1], \\ \tilde{a} \succeq \tilde{b} \; iff (a^L_\lambda, a^R_\lambda)^T \geq (b^L_\lambda, b^R_\lambda)^T, \forall\lambda \in [0,1], \\ \tilde{a} \succ \tilde{b} \; iff (a^L_\lambda, a^R_\lambda)^T > (b^L_\lambda, b^R_\lambda)^T, \forall\lambda \in [0,1], \end{aligned}$$

are called binary relations $\gtrsim, \succeq$ *and* $\succ$ *a fuzzy max order, a strict fuzzy max order and a strong fuzzy max order, respectively.*

Dubois and Prade [2] proposed four indices based on the possibility measure and the necessity measure.

Definition 2.5. *Let* $\tilde{a}, \tilde{b}$ *be any fuzzy intervals. The inequality relations are defined as follows:*

(i) $Pos(\tilde{a} \geqq \tilde{b}) \equiv sup\{min(\mu_{\tilde{a}}(x), \mu_{\tilde{b}}(y)) \,|\, x \geqq y\,\}$,
(ii) $Pos(\tilde{a} > \tilde{b}) \equiv sup_x\{inf_y\{min(\mu_{\tilde{a}}(x), 1 - \mu_{\tilde{b}}(y)) \,|\, x \leqq y\,\}\}$,
(iii) $Nes(\tilde{a} \geqq \tilde{b}) \equiv inf_x\{sup_y\{max(1 - \mu_{\tilde{a}}(x), \mu_{\tilde{b}}(y)) \,|\, x \geqq y\,\}\}$,
(iv) $Nes(\tilde{a} > \tilde{b}) \equiv inf\{max(\mu_{1-\tilde{a}}(x), 1 - \mu_{\tilde{b}}(y)) \,|\, x \leqq y\,\}$.

Theorem 2.1. *Let* $\tilde{a}$, $\tilde{b}$ *be any LR fuzzy intervals and let* $\lambda \in [0,1]$ *be any real number. Then the following relationships hold:*

(i) $Pos(\tilde{a} \geqq \tilde{b}) \geqq \lambda \Leftrightarrow a^R_\lambda \geqq b^L_\lambda$,
(ii) $Pos(\tilde{a} > \tilde{b}) \geq \lambda \Leftrightarrow a^R_\lambda \geqq b^R_{1-\lambda}$,
(iii) $Nes(\tilde{a} \geqq \tilde{b}) \geq \lambda \Leftrightarrow a^L_{1-\lambda} \geqq b^L_\lambda$,
(iv) $Nes(\tilde{a} > \tilde{b}) \geq \lambda \Leftrightarrow a^L_{1-\lambda} \geqq b^R_{1-\lambda}$.

3 Fuzzy Linear Programming Problem and Fuzzy Max Order

In the section, we will discuss FLP problems which can be stated as:

$$(\text{FLP}) \begin{cases} \text{maximize } < \tilde{c}, x > = \sum_{i=1}^n \tilde{c}_i x_i, \\ \text{subject to } Ax \leqq b, \;\; x \geqq 0, \end{cases} \tag{1}$$

where $\tilde{c} = (\tilde{c}_1, \tilde{c}_2, \cdots, \tilde{c}_n)^T$, $\tilde{c}_i \in \mathscr{F}^*_{LR}(R)$ $(i = 1, \cdots, n)$, A is an $m \times n$ matrix and $b \in R^m$.

(FLP) has no absolute optimum because of the fuzziness of coefficients in the objective function. Therefore, we will introduce the concepts of optimal solution to FLP problems proposed by Maeda [5]. For the sake of simplicity, we set $X \equiv \{ x \in R^n \mid AX \leqq b, x \geqq 0 \}$ and assume that X is compact.

Definition 3.1. *A point $x^* \in X$ is said to be an optimal solution to FLP problems if it holds that $< \tilde{c}, x^* > \succsim\!\!\approx < \tilde{c}, x >$ for all $x \in X$.*

Definition 3.2. *A point $x \in X$ is said to be a nondominated solution to FLP problems if there does not exist $x \in X$ such that $< \tilde{c}, x > \succeq < \tilde{c}, x^* >$ holds.*

Definition 3.3. *A point $x \in X$ is said to be a weak nondominated solution to FLP problems if there does not exist $x \in X$ such that $< \tilde{c}, x > \succ < \tilde{c}, x^* >$ holds.*

We denote the sets of all nondominated solutions and all weak nondominated solutions to (FLP) by X^F and X^{wF} , respectively. Then, by definition, it holds that $X^F \subseteq X^{wF}$.

Associated with (FLP), we consider the following MOLP problem:

$$\text{(MOLP)} \begin{cases} \text{maximize } \ (< c_0^L, x >, < c_1^L, x >, < c_0^R, x >, < c_1^R, x >)^T, \\ \text{subject to } Ax \leqq b, \quad x \geqq 0, \end{cases} \tag{2}$$

where $c_i^L = (c_{i1}^L, c_{i2}^L, \cdots, c_{in}^L)^T$, $c_i^R = (c_{i1}^R, c_{i2}^R, \cdots, c_{in}^R)^T \in R^n$, $i = 0, 1$.

The concepts of optimal solutions of (MOLP) can be seen in [5]. Now, we consider reducing (FLP) to (MOLP) based on fuzzy max order. In [9], the following lemma is claimed:

Lemma 3.1 ([9])**.** For any $a, b, c \in R^n$ and $a \leqq b \leqq c$. If there exist $x^*, x \in R^n$ such that $< a, x > \leqq < a, x^* >$ and $< c, x > \leqq < c, x^* >$, then $< b, x > \leqq < b, x^* >$.

As a counterexample, consider the following instance:

Let $a = (-2, -3)$, $b = (-2, -1)$, $c = (-1, -1)$, $x^* - x = (11, -20)$, we have $a \leqq b \leqq c$, $< a, x^* - x > = 38 \geqq 0$ and $< c, x^* - x > = 9 \geqq 0$, $< b, x^* - x > = -2 \leqq 0$, that is, $< a, x > \leqq < a, x^* >$ and $< c, x > \leqq < c, x^* >$, but $< b, x > \geqq < b, x^* >$.

Theorefore, we propose the following lemma:

Lemma 3.1. *Let $\tilde{a} = (m_1, n_1, \alpha_1, \beta_1)$, $\tilde{b} = (m_2, n_2, \alpha_2, \beta_2) \in \mathscr{F}^*_{LR}$, if*

$$a_0^L \geqq b_0^L,\ a_0^R \geqq b_0^R,\ a_1^L \geqq b_1^L,\ a_1^R \geqq b_1^R, \tag{3}$$

then

$$\tilde{a} \succsim\!\!\approx \tilde{b}.$$

Proof. For any $\lambda \in [0,1]$, let $\mu_{\tilde{a}}(x) = \mu_{\tilde{b}}(y) = \lambda$, that is

$$\lambda = L(\frac{m_1 - x}{\alpha_1}) = L(\frac{m_2 - y}{\alpha_2}). \tag{4}$$

Let $l_1 = \inf\{x \,|L(x) = \lambda\,\}$, we have $-1 \leq l_1 \leq 0$. Since $a_\lambda^L = \inf\{x \,|L\,(\frac{m_1-x}{\alpha_1}) = \lambda\,\}$, $b_\lambda^L = \inf\{y \,|L\,(\frac{m_2-y}{\alpha_2}) = \lambda\,\}$, and $L(\,\cdot\,)$ is nonincreasing on [0,1], we have

$$\frac{m_1 - a_\lambda^L}{\alpha_1} = \frac{m_2 - b_\lambda^L}{\alpha_2} = -l_1. \tag{5}$$

Since $a_0^L \geqq b_0^L, a_1^L \geqq b_1^L$, we have $m_1 - \alpha_1 \geqq m_2 - \alpha_2, m_1 \geqq m_2$, that is $\alpha_2 \geqq m_2 - m_1 + \alpha_1$, therefore, $b_\lambda^L = m_2 + l_1\alpha_2 \leqq m_2 + l_1(m_2 - m_1 + \alpha_1) = (m_2 - m_1)(1 + l_1) + a_\lambda^L$. Since $-1 \leqq l_1 \leqq 0$, we have $a_\lambda^L \geqq b_\lambda^L$. In the same way, we can proof $a_\lambda^R \geqq b_\lambda^R$, then by Definition 2.4, we have $\tilde{a} \gtrapprox \tilde{b}$.

According Lemma 3.1, we obtain the solution relationship between (FLP) and (MOLP) by the following theorems for fuzzy intervals:

Theorem 3.1. *Let a point $x^* \in X$ be a feasible solution to (FLP). Then x^* is an optimal solution to the problem if and only if x^* is a complete optimal solution to (MOLP).*

Proof. If x^* is an optimal solution to (1), for all $x \in X$, we have $< \tilde{c}, x^* > \gtrapprox < \tilde{c}, x >$. Therefore, for all $\lambda \in [0,1]$, we have $< \tilde{c}, x^* >_\lambda^L \geqq < \tilde{c}, x >_\lambda^L$ and $< \tilde{c}, x^* >_\lambda^R \geqq < \tilde{c}, x >_\lambda^R$. If $\lambda = 0, 1$, then

$$(< c_0^L, x^* >, < c_1^L, x^* >, < c_0^R, x^* >, < c_1^R, x^* >)^T \geqq$$
$$(< c_0^L, x >, < c_1^L, x >, < c_0^R, x >, < c_1^R, x >)^T.$$

Therefore, x^* is a complete optimal solution to (MOLP).

If x^* is a complete optimal solution to (MOLP), for all $x \in X$, we have

$$(< c_0^L, x^* >, < c_1^L, x^* >, < c_0^R, x^* >, < c_1^R, x^* >)^T \geqq$$
$$(< c_0^L, x >, < c_1^L, x >, < c_0^R, x >, < c_1^R, x >)^T.$$

Since $\tilde{c}_i \,(i = 1, \cdots, n)$ are LR fuzzy intervals or fuzzy numbers with the same reference functions, $< \tilde{c}, x^* >$ and $< \tilde{c}, x >$ are also LR fuzzy intervals or fuzzy numbers with the same reference functions. According to Lemma 3.1, we have

$$< \tilde{c}, x^* > \gtrapprox < \tilde{c}, x > .$$

Therefore, x^* is an optimal solution to (1).

Theorem 3.2. *Let a point $x \in X$ be any feasible solution to (FLP). Then x^* is a nondominated (resp. weak nondominated) solution to the problem if and only if x^* is a Pareto (resp. weak Pareto) optimal solution to (MOLP).*

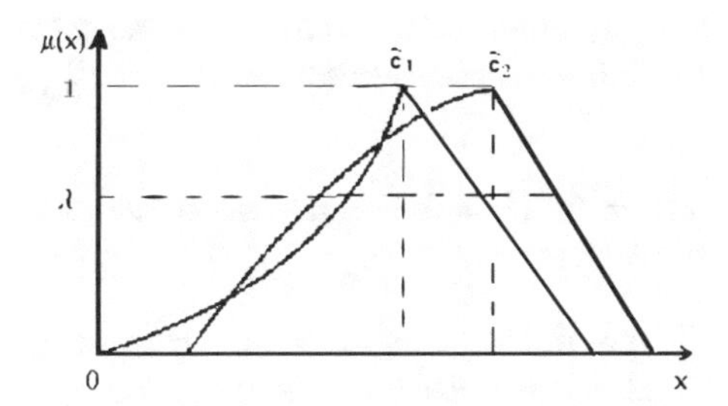

Fig. 1. Comparison between two fuzzy numbers

In [9], it is claimed that Theorem 3.1 (Theorem 3.1 in[9]) is valid for any form fuzzy numbers. We present an example in Fig.1 to say that the sufficient condition for Theorem 3.1 does not hold if the reference functions of fuzzy coefficients do not have the same shape.

In Fig.1, the left of $\tilde{c}_1$ and $\tilde{c}_2$ have different reference function and satisfy $c_{01}^L \le c_{02}^L$, $c_{01}^R \le c_{02}^R$, $c_{11}^L = c_{11}^R \le c_{12}^L = c_{12}^R$. Let $x = (1,0)^T$, $x^* = (0,1)^T$, we have $< \tilde{c}, x >= \tilde{c}_1$ and $< \tilde{c}, x^* >= \tilde{c}_2$, then $< \tilde{c}, x >$ and $< \tilde{c}, x^* >$ satisfy $< \tilde{c}, x >_0^L \le < \tilde{c}, x^* >_0^L$, $< \tilde{c}, x >_0^R \le < \tilde{c}, x^* >_0^R$, $< \tilde{c}, x >_1^L \le < \tilde{c}, x^* >_1^L$, $< \tilde{c}, x >_1^R \le < \tilde{c}, x^* >_1^R$. However, they do not satisfy $< \tilde{c}, x >_\lambda^L \le < \tilde{c}, x^* >_\lambda^L$ and $< \tilde{c}, x >_\lambda^R \le < \tilde{c}, x^* >_\lambda^R$ for any $\lambda \in [0,1]$. Therefore, $< \tilde{c}, x >$ and $< \tilde{c}, x^* >$ can not compare by using fuzzy max order if $\tilde{c}_1$ and $\tilde{c}_2$ have different reference function. In practice, this result still holds if the fuzzy coefficients belong to the same reference function family with different parameters. For example, construct $\tilde{c}_1$, $\tilde{c}_2$ as follows:

$$\mu_{\tilde{c}_1} = \begin{cases} x^4, & 0 \le x \le 1, \\ -x+2, & 1 \le x \le 2, \end{cases} \quad and \quad \mu_{\tilde{c}_2} = \begin{cases} (x-0.5)^{\frac{1}{2}}, & 0.5 \le x \le 1.5, \\ -x+2.5, & 1.5 \le x \le 2.5, \end{cases}$$

where the reference functions belong to the power family. For that reason, it is impossible that Theorem 3.1 is valid for any form fuzzy numbers.

Let

$$\begin{aligned} \Lambda^{+} &= \{\, w = (w_1, \cdots, w_4)^T \mid \textstyle\sum_{i=1}^4 w_i = 1,\ w_i \ge 0\}, \\ \Lambda^{++} &= \{\, w = (w_1, \cdots, w_4)^T \mid \textstyle\sum_{i=1}^4 w_i = 1,\ w_i > 0\}, \end{aligned}$$

in order to find all nondominated or all weak nondominated solutions to (FLP), it suffices to find all Pareto or weak Pareto optimal solution to (MOLP). Now, consider the following parameter linear programming problem:

$$(\mathrm{LP})_w \begin{cases} \text{maximize } w_1 < c_0^L, x > + w_2 < c_1^L, x > \\ \qquad\qquad + w_3 < c_0^R, x > + w_4 < c_1^R, x >, \\ \text{subject to } Ax \leqq b, \quad x \geqq 0, \end{cases} \tag{6}$$

where $w_i \in R\,(i = 1, \cdots, 4)$.

Based on Theorem 3.1-Theorem 3.2 and the relationships between (MOLP) and $(\mathrm{LP})_w$, the (weak) nondominated optimality concepts of (FLP) can be characterized by the following theorems:

Theorem 3.3. *x^* is a nondominated (reps. weak nondominated) solution to (FLP) if and only if x^* is optimal solution to the $(LP)_w$ for some $w \in \Lambda^{++}$ (reps. $w \in \Lambda^{+}$).*

Assume that $c_{1i}^L = c_{1i}^R = c_i$, $(i = 1, \cdots, n)$, we consider the following MOLP problem with three objectives:

$$(\mathrm{MOLP})_3 \begin{cases} \text{maximize } (< c_0^L, x >, < c, x >, < c_0^R, x >)^T, \\ \text{subject to } Ax \leqq b, \quad x \geqq 0, \end{cases} \tag{7}$$

where $c_0^L = (c_{01}^L, c_{02}^L, \cdots, c_{0n}^L)^T, c_0^R = (c_{01}^R, c_{02}^R, \cdots, c_{0n}^R)^T, c = (c_1, c_2, \cdots, c_n)^T \in R^n$.

Similar to (FLP) with fuzzy intervals, the nondominated optimal solutions to (FLP) with fuzzy numbers have the same relationships with the optimal solutions to the following parameter linear programming:

$$(\mathrm{LP})_{w_1, w_2} \begin{cases} \text{maximize } < c, x > -w_1 < \alpha, x > +w_2 < \beta, x >, \\ \text{subject to } Ax \leqq b, \quad x \geqq 0, \end{cases} \tag{8}$$

where $\alpha = c - c_0^L$, $\beta = c_0^R - c$, $w_i \in [0, 1]$, $i = 1, 2$.

Assume that $\alpha = \beta$, the solution of (FLP) can be obtained by considering the following parameter linear programming problem:

$$(\mathrm{LP})_\omega \begin{cases} \text{maximize } < c, x > +\omega < \alpha, x >, \\ \text{subject to } Ax \leqq b, \quad x \geqq 0, \end{cases} \tag{9}$$

where $\omega \in [-1, 1]$.

Theorem 3.4. *Let $c_1^L = c_1^R = c$, $\alpha = \beta$, Then,*

(i) x^* is a nondominated solution to (FLP) if and only if x^* is optimal solution to the $(LP)_\omega$ for some $\omega \in (-1, 1)$;

(ii)x^* is a weak nondominated solution to (FLP) if and only if x^* is optimal solution to the $(LP)_\omega$ for some $\omega \in [-1, 1]$.

Proof. (i) Let x^* is a nondominated solution to (FLP). Then, from Theorem 3.3, there exists $w_1, w_2 \in (0, 1)$ such that x^* is the optimal solution to $(\mathrm{LP})_{w_1, w_2}$. Assuming $\omega = w_2 - w_1$, we have $\omega \in (-1, 1)$. Conversely, if x^* is optimal solution to the $(\mathrm{LP})_\omega$ for some $\omega \in (-1, 1)$, we get the value of w_1, $w_2 \in (0, 1)$ which satisfy $w_2 - w_1 = \omega$. Then, from Theorem 3.3, x^* is a nondominated solution to (FLP).

(ii)Similar to the proof of (i).

4 Possibility and Necessity Maximization Problems

Maeda [5] proposed possibility and necessity maximization problems and characterize the membership function of optimal value to FLP problems with triangular fuzzy numbers.

Therefore, in this section, we shall discuss the relationship between the Pareto optimal solution to possibility and necessity maximization problems and the nondominated optimal solution to FLP problems with LR-fuzzy interval or fuzzy numbers. The possibility maximization problem is defined by

$$\text{(P)} \begin{cases} \text{maximize} & (Pos(<\tilde{c}, x> \geqq v),\, v)^T, \\ \text{subject to} & Ax \leqq b, \quad x \geqq 0,\, v \in V, \end{cases} \tag{10}$$

where $V = [v_1, v_0]$, and $v_0 = max_{x \in X} < c_0^R, x >$, $v_1 = max_{x \in X} < c_1^R, x >$. We denote that X^P is the set of all Pare to optimal solution to problem (P).

First, we show the relationships between the Pareto optimal solution to (P) and the optimal solution to the following parameter programming:

$$\text{(LP)}_{Sup\,\{R^{-1}(\lambda)\}} \begin{cases} \text{maximize} & < c_1^R, x > + \sup\{R^{-1}(\lambda)\} < \beta, x >, \\ \text{subject to} & Ax \leqq b, \quad x \geqq 0, \end{cases} \tag{11}$$

where $\beta = c_0^R - c_1^R$, $\lambda \in [0,1]$, $sup\,\{R^{-1}(1)\} = 0$ and $sup\,\{R^{-1}(0)\} = 1$. The relationships can be characterized by the following theorems:

Theorem 4.1. *$(x^*, v^*) \in X \times V$ be a Pareto optimal solution to Problem (P) if and only if x^* is an optimal solution to the $(LP)_{Sup\,\{R^{-1}(\lambda)\}}$ with $\lambda \in [0,1]$, where $\lambda \equiv Pos(<\tilde{c}, x^* > \geqq v^*)$ and $v^* = < c_1^R + sup\,\{R^{-1}(\lambda)\}\,\beta, x^* >$.*

Proof. Let $(x^*, v^*) \in X \times V$ be a Pareto optimal solution to Problem (P) and $\lambda \equiv Pos(<\tilde{c}, x^* > \geq v^*)$. By Theorem 2.1, $< \tilde{c}, x^* >_\lambda^R = v^*$ holds, that is, $< c_\lambda^R, x^* > = v^*$ where $c_\lambda^R = c_1^R + sup\,\{R^{-1}(\lambda)\}\,\beta$. If x^* is not a Pareto optimal solution of the $\text{(LP)}_{\text{Sup}\,\{R^{-1}(\lambda)\}}$, then exists $\bar{x} \in X$ such that $< c_\lambda^R, \bar{x} > > < c_\lambda^R, x^* > = v^*$. If $\lambda = 1$, then $v_1 \geq < c_1^R, \bar{x} > > < c_1^R, x^* > = v^* \geq v_1$. This is a contradiction. This implies $\lambda < 1$. Since $< c_\lambda^R, \bar{x} > > < c_\lambda^R, x^* > = v^*$, there exists positive numbers $\theta > 0$ and $\rho > 0$ such that $< c_{\lambda+\theta}^R, \bar{x} > \geq v^* + \rho$. By Theorem 2.1, we have $Pos(<\tilde{c}, \bar{x} > \geqq v^* + \rho) \geq \lambda + \theta$. Let $\bar{v} = v^* + \rho$, we have $(Pos(<\tilde{c}, \bar{x} > \geq \bar{v}), \bar{v})^T > (Pos(<\tilde{c}, x^* > \geq v^*), v^*)^T$. This contradicts the assumption that (x^*, v^*) is a Pareto optimal solution to Problem (P).

Conversely, let $x^* \in X$ is an optimal solution to the $(LP)_{Sup\{R^{-1}(\lambda)\}}$ and $v^* = < c_1^R + sup\{R^{-1}(\lambda)\}\,\beta, x^* >$, that is, $v^* = < c_\lambda^R, x^* >$, then $\lambda = Pos(<\tilde{c}, x^* > \geqq v^*)$. If (x^*, v^*) is not a Pareto solution of the (P), then there exists $< \bar{x}, \bar{v} >$ such that

$$(Pos(<\tilde{c}, \bar{x} > \geq \bar{v}), \bar{v})^T \geq (Pos(<\tilde{c}, x^* > \geq v^*), v^*)^T. \tag{12}$$

First, suppose that $\lambda = 1$. From (10), we have $Pos(<\tilde{c}, \bar{x} > \geq \bar{v}) = 1$. By Theorem 2.1 and (10), we have $v_1 \geq < c_1^R, \bar{x} > = \bar{v} > v^* = v_1$. This is a contradiction. Therefore, we have $\lambda < 1$. If $\bar{v} > v^*$, then $Pos(<\tilde{c}, \bar{x} > \geq \bar{v}) \geq \lambda$. By Theorem 2.1, we have $< c_\lambda^R, \bar{x} > \geq \bar{v} > v^*$. This contradicts the assumption that $x^* \in X$ is an optimal solution to the $(LP)_{Sup\,\{R^{-1}(\lambda)\}}$. Therefore, we have $\bar{v} = v^*$ and $Pos(<\tilde{c}, \bar{x} > > \bar{v}) > \lambda$. This implies that $< c_\lambda^R, \bar{x} > > \bar{v} = v^*$. Similarly, this contradicts the assumption that $x^* \in X$ is an optimal solution to the $(LP)_{Sup\,\{R^{-1}(\lambda)\}}$.

The necessity maximization problems is defined by

$$(\mathrm{N})\begin{cases}\text{maximize} & (Nes(<\tilde{c},x>\geqq w),\, w)^T,\\ \text{subject to} & Ax\leqq b,\quad x\geqq 0,\, w\in W,\end{cases}\tag{13}$$

where $W=[w_0,\, w_1]$, and $w_0=max_{x\in X}<c_0^L,x>$, $w_1=max_{x\in X}<c_1^L,x>$. We denote that X^N is the set of all Pareto optimal solution to problem (N).

Next, we show the relationships between the Pareto optimal solution to (N) and the optimal solution to the following parameter programming:

$$(LP)_{\inf\{L^{-1}(\lambda)\}}\begin{cases}\text{maximize} & <c_1^L,x>+\inf\{L^{-1}(\lambda)\}<\alpha,x>,\\ \text{subject to} & Ax\leqq b,\quad x\geqq 0,\end{cases}\tag{14}$$

where $\alpha=c_1^L-c_0^L$, $\lambda\in[0,1]$, $inf\,\{L^{-1}(1)\}=0$ and $inf\,\{R^{-1}(0)\}=-1$. In the same way, the relationships between (13) and (14) can be characterized by the following theorems:

Theorem 4.2. *$(x^*,\, w^*)\in X\times W$ be a Pareto optimal solution to Problem (N) if and only if x^* is an optimal solution to the $(LP)_{inf\,\{L^{-1}(\lambda)\}}$ with $\lambda\in[0,1]$, where $\lambda=1-Nes(<\tilde{c},x^*>\geqq w^*)$ and $w^*=<c_1^L+inf\,\{L^{-1}(\lambda)\}\,\alpha,\, x^*>$.*

Then, we show the relationship between the Pareto optimal solution to possibility and necessity maximization problems and the nondominated optimal solution to (FLP) with LR-fuzzy intervals or fuzzy numbers.

Theorem 4.3. *If $(x^*,\, v^*)\in X\times V$ (resp. $(x^*,\, w^*)\in X\times W$) is a Pareto optimal solution to Problem (P) (resp. (N)), then x^* is a weak nondominated optimal solution to (FLP).*

Proof. If $(x^*,\, v^*)\in X\times V$ is a Pareto optimal solution to Problem (P), by Theorem 4.1, x^* is an optimal solution to the $(LP)_{Sup\,\{R^{-1}(\lambda)\}}$ with $\lambda\in[0,1]$, where $\lambda=\mathrm{P}os(<\tilde{c},x^*>\geqq v^*)$. Then, x^* is an optimal solution to the $(\mathrm{LP})_w$, where $w_1=w_3=0$, $w_2=\frac{1}{1+Sup\{R^{-1}(\lambda)\}}$, $w_4=\frac{Sup\{R^{-1}(\lambda)\}}{1+Sup\{R^{-1}(\lambda)\}}$, $\lambda\in[0,1]$. By Theorem 3.3, x^* is a weak nodominated optimal solution to (FLP).

In the same way, we can proof if $(x^*,\, w^*)\in X\times W$ be a Pareto optimal solution to Problem (N), then x^* is also a weak nodominated optimal solution to (FLP).

In the following, we present some theorems which are appropriate to fuzzy numbers:

Theorem 4.4. *Let $c_1^L=c_1^R=c$, $\alpha=\beta$, $(x^*,\, v^*)\in X\times V$ (resp. $(x^*,\, w^*)\in X\times W$) be a Pareto optimal solution to Problem (P) (resp. (N)). If $\lambda=Pos(<\tilde{c},x^*>\geqq v^*)>0$ (resp. $\lambda=1-Nes(<\tilde{c},x^*>\geqq w^*)>0$), then x^* is a nondominated optimal solution to (FLP).*

Theorem 4.5. *Let $c_1^L=c_1^R=c$, $\alpha=\beta$, L and R be strictly monotone. If x^* is a weak nondominated optimal (resp. nondominated optimal)solution to (FLP), then there exists a real number $\omega\in[-1,1]$ (resp. $\omega\in(-1,1)$) such that $<c+\omega\alpha,x^*>$ is a Pareto optimal solution to (P) or (N).*

Proof. Let $c_1^L = c_1^R = c$, $\alpha = \beta$. If x^* is a weak nondominated optimal solution to (FLP), by Theorem 3.7, there exists a real number $\omega \in [-1,1]$ such that x^* is the optimal solution to the $(\mathrm{LP})_\omega$. If $\omega \in [-1,0]$, we assume that $w^* =< c+\omega\,\alpha, x^* >$. Since L is strictly monotone, we assume that $\lambda = L(\omega)$. Then, by Theorem 4.2, (x^*, w^*) is a Pareto optimal solution to (N). If $\omega \in [0,1]$, we assume that $v^* =< c+\omega\,\alpha, x^* >$. Since R is strictly monotone, we assume that $\lambda = R(\omega)$. By Theorem 4.1, (x^*, v^*) is a Pareto optimal solution to (P).

According to the above theorems, we have the following corollary:

Corollary 4.1. *In the (FLP), it holds that*

$$\{x \in X \mid (x,v) \in X^P\} \cup \{x \in X \mid (x,w) \in X^N\} \subseteq X^{wF}. \tag{15}$$

Corollary 4.2. *Let $c_1^L = c_1^R = c$, $\alpha = \beta$, L and R be strictly monotone. In (FLP), it holds that*

$$\begin{aligned} X^{wF} &= \{x \in X \mid (x,v) \in X^P\} \cup \{x \in X \mid (x,w) \in X^N\}, \\ X^F &= \{x \in X \mid (x,v) \in X^P,\ Pos(< \tilde{c}, x > \geqq v) > 0\} \\ &\quad \cup \{x \in X \mid (x,w) \in X^N,\ Nes(< \tilde{c}, x > \geqq w) < 1\}. \end{aligned} \tag{16}$$

From the above corollaries, all Pareto optimal solutions of the possibility-necessity approaches are the subset of the weak dominated solutions of (FLP). Only under a strictly restrictive condition that all fuzzy coefficients are fuzzy numbers, the left and right spreads are equal and L, R reference functions are strictly monotone, the possibility-necessity approaches are equivalent to fuzzy max order approach. Therefore, Maeda's conclusion (See [5]) is a special case of LR fuzzy numbers.

5 Numerical Examples

Example 1

$$(FLP)\begin{cases} \text{maximize } f(x,y) = \tilde{c}_1 x + \tilde{c}_2 y, \\ \text{subject to } 3x + y \le 12, \\ \qquad\qquad\quad x + 2y \le 12, \\ \qquad\qquad\quad x \ge 0,\ y \ge 0, \end{cases} \tag{17}$$

where $\tilde{c}_1 = (2,3,1,2)$, $\tilde{c}_2 = (4,5,1,2)$, with the following reference functions:

$$L(x) = \begin{cases} 0, & x < -1, \\ (1+x)^{\frac{1}{2}}, & -1 \le x < 0, \end{cases} \quad \text{and} \quad R(x) = \begin{cases} 1 - x^2, & 0 \le x < 1, \\ 0, & x > 1. \end{cases} \tag{18}$$

Associated with (FLP) , consider the following MOLP:

$$(\mathrm{MOLP})\begin{cases} \text{maximize } \{2x+4y, 3x+5y, x+4y, 5x+7y\}, \\ \text{subject to } 3x + y \le 12, \\ \qquad\qquad\quad x + 2y \le 12, \\ \qquad\qquad\quad x \ge 0,\ y \ge 0. \end{cases} \tag{19}$$

By Theorem 3.1, the complete optimal solution of (19) is the optimal solution of the (17). In order to solve (19), consider the following linear programming problem:

$$(\mathrm{MOLP})_w \begin{cases} \text{maximize } w_1(2x+4y)+w_2(3x+5y) \\ \qquad +w_3(x+4y)+w_4(5x+7y), \\ \text{subject to } 3x+y \le 12, \\ \qquad x+2y \le 12, \\ \qquad x \ge 0,\ y \ge 0. \end{cases} \tag{20}$$

where $w \in \Lambda^+$. By Theorem 3.3, for $w \in \Lambda^{++}$, the optimal solution of (20) is the nondominated solution of (17).

Consider the following parameter programming problems:

$$(LP)_\omega \begin{cases} \text{maximize } (3x+5y)+\omega(2x+2y), \\ \text{subject to } 3x+y \le 12, \\ \qquad x+2y \le 12, \\ \qquad x \ge 0,\ y \ge 0, \end{cases} \tag{21}$$

where $\omega \in [0,1]$. Let $\omega = \frac{1}{2}$, the optimal solution of (21) is $(x^*, y^*) = (2.4, 4.8)$. By Theorem 3.3, (x^*, y^*) is a weak nodominated solution to (FLP). Let $v^* = 3x^* + 5y^* + R(\frac{1}{2})(2x^* + 2y^*) = 42$, then (2.4, 4.8, 42) is a Pareto optimal solution to the (P).

Similarly, consider the following parameter programming:

$$(LP)_\omega \begin{cases} \text{maximize } (2x+4y)+\omega(x+y), \\ \text{subject to } 3x+y \le 12, \\ \qquad x+2y \le 12, \\ \qquad x \ge 0,\ y \ge 0, \end{cases} \tag{22}$$

where $\omega \in [-1,0]$. Let $\omega = -\frac{3}{4}$, the optimal solution of (22) is $(x^*, y^*) = (2.4, 4.8)$. Let $w^* = 2x^* + 4y^* + L(-\frac{3}{4})(x^* + y^*) = 27.6$, then (2.4,4.8,27.6) is a Pareto optimal solution to the (N).

Acknowledgements

This work described here is supported by the grants from the National Natural Science Foundation of China (NSFC No. 60433020, 10471045).

References

1. Dubois D, Kerre E, Mesiar R, Prade H (2000) Fuzzy interval analysis, in: Dubois D, Prade H (eds) Fundamentals of Fuzzy Sets. Kluwer Academic, Dordrecht
2. Dubois D, Prade H (1983) Ranking fuzzy numbers in the setting of possibility theory, Information Science 30: 183–224

3. Lai YJ, Hwang CL (1992) Fuzzy mathematical programming. Springer-Verlag
4. Luhandjula MK (1987)Multiple objective programming with possibilitic coefficients. Fuzzy Sets and Systems 21: 135–146
5. Maeda T (2001) Fuzzy linear programming problems as bi-criteria optimization problems. Applied Mathematics and Computation 120: 109–121
6. Rommelfanger H, Hanuscheck R, Wolf J (1989) Linear programming with fuzzy objectives. Fuzzy Sets and Systems 29: 31–48
7. Sakawa M, Yano H (1989) Interactive fuzzy satisficing method for multiobjective nolinear programming problems with fuzzy parameters. Fuzzy Sets and Systems 30: 221–238
8. Tanaka H, Ichihashi H, Asai K (1984) A formulation of linear programming problems based on comparision of fuzzy numbers. Control and Cybernetics 13: 185–194
9. Zhang GQ, Wu YH, Remias M, Lu J (2003) Formulation of fuzzy linear programming problems as four-objective constrained optimization problems. Applied Mathematics and Computation 139: 383–399

On Relationships of Filters in Lattice Implication Algebra

Jia-jun Lai[1], Shu-wei Chen[1], Yang Xu[1], Ke-yun Qin[1], and Li Chen[2]

[1] Intelligent Control Development Center, Southwest Jiaotong University, Chengdu 610031, P.R. China
eagles_laijiajun7309@126.com, swchen@zzu.edu.cn, yxu@home.swjtu.edu.cn
[2] Department of Mathematics, JianXi Agricultural University, NanChang 330045, P.R. China

Abstract. In this paper, we introduced the notions of fuzzy lattice filter (briefly, *FL*-filter), and normal fuzzy filters (briefly, *NF*-filter) of lattice implication algebras. And we also studied the properties of *FL*-filters, *NF*-filters in lattice implication algebra. Several characteristic of *FL*-filters and *NF*-filters are given. Respectively the relation between *FL*-filters and *NF*-filters, between fuzzy filters and *FL*-filters are investigated. This article aims at discussing new development of fuzzy filters and its properties.

Keywords: FL-filter, NF-filter, Lattice Implication Algebra.

1 Introduction

Non-classical logic has become a considerable formal tool for computer science and artificial intelligence to deal with fuzzy information and uncertain information. Many-Valued logic, a great extension and development of classical logic [1], has always been a crucial direction in non-classical logic. In order to research the many-valued logical system whose propositional value is given in a lattice, in 1990 Xu [2, 3] proposed the concept of lattice implication algebras and discussed its some properties. Since then this logical algebra has been extensively investigated by several researchers. In [4], Xu and Qin introduced the notion of lattice H implication algebras, which is an important class of lattice implication algebras. In [5], Xu and Qin introduced filters and implicative filters in lattice implication algebras, and investigated their some properties. In [6], Xu and Qin defined the concepts of fuzzy filters and fuzzy implicative filters and investigated its some properties. In [7], Liu and Xu researched the structures of the filters of lattice implication algebras. In [8], Qin and Xu proposed the notion of *Ultra*-filters and discussed some of their properties. In [9], Wang et al. introduced obstinate filters, and obtained its some properties. In [10], Jun, Xu and Qin proposed the notions of positive implicative and associative filters, and investigated its some properties. In [11], Zhao and zhu introduced the concept of primary filter in lattice implication algebras, and given three characterizations of primary filters in lattice implication algebras. In this article, as an extension of above-mention work we introduce the notions of *FL*-filters, *NF*-filters and *CNF*-filters in lattice implication algebras, and investigate the properties of NF-filters, *CNF*-filters and *FL*-filters,

B.-Y. Cao (Ed.): Fuzzy Information and Engineering (ICFIE), ASC 40, pp. 1000–1008, 2007.
springerlink.com

respectively. In Section 2, we list some basic information on the lattice implication algebras which is needed for development of this topic. In Section 3, the notions of *FL*-filter in lattice implication algebra is introduce, we give some characteristics of *FL*-filter. In Section 4, we introduce the notions of *NF*-filters in lattice implication algebra, and the properties of NF-filter are investigated. The relations between *NF*-filter and *FL*-filter are investigated.

2 Preliminaries

Definition 2.1. [2] A bounded lattice $(L, \vee, \wedge, ', O, I)$ with ordered-reversing involution $'$ and a binary operation $\to$ is called a lattice implication algebra if it satisfies the following axioms:

(L_1) $x \to (y \to z) = y \to (x \to z)$,

(L_2) $x \to x = I$,

(L_3) $x \to y = y' \to x'$,

(L_4) $x \to y = y \to x = I$ imply $x = y$,

(L_5) $(x \to y) \to y = (y \to x) \to x$,

(L_6) $(x \vee y) \to z = (x \to z) \wedge (y \to z)$,

(L_7) $(x \wedge y) \to z = (x \to z) \vee (y \to z)$.

Definition 2.2. [4] A lattice implication algebra L is said to be a lattice H implication algebra if it satisfies

$$x \vee y \vee ((x \wedge y) \to z) = I \text{ for all } x, y, z \in L.$$

Theorem 2.1. [4] $(L, \vee, \wedge, ', \to)$ is a lattice H implication algebra if and only if $(L, \vee, \wedge, ')$ is a Boolean lattice, x' is the complement of x and $x \to y = x' \vee y$ for any $x, y \in L$.

Definition 2.3. [2, 5] Let L be a lattice implication algebra. A subset $A \subseteq L$ is called a filter of L if it satisfies the following statement conditions:

(1) $I \in A$; (2) if $x \in A$ and $x \to y \in A$ then $y \in A$ for all $x, y \in L$.

Theorem 2.4. [2] Let L_1 is a filter of a lattice implication algebra L. If $x \le y$ and $x \in L_1$ then $y \in L_1$.

Next we give a property of filter which will be needed in the sequel.

Theorem 2.5. Let L_1 be a non-empty subset of a lattice implication algebra *L*. Then L_1 is a filter of *L* if and only if it satisfies for all $x, y \in L_1, z \in L$, $y \le x \to z$ implies $z \in L_1$.

Proof. Suppose that L_1 is a filter and $x, y \in L_1, z \in L$. If $y \le x \to z$, then $x \to z \in L_1$ by Theorem 2.2. Using the defintion2.3 we obtain $z \in L_1$. Conversely, suppose that for all $x, y \in L_1$ and $z \in L$, $y \le x \to z$ implies $z \in L_1$. Since L_1 is a non-empty subset of *L*, we assume $x \in L_1$. Because $x \le x \to I$ we have $I \in L_1$. Let $x \to y \in L_1, x \in L_1$, since $x \to y \le x \to y$, we have $y \in L_1$. Hence L_1 is a filter of *L*.

Definition 2.6. [6] Let *L* be a lattice implication algebra. $B_{(L)}$ is the collections of fuzzy sets of *L*, $A \in B_{(L)}$ and $A \neq \varnothing$. A is called a fuzzy filter of *L* if it satisfies: *(1)* $A(x) \le A(1)$; *(2)* $A(y) \ge \min\{A(x \to y), A(x)\}$ for any $x, y \in L$.

Theorem 2.7. [2, 6] Let *L* be a lattice implication algebra and *A* a fuzzy filter of *L*. Then for any $x, y \in L$, $x \le y$ imply $A(x) \le A(y)$.

Theorem 2.8. [2] Let *L* be a lattice implication algebra,

(1) if *A* is fuzzy filter of *L*, then $\{x : x \in L, A(x) = A(I)\}$ is a filter of *L*.

(2) if A is filter of *L*, then x_A is a fuzzy filter of *L*, where x_A is the characteristic function of *A*.

3 *FL*-Filter of Lattice Implication Algebra

Definition 3.1. *A fuzzy subset B of a bounded lattice (L,* $\vee$ *,* $\wedge$*, O, I) is called fuzzy lattice filter (briefly, FL-filter) if for all* $x, y \in L$ *(1) if* $x \le y$*, then* $B(x) \le B(y)$*; (2)* $B(x \wedge y) \ge \min\{B(x), B(y)\}$ *hold.*

Theorem 3.2. *Let L be a lattice implication algebra. Every fuzzy filter of L is a FL-filter.*

Proof. Suppose *A* is a fuzzy filter of *L*. By Theorem 2 *A* is order-preserving, obviously, the condition *(1)* of definition 3.1 holds. By

$$x \to (x \wedge y) = (x \to x) \wedge (x \to y) = (x \to y) \ge y,$$

We can get

$$A(x \wedge y) \geq \min\{A(x \rightarrow (x \wedge y)), A(x)\}$$
$$\geq \min\{A(y), A(x)\}.$$

Hence, the condition *(2)* of definition 3.1 is holds. That said that proposition is true.

Theorem 3.3. *In a lattice H implication algebra L, every FL-filter A of L is a fuzzy filter.*

Proof. Since $x \leq I$, it follows that $A(x) \leq A(I)$ for any $x \in L$. Let $x, y \in L$, we have

$$A(y) \geq A(x \wedge y) = A(x \wedge (x' \vee y))$$
$$= A(x \wedge (x \rightarrow y)) \geq \min\{A(x), A(x \rightarrow y)\}.$$

So, *A* is a fuzzy filter.

Theorem 3.4. *Let A be a fuzzy subset of a lattice implication algebra L. Then A is a fuzzy filter if and only if* A_t *is a fuzzy filter when* $A_t \neq \varnothing$, $t \in [0, 1]$.

Proof. Assume that *A* is a fuzzy filter of *L* and $t \in [0, 1]$. Such that $A_t \neq \varnothing$, clearly $I \in A_t$. Suppose $x, y \in L$, $x \rightarrow y \in A_t$, $x \in A_t$. Then $A(x \rightarrow y) \geq t$ and $A(x) \geq t$. It follows that

$$A(y) \geq \min\{A(x \rightarrow y), A(x)\} \geq t.$$

So that $y \in A_t$. A_t is a fuzzy filter of *L*.

Conversely, suppose $A_t (t \in [0, 1])$ is a fuzzy filter of *L* when $A_t \neq \varnothing$. For any $y \in L$, $y \in A_{A(y)}$, it follows that $A_{A(x)}$ is a fuzzy filter of *L* and hence $I \in A_{A(x)}$, that is $A(x) \leq A(I)$. For any $x, y \in L$, let $t = \min\{A(x \rightarrow y), A(x)\}$, it follows that A_t is a fuzzy filter and $x \rightarrow y \in A_t$, $x \in A_t$, this implies that $y \in A_t$ and $A(y) \geq t = \min\{A(x \rightarrow y), A(x)\}$, i.e. *A* is a fuzzy filter.

Theorem 3.5. *Let lattice L be a completely, In what follows we denoted by* $F_{(L)}$ *the set of all fuzzy subsets of L,* $A \in F_{(L)}$ *and* $A \neq \varnothing$, $a \in L$. *If A is a fuzzy lattice filter of L, then* $B = \{x : x \in L, A(a) \leq A(x)$ *is a lattice filter of L.*

Proof. Let $x \in B$, $y \in L$, $x \leq y$, then $A(a) \leq A(x)$. Since A is a fuzzy lattice filter, hence $A(a) \leq A(x) \leq A(y)$, i.e. $y \in B$. Suppose $x, y \in B$, then

$$A(a) \leq A(x),\ A(a) \leq A(y).$$

It follows that $A(x \wedge y) \geq \min\{A(x), A(y)\} \geq A(a)$, i.e. $x \wedge y \in B$. Moreover, B is a lattice filter of L.

Theorem 3.6. *Let L be a lattice H implication algebra. (1). every fuzzy filter of L is a fuzzy lattice filter; (2). Every fuzzy implicative filter of L is a fuzzy lattice filter of L.*

Proof. (1). Suppose A is a fuzzy filter of L. A is order-preserving by theorem 2.1, hence for all $x, y \in L$, $x \leq y$ implies $A(x) \leq A(y)$ holds. Since

$$x \to (x \wedge y) = x' \vee (x \wedge y) = (x' \vee x) \wedge (x' \vee y)$$

$$= I \wedge (x' \vee y) = x' \vee y \geq y$$

Hence $A(x \to (x \wedge y) \geq A(y)$, we get

$$A(x \wedge y) \geq \min\{A(x \to (x \wedge y)), A(x)\} \geq \min\{A(x), A(y)\}.$$

Moreover A is a fuzzy lattice filter.

(2). Since, fuzzy implicative filter of L is a fuzzy filter. According to (1), we obtain (2) holds.

4 On FL-Filter and NF-Filter of Lattice Implication Algebra

Definition 4.1. [12] *A fuzzy filter A of a lattice implication algebra L is called a normal fuzzy filters (briefly, NF-filters) if there exists* $x \in L$ *such that* $A(x) = 1$.

Corollary 4.2. *Every NF-filter of lattice implication algebra is a FL-filter.*

Corollary 4.3. *A is NF-filter if and only if* $A(1) = 1$.

Let L be a lattice implication algebra, $A \in B(L)$,

$$L_A \overset{\Delta}{=} \{x \in L : A(x) = A(1)\}.$$

Theorem 4.4. *Given a fuzzy filter A of L, let* A^* *be a fuzzy subset of L defined by* $A^*(x) = A(x) + 1 - A(1)$, *for any* $x \in L$. *Then* A^* *is a FL-filter of L containing A.*

Proof. Notice that for any filters A_1 of L the characteristic function x_A of A_1 is a normal fuzzy filter of L. It is clear that A is normal if and only if $A = A^*$. If A is a fuzzy filter of L, then $A^* = (A^*)^*$. Hence, if A is normal, then $A = (A^*)^*$.

According to Definitions 2.4 and 4.1, we know that if A and A_0 are fuzzy filters of L such that $A \subseteq A_0$ and $A(1) = A_0(1)$, then $L_A \subseteq L_{A_0}$. If A is a fuzzy filter of L and there exists a fuzzy filter of L such that $A_0^* \subseteq A$, then A is normal. By Definition 4.1 and Theorem 3.2, we have A^* is a *FL*-filter of L containing A. This means that the propositional holds.

Theorem 4.5. *Let A be a fuzzy filter of L and a map f: [0, A (I)]* $\rightarrow$ *[0, 1] is an increasing function. Then the fuzzy subset* $A_f(x) = f(A(x))$ *for any* $x \in L$ *is a FL-filter of L.*

Proof. Notice that $A(x) \leq A(I)$ for all $x \in L$, since *f* is an increasing function it follows that

$$A_f(I) = f(A(I)) \geq f(A(x)) = A_f(x)$$

for any $x \in L$. Let $x, y \in L$, then

$$\begin{aligned}\min\{A_f(x \rightarrow y), A_f(x)\} &= \min\{f(A(x \rightarrow y)), f(A(x))\} \\ &= f(\min\{A(x \rightarrow y), A(x)\}) \\ &\leq f(A(y)) = A_f(y),\end{aligned}$$

i.e. , $A_f(y) \geq \min\{A_f(x \rightarrow y), A_f(x)\}$.

Moreover, A_f is a fuzzy filter of L. According Theorem 3.2, we get the fuzzy subset $A_f(x) = f(A(x))$ for any $x \in L$ is a *FL*-filter of *L*.

Theorem 4.6. *If* $A\ (\neq \varnothing)$ *and* $B\ (\neq \varnothing)$ *are fuzzy filters of L such that for all* $x, y \in L$, $A(x) \leq A(y)$ *if and only if* $B(x) \leq B(y)$, *then* $(A \circ B)$ *is also a fuzzy filter of L and* $L_{A \circ B} = L_A \cap L_B$, *where* $(A \circ B)(x) = A(x)B(x)$ *for any* $x \in L$.

Proof. Notice that for any $x \in L$, $(A \circ B)(x) = A(x)B(x)$

$$\leq A(I)B(I) = (A \circ B)(I),$$

i.e. $(A \circ B)(x) \leq (A \circ B)(I)$; for any $x, y \in L$,

$$\begin{aligned}(A \circ B)(y) &= A(y)B(y) \\ &\geq (\min\{A(x \rightarrow y), A(x)\})(\min\{B(x \rightarrow y), B(x)\}) \\ &= \min(\{A(x \rightarrow y), A(x)\}\{B(x \rightarrow y), B(x)\}) \\ &= \min\{(A \circ B)(x \rightarrow y), (A \circ B)(x)\}\end{aligned}$$

i.e., $(A \circ B)(y) \geq \min\{(A \circ B)(x \rightarrow y), (A \circ B)(x)\}$.

Hence, $A \circ B$ is a fuzzy filter of *L*.

For all $x \in L$, if $x \in L_A \cap L_B$, then $A(x) = A(I), B(x) = B(I)$.

So, $(A \circ B)(x) = (A \circ B)(I)$, i.e., $x \in L_{A \circ B}$. Therefore, we have $L_A \cap L_B \subseteq L_{A \circ B}$.

Let $x \in L_{A \circ B}$ then $(A \circ B)(x) = (A \circ B)(I) = A(I)B(I)$.

Also, $A(x) = A(I)$ and $B(x) = B(I)$. If $A(x) \prec A(I)$, then

$A(x)B(x) \prec A(I)B(I)$.

This is contradiction.

Similarly, it is also a contradiction when $B(x) \prec B(I)$. Moreover, $x \in L_A \cap L_B$, i.e. , $L_{A\circ B} \subseteq L_A \cap L_B$. Hence, we obtain $L_{A\circ B} = L_A \cap L_B$.

Definition 4.7. [12] *A fuzzy filter is called a maximal fuzzy filter if it is not L, and it is a maximal element of the set of all fuzzy filters with respect to fuzzy set inclusion.*

Theorem 4.8. *If A is a maximal fuzzy filter of L, then*

(1) A is a FL-filter;
(2) A taker only the values 0 and 1;
(3) L_A is a maximal filter of L.

Proof. (1). If A is not FL-filter, then A is not NF-filter for any $x \in L$,

$A(x) \leq A(x)+1-A(I)=A^*(x)$,

$A(I) \prec 1=A(I)+1-A(I)=A^*(I)$,

It follows that A is not a maximal fuzzy filter of L, which is a contradiction.

(2). Note that $A(I)=1$ since A is normal. Let $t \in L$ and $A(t) \neq 1$, we claim that $A(t)=0$. If not, then

$$0 \prec A(t) \prec 1.$$

Let B be a fuzzy subset of L defined by $B(x)=\frac{1}{2}(A(x)+A(t))$ for all $x \in L$, we obtain

$$B(I)= \frac{1}{2}(A(I)+ A(t))$$
$$= \frac{1}{2}(1+A(t)) \geq \frac{1}{2}(A(x)+A(t))=B(x).$$

For all $x, y \in L$, then

$$B(y)=\frac{1}{2}(A(y)+A(t)) \geq \frac{1}{2}(\min\{A(x \to y), A(x)\}+A(t))$$
$$=\min (\frac{1}{2}\{A(x \to y)+A(t), A(x)+A(t)\})$$
$$=\min \{\frac{1}{2}(A(x \to y)+A(t), \frac{1}{2}(A(x)+A(t))\}$$
$$=\min \{B(x \to y), B(x)\}.$$

Hence B is a fuzzy filter of L. It follows from Theorem 4.1 that for any $x \in L$,

$$B^*(x)=B(x)+1-B(I)= \frac{1}{2}(A(x)+A(t))+1-\frac{1}{2}(A(I)+A(t))$$

$$=\frac{1}{2}A(x)+\frac{1}{2}=\frac{1}{2}(A(x)+1)\geq A(x)$$

Note that $B^*(t)=B(t)+1-B(I)=\frac{1}{2}(A(t)+1)\prec 1=B^*(I)$. Hence B^* is non-constant and A is not a maximal fuzzy filter. This is a contradiction.

(3). L_A is a filter of L. If $L_A\subseteq A_O\neq L$ and A_O is a filter, then the Characteristic functions x_{A_O} and x_{L_A} are fuzzy filters and $A=x_{L_A}\subseteq x_{A_O}$, it follows that $A=x_{A_O}$ because A is a maximal fuzzy filter and hence $L_A=A_O$. This proved that L_A is a maximal filter of L.

Theorem 4.9. Every maximal fuzzy filter of L is a *FL*-filter.

5 Conclusions

In order to research the many- valued logical system whose propositional value is given in a lattice, Xu initiated the notion of lattice implication algebras. Moreover, for development of this many-valued logical system, it is needed to make clear the structure of lattice implication algebras. It is well known that the filters with special properties play an important role in investigating the structure of logical system. In this article, we proposed the notions of *FL*-filter and *NF*-filter in lattice implication algebras, discussed their properties. We hope above work would serve as a foundation for further on study the structure of lattice implication algebras and develop corresponding many-valued logical system.

Acknowledgements

Authors would like to express their sincere thanks to the referees for their valuable suggestions and comments.

Remark: This work was supported by the National Science Foundation of P.R. China (Grant No. 60474022).

References

[1] D. W. Borns, J. M. Mack, *An algebraic Introduction on Mathematical Logic*, Springer,Berlin,1975.
[2] Y. Xu, Lattice implication algebras, *J. Southwest Jiaotong Univ.* 28(1), (1993)20-27(in Chinese).
[3] Y. Xu, D. Ruan, K. Y. Qin, J. Liu, *Lattice-Valued Logic*, Springer, Berlin, 2003.
[4] Y. Xu, K. Y. Qin, Lattice H implication algebras and lattice implication algebra classes, *J. Hebei Mining Civil Engineering Institute*, 3 (1992)139-143 (in Chinese).

[5] Y. Xu, K. Y. Qin, On filters of lattice implication algebras, *J. Fuzzy Math.* 1(1993)251-260 (in Chinese).
[6] Y. Xu, K. Y. Qin, Fuzzy lattice implication algebras, *J. Southwest Jiaotong Univ.* 2(1995)121-127(in Chinese).
[7] J. Liu, Y. Xu, On filters and structures of lattice implication algebras, *Chinese Science Bulletin*, Vol.42, No.10, (May 1997)1049-1052(in Chinese).
[8] K. Y. Qin, Y. Xu, The Ultra-filter of lattice implication algebras, *J. Southwest Jiaotong Univ.* Vol.34, No.1, (Feb.1999)51-54(in Chinese).
[9] X. F. Wang, Y. Xu, Z. M. Song, Some of properties of filters in lattice implication algebras, *J. Southwest Jiaotong Univ.* 5(2001)536-539(in Chinese).
[10] Y. B. Jun, Y.Xu, K.Y.Qin, Positive implicative and associative filters of lattice implication algebras, *J. Bull. Korean Math.* Soc. 35(1) (1998) 53-61(in Chinese).
[11] J. B. Zhao, H. Zhu, Primary filter of residuated lattice implication algebras, *J.Nanyang Teacher' s College*, Vol.5, No.3 (Mar. 2006) 14-16 (in Chinese).
[12] S. W. Chen, Y. Xu, The normal fuzzy filter of lattice implication algebra, *J. Fuzzy Math. , Vol. 14 No. 2 (2006).*

Study on Adaptive Fuzzy Control System Based on Gradient Descent Learning Algorithm

Jundi Xiong[1], Taifu Li[2], Huihui Xiao[1], and Renming Deng[3]

[1] College of Information and Automation, Chongqing Institute of Technology, Chongqing, 400050, China
xiongjundi@126.com
[2] College of Electronic Information Engineering, Chongqing University of Science and Technology, Chongqing, 400050, China
litaifu@tom.com
[3] College of Automation, Chongqing University, Chongqing, 400044, China

Abstract. For the general two-dimensional fuzzy controller, the output control value is a rigid function with respect to the error and the change in error, so it is difficult to obtain the desired effects for the plant that is uncertain and time-varying. Aimed at the problem, the paper applies gradient descent learning algorithm to correct the mean and variance of Gaussian membership functions of all fuzzy sets in the input and output universes, in this way, the fuzzy control system is adaptive. Selecting input signals as step, ramp, acceleration and sine signals, respectively, all simulation studies were carried out. The results demonstrate that the control algorithm is feasible, and its effect is better than that of the fuzzy control system without adaptability.

Keywords: Fuzzy control, Adaptive, Gradient descent, Simulation.

1 Introduction

Fuzzy control rules are based on people's experience that is formed by learning, trying and accumulating. Designing a general fuzzy control system includes five major steps that are to establish fuzzy control rules, confirm the membership functions of the input and output universes, select fuzzy inference engine, choose fuzzification and defuzzification methods, respectively. Input signals generally select the error and the change in error in this kind of fuzzy control system. Once above-mentioned five steps have been confirmed, the control value of the fuzzy control system should be a rigid function with respect to the error and the change in error. Since the plant is complex, uncertain and time-varying, it is difficult to obtain the desired control effect through the rigid function[1]. On the other hand, since the expert experience isn't always perfect, it is possible to lead to a bad control performance because the definition of the membership function is inaccurate, or the rule-base is not complete. Therefore, the paper applies gradient descent learning algorithm to modify the mean and variance of Gaussian membership functions of all fuzzy sets in the input and output universe, in this way, the fuzzy control system is adaptive for parameters varying and various disturbances.

B.-Y. Cao (Ed.): Fuzzy Information and Engineering (ICFIE), ASC 40, pp. 1009–1020, 2007.
springerlink.com

2 How to Design a General Fuzzy Control System

For a general fuzzy control system without self-learning and adaptability, the control engineer typically designs the fuzzy control system by doing the following:

(1) Deciding the structure of a fuzzy controller, i.e., selecting input and output variables.

(2) summarizing the fuzzy control rules (which is formed by the human's learning and trying).

(3) confirming membership function for all fuzzy sets in the input and output universes.

(4) choosing fuzzy inference engine, fuzzification and defuzzification. In this way, a fuzzy control system actually may be replaced with a accurate mathematical expression.

The rule-base for the fuzzy controller has rules of the form

R^l:IF e is E^l and ce is CE^l Then u is U^l

The corresponding mathematical expression of fuzzy controller is

$$u = f(e, ce) = \frac{\sum_{l=1}^{m} \bar{u}^l \mu_{E^l}(e) \mu_{CE^l}(ce)}{\sum_{l=1}^{m} \mu_{E^l}(e) \mu_{CE^l}(ce)} \tag{1}$$

Above-mentioned fuzzy controller can get the output control signal value while the error and the change in error are both known. The fuzzy controller is quite dependent on five design steps above-mentioned, once five steps have been confirmed, the fuzzy control system should be a rigid function with respect to the error and the change in error, shown in Figure 1. For the different plant, it is difficult to obtain the desired control effect for the invariable rigid function.

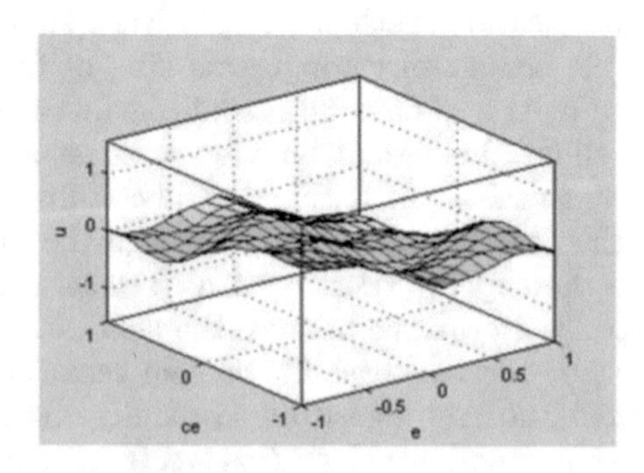

Fig. 1. The rigid function surface of general two-dimensional fuzzy controller

3 Adaptive Fuzzy Control Algorithm

3.1 System Structure

In this paper, the suggested adaptive fuzzy control system based on gradient descent method is shown in Figure 2. The adaptive fuzzy controller selects the error between the reference input and actual output signal as learning criterion, and modifies parameters in fuzzy controller, in this way, the actual output value could follow the reference input, and can adapt all kinds of disturbances, and the structure-varying and parameter-varying in the controlled process.

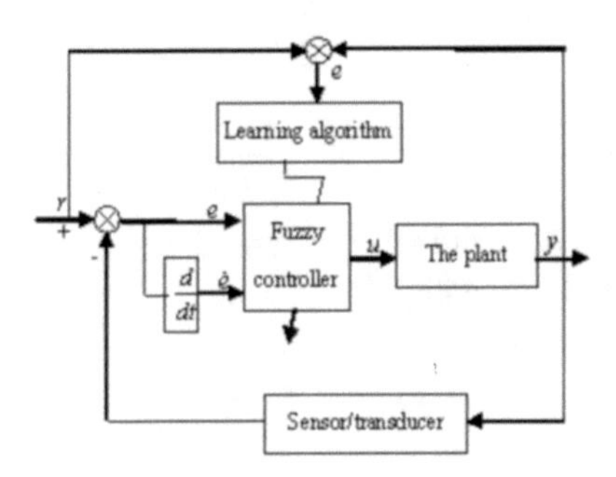

Fig. 2. Block diagram of adaptive fuzzy control system

3.2 Adaptive Control Algorithm

The fuzzy controller chooses Mamdani product inference engine, singleton fuzzification, center average defuzzification and Gaussian membership function, the output value of the fuzzy controller is

$$\mathrm{u} = f(e, ce) = \frac{\sum_{l=1}^{M} \bar{u}^l [\exp(-(\frac{e-\bar{e}^l}{\sigma_e^l})^2) \exp(-(\frac{ce-\bar{ce}^l}{\sigma_{ce}^l})^2)]}{\sum_{l=1}^{M} [\exp(-(\frac{e-\bar{e}^l}{\sigma_e^l})^2) \exp(-(\frac{ce-\bar{ce}^l}{\sigma_{ce}^l})^2)]} \tag{2}$$

where, M represents the number of the rules, $\bar{u}^l$ stands for the center value of the membership function of some fuzzy set at the output control universe in the lth rule, $\overline{e}^l$ stands for the mean of the Gaussian membership function of some fuzzy set at the error universe in the lth rule, $\overline{ce}^l$ stands for the mean of the membership function of some fuzzy set at the error change universe in the lth rule, σ_e^lstands for the variance of the Gaussian membership function of some fuzzy set at the error universe in the lth rule, σ_{ce}^l stands for the variance of the Gaussian membership function of some fuzzy set at the error change universe in the lth rule. Shown as Figure 2, defining the error power function as the following:

$$E = \frac{1}{2}e^2 = \frac{1}{2}(r-y)^2 \tag{3}$$

Simplifying the equation (2) as the following[2]:

$$u = \frac{a}{b} \tag{4}$$

$$a = \sum_{l=1}^{M} (\bar{u}^l \bar{z}^l) \tag{5}$$

$$b = \sum_{l=1}^{M} \bar{z}^l \tag{6}$$

$$\bar{z}^l = \mu_{E^l}(e)\mu_{CE^l}(ce) \ = \exp(-(\frac{e-\bar{e}^l}{\sigma_e^l})^2)\exp(-(\frac{ce-\bar{ce}^l}{\sigma_{ce}^l})^2) \tag{7}$$

For fuzzy sets in the output control universe, defining negative gradient descent learning algorithm listed below[3], [4]:

$$\bar{u}^l(k+1) = \bar{u}^l(k) - \alpha\frac{\partial E}{\partial \bar{u}^l}|_k \tag{8}$$

$$\frac{\partial E}{\partial \bar{u}^l}|_k = \frac{\partial E(k)}{\partial y(k)}\frac{\partial y(k)}{\partial u(k)}\frac{\partial u(k)}{\partial a(k)}\frac{\partial a(k)}{\partial \bar{u}^l(k)} \tag{9}$$

Because the model of the controlled process is unknown, hypothetically, $y = g(u)$, defining $\frac{\partial y}{\partial u}|_k$ as the following:

$$\left.\frac{\partial y}{\partial u}\right|_k = \lim_{\Delta u \to 0}\frac{\Delta y}{\Delta u} = \frac{y(k)-y(k-1)}{u(k)-u(k-1)} \tag{10}$$

thus,

$$\begin{aligned}\frac{\partial E}{\partial \bar{u}^l}|_k &= [y(k)-r(k)]\frac{y(k)-y(k-1)}{u(k)-u(k-1)}\frac{1}{b(k)} \\ &\exp(-(\frac{e(k)-\bar{e}^l(k)}{\sigma_e^l(k)})^2)\exp(-(\frac{ce(k)-\bar{ce}^l(k)}{\sigma_{ce}^l(k)})^2)\end{aligned} \tag{11}$$

alike,

$$\bar{e}^l(k+1) = \bar{e}^l(k) - \alpha\frac{\partial E}{\partial \bar{e}^l}|_k \tag{12}$$

$$\begin{aligned}\frac{\partial E}{\partial \bar{e}^l}|_k &= \frac{\partial E(k)}{\partial y(k)}\frac{\partial y(k)}{\partial u(k)}\frac{\partial u(k)}{\partial z^l(k)}\frac{\partial z^l(k)}{\partial \bar{e}^l(k)} \\ &= \frac{\partial E(k)}{\partial y(k)}\frac{\partial y(k)}{\partial u(k)}\left(\frac{\partial u(k)}{\partial a(k)}\frac{\partial a(k)}{\partial z^l(k)} + \frac{\partial u(k)}{\partial b(k)}\frac{\partial b(k)}{\partial z^l(k)}\right)\frac{\partial z^l(k)}{\partial \bar{e}^l(k)}\end{aligned} \tag{13}$$

$$\overline{ce}^l(k+1) = \overline{ce}^l(k) - \alpha\frac{\partial E}{\partial \overline{ce}^l}|_k \tag{14}$$

$$\begin{aligned}\frac{\partial E}{\partial \bar{ce}^l}|_k &= \frac{\partial E(k)}{\partial y(k)}\frac{\partial y(k)}{\partial u(k)}\frac{\partial u(k)}{\partial z^l(k)}\frac{\partial z^l(k)}{\partial \bar{ce}^l(k)} \\ &= \frac{\partial E(k)}{\partial y(k)}\frac{\partial y(k)}{\partial u(k)}\left(\frac{\partial u(k)}{\partial a(k)}\frac{\partial a(k)}{\partial z^l(k)} + \frac{\partial u(k)}{\partial b(k)}\frac{\partial b(k)}{\partial z^l(k)}\right)\frac{\partial z^l(k)}{\partial \bar{ce}^l(k)}\end{aligned} \tag{15}$$

$$\sigma_e^l(k+1) = \sigma_e^l(k) - \alpha \frac{\partial E}{\partial \sigma_e^l}|_k \tag{16}$$

$$\begin{aligned} \frac{\partial E}{\partial \sigma_e^l}|_k &= \frac{\partial E(k)}{\partial y(k)} \frac{\partial y(k)}{\partial u(k)} \frac{\partial u(k)}{\partial z^l(k)} \frac{\partial z^l(k)}{\partial \sigma_e^l(k)} \\ &= \frac{\partial E(k)}{\partial y(k)} \frac{\partial y(k)}{\partial u(k)} \left(\frac{\partial u(k)}{\partial a(k)} \frac{\partial a(k)}{\partial z^l(k)} + \frac{\partial u(k)}{\partial b(k)} \frac{\partial b(k)}{\partial z^l(k)} \right) \frac{\partial z^l(k)}{\partial \sigma_e^l(k)} \end{aligned} \tag{17}$$

$$\sigma_{ce}^l(k+1) = \sigma_{ce}^l(k) - \alpha \frac{\partial E}{\partial \sigma_{ce}^l}|_k \tag{18}$$

$$\begin{aligned} \frac{\partial E}{\partial \sigma_{ce}^l}|_k &= \frac{\partial E(k)}{\partial y(k)} \frac{\partial y(k)}{\partial u(k)} \frac{\partial u(k)}{\partial z^l(k)} \frac{\partial z^l(k)}{\partial \sigma_{ce}^l(k)} \\ &= \frac{\partial E(k)}{\partial y(k)} \frac{\partial y(k)}{\partial u(k)} \left(\frac{\partial u(k)}{\partial a(k)} \frac{\partial a(k)}{\partial z^l(k)} + \frac{\partial u(k)}{\partial b(k)} \frac{\partial b(k)}{\partial z^l(k)} \right) \frac{\partial z^l(k)}{\partial \sigma_{ce}^l(k)} \end{aligned} \tag{19}$$

4 Fuzzy Control Rules

For above-mentioned adaptive fuzzy control algorithm, because the parameters that will be modified is more, so computation is very large, therefore, only selecting five fuzzy sets in the error universe, i.e., NB, NS, O, PS, PB, three fuzzy sets in the error change universe, i.e., N, O, P, five fuzzy sets in the output control universe, i.e., NB, NS, O, PS, PB. The fuzzy control rules is shown in Table 1.

Table 1. The fuzzy control rules for u

e \ ce	N	O	P
NB	PB	PB	O
NS	PB	PS	NS
O	PS	O	NS
PS	PS	NS	NB
PB	O	NB	NB

Above fuzzy rule table could written as the following form:

R^1: IF e is NB and ce is N, then u is PB
R^2: IF e is NB and ce is O, then u is PB
R^3: IF e is NB and ce is P, then u is O
R^4: IF e is NS and ce is N, then u is PB
R^5: IF e is NS and ce is O, then u is PS
R^6: IF e is NS and ce is P, then u is NS
R^7: IF e is O and ce is N, then u is PS
R^8: IF e is O and ce is O, then u is O
R^9: IF e is O and ce is P, then u is NS
R^{10}: IF e is PS and ce is N, then u is PS
R^{11}: IF e is PS and ce is O, then u is NS
R^{12}: IF e is PS and ce is P, then u is NB

R^{13}: IF e is PB and ce is N, then u is O
R^{14}: IF e is PB and ce is O, then u is NB
R^{15}: IF e is PB and ce is P, then u is NB

There are 15 pieces of rules in all, based on equation (8), (9), (12), (14), (16) and (18), the parameters that would be modified are $15 \times 5 = 75$. Actually, alterable parameters are the mean and the variance in Gaussian membership function. In the output control universe, it is only to adjust the mean in Gaussian membership function because the defuzzification is central average method. Therefore, all alterable parameters are $5 \times 2 + 3 \times 2 + 5 = 21$. Detailed approach, that corrects alterable parameters in membership functions in all kinds of fuzzy sets, will be explored with several examples listed below. For example

$$\mu_{PB}(e) = \exp(-(\frac{e - \bar{e}_{PB}}{\sigma_{PBe}})^2) \tag{20}$$

For 15 pieces of rules in equation (2), there are 3 pieces of rules containing $\mu_{PB}(e)$, i.e., R10, R11, and R12, thus, the mathematical expressions correcting parameters $\bar{e}_{PB}$ and σ_{PBe} are

$$\bar{e}_{PB}(k+1) = \bar{e}_{PB}(k) - \alpha(\frac{\partial E}{\partial \bar{e}^{10}}|_k + \frac{\partial E}{\partial \bar{e}^{11}}|_k + \frac{\partial E}{\partial \bar{e}^{12}}|_k) \tag{21}$$

$$\sigma_{PBe}(k+1) = \sigma_{PBe}(k+1) - \alpha(\frac{\partial E}{\partial \sigma_e^{10}}|_k + \frac{\partial E}{\partial \sigma_e^{11}}|_k + \frac{\partial E}{\partial \sigma_e^{12}}|_k) \tag{22}$$

Other example

$$\mu_N(ce) = \exp(-(\frac{ce - \overline{ce}_N}{\sigma_{Nce}})^2) \tag{23}$$

For 15 pieces of rules in equation(2) , there are 5 pieces of rules containing $\mu_N(ce)$, i.e., R1, R4, R7, R10, and R13, thus, the mathematical expressions correcting parameters $\overline{ce}_N$ and σ_{Nce} are

$$\overline{ce}_N(k+1) = \overline{ce}_N(k) - \alpha(\frac{\partial E}{\partial \overline{ce}^1}|_k + \frac{\partial E}{\partial \overline{ce}^4}|_k + \frac{\partial E}{\partial \overline{ce}^7}|_k + \frac{\partial E}{\partial \overline{ce}^{10}}|_k + \frac{\partial E}{\partial \overline{ce}^{13}}|_k) \tag{24}$$

The third example

$$\mu_{PS}(u) = \exp(-(\frac{u - \bar{u}_{PS}}{\sigma_{PSu}})^2) \tag{25}$$

For 15 pieces of rules in equation (2), there are 3 pieces of rules containing $\mu_{PS}(u)$, i.e., R5, R7, and R10, thus, the mathematical expression correcting parameters $\bar{u}_{PS}$ is

$$\bar{u}_{PS}(k+1) = \bar{u}_{PS}(k) - \alpha(\frac{\partial E}{\partial \bar{u}^5}|_k + \frac{\partial E}{\partial \bar{u}^7}|_k + \frac{\partial E}{\partial \bar{u}^{10}}|_k) \tag{26}$$

5 Simulation Results and Discussions

Above-mentioned adaptive fuzzy control algorithm had been carried out through simulation with MATLAB. In addition, in order to verify influences of adaptive fuzzy control algorithm, the same fuzzy controller without adaptive algorithm also had been carried out simulation. The controlled process is the third-order system as listed below.

$$\frac{523500}{s^3 + 87.35s^2 + 10470s} \tag{27}$$

The universe of error was selected as [-1, 1], the initial parameters of membership function of all fuzzy sets are shown in Table 2.

Table 2. The initial parameters for e

	NB	NS	O	PS	PB
μ	-1	-0.5	0	0.5	1
σ	0.2	0.2	0.2	0.2	0.2

The universe of the change in error ce was selected as [-1, 1], the initial parameters of membership function of all fuzzy sets are shown in Table 3.

Table 3. The initial parameters for ce

	N	O	P
μ	-1	0	1
σ	0.4	0.4	0.4

The universe of the output control u was selected as [-2, 2], the initial parameters of membership function of all fuzzy sets are shown in Table 4.

Table 4. The initial parameters for u

	NB	NS	O	PS	PB
μ	-2	-1	0	1	2
σ	0.2	0.2	0.2	0.2	0.2

5.1 Step Response

The simulation results of step responses are shown in Figure 3.

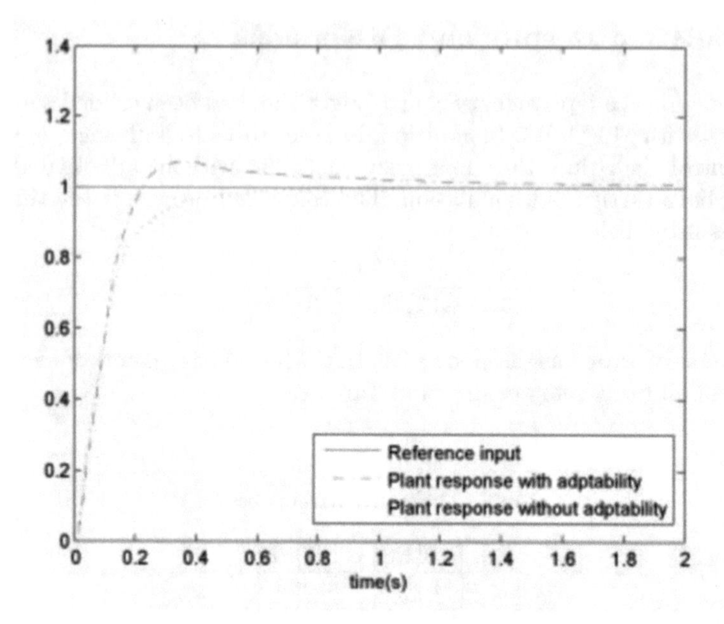

Fig. 3. Comparison of step responses

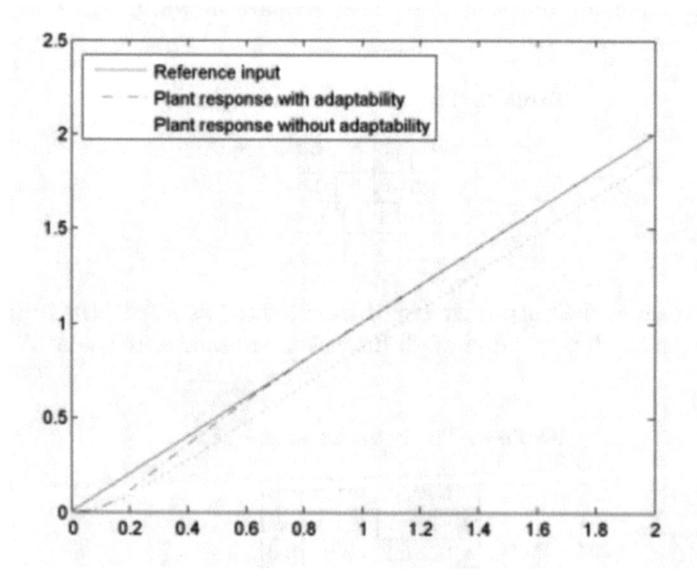

Fig. 4. Comparison of ramp responses

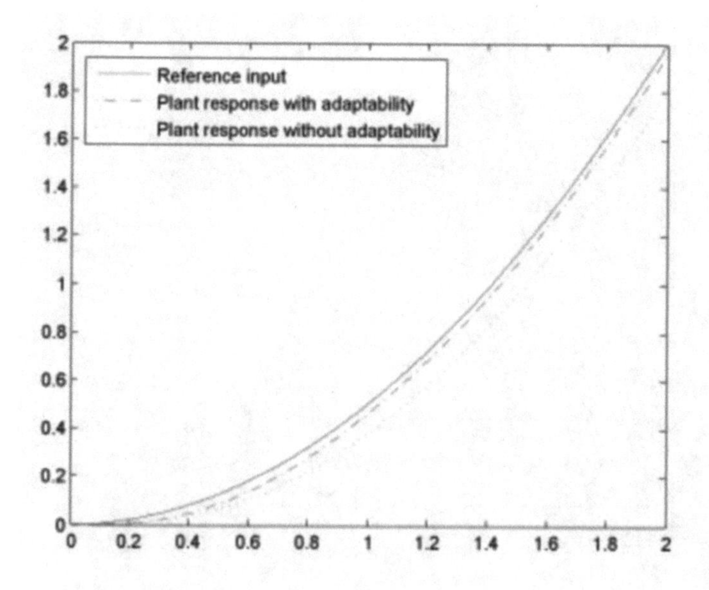

Fig. 5. Comparison of acceleration responses

5.2 Ramp Response

The simulation results of ramp responses are shown in Figure 3 with input signal "rin(k)=k*ts".

5.3 Acceleration Response

The simulation results of acceleration responses are shown in Figure 5 with input signal rin(k) $=\frac{(k\times ts)^2}{2}$, the changed parameters of all membership function curves are shown in Table 5, 6 and 7 after simulation finished.

Table 5. The changed parameters for e

	NB	NS	O	PS	PB
μ	-1	-0.486	0.056	0.505	1
σ	0.2	0.229	0.181	0.185	0.2

5.4 Sine Response

The simulation results of sine responses are shown in Figure 11 with input signal "rin(k)=0.5*sin(0.5*2*pi*k*ts)".

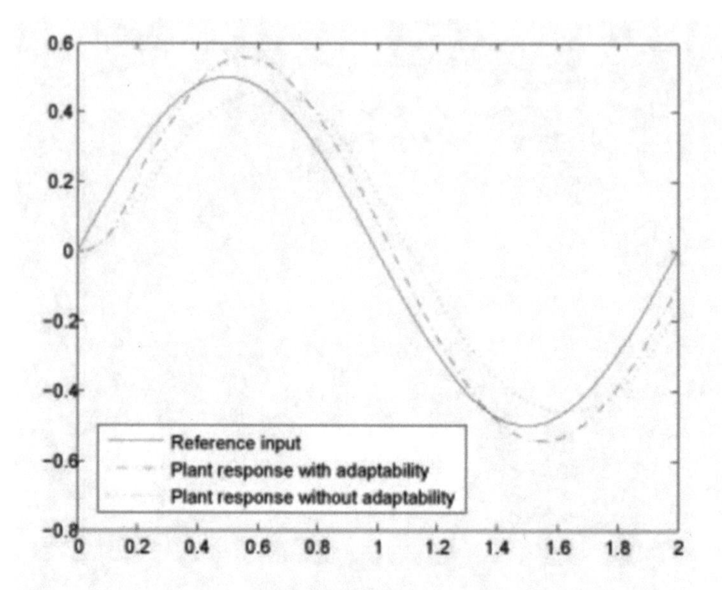

Fig. 6. Comparison of sine responses

Table 6. The changed parameters for ce

	N	O	P
μ	-0.9964	2.269e-006	1.006
σ	0.4089	0.4	0.3857

Table 7. The changed parameters for u

	NB	NS	O	PS	PB
μ	-2	-0.999	-0.1866	1.001	2
σ	0.2	0.2	0.2	0.2	0.2

5.5 Discussions

The simulation results of step responses show that the fuzzy controller without adaptive algorithm is better than the adaptive fuzzy controller. We thought that the requirement of control is relatively low, and adaptive fuzzy control system is larger in computation than the general non-adaptive fuzzy controller. The simulation results of ramp responses show that the adaptive fuzzy controller is better than the general non-adaptive fuzzy controller, its error rapidly approaches zero, but the error of the general non-adaptive fuzzy controller will approach a constant that isn't small. The simulation results of acceleration responses show that the advantage of the adaptive fuzzy control system is more obvious than that of the general non-adaptive fuzzy controller, though its error rapidly approaches a smaller constant, but the general non-adaptive fuzzy controller is unstable, the error will be divergent. The simulation results of sine response show that the adaptive fuzzy control system is still successful, and can meet engineering application at some tolerance, but the result of the general non-adaptive fuzzy controller is obviously bad. In addition, after simulation finished, the alterable parameters had changed, it also shows that the suggested adaptive fuzzy control algorithm is not an ornaments.

6 Stability Discussion

The criterion function of adaptive algorithm is

$$\min E(k) = \frac{1}{2}\left[r(k) - y(k)\right]^2 = \frac{1}{2}\,e^2(k) \tag{28}$$

If

$$\frac{1}{2}\,e^2(k) \le \varepsilon \tag{29}$$

or

$$e(k) \to 0 \tag{30}$$

Then

$$y(k) \to r(k) \tag{31}$$

It is very important to choose initial parameters for the convergence in gradient descent learning algorithm. If the chosen initial parameters are very close to optimal parameters, it is possible to converge at the global optimization solutions for the gradient descent learning algorithm. Otherwise, the algorithm may converge at local optimization solutions, even can't converge. There is an advantage to utilize fuzzy control, i.e., the fuzzy controller summarized from effective control experience, actually it has chosen better initial parameters. Therefore, the adaptive fuzzy control algorithm suggested in this paper can normally converge at the global optimization solutions.

7 Conclusions

Aimed at the problem that many controlled processes are complex, uncertain and time-varying, the paper analyzed the shortage of general two-dimensional fuzzy controller, then suggested a kind of adaptive fuzzy control algorithm based on gradient descent method. In order to verify the suggested theory, the paper had carried out simulation for the suggested adaptive fuzzy control system and the general non-adaptive fuzzy control system by four kinds of input signals such as step, ramp, acceleration and sine. The simulation results show that the suggested adaptive fuzzy control algorithm is feasible and effective.

Acknowledgment

The supports from Chongqing Educational Committee Science and Technology Research Project (No.040603) and Chongqing Natural Science Foundation of China (No.2004-8661) are gratefully acknowledged.

References

1. Li, Shiyong.(1999). *Fuzzy Control, Neural Control and Intelligent Cybernetics.* Haierbin: Press of Haierbin Institute of Technology.
2. Wang, L.X.(1994). *Adaptive Fuzzy Systems and Control: Design and Stability Analysis.* NJ: Prentice Hall, Englewood Cliffs.
3. Jang, J.R.(1993)."ANFIS: Adaptive-network-based fuzzy inference system," *IEEE Transactions on Systems, Man, and Cybernetics,* 23, 665-685.
4. Lin, C.T.(1994). *Neural Fuzzy Control Systems with Structure and Parameter Learning.* Singapore: World Scientific.

Uncertainty Measure of Fuzzy Rough Set

Ying Han[1] and Hong-tao Zhang[2]

[1] Department of Mathematics and Computer, Chaoyang Teacher's College, Chaoyang, 122000, P.R. China
hanying67@126.com
[2] School of Science, Shenyang University of Technology, Shenyang, 110023, P.R. China
victory19792005@163.com

Abstract. In this paper, by the use of the concept of cut set and information entropy, uncertainty and rough entropy about fuzzy knowledge are described. A new definition of rough entropy of a set X about knowledge R is given, and definition of rough entropy of fuzzy rough set is proposed.

Keywords: Rough set, Information entropy, Rough entropy, Uncertainty.

1 Introduction

Rough set theory, introduced by Z. Pawlak[1, 8], is a relatively new mathematical tool to deal with imprecise, uncertain and incomplete information and uncertainty is one of the important topics in the research on the rough set theory. Rough set theory inherently models two kinds of uncertainties. The first type of uncertainty, called set-uncertainty, arises from the boundary region of rough set on the approximation space. The other type of uncertainty, named knowledge-uncertainty, is modeled through knowledge granulation which is derived from binary relation on the universe. Roughness is proposed to measure set-uncertainty[1] and knowledge-uncertainty is measured by the use of information entropy and rough entropy[2, 3, 4]. The above results are obtained on the basis of rough set theory[1]. Nanda and Majumdar[6] presented fuzzy rough set. However, how to measure the uncertainty of fuzzy rough set? In this paper, fuzzy knowledge, i.e., fuzzy set, is locally made accurate by the use of cut set, and information entropy and rough entropy of fuzzy knowledge is obtained by the use of series of information entropies and rough entropies of locally accurate knowledge. Finally, uncertainty measure of fuzzy rough set is presented and it is proved that the measure of uncertainty is more accurate.

2 Prelimilary

Let $K = (U, R)$ be an approximation space, where U, a non-empty, finite set, is called the universe. R is a partition of U, or a classical equivalence relation. We

B.-Y. Cao (Ed.): Fuzzy Information and Engineering (ICFIE), ASC 40, pp. 1021–1027, 2007.
springerlink.com

assume that $R = \{R^1, R^2, \cdots, R^n\}$, where $R^i\,(1 \leqslant i \leqslant n)$ is an equivalence class and $\hat{R} = \big\{\{x\} \mid x \in U\big\}$ and $\check{R} = \{U\}$.

Two numerical characterizations of imprecision of a rough set X, called accuracy and roughness, are discussed[1]. Accuracy is defined as $\alpha_R(X)$ and

$$\alpha_R(X) = \frac{|\underline{apr}_R(X)|}{|\overline{apr}_R(X)|}, \text{ where } 0 \leqslant \alpha_R(X) \leqslant 1.$$

The second measure, roughness, is defined as $\beta_R(X)$ and

$$\beta_R(X) = 1 - \alpha_R(X).$$

Definition 1[2]. *Let (U, R) be an approximation space, and R a partition of U. A measure of uncertainty in rough set theory is defined by*

$$G(R) = -\sum_{i=1}^{n} \frac{|R_i|}{|U|} \log_2 \frac{|R_i|}{|U|}$$

If $R = \hat{R}$, then information measure of knowledge R achieves maximum value $\log_2 |U|$. If $R = \check{R}$, then information measure of knowledge R achieves minimum value 0. Obviously, $0 \leqslant G(R) \leqslant \log_2 |U|$.

Let P and Q be partitions of a finite set U, and we define

$$P \preceq Q \Leftrightarrow \forall P_i \in P, \exists Q_j \in Q \text{ such that } P_i \subseteq Q_j.$$

If $P \preceq Q$ and $P \neq Q$ then we say that Q is strictly coarser than P (or P is strictly finer than Q) and write $P \prec Q$.

Theorem 1[2]. *Let P and Q be two partitions of finite set U. If $P \prec Q$, then $G(Q) < G(P)$.*

Definition 2[4]. *Let $K = (U, R)$ be an approximate space, and R a partition of U. The rough entropy C of knowledge R is defined by*

$$C(R) = -\sum_{i=1}^{m} \frac{|R_i|}{|U|} \log_2 \frac{1}{|R_i|}$$

If $R = \hat{R}$, then the rough entropy of knowledge R achieves minimum value 0. If $R = \check{R}$, then the rough entropy of knowledge R achieves maximum value $\log_2 |U|$. Obviously, $0 \leqslant C(R) \leqslant \log_2 |U|$.

Theorem 2[4]. *Let P and Q be two partitions of finite set U. If $P \prec Q$, then $C(P) < C(Q)$.*

A better measure of uncertainty of rough set X is given as follows:

Definition 3[5]. *Let $K = (U, R)$ be an approximate space, and R a partition of U. The rough entropy C of knowledge X about R is defined by*

$$C_R(X) = C(R) \cdot \beta_R(X)$$

Definition 4[7]. *Let U be an universe. A real function $e : \mathscr{F}(U) \to R^+$ is called a fuzzy entropy on $\mathscr{F}(U)$, if e has the following properties:*

(1) $e(D) = 0, \forall D \in \mathscr{P}(U)$.
(2) $e([\frac{1}{2}]_U) = \max_{\mathscr{A} \in \mathscr{F}(U)} e(\mathscr{A})$.
(3) $\forall \mathscr{A}, \mathscr{B} \in \mathscr{F}(U)$, *if* $\mu_{\mathscr{B}}(x) \geqslant \mu_{\mathscr{A}}(x) \geqslant \frac{1}{2}$ *or* $\mu_{\mathscr{B}}(x) \leqslant \mu_{\mathscr{A}}(x) \leqslant \frac{1}{2}$, *then* $e(\mathscr{A}) \geqslant e(\mathscr{B})$.
(4) $e(\mathscr{A}^c) = e(\mathscr{A}), \forall \mathscr{A} \in \mathscr{F}(U)$.
where $[\frac{1}{2}]_U$ *indicates a fuzzy set on* U *and its membership in every point is* $\frac{1}{2}$.

If $e([\frac{1}{2}]_U) = 1$, we call e a normal entropy on $\mathscr{F}(U)$. We assume that fuzzy entropy which is mentioned is normal.

3 Rough Entropy of Fuzzy Rough Set

3.1 Information Entropy and Rough Entropy of Fuzzy Knowledge $\mathscr{R}$

Let $\mathscr{R}$ be fuzzy equivalence relation on U. For any $\alpha \in [0,1], \mathscr{R}_\alpha = \{(x,y) \mid \mathscr{R}(x,y) \geqslant \alpha\}$ is a classical equivalence relation on U. We suppose $\mathrm{Im}(\mathscr{R}) = \{\lambda_1, \lambda_2, \cdots, \lambda_k\}$, $\lambda_1 < \lambda_2 < \cdots < \lambda_k$. Then $\mathscr{R}_{\lambda_i} = \{\mathscr{R}^1_{\lambda_i}, \mathscr{R}^2_{\lambda_i}, \cdots, \mathscr{R}^m_{\lambda_i}\} (i = 1, 2, \cdots, k)$ is a knowledge block on U. When $\lambda_1 < \lambda_2$, $\mathscr{R}_{\lambda_1} \supset \mathscr{R}_{\lambda_2}$. If x and y are in the same equivalence class of $\mathscr{R}_{\lambda_2}$, then x and y are also in the same equivalence class of $\mathscr{R}_{\lambda_1}$. Therefore, $\forall \mathscr{R}^j_{\lambda_2}, \exists \mathscr{R}^i_{\lambda_1}$, satisfies $\mathscr{R}^j_{\lambda_2} \subseteq \mathscr{R}^i_{\lambda_1}$, i.e., $\mathscr{R}_{\lambda_2} \prec \mathscr{R}_{\lambda_1}$. So we have $\mathscr{R}_{\lambda_k} \prec \mathscr{R}_{\lambda_{k-1}} \prec \cdots \prec \mathscr{R}_{\lambda_2} \prec \mathscr{R}_{\lambda_1}$. Furthermore, we have $G(\mathscr{R}_{\lambda_1}) < G(\mathscr{R}_{\lambda_2}) < \cdots < G(\mathscr{R}_{\lambda_{k-1}}) < G(\mathscr{R}_{\lambda_k})$. Thus, we obtain a function as follows:

$$f_{\mathscr{R}} : [0,1] \to [0, \log_2 |U|], \qquad \lambda \to G(\mathscr{R}_\lambda)$$

Because of finiteness of the universe U, $\mathrm{Im}(\mathscr{R})$ is a finite set. Thus $\mathscr{R}_\lambda = \check{R}, 0 \leqslant \lambda \leqslant \lambda_1. \mathscr{R}_\lambda = \mathscr{R}_{\lambda_{i+1}}, \lambda_i < \lambda \leqslant \lambda_{i+1}, i = 1, 2, \cdots, n-1$. Obviously, the function $f_{\mathscr{R}} = G(\mathscr{R}_\lambda)$ is piecewise-smooth and increasing in the interval of $[0,1]$. For the same reason, the function $C(\mathscr{R}_\lambda)$ is piecewise-smooth and decreasing in [0,1].

Definition 5. *Let* $K = (U, \mathscr{R})$ *be an approximate space,* e *is a normal fuzzy entropy on* $\mathscr{F}(U \times U)$. $\mathscr{R}$ *is a fuzzy equivalence relation on* U. *The information entropy of knowledge* $\mathscr{R}$ *is defined by:*

$$G(\mathscr{R}) = 2^{e(\mathscr{R})} \int_0^1 G(\mathscr{R}_\lambda) \mathrm{d}\lambda$$

Theorem 3. *Let* R *be a classical equivalence relation and* $\mathscr{R}$ *a fuzzy equivalence relation. If* $\mathscr{R} = R$, *then* $G(\mathscr{R}) = G(R)$.

Proof. If $\mathscr{R} = R$, then $\mathscr{R}_0 = \check{R}$, $\mathscr{R}_\lambda = R, 0 < \lambda \leqslant 1$. Thus $G(\mathscr{R}_0) = 0, G(\mathscr{R}_\lambda) = G(R)$. Furthermore, $\int_0^1 G(\mathscr{R}_\lambda)\mathrm{d}\lambda = \int_0^1 G(R)\mathrm{d}\lambda = G(R)$. Note that $e(\mathscr{R}) = 0$, therefore $G(\mathscr{R}) = G(R)$.

Theorem 3 indicates Definition 1 is a special case of Definition 5.

Let $\mathscr{P}$ and $\mathscr{Q}$ be fuzzy equivalence relations on U. We write $\mathscr{P} \preceq \mathscr{Q}$, if $\mathscr{P} \subseteq \mathscr{Q}$. If $\mathscr{P} \preceq \mathscr{Q}$ and $\mathscr{P} \neq \mathscr{Q}$, then we write $\mathscr{P} \prec \mathscr{Q}$.

When $\mathscr{P} \preceq \mathscr{Q}$ and $\mathscr{P} \subseteq \mathscr{Q} \subseteq [\frac{1}{2}]_{U \times U}$, write $\mathscr{P} \preccurlyeq \mathscr{Q}$. when $\mathscr{P} \preceq \mathscr{Q}$ and $[\frac{1}{2}]_{U \times U} \subseteq \mathscr{P} \subseteq \mathscr{Q}$, write $\mathscr{P} \precneqq \mathscr{Q}$.

Theorem 4. *If $\mathscr{P} \preccurlyeq \mathscr{Q}$, then $G(\mathscr{P}) \geqslant G(\mathscr{Q})$.*

Proof. Since $\mathscr{P} \subseteq \mathscr{Q}$, then $\mathscr{P}_\lambda \subseteq \mathscr{Q}_\lambda, \lambda \in [0,1]$. Furthermore, $\mathscr{P}_\lambda \preceq \mathscr{Q}_\lambda$, then $G(\mathscr{P}_\lambda) \geqslant G(\mathscr{Q}_\lambda)$. Hence $\int_0^1 G(\mathscr{P}_\lambda)\mathrm{d}\lambda \geq \int_0^1 G(\mathscr{Q}_\lambda)\mathrm{d}\lambda$. Note that $\mathscr{P} \preccurlyeq \mathscr{Q}$, so $e(\mathscr{P}) \geqslant e(\mathscr{Q})$. Therefore $G(\mathscr{P}) \geqslant G(\mathscr{Q})$.

Theorem 4 indicates that when $\mathscr{P} \preccurlyeq \mathscr{Q} \preccurlyeq [\frac{1}{2}]_{U\times U}$, the information entropy G of $\mathscr{P}$ is larger than or equal to the information entropy of $\mathscr{Q}$.

Definition 6. *Let $K = (U, \mathscr{R})$ be an approximate space, $\mathscr{R}$ is a fuzzy equivalence relation on U. The rough entropy C of knowledge $\mathscr{R}$ is defined by:*

$$C(\mathscr{R}) = 2^{e(\mathscr{R})} \int_0^1 C(\mathscr{R}_\lambda)\mathrm{d}\lambda$$

Theorem 5. *Let $\mathscr{R}$ be a fuzzy equivalence relation and R a classical equivalence relation. If $\mathscr{R} = R$, then $C(\mathscr{R}) = C(R)$.*

Proof. Note that $\mathscr{R} = R$, then $\mathscr{R}_0 = \check{R}, \mathscr{R}_\lambda = R, 0 < \lambda \leqslant 1$. Hence $C(\mathscr{R}_0) = \log_2 |U|, C(\mathscr{R}_\lambda) = C(R)$. Furthermore, $\int_0^1 C(\mathscr{R}_\lambda)\mathrm{d}\lambda = \int_0^1 C(R)\mathrm{d}\lambda = C(R)$. Note also that $e(\mathscr{R}) = 0$, thus $C(\mathscr{R}) = C(R)$.

Theorem 5 also indicates that in a special case, Definition 2 is the same with Definition 6.

Theorem 6. *If $\mathscr{P} \preccurlyeq \mathscr{Q}$, then $C(\mathscr{P}) \leqslant C(\mathscr{Q})$.*

Proof. If $\mathscr{P} \subseteq \mathscr{Q}$, then $\mathscr{P}_\lambda \subseteq \mathscr{Q}_\lambda, \lambda \in [0,1]$. Hence $\mathscr{P}_\lambda \preceq \mathscr{Q}_\lambda$, furthermore, $C(\mathscr{P}_\lambda) \leqslant C(\mathscr{Q}_\lambda)$. Thus $\int_0^1 C(\mathscr{P}_\lambda)\mathrm{d}\lambda \leqslant \int_0^1 C(\mathscr{Q}_\lambda)\mathrm{d}\lambda$.

Note that $\mathscr{P} \preccurlyeq \mathscr{Q}$, therefore $e(\mathscr{P}) \leqslant e(\mathscr{Q})$. So $C(\mathscr{P}) \leqslant C(\mathscr{Q})$.

3.2 Rough Entropy of Fuzzy Rough Set

In Pawlak rough set theory, uncertainty inherently includes set-uncertainty and knowledge-uncertainty. In order to fully characterize the uncertainty of rough set, we have to compose the measures of two types of uncertainties in some way. A composition of measures of two different uncertainties is given[5]. In the following example, we will measure uncertainty of rough set using the composition way[5].

Example 1. Let $U = \{x_1, x_2, x_3, x_4, x_5, x_6, x_7, x_8, x_9, x_{10}\}$ be a finite universe and knowledge R of information system $K : R_1 = \{\{x_1, x_2, x_3\}, \{x_4, x_5, x_6\}, \{x_7, x_8, x_9, x_{10}\}\}$, $R_2 = \{\{x_1\}, \{x_2, x_3\}, \{x_4, x_5\}, \{x_6\}, \{x_7, x_8, x_9, x_{10}\}\}$, $R_3 = \{\{x_1\}, \{x_2, x_3\}, \{x_4, x_5\}, \{x_6, x_7\}, \{x_8, x_9, x_{10}\}\}$, $R_4 = \{\{x_1, x_2, x_3\}, \{x_4, x_5, x_6, x_7\}, \{x_8, x_9, x_{10}\}\}$. Then the lower approximation and upper approximation operators of $X = \{x_1, x_2, x_4, x_5, x_6, x_7\}$ about knowledge R_1, R_2, R_3, R_4 are as follows:

$$\underline{apr}_{R_1}(X) = \{x_1, x_2, x_3, x_4, x_5, x_6\},$$
$$\overline{apr}_{R_1}(X) = \{x_1, x_2, x_3, x_4, x_5, x_6, x_7, x_8, x_9, x_{10}\};$$
$$\underline{apr}_{R_2}(X) = \{x_1, x_2, x_3, x_4, x_5, x_6\},$$
$$\overline{apr}_{R_2}(X) = \{x_1, x_2, x_3, x_4, x_5, x_6, x_7, x_8, x_9, x_{10}\};$$

$$\underline{apr}_{R_3}(X) = \{x_1, x_2, x_3, x_4, x_5, x_6, x_7\},$$
$$\overline{apr}_{R_3}(X) = \{x_1, x_2, x_3, x_4, x_5, x_6, x_7\};$$
$$\underline{apr}_{R_4}(X) = \{x_1, x_2, x_3, x_4, x_5, x_6, x_7\},$$
$$\overline{apr}_{R_4}(X) = \{x_1, x_2, x_3, x_4, x_5, x_6, x_7\}.$$

Then roughnesses β of X about knowledge R_1, R_2, R_3, R_4 are as follows:

$$\beta_{R_1}(X) = 0.4,\ \beta_{R_2}(X) = 0.4,\ \beta_{R_3}(X) = 0,\ \beta_{R_4}(X) = 0.$$

From Definition 2, we have that rough entropies C of knowledge R_1, R_2, R_3, R_4 are: $C(R_1) = 1.7510$, $C(R_2) = 1.2000$, $C(R_3) = 1.0755$, $C(R_4) = 1.7510$. Furthermore, from Definition 3, we obtain that rough entropies of X about R_1, R_2, R_3, R_4 are: $C_{R_1}(X) = 0.7004$, $C_{R_2}(X) = 0.4800$, $C_{R_3}(X) = 0$, $C_{R_4}(X) = 0$.

From Example 1, we find that for rough entropies C of X about R_1, R_2, Definition 3 fully shows the knowledge-uncertainty. However, we also find that for rough entropies C of X about R_3, R_4, Definition 3 can not fully reflect the knowledge-uncertainty. Therefore, we improve the construction of rough entropy of X about R to measure uncertainty of rough set.

Definition 7. *Let $K = (U, R)$ be an approximation space, $X \subseteq U$. Rough entropy of X about R is defined by:*

$$C_R(X) = C(R) \cdot \big(1 + \beta_R(X)\big)$$

From Example 1, we find that using Definition 7, rough entropies of X about R_1, R_2, R_3, R_4 are obtained: $C_{R_1}(X) = 2.4514$, $C_{R_2}(X) = 1.6800$, $C_{R_3}(X) = 1.0755$, $C_{R_4}(X) = 1.7510$. Obviously, Definition 7 not only fully measures uncertainty of rough set, but also better reflects the relation between knowledge-uncertainty and set-uncertainty.

In order to study roughness and rough entropy of fuzzy rough set, the following definition is first given.

Definition 8. *Let $K = (U, \mathscr{R})$ be an approximation space, $\mathscr{R}$ is a fuzzy equivalence relation on U , $\mathscr{A} \in \mathscr{F}(U)$. Lower approximation operator and upper approximation operator of $\mathscr{A}$ about information system K are defined by:*

$$\underline{apr}_{\mathscr{R}}(\mathscr{A})(x) = \bigwedge_{y \in X} \big((1 - \mathscr{R}(x, y)) \vee \mathscr{A}(y)\big)$$
$$\overline{apr}_{\mathscr{R}}(\mathscr{A})(x) = \bigvee_{y \in X} \big(\mathscr{R}(x, y) \wedge \mathscr{A}(y)\big)$$

where $\big(\underline{apr}_{\mathscr{R}}(\mathscr{A}), \overline{apr}_{\mathscr{R}}(\mathscr{A})\big)$ is called fuzzy rough set of $\mathscr{A}$.

For roughness of fuzzy rough set, we present the following definition:

Definition 9. *Let $K = (U, \mathscr{R})$ be an approximation space, $\mathscr{A} \in \mathscr{F}(U)$, $\mathscr{R}$ is a fuzzy equivalence relation on U. Accuracy of $\mathscr{A}$ about $\mathscr{R}$ is defined by:*

$$\alpha_{\mathscr{R}}(\mathscr{A})=\frac{\sum_{i=1}^{|U|}\underline{apr}_{\mathscr{R}}(\mathscr{A})(x_i)}{\sum_{i=1}^{|U|}\overline{apr}_{\mathscr{R}}(\mathscr{A})(x_i)},\ \textit{where}\ 0\leqslant\alpha_{\mathscr{R}}(\mathscr{A})\leqslant 1.$$

Roughness of $\mathscr{A}$ *about* $\mathscr{R}$ *is defined by:* $\beta_{\mathscr{R}}(\mathscr{A})=1-\alpha_{\mathscr{R}}(\mathscr{A})$.

It is proved that Definition 9 is a generalization of roughness of Pawlak rough set theory[1].

Based on the above discussion , we may obtain the definition of uncertainty measure of fuzzy rough set—rough entropy:

Definition 10. *Let* $K=(U,\mathscr{R})$ *be an approximation space,* $\mathscr{R}$ *is a fuzzy equivalence relation on* U. $\mathscr{A}\in\mathscr{F}(U)$. *Rough entropy of knowledge* $\mathscr{A}$ *about* $\mathscr{R}$ *is defined by:*

$$C_{\mathscr{R}}(\mathscr{A})=C(\mathscr{R})\cdot\big(1+\beta_{\mathscr{R}}(\mathscr{A})\big)$$

4 Conclusions

Uncertainty of rough set includes set-uncertainty and knowledge-uncertainty. However, fuzzy rough set not only inherently models set-uncertainty and knowledge-uncertainty, but also has the uncertainty of set and knowledge which is caused by their fuzziness. In the paper, fuzzy entropy is used to measure fuzzy uncertainty. By the use of cut set, fuzzy knowledge is first made locally accurate, and at the same time information entropy and rough entropy of fuzzy knowledge is given on the basis of the above content. We also improve the composition of measures of different uncertainties so that characterization of uncertainty of rough set becomes more accurate.

References

1. Pawlak Z.(1991) Rough Sets: Theoretecal Aspects of Reasoning about Data. Dordrecht: Kluwer Academic Publishers.
2. M. J. Wierman (1999) Measuring uncertainty in rough set theory. International Journal of General Systems 28: 283-197.
3. J. Y. Liang, K. S. Chin, C. Y. Dang, and C. M. YAM Richard (2002) A new method for measuring uncertainty and fuzziness in rough set theory. International Journal of General Systems 31: 331-342.
4. J. Y. Liang, Z. Z. Shi (2004) The information entropy,rough entropy and knowledge granulation in rough set theory. International Journal of Uncertainty,Fuzziness and Knowledge-Based Systems 12: 37-46.
5. Beaubouef T, Petry F, and Arora G (1998) Information-theoretic measures of uncertainty for rough sets and rough relational databases. Information Sciences 109: 185-195.
6. Nanda S, S. Majumdar (1992) Fuzzy rough sets. Fuzzy Sets and Systems 45: 157-160.

7. X. C. Liu (1992) Entropy, distance measure and similarity measure of fuzzy sets and their relations. Fuzzy Sets and Systems 52: 305-318.
8. Pawlak Z.(1982) Rough sets. International Journal of Computer and Information-Sciences 11: 341-356.

Author Index

Printing: Mercedes-Druck, Berlin
Binding: Stein+Lehmann, Berlin

Printed in the United States
By Bookmasters